AQUIFER HYDRAULICS

AQUIFER HYDRAULICS

A Comprehensive Guide to Hydrogeologic Data Analysis

Vedat Batu, Ph.D., P.E.

A Wiley-Interscience Publication

JOHN WILEY & SONS, INC.

New York / Chichester / Weinheim / Brisbane / Singapore / Toronto

This text is printed on acid-free paper. ∞

Library of Congress Cataloging-in-Publication Data:

Batu, Vedat.
Aquifer hydraulics : a comprehensive guide to hydrogeologic data analysis / Vedat Batu.
p. cm.
Includes index.
ISBN 0-471-18502-7 (alk. paper)
1. Groundwater flow. 2. Aquifers. I. Title.
GB1197.7.B37 1998
551.49—dc21 97-17107

Printed in the United States of America

10 9 8 7 6 5 4 3 2 1

This book is dedicated to members of my family: past, present, and future.

Bu kitap, ailemin geçmişteki, şimdiki, ve gelecekteki bireylerine ithaf edilmiştir.

CONTENTS

PREFACE xv

ACKNOWLEDGMENTS xvii

I INTRODUCTION AND FUNDAMENTALS OF AQUIFER HYDRAULICS

1 INTRODUCTION 3

1.1 Earth's Vital Natural Resources: Aquifers / 3
1.2 Purpose of This Book / 4
1.3 Scope and Organization / 6
1.4 Key Features of This Book / 10
1.5 How to Use This Book / 11

2 FUNDAMENTALS OF AQUIFER HYDRAULICS 21

2.1 Introduction / 21
2.2 Types of Aquifers / 22
2.2.1 Unconfined and Perched Aquifers / 22
2.2.2 Confined Aquifers / 24
2.2.3 Leaky Aquifers / 24
2.2.4 Multiple Aquifers / 24
2.3 Hydraulic and Hydrogeologic Characteristics of Aquifers / 25

2.3.1 Hydraulic Head and Drawdown / 26
2.3.2 Porosity and Effective Porosity / 27
2.3.3 Discharge Velocity and Ground Water Velocity / 28
2.3.4 Darcy's Law / 29
2.3.5 Hydraulic Conductivity, Transmissivity, and Permeability / 34
2.3.6 Estimation of Hydraulic Conductivity from Grain Size / 40
2.3.7 Classification of Aquifers with Respect to Hydraulic Conductivity / 46
2.3.8 Capillarity and Capillary Fringe / 51

2.4 Compressibility and Elasticity of Aquifers / 53
2.4.1 Specific Storage / 53
2.4.2 Storage Coefficient / 61
2.4.3 Specific Yield / 65
2.4.4 Specific Retention / 66
2.4.5 Barometric Efficiency / 67

2.5 Equations of Motion in Aquifers: Generalization of Darcy's Law / 80
2.5.1 Equations of Motion: Principal Directions of Anisotropy Coincide with the Directions of Coordinate Axes / 80
2.5.2 Equations of Motion: Principal Directions of Anisotropy Do Not Coincide with the Directions of Coordinate Axes / 83

2.6 Principal Hydraulic Conductivities / 84
2.6.1 Principal Hydraulic Conductivities and Transmissivities for the Two-Dimensional Case / 85
2.6.2 Mohr's Circle for Principal Hydraulic Conductivities and Transmissivities for the Two-Dimensional Case / 87

2.7 Directional Hydraulic Conductivities / 88
2.7.1 Directional Hydraulic Conductivities and Tranmissivities for Two-Dimensional Cases / 89
2.7.2 Directional Hydraulic Conductivities for Three-Dimensional Cases / 91

2.8 Differential Equations of Flow in Aquifers / 92
2.8.1 Differential Equations in Cartesian Coordinates / 92
2.8.2 Differential Equations in Polar Coordinates / 95
2.8.3 Differential Equations Under Steady-State Conditions / 96

2.9 Initial and Boundary Conditions / 97
2.9.1 Introduction / 97
2.9.2 Initial Conditions / 97
2.9.3 Types of Boundary Conditions / 97

II HYDRAULICS OF AQUIFERS UNDER STEADY PUMPING CONDITIONS FROM WELLS AND HYDROGEOLOGIC DATA ANALYSIS METHODS

3 FULLY PENETRATING PUMPING WELLS IN HOMOGENEOUS AND ISOTROPIC CONFINED AND UNCONFINED AQUIFERS 113

3.1 Introduction / 113
3.2 Well Hydraulics in Confined Aquifers / 113
3.2.1 Nonleaky Aquifer Case: Thiem Equation / 113
3.2.2 Leaky Aquifer Case / 116
3.3 Well Hydraulics in Unconfined Aquifers / 123
3.3.1 Dupuit–Forchheimer Well Discharge Formula / 123
3.4 Methods of Analysis for Steady-State Pumping Test Data / 128
3.4.1 Confined Nonleaky Aquifers: Thiem's Methods / 128
3.4.2 Confined Leaky Aquifers / 131
3.4.3 Unconfined Aquifers / 137

III HYDRAULICS OF AQUIFERS UNDER TRANSIENT PUMPING CONDITIONS FROM WELLS AND HYDROGEOLOGIC DATA ANALYSIS METHODS

4 FULLY PENETRATING PUMPING WELLS IN HOMOGENEOUS AND ISOTROPIC NONLEAKY CONFINED AQUIFERS 145

4.1 Introduction / 145
4.2 Small-Diameter Wells: The Theis Model / 145
4.2.1 Problem Statement and Assumptions / 145
4.2.2 Governing Differential Equation / 147
4.2.3 Initial and Boundary Conditions / 148
4.2.4 Solution / 148
4.2.5 Solution for a Finite Confined Aquifer and Criterion for the Infinite Aquifer Assumption / 154
4.2.6 Approximated Forms of the Theis Equation / 156
4.2.7 Methods of Analysis for Pumping Test Data / 158
4.3 Large-Diameter Well (Dug Well): Papadopulos and Cooper Model / 187
4.3.1 Problem Statement and Assumptions / 187
4.3.2 Governing Differential Equation / 187
4.3.3 Initial and Boundary Conditions / 188
4.3.4 Solution / 189
4.3.5 Methods of Analysis for Pumping Test Data / 195

5 FULLY PENETRATING PUMPING WELLS IN HOMOGENEOUS AND ANISOTROPIC CONFINED NONLEAKY AQUIFERS 206

5.1 Introduction / 206
5.2 Papadopulos-Model-Based Methods / 207
5.2.1 Papadopulos Model / 207
5.3 Hantush's Methods / 231
5.3.1 Drawdown Equations for Isotropic Confined Aquifers / 231
5.3.2 Drawdown Equations for Anisotropic Confined Aquifers / 232
5.3.3 Analysis of Drawdown Equations / 232
5.3.4 Pumping Test Data Analysis Methods / 235
5.4 Hantush and Thomas Method / 242
5.4.1 Equations for the Equal Drawdown Curves / 242
5.4.2 Equations for Residual Equal Drawdown Curves / 244
5.4.3 Pumping Test Data Analysis Methods / 245

6 FULLY PENETRATING PUMPING WELLS IN HOMOGENEOUS AND ISOTROPIC CONFINED LEAKY AQUIFERS WITHOUT THE STORAGE OF THE CONFINING LAYER 248

6.1 Introduction / 248
6.2 Hantush and Jacob Model for Homogeneous and Isotropic Leaky Confined Aquifers / 249
6.2.1 Problem Statement and Assumptions / 249
6.2.2 Governing Differential Equation / 250
6.2.3 Initial and Boundary Conditions / 252
6.2.4 Solution / 252
6.3 Pumping Test Data Analysis Methods for Homogeneous and Isotropic Aquifers / 264
6.3.1 Walton's Type-Curve Method / 264
6.3.2 Hantush's Inflection-Point Methods / 268
6.3.3 Hantush's Type-Curve Method / 281
6.4 Pumping Test Data Analysis Methods for Homogeneous and Anisotropic Leaky Confined Aquifers / 285
6.4.1 Equations for Anisotropic Leaky Confined Aquifers / 286
6.4.2 Hantush's Pumping Test Data Analysis Methods / 287

7 FULLY PENETRATING PUMPING WELLS IN HOMOGENEOUS AND ISOTROPIC CONFINED LEAKY AQUIFERS WITH THE STORAGE OF THE CONFINING LAYERS 293

7.1 Introduction / 293
7.2 Hantush's One Confined Aquifer and Two Confining Layers Model: Asymptotic Solutions / 294
7.2.1 Problem Statement and Assumptions / 294

7.2.2 Governing Equations / 296
7.2.3 Initial and Boundary Conditions / 297
7.2.4 Solutions / 299
7.2.5 Solutions for Special Cases / 325
7.2.6 Evaluation of the Solutions / 326
7.2.7 Application to Pumping Test Data Analysis and Leakage Rates / 327

7.3 Neuman and Witherspoon's Two/One Confined Aquifers and One Aquiclude Models: Complete Solutions / 345
7.3.1 Two Confined Aquifers and One Aquiclude Model / 347
7.3.2 One Confined Aquifer and One Aquiclude of Infinite Thickness Model / 351
7.3.3 Evaluation of the Models and Their Potential Usage / 353

7.4 Neuman and Witherspoon's Two Confined Aquifers and One Aquitard Model: Complete Solution / 358
7.4.1 Problem Statement and Assumptions / 358
7.4.2 Governing Equations / 359
7.4.3 Initial and Boundary Conditions / 359
7.4.4 General Solution / 360
7.4.5 Solutions for Special Cases / 362
7.4.6 Evaluation of the Solutions / 365
7.4.7 Evaluation of the Two Key Assumptions of the Hantush and Jacob (1955) and Hantush (1960b) Models / 368
7.4.8 Evaluation of the Effect of Neglecting Storage in Aquitard on Drawdown in Pumped and Unpumped Aquifers / 371
7.4.9 Application to Pumping Test Data Analysis for Leaky Aquifer Conditions: Neuman and Witherspoon's Ratio Method for Aquitards / 372

8 PARTIALLY PENETRATING PUMPING AND OBSERVATION WELLS IN HOMOGENEOUS AND ANISOTROPIC CONFINED AQUIFERS 386

8.1 Introduction / 386

8.2 Hantush's Leaky Aquifer Model: The Screen of the Pumped Well Extends to the Top of the Aquifer / 387
8.2.1 Problem Statement and Assumptions / 387
8.2.2 Governing Equations / 388
8.2.3 Initial and Boundary Conditions / 388
8.2.4 Solution / 389
8.2.5 Special Solutions / 390
8.2.6 Evaluation of Solutions / 392

8.3 Hantush's Leaky Aquifer Model: The Screen of the Pumped Well Does Not Extend to the Top of the Aquifer (General Case) / 394
8.3.1 Problem Statement and Assumptions / 394

8.3.2 Governing Equations / 395
8.3.3 Initial and Boundary Conditions / 395
8.3.4 Solution for Drawdown in Piezometers / 396
8.3.5 Solution for Average Drawdown in Observation Wells / 408
8.3.6 Drawdown in Piezometers or Wells for $r/b > 1.5$ / 411
8.3.7 Recovery Equations / 412
8.3.8 Evaluation of Solutions / 413
8.4 Application to Pumping Test Data Analysis / 415
8.4.1 Introduction / 415
8.4.2 Type-Curve Methods for Nonleaky Confined Aquifers / 416
8.4.3 Straight-Line and Type-Curve Methods for Nonleaky Confined Aquifers / 425
8.4.4 Hantush's Method for Recovery Test Data / 451

9 FULLY AND PARTIALLY PENETRATING PUMPING AND OBSERVATION WELLS IN HOMOGENEOUS AND ANISOTROPIC UNCONFINED AQUIFERS 458

9.1 Introduction / 458
9.2 Physical Mechanism of Flow to a Well in an Unconfined Aquifer / 459
9.3 Boulton's Model for a Fully Penetrating Well in an Unconfined Aquifer Without Delayed Response / 461
9.3.1 Introduction / 461
9.3.2 Problem Statement and Assumptions / 461
9.3.3 Governing Equations / 462
9.3.4 Initial and Boundary Conditions / 463
9.3.5 Solution / 464
9.3.6 Approximate Formulas for $V(\rho, \tau)$ / 465
9.3.7 Accuracy and Limitations of the Drawdown Equation: Method of Correction / 468
9.3.8 Recovery Equation / 469
9.3.9 Examples / 470
9.4 Boulton's Model for a Fully Penetrating Well in Unconfined Aquifers with the Delayed Response / 472
9.4.1 Problem Statement and Assumptions / 472
9.4.2 Governing Equations / 472
9.4.3 Initial and Boundary Conditions / 473
9.4.4 Solution / 474
9.4.5 Boulton's Well Function for Unconfined Aquifers / 475
9.4.6 Application to Pumping Test Data Analysis / 476
9.5 Neuman's Models for Fully and Partially Penetrating Wells in Unconfined Aquifers with the Delayed Response / 492
9.5.1 Introduction / 492

9.5.2 Neuman's Fully Penetrating Wells Model / 492
9.5.3 Neuman's Partially Penetrating Wells Model / 504
9.5.4 Application to Pumping Test Data Analysis / 513

10 FULLY PENETRATING PUMPING WELLS IN HOMOGENEOUS AND ISOTROPIC BOUNDED NONLEAKY CONFINED AQUIFERS 555

10.1 Introduction / 555
10.2 One or Two Impervious Boundaries / 556
10.2.1 Theory with the Application of the Theis Equation / 556
10.2.2 Boundary Detection Methods for Confined Aquifers / 561
10.3 One Recharge Boundary / 570
10.3.1 Theory / 570
10.3.2 Hantush's Methods for One Recharge Boundary / 572

IV WELL EFFICIENCY AND HYDROGEOLOGIC DATA ANALYSIS METHODS

11 FULLY PENETRATING PUMPING WELLS IN HOMOGENEOUS AND ISOTROPIC NONLEAKY CONFINED AQUIFERS 599

11.1 Introduction / 599
11.2 Drawdown in Discharging Wells / 600
11.3 Specific Capacity of Wells / 601
11.3.1 Specific Capacity of Wells Under Steady-State Conditions / 601
11.3.2 Specific Capacity of Wells Under Unsteady-State Conditions / 604
11.3.3 Properties of the Well Loss Constant / 605
11.4 Step-Drawdown Test Data Analysis Methods / 607
11.4.1 Straight-Line Method / 607
11.4.2 Rorabaugh's Method / 610
11.4.3 Miller and Weber Method / 615
11.4.4 Labadie and Helweg's Computer Method / 621

V HYDRAULICS OF SLUG TESTS AND HYDROGEOLOGIC DATA ANALYSIS METHODS

12 FULLY AND PARTIALLY PENETRATING WELLS IN AQUIFERS 627

12.1 Introduction / 627
12.2 Bouwer and Rice Slug Test Model / 628
12.2.1 Theory / 628

12.2.2 Evaluation of R_e / 631
12.2.3 Data Analysis Methodology / 632
12.2.4 Guidelines for the Applicability of the Bouwer and Rice Method / 633
12.2.5 Evaluation of the Bouwer and Rice Model with Other Models and Cautions / 639
12.3 Hvorslev Slug Test Model / 648
12.3.1 Theory for a Fully-Penetrated Well in a Confined Aquifer / 648
12.3.2 Hvorslev Equations / 651
12.3.3 Comparison of the Hvorslev Equations with the Theoretically Derived Equation / 653
12.3.4 Data Analysis Methodology / 654
12.3.5 Guidelines for the Applicability of the Hvorslev Method / 654
12.3.6 Application / 654
12.4 Cooper, Bredehoeft, and Papadopulos Slug Test Model / 662
12.4.1 Theory / 662
12.4.2 Type-Curve Method / 664
12.4.3 Guidelines for the Type-Curve Method / 666
12.4.4 Application / 667

VI HYDRAULICS OF PRESSURE PULSE AND CONSTANT HEAD INJECTION TESTS FOR TIGHT FORMATIONS AND HYDROGEOLOGIC DATA ANALYSIS METHODS

13 FULLY PENETRATING WELLS IN CONFINED AQUIFERS 675

13.1 Introduction / 675
13.2 Pressure Pulse Test Method / 676
13.2.1 Introduction / 676
13.2.2 Bredehoeft and Papadopulos Method / 676
13.3 Constant Head Injection Tests / 692
13.3.1 Introduction / 692
13.3.2 Steady-State Flow Model / 692
13.3.3 Unsteady-State Flow Model: The Jacob And Lohman Equation / 694
13.3.4 Data Analysis Methods / 698

REFERENCES 705
ABOUT THE AUTHOR 713
ABOUT THE DISK 715
AUTHOR INDEX 717
SUBJECT INDEX 721

PREFACE

In almost every region of the world, aquifers are vital natural resources—for irrigation and for industrial and potable water supply. Unfortunately, many aquifers have been contaminated by human activities, and the contamination rate has been increasing, especially during the past half century. In some aquifers, man-made contamination dates to the beginning of the industrial revolution or even earlier. Both industrial and developing countries have been trying to clean up and protect these resources, and expenditures toward this end have increased significantly during the past two decades. Until the 1960s or 1970s, aquifer problems were primarily related to adequate water supply. Since the 1970s, however, the following issues have been added to the study of aquifers: (1) identification of contamination levels, (2) cleanup, and (3) protection from further contamination. Investigation of water supply or aquifer contamination requires the measurement of aquifer hydraulic parameters, in addition to geologic, hydrogeologic, hydrologic, geochemical, and other information. Without these hydraulic parameters, predictions as to the quantity and quality of ground water cannot be made.

This book is entitled *Aquifer Hydraulics: A Comprehensive Guide to Hydrogeologic Data Analysis* because the hydraulics of aquifers and the determination of their hydrogeologic parameters based on field tests are the main focus. This book started with a series of lecture notes that I used when I taught at Karadeniz Technical University, Trabzon, Turkey, and several universities in the United States at which I had visiting positions. The majority of the subjects in this book were inspired, developed, and enriched during my past 14 years of industrial practice on real world projects. The feedback I received from many practicing hydrologists, hydrogeologists, geologists, engineers, and others has helped me greatly in selecting the scope and shaping the organization of this book.

Characterization of a ground-water system is a critically important first step in solving aquifer problems. Designing a hydrogeologic test, conducting the test in the field, and analyzing the data require a thorough knowledge of aquifer hydraulics, along with familiarity with the existing models in the literature as well as data analysis methods, especially those published since the 1930s. It is not easy for practitioners and students in related disciplines to obtain these models through libraries to use in practical applications. Moreover, most papers and reports on these subjects do not include examples that practitioners and students can easily follow in a timely manner and apply to their work. For most practitioners, time and funds are limited. The above-mentioned factors and the feedback I received from my many practicing colleagues in various disciplines in different corporations, as well as from researchers in various universities and other organizations, played a pivotal role in helping me select the subjects and examples for this book.

Even though the world is in the process of changing to the usage of a single international system of units, in this book both the metric S.I. (Système International d'Unités) and the English FSS (foot–slug–second) system (or U.S. customary system) have been used. For the graphs of this book both the regular and scientific notation (E format) are used as appropriate. For example, 1.0E−08 is equivalent to 1.0×10^{-8}, and 3.5E+03 is equivalent to 3.5×10^{3}. In some tables, the same format is used.

The examples presented in this book are one of its most important strengths. The book includes around 100 example problems, and their solutions are presented in a systematic manner. In the examples both the metric S.I. and English FSS units are used. However, only one unit system is used in each example.

Computers have been playing a significant role in the areas of aquifer hydraulics and hydrogeology, as they have in many other fields. As a result, during the past decade a number of companies have released software packages for aquifer test data analysis. Software packages generally assume that the user has a certain level of experience to be able to select the appropriate model and data analysis method, analyze a given data set, and interpret the results. Therefore, a certain level of familiarity with the fundamentals of aquifer hydraulics is necessary for proper usage of a software package. The aforementioned points have played a significant role in selecting the subjects and examples and in shaping the presentation of this book.

This book is designed to be a self-contained document with (a) some data generation programs for partially penetrating wells and (b) some graph files for type curves for fully penetrating wells. Some key features of this book are as follows: (1) This book presents sufficient theory with enriched practical applications to determine aquifer hydrogeologic parameters. (2) The book covers almost all practical well-known methods published until the 1990s. (3) A "step-by-step methodology" is adopted in presenting the methods as well as the examples. (4) For almost all examples, the data are presented in both digitized and graphical forms so the reader has the opportunity to solve an example problem from beginning to end and compare his or her results with the ones in the book. (5) The book can save the reader a significant amount of time thanks to a relatively simple way of presenting cases—including complex ones.

This book can be used by practitioners, undergraduate and graduate students, researchers, and instructors in related disciplines.

VEDAT BATU

Naperville, Illinois
October 1997

ACKNOWLEDGMENTS

My association with ground water and aquifers started with my Ph.D. studies in the early 1970s at the Department of Hydraulic Engineering, Civil Engineering Faculty, Istanbul Technical University, Istanbul, Turkey. I would like to acknowledge my former teacher and Ph.D. supervisor the late Professor Dr. Kâzım Çeçen and my former teacher Professor Dr. Eren Omay, who inspired and encouraged me to study these subjects.

I am grateful to Rust Environment & Infrastructure, Inc., as well as its parent company, Waste Management, Inc., for their continuous support to employees who are willing to publish.

I owe special thanks to Dr. James J. Butler, Jr. of Kansas Geological Survey for his invaluable contributions on the slug test data analysis methods, providing me with the original data for some figures in one of his coauthored papers; I also thank him for his comments on the manuscript of this book.

I am grateful to Dr. M. Yavuz Çorapçıoğlu of Texas A&M University for his continuous support by discussing various aquifer hydraulics problems and providing me with some pertinent literature; I also thank him for his comments on the manuscript of this book.

I would like to thank Drs. John W. Labadie and O. J. Helweg of Colorado State University for their permission to include a revised version of their original FASTEP program for the analysis of step-drawdown test data.

My special thanks go to Peter A. Mock, of Peter A. Mock and Associates, Phoenix, Arizona, who provided me with the data set for partially penetrating wells in an unconfined aquifer.

I owe special thanks to Dr. Allen F. Moench of the U.S. Geological Survey for his generous and timely response regarding some aquifer hydraulics problems and for the documents he provided me.

I would like to thank Dr. Shlomo P. Neuman of the University of Arizona for discussing some problems in aquifer hydraulics and for his permission to include a revised version of his original DELAY2 program for generation of type curve data for unconfined aquifers under partially penetrating well conditions.

My special thanks go to Dr. Cristopher E. Neuzil of the U.S. Geological Survey for taking the time to discuss some problems regarding pressure pulse test techniques and the data of some example problems.

I am indebted to Dr. Stavros S. Papadopulos, president of S. S. Papadopulos & Associates, Inc., who gave his time generously in discussing many problems regarding aquifer hydraulics and who provided me with most of the papers of the late Dr. Mahdi S. Hantush and some other pertinent literature.

I would like to thank Dr. Paul A. Witherspoon of the University of California–Berkeley for his quick responses to my phone messages and for sending me one of his coauthored original reports published in the 1960s.

I owe special thanks to my colleague Martin N. Sara at Waste Management, Inc., for his continuous support in discussing many problems related to aquifers. Special thanks are also extended to my colleagues Jean A. Crowley, Edward J. Doyle, Patrick W. Dunne, John C. Geiger, Daniel L. Kelleher, Daniel P. Korth, Dr. E. Sabri Motan, Michael C. Ray, and others for their help on many occasions. My special thanks also go to my former colleagues Dr. Sirous Haji-Djafari and Dr. Neville F. Allen, as well as Harry H. Morris, for similar help.

My special thanks go to Sharon M. Guest, who drew all the schematic figures and some of the graphical figures for this book.

I am grateful to John Wiley & Sons, Inc., for undertaking this project. Special thanks are due Daniel R. Sayre, Neil Levine, David Sassian, and Rose L. Kish.

Finally I thank my wife, Nevin, and my sons, Özer and Eren, for their patience during the long hours I spent daily writing this book.

PART I

INTRODUCTION AND FUNDAMENTALS OF AQUIFER HYDRAULICS

CHAPTER 1

INTRODUCTION

1.1 EARTH'S VITAL NATURAL RESOURCES: AQUIFERS

Life on earth is dependent on its natural resources, which are mainly composed of air, surface water, ground water, and soil. Ground water exists mainly in aquifers, which are capable of yielding significant amounts of water. Aquifers are vital natural resources of the earth. In general, beneath the ground surface an unsaturated zone exists, and at most locations the unsaturated zone ends with the water table of an unconfined aquifer. Although the physical and mathematical foundations of flow through unsaturated and saturated zones are similar or the same in many respects, the reasons they are investigated (their research and practical applications) are generally different. For example, the hydraulics of unsaturated zones is especially important in the areas of agriculture and artificial recharge and for some hydraulic structures; hydraulic engineers, hydrologists, soil scientists, and others in different disciplines have made significant contributions in solving many associated problems. Saturated zones, or aquifers, are being used as vital potable, industrial, and irrigation water sources—as well as other sources—in many countries around the world. The study of aquifers is germane to agricultural engineers, chemical engineers, civil engineers, forestry engineers, ecologists, geographers, geologists, geotechnical engineers, hydraulic engineers, hydrologists, hydrogeologists, irrigation engineers, mining engineers, petroleum engineers, soil scientists, and others.

In almost every country of the world, aquifers are part of the ground-water system and are vital natural resources for water supply. Unfortunately, these important resources are being contaminated by human activities, and the contamination rate has been increasing in most aquifers of many industrial and developing countries along with industrial and agricultural developments. Many countries have been trying to clean up and protect these important resources, and the expenditure in this effort has increased significantly since the 1970s. As a result, aquifers are being studied by many researchers and practitioners in different scientific disciplines, including geology, hydrogeology, hydrology, civil engineering, and agricultural engineering. This focus will increase in the future. Problems regarding aquifers may be categorized in three groups: (1) water supply, (2) cleanup of contaminated aquifers, and (3) protection of aquifers from potential contamination.

It is a well-known fact that most aquifers have been being contaminated since the beginning of the industrial revolution, but the importance of the contamination of aquifers, or subsurface flow systems, was not recognized until the 1960s or 1970s. The formation of the Environmental Protection Agency in 1972 in the United States was a recognition of the importance of environmental contamination, including that of aquifers. During the past two decades, many industrial as well as developing countries have formed similar organizations at the ministry or equivalent level as a result of environmental concerns. At the same time, many universities and colleges in different countries have developed new programs and formed new departments or expanded some of their existing departments regarding environment-related teaching and research. There is no doubt that our planet would be much cleaner had these activities started at the beginning of the twentieth century.

During the past century, aquifers have been studied especially for ground-water resource evaluation purposes. However, because of environmental concerns, studies have been intensified especially in many industrial countries to clean up contaminated aquifers and to take preventive actions to reduce or eliminate potential contamination of clean aquifers. The contaminant migration rate in aquifers is typically much slower than that in the four natural resources (air, surface water, ground water, and soil) mentioned above. However, identification of contamination levels in an aquifer with potential contaminant sources and receptors (remedial investigation and feasibility study), along with cleanup of the aquifer (remediation), is a much more expensive and time-consuming process depending on the hydrogeologic characteristics of the aquifer. For example, the velocity of water in a typical river or stream is around 1 meter per second. In aquifers, even 1 meter per day is a high velocity. In some water-bearing formations the velocity of water is less than 1 meter per year. The above-mentioned factors make the cleaning-up process of aquifers more difficult and expensive. During the past two decades, many industrial countries, especially the United States, have been spending a tremendous amount of money to clean up the contaminated aquifers. The starting dates of contamination in some aquifers go back to the nineteenth and eighteenth centuries, or even earlier. The big challenge is to protect the existing aquifers from further contamination in order to reduce the cleanup cost and improve the environmental quality.

1.2 PURPOSE OF THIS BOOK

With the general conditions outlined above in mind, this book will first characterize the subject subsurface flow system, or aquifer. *Characterization* is a general term and covers geologic, hydrogeologic, hydraulic, geochemical, and other site features. The main focus of this book is the hydrogeologic characterization of aquifers based on field data, and that is why the title of the book is *Aquifer Hydraulics: A Comprehensive Guide to Hydrogeologic Data Analysis*. Other characterization subjects and their field test procedures are not included in this book. The hydraulic parameters of an aquifer are the integral part of the hydrogeology of the aquifer and are heavily related to aquifer hydraulics. For instance, without quantification of these parameters, predictions for a water supply well or future migration of a contaminant plume cannot be made. Determination of aquifer hydrogeologic parameters generally require field testing and the analysis of field test data with the appropriate model or models in the subject of aquifer hydraulics. To properly select the appropriate model or models for a given aquifer or a system of aquifers, a thorough knowledge of aquifer hydraulics is required.

The subjects regarding aquifer hydraulics and data analysis methods presented in this book are all based on the *porous media flow theory*. Therefore, subjects in the *fractured media flow theory* are not included. There are several reasons: First, data analysis methods for fractured media are still in the developmental process in the literature, although a limited number of methods for confined aquifers with some relatively simple cases are available. The available data analysis methods do not have widespread practical application, and models based on porous media flow theory may also be used for most fractured aquifers depending on the sizes and frequencies of fractures.

Research activities regarding aquifers by many universities and by some public and private organizations have been intensified in the United States and in some other countries especially during the past three decades. As a result, new approaches and methods have emerged for solving various aquifer problems in different scientific disciplines. A number of books (Bear, 1972, 1979; Bouwer, 1978; Davis and DeWiest, 1966; Dawson and Istok, 1991; de Marsily, 1986; Domenico, 1972; Domenico and Schwartz, 1990; Driscoll, 1986; Fetter, 1994; Freeze and Cherry, 1979; Gelhar, 1993; Hall and Chen, 1996; Kashef, 1986; Kruseman and De Ridder, 1991; Mariño and Luthin, 1982; Polubarinova-Kochina, 1962; Raghunath, 1987; Rushton and Redshaw, 1979; Şen, 1994; Todd, 1980; Walton, 1970; and others) and reports by various organizations have been published during the past half century on the subject of aquifers. Most of these books deal generally with theory, although a small number of them with practical applications regarding hydraulics and testing of aquifers have been published during the past three decades. Also during the past half century, coverage of aquifer hydraulics problems has intensified in journals and symposia proceedings. As a result, many important contributions have been presented in numerous papers and reports on the topic of solving different aquifer problems. Many practitioners as well as students in related disciplines may not have easy access to these documents. Even if they do, it may be time-consuming to learn and apply a methodology in the subject of aquifer hydraulics from these documents. Moreover, most papers published on this subject do not include examples that practitioners and students can easily follow. Most practitioners do not have enough time or funds to search for and study pertinent papers or documents and apply them to their projects. This is also the case for undergraduate students and most graduate students.

During the past decade a number of software packages have been released by different private and public organizations for the analysis of aquifer test data. Almost all of these packages assume that the user has a certain level of experience to select the appropriate model or models, analyze the data, and interpret the results for a given aquifer. A certain level of familiarity with the fundamentals of aquifer hydraulics and models is required to properly analyze an aquifer data set either by hand or by using a software package. The following key points illustrate the importance of a certain level of knowledge of aquifer hydraulics and data analysis methods.

1. The first step is to select the most appropriate analytical aquifer model or models and data analysis methods to analyze a given data set for an aquifer for which the hydrogeologic model is generated based on field investigation. A given data set generally includes a site conceptual hydrogeologic model, formation thicknesses, horizontal and vertical geometries of the wells, and other qualitative and quantitative information. It is generally rare to have

this information package be complete. For example, the saturated thickness of the subject unconfined aquifer or the thickness of the subject confined aquifer may not be known. The hydraulics of wells in relatively thin and relatively thick confined and unconfined aquifers has certain features. Knowledge of these features can help the user significantly in interpreting the field data and in selecting the most appropriate models and hydrogeologic data analysis methods. Practicing hydrogeologists often face cases where the thickness of a confined aquifer or an unconfined aquifer is unknown under partially penetrating pumping and observation well conditions. There are procedures to analyze aquifer test data under these conditions. For instance, with one of the inflection point methods in the literature, the thickness of a confined aquifer can be estimated.

2. Data reduction and interpretation are important steps to take before attempting to analyze aquifer test data. For example, there may be relatively few pervious formations or bedrock outcrops near a pumping test site. If such a formation is close to the center of the test site and the test period is long enough, the drawdown versus time curves at observation wells may reflect its effects. Sometimes a surface water body (river, lake, pond, etc.) may exist around the test area, and the observed drawdowns may be significantly affected under pumping conditions.

3. Sometimes a sophisticated type-curve matching technique either by hand or by a software package for given pumping test data may not be warranted because of (1) time constraints, (2) budgetary limitations, (3) limited data, (4) purpose of the project, or (5) all or several of the previous items.

To properly analyze an aquifer data set either manually or with a software package, one must have a certain level of familiarity with the fundamentals of aquifer hydraulics, available models for well hydraulics in the literature with their key features, and potential data analysis methods. It is the main purpose of this book to address potential cases, in a systematic manner and to save the reader a significant amount of time during learning and application.

1.3 SCOPE AND ORGANIZATION

This book is composed of six parts and has a total of 13 chapters. Part I is composed of Chapter 2, which presents the fundamentals of aquifer hydraulics. Part II consists of Chapter 3, in which the aquifer hydraulics and data analysis methods under steady-state conditions are presented. Part III is composed of Chapter 4 through Chapter 10, in which the aquifer hydraulics and data analysis methods under transient conditions for fully and partially penetrating well cases in single and multiple aquifers are presented. Part IV consists of Chapter 11, which is related to well efficiency and methods of determining well characteristics. In Part V, which is composed of Chapter 12, the hydraulics and data analysis methods for slug tests in aquifers are presented. Finally, in Part VI, Chapter 13 presents the hydraulics and data analysis methods for tests in tight formations. Each chapter is briefly described below.

In Chapter 2, the fundamentals of aquifer hydraulics are presented. This is a stand-alone chapter and presents the general laws of aquifer hydraulics and physical meaning of various parameters, derivation of governing equations, and initial and boundary conditions commonly used in solving aquifer hydraulics problems. Some general methods, which are applicable to almost all aquifer test data, are also included in Chapter 2. The *barometric efficiency* is one of them. The estimation procedures of hydraulic conductivity based on grain-size data are also included in Chapter 2. These procedures are helpful for designing a pumping test for an aquifer if available data are very limited or only grain-size distribution data for granular materials are available.

Chapter 3 presents the aquifer hydraulics under steady-state conditions for fully penetrating pumping and observation wells in homogeneous and isotropic (nonleaky and leaky) confined and unconfined aquifers. Equations included in this chapter are the well-known Thiem equation for nonleaky confined aquifers, the De Glee equation for leaky confined aquifers, and the *Dupuit–Forchheimer well discharge formula* for unconfined aquifers. The corresponding type-curve and straight-line methods are included, with examples.

Chapter 4 presents the aquifer hydraulics under transient conditions for fully penetrating small- and large-diameter pumping and observation wells in homogeneous and isotropic nonleaky confined aquifers. In the first part, the well-known *Theis equation* and its approximate form, known as the *Cooper and Jacob approximation*, for small-diameter wells are presented. Also, the *Theis type-curve method* and the *Cooper and Jacob straight-line method* are included, with examples. The corresponding type-curve and straight-line data analysis methods are also included. In the second part, the *Papadopulos and Cooper equation* for large-diameter wells is presented, as well as two type-curve methods, with examples.

Chapter 5 presents the aquifer hydraulics under transient conditions for fully penetrating pumping and observation wells in homogeneous and anisotropic nonleaky confined aquifers, along with hydrogeologic data analysis methods. Papadopulos' model is presented with its associated type-curve and straight-line methods, based on four wells and three wells. With these methods, the principal directions of anisotropy can also be specified. Also given are Hantush's methods under the conditions of known and unknown principal directions of anisotropy. Finally, the Hantush and Thomas method is presented.

Chapter 6 presents aquifer hydraulics under transient conditions for fully penetrating pumping and observation wells in homogeneous and isotropic/anisotropic confined leaky aquifers without the storage effects of the underlain or overlain confining layer. Then the *Hantush and Jacob formula* for leaky confined aquifers and its associated hydrogeologic data analysis methods are presented, with examples. Included methods are *Walton's type-curve method*, *Hantush's inflection-point methods*, and *Hantush's type-curve method.*

Chapter 7, first presents aquifer hydraulics under transient conditions for fully penetrating pumping and observation wells in homogeneous and isotropic confined leaky aquifers with the storage effects of the confining layers. Chapter 7 is composed of three parts. In the first part, Hantush's leaky confined aquifer models, which take into account the storage effects of the confining layers, are presented with examples based on a type-curve method. In the second part, Neuman and Witherspoon's two/one confined aquifer and one aquiclude models are presented. In the third part, Neuman and Witherspoon's two confined aquifer and one aquitard model (and its practical application regarding the measurement of the hydraulic conductivity of aquitards, *Neuman and Witherspoon's ratio method*) are presented, with an example.

Chapter 8 presents the aquifer hydraulics under transient conditions for partially penetrating pumping and observation wells in homogeneous and anisotropic confined aquifers and hydrogeologic data analysis methods, based on Hantush's original works. First, Hantush's initial model regarding a nonleaky confined aquifer in which the screen of the pumped well extends to the top of the aquifer is presented. Then, Hantush's generalized leaky aquifer model, for which the screen of the pumped well does not extend to the top of the aquifer, is presented. Evaluation of the solutions including special cases are also included in Chapter 8. Later, application of these models to hydrogeologic data analysis under partially penetrating wells conditions are presented, with examples. Included methods are the standard type-curve method, Hantush's inflection-point method, Hantush's type-curve method, Hantush's method for recovery test data, and others. Also, an IBM PC (Personal Computer) DOS (Disk Operating System) version menu-driven type-curve data generation computer program, called HTYCPC, which is a modified version of a program extracted from the mainframe FORTRAN computer program written for several aquifer models (called TYPCRV) by the U.S. Geological Survey (Reed, 1980) for Hantush's general model, and its application to hydrogeologic data analysis, are included. Modifications in HTYCPC are only related with the input data instructions and output structure; the rest of part of the original program, including its algorithm, was kept the same.

Chapter 9 presents aquifer hydraulics under transient conditions for fully and partially penetrating pumping and observation wells in homogeneous and anisotropic unconfined aquifers and hydrogeologic data analysis methods, based on Boulton's and Neuman's original works.

First, the physical mechanism of flow to a well in an unconfined aquifer is presented. Then, Boulton's models for fully penetrating pumping and observation wells without and with delayed response and *Boulton's type-curve method* are presented, with examples. Later, Neuman's models for fully and partially penetrating pumping and observation wells with the delayed response, *Neuman's type-curve method*, and *Neuman's semilogarithmic method* are presented, with examples. Evaluation of the solutions, with special cases including the unique features of Boulton's and Neuman's approaches, is also included. Also, an IBM PC DOS version menu-driven type-curve data generation program, called DELAY2PC, under partially penetrating well conditions, which is a modified version of the mainframe computer program (called DELAY2) developed by Shlomo P. Neuman in 1974 (Neuman, 1974) in FORTRAN, and its application to hydrogeologic data analysis are included. Modifications in DELAY2PC are only related with the input data instructions and output structure; all the rest of part of the original DELAY2 program, including its algorithm, was kept the same.

Chapter 10 presents aquifer hydraulics under transient conditions for fully penetrating pumping and observation wells in homogeneous and isotropic bounded nonleaky confined aquifers. Included cases involve one or two impervious boundaries, and one recharge boundary, which represents a stream or a river. Then, boundary detection methods and Hantush's hydrogeologic data analysis methods for one recharge boundary are presented, with examples.

Chapter 11 presents well efficiency and methods for determining well characteristics. First, specific capacity of wells under steady and transient cases is included. Then, the step-drawdown test data analysis methods are presented. Included methods are the *straight-line method*, *Rorabaugh's method*, the *Miller and Weber method*, and *Labadie and Helweg's computer method*. Also, an IBM PC DOS version menu-driven computer program, called FASTEPPC, which is a modified version of the mainframe computer program (called FASTEP) developed by Labadie and Helweg (1975), and its application to step-drawdown test data analysis are included. The FASTEPPC program is the same as the original FASTEP program, with the exception of its adapted portions for IBM PC DOS computers.

Chapter 12 presents the hydraulics of slug tests in aquifers and the hydrogeologic data analysis methods. First, the *Bouwer and Rice slug test model* for partially penetrating wells for both confined and unconfined aquifers with application to slug test data analysis are presented. In this section, guidelines for the applicability of the method and its evaluation against other models, including the partially penetrating slug test model of the Kansas Geological Survey (KGS), are also presented. Then, the *Hvorslev slug test model* for fully penetrating wells in confined aquifers and its application to slug test data analysis are presented. Finally, the *Cooper, Bredehoeft, and Papadopulos slug test model* for fully penetrating wells in confined aquifers, along with its associated type-curve method, are presented with application to slug test data analysis.

Chapter 13 presents the hydraulics of tests in tight formations and hydrogeologic data analysis methods. First, the *Bredehoeft and Papadopulos pressure pulse test method* for fully penetrating wells in confined formations and its associated data analysis methods are presented. In the second part, models for constant head injection tests in confined aquifers and formations are presented for steady and transient cases. First, the *Thiem equation* and its application to steady-state constant head injection data are presented. Then, *Jacob and Lohman's model* for transient constant head injection tests and its application to data analysis are presented. Included methods are the *Jacob and Lohman type-curve method* and the *Jacob and Lohman straight-line method.*

In each chapter, first, the corresponding analytical model or models with their theoretical foundations are presented. Then, evaluation of the model, including the limitations and cautions, and hydrogeologic data analysis methods are included. The following format is followed in each chapter as appropriate:

Problem Statement and Assumptions: First, the statement of each analytical model is presented with schematic pictures and corresponding assumptions. Some assumptions are critically important for certain models. Therefore, critical evaluation of those assumptions as well as their limitations is included. A typical example is the vertical flow assumption in a confining layer that is underlain or overlain by an aquifer (see Chapter 8).

Governing Differential Equation(s): Every initial and boundary value problem requires a representative governing differential equation or a group of differential equations. The corresponding governing differential equation(s) of a model is presented by referencing the corresponding section of Chapter 2, which includes the fundamentals of aquifer hydraulics.

Initial and Boundary Conditions: All models require the specification of initial and boundary conditions, with the exception of the models under steady-state conditions in Chapter 3 for which only boundary conditions are required. First, the mathematical expressions of each initial and boundary condition are presented. Then, the physical meaning of each condition is explained.

Solution: The solution based on the corresponding initial and boundary conditions is presented for each case without including the intermediate steps, which are generally related with various solution methods (Laplace, Fourier, and Hankel transform techniques and others) in solving partial differential equations. Special cases of each solution and their available approximate forms are also presented as appropriate. Then, evaluation of the solution is presented in tables and graphs as appropriate.

Application to Hydrogeologic Data Analysis: In each chapter, the appropriate hydrogeologic data analysis methods (for pumping tests, slug tests, pulse pressure tests, or constant head injection tests) are presented based on the corresponding solution. For some models, several methods are presented including type-curve, straight-line, inflection-point methods, and others. A step-by-step methodology, which is one of the most important features of the book, is adopted in presenting each method. Then, examples are presented in the same step-by-step manner. The data of examples are mostly taken from the pertinent literature, and for each example the corresponding data are presented in both tabular and graphical forms. If only the graphical form of a data set is available in the pertinent literature, the graph is digitized and both its tabular and graphical forms are included in this book. As a result, each example is a complete set. The reader will thus be able to solve each example problem and compare his or her own generated results with the ones presented in the book. The above procedures are applied from the simplest to the most complex example problems. The main idea behind this approach is to try to minimize the reader's learning time for the data analysis methods. This is extremely important, especially for practitioners who are often required to analyze an aquifer data set in a relatively tight time period and with a limited budget. There is no doubt that the aforementioned procedure will also minimize the learning efforts of students in the corresponding scientific disciplines.

Figures and Tables: This book contains close to 300 figures and over 100 tables. The figures are composed of schematics and x–y-type graphs. The type-curve figures in the book are also presented as LOTUS 123 graph files (a total of 14) in the attached diskette. In the graphs of the book the regular and scientific notation (E format) are used simultaneously as appropriate. For example, 1.0E−08 is equivalent to 1.0×10^{-8}, and $3.5E + 03$ is equivalent to 3.5×10^3. In some tables, the same formats are used in the same manner.

Units: The world is in the process of changing to the usage of a single international system of units. The metric system, called Système International d'Unités (S.I.), has already been adopted by almost all countries around the world. However, in the United States the English system (or U.S. customary system) is still being used in practice; the S.I. system has been partially adopted by some research organizations. In this book, both the metric S.I. system and the English FSS (foot–slug–second) system of units are used.

1.4 KEY FEATURES OF THIS BOOK

As mentioned above, this book is designed to be a self-contained document with some type-curve data generation computer programs for partially penetrating well cases and some type-curve data and curves for fully penetrating well cases. The major features of this book are as follows:

1. This book is a self-contained document for aquifer hydraulics and analysis of aquifer tests to determine the hydrogeologic parameters.
2. This book covers almost all practical well-known methods published through the mid 1990s.
3. This book presents sufficient theory with enriched practical applications to hydrogeologic data analysis.
4. A *step-by-step methodology* is adopted in presenting the methods as well as the examples.
5. This book can save readers (practitioners, undergraduate and graduate students) a significant amount of time thanks to a relatively simple way of presenting the cases, including the complex ones.

6. For almost all examples, the data are presented in digitized form as well as graphical form. The reader will thus have the opportunity to solve an example problem from the beginning to end and compare his or her results with the ones presented in this book.

1.5 HOW TO USE THIS BOOK

As previously mentioned, this book has basically been written to present information on aquifer hydraulics and analysis of aquifer test data. The book is designed to be a self-contained document. If the reader does not have sufficient background regarding the fundamentals of aquifer hydraulics, it is suggested he or she read Chapter 2 first. If the reader has sufficient background in the fundamentals of aquifer hydraulics and wants to learn about a specific aquifer and well hydraulics case or analyze a given data set, he or she must first find the appropriate chapter and section including the most appropriate method or methods for the analysis of aquifer test data. Table 1-1 through Table 1-12 can be used as guides to select the most appropriate methods for a given data set. To find the most appropriate data analysis methods, the reader is advised to take the following steps: (1) Generate the conceptual model of the hydrogeologic system using the available qualitative and quantitative information for the test site. (2) Assign thickness values of the formations. If data are not available, make the best estimates. (3) Locate existing nearby surface water bodies and less pervious formations around the test site if information is available. (4) Locate the screen intervals of the pumping and observation wells, and assign the geometric dimensions. If data are not available, make the best estimates. (5) Based on the information of the previous steps, select the most appropriate aquifer model and data analysis methods from Table 1-1 through Table 1-12 and apply them.

The attached three computer programs are menu-driven and are relatively easy to use. These programs can be run on almost every IBM PC DOS-compatible computer. The purpose of FASTEPPC is to calculate some well parameters from a given step-drawdown data set.The purpose of the HTYCPC and DELAY2PC programs is to generate type curve data for partially penetrating wells in confined and unconfined aquifers, respectively. These

TABLE 1-1 Hydrogeologic Data Analysis Methods Under Steady-State Conditions for Homogeneous and Isotropic Confined and Unconfined Aquifers with Fully Penetrating Wells

Type of Aquifer	Penetration	Type of Flow	Conceptual Model	Method	Section	Example	Available Aquifer Parameters
CON	FP	SS	Figure 3-1	SC	3.4.1	3-4	$T(K_h)$
CON	FP	SS	Figure 3-1	SL	3.4.1	3-5	$T(K_h)$
CL	FP	SS	Figure 3-2	TC	3.4.2.1	3-6	$T(K_h)$ b'/K'
CL	FP	SS	Figure 3-2	SL	3.4.2.2	3-7	$T(K_h)$
UC	FP	SS	Figure 3-3	SL	3.4.3.3	3-8	$T(K_h)$

Notation: CON, confined; CL, confined leaky; FP, full penetration; SC, straight calculation; SS, steady state; SL, straight line; TC, type curve; UC, unconfined; b', thickness of confining layer; K_h, horizontal hydraulic conductivity; K', vertical hydraulic conductivity of confining layer; T, transmissivity.

TABLE 1-2 Hydrogeologic Data Analysis Methods Under Transient Conditions for Nonleaky Homogeneous and Isotropic Confined Aquifers with Fully Penetrating Small- and Large-Diameter Wells

Type of Aquifer	Penetration	Type of Flow	Conceptual Model	Method	Section	Example	Available Aquifer Parameters
CON	FP(SDW)	TS	Figure 4-1	TC	4.2.7.1	4-6	$T(K_h)$ S
CON	FP(SDW)	TS	Figure 4-1	SL	4.2.7.2.2	4-7	$T(K_h)$ S
CON	FP(SDW)	TS	Figure 4-1	SL	4.2.7.2.2	4-8	$T(K_h)$ S
CON	FP(SDW)	TS	Figure 4-1	SL	4.2.7.2.2	4-9	$T(K_h)$ S
CON	FP(SDW)	TS	Figure 4-1	SL	4.2.7.3.2	4-10	$T(K_h)$ S
CON	FP(SDW)	TS	Figure 4-1	SLR	4.2.7.4.2	4-11	$T(K_h)$
CON	FP(SDW)	TS	Figure 4-1	SLR	4.2.7.4.3	4-12	$T(K_h)$ S
CON	FP(LDW)	TS	Figure 4-21	TC	4.3.5.1 4.3.5.3	4-13	$T(K_h)$ S
CON	FP(LDW)	TS	Figure 4-21	TC	4.3.5.2 4.3.5.3	4-14	$T(K_h)$ S

Notation: CON, confined; FP, full penetration; LDW, large-diameter well; SDW, small-diameter well; SL, straight line; SLR, straight-line recovery; TS, transient state; TC, type-curve; K_h, horizontal hydraulic conductivity; S, storage coefficient; T, transmissivity.

programs are not capable of generating type curves and type-curve matching for a given data set. The user is strongly advised to read the corresponding chapters before starting to run these programs. Neither the author of this book nor the original developers of these programs nor John Wiley & Sons, Inc., has responsibility for the results generated by the user. Also, no technical support is available for these programs by the author of this book, the original developers of these programs, or John Wiley & Sons, Inc.

TABLE 1-3 Hydrogeologic Data Analysis Methods Under Transient Conditions for Nonleaky Homogeneous and Anisotropic Confined Aquifers with Fully Penetrating Wells

Type of Aquifer	Penetration	Type of Flow	Conceptual Model	Method	Section	Example	Available Aquifer Parameters
CON	FP	TS	Figure 4-1 Figure 5-1	TC(FW)	5.2.1.5.1.1	5-1	$T_{xx}(K_{xx})$ $T_{yy}(K_{yy})$ $T_{xy}(K_{xy})$ $T_{\xi\xi}(K_{\xi\xi})$ $T_{\eta\eta}(K_{\eta\eta})$ θ

TABLE 1-3 (*continued*)

Type of Aquifer	Penetration	Type of Flow	Conceptual Model	Method	Section	Example	Available Aquifer Parameters
CON	FP	TS	Figure 4-1 Figure 5-1	SL(FW)	5.2.1.5.1.2	5-2	$T_{xx}(K_{xx})$ $T_{yy}(K_{yy})$ $T_{xy}(K_{xy})$ $T_{\xi\xi}(K_{\xi\xi})$ $T_{\eta\eta}(K_{\eta\eta})$ θ
CON	FP	TS	Figure 4-1 Figure 5-1	TC(FW)	5.2.1.5.2.1	—	$T_{xx}(K_{xx})$ $T_{yy}(K_{yy})$ $T_{xy}(K_{xy})$ $T_{\xi\xi}(K_{\xi\xi})$ $T_{\eta\eta}(K_{\eta\eta})$ θ
CON	FP	TS	Figure 4-1 Figure 5-1	SL(FW)	5.2.1.5.2.6	5-3 5-4	$T_{xx}(K_{xx})$ $T_{yy}(K_{yy})$ $T_{xy}(K_{xy})$ $T_{\xi\xi}(K_{\xi\xi})$ $T_{\eta\eta}(K_{\eta\eta})$ θ
CON	FP	TS	Figure 4-1 Figure 5-1	TC(PK)	5.3.4.1	—	$T_{xx}(K_{xx})$ $T_{yy}(K_{yy})$ $T_{xy}(K_{xy})$ $T_{\xi\xi}(K_{\xi\xi})$ $T_{\eta\eta}(K_{\eta\eta})$ θ
CON	FP	TS	Figure 4-1 Figure 5-1	SL(PK)	5.3.4.1	—	$T_{xx}(K_{xx})$ $T_{yy}(K_{yy})$ $T_{xy}(K_{xy})$ $T_{\xi\xi}(K_{\xi\xi})$ $T_{\eta\eta}(K_{\eta\eta})$ θ
CON	FP	TS	Figure 4-1 Figure 5-1	TC(PU)	5.3.4.2	5-5 5-6	$T_{xx}(K_{xx})$ $T_{yy}(K_{yy})$ $T_{xy}(K_{xy})$ $T_{\xi\xi}(K_{\xi\xi})$ $T_{\eta\eta}(K_{\eta\eta})$ θ
CON	FP	S	Figure 4-1 Figure 5-1	SL(PU)	5.3.4.2	5-5 5-6	$T_{xx}(K_{xx})$ $T_{yy}(K_{yy})$ $T_{xy}(K_{xy})$ $T_{\xi\xi}(K_{\xi\xi})$ $T_{\eta\eta}(K_{\eta\eta})$ θ

TABLE 1-3 (*continued*)

Type of Aquifer	Penetration	Type of Flow	Conceptual Model	Method	Section	Example	Available Aquifer Parameters
CON	FP	TS	Figure 4-1 Figure 5-16	TC(FW)	5.4.3.1	—	$T_{xx}(K_{xx})$ $T_{yy}(K_{yy})$ $T_{xy}(K_{xy})$ $T_{\xi\xi}(K_{\xi\xi})$ $T_{\eta\eta}(K_{\eta\eta})$ θ
CON	FP	TS	Figure 4-1 Figure 5-16	SLR	5.4.3.2	—	$T_{xx}(K_{xx})$ $T_{yy}(K_{yy})$ $T_{xy}(K_{xy})$ $T_{\xi\xi}(K_{\xi\xi})$ $T_{\eta\eta}(K_{\eta\eta})$ θ

Notation: CON, confined; FW, four wells; FP, full penetration; PK, principal directions of anisotropy are known; PU, principal directions of anisotropy are unknown; SL, straight line; SLR, straight-line recovery; TC, type curve; TW, three wells; TS, transient state; $T_{xx}(K_{xx})$, transmissivity (hydraulic conductivity) tensor component; $T_{yy}(K_{yy})$, transmissivity (hydraulic conductivity) tensor component; $T_{xy}(K_{xy})$, transmissivity (hydraulic conductivity) tensor component; $T_{\xi\xi}(K_{\xi\xi})$, principal transmissivity (hydraulic conductivity) in $\xi\xi$ direction; $T_{\eta\eta}(K_{\eta\eta})$, principal transmissivity (hydraulic conductivity) in $\eta\eta$ direction; θ, the angle between the x axis and the ξ axis.

TABLE 1-4 Hydrogeologic Data Analysis Methods Under Transient Conditions for Leaky Homogeneous and Isotropic Confined Aquifers for Fully Penetrating Wells Without the Storage Effects of the Confining Layer

Type of Aquifer	Penetration	Type of Flow	Conceptual Model	Method	Section	Example	Available Aquifer Parameters
CL(WHS)	FP	TS	Figure 6-1	TC	6.3.1	6-1	$T(K_h)$ b'/K'
CL(WHS)	FP	TS	Figure 6-1	IP(OW)	6.3.2.2	6-2 6-3	$T(K_h)$ b'/K'
CL(WHS)	FP	TS	Figure 6-1	IP(MOW)	6.3.2.3	6-4	$T(K_h)$ b'/K'
CL(WHS)	FP	TS	Figure 6-1	TC	6.3.3	6-5	$T(K_h)$ b'/K'
CL(WHS)	FP	TS	Figure 6-1	TC(PK)	6.4.2.1	—	$T_{xx}(K_{xx})$ $T_{yy}(K_{yy})$ $T_{xy}(K_{xy})$ $T_{\xi\xi}(K_{\xi\xi})$ $T_{\eta\eta}(K_{\eta\eta})$ θ
CL(WHS)	FP	TS	Figure 6-1	IP(PK)	6.4.2.1	—	$T_{xx}(K_{xx})$ $T_{yy}(K_{yy})$ $T_{xy}(K_{xy})$ $T_{\xi\xi}(K_{\xi\xi})$ $T_{\eta\eta}(K_{\eta\eta})$ θ
CL(WHS)	FP	TS	Figure 6-1	TC(PU)	6.4.2.2	6-6	$T_{xx}(K_{xx})$ $T_{yy}(K_{yy})$ $T_{xy}(K_{xy})$ $T_{\xi\xi}(K_{\xi\xi})$ $T_{\eta\eta}(K_{\eta\eta})$ θ
CL(WHS)	FP	TS	Figure 6-1	IP(PU)	6.4.2.2	6-6	$T_{xx}(K_{xx})$ $T_{yy}(K_{yy})$ $T_{xy}(K_{xy})$ $T_{\xi\xi}(K_{\xi\xi})$ $T_{\eta\eta}(K_{\eta\eta})$ θ

Notation: CL, confined leaky; FW, four wells; FP, full penetration; IP, inflection points; MOW, more than one observation well; OW, one observation well; PK, principal directions of anisotropy are known; PU, principal directions of anisotropy are unknown; SL, straight line; SLR, straight-line recovery; TC, type curve; TW, three wells; TS, transient state; WHS, without storage effects of the confining layers; b', thickness of confining layer; K', vertical hydraulic conductivity of confining layer; $T_{xx}(K_{xx})$, transmissivity (hydraulic conductivity) tensor component; $T_{yy}(K_{yy})$, transmissivity (hydraulic conductivity) tensor component; $T_{xy}(K_{xy})$, transmissivity (hydraulic conductivity) tensor component; $T_{\xi\xi}(K_{\xi\xi})$, principal transmissivity (hydraulic conductivity) in $\xi\xi$ direction; $T_{\eta\eta}(K_{\eta\eta})$, principal transmissivity (hydraulic conductivity) in $\eta\eta$ direction; θ, the angle between the x axis and the ξ axis.

TABLE 1-5 Hydrogeologic Data Analysis Methods Under Transient Conditions for Leaky Homogeneous and Isotropic Confined Aquifers for Fully Penetrating Wells with the Storage Effects of the Confining Layer

Type of Aquifer	Penetration	Type of Flow	Conceptual Model	Method	Section	Example	Available Aquifer Parameters
CLM(WS)	FP	TS(SVT)	Figure 7-1 Figure 7-2 Figure 7-3	TC	7.2.7.1.1	7-1	$T(K_h)$ S $K'S$ $K''S''$
CLM(WS)	FP	TS(LVT)	Figure 7-1	TC	7.2.7.1.2.1	—	$T(K_h)$ $S + S'/3 + S''/3$
CLM(WS)	FP	TS(LVT)	Figure 7-2	TC	7.2.7.1.2.2	—	$T(K_h)$ $S + S' + S''$
CLM(WS)	FP	TS(LVT)	Figure 7-3	TC	7.2.7.1.2.3	—	$T(K_h)$ $S + S'/3 + S''$
CLM(WS)	FP	TS(LVT)	Figure 7-19	RM	7.4.9.3	7-3	α'

Notation: CLM, confined leaky multiple aquifers; SVT, small values of time; LVT, large values of time; TS, transient state; WS, with storage effects of the confining layers; $T(K_h)$, transmissivity (horizontal hydraulic conductivity) of aquifer; S, storage coefficient of aquifer; S', storage coefficient of the overlain confining layer; S'', storage coefficient of the overlain confining layer; α', diffusivity.

TABLE 1-6 Hydrogeologic Data Analysis Methods Under Transient Conditions for Nonleaky Homogeneous and Anisotropic Confined Aquifers with Partially Penetrating Wells

Type of Aquifer	Penetration	Type of Flow	Conceptual Model	Method	Section	Example	Available Aquifer Parameters
CON	PP	TS	Figure 8-5	TC(CP)	8.4.2.2.2	8-1	K_h K_z S
CON	PP	TS	Figure 8-5	IP	8.4.3.2.1	8-2 8-3	K_h S
CON	PP	TS	Figure 8-5	TC	8.4.3.2.2	8-2 8-3	K_h S
CON	PP	TS	Figure 8-5	TCJMB	8.4.3.2.3	8-3	K_h S
CON	PP	TS	Figure 8-5	CJMPP	8.4.3.2.4	—	$T(K_h)$ S
CON	PP	TS	Figure 8-5	R	8.4.4.2	8-4	K_h S

Notation: CON, confined; CJMPP, Cooper and Jacob method adjusted for partial penetration; CP, requires to run a computer program for Eq. (8–102); IP, inflection point; PP, partial penetration; R, recovery; TC, type curve; TCJO, Theis' or Cooper and Jacob, or both; TS, transient state; K_h, horizontal hydraulic conductivity; S, storage coefficient; T, transmissivity.

TABLE 1-7 Hydrogeologic Data Analysis Methods Under Transient Conditions for Nonleaky Homogeneous and Anisotropic Unconfined Aquifers with Fully and Partially Penetrating Wells

Type of Aquifer	Penetration	Type of Flow	Conceptual Model	Method	Section	Example	Available Aquifer Parameters
UC	FP	TS	Figure 9-3	TC	9.4.6.3	9-3 9-4	K_h S S_y
UC	FP	TS	Figure 9-15	TC	9.5.4.1.1	9-5	K_h K_z S S_y
UC	FP	TS	Figure 9-15	SL	9.5.4.1.2	9-6	K_h S S_y
UC	FP	TS	Figure 9-15	R	9.5.4.1.3	—	K_h S
UC	FP	TS	Figure 9-22	TC(CP)	9.5.4.2.1	9-7	K_h K_z S S_y
UC	FP	TS	Figure 9-22	R	9.5.4.1.3	—	K_h S

Notation: UC, unconfined; CP, requires to run a computer program for Eq. (9–115); FP, full penetration; PP, partial penetration; R, recovery; SL, semilogarithmic; TC, type curve; TS, transient state; K_h, horizontal hydraulic conductivity; K_z, vertical hydraulic conductivity; S, storage coefficient; S_y, specific yield.

TABLE 1-8 Hydrogeologic Data Analysis Methods Under Transient Conditions for Nonleaky Homogeneous and Isotropic Confined Aquifers with Fully Penetrating Wells with One Recharge Boundary

Type of Aquifer	Penetration	Type of Flow	Conceptual Model	Method	Section	Example	Available Aquifer Parameters
CON	FP	TS	Figure 10-9	OOW	10.3.2.5.1	10-3 10-4	$T(K_h)$ S
CON	FP	TS	Figure 10-9	TMOW	10.3.2.5.2	10-5	$T(K_h)$ S

Notation: CON, confined; FP, full penetration; OOW, one observation well; TMOW, two or more observation wells; K_h, horizontal hydraulic conductivity; T, transmissivity; S, storage coefficient.

TABLE 1-9 Methods for Determining Well Characteristics Based on Step-Drawdown Test Data

Type of Aquifer	Penetration	Type of Flow	Conceptual Model	Method	Section	Example	Available Parameters
CON	FP	TS	Figure 11-1	SL	11.4.1	11-3	B
			Figure 11-2			11-4	C
CON	FP	TS	Figure 11-1	G	11.4.2	11-5	B
			Figure 11-2			11-6	C
							n
CON	FP	TS	Figure 11-1	I	11.4.3	11-7	B
			Figure 11-2			11-8	C
							n
CON	FP	TS	Figure 11-1	COM	11.4.4	11-9	B
			Figure 11-2				C
							n

CON, confined; COM, computer; FP, full penetration; G, graphical; I, iteration; TS, transient state; B, parameter; C, parameter; n, parameter.

TABLE 1-10 Slug Test Data Analysis Methods

Type of Aquifer	Penetration	Type of Flow	Conceptual Model	Method	Section	Example	Available Aquifer Parameters
CON	PP[a]	TS	Figure 12-1	SL	12.2.3	12-1	$T(K_h)$
UC							
CON	FP[b]	TS	Figure 12-17	SL	12.3.4	12-2	$T(K_h)$
UC						12-3	
						12-4	
CON	FP[b]	TS	Figure 12-17	TC	12.4.2	12-5	$T(K_h)$
UC						12-6	S

[a]The method is also applicable to confined aquifers.
[b]The method is also applicable to confined and unconfined aquifers under partially penetrating well conditions.
Notation: CON, confined; FP, full penetration; PP, partial penetration; TS, transient state; TC, type curve; K_h, horizontal hydraulic conductivity; S, storage coefficient; T, transmissivity.

TABLE 1-11 Data Analysis Methods for Tight Formations

Type of Aquifer	Penetration	Type of Flow	Conceptual Model	Method	Section	Example	Available Aquifer Parameters
CON	FP[(a)]	PPT(TS)	Figure 13-1	TC	13.2.2.4	13-2	$T(K_h)$
UC			Figure 13-3				S
CON	FP[(a)]	CHI(SS)	Figure 13-10	SC	13.3.2.2	13-3	$T(K_h)$
UC							
CON	FP[(a)]	CHI(TS)	Figure 13-10	TC	13.3.4.1	13-4	$T(K_h)$
UC							S
CON	FP[(a)]	CHI(TS)	Figure 13-10	SL	13.3.4.2	13-5	$T(K_h)$
UC							S

[a]The method is also applicable under partially penetrating well conditions.
Notation: CHI, constant head injection test; CON, confined; FP, full penetration; SC, straight calculation; PP, partial penetration; PPT, pulse pressure test; SS, steady state; TS, transient state; TC, type curve; K_h, horizontal hydraulic conductivity; S, storage coefficient; T, transmissivity.

TABLE 1-12 Boundary Detection Methods Under Transient Conditions for Nonleaky Homogeneous and Isotropic Bounded Confined Aquifers with Fully Penetrating Wells

Type of Aquifer	Penetration	Type of Flow	Conceptual Model	Method	Section	Example	Boundarics to Be Detectable
CON	FP	TS	Figure 10-1	SL	10.2.2.1	10-1	OIB
			Figure 10-2			10-2	
CON	FP	TS	Figure 10-8	SL	10.2.2.2	—	TIB

CON, confined; FP, full penetration; OIB, one impervious boundary; SL, semilogarithmic; TIB, two impervious boundaries; TS, transient state.

CHAPTER 2

FUNDAMENTALS OF AQUIFER HYDRAULICS

2.1 INTRODUCTION

Any project for an aquifer or a system of aquifers, whose geology and hydrogeology are well defined by field testing methods, requires the determination of aquifer hydraulic characteristics in the field. Horizontal and vertical hydraulic conductivities, and specific storage (or storage coefficient), are the main aquifer hydraulic characteristics. Almost all field testing procedures are based on the vertical well methods (theories of horizontal wells are still in the development process, and no field testing procedure is available yet). Pumping tests, slug tests, pressure pulse tests, and constant head injection tests are routinely used on vertical wells for the determination of aquifer hydraulic parameters. For example, pumping tests are mostly performed by withdrawing water at a constant rate from one well and observing the temporal variation of the water levels in the pumping well and in the nearby observation wells. The temporal variation of the water level in an observation well depends on the aquifer permeability, the magnitude of the pumping rate, and the well and aquifer geometries. Obviously, conducting a pumping test is not sufficient to determine aquifer hydraulic parameters. The next step is to identify the most appropriate analytical model that fits the conditions of the aquifer under question. Properly identifying an analytical model requires familiarization with the initial and boundary conditions upon which the potential analytical models are based.

The determination of aquifer hydraulic characteristics by means of well tests requires a thorough understanding of the hydraulics of ground-water flow in aquifers under the hydraulic conditions created by wells. As a result, this chapter is devoted to fundamental principles of aquifer hydraulics. Types of aquifers and the definitions of hydraulic/hydrogeologic parameters of aquifers are described in detail in this chapter. Before attempting to analyze the results of a pumping test in terms of aquifer hydraulic parameters, it is important to consider the effects of changes in barometric pressure and earth tides on temporal water-level variations in wells. Therefore, the mechanics associated with these effects as well as compressibility and elasticity of aquifers are all covered in detail, with examples. The Darcy equation was originally established as a one-dimensional empirical equation (Darcy, 1856). In the subsequent decades, the one-, two-, and three-dimensional differential forms of Darcy's equation under isotropic and anisotropic conditions were formed to derive the general governing partial differential equations of ground-water flow based on the mass conservation principle. Understanding of the foundations of these equations is a critically important step in aquifer hydraulics; they are presented in detail in this chapter. The analytical solutions of well methods are based on the governing equations and initial and boundary conditions. Also in this chapter, the initial and boundary conditions encountered in aquifer hydraulics are included in detail, with examples.

2.2 TYPES OF AQUIFERS

The water body below the ground surface is called *subsurface water*. The subsurface water system is composed of unsaturated and saturated zones. The term *aquifer* is used for saturated formations. Etymologically, *aquifer* means "water-bearing formation." The word *aquifer* is produced from two Latin words: *aqua* (water) and *ferre* (to bear). In subsurface hydrology, an aquifer is defined as a single geologic formation or a group of geologic formations that transmits and yields a significant amount of water.

If a geologic formation does not have the capability to transmit a significant amount of water, it is called an *aquiclude*.

An *aquitard*, or a *confining layer*, is defined as a geologic formation that can transmit water at a relatively low rate compared with aquifers.

If a geologic formation neither absorbs nor transmits water, it is called an *aquifuge*.

Figure 2-1 is a generalized cross section of a subsurface flow system that includes different types of aquifers as well as the unsaturated zone. Based on Figure 2-1, in the following sections the definitions of different types of aquifers will be presented.

2.2.1 Unconfined and Perched Aquifers

Unconfined aquifers, which are also called *water-table aquifers* or *phreatic aquifers*, are bounded by a free surface at the upper boundary. As a result, the free surface of an unconfined aquifer is under atmospheric pressure. In reality, there is a *capillary fringe* above the water table.

The term *phreatic water* was used in French by Daubré (1887), who included under this term only the water in the upper part of the saturated zone (Meinzer, 1923). The term *water-table aquifers* has widely been used for unconfined aquifers as well.

In Figure 2-1, Aquifer 1 is an unconfined aquifer because it is under water-table conditions. The far left portions of Aquifer 2 and Aquifer 3 are under water-table or unconfined aquifer conditions as well. A typical indicator of an unconfined aquifer is that the water level in a well penetrating the unconfined aquifer is equal to the water-table elevation at the same location of the aquifer. In other words, water does not rise above the water table. For example, the water level in Well W4 in Figure 2-1 is equal to the water-table elevation at that location.

A *perched aquifer* is a special form of an unconfined aquifer and is under water-table conditions as well (see at the far right of Figure 2-1). A perched water-table aquifer occurs above the main water table if a relatively small aquitard exists between the water table and the ground surface.

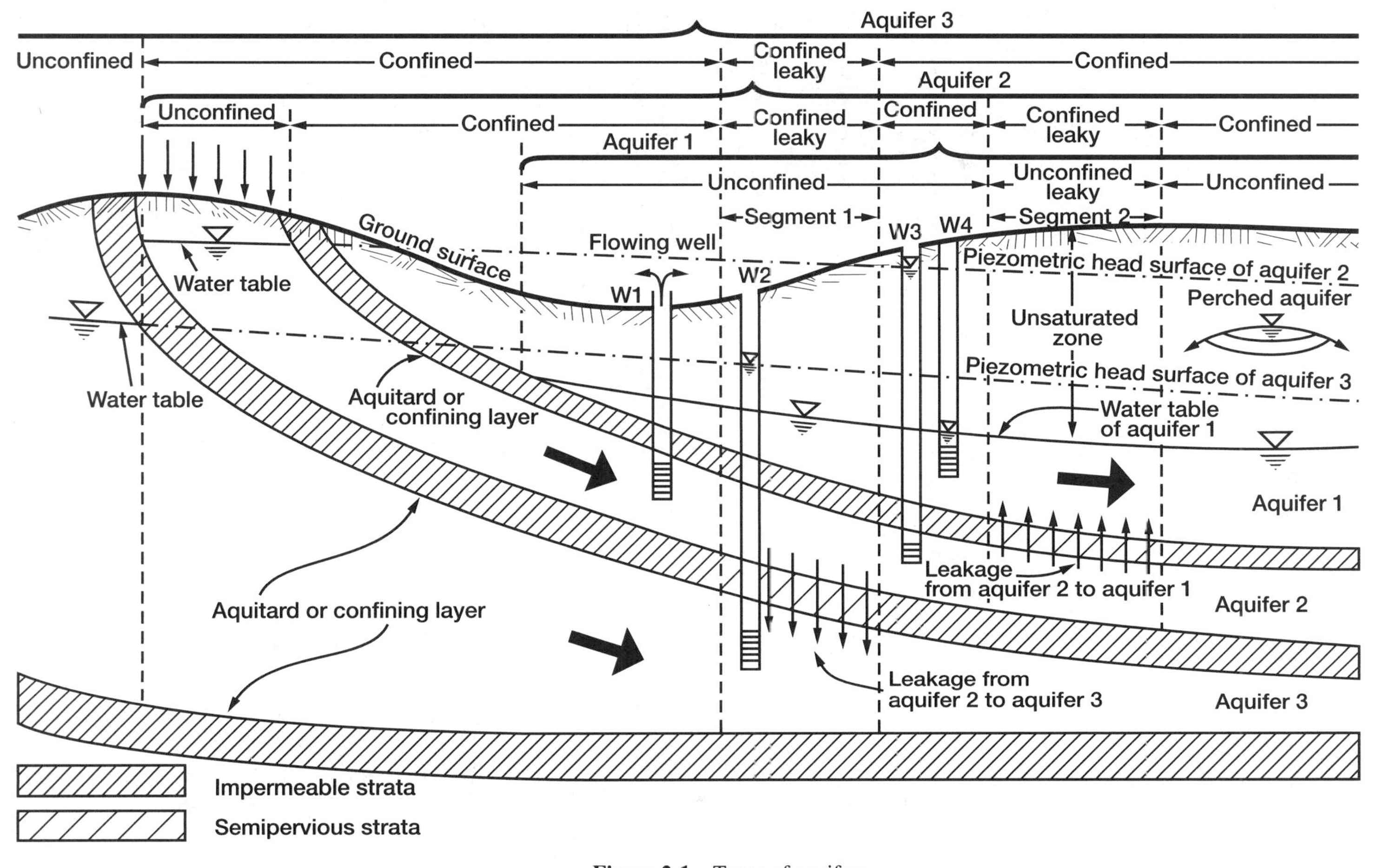

Figure 2-1 Types of aquifers.

A *capillary fringe* generally exists immediately above the water table in an unconfined aquifer. The thickness of a capillary fringe depends on the texture of the aquifer materials and ranges from a small fraction of a centimeter (cm) in coarse gravel materials to more than 1 meter (m) in silty materials. The lower part of a capillary fringe is completely saturated, as is the case for the aquifer portion below the water table, but the pressure of its water is negative, namely less than the atmospheric pressure. Under extraction conditions from a well, this water is normally not extracted through the well. Generally, the effects of the capillary fringe are neglected in most aquifer hydraulics studies.

2.2.2 Confined Aquifers

Confined aquifers, which are also called *artesian aquifers* or *pressure aquifers*, are bounded by impervious or semipervious strata. In Figure 2-1, Aquifer 2 and Aquifer 3 are under confined flow conditions, with the exception of their far left portions. Water level in a well penetrating a confined aquifer will be above the base of the upper confining layer or the upper boundary of the confined aquifer (Well 3 in Aquifer 2 and Well 2 in Aquifer 3). There may be cases where the piezometric surface of a confined aquifer is above the ground surface. For example, in Figure 2-1 the piezometric surface of Aquifer 2 is located above the ground surface; that is why Well W1 is a *flowing well*. The *piezometric surface* is an imaginary surface to which the water level would reach in tightly cased wells from a given point in a confined aquifer. The term *potentiometric surface* is often used instead of the term *piezometric surface*.

The term *piezometric surface* is obtained from the French (Meinzer, 1923). A *piezometric surface* may be very different from a *water table*, but they have the same general behavior in many respects.

The term *artesian aquifer* is also widely used for the term *confined aquifer*; the word *artesian* comes from the name of a place that had *flowing wells*. As mentioned above, this term is widely understood to refer only to water that will rise beyond the ground surface. According to Meinzer (1923), the word *artesian* was adopted by both Norton (1897) and Fuller (1906).

2.2.3 Leaky Aquifers

If an aquifer (confined or unconfined) loses or gains water through adjacent semipervious layers, it is called a *leaky aquifer*. For example, in Figure 2-1, Segment 2 of Aquifer 1 is a leaky unconfined aquifer because of the leakage from Aquifer 2 to Aquifer 1. As seen in Figure 2-1, Aquifer 2 is losing water to Aquifer 1 and Aquifer 3 through Segment 1 and Segment 2, respectively. Therefore, Segment 1 and Segment 2 of Aquifer 2 are under leaky confined aquifer conditions. Segment 1 also makes Aquifer 3 locally leaky along its length.

The term *nonleaky* is also used to describe the status of an aquifer. For example, Aquifer 1 is a nonleaky unconfined aquifer, with the exception of Segment 2. Likewise, Aquifer 2 is a nonleaky confined aquifer, with the exception of Segment 1 and Segment 2 as well as its far left unconfined portion.

2.2.4 Multiple Aquifers

Hydraulically, single aquifers seldom exist in nature. An aquifer is generally part of a system of aquifers. For example, Figure 2-1 displays a cross section of multiple aquifers that is composed of Aquifer 1, Aquifer 2, and Aquifer 3. A system of multiple aquifers consists of a series of aquifers separated from each other by confining layers that are relatively less permeable as compared with aquifers. The mechanics of flow through a system of multiple aquifers may become complex depending on the degree of the hydraulic communication between the individual aquifers. Quite often, when water is extracted from one aquifer, how much of the extraction comes from the aquifer in which the well is completed and how much comes through the underlain and overlain confining layers becomes important. For example, Well W2 in Figure 2-1 is completed in Aquifer 3, and under extraction conditions the leakage through Segment 1 will be increased because of the increased hydraulic gradicnts. In othcr words, cxtraction from Aquifer 3 will affect Aquifer 2 as well as Aquifer 1, depending on the distances of Segment 1 and Segment 2 to the pumping well.

De Glee (1930) was apparently the first to recognize the fact that aquifers are part of a more complex hydrogeologic system, and he developed a mathematical approach for steady-state leakage of water through confining layers into an aquifer that is being pumped. Jacob (1946) and Hantush and Jacob (1955) extended the mathematical approach to include the transient effects of leakage. Later, Hantush (1960b) modified this approach by including the storage effects of confining layers, Neuman and Witherspoon (1968, 1969a–c, 1972) extended the existing theories to multiple aquifer systems. All of the aforementioned approaches and their associated analytical models and data analysis methods will be presented in detail in other chapters.

Predictions made by multiple-aquifer-theory-based models are closer to reality than the single-aquifer models. For example, as will be seen in detail in Chapter 3, the Theis model (Theis, 1935) for fully penetrating wells in confined aquifers is based on the assumption that the upper and lower boundaries of the aquifer are impermeable. This means that the hydraulic interactions between its underlain and/or overlain less permeable formations (aquitards) are neglected. A similar situation is also valid for a single unconfined aquifer model for which the aquifer bottom is assumed to be impermeable. Under the framework of these models, there is a tendency to focus attention on the more permeable aquifers of an aquifer system in evaluating potential water supplies. However, sedimentary aquifer systems usually consist of a series of aquifers separated by less permeable confining layers, which may cause vertical hydraulic interactions from one aquifer to another. Because fine-grained sediments are generally much more compressible than underlain and/or overlain coarse-grained aquifer materials, they can potentially release large quantities of water from storage, resulting in a situation that the supply available to the aquifer increases. As a result, it can be concluded that single aquifer theories may underestimate the potential supply capacity of an aquifer (Neuman and Witherspoon, 1972).

2.3 HYDRAULIC AND HYDROGEOLOGIC CHARACTERISTICS OF AQUIFERS

Hydraulic head; drawdown; porosity; discharge velocity and ground-water velocity; hydraulic conductivity, permeability, and transmissivity; capillarity and capillary fringe are the key hydraulic and hydrogeological quantities used to characterize aquifers. These quantities are used for almost every aquifer problem. In the following sections, definitions as well as some key aspects of these quantities are presented.

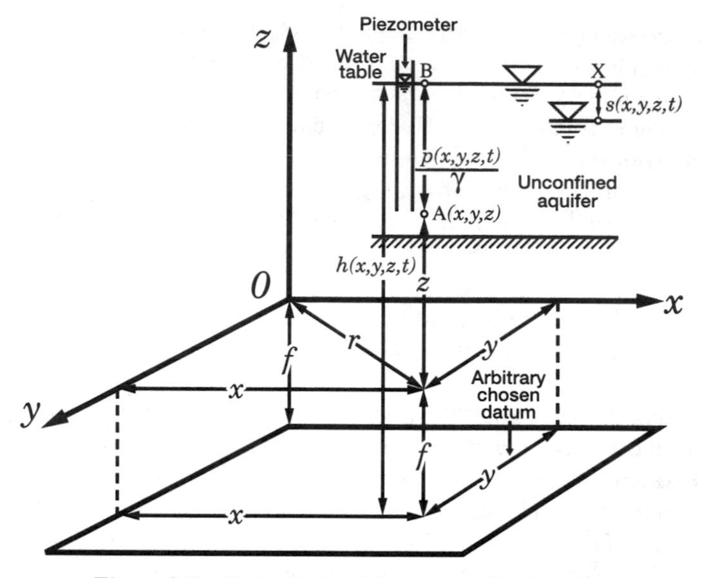

Figure 2-2 Hydraulic head in an unconfined aquifer.

2.3.1 Hydraulic Head and Drawdown

2.3.1.1 Hydraulic Head. *Hydraulic head*, which is also called *head* or *piezometric head*, is defined as the fluid pressure of formation water produced by the height above a given point. Geometrically, the hydraulic head at a point of an unconfined aquifer and a confined aquifer is shown in Figure 2-2 and Figure 2-3, respectively. The water level in a piezometer, which is a small tube that is open only at its bottom, in an unconfined aquifer

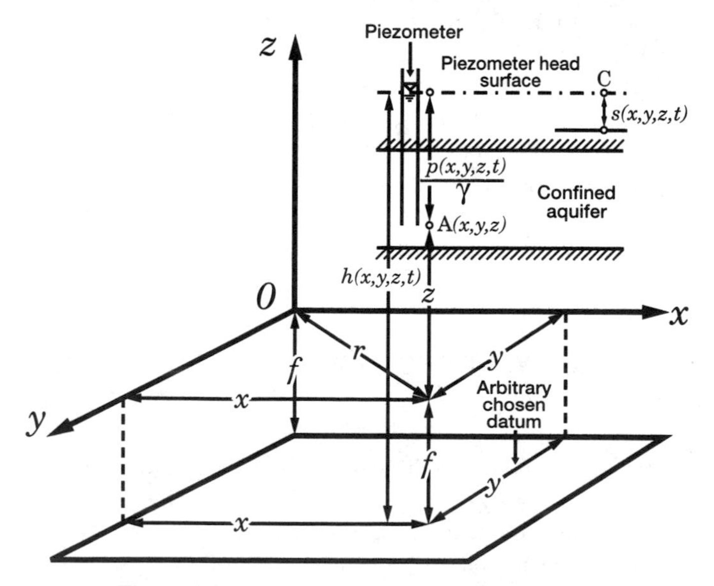

Figure 2-3 Hydraulic head in a confined aquifer.

is the same as the water-table elevation (see Figure 2-2), whereas for a confined aquifer case the water level in a piezometer is the same as the piezometric surface elevation of the aquifer at the same point (see Figure 2-3). As shown in Figures 2-2 and 2-3, the *hydraulic head*, $h(x, y, z, t)$ at point $A(x, y, z)$, consists of the *pressure head*, $p(x, y, z, t)/\gamma$, and *the elevation of the bottom of the piezometer*, z, above an arbitrarily chosen datum elevation. The pressure head, $p(x, y, z, t)/\gamma$, is the height of the column of water in the piezometer above its bottom or above point $A(x, y, z)$. As a result, the hydraulic head is defined as (Hantush, 1964)

$$h(x, y, z, t) = \frac{p(x, y, z, t)}{\gamma} + z + f \tag{2-1}$$

in which x, y, and z are the Cartesian coordinates, γ is the unit weight of water, f is the elevation of the x–y plane above an arbitrarily chosen datum elevation, and t is time observation since a reference time.

The dimension of hydraulic head is *length* $[L]$. The hydraulic head is generally expressed as "meters of water" or "feet of water."

2.3.1.2 Drawdown. Under extraction or injection conditions from a well the initial hydraulic head at a point in an aquifer changes. Under extraction conditions from a well, the vertical distance between the initial hydraulic head at a point in the aquifer and the lowered position of the hydraulic head for the same point is called *drawdown*. For an unconfined aquifer the water level at point C on the water table is dropped by a vertical distance $s(x, y, z, t)$, which is the drawdown at the same point (see Figure 2-2). For a confined aquifer case, the drawdown is defined with respect to the piezometric surface (see Figure 2-3).

The dimension of drawdown is *length* $[L]$. The drawdown is generally expressed in "meters of water" or "feet of water."

2.3.2 Porosity and Effective Porosity

Porosity of an aquifer, or porous medium, n, is defined as the percentage of void spaces:

$$n = \frac{\text{volume of voids of a sample}}{\text{total volume of the sample}} \tag{2-2}$$

The *effective porosity* is the portion of pore space in a saturated porous material in which water flow occurs. In other words, it is the volume of the interconnected voids of a sample. This definition is based on the fact that all the pore space of a porous medium filled with water is not open for water flow. Likewise, Eq. (2-2), the definition of *effective porosity*, n_e, is

$$n_e = \frac{\text{volume of the interconnected voids of a sample}}{\text{total volume of the sample}} \tag{2-3}$$

Obviously, effective porosity of a porous medium is smaller than its porosity.

Table 2-1, based on the original data of Morris and Johnson (1967), presents representative porosity (n) ranges as well as arithmetic mean (n_{av}) for various aquifer materials. In

TABLE 2-1 Porosity of Various Porous Materials

Group	Porous Material	Number of Analyses	Range of Porosity (n)	Arithmetic Mean Average Porosity (n_{av})
Igneous rocks	Weathered granite	8	0.34–0.57	0.45
	Weathered gabbro	4	0.42–0.45	0.43
	Basalt	94	0.03–0.35	0.17
Sedimentary materials	Sandstone	65	0.14–0.49	0.34
	Siltstone	7	0.21–0.41	0.35
	Sand (fine)	243	0.26–0.53	0.43
	Sand (coarse)	26	0.31–0.46	0.39
	Gravel (fine)	38	0.25–0.38	0.34
	Gravel (coarse)	15	0.24–0.36	0.28
	Silt	281	0.34–0.61	0.46
	Clay	74	0.35–0.57	0.42
	Limestone	74	0.07–0.56	0.30
Metamorphic rocks	Schist	18	0.04–0.49	0.39

Source: After Morris and Johnson (1967) and Mercer et al. (1982).

general, clays have smaller pores than sandy materials but may have much greater porosity values.

2.3.3 Discharge Velocity and Ground Water Velocity

The *discharge velocity* is defined as the quantity of fluid that passes through a unit of total area of the porous medium in a unit of time. To better define these velocities, let us consider a volume element in a porous medium as shown in Figure 2-4. The dimensions of the parallelepiped volume element are Δx, Δy, and Δz in the x, y, and z directions, respectively. Now consider a horizontal section at z elevation from the x–y plane. The total area and the area of the pores of the horizontal section are A and A_p, respectively. And let us define m as

$$m = \frac{\text{the area of pores}}{\text{the total area}} \tag{2-4}$$

Let Q_z be the vertical component of Q discharge passing the area A_p. Therefore,

$$Q_z = mAv \tag{2-5}$$

is the vertical component of the Q discharge vector. Here, mv and v are the *discharge velocity* and *ground-water velocity* (seepage velocity), respectively.

In general, the area of pores changes with the z coordinate. Therefore, Eq. (2-4) takes the form

$$m_z = \frac{A_p(z)}{A} \tag{2-6}$$

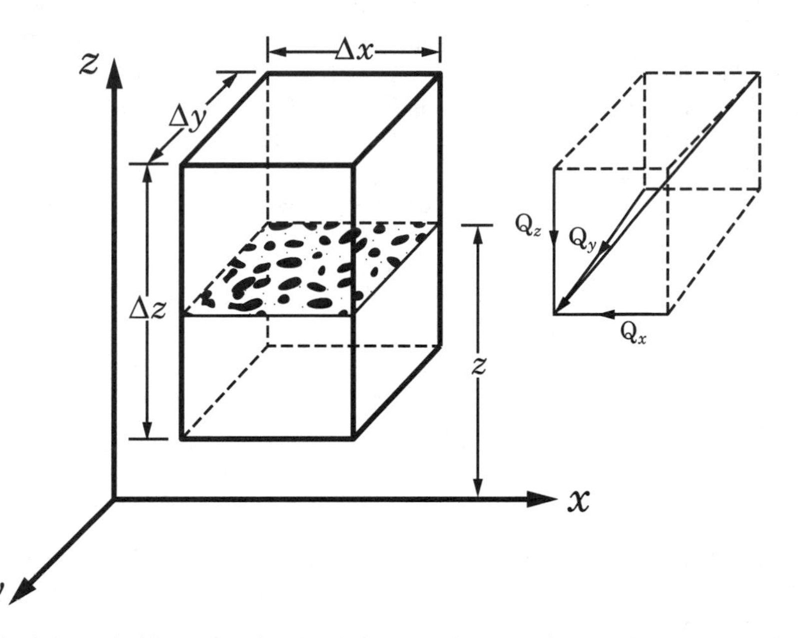

Figure 2-4 Schematic illustration for the derivation of the relationship between the discharge velocity and ground-water velocity.

The average value of m over the parallelepiped of height Δz gives

$$m = \frac{1}{\Delta z}\int_0^{\Delta z} m(z)\,dz = \frac{1}{(\Delta x)(\Delta y)(\Delta z)}\int_0^{\Delta z} A_p(z)\,dz \tag{2-7}$$

or

$$m = \frac{1}{V}\int_0^{\Delta z} A_p(z)\,dz \tag{2-8}$$

where V is the total volume and the integral is the volume of voids. As a result, the average value of m is the volume-based effective porosity (n_e). Therefore, Eq. (2-5) takes the form

$$Q_z = n_e A v = qA \tag{2-9}$$

and

$$q = n_e v \tag{2-10}$$

is the relationship between the *discharge velocity* and *ground-water velocity*. The term *seepage velocity* is also used for ground-water velocity.

2.3.4 Darcy's Law

Darcy's law was established in 1856 by Henry Darcy based on experimental study for flow of water through sands (Darcy, 1856). Here, the one-dimensional form of Darcy's law will be presented in light of the fundamental principles of hydromechanics. The concepts regarding the components of hydraulic head will be presented before Darcy's equation.

The schematic picture of the experimental apparatus shown in Figure 2-5 is equivalent to that used by Darcy in his original experiments, except that Darcy's apparatus was vertical. The pipe shown in Figure 2-5 is filled with a porous material, and water flow occurs in the arrow direction. Two piezometers are located at points 1 and 2 in the pipe. The geometric elevations of these two points are z_1 and z_2 with respect to a datum. During water flow, head losses occur between points 1 and 2. As a result, the hydraulic head at point 2, h_2, is less than the hydraulic head h_1 by the amount of Δh. The pressure heads at points 1 and 2 are p_1/γ and p_2/γ, respectively.

For an incompressible inviscid (nonviscous) fluid flowing in a pipe, horizontal, vertical, or inclined, with smooth walls, the steady-state Bernoulli equation is

$$\frac{p}{\rho g} + z + \frac{v^2}{2g} = \text{constant} = h \tag{2-11}$$

where ρ is the density of the fluid, g is the acceleration of gravity, p is the pressure, z is the geometric elevation from the datum, v is the ground-water (or seepage) velocity, and h is the total head. Also,

$$\frac{p}{\rho g} = \frac{p}{\gamma} = \text{piezometric head}$$

$$\frac{v^2}{2g} = \text{velocity head}$$

where γ is the unit weight of the fluid (water). The *Bernoulli equation*, Eq. (2-11), states that for all points of the pipe, the sum of the *piezometric head* (p/γ), *geometric elevation* (z), and *velocity head* $(v^2/2g)$ remains a constant value.

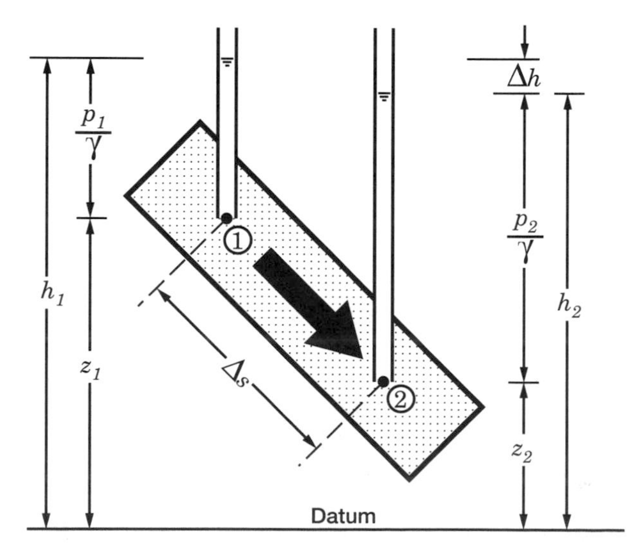

Figure 2-5 Schematic illustration of Darcy's experiment.

Now let us look closely at how the flow in a porous medium develops. As shown in Figure 2-5, consider two points, shown by 1 and 2, along the axis of the pipe, which are separated by a distance Δs from each other. Piezometers are inserted at these points, and the water levels in the piezometers are at h_1 and h_2 elevations from the datum. In ground-water flow, due to resistance against the flow within pores, the Bernoulli equation takes the following form for points 1 and 2:

$$\frac{p_1}{\gamma} + z_1 + \frac{v_1^2}{2g} = \frac{p_2}{\gamma} + z_2 + \frac{v_2^2}{2g} + \Delta h \tag{2-12}$$

where Δh is the head loss over the distance Δs.

The *hydraulic gradient*, I, or the *gradient of piezometric head*, is defined as

$$I = -\lim_{\Delta s \to 0} \frac{\Delta h}{\Delta s} = -\frac{dh}{ds} \tag{2-13}$$

In porous media flow, velocities are relatively low. As a result, the velocity heads in Eq. (1-12) may be neglected without appreciable error. Therefore, from Eq. (2-12) one can write

$$\Delta h = \left(\frac{p_1}{\gamma} + z_1\right) - \left(\frac{p_2}{\gamma} + z_2\right) = h_1 - h_2 \tag{2-14}$$

which means that the expression for the hydraulic head at any point in the flow domain is

$$h = \frac{p}{\gamma} + z \tag{2-15}$$

Equation (2-15) is one of the key expressions in ground-water hydraulics, whereas Eq. (2-11) is the hydraulic head expression for surface-water flows such as rivers, streams, canals, and so on. In a typical river or stream the average velocity is around 1 m/sec. However, in most aquifers, even 1 m/day is a very high velocity. In some tight formations, the velocity of water is less than 1 m/year. Therefore, the velocity heads in porous media flow are extremely low; this is shown in the following example.

Example 2-1: The average velocity in a river is around 1.5 m/sec. Based on hydrogeologic investigations, the average ground-water velocity in a nearby aquifer is around 1 m/day. Determine the velocity heads for each flow case and make a comparative evaluation.

Solution: For river flow case, the velocity head in Eq. (2-11) has the value

$$h_{swv} = \frac{v^2}{2g} = \frac{(1.5 \text{ m/s})^2}{2(9.81 \text{ m/s}^2)} = 0.11 \text{ m}$$

The ground-water velocity in the aquifer is

$$v = 1 \text{ m/day} = 1.16 \times 10^{-5} \text{ m/s}$$

Therefore, the velocity head for ground-water flow is

$$h_{gwv} = \frac{v^2}{2g} = \frac{(1.16 \times 10^{-5}\ \text{m/s})^2}{2(9.81\ \text{m/s}^2)} = 6.83 \times 10^{-12}\ \text{m}$$

And the ratio between these velocities is

$$\frac{h_{gwv}}{h_{swv}} = \frac{6.83 \times 10^{-12}\ \text{m}}{0.11\ \text{m}} = 6.21 \times 10^{-11}$$

The above values clearly show that the ground-water velocity head is extremely small as compared with the other hydraulic head components. Therefore, the velocity head term in Eq. (2-11) can safely be neglected for flow in porous media.

2.3.4.1 Darcy Experiment. To generate the flow equation in porous media, a relationship is needed between the ground-water velocity and the hydraulic gradient. In 1856, Henry Darcy found experimentally that the discharge, Q, is proportional to the difference of water levels, Δh, and inversely proportional to Δs (see Figure 2-5):

$$Q \propto \Delta h \qquad Q \propto \frac{1}{\Delta s} \tag{2-16}$$

Introducing a proportionality constant, K, leads to the equation

$$Q = -KA\frac{h_2 - h_1}{\Delta h} = -KA\frac{\Delta h}{\Delta s} \tag{2-17}$$

or

$$Q = -KA\frac{dh}{ds} \tag{2-18}$$

where A is the cross-sectional area of the column. From Eq. (2-18) the discharge velocity can be expressed as

$$q = \frac{Q}{A} = -K\frac{dh}{ds} = -KI \tag{2-19}$$

where q is the *discharge velocity* or *Darcy velocity* (*Darcy flux* and *specific discharge* terms are also used), K is the *hydraulic conductivity*, and $I = dh/ds$ is the *hydraulic gradient*. Equation (2-19) is commonly called *Darcy's law*, which states that there is a linear relationship between the hydraulic gradient, I, and the discharge velocity, q. The negative sign in Eq. (2-19) indicates that the flow of water is in the direction of decreasing head. It must be emphasized that Darcy's law does not describe the state within individual pores. Darcy's law represents the statistical macroscopic equivalent of the Navier–Stokes equations of motion for viscous flow of water in porous media. Equation (2-19) is the simple form of Darcy's law, and its two- and three-dimensional forms are presented in Section 2.5.

The head loss, Δh, is independent of the inclination of the column. Figure 2-6 presents two schemes of experiments to check Darcy's law. In Figure 2-6a, the column is horizontal, and the Δh head difference occurs over the Δs horizontal distance in the porous medium.

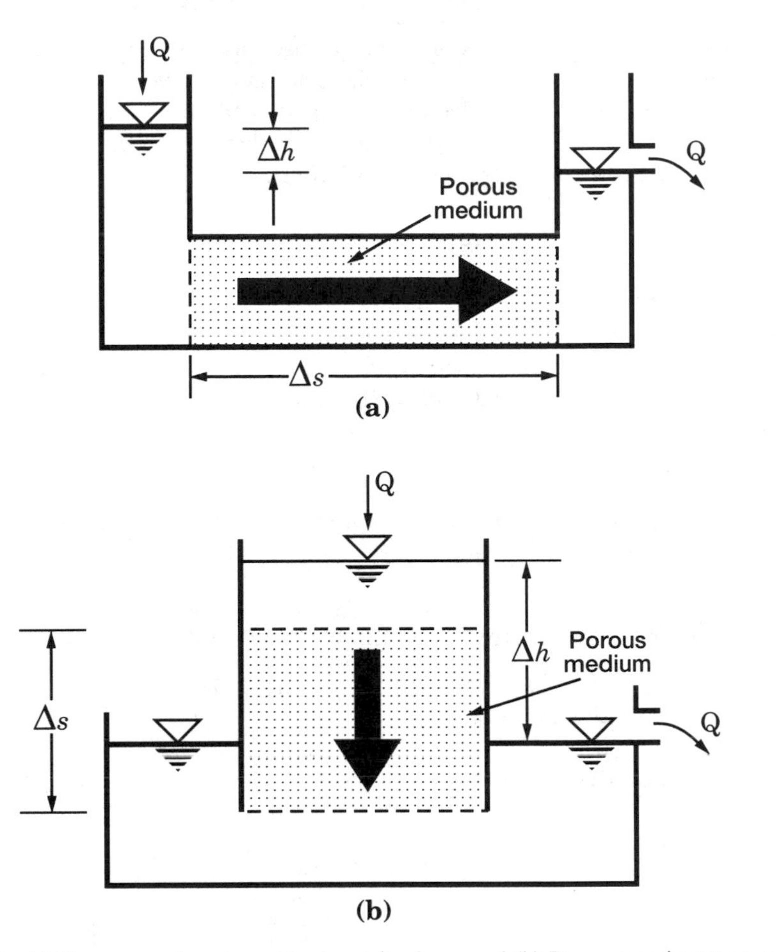

Figure 2-6 (a) Darcy experiment on a horizontal column and (b) Darcy experiment on a vertical column.

In Figure 2-6b, the column is vertical and the Δh head difference occurs over the vertical Δs distance in the porous medium. Different forms of Darcy's law as given by Eqs. (2-17), (2-18), and (2-19) are also valid for these experimental cases.

2.3.4.2 Validity of Darcy's Law. According to Eq. (2-19), Darcy's law states that the discharge velocity is proportional to the first power of the hydraulic gradient. Because the velocity in laminar flow is also proportional to the first power of the hydraulic gradient, it can be inferred that the flow in a porous medium must be laminar for Darcy's law to be valid. In hydrodynamics, the usual criterion to determine whether the flow is laminar or turbulent is the Reynolds number. Experiments have shown that Darcy's equation has definite limitations. For homogeneous aquifer materials, Darcy's law is valid only under the condition of the inequality for the Reynolds number:

$$N_R = \frac{\rho q d}{\mu} = \frac{q d}{\nu} \leq N_L \tag{2-20}$$

where ρ is the density of water, μ is the dynamic viscosity, q is the discharge velocity, d is the average diameter of soil particles, ν is the kinematic viscosity of water, and N_L is a number for which various authors give different values between 3 and 10 (Polubarinova–Kochina, 1962; Bear, 1979). With $\nu = 0.018\ \text{cm}^2/\text{s}$, the kinematic viscosity of water under average temperature conditions for water, Eq. (2-20) takes the form

$$qd \leq 0.054 - 0.180 \tag{2-21}$$

where q is expressed in centimeters per second and d is expressed in centimeters. For large grained materials or relatively fast water flow in aquifers, other relationships between q and I have been established based on experimental results. For most turbulent flows in porous media a polynomial of the second degree has been considered (Jaeger, 1956):

$$I = Aq + Bq^2 \tag{2-22}$$

If turbulence is fully developed, the term Aq in Eq. (2-22) is negligible. The constants A and B are determined experimentally.

2.3.5 Hydraulic Conductivity, Transmissivity, and Permeability

These quantities are widely used in characterizing aquifers. Their definitions with the units used in practice are presented below.

Hydraulic Conductivity: Definition of *hydraulic conductivity* comes from Darcy's law. Solving Eq. (2-19) for K gives

$$K = \frac{q}{I} = -\frac{Q}{A\left(\dfrac{dh}{ds}\right)} \tag{2-23}$$

It is apparent from Eq. (2-23) that K has the dimensions of "length/time" (L/T) or velocity. This coefficient is called *hydraulic conductivity* or *saturated hydraulic conductivity*, and it is a function of properties of both the porous medium and the fluid flowing through it. The saturated hydraulic conductivity of a given aquifer material can be determined in the laboratory by one of the schematical devices shown in Figure 2-6. Tables and nomographs regarding saturated hydraulic conductivity values for different formations have been presented in a number of publications especially during the past century. It is important to note that these tables mostly use the term "hydraulic conductivity" rather than making it specific for horizontal and vertical hydraulic conductivity values. By default, these values should be viewed as tables for "horizontal hydraulic conductivity" unless otherwise stated. It is a well-known fact that vertical hydraulic conductivities may be one to three orders of magnitude less than horizontal hydraulic conductivities, depending on the type of formation. For practical usage, two different forms of tables are presented that are commonly used in some well-known literature. The first one presents average ranges of the saturated hydraulic conductivity for some typically encountered soils (Table 2-2). From another perspective, Table 2-3 (Davis, 1969; Freeze and Cherry, 1979) presents the range of values of hydraulic conductivity and intrinsic permeability (or permeability) in a nomographic format. Different units are used in practice for hydraulic conductivity, and

TABLE 2-2 Hydraulic Conducitivity of Porous Materials

Group	Porous Material	Number of Analysis	Range of Hydraulic Conductivity, K (cm/s)	Arithmetic Mean Average Hydraulic Conductivity, K_{av} (cm/s)
Igneous rocks	Weathered granite	7	$(3.3–52) \times 10^{-4}$	1.65×10^{-3}
	Weathered gabbro	4	$(0.5–3.8) \times 10^{-4}$	1.98×10^{-4}
	Basalt	93	$(0.2–4250) \times 10^{-6}$	9.45×10^{-6}
Sedimentary materials	Sandstone (fine)	20	$(0.5–2270) \times 10^{-6}$	3.31×10^{-4}
	Siltstone	8	$(0.1–142) \times 10^{-8}$	1.9×10^{-7}
	Sand (fine)	159	$(0.2–189) \times 10^{-4}$	2.28×10^{-3}
	Sand (medium)	255	$(0.9–567) \times 10^{-4}$	1.65×10^{-2}
	Sand (coarse)	158	$(0.9–6610) \times 10^{-4}$	5.20×10^{-2}
	Gravel	40	$(0.3–31.2) \times 10^{-1}$	4.03×10^{-1}
	Silt	39	$(0.09–7090) \times 10^{-7}$	2.83×10^{-5}
	Clay	19	$(0.1–47) \times 10^{-8}$	9.00×10^{-8}
Metamorphic rocks	Schist	17	$(0.002–1130) \times 10^{-6}$	1.90×10^{-4}

Source: After Morris and Johnson (1967) and Mercer et al. (1982).

their conversion factors are included in Table 2-4. In Table 2-5, units and conversion factors for transmissivity are included.

In the geotechnical engineering literature the term *coefficient of permeability* is mostly used instead of *hydraulic conductivity*. Based on the aforementioned expressions, hydraulic conductivity can be defined as the rate of discharge of water under laminar flow conditions through a unit cross-sectional area of a porous medium under a unit hydraulic gradient and standard temperature conditions (usually 20°C).

Transmissivity: For a confined aquifer of thickness b, the *transmissivity* (or *transmissibility*) T is defined as

$$T = Kb \tag{2-24}$$

The transmissivity concept can also be used for unconfined aquifers. In Eq. (2-24), b is now the saturated thickness of the aquifer or the height of the water table above the underlying aquitard or impermeable boundary. Different units are used in practice for transmissivity and its conversion factors are included in Table 2-5.

Permeability: As mentioned above, the term *coefficient of permeability* is used instead of *hydraulic conductivity*, especially in the geotechnical engineering literature. This is inappropriate because K depends not only on the permeability of the porous medium, but also on the properties of the fluid flowing through the porous medium. The fluid properties affecting the flow are its *dynamic viscosity* (μ) and *specific weight* (γ). The dynamic viscosity μ of the fluid is a measure of the resistive force within the pores of the porous medium. The specific weight γ may be considered as the driving force exerted by gravity. On the other hand, the size of a typical pore is proportional to the mean grain diameter d.

TABLE 2-3 Hydraulic Conductivity and Permeability Range of Values in Different Units

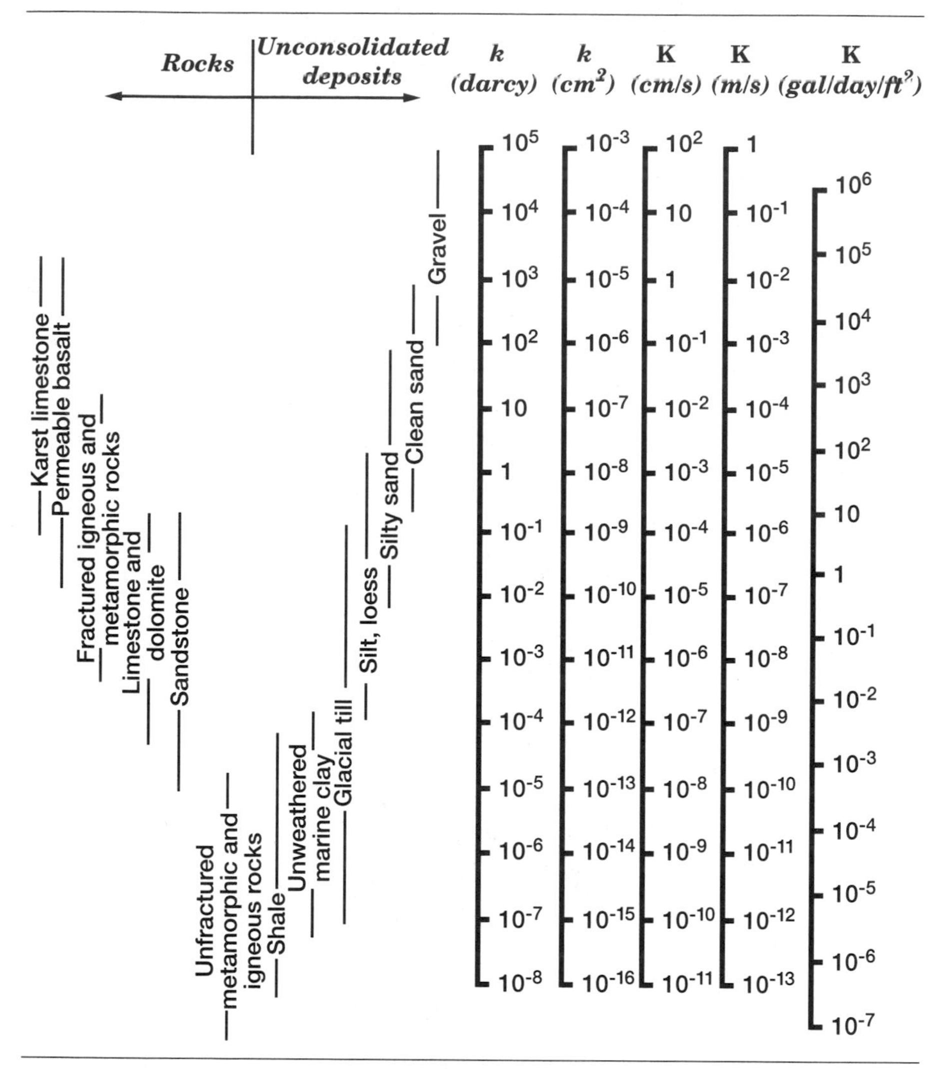

Source: After Davis (1969) and Freeze and Cherry (1979).

Using the method of dimensional analysis, the relationship between K, μ, γ, and d is determined as (Jacob, 1950)

$$K = \frac{Cd^2\gamma}{\mu} = k\frac{\gamma}{\mu} = k\frac{\rho g}{\mu} = k\frac{g}{\nu} \tag{2-25}$$

where

$$k = Cd^2 \tag{2-26}$$

TABLE 2-4 Conversions of Hydraulic Conducitivity (K) and Intrinsic Permeability (k)

	Hydraulic Conductivity (K) Units								Intrinsic Permeability (k) Units		
Unit	cm/s	m/s	m/day	ft/s	ft/day	ft/yr	USgpd/ft^2	UKgpd/ft^2	darcy	cm^2	ft^2
cm/s	1	1.00E−2	8.64E2	3.28E−2	2.83E3	1.03E6	2.12E4	1.77E4	1.16E3	1.15E−5	1.24E−8
m/s	1.00E2	1	8.64E4	3.28	2.83E5	1.03E8	2.12E6	1.77E6	1.16E5	1.15E−3	1.24E−6
m/day	1.16E−3	1.16E−5	1	3.80E−5	3.28	1.20E3	2.45E1	2.04E1	1.35	1.33E−8	1.43E−11
ft/s	3.05E1	0.305	2.63E4	1	8.64E4	3.15E7	6.46E5	5.38E5	3.55E4	3.50E−4	3.77E−7
ft/day	3.53E−4	3.53E−6	0.305	1.16E−5	1	3.65E2	7.48	6.23	0.411	4.06E−9	4.36E−12
ft/yr	9.66E−7	9.66E−9	8.35E−4	3.17E−8	2.74E−3	1	2.05E−2	1.71E−2	1.13E−3	1.11E−11	1.20E−14
USgpd/ft^2	4.72E−5	4.72E−7	4.07E−2	1.55E−6	0.134	4.88E1	1	0.833	5.49E−2	5.42E−10	5.83E−13
UKgpd/ft^2	5.66E−5	5.66E−7	4.89E−2	1.86E−6	0.161	5.86E1	1.20	1	6.60E−2	6.51E−10	7.01E−13
darcy	8.58E−4	8.58E−6	7.42E−1	2.82E−5	2.43	8.88E2	1.82E1	1.52E1	1	9.87E−9	1.06E−11
cm^2	8.70E4	8.70E2	7.51E7	2.85E3	2.47E8	9.00E10	1.84E9	1.54E9	1.01E8	1	1.08E−3
ft^2	8.08E7	8.08E5	6.98E10	2.65E6	2.29E11	8.36E13	1.71E12	1.43E12	9.41E10	9.29E2	1

Notes: 1. The relation between units K and k is temperature-dependent: these factors are for 15.5°C (60°F).

2. Enter table at the left with the given unit; move right to the column of the unit to be derived; read the conversion factor as a multiplier.

3. Example: To convert 2.1 ft/day (hydraulic conductivity) to cm/s: $2.1 \times 3.53E{-4} = 7.4E{-4}$ cm/s.

4. Conversion factors are given in FORTRAN/BASIC notation (or scientific notation); thus $3.53E{-4} = 3.53 \times 10^{-4}$.

Source: After Ground Water Publishing Company (1996).

TABLE 2-5 Conversions of Transmissivity (T)

	Transmissivity (T) Units						
Unit	m^2/s	m^2/min	m^2/day	ft^2/s	ft^2/day	USgpd/ft	UKgpd/ft
m^2/s	1	6.00E1	8.64E4	1.08E1	9.30E5	6.96E6	5.79E6
m^2/min	1.67E−2	1	1.44E3	1.79E−1	1.55E4	1.16E5	9.65E4
m^2/day	1.16E−5	6.94E−4	1	1.25E−4	1.08E1	8.05E1	6.70E1
ft^2/s	9.29E−2	5.57	8.03E3	1	8.64E4	6.46E5	5.38E5
ft^2/day	1.08E−6	6.45E−5	9.29E−2	1.16E−5	1	7.48	6.23
USgpd/ft	1.44E−7	8.62E−6	1.24E−2	1.55E−6	1.34E−1	1	0.833
UKgpd/ft	1.73E−7	1.04E−5	1.49E−2	1.86E−6	1.61E−1	1.20	1

Notes: 1. Enter table at the left with the given unit; move right to the column of the unit to be derived; read conversion factor as a multiplier.
2. Example: to convert 63.0 ft^2/day (transmissivity) to m^2/day: 63.0 × 9.29E−2 = 585.27E−2 m^2/day.
3. Conversion factors are given in FORTRAN/BASIC notation (or scientific notation); thus 9.29E−2 = 9.20 × 10^{-2}.

Source: After Ground Water Publishing Company (1996).

and

$$\nu = \frac{\mu}{\rho} \tag{2-27}$$

is the kinematic viscosity whose dimensions are $[L^2/T]$ and k depends on the character of thc porous mcdium only. The shape factor C depends on the porosity of the medium, the grain-size distribution, the shape of the grains, and their orientation and arrangement. The dimension of k is, obviously, $[L^2]$. Since k depends on only the character of the porous medium, it is called the *coefficient of permeability* or just *permeability*. The term *intrinsic permeability* is also used by some authors. Different units are used in practice for permeability, and their conversion factors are included in Table 2-4.

In Eq. (2-25), both γ and μ vary with temperature (see Tables 2-6a and 2-6b). Therefore, the ratio γ/μ depends on the temperature as well. However, in many cases for aquifer problems, the temperature does not vary significantly, and the hydraulic conductivity (K) may then be considered as constant in the associated equations.

Units of permeability (k) can be in square centimeters (cm^2) or square feet (ft^2). In the petroleum industry the unit of permeability is *darcy*, according to the relationship

$$k = \frac{\dfrac{\mu Q}{A}}{\dfrac{\Delta p}{\Delta s}} \tag{2-28}$$

which is obtained from Eqs. (2-17) and (2-25) by replacing $h = p/\gamma$ where p is the pressure. Based on Eq. (2-28), one darcy is defined by taking $\mu = 1$ centipoise $= 0.01$ dyne-s/cm^2, $Q = 1\ cm^3$/s, $A = 1\ cm^2$, $\Delta p = 1$ atmosphere $= 1.0132 \times 10^6$ dynes/cm^2, and $\Delta s = 1$ cm. Therefore, substitution of the above values into Eq. (2-28) gives the definition of darcy:

TABLE 2-6a Specific Weight, Density, Dynamic Viscosity, and Kinematic Viscosity of Water as Function of Temperature in S.I. Units

Temperature (°C)	Specific Weight, γ (kN/m^3)	Density, ρ (kg/m^3)	Dynamic Viscosity, $\mu \times 10^3$ (Pa · s)	Kinematic Viscosity, $\nu \times 10^6$ (m^2/s)
0	9.805	999.8	1.781	1.785
5	9.807	1000.0	1.518	1.518
10	9.804	999.7	1.307	1.306
15	9.798	999.1	1.139	1.139
20	9.789	998.2	1.002	1.003
25	9.777	997.0	0.890	0.893
30	9.764	995.7	0.798	0.800
40	9.730	992.2	0.653	0.658
50	9.689	988.0	0.547	0.553
60	9.642	983.2	0.466	0.474
70	9.589	977.8	0.404	0.413
80	9.530	971.8	0.354	0.364
90	9.466	965.3	0.315	0.326
100	9.399	958.4	0.282	0.294

Example: If $\mu \times 10^3 = 1.781$ Pa · s, then $\mu = 1.781 \times 10^{-3}$ Pa · s.

Source: Adapted from Vennard and Street (1982).

TABLE 2-6b Specific Weight, Density, Dynamic Viscosity, and Kinematic Viscosity of Water as Function of Temperature in the English FSS Units

Temperature (°F)	Specific Weight, γ (lb/ft^3)	Density, ρ (slug/ft^3)	Dynamic Viscosity, $\mu \times 10^5$ (lb · s/ft^2)	Kinematic Viscosity, $\nu \times 10^5$ (ft^2/s)
32	62.42	1.940	3.746	1.931
40	62.43	1.940	3.229	1.664
50	62.41	1.940	2.735	1.410
60	62.37	1.938	2.359	1.217
70	62.30	1.936	2.050	1.059
80	62.22	1.934	1.799	0.930
90	62.11	1.931	1.595	0.826
100	62.00	1.927	1.424	0.739
110	61.86	1.923	1.284	0.667
120	61.71	1.918	1.168	0.609
130	61.55	1.913	1.069	0.558
140	61.38	1.908	0.981	0.514
150	61.20	1.902	0.905	0.476
160	61.00	1.896	0.838	0.442
170	60.80	1.890	0.780	0.413
180	60.58	1.883	0.726	0.385
190	60.36	1.876	0.678	0.362
200	60.12	1.868	0.637	0.341
212	59.83	1.860	0.593	0.319

Example: If $\mu \times 10^5 = 3.746$ lb · s/ft^2, then $\mu = 3.746 \times 10^{-5}$ lb · s/ft^2.

Source: Adapted from Vennard and Street (1982).

$$1 \text{ darcy} = \frac{\dfrac{(1 \text{ centipoise})(1 \text{ cm}^3/\text{s})}{(1 \text{ cm}^2)}}{\dfrac{(1 \text{ atmosphere})}{(1 \text{ cm})}} = 9.87 \times 10^{-9} \text{ cm}^2$$

Example 2-2: The kinematic viscosity and permeability of an aquifer are given as $\nu = 1.12 \text{ cm}^2/\text{s}$ and $k = 7.5$ darcys. Determine the value of the hydraulic conductivity of the aquifer.

Solution: From Eq. (2-25) we have

$$K = k\frac{g}{\nu} = \left(7.5 \text{ darcy} \times 9.87 \times 10^{-9} \frac{\text{cm}^2}{\text{darcy}}\right) \frac{(981 \text{ cm/s}^2)}{(1.12 \text{ cm}^2/\text{s})}$$

$$K = 6.48 \times 10^{-5} \text{ cm/s}$$

2.3.6 Estimation of Hydraulic Conductivity from Grain Size

It is evident from Eq. (2-26) that hydraulic conductivity is related to the grain size of granular porous media. Based on this relationship, during the past century, numerous investigators have studied the correlation between hydraulic conductivity or permeability and grain-size distribution of saturated materials to generate practical formulas for the estimation of these quantities. These formulas are useful especially during the initial stages of many aquifer studies, such as designing aquifer pump tests and estimation of some other quantities, if measured hydraulic conductivities under field conditions are not available. However, grain-size distribution data can be available in a relatively cheaper and faster manner. As mentioned in Section 2.5, it is important to note that the term "hydraulic conductivity," rather than making it specific for "horizontal hydraulic conductivity" and "vertical hydraulic conductivity," is used in the literature. And by default, the values determined from grain-size distribution should be viewed as "horizontal hydraulic conductivity" unless otherwise stated. In the following sections, some of the well-known equations as well as the new approaches developed during the past decade are presented, with examples.

2.3.6.1 Hazen Equation. Hazen's approximation (Hazen, 1892), which is a simple relationship, is based on Eq. (2-26). In Eq. (2-26), C is dimensionless and, therefore, the permeability k is in dimensions of "length squared" and d is "length" parameter, making the equation dimensionally consistent. However, range of values C and definition of d vary from author to author. For example, Bear (1979, p. 67) states that C is a coefficient in the range between 45 for clayey sand and 140 for pure sand (often the value of $C = 100$ is used as an average), and d is the effective grain diameter, d_{10}, which is defined as 10% of soil particles are finer and 90% are coarser. Todd (1980, p. 71) defines d as a "characteristic grain diameter" without mentioning sieve-analysis-based value. Freeze and Cherry (1979) give the following empirical relation due to Hazen (1892) for hydraulic conductivity estimates:

$$K = Ad_{10}^2 \tag{2-29}$$

where the units of K and d_{10} are cm/s and mm, respectively, and $A = 1.0$.

2.3.6.2 Shepherd Equations. Shepherd (1989) performed statistical power regression analyses on 19 sets of published data on hydraulic conductivity versus grain size using the following Eq. (2-26)-type simple power equation:

$$K = ad^b \tag{2-30}$$

where K is hydraulic conductivity, d is grain diameter, and a and b are some parameters. According to the results of Shepherd, values of the coefficient a ranged from 1014 to 208,808 gpd/ft^2 (gallons per day per foot square) (4.79×10^{-2} to 9.86 cm/s), but only two values were over 25,000 gpd/ft^2 (1.18 cm/s). The exponent b (dimensionless) ranged from 1.11 to 2.05, with an average of 1.72. The results of Shepherd are presented in Figure 2-7 for different sets of materials with the regression-analyses-derived equations. As can be seen

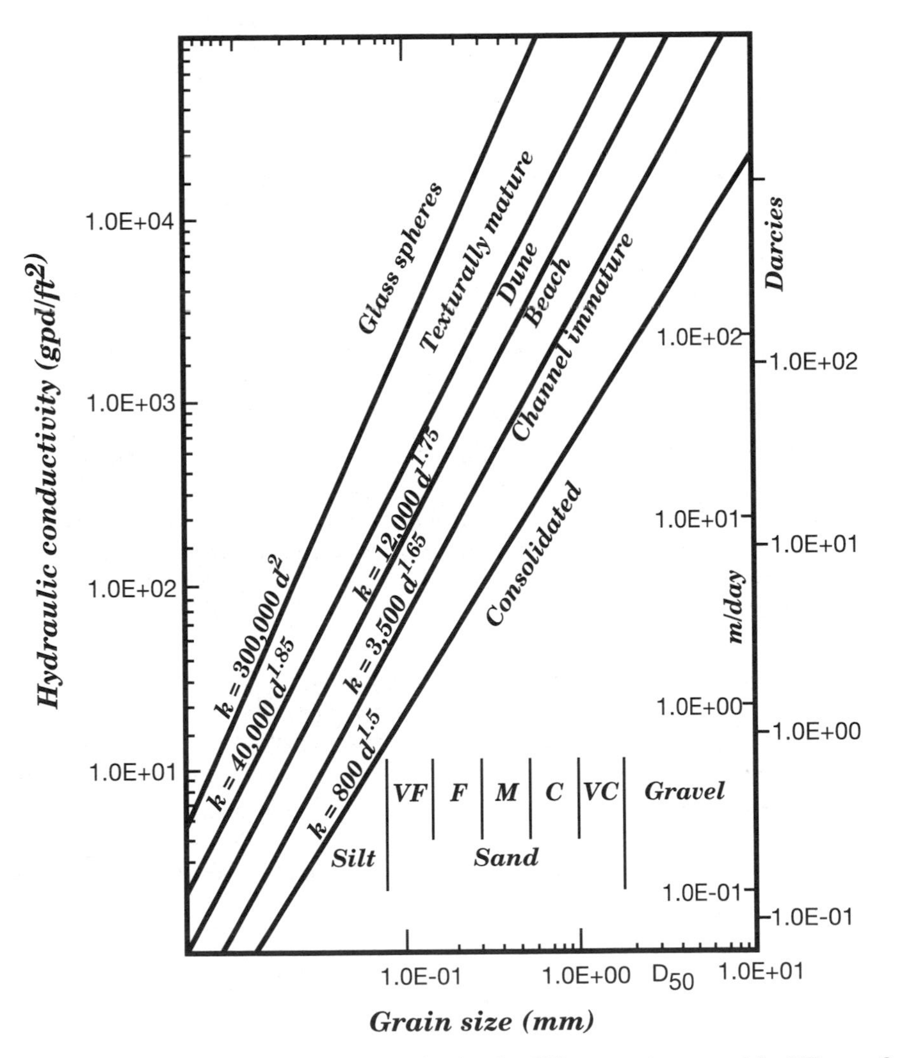

Figure 2-7 Hydraulic conductivity versus grain size for different porous materials: VF, very fine; F, fine; M, medium; C, coarse; VC, very coarse. (After Shepherd, 1989.)

from Figure 2-7, (1) higher values of coefficient a correspond to more texturally mature samples, (2) the lowest values of exponent b correspond to texturally immature sediments, and (3) all the lines representing natural sediments are within a relatively narrow field.

2.3.6.3 Kozeny–Carman Equation. One of the most widely accepted derivations of permeability as a function of the characteristics of the medium was proposed by Kozeny (1927) and later modified by Carman (1937, 1956). As a result, in the literature it is called the *Kozeny–Carman equation*. The *Kozeny–Carman equation* (Bear, 1972) is

$$K = \left(\frac{\rho g}{\mu}\right)\left[\frac{n^3}{(1-n)^2}\right]\left(\frac{d_m^2}{180}\right) \tag{2-31}$$

where K is hydraulic conductivity, ρ is the density of water, g is the acceleration of gravity, μ is the dynamic viscosity, n is porosity, and d_m is a representative grain size. Equation (2-30) is dimensionally correct and suitable for application with any consistent set of units.

2.3.6.4 Fair and Hatch Equation. Based on dimensional considerations and experimental verification, Fair and Hatch (1933) developed the following equation for the estimation of hydraulic conductivity (Bear, 1979):

$$K = \left(\frac{\rho g}{\mu}\right)\left[\frac{n^3}{(1-n)^2}\right]\left[\frac{1}{\beta\left(\frac{\alpha}{100}\sum_{m=0}^{N}\frac{P_m}{d_m}\right)^2}\right] \tag{2-32}$$

where K is hydraulic conductivity; ρ is the density of water; g is the acceleration of gravity; μ is the dynamic viscosity; n is porosity; β is a packing factor, found experimentally to be about 5; α is a sand shape factor, varying from 6.0 for spherical grains to 7.7 for angular grains; P_m is percentage of sand held between adjacent sieves; d_m is the geometric mean diameter of the adjacent sieves; m is a dummy variable; and N is the number of the percentages held between adjacent sieves.

2.3.6.5 Alyamani and Şen Equation. The equations presented above, which relate the hydraulic conductivity to grain-size distributions, are based on a single parameter, such as the *effective grain diameter* for the Hazen equation, *representative grain diameter* for the Kozeny–Carman equation, and the *geometric mean diameter* for the Fair and Hatch equation. In the equations presented by Shepherd, the *grain diameter* is a variable. As an alternative, Alyamani and Şen (1993) proposed the following equation based on analysis of 32 samples from Saudi Arabia and Australia that incorporates the initial slope and the intercept of the grain-size distribution curve:

$$K = 1300[I_0 + 0.025(d_{50} - d_{10})]^2 \tag{2-33}$$

where K (units m/day) is hydraulic conductivity and I_0 (unit mm) is the x intercept of the slope of the line formed by d_{50} and d_{10} of the grain-size distribution curve. Here, d_{50} is the effective grain size where 50% of particles are finer than d (mm). The linear x and y

coordinates of the grain-size distribution curve are the "grain size" in mm and "percent finer," respectively (see Figure 2-8). Some of the key aspects of Eq. (2-33) are presented below (Alyamani and Şen, 1993):

Intercept (I_0). The intercept occurs where the observed straight line crosses the horizontal axis. The value of the x intercept is expected to be very close to zero or to the effective grain diameter. Physically, this means that there is no passing material from the set of sieves. The higher the x-intercept values, the higher the hydraulic conductivities.

Properties of Eq. (2-33). (1) In general, I_0 is very close to the value of d_{10}. (2) From Eq. (2-33), it can be said that the hydraulic conductivity, K, is proportional to d_{10}, which is the base of the Hazen equation as given by Eq. (2-29). In other words, the Hazen equation, Eq. (2-29), is a special case of the Alyamani and Şen equation, Eq. (2-33). (3) The appearance of difference ($d_{50} - d_{10}$) in Eq. (2-33) implies that the hydraulic conductivity is proportional to the dispersion of grain size.

2.3.6.6 Evaluation of the Equations. Sperry and Peirce (1995) performed an evaluation by comparing the measured K values of different porous materials with those determined from their own equations as well as the equations of Hazen, Kozeny–Carman, and Alyamani and Şen. Sperry and Peirce (1995) reached the following conclusions for the tested materials: (1) Overall, the Hazen equation provides the best estimate of the hydraulic conductivity of the media studied, except for irregularly shaped particles. The

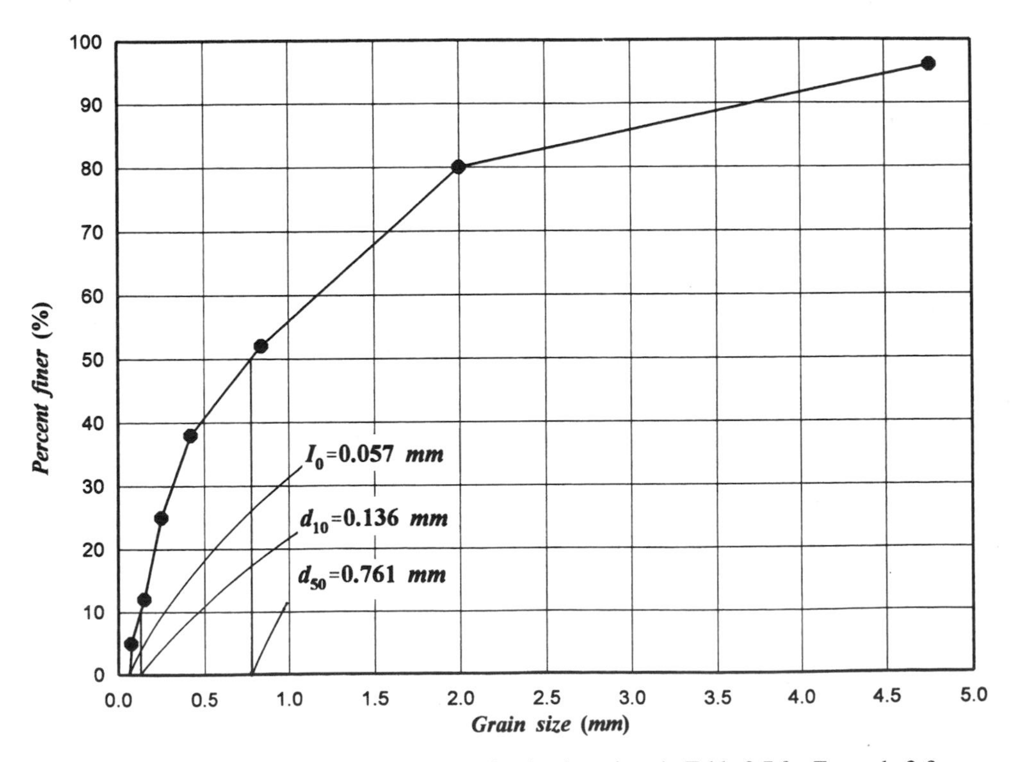

Figure 2-8 Grain-size distribution curve for the data given in Table 2-7 for Example 2-3.

values determined from the Hazen model are within a factor or two of the experimental values, except for irregularly shaped particles for which the values are within 330% of the experimental values. This reflects the fact that the Hazen equation is not convenient for irregularly shaped particles. (2) The Kozeny–Carman equation estimates are 73% to 83% lower than the measured hydraulic conductivity for the filter pack sands. (3) The Alyamani and Şen equation estimates are 30% to 36% greater for the same media.

Example 2-3: The grain-size characteristics of a soil are given in Table 2-7, which is taken from McCarthy (1988, p. 84, Problem 4-2). Assumed porosity value is $n = 0.40$. Also, $\gamma = 9.789$ kN/m^3 and $\mu = 1.002 \times 10^{-3}$ N · s/m^2, which correspond to 20°C (see Table 2-6a). Using the above-given five methods of estimation, determine the values of hydraulic conductivity (K) and make a comparative evaluation.

Solution: The corresponding grain-size distribution curve for the data in Table 2-7 is shown in Figure 2-8 in the linear coordinates system. As shown in Figure 2-8, $d_{10} = 0.136$ mm and $d_{50} = 0.761$ mm. Based on the above values, results of each method are given below.

Hazen Equation. From Eq. (2-29), with $A = 1.0$ and $d_{10} = 0.136$ mm we obtain

$$K = (1.0)(0.136 \text{ mm})^2 = 1.85 \times 10^{-2} \text{ cm/s}$$

Shepherd Equations. As can be seen from the equations for various materials given in Figure 2-7, the hydraulic conductivity of a type of material is a function of its grain size, d. To calculate a K value, an average d value should be selected. Let us consider $d = d_{50} = 0.136$ mm as an average value for the material under question (see Figure 2-8). Because it is a natural material, calculations will be made with the formulas corresponding to "channel immature" (K_{CI}) and "consolidated" (K_{C}) materials given in Figure 2-7:

$$K_{\text{CI}} = 3500d^{1.65} = 3500(0.136 \text{ mm})^{1.65} = 130.14 \text{ gpd/ft}^2 = 6.14^{-3} \text{ cm/s}$$

$$K_{\text{C}} = 800d^{1.5} = 800(0.136 \text{ mm})^{1.5} = 40.12 \text{ gpd/ft}^2 = 1.89 \times 10^{-3} \text{ cm/s}$$

Kozeny–Carman Equation. Likewise in the Shepherd method, by selecting $d_m = 0.136 \text{ mm} = 0.136 \times 10^{-3}$ m, $n = 0.40$, and remembering that $\rho g = \gamma$, Eq. (2-31)

TABLE 2-7 The Grain-size Characteristics of a Soil for Example 2-3

Grain size, d (mm)	Percent finer
4.760	96.0
2.000	80.0
0.840	52.0
0.420	38.0
0.250	25.0
0.149	12.0
0.074	5.0

gives

$$K = \frac{9.789\ \text{kN/m}^3}{1.002 \times 10^{-3}\ \text{N}\cdot\text{s/m}^2}\left[\frac{(0.40)^2}{(1-0.40)^2}\right]\frac{(0.136\times 10^{-3}\ \text{m})^2}{180}$$

$$= 1.78 \times 10^{-4}\ \text{m/s}$$

$$K = 1.78 \times 10^{-2}\ \text{cm/s}$$

If porosity (n) is selected to be 0.35, the resulting value would be $K = 1.02 \times 10^{-2}$ cm/sec.

Fair and Hatch Equation. For this method, $\beta = 5$ and $\alpha = 6.5$ were selected as the packing factor and shape factor in Eq. (2-32), respectively. In Table 2-8, the calculated values of P_m and d_m are listed using the values in Table 2-7. For example, $P_1 = 96.0 - 80.0 = 16.0$ and $d_1 = [(4.76\ \text{mm})(2.00\ \text{mm})]^{1/2} = 3.085$ mm correspond to $m = 1$. Similarly, $P_2, P_3, \ldots, P_N$ and $d_2, d_3, \ldots, d_N$ were calculated and are listed in Table 2-8. As seen, $N = 6$. Using the values in Table 2-8, the series term in Eq. (2-32) can be calculated as

$$\sum_{m=1}^{N=8}\frac{P_m}{d_m} = \frac{P_1}{d_1}+\frac{P_2}{d_2}+\frac{P_3}{d_3}+\frac{P_4}{d_4}+\frac{P_5}{d_5}+\frac{P_6}{d_6}$$

$$= \frac{16.0}{3.085\ \text{mm}}+\frac{28.0}{1.296\ \text{mm}}+\frac{14.0}{0.594\ \text{mm}}+\frac{13.0}{0.324\ \text{mm}}+\frac{13.0}{0.193\ \text{mm}}+\frac{7.0}{0.105\ \text{mm}}$$

$$= 224.508\ \text{mm}^{-1} = 224.508 \times 10^3\ \text{m}^{-1}$$

And substituting this and the rest of the values in Eq. (2-32) gives

$$K = \frac{9.789\ \text{kN/m}^3}{1.002 \times 10^{-3}\ \text{N}\cdot\text{s/m}^2}\left[\frac{(0.40)^3}{(1-0.40)^2}\right]\frac{1}{5\left[\left(\frac{6.5}{100}\right)(224.508\times 10^3\ \text{m}^{-1})\right]^2}$$

$$= 1.63 \times 10^{-3}\ \text{m/s} = 1.63 \times 10^{-1}\ \text{cm/s}$$

TABLE 2-8 Generated P_m and d_m Data Set from the Data in Table 2-7 for the Fair and Hatch Method for Example 2-3

Grain Size, d (mm)	Percent Finer	Percentage Held Between Adjacent Sieves (P_m)	Geometric Mean of the Sizes of the Adjacent Sieves, d_m (mm)	Dummy Variable, m
4.760	96.0	—	—	—
2.000	80.0	16.0	3.085	1
0.840	52.0	28.0	1.296	2
0.420	38.0	14.0	0.594	3
0.250	25.0	13.0	0.324	4
0.149	12.0	13.0	0.193	5
0.074	5.0	7.0	0.105	6

If porosity (n) is selected to be 0.35, Eq. (2-32) gives

$$K = \frac{9.789 \text{ kN/m}^3}{1.002 \times 10^{-3} \text{ N} \cdot \text{s/m}^2} \left[\frac{(0.35)^3}{(1-0.35)^2} \right] \frac{1}{5\left[\left(\frac{6.5}{100}\right)(224.508 \times 10^3 \text{ m}^{-1})\right]^2}$$

$$= 9.31 \times 10^{-4} \text{ m/s}$$

$$K = 9.31 \times 10^{-2} \text{ cm/s}$$

Alyamani and Şen Equation. This method requires the value of I_0 other than d_{10} and d_{50} values. Substituting the values in Eq. (2-33) gives:

$$K = 1300[0.057 \text{ mm} + 0.025(0.761 \text{ mm} - 0.136 \text{ mm})]^2$$

$$= 6.86 \text{ m/d}$$

$$K = 7.94 \times 10^{-3} \text{ cm/s}$$

Comparison of the Results. The hydraulic conductivity values estimated from different methods are listed in Table 2-9. The equations of Hazen, Kozeny and Carman, and Fair and Hatch generate somewhat higher K values than the other equations. Two different porosity values do not make a significant difference for Kozeny and Carman equations and for Fair and Hatch equations. The K values determined from Shepherd equations and from Alyamani and Şen equations are close to each other.

2.3.7 Classification of Aquifers with Respect to Hydraulic Conductivity

As seen in Section 2.3.4, hydraulic conductivity of a porous medium is a measure of its fluid transmitting capability. In aquifers, hydraulic conductivity values usually vary from location to location; in other words, it is a spatially variable quantity. Also, hydraulic conductivity values show variation from direction to direction at a point or location. As a result, the flow rate and discharge velocity will vary from point to point as well as from direction to direction at a point. Aquifer problems associated with these kinds of situations obviously cannot be handled with the simplistic definition of hydraulic conductivity given

TABLE 2-9 Comparisons of *K* Values Determined from Different Methods for Example 2-3

Method	Porosity, n	Hydraulic Conductivity, K (cm/s)
Hazen	Not required	1.85×10^{-2}
Shepherd	Not required	6.14×10^{-3} [a]
		1.89×10^{-3} [b]
Kozeny and Carman	0.40	1.78×10^{-2}
	0.35	1.02×10^{-2}
Fair and Hatch	0.40	16.3×10^{-2}
	0.35	9.31×10^{-2}
Alyamani and Şen	Not required	7.94×10^{-3}

[a] Corresponds to the equation of "channel immature" in Figure 2-7.
[b] Corresponds to the equation of "consolidated" in Figure 2-7.

by the original form of Darcy's law [Eq. (2-19)]. For this reason, aquifers are categorized with respect to the variation of their hydraulic conductivity, and in the following sections the definitions of some cases are presented.

2.3.7.1 Basic Definitions

Isotropic and Anisotropic Aquifers. If the hydraulic conductivity, K, at a point in an aquifer is the same at every direction, the aquifer is called an *isotropic aquifer*. If the hydraulic conductivity, K, at a point in an aquifer varies with the direction, the aquifer is called an *anisotropic aquifer*.

Principal Directions of Anisotropy. For the sake of simplicity, the principal directions of anisotropy will be explained in a plane perpendicular to the z axis, the x–y plane (see Figure 2-9). Let θ be the angle between the x axis and the axis at which its value is determined, the ξ axis. Because K changes with θ on the x–y plane, the value of K varies between its minimum and maximum values from $\theta = 0$ to $\theta = 2\pi$ radians. The axes, on which the minimum and maximum values are perpendicular to each other, will be shown in Section 2.6. These axes are called the *principal directions of anisotropy*. For example, if these K values are located on the ξ and η axes in Figure 2-9, these axes are the principal directions of anisotropy. For the special case in Figure 2-9, the third axis, the ζ axis, coincides with the z-axis. If one considers a plane perpendicular to either the x axis or the y axis, the principal directions of anisotropy in that plane can similarly be defined.

2.3.7.2 Classification of Aquifers.

Consider that the principal directions of anisotropy of an aquifer are in the directions of the x, y, and z coordinates in a Cartesian coordinate system. Four cases can be defined for heterogeneity and anisotropy (Freeze and Cherry, 1979). These cases are defined below.

Homogeneous and Isotropic Aquifer. This is the simplest case. A homogeneous and isotropic aquifer is schematically defined in Figure 2-10. As shown in Figure 2-10, consider two points in the aquifer whose coordinates are (x_1, y_1, z_1) for point 1 and (x_2, y_2, z_2) for point 2. If, as shown in Figure 2-10, the hydraulic conductivities satisfy for points 1 and 2,

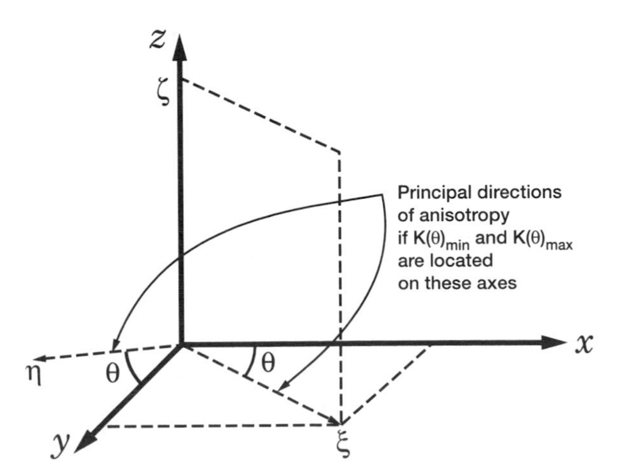

Figure 2-9 Principal directions of anisotropy in the x–y plane.

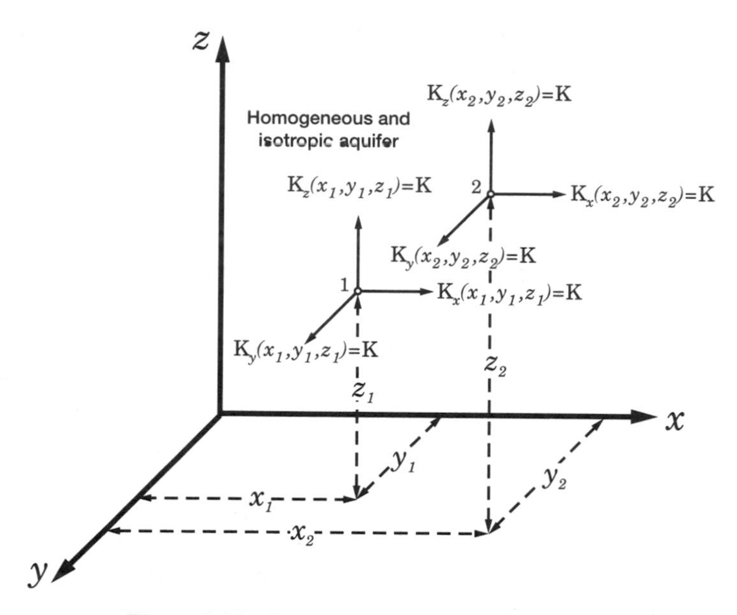

Figure 2-10 Homogeneous and isotropic aquifer.

respectively, that

$$\begin{aligned} K_x(x_1, y_1, z_1) &= K_y(x_1, y_1, z_1) = K_z(x_1, y_1, z_1) \\ &= K_x(x_2, y_2, z_2) = K_y(x_2, y_2, z_2) = K_z(x_2, y_2, z_2) = K \end{aligned} \tag{2-34}$$

the aquifer is said to be a *homogeneous and isotropic aquifer*. The expressions in Eq. (2-34) mean that the hydraulic conductivity does not vary from point to point as well as with direction.

Homogeneous and Anisotropic Aquifer. A homogeneous and anisotropic aquifer is schematically defined in Figure 2-11. Consider points 1 and 2 in the aquifer whose coordinates are (x_1, y_1, z_1) and (x_2, y_2, z_2) for points 1 and 2, respectively. If the hydraulic conductivities satisfy for points 1 and 2, respectively, that

$$\begin{aligned} K_x(x_1, y_1, z_1) &= K_x(x_2, y_2, z_2) = K_x \\ K_y(x_1, y_1, z_1) &= K_y(x_2, y_2, z_2) = K_y \\ K_z(x_1, y_1, z_1) &= K_z(x_z, y_2, z_2) = K_z \end{aligned} \tag{2-35}$$

the aquifer is said to be a *homogeneous and anisotropic aquifer*.

Heterogeneous and Isotropic Aquifer. A heterogeneous and isotropic aquifer is schematically defined in Figure 2-12. Let the coordinates of points 1 and 2 be (x_1, y_1, z_1) and (x_2, y_2, z_2), respectively. If the hydraulic conductivities satisfy for points 1 and 2,

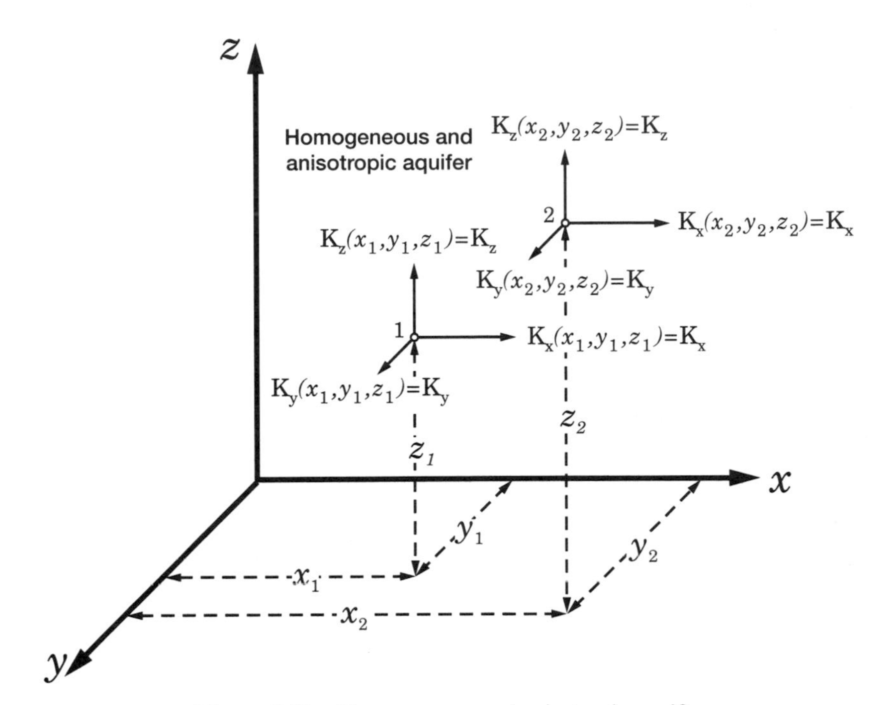

Figure 2-11 Homogeneous and anisotropic aquifer.

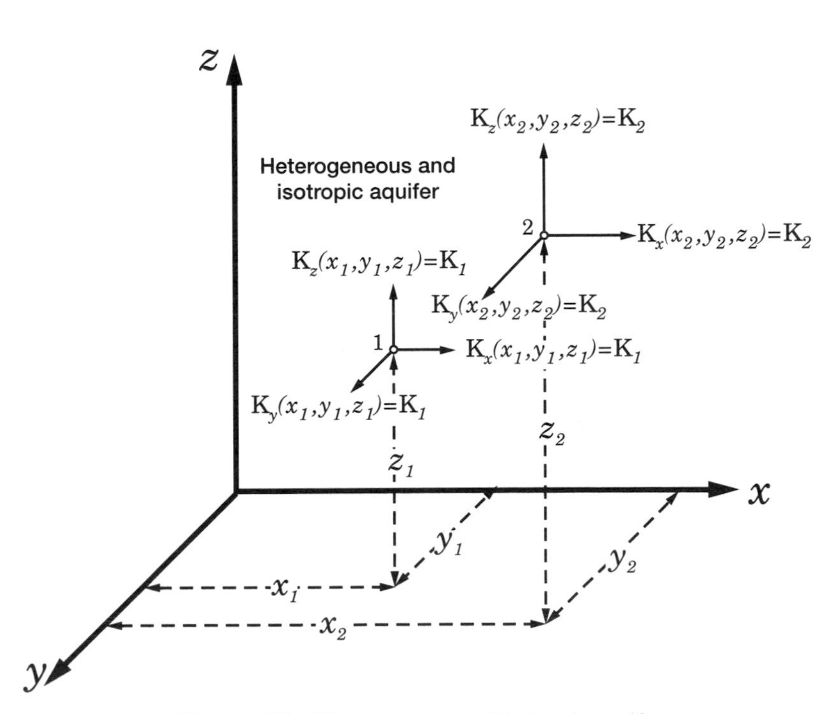

Figure 2-12 Heterogeneous and isotropic aquifer.

respectively, that

$$K_x(x_1, y_1, z_1) = K_y(x_1, y_1, z_1) = K_z(x_1, y_1, z_1) = K_1$$
$$K_x(x_2, y_2, z_2) = K_y(x_2, y_2, z_2) = K_z(x_2, y_2, z_2) = K_2 \tag{2-36}$$

the aquifer is said to be a *heterogeneous and isotropic aquifer*. If, as a special case of Eq. (2-36), we have

$$K_x(x_1, y_1, z_1)=K_y(x_1, y_1, z_1) = K_x(x_2, y_2, z_2) = K_y(x_2, y_2, z_2) = K_h$$
$$K_z(x_1, y_1, z_1)=K_z(x_2, y_2, z_2) = K_z = K_v \tag{2-37}$$

the aquifer is said to be a *transversely isotropic aquifer*. Even though the case defined by the expressions in Eq. (2-37) is a somewhat simplified form, it is widely used in deriving analytical solutions for well hydraulics under fully and partially penetrating well conditions; these will be covered in subsequent chapters.

Heterogeneous and Anisotropic Aquifer. This is the most general case as shown in Figure 2-13. Likewise, consider points 1 and 2, whose coordinates are (x_1, y_1, z_1) and (x_2, y_2, z_2), respectively. If the hydraulic conductivities satisfy for points 1 and 2, respectively, that

$$K_x(x_1, y_1, z_1) \neq K_y(x_1, y_1, z_1) \neq K_z(x_1, y_1, z_1)$$
$$K_x(x_2, y_2, z_2) \neq K_y(x_2, y_2, z_2) \neq K_z(x_2, y_2, z_2) \tag{2-38}$$

the aquifer is said to be a *heterogeneous and anisotropic aquifer*.

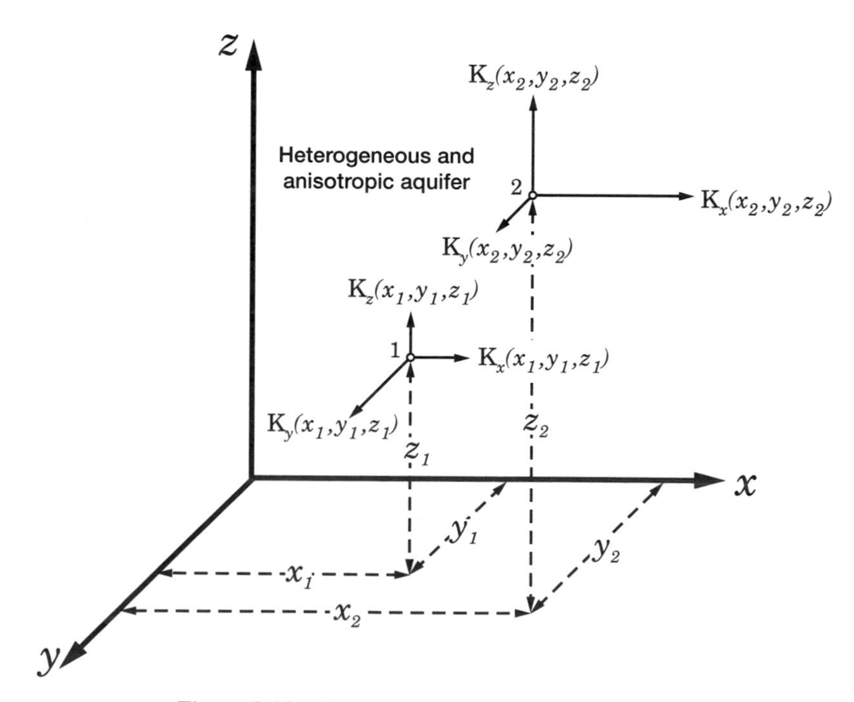

Figure 2-13 Heterogeneous and anisotropic aquifer.

2.3.8 Capillarity and Capillary Fringe

2.3.8.1 Definitions. *Capillarity* is defined as the rise or motion of water in the interstices of soils or rocks due to (1) the molecular attraction (or adhesion) between the solid particles and fluid and (2) the surface tension of the fluid, which is the result of the attraction (cohesion) between the molecules of the fluid. The molecular attraction between the fluid and solid particles depends on the composition of fluid and solid particles and its magnitude varies from soil to soil. A typical example is the water rise in a relatively small-diameter tube filled with porous material as shown schematically in Figure 2-14. This rise (h_c in Figure 2-14) is the result of the aforementioned forces and is called *capillary rise*. One should bear in mind that h_c represents the average in the pipe.

Capillary rise occurs in a limited zone above the water table of an unconfined aquifer, which is shown in a vertical distribution of subsurface diagram in Figure 2-15. In the saturation zone all pores are filled with water under hydrostatic pressure, whereas in the unsaturated zone the pores are filled partially by water and partially by air. The *capillary zone*, or *capillary fringe*, starts from the water table and extends up to the limit of capillary rise of water, which is dependent on the soil texture and its hydraulic conductivity. The *intermediate vadose zone* extends from the upper edge of the capillary zone to the lower edge of the soil–water zone, and its thickness may vary from zero up to a relatively high value, more than 100 m under deep water-table conditions. The *soil–water zone* extends from the upper edge of the intermediate vadose zone to the ground surface and includes the roots of vegetation. Its thickness depends on the soil type and vegetation (Todd, 1980).

2.3.8.2 Quantification of the Capillary Rise. Based on the equilibrium of a column of water in a capillary tube of radius r (Figure 2-14), the expression for capillary rise is (e.g., Bear, 1979)

$$h_c = \frac{2T \cos \theta}{\gamma r} \tag{2-39}$$

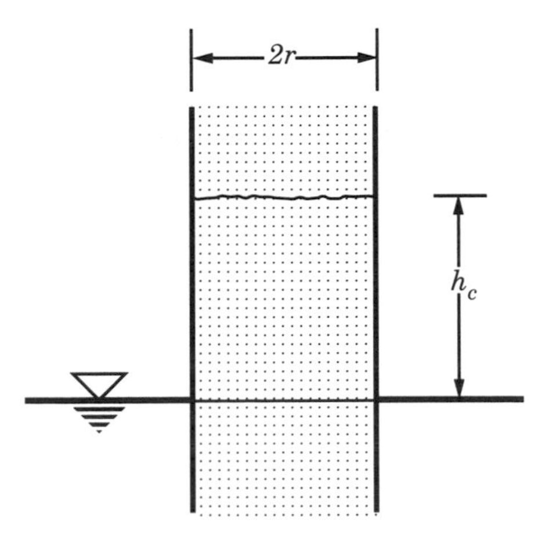

Figure 2-14 Capillary rise in a porous tube.

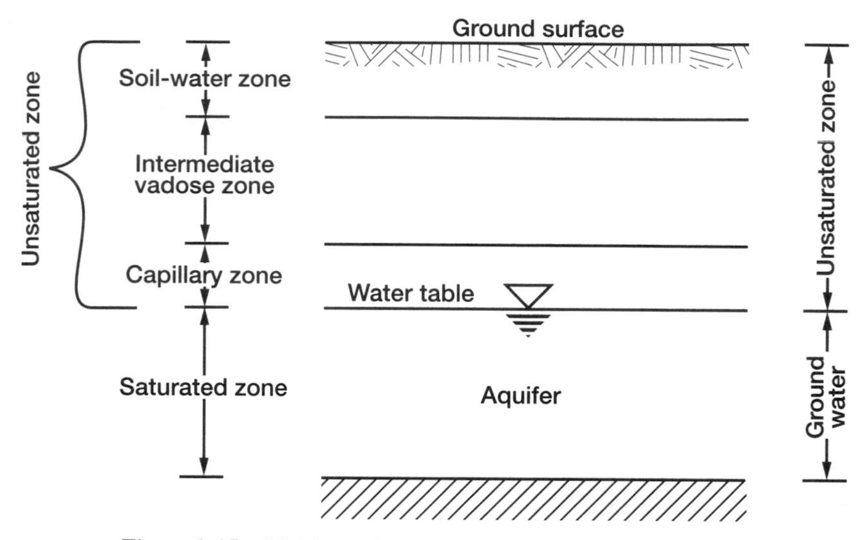

Figure 2-15 Divisions of subsurface water. (After Todd, 1980.)

where T (units L/T) is the *surface tension* of water, θ is the contact angle and is dependent on the chemical properties of both the tube and water, γ is the specific weight of water, and r is the radius of the tube. The values of surface tension do not change significantly with temperature, and its values at 0°C (32°F), 20°C (68°F), and 100°C (212°F) are 0.0756 N/m (0.00518 lb/ft), 0.0728 N/m (0.00498 lb/ft), and 0.0589 N/m (0.00404 lb/ft), respectively (Vennard and Street, 1982). For clean glass and pure water, $\theta = 0$.

Example 2-4: For water at 20°C, $T = 0.0756$ N/m and $\gamma = 9.789$ kN/m^3. Taking $\theta = 0$, determine the capillary rise value for $r = 0.001$ m.

Solution: Substituting the values in Eq. (2-39) gives

$$h_c = \frac{2(0.0756 \text{ N/m})(\cos\theta)}{r(9.789 \text{ kN/m}^3)} = \frac{1.545 \times 10^{-5} \text{ m}^2}{\text{r}} \tag{2-40}$$

TABLE 2-10 Capillary Rise in Samples of Unconsolidated Materials[a]

Material	Grain size (mm)	Capillary rise (cm)
Fine gravel	5–2	2.5
Very coarse sand	2–1	6.5
Coarse sand	1–0.5	13.5
Medium sand	0.5–0.2	24.6
Fine sand	0.2–0.1	42.8
Silt	0.1–0.05	105.5
Silt	0.05–0.02	200[b]

[a]Capillary rise after 72 days; all samples have virtually the same porosity of 41%, after 72 days.
[b]Still rising after 72 days.

Source: After Lohman (1972).

For $r = 0.001$ m, Eq. (2-40) gives

$$h_c = \frac{1.545 \times 10^{-5}\ \text{m}^2}{0.001\ \text{m}} = 0.015\ \text{m}$$

It is evident from Eq. (2-40) that the thickness of the capillary zone is inversely related to the pore size of a porous medium, such as soil or rock. Measured capillary rise values are given in Table 2-10. Notice that the values in Table 2-10 indicate that the capillary rise is nearly inversely proportional to the grain size.

2.4 COMPRESSIBILITY AND ELASTICITY OF AQUIFERS

Aquifers are compressible and elastic mediums like all other solid mediums. However, the main difference between aquifers and solid mediums is that aquifers are composed partly of solid grains and partly of pores that are filled with water. As a result, the concept of *volume elasticity of aquifers* has been established based on laboratory and field observations (Meinzer, 1928). Good evidence of compressibility and elasticity of aquifers is fluctuations of water levels in wells as a response to barometric pressure changes, tidal effects, earthquake effects, ground surface loads for very shallow water-table cases, and subsurface subsidence.

Compressibility and elasticity are the key aspects of transient saturated ground-water flow in confined and unconfined aquifers as well as in confining layers, such as aquitards. The compressibility of an aquifer is more dominant when the aquifer is confined and completely saturated. These concepts and their associated quantities are presented in the following sections (e.g., Jacob, 1950; Hantush, 1964; Cooper, 1966; Bear, 1979).

2.4.1 Specific Storage

The definition of this important quantity will be presented by deriving some important expressions based on the *effective stress concept.*

2.4.1.1 Relation Between Water Presssure and Compressive Stress. Consider an elementary volume $\Delta V = (\Delta A)(\Delta z)$ in a compressible aquifer in which ΔA is the horizontal cross-sectional area and Δz is the height of the elementary volume. The relationship between the water pressure (p) and compressive stress (σ) of the solid skeleton of the aquifer using the effective stress concept was first introduced by Terzaghi (1925) and was based on the assumptions that (1) the water pressure acts effectively throughout the elementary volume (ΔV), (2) the compressive stress (σ) of the solid skeleton of the aquifer is considered to act over the entire horizontal cross-sectional area (ΔA), and (3) the atmospheric pressure is constant. A cross section through a confined aquifer (also for an unconfined aquifer) and the details at any internal horizontal plane (AB), at which ΔV is located, are shown in Figures 2-16a and 2-16b, respectively. The stress resulting from the total weight of soil and water above ΔV in the aquifer and the load at the ground surface including the atmospheric pressure (σ_T) is balanced by the water pressure (p) and the compressive stress of the solid skeleton of the aquifer (σ):

$$\sigma_T = \sigma + p \tag{2-41}$$

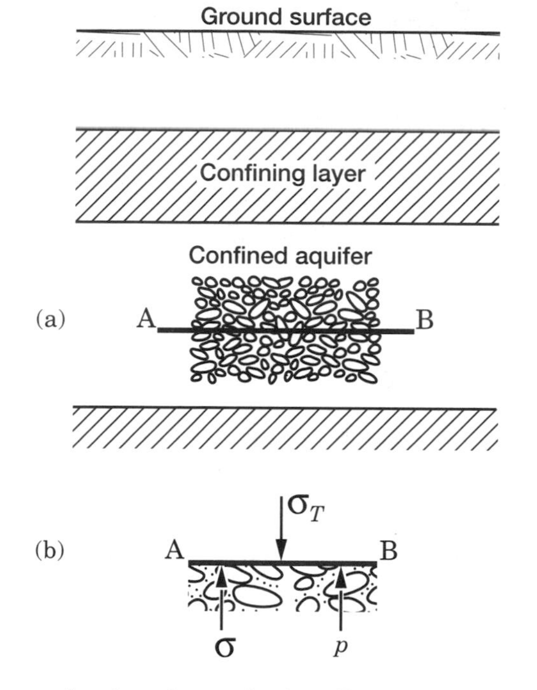

Figure 2-16 (a) Cross section through a confined aquifer. (b) Details of any internal horizontal elemental plane AB in an aquifer.

In Eq. (2-41), σ_T can safely be assumed to be constant because it is the result of the overburden load including the atmospheric pressure. As a result, $d\sigma_T = 0$; and by differentiation, Eq. (2-41) gives

$$d\sigma = -dp \tag{2-42}$$

Equation (2-42) is the relation between the water pressure and compressive stress and means that a decrease of water pressure (p) results in a corresponding increase of solid compressive stress.

2.4.1.2 Stress–Strain Relationships. Aquifers follow Hooke's law of deformation. Consider an elementary volume in an aquifer as shown in Figure 2-17. If the original height is H_0, $\Delta\sigma$ is change in effective stress, and ΔH is the change in height, then from Hooke's law the expression for strain (ϵ) can be expressed as

$$\epsilon = \frac{\Delta H}{H_0} = \frac{\Delta\sigma}{E_s} \tag{2-43}$$

where E_s is the *modulus of elasticity* of the solid skeleton. In aquifer hydraulics, the reciprocal of the modulus of elasticity

$$\alpha = \frac{1}{E_s} \tag{2-44}$$

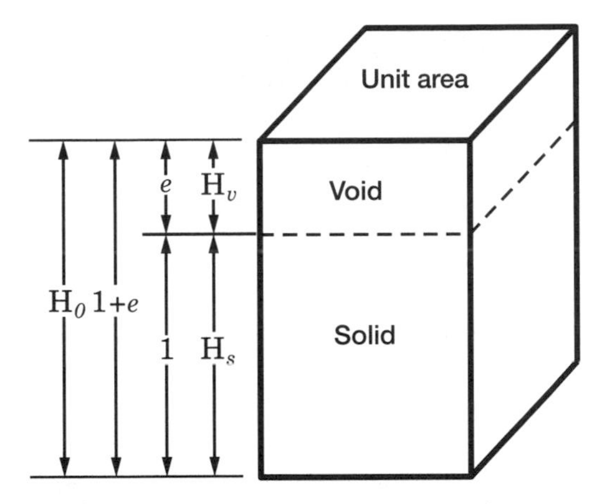

Figure 2-17 An elementary aquifer volume for definition of stress–strain relationships.

is defined as the *vertical compressibility of the solid skeleton*. In soil mechanics (e.g., Lambe and Whitman, 1969), α (denoted by m_v) is called the *coefficient of volume change*.

Figure 2-17 is also a diagrammatic representation of the void ratio of a unit aquifer element. The height of the element consists of the height of solids (H_s) and the height of voids (H_v). The void ratio (e) is defined as the ratio between the volume of voids (V_v) and the volume of solids (V_s):

$$e = \frac{V_v}{V_s} \tag{2-45}$$

Since the solids are assumed to be incompressible, changes in the height of element (ΔH_0) are proportional to changes in void ratio (Δe). Therefore, the relative compression of the element or the relative amount of water released per unit height can be expressed as

$$\frac{\Delta H}{H_0} = \frac{\Delta e}{1 + e} \tag{2-46}$$

The change in void ratio, a_v, is defined as (e.g., Lambe and Whitman, 1969)

$$\Delta e = a_v \Delta \sigma \tag{2-47}$$

where a_v is the *coefficient of compressibility* or rate of *change of void ratio* with respect to rate of change of effective pressure causing consolidation. In soil mechanics, a_v is the slope of the line determined by plotting void ratio versus pressure for test specimens (Terzaghi and Peck, 1967). From Eqs. (2-43), (2-46), and (2-47) one can write

$$\alpha = \frac{1}{E_s} = \frac{a_v}{1 + e} = \frac{\Delta e}{\Delta \sigma (1 + e)} \tag{2-48}$$

2.4.1.3 Derivation of the Specific Storage Equation. Again, let us consider an elemental volume $\Delta V = (\Delta A)(\Delta z)$ in a compressible aquifer. If the solid skeleton of ΔV is compressed, its porosity n and its height Δz will concurrently change. The specific storage equation will be derived on the assumption that the compressive force acts in a vertical direction over a large areal extent, so that changes in the horizontal direction are negligible.

When ΔV is compressed, from Eqs. (2-42) and (2-43) one can write

$$\frac{d(\Delta z)}{\Delta z} = -\alpha d\sigma = \alpha dp \tag{2-49}$$

The negative sign in Eq. (2-49) indicates a decrease in Δz with increasing stress. The volume of solids, ΔV_s in ΔV, can be expressed as

$$\Delta V_s = (1 - n)\Delta V = (1 - n)(\Delta A)(\Delta V) \tag{2-50}$$

As the compressibility of the individual solid grains is small compared with the change in porosity (n), the volume of the solid material, ΔV_s, may be assumed to be constant. Therefore,

$$\Delta V_s = (1 - n)(\Delta A)(\Delta z) = \text{ constant} \tag{2-51}$$

In Eq. (2-51), ΔA is constant; by differentiation we obtain

$$\frac{d(\Delta V_s)}{\Delta A} = (1 - n)d(\Delta z) - (\Delta z)(dn) = 0 \tag{2-52}$$

Using Eq. (2-49) and after simplification, one gets

$$dn = \alpha(1 - n)dp \tag{2-53}$$

Based on the definition of the bulk modulus of elasticity, the density of water, ρ, increases as p increases in accordance with (Jacob, 1950)

$$d\rho = \beta\rho_0 dp \tag{2-54}$$

where β is the compressibility of water or the reciprocal of its bulk modulus of elasticity, E_w:

$$\beta = \frac{1}{E_w} = \frac{-\dfrac{\Delta V}{V_0}}{\Delta p} \tag{2-55}$$

ρ_0 is the reference density, conveniently taken at atmospheric pressure. Variation of E_w and β with other parameters in the S.I. and English FSS units are given in Table 2-11 and Table 2-12, respectively.

The mass ΔM of the volume of water ΔV_w contained in the elementary volume ΔV is given by

$$\Delta M = \rho\Delta V_w = \rho n(\Delta A)(\Delta z) \tag{2-56}$$

TABLE 2-11 Modulus of Elasticity and Compressibility of Water as a Function of Temperature, Specific Yield, and Density in the S.I. Units

Temperature °C	Specific Weight, γ (kN/m^3)	Density, ρ (kg/m^3)	Modulus of Elasticity, E_W (N/m^2)	Compressibility, $\beta = 1/E_w$ (m^2/N)
0	9.805	999.8	1.98×10^9	5.05×10^{-10}
5	9.807	1000.0	2.05×10^9	4.88×10^{-10}
10	9.804	999.7	2.10×10^9	4.76×10^{-10}
15	9.798	999.1	2.15×10^9	4.65×10^{-10}
20	9.789	998.2	2.17×10^9	4.61×10^{-10}
25	9.777	997.0	2.22×10^9	4.50×10^{-10}
30	9.764	995.7	2.25×10^9	4.44×10^{-10}
40	9.730	992.2	2.28×10^9	4.39×10^{-10}
50	9.689	988.0	2.29×10^9	4.37×10^{-10}
60	9.642	983.2	2.28×10^9	4.39×10^{-10}
70	9.589	977.8	2.25×10^9	4.44×10^{-10}
80	9.530	971.8	2.20×10^9	4.55×10^{-10}
90	9.466	965.3	2.14×10^9	4.67×10^{-10}
100	9.399	958.4	2.07×10^9	4.83×10^{-10}

Source: Adapted from Vennard and Street (1982).

TABLE 2-12 Modulus of Elasticity and Compressibility of Water as a Function of Temperature, Specific Yield, and Density in the English FSS Units

Temperature °F	Specific Weight, γ (lb/ft^3)	Density, ρ (slug/ft^3)	Modulus of Elasticity, E_W(psi)	Compressibility, $\beta = 1/E_w$ (in.^{2}/lb)
32	62.42	1.940	287×10^3	3.48×10^{-6}
40	62.43	1.940	296×10^3	3.38×10^{-6}
50	62.41	1.940	305×10^3	3.28×10^{-6}
60	62.37	1.938	313×10^3	3.19×10^{-6}
70	62.30	1.936	319×10^3	3.13×10^{-6}
80	62.22	1.934	324×10^3	3.09×10^{-6}
90	62.11	1.931	328×10^3	3.05×10^{-6}
100	62.00	1.927	331×10^3	3.02×10^{-6}
110	61.86	1.923	332×10^3	3.01×10^{-6}
120	61.71	1.918	332×10^3	3.01×10^{-6}
130	61.55	1.913	331×10^3	3.02×10^{-6}
140	61.38	1.908	330×10^3	3.03×10^{-6}
150	61.20	1.902	328×10^3	3.05×10^{-6}
160	61.00	1.896	326×10^3	3.07×10^{-6}
170	60.80	1.890	322×10^3	3.11×10^{-6}
180	60.58	1.883	318×10^3	3.14×10^{-6}
190	60.36	1.876	313×10^3	3.19×10^{-6}
200	60.12	1.868	308×10^3	3.25×10^{-6}
212	59.83	1.860	300×10^3	3.33×10^{-6}

Source: Adapted from Vennard and Street (1982).

In Eq. (2-56), when ΔV is compressed, ΔM will change only with ρ, n, and Δz, and ΔA will remain constant. Therefore, by differentiation, from Eq. (2-56) one gets

$$d(\Delta M) = nd\rho(\Delta A)(\Delta z) + \rho dn(\Delta A)(\Delta z) + \rho n(\Delta A)d(\Delta z) \tag{2-57}$$

Remembering that $\Delta V = (\Delta A)(\Delta z)$ and dividing both sides of Eq. (2-51) by ΔV gives

$$\frac{d(\Delta M)}{\Delta V} = nd\rho + \rho dn + \rho n\frac{d(\Delta z)}{\Delta z} \tag{2-58}$$

Using the equivalents of $d(\Delta z)/\Delta z$, dn, and $d\rho$ from Eqs. (2-49), (2-53), and (2-54), respectively, Eq. (2-58) takes the form

$$\frac{d(\Delta M)}{\Delta V} = (n\beta\rho_0 + \alpha\rho)dp \tag{2-59}$$

If the small differences between ρ and ρ_0 are neglected, the change in the volume of water, which is released from storage, per unit bulk volume, will be expressed as

$$\frac{d(\Delta M)}{\rho\Delta V} = \frac{d(\Delta V_w)}{\Delta V} = (\alpha + n\beta)dp \tag{2-60}$$

From Eq. (2-1), it can be seen that the pressure head, p/γ, in the elemental volume, ΔV, changes directly with the hydraulic head, h. Therefore, Eq. (2-1), by differentiation, gives

$$dp = \gamma dh \tag{2-61}$$

and substituting this into Eq. (2-60) gives

$$\frac{d(\Delta M)}{\rho\Delta V} = \frac{d(\Delta V_w)}{\Delta V} = (\alpha + n\beta)\gamma dh = S_s dh \tag{2-62}$$

where

$$S_s = (\alpha + n\beta)\gamma = \gamma n\beta\left(1 + \frac{\alpha}{n\beta}\right) = \frac{\gamma}{E_s} + \frac{n\gamma}{E_w} \tag{2-63}$$

The term S_s, which has a dimension of L^{-1}, has been called *specific storage*. From Eq. (2-62) it may be defined as *the volume of water that is released from or taken into storage per unit volume of a confined aquifer or a confined aquifer layer per unit change in hydraulic head*. The factor $\gamma n\beta$ of Eq. (2-63) gives the fraction of storage derived from expansion of water and the product $\gamma\alpha$ gives the fraction derived from compression of aquifer. Specific storage values for a variety of materials are given in Table 2-13.

2.4.1.4 Specific Storage Equations Based on Consolidation Coefficient. The specific storage can also be estimated based on consolidation coefficient of soils, which is widely used in the field of soil mechanics. In soil mechanics, consolidation is the adjustment of a saturated soil in response to increased load, which involves the squeezing of water from the pores and a decrease in void ratio. Although physically the same quantities

TABLE 2-13 Range of Values of Specific Storage for Various Porous Materials

Porous Material	Specific Storage, S (m^{-1})
Plastic clay	2.0×10^{-2}–2.6×10^{-3}
Stiff clay	2.6×10^{-3}–1.3×10^{-3}
Medium-hard clay	1.3×10^{-3}–9.2×10^{-4}
Loose sand	1.0×10^{-3}–4.9×10^{-4}
Dense sand	2.0×10^{-4}–1.3×10^{-4}
Dense sandy gravel	1.0×10^{-4}–4.9×10^{-5}
Rock, fissured, jointed	6.9×10^{-5}–3.3×10^{-6}
Rock, sound	Less than 3.3×10^{-6}

Source: After Domenico (1972) and Mercer et al. (1982).

are used in soil mechanics, the terminology in the soil mechanics literature is somewhat different as compared in the ground-water hydraulics or hydrogeology areas. Moreover, in soil mechanics, the specific storage-related subjects are generally included under the *theory of consolidation*, and the corresponding expressions are presented below. Before the corresponding approach is addressed, some important governing equations used in the area of soil mechanics will be presented.

It is useful to start with the governing equations of the consolidation phenomenon (the corresponding forms of these equations used in ground-water hydraulics and hydrogeology areas are given in Section 2.7). The three-dimensional consolidation equation in soil mechanics is given as (e.g., Lambe and Whitman, 1969; Kézdi, 1974)

$$\frac{\partial u}{\partial t} = \frac{1 + e}{a_v \gamma} \left(K_x \frac{\partial^2 u}{\partial x^2} + K_y \frac{\partial^2 u}{\partial y^2} + K_z \frac{\partial^2 u}{\partial z^2} \right) \tag{2-64}$$

where u is the pore pressure and is expressed as follows:

$$u = \gamma h \tag{2-65}$$

e is the void ratio defined by Eq. (2-45), and a_v is the coefficient of compressibility defined as (Lambe and Whitman, 1969)

$$a_v = -\frac{de}{d\sigma_v} \tag{2-66}$$

where σ_v is the vertical stress component.

The relation between the *coefficient of consolidation*, c_v, and the vertical hydraulic conductivity, K_z, for compressible fine-grained materials (e.g., see Lambe and Whitman, 1969; Kézdi, 1974) is given by

$$c_v = \frac{K_z(1 + e)}{\gamma a_v} = \frac{K_z}{\gamma \alpha} \tag{2-67}$$

where α, which, defined by Eq. (2-44), can also be expressed as (e.g., see Lambe and Whitman, 1969)

$$\alpha = \frac{\Delta \epsilon_v}{\Delta \sigma_v} \tag{2-68}$$

where ϵ is the strain in the vertical direction defined as

$$\epsilon_v = \frac{\sigma_v}{E_s} \tag{2-69}$$

where E_s is the *modulus of elasticity* of the solid skeleton. Introducing the first part of Eq. (2-67) into Eq. (2-64) gives

$$S_s \frac{\partial h}{\partial t} = K_x \frac{\partial^2 h}{\partial x^2} + K_y \frac{\partial^2 h}{\partial y^2} + K_z \frac{\partial^2 h}{\partial z^2} \tag{2-70}$$

where

$$S_s = \frac{K_z}{c_v} = \alpha \gamma \tag{2-71}$$

Comparing Eq. (2-71) with Eq. (2-63) shows that the $n\beta\gamma$ term in Eq. (2-63), which is the fraction of storage derived from expansion of water, is zero. In highly compressible materials, this term is generally negligible as compared with the compression of the skeleton. In other words, in highly compressible confining layers, the volume of water released from expansion of water is negligible compared with that released through a change in porosity (Domenico and Mifflin, 1965).

The one-dimensional form of Eq. (2-70) is

$$\frac{S_s}{K_z} \frac{\partial h}{\partial t} = \frac{\partial^2 h}{\partial z^2} \tag{2-72}$$

which has a special importance in the soil mechanics literature. However, in soil mechanics, the term p/γ is used instead of h. Therefore, Eq. (2-72) is a special form of Eq. (2-64) as well. As mentioned in Section 7.2 of Chapter 7, the flow direction in confining layers (or aquitards) is predominantly vertical and horizontal in the aquifers. Combination of Eqs. (2-48) and (2-71) gives

$$S_s = \alpha \gamma = \frac{\gamma}{E_s} = \frac{a_v \gamma}{1 + e} = \frac{\gamma \Delta e}{\Delta \sigma (1 + e)} \tag{2-73}$$

In soil mechanics, to reduce the number of variables, the *consolidation coefficient* (c_v) has been introduced to combine the effects of compressibilty and vertical hydraulic conductivity, and its value is determined in the laboratory. As a result, in soil mechanics literature, Eq. (2-72) is generally expressed in terms of pressure. Using Eq. (2-65) in Eq. (2-72) gives (e.g., see Lambe and Whitman, 1969)

$$\frac{1}{c_v} \frac{\partial u}{\partial t} = \frac{\partial^2 u}{\partial z^2} \tag{2-74}$$

2.4.1.5 Specific Storage Values of Natural Sediments Based on Consolidation Tests. If the compressibility of water can be neglected, the specific storage of natural sediments may be estimated using Eq. (2-71) provided that the compressibility of soil (α) or modulus of elasticity (E_s) is known. If the vertical hydraulic conductivity (K_z) and consolidation coefficient (c_v) are known from laboratory measurements, the value of α can be determined from Eq. (2-67). Based on this approach, Domenico and Mifflin (1965) presented a range of values for the modulus of elasticity (E_s) and specific storage ($S_s = \gamma/E_s$), and their values are presented in Tables 2-14 and 2-15 in the S.I. units and English FSS units, respectively.

2.4.2 Storage Coefficient

2.4.2.1 Storage Coefficient of a Single Layer. The *storage coefficient* (or *storativity*) S of a saturated confined aquifer or confined layer of thickness b is defined as

$$S = S_s b \tag{2-75}$$

TABLE 2-14 Range of Values for the Modulus of Elasticity (E_s) and Specific Storage ($S_s = \gamma/E_s$) of Various Porous Materials in the S.I. Units

Porous Material	Modulus of Elasticity, E_s (N/m^2)	Specific Storage, S_s (m^{-1})
Plastic clay	4.78×10^5–3.82×10^6	2.03×10^{-3}–2.56×10^{-3}
Stiff clay	3.82×10^6–7.65×10^6	2.56×10^{-3}–1.28×10^{-3}
Medium hard clay	7.65×10^6–1.43×10^7	1.28×10^{-3}–9.19×10^{-4}
Loose sand	9.56×10^5–1.91×10^7	1.02×10^{-3}–4.92×10^{-4}
Dense sand	4.78×10^7–7.65×10^7	2.03×10^{-4}–1.28×10^{-4}
Dense sandy gravel	9.56×10^7–1.91×10^8	1.02×10^{-4}–4.92×10^{-5}
Rock, fissured, jointed	1.43×10^8–2.99×10^9	6.89×10^{-5}–3.28×10^{-6}
Rock, sound	Greater than 2.99×10^9	Less than 3.28×10^{-6}

Source: Adapted from Domenico and Mifflin (1965).

TABLE 2-15 Range of Values for the Modulus of Elasticity (E_s) and Specific Storage ($S_s = \gamma/E_s$) of Various Porous Materials in the English FSS Units

Porous Material	Modulus of Elasticity, E_s (fb/ft^2)	Specific Storage, S_s (ft^{-1})
Plastic clay	1.00×10^4–8.00×10^4	6.20×10^{-3}–7.80×10^{-4}
Stiff clay	8.00×10^4–1.60×10^5	7.80×10^{-4}–3.90×10^{-4}
Medium hard clay	1.60×10^5–3.00×10^5	3.90×10^{-4}–2.80×10^{-4}
Loose sand	2.00×10^5–4.00×10^5	3.10×10^{-4}–1.50×10^{-4}
Dense sand	1.00×10^6–1.60×10^6	6.20×10^{-5}–3.90×10^{-5}
Dense sandy gravel	2.00×10^6–4.00×10^6	3.10×10^{-5}–1.50×10^{-5}
Rock, fissured, jointed	3.00×10^6–6.25×10^7	2.10×10^{-5}–1.00×10^{-6}
Rock, sound	Greater than 6.25×10^7	Less than 1.00×10^{-6}

Source: Adapted from Domenico and Mifflin (1965).

here S_s is the *specific storage* and is given by Eq. (2-63). Substitution of Eq. (2-63) into Eq. (2-75) gives

$$S = (\alpha + n\beta)\gamma b = \gamma n\beta\left(1 + \frac{\alpha}{n\beta}\right) = \frac{\gamma b}{E_s} + \frac{n\gamma b}{E_w} = S_k + S_w \tag{2-76}$$

where S_k is the *storage coefficient of soil* and is defined as

$$S_k = \frac{\gamma b}{E_s} = \alpha\gamma b \tag{2-77}$$

and S_w is the *storage coefficient of water* and is defined as

$$S_w = \frac{n\gamma b}{E_w} = n\beta\gamma b \tag{2-78}$$

Unlike the specific storage (S_s), which has dimension of L^{-1}, the storage coefficient is dimensionless. From Eq. (2-76) it may be defined as *the volume of water that is released from or taken into storage per unit surface area of a confined aquifer or a confined aquifer layer per unit change in hydraulic head.*

The definition of storage coefficient is further clarified on a confined aquifer (Figure 2-18), which has uniform thickness and whose upper and lower boundaries are horizontal for the sake of convenience (Ferris et al., 1962). If the hydraulic head in the confined aquifer is decreased, some volume of water will be released proportionally to the change in hydraulic head. Let us consider a representative parallelepiped extending from the base of the aquifer to its piezometric surface as shown in Figure 2-18. The ratio between the volume of water released from storage in the parallelepiped (ΔV) and $(\Delta A)(\Delta h)$ is the storage coefficient

$$S = \frac{\Delta V}{(\Delta A)(\Delta h)} \tag{2-79}$$

where ΔA and Δh are the horizontal cross-sectional area of the parallelepiped and the change in hydraulic head, respectively. Although the storage coefficient definition is given for a confined aquifer, the concept can also be applied to unconfined aquifers.

It is clear from Eq. (2-76) that, besides the compressibilities of soil and water, the storage coefficient (S) is a function of formation thickness (b), which is a site-specific quantity. Therefore, it may not be appropriate to present tables like the tables for the specific storage. However, for most aquifers, confined or unconfined, the storage coefficient values generally fall in the range from 1.0×10^{-5} to 5.0×10^{-3} (Ferris et al., 1962). The storage coefficient of an aquifer can be measured by pumping and slug tests. The data analysis methods in determining storage coefficient are presented in subsequent chapters for various water-bearing formations, including aquitards.

Example 2-5: The characteristics of a saturated clay are given as $c_v = 2.94 \times 10^{-3}$ cm^2/s (consolidation coefficient), $b = 10.0$ m (clay thickness); $K_z = 10^{-7}$ cm/s (vertical hydraulic conductivity), and $n = 0.35$ (total porosity). Determine the storage coefficient of the clay (S) and compare it with the storage coefficient of water (S_w) using the properties of water at 5°C.

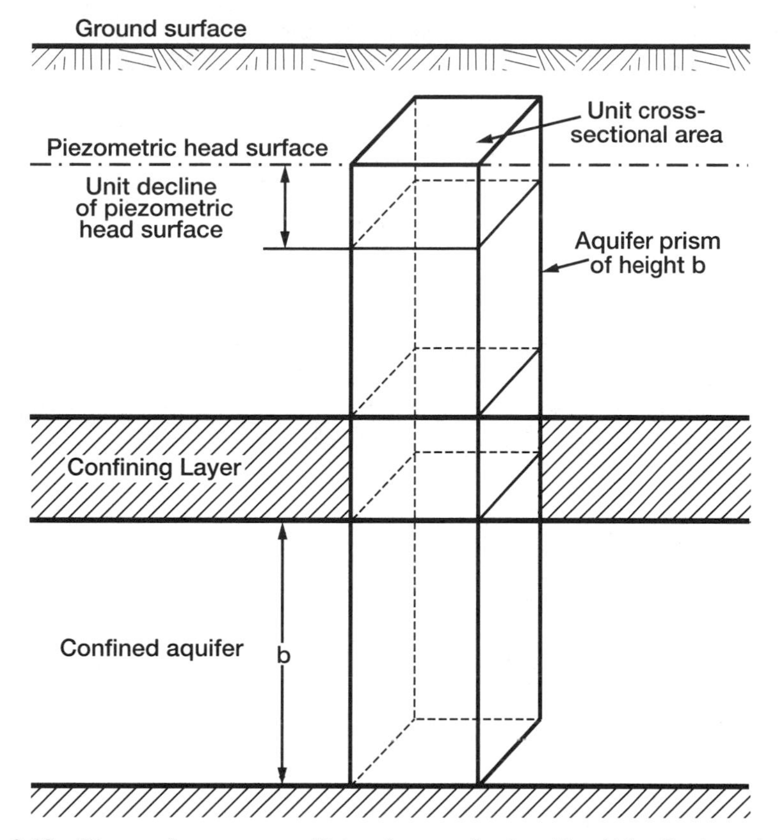

Figure 2-18 Diagram for storage coefficient for a confined aquifer. (After Ferris et al., 1962.)

Solution: From Table 2-11, the specific weight and compressibility of water at 5°C are $\gamma = 9.807\ \text{kN/m}^3$ and $\beta = 4.88 \times 10^{-10}\ \text{m}^2/\text{N}$, respectively. From Eq. (2-71), with the unit conversion, the value of specific storage is

$$S_s = \frac{K_z}{c_v} = \frac{10^{-9}\ \text{m/s}}{2.94 \times 10^{-7}\ \text{m}^2/\text{s}} = 3.40 \times 10^{-3}\ \text{m}^{-1}$$

And from Eq. (2-75), the storage coefficient value is

$$S = S_s b = (3.40 \times 10^{-3}\ \text{m}^{-1})(10\ \text{m}) = 3.4 \times 10^{-2}$$

From Eq. (2-78), the water storage coefficient value is

$$S_w = n\beta\gamma b = (0.35)(4.88 \times 10^{-10}\ \text{m}^2/\text{N})(9.807\ \text{kN/m}^3)(10.0\ \text{m}) = 1.68 \times 10^{-5}$$

Furthermore,

$$\frac{S_w}{S} = \frac{1.68 \times 10^{-5}}{3.4 \times 10^{-2}} = 4.94 \times 10^{-4}$$

which shows that the storage coefficient of water is a relatively small fraction of the storage coefficient of clay ($4.94 \times 10^{-2}\%$). Therefore, the storage coefficient of water is negligible.

Example 2-6: The total porosity (n) of an unconsolidated sandy aquifer is 0.40, and its thickness is 40 m. Determine the storage coefficient of water (S_w) at 15°C.

Solution: From Table 2-11, the specific weight and compressibility of water at 15°C are $\gamma = 9.798$ kN/m^3 and $\beta = 4.65 \times 10^{-10}$ m^2/N, respectively. From Eq. (2-78) the value of specific storage for water is

$$S_w = n\beta\gamma b = (0.40)(4.65 \times 10^{-10}\ \text{m}^2/\text{N})(9.798\ \text{kN/m}^3)(40\ \text{m}) = 7.29 \times 10^{-5}$$

Example 2-7: Pumping test was conducted for the aquifer described in Example 2-6, and the storage coefficient was found to be $S = 2.0 \times 10^{-5}$. Is this value correct?

Solution: The storage coefficient of water (S_w) cannot be higher than the storage coefficient (S) of the aquifer. Because $S = 2.0 \times 10^{-5} < S_w = 7.29 \times 10^{-5}$, the aquifer test value is in error.

Example 2-8: A confined aquifer has the following characteristics: $A = 6.5 \times 10^5$ m^2 (area), $b = 40.0$ m (thickness), and $S = 2.0 \times 10^{-3}$ (storage coefficient). By assuming no leakage through the underlain and overlain confining layers, estimate the volume of recovered water by reducing the hydraulic head, $\Delta h = 10$ m, in the aquifer.

Solution: The specific storage value is needed to carry out the calculation. From Eq. (2-75), the specific storage value is

$$S_s = \frac{S}{b} = \frac{2 \times 10^{-3}}{40.0\ \text{m}} = 5.0 \times 10^{-5}\ \text{m}^{-1}$$

The aquifer volume is

$$V = Ab = (6.5 \times 10^5\ \text{m}^2)(40.0\ \text{m}) = 2.6 \times 10^7\ \text{m}^3$$

and

$$\text{Volume recovered} = AS_s\Delta h = (2.6 \times 10^7\ \text{m}^3)(5.0 \times 10^{-5}\ \text{m}^{-1})(10\ \text{m}) = 1.3 \times 10^4\ \text{m}^3$$

2.4.2.2 Storage Coefficient of Layered Systems. The equations for the storage coefficient developed in Section 2.4.2.1 are for single layers. However, most systems encountered in nature are composed of more than one layer, and each layer has its own hydrogeologic characteristics. An approach for this kind of systems is presented below (Hantush, 1960b; Jorgensen, 1980).

The equations presented by Jorgensen (1980) are based on the results of Hantush's modified theory for leaky aquifers (Hantush, 1960b), which is presented in detail in Chapter 7. The analysis of Hantush under pumping conditions from a confined aquifer, underlain and/or overlain by confining layers, for a relatively long period of time showed that the drawdown equations are the function of the sum of the storage coefficients of the individual layers [see Eqs. (7-49), (7-57), and (7-61) in Chapter 7]. Based on these results,

Jorgensen (1980) concluded that, for some purposes, the storage coefficient of a layered system, called the *effective storage coefficient*, can be considered to be equal to the sum of the storage coefficients of the individual layers. Therefore, the effective storage coefficient of the system can be written as

$$S_{\text{system}} = S_1 + S_2 + \cdots + S_n \tag{2-80}$$

and from Eq. (2-75), in terms of specific storage, one gets

$$S_{\text{system}} - S_{s_1} b_1 + S_{s_2} b_2 + \cdots + S_{s_n} b_n \tag{2-81}$$

where S_{s_n} and n_n are the specific storage and thickness of the nth layer, respectively. Thus, the storage coefficient for the system is

$$S_{\text{system}} = S_{s_{\text{system}}} b_T \tag{2-82}$$

where b_T is the total thickness of the aquifer:

$$b_T = b_1 + b_2 + \cdots + b_n \tag{2-83}$$

The specific storage of the system, which is composed of n layers, is the weighted mean of the specific storage of the individual layers:

$$S_{s_{\text{system}}} = \frac{S_{s_1} b_1 + S_{s_2} b_2 + \cdots + S_{s_n} b_n}{b_T} \tag{2-84}$$

2.4.3 Specific Yield

The *specific yield*, S_y, is the measure of water yielded by gravity drainage when the water table of an unconfined aquifer declines. And, in general terms, it is defined as *the volume of water released from or taken into a unit area of an unconfined aquifer when the water table moves by one unit of height* (Meinzer, 1923; Ferris et al., 1962).

The definition of specific yield is further illuminated on an unconfined aquifer (Figure 2-19), which has uniform thickness and whose lower boundary is horizontal for the sake of convenience (Ferris et al., 1962). When the hydraulic head in the unconfined aquifer is dropped, some volume of water will be released by gravity drainage. Let us consider a representative parallelepiped extending from the base of the aquifer to its water table as shown in Figure 2-19. The ratio between the volume of water involved in the gravity drainage or refilling in the parallelepiped (ΔV_g) and $(\Delta A)(\Delta h)$ is the *specific yield*

$$S_y = \frac{\Delta V_g}{(\Delta A)(\Delta h)} \tag{2-85}$$

where ΔA and Δh are the horizontal cross-sectional area of the parallelepiped and the change in water-table elevation, respectively. Like the storage coefficient, the specific yield is a dimensionless quantity.

The specific yields of unconfined aquifers (S_y) are much higher than the storage coefficients (S) of confined aquifers. The specific yields of unconfined aquifers generally range

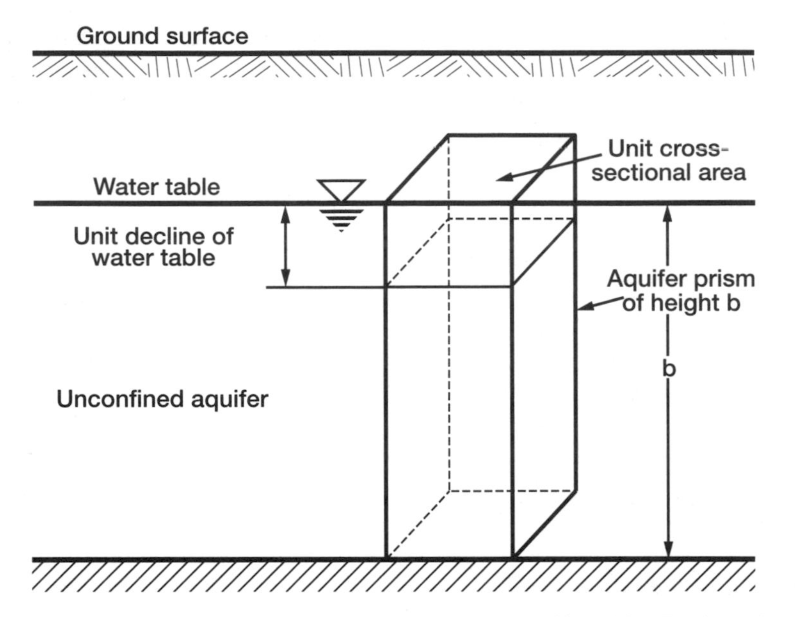

Figure 2-19 Diagram for specific yield for an unconfined aquifer. (After Ferris et al., 1962.)

from about 0.01 to 0.30, depending on the type of porous materials. For fine-grained porous materials the specific yield values are significantly different from their porosity values. For coarse-grained materials the specific yield values are very close to their porosity values. Typical values for various porous materials are given in Table 2-16. Methods for the determination of specific yield include the analysis of time versus drawdown data from pumping tests in unconfined aquifers and are included in Chapter 9.

2.4.4 Specific Retention

The *specific retention*, S_r, is the measure of water retained in the soil against gravity by capillarity and hygroscopic forces when the water table of an unconfined aquifer drops. It is defined as *the volume of water retained against gravity in a unit area of an unconfined aquifer when the water table drops by one unit of height* (Meinzer, 1923; Lohman, 1972).

The definition of specific retention can further be illuminated on an unconfined aquifer (Figure 2-19), which has uniform thickness and whose lower boundary is horizontal for the sake of convenience. As mentioned above, when the hydraulic head in the unconfined aquifer is dropped, some volume of water will be retained by capillarity and hygroscopic forces against gravity. Let us consider a representative parallelepiped extending from the base of the aquifer to its water table, as shown in Figure 2-19. The ratio between the volume of water involved in the retention in the parallelepiped (ΔV_r) and $(\Delta A)(\Delta h)$ is the *specific retention*

$$S_r = \frac{\Delta V_r}{(\Delta A)(\Delta h)} = n - S_y \tag{2-86}$$

where ΔA and Δh are the horizontal cross-sectional area of the parallelepiped and the change in water-table elevation, respectively.

TABLE 2-16 Specific Yield of Various Porous Materials

Group	Porous Material	Number of Analyses	Range of Specific Yield (S_y)	Arithmetic Mean Average Specific Yield ($S_{y_{av}}$)
Sedimentary materials	Sandstone (fine)	47	0.02–0.40	0.21
	Sandstone (medium)	10	0.12–0.41	0.27
	Siltstone	13	0.01–0.33	0.12
	Sand (fine)	287	0.01–0.46	0.33
	Sand (medium)	297	0.16–0.46	0.32
	Sand (coarse)	143	0.18–0.43	0.30
	Gravel (fine)	33	0.13–0.40	0.28
	Gravel (medium)	13	0.17–0.44	0.24
	Gravel (coarse)	9	0.13–0.25	0.21
	Silt	299	0.01–0.39	0.20
	Clay	27	0.01–0.18	0.06
	Limestone	32	0–0.36	0.14
Wind-laid materials	Loess	5	0.14–0.22	0.18
	Eolian sand	14	0.32–0.47	0.38
	Tuff	90	0.02–0.47	0.21
Metamorphic rocks	Schist	11	0.22–0.33	0.26

Source: After Morris and Johnson, (1967), as presented in Mercer et al. (1982).

Like the storage coefficient (S) and specific yield (S_y), the specific retention is a dimensionless quantity. To determine the values of specific retention (S_r), the values in Table 2-16 can be subtracted from the porosity values in Table 2-1 according to Eq. (2-86).

2.4.5 Barometric Efficiency

Water levels observed in wells under both confined and unconfined flow conditions are commonly affected by changes in atmospheric pressure, which is also called the *barometric pressure*. Barometric pressure effects on water levels in wells were first noted by the French mathematician and philosopher Blaise Pascal in 1663 (Crawford, 1994). These changes sometimes become significant, especially during relatively long aquifer pumping tests. Typically, a pumping test period may range from a couple of hours to a couple of days, depending on the purpose and type of aquifer materials. If the barometric pressure changes significantly and causes sizable fluctuations during the test period, the drawdown data collected from an observation well distant from the pumping well will be inadequate for interpreting the test. As a result, to estimate the undisturbed hydraulic head, water-level fluctuations in observation wells resulting from the barometric pressure changes must be quantified, and the necessary changes must be made on the disturbed drawdown data.

Barometric efficiency (BE) is defined as the ratio of change in hydraulic head to the change in barometric pressure head. Barometric efficiency can be used for the correction of drawdown data. Moreover, barometric efficiency can also be used to estimate the storage coefficient of a confined aquifer (Jacob, 1940) and bulk elastic properties (Domenico, 1983; Rojstaczer and Agnew, 1989).

According to some pertinent literature, the barometric pressure fluctuations are only important for confined aquifers; in unconfined aquifers, water levels are not affected by the barometric pressure fluctuations (e.g., Ferris et al., 1962; Todd, 1980). Some studies conducted at the beginning of the nineteenth century showed that barometric pressure fluctuations may be important for the water-level fluctuations in deep unconfined aquifers as well (Weeks, 1979).

In this section, the theoretical foundations of the effects of barometric fluctuations on the hydraulic heads in both unconfined and confined aquifers as well as the estimation method of the undisturbed hydraulic head will be presented, with examples.

2.4.5.1 Barometric Fluctuations in Wells in Unconfined Aquifers. As mentioned above, some studies from the beginning of the nineteenth century showed that barometric pressure fluctuations can be important for the water-level fluctuations in deep unconfined aquifers. A theoretical explanation regarding this situation is presented below based on Weeks (1979).

The water level in a well screened only below the water table in an unconfined aquifer fluctuates in response to barometric changes at land surface. The magnitude of these fluctuations may be significant if the overlying materials in the unsaturated zone are thick or have relatively low permeability to air at their prevailing water content. As a result, it is sometimes necessary to correct observed water-level data collected during tests on unconfined aquifers for such effects.

Water levels in wells screened below the water table of unconfined aquifers are affected by changes in barometric pressure because of the fact that air must move into or out of the unsaturated zone above the water table to transmit the pressure change to the water table. This movement is relatively slow because of the relatively low permeability of the unsaturated materials and because of their capacity to store or release soil gas as the pressure changes. As a result, the change in soil gas pressure at the water table lags behind that at ground surface. But barometric pressure changes are transmitted essentially instantaneously. This situation results in a pressure imbalance between the water level in the well and the water level in the adjacent aquifer. Consequently, the pressure difference creates fluctuation of water level in the well.

The aforementioned phenomenon is illustrated schematically in Figure 2-20 by considering the effects of a step change in barometric pressure on the water level in a well whose screen interval is below the water table. After the atmospheric pressure changes at the ground surface by an amount $\Delta(p_a/\gamma)$, only a fraction, $\Delta(p_t/\gamma)$, of the change is transmitted through the unsaturated zone to the water table. This generally takes place in a relatively short period of time. The same atmospheric pressure change is transmitted unattenuated to the water surface in the well. As a result, a temporary pressure imbalance is created between the water in the well and that in the aquifer. This results in a water-level decline that is equal to the pressure head difference of $[\Delta(p_a/\gamma) - \Delta p_t/\gamma)]$. As time goes on, the whole pressure change, $\Delta(p_a/\gamma)$, is transmitted through the unsaturated zone, and the water level recovers to its initial position.

2.4.5.2 Barometric Fluctuations in Wells in Confined Aquifers. Barometric pressure fluctuations can cause significant effects on the water level in a well tapping a confined aquifer. A theoretical explanation for this phenomenon is presented below based on Ferris et al. (1962) and Weeks (1979).

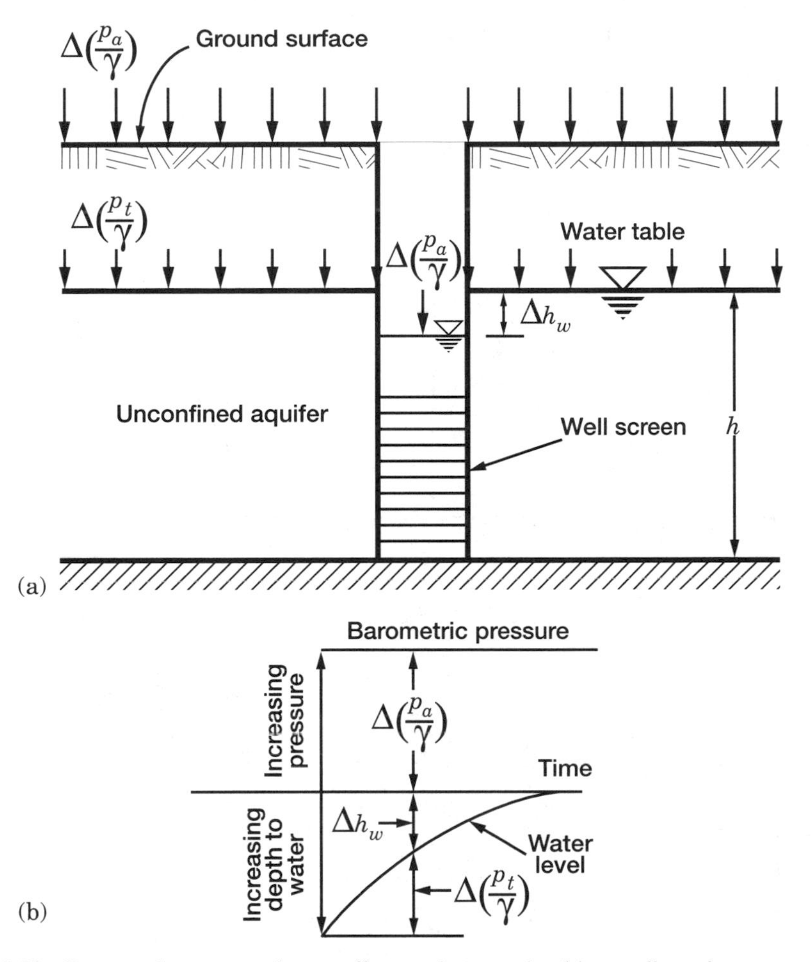

Figure 2-20 Barometric pressure change effect on the water level in a well tapping an unconfined aquifer: (a) Idealized section of an unconfined aquifer. (b) Idealized barograph and hydrograph showing water-level response with time. (After Weeks, 1979.)

A well tapping a confined aquifer is shown in Figure 2-21. Based on the model of Jacob (1940) for barometric pressure effects on confined aquifers, the pressure change is transmitted instantaneously without attenuation to the interface between the confining layer and the confined aquifer. In Figure 2-21, the force $\Delta(p_a/\gamma)$, which represents the change in atmospheric pressure, acts on the free water surface in the well as well as on the interface between the confining layer and aquifer. According to the basis of the theory of Jacob (1940), the fluctuations in a well in a confined aquifer are an index of the elasticity of the aquifer. In other words, the confining layer, which is viewed as a unit, has no resistance to withstand or contain any sensible part of an applied load. This means that the whole atmospheric pressure head change, $\Delta(p_a/\gamma)$, is transmitted to the interface without any reduction. This force as well as the other forces acting at a point at the interface between the aquifer and confining layer are shown in the inset sketch in Figure 2-21. As shown in this sketch, two opposite stress increments, $b\Delta p_w$ (b is the fraction of the interface between

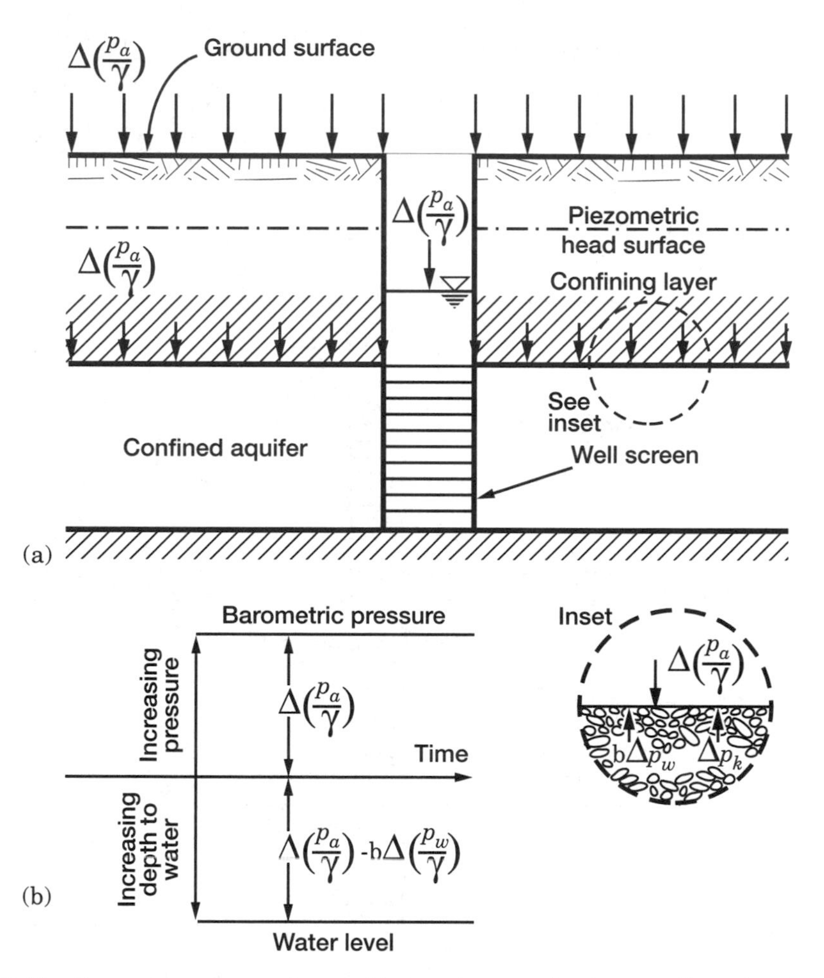

Figure 2-21 Barometric pressure change effect on the water level in a well tapping a confined aquifer: (a) Idealized section of a confined aquifer. (b) Idealized barograph and hydrograph showing water-level response with time. (After Weeks, 1979.)

the confined aquifer and confining layer that is in contact with the water in the aquifer, and Δp_w is the portion of the atmospheric pressure change borne by the water in the confined aquifer) and Δp_k (the portion of the atmospheric pressure change borne by the aquifer skeleton in the confined aquifer), plotted against $\Delta(p_a/\gamma)$, are created at the interface. As shown in Figure 2-21, within the well, the pressure change is borne entirely by the water in the well. As a result, a pressure head imbalance equal to $[\Delta(p_a/\gamma) - \Delta(p_w/\gamma)]$ is created between the water in the well and the pore water. This pressure head imbalance results in a water-level change in the well. Therefore, under confined aquifer conditions, water-level fluctuation in a well will be observed as a result of barometric pressure fluctuations.

Relationship Between Barometric Efficiency and Confined Aquifer Properties. As mentioned above, *barometric efficiency* (BE) is defined as the ratio of change in hydraulic head to the change in barometric pressure head:

$$BE = \frac{\Delta h}{\Delta \left(\dfrac{p_a}{\gamma} \right)} \tag{2-87}$$

The barometric efficiency of a confined aquifer is related to its properties. The derivation of this relationship is presented below (Jacob, 1940; Hantush, 1964).

With the assumption that the whole atmospheric pressure head change, $\Delta(p_a/\gamma)$, is transmitted to the interface without any reduction, Eq. (2-41) takes the form

$$p + \sigma = \sigma_T = p_a + \text{ constant} \tag{2-88}$$

where p_a is atmospheric pressure. By differentiation, Eq. (2-88) gives

$$dp - dp_a = -d\sigma \tag{2-89}$$

The weight of the column of water in the well above the upper boundary of the aquifer plus the weight of the atmosphere is balanced by the water pressure in the aquifer. Therefore, the change of water pressure is given by

$$dp = dp_a + \gamma dh \tag{2-90}$$

where h is the elevation of the water surface in the well. From Eqs. (2-89) and (2-90), equations can be written, respectively, as

$$\frac{dp_a}{\gamma} = \frac{dp}{\gamma} + \frac{d\sigma}{\gamma} \tag{2-91}$$

$$dh = \frac{dp}{\gamma} - \frac{dp_a}{\gamma} \tag{2-92}$$

Dividing both sides of Eq. (2-92) by the corresponding sides of Eq. (2-91) and using Eq. (2-89), one gets

$$\frac{dh}{\dfrac{dp_a}{\gamma}} = -\frac{1}{\dfrac{dp}{d\sigma} + 1} \tag{2-93}$$

The left-hand side of Eq. (2-93) is nothing but the definition of *barometric efficiency* (BE) as defined by Eq. (2-87). Therefore,

$$\text{BE} = \frac{dh}{\dfrac{dp_a}{\gamma}} = -\frac{1}{1 + \dfrac{dp}{d\sigma}} \tag{2-94}$$

If the aquifer grains are assumed to be incompressible, then as the aquifer is compressed the change in the bulk volume $d(\Delta V)$ is equal to the change of water volume $d(\Delta V_w)$.

Therefore, the following relations can be written:

$$\Delta V_w = n\Delta V, \qquad d(\Delta V_w) = d(\Delta V) \tag{2-95}$$

And dividing their corresponding sides gives

$$\frac{d(\Delta V_w)}{\Delta V_w} = \frac{d(\Delta V)}{d\Delta V} \tag{2-96}$$

From the definition of the bulk modulus of elasticity of water, as given by Eq. (2-55) in Section 2.4.1, one can write

$$\frac{d(\Delta V_w)}{\Delta V_w} = -\beta dp \tag{2-97}$$

where β is the compressibility of water, which is the reciprocal of its bulk modulus of elasticity, E_w. For the solid skeleton, from Eq. (2-49) one can write

$$\frac{d(\Delta V)}{\Delta V} = -\alpha d\sigma \tag{2-98}$$

where α is the vertical compressibility of the solid skeleton, which is the reciprocal of the modulus of elasticity, E_s, as defined by Eq. (2-44). The negative signs in Eqs. (2-97) and (2-98) indicate decrease in volume accompanying increase in stress. Substituting Eqs. (2-97) and (2-98) into Eq. (2-96) gives

$$\frac{dp}{d\sigma} = \frac{\alpha}{n\beta} \tag{2-99}$$

and introducing this into Eq. (2-94) yields the following equation:

$$\text{BE} = \frac{dh}{d\left(\dfrac{p_a}{\gamma}\right)} = -\frac{1}{1 + \dfrac{\alpha}{n\beta}} \tag{2-100}$$

The negative sign in Eq. (2-100) indicates that a rise in barometric pressure is accompanied by depression of water level in the well. From Eqs. (2-63), (2-75), and (2-100) one finally obtains

$$S = \frac{\gamma n\beta b}{\text{BE}} \tag{2-101}$$

Equation (2-101) shows that if the barometric efficiency (BE) is known, the storage coefficient (S) can be estimated using the estimated thickness of the aquifer (b), the specific weight of water (γ), the porosity (n) of the aquifer, and the compressibility of water (β).

2.4.5.3 Estimation of Barometric Efficiency and Drawdown Correction. As mentioned above, the *barometric efficiency* is defined as the ratio of change in hydraulic head to the change in barometric pressure head and is expressed by Eq. (2-87). This equation is the foundation of the estimation methods, and two methods will be presented based on this equation. As mentioned above, significant barometric pressure fluctuations are especially important in evaluating drawdown data collected from discharging well tests. It is advised to measure both the water levels in the wells and local barometric pressure for several days prior to the start of the test to determine the influence of pressure changes on water levels. The methods presented below are based on a constant barometric efficiency parameter, which is based on the assumption that the aquifer is confined, elastic, homogeneous, and uniform in thickness and infinite in areal extent and that borehole storage effects are minimal (Davis and Rasmussen, 1993). The first method (Method I) is based on the assumption that the changes in water level were due only to changes in the atmospheric pressure. The second method (Method II), which was developed by Clark (1967), takes into account the effects other than the atmospheric pressure. These methods are presented below.

Method I. This method is based on a constant barometric efficiency parameter. It is also based on the assumption that the changes in water level were due only to changes in the atmospheric pressure. Using the classical method (U.S. Department of the Interior, 1981), the value of barometric efficiency can be estimated by plotting the water-level changes as ordinates and barometric pressure changes as abscissas in the Cartesian coordinate system. The slope of the straight line fitted through the plotted points is the barometric efficiency (BE), which may be as high as 80%. The method is explained in Example 2.9.

Example 2-9: This example is adapted from the *Ground Water Manual* of the U.S. Department of the Interior (1981). Barometric pressure head and the observed head values in a well in an aquifer are given in Table 2-17. The measurements started at 08 00 hour and continued until 20 00 hour with 1-hour increments. Using these data, estimate the barometric efficiency. Then, with the estimated barometric efficiency value, determine the undisturbed head values in the well during the measurement period.

TABLE 2-17 Observed Barometric Pressure and Water-Level Head Data for Example 2-9

Time, t (hours)	p_a/γ (m)	h (m)
08 00	10.256	1604.254
09 00	10.247	1604.257
10 00	10.244	1604.260
11 00	10.241	1604.260
12 00	10.232	1604.266
13 00	10.223	1604.269
14 00	10.217	1604.275
15 00	10.196	1604.284
16 00	10.208	1604.278
17 00	10.214	1604.275
18 00	10.217	1604.275
19 00	10.233	1604.269
20 00	10.232	1604.260

TABLE 2-18 Determination of $\Sigma\Delta(p_a/\gamma)$ Versus $\Sigma\Delta h$ for Example 2-9

Time, t (hours)	p_a/γ (m)	$\Delta(p_a/\gamma)$ (m)	$\Sigma\Delta(p_a/\gamma)$ (m)	h (m)	Δh (m)	$\Sigma\Delta h$ (m)
08 00	10.256	0.000	0.000	1604.254	0.000	0.000
09 00	10.247	−0.009	−0.009	1604.257	+0.003	+0.003
10 00	10.244	−0.003	−0.012	1604.260	+0.003	+0.006
11 00	10.241	−0.003	−0.015	1604.260	+0.000	+0.006
12 00	10.232	−0.009	−0.024	1604.266	+0.006	+0.012
13 00	10.223	−0.009	−0.033	1604.269	+0.003	+0.015
14 00	10.217	−0.006	−0.039	1604.275	+0.006	+0.021
15 00	10.196	−0.021	−0.060	1604.284	+0.009	+0.030
16 00	10.208	+0.012	−0.048	1604.278	−0.006	+0.024
17 00	10.214	+0.006	−0.042	1604.275	−0.003	+0.021
18 00	10.217	+0.003	−0.039	1604.275	−0.000	+0.021
19 00	10.233	+0.006	−0.033	1604.269	−0.006	+0.015
20 00	10.232	+0.009	−0.024	1604.260	−0.009	+0.006

Solution: Using the values in Table 2-17, the calculated data for $\Sigma\Delta(p_a/\gamma)$ versus $\Sigma\Delta h$ are given in Table 2-18. And the corresponding graph is given in Figure 2-22, where the geometric slope of the fitted line is shown. Therefore, application of Eq. (2-87) gives

$$\text{BE} = \frac{\Delta(\Sigma\Delta h)}{\Delta\left[\Sigma\Delta(dp_a/\gamma)\right]} = \frac{0.01\ \text{m}}{0.02\ \text{m}} = 0.50$$

In other words, BE is the geometric slope of the $\Sigma\Delta h$ versus $\Sigma\Delta(p_a/\gamma)$ line as shown in Figure 2-22.

With this BE value, the hydraulic heads are corrected and listed in Table 2-19. Negative $\Delta(p_a/\gamma)$ values correspond to higher heads. Therefore, in Table 2-19, Δh values are subtracted from the observed head values. For example,

$$h = 1604.257\ \text{m}$$

$$\Delta h = \text{BE}\ \Delta\left(\frac{p_a}{\gamma}\right) = (0.50)(-0.009\ \text{m}) = -0.0045\ \text{m}$$

$$h_c = h - \Delta h = 1604.257\ \text{m} - 0.0045\ \text{m} = 1604.2525\ \text{m}$$

For positive Δh values they are added to the h values.

Example 2-10: The observed barometric efficiency (BE = 0.50) determined in Example 2-9 belongs to a confined aquifer whose average thickness (b) is 45 m and whose average porosity (n) is 0.32. The specific weight (γ) and compressibility of water (β) are given as 9799.74 N/m^3 and 4.786×10^{-10} m^2/N, respectively. Estimate the storage coefficient of the aquifer.

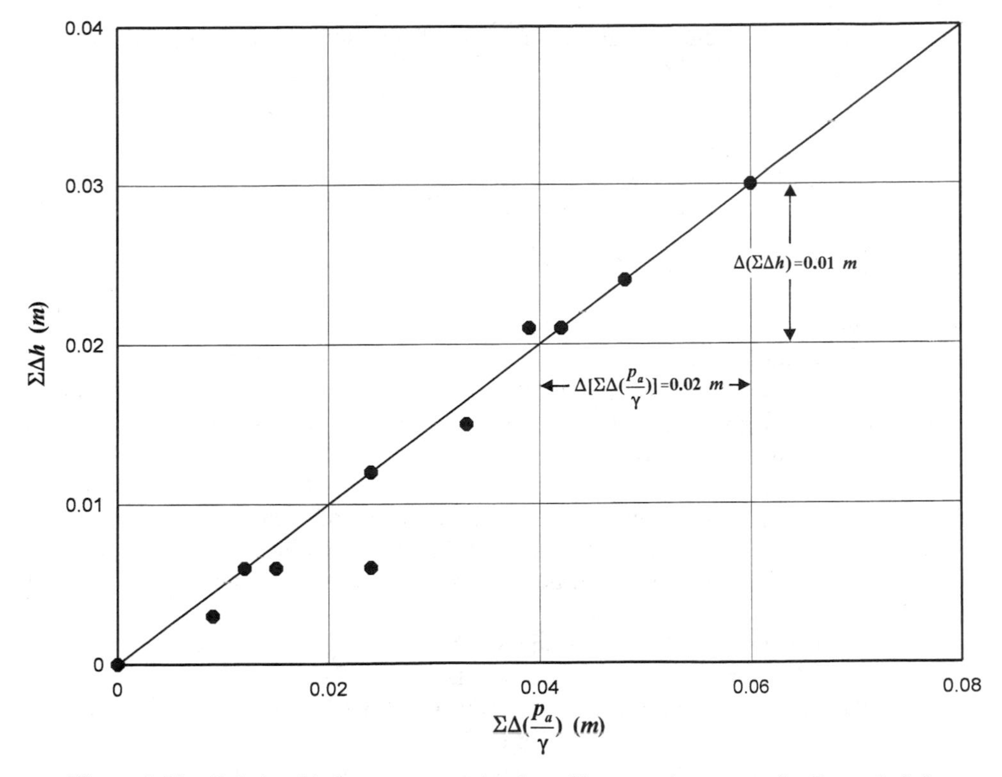

Figure 2-22 Relationship between water levels and barometric pressure for Example 2-9.

TABLE 2-19 Corrected Hydraulic Heads for Example 2-9

Time, t (hours)	h (m)	$\Delta p_a/\gamma$ (m)	$\Delta h = \text{BE}\Delta(p_a/\gamma)$ (m)	Corrected Head, h_c (m)
08 00	1604.254	0.000	0.0000	1604.2540
09 00	1604.257	−0.009	−0.0045	1604.2525
10 00	1604.260	−0.003	−0.0015	1604.2585
11 00	1604.260	−0.003	−0.0015	1604.2585
12 00	1604.266	−0.009	−0.0045	1604.2615
13 00	1604.269	−0.009	−0.0045	1604.2645
14 00	1604.275	−0.006	−0.0030	1604.2720
15 00	1604.284	−0.021	−0.0105	1604.2735
16 00	1604.278	+0.012	+0.0060	1604.2840
17 00	1604.275	+0.006	+0.0030	1604.2780
18 00	1604.275	+0.003	+0.0015	1604.2765
19 00	1604.269	+0.006	+0.0030	1604.2720
20 00	1604.260	+0.009	+0.0045	1604.2645

Solution: Substituting the given values in Eq. (2-101) gives the storage coefficient value:

$$S = \frac{\gamma n \beta b}{\text{BE}}$$

$$= \frac{(9799.74\ \text{N/m}^3)(0.32)(4.786 \times 10^{-10}\ \text{m}^2\text{/N})(45\ \text{m})}{0.50}$$

$$S = 1.35 \times 10^{-4}$$

Method II: Clark's Method. When the water level in a confined aquifer is fluctuating due to causes other than changes in barometric pressure (such as temporal variations in regional recharge and discharge rates) and earth tides, estimation of barometric efficiency from a data set of hydraulic head versus barometric pressure is not straightforward. To eliminate the influence of effects other than the atmospheric pressure, Clark (1967) developed a method for estimating the barometric efficiency. Clark's method is based on the assumption that the barometric efficiency is a constant and that rapid equilibration occurs between barometric pressure perturbations and water-level responses in wells (Davis and Rasmussen, 1993).

Clark's method employs observed changes in barometric pressure head, $\Delta(p_a/\gamma)$, and hydraulic head, Δh, for constant time increments to estimate the barometric efficiency. Clark's method assigns a positive sign to the value of Δh when the water level in the well is rising, and it also assigns a positive sign to $\Delta(p_a/\gamma)$ when the barometric pressure is decreasing. To estimate the barometric efficiency, two sums are made, $\Sigma\Delta(p_a/\gamma)$ and $\Sigma\Delta h$, in accordance with the following rules (Clark, 1967): (1) When $\Delta(p_a/\gamma)$ is zero, neglect the corresponding value of Δh in determining $\Sigma\Delta h$. (2) When Δh and $\Delta(p_a/\gamma)$ have the same signs, subtract the absolute value of Δh in determining $\Sigma\Delta h$; when Δh and $\Delta(p_a/\gamma)$ have opposite signs, subtract the absolute value of Δh in determining $\Sigma\Delta h$. (3) $\Sigma\Delta(p_a/\gamma)$ is the sum of the absolute values of $\Delta(p_a/\gamma)$.

Application of Clark's method and clarification of the aforementioned rules are shown with an example taken from Clark (1967) and illustrated in Figure 2-23. The reason for Rule 1 is clear. The reason for Rule 2 is clarified in Figure 2-23, which shows a hypothetical example regarding water level and atmospheric pressure head records and graphs. In constructing the graphs, Clark (1967) assumed that in the absence of any change in atmospheric pressure the water level would have risen steadily, as shown by the dashed lines in Figure 2-23. The water level that would have been observed in the well is shown in the upper graphs. These graphs were obtained by superposing the atmospheric pressure multiplied by the 100% assumed barometric efficiency on the trend of the water level.

In Table A of Figure 2-23, the values of $\Sigma\Delta h$ and $\Delta(p_a/\gamma)$ have the same sign. Graph A is the corresponding graph. As shown in Figure 2-23, the components resulting from the trend of the water level are included in Δh in time intervals 1 and 3 and are subtracted from Δh in time intervals 2 and 4. In Table B of Figure 2-23, the values of Δh and $\Delta(p_a/\gamma)$ have the same signs and opposite signs in time intervals 2 and 4. In time intervals 1 and 3, Δh is composed of the component yielded by the trend plus the component yielded by the change in atmospheric pressure head. In time intervals 2 and 4, Δh is composed of the component yielded by the trend minus the component yielded by the change in atmospheric pressure head. When Δh and $\Delta(p_a/\gamma)$ have opposite signs, subtracting one from the other eliminates the component of the fluctuation resulting from the straight-line trend.

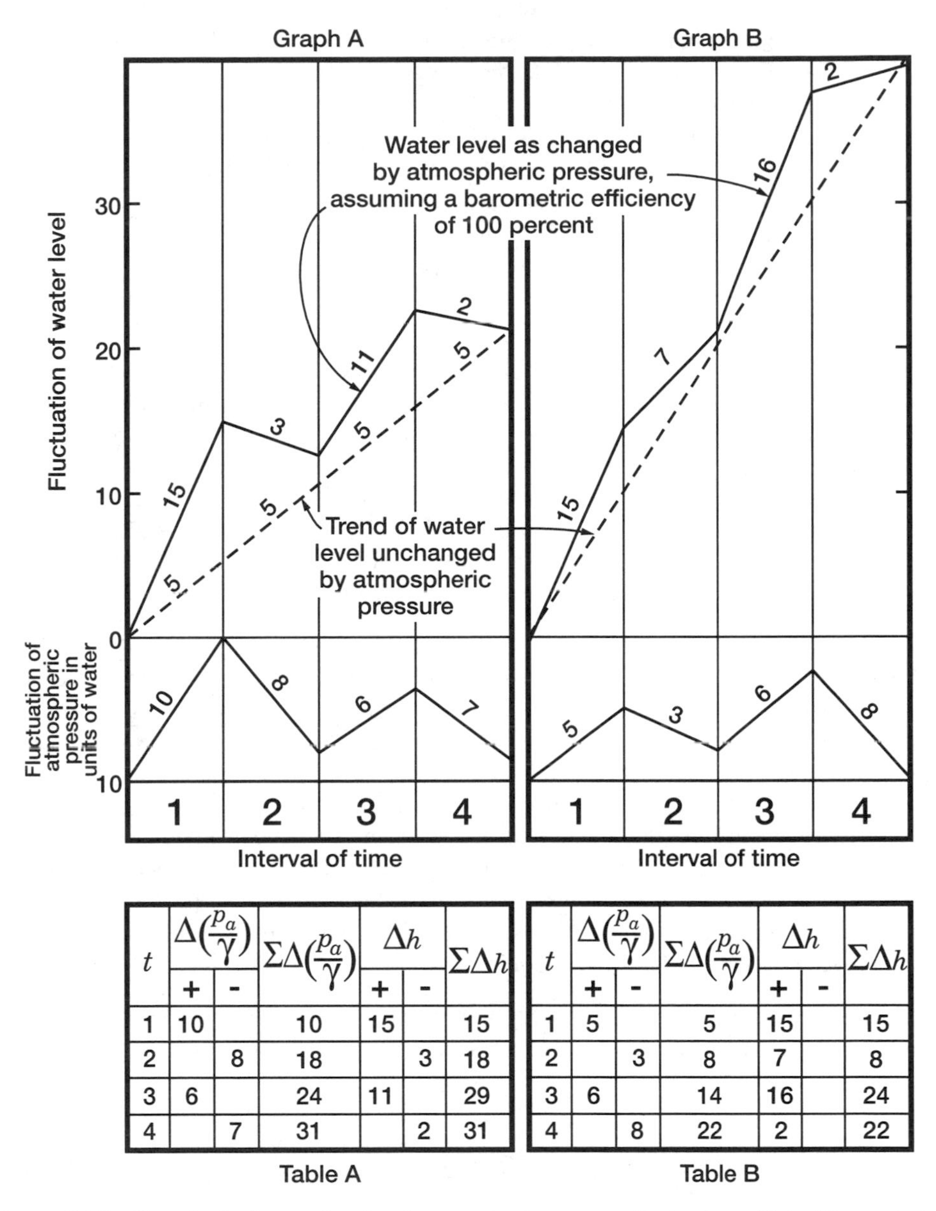

t	$\Delta\left(\frac{p_a}{\gamma}\right)$ +	$\Delta\left(\frac{p_a}{\gamma}\right)$ -	$\Sigma\Delta\left(\frac{p_a}{\gamma}\right)$	Δh +	Δh -	$\Sigma\Delta h$
1	10		10	15		15
2		8	18		3	18
3	6		24	11		29
4		7	31		2	31

Table A

t	$\Delta\left(\frac{p_a}{\gamma}\right)$ +	$\Delta\left(\frac{p_a}{\gamma}\right)$ -	$\Sigma\Delta\left(\frac{p_a}{\gamma}\right)$	Δh +	Δh -	$\Sigma\Delta h$
1	5		5	15		15
2		3	8	7		8
3	6		14	16		24
4		8	22	2		22

Table B

Figure 2-23 Hypothetical water level and atmospheric pressure fluctuations. *Note:* Numerals indicated number of units of change in water level or atmospheric pressure during increment of time. (Adapted from Clark, 1967.)

Once the aforementioned rules are used for generation of the values of Δh and $\Sigma\Delta(p_a/\gamma)$, the barometric efficiency is estimated from the following equation:

$$\text{BE} = \frac{\Sigma\Delta h}{\Sigma\Delta(p_a/\gamma)} \tag{2-102}$$

It is important to point out that the absolute value of $\Delta(p_a/\gamma)$ is always used to determine the $\Sigma\Delta(p_a/\gamma)$ term, while the term $\Sigma\Delta h$ can be positive or negative depending upon the corresponding value of $\Sigma\Delta(p_a/\gamma)$.

Davis and Rasmussen (1993) compared the results of Clark's method with the method of linear regression for estimating barometric efficiency of confined aquifers. The authors concluded that (1) Clark's method provides a robust estimate of the time-averaged barometric efficiency for a wide variety of confined aquifer conditions and parameters and (2) the estimate is consistent and unbiased when positive and negative changes in barometric pressure are equally likely.

2.4.5.4 Tidal Efficiency. It is well known that lunar and solar tide-generating forces can cause periodic fluctuations to the water level in a well in an aquifer. Studies have shown that certain water-level changes in wells in various parts of the world were the results of tidal effects (Robinson and Bell, 1971). Also, the aforementioned forces generate significant water-level fluctuations in surface water bodies, such as oceans, lakes, streams, and so on, depending on the location on the earth. In aquifers in contact with these kinds of water bodies, sinusoidal fluctuations of ground-water levels occur in response to tides. In the following sections, tidal fluctuations in wells in aquifers will be analyzed in coastal aquifers under tidal effects.

Tidal Fluctuation in Aquifers. In coastal areas of certain parts of the earth the periodic rise and fall of water level in the adjacent ocean or lake, hydraulically connected streams, and tidal marshes produce sinusoidal ground-water head fluctuations in adjacent aquifers. Ground-water head fluctuations produced by ocean tides in confined and unconfined aquifers are shown schematically in Figure 2-24. In confined aquifers, which are effectively separated from the body of surface water by an extensive confining layer, the response in hydraulic head is due to the changing load on the aquifer, transmitted through the confining layer with the changing surface water level because of tidal effects. In unconfined aquifers, the water-level response to tidal fluctuations in the surface water body is actual movement of water level in the aquifer.

Relationship Between Tidal Efficiency and Confined Aquifer Properties. *Tidal efficiency* (TE) of a confined aquifer is defined as the ratio of change in hydraulic head (h) to the change in tide stage (H):

$$\mathrm{TE} = \frac{\Delta h}{\Delta H} \tag{2-103}$$

As with the barometric efficiency case, the tidal efficiency of a confined aquifer is also related to its properties. The derivation of this relationship is presented below (Jacob, 1940; Hantush, 1964).

Because of the tidal effects, the vertical load on the aquifer changes with the stage of the tide. Therefore, the relation given by Eq. (2-41) takes the form

$$p + \sigma = \gamma H + \text{constant} \tag{2-104}$$

where H is the stage of the tide, corrected for density when necessary. By differentiation, Eq. (2-104) can be written as

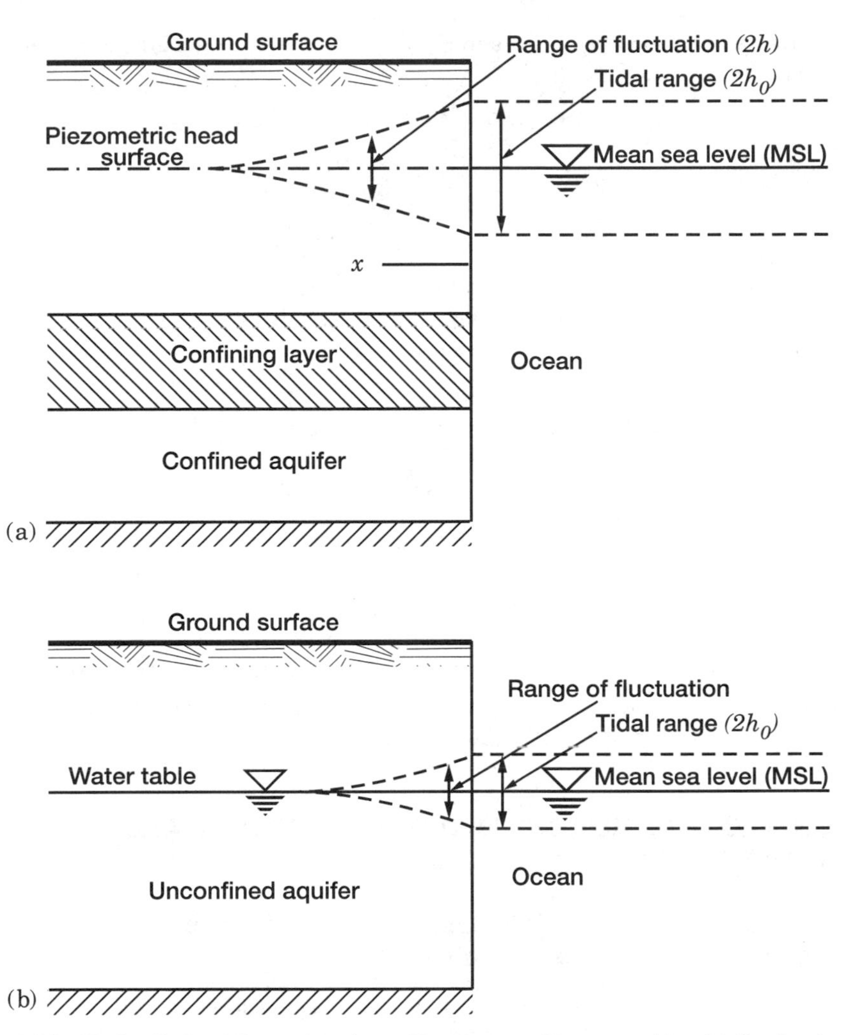

Figure 2-24 Hydraulic head fluctuations in aquifers generated by ocean tides: (a) Confined aquifer case. (b) Unconfined aquifer case. (Adapted from Todd, 1980.)

$$dp + d\sigma = \gamma dH \tag{2-105}$$

The change of the pressure of the water in the aquifer is given by

$$dp = \gamma dh \tag{2-106}$$

From Eqs. (2-105) and (2-106) one can write

$$\frac{dh}{dH} = \frac{dp}{\gamma dH} = \frac{\dfrac{dp}{d\sigma}}{1 + \dfrac{dp}{d\sigma}} \tag{2-107}$$

Substituting $dp/d\sigma$ from Eq. (2-99) into Eq. (2-107) gives the tidal efficiency defined by Eq. (2-103):

$$\mathrm{TE} = \frac{dh}{dH} = \frac{\dfrac{\alpha}{n\beta}}{1 + \dfrac{\alpha}{n\beta}} \tag{2-108}$$

The sum of the absolute values of the barometric efficiency (BE) as given by Eq. (2-100) and tidal efficiency as given by Eq. (2-108) equals unity; that is,

$$|\mathrm{BE}| + |\mathrm{TE}| = \left| \frac{-1}{1 + \dfrac{\alpha}{n\beta}} \right| + \left| \frac{\dfrac{\alpha}{n\beta}}{1 + \dfrac{\alpha}{n\beta}} \right| = 1 \tag{2-109}$$

Thus, from Eqs. (2-101) and (2-109)

$$TE = 1 - BE = 1 - \frac{\gamma n \beta b}{S} \tag{2-110}$$

Equation (2-110) means that for a perfectly elastic aquifer the percentage of storage attributable to expansion of water is equal to the barometric efficiency (BE), and that attributable to compression of aquifer is equal to the tidal efficiency (TE) (Hantush, 1964).

2.5 EQUATIONS OF MOTION IN AQUIFERS: GENERALIZATION OF DARCY'S LAW

2.5.1 Equations of Motion: Principal Directions of Anisotropy Coincide with the Directions of Coordinate Axes

Both Cartesian and cylindrical coordinates have wide applications in solving aquifer hydraulics problems. To derive the governing partial differential equations of flow in aquifers and to solve these equations, expressions of Darcy velocities in these coordinate systems are required. In the following sections these equations are presented.

2.5.1.1 Darcy Velocities in Two- and Three-Dimensional Cartesian Coordinates. Equations of motion will be written in the Cartesian coordinates for the special case where the principal directions of anisotropy coincide with x, y, and z directions of the coordinate axes. In other words, the Darcy velocity vector has three components that are aligned with the x, y, and z coordinate axes. These components are shown on an infinitesimal parallelepiped volume, which represents a point in the flow field (Figure 2-25).

Equation (2-19) is the one-dimensional differential form of Darcy's law; it states that the flow rate through porous materials in any direction is proportional to the negative rate of change of the hydraulic head in that direction. The meaning of the negative sign is that the fluid moves in the direction of decreasing hydraulic head.

Darcy's law was extended to three-dimensional cases (e.g., Polubarinova-Kochina, 1962; Hantush, 1964; Bear, 1972, 1979; Freeze and Cherry, 1979). The general form of Darcy's law in a nonhomogeneous and anisotropic porous medium with the principal axes of the

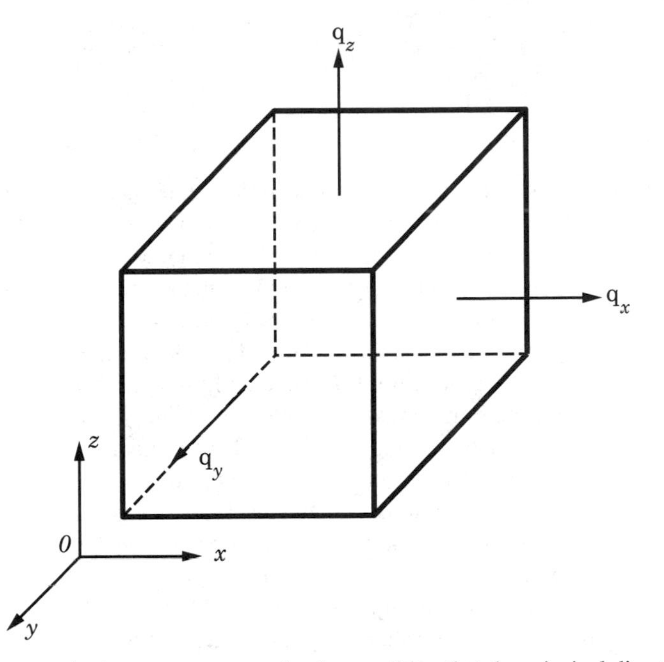

Figure 2-25 Darcy velocity components under the condition that the principal directions of anisotropy coincide with x, y, and z directions of the coordinate axes.

hydraulic conductivity tensor parallel to the coordinate of the Cartesian coordinate axes in vector form is

$$\mathbf{q} = q_x\mathbf{i} + q_y\mathbf{j} + q_z\mathbf{k} \tag{2-111}$$

where: $\mathbf{q}$ is the Darcy velocity vector; q_x, q_y, and q_z are the Darcy velocity components in the x, y, and z directions, respectively; and $\mathbf{i}$, $\mathbf{j}$, and $\mathbf{k}$ are the unit vectors in the same directions, respectively. The general spatially and temporarily variable forms of the Darcy velocity components in the x, y, and z directions, respectively, are

$$q_x = q_x(x, y, z, t) = -K_x(x, y, z)\frac{\partial h(x, y, z, t)}{\partial x} \tag{2-112a}$$

$$q_y = q_y(x, y, z, t) = -K_y(x, y, z)\frac{\partial h(x, y, z, t)}{\partial y} \tag{2-112b}$$

$$q_z = q_z(x, y, z, t) = -K_z(x, y, z)\frac{\partial h(x, y, z, t)}{\partial z} \tag{2-112c}$$

For the two-dimensional case, Eqs. (2-112a), (2-112b), and (2-112c), respectively, take the forms

$$q_x = q_x(x, y, t) = -K_x(x, y)\frac{\partial h(x, y, t)}{\partial x} \tag{2-113a}$$

$$q_y = q_y(x, y, t) = -K_y(x, y)\frac{\partial h(x, y, t)}{\partial y} \tag{2-113b}$$

$$q_z = 0 \tag{2-113c}$$

In reality, the hydraulic conductivity components of an aquifer are spatially variable quantities as indicated in the above equations. As a result, the Darcy velocity components will be spatially variable quantities as well. As will be shown in the remaining chapters, aquifer test analysis methods are based on analytical and semianalytical solutions that assume that the principal hydraulic conductivities do not vary with the coordinates. This assumption generally produces practical, reasonable results for most cases encountered in various aquifers. Recent studies based on finite-difference and finite-element numerical methods of solving ground-water problems showed that spatial variability of hydraulic conductivity corresponds to a lower level of spatially variable Darcy velocity components. In other words, the variability of Darcy velocity components is significantly less than the variability of hydraulic conductivity. For this purpose, Batu (1984) developed a finite-element method to determine Darcy velocities in any kind of nonhomogeneous and anisotropic aquifers, and the results on some hypothetical space-dependent hydraulic conductivity functions confirmed the above statements [see Figure 4 of Batu (1984)].

With the assumption of constant principal hydraulic conductivities, in aquifer hydraulics practice, it is further assumed that the principal hydraulic conductivities in the horizontal plane are the same; that is, $K_x(x, y, z) = K_y(x, y, z) = K_h$ and $K_z(x, y, z) = K_z$, Eqs. (2-112a), (2-112b), and (2-112c), respectively, take the following form:

$$q_x = q_x(x, y, z, t) = -K_h\frac{\partial h(x, y, z, t)}{\partial x} \tag{2-114a}$$

$$q_y = q_y(x, y, z, t) = -K_h\frac{\partial h(x, y, z, t)}{\partial y} \tag{2-114b}$$

$$q_z = q_z(x, y, z, t) = -K_z\frac{\partial h(x, y, z, t)}{\partial z} \tag{2-114c}$$

In Eq. (2-114c), the symbol of K_v is often used instead of K_z.

2.5.1.2 Darch Velocities in Two- and Three-Dimensional Cylindrical Coordinates. Cylindrical coordinates system has wide applications in deriving analytical solutions for circular vertical well hydraulics problems because of the geometric shape of wells; these solutions will be presented in subsequent chapters. A circular well generally has a finite radius, and it is assumed that the central axis of the well coincides with the z coordinate axis. The radial distance in a horizontal plane is measured from the vertical axis. In Figure 2-26 both the Cartesian and cylindrical coordinates are shown. Under extraction or injection conditions from or to a circular well, a radial flow to the well or from the well occurs. Under this condition, the hydraulic conductivity K_r is the only horizontal hydraulic conductivity instead of K_x and K_y, in the x and y directions, respectively. The vertical hydraulic conductivity K_z remains the same. Therefore, the equivalent forms of Eqs. (2-114) take the following forms:

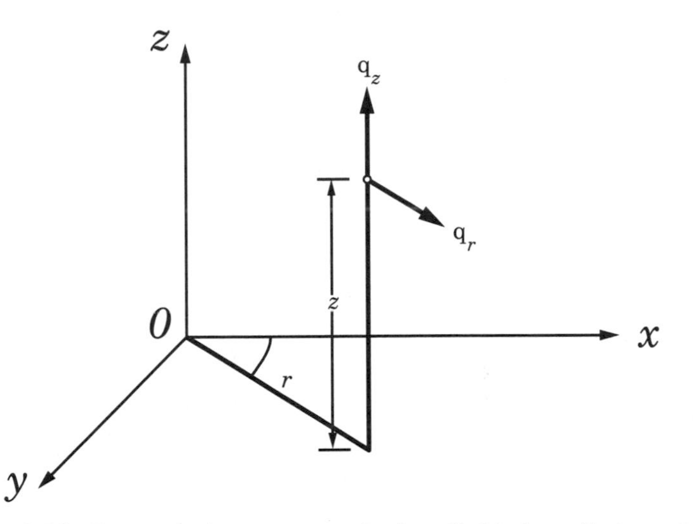

Figure 2-26 Darcy velocity components for the cylindrical coordinates system.

$$q_r = q_r(r, z, t) = -K_r \frac{\partial h(r, z, t)}{\partial r} \tag{2-115a}$$

$$q_z = q_z(r, z, t) = -K_z \frac{\partial h(r, z, t)}{\partial z} \tag{2-115b}$$

It is apparent that the equivalent forms of Eq. (2-115a) in the Cartesian coordinates are Eqs. (2-114a) and (2-114b). Equation (2-115a) is based on the assumption that the principal direction of the horizontal hydraulic conductivity is in the radial direction (K_r), and its value is the same in all directions around the well. The direction of the vertical hydraulic conductivity is in the direction of the z coordinate.

2.5.2 Equations of Motion: Principal Directions of Anisotropy Do Not Coincide with the Directions of Coordinate Axes

Equations of motion will be written for the general case where the principal directions of anisotropy do not coincide with x, y, and z directions of the coordinate axes. These components are shown on an infinitesimal parallelepiped volume, which represents a point in the flow field shown in Figure 2-27. The Darcy velocity components of the vector **q** given by Eq. (2-111) are (e.g., Bear, 1972; Domenico and Schwartz, 1990)

$$q_{xx} = -K_{xx} \frac{\partial h}{\partial x} - K_{xy} \frac{\partial h}{\partial y} - K_{xz} \frac{\partial h}{\partial z} \tag{2-116a}$$

$$q_{yy} = -K_{yx} \frac{\partial h}{\partial x} - K_{yy} \frac{\partial h}{\partial y} - K_{yz} \frac{\partial h}{\partial z} \tag{2-116b}$$

$$q_{zz} = -K_{zx} \frac{\partial h}{\partial x} - K_{zy} \frac{\partial h}{\partial y} - K_{zz} \frac{\partial h}{\partial z} \tag{2-116c}$$

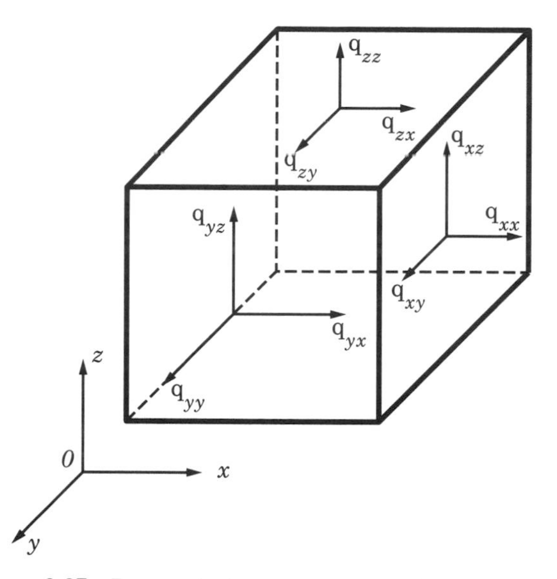

Figure 2-27 Darcy velocity components for the general case.

where x, y, and z are the Cartesian coordinates and $K_{xx}, K_{xy}, \ldots, K_{zz}$ are nine constant components of the hydraulic conductivity tensor in the most general case. The nine components in matrix form display a second-rank symmetric tensor known as the *hydraulic conductivity tensor* (Bear, 1972). In Eqs. (2-116), the first subscript indicates the direction perpendicular to the plane upon which the Darcy velocity vector acts, and the second subscript indicates the direction of the Darcy velocity vector in that plane. If the principal directions of anisotropy coincide with x, y, and z directions of the coordinate axes, the six components K_{xy}, K_{xz}, K_{yx}, K_{yz}, K_{zx}, and K_{zy} all become equal to zero and Eqs. (2-116) reduce to Eqs. (2-112) for $K_x(x, y, z) = K_x$, $K_y(x, y, z) = K_y$ and $K_z(x, y, z) = K_z$.

2.6 PRINCIPAL HYDRAULIC CONDUCTIVITIES

As mentioned in Section 2.5.1.1, if the principal directions of anisotropy coincide with x, y, and z directions of the coordinate axes, the Darcy velocity vector has three components expressed by Eqs. (2-114) that are aligned with the x, y, and z coordinate axes. In other words, the six components K_{xy}, K_{xz}, K_{yx}, K_{yz}, K_{zx}, and K_{zy} of the hydraulic conductivity tensor all become equal to zero. Under these conditions, K_{xx}, K_{yy}, and K_{zz} are called *principal hydraulic conductivities*. And the x, y, and z coordinate directions are called *principal directions*. Now let us consider the situation with Eqs. (2-116) and Figure 2-27 for which the aforementioned six components are not equal to zero. The question is: What is the orientation of the principal directions of anisotropy axes corresponding to Eqs. (2-116)? The answer is that there are three principal directions of anisotropy (shown by ξ, η, and ζ) on which the aforementioned six hydraulic conductivity components are zero. The ξ, η, and ζ coordinates have the same origin as the x, y, and z coordinates, but their orientation as well as the corresponding hydraulic conductivities on these axes are not known. The solution of the two-dimensional version of this problem has significantly practical importance, and its details are presented below.

2.6.1 Principal Hydraulic Conductivities and Transmissivities for the Two-Dimensional Case

Let us consider the two-dimensional form of Eqs. (2-116). For two-dimensional flow, Eqs. (2-116) take the form

$$q_x = -K_{xx}\frac{\partial h}{\partial x} - K_{xy}\frac{\partial h}{\partial y} \quad \text{(2-117a)}$$

$$q_y = -K_{yx}\frac{\partial h}{\partial x} - K_{yy}\frac{\partial h}{\partial y} \quad \text{(2-117b)}$$

The x–y and ξ–η coordinates are shown in Figure 2-28. The orientation angle θ and the hydraulic conductivities or transmissivities on the ξ and η coordinates are not known. Now we will derive the corresponding expressions in terms of K_{xx}, K_{yy}, K_{xy}, and K_{yx} regarding these quantities.

As can be seen from Figure 2-28, the Darcy velocity vector is the same in x–y and ξ–η coordinate systems. Therefore, one can write

$$q_x = q_\xi \cos\theta - q_\eta \sin\theta \quad \text{(2-118a)}$$

$$q_y = q_\eta \cos\theta + q_\xi \sin\theta \quad \text{(2-118b)}$$

From the geometry in Figure 2-28, the coordinates ξ–η are related to x–y coordinates by

$$x = \xi\cos\theta - \eta\sin\theta \quad \text{(2-119a)}$$

$$y = \eta\cos\theta + \xi\sin\theta \quad \text{(2-119b)}$$

On the other hand, from the rules of derivatives and Eqs. (2-119), the following equations can be written:

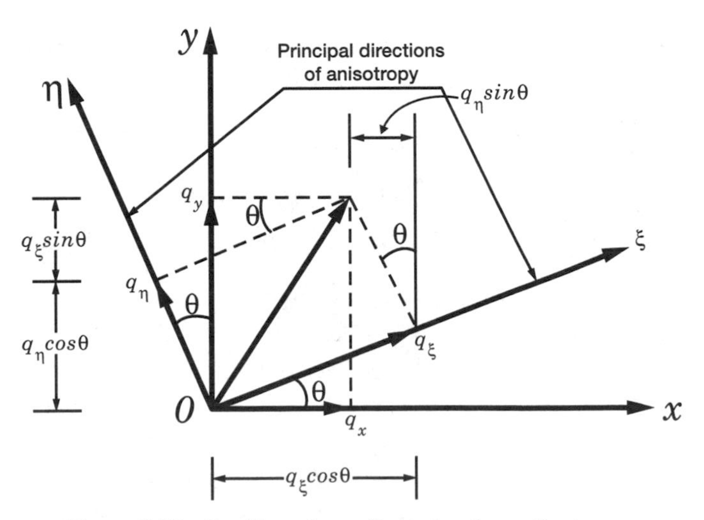

Figure 2-28 Rotation of two-dimensional coordinate axes.

$$\frac{\partial h}{\partial \xi} = \frac{\partial h}{\partial x}\frac{\partial x}{\partial \xi} + \frac{\partial h}{\partial y}\frac{\partial y}{\partial \xi} = \frac{\partial h}{\partial x}\cos\theta + \frac{\partial h}{\partial y}\sin\theta \tag{2-120a}$$

$$\frac{\partial h}{\partial \eta} = \frac{\partial h}{\partial x}\frac{\partial x}{\partial \eta} + \frac{\partial h}{\partial y}\frac{\partial y}{\partial \eta} = -\frac{\partial h}{\partial x}\sin\theta + \frac{\partial h}{\partial y}\cos\theta \tag{2-120b}$$

From the definition of principal directions of anisotropy, the Darcy velocity components on the ξ and η coordinates in Figure 2-28, respectively, are

$$q_\xi = -K_{\xi\xi}\frac{\partial h}{\partial \xi} \tag{2-121a}$$

$$q_\eta = -K_{\eta\eta}\frac{\partial h}{\partial \eta} \tag{2-121b}$$

From Eqs. (2-118), (2-120), and (2-121), the Darcy velocity components on the x- and y-coordinate directions can be written, respectively, as

$$q_x = -(K_{\xi\xi}\cos^2\theta + K_{\eta\eta}\sin^2\theta)\frac{\partial h}{\partial x} + (K_{\eta\eta} - K_{\xi\xi})\sin\theta\cos\theta\frac{\partial h}{\partial y} \tag{2-122a}$$

$$q_y = -(K_{\xi\xi}\sin^2\theta + K_{\eta\eta}\cos^2\theta)\frac{\partial h}{\partial y} + (K_{\eta\eta} - K_{\xi\xi})\sin\theta\cos\theta\frac{\partial h}{\partial x} \tag{2-122b}$$

From Eqs. (2-117a) and (2-122a) one can write

$$K_{xx} = K_{\xi\xi}\cos^2\theta + K_{\eta\eta}\sin^2\theta \tag{2-123a}$$

$$K_{xy} = (K_{\eta\eta} - K_{\xi\xi})\sin\theta\cos\theta \tag{2-123b}$$

Similarly, from Eqs. (2-117b) and (2-122b) one gets

$$K_{yy} = K_{\xi\xi}\sin^2\theta + K_{\eta\eta}\cos^2\theta \tag{2-124a}$$

$$K_{yx} = (K_{\eta\eta} - K_{\xi\xi})\sin\theta\cos\theta \tag{2-124b}$$

It is interesting to see from Eqs. (2-123b) and (2-124b) that $K_{xy} = K_{yx}$. Finally, substituting the equivalents of $\sin^2\theta$, $\cos^2\theta$, and $\sin\theta\cos\theta$ in Eqs. (2-123) and (2-124) gives

$$K_{xx} = \tfrac{1}{2}(K_{\xi\xi} + K_{\eta\eta}) + \tfrac{1}{2}(K_{\xi\xi} - K_{\eta\eta})\cos(2\theta) \tag{2-125}$$

$$K_{yy} = \tfrac{1}{2}(K_{\xi\xi} + K_{\eta\eta}) - \tfrac{1}{2}(K_{\xi\xi} - K_{\eta\eta})\cos(2\theta) \tag{2-126}$$

$$K_{xy} = K_{yx} = -\tfrac{1}{2}(K_{\xi\xi} - K_{\eta\eta})\sin(2\theta) \tag{2-127}$$

Since according to Eq. (2-24) the transmissivity is a product of the hydraulic conductivity and aquifer thickness, by multiplying both sides of Eqs. (2-125), (2-126), and (2-127) by the aquifer thickness (b) the corresponding equations for transmissivity can be obtained as

$$T_{xx} = \tfrac{1}{2}(T_{\xi\xi} + T_{\eta\eta}) + \tfrac{1}{2}(T_{\xi\xi} - T_{\eta\eta})\cos(2\theta) \tag{2-128}$$

$$T_{yy} = \tfrac{1}{2}(T_{\xi\xi} + T_{\eta\eta}) - \tfrac{1}{2}(T_{\xi\xi} - T_{\eta\eta})\cos(2\theta) \tag{2-129}$$

$$T_{xy} = T_{yx} = -\tfrac{1}{2}(T_{\xi\xi} - T_{\eta\eta})\sin(2\theta) \tag{2-130}$$

2.6.2 Mohr's Circle for Principal Hydraulic Conductivities and Transmissivities for the Two-Dimensional Case

In the mechanics of materials, *Mohr's circle* is a pictorial interpretation of the formulas for determining the principal stresses and the maximum shearing stresses at a point in a stressed rigid body. The German engineer Otto Mohr (1835–1918) developed an useful graphic interpretation for the stress equations. The particular case for plane stress was presented by the German engineer K. Culmann (1829–1881) about 16 years before Mohr's paper was published. This method, commonly known as *Mohr's circle* in solid mechanics, involves the construction of a circle such that the coordinates of each point on the circle represent the normal and shearing stresses on one plane through the stressed point, and the angular position of the radius to the point gives the orientation of the plane. Mohr's circle is also used as a pictorial interpretation of the formulas in determining maximum and minimum second moments of areas and maximum products of inertia (e.g., see Terzaghi, 1943; Highdon et al., 1976).

The Mohr's circle approach also provides a convenient way to interpret Eqs. (2-125), (2-126), and (2-127) to determine the principal hydraulic conductivities and their orientation angle (e.g. Verruijt, 1970; Bear, 1972). The Mohr's circle corresponding to Eqs. (2-125), (2-126), and (2-127) is shown in Figure 2-29, where K_{xx} and K_{yy} are plotted as horizontal coordinates and K_{xy} and K_{yx} are plotted as vertical coordinates. The circle is centered on the K_{xx} and K_{yy} axis at a distance $(K_{xx} + K_{yy})/2$ from the K_{yx} and $-K_{xy}$ axis, and the radius of the circle is given by

$$r = \left[\left(\frac{K_{xx} - K_{yy}}{2}\right)^2 + K_{xy}^2\right] \tag{2-131}$$

From the horizontal coordinates of points E and D in Figure 2-29 the expressions for $K_{\xi\xi}$ and $K_{\eta\eta}$ can be determined as

$$K_{\xi\xi} = \tfrac{1}{2}(K_{xx} + K_{yy}) + \left[\tfrac{1}{4}(K_{xx} - K_{yy})^2 + K_{xy}^2\right]^{1/2} \tag{2-132}$$

$$K_{\eta\eta} = \tfrac{1}{2}(K_{xx} + K_{yy}) - \left[\tfrac{1}{4}(K_{xx} - K_{yy})^2 + K_{xy}^2\right]^{1/2} \tag{2-133}$$

The angle 2θ from CV to CD is counterclockwise or positive, and from Figure 2-29 we have

$$\tan(2\theta) = \frac{2K_{xy}}{K_{xx} - K_{yy}} \tag{2-134}$$

Since $\beta = \theta$ in Figure 2-29, the following equation can also be written:

$$\tan\theta = \tan\beta = \frac{K_{\xi\xi} - K_{xx}}{K_{xy}} \tag{2-135}$$

According to Eq. (2-24), the transmissivity is a product of the hydraulic conductivity and aquifer thickness. Therefore, by multiplying both sides of Eqs. (2-132) and (2-133) by the

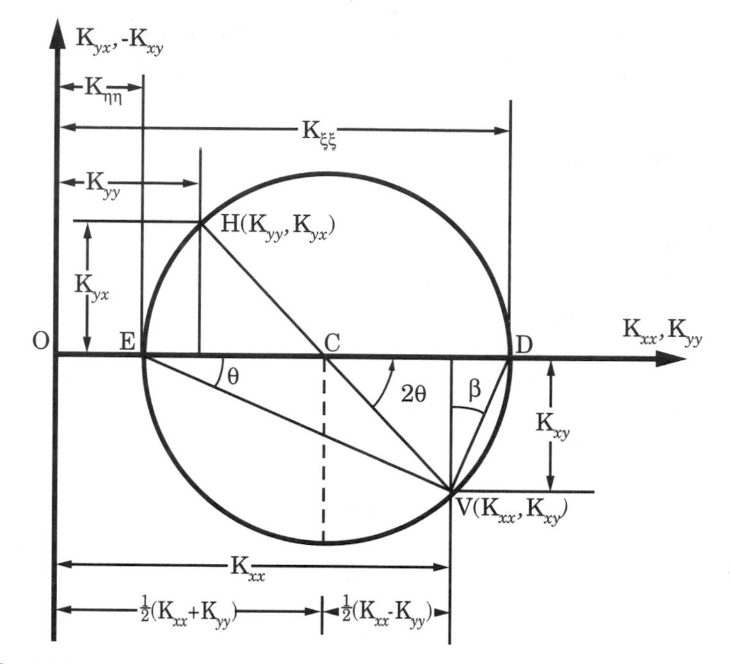

Figure 2-29 Mohr's circle for hydraulic conductivity and transmissivity in two-dimensional flow cases in aquifers.

aquifer thickness (b), the expressions for $T_{\xi\xi}$ and $T_{\eta\eta}$ can be obtained:

$$T_{\xi\xi} = \tfrac{1}{2}(T_{xx} + T_{yy}) + \left[\tfrac{1}{4}(T_{xx} - T_{yy})^2 + T_{xy}^2\right]^{1/2} \tag{2-136}$$

$$T_{\eta\eta} = \tfrac{1}{2}(T_{xx} + T_{yy}) - \left[\tfrac{1}{2}(T_{xx} - T_{yy})^2 + T_{xy}^2\right]^{1/2} \tag{2-137}$$

Likewise, Eqs. (2-134) and (2-135) can, respectively, be written as

$$\tan(2\theta) = \frac{2T_{xy}}{T_{xx} - T_{yy}} \tag{2-138}$$

$$\tan\theta = \tan\beta = \frac{T_{\xi\xi} - T_{xx}}{T_{xy}} \tag{2-139}$$

Application of the above equations is given in Section 5.2.1.5 of Chapter 5.

2.7 DIRECTIONAL HYDRAULIC CONDUCTIVITIES

In anisotropic aquifers, hydraulic conductivity varies from direction to direction. In Section 2.6, for a two-dimensional anisotropic aquifer case, the orientation of the principal directions as well as the expressions of hydraulic conductivities ($K_{\xi\xi}$ and $K_{\eta\eta}$) on these axes in terms of the hydraulic conductivity components (K_{xx}, K_{yy}, K_{xy}, and K_{yx}) are given.

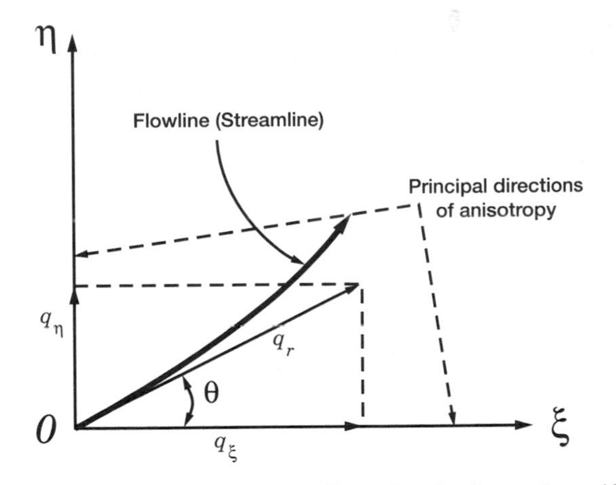

Figure 2-30 Flow line in a two-dimensional anisotropic aquifer.

As shown in Figure 2-30, let us consider a two-dimensional flow system in which the (ξ, η) coordinate axes are the principal axes on which the principal hydraulic conductivities ($K_{\xi\xi}$ and $K_{\eta\eta}$) are known. The problem is to find the hydraulic conductivity expression on the r direction that makes a θ angle with the ξ coordinate axis. The corresponding expressions for three-dimensional flow systems are also presented.

2.7.1 Directional Hydraulic Conductivities and Tranmissivities for Two-Dimensional Cases

Consider a two-dimensional homogeneous and anisotropic flow system in which the principal hydraulic conductivities ($K_{\xi\xi}$ and $K_{\eta\eta}$) or transmissivities ($T_{\xi\xi}$ and $T_{\eta\eta}$) are aligned with the ξ and η axes. The tangent to the flow line passing through the origin makes a θ angle with the ξ axis. Therefore, the Darcy velocity along the streamline is

$$q_r = -K_r \frac{\partial h}{\partial r} \tag{2-140}$$

and its components in the ξ and η directions, respectively, are

$$q_\xi = -K_{\xi\xi} \frac{\partial h}{\partial \xi} = q_r \cos\theta \tag{2-141}$$

$$q_\eta = -K_{\eta\eta} \frac{\partial h}{\partial \eta} = q_r \sin\theta \tag{2-142}$$

Since $h = h(\xi, \eta)$, $\xi = \xi(r)$, and $\eta = \eta(r)$, we obtain

$$\frac{\partial h}{\partial r} = \frac{\partial h}{\partial \xi}\frac{\partial \xi}{\partial r} + \frac{\partial h}{\partial \eta}\frac{\partial \eta}{\partial r} \tag{2-143}$$

and from Figure 2-30 one can write

$$\frac{\partial \xi}{\partial r} = \cos\theta, \qquad \frac{\partial \eta}{\partial r} = \sin\theta \tag{2-144}$$

When we substitute Eqs. (2-140), (2-141), (2-142), and (2-144) into Eq. (2-143) and then simplify, the following equation can be obtained:

$$\frac{1}{K_r} = \frac{\cos^2\theta}{K_{\xi\xi}} + \frac{\sin^2\theta}{K_{\eta\eta}} \tag{2-145}$$

Solving Eq. (2-145) for K_r, one gets

$$K_r = \frac{K_{\xi\xi}}{\cos^2\theta + \dfrac{K_{\xi\xi}}{K_{\eta\eta}}\sin^2\theta} \tag{2-146}$$

According to Eq. (2-24), the transmissivity is a product of the hydraulic conductivity and aquifer thickness. Therefore, the principal transmissivities become

$$T_{\xi\xi} = K_{\xi\xi}b, \qquad T_{\eta\eta} = K_{\eta\eta}b \tag{2-147}$$

and with Eq. (2-146) the transmissivity ($T_r = K_r b$) in the direction of s becomes

$$T_r = \frac{T_{\xi\xi}}{\cos^2\theta + \dfrac{T_{\xi\xi}}{T_{\eta\eta}}\sin^2\theta} \tag{2-148}$$

Equations (2-146) and (2-148) give the directional hydraulic conductivity (K_r) and directional transmissivity (T_r), respectively, in any angular direction θ. Applications of Eq. (2-148) regarding aquifer data analysis are given in Section 5.3.3 of Chapter 5.

Hydraulic Conductivity and Transmissivity Ellipses. With $\xi = r\cos\theta$ and $\eta = r\sin\theta$, Eq. (2-145) takes the form

$$\frac{r^2}{K_r} = \frac{\xi^2}{K_{\xi\xi}} + \frac{\eta^2}{K_{\eta\eta}} \tag{2-149}$$

By drawing a segment of length $r = (K_r)^{1/2}$ in the direction of q_r, Eq. (2-149) takes the form (Maasland, 1957; Bear, 1972)

$$\frac{\xi^2}{K_{\xi\xi}} + \frac{\eta^2}{K_{\eta\eta}} = 1 \tag{2-150}$$

Equation (2-150) is the *hydraulic conductivity ellipse* in the ξ and η coordinate system. The semiaxes of the ellipse in the ξ and η directions are $(K_{\xi\xi})^{1/2}$ and $(K_{\eta\eta})^{1/2}$, respectively (see Figure 2-31). The ellipse gives the directional hydraulic conductivity in the direction of the flow. The directional hydraulic conductivity K_r at angular direction θ can also be determined from Eq. (2-145) using the known values of $K_{\xi\xi}$ and $K_{\eta\eta}$ on the principal axes.

The equations presented above can also be applied to directional transmissivities and transmissivity ellipse using Eqs. (2-147) with $T_r = K_r b$.

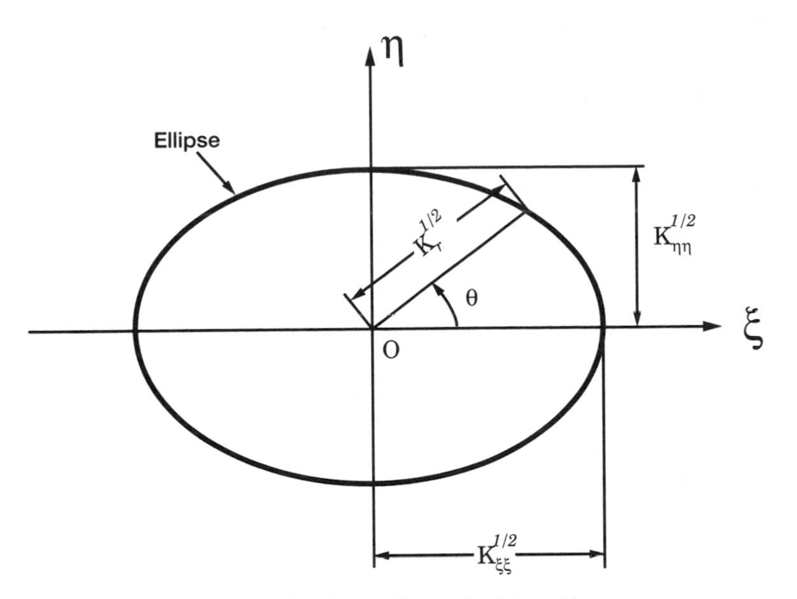

Figure 2-31 Hydraulic conductivity ellipse.

2.7.2 Directional Hydraulic Conductivities for Three-Dimensional Cases

For three-dimensional cases equations similar to those in Section 2.7.1 can be developed (Maasland, 1957; Bear, 1972). The corresponding equation to Eq. (2-145) is

$$\frac{1}{K_r} = \frac{\cos^2 \alpha_x}{K_{\xi\xi}} + \frac{\cos^2 \alpha_y}{K_{\eta\eta}} + \frac{\cos^2 \alpha_z}{K_{\zeta\zeta}} \tag{2-151}$$

where α_x, α_y, and α_z are the angles between the direction of q_r and the three principal axes ξ, η, and ζ, respectively.

Hydraulic Conductivity Ellipsoid. With $\xi = r \cos \alpha_x$, $\eta = r \cos \alpha_y$, and $\zeta = r \cos z$, Eq. (2-151) takes the form

$$\frac{r^2}{K_r} = \frac{\xi^2}{K_{\xi\xi}} + \frac{\eta^2}{K_{\eta\eta}} + \frac{\zeta^2}{K_{\zeta\zeta}} \tag{2-152}$$

By drawing a segment of length $r = (K_r)^{1/2}$ in the direction of q_r, Eq. (2-152) takes the form (Maasland, 1957; Bear, 1972)

$$\frac{\xi^2}{K_{\xi\xi}} + \frac{\eta^2}{K_{\eta\eta}} + \frac{\zeta^2}{K_{\zeta\zeta}} = 1 \tag{2-153}$$

Eq. (2-153) is the *hydraulic conductivity ellipsoid* in the ξ-, η-, and ζ-coordinate system. The semiaxes of the ellipse in the ξ, η, and ζ directions are $(K_{\xi\xi})^{1/2}$, $(K_{\eta\eta})^{1/2}$, and $(K_{\zeta\zeta})^{1/2}$, respectively. Figure 2-31 can be viewed as the horizontal cross-section of the hydraulic conductivity ellipsoid. The ellipse gives the directional hydraulic conductivity in the direction of the flow. The directional hydraulic conductivity K_r at a direction defined by α_x, α_y, and α_z angles can also be determined from Eq. (2-151) using the known values of $K_{\xi\xi}$, $K_{\eta\eta}$, and $K_{\zeta\zeta}$ on the principal axes.

2.8 DIFFERENTIAL EQUATIONS OF FLOW IN AQUIFERS

Analytical as well as numerical solution of an aquifer problem is based on its governing differential equation and the corresponding initial and boundary conditions. Initial and boundary conditions typically encountered in most aquifer problems are included in Section 2.9. Here, the governing equations of flow in aquifers are presented in both Cartesian and cylindrical coordinates (e.g., Hantush, 1964; Bear, 1972, 1979).

2.8.1 Differential Equations in Cartesian Coordinates

2.8.1.1 General Case: The Principal Directions of Anisotropy Are not the Coordinate Axes

Three-Dimensional Flow Equation. Let us consider a volume element ΔV in an aquifer whose mass is ΔM. According to the conservation of mass principle, the sum of the net inward flux through ΔV and the amount of mass generated within the element per unit time must be equal to the rate of mass accumulating within ΔV. Then

$$\text{rate of change of } \Delta M \text{ per unit time per unit volume} = \frac{1}{\Delta V}\frac{\partial(\Delta M)}{\Delta V} \qquad \text{(2-154)}$$

If $\mathbf{q}$ is the bulk velocity vector and ρ is the density of the fluid, then the mass flow per unit area will be $\rho\mathbf{q}$. The mathematical expression of the aforementioned conservation of mass principle is (Hantush, 1964)

$$-\int\!\!\int(\rho\mathbf{q})\cdot d\mathbf{S} = \int\!\!\int\!\!\int\left[\frac{1}{\Delta V}\frac{\partial(\Delta M)}{\partial t}\right]dV \qquad \text{(2-155)}$$

in which the left-hand side is the surface integral taken over the closed surface of the volume of ΔV and the right-hand side is the volume integral over the volume element. The dot on the left-hand side means that the *scalar product* or *dot product* of the two vector quantities $\rho\mathbf{q}$ and $d\mathbf{S}$ is taken (e.g., see Spiegel, 1971). From Eq. (2-62) one can write

$$\frac{\partial(\Delta M)}{\Delta V} = \rho S_s\,\partial h \qquad \text{(2-156)}$$

The surface integral on the left-hand side of Eq. (2-155) can be converted to a volume integral by applying the *divergence theorem* (e.g., Spiegel, 1971). Using this theorem and substituting Eq. (2-156) into Eq. (2-155), we obtain

$$-\int\!\!\int\!\!\int\left[\nabla\cdot(\rho\mathbf{q}) + \rho S_s\frac{\partial h}{\partial t}\right]dV = 0 \qquad \text{(2-157)}$$

where ∇ is the *gradient operator* and is defined as (e.g., see Spiegel, 1971)

$$\nabla = \frac{\partial}{\partial x}\mathbf{i} + \frac{\partial}{\partial y}\mathbf{j} + \frac{\partial}{\partial z}\mathbf{k} \tag{2-158}$$

where **i**, **j**, and **k** are the unit vectors in the x, y, and z directions, respectively. Since the volume element is arbitrary, the integral in Eq. (2-157) is valid for any volume element. As a result, with the assumption that the density of water ρ is constant, the following equation can be written:

$$-\nabla \cdot \mathbf{q} = S_s \frac{\partial h}{\partial t} \tag{2-159}$$

where **q** is the Darcy velocity in general form and its components are given by Eqs. (2-116). Using tensor notation, Eqs. (2-116) can be expressed in a concise form:

$$\mathbf{q} = -K\nabla h \tag{2-160}$$

where **K** is the *hydraulic conductivity tensor* and is defined as

$$[K] = \begin{bmatrix} K_{xx} & K_{xy} & K_{xz} \\ K_{yx} & K_{yy} & K_{yz} \\ K_{zx} & K_{zy} & K_{zz} \end{bmatrix} \tag{2-161}$$

Substitution of Eq. (2-160) into Eq. (2-159) gives

$$\nabla \cdot (K\nabla h) = S_s \frac{\partial h}{\partial t} \tag{2-162}$$

Equation (2-162) can also be expressed in scalar form by using Eqs. (2-158) and (2-161). Since its scalar form is too lengthy, only its two-dimensional form under some special conditions is presented below.

Two-Dimensional Flow Equation Under Special Conditions. The two-dimensional form of Eq. (2-161) for $K_{xy} = K_{yx}$ is

$$[K] = \begin{bmatrix} K_{xx} & K_{xy} \\ K_{xy} & K_{yy} \end{bmatrix} \tag{2-163}$$

which represents a symmetric hydraulic conductivity tensor. Combination of Eqs. (2-117), (2-162), and (2-163) gives

$$K_{xx}\frac{\partial^2 h}{\partial x^2} + 2K_{xy}\frac{\partial^2 h}{\partial x\, \partial y} + K_{yy}\frac{\partial^2 h}{\partial y^2} = S_s \frac{\partial h}{\partial t} \tag{2-164}$$

2.8.1.2 Special Case: The Principal Directions of Anisotropy Are the Coordinate Axes.

As mentioned in Section 2.5.1.1, if the principal directions of anisotropy coincide with x, y, and z directions of the coordinate axes, the Darcy velocity vector has three components expressed by Eqs. (2-112) that are aligned with the x, y, and z coordinate axes. In other words, the six components K_{xy}, K_{xz}, K_{yx}, K_{yz}, K_{zx}, and K_{zy} of the hydraulic conductivity tensor all become equal to zero. Under this condition, Eq. (2-161) takes the form

$$[K] = \begin{bmatrix} K_{xx} & 0 & 0 \\ 0 & K_{yy} & 0 \\ 0 & 0 & K_{zz} \end{bmatrix} \tag{2-165}$$

and the corresponding forms of Darcy velocity components are given by Eqs. (2-112). In the following sections the differential equations of several cases in accordance with Section 2.3.7.2 will be presented.

Homogeneous and Isotropic Aquifer. This is the simplest case, and its definition is given in Figure 2-10. When we substitute Eqs. (2-34) into Eqs. (2-112), Eq. (2-162) takes the form

$$\frac{\partial^2 h}{\partial x^2} + \frac{\partial^2 h}{\partial y^2} + \frac{\partial^2 h}{\partial z^2} = \frac{S_s}{K}\frac{\partial h}{\partial t} \tag{2-166}$$

which is the differential equation of flow in a homogeneous and isotropic aquifer.

Homogeneous and Anisotropic Aquifer. A homogeneous and anisotropic aquifer is schematically defined in Figure 2-11. Substitution of Eqs. (2-35) into Eqs. (2-112) and then into Eq. (2-162) gives

$$K_x\frac{\partial^2 h}{\partial x^2} + K_y\frac{\partial^2 h}{\partial y^2} + K_z\frac{\partial^2 h}{\partial z^2} = S_s\frac{\partial h}{\partial t} \tag{2-167}$$

which is the differential equation of flow in a homogeneous and anisotropic aquifer.

Heterogeneous and Isotropic Aquifer. A heterogeneous and isotropic aquifer is schematically defined in Figure 2-12. According to Eqs. (2-36) the three principal hydraulic conductivities at a point in the aquifer are equal, but their values vary from point to point. Therefore, with Eqs. (2-112), Eq. (2-162) takes the form

$$\frac{\partial}{\partial x}\left[K(x,y,z)\frac{\partial h}{\partial x}\right] + \frac{\partial}{\partial y}\left[K(x,y,z)\frac{\partial h}{\partial y}\right] + \frac{\partial}{\partial z}\left[K(x,y,z)\frac{\partial h}{\partial z}\right] = S_s\frac{\partial h}{\partial t} \tag{2-168}$$

which is the differential equation of flow in a heterogeneous and isotropic aquifer.

Transversely Isotropic Aquifer. From Eqs. (2-37), (2-112), and (2-162) one gets

$$K_h\left(\frac{\partial^2 h}{\partial x^2} + \frac{\partial^2 h}{\partial y^2}\right) + K_z\frac{\partial^2 h}{\partial z^2} = S_s\frac{\partial h}{\partial t} \tag{2-169}$$

which is the differential equation of flow in a transversely isotropic aquifer.

Heterogeneous and Anisotropic Aquifer. A heterogeneous and anisotropic aquifer is schematically defined in Figure 2-13, and from the conditions expressed by Eqs. (2-38) and the combination of Eqs. (2-112) and (2-162) we obtain the corresponding differential equation of flow:

$$\frac{\partial}{\partial x}\left[K_x(x,y,z)\frac{\partial h}{\partial x}\right] + \frac{\partial}{\partial y}\left[K_y(x,y,z)\frac{\partial h}{\partial y}\right] + \frac{\partial}{\partial z}\left[K_z(x,y,z)\frac{\partial h}{\partial z}\right] = S_s\frac{\partial h}{\partial t} \tag{2-170}$$

The conditions for Eqs. (2-168) and (2-170) are that the hydraulic conductivity functions must be continuous and have continuous first derivative everywhere in the flow domain. The one-, two-, and three-dimensional forms of Eqs. (2-164), (2-166), (2-167), and (2-169) are widely used in developing analytical solutions for aquifer problems. Equations (2-168) and (2-170) can only be solved numerically with the exception of some special hydraulic conductivity distribution functions.

2.8.2 Differential Equations in Polar Coordinates

A vertical extraction or injection well causes *radial flow* in the aquifer toward the well or from the well. It is easier to work with partial differential equations of flow in aquifers written in *polar coordinates*. Both Cartesian and polar coordinates are shown in Figure 2-32, in which the well is located at the origin of the coordinate system and the axis of the well is perpendicular to the x–y plane. This implicitly means that only the x and y Cartesian coordinates must be converted to polar coordinates because the vertical coordinate z remains the same. Because flow is radial, hydraulic conductivity in the radial direction or *radial hydraulic conductivity* (K_r) is used instead of K_x and K_y. This concept is based on the assumption that both the hydraulic conductivity and the hydraulic head do not change with the angular direction θ. The hydraulic head changes only with the radial distance r from the well. In the following sections, the differential equations of ground-water flow in three- and two-dimensional cases will be presented.

Differential Equation in Three Dimensions for a Transversely Isotropic Aquifer Case. The definition of a transversely isotropic aquifer is given in Section 2.3.7.2, and the corresponding partial differential equation in Cartesian coordinates is given by Eq. (2-169). Because of the reasons mentioned above, only the derivatives of h with respect to x and y

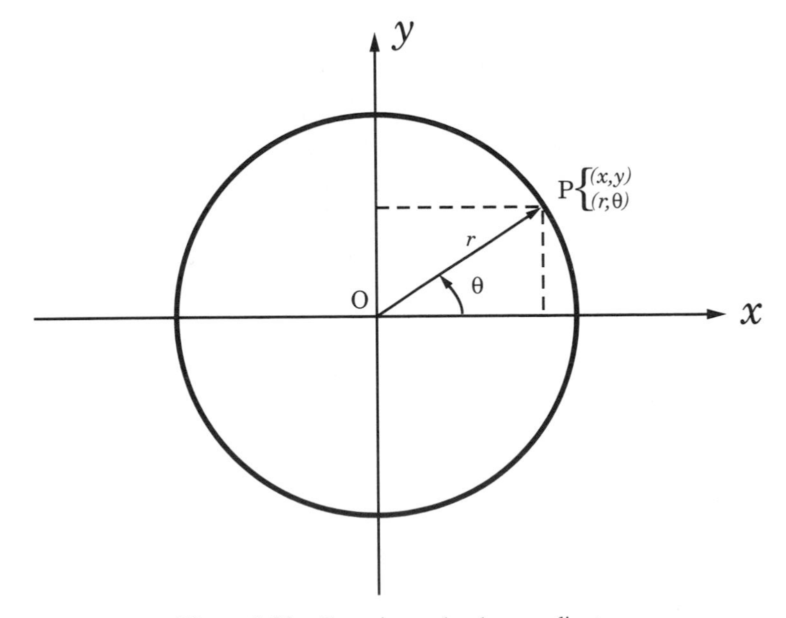

Figure 2-32 Cartesian and polar coordinates.

must be converted into polar coordinates. From the geometry of Figure 2-32 one can write

$$
\begin{aligned}
r &= r\cos\theta, \qquad & y &= r\sin\theta \\
r &= (x^2 + y^2)^{1/2}, \qquad & \theta &= \arctan\left(\frac{y}{x}\right)
\end{aligned}
\tag{2-171}
$$

By applying the *chain rule of differentiation* to the derivatives of h, one obtains (Verruijt, 1970)

$$\frac{\partial^2 h}{\partial x^2} = \frac{x^2}{r^2}\frac{\partial^2 h}{\partial r^2} + \frac{y^2}{r^3}\frac{\partial h}{\partial r} + \frac{y^2}{r^4}\frac{\partial h}{\partial \theta^2} - \frac{2xy}{r^4}\frac{\partial h}{\partial \theta} \tag{2-172}$$

$$\frac{\partial^2 h}{\partial y^2} = \frac{y^2}{r^2}\frac{\partial^2 h}{\partial r^2} + \frac{x^2}{r^3}\frac{\partial h}{\partial r} + \frac{x^2}{r^4}\frac{\partial^2 h}{\partial \theta^2} + \frac{2xy}{r^4}\frac{\partial h}{\partial \theta} \tag{2-173}$$

If we substitute Eqs. (2-172) and (2-173) into Eq. (2-169), replace K_r instead of K_h, and remembering that h does not change with θ, after simplification we obtain

$$K_r\frac{\partial^2 h}{\partial r^2} + \frac{K_r}{r}\frac{\partial h}{\partial r} + K_z\frac{\partial^2 h}{\partial z^2} = S_s\frac{\partial h}{\partial t} \tag{2-174}$$

Equation (2-174) has wide application for circular fully and partially penetrating wells in aquifers.

Differential Equation in Two Dimensions. For two-dimensional radial flow cases the third term on the left-hand side becomes zero, and it is appropriate to use transmissivity instead of hydraulic conductivity using the aquifer thickness (b). Therefore, using Eqs. (2-24) and (2-75), the differential equation in two dimensions from Eq. (2-174) takes the form

$$\frac{\partial^2 h}{\partial r^2} + \frac{1}{r}\frac{\partial h}{\partial r} = \frac{S}{T}\frac{\partial h}{\partial t} \tag{2-175}$$

where $T = K_r b$. Another derivation of Eq. (2-175) is given in Section 4.2.2 of Chapter 4.

2.8.3 Differential Equations Under Steady-State Conditions

The term $S_s\partial h/\partial t$ or $(S/T)\partial h/\partial t$ in Eqs. (2-159) through (2-175) is the one that makes the cases transient. Therefore, if this term is zero, all of those equations correspond to the steady-state case. For example, under this condition, Eq. (2-167), which corresponds to a *homogeneous and anisotropic aquifer*, takes the form

$$K_x\frac{\partial^2 h}{\partial x^2} + K_y\frac{\partial^2 h}{\partial y^2} + K_z\frac{\partial^2 h}{\partial z^2} = 0 \tag{2-176}$$

For a *homogeneous and isotropic aquifer*, Eq. (2-166) reduces to the well-known Laplace equation:

$$\nabla^2 h \equiv \frac{\partial^2 h}{\partial x^2} + \frac{\partial^2 h}{\partial y^2} + \frac{\partial^2 h}{\partial z^2} = 0 \tag{2-177}$$

where ∇^2 is the *Laplacian operator*.

2.9 INITIAL AND BOUNDARY CONDITIONS

2.9.1 Introduction

The type of differential equations presented by Eqs. (2-159) through (2-175) is commonly encountered in various branches of mathematical physics such as fluid dynamics, electricity, magnetism, heat flow, and others. Even though these equations are written for water flow in aquifers, they do not include any information regarding any specific case of flow or the shape and boundaries of the flow domain. As a result, each of the aforementioned equations has infinite solutions and each corresponds to a particular flow case in an aquifer.

The shape and boundaries as well as initial conditions for a particular aquifer problem correspond to one of the solutions of the corresponding differential equation for the aquifer. Therefore, the specification of appropriate initial and boundary conditions for an aquifer problem is a required step in determining the corresponding solution. Based on numerous initial and boundary conditions, a number of solutions were developed for the steady-state and unsteady-state equations of flow in aquifers in the nineteenth and twentieth centuries. A significant number of solutions are covered in this book. Understanding of the physical meaning of different initial and boundary conditions is important not only for the development of mathematical solutions, but also for the selection of a solution from a set of solutions to be used for a particular case faced in practice. This particular case may be a solution to be used for predictions of hydraulic head or drawdown or a solution to be used for the analysis of pumping or slug test data or for other purposes. The solution of a specific problem is one of the special solutions of the ground-water flow differential equations under steady- and unsteady-state conditions. Because the specific solution is dependent on the boundary conditions, the problem is called *boundary value problem.*

Based on the above-mentioned facts, in the following sections the initial and boundary conditions typically encountered in aquifer hydraulics problems will be presented, with examples.

2.9.2 Initial Conditions

Specification of initial conditions for a particular problem is only required for unsteady-state aquifer problems. For steady-state flow problems, initial conditions are not required. For unsteady-state flow problems, the distribution of hydraulic head for the flow system at a particular time is assumed to be known. Initial conditions are generally specified when time is zero ($t = 0$). The mathematical expression for this condition is

$$h = f(x, y, z, 0) \tag{2-178}$$

for all points inside the flow domain at $t = 0$.

2.9.3 Types of Boundary Conditions

After defining the problem for a particular aquifer case, the first step is to convert the physical problem to a mathematical problem. The second step is to establish the framework of the boundary value problem. In aquifer flow problems, the final solution is usually expressed in terms of head (h) or drawdown (s), and sometimes flow rate (Q).

The selection of boundary conditions is a critically important step for conceptualizating and developing a model for an aquifer flow system. Even if one of the boundary conditions is inappropriate for a defined aquifer problem, it is most likely that the final expression will be in error or will produce ambiguous results.

There are several boundary conditions used in solving flow problems in aquifers. As a matter of fact, the same boundary conditions are also used in solving problems for similar phenomena such as fluid dynamics, electricity, magnetism, heat flow, and others. The boundary conditions often used in solving aquifer problems are described below, with examples.

2.9.3.1 Constant Head Boundary.

This boundary condition is also known as the *Dirichlet boundary condition* or the *first-type boundary condition*. The frequently used term is *constant head boundary condition*. According to Eq. (2-1), the hydraulic head is the sum. The general mathematical form of the steady constant head boundary condition is

$$h = f_1(x, y, z) \tag{2-179}$$

If the boundary condition changes with time, it takes the form

$$h = f_2(x, y, z, t) \tag{2-180}$$

Here, f_1 and f_2 are known functions. These boundary conditions may be valid for the whole flow domain or for a portion of it. The constant head boundary condition occurs when the aquifer flow domain is adjacent to a surface water body. Several examples are presented below for this boundary condition.

Example 2-11: Describe the constant head boundaries for an unconfined aquifer bounded by two parallel rivers as shown in Figure 2-33, which is a cross section perpendicular to the axes of rivers.

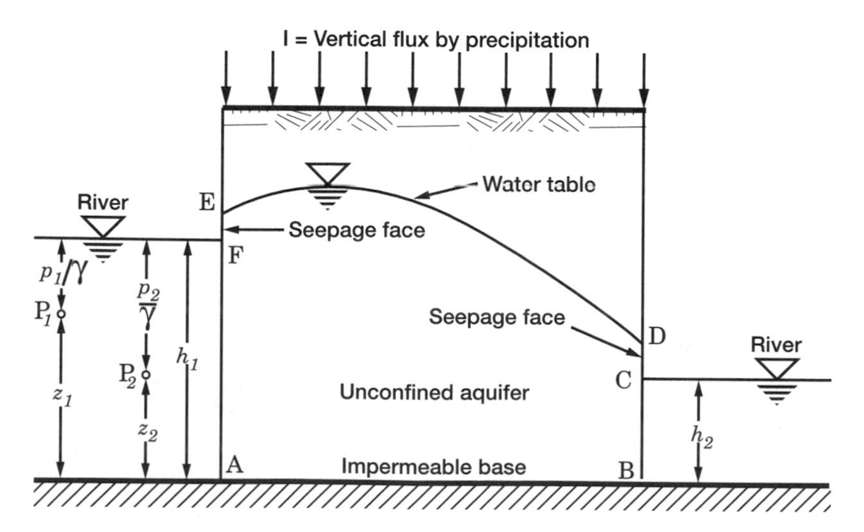

Figure 2-33 An unconfined aquifer between two parallel rivers.

Solution: In Figure 2-33, there are six boundary segments (AB, BC, CD, DE, EF, and FA). Of these six boundaries, only two (AF and BC) are constant head boundaries. Let us consider that the impermeable horizontal base (AB) is the chosen datum. Consider two points (P_1 and P_2) in the river on the left-hand side of Figure 2-33. For these points, application of Eq. (2-1) gives

$$h_1 = \frac{p_1}{\gamma} + z_1 = \frac{p_2}{\gamma} + z_2$$

which means that the heads of all points of AF are the same and equal to h_1. Similarly, the heads on all points of BC are equal to h_2. Therefore, in mathematical terms the constant head boundary conditions take the forms

$$h = h_1 \qquad \text{on AF}$$

$$h = h_2 \qquad \text{on BC}$$

If h_1 and h_2 change with time, the boundary conditions take the forms

$$h = h_1(t) \qquad \text{on AF}$$

$$h = h_2(t) \qquad \text{on BC}$$

Because on AF and BC h is constant, they are equipotential boundaries.

Example 2-12: Describe the constant head boundaries for the hydraulic structures shown in Figures 2-34, 2-35, and 2-36.

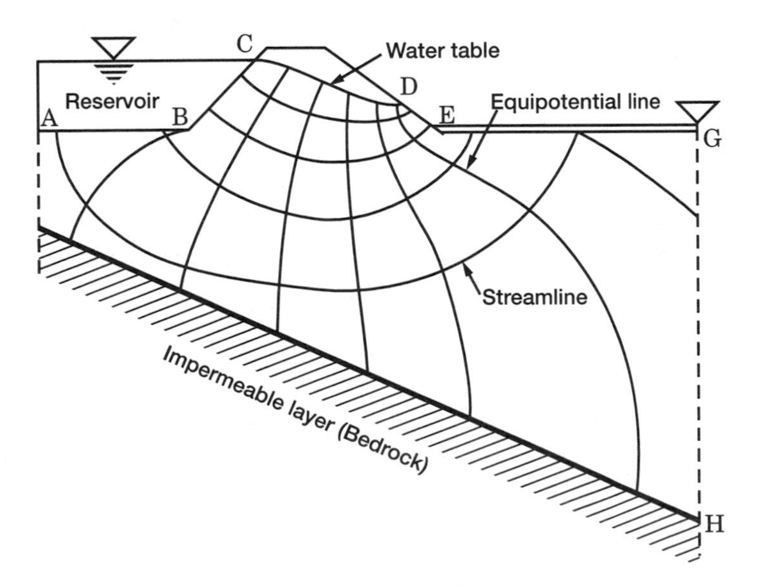

Figure 2-34 Flow through and beneath an earth dam underlain by sloping bedrock. (After Franke et al., 1984.)

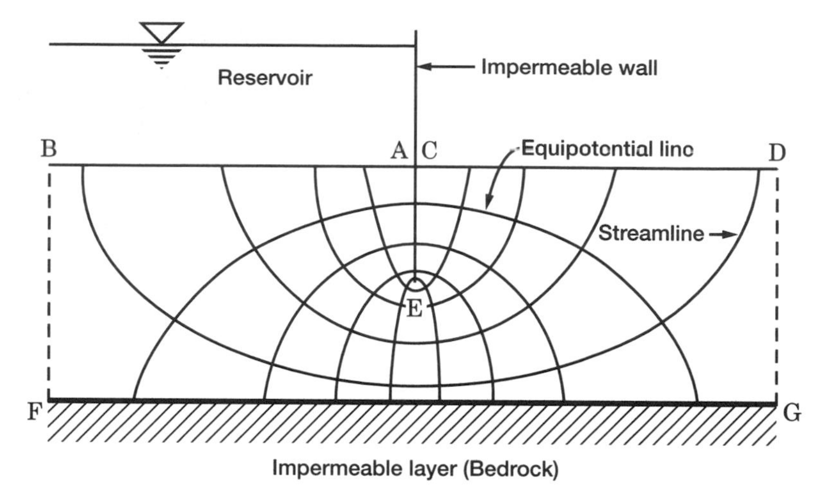

Figure 2-35 Flow beneath a vertical impermeable wall underlain by horizontal bedrock. (After Franke et al., 1984.)

Solution: Figure 2-34 is a cross section of an earth dam underlain by sloping bedrock. Here, the segments ABC and EG are constant head boundaries. The inclined segment is a somewhat general form of the segments AF and BC in Figure 2-33. However, the same principle can be applied, as shown in Example 2-11, to see that the head is indeed the same at different points on the inclined BC segment.

Figure 2-35 is a cross section of a reservoir behind a vertical impermeable wall. With the same principle, as applied in Example 2-11, it can be shown that the segments BA and CD are constant head boundaries.

Figure 2-36 is a cross section of an impermeable dam with a vertical impermeable wall above bedrock. Likewise, it can be shown that the segments *AB* and *CD* are constant head boundaries.

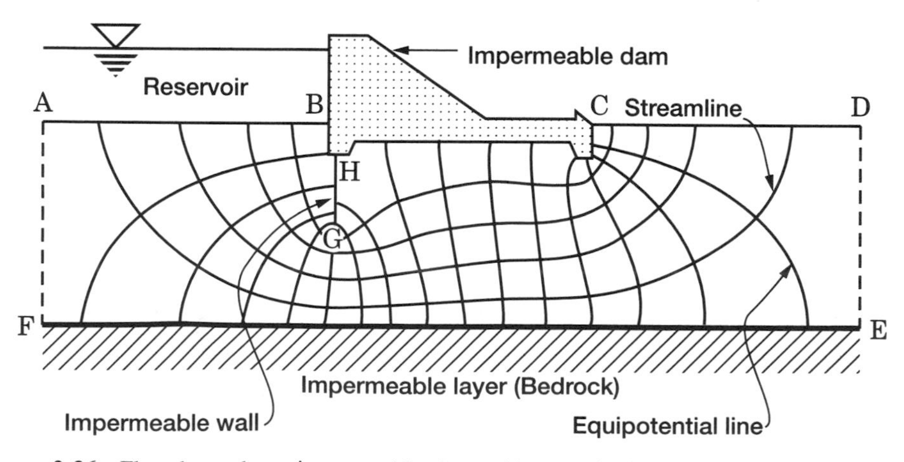

Figure 2-36 Flow beneath an impermeable dam with a vertical impermeable wall and above horizontal bedrock. (After Franke et al., 1984.)

2.9.3.2 Specified Head Boundary. The *specified head boundary condition* (another term is *prescribed head boundary condition*) is used in numerical models (finite-difference, finite-element, or others). The common practice is to assign the known boundary head values determined by field observations to the corresponding boundary segments. In reality, the constant head boundary segments act as though they are a portion of adjacent surface water bodies (see Section 2.9.3.1). Therefore, special care is required when using this type of boundary condition. The main criterion in using this type of boundary condition is to check the head values near the boundaries and to make sure that they are not affected under different stress conditions (extraction, injection, and others) in the aquifer. For example, as long as the cone of depression of an extraction well does not reach to the specified head boundary segments, they can be used safely. If the cone of depression of an extraction well reaches to the specified boundary segments, they act as though some real surface water bodies exist there. As a result, the solution will be in error.

2.9.3.3 Specified Flux Boundary. The term *flux* means the volume of fluid per unit time passing through a unit surface area. If the flux normal on a given part of the boundary surface can be specified as a function of space and time, that boundary is a *specified boundary*. This boundary condition is also known as *Neuman boundary condition* or *second-type boundary condition*. The mathematical expression of the flux boundary condition is

$$q_n = f(x, y, z, t) \tag{2 181}$$

where q_n is the flux component normal to the boundary surface and $f(x, y, z, t)$ is a known function.

The simplest form of the specified flux boundary is the constant areal net precipitation recharge (I) as shown in Figure 2-33. The flux can also be specified using Darcy's equation as given in Section 2.5 in various forms:

$$q_n = -K_n \frac{\partial h}{\partial n} \tag{2-182}$$

where n represents the direction perpendicular to boundary. This boundary condition is often used in solving well hydraulics problems (see, for example, Section 4.2 in deriving the Theis equation for a discharging fully penetrating well in a confined aquifer).

2.9.3.4 No-Flow Boundary. There are several forms of the no-flow boundary condition. These are described below.

Impermeable Boundary Layer. Relatively less pervious layers or formations act as impermeable boundary layers. For example, the interfaces between the aquitards and aquifers shown in Figure 2-1 act as impermeable boundaries. The Darcy velocity normal to this kind of boundaries is equal to zero. Therefore, from Eq. (2-182) we obtain

$$q_n = -K_n \frac{\partial h}{\partial n} = 0 \tag{2-183}$$

Since K_n cannot be zero, the no-flow boundary condition takes the form

$$\frac{\partial h}{\partial n} = 0 \qquad (2\text{-}184)$$

which is a special form of the Neuman boundary condition given in Section 2.9.3.3.

Example 2-13: Describe the no-flow boundaries for the hydraulic structures shown in Figures 2-34, 2-35, and 2-36.

Solution: In Figure 2-34, the bedrock acts as an impermeable layer. Therefore, the upper end of the bedrock (line HI) is a no-flow boundary. Mathematically speaking, on this boundary we have $\partial h/\partial n = 0$, where n is the direction normal to HI.

In Figure 2-35, the buried portion of the impermeable wall (line AEC) is a no-flow boundary of the flow domain. Likewise, the upper boundary of the bedrock (line FG) is a no-flow boundary as well.

In Figure 2-36, lines BGHC and EF are no-flow boundaries.

Water Divide. This *water divide* boundary is a special case of the no-flow boundary and will be explained with an example. Let us consider the flow lines in a homogeneous and isotropic unconfined aquifer with constant areal recharge and discharge to two symmetrically located streams (see Figure 2-37). In Figure 2-37, because of symmetry, lines BD, GE, and CF are flow lines. Therefore, the velocities perpendicular to these lines are zero. As a result, these lines are no-flow boundaries, and Eq. (2-184) is valid for these boundaries as well. In other words, the lines BD, GE, and CF create separate flow zones in the flow domain. That is why each one is called a *water divide*, a term that is often used by hydrologists. Therefore, a water divide is a no-flow boundary. Usage of this type of boundary significantly simplifies a relatively complex aquifer problem.

2.9.3.5 Seepage Face Boundary. This type of boundary occurs at the interface between an unconfined aquifer and an adjacent water body such that the water table ends at a point that is above the free surface of the water body. The segment of the boundary above

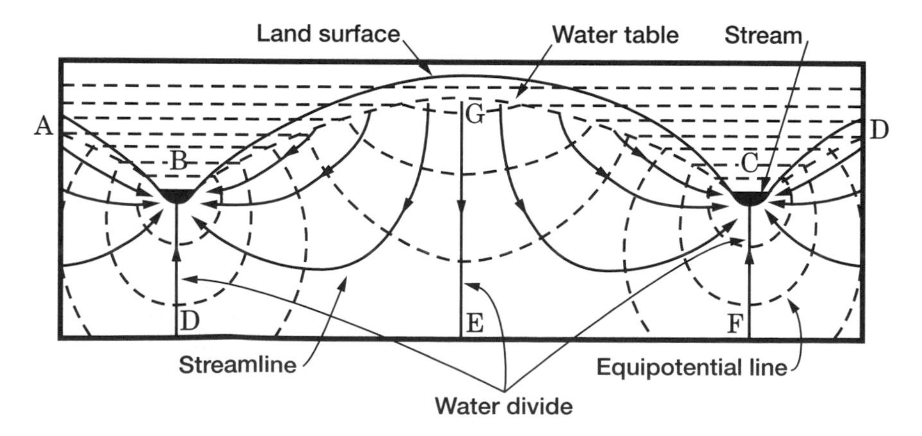

Figure 2-37 Flow lines in a homogeneous and isotropic unconfined aquifer having two symmetrically located streams with uniform areal recharge. (After Franke et al., 1984.)

the water surface and below the water table of the unconfined aquifer is called *seepage face*. For example, in Figure 2-33, the segments DC and EF are seepage faces. Water comes out from the unconfined aquifer along the seepage faces and follows the vertical lines DC and EF. This type of boundary also occurs along the internal face of the screen interval of a well in an unconfined aquifer.

The mathematical expression of a seepage face boundary can easily be written from Eq. (2-1). Since a seepage face boundary is open to atmosphere, the pressure along the seepage face is atmospheric pressure. The atmospheric pressure is the same everywhere on the free surface; and, therefore, for the sake of simplicity, the pressure (p) can be taken to be equal to zero. Then Eq. (2-1) becomes

$$h = z \tag{2-185}$$

which is the boundary condition at the seepage face. The geometry of a seepage face matches the boundary of the aquifer. However, the upper end of a seepage face (points D and E in Figure 2-33) is unknown. Therefore, determination of the upper ends of the seepage faces is part of the solution. Under unsteady-state flow conditions, the upper point of a seepage face varies with time.

2.9.3.6 Head-Dependent Flux Boundary. This type of boundary condition occurs when an aquifer is adjacent to a less pervious layer such as an aquitard or a confining layer. For example, in Figure 2-38 the confined aquifer is overlain by a confining layer. For this

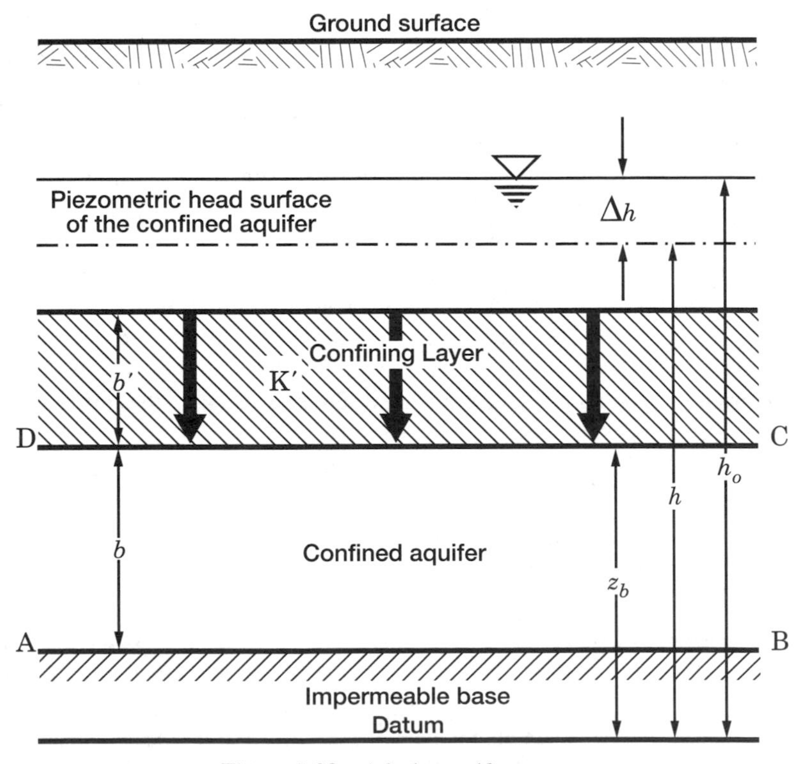

Figure 2-38 A leaky aquifer system.

kind of system, it has been customary to simplify the mathematics by assuming that flow is essentially horizontal in the aquifers and vertical in the confining layers. The validity of this assumption has been investigated by Neuman and Witherspoon (1969b) for the case of two aquifers and one aquitard (see also Section 7.2.1 of Chapter 7). Neuman and Witherspoon found that when the hydraulic conductivities of the aquifers are more than two orders of magnitude greater than that of the aquitard, the errors introduced by this assumption are usually less than 5%. For most cases the hydraulic conductivity contrast between aquifer and aquitard is greater than three orders of magnitude. Therefore, it would appear that the above assumption can safely be used. As can be seen from Figure 2-38, there is a Δh head difference between the water level of the unconfined aquifer and piezometric head level of the confined aquifer. The hydraulic conductivity contrast is responsible for this head difference.

The vertical flux through the upper boundary of the confined aquifer (DC in Figure 2-38) will be dependent on the Δh head difference and the aquitard hydraulic conductivity (K'). From the one-dimensional form of Darcy's law, as given by Eq. (2-19), one can write

$$q_v(h) = -K'\frac{h - h_0}{b'} = -K'\frac{\Delta h}{b} \tag{2-186}$$

The vertical flux is either downward or upward, depending on the head levels. Here, K' is the vertical hydraulic conductivity of the confining layer, b' is the thickness of the confining layer, h_0 is the fixed head above the confining layer, and h is the head in the confined aquifer. Since $q_v(h)$ varies with the head, it is called the *head-dependent flux*. In mathematical physics, it is called the *third-type boundary condition* or the *Cauchy boundary condition*. In some literature the term *mixed boundary condition* is also used. In Eq. (2-186), the flux is a linear function of the head in the aquifer. As the head falls, the flux through the confining layer increases; and as h rises, the flux decreases. If the head (h) level of the confined aquifer falls below the top boundary (DC in Figure 2-38), the aquifer is no longer confined and is under unconfined flow condition. However, the confining layer remains saturated from its top to its bottom. Under these conditions, the bottom of the confining layer becomes a *seepage face* (see Section 2.9.3.5), and the pressure acting there becomes equal to the atmospheric pressure. Since the atmospheric pressure acts to the top and bottom of the confining layer, for the sake of simplicity the atmospheric pressure (p) can be taken to be virtually equal to zero (see also Section 2.9.3.5). As a result, with the atmospheric pressure considered to be zero, the head at the bottom of the confining layer becomes equal to the geometric elevation (z_b) of its bottom. And, Eq. (2-186) takes the form

$$q_v = -K'\frac{z_b - h_0}{b'} \tag{2-187}$$

Therefore, the vertical flux remains constant.

2.9.3.7 *Interface Boundary.* Let us consider two adjacent aquifers (see Figure 2-39) having uniform different hydraulic conductivities. The change of hydraulic conductivity may be gradual or abrupt. The case in Figure 2-39 is idealized since the two aquifers form a common boundary. Let us designate the heads of Aquifer 1 and Aquifer 2 as $h_1(x, y, z, t)$ and $h_2(x, y, z, t)$, respectively. The hydraulic conductivities of the aquifers are K_1 and K_2.

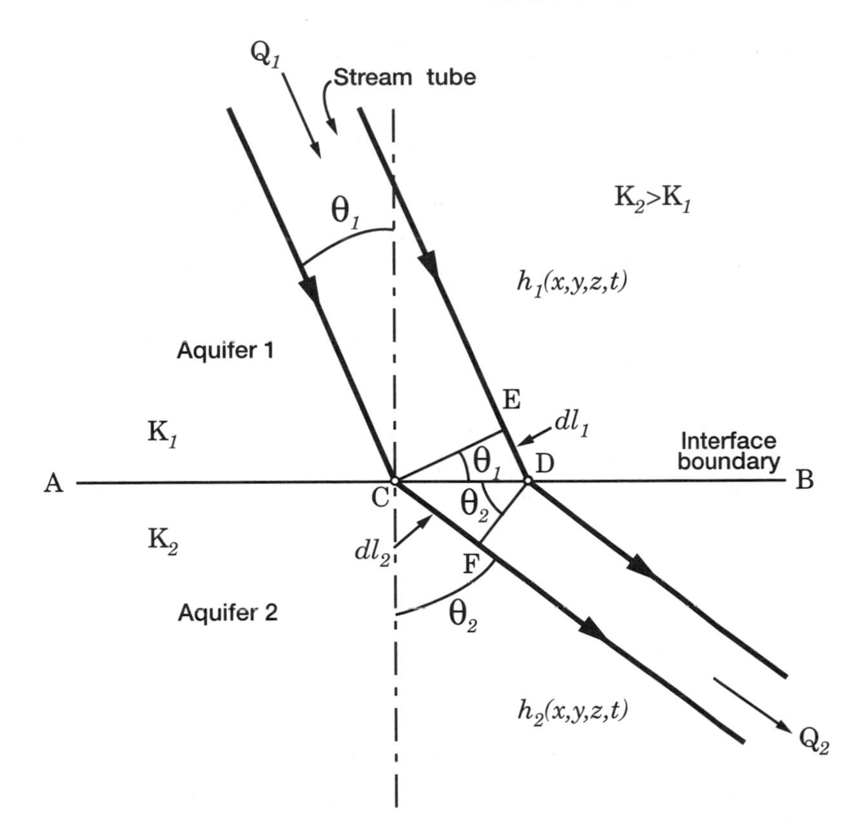

Figure 2-39 Two adjacent aquifers having uniform but different hydraulic conductivities.

Since the head at the common boundary is the same, one can write

$$h_1(x, y, z, t) = h_2(x, y, z, t) \qquad \text{at the interface AB} \tag{2-188}$$

The other condition is that whatever enters the boundary at one side must come out on the other side of the boundary. Now let us consider a stream tube as shown in Figure 2-39. The stream tube has a unit width perpendicular to the page. Since the inflow rate Q_1 must be equal to the outflow rate Q_2, from Darcy's law one can write

$$K_1 \frac{dh_1}{dl_1}(\overline{\text{CE}}) = K_2 \frac{dh_2}{dl_2}(\overline{\text{DF}}) \tag{2-189}$$

where dh_1 is the head drop across the distance dl_1, and dh_2 is the head drop across the distance dl_2. From the geometry in Figure 2-39, we obtain

$$\overline{\text{CE}} = \overline{\text{CD}} \cos \theta_1, \qquad \overline{\text{DF}} = \overline{\text{CD}} \cos \theta_2$$

and

$$dl_1 = \overline{\text{CD}} \sin \theta_1 \qquad dl_2 = \overline{\text{CD}} \sin \theta_2$$

Substituting these into Eq. (2-189) and making $dh_1 = dh_2$ gives

$$\frac{K_1}{K_2} = \frac{\tan\theta_1}{\tan\theta_2} \tag{2-190}$$

Equation (2-190) is the *tangent law* for the refraction of aquifer flow.

2.9.3.8 Free Surface Boundary. If the pressure on a stream surface is uniform, that stream surface is called the *free surface*. For unconfined flow cases, for which the capillary fringe effects may be neglected, the water table may be regarded as the upper bounding surface of the flow. Under these conditions, the water table is a free surface and the uniform pressure on it is equal to the atmospheric pressure. The atmospheric pressure is taken to be zero because of the reasons mentioned in Sections 2.9.3.5 and 2.9.3.6.

The location and shape of the free surface are unknown, and they are parts of the solution. Therefore, from Eq. (2-1), the head on a free surface is expressed as

$$h(x, y, z, t) = \frac{p_a}{\gamma} + z + f \tag{2-191}$$

where p_a is the atmospheric pressure (taken to be zero) and z is the vertical coordinate at any point on the free surface. Referring to Figure 2-2, z is the vertical coordinate of a point B located on the water table.

Differential Equation of the Free Surface. Under the aforementioned conditions the differential equation of the free surface will be obtained using vector notation (Hantush, 1964). Equation (2-1) can also be written as

$$h(\mathbf{r}, t) = \frac{p(\mathbf{r}, t)}{\gamma} + z + f \tag{2-192}$$

where $\mathbf{r}$ is the position vector and defined as

$$\mathbf{r} = x\mathbf{i} + y\mathbf{j} + z\mathbf{k} \tag{2-193}$$

where $\mathbf{i}$, $\mathbf{j}$, and $\mathbf{k}$ are the unit vectors in the directions of x, y, and z coordinates, respectively. From Eq. (2-192), the total change in $p(\mathbf{r}, t)$ with time is

$$\frac{1}{\gamma}\frac{d[p(\mathbf{r}, t)]}{dt} = \frac{d[h(\mathbf{r}, t) - z - f]}{dt} = \frac{\partial h}{\partial t} + \nabla \cdot \frac{d\mathbf{r}}{dt} - \frac{dz}{dt} \tag{2-194}$$

and from Darcy's law (see Section 2.5.1.1) one can write

$$\frac{d\mathbf{r}}{dt} = \frac{\mathbf{q}}{n_e} = -\frac{1}{n_e}\left[K_x\frac{\partial h}{\partial x}\mathbf{i} + K_y\frac{\partial h}{\partial y}\mathbf{j} + K_z\frac{\partial z}{\partial z}\mathbf{k}\right] \tag{2-195}$$

and

$$\frac{dz}{dt} = \frac{q_z}{n_e} = -\frac{1}{n_e}K_z\frac{\partial h}{\partial z} \tag{2-196}$$

where $\mathbf{q}$ is Darcy velocity vector given by Eq. (2-111) and n_e is the effective porosity (see Section 2.3.2). Since the pressure is constant at the free surface, $dp/dt = 0$, and then Eq. (2-194) takes the form

$$\frac{\partial h}{\partial t} = -\nabla h \cdot \frac{d\mathbf{r}}{dt} + \frac{dz}{dt} \tag{2-197}$$

Substituting Eqs. (2-111), (2-195), and (2-196) into Eq. (2-197) gives

$$\frac{\partial h}{\partial t} = -\nabla h \cdot \left[-\frac{1}{n_e} \left(K_x \frac{\partial h}{\partial x}\mathbf{i} + K_y \frac{\partial h}{\partial y}\mathbf{j} + K_z \frac{\partial h}{\partial z}\mathbf{k} \right) \right] - \frac{1}{n_e} K_z \frac{\partial h}{\partial z} \tag{2-198}$$

or after performing the dot product one obtains

$$\frac{\partial h}{\partial t} = \frac{1}{n_e} \left[K_x \left(\frac{\partial h}{\partial x} \right)^2 + K_y \left(\frac{\partial h}{\partial y} \right)^2 + K_z \left(\frac{\partial h}{\partial z} \right)^2 - K_z \frac{\partial h}{\partial z} \right] \tag{2-199}$$

This nonlinear differential equation is the boundary condition to be satisfied on the transient free surface of the unconfined aquifer.

With a constant recharge rate (I) at the surface, Eq. (2-199) takes the form (Bear, 1979)

$$\frac{\partial h}{\partial t} = \frac{1}{n_e} \left[K_x \left(\frac{\partial h}{\partial x} \right)^2 + K_y \left(\frac{\partial h}{\partial y} \right)^2 + K_z \left(\frac{\partial h}{\partial z} \right)^2 - \frac{\partial h}{\partial z}(K_z + I) + I \right] \tag{2-200}$$

For a homogeneous and isotropic unconfined aquifer case, $K = K_x = K_y = K_z$, and then Eq. (2-199) takes the form

$$\frac{n_e}{K} \frac{\partial h}{\partial t} = \left(\frac{\partial h}{\partial x} \right)^2 + \left(\frac{\partial h}{\partial y} \right)^2 + \left(\frac{\partial h}{\partial z} \right)^2 - \frac{\partial h}{\partial z} \tag{2-201}$$

2.9.3.9 Illustrative Example. With the governing differential equations presented in Section 2.8 and initial and boundary conditions presented in Section 2.9, aquifer hydraulics problems in one, two, and three dimensions can be solved using analytical and numerical solution methods. However, the main focus of this book is to present data analysis methods for the determination of aquifer parameters. These methods depend on analytical solutions based on the aforementioned equations. Therefore, in the remaining chapters of this book, a number of analytical solutions regarding well hydraulics problems will be presented. Below, just to show an application for the solution of aquifer differential equations, a sample analytical solution is presented for a hypothetical aquifer problem.

Example 2-14: This example is taken from Batu (1984). The vertical cross section of a confined aquifer is shown in Figure 2-40. Two surface water bodies exist at the left side of the y–z plane and on the right side of the plane parallel to the $y - z$ plane at distance L. The hydraulic conductivity (K_x and K_z) are assumed to be given by the functions

$$K_x(x, z) = a \exp(bx + cz + d) \qquad K_z(x, z) = 0$$

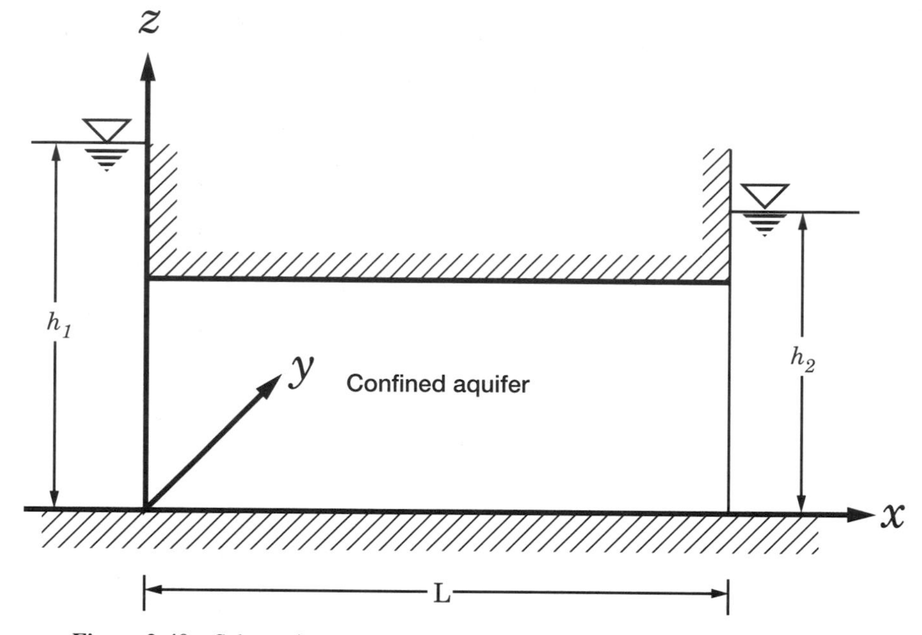

Figure 2-40 Schematic representation of the confined aquifer for Example 2-14.

where $a(L/T)$, $b(1/L)$, $c(1/L)$, and d (dimensionless) are some constants. Determine the head (h) and Darcy velocity distributions in the aquifer under steady state conditions.

Solution: The given hydraulic conductivity function corresponds to a heterogeneous and isotropic aquifer case (see Section 2.3.7.2). Since the ground-water flow in the aquifer is one-dimensional, the differential equation to be used is the one-dimensional form of Eq. (2-170) under steady-state conditions. Therefore, from Eq. (2-170) one can write

$$\frac{\partial}{\partial x}\left[K_x(x,z)\frac{\partial h}{\partial x}\right] = 0$$

or

$$\frac{\partial K_x(x,z)}{\partial x}\frac{\partial h}{\partial x} + K_x(x,z)\frac{\partial^2 h}{\partial x^2} = 0$$

The general solution of this equation is

$$h(x) = C_1 \exp(m_1 x) + C_2 \exp(m_2 x)$$

where C_1 and C_2 are constants and m_1 and m_2 are the solutions of the characteristic equation of the differential equation. They are $m_1 = 0$ and $m_2 = -b$. Introducing these in the above solution, one obtains

$$h(x) = C_1 + C_2 \exp(-bx)$$

The boundary conditions in Figure 2-40 are

$$h = h_1 \qquad \text{at } x = 0$$

and

$$h = h_2 \qquad \text{at } x = L$$

From these boundary conditions, C_1 and C_2 can be determined. The final result for head distribution is

$$h(x) = \frac{h_1 \exp(-bL) - h_2 + (h_2 - h_1)\exp(-bx)}{\exp(-bL) - 1}$$

Using this equation along with Eqs. (2-112a) and (2-112c) gives the Darcy velocities in the x and z directions:

$$q_x(x, z) = \frac{ab(h_1 - h_2)\exp(cz + d)}{1 - \exp(-bL)}, \qquad q_z(x, z) = 0$$

The above equations were verified by comparing their numerical results with the finite-element method (Batu, 1984, Figure 9).

PART II

HYDRAULICS OF AQUIFERS UNDER STEADY PUMPING CONDITIONS FROM WELLS AND HYDROGEOLOGIC DATA ANALYSIS METHODS

CHAPTER 3

FULLY PENETRATING PUMPING WELLS IN HOMOGENEOUS AND ISOTROPIC CONFINED AND UNCONFINED AQUIFERS

3.1 INTRODUCTION

After a long period of pumping from a well or recharging to a well, the ground-water flow around the well approaches steady state. This means that the potentiometric head at any point in the ground-water flow field does not change with time. The time required to reach steady state depends on the hydraulic characteristics of the aquifer. For less permeable aquifers the time period is longer than for highly permeable aquifers.

Steady-state solutions for confined and unconfined aquifers play a very important role in analyzing the drawdown data for the determination of aquifer hydraulic characteristics and in making a quick assessment of the influence zone of a well or combination of wells.

In this chapter, the governing equations and solutions for a well in confined and unconfined aquifers are presented in detail. Derived equations include the well-known *Thiem equation* and *Dupuit–Forchheimer well discharge formula*. Application of the solutions for pumping test data analysis are presented in Section 3.4.

3.2 WELL HYDRAULICS IN CONFINED AQUIFERS

3.2.1 Nonleaky Aquifer Case: Thiem Equation

Thiem (1906) was the first to derive the solution for a well in a confined aquifer under steady-state conditions. This equation is the simplest with regard to its mathematical form and its derivation process. In the following sections the details for the *Thiem equation* are presented.

3.2.1.1 Problem Statement and Assumptions. Figure 3-1 shows a fully penetrating well in a confined aquifer. A solution for the piezometric head will be presented with the following assumptions:

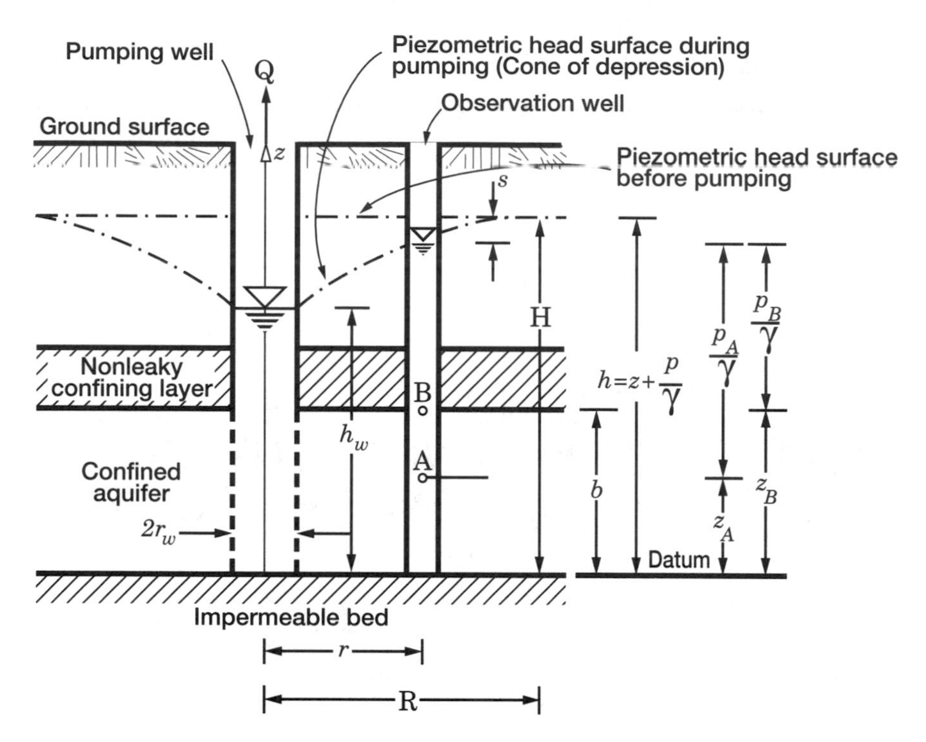

Figure 3-1 Fully penetrating well in a confined aquifer

1. The aquifer is horizontal and has a constant thickness.
2. The aquifer is homogeneous and isotropic and its boundaries go to infinity.
3. The potentiometric or piezometric head surface is horizontal before pumping.
4. Darcy's law is valid for the flow in the aquifer.
5. Water is instantaneously removed from storage as the piezometric head declines.
6. The pumping rate of the well is constant.
7. The flow is symmetric with respect to the axis of the well.

3.2.1.2 Governing Differential Equation. For two-dimensional radial flow, the governing differential equation under transient conditions is given by Eq. 2-175 of Chapter 2. Under steady-state conditions the term on the right-hand side becomes zero. Therefore, for two-dimensional radial flow under steady-state conditions it takes the form

$$\frac{1}{r}\frac{\partial h}{\partial r} + \frac{\partial^2 h}{\partial r^2} = 0 \tag{3-1}$$

3.2.1.3 Boundary Conditions. The boundary conditions are

$$h = h_w, \qquad r = r_w \tag{3-2}$$

$$h = H, \qquad r = R \tag{3-3}$$

where h_w is the piezometric head at the well boundary, r_w is the well radius, H is the piezometric head level before pumping, and R is the well influence radius at which the drawdown is zero (see Figure 3-1). The boundary conditions given by Eqs. (3-2) and (3-3) are the Dirichlet or first-type boundary conditions (see Section 2.9.3.1 of Chapter 2).

The boundary condition described by Eq. (3-2) states that the piezometric head at the well boundary is h_w. The second condition described by Eq. (3-3) states that at the influence radius R the piezometric head is equal to the initial piezometric head (H).

3.2.1.4 Solution. The potentiometric head distribution, $h(r)$, can be determined from the governing differential equation and the boundary conditions given above. The same distribution can also be determined directly from the continuity equation. The continuity equation is

$$-Q = 2\pi r b q_r \tag{3-4}$$

where q_r is the radial velocity that is given by Eq. (2-115a) of Chapter 2. According to the assumed sign convention $-Q$ corresponds to withdrawal and $+Q$ corresponds to injection. Substitution of the q_r expression into Eq. (3-4) with $K_r = K$ gives

$$K\frac{\partial h}{\partial r} = \frac{Q}{2\pi r b} \tag{3-5}$$

From Eq. (3-5) one can write

$$h(r) = \frac{Q}{2\pi T}\ln r + C \tag{3-6}$$

where T is transmissivity and is defined by Eq. (2-24) of Chapter 2. One can verify that Eq. (3-6) satisfies Eq. (3-1), which is the governing differential equation. From Eqs. (3-2) and (3-6), C can be determined. Then, introducing C in Eq. (3-6) gives

$$h(r) - h_w = \frac{Q}{2\pi T}\ln\left(\frac{r}{r_w}\right) \tag{3-7}$$

The properties of Eq. (3-7) are as follows: (1) The piezometric head h increases indefinitely with increasing radial distance r. (2) It is clear that the piezometric surface cannot rise above $h(R)$. For this reason, steady radial flow does not exist in an areally extensive aquifer. (3) Equation (3-7) is valid only in the close proximity of a well where steady flow has been established.

From Eqs. (3-3) and (3-6) we obtain

$$H - h_w = h(R) - h(r_w) = \frac{Q}{2\pi T}\ln\left(\frac{R}{r_w}\right) \tag{3-8}$$

Between any two distances (r_1 and r_2, $r_2 > r_1$), Eq. (3-8) takes the form

$$h(r_2) - h(r_1) = s(r_1) - s(r_2) = \frac{Q}{2\pi T}\ln\left(\frac{r_2}{r_1}\right) \tag{3-9}$$

Equation (3-9) is known as the *Thiem equation* (Thiem, 1906). For $r_1 = r$, and $r_2 = R$, Eq. (3-9) takes the following form:

$$h(R) - h(r) = \frac{Q}{2\pi T} \ln\left(\frac{R}{r}\right) \tag{3-10}$$

Some important aspects of Eqs. (3-8), (3-9), and (3-10) are as follows: (1) The distance R, for which the drawdown is zero is the *influence radius of the well.* (2) The parameter R has to be estimated before the prediction of drawdowns. (3) In Eq. (3-10), R is in the form of $\ln R$. For this reason, even a large error in estimating R does not significantly affect the drawdown determined by Eq. (3-10).

Example 3-1: For the aquifer shown in Figure 3-1, the following values are given: $Q =$ 5 liters/s, $r_w = 0.1$ m, $b = 6.0$ m, $K = 0.0001$ m/s. Select the range of R and compute the drawdown at the well using different values of R.

Solution: Equation (3-8) may be used. $Q/(2\pi T) = 1.33$ m. For different values of R the drawdown values are given in Table 3-1.

As can be seen from the results in Table 3-1, the value of R is increased from $R = 50$ m to 300 m, but the drawdown is inreased by 29%.

3.2.2 Leaky Aquifer Case

The type of aquifer described in Section 2.1 will be analyzed under leakege conditions from an overlain or underlain confined aquifer. In the following sections the details are presented.

3.2.2.1 Problem Statement and Assumptions. Figure 3-2 shows a fully penetrating well in a confined aquifer through which leakage occurs from an overlain aquifer. The solution for piezometric head distribution will be presented with the following assumptions:

1. The leaky confined aquifer is underlain by an impereable bed and overlain by a cofining layer.

TABLE 3-1 Drawdown as a Function of the Influence Radius for Example 3-1

Influence Radius, R (m)	Drawdown, s (m)
50	8.27
100	9.19
150	9.73
200	10.11
250	10.41
300	10.65

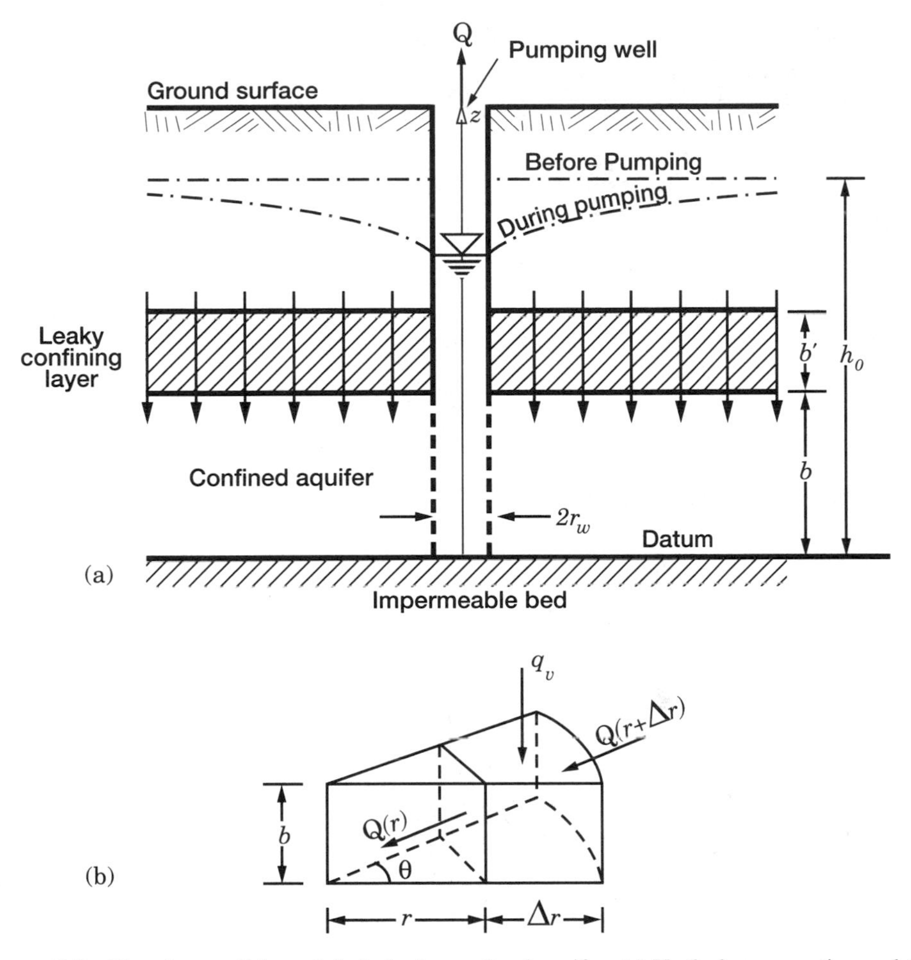

Figure 3-2 Flow to a well in an infinite leaky confined aquifer: (a) Vertical cross section and (b) aquifer element with the flux components.

2. Above the confining layer, an unconfined aquifer exists. The unconfined aquifer has a horizontal water table whose hydraulic head (h_0) is constant. The supply of water to the unconfined aquifer is sufficient to maintain h_0 constant.
3. The flow in the confining layer is vertical (see Section 2.9.3.6 of Chapter 2).
4. The rest of the assumptions are the same as in Section 3.2.1.

3.2.2.2 Governing Differential Equation. The governing differential equation can be derived by applying the continuity principle. Applying the principle of continuity, from Figure 3-2b one can write

$$Q(r + \Delta r) - Q(r) + (2\pi r \Delta r)q_v = 0 \tag{3-11}$$

Dividing both sides by Δr and, as $\Delta r \to 0$, Eq. (3-11) yields

$$\frac{\partial Q}{\partial r} + 2\pi r q_v = 0 \tag{3-12}$$

On the other hand, the discharge $Q(r)$ can be expressed as

$$Q(r) = (2\pi r b)K\frac{\partial h}{\partial r} = 2\pi r T\frac{\partial h}{\partial r} \tag{3-13}$$

From Darcy's law, q_v is [see Eq. (2-186) of Chapter 2]

$$q_v = K'\frac{h_0 - h}{b'} \tag{3-14}$$

Introducing Eqs. (3-13) and (3-14) into Eq. (3-12) gives

$$\frac{1}{r}\frac{\partial}{\partial r}\left(\frac{r\partial h}{\partial r}\right) + \frac{h_0 - h}{B^2} = 0 \tag{3-15}$$

where

$$B^2 = \frac{bb'K}{K'} \tag{3-16}$$

B is a characteristic length of the leaky aquifer and is called *leakage factor*. In the same equation, b'/K' is known as *hydraulic resistance.*

Equation (3-15) can be written as an ordinary differential equation since h is only a function of r:

$$r^2\frac{d^2s}{dr^2} + r\frac{ds}{dr} - \frac{r^2}{B^2}s = 0 \tag{3-17}$$

where $s = h_0 - h$ is the drawdown. With

$$\frac{r}{B} = x, \qquad dr = Bdx \tag{3-18}$$

Eq. (3-17) takes the form

$$x^2\frac{d^2s}{dx^2} + x\frac{ds}{dx} - x^2 s = 0 \tag{3-19}$$

3.2.2.3 General and Special Solutions

General Solution. Equation (3-19) is a modified Bessel differential equation, and its general solution is [Beyer, 1986, p. 352, Eq. (14)]

$$s = h_0 - h(r) = C_1 I_0\left(\frac{r}{B}\right) + C_2 K_0\left(\frac{r}{B}\right) \tag{3-20}$$

where C_1 and C_2 are constants, $I_0(r/B)$ is a zero-order modified Bessel function of the first kind, and $K_0(r/B)$ is a zero-order modified Bessel function of the second kind. The values of $K_0(x)$ are tabulated in Table 3-2.

3.2.2.4 Special Solution for an Infinite Aquifer.

For this case, the boundary condition at infinity is

$$h = h_0, \qquad r = \infty \tag{3-21}$$

Because $I_0(\infty) = \infty$ and $K_0(\infty) = 0$, it follows from Eq. (3-20) that $C_1 = 0$. On the other hand, from continuity we obtain

$$Q = 2\pi r_w bK \frac{\partial h}{\partial r} \qquad \text{at } r = r_w \tag{3-22}$$

and

$$\frac{\partial h}{\partial r} = -C_2 \frac{\partial}{\partial r}\left[K_0\left(\frac{r}{B}\right)\right] = C_2 \frac{1}{B} K_1\left(\frac{r}{B}\right) \tag{3-23}$$

where $K_1(r/B)$ is a first-order modified Bessel function of the second kind. Introducing Eq. (3-23) into Eq. (3-22) with $r = r_w$ we obtain

$$Q = 2\pi r_w bKC_2 \frac{1}{B} K_1\left(\frac{r_w}{B}\right) \tag{3-24}$$

and solving for C_2 gives

$$C_2 = \frac{Q}{2\pi T \dfrac{r_w}{B} K_1\left(\dfrac{r_w}{B}\right)} \tag{3-25}$$

With the values of C_1 and C_2, Eq. (3-20) takes the following form:

$$s(r) = h_0 - h(r) = \frac{Q}{2\pi T} \frac{K_0\left(\dfrac{r}{B}\right)}{\dfrac{r_w}{B} K_1\left(\dfrac{r_w}{B}\right)} \tag{3-26}$$

Equation (3-26) can be simplified by considering the physical range of r_w/B. From Eq. (3-16) we obtain

$$\frac{r_w}{B} = r_w \left(\frac{K'}{bb'K}\right)^{1/2} \tag{3-27}$$

In practice (Mariño and Luthin, 1982) we have

$$\frac{r_w}{B} < 0.01 \tag{3-28}$$

TABLE 3-2 Values of the Functions e^x, $K_0(x)$, $e^xK_0(x)$, $-\mathrm{Ei}(-x)$, and $-\mathrm{Ei}(-x)e^x$

x	e^x	$K_0(x)$	$e^xK_0(x)$	$-\mathrm{Ei}(-x)$	$-\mathrm{Ei}(-x)e^x$	x	e^x	$K_0(x)$	$e^xK_0(x)$	$-\mathrm{Ei}(-x)$	$-\mathrm{Ei}(-x)e^x$	x	e^x	$K_0(x)$	$e^xK_0(x)$	$-\mathrm{Ei}(-x)$	$-\mathrm{Ei}(-x)e^x$
0.010	1.0101	4.7212	4.7687	4.0379	4.0787	0.10	1.1052	2.4271	2.6823	1.8229	2.0147	1.0	2.7183	0.4210	1.1445	.2194	.5964
11	1.0111	4.6260	4.6771	3.9436	3.9874	11	1.1163	2.333	2.6046	1.7371	1.9391	1.1	3.0042	.3656	1.0983	.1860	.5588
12	1.0121	4.5390	4.5938	3.8576	3.9044	12	1.1275	2.2479	2.5345	1.6595	1.8711	1.2	3.3201	.3185	1.0575	.1584	.5259
13	1.0131	4.4590	4.5173	3.7785	3.8282	13	1.1388	2.1695	2.4707	1.5889	1.8094	1.3	3.6693	.2782	1.0210	.1355	.4972
14	1.0141	4.3849	4.4467	3.7054	3.7578	14	1.1503	2.0972	2.4123	1.5241	1.7532	1.4	4.0552	.2437	0.9881	.1162	.4712
15	1.0151	4.3159.	4.3812	3.6374	3.6925	15	1.1618	2.0300	2.3585	1.4645	1.7015	1.5	4.4817	.2138	.9582	.1000	.4482
16	1.0161	4.2514	4.3200	3.5739	3.6317	16	1.1735	1.9674	2.3088	1.4092	1.6537	1.6	4.9530	.1880	.9309	.0863	.4275
17	1.0171	4.1908	4.2627	3.5143	3.5746	17	1.1853	1.9088	2.2625	1.3578	1.6094	1.7	5.4739	.1655	.9059	.0747	.4086
18	1.0182	4.1337	4.2088	3.4581	3.5209	18	1.1972	1.8537	2.2193	1.3098	1.5681	1.8	6.0496	.1459	.8828	.0647	.3915
19	1.0192	4.0797	4.1580	3.4050	3.4705	19	1.2093	1.8018	2.1788	1.2649	1.5295	1.9	6.6859	.1288	.8614	.0562	.3758
0.020	1.0202	4.0285	4.1098	3.3547	3.4225	0.20	1.2214	1.7527	2.1408	1.2227	1.4934	2.0	7.3891	.1139	.8416	.0489	.3613
21	1.0212	3.9797	4.0642	3.3069	3.3771	21	1.2337	1.7062	2.1049	1.1829	1.4593	2.1	8.1662	.1088	.8230	.0426	.3480
22	1.0222	3.9332	4.0207	3.2614	3.3340	22	1.2461	1.6620	2.0710	1.1454	1.4273	2.2	9.0250	.0893	.8057	.0372	.3356
23	1.0233	3.8888	3.9793	3.2179	3.2927	23	1.2586	1.6199	2.0389	1.1099	1.3969	2.3	9.9742	.0791	.7894	.0325	.3242
24	1.0243	3.8463	3.9398	3.1763	3.2535	24	1.2713	1.5798	2.0084	1.0762	1.3681	2.4	11.0232	.0702	.7740	.0284	.3135
25	1.0253	3.8056	3.9019	3.1365	3.2159	25	1.2840	1.5415	1.9793	1.0443	1.3409	2.5	12.1825	.0623	.7596	.0249	.3035
26	1.0263	3.7664	3.8656	3.0983	3.1799	26	1.2969	1.5048	1.9517	1.0139	1.3149	2.6	13.4637	.0554	.7459	.0219	.2942
27	1.0274	3.7287	3.8307	3.0615	3.1452	27	1.3100	1.4697	1.9253	.9849	1.2902	2.7	14.8797	.4093	.7329	.0192	.2854
28	1.0284	3.6924	3.7972	3.0261	3.1119	28	1.3231	1.4360	1.9000	.9573	1.2666	2.8	16.4446	.0438	.7206	.0169	.2773
29	1.0294	3.6574	3.7650	2.9920	3.0800	29	1.3364	1.4036	1.8758	.9309	1.2441	2.9	18.1742	.0390	.7089	.0148	.2693
0.030	1.0305	3.6235	3.7339	2.9591	3.0494	0.30	1.3499	1.3725	1.8526	.9057	1.2226	3.0	20.0855	.0347	.6978	.0131	.2621
31	1.0315	3.5908	3.7039	2.9273	3.0196	31	1.3634	1.3425	1.8304	.8815	1.2018	3.1	22.1980	.0310	.6871	.0115	.2551
32	1.0325	3.5591	3.6749	2.8965	2.9908	32	1.3771	1.3136	1.8089	.8583	1.1820	3.2	24.5325	.0276	.6770	.0101	.2485
33	1.0336	3.5284	3.6468	2.8668	2.9631	33	1.3910	1.2857	1.7883	.8361	1.1630	3.3	27.1126	.0246	.6773	.0089	.2424
34	1.0346	3.4986	3.6196	2.8379	2.9362	34	1.4050	1.2587	1.7685	.8147	1.1446	3.4	29.9641	.0220	.6580	.0079	.2365
35	1.0356	3.4697	3.5933	2.8099	2.9101	35	1.4191	1.2327	1.7493	.7942	1.1270	3.5	33.1155	.0196	.6490	.0070	.2308
36	1.0367	3.4416	3.5678	2.7827	2.8848	36	1.4333	1.2075	1.7308	.7745	1.1101	3.6	36.5982	.0175	.6405	.0062	.2254
37	1.0377	3.4143	3.5430	2.7563	2.8603	37	1.4477	1.1832	1.7129	.7554	1.0936	3.7	40.4473	.0156	.6322	.0055	.2204
38	1.0387	3.3877	3.5189	2.7306	2.8364	38	1.4623	1.1596	1.6956	.7371	1.0779	3.8	44.7012	.0140	.6243	.0048	.2155
39	1.0398	3.3618	3.4955	2.7056	2.8133	39	1.4770	1.1367	1.6789	.7194	1.0626	3.9	49.4025	.0125	.6166	.0043	.2108
0.040	1.0408	3.3365	3.4727	2.6813	2.7907	0.40	1.4918	1.1145	1.6627	.7024	1.0478	4.0	54.5982	.0112	.6093	.0038	.2063
41	1.0419	3.3119	3.4505	2.6576	2.7688	41	1.5068	1.0930	1.6470	.6859	1.0335	4.1	60.3403	.0100	.6022	.0033	.2021
42	1.0429	3.2879	3.4289	2.6344	2.7474	42	1.5220	1.0721	1.6317	.6700	1.0197	4.2	66.6863	.0089	.5953	.0030	.1980
43	1.0439	3.2645	3.4079	2.6119	2.7267	43	1.5373	1.0518	1.6169	.6546	1.0063	4.3	73.6998	.0080	.5887	.0026	.1941
44	1.0450	3.2415	3.3874	2.5899	2.7064	44	1.5527	1.0321	1.6025	.6397	.9933	4.4	81.4509	.0071	.5823	.0023	.1903
45	1.0460	3.2192	3.3673	2.5684	2.6866	45	1.5683	1.0129	1.5886	.6253	.9807	4.5	90.0171	.0064	.5761	.0021	.1866
46	1.0471	3.1973	3.3478	2.5474	2.6672	46	1.5841	0.9943	1.5750	.6114	.9685	4.6	99.4843	.0057	.5701	.0018	.1832
47	1.0481	3.1758	3.3287	2.5268	2.6483	47	1.6000	.9761	1.5617	.5979	.9566	4.7	109.9472	.0051	.5643	.0016	.1798
48	1.0492	3.1549	3.3100	2.5068	2.6300	48	1.6161	.9584	1.5489	.5848	.9451	4.8	121.5104	.0046	.5586	.0014	.1766
49	1.0502	3.1343	3.2918	2.4871	2.6120	49	1.6323	.9412	1.5363	.5721	.9338	4.9	134.2898	.0041	.5531	.0013	.1734

0.050	1.0513	3.1142	3.2739	2.4679	2.5945
51	1.0523	3.0945	3.2564	2.4491	2.5773
52	1.0534	3.0752	3.2393	2.4306	2.5604
53	1.0544	3.0562	3.2226	2.4126	2.5440
54	1.0555	3.0376	3.2062	2.3948	2.5278
55	1.0565	3.0194	3.1901	2.3775	2.5120
56	1.0576	3.0015	3.1744	2.3604	2.4964
57	1.0587	2.9839	3.1589	2.3437	2.4811
58	1.0597	2.9666	3.1437	2.3273	2.4663
59	1.0608	2.9496	3.1288	2.3111	2.4516
0.060	1.0618	2.9329	3.1142	2.2953	2.4371
61	1.0629	2.9165	3.0999	2.2797	2.4230
62	1.0640	2.9003	3.0058	2.2645	2.4092
63	1.0650	2.8844	3.0719	2.2494	2.3956
64	1.0661	2.8688	3.0584	2.2346	2.3822
65	1.0672	2.8534	3.0450	2.2201	2.3691
66	1.0682	2.8362	3.0319	2.2058	2.3562
67	1.0693	2.8233	3.0189	2.1917	2.3434
68	1.0704	2.8086	3.0062	2.1779	2.3310
69	1.0714	2.7941	2.9937	2.1643	2.3188
0.070	1.0725	2.7798	2.9814	2.1508	2.3067
71	1.0736	2.7657	2.9693	2.1376	2.2949
72	1.0747	2.7519	2.9573	2.1246	2.2832
73	1.0757	2.7382	2.9455	2.1118	2.2717
74	1.0768	2.7247	2.9340	2.0991	2.2603
75	1.0779	2.7114	2.9226	2.0867	2.2492
76	1.0790	2.6983	2.9113	2.0744	2.2381
77	1.0800	2.6853	2.9002	2.0623	2.2273
78	1.0811	2.6726	2.8894	2.0503	2.2165
79	1.0822	2.6589	2.8786	2.0386	2.2062
0.080	1.0833	2.6475	2.8680	2.0269	2.1957
81	1.0844	2.6352	2.8575	2.0155	2.1856
82	1.0855	2.6231	2.8472	2.0042	2.1754
83	1.0865	2.6111	2.8370	1.9930	2.1655
84	1.0876	2.5992	2.8270	1.9820	2.1557
85	1.0887	2.5875	2.8171	1.9711	2.1460
86	1.0898	2.5759	2.8073	1.9604	2.1364
87	1.0909	2.5645	2.7976	1.9498	2.1270
88	1.0920	2.5532	2.7881	1.9393	2.1176
89	1.0931	2.5421	2.7787	1.9290	2.1086
0.090	1.0942	2.5310	2.7694	1.9187	2.0994
91	1.0953	2.5201	2.7602	1.9087	2.0906
92	1.0964	2.5093	2.7511	1.8987	2.0818
93	1.0975	2.4986	2.7421	1.8888	2.0729
94	1.0986	2.4881	2.7333	1.8791	2.0643
95	1.0997	2.4776	2.7246	1.8695	2.0558
96	1.1008	2.4673	2.7159	1.859	2.0473
97	1.1019	2.4571	2.7074	1.8505	2.0390
98	1.1030	2.4470	2.6989	1.8412	2.0307
99	1.1041	2.4370	2.6906	1.8320	2.0227
0.100	1.1052	2.4271	2.6823	1.8229	2.0147

0.50	1.6487	.9244	1.5241	.5598	.9229
51	1.6653	.9081	1.5122	.5478	.9123
52	1.6820	.8921	1.5006	.5362	.9019
53	1.6989	.8766	1.4892	.5250	.8919
54	1.7160	.8614	1.4781	.5140	.8820
55	1.7333	.8466	1.4673	.5034	.8725
56	1.7507	.8321	1.4567	.4930	.8631
57	1.7683	.8180	1.4464	.4830	.8541
58	1.7860	.8042	1.4363	.4732	.8451
59	1.8040	.7907	1.4264	.4637	.8365
0.60	1.8221	.7775	1.4167	.4544	.8280
61	1.8404	.7646	1.4073	.4454	.8197
62	1.8589	.7520	1.3980	.4366	.8116
63	1.8776	.7397	1.3889	.4280	.8036
64	1.8965	.7277	1.3800	.4197	.7960
65	1.9155	.7159	1.3713	.4115	.7882
66	1.9348	.7043	1.3627	.4036	.7809
67	1.9542	.6930	1.3543	.3959	.7737
68	1.9739	.6820	1.3461	.3883	.7665
69	1.9937	.6711	1.3380	.3810	.7596
0.70	2.0138	.6605	1.3301	.3738	.7528
71	2.0340	.6501	1.3223	.3668	.7461
72	2.0544	.6399	1.3147	.3599	.7394
73	2.0751	.6300	1.3072	.3532	.7329
74	2.0959	.6202	1.2998	.3467	.7266
75	2.1170	.6106	1.2926	.3403	.7204
76	2.1383	.6012	1.2855	.3341	.7144
77	2.1598	.5920	1.2785	.3280	.7084
78	2.1815	.5829	1.2716	.3221	.7027
79	2.2034	.5740	1.2649	.3163	.6969
0.80	2.2255	.5653	1.2582	.3106	.6912
81	2.2479	.5568	1.2517	.3050	.6856
82	2.2705	.5484	1.2452	.2996	.6802
83	2.2933	.5405	1.2389	.2943	.6749
84	2.3164	.5321	1.2326	.2891	.6697
85	2.3397	.5242	1.2265	.2840	.6644
86	2.3632	.5165	1.2205	.2790	.6593
87	2.3869	.5088	1.2145	.2742	.6545
88	2.4109	.5013	1.2086	.2694	.6495
89	2.4351	.4940	1.2029	.2647	.6446
0.90	2.4596	.4867	1.1972	.2602	.6400
91	2.4843	.4796	1.1916	.2557	.6352
92	2.5093	.4727	1.1860	.2513	.6306
93	2.5345	.4658	1.1806	.2470	.6260
94	2.5600	.4591	1.1752	.2429	.6218
95	2.5857	.4524	1.1699	.2387	.6172
95	2.6117	.4459	1.1647	.2347	.6130
97	2.6379	.4396	1.1595	.2308	.6088
98	2.6645	.4333	1.1544	.2269	.6046
99	2.6912	.4271	1.1494	.2231	.6004
1.00	2.7183	.4210	1.1445	.2194	.5964

5.0	148.4132	.0037	.5478	.0011	.1704

Source: After Hantush (1956)

and because of this we obtain

$$\frac{r_w}{B} K_1 \left(\frac{r_w}{B} \right) \cong 1 \tag{3-29}$$

and finally Eq. (3-26) takes the form

$$s(r) \cong \frac{Q}{2\pi T} K_0 \left(\frac{r}{B} \right) \tag{3-30}$$

Under the conditions described above, $s(r)$ is independent of r_w. Equation (3-30) gives the drawdown distribution under steady-state conditions. Therefore, it describes the maximum drawdown distribution in the vicinity of a well.

Equation (3-30) was first derived by De Glee (1930), which is referenced in Kruseman and DeRidder (1991). The same equation was also derived by Jacob (1946) and Hantush and Jacob (1955) being unaware of the publication of De Glee around a quarter century ago.

Hantush (1964) noted that when r/B is small ($r/B \leq 0.05$), Eq. (3-30) may, for practical purposes, be approximated by the following equation:

$$s(r) \cong \frac{2.303Q}{2\pi T} \log \left(\frac{1.12B}{r} \right) \tag{3-31}$$

Example 3-2: For the confined aquifer shown in Figure 3-2, the following values are given: $b = 10$ m, $b' = 4$ m, $K = 0.02$ cm/s $= 17.28$ m/day, $K' = 2 \times 10^{-6}$ cm/s $=$ 17.28×10^{-4} m/day, $r_w = 0.2$ m, and $Q = 5$ liters/s $= 432$ m^3/day. Determine the drawdown under steady-state conditions at a 400-m distance from the well and at the face of the well.

Solution: Eq. (3-30) is the equation to be used for steady-state drawdown calculation. From Eq. (3-16), the leakage factor is

$$B = \left[\frac{(10 \text{ m})(4 \text{ m})(17.28 \text{ m/day})}{0.001728 \text{ m/day}} \right]^{1/2} = 632 \text{ m}$$

The value of $Q/(2\pi T)$ is

$$\frac{Q}{2\pi T} = \frac{432 \text{ m}^3/\text{day}}{2\pi(17.28 \text{ m/day})(10 \text{ m})} = 0.40 \text{ m}$$

$r/B = 400 \text{ m}/632 \text{ m} = 0.63$. From Table 3-1, $K_0(0.63) = 0.7397$. When we substitute the values in Eq. (3-30), the drawdown at 400 m is

$$s(r = 400 \text{ m}) = (0.40 \text{ m})(0.7397) = 0.30 \text{ m}$$

$r/B = 0.2 \text{ m}/632 \text{ m} = 0.00032$ at the face of the well. Because this value is not included in Table 3-2 for $K_0(r/B)$, the approximate expression given by Eq. (3-31) will be used to calculate the drawdown at the face of the well:

$$s_w(r = 0.2\text{ m}) = (2.303)(0.40\text{ m})\log\left[\frac{(1.12)(632\text{ m})}{0.2\text{ m}}\right] = 3.27\text{ m}$$

3.3 WELL HYDRAULICS IN UNCONFINED AQUIFERS

3.3.1 Dupuit–Forchheimer Well Discharge Formula

Dupuit and Forchheimer derived the subject equation without being aware of each other. For this reason, the equation bears their names. This equation is the simplest equation for unconfined aquifers. In the following sections the details for the *Dupuit–Forchheimer equation* are presented.

3.3.1.1 Problem Statement and Assumptions. Figure 3-3 shows a fully penetrating well in an unconfined aquifer. A solution for the piezometric head will be presented with the following assumptions:

1. The aquifer is homogeneous and isotropic and its boundaries go to infinity.
2. The water table is horizontal before pumping.
3. Darcy's law is valid for the flow in the aquifer.
4. Water is instantaneously removed from storage as the piezometric head declines.
5. The pumping rate of well is constant.

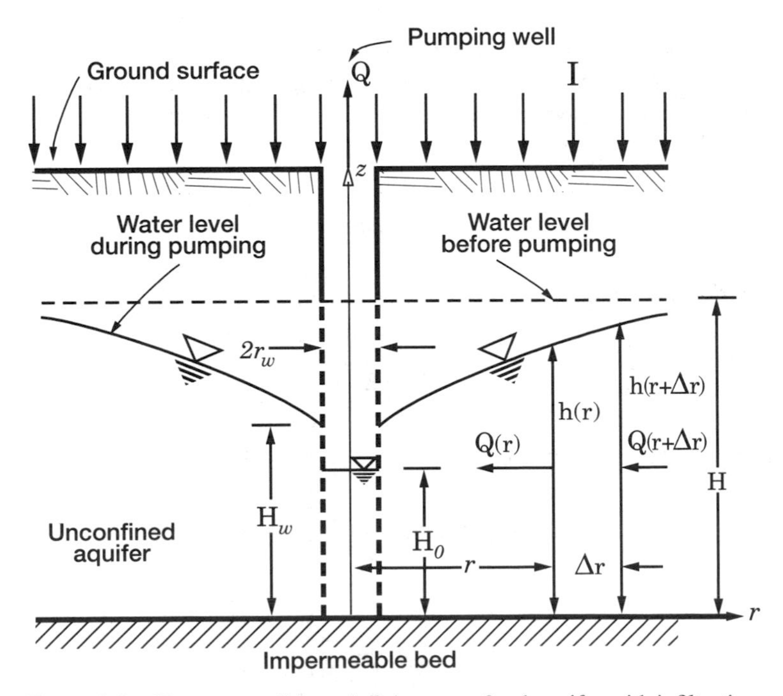

Figure 3-3 Flow to a well in an infinite unconfined aquifer with infiltration.

6. The Dupuit assumptions are valid.
7. The flow is symmetric with respect to the axis of the well.
8. The seepage face of the well is negligible and the aquifer receives a constant recharge rate (I).

Of these eight assumptions, the one regarding the Dupuit assumptions is critically important. The assumptions are discussed below.

Dupuit Assumptions. Dupuit (1863) pointed out that the slope of the water table of an unconfined aquifer under no extraction conditions along a vertical cross section is very small. Typical slopes range from 1/1000 to 10/1000. Around an extraction well in an unconfined aquifer, this slope is higher, with the decreasing radial distance from the well depending on the horizontal and vertical hydraulic conductivities of the aquifer. The assumption of small geometric slope means that the flow is essentially horizontal and the hydraulic head (h) is equal to the water table elevation (Bear, 1979).

For unconfined aquifers, the height of water in the well (H_0) is always less than the thickness of saturation at the face of well (H_w). As mentioned above, in this analysis, the seepage face is neglected and it is assumed that H_0 and H_w are the same.

3.3.1.2 Governing Differential Equation. With the assumptions given above, the derivation of the governing differential equation will be derived. The geometry of the problem is given in Figure 3-3. The continuity principle for the portion of aquifer between two cylinders of radius r and $r + \Delta r$ gives

$$Q(r + \Delta r) - Q(r) + (2\pi r \Delta r)I = 0 \tag{3-32}$$

where $I(L/T)$ represents the volume of water entering in a unit horizontal area of the aquifer per unit time, due to precipitation recharge. Positive and negative values of I represent recharge and evaporation, respectively. Dividing both sides of Eq. (3-32) by Δr, as $\Delta r \to 0$, yields

$$\frac{\partial Q}{\partial r} + 2\pi r I = 0 \tag{3-33}$$

From Eq. (2-115a) of Chapter 2, the radial Darcy velocity is

$$q_r = -K\frac{\partial h}{\partial r} \tag{3-34}$$

where K is the radial or horizontal hydraulic conductivity. From Figure 3-3 one can write

$$Q(r) = -(2\pi r h)q_r = -(2\pi r h)K\frac{\partial h}{\partial r} \tag{3-35}$$

Equation (3-35) can be expressed as

$$Q(r) = 2\pi r K\frac{1}{2}\frac{\partial(h^2)}{\partial r} \tag{3-36}$$

Substituting Eq. (3-36) into Eq. (3-33) gives the governing differential equation:

$$\frac{1}{r}\frac{d}{dr}\left[r\frac{d(h^2)}{dr}\right] + \frac{2I}{K} = 0 \tag{3-37}$$

3.3.1.3 Boundary Conditions. The boundary conditions are

$$h = H_w, \qquad r = r_w \tag{3-38}$$

$$h = H, \qquad r = R \tag{3-39}$$

where H_w is the piezometric head at the well face, r_w is the well radius, H is the piezometric head before pumping, and R is the influence radius at which the drawdown is zero (see Figure 3-3). The boundary conditions given above are the *Dirichlet boundary conditions* or *first-type boundary conditions* (see Section 2.9.3.1 of Chapter 2). The boundary condition described by Eq. (3-38) states that the piezometric head at the well face is h_w. The second condition described by Eq. (3-39) states that at the influence radius R the piezometric head is equal to the initial piezometric head H (the piezometric head of aquifer before pumping).

3.3.1.4 Solution. The piezometric head distribution, $h(r)$, can be determined from the governing differential equation and the boundary conditions given above. The general solution of Eq. (3-37) is

$$h(r)^2 = -\frac{I}{2K}r^2 + C_1 \ln r + C_2 \tag{3-40}$$

where C_1 and C_2 are some constants.

Using the boundary conditions, described by Eqs. (3-38) and (3-39), the constants C_1 and C_2 can be determined. After some manipulations the solution takes the form

$$h(r)^2 = H^2 + \frac{I}{2K}(R^2 - r^2) - \left[H^2 - H_w^2 + \frac{I}{2K}(R^2 - r_w^2)\right]\frac{\ln\left(\frac{r}{R}\right)}{\ln\left(\frac{r_w}{R}\right)} \tag{3-41}$$

The discharge from the well, Q, can be determined from

$$Q = -2\pi r_w h q_r = 2\pi r_w h K\frac{\partial h}{\partial r} = \pi K r_w \frac{d(h^2)}{dr}, \qquad r = r_w \tag{3-42}$$

From Eqs. (3-41) and (3-42) one can write

$$Q = -I\pi r_w^2 - \left[H^2 - H_w^2 + \frac{I}{2K}(R^2 - r_w^2)\right]\frac{\pi K}{\ln\left(\frac{r_w}{R}\right)} \tag{3-43}$$

The first term on the right-hand side corresponds to the recharge quantity into the well itself and is neglegibly small compared with the other terms. Neglecting this term gives

$$Q = -\left[H^2 - H_w^2 + \frac{I}{2K}(R^2 - r_w^2)\right] \frac{\pi K}{\ln\left(\frac{r_w}{R}\right)} \tag{3-44}$$

Introducing Eq. (3-44) into Eq. (3-41) yields the following equation:

$$h^2 = H^2 + \frac{I}{2K}(R^2 - r^2) + \frac{Q}{\pi K} \ln\left(\frac{r}{R}\right) \tag{3-45}$$

With $I = 0$, Eq. (3-45) takes the form

$$h^2 = H^2 + \frac{Q}{\pi K} \ln\left(\frac{r}{R}\right) \tag{3-46}$$

Equations (3-45) and (3-46) represent the piezometric head distributions with recharge and without recharge cases, respectively. For the condition described by Eq. (3-38), the well discharge rate, Q, can be expressed as

$$Q = \frac{\pi K(H^2 - H_w^2)}{\ln\left(\frac{R}{r_w}\right)} \tag{3-47}$$

Between any two distances (r_1 and r_2, $r_2 > r_1$), Eq. (3-46) takes the form

$$h_1^2 = h_2^2 + \frac{Q}{\pi K} \ln\left(\frac{r_1}{r_2}\right) \tag{3-48}$$

Equation (3-47) is known as the *Dupuit–Forchheimer well discharge formula*. Equations (3-46) and (3-48) are other forms of the Dupuit–Forchheimer formula.

Equation (3-47) is obtained on the basis of Dupuit's assumptions. These assumptions do not take into account the curvilinear shape of flow in the radial plane. The vertical flow components as well as the variation of the horizontal velocity in the vertical plane are neglected. Equation (3-47) provides results with reasonable accuracy if the radial distance (r) is sufficiently great so that the curvilinear effects are negligible. Later, numerical methods (Boulton, 1951) and experimental investigations (Babbitt and Caldwell, 1948; Peterson et al., 1952) showed that Eq. (3-47) closely represents the free-surface curve for values of $r \geq 1.5h$, even when the water level in the well (H_0) is zero in Figure 3-3. Agreement occurs for smaller values of r as H_0 is increased. These investigations also showed that the free surface near and at some distance around the well is not correctly represented by the Dupuit–Forchheimer theory and that the free surface joins the well face at some distance above the water level in the well. Hantush (1962) analyzed the validity of the Dupuit–Forchheimer well discharge formula by taking into account both the water level in the well (H_0) and the water level at the well face (H_w) and obtained Eq. (3-47) with the exception that $H_w = H_0$. This means that the curvilinear nature of the flow, along with the seepage face, was taken into account and virtually the same equation was obtained.

Example 3-3: For the unconfined aquifer shown in Figure 3-3, the following values are given: $H = 20$ ft, $r_w = 0.5$ ft, and $K = 30$ ft/day. Determine the approximate maximum pumping rate to create a 2-ft drawdown at the well. Repeat the same evaluation for $K = 50$ ft/day.

Solution: The Dupuit–Forchheimer well discharge formula, Eq. (3-47), gives the drawdown distribution under steady-state conditions which represents the maximum drawdown distribution for a given pumping rate.

The drawdown at the well is $s_w = H - H_w = 2$ ft. Because $H = 20$ ft, $H_w = 18$ ft. Substituting the values into Eq. (3-47) gives

$$Q = \frac{\pi(30\ \text{ft/day})(20^2 - 18^2)\ \text{ft}^2}{\ln\left(\dfrac{R}{0.5\ \text{ft}}\right)} = \frac{7163}{\ln(2R)}$$

In the above expression, R is the radius of influence and its value is not known and Q varies with R. Using this equation, the variation of Q versus R for $K = 30$ ft/day is determined and shown in Figure 3-4. As can be observed from the figure, the pumping rate (Q) is not too sensitive to the expected range of R. As shown in Figure 3-4, for $R = 150$ ft to 500 ft range, $Q_{\text{avg}} \sim 6$ gpm. This means that both $R = 150$ ft and 500 ft correspond to approximately the same pumping rate.

As shown in Figure 3-4, for $K = 50$ ft/day, $Q_{\text{avg}} \sim 10$ gpm for the same range of R.

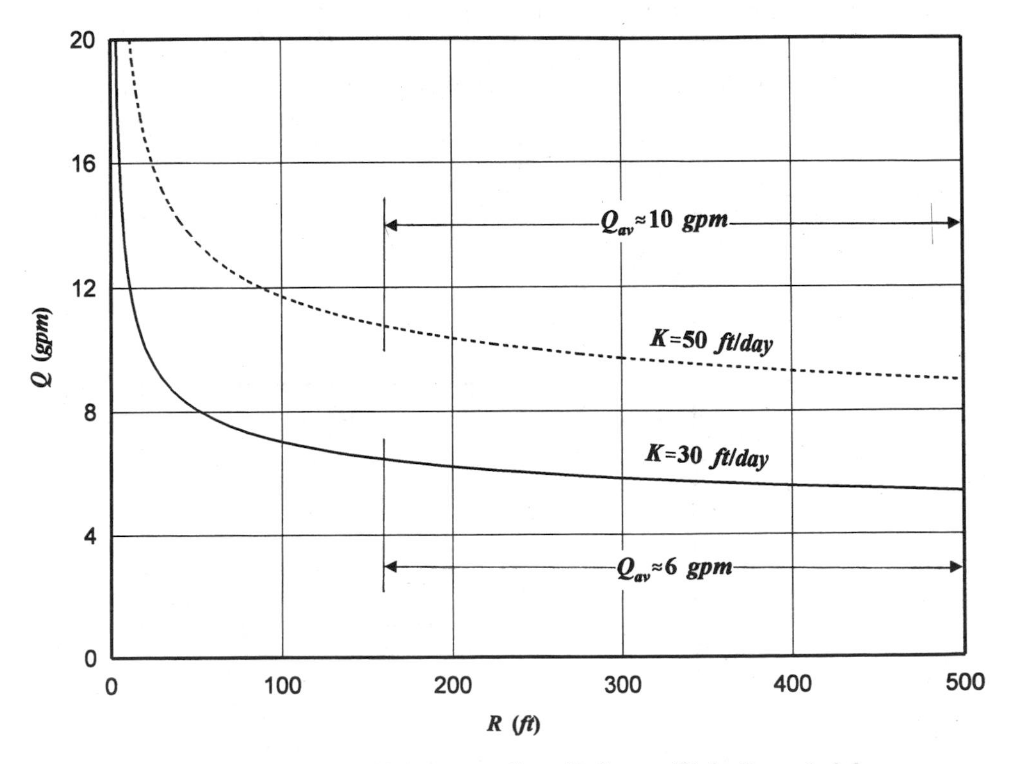

Figure 3-4 Flow rate (Q_w) versus radius of influence (R) for Example 3-3.

3.4 METHODS OF ANALYSIS FOR STEADY-STATE PUMPING TEST DATA

In the previous sections, the basic theories of the Thiem equation for confined aquifers, the De Glee equation for leaky-confined aquifers, and the Dupuit–Forchheimer equation for unconfined aquifers are presented in detail. These equations can be used for pumping test data analysis to determine aquifer transmissivity if the data correspond to steady-state (or equilibrium) conditions. In the following sections, the methods based on these equations are presented with examples.

3.4.1 Confined Nonleaky Aquifers: Thiem's Methods

If the piezometric head (h) values are measured in at least two piezometers or observation wells under steady-state conditions, it is possible to determine the transmissivity of the aquifer using Thiem's method (Thiem, 1906). The schematic cross section of pumping and observation wells is shown in Figure 3-5. Because $h_1 + s_1 = h_2 + s_2 = h_0$, Eq. (3-9) can also be expressed as

$$Q = \frac{2\pi T(s_1 - s_2)}{\ln\left(\frac{r_2}{r_1}\right)} \tag{3-49}$$

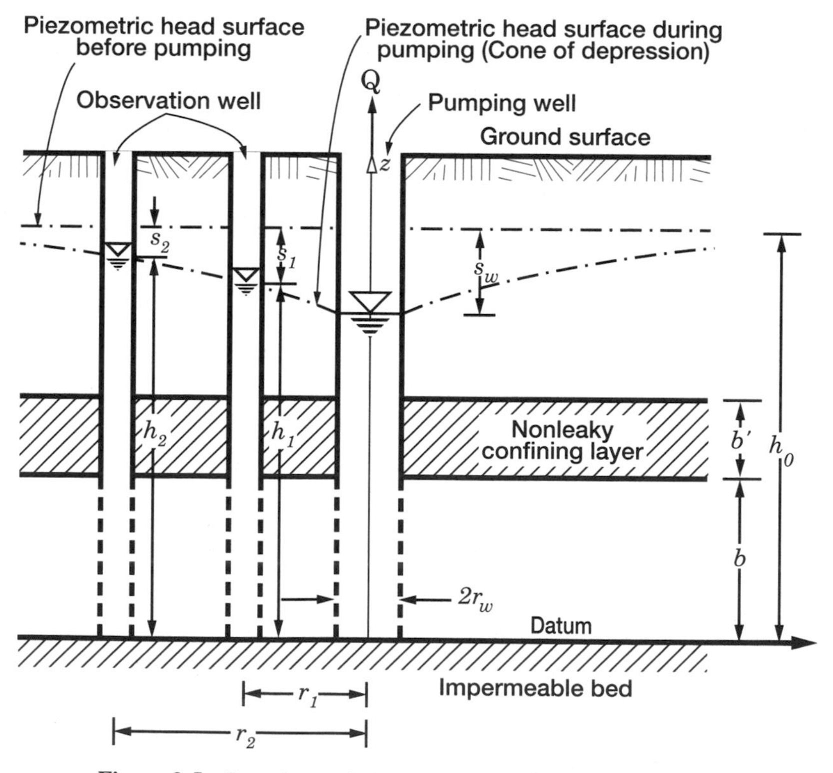

Figure 3-5 Pumping and observation wells in a confined aquifer.

where h_0 is the piezometric head before pumping and s_1 and s_2 are the drawdowns. If only one observation is available at a distance r_1 from the pumping well, Eq. (3-49) takes the form

$$Q = \frac{2\pi T(s_w - s_1)}{\ln\left(\frac{r_1}{r_w}\right)} \tag{3-50}$$

where s_w is the steady-state drawdown at the pumping well and r_w is the radius of the pumping well.

Some important aspects of use of Eq. (3-50) are as follows: (1) Eq. (3-50) has limited use because well losses by the well screen and the flow inside the well to the pump intake affect the value of s_w; (2) Eq. (3-50) should be used whenever additional data are not available; and (3) The best way is to use piezometer tubes in the close proximity of the well to specify the drawdown at the well.

Using the equations given above, two methods can be used to determine the transmissivity of aquifers.

Thiem's Method I. The steps of the method are given below:

Step 1: Plot the measured drawdown versus time for each observation well in linear coordinates. Draw the drawdown–time curve for each observation well. If the drawdown curves tend to become parallel to the time axis for the later time, the ground-water flow can be considered to be steady state.

Step 2: Substitute the values of steady-state drawdowns (s_1 and s_2) of two observation wells with their corresponding values of radial distances from the pumping well (r_1 and r_2) and the known value of Q into Eq. (3-49) and solve for T. If the number of observation wells are more than two, repeat the same calculation for all possible combinations of the drawdowns measured at different observation wells.

Step 3: The transmissivity values determined for each combination may show a close agreement. If this is the case, calculate the arithmetic mean.

Example 3-4: Measured data under steady-state conditions from a pumping test having four monitoring wells are given Table 3-3. The steady flow rate is 540 m^3/day. Using Thiem's method I, determine the transmissivity value of the aquifer.

TABLE 3-3 Distance and Drawdowns for Example 3-4

Well No.	Distance (m)	Drawdown (m)
1	1.2	1.85
2	25.0	0.78
3	75.0	0.56
4	225.0	0.18

Solution:

Step 1: The given data represent steady-state conditions. Therefore, the next step can be proceeded.

Step 2: Different combinations can be used to calculate the transmissivity. For example, substitution of the values for wells No. 1 and No. 2 in Table 3-3 into Eq. (3-49) gives

$$T = \frac{540\ \text{m}^3/\text{day}}{2\pi(0.78\ \text{m} - 0.56\ \text{m})} \ln\left(\frac{75\ \text{m}}{25\ \text{m}}\right) = 429\ \text{m}^2/\text{day}$$

The same equation can be used for other combinations of the measured drawdowns. The results are given Table 3-4.

Step 3: The transmissivity has values between 244.0 m^2/day and 429.0 m^2/day. The arithmetic mean of these values is 297.0 m^2/day.

Thiem's Method II. The steps of the method are given below:

Step 1: On a single logarithmic paper, plot distance (logarithmic) versus steady-state drawdown of each observation well. Then draw the best straight line through the points.

Step 2: Determine the geometric slope of the straight line. This can be achieved in an easier way by taking the difference of drawdown per log cycle of the distance (r). With this, $r_2/r_1 = 10$ or $\ln(r_2/r_1) = 2.303 \log(r_2/r_1) = 2.303$. Introducing this into Eq. (3-49) gives

$$Q = 2.728T\Delta s \tag{3-51}$$

where Δs is the difference of drawdowns per log cycle. Introduce the values of Q and Δs in Eq. (3-51) and solve for transmissivity (T).

Example 3-5: Apply Thiem's Method II to Example 3-4.

Solution:

Step 1: Using the given data in Table 3-3, a distance versus steady-state drawdown diagram is given in Figure 3-6.

TABLE 3-4 Results of Calculations for Different Combinations of the Measured Data for Example 3-4

r_1 (m)	r_2 (m)	s_1 (m)	s_2 (m)	T (m^2/day)
25.0	75.0	0.78	0.56	429.0
1.2	25.0	1.85	0.78	244.0
1.2	75.0	1.85	0.56	276.0
1.2	225.0	1.85	0.18	269.0
25.0	225.0	0.78	0.18	315.0
75.0	225.0	0.56	0.18	249.0

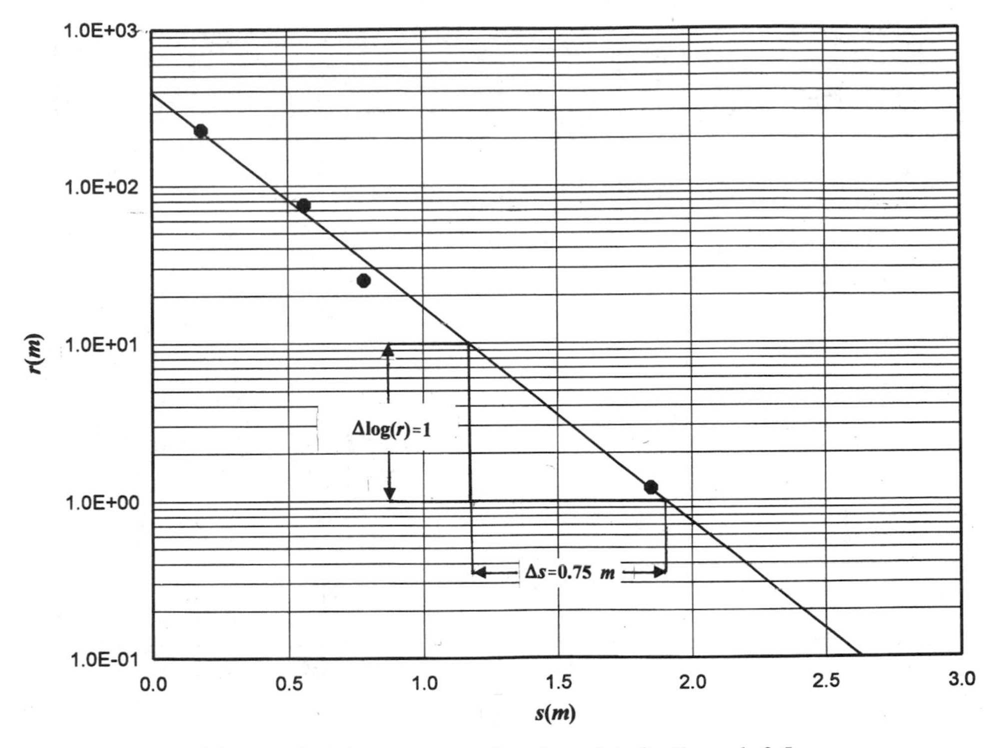

Figure 3-6 Distance versus drawdown data for Example 3-5.

Step 2: From Figure 3-6, $\Delta s = 0.75$ m. Introducing this value and the value of Q in Eq. (3-51) and solving for T gives

$$T = \frac{Q}{2.728\Delta s} = \frac{540.0 \text{ m}^3/\text{day}}{(2.728)(0.75 \text{ m})} = 264 \text{ m}^2/\text{day}$$

This value agrees reasonable well with the average value obtained from Thiem's Method I (see Example 3-4).

3.4.2 Confined Leaky Aquifers

Methods were developed to determine the transmissivity and leakage factor for confined leaky aquifers using the drawdown data under steady-state conditions. The methods that are called the *type-curve method* and the *straight-line method* are given in the following sections, with examples.

3.4.2.1 Type Curve Method. The method is based on the steady-state drawdown distribution for a confined leaky aquifer and is given by Eq. (3-30) [De Glee, 1930; Hantush, 1964; Kruseman and De Ridder, 1990]. The steps of the method are given below:

Step 1: Plot the type curve of r/B versus $K_0(r/B)$ on log–log paper using the data in Table 3-2. The type curve is shown in Figure 3-7.

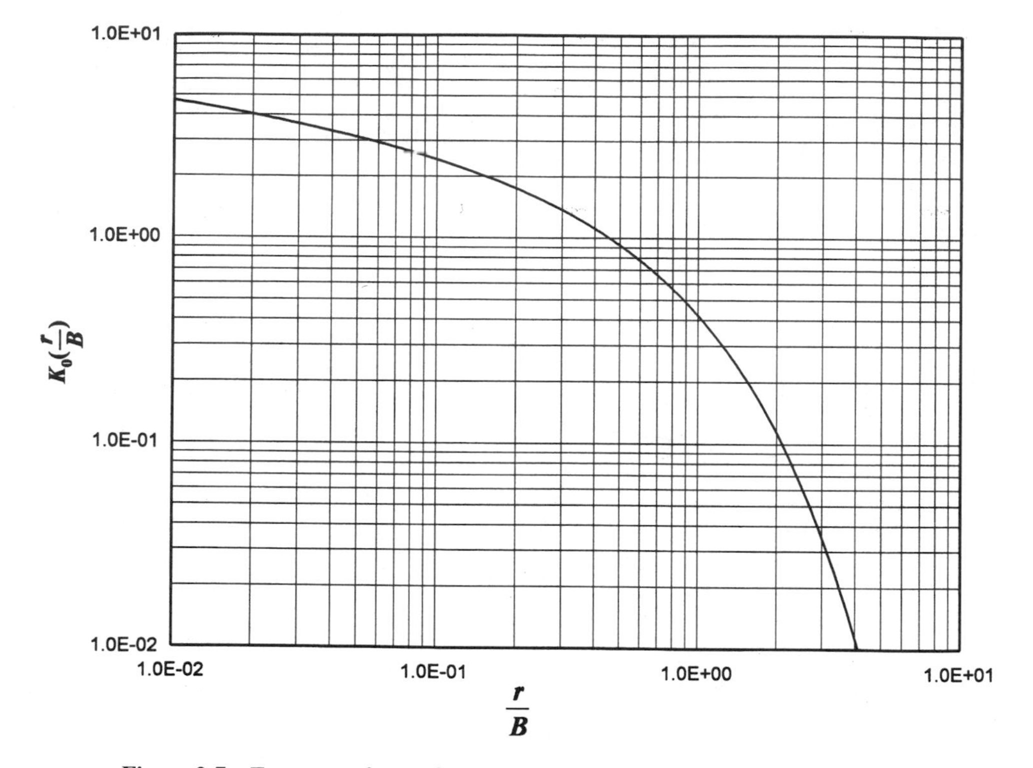

Figure 3-7 Type curve for confined leaky aquifers under steady-state conditions.

Step 2: On another sheet of log–log paper with the same scale as the one in Step 1, plot the measured drawdown (s) data versus radial distance (r).

Step 3: Keeping the coordinate axes parallel at all times, superimpose the two sheets until the best fit of the data curve and the type curve is obtained.

Step 4: Select a common point, the match point, arbitrarily chosen on the overlapping part of the curve or even anywhere on the overlapping portion of the sheets. Read the corresponding values of s, r, $K_0(r/B)$, and r/B.

Step 5: From Eq. (3-30) we obtain

$$T = \frac{Q}{2\pi s} K_0\left(\frac{r}{B}\right) \tag{3-52}$$

and from Eq. (3-16) one can write

$$\frac{b'}{K'} = \frac{r^2}{T\left(\frac{r}{B}\right)^2} \tag{3-53}$$

Substitute the values of Step 4 in Eq. (3-52) and determine the value of $T = Kb$ (transmissivity). Then, from Eq. (3-53) determine the value of b'/K' (hydraulic resistance).

TABLE 3-5 Steady-State Pumping Test Data for Example 3-6

Distance, (r) (m)	Drawdown, (s) (m)
8.0	0.220
25.0	0.172
50.0	0.125
100.0	0.106
125.0	0.094
350.0	0.047

Example 3-6: Pumping test was conducted for a leaky confined aquifer under steady-state conditions. Measured drawdown values are given in Table 3-5. The steady-state pumping rate is 350 m^3/day. Using the type-curve method, determine the transmissivity (T) and hydraulic resistance (b'/K') of the aquifer.

Solution:

Step 1: Using Table 3-2, the type curve r/B versus $K_0(r/B)$ is given in Figure 3-7.
Step 2: Using Table 3-6, the observation data graph r versus s is given in Figure 3-8.

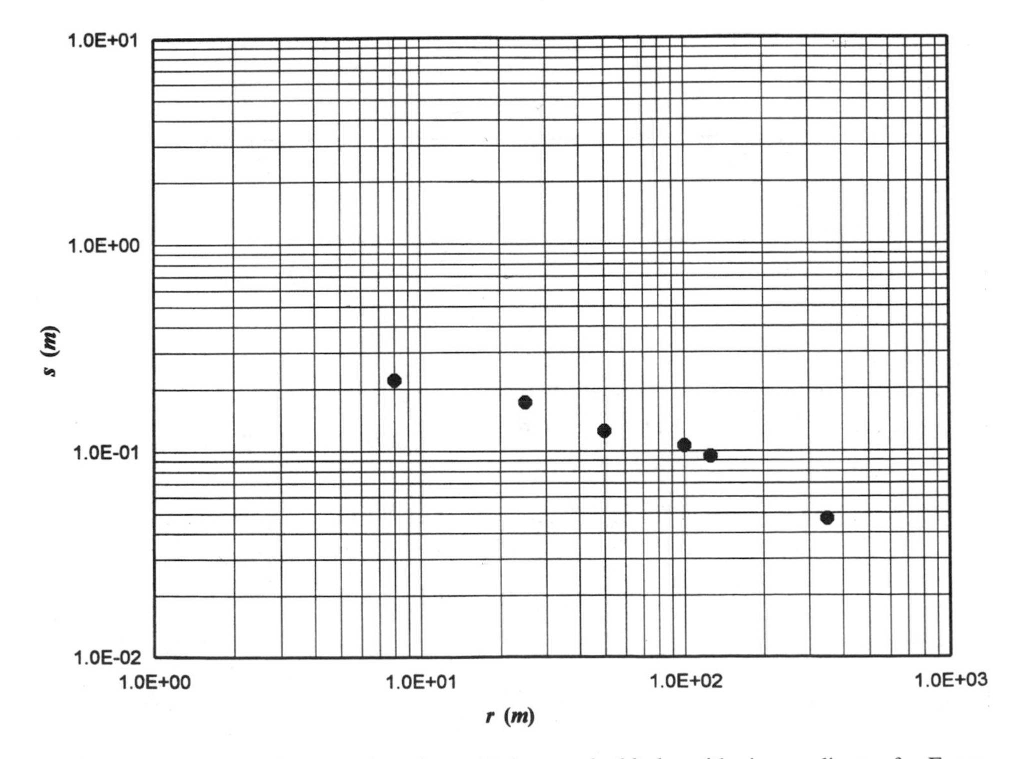

Figure 3-8 Distance (r) versus drawdown (s) data on double logarithmic coordinates for Example 3-6.

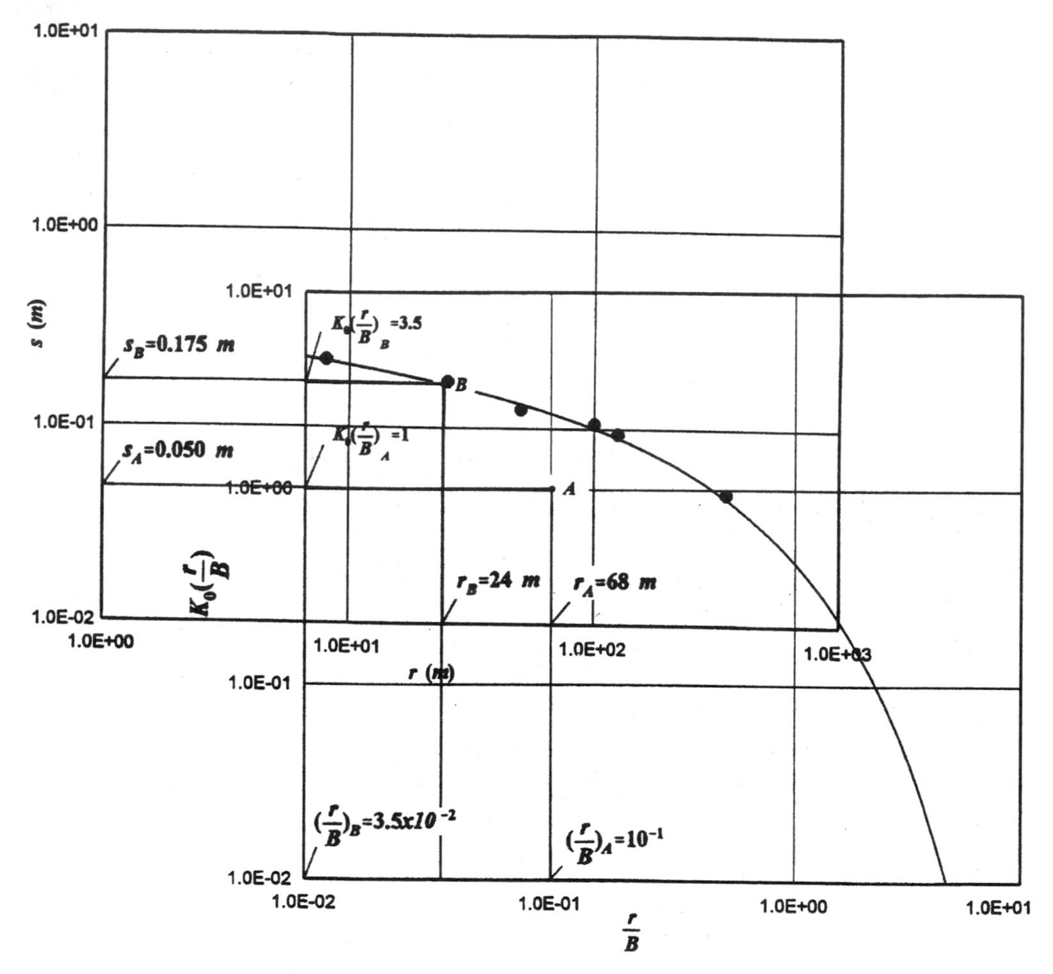

Figure 3-9 Type-curve matching for Example 3-6.

Step 3: The superimposed graphs are shown in Figure 3-9.

Step 4: In Figure 3-9, A is an arbitrarily chosen point; and B is a point on the overlapping part of the curve. The coordinates of A are

$$\left(\frac{r}{B}\right)_A = 0.1, \qquad K_0\left(\frac{r}{B}\right)_A = 1$$

and

$$r_A = 68 \text{ m}, \qquad s_A = 0.050 \text{ m}$$

The coordinates of B are

$$\left(\frac{r}{B}\right)_B = 0.035, \qquad K_0\left(\frac{r}{B}\right)_B = 3.5$$

and

$$r_B = 24 \text{ m}, \qquad s_B = 0.175 \text{ m}$$

Step 5: Using the match-point values for point A, the value of T from Eq. (3-52) is:

$$T = \frac{(350 \text{ m}^3/\text{day})(1)}{(2\pi)(0.050 \text{ m})} = 1114 \text{ m}^2/\text{day}$$

and from Eq. (3-53) the value of b'/K' is

$$\frac{b'}{K'} = \frac{(68 \text{ m})^2}{\left(1114 \text{ m}^2/\text{day}\right)(0.1)^2} = 415 \text{ days}$$

From the values of point B, $T = 1114 \text{ m}^2/\text{day}$ and $b'/K' = 422$ days. The two matching points give approximately the same values. The differences are reading errors from the graph.

In reality, one matching point is enough. The second match point is included just for demonstration purpose. Theoretically, all match points in the overlapping portion of the figures should give the same results. As mentioned above, because of reading errors, the results may be slightly different, but they are the same for all practical purposes.

3.4.2.2 Straight-Line Method. The method is similar to that of Cooper and Jacob method (see Section 4.2.7.2 of Chapter 4). The method is based on Eq. (3-31), which is an approximated version of Eq. (3-30). A semilogarithmic data plot of s versus r, with r on the logarithmic scale, will define a straight-line variation if the observed points fall in the range $r/B < 0.05$. If $r/B > 0.05$, the points fall on a curve that approaches the zero drawdown asymptotically. Equation (3-31) can also be expressed as

$$s \cong \frac{2.303Q}{2\pi T} \log(1.12B) - \frac{2.303Q}{2\pi T} \log r \tag{3-54}$$

The first term on the right-hand side of Eq. (3-54) is a constant. By differential calculus, remembering that the derivative of a constant is zero, one may obtain

$$\Delta s = \left| \frac{2.303Q}{2\pi T} \Delta(\log r) \right| \tag{3-55}$$

The straight-line portion of the curve intercepts the r-axis where the drawdown is zero at the point r_0. Consequently, the interception point has the coordinates $s = 0$, and $r = r_0$. Introducing these values into Eq. (3-31) gives

$$0 = \frac{2.303Q}{2\pi T} \log\left(\frac{1.12B}{r_0}\right) \tag{3-56}$$

Because $2.303Q/(4\pi T)$ cannot be zero, it follows that

$$\frac{1.12B}{r_0} = 1 \tag{3-57}$$

or using Eq. (3-16)

$$\frac{b'}{K'} = \frac{1}{T}\left(\frac{r_0}{1.12}\right)^2 \tag{3-58}$$

The steps of the method are given below:

Step 1: Plot the data of all observation wells on a single logarithmic paper between s versus r (logarithmic). Then, draw a straight line through the points.

Step 2: Determine the geometric slope of the straight line—that is, the drawdown difference Δs per log cycle of r.

Step 3: Find the interception point of the straight line with the r-axis where $s = 0$. Read the value of r_0.

Step 4: With the known values of Δs, Q, $\Delta(\log r)$, and r_0, calculate the values of T and b'/K' using Eqs. (3-55) and (3-58), respectively.

Example 3-7: Using the straight-line method, determine the transmissivity (T) and hydraulic resistance (b'/K') for the observation data of Example 3-6.

Solution:

Step 1: The graph for s versus r (logarithmic) is given in Figure 3-10.

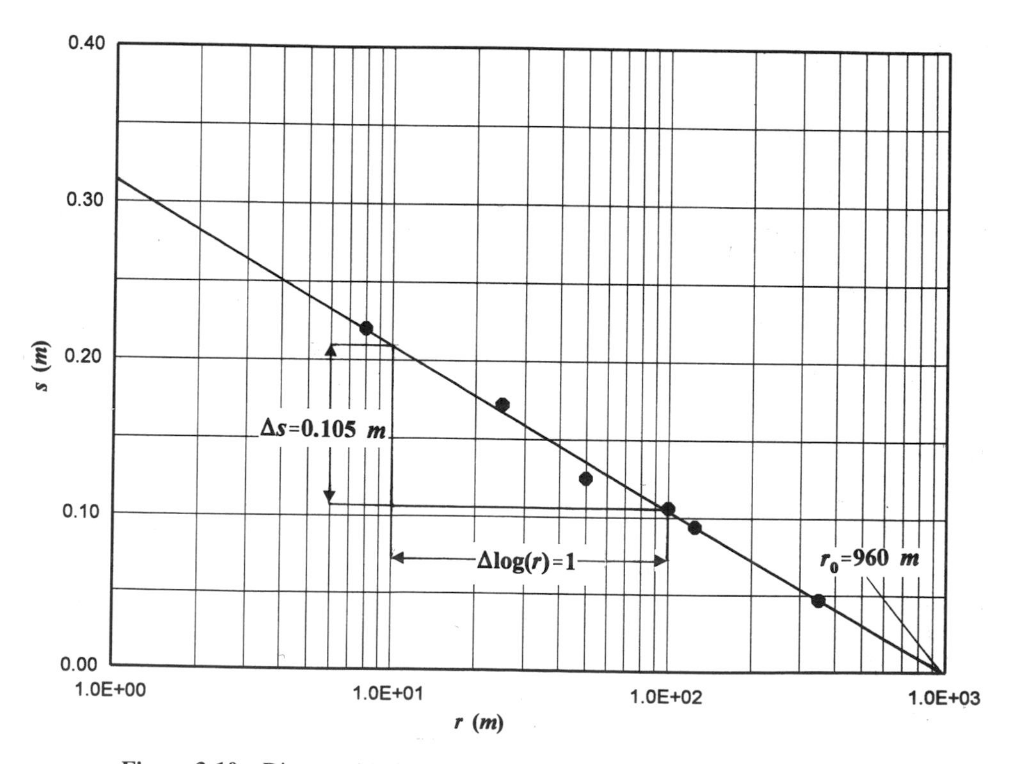

Figure 3-10 Distance (r) (logarithmic) versus drawdown (s) for Example 3-7.

Step 2: From Figure 3-10, the corresponding values for the geometric slope are Δs = 0.105 m and $\Delta(\log r) = 1$.

Step 3: The interception point of the straight line with the r axis, where $s = 0$, is r_0 = 960 m.

Step 4: From Eq. (3-55), with the values given above and $Q = 350\ \text{m}^3/\text{day}$, we obtain

$$T = \frac{(2.303)(350\ \text{m}^3/\text{day})}{(2\pi)(0.105\ \text{m})}(1) = 1222\ \text{m}^2/\text{day}$$

and from Eq. (3-58) we obtain

$$\frac{b'}{K'} = \frac{1}{(1222\ \text{m}^2/\text{day})}\left(\frac{960\ \text{m}}{1.12}\right)^2 = 601\ \text{days}$$

3.4.3 Unconfined Aquifers

Equation (3-48) is the main equation for unconfined aquifer data analysis under steady-state pumping conditions. Equation (3-48) was derived on the assumption that the drawdown of the water table (s) remains small as compared with the original saturated thickness (b) of the aquifer (Figure 3-11). In thin unconfined aquifers, s may be a large fraction of b. For

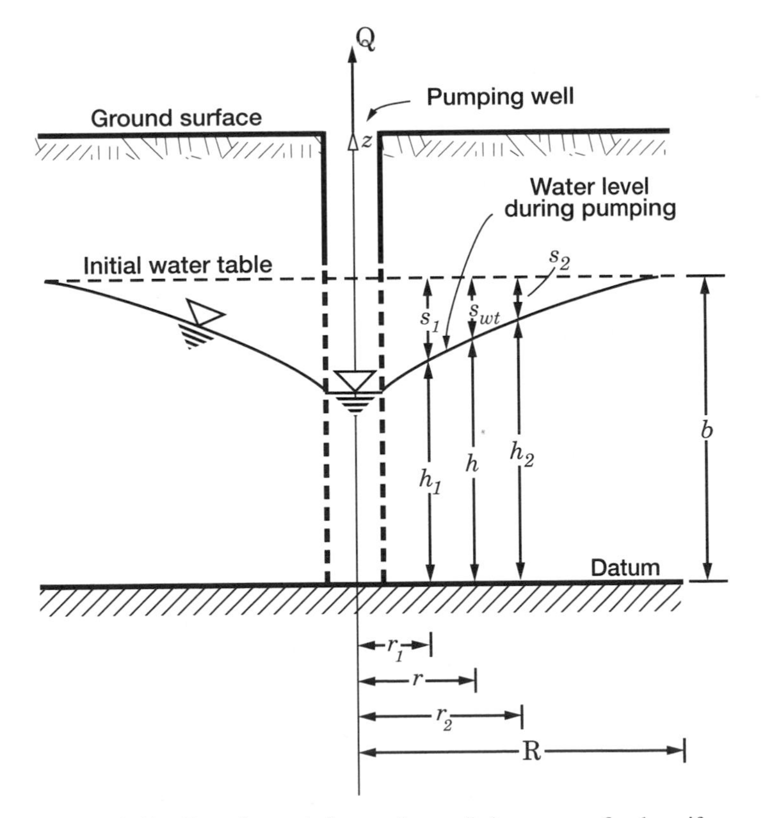

Figure 3-11 Pumping and observation wells in an unconfined aquifer.

this condition, Jacob (1944) recommended that prior to analyzing the pumping test data the drawdown data should be corrected. In the following sections, first the theory behind the *Jacob's drawdown correction scheme* is presented and then the methods of data analysis using the *Dupuit–Forchheimer formula* are given for relatively thick and thin aquifers.

3.4.3.1 Jacob's Drawdown Correction Scheme. Jacob's correction scheme (Jacob, 1944) adopted the Dupuit assumptions, particularly by assuming that the drawdown along any vertical section are always equal to the drawdown of the water table (s_{wt}). The flow rate across a cylindrical surface of radius r centered around the pumping well (F) can be expressed as (see Figure 3-11)

$$F = -q_r 2\pi r(b - s_{wt}) \tag{3-59}$$

where q_r is the Darcy velocity in the radial direction, whose expression is given by Eq. (2-115a) of Chapter 2:

$$q_r = -K\frac{\partial h}{\partial r} = -K\frac{\partial(b - s_{wt})}{\partial r} \tag{3-60}$$

where K is the hydraulic conductivity in the radial direction. Introducing this in Eq. (3-59) gives

$$F = 2\pi rK(b - s_{wt})\frac{\partial(b - s_{wt})}{\partial r} = \pi rK\frac{\partial(b - s_{wt})^2}{\partial r} \tag{3-61}$$

Equation (3-61) can also be expressed as

$$F = \pi rK\frac{\partial(b^2 - 2bs_{wt} + s_{wt}^2)}{\partial r}\left(\frac{2b}{2b}\right) = 2\pi rbK\frac{\partial\left(\frac{b}{2} - s_{wt} + \frac{s_{wt}^2}{2b}\right)}{\partial r} \tag{3-62}$$

Finally, since the derivative of $b/2$ with respect to r is zero, Eq. (3-62) takes the form

$$F = -2\pi rbK\frac{\partial s_c}{\partial r} \tag{3-63}$$

where

$$s_c = s_{wt} - \frac{s_{wt}^2}{2b} \tag{3-64}$$

The second term on the right-hand side of Eq. (3-64) is the *drawdown correction term.*

3.4.3.2 Equations for Data Analysis. The equations for thick and thin unconfined aquifers that are used for data analysis are presented in the following paragraphs.

Thick Unconfined Aquifers. Converting natural logarithm (ln) to common logarithm (log), Eq. (3-48) gives

$$K = \frac{2.303Q \log\left(\frac{r_2}{r_1}\right)}{\pi(h_2^2 - h_1^2)} \tag{3-65}$$

In thick unconfined aquifers where the drawdown (s) is negligible compared with the original thickness (b), $h_1 + h_2 \approx 2b$. Then, as $h_2^2 - h_1^2 = (h_2 + h_1)(h_2 - h_1)$, $h_2 - h_1 = s_1 - s_2$, and $T = Kb$, Eq. (3-65) takes the following form:

$$T = \frac{2.303Q \log\left(\frac{r_2}{r_1}\right)}{2\pi(s_1 - s_2)} \tag{3-66}$$

Equation (3-66) is the same form of the Thiem equation [Eq. (3-49)], with the exception that it is expressed with the common logarithms.

Thin Unconfined Aquifers. In thin unconfined aquifers the drawdown (s) may be a significant portion of the aquifer thickness (b). If this is the case, the drawdowns need to be corrected in accordance with Eq. (3-64). With this correction, Eq. (3-66) takes the following form:

$$T = \frac{2.303Q \log\left(\frac{r_2}{r_1}\right)}{2\pi\left[\left(s_1 - \frac{s_1^2}{2b}\right) - \left(s_2 - \frac{s_2^2}{2b}\right)\right]} \tag{3-67}$$

with

$$\log\left(\frac{r_2}{r_1}\right) = \log(r_2) - \log(r_1) = \Delta(\log r) \tag{3-68}$$

and

$$\left(s_1 - \frac{s_1^2}{2b}\right) - \left(s_2 - \frac{s_2^2}{2b}\right) = \Delta\left(s - \frac{s^2}{2b}\right) \tag{3-69}$$

Substituting Eqs. (3-68) and (3-69) into Eq. (3-67) gives

$$T = \frac{2.303Q\Delta(\log r)}{2\pi\Delta\left(s - \frac{s^2}{2b}\right)} \tag{3-70}$$

3.4.3.3 Method of Data Analysis.

Measured data may be analyzed using the distance versus drawdown method (Lohman, 1972). Based on the equations given in Section 3.4.3.2, the steps of the method are given below:

Step 1: Using the measured drawdown data, calculate the corrected drawdown (s_c) from Eq. (3-64) for each observation well and create a table of radial distance (r) versus s_c.

Step 2: On a single logarithmic paper, plot distance (logarithmic) versus corrected drawdown for each observation well. Then, draw the best straight line through the points.

Step 3: Determine the geometric slope of the straight line. This can be achieved in an easier way by taking the difference of corrected drawdown over one log cycle of radial distance—that is, $\Delta(s - s^2/2b)$ versus $\Delta(\log r)$.

Step 4: Using the values of Q, $\Delta(s - s^2/2b)$, and $\Delta(\log r)$ calculate T from Eq. (3-70).

Example 3-8: The data of this example is taken from Lohman (1972). The period of a pumping test was 18 days during which the pumping rate was kept approximately 1000 gpm (gallons per minute). The aquifer was unconfined (refer to Figures 3-11 and 3-12) and the initial saturated thickness (b) was 26.8 ft. Table 3-6 presents the observation data. Determine the transmissivity (T) of the aquifer.

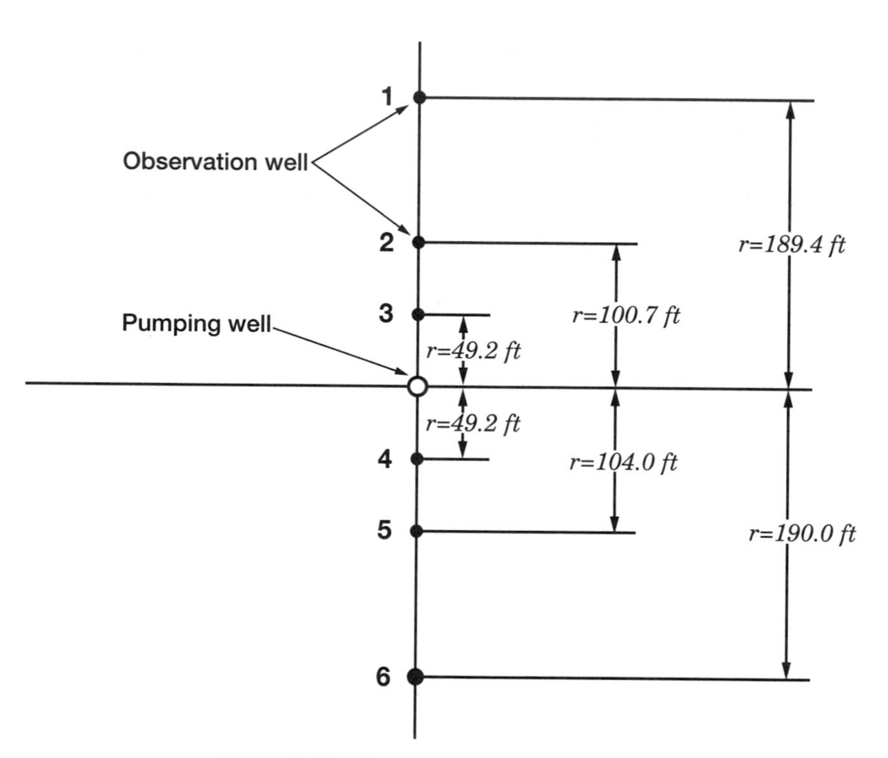

Figure 3-12 Well locations for Example 3-8.

TABLE 3-6 Steady-State Pumping Test Data for Example 3-8

Well No.	Distance, (r) (ft)	Drawdown, (s) (ft)
1	189.4	3.42
2	100.7	4.58
3	49.2	5.91
4	49.0	5.48
5	100.4	4.31
6	190.0	3.19

TABLE 3-7 Corrected Drawdowns for Example 3-8

Well No.	r (ft)	s (ft)	$s^2/2b$ (ft)	$s - s^2/2b$ (ft)
1	189.4	3.42	0.22	3.20
2	100.7	4.58	0.39	4.19
3	49.2	5.91	0.65	5.26
4	49.0	5.48	0.56	4.92
5	100.4	4.31	0.35	3.96
6	190.0	3.19	0.19	3.00

Solution:

Step 1: The calculated radial distance (r) versus corrected drawdown (s_c) data is given in Table 3-7.

Step 2: Using the data in Table 3-7, distance (logarithmic) versus corrected drawdown diagram is shown in Figure 3-13.

Step 3: From Figure 3-13, $\Delta(s - s^2/2b) = 3.38$ ft and $\Delta(\log r) = 1$.

Step 4: Using the values of the previous step and $Q = 1000$ gpm in Eq. (3-70) gives

$$T = \frac{(2.303)(1000 \text{ gal/min})(1440 \text{ min/day})}{(2\pi)(7.48 \text{ gal/ft}^3)(3.38 \text{ ft})} = 20{,}877 \text{ ft}^2\text{/day}$$

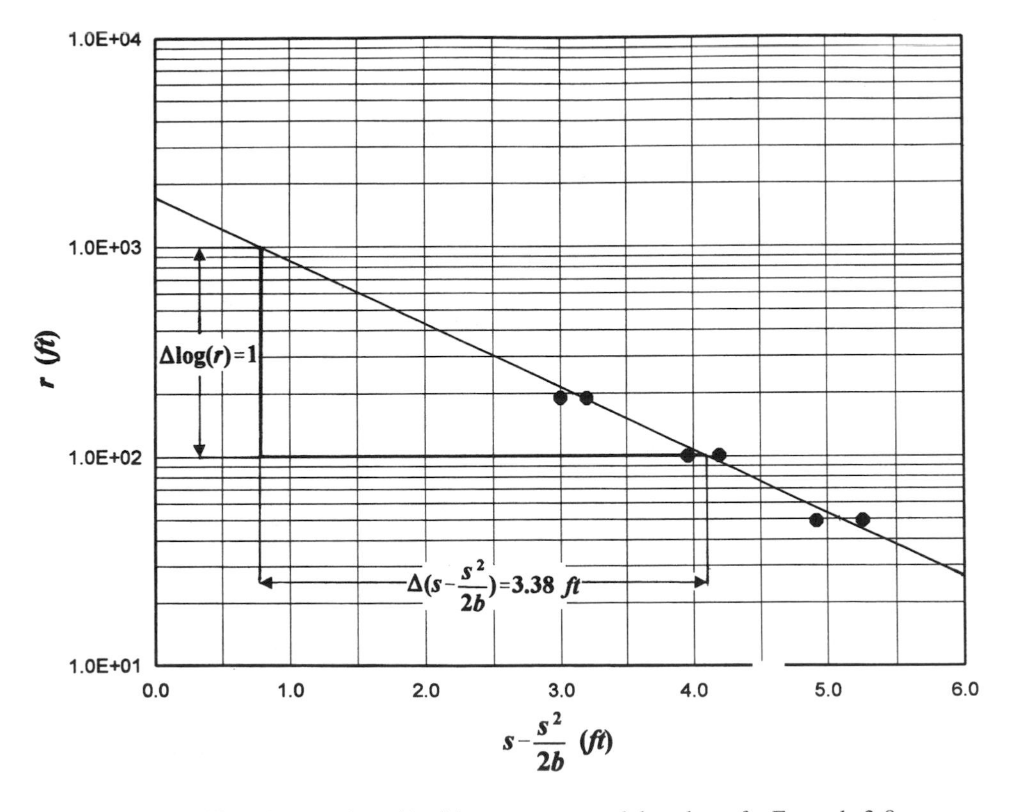

Figure 3-13 Distance (logarithmic) versus corrected drawdown for Example 3-8.

PART III

HYDRAULICS OF AQUIFERS UNDER TRANSIENT PUMPING CONDITIONS FROM WELLS AND HYDROGEOLOGIC DATA ANALYSIS METHODS

CHAPTER 4

FULLY PENETRATING PUMPING WELLS IN HOMOGENEOUS AND ISOTROPIC NONLEAKY CONFINED AQUIFERS

4.1 INTRODUCTION

Transient well hydraulics in a single homogeneous and isotropic nonleaky confined aquifer is the simplest case in both the practical and the mathematical sense. The theory was developed by Theis (1935); his famous formula, known as the *Theis equation*, is the first equation in the literature describing transient flow in aquifers under extraction conditions from a well in a confined nonleaky aquifer. Despite its well-known limitations, the Theis equation has a special importance in aquifer hydraulics. The Theis equation, like many other equations in well hydraulics, treats well as a line source and, therefore, does not take into consideration the water derived from storage within the well. Later, Papadopulos and Cooper (1967) generalized the Theis model by taking into consideration the storage effects in the well.

In this chapter, assumptions, governing equations, solutions with regard to the Theis equation, and the Papadopulos and Cooper equation for confined aquifers are presented in detail. Also, applications of the solutions for pumping test data analysis are presented.

4.2 SMALL-DIAMETER WELLS: THE THEIS MODEL

In the following sections the theoretical foundations and the practical implications of the Theis model are presented.

4.2.1 Problem Statement and Assumptions

Figure 4-1 shows areally and cross-sectionally a fully penetrating well in a nonleaky confined aquifer. Theis (1935) was the first to analyze the aquifer dynamics under transient pumping conditions with the following assumptions:

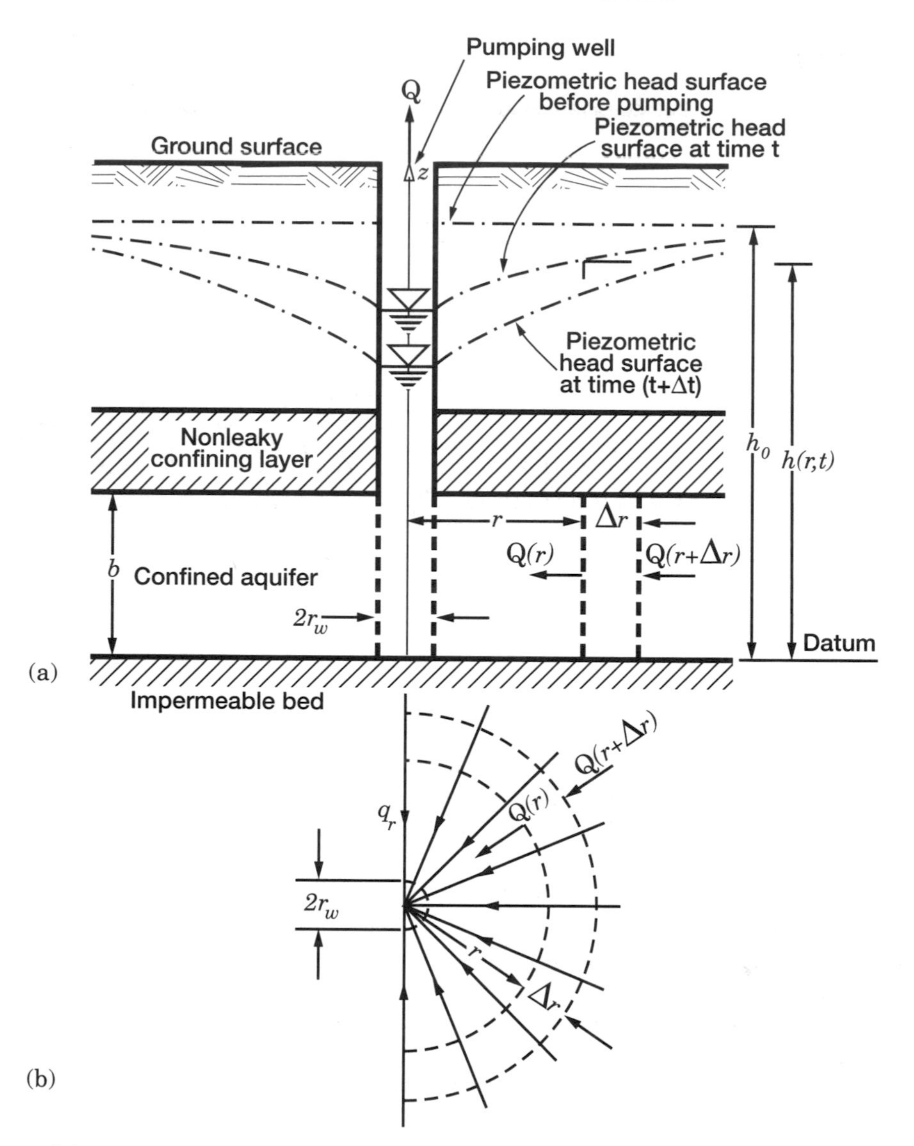

Figure 4-1 Unsteady flow to a fully penetrating well in a nonleaky confined aquifer: (a) Vertical cross section. (b) Plan.

1. The aquifer is homogeneous and isotropic.
2. The aquifer is horizontal and has a constant thickness (b).
3. Well discharges at a constant rate (Q).
4. The aquifer is not leaky.
5. The aquifer is infinite in horizontal extent.
6. The diameter of the well is infinitesimally small; that is, the storage in the well is negligible.
7. The well penetrates the entire aquifer.

8. Before pumping, the piezometric head in the aquifer is the same at every point in the aquifer.
9. Discharge from the well is derived exclusively from storage in the aquifer.
10. Water is immediately released from storage upon decline of the potentiometric head.
11. Storage in the aquifer is proportional to the head.

4.2.2 Governing Differential Equation

From Eq. (2-169) of Chapter 2 and using Eqs. (2-24) and (2-75) of the same chapter, the partial differential equation describing the flow in a horizontal confined aquifer in the two-dimensional x–y Cartesian coordinate system is

$$\frac{\partial^2 h}{\partial x^2} + \frac{\partial^2 h}{\partial y^2} = \frac{S}{T}\frac{\partial h}{\partial t} \tag{4-1}$$

where S and T are the storage coefficient and transmissivity, respectively. Because of cylindrical symmetry (see Figure 4-1), it is much easier to work with the polar coordinates. As mentioned in Section 2.8.2 of Chapter 2, the conversion from Cartesian to polar coordinates can be accomplished through the relation $r = (x^2 + y^2)^{1/2}$, which was originally proposed by Jacob (1950). To give more insight to the reader, another derivation method of Eq. (4-1) in polar coordinates will be presented here (Bear, 1979).

Consider an aquifer ring as shown in Figure 4-1. From Eq. (2-115a) of Chapter 2, the Darcy velocity in the radial direction is

$$q_r = -K\frac{\partial h}{\partial r} \tag{4-2}$$

where K is the hydraulic conductivity in the radial direction. The total flow rate at the distance r from the well is

$$Q(r) = -2\pi r b q_r = 2\pi r T\frac{\partial h}{\partial r} \tag{4-3}$$

where $T = Kb$. The piezometric head at the distance r is $h(r,t)$. After Δt time interval, the piezometric head will be $h(r, t + \Delta t)$. The decline of the piezometric head is

$$\Delta h = h(r,t) - h(r, t + \Delta t) \tag{4-4}$$

Using the continuity principle, the following equation can be written for the ring shown in Figure 4-1:

$$[Q(r) - Q(r + \Delta r)]\Delta t = 2\pi r \Delta r \Delta h S \tag{4-5}$$

or as $\Delta r \to 0$ and $\Delta t \to 0$

$$\frac{\partial Q}{\partial r} = 2\pi r S\frac{\partial h}{\partial t} \tag{4-6}$$

Introducing this in Eq. (4-3) gives

$$\frac{1}{r}\frac{\partial}{\partial r}\left(r\frac{\partial h}{\partial r}\right) = \frac{S}{T}\frac{\partial h}{\partial t} \tag{4-7}$$

or

$$\frac{\partial^2 h}{\partial r^2} + \frac{1}{r}\frac{\partial h}{\partial r} = \frac{S}{T}\frac{\partial h}{\partial t} \tag{4-8}$$

which is the same as Eq. (2-175) of Chapter 2. With the drawdown expression

$$s = h_0 - h \tag{4-9}$$

Eq. (4-8) finally can be expressed as

$$\frac{\partial^2 s}{\partial r^2} + \frac{1}{r}\frac{\partial s}{\partial r} = \frac{S}{T}\frac{\partial s}{\partial t} \tag{4-10}$$

4.2.3 Initial and Boundary Conditions

The initial condition for the drawdown $s(r, t)$ is

$$s(r, 0) = 0 \qquad \text{for all } r \tag{4-11}$$

The boundary condition at infinity ($r = \infty$) is

$$s(\infty, t) = 0 \qquad \text{for all } t \tag{4-12}$$

The discharge $Q(L^3/T)$ is

$$\begin{aligned} Q &= 0 && \text{for } t < 0 \\ Q &= \text{constant} && \text{for } t \geq 0 \end{aligned} \tag{4-13}$$

From Eqs. (4-3) and (4-9), the constant pumping rate (Q) at the well is

$$\lim_{r \to 0}\left(r\frac{\partial s}{\partial r}\right) = -\frac{Q}{2\pi T} \qquad \text{for } t \geq 0 \tag{4-14}$$

Equation (4-11) states that initially the drawdown is zero everywhere in the aquifer. The boundary condition described by Eq. (4-12) assumes zero drawdown when the value of r goes to infinity. Equation (4-13) states that the discharge of the pumping well is constant throughout the pumping period. Equation (4-14) states that near the pumping well the flow toward the well is equal to its discharge.

4.2.4 Solution

4.2.4.1 Theis Equation. The drawdown distribution function, $s(r, t)$, can be determined from the governing differential equation, Eq. (4-10), and its associated initial and boundary conditions. Theis (1935) was the first to solve this problem by utilizing an analogy to heat-flow theory in solids to find an analytical solution. Here, because of its importance, the same solution will be presented in detail. The solution (Bear, 1979)

$$s(r,t) = \frac{A}{t}\exp(-u), \qquad A = \text{constant}, \qquad u = \frac{r^2 S}{4Tt} \tag{4-15}$$

satisfies Eq. (4-10). For $t > 0$, the total volume of water taken from the aquifer is

$$V_0 = \int_0^\infty 2\pi r s S\, dr \tag{4-16}$$

or introducing Eq. (4-15) into Eq. (4-16) gives

$$V_0 = 4\pi T A, \qquad A = \frac{V_0}{4\pi T} \tag{4-17}$$

Then Eq. (4-15) takes the form

$$s(r,t) = \frac{V_0}{4\pi T t}\exp\left(-\frac{r^2 S}{4Tt}\right) \tag{4-18}$$

The solution given by Eq. (4-18) may be interpreted as the drawdown in an infinite confined aquifer due to the volume of water V_0 instantaneously withdrawn at $t = 0$ through a point at the origin. In other words, V_0 is removed during the time period (dt). Therefore, $V_0 = Q\,dt$. If water is pumping at the rate of Q per unit time from $t = 0$ to $t = t$ at the origin, by integrating Eq. (4-18) (Carslaw and Jaeger, 1959, p. 261) we obtain

$$s(r,t) = \frac{1}{4\pi T}\int_0^t \frac{Q(t')}{t-t'}\exp\left[-\frac{r^2 S}{4T(t-t')}\right] dt' \tag{4-19}$$

Equation (4-19) can be used for any functional distribution of $Q(t')$. If $Q(t) = Q = \text{constant}$, it takes the form

$$s(r,t) = \frac{Q}{4\pi T}\int_0^t \frac{1}{t-t'}\exp\left[-\frac{r^2 S}{4T(t-t')}\right] dt' \tag{4-20}$$

With the transformation

$$y = \frac{r^2 S}{4T(t-t')} \tag{4-21}$$

Eq. (4-20) yields the Theis equation:

$$s(r,t) = h_0 - h(r,t) = \frac{Q}{4\pi T}\int_{y=u}^\infty \frac{e^{-y}}{y}\, dy \tag{4-22}$$

where

$$u = \frac{r^2 S}{4Tt} \tag{4-23}$$

4.2.4.2 Theis Well Function and Type Curve. Equation (4-22) may be expressed as

$$s(r, t) = \frac{Q}{4\pi T} W(u) \tag{4-24}$$

where $W(u)$ is called the *Theis well function* and is defined as

$$W(u) = \int_u^\infty \frac{e^{-y}}{y} dy = -\mathrm{Ei}(-u) \tag{4-25}$$

In mathematics, $-\mathrm{Ei}(-u)$ is called the *exponential integral*. This exponential integral function, $W(u)$, can be expressed by expanding the term to be integrated into series:

$$\begin{aligned} W(u) &= -0.5772 - \ln u + u - \frac{u^2}{2.2!} + \frac{u^3}{3.3!} - \frac{u^4}{4.4!} + \cdots \\ &= -0.5772 - \ln u - \sum_{n=1}^{\infty} (-1)^n \frac{u^n}{n.n!} \end{aligned} \tag{4-26}$$

Values of the well function are widely available. Table 4-1 provides values of $W(u)$ versus u from 10^{-15} to 9.9 (Wenzel, 1942; Ferris, et al., 1962). The notation $\mathrm{Ei}(-u)$ for the well function is primarily used in mathematical literature (e.g., Beyer, 1986).

In Figure 4-2, $W(u)$ versus u is presented on a logarithmic coordinate system using selected values from Table 4-1. This curve is called *Theis type curve* and is used to determine the hydrogeologic parameters of confined aquifers using pumping test data (see Section 4.2.7.1).

4.2.4.3 Evaluation of the Theis Equation and its Application. The variation of drawdown (s) with radial distance (r) for two values of t (time), where the second time is twice the value of the first, is shown in Figure 4-3a. Shown in Figure 4-3b is the variation of s with t for two different values of r, where the second one is twice the first (Jacob, 1950).

The assumptions in Section 4.2.1 are applicable to confined (artesian) aquifers, shown in Figure 4-1. However, the solution may be applied to unconfined aquifers if drawdown is small compared with the saturated thickness (b) of the aquifer.

According to assumption 6 in Section 4.2.1, the solution does not take into account the change in storage within the pumping well itself. This assumption is applicable under the following condition (Papadopulos and Cooper, 1967):

$$t > 250 \frac{r_c^2}{T} \tag{4-27}$$

which is also given by Eq. (4-117) of Section 4.3. Here, r_c is the radius of the well casing in the interval over which the water level declines, and other symbols are defined previously.

If the aquifer parameters K, b, and S and the pumping rate, Q, are known, then the drawdown in a potentiometric head in a confined aquifer, as well as the water level or hydraulic head in an unconfined aquifer, at any distance r from a well at any time can be determined. It is simply necessary to calculate u from Eq. (4-23): Look up the corresponding value of well function, $W(u)$, in Table 4-1, and calculate the drawdown from Eq. (4-24).

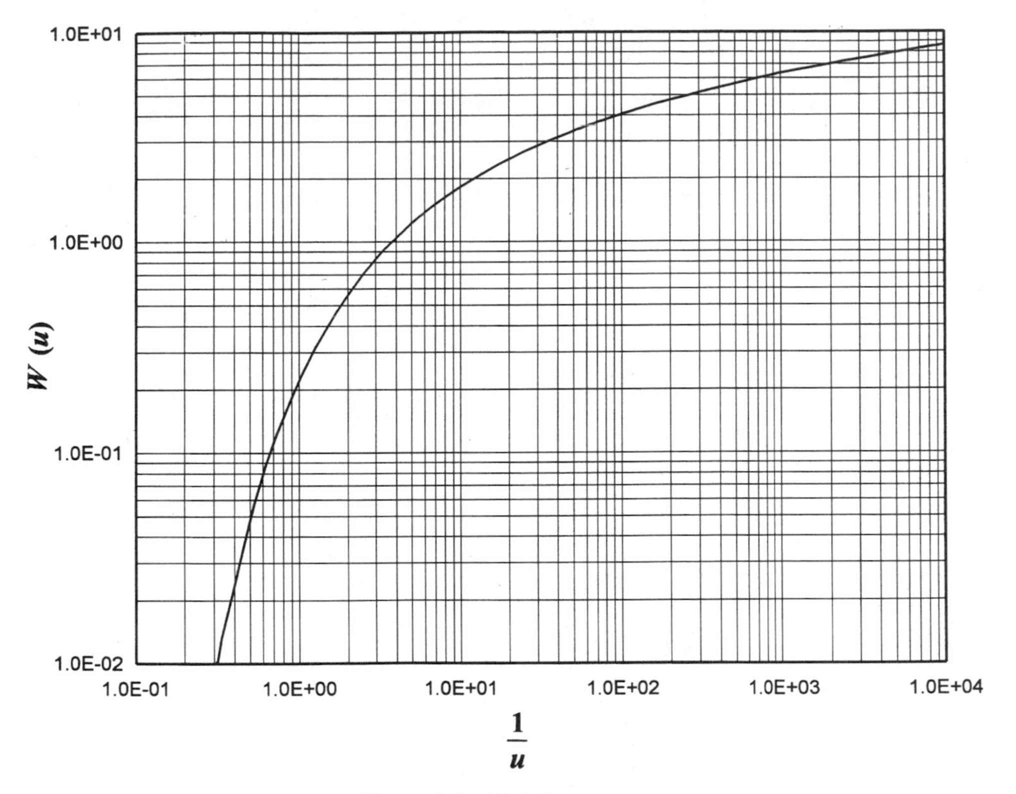

Figure 4-2 Theis type curve.

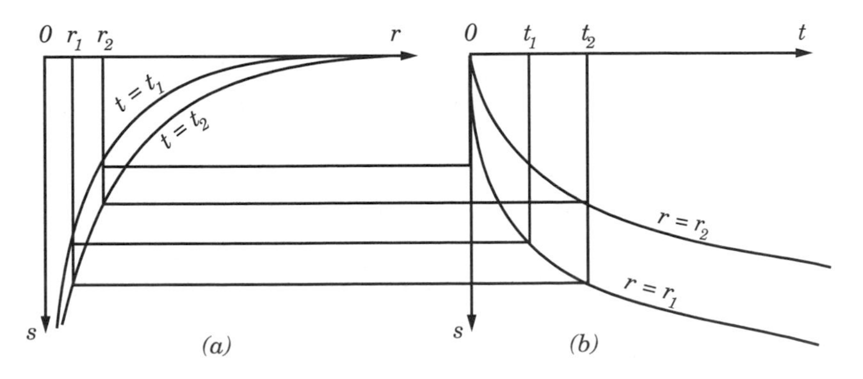

Figure 4-3 Variation of drawdown with time around a fully penetrating well in a confined aquifer: (a) Drawdown versus radial distance at different times. (b) Drawdown versus time at different radial distances. (After Jacob, 1950).

TABLE 4-0 Values of $W(u)$ for Values of u Between 10^{-15} and 9.9

N/u	$N \times 10^{-15}$	$N \times 10^{-14}$	$N \times 10^{-13}$	$N \times 10^{-12}$	$N \times 10^{-11}$	$N \times 10^{-10}$	$N \times 10^{-9}$	$N \times 10^{-8}$	$N \times 10^{-7}$	$N \times 10^{-6}$	$N \times 10^{-5}$	$N \times 10^{-4}$	$N \times 10^{-3}$	$N \times 10^{-2}$	$N \times 10^{-1}$	N
1.0	33.9616	31.6590	29.3564	27.0538	24.7512	22.4486	20.1460	17.8435	15.5409	13.2383	10.9357	8.6332	6.3315	4.0379	1.8229	0.2194
1.1	33.8662	31.5637	29.2611	26.9585	24.6559	22.3533	20.0507	17.7482	15.4456	13.1430	10.8404	8.5379	6.2363	3.9436	1.7371	.1860
1.2	33.7792	31.4767	29.1741	26.8715	24.5689	22.2663	19.9637	17.6611	15.3586	13.0560	10.7534	8.4509	6.1494	3.8576	1.6595	.1584
1.3	33.6992	31.3966	29.0940	26.7914	24.4889	22.1863	19.8837	17.5811	15.2785	12.9759	10.6734	8.3709	6.0695	3.7795	1.5889	.1355
1.4	33.6251	31.3225	29.0199	26.7173	24.4147	22.1122	19.8096	17.5070	15.2044	12.9018	10.5993	8.2968	5.9955	3.7054	1.5241	.1162
1.5	33.5561	31.2535	28.9509	26.6483	24.3458	22.0432	19.7406	17.4380	15.1354	12.8328	10.5303	8.2278	5.9266	3.6374	1.4645	.1000
1.6	33.4916	31.1890	28.8864	26.5838	24.2812	21.9786	19.6760	17.3735	15.0709	12.7683	10.4657	8.1634	5.8621	3.5739	1.4092	.08631
1.7	33.4309	31.1283	28.8258	26.5232	24.2206	21.9180	19.6154	17.3128	15.0103	12.7077	10.4051	8.1027	5.8016	3.5143	1.3578	.07465
1.8	33.3738	31.0712	28.7686	26.4660	24.1634	21.8608	19.5583	17.2557	14.9531	12.6505	10.3479	8.0455	5.7446	3.4581	1.3089	.06471
1.9	33.3197	31.0171	28.7145	26.4119	24.1094	21.8068	19.5042	17.2016	14.8990	12.5964	10.2939	7.9915	5.6906	3.4050	1.2649	.05620
2.0	33.2684	30.9658	28.6632	26.3607	24.0581	21.7555	19.4529	17.1503	14.8477	12.5451	10.2426	7.9402	5.6394	3.3547	1.2227	.04890
2.1	33.2196	30.9170	28.6145	26.3119	24.0093	21.7067	19.4041	17.1015	14.7989	12.4964	10.1938	7.8914	5.5907	3.3069	1.1829	.04261
2.2	33.1731	30.8705	28.5679	26.2653	23.9628	21.6602	19.3576	17.0550	14.7524	12.4498	10.1473	7.8449	5.5443	3.2614	1.1454	.03719
2.3	33.1286	30.8261	28.5235	26.2209	23.9183	21.6157	19.3131	17.0106	14.7080	12.4054	10.1028	7.8004	5.4999	3.2179	1.1099	.03250
2.4	33.0861	30.7835	28.4809	26.1783	23.8758	21.5732	19.2706	16.9680	14.6654	12.3628	10.0603	7.7579	5.4575	3.1763	1.0762	.02844
2.5	33.0453	30.7427	28.4401	26.1375	23.8349	21.5323	19.2298	16.9272	14.6246	12.3220	10.0194	7.7172	5.4167	3.1365	1.0443	.02491
2.6	33.0060	30.7035	28.4009	26.0983	23.7957	21.4931	19.1905	16.8880	14.5854	12.2828	9.9802	7.6779	5.3776	3.0983	1.0139	.02185
2.7	32.9683	30.6657	28.3631	26.0606	23.7580	21.4554	19.1528	16.8502	14.5476	12.2450	9.9425	7.6401	5.3400	3.0615	.9849	.01918
2.8	32.9319	30.6294	28.3268	26.0242	23.7216	21.4190	19.1164	16.8138	14.5113	12.2087	9.9061	7.6038	5.3037	3.0261	.9573	.01686
2.9	32.8968	30.5943	28.2917	25.9891	23.6865	21.3839	19.0813	16.7788	14.4762	12.1736	9.8710	7.5687	5.2687	2.9920	.9309	.01482
3.0	32.8629	30.5604	28.2578	25.9552	23.6526	21.3500	19.0474	16.7449	14.4423	12.1397	9.8371	7.5348	5.2349	2.9591	.9057	.01305
3.1	32.8302	30.5276	28.2250	25.9224	23.6198	21.3172	19.0146	16.7121	14.4095	12.1069	9.8043	7.5020	5.2022	2.9273	.8815	.01149
3.2	32.7984	30.4958	28.1932	25.8907	23.5880	21.2855	18.9829	16.6803	14.3777	12.0751	9.7726	7.4703	5.1706	2.8965	.8583	.01013
3.3	32.7676	30.4651	28.1625	25.8599	23.5573	21.2547	18.9521	16.6495	14.3470	12.0444	9.7418	7.4395	5.1399	2.8668	.8361	.008939
3.4	32.7378	30.4352	28.1326	25.8300	23.5274	21.2249	18.9223	16.6197	14.3171	12.0145	9.7120	7.4097	5.1102	2.8379	.8147	.007971
3.5	32.7088	30.4062	28.1036	25.8010	23.4985	21.1959	18.8933	16.5907	14.2881	11.9855	9.6830	7.3807	5.0813	2.8099	.7942	.006970
3.6	32.6806	30.3780	28.0755	25.7729	23.4703	21.1677	18.8651	16.5625	14.2599	11.9574	9.6548	7.3526	5.0532	2.7827	.7745	.006160
3.7	32.6532	30.3506	28.0481	25.7455	23.4429	21.1403	18.8377	16.5351	14.2325	11.9300	9.6274	7.3252	5.0259	2.7563	.7554	.005448
3.8	32.6266	30.3240	28.0214	25.7188	23.4162	21.1136	18.8110	16.5085	14.2059	11.9033	9.6007	7.2985	4.9993	2.7306	.7371	.004820
3.9	32.6006	30.2980	27.9954	25.6928	23.3902	21.0877	18.7851	16.4825	14.1799	11.8773	9.5748	7.2725	4.9735	2.7056	.7194	.004267
4.0	32.5753	30.2727	27.9701	25.6675	23.3649	21.0623	18.7598	16.4572	14.1546	11.8520	9.5495	7.2472	4.9482	2.6813	.7024	.003779
4.1	32.5506	30.2480	27.9454	25.6428	23.3402	21.0376	18.7351	16.4325	14.1299	11.8273	9.5248	7.2225	4.9236	2.6576	.6859	.003349
4.2	32.5265	30.2239	27.9213	25.6187	23.3161	21.0136	18.7110	16.4084	14.1058	11.8032	9.5007	7.1985	4.8997	2.6344	.6700	.002969
4.3	32.5029	30.2004	27.8978	25.5952	23.2926	20.9900	18.6874	16.3884	14.0823	11.7797	9.4771	7.1749	4.8762	2.6119	.6546	.002633
4.4	32.4800	30.1774	27.8748	25.5722	23.2696	20.9670	18.6644	16.3619	14.0593	11.7567	9.4551	7.1520	4.8533	2.5899	.6397	.002336
4.5	32.4575	30.1519	27.8523	25.5497	23.2471	20.9446	18.6420	16.3394	14.0368	11.7342	9.4317	7.1295	4.8310	2.5684	.6253	.002073
4.6	32.4355	30.1329	27.8303	25.5277	23.2252	20.9226	18.6200	16.3174	14.0148	11.7122	9.4097	7.1075	4.8091	2.5474	.6114	.001841
4.7	32.4140	30.1114	27.8088	25.5062	23.2037	20.9011	18.5985	16.2959	13.9933	11.6907	9.3882	7.0860	4.7877	2.5268	.5979	.001635
4.8	32.3929	30.0904	27.7878	25.4852	23.1826	20.8800	18.5774	16.2748	13.9723	11.6697	9.3671	7.0650	4.7667	2.5068	.5848	.001453
4.9	32.3723	30.0697	27.7672	25.4646	23.1620	20.8594	18.5568	16.2542	13.9516	11.6491	9.3465	7.0444	4.7462	2.4871	.5721	.001291
5.0	32.3521	30.0495	27.7470	25.4444	23.1418	20.8392	18.5366	16.2340	13.9314	11.6289	9.3263	7.0242	4.7261	2.4679	.5598	.001148
5.1	32.3323	30.0297	27.7271	25.4246	23.1220	20.8194	18.5168	16.2142	13.9116	11.6091	9.3065	7.0044	4.7064	2.4491	.5478	.001021
5.2	32.3129	30.0103	27.7077	25.4051	23.1026	20.8000	18.4974	16.1948	13.8922	11.5896	9.2871	6.9850	4.6871	2.4306	.5362	.0009086
5.3	32.2939	29.9913	27.6887	25.3861	23.0835	20.7809	18.4783	16.1758	13.8732	11.5706	9.2681	6.9659	4.6681	2.4126	.5250	.0008086

5.4	32.2752	29.9726	27.6700	25.3674	23.0648	20.7622	18.4596	16.1571	13.8545	11.5519	9.2494	6.9473	4.6495	2.3948	.5140	.0007198
5.5	32.2568	29.9542	27.6516	25.3491	23.0465	20.7439	18.4413	16.1387	13.8361	11.5336	9.2310	6.9289	4.6313	2.3775	.5034	.0006409
5.6	32.2388	29.9362	27.6336	25.3310	23.0285	20.7259	18.4233	16.1207	13.8181	11.5155	9.2130	6.9109	4.6134	2.3604	.4930	.0005708
5.7	32.2211	29.9185	27.6159	25.3133	23.0108	20.7082	18.4056	16.1030	13.8004	11.4978	9.1953	6.8932	4.5958	2.3437	.4830	.0005085
5.8	32.2037	29.9011	27.5985	25.2959	22.9934	20.6908	18.3882	16.0855	13.7830	11.4804	9.1779	6.8758	4.5785	2.3273	.4732	.0004532
5.9	32.1866	29.8840	27.5814	25.2789	22.9763	20.6737	18.3711	16.0685	13.7659	11.4633	9.1608	6.8588	4.5615	2.3111	.4637	.0004039
6.0	32.1698	29.8672	27.5646	25.2620	22.9595	20.6569	18.3543	16.0517	13.7491	11.4465	9.1440	6.8420	4.5448	2.2953	.4544	.0003601
6.1	32.1533	29.8507	27.5481	25.2455	22.9429	20.6403	18.3378	16.0352	13.7326	11.4300	9.1275	6.8254	4.5283	2.2797	.4454	.0003211
6.2	32.1370	29.8344	27.5318	25.2293	22.9267	20.6241	18.3215	16.0189	13.7163	11.4138	9.1112	6.8092	4.5122	2.2645	.4366	.0002864
6.3	32.1210	29.8184	27.5158	25.2133	22.9107	20.6081	18.3055	16.0029	13.7003	11.3978	9.0952	6.7932	4.4963	2.2494	.4280	.0002555
6.4	32.1053	29.8027	27.5001	25.1975	22.8949	20.5923	18.2898	15.9872	13.6846	11.3820	9.0795	6.7775	4.4806	2.2346	.4197	.0002279
6.5	32.0898	29.7872	27.4846	25.1820	22.9794	20.5768	18.2742	15.9717	13.6691	11.3665	9.0640	6.7660	4.4652	2.2201	.4115	.0002034
6.6	32.0745	29.7719	27.4693	25.1667	22.8641	20.5616	18.2590	15.9564	13.6538	11.3512	9.0487	6.7467	4.4501	2.2058	.4036	.0001816
6.7	32.0595	29.7569	27.4543	25.1517	22.8491	20.5465	18.2439	15.9414	13.6388	11.3362	9.0337	6.7317	4.4351	2.1917	.3959	.0001621
6.8	32.0446	29.7421	27.4395	25.1369	22.8343	20.5317	18.2291	15.9265	13.6240	11.3214	9.0189	6.7169	4.4204	2.1779	.3883	.0001448
6.9	32.0300	29.7275	27.4249	25.1223	22.8197	20.5171	18.2145	15.9119	13.6094	11.3608	9.0043	6.7023	4.4059	2.1643	.3810	.0001293
7.0	32.0156	29.7131	27.4105	25.1079	22.8053	20.5027	18.2001	15.8976	13.5950	11.2924	8.9899	6.6879	4.3916	2.1508	.3738	.0001155
7.1	32.0015	29.6989	27.3963	25.0937	22.7911	20.4885	18.1860	15.8834	13.5808	11.2782	8.9757	6.6737	4.3775	2.1376	.3668	.0001032
7.2	31.9875	29.6849	27.3823	25.0797	22.7771	20.4746	18.1720	15.8694	13.5668	11.2642	8.9617	6.6598	4.3636	2.1246	.3599	.00009219
7.3	31.9737	29.6711	27.3685	25.0659	22.7633	20.4608	18.1582	15.8556	13.5530	11.2504	8.9479	6.6460	4.3500	2.1118	.3532	.00008239
7.4	31.9601	29.6575	27.3549	25.0523	22.7497	20.4472	18.1446	15.8420	13.5394	11.2368	8.9343	6.6324	4.3364	2.0991	.3467	.00007364
7.5	31.9467	29.6441	27.3415	25.0389	22.7363	20.4337	18.1311	15.8286	13.5260	11.2234	8.9209	6.6190	4.3231	2.0867	.3403	.00006583
7.6	31.9334	29.6308	27.3282	25.0257	22.7231	20.4205	18.1179	15.8153	13.5127	11.2102	8.9076	6.6057	4.3100	2.0744	.3341	.00005886
7.7	31.9203	29.6178	27.3152	25.0126	22.7100	20.4074	18.1048	15.8022	13.4997	11.1971	8.8946	6.5927	4.2970	2.0623	.3280	.00005263
7.8	31.9074	29.6048	27.3023	24.9997	22.6971	20.3945	18.0919	15.7893	13.4868	11.1842	8.8817	6.5798	4.2842	2.0503	.3221	.00004707
7.9	31.8947	29.5921	27.2895	24.9869	22.6844	20.3818	18.0792	15.7766	13.4740	11.1714	8.8689	6.5671	4.2716	2.0386	.3163	.00004210
8.0	31.8821	29.5795	27.2769	24.9744	22.6718	20.3692	18.0666	15.7640	13.4614	11.1589	8.8563	6.5545	4.2591	2.0269	.3106	.00003767
8.1	31.8697	29.5671	27.2645	24.9619	22.6594	20.3568	18.0542	15.7516	13.4490	11.1464	8.8439	6.5421	4.2468	2.0155	.3050	.00003370
8.2	31.8574	29.5548	27.2523	24.9497	22.6471	20.3445	18.0419	15.7393	13.4367	11.1342	8.8317	6.5298	4.2346	2.0042	.2996	.00003015
8.3	31.8453	29.5427	27.2401	24.9375	22.6350	20.3324	18.0298	15.7272	13.4246	11.1220	8.8195	6.5177	4.2226	1.9930	.2943	.00002699
8.4	31.8333	29.5307	27.2282	24.9256	22.6230	20.3204	18.0178	15.7152	13.4126	11.1101	8.8076	6.5057	4.2107	1.9820	.2891	.00002415
8.5	31.8215	29.5189	27.2163	24.9137	22.6112	20.3086	18.0060	15.7034	13.4008	11.0982	8.7957	6.4939	4.1990	1.9711	.2840	.00002162
8.6	31.8098	29.5072	27.2046	24.9020	22.5995	20.2969	17.9943	15.6917	13.3891	11.0865	8.7840	6.4822	4.1874	1.9604	.2790	.00001936
8.7	31.7982	29.4957	27.1931	24.8905	22.5879	20.2853	17.9827	15.6801	13.3776	11.0750	8.7725	6.4707	4.1759	1.9498	.2742	.00001733
8.8	31.7868	29.4842	27.1816	24.8790	22.5765	20.2739	17.9713	15.6687	13.3661	11.0635	8.7610	6.4592	4.1646	1.9393	.2694	.00001552
8.9	31.7755	29.4729	27.1703	24.8678	22.5652	20.2626	17.9600	15.6574	13.3548	11.0523	8.7497	6.4480	4.1534	1.9290	.2647	.00001390
9.0	31.7643	29.4618	27.1592	24.8566	22.5540	20.2514	17.9488	15.6462	13.3437	11.0411	8.7386	6.4368	4.1423	1.9187	.2602	.00001245
9.1	31.7533	29.4507	27.1481	24.8455	22.5429	20.2404	17.9378	15.6352	13.3326	11.0300	8.7275	6.4258	4.1313	1.9087	.2557	.00001115
9.2	31.7424	29.4398	27.1372	24.8346	22.5320	20.2294	17.9268	15.6243	13.3217	11.0191	8.7166	6.4148	4.1205	1.8987	.2513	.000009988
9.3	31.7315	29.4290	27.1264	24.8238	22.5212	20.2186	17.9160	15.6135	13.3109	11.0084	8.7058	6.4040	4.1098	1.8888	.2470	.000008948
9.4	31.7208	29.4183	27.1157	24.8131	22.5105	20.2079	17.9053	15.6028	13.3002	10.9976	8.6951	6.3934	4.0992	1.8791	.2429	.000008018
9.5	31.7103	29.4077	27.1051	24.8025	22.4999	20.1973	17.8948	15.5922	13.2896	10.9870	8.6845	6.3828	4.0887	1.8695	.2387	.000007185
9.6	31.6998	29.3972	27.0946	24.7920	22.4895	20.1869	17.8843	15.5817	13.2791	10.9765	8.6740	6.3723	4.0784	1.8599	.2347	.000006439
9.7	31.6894	29.3868	27.0843	24.7817	22.4791	20.1765	17.8739	15.5713	13.2688	10.9662	8.6637	6.3620	4.0681	1.8505	.2308	.000005771
9.8	31.6792	29.3766	27.0740	24.7714	22.4688	20.1663	17.8637	15.5611	13.2585	10.9559	8.6534	6.3517	4.0579	1.8412	.2269	.000005173
9.9	31.6690	29.3664	27.0639	24.7613	22.4587	20.1561	17.8535	15.5509	13.2483	10.9458	8.6433	6.3416	4.0479	1.8320	.2231	.000004637

Source: After Ferris et al. (1962).

Example 4-1: For the confined aquifer shown in Figure 4-1, the following values are given: $b = 12\,\text{m}, K = 0.01\,\text{cm/s} = 8.64\,\text{m/day}, S = 0.001$, and $Q = 0.004\,\text{m}^3/\text{s} = 345.6\,\text{m}^3/\text{day}$. Determine the drawdown after 8 hours at a 25-m distance from the well.

Solution: From Eq. (4-23) we obtain

$$u = \frac{(25\text{ m})^2(0.001)}{4(8.64\text{ m/day})(12\text{ m})\left(\dfrac{8}{24}\text{ day}\right)} = 4.52 \times 10^{-3}$$

From Table 4-1, $W(u) = 4.83$. From Eq. (4-24) the drawdown is

$$s = \frac{345.6\text{ m}^3/\text{day}}{4\pi\,(8.64\text{ m/day})\,(12\text{ m})}(4.83) = 1.28\text{ m}$$

4.2.5 Solution for a Finite Confined Aquifer and Criterion for the Infinite Aquifer Assumption

Chen (1984) reinvestigated the Theis problem by using the following boundary condition instead of the boundary condition described by Eq. (4-12):

$$s(R, t) = 0 \tag{4-28}$$

where R is the radial distance at which the drawdown is zero. The rest of initial and boundary conditions are the same as described in Section 4.2.3. Equation (4-28) states that the zero-drawdown boundary is prescribed at a finite distance R, instead of at an infinite distance, as in the Theis problem.

By solving the modified boundary-value problem, Chen (1984) obtained the following solution:

$$s(r, t) = \frac{Q}{4\pi T}\,[W(u) - W(U) + 2I] \tag{4-29}$$

where

$$U = \frac{R^2 S}{4Tt} \tag{4-30}$$

$$I = \sum_{n=0}^{\infty} \frac{J_0\left[\left(\dfrac{u}{U}\right)^{1/2} \chi_n\right]}{\chi_n J_1(\chi_n)} \int_0^1 \exp\left[-\frac{U}{x} - \frac{\chi_n^2(1-x)}{4U}\right]\frac{dx}{x} \tag{4-31}$$

$$\chi_n = R\beta_n \tag{4-32}$$

and u and W are defined by Eqs. (4-23) and (4-25), respectively. J_0 and J_1 are the Bessel functions of zero order and first order, respectively, and β_n is the nth root satisfying $J_0(R\chi_n) = 0$.

The results of Eq. (4-29) are presented in Figure 4-4 as a type curve of the u versus $4\pi Ts/Q$ coordinate system. As can be seen from Figure 4-4, for $U \geq 4$, Eq. (4-29) and the

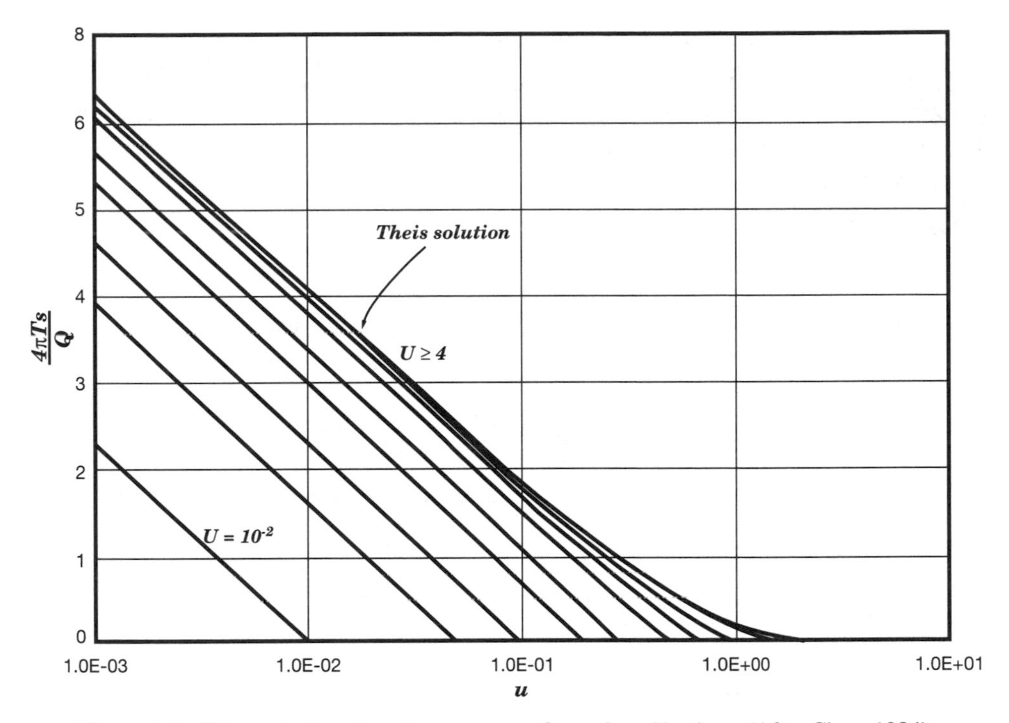

Figure 4-4 Dimensionless drawdown versus u for various U values. (After Chen, 1984).

Theis solution, Eq. (4-24), are practically the same; the maximum difference is less than 1% when $U = 4$ and smaller when $U > 4$. Therefore, Chen established the following time criterion during which an aquifer can be analyzed as an infinite aquifer:

$$t \leq \frac{R^2 S}{16T} \tag{4-33}$$

This criterion can be used for those cases where exterior boundaries exist around the close proximity of the well. When the above criterion is applied to such a field case, the approximate minimum distance from the pumping well to the boundary can be used for R.

Example 4-2: The confined aquifer described in Example 4-1 has an exterior boundary at approximately an 80-m distance. Determine the time period during which the aquifer can be analyzed as an infinite one.

Solution: From Eq. (4-33) we obtain

$$t \leq \frac{(80\text{ m})^2(0.001)}{16(8.64\text{ m/day})(12\text{m})} = 38.58 \times 10^{-3}\text{ days} = 5.56\text{ min}$$

Physically, this means that the aquifer can be analyzed as an infinite aquifer during an approximately 6-minute period of time since pumping started.

4.2.6 Approximated Forms of the Theis Equation

4.2.6.1 Drawdown Equation. For small values of u, $u <$ about 0.01, the sum of the terms beyond $\ln(u)$ in Eq. (4-26) is not significant (Cooper and Jacob, 1946; Jacob, 1950). The value of u decreases as the time (t) increases and as the radial distance (r) decreases. Under these conditions, Eq. (4-24) may be approximated by

$$s(r,t) \cong \frac{Q}{4\pi T}[-0.5772 - \ln(u)] \tag{4-34}$$

Noting that $\ln(1.781) = 0.5772$, Eq. (4-34) takes the following form (see also Section 4.2.7.2):

$$s(r,t) \cong \frac{Q}{4\pi T}[-\ln(1.781) - \ln(u)] \tag{4-35}$$

4.2.6.2 Radius of Influence Zone. Equation (4-35) can also be expressed as

$$s = \frac{Q}{2\pi T} \ln\left[\frac{1.5}{r}\left(\frac{Tt}{S}\right)^{1/2}\right] \tag{4-36}$$

By comparing Eq. (4-36) with Eq. (3-10) of Chapter 3, one may define a *radius of influence zone* (R), a circle over which the drawdown is zero, by (Bear, 1979)

$$R(t) = 1.5\left(\frac{Tt}{S}\right)^{1/2} \tag{4-37}$$

which is a time-dependent quantity.

Example 4-3: For the confined aquifer described in Example 4-1, determine the radius of influence zone at 8, 16, and 24 hours.

Solution: From Eq. (4-37), at t = 8 hours = 0.33 days

$$R = 1.5\left[\frac{(8.64 \text{ m/days})(12 \text{ m})(0.333 \text{ day})}{0.001}\right]^{1/2} = 279 \text{ m}$$

At the end of 16- and 24-hour time periods, the radius of influence zone can be determined in the same way as 394 m and 483 m, respectively.

4.2.6.3 Equations for Specific Capacity and Transmissivity Estimation. The *specific capacity* of a well is defined as the ratio of its discharge to its total drawdown—that is, discharge per unit drawdown. Simple equations can be developed from the approximated form of the Theis equation to estimate the specific capacity and transmissivity of a well (Driscoll, 1986). The derivation of these equations are based on an average well diameter, average pumping period, and typical values of the applicable storage coefficient and specific yield.

For a confined aquifer, Driscoll (1986) assumed the following typical values: t = 1 day, r_w (well diameter) = 0.5 ft = 0.152 m, T = 30,000 gpd/ft = 4011 ft^2/day = 373 m^2/day, and S = 0.001. Substituting these values into Eq. (4-35), the following equations can be obtained for a confined aquifer in two different unit systems:

$$\frac{Q\ (\mathrm{m^3/day})}{s_w\ (m)} = \frac{T}{1.385}(\mathrm{m^2/day}),\ \frac{Q\ (\mathrm{gpm})}{s_w\ (\mathrm{ft})} = \frac{T\ (\mathrm{gpd/ft})}{2000} \tag{4-38}$$

For an unconfined aquifer, using $S_y = 0.075$ as a typical specific yield value and the rest of above-given values, Eq. (4-35) yields the following equations in the same unit systems:

$$\frac{Q\ (\mathrm{m^3/day})}{s_w\ (\mathrm{m})} = \frac{T\ (\mathrm{m^2/day})}{1.042},\quad \frac{Q\ (\mathrm{gpm})}{s_w\ (\mathrm{ft})} = \frac{T\ (\mathrm{gpd/ft})}{1500} \tag{4-39}$$

Example 4-4: For an unconfined aquifer the following values are given: $Q = 10$ gpm and $s_w = 4$ ft. This drawdown corresponds to a relatively long period of time. Estimate the transmissivity of the aquifer in the two unit systems.

Solution: From Eq. (4-39) we obtain

$$T = 1500\frac{10\ \mathrm{gpm}}{4\ \mathrm{ft}} = 3750\ \mathrm{gpd/ft} = 501\ \mathrm{ft^2/day} = 47\ \mathrm{m^2/day}$$

$Q = 10$ gpm $= 54.51$ m^3/day and $s_w = 4$ ft $= 1.22$ m. The transmissivity in the metric system is

$$T = 1.042\frac{54.51\ \mathrm{m^3/day}}{1.22\ \mathrm{m}} = 47\ \mathrm{m^2/day}$$

Notice that both equations yielded the same value.

Example 4-5: Water is pumping from four partially penetrated production wells in a confined aquifer which is composed of sand and gravel. Pumping rates, well drawdowns, and screening intervals of the wells are given in Table 4-2. Using these data, estimate average hydraulic conductivity of the aquifer.

Solution: Because the aquifer is confined, Eq. (4-38) should be used. Using the values in Table 4-2, the calculated transmissivities and hydraulic conductivities are presented in Table 4-3. Hydraulic conductivities are determined by dividing the transmissivity values by the length of the screening interval. The arithmetic average of the hydraulic conductivities

TABLE 4-1 Observed Data for Example 4-5

Well No.	Pumping Rate, Q (gpm)	Drawdown at the Well, s_W (ft)	Screen Interval, L (ft)
1	1298	14.7	144
2	1382	41.0	16
3	1164	37.0	53
4	1676	9.0	100

TABLE 4-2 Estimated Hydraulic Conductivities for Example 4-5

Well No.	Transmissivity, T (gpd/ft)	Hydraulic Conductivity, K (ft/day)
1	176,599	164
2	67,415	563
3	62,919	159
4	372,444	498

is 346 ft/day (0.122 cm/s). The weighted average (K_{wav}) can be calculated by the following formula:

$$K_{wav} = \frac{\sum_{n=1}^{4} K_n L_n}{\sum_{n=1}^{4} L_n}$$

Using the L values in Table 4-2 and the K values in Table 4-3, the weighted average hydraulic conductivity is 290 ft/day (0.102 cm/s). As can be seen from these values, both arithmetic and weighted average hydraulic conductivities are in the same order of magnitude and they correspond to sand and gravel (see Tables 2-2 and 2-3 of Chapter 2).

4.2.7 Methods of Analysis for Pumping Test Data

A number of methods have been developed in analyzing pumping test data for aquifers. In this section some of the well-known methods will be described in detail, with examples. The Theis solution, which is described in detail in the previous sections, is the basis of these methods.

4.2.7.1 Theis Type-Curve Method

Steps for the Method. Theis (1935) established a graphical method to determine the aquifer transmissivity (T) and storage coefficient (S) from Eqs. (4-23) and (4-24) using the constant pumping rate (Q), measured drawdowns (s) for one value of radial distance (r), and several values of time (t); or for one value of t and several values of r (Ferris et al., 1962).

Equation (4-23) may be rearranged to obtain

$$\frac{t}{r^2} = \left(\frac{S}{4T}\right)\frac{1}{u} \tag{4-40}$$

or

$$\log\left(\frac{t}{r^2}\right) = \left[\log\left(\frac{S}{4T}\right)\right] + \log\left(\frac{1}{u}\right) \tag{4-41}$$

Similarly, from Eq. (4-24) we obtain

$$s = \left(\frac{Q}{4\pi T}\right) W(u) \tag{4-42}$$

or

$$\log(s) = \left[\log\left(\frac{Q}{4\pi T}\right)\right] + \log W(u) \tag{4-43}$$

Because Q is constant, the bracketed parts of Eqs. (4-41) and (4-43) are constant as well. Therefore, $W(u)$ is related to u in such a way that s is related to r^2/t as shown in Figure 4-5. If values of s are plotted against r^2/t on double logarithmic paper to the same scale as the type curve, the data curve should be similar to the type curve. Keeping the coordinate axes of the two curves parallel, the data curve is then superimposed on the type curve and translated to a position that establishes the best fit of the field data to the type curve.

An alternative method is to plot $W(u)$ versus $1/u$ as the type curve. For the data curve, s may be plotted against t/r^2 (or t, if only one observation well is used). The corresponding coordinates of the measured data are s versus t or t/r^2. This method eliminates the necessity of computing $1/t$ values for s values.

The process is called *type-curve matching*, and its associated steps are given below in a stepwise manner:

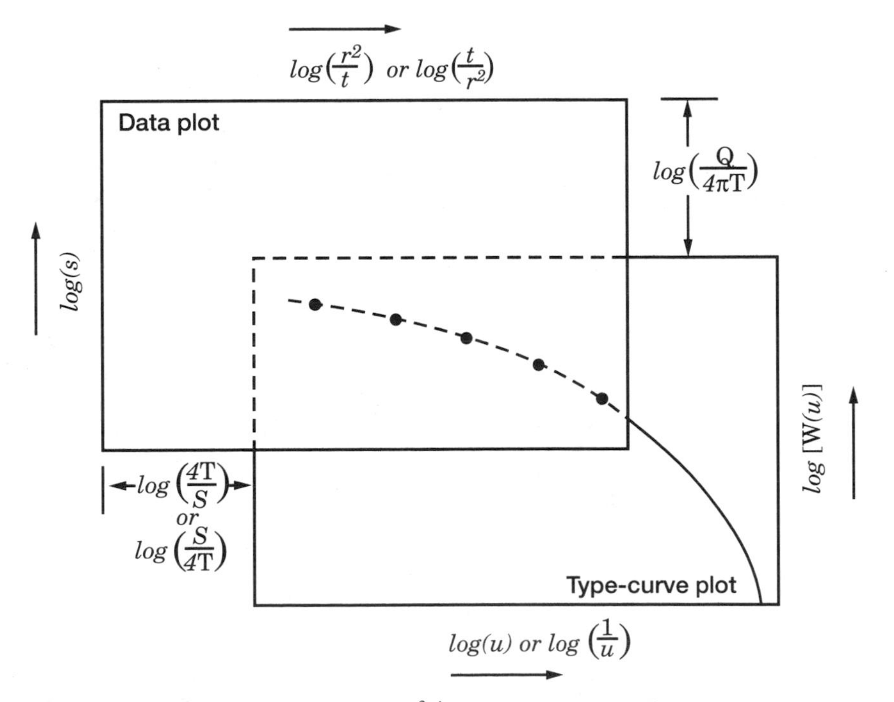

Figure 4-5 Relation of $W(u)$ and u to s and r^2/t and displacements of graph scales by amounts of constants shown. (After Ferris et al., 1962).

Step 1: Plot the *type curve* of $1/u$ versus $W(u)$ or u versus $W(u)$ (Figure 4-2) on log–log paper using the data in Table 4-1.

Step 2: On another sheet of log–log paper with the same as in Step 1, plot the observation data of s versus t/r^2 or s versus r^2/t.

Step 3: Keeping the coordinate axes parallel at all times, superimpose the two sheets until the best fit of data curve and the type curve is obtained.

Step 4: Select a common point, the match point, arbitrarily chosen on the overlapping part of the curve, or even anywhere on the overlapping portion of the sheets. Read the corresponding values of u, $W(u)$, s, and t/r^2 or s and r^2/t.

Step 5: Substitute the values of $W(u)$ and s into Eq. (4-24) and calculate the transmissivity (T). Then, substitute the values of u, T, and t/r^2 (or r^2/t) into Eq. (4-23) to calculate the storage coefficient.

Cautions and Limitations of the Method. Historically, the Theis method is the first method that was established in the 1930s to analyze pumping test data for aquifers. After the 1950s, a number of models were developed for pumping test data analysis that can treat much more complex situations than the Theis model. These models are all included in subsequent chapters. Because of their complexities, one may prefer to use the Theis model to determine aquifer parameters from pumping test data. However, the hydrologist should be aware of the assumptions and limitations of the model. Some guidelines are presented below (Ferris et al., 1962; Hantush, 1964):

1. First of all, one has to be aware of the assumptions of the Theis model (see Section 4.2.1). The key assumptions of the model are that the aquifer is a nonleaky confined aquifer and that the pumping and observation wells fully penetrate the aquifer.
2. The Theis type-curve method is a convenient method of analysis if the log–log plot of most of the measured data shows a well-defined curvature in a way that the type curve is in the range of $u < 0.01$ or $1/u < 100$.
3. Initially, the well discharge may be variable because of pump adjustment. This situation may result in an initial disagreement between the theory and real flow conditions. As the time of pumping becomes large, such effects are minimized and closer agreement between the type curve and measured results may be obtained.
4. Theoretically, the Theis equation is only valid for confined aquifers. However, the same equation can also be used in approximating the conditions for unconfined aquifers as well. If the drawdowns for an unconfined aquifer are large compared with the initial thickness of the aquifer, then it is necessary to adjust the observed drawdowns in accordance of Eq. (3-64) of Chapter 3. According to Jacob (1944), if the drawdowns are adjusted by the quantity $s^2/2b$, the value of transmissivity (T) will correspond to an equivalent confined aquifer, and the value of storage coefficient (S) will more closely approximate the true value.

Examples

Example 4-6: Measured data collected from a pumping test for a confined aquifer are given in Tables 4-4a and 4-4b. Both the pumping well and observation wells fully penetrate the aquifer. The drawdowns were measured at two observation wells having 25- and 75-m radial distances from the pumping well. The steady-state flow rate was 540 m^3/day. Using

TABLE 4-4a Observed Drawdown Versus Time Data at r = 25 m for Example 4-6

t (min)	s (m)	t/r^2 (min/m^2)
1.0	0.18	1.60×10^{-3}
1.4	0.25	2.24×10^{-3}
2.0	0.32	3.20×10^{-3}
2.8	0.40	4.48×10^{-3}
3.8	0.46	6.08×10^{-3}
4.7	0.50	7.52×10^{-3}
5.6	0.55	8.96×10^{-3}
6.7	0.60	1.07×10^{-2}
8.0	0.65	1.28×10^{-2}
10.7	0.70	1.71×10^{-2}
13.6	0.76	2.18×10^{-2}
16.5	0.80	2.64×10^{-2}
17.5	0.82	2.80×10^{-2}
21.0	0.84	3.36×10^{-2}
26.3	0.90	4.21×10^{-2}
36.0	0.96	5.76×10^{-2}
54.0	1.04	8.64×10^{-2}
65.0	1.05	1.04×10^{-1}
80.0	1.09	1.28×10^{-1}
97.0	1.11	1.55×10^{-1}
119.0	1.15	0.90×10^{-1}
162.0	1.20	2.59×10^{-1}
190.0	1.22	3.04×10^{-1}
280.0	1.28	4.48×10^{-1}
364.0	1.31	5.82×10^{-1}
490.0	1.35	7.84×10^{-1}
610.0	1.40	9.76×10^{-1}
730.0	1.42	1.17
950.0	1.47	1.52
1220.0	1.48	1.95
1480.0	1.50	2.37

TABLE 4-4b Observed Drawdown Versus Time Data at r = 75 m for Example 4-6

t (min)	s (m)	t/r^2 (min/m^2)
3.0	0.022	5.33×10^{-4}
4.0	0.030	7.11×10^{-4}
4.3	0.033	7.64×10^{-4}
5.3	0.062	9.42×10^{-4}
6.0	0.076	1.07×10^{-3}
7.0	0.105	1.24×10^{-3}
8.0	0.126	1.42×10^{-3}
8.7	0.146	1.55×10^{-3}
11.0	0.186	1.96×10^{-3}
12.0	0.214	2.13×10^{-3}
15.0	0.250	2.67×10^{-3}
18.0	0.288	3.20×10^{-3}
27.0	0.350	4.80×10^{-3}
31.0	0.385	5.51×10^{-3}
35.0	0.427	6.22×10^{-3}
50.0	0.487	8.89×10^{-3}
60.0	0.510	1.07×10^{-2}
80.0	0.566	1.42×10^{-2}
106.0	0.602	1.88×10^{-2}
120.0	0.622	2.13×10^{-2}
150.0	0.654	2.67×10^{-2}
180.0	0.690	3.20×10^{-2}
220.0	0.710	3.91×10^{-2}
245.0	0.740	4.36×10^{-2}
300.0	0.774	5.33×10^{-2}
360.0	0.795	6.40×10^{-2}
500.0	0.830	8.89×10^{-2}
610.0	0.860	1.08×10^{-1}
730.0	0.890	1.30×10^{-1}
850.0	0.920	1.51×10^{-1}
1100.0	0.950	1.96×10^{-1}

the Theis type-curve method, determine the transmissivity (T) and storage coefficient (S) of the aquifer.

Solution:

Step 1: Using Table 4-1, the type curve $1/u$ versus $W(u)$ is given in Figure 4-2.

Step 2: First, using the values in Tables 4-4a and 4-4b, t/r^2 is calculated and listed in the same tables. Then, with these tables, the graphs in Figure 4-6 are plotted.

Step 3: The superimposed graphs are shown in Figure 4-7.

Step 4: In Figure 4-7, A is an arbitrarily chosen point, and B is a point on the overlapping part of the curve. The coordinates of A are

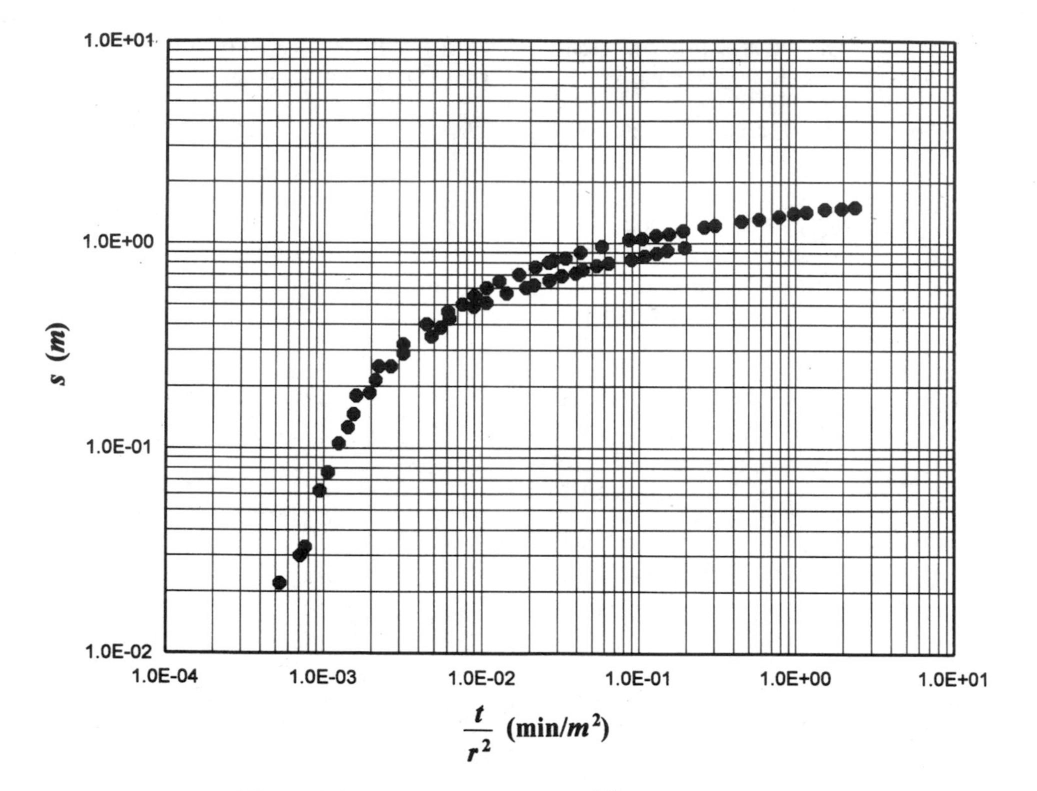

Figure 4-6 Drawdown (s) versus t/r^2 for Example 4-6.

$$\left(\frac{1}{u}\right)_A = 10, \qquad W_A(u) = 1$$

$$s_A = 0.23\text{m}, \qquad \left(\frac{t}{r^2}\right)_A = 8.2 \times 10^{-3} \text{ min/m}^2$$

The coordinates of B are

$$\left(\frac{1}{u}\right)_B = 100, \qquad W_B(u) = 3.9$$

$$s_B = 0.9 \text{ m}, \qquad \left(\frac{t}{r^2}\right)_B = 8.4 \times 10^{-2} \text{ min/m}^2$$

Step 5: Using the match point values for point A, from Eqs. (4-24) and (4-23) we obtain

$$T = \frac{(540 \text{ m}^3/\text{day})(1)}{(4\pi)(0.23 \text{ m})} = 187 \text{ m}^2/\text{day}$$

$$S = \frac{4\left(187 \dfrac{\text{m}^2}{24 \times 60 \text{ min}}\right)(8.2 \times 10^{-3} \text{ min/m}^2)}{(10)} = 4.26 \times 10^{-4}$$

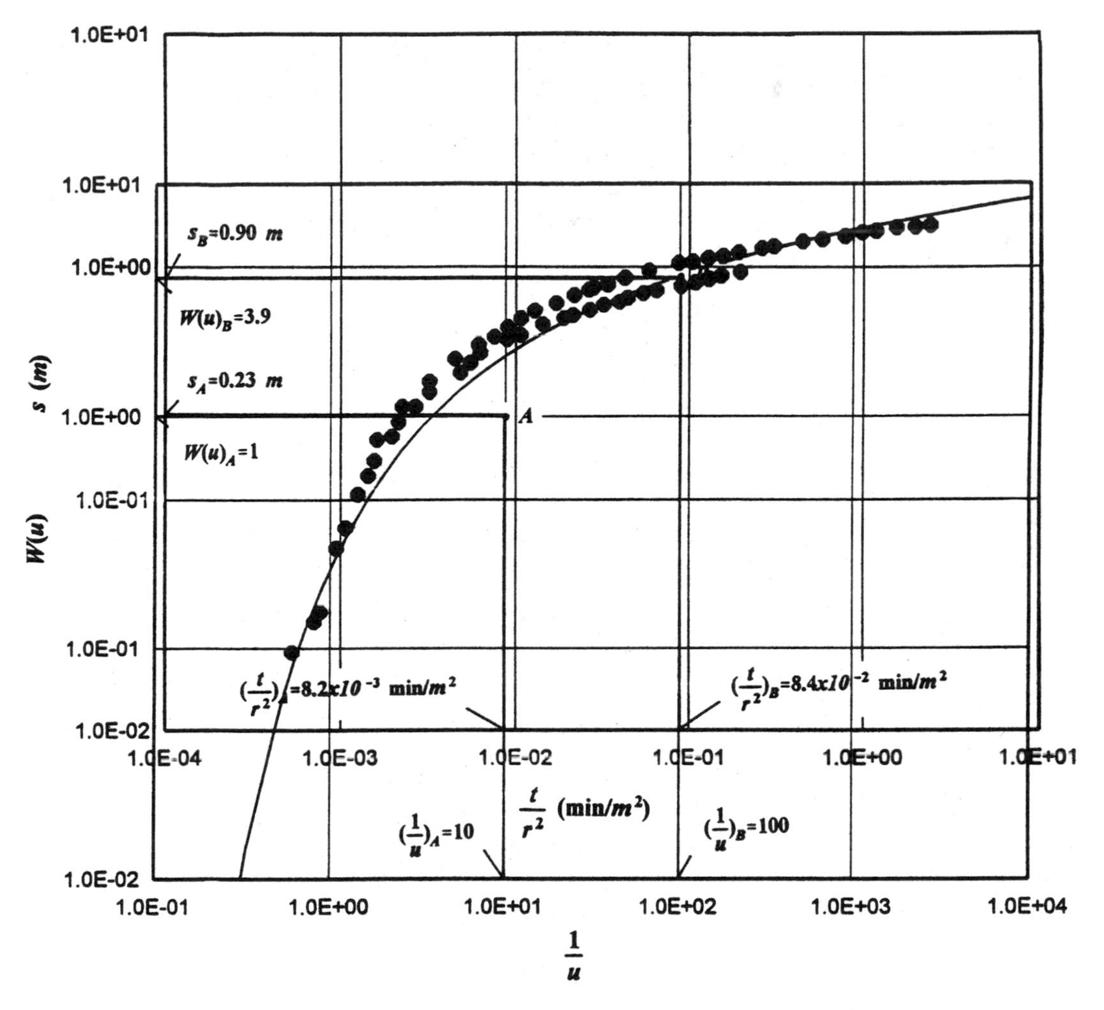

Figure 4-7 Theis type-curve matching for the data of Example 4-6.

From the values of point B we obtain

$$T = \frac{(540\ \mathrm{m^3/day})(3.9)}{(4\pi)(0.9\ \mathrm{m})} = 186\ \mathrm{m^2/day}$$

$$S = \frac{4\left(186\dfrac{\mathrm{m^2}}{24 \times 60\ \mathrm{min}}\right)(8.4 \times 10^{-2}\ \mathrm{min/m^2})}{(100)} = 4.34 \times 10^{-4}$$

As can be seen from the above values, the two matching points A and B give approximately the same values of T and S.

4.2.7.2 Cooper and Jacob Method. Cooper and Jacob (1946) developed an approximation for the Theis equation and a data analysis method which does not require type-curve matching. In the following sections the approximation method and the associated data analysis methods are presented.

Approximation for the Theis Equation. As mentioned in Section 4.2.6.1, Cooper and Jacob (1946) found that in the series of Eq. (4-26) the sum of the terms with the exception of the first two terms in Eq. (4-26) is insignificant for small values of u. According to Eq. (4-23), the value of u decreases as the time (t) increases and as the radial distance (r) decreases. Therefore, for large values of t and reasonably small values of r, the terms with the exeption of the first two in Eq. (4-26) may be neglected. Thus, the Theis equation approximated by Eq. (4-35) takes the form

$$s \cong \frac{Q}{4\pi T}\left[-\ln(1.781) - \ln\left(\frac{r^2 S}{4Tt}\right)\right] \tag{4-44}$$

Eq. (4-44) may be written as

$$s \cong \frac{Q}{4\pi T}\left[\ln\left(\frac{1}{u}\right) - \ln(1.781)\right] \tag{4-45}$$

Converting natural logarithms (ln) to common logarithms (log) [i.e., $\ln(u) = 2.303\log(u)$] Eq. (4-45) may be rewritten as

$$s \cong \frac{2.303Q}{4\pi T}\log\left(\frac{2.25Tt}{r^2 S}\right) \tag{4-46}$$

Equation (4-46) is an asymptotic expression which predicts the drawdown very closely under the condition that a sufficient time has elapsed since the starting of pumping.

Note that in Eq. (4-46), Q, T, and S are constants, and only t varies. Equation (4-46) may be rewritten as

$$s = \frac{2.303Q}{4\pi T}\log\left(\frac{2.25T}{r^2 S}\right) + \frac{2.303Q}{4\pi T}\log(t) \tag{4-47}$$

Equation (4-47) is the equation of a straight line on a semilogarithmic coordinate system s versus $\log(t)$. The slope of the straight line is equal to $2.303Q/(4\pi T)$, and this line intercepts the time-axis where $s = 0$. Consequently, the interception point has the coordinates $s = 0$ and $t = t_0$. Introducing these values into Eq. (4-46) gives

$$0 = \frac{2.303Q}{4\pi T}\log\left(\frac{2.25Tt_0}{r^2 S}\right) \tag{4-48}$$

Because $2.303Q/(4\pi T)$ cannot be zero, it follows that

$$\frac{2.25Tt_0}{r^2 S} = 1 \tag{4-49}$$

and solving for S gives

$$S = \frac{2.25Tt_0}{r^2} \tag{4-50}$$

If the units of T, t_0, and r are gpd/ft, day, and ft, respectively, Eq. (4-50) takes the form

$$S = \frac{0.3Tt_0}{r^2} \tag{4-51}$$

Because the derivative of a constant is zero, the derivation of Eq. (4-47) with respect to $\log(t)$ is

$$\Delta s = \frac{2.303Q}{4\pi T}\Delta \log(t) \tag{4-52}$$

and solving it for T gives

$$T = \frac{2.303Q}{4\pi\Delta s}\Delta \log(t) \tag{4-53}$$

If the units of T, Q, and Δs are gpd/ft, gpm, and ft, respectively, Eq. (4-53) takes the following form:

$$T = \frac{264Q}{\Delta s}\Delta \log(t) \tag{4-54}$$

In Eq. (4-54), t can be in any unit.

To apply the Cooper and Jacob method, the following conditions should be satisfied (Ferris et al., 1962): (1) the assumptions for the Theis model (Section 4.2.1) and (2) small values of u (generally less than 0.01). In other words, r is small and t is large.

Methods of Analysis. Using the equations given in the previous section, three methods can be used to determine the transmissivity and storage coefficient of a confined aquifer (Kruseman and De Ridder, 1991). The term *straight-line methods* is also used for these methods.

METHOD I. The steps are given below:

Step 1: Plot the data of one of the observation wells on single logarithmic paper (t on logarithmic scale) and draw a line through the points.

Step 2: Find the interception point of the line with the time axis where the drawdown (s) is zero. Read the value of t_0.

Step 3: Determine the geometric slope of the line—that is, the drawdown difference Δs per log cycle of time.

Step 4: Introduce the values of Q, Δs, and $\Delta(\log t)$ into Eq. (4-53) to determine the value of T. For one log cycle of time, $\Delta \log(t) = 1$ and the corresponding Δs should be taken from the graph. More than one log cycle can also be used.

Step 5: With the known values of T, t_0, and r, calculate the storage coefficient (S) from Eq. (4-50).

Step 6: Check the condition if $u < 0.01$.

Cautions and Limitations of the Method

1. The method should be repeated for all available data of the observation wells having different radial distances from the pumped well. The calculated transmissivity (T) and storage coefficient (S) values for each observation well should be in close agreement.
2. After determination of the values of T and S, they should be introduced into the equation $u = r^2S/(4Tt)$ to check if $u < 0.01$, which is the key condition for the applicability of the Jacob method.
3. Before using Eqs. (4-50) and (4-53), all quantities must be expressed in the same set of units. For example, the units of time and transmissivity may be expressed as minutes and m^2/day (or ft^2/day), respectively.

Example 4-7: Using the data of Example 4-6, determine the transmissivity (T) and storage coefficient (S) of the aquifer with the Cooper and Jacob method. The measured data for $r = 25$ m and 75 m are given in Table 4-4a and Table 4-4b, respectively. The steady-state flow rate is 540 m^3/day.

Solution:

Step 1: The graphs for the data of the the observation wells having radial distances r = 25 m and 75 m are shown in Figures 4-8 and 4-9, respectively (t on logarithmic scale). Straight lines are drawn through the points of each observation well.

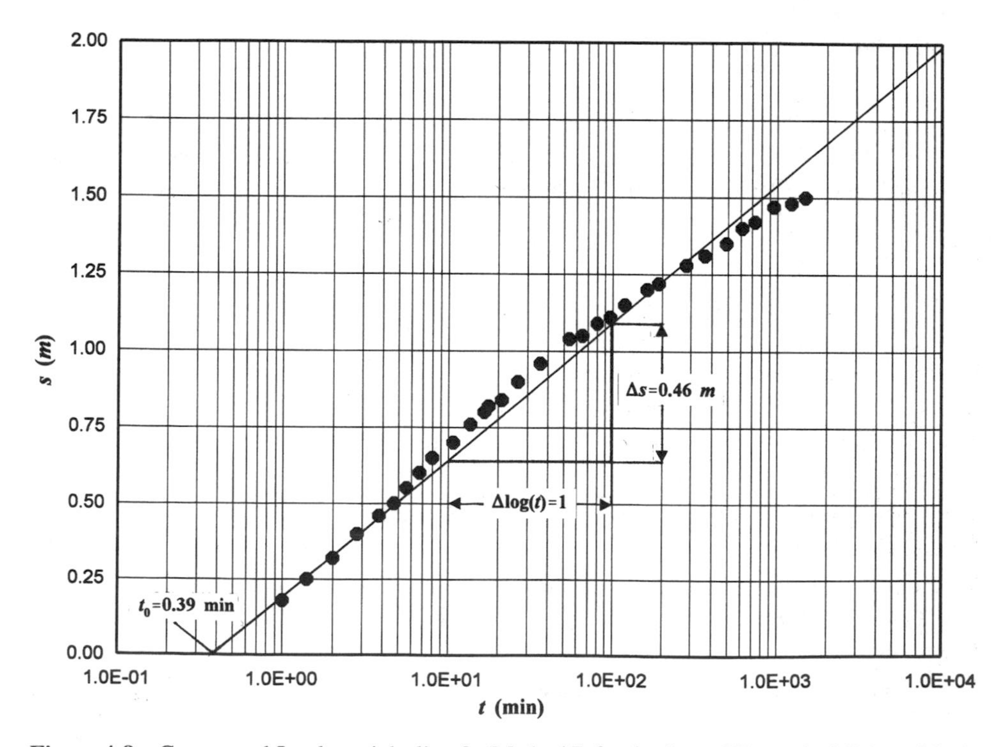

Figure 4-8 Cooper and Jacob straight-line fit (Method I) for the data of Example 4-7 (r = 25 m).

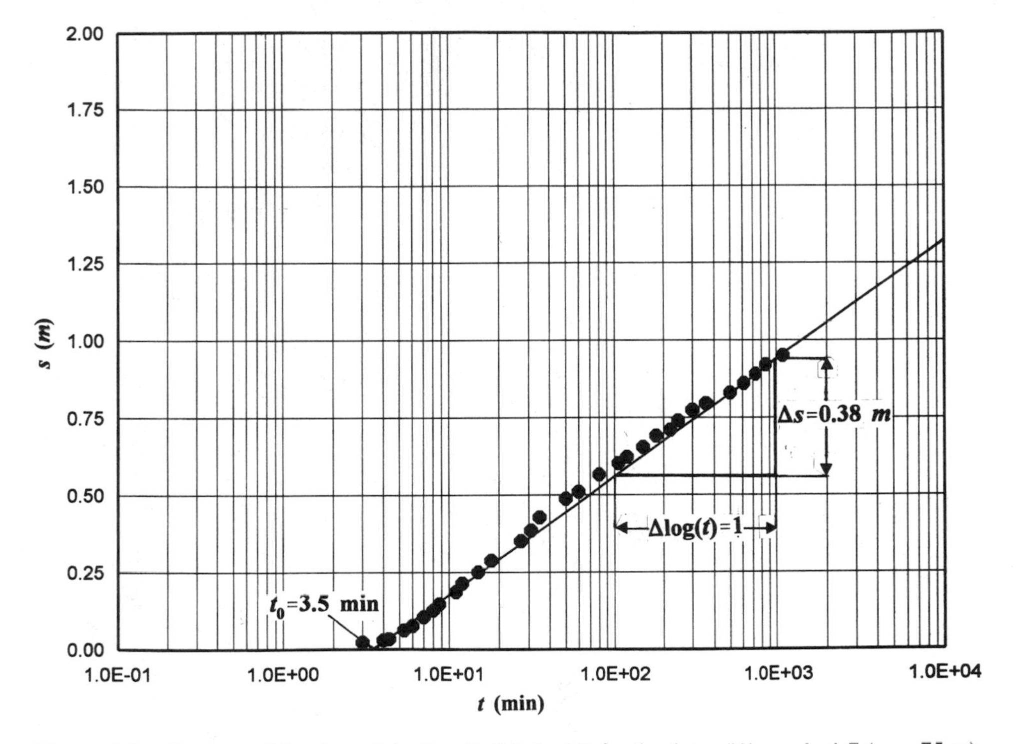

Figure 4-9 Cooper and Jacob straight-line fit (Method I) for the data of Example 4-7 ($r = 75$ m).

Step 2: The interception points of the straight lines with the time axis are

$$t_0 = 0.39 \text{ min} \qquad \text{for } r = 25 \text{ m}$$
$$t_0 = 3.50 \text{ min} \qquad \text{for } r = 75 \text{ m}$$

Step 3: The corresponding values for geometric slopes are:

$$\Delta s = 0.46 \text{ m}, \qquad \Delta \log(t) = 1 \qquad \text{for } r = 25 \text{ m}$$
$$\Delta s = 0.38 \text{ m}, \qquad \Delta \log(t) = 1 \qquad \text{for } r = 75 \text{ m}$$

Step 4: Substitution of the values of Step 3 into Eq. (4-53) with $Q = 540\text{m}^3/\text{day}$ gives

$$T = \frac{(2.303)(540 \text{ m}^3/\text{day})}{(4\pi)(0.46 \text{ m})}(1) = 215 \text{ m}^2/\text{day} \qquad \text{for } r = 25 \text{ m}$$

$$T = \frac{(2.303)(540 \text{ m}^3/\text{day})}{(4\pi)(0.38 \text{ m})} = 260 \text{ m}^2/\text{day} \qquad \text{for } r = 75 \text{ m}$$

Step 5: Introducing the values of T, t_0, and r in Eq. (4-50) gives

$$S = \frac{(2.25)(215 \text{ m}^2/\text{day})}{(25 \text{ m})^2}\left(\frac{0.39}{24 \times 60} \text{ day}\right) = 2.1 \times 10^{-4} \qquad \text{for } r = 25 \text{ m}$$

$$S = \frac{(2.25)(260\ \text{m}^2/\text{day})}{(75\ \text{m})^2}\left(\frac{3.50}{24 \times 60}\ \text{day}\right) = 2.5 \times 10^{-4} \qquad \text{for } r = 75\ \text{m}$$

Step 6: Substitution of the values of T, S, and r in $u = r^2S/(4Tt)$ gives

$$u = \frac{(25\ \text{m})^2(2.1 \times 10^{-4})}{(4)(215\ \text{m}^2/\text{day})t} = \frac{1.53 \times 10^{-4}}{t} \qquad \text{for } r = 25\ \text{m}$$

$$u = \frac{(75\ \text{m})^2(2.5 \times 10^{-4})}{(4)(260\ \text{m}^2/\text{day})t} = \frac{13.52 \times 10^{-4}}{t} \qquad \text{for } r = 75\ \text{m}$$

Hence, $t > 1.53 \times 10^{-2}$ day or $t > 22$ min, which satisfies $u < 0.01$ for $r = 25$ m; $t > 195$ min satisfies the same criterion for $r = 75$ m.

METHOD II. Equation (4-46) can also be expressed as

$$s = \frac{2.303Q}{4\pi T}\log\left(\frac{2.25Tt}{S}\right) - \frac{2.303Q}{2\pi T}\log(r) \tag{4-55}$$

Taking its derivative with respect to r gives

$$\Delta s = \left|\frac{2.303Q}{2\pi T}\Delta \log(r)\right| \tag{4-56}$$

or

$$T = \frac{2.303Q}{2\pi \Delta s}\Delta \log(r) \tag{4-57}$$

The storage coefficient (S) can be determined from Eq. (4-50) by replacing t for t_0:

$$S = \frac{2.25Tt_0}{r_0^2} \tag{4-58}$$

The steps of the method are given below:

Step 1: Plot the data on single logarithmic paper s versus r (r on logarithmic scale) for a constant value of time (t). Then, draw a straight line through the points.

Step 2: Find the interception point of the straight line with the r axis where $s = 0$. Read the value of r_0.

Step 3: Determine the geometric slope of the straight line—that is, the drawdown difference Δs per log cycle of r.

Step 4: Using the known values of Q, Δs, and $\Delta \log(r)$, determine T from Eq. (4-57). Then, using T, t, and r_0, determine S from Eq. (4-58).

Cautions and Limitations of the Method.

1. Special attention shoud be paid in determining Δs in Eq. (4-57).
2. At least two observation wells of data are necessary in order to get reliable results.

3. If the drawdowns measured at different observation wells do not correspond to the same time, the drawdowns at selected times should be determined from the drawdown versus time relationship used in Method I.
4. The method should be applied for several values of t. The values of T and S should agree with a reasonable approximation.

Example 4-8: Using the data of Example 4-6, determine the transmissivity (T) and storage coefficient (S) of the aquifer. Use $t = 150$ min and 300 min as the constant values for the time.

Solution:

Step 1: For plotting the data, the corresponding drawdowns should be determined. The determined values are given below:

$r = 25$ m, $t = 150$ min: $s = 1.17$ m (from Figure 4-8)

$r = 75$ m, $t = 150$ min: $s = 0.654$ m (from Table 4-4b)

$r = 25$ m, $t = 300$ min: $s = 1.31$ m (from Figure 4-8)

$r = 75$ m, $t = 300$ min: $s = 0.774$ m (from Table 4-4b)

The graphs are given in Figure 4-10.

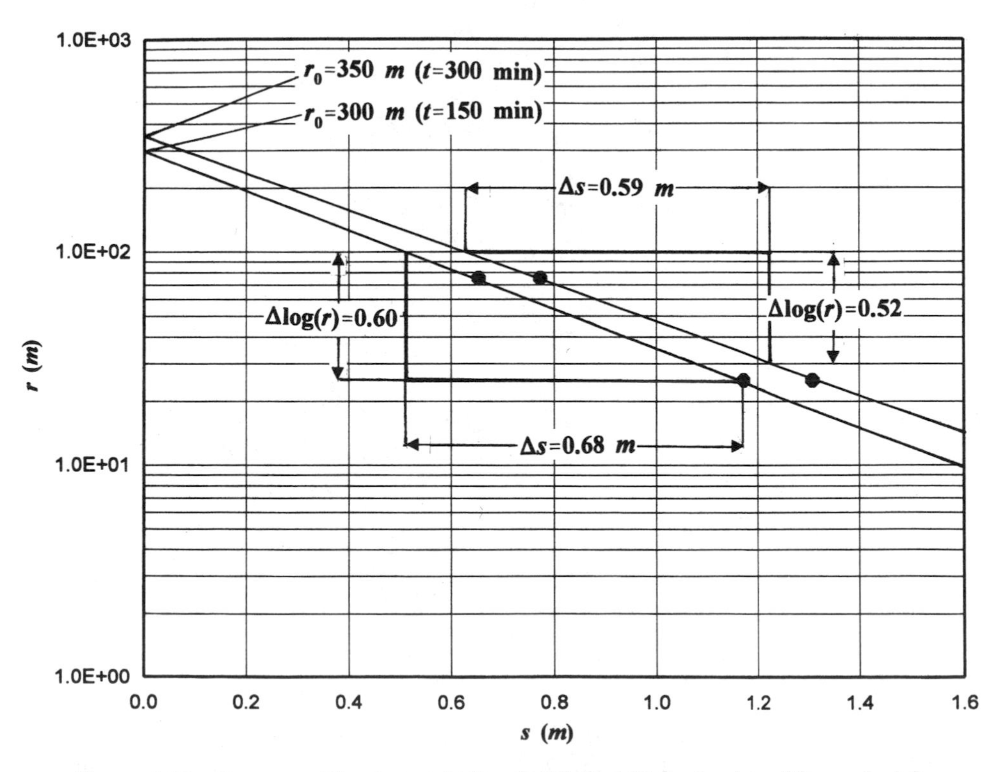

Figure 4-10 Cooper and Jacob straight-line fit (Method II) for the data of Example 4-8.

Step 2: From Figure 4-10, the interception points of the straight lines with the r axis are as follows:

$$r_0 = 300 \text{ m} \qquad \text{for } t = 150 \text{ min}$$

$$r_0 = 350 \text{ m} \qquad \text{for } t = 300 \text{ min}$$

Step 3: The geometric slopes of the straight lines can be determined as

$$\Delta s = 0.68 \text{ m}, \quad \Delta \log(r) = 0.60 \qquad \text{for } t = 150 \text{ min}$$

$$\Delta s = 0.59 \text{ m}, \quad \Delta \log(r) = 0.52 \qquad \text{for } t = 300 \text{ min}$$

Step 4: Substitution of the values of Step 3 in Eqs. (4-57) and (4-58) with $Q = 540 \text{ m}^3/\text{day}$ gives

$$T = \frac{(2.303)(540 \text{ m}^3/\text{day})}{(2\pi)(0.68 \text{ m})}(0.60) = 182 \text{ m}^2/\text{day} \qquad t = 150 \text{ min}$$

$$T = \frac{(2.303)(540 \text{ m}^3/\text{day})}{(2\pi)(0.59 \text{ m})}(0.52) = 174 \text{ m}^2/\text{day} \qquad t = 300 \text{ min}$$

$$S = \frac{(2.25)(182 \text{ m}^2/\text{day})}{(300 \text{ m})^2} \frac{150}{(24)(60)} \text{ day} = 4.74 \times 10^{-4} \qquad t = 150 \text{ min}$$

$$S = \frac{(2.25)(174 \text{ m}^2/\text{day})}{(350 \text{ m})^2} \frac{300}{(24)(60)} \text{ day} = 6.66 \times 10^{-4} \qquad t = 300 \text{ min}$$

METHOD III. Equation (4-47) may be rewritten as

$$s = \frac{2.303Q}{4\pi T}\left[\log\left(\frac{2.25T}{S}\right) + \log\left(\frac{t}{r^2}\right)\right] \tag{4-59}$$

Equation (4-59) represents a straight line on single logarithmic paper s versus $\log(t/r^2)$. From Eq. (4-59) we obtain

$$\Delta s = \frac{2.303Q}{4\pi T}\Delta\left[\log\left(\frac{t}{r^2}\right)\right] \tag{4-60}$$

Per log cycle of t/r^2, Eq. (4-60) takes the form

$$\Delta s = \frac{2.303Q}{4\pi T}, \Delta\left[\log\left(\frac{t}{r^2}\right)\right] = 1 \tag{4-61}$$

From Eq. (4-60), the transmissivity (T) is

$$T = \frac{2.303Q}{4\pi \Delta s}\Delta\left[\log\left(\frac{t}{r^2}\right)\right] \tag{4-62}$$

The straight line intercepts the time axis at $s = 0$. Consequently, the interception point has the coordinates $s = 0$ and $t/r^2 = (t/r^2)_0$. Introducing these values into Eq. (4-46) gives

$$0 = \frac{2.303Q}{4\pi T}\log\left[\left(\frac{2.25T}{S}\right)\left(\frac{t}{r^2}\right)_0\right] \tag{4-63}$$

Because $2.303Q/(4\pi T)$ is not equal to zero, it follows that

$$\left(\frac{2.25T}{S}\right)\left(\frac{t}{r^2}\right)_0 = 1 \tag{4-64}$$

or solving it for S gives

$$S = 2.25T\left(\frac{t}{r^2}\right)_0 \tag{4-65}$$

This method uses all data points of the observation wells on single paper. The steps of the method are given below:

Step 1: Plot all observation wells data on a single logarithmic paper between s and t/r^2 (logarithmic). Then draw a straight line through the points.
Step 2: Find the interception point of the straight line with the t/r^2 axis where $s = 0$. Read the coordinates of the interception point: $t/r^2 = (t/r^2)_0$, $s = 0$.
Step 3: Determine the geometric slope of the straight line.
Step 4: Calculate T from Eqs. (4-61) or (4-62). Then, calculate S from Eq. (4-65).

Example 4-9: Using the data of Example 4-6, determine the transmissivity (T) and storage coefficient of the aquifer.

Solution:

Step 1: The data, drawdown (s) versus t/r^2, are given in Tables 4-4a and 4-4b for $r = 25$ m and $r = 75$ m, respectvely. The graphical form of these data is given in Figure 4-11. A straight line is drawn through these points.
Step 2: The interception point of the line with the t/r^2 axis is $(t/r^2)_0 = 6 \times 10^{-4}$ min/m^2.
Step 3: From Figure 4-11, the corresponding values for the geometric slope of the fitted straight line are

$$\Delta s = 0.43 \text{ m}, \qquad \Delta[\log(t/r^2)] = 1$$

Step 4: Substitution of the values of Step 3 into Eq. (4-62) with $Q = 540$ m^3/day gives

$$T = \frac{(2.303)(540 \text{ m}^3/\text{day})}{(4\pi)(0.43 \text{ m})} = 230 \text{ m}^2/\text{day}$$

Introducing the values in Eq. (4-65) gives the storage coefficient:

$$S = (2.25)(230 \text{ m}^2/\text{day})\left(\frac{6 \times 10^{-4}}{24 \times 60} \text{ m}^2/\text{day}\right) = 2.16 \times 10^{-4}$$

4.2.7.3 Chow Method. The method was established by Chow (1952) based on the Theis model. The advantages of the Chow method can be summarized as follows: (1) The method does not require curve fitting as the Theis method, and (2) This method does not have restrictions for the values of r and t as the Jacob method. In the following sections, its theory and the associated data analysis method are given.

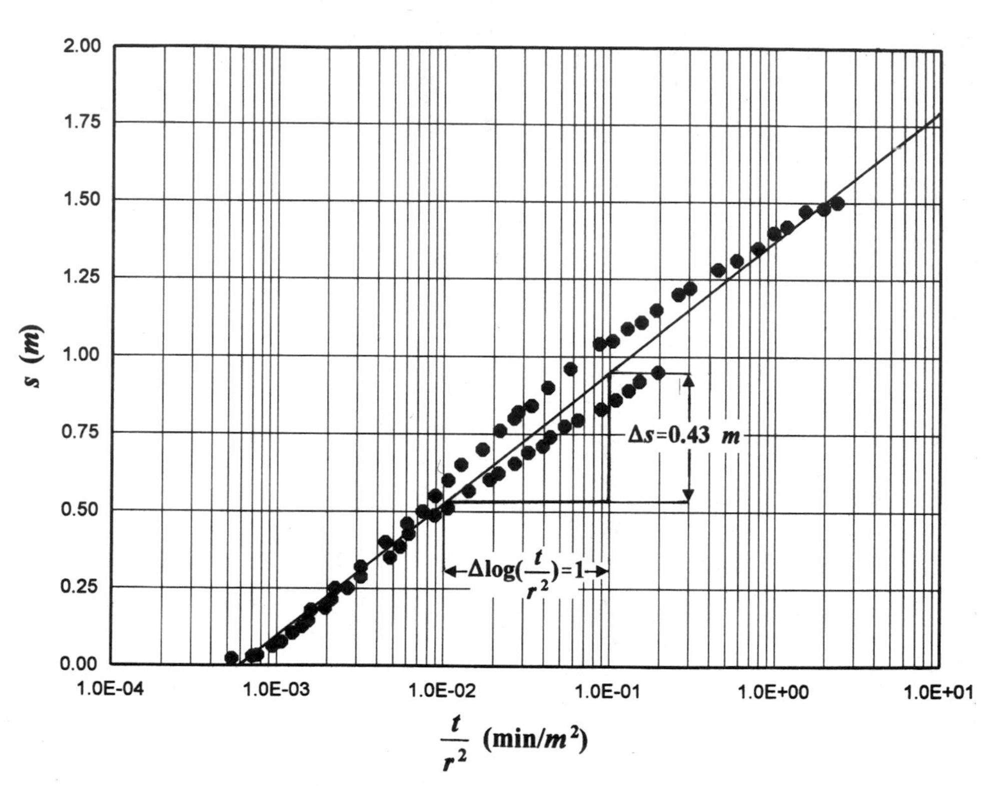

Figure 4-11 Cooper and Jacob straight-line fit (Method III) for the data of Example 4-9.

Theory. The main equations of the method are derived from the Theis equations [Eqs. (4-23), (4-24), and (4-25)]. From Eq. (4-23) we obatin

$$\ln(u) = \ln\left(\frac{r^2 S}{T}\right) - \ln(t) \tag{4-66}$$

Taking the derivatives of u, $\ln u$, and $W(u)$ of Eqs. (4-23), (4-66), and (4-26), respectively, the resulting expressions are

$$\frac{du}{d[\ln(t)]} = -u \tag{4-67}$$

$$\frac{d[\ln(u)]}{d[\ln(t)]} = -1 \tag{4-68}$$

$$\frac{d[W(u)]}{d[\ln(t)]} = -\frac{d[\ln(u)]}{d[\ln(t)]} + \frac{du}{d[\ln(t)]}\left(1 - \frac{u}{2!} + \frac{u^2}{3!} - \cdots\right) \tag{4-69}$$

Introducing Eqs. (4-67) and (4-68) into Eq. (4-69) gives

$$\begin{aligned}\frac{d[W(u)]}{d[\ln(t)]} &= 1 - u\left(1 - \frac{u}{2!} + \frac{u^2}{3!} - \cdots\right)\\ &= 1 - u + \frac{u^2}{2!} - \frac{u^3}{3!} + \cdots \\ &= e^{-u}\end{aligned} \tag{4-70}$$

The derivative of s of Eq. (4-24) with respect to $\ln t$ is

$$\frac{ds}{d[\ln(t)]} = \frac{Q}{4\pi T}\frac{d[W(u)]}{d\ln t} \tag{4-71}$$

or, from Eq. (4-70),

$$\frac{ds}{d[\ln(t)]} = \frac{Q}{4\pi T}e^{-u} \tag{4-72}$$

When we convert natural logarithms (ln) to common logarithms (log) [i.e., $\ln u = 2.303 \log u$], Eq. (4-72) becomes

$$\frac{ds}{d[\log(t)]} = \frac{2.303Q}{4\pi T}e^{-u} \tag{4-73}$$

Dividing Eq. (4-24) by Eq. (4-73) gives

$$\frac{s}{\dfrac{ds}{d[\log(t)]}} = W(u)\frac{e^u}{2.303} \tag{4-74}$$

Let

$$F(u) = \frac{s}{\dfrac{ds}{d[\log(t)]}} \tag{4-75}$$

Then

$$F(u) = W(u)\frac{e^u}{2.303} \tag{4-76}$$

Therefore, for a given value of u, the values of $W(u)$ and $F(u)$ can be computed by Eqs. (4-25) and (4-76), respectively. Consequently, a relation between $F(u)$ and $W(u)$ can be plotted as shown by Figure 4-12.

Data Analysis Method. Based on the equations presented above, the steps of the Chow method are given below (Chow, 1952):

Step 1: Plot the measured data for a well on single logarithmic paper (t on logarithmic scale) as drawdown (s) versus time (t). Draw a curve through the points.

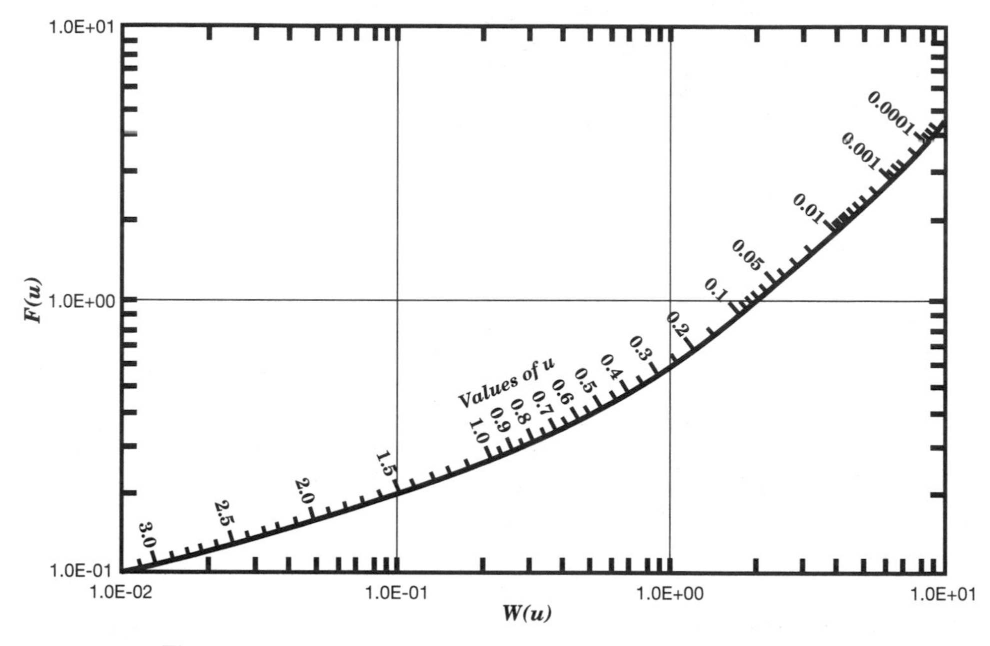

Figure 4-12 Relation among $F(u)$, $W(u)$, and u. (After Chow, 1952).

Step 2: Select an arbitrary point A on the curve and draw a tangent to the curve at that point.

Step 3: Read the drawdown for point $A(s_A)$ and the geometric slope of the tangent line—that is, the drawdown difference per log cycle of time, Δs.

Step 4: With the values of s_A and Δs, calculate the value of $F(u)$ using the finite-difference form of Eq. (4-75)

$$F(u) = \frac{s_A}{\dfrac{\Delta s}{\Delta[\log(t)]}} \tag{4-77}$$

or, with the drawdown difference per log cycle [i.e., $\Delta[\log(t)] = 1$],

$$F(u) = \frac{s_A}{\Delta s} \tag{4-78}$$

Step 5: With the known value of $F(u)$, find the corresponding value of u from Figure 4-12 and $W(u)$ from Figure 4-12 or Table 4-1.

Step 6: Read the value of t_A from the data curve and substitute the values in Eqs. (4-24) and (4-23) to solve for the transmissivity (T) and storage coefficient (S), respectively.

Example 4-10: Using the Chow method, determine the transmissivity (T) and storage coefficient (S) for the data at $r = 25$ m of Example 4-6 (Table 4-4a).

Solution:

Step 1: The graph for the data is presented in Figure 4-13.

Step 2: On the curve in Figure 4-13, a point A is chosen arbitrarily and a tangent to the curve at this point is drawn.

Step 3: From Figure 4-13, the corresponding values are

$$s_A = 1.0 \text{ m}, \qquad \Delta s = 0.44 \text{ m}, \qquad \Delta \log(t) = 1$$

Step 4: From Eqs. (4-77) or (4-78) we obtain

$$F(u) = \frac{1.0 \text{ m}}{0.44 \text{ m}} = 2.27$$

Step 5: From Figure 4-12, it can be found that $F(u) = 2.27$ corresponds to $u = 0.003$ and from Figure 4-12 or Table 4-1 $W(u) = 5.23$.

Step 6: From Figure 4-13, $t_A = 45$ min. Substituting the values of Step 3, $Q = 540 \text{ m}^3/\text{day}$, and t_A in Eqs. (4-24) and (4-23) gives

$$T = \frac{(540 \text{ m}^3/\text{day})}{(4\pi)(1.0 \text{ m})}(5.23) = 225 \text{ m}^2/\text{day}$$

$$S = \frac{(4)(225 \text{ m}^2/\text{day})(0.003)}{(25 \text{ m})^2}\left(\frac{45}{24 \times 60} \text{ day}\right) = 1.35 \times 10^{-4}$$

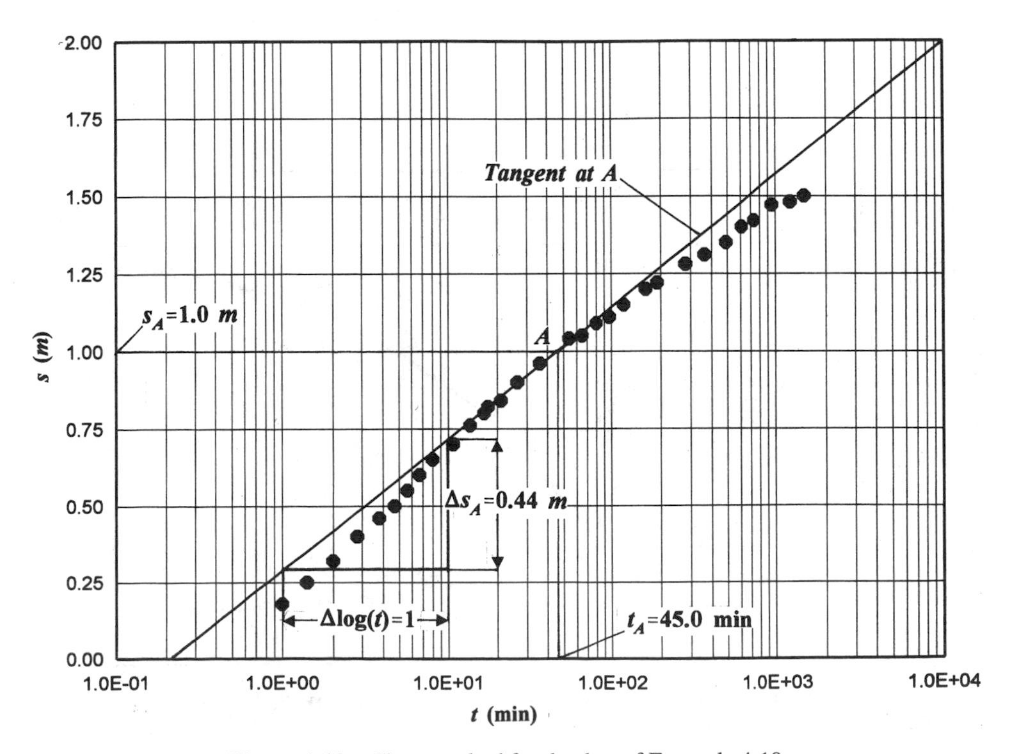

Figure 4-13 Chow method for the data of Example 4-10.

4.2.7.4 Theis Recovery Method. After pumping of water from a confined aquifer is stopped, the observed drawdown at an observation well starts to decrease and eventually approaches zero. This decreasing drawdown is called *residual drawdown*. Residual drawdown data contain valid information regarding the hydrogeological behavior of the aquifer. It is a common practice to record the residual drawdown data once the extraction of water from the aquifer is stopped. Sometimes, mulfunction of the recording equipment or accidental stopping of the pumping equipment may impair the value of the drawdown data. Consequently, the success of the aquifer test may then depend on recovery data collected after the pumping stopped.

The theory of the recovery methods for nonleaky confined aquifers are based on Theis (1935) and Cooper and Jacob (1946). In the following sections the theory of the Theis recovery method will be presented first. Then, two methods will be presented to determine aquifer hydrogeological parameters. With the first method, only the transmissivity can be determined. Both transmissivity and storage coefficient can be determined with the second method.

Theory. If a well is pumped at a constant rate for a known period of time and then allowed to recover, the residual drawdown at any instant will be the same as if the discharge of the well had been continued but a recharge well with the same flow rate had been introduced at the same point at the instant the discharge stopped (Theis, 1935; Ferris et al., 1962). The schematic time-drawdown and time-residual drawdown diagram is shown in Figure 4-14. Some explanatory information about this figure and the derivation of residual drawdown equation are given in the following paragraphs.

In Figure 4-14, t is the time since pumping started, and t' is the time since pumping stopped. h_0 is the initial value of the hydraulic head before pumping started, then $h_0 - h' = s'$

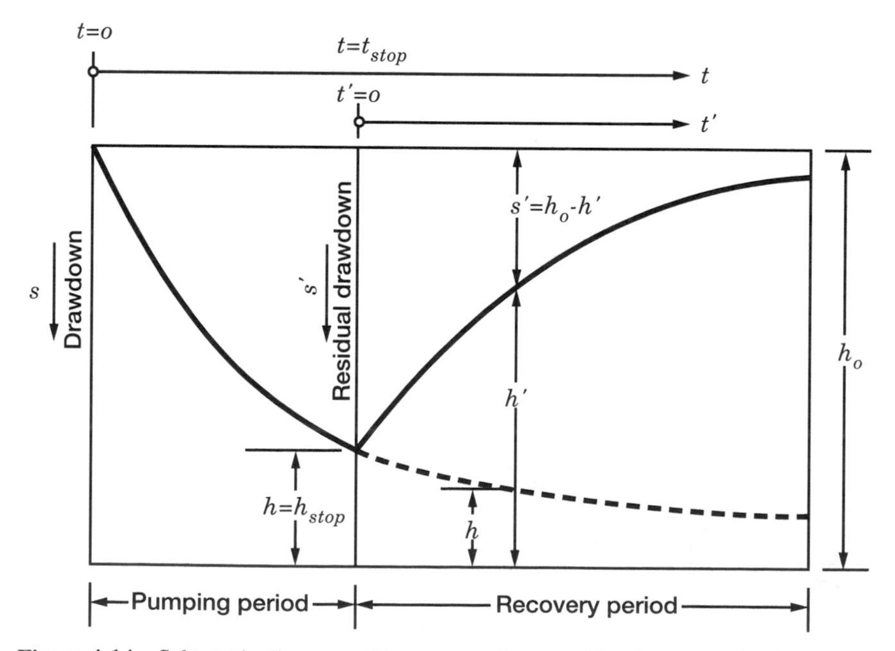

Figure 4-14 Schematic diagram of the recovery in an aquifer after extraction is stopped.

is called the *residual drawdown*. The hydraulic head will recover from its lowest value $h = h_{stop}$ when pumping stopped. The residual drawdown equation can be derived with the application of the superposition principle. This means that an imaginary injection well with the same constant flow rate is superimposed on the same well. The two flow rates represent an idle well because they cancel each other. Thus, the residual drawdown can be expressed as

$$s' = h_0 - h - (h - h') \tag{4-79}$$

or, using the Theis equation as given by Eq. (4-24),

$$s' = \frac{Q}{4\pi T}[W(u) - W(u')] \tag{4-80}$$

where u is given by Eq. (4-23) and

$$u' = \frac{r^2 S}{4Tt'} \tag{4-81}$$

For u' sufficiently small, the approximation made in the derivation of Eq. (4-46) is also valid here, and therefore

$$\begin{aligned} s' &= \frac{2.303Q}{4\pi T}\left[\log\left(\frac{2.25Tt}{r^2 S}\right) - \log\left(\frac{2.25Tt'}{r^2 S}\right)\right] \\ &= \frac{2.303Q}{4\pi T}\log\left(\frac{t}{t'}\right) \end{aligned} \tag{4-82}$$

Data Analysis Method for Transmissivity. Equation (4-82) is similar to the modified Theis equation, Eq. (4-46), developed by Cooper and Jacob. This equation permits the determination of transmissivity (T) of an aquifer from the observation of the rate of recovery of water level in a pumped well, or in an observation well close to the pumping well. The steps of the method are given below (Ferris et al., 1962; Kruseman and De Ridder, 1991):

Step 1: Plot the data of the observation well as residual drawdown (s') versus time (t/t') on a single sheet of logarithmic paper (t/t' on logarithmic scale). Draw a straight line through the points.

Step 2: From Eq. (4-82), the slope of the line is

$$\frac{\Delta s'}{\Delta\left[\log\left(\frac{t}{t'}\right)\right]} = \frac{2.303Q}{4\pi T} \tag{4-83}$$

and solving for T gives

$$T = \frac{2.303Q}{4\pi \Delta s'}\Delta\left[\log\left(\frac{t}{t'}\right)\right] \tag{4-84}$$

Determine the slope of the straight line (i.e., the residual drawdown difference, $\Delta s'$), per log cycle—that is, $\Delta \log(t/t') = 1$. When we use the English system, Eq. (4-84) takes the following form:

$$T = \frac{264Q}{\Delta s'} \tag{4-85}$$

where T is the transmissivity (gpd/ft), Q is the pumping rate (gpm), and $\Delta s'$ is the geometric slope of the straight line expressed as the change in residual drawdown over one log cycle.

Substitute the values either in Eqs. (4-84) or (4-85) to calculate the transmissivity (T).

Cautions and Limitations of the Method. There are some cautions and limitations for the application of the Theis recovery method. Some important points regarding this method are given below (Ferris et al., 1962; Driscoll, 1986; Kruseman and De Ridder, 1991).

1. If observation wells are not available, the recovery data from the pumped well is the only way for determining the transmissivity of the aquifer under both confined and unconfined aquifer conditions (see Section 9.5.4.1 of Chapter 9 for unconfined aquifers).
2. The residual drawdown plot cannot be used for the determination of the storage coefficient, despite the fact that the same plot is valid for calculating the transmissivity.
3. If measurements are made in at least one observation well during the recovery period, the storage coefficient can be determined from portions of these data. The data must be plotted as $(s - s')$ versus t'.
4. If a geologic boundary has been intercepted by the cone of depression during pumping, it may be reflected in the rate of recovery of the pumped well, and the value of transmissivity (T) determined by the recovery formula could be in error.

Example 4-11: Recovery data for a well is given in Table 4-5. The flow rate is 1123 m^3/day. Using the Theis recovery method, determine the transmissivity of the aquifer.

Solution:

Step 1: The graph for the data is given in Figure 4-15.
Step 2: From Figure 4-15, the geometric slope values are as follows:

$$\Delta s' = 1.07 \text{ m}, \qquad \Delta \log(t/t') = 1$$

Substitution of these values into Eq. (4-84) gives

$$T = \frac{(2.303)(1123 \text{ m}^3/\text{day})}{(4\pi)(1.07 \text{ m})}(1) = 192 \text{ m}^2/\text{day}$$

Data Analysis Method for Transmissivity and Storage Coefficient. In the previous method (Section 4.2.7.4.2), only the transmissivity (T) can be determined. In this method, an equation for the storage coefficient (S) does not exist. Here, both the storage coefficient and transmissivity equations will be derived based on the projected drawdown data after the extraction of water is stopped (Chapius, 1992a).

TABLE 4-4 Recovery Data for Example 4-11

Time Since Pumping Began, t (min)	Time Since Pumping Stopped, t' (min)	Ratio of Times, t/t'	Residual Drawdown s' (m)
0	—	—	—
44	0	—	2.44
45	1	45.0	1.69
46	2	23.0	1.30
47	3	15.7	1.14
48	4	12.0	1.01
49	5	9.8	0.91
50	6	8.3	0.84
51	7	7.3	0.78
52	8	6.5	0.73
54	10	5.4	0.63
56	12	4.7	0.56
58	14	4.1	0.51
61	17	3.6	0.43
64	20	3.2	0.38
67	23	2.9	0.33
71	26	2.7	0.29
75	31	2.4	0.24
80	36	2.2	0.21
88	44	2.0	0.15

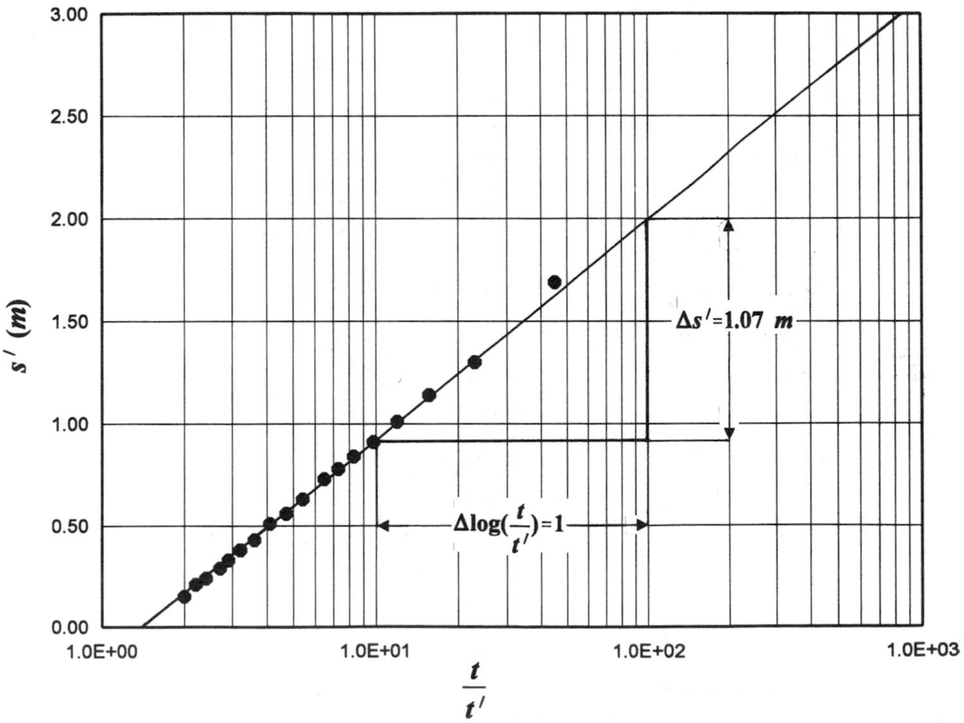

Figure 4-15 Theis recovery method straight-line fit for the data of Example 4-11.

As shown previously, if the times t and t' are large enough, the residual drawdown (s') can be expressed by Eq. (4-82). As can be seen from Eq. (4-82), the residual drawdown (s') is only function of Q, T, and t/t' and is independent of S. This means that S cannot be determined by this method. An equation for S is presented below based on the *projected drawdown*.

A schematic representation of the projected drawdown is shown in Figure 4-16. In Figure 4-16, s_p is the pumping period drawdown projected to time t' and is defined as

$$s_p = \frac{Q}{4\pi T} \ln\left(\frac{t}{t_0}\right) = \frac{2.303Q}{4\pi T} \log\left(\frac{t}{t_0}\right) \tag{4-86}$$

where t_0 is the intercept on the time axis for the pumping period. Equation (4-86) is obtained at any time t' by extrapolating the Cooper–Jacob-type straight-line graph. Since

$$\frac{2.303Q}{4\pi T} = \frac{\Delta s}{\Delta\left[\log\left(\frac{t}{t_0}\right)\right]} \tag{4-87}$$

Eq. (4-86) can be written as

$$s_p = \Delta s \frac{\log\left(\frac{t}{t_0}\right)}{\Delta\left[\log\left(\frac{t}{t_0}\right)\right]} \tag{4-88}$$

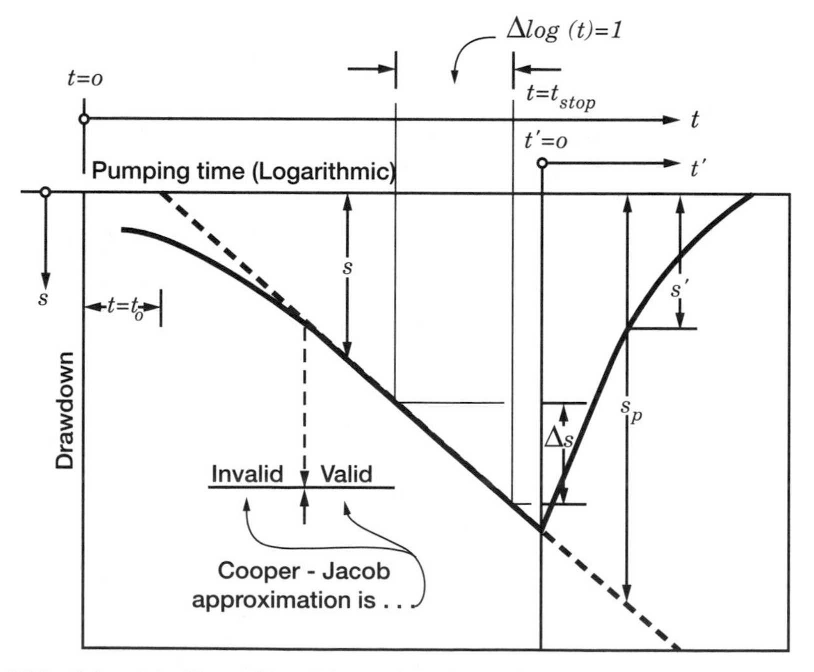

Figure 4-16 Schematic illustration of the projected drawdown (s_p) after extraction is stopped. (Adapted from Chapius, 1992a.)

or, with $\Delta[\log(t/t_0)] = 1$ per log cycle,

$$s_p = \Delta s \log\left(\frac{t}{t_0}\right) \tag{4-89}$$

Then, the difference $(s_p - s')$ from Eqs. (4-82) and (4-86) is

$$s_p - s' = \frac{2.303Q}{4\pi T} \log\left(\frac{t'}{t_0}\right) \tag{4-90}$$

Similarly, from Eq. (4-90) one can write

$$\Delta(s_p - s') = \frac{2.303Q}{4\pi T} \Delta\left[\log\left(\frac{t'}{t_0}\right)\right] \tag{4-91}$$

and combination with Eq. (4-90) gives

$$s_p - s' = \Delta(s_p - s') \frac{\log\left(\frac{t'}{t_0}\right)}{\Delta\left[\log\left(\frac{t'}{t_0}\right)\right]} \tag{4-92}$$

and for one log cycle we have

$$s_p - s' = \Delta(s_p - s') \log\left(\frac{t'}{t_0}\right) \tag{4-93}$$

Equation (4-93) indicates a straight-line relationship between $(s_p - s')$ and $\log(t')$ with a slope $\Delta(s_p - s')$ over one log cycle and a time intercept $t = t_0$ when $(s_p - s') = 0$. Equation (4-93) is equivalent to the Cooper and Jacob equation, as given by Eq. (4-82), for the pumping period drawdown, and it is valid for similar conditions on either u', given by Eq. (4-81), or t'.

Consequently, when $(s_p - s')$ values are plotted against $\log(t')$, as shown schematically in Figure 4-16, the transmissivity (T) can be obtained from Eq. (4-91) as

$$T = \frac{2.303Q}{4\pi\Delta(s_p - s')} \Delta\left[\log\left(\frac{t'}{t_0}\right)\right] \tag{4-94}$$

or, for one log cycle,

$$T = \frac{2.303Q}{4\pi\Delta(s_p - s')} \tag{4-95}$$

Then, the storage coefficient (S) can be obtained from Eq. (4-50) as

$$S = \frac{2.25Tt_0'}{r^2} \tag{4-96}$$

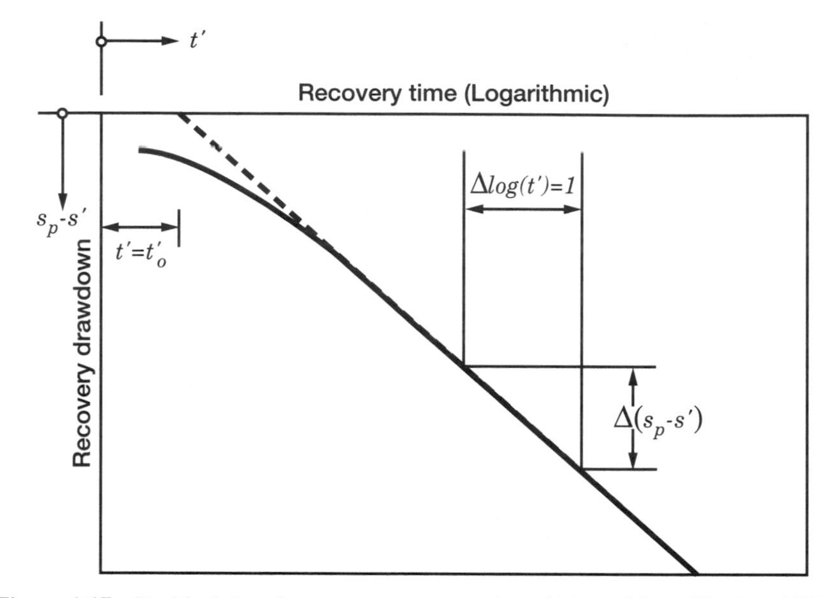

Figure 4-17 Residual drawdown versus recovery time. (Adapted from Chapius, 1992a.)

The schematic representation shown in Figure 4-17 is similar to that of the Cooper and Jacob (1946) representation in Figure 4-14. As a result, the equations for both T and S are similar.

The equation for storage coefficient (S) in the U.S. Department of the Interior (USDI) *Ground Water Manual* [1981, p. 115, Eq. (24)] has a different form than Eq. (4-50). It can be derived from Eqs. (4-50) and (4-93) as

$$s_p - s' = \Delta(s_p - s') \log\left(\frac{2.25Tt'}{r^2 S}\right) \tag{4-97}$$

or

$$\frac{2.25Tt'}{r^2 S} = \log^{-1}\left[\frac{s_p - s'}{\Delta(s_p - s')}\right] \tag{4-98}$$

which yield the the aforementioned USDI *Ground Water Manual* equation:

$$S = \frac{2.25Tt'}{r^2} \frac{1}{\log^{-1}\left[\dfrac{s_p - s'}{\Delta(s_p - s')}\right]} \tag{4-99}$$

The steps of the method are as follows:

Step 1: Plot the data of one of the observation wells on single logarithmic paper (t on logarithmic scale) in accordance to the method (Method I) outlined in Section 4.2.7.2. Draw a straight line through the points.

Step 2: Plot the residual drawdown (s') versus time (t/t') for the same observation well on a single sheet of logarithmic paper (t/t' on logarithmic scale) in accordance of the Theis recovery method as outlined previously. Draw a straight line through the points.

Step 3: Determine the geometric slope of the straight line and the interception point of the same line in Step 1 with the time axis where the drawdown (s) is zero in accordance to the method (Method I) outlined in Section 4.2.7. Then, calculate T and S values from Eqs. (4-53) and (4-50), respectively.

Step 4: Determine the geometric slope of the straight line in Step 2 as outlined previously for the Theis recovery method. Then, calculate T from Eq. (4-84).

Step 5: Using the Δs and t_0 values from Step 3, determine the s_p versus t equation from Eq. (4-88). Then, calculate the s_p values as well as $s_p - s'$ values for each time during the recovery period and tabulate them in the table of the recovery data.

Step 6: Plot the graph ($s_p - s'$) versus t' (on logarithmic scale) and determine its geometric slope and the t'_0 intercept on the time axis.

Step 7: Using the values from the previous steps, calculate the value of S from Eq. (4-96).

Example 4-12: The data for a pumping and recovery tests are given in Table 4-6 and Table 4-7, respectively, which are taken from Todd (1980). The steady-state extraction rate was 2500 m^3/day (28.94 liters/s), and the observation well was 60 m from the pumping well. The pumping well was shut down after 240 min of pumping.

Solution:

Step 1: The graph for the pumping test data, given in Table 4-6, is given in Figure 4-18 (t on logarithmic scale). A straight line is drawn through the points.

Step 2: The residual drawdown versus time graph for the recovery period is given in Figure 4.19. A straight line is drawn through the points.

Step 3: From Figure 4-18 we have

$$\Delta s = 0.40 \text{ m}, \qquad \Delta \log(t) = 1, \qquad t_0 = 0.39 \text{ min}$$

TABLE 4-5 Pumping Test Data for Example 4-12

t (min)	s (m)	t (min)	s (m)
1.0	0.20	24	0.72
1.5	0.27	30	0.76
2.0	0.30	40	0.81
2.5	0.34	50	0.85
3.0	0.37	60	0.90
4	0.41	80	0.93
5	0.45	100	0.96
6	0.48	120	1.00
8	0.53	150	1.04
10	0.57	180	1.07
12	0.60	210	1.10
14	0.63	240	1.12
18	0.67		

TABLE 4-6 Recovery Data and Calculated Values for Example 4-12

t (min)	t' (min)	t/t'	s' (m)	s_p (m)	$s_p - s'$ (m)
241	1	241	0.89	1.12	0.23
242	2	121	0.81	1.12	0.31
243	3	81	0.76	1.12	0.36
245	5	49	0.68	1.12	0.44
247	7	35.3	0.64	1.12	0.48
250	10	25	0.56	1.12	0.56
255	15	17	0.49	1.13	0.64
260	20	13	0.45	1.13	0.68
270	30	9	0.38	1.14	0.76
280	40	7	0.34	1.14	0.80
300	60	5	0.28	1.15	0.87
320	80	4	0.24	1.17	0.93
340	100	3.4	0.21	1.18	0.97
380	140	2.7	0.17	1.20	1.03
420	180	2.3	0.14	1.21	1.07

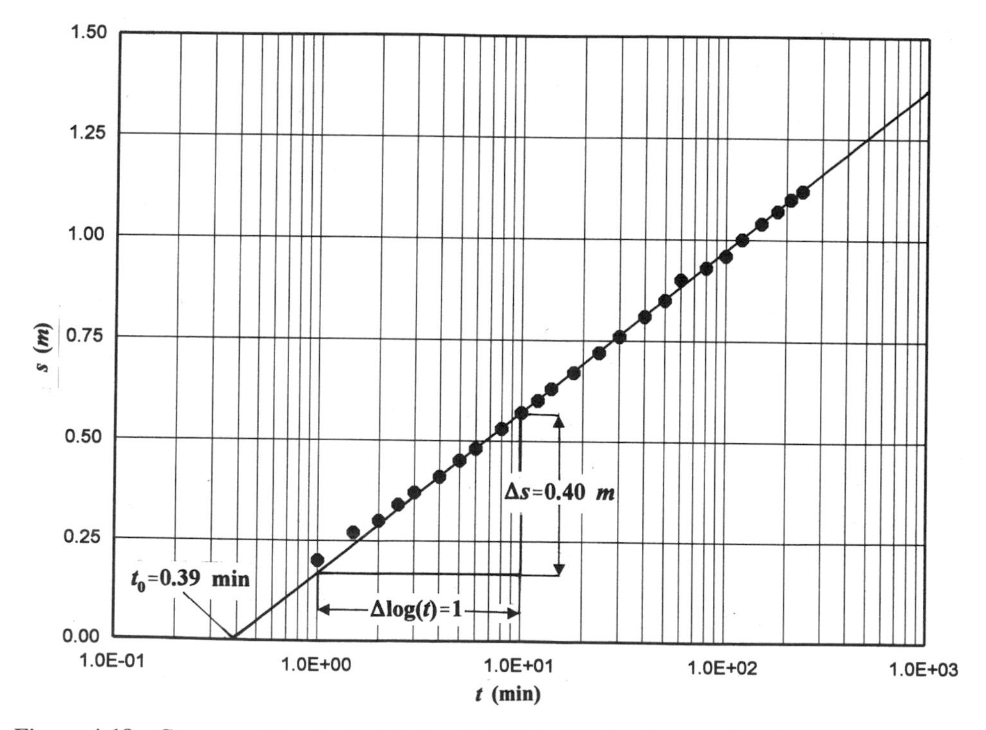

Figure 4-18 Cooper and Jacob straight-line fit (Method I) for the pumping test data of Example 4-12.

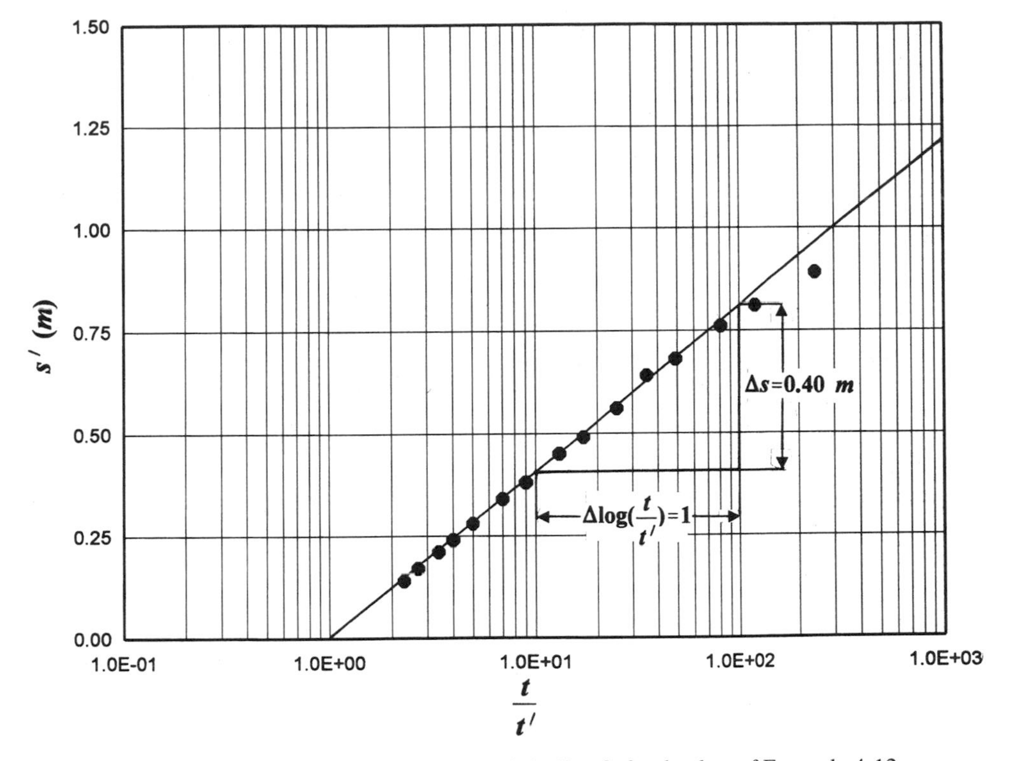

Figure 4-19 Theis recovery method straight-line fit for the data of Example 4-12.

Substitution of these values into Eq. (4-53) gives the value of transmissivity (T) for the pumping test portion:

$$T = \frac{(2.303)(2500\ \text{m}^3/\text{day})}{(4\pi)(0.40\ \text{m})}(1) = 1145\ \text{m}^2/\text{day}$$

And from Eq. (4-50) the value of storage coefficient (S) is

$$S = \frac{(2.25)(1145\ \text{m}^2/\text{day})\left(\dfrac{0.39}{24 \times 60}\ \text{day}\right)}{(60\ \text{m})^2} = 1.94 \times 10^{-4}$$

Step 4: From Figure 4-19 we obtain

$$\Delta s' = 0.40\ \text{m}, \qquad \Delta \log\left(\frac{t}{t'}\right) = 1$$

and substitution of these values into Eq. (4-84) gives the value of T:

$$T = \frac{(2.303)(2500\ \text{m}^3/\text{day})}{(4\pi)(0.40\ \text{m})}(1) = 1145\ \text{m}^2/\text{day}$$

Step 5: Introducing Δs and t_0 values of Step 3 into Eq. (4-89) gives

$$s_p = (0.40 \text{ m}) \log\left(\frac{t}{0.39 \text{ min}}\right)$$

and from this equation the calculated s_p values using the t values in Table 4-7 are included in the same table. The $s_p - s'$ values are included in Table 4-7 as well.

Step 6: The graph of $(s_p - s')$ versus $\log(t')$ is shown in Figure 4-20. This recovery graph yields exactly the same slope $\Delta(s_p - s') = 0.40$ m and the same time intercept $t'_0 = 0.39$ min as the pumping drawdown graph presented in Figure 4-18. Thus, the T and S values are exactly identical for pumping and recovery period.

Step 7: Substitution of the above values into Eq. (4-96) gives the value of storage coefficient (S):

$$S = \frac{(2.25)(1145 \text{ m}^2/\text{day})\left(\frac{0.39}{24 \times 60}\text{ day}\right)}{(60 \text{ m})^2} = 1.94 \times 10^{-4}$$

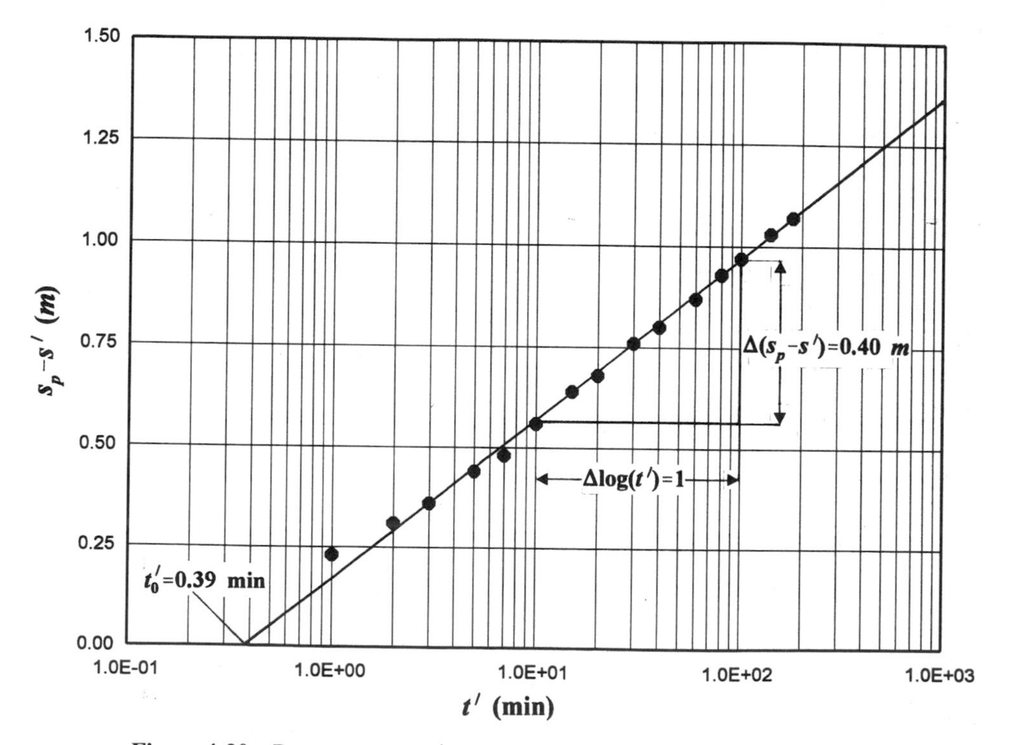

Figure 4-20 Recovery $(s_p - s')$ versus recovery time (t') for Example 4-12.

4.3 LARGE-DIAMETER WELL (DUG WELL): PAPADOPULOS AND COOPER MODEL

The radius of small-diameter wells generally vary between 0.05 m and 0.25 m. As mentioned in Section 4.2, these are represented as line sources in the mathematical models. Although this approach is practically enough for most wells having radii in the above range, it may not be appropriate for large-diameter wells which are in use especially in the Middle Eastern and Asian countries. The radius of these wells generally vary between 0.5 m and 2 m, and may even be higher; these wells are called *dug wells*.

The theory of the Theis equation assumes that the pumped well is a line source. As a matter of fact, almost all well hydraulics models are based on this assumption. This assumption may not be valid if the well bore storage effects are significant. Effects of well bore storage may become important when the aquifer transmissivity and storage coefficient are small or when the pumped well diameter is large. Papadopulos and Cooper (1967) and subsequently Papadopulos (1967) developed analytical solutions and type curves in and around a large-diameter well in a homogeneous and isotropic nonleaky confined aquifer by taking into account the effects of well bore storage in the pumped well. Later, Moench (1985) presented mathematical models that combine the leaky aquifer theory of Hantush (1960b) (see Chapter 7) with the aforementioned theory of flow to a large-diameter well.

In this section, the theory as well as practical application of the Papadopulos and Cooper model will be presented.

4.3.1 Problem Statement and Assumptions

Figure 4-21 shows cross-sectionally a fully penetrating large-diameter well in a nonleaky confined aquifer. Papadopulos and Cooper (1967) developed an analytical solution under transient conditions with the following assumptions:

1. The aquifer is homogeneous and isotropic.
2. The aquifer is horizontal and has a constant thickness (b).
3. Well discharges at a constant rate (Q).
4. The aquifer is not leaky and is infinite in horizontal extent.
5. The well penetrates the entire aquifer thickness.
6. The well losses are negligible.
7. Before pumping, the potentiometric head in the aquifer is the same at every point in the aquifer.
8. Discharge from the well is derived exclusively from storage in the aquifer.
9. Water is immediately released from storage upon the decline of the piezometric head.
10. Storage in the aquifer is proportional to head.

4.3.2 Governing Differential Equation

The governing differential equation is the same as Eq. (4-10) conditionally with the radius of the well as

$$\frac{\partial^2 s}{\partial r^2} + \frac{1}{r}\frac{\partial s}{\partial r} = \frac{S}{T}\frac{\partial s}{\partial t}, \qquad r \geq r_w \tag{4-100}$$

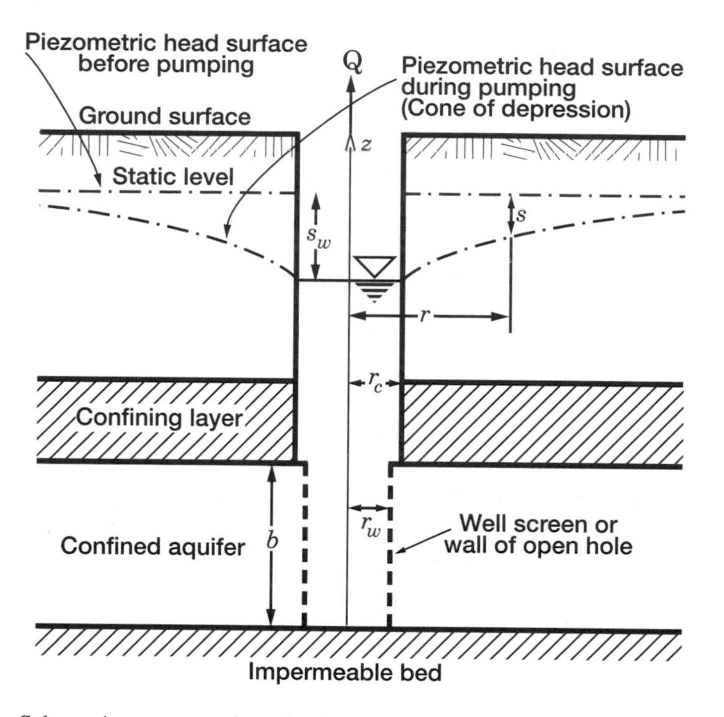

Figure 4-21 Schematic representation of a large-diameter well in a confined aquifer. (After Papadopulos and Cooper, 1967.)

where s is the drawdown in the aquifer at distance r and time t; r is the radial distance from the center of the well; S is the storage coefficient of aquifer; T is the transmissivity which is defined by Eq. (2-24) of Chapter 2, and r_w is the effective radius of well screen or open hole.

4.3.3 Initial and Boundary Conditions

The initial conditions in the aquifer and well itself, respectively, are

$$s(r, 0) = 0, \qquad r \geq r_w \tag{4-101}$$

$$s_w(0) = 0 \tag{4-102}$$

The boundary conditions are

$$s(r_w, t) = s_w(t) \tag{4-103}$$

$$s(\infty, t) = 0 \tag{4-104}$$

$$2\pi r_w T \frac{\partial s(r_w, t)}{\partial t} - \pi r_c^2 \frac{\partial s_w(t)}{\partial t} = -Q \qquad t \geq 0 \tag{4-105}$$

where $s_w(t)$ is the drawdown in the well at time t and r_c is the radius of well in the interval over which the water level declines. Equation (4-101) states that initially the drawdown in the aquifer is zero. Equation (4-102) states that initially the drawdown in the well is zero.

Equation (4-103) states that at any time the drawdown in the aquifer at the face of the well is equal to that in the well. Eq. (4-104) states that at infinity—that is, at large distances from the well—the drawdown is zero. Finally, Eq. (4-105) expresses the fact that the rate of discharge of the well is equal to the sum of the rate of flow of water into the well and the rate of decrease in the volume of water within the well.

4.3.4 Solution

The above-described problem was solved by the Laplace transform method and the following solution was obtained (Papadopulos and Cooper, 1967; Papadopulos, 1967; Reed, 1980):

$$s(r,t) = \frac{Q}{4\pi T} F(u, \alpha, \rho) \tag{4-106}$$

where

$$F(u, \alpha, \rho) = \frac{8\alpha}{\pi} \int_0^{\infty} \frac{C(\beta)}{D(\beta)\beta^2} \, d\beta \tag{4-107}$$

and

$$C(\beta) = \left[1 - \exp\left(-\beta^2 \frac{\rho^2}{4u}\right)\right] [J_0(\beta\rho)A(\beta) - Y_0(\beta\rho)B(\beta)] \tag{4-108}$$

$$A(\beta) = \beta Y_0(\beta) - 2\alpha Y_1(\beta) \tag{4-109}$$

$$B(\beta) = \beta J_0(\beta) - 2\alpha J_1(\beta) \tag{4-110}$$

$$D(\beta) = [A(\beta)]^2 + [B(\beta)]^2 \tag{4-111}$$

$$u = \frac{r^2 S}{4Tt}, \qquad \alpha = \frac{r_w^2 S}{r_c^2}, \qquad \rho = \frac{r}{r_w} \tag{4-112}$$

and J_0 (and Y_0), and Y_1 are zero-order and first-order Bessel functions of the first and second kind, respectively.

Drawdown Inside the Pumped Well. The drawdown inside the pumped well is obtained at $r = r_w$ and can be expressed as

$$s_w(t) = \frac{Q}{4\pi T} F(u_w, \alpha) \tag{4-113}$$

where

$$F(u_w, \alpha) = F(u, \alpha, 1) \tag{4-114}$$

and

$$u_w = \frac{r_w^2 S}{4Tt} \tag{4-115}$$

4.3.4.1 Type Curves. Values of $F(u, \alpha, \rho)$ computed by numerical integration from Eq. (4-107) are given in Table 4-8. Table 4-9 presents values for $F(u_w, \alpha)$ which correspond to $\rho = 1$. In Figure 4-22, the values from Table 4-9 are represented as a family of five curves of $s_w/[Q/(4\pi T)]$ versus $1/u_w$, one curve for each of the five values of the parameter α. The Theis curve corresponding Eq. (4-24) and illustrated in Figure 4-2 is also shown in Figure 4-22.

Properties of the Function $F(u, \alpha, \rho)$. The function $F(u, \alpha, \rho)$ has some important properties and they are presented below (Papadopulos and Cooper, 1967; Papadopulos, 1967).

It is apparent from Figure 4-22 that the drawdown predicted by the Theis solution approximate the drawdown in a well of finite diameter only for relatively large values of time. Papadopulos (1967) stated that upon comparison of Cooper and Papadopulos type-curve data for large-diameter wells (Table 4-8) with the table for Theis type-curve data (Table 4-1) the function $F(u, \alpha, \rho)$ can closely be approximated by

$$F(u, \alpha, \rho) \simeq W(u) \qquad \text{for } t > 2.5\frac{10^3 r_c}{T}, \quad \frac{\alpha \rho^2}{u} > 10^4 \tag{4-116}$$

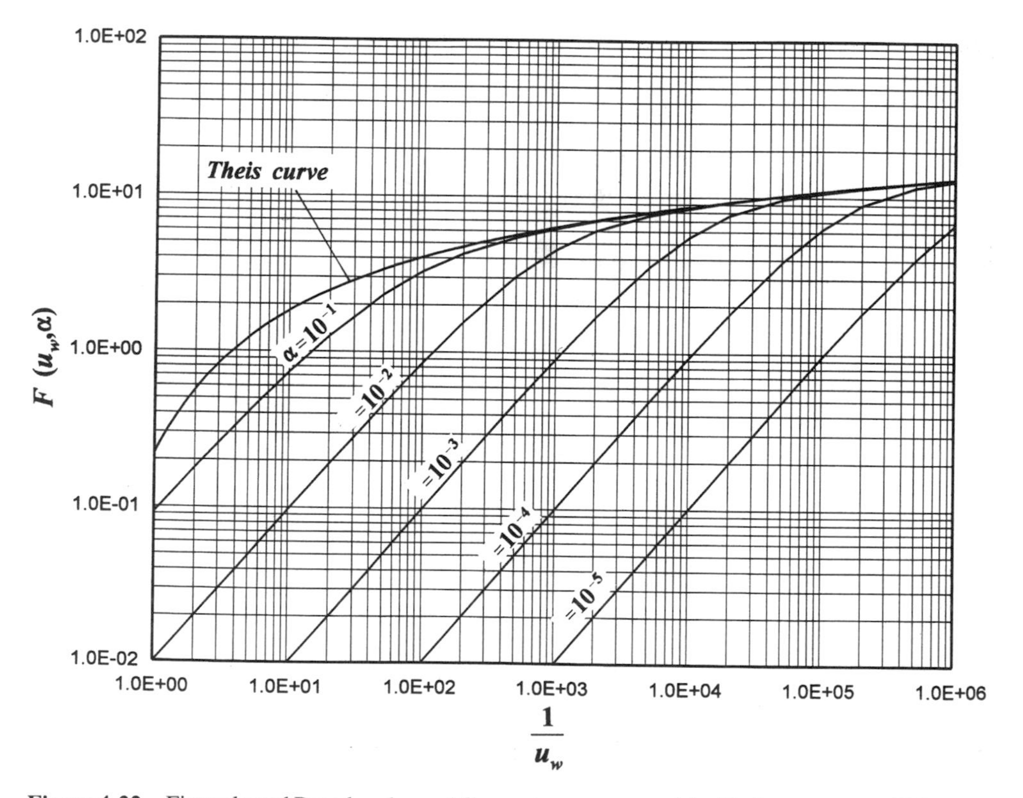

Figure 4-22 Five selected Papadopulos and Cooper type curves and the Theis type curve of $F(u_w, \alpha)$ versus $1/u_w$. (After Papadopulos and Cooper, 1967.)

TABLE 4-7 Values of the Function $F(u, \alpha, p)$

u	p = 1	2	5	10	20	50	100	200
				For $\alpha = 10^{-1}$				
2×10^{0}	4.88×10^{-2}	1.96×10^{-2}	1.75×10^{-2}	2.41×10^{-2}	3.48×10^{-2}	4.24×10^{-2}	4.48×10^{-2}	4.50×10^{-2}
1	9.19	7.01	9.55	1.41×10^{-1}	1.85×10^{-1}	2.09×10^{-1}	2.14×10^{-1}	2.15×10^{-1}
5×10^{-1}	1.77×10^{-1}	1.95×10^{-1}	3.21×10^{-1}	4.44	5.20	5.49	5.55	5.59
2	4.06	5.78	9.42	1.13×10^{0}	1.19×10^{0}	1.22×10^{0}		
1	7.34	1.11×10^{0}	1.60×10^{0}	1.76	1.80			
5×10^{-2}	1.26×10^{0}	1.84	2.33	2.43	2.46			
2	2.30	2.97	3.28	3.34	3.35			
1	3.28	3.81	4.00	4.03				
5×10^{-3}	4.26	4.60	4.70	4.72				
2	5.42	5.58	5.63	5.64				
1	6.21	6.30	6.33					
5×10^{-4}	6.96	7.01						
2	7.87	7.93						
1	8.57	8.63						
5×10^{-5}	9.32							
2	10.24							
				For $\alpha = 10^{-2}$				
2×10^{0}	4.99×10^{-3}	2.13×10^{-3}	2.11×10^{-3}	3.52×10^{-3}	7.47×10^{-3}	2.03×10^{-2}	3.44×10^{-2}	4.35×10^{-2}
1	9.91	7.99	1.32×10^{-2}	2.69×10^{-2}	6.12×10^{-2}	1.42×10^{-1}	1.91×10^{-1}	2.11×10^{-1}
5×10^{-1}	1.97×10^{-2}	2.40×10^{-2}	5.40	1.21×10^{-1}	2.63×10^{-1}	4.65	5.31	5.51
2	4.89	8.34	2.33×10^{-1}	5.12	9.15	1.16×10^{0}	1.20×10^{0}	1.22×10^{0}
1	9.67	1.93×10^{-1}	5.67	1.12×10^{0}	1.58×10^{0}	1.78	1.81	
5×10^{-2}	1.90×10^{-1}	4.16	1.18×10^{0}	1.95	2.32	2.44	2.46	
2	4.53	1.03×10^{0}	2.42	3.11	3.29	3.34	3.35	
1	8.52	1.87	3.48	3.90	4.00	4.03		
5×10^{-3}	1.54×10^{0}	3.05	4.43	4.65	4.71	4.72		
2	3.04	4.78	5.52	5.61	5.63	5.64		
1	4.55	5.90	6.27	6.31	6.33			
5×10^{-4}	6.03	6.81	6.99	7.01				
2	7.56	7.95	7.92	7.94				
1	8.44	8.59	8.63					
5×10^{-5}	9.23	9.30						
2	10.20	10.23						
1	10.87	10.93						
5×10^{-6}	11.62	11.63						
2	12.54							
1	13.24							

(continued)

TABLE 4-7 *(continued)*

u	ρ = 1	2	5	10	20	50	100	200
				For $\alpha = 10^{-3}$				
2×10^{0}	5.00×10^{-4}	2.15×10^{-4}	2.15×10^{-4}	3.70×10^{-4}	8.35×10^{-3}	3.05×10^{-3}	8.38×10^{-3}	1.50×10^{-2}
1	9.99	8.11	1.37×10^{-3}	2.95×10^{-3}	7.58×10^{-3}	2.81×10^{-2}	7.56×10^{-2}	1.47×10^{-1}
5×10^{-1}	2.00×10^{-3}	2.45×10^{-3}	5.77	1.42×10^{-2}	3.90×10^{-2}	1.54×10^{-1}	3.23×10^{-1}	4.78
2	4.99	8.71	2.67×10^{-2}	7.24	2.03×10^{-1}	6.59	1.02×10^{0}	1.17×10^{0}
1	9.97	2.07×10^{-2}	7.16	2.01×10^{-1}	5.41	1.38×10^{0}	1.70	1.79
5×10^{-2}	1.99×10^{-2}	4.66	1.74×10^{-1}	4.87	1.19×10^{0}	2.27	2.40	2.45
2	4.95	1.29×10^{-1}	5.05	1.31×10^{0}	2.52	3.22	3.32	3.35
1	9.83	2.70	1.04×10^{0}	2.38	3.59	3.96	4.02	
5×10^{-3}	1.95×10^{-1}	5.47	1.96	3.68	4.50	4.69	4.72	
2	4.73	1.31×10^{0}	3.81	5.23	5.55	5.63	5.64	
1	9.07	2.39	5.34	6.13	6.28	6.32		
5×10^{-4}	1.69×10^{-1}	3.98	6.57	6.92	7.00	7.02		
2	3.52	6.44	7.77	7.90	7.93			
1	5.53	7.95	8.55	8.61	8.63			
5×10^{-5}	7.63	9.02	9.28	9.31				
2	9.68	10.12	10.22	10.24				
1	10.68	10.88	10.93					
5×10^{-6}	11.50	11.59	11.62					
2	12.49	12.53	12.54					
1	13.21	13.23	13.24					
5×10^{-7}	13.92	13.93						
2	14.84							
1	15.54							

(continued)

TABLE 4-7 (*continued*)

u	ρ = 1	2	5	10	20	50	100	200
				For $\alpha = 10^{-4}$				
2×10^{0}	5.00×10^{-5}	2.17×10^{-5}	2.18×10^{-5}	3.73×10^{-5}	8.46×10^{-5}	3.16×10^{-4}	9.56×10^{-4}	3.83×10^{-3}
1	1.00×10^{-4}	8.15	1.38×10^{-4}	2.98×10^{-4}	7.77×10^{-4}	3.23×10^{-3}	1.01×10^{-2}	3.42×10^{-2}
5×10^{-1}	2.00	2.47×10^{-4}	5.81	1.45×10^{-3}	4.10×10^{-3}	1.80×10^{-2}	5.62	1.75×10^{-1}
2	5.00	8.76	2.71×10^{-3}	7.54	2.27×10^{-2}	1.03×10^{-1}	3.04×10^{-1}	7.10
1	1.00×10^{-3}	2.09×10^{-3}	7.34	2.16×10^{-2}	6.69	2.97	7.92	1.43×10^{0}
5×10^{-2}	2.00	4.72	1.82×10^{-2}	5.55	1.74×10^{-1}	7.30	1.62×10^{0}	2.24
2	5.00	1.32×10^{-2}	5.56	1.74×10^{-1}	5.36	1.87×10^{-0}	2.95	3.28
1	9.98	2.81	1.23×10^{-1}	3.86	1.14×10^{0}	3.08	3.84	4.02
5×10^{-3}	1.99×10^{-2}	5.88	2.64	8.13	2.17	4.25	4.63	4.71
2	4.97	1.53×10^{-1}	6.89	1.97×10^{0}	4.41	5.47	5.60	5.63
1	9.90	3.10	1.36×10^{0}	3.44	5.61	6.24	6.31	6.33
5×10^{-4}	1.97×10^{-1}	6.18	2.53	5.26	6.71	6.98	7.01	
2	4.81	1.48×10^{0}	4.95	7.33	7.82	7.92	7.94	
1	9.34	2.72	7.03	8.37	8.57	8.62		
5×10^{-5}	1.77×10^{0}	4.65	8.65	9.20	9.29	9.32		
2	3.83	7.87	10.02	10.19	10.23	10.24		
1	6.25	9.92	10.83	10.91	10.93			
5×10^{-6}	8.99	11.23	11.57	11.62	11.63			
2	11.74	12.40	12.52	12.54				
1	12.91	13.17	13.23	13.24				
5×10^{-7}	13.78	13.90	13.93					
2	14.79	14.83						
1	15.51	15.53						
5×10^{-8}	16.22	16.23						
2	17.14							
1	17.84							

(*continued*)

TABLE 4-7 *(continued)*

u	ρ = 1	2	5	10	20	50	100	200
				For $\alpha = 10^{-5}$				
2×10^{0}	5.00×10^{-6}	2.27×10^{-6}	2.48×10^{-6}	4.19×10^{-6}	9.00×10^{-6}	3.21×10^{-5}	9.77×10^{-5}	3.15×10^{-4}
1	1.00×10^{-5}	8.36	1.44×10^{-5}	3.07×10^{-5}	7.89×10^{-5}	3.27×10^{-4}	1.04×10^{-3}	3.44×10^{-3}
5×10^{-1}	2.00	2.51×10^{-5}	5.94	1.47×10^{-4}	4.14×10^{-4}	1.84×10^{-3}	6.02	2.00×10^{-2}
2	5.00	8.87	2.74×10^{-4}	7.61	2.31×10^{-3}	1.08×10^{-2}	3.61×10^{-2}	1.19×10^{-1}
1	1.00×10^{-4}	2.11×10^{-4}	7.42	2.18×10^{-3}	6.85	3.30	1.10×10^{-1}	3.50
5×10^{-2}	2.00	4.77	1.84×10^{-3}	5.65	1.82×10^{-2}	8.90	2.92	8.57
2	5.00	1.34×10^{-3}	5.64	1.80×10^{-2}	5.92	2.89×10^{-1}	8.91	2.12×10^{0}
1	1.00×10^{-3}	2.84	1.26×10^{-2}	4.09	1.36×10^{-1}	6.49	1.80×10^{0}	3.34
5×10^{-3}	2.00	5.96	2.74	9.03	3.01	1.35×10^{0}	3.14	4.40
2	5.00	1.56×10^{-2}	7.43	2.47×10^{-1}	8.06	3.03	5.01	5.52
1	9.99	3.20	1.55×10^{-1}	5.15	1.60×10^{0}	4.75	6.06	6.27
5×10^{-4}	2.00×10^{-2}	6.54	3.20	1.04×10^{0}	2.96	6.31	6.90	6.99
2	4.98	1.66×10^{-1}	8.08	2.45	5.58	7.71	7.89	7.93
1	9.93	3.34	1.58×10^{0}	4.28	7.54	8.52	8.61	8.63
5×10^{-5}	1.98×10^{-1}	6.62	2.93	6.63	8.90	9.21	9.31	
2	4.86	1.59×10^{0}	5.86	9.36	10.10	10.22	10.24	
1	9.49	2.95	8.53	10.60	10.86	10.92		
5×10^{-6}	1.82×10^{0}	5.15	10.67	11.48	11.59	11.62		
2	4.03	9.08	12.28	12.49	12.53	12.54		
1	6.78	11.76	13.12	12.49	13.23	13.24		
5×10^{-7}	10.13	13.41	13.88	13.92	13.93			
2	13.71	14.68	14.83	14.85				
1	15.13	15.46	15.54					
5×10^{-8}	16.05	16.20						
2	17.08	17.14						
1	17.81	17.84						
5×10^{-9}	18.51							
2	19.40							
1	20.15							

Source: From Papadopulos (1967), as presented in Reed (1980).

Papadopulos and Cooper (1967) stated that

$$F(u_w, \alpha) \simeq W(u_w) \qquad \text{for } t > 2.5\frac{10^2 r_c^2}{T}, \quad \frac{\alpha}{u_w} > 10^3 \tag{4-117}$$

The approximations in Eqs. (4-116) and (4-117) are valid for both conditions. For wells of having small diameter and/or aquifers of relatively high transmissivity, the period defined in the above equations is very small. However, for wells having large diameter and/or aquifers of relatively low transmissivity, this period is considerably larger.

Figure 4-22 also shows that, as $1/u_w$ becomes sufficiently small, the curves approach straight lines that satisfy the equation

$$s_w = \frac{Qt}{\pi r_c^2} = \frac{\text{Volume of water discharged}}{\text{Area of well casing}} = \frac{Q}{4\pi T}\frac{\alpha}{u_w} \tag{4-118}$$

or

$$F(u_w, \alpha) = \frac{\alpha}{u_w} \tag{4-119}$$

The early parts (short time) of the curves in Figure 4-22 are straight lines that represent conditions under which all the water pumped is derived from storage within the well. As a result, as pointed out by Papadopulos and Cooper (1967), data that fall on this straight part of the type curves do not indicate information about the aquifer hydrogeologic characteristics.

4.3.5 Methods of Analysis for Pumping Test Data

Papadopulos and Cooper (1967) evaluated the possibilities of determination of aquifer parameters based on type-curve analysis without presenting a numerical example. Some of the points of the authors are presented below.

Using the observed drawdown within the pumped well (s_w) versus time data, a family of type curves plotted on double logarithmic paper, as shown in Figure 4-22, permits the determination aquifer transmissivity and storage coefficient. The method is similar to the Theis type-curve method (see Section 4.2.7.1) except that it involves more than a single type curve. In order to make a productive analysis, most of the drawdown data should fall on the curved part of the type curves. As mentioned above, the almost-straight-line portion of the type curves corresponds to periods when most of the water is derived from storage inside the well. Therefore, the data points falling on this portion of the type curves do not adequately reflect the aquifer characteristics. As a result, data that completely fall on the almost-straight-line part of the type curves cannot be analyzed by this method or, in fact, by any other method.

In principle, the storage coefficient (S) can also be determined using the type curve proposed by Papadopulos and Cooper. However, as the authors stated, determination of the values of S has questionable reliability. This is due to the fact that the matching of data curve to the type curves depends on the shape of the type curves, which differ only slightly when α differs by an order of magnitude. The determination of transmissivity (T) is not so sensitive as the case for storage coefficient. For example, when the data curve is moved from one type curve to another, T will change only slightly, whereas S will change by an order of magnitude. In order to overcome these difficulties, Wikramaratna (1985) proposed an alternative type-curve method based on the original Papadopulos and Cooper (1967) large-diameter well model.

In the following sections the aforementioned type-curve methods are presented with examples.

4.3.5.1 Papadopulos and Cooper Type Curve Method. The steps of the method are as follows:

Step 1: Plot the type curves of $1/u_w$ versus $F(u_w, \alpha)$ on log–log paper using the data in Table 4-9 as shown in Figure 4-22.

Step 2: On another sheet of log–log paper with the same as Step 1, plot observation data of s_w versus time (t).

Step 3: Keeping the coordinate axes parallel at all times, superimpose the two sheets until the best fit of data curve and one of the type curves is obtained.

Step 4: Select a common point, the match point, arbitrarily chosen on the overlapping part of the curves, or even anywhere on the overlapping portion of the sheets. Read the corresponding coordinate values of $1/u_w$ and $F(u_w, \alpha)$ from the type-curve sheet, and read t and s_w from the data sheet.

Step 5: Using the match-point coordinates, calculate the value of T from Eq. (4-113).

Step 6: Using the match-point coordinates, calculate the values of S from Eq. (4-115) as well as from the second expression of Eqs. (4-112), respectively:

$$S = \frac{4Ttu_w}{r_w^2} \tag{4-120}$$

and

$$S = \alpha \frac{r_c^2}{r_w^2} \tag{4-121}$$

4.3.5.2 Wikramaratna Type-Curve Method. As mentioned previously, determination of the value of storage coefficient (S), based on the Papadopulos and Cooper type-curve method, has questionable reliability. In order to overcome this limitation, Wikramaratna (1985) proposed to use a set of type curves $F(u_w, \alpha)$ versus α/u_w on double logarithmic paper, together with the match line, according to Eq. (4-119):

$$\log F(u_w, \alpha) = \log\left(\frac{\alpha}{u_w}\right) \tag{4-122}$$

The modified Papadopulos and Cooper (1967) type-curve by Wikramaratna (1985) is given in Figure 4-23. As seen from Figure 4-23, each of the type curves is asymptotic to the match line as α/u_w approaches zero.

From Eq. (4-118), taking logarithms, we obtain

$$\log(s_w) = \log(t) + \log\left(\frac{Q}{\pi r_c^2}\right) \tag{4-123}$$

which represents the asymptotic well response as the value of t approaches zero. At sufficiently small times, all the water is extracted from storage inside the well. Therefore,

TABLE 4-8 Values of the Function $F(u_w, \alpha)$

u_w	$\alpha = 10^{-1}$	$\alpha = 10^{-2}$	$\alpha = 10^{-3}$	$\alpha = 10^{-4}$	$\alpha = 10^{-5}$
10	9.755×10^{-3}	9.976×10^{-4}	9.998×10^{-5}	1.000×10^{-5}	1.000×10^{-6}
1	9.192×10^{-2}	9.914×10^{-3}	9.991×10^{-4}	1.000×10^{-4}	1.000×10^{-5}
5×10^{-1}	1.767×10^{-1}	1.974×10^{-2}	1.997×10^{-3}	2.000	2.000
2	4.062	4.890	4.989	4.999	5.000
1	7.336	9.665	9.966	9.997	1.000×10^{-4}
5×10^{-2}	1.260×10^{0}	1.896×10^{-1}	1.989×10^{-2}	1.999×10^{-3}	2.000
2	2.303	4.529	4.949	4.995	5.000
1	3.276	8.520	9.834	9.984	$1.000 \times 10^{\ 3}$
5×10^{-3}	4.255	1.540×10^{0}	1.945×10^{-1}	1.994×10^{-2}	2.000
2	5.420	3.043	4.725	4.972	4.998
1	6.212	4.545	9.069	9.901	9.992
5×10^{-4}	6.960	6.031	1.688×10^{0}	1.965×10^{-1}	1.997×10^{-2}
2	7.866	7.557	3.523	4.814	4.982
1	8.572	8.443	5.526	9.340	9.932
5×10^{-5}	9.318	9.229	7.631	1.768×10^{0}	1.975×10^{-1}
2	1.024×10^{1}	1.020×10^{1}	9.676	3.828	4.861
1	1.093	1.087	1.068×10^{1}	6.245	9.493
5×10^{-6}	1.163	1.162	1.150	8.991	1.817×10^{0}
2	1.255	1.254	1.249	1.174×10^{1}	4.033
1	1.324	1.324	1.321	1.291	6.779
5×10^{-7}	1.393	1.393	1.392	1.378	1.013×10^{1}
2	1.485	1.485	1.484	1.479	1.371
1	1.554	1.554	1.554	1.551	1.513
5×10^{-8}	1.623	1.623	1.623	1.622	1.605
2	1.705	1.705	1.705	1.714	1.708
1	1.784	1.784	1.784	1.784	1.781
5×10^{-9}	1.854	1.854	1.854	1.854	1.851
2	1.945	1.945	1.945	1.945	1.940
1	2.015	2.015	2.015	2.015	2.015

Source: After Papadopulos and Copper (1967).

the well response is independent of the aquifer parameters given by Eq. (4-118). It is evident from Eq. (4-123) that when plotted on double logarithmic paper it gives a straight line with geometric slope equal to unity passing through the point with coordinates $t = 1$ and $s_w = Q/(\pi r_c^2)$.

Two formulas are available for transmissivity (T). From Eq. (4-113) one can write

$$T = \frac{Q}{4\pi s_w} F(u_w, \alpha) \tag{4-124}$$

Combination of the second expression of Eqs. (4-112) with Eq. (4-115) gives the second alternative equation:

$$T = \frac{r_c^2}{4t} \frac{\alpha}{u_w} \tag{4-125}$$

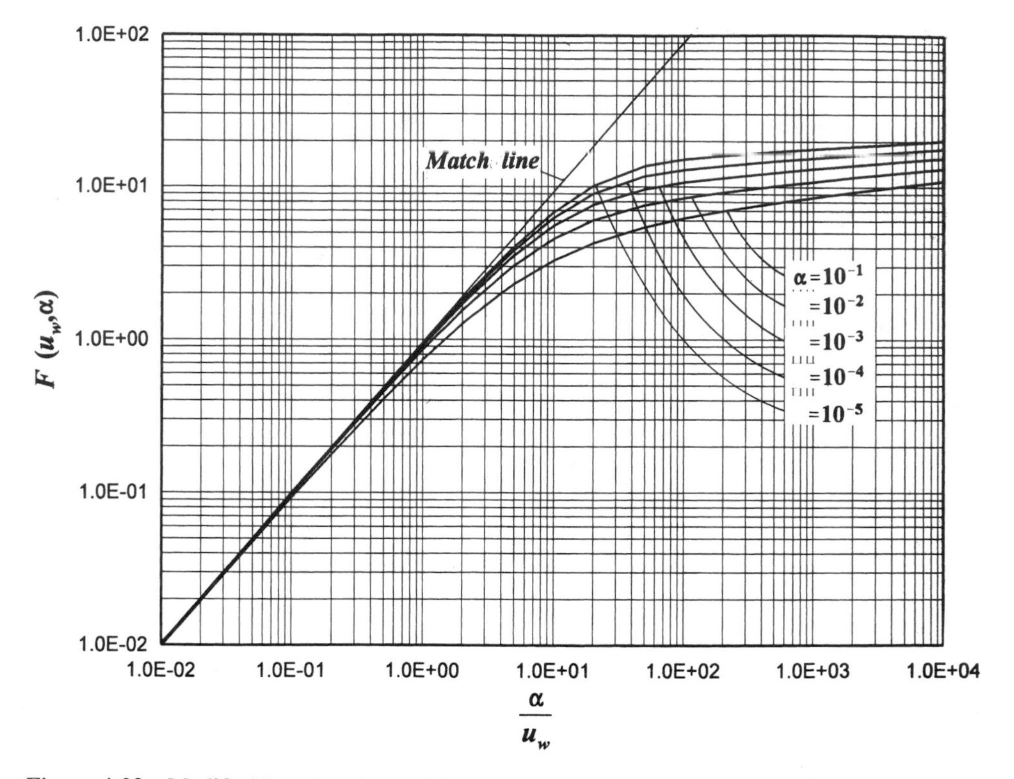

Figure 4-23 Modified Papadopulos and Cooper type curves $F(u_w, \alpha)$ versus α/u_w. (After Wikramaratna, 1985.)

If the match line on the observed response has been chosen correctly and the two match lines coincide, the two values of T determined from Eqs. (4-124) and (4-125) will be the same. This, obviously, provides a useful check that the correct match line has been chosen for tha data.

The match point can be chosen at any place on the overlapping portions of the type curves and observed data sheets. However, it is convenient to choose a point on the match line whose equation is given by Eq. (4-119). If this is chosen to be an exact power of 10, the subsequent calculation of T using either Eqs. (4-124) or (4-125) becomes particularly simple. In this case, Eq. (4-118) will also hold at the match point. It is obvious from the inspection of Eqs. (4-124) and (4-125) that if both Eqs. (4-118) and (4-119) hold, the two determined transmissivity values will be equal. On the other hand, if one of Eqs. (4-118) and (4-119) holds but the other does not, then the two determined values of transmissivity will be different.

Based on the foregoing analysis, the steps of the method are as follows:

Step 1: Plot the type curves of α/u_w versus $F(u_w, \alpha)$ on log–log paper using the data in Table 4-9 and the match line defined by Eq. (4-119) as shown in Figure 4-23.

Step 2: On another sheet of log–log paper with the same as Step 1, plot observation data of s_w versus time (t). Plot also the match line.

Step 3: Keeping the coordinate axes parallel at all times and keeping the match lines matched, superimpose the two sheets until the best-fit-of-data curve and one of the type curves is obtained.

Step 4: Select a common point on the match line. (The match point can be chosen arbitrarily on the overlapping portion of the sheets. However, it is convenient to choose a point on the match line.) Read the corresponding coordinate values of $1/u_w$ and $F(u_w, \alpha)$ from the type-curve sheet, and read t and s_w from the data sheet.

Step 5: Using the match-point coordinates, calculate the value of T from Eqs. (4-124) and (4-125).

Step 6: Using the match-point coordinates, calculate the values of S from the second expression of Eqs. (4-112).

4.3.5.3 *Examples.* Application of the two methods presented above are given below.

Example 4-13: The drawdown within the well (s_w) versus time (t) data of this example (Table 4-10) is taken from Wikramaratna (1985) using the following values from the Papadopulos and Cooper equation, as given by Eq. (4-113): $S = 0.01$, $T = 86.4$ m^2/day, $Q = 432.0$ m^3/day, and $r_w = r_c = 0.2$ m. By treating these data as the results of an idealized pumping test and assuming that Q, r_w, and r_c are known, determine the values of S and T using (a) the Cooper and Papadopulos type-curve method and (b) the Wikramaratna type-curve method.

TABLE 4-9 Drawdown Versus Time Data for Example 4-13

t		
Seconds	Minutes	s_w (m)
10	0.167	0.34
30	0.5	0.84
60	1	1.36
90	1.5	1.71
120	2	1.97
180	3	2.32
300	5	2.69
600	10	3.11
900	15	3.31
1200	20	3.44
1800	30	3.63
2400	40	3.75
3000	50	3.85
3600	60	3.92
5400	90	4.09
7200	120	4.21
9000	150	4.30
10800	180	4.37
14400	250	4.49
18000	300	4.58

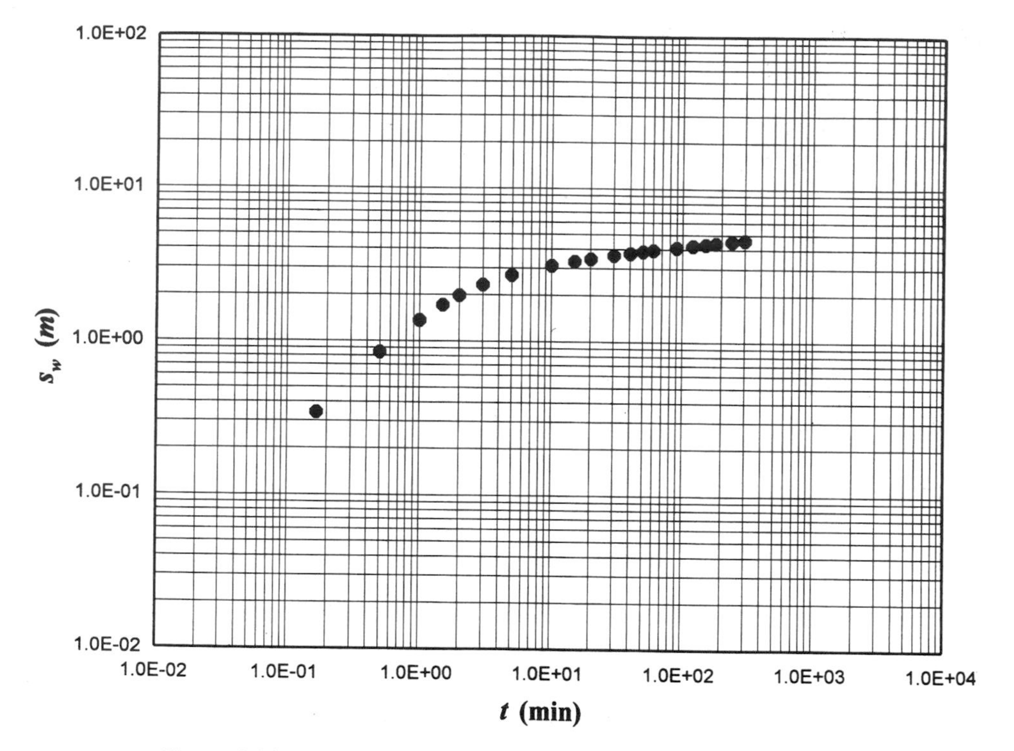

Figure 4-24 Observed drawdown versus time data for Example 4-13.

Solution:

(a) The Papadopulos and Cooper type-curve method is utilized as follows:

Step 1: Using the values in Table 4-9, the type curves are given in Figure 4-22.

Step 2: Using the values in Table 4-10, the observed drawdown (s_w) versus time (t) are shown in Figure 4-24.

Step 3: As mentioned in previously, when the data curve is moved from one type curve to another, T will change only slightly, whereas S will change by an order of magnitude. In order to show this situation, three type-curve matchings are presented with the type curves corresponding $\alpha = 10^{-1}$, 10^{-2}, and 10^{-3} in Figures 4-25, 4-26, and 4-27, respectively.

Step 4: The coordinates of the match-point data are given in Table 4-11.

Step 5: for $\alpha = 10^{-1}$, the value of T from Eq. (4-124) is

$$T = \frac{432 \text{ m}^3/\text{day}}{(4\pi)(5.5 \text{ m})}(10) = 62.50 \text{ m}^2/\text{day}$$

Similarly, the values of T for $\alpha = 10^{-2}$ and 10^{-3} can be calculated. The values are included in Table 4-11.

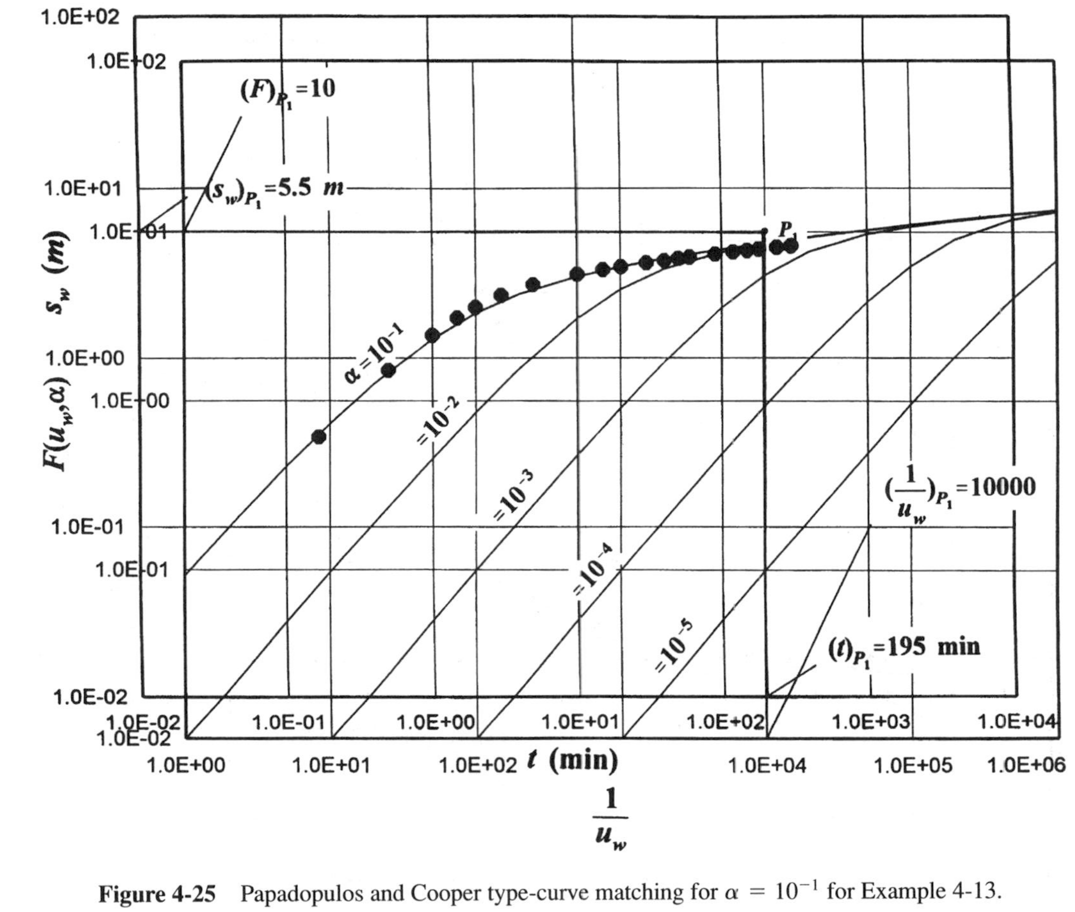

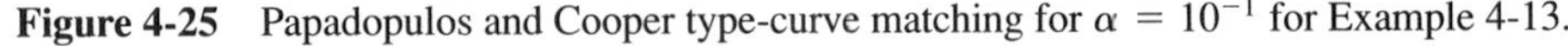

Figure 4-25 Papadopulos and Cooper type-curve matching for $\alpha = 10^{-1}$ for Example 4-13.

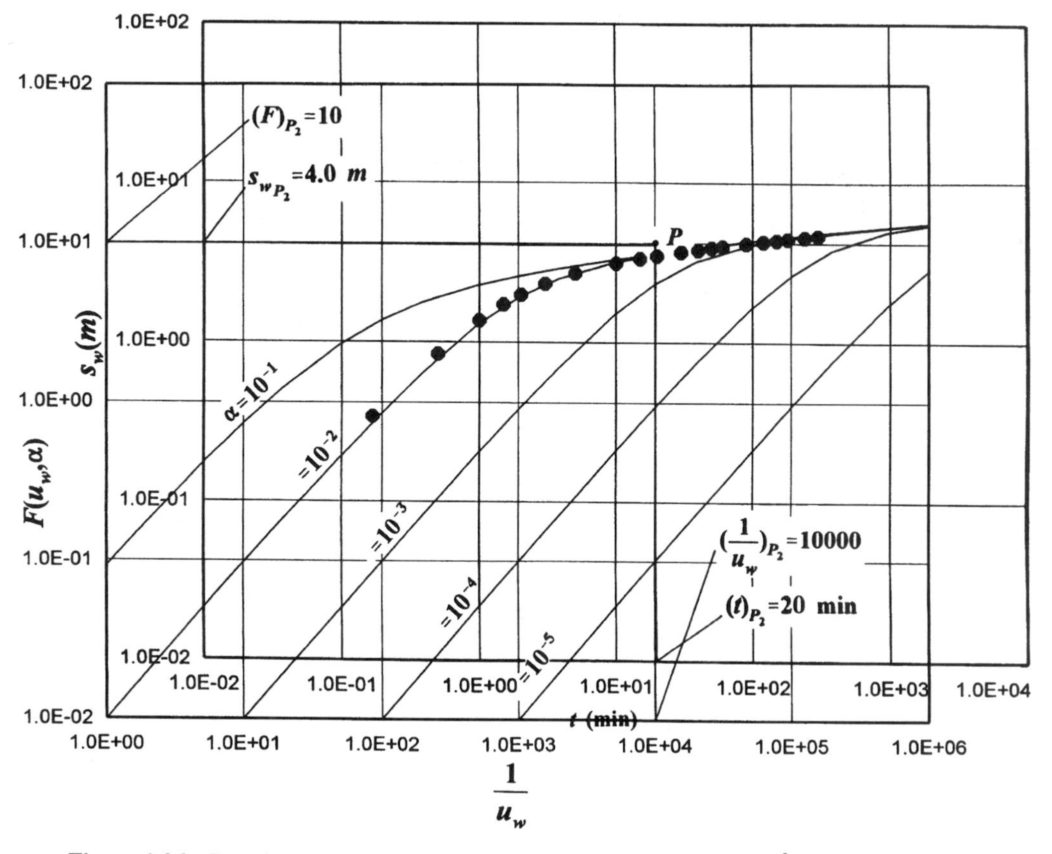

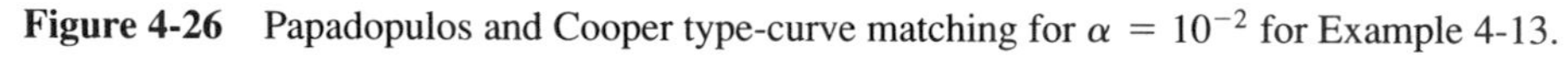

Figure 4-26 Papadopulos and Cooper type-curve matching for $\alpha = 10^{-2}$ for Example 4-13.

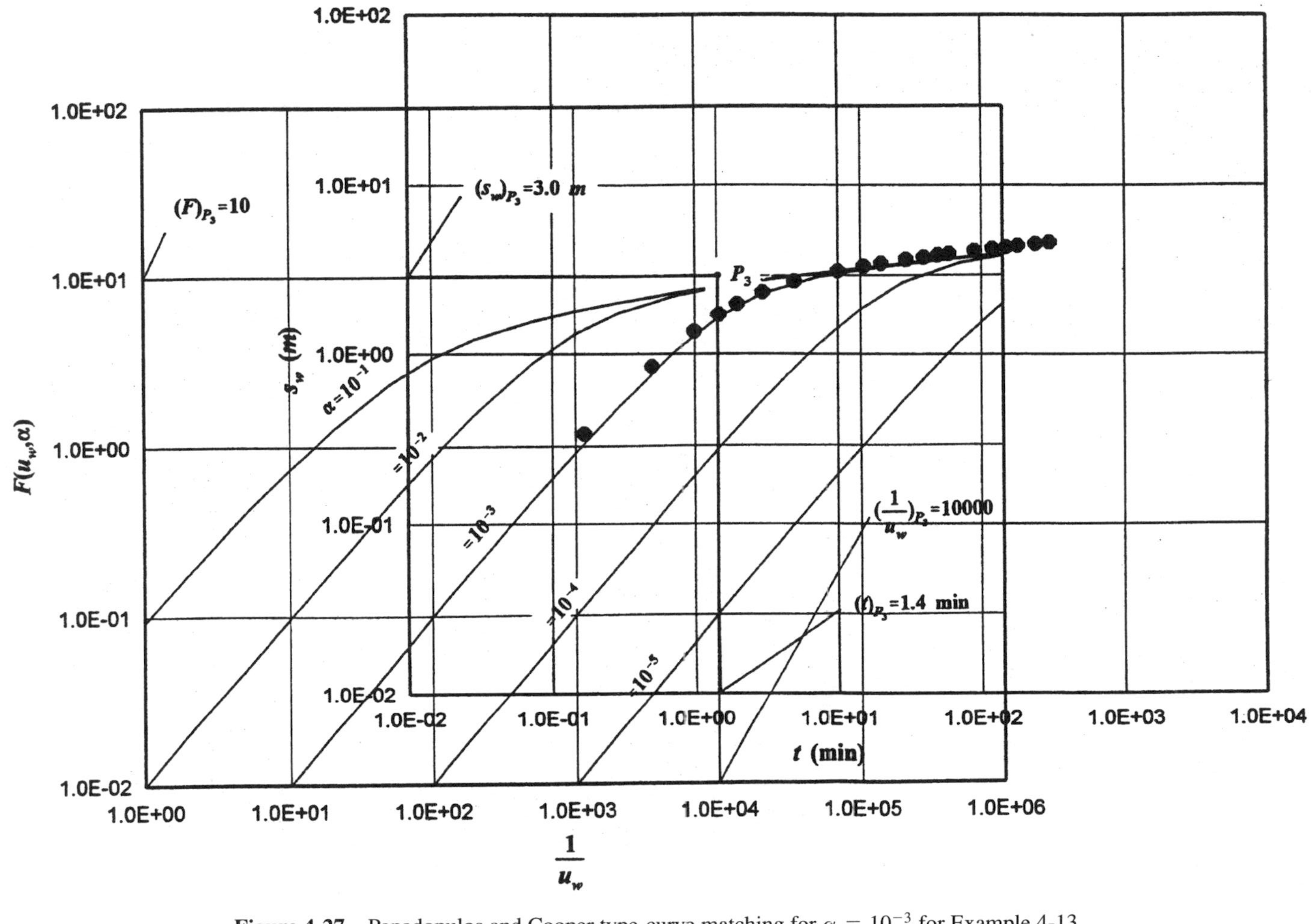

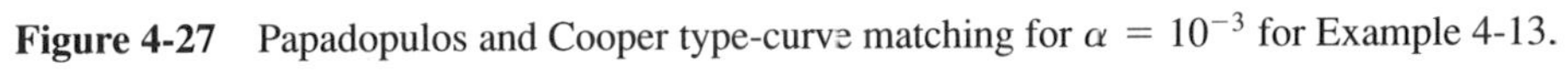

Figure 4-27 Papadopulos and Cooper type-curve matching for $\alpha = 10^{-3}$ for Example 4-13.

TABLE 4-10 Coordinates of Possible Match Points and Resulting Hydrogeologic Parameters for the Papadopulos and Cooper Method

Point	$F(u_w, \alpha)$	α	$1/u_w$	s_w (m)	t (min)	T (m^2/day)	S^a	S^b
P_1	10	10^{-1}	10^4	5.5	195	62.50	0.0846	0.1
P_2	10	10^{-2}	10^4	4.0	20	85.94	0.0119	0.01
P_3	10	10^{-3}	10^4	3.0	1.4	114.59	0.0011	0.001

[a]From Eq. (4-115).
[b]From the second expression of Eqs. (4-112).

Step 6: For $\alpha = 10^{-1}$, the value of S from Eq. (4-120) is

$$S = \frac{4(62.50\ \text{m}^2/\text{day})\left(\dfrac{195}{24 \times 60}\ \text{days}\right)}{(0.2\ \text{m})^2}\frac{1}{10000} = 0.0846$$

Similarly, the values of S for $\alpha = 10^{-2}$ and 10^{-3} can be calculated. The values are included in Table 4-11. Since $r_w = r_c$, the second expression of Eqs. (4-112) gives $S = \alpha$. The values are included in Table 4-11.

(b) The Wikramaratna type-curve method is utilized as follows:

Step 1: Using the values in Table 4-9, the type curves are given in Figure 4-23.
Step 2: Using the values in Table 4-10, the observed drawdown (s_w) versus time (t) are shown in Figure 4-24.
Step 3: The matched curves and match lines are shown in Figure 4-28.
Step 4: The coordinates of the match-point data are

$$\frac{\alpha}{u_w} = 1, \qquad F(u_w, \alpha) = 1$$

$$t = 0.17\ \text{min}, \qquad s_w = 0.4\ \text{m}$$

Step 5: From Eq. (4-124) the value of T is

$$T = \frac{432\ \text{m}^3/\text{day}}{(4\pi)(0.4\ \text{m})}(1) = 85.94\ \text{m}^2/\text{day}$$

From Eq. (4-125) the value of T is

$$T = \frac{(0.2\ \text{m})^2}{4(0.17\ \text{min})}(1) = 0.0588\ \text{m}^2/\text{min} = 84.67\ \text{m}^2/\text{day}$$

The two values of T are very close to each other.
Step 6: The value of S from the second expression of Eqs. (4-112) is

$$S = 10^{-2}\frac{(0.2\ \text{m})^2}{(0.2\ \text{m})} = 0.01$$

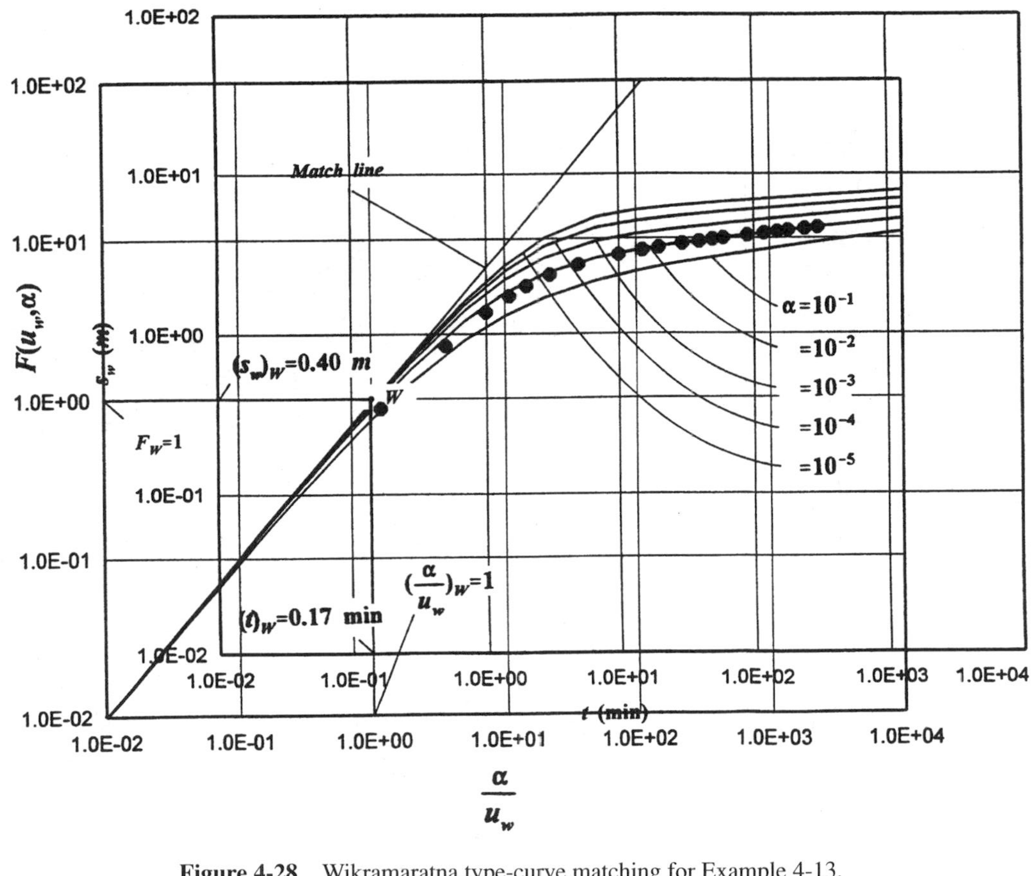

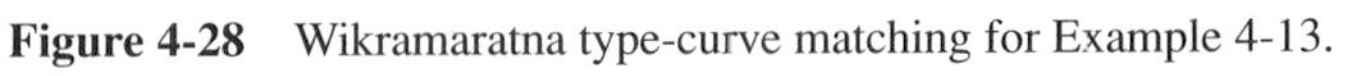

Figure 4-28 Wikramaratna type-curve matching for Example 4-13.

CHAPTER 5

FULLY PENETRATING PUMPING WELLS IN HOMOGENEOUS AND ANISOTROPIC CONFINED NONLEAKY AQUIFERS

5.1 INTRODUCTION

The Theis model (see Chapter 4, Section 4.2) is based on the assumption that the aquifer is homogeneous and isotropic and that the velocity vector and its corresponding hydraulic gradient are parallel to each other. In anisotropic aquifers (for definitions refer to Chapter 2, Section 2.3.7) the velocity vector and its corresponding hydraulic gradient are not generally parallel to each other, and the deficiency depends on the degree of anisotropy. In other words, for anisotropic flow the streamlines are no longer orthogonal to the equipotential lines. By taking into account the anisotropy, Papadopulos (1965) developed an analytical model for the drawdown distribution around a well discharging at a constant rate from an infinite anisotropic aquifer and developed methods for the analysis of drawdown data to determine the aquifer parameters. In the following year, Hantush's generalized theory for homogeneous and anisotropic aquifers was published (Hantush, 1966a). Based on his general theory, Hantush (1966b) developed methods for the analysis of drawdown data in anisotropic aquifers for different cases. Apart from these, there are some additional studies regarding anisotropic confined aquifers (Hantush and Thomas, 1966; Neuman et al., 1984).

There are certain requirements for the design of pumping tests in anisotropic aquifers. For this reason, the reader is strongly advised to be familiar with the methods of data analysis for anisotropic aquifers before attempting to design the pumping test itself; that is, locate the pumping and observation wells.

In this chapter, assumptions, governing equations, and solutions for wells in homogeneous and anisotropic confined aquifers, as well as pumping test data analysis methods, are presented in detail, with examples.

5.2 PAPADOPULOS-MODEL-BASED METHODS

5.2.1 Papadopulos Model

In the following sections the theoretical foundations and practical implications of the Papadopulos model are presented.

5.2.1.1 Problem Statement and Assumptions. Papadopulos (1965) extended the Theis solution to an anisotropic, nonleaky, fully penetrating confined aquifer (refer to Figure 4-1 of Chapter 4) with the following assumptions:

1. The aquifer is homogeneous and anisotropic, and it has an infinite areal extent and a constant thickness (b).
2. The transmissivity of the aquifer (T) is a two-dimensional symmetric tensor.

The rest of the assumptions are the same as those in the Theis model (Chapter 4, Section 4.2.1).

5.2.1.2 Governing Differential Equation. If the hydraulic conductivity is dependent on the direction of flow in an aquifer, the aquifer is said to be anisotropic with regard to hydraulic conductivity. In other words, a horizontal anisotropic aquifer is characterized by its hydraulic conductivity or transmissivity tensor, which is dependent on flow direction in the horizontal plane (for definitions refer to Section 2.3.7 of Chapter 2). Anisotropy exists in all naturally deposited formations and is negligible for most of them. But, in sedimentary deposits, flow may become more dominant along the plane of deposition as compared with the direction that is perpendicular to the deposition direction. Anisotropy depends on different controlling factors, including sediment deposition rate, depositional environment, shape, size, and orientation of particles (Quiñones-Aponte, 1989).

The transmissivity tensor (T) is defined as the multiplication of the two-dimensional hydraulic conductivity tensor and the thickness (b) of the aquifer. In matrix notation, the two-dimensional symmetric transmissivity tensor can be expressed using Eqs. (2-24) and (2-163) as (see Section 2.8.1.1 of Chapter 2)

$$T = \begin{bmatrix} T_{xx} & T_{xy} \\ T_{xy} & T_{yy} \end{bmatrix} \tag{5-1}$$

where x and y are the Cartesian coordinates (Figure 5-1). Equation (5-1) corresponds to Eq. (2-163) of Chapter 2. If the principal directions of anisotropy are shown by ξ and η, Eq. (5-1) reduces to (see Section 2.6 and 2.8 of Chapter 2)

$$T = \begin{bmatrix} T_{\xi\xi} & 0 \\ 0 & T_{\eta\eta} \end{bmatrix} \tag{5-2}$$

where $T_{\xi\xi}$ and $T_{\eta\eta}$ are the principal transmissivities. Using Eq. (5-1), we obtain the governing partial differential equation describing a well in a horizontal confined aquifer in two-dimensional Cartesian coordinates system (Papadopulos, 1965) (see Section 2.8.1.1 of Chapter 2):

$$T_{xx}\frac{\partial^2 s}{\partial x^2} + 2T_{xy}\frac{\partial^2 s}{\partial x \partial y} + T_{yy}\frac{\partial^2 s}{\partial y^2} + Q\delta(x)\delta(y) = S\frac{\partial s}{\partial t} \tag{5-3}$$

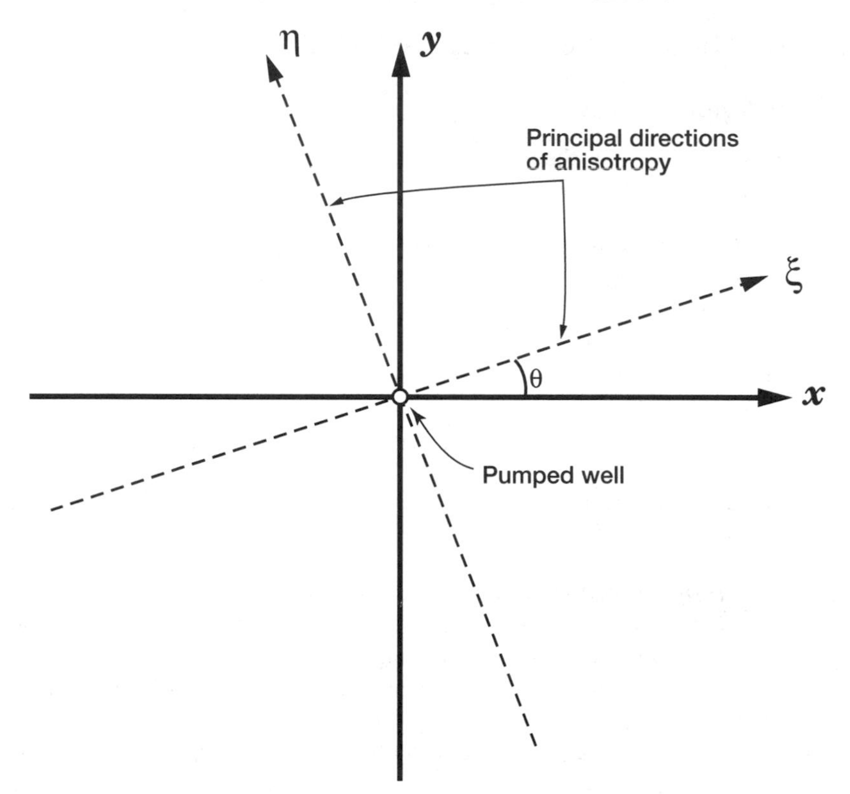

Figure 5-1 Plan view of the coordinate axes.

where s is the drawdown, T_{xx}, T_{yy}, and T_{xy} are the transmissivity tensor components, S is the storage coefficient, Q is the well discharge, δ is the Dirac delta function, x and y are the Cartesian coordinates, and t is the time since the pumping started. Equation (5-3) describes the unsteady flow in a homogeneous and anisotropic aquifer with a constantly discharging well at the origin ($x = 0, y = 0$) in Figure 5-1.

5.2.1.3 Initial and Boundary Conditions. The initial and boundary conditions for the drawdown $s(x, y, t)$ are as follows:

$$s(x, y, 0) = 0 \tag{5-4}$$

$$s(\pm\infty, y, t) = 0 \tag{5-5}$$

$$s(x, \pm\infty, t) = 0 \tag{5-6}$$

Equation (5-4) states that, initially, the drawdown is zero everywhere in the aquifer. Equations (5-5) and (5-6) state that the drawdown approaches zero as distance from the discharging well approaches infinity.

5.2.1.4 Solution

Papadopulos Equation. Papadopulos (1965) used the theory of Laplace and Fourier transforms in solving the problem and gave the solution as

$$s(x, y, t) = \frac{Q}{4\pi(T_{xx}T_{yy} - T_{xy}^2)^{1/2}} W(u_{xy}) \tag{5-7}$$

where W is the well-known *Theis well function* as defined in Chapter 4 [Eq. (4-25)] and

$$u_{xy} = \frac{S}{4t}\left(\frac{T_{xx}y^2 + T_{yy}x^2 - 2T_{xy}xy}{T_{xx}T_{yy} - T_{xy}^2}\right) \tag{5-8}$$

If the coordinate axes x and y coincide with the principal axes ξ and η (Figure 5-1) of the transmissivity tensor, Eq. (5-7) reduces to

$$s(\xi, \eta, t) = \frac{Q}{4\pi(T_{\xi\xi}T_{\eta\eta})^{1/2}} W(u_{\xi\eta}) \tag{5-9}$$

where

$$u_{\xi\eta} = \frac{S}{4t}\left(\frac{T_{\xi\xi}\eta^2 + T_{\eta\eta}\xi^2}{T_{\xi\xi}T_{\eta\eta}}\right) \tag{5-10}$$

Notice that if $T_{xx} = T_{yy} = T$ and $T_{xy} = 0$, Eqs. (5-7) and (5-8) turn out to be exactly the same form of Theis equation as given by Eqs. (4-22) and (4-23) of Chapter 4.

Evaluation of the Papadopulos Solution. To demonstrate the nature of the solution, Papadopulos (1965) presented some numerical results for $T_{\xi\xi} = 12$ cm^2/s, $T_{\eta\eta} = 3$ cm^2/s, $S = 0001$, $Q = 5.46$ liters/s, and $t = 400$ hours and are presented in Figure 5-2. As can be seen from Figure 5-2, the drawdown contours around a pumping well for an anisotropic aquifer are in the form of concentric ellipses and the longitudinal axis of ellipse (ξ axis) corresponds to the higher transmissivity direction.

The analytical solution as given by Eq. (5-7) was verified by an electric analog model (Papadopulos, 1965). The results of the Papadopulos solution were also checked against the finite-difference method results (Ward et al., 1984).

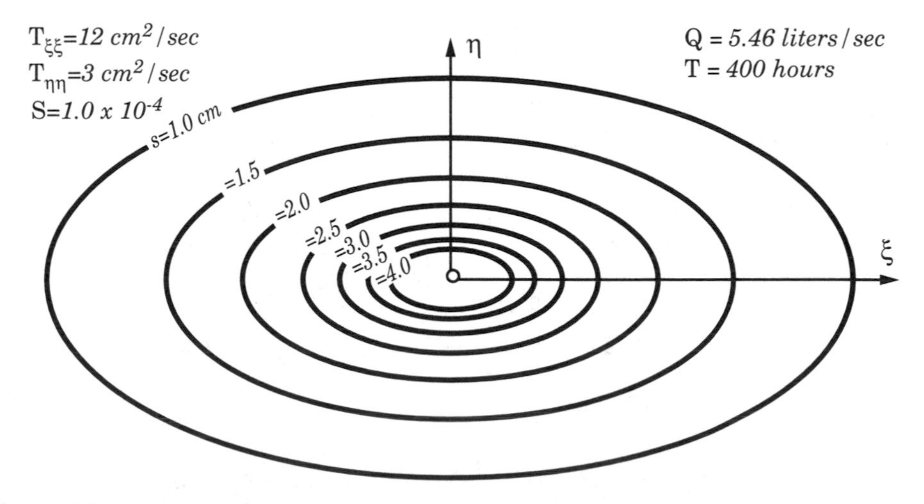

Figure 5-2 Drawdown distribution in an anisotropic confined aquifer. (After Papadopulos, 1965.)

Equations for the Principal Transmissivities. If T_{xx}, T_{yy}, and T_{xy} are the components of the transmissivity tensor in the x and y coordinate system, the principal transmissivities in another coordinate system (ξ and η) whose $O\xi$ axis makes a θ angle with the Ox axis, the principal transmissivities $T_{\xi\xi}$ and $T_{\eta\eta}$ along the ξ and η axes, respectively, and the angle θ can be determined using the equations given in Section 2.6 of Chapter 2 (refer to Figure 2-28).

From Eq. (2-136) of Chapter 2, the expression for the first principal transmissivity is

$$T_{\xi\xi} = \frac{1}{2}\left\{(T_{xx} + T_{yy}) + [(T_{xx} - T_{yy})^2 + 4T_{xy}^2]^{1/2}\right\} \tag{5-11}$$

From Eq. (2-137) of Chapter 2, the expression for the second principal transmissivity is

$$T_{\eta\eta} = \frac{1}{2}\left\{(T_{xx} + T_{yy}) - [(T_{xx} - T_{yy})^2 + 4T_{xy}^2]^{1/2}\right\} \tag{5-12}$$

The angle is [Chapter 2, Eq. (2-139)]

$$\theta = \arctan\left(\frac{T_{\xi\xi} - T_{xx}}{T_{xy}}\right) \tag{5-13}$$

or from Eq. (2-138) of Chapter 2

$$\theta = \frac{1}{2}\arctan\left(\frac{2T_{xy}}{T_{xx} - T_{yy}}\right) \tag{5-14}$$

The angle θ is between the x axis and the ξ axis and is positive in a counterclockwise direction from the x axis. For convenience, the angle θ is restricted to the interval $0 \leq \theta < \pi$.

Approximated Forms of the Papadopulos Equation. For small values of u, $u < {\sim}0.01$, the well function appearing in Eqs. (5-7) and (5-9) can be closely approximated by (see Section 4.2.7.2, Chapter 4)

$$W(u) \cong -0.5772 - \ln(u) = 2.303 \log\left(\frac{2.25}{4u}\right) \tag{5-15}$$

Introducing Eq. (5-15) into Eq. (5-7) gives the drawdown in (x, y) coordinates system for relatively large values of time as

$$s(x, y, t) = \frac{2.303Q}{4\pi(T_{xx}T_{yy} - T_{xy}^2)^{1/2}} \log\left[\frac{2.25t}{S}\left(\frac{T_{xx}T_{yy} - T_{xy}^2}{T_{xx}y^2 + T_{yy}x^2 - 2T_{xy}}\right)\right] \tag{5-16}$$

Similarly, substitution of Eq. (5-15) in Eq. (5-9) gives the drawdown in (ξ, η) coordinates:

$$s(\xi, \eta, t) = \frac{2.303Q}{4\pi(T_{\xi\xi}T_{\eta\eta})^{1/2}} \log\left[\frac{2.25t}{S}\left(\frac{T_{\xi\xi}T_{\eta\eta}}{T_{\xi\xi}\eta^2 + T_{\eta\eta}\xi^2}\right)\right] \tag{5-17}$$

5.2.1.5 Methods of Analysis for Pumping Test Data. Based on the mathematical model developed by Papadopulos (1965), some methods were developed for the determination of the hydrogeologic parameters of anisotropic aquifers. These methods are the extensions of the type-curve and straight-line methods for isotropic aquifers, originally developed by Theis (1935) and Cooper and Jacob (1946), respectively. These methods are included in detail in Section 4.2.7.1 and 4.2.7.2 of Chapter 4, respectively. The methods that were developed by Papadopulos (1965) and Neuman et al. (1984) are presented in detail in the following sections.

Four-Well Method. The method originally was developed by Papadopulos (1965). For anisotropic aquifers, Eqs. (5-7) or (5-16) are the appropriate equations if the principal axes (ξ and η) are not known. If the principal axes are known, Eq. (5-9) or (5-17) are the appropriate equations to be used. For an anisotropic confined aquifer, there are four constants (T_{xx}, T_{yy}, T_{xy}, and S) to be determined. Therefore, a minimum of three observation wells at different distances and directions from the pumped well are necessary. Because the method requires four wells, which are shown in Figure 5-3—one for water withdrawal and three for drawdown observations—the method is called the *four-well method*. If data are available for more than three observation wells, the same method can be applied by grouping them into sets of three.

TYPE-CURVE METHOD. The method is similar in many respects to the Theis type-curve method (see Chapter 4, Section 4.2.7.1), and its associated steps are given below in stepwise manner:

Step 1: Select an appropriate x and y Cartesian coordinate system having the pumped well at the origin and determine the x and y coordinates of each observation well.

Step 2: Plot the type curve of $1/u_{xy}$ versus $W(u_{xy})$ or u_{xy} versus $W(u_{xy})$ on log–log paper using the data in Table 4-1 of Chapter 4.

Step 3: On another sheet of log–log paper with the same scale as the one in Step 2, plot the observation data of drawdown, s, against time, t, for each of the three observation wells.

Step 4: Keeping the coordinate axes parallel at all times, superimpose the two sheets until the best fit of the data curve and the type curve is obtained. Repeat this process for each well. Choose a match point for each well and record the dual coordinates $W(u_{xy})$, s, $1/u_{xy}$, and t of each match point.

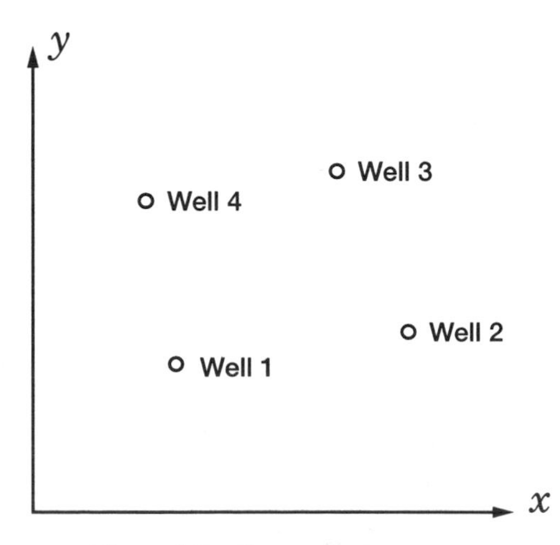

Figure 5-3 Four-well arrangement.

Step 5: Substitute the values of $W(u_{xy})$ and s from each match point into Eq. (5-7) and determine $(T_{xx}T_{yy} - T_{xy}^2)$. The three match points should yield approximately the same value for $(T_{xx}T_{yy} - T_{xy}^2)$. If they are not, an average value may be determined based on judgment.

Step 6: Substitute the values of u_{xy}, t (obtained in Step 4), $(T_{xx}T_{yy} - T_{xy}^2)$ (obtained in Step 5), and the x and y observation well coordinates into Eq. (5-8), and solve the resulting three equations for the products ST_{xx}, ST_{yy}, and ST_{xy}.

Step 7: Determine T_{xx}, T_{yy}, and T_{xy} in terms of S, and substitute these values into the expression $(T_{xx}T_{yy} - T_{xy}^2)$, which has known value from Step 5, to determine S.

Step 8: With the known value of S, determine T_{xx}, T_{yy}, and T_{xy} from the products in Step 7.

Step 9: Determine the principal transmissivities $T_{\xi\xi}$ and $T_{\eta\eta}$ and the angle θ between the x and the ξ from Eqs. (5-11), (5-12), and (5-13) or (5-14), respectively.

Example 5-1: The data for this example are taken from Papadopulos (1965). A 12-hour pumping test was conducted for the determination of the hydrogeologic characteristics of an anisotropic aquifer. The well PW was pumped at a rate of 12.57 liters/s, and the drawdown was observed at three observation wells OW-1, OW-2, and OW-3. The pumped and observation wells along with their coordinates are shown in Figure 5-4. The observed drawdown data are given in Table 5-1. Using the type-curve method determine the values of

TABLE 5-1 Drawdown Data for Example 5-1

	s (m)		
t (min)	OW-1	OW-2	OW-3
0.5	0.335	0.153	0.492
1	0.591	0.343	0.762
2	0.911	0.611	1.089
3	1.082	0.762	1.284
4	1.215	0.911	1.419
6	1.405	1.089	1.609
8	1.549	1.225	1.757
10	1.653	1.329	1.853
15	1.853	1.531	2.071
20	2.019	1.677	2.210
30	2.203	1.853	2.416
40	2.344	2.019	2.555
50	2.450	2.123	2.670
60	2.541	2.210	2.750
90	2.750	2.416	2.963
120	2.901	2.555	3.118
150	2.998	2.670	3.218
180	3.075	2.750	3.310
240	3.235	2.901	3.455
300	3.351	2.998	3.565
360	3.438	3.118	3.649
480	3.587	3.247	3.802
720	3.784	3.455	3.996

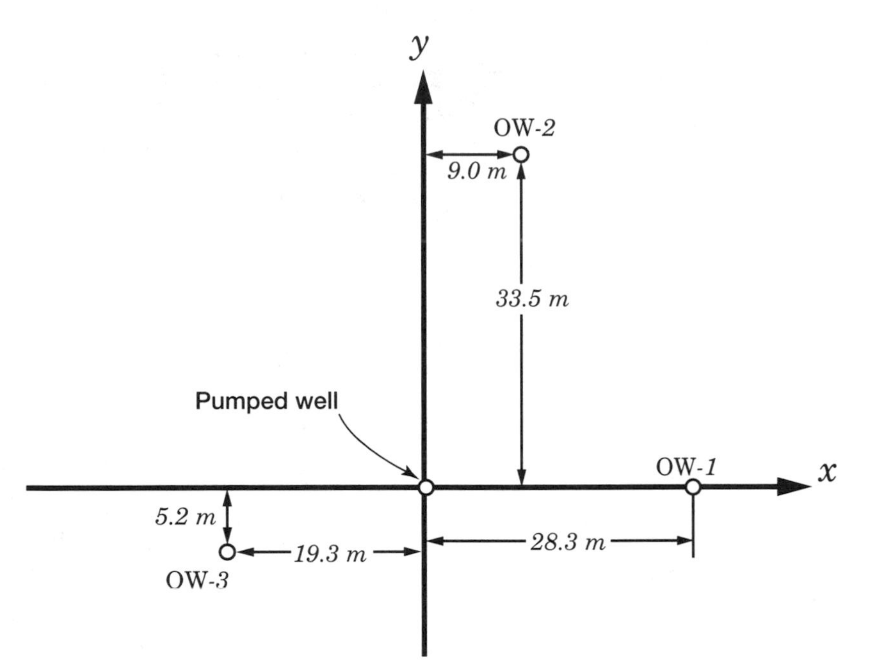

Figure 5-4 Well locations and coordinate system for Example 5-1.

storage coefficient (S), the components of transmissivity tensor (T_{xx}, T_{yy}, and T_{xy}), principal transmissivities of the aquifer ($T_{\xi\xi}$ and $T_{\eta\eta}$), and the direction of the principal axes (the angle θ between the x-axis and ξ-axis).

Solution: The geometry of the pumping system is shown in Figure 5-4. The drawdown data are given in Table 5-1. The type-curve steps in accordance with the above steps are given below.

Step 1: The pumped well (PW) is selected as the origin of the x and y Cartesian coordinate system, and the coordinates of the observation wells are determined as follows: Observation Well OW-1: $x = 28.3$ m, $y = 0$; Observation Well OW-2: $x = 9.0$ m, $y = 32.5$ m; and Observation Well OW-3: $x = -19.3$ m, $y = -5.2$ m.

Step 2: The type curve $1/u_{xy}$ versus $W(u_{xy})$ is the same as Figure 4-2 of Chapter 2, which is drawn using the values in Table 4-1.

Step 3: Observed values of drawdown (s) against time (t) for each of the three observation wells are given in Figure 5-5.

Step 4: The superimposed type-curve and drawdown data for the observation wells OW-1, OW-2, and OW-3 are given Figures 5-6, 5-7, and 5-8, respectively. The coordinates of the match point A from Figure 5-6 are

$$\left(\frac{1}{u_{xy}}\right) = 88, \qquad W(u_{xy})_A = 2$$

$$t_A = 20 \text{ min}, \qquad s_A = 1.0 \text{ m}$$

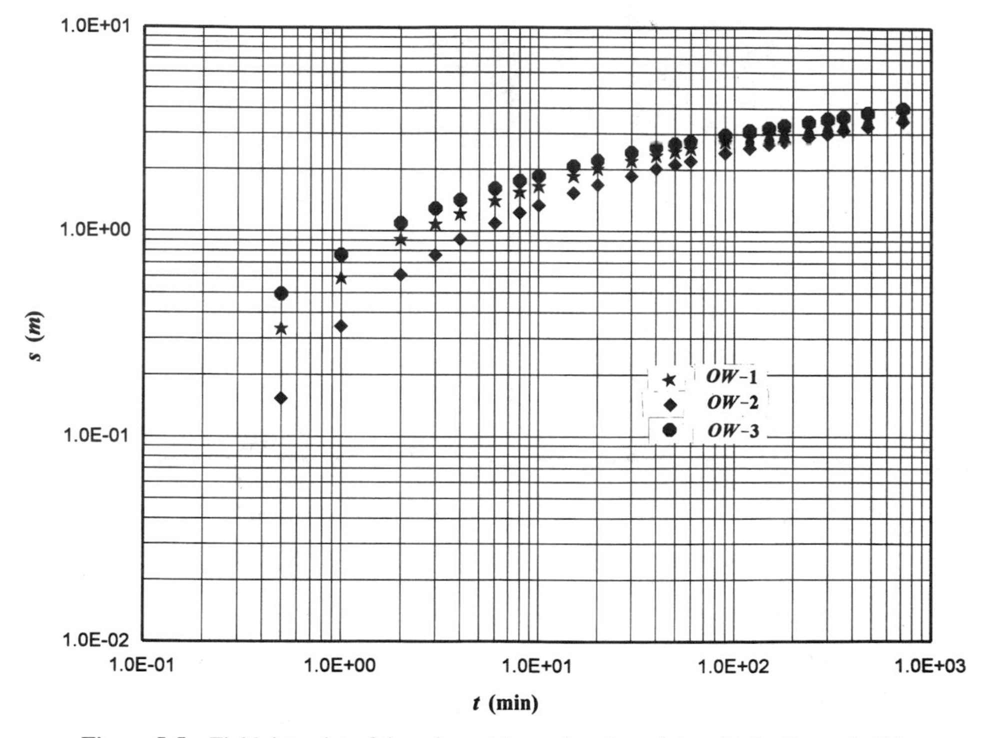

Figure 5-5 Field data plot of drawdown (s) as a function of time (t) for Example 5-1.

The coordinates of the match point B from Figure 5-7 are

$$\left(\frac{1}{u_{xy}}\right)_B = 48, \qquad W(u_{xy})_B = 2$$

$$t_B = 20 \text{ min}, \qquad s_B = 1.0 \text{ m}$$

The coordinates of the match point C from Figure 5-8 are

$$\left(\frac{1}{u_{xy}}\right)_C = 135, \qquad W(u_{xy}) = 2$$

$$t_C = 20 \text{ min}, \qquad s_C = 1.0 \text{ m}$$

Step 5: Substituting the values in Step 4 into Eq. (5-7) gives:

$$\begin{aligned} T_{xx}T_{yy} - T_{xy}^2 &= \left[\frac{QW(u_{xy})}{4\pi s}\right]^2 \\ &= \left[\frac{(12.57 \text{ liters/s})(2.0)}{(1000 \text{ liters/m}^3)(4\pi)(1.0m)}\right]^2 \\ &= 4.002 \times 10^{-6} \text{ m}^4/\text{s}^2 \end{aligned}$$

All three match points yield the same results because the values of $W(u_{xy})$ and s are the same.

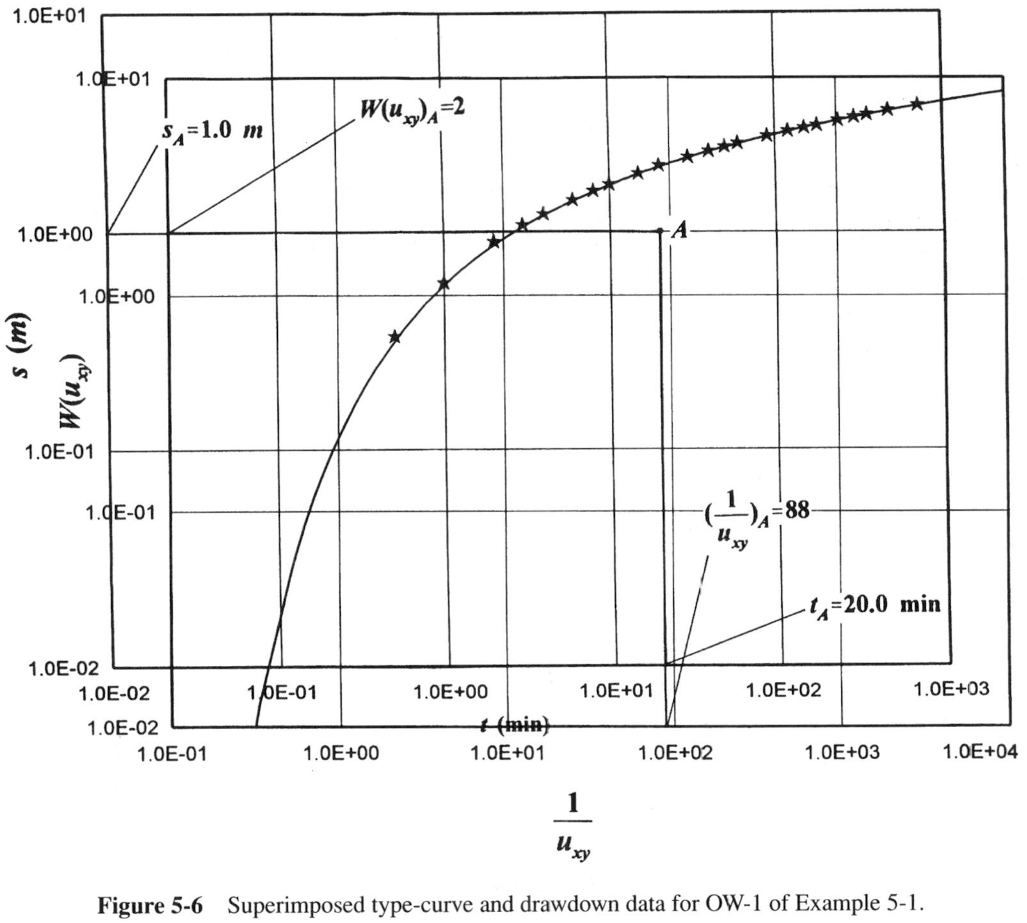

Figure 5-6 Superimposed type-curve and drawdown data for OW-1 of Example 5-1.

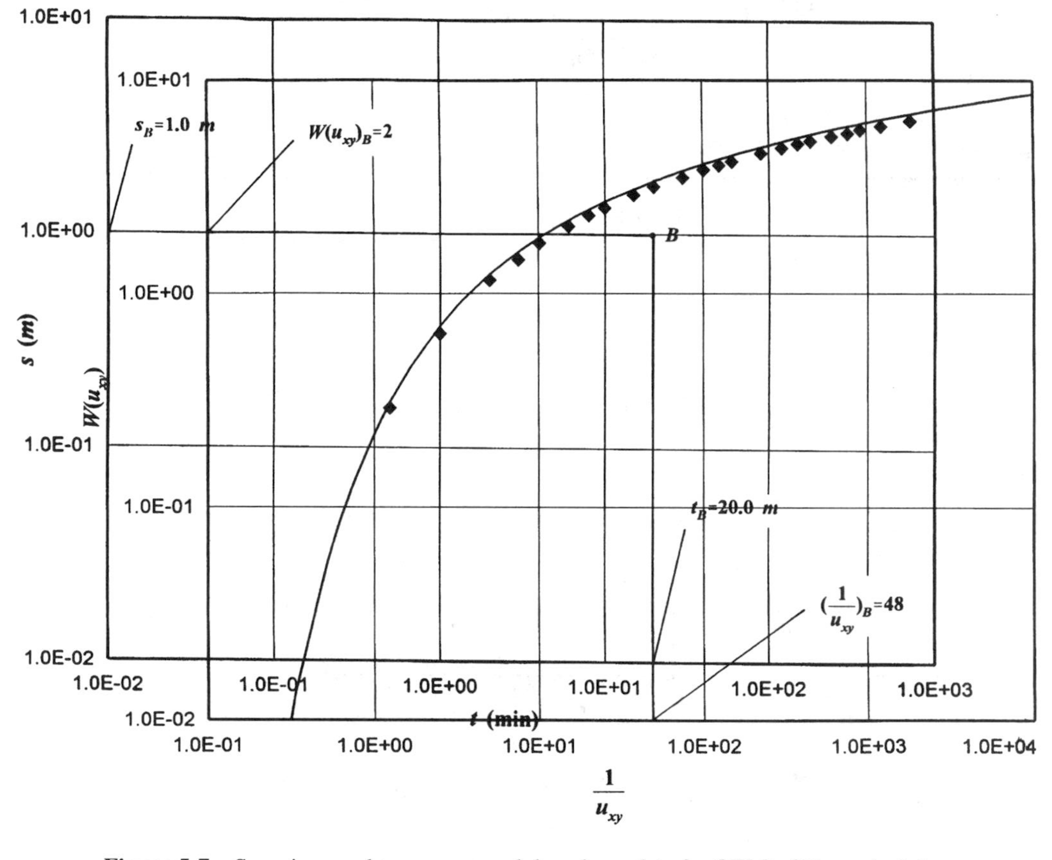

Figure 5-7 Superimposed type-curve and drawdown data for OW-2 of Example 5-1.

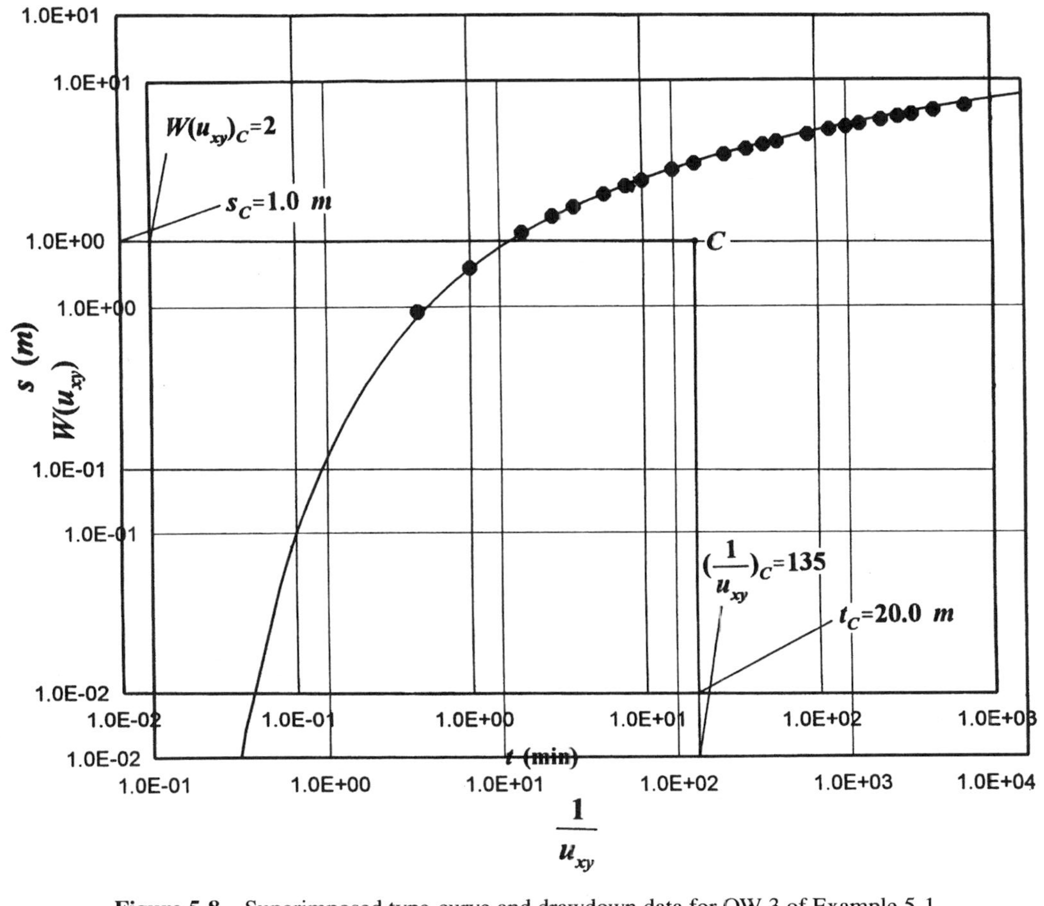

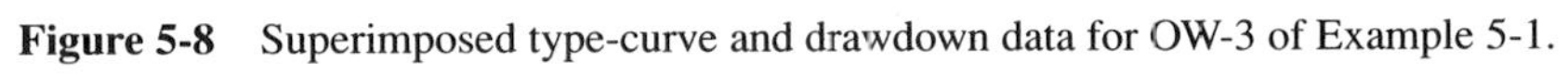

Figure 5-8 Superimposed type-curve and drawdown data for OW-3 of Example 5-1.

Step 6: Substituting the values in Step 1, Step 4, and Step 5 into Eq. (5-8) yields

$$ST_{yy} = (4)(20 \times 60\ \text{s})(4.002 \times 10^{-6}\ \text{m}^4/\text{s}^2)/(28.3\text{m})^2(88) = 2.726 \times 10^{-7}\ \text{m}^2/\text{s}$$

$$(33.5\ \text{m})^2 ST_{xx} + (9.0\ \text{m})^2 ST_{yy} - 2(33.5\ \text{m})(9.0\ \text{m}) ST_{xy} = 4.002 \times 10^{-4}\ \text{m}^4/\text{s}$$

$$(5.2\ \text{m})^2 ST_{xx} + (19.3\ \text{m})^2 ST_{yy} - 2(5.2\ \text{m})(19.3\ \text{m}) ST_{xy} = 1.423 \times 10^{-4}\ \text{m}^4/\text{s}$$

Solving these equations simultaneously gives the following results:

$$T_{xx} = \frac{2.455 \times 10^{-7}}{S}\ \text{m}^2/\text{s}$$

$$T_{yy} = \frac{2.726 \times 10^{-7}}{S}\ \text{m}^2/\text{s}$$

$$T_{xy} = \frac{1.702 \times 10^{-7}}{S}\ \text{m}^2/\text{s}$$

Step 7: Introducing the values of T_{xx}, T_{yy}, and T_{xy} given in Step 6 into the equation of Step 5 gives the value of S:

$$\frac{(2.455\ \text{m}^2/\text{s})(2.726\text{m}^2/\text{s})(10^{-14}) - (-1.702)^2(10^{-14})}{S^2} = 4.002 \times 10^{-6}\ \text{m}^4/\text{s}^2$$

solving for S gives

$$S = 0.974 \times 10^{-4}$$

Step 8: Substituting the values of S in the equations given in Step 6 gives the components of transmissivity tensor:

$$T_{xx} = 2.52 \times 10^{-3}\ \text{m}^2/\text{s} = 25.2\ \text{cm}^2/\text{s}$$

$$T_{yy} = 2.80 \times 10^{-3}\ \text{m}^2/\text{s} = 28.0\ \text{cm}^2/\text{s}$$

$$T_{xy} = -1.75 \times 10^{-3}\ \text{m}^2/\text{s} = -17.5\ \text{cm}^2/\text{s}$$

Step 9: With the known transmissivity tensor components, the principal transmissivities can be determined from Eqs. (5-11) and (5-12) as

$$\begin{aligned} T_{\xi\xi} &= \tfrac{1}{2}\Big\{(2.52 + 2.80)(10^{-3})\ \text{m}^2/\text{s} + [(2.52 - 2.80)^2(10^{-3})^2\ \text{m}^4/\text{s}^2 \\ &\quad + (4)(-1.75)^2(10^{-3})^2\ \text{m}^4/\text{s}^2]^{1/2}\Big\} \\ &= 4.42 \times 10^{-3}\ \text{m}^2/\text{s} \\ T_{\eta\eta} &= \frac{1}{2}\Big\{(2.52 + 2.80)(10^{-3})\ \text{m}^2/\text{s} - [(2.52 - 2.80)^2(10^{-3})^2\ \text{m}^4/\text{s}^2 \\ &\quad + (4)(-1.75)^2(10^{-3})^2\ \text{m}^4/\text{s}^2]^{1/2}\Big\} \\ &= 0.90 \times 10^{-3}\ \text{m}^2/\text{s} \\ &= 9.0\ \text{cm}^2/\text{s} \end{aligned}$$

The angle θ between the x and ξ axes can be determined from Eq. (5-14) as follows:

$$\tan(2\theta) = \frac{2(-1.75 \times 10^{-3}\ \mathrm{m^2/s})}{(2.52 - 2.80)(10^{-3})\ \mathrm{m^2/s}} = 12.5$$

$$2\theta = 85^\circ\ 26'$$

finally,

$$\theta = 42^\circ\ 43'$$

STRAIGHT-LINE METHOD. The straight-line method of analysis is also known as the *Cooper and Jacob method* (Chapter 4, Section 4.2.7.2). Equation (5-16) is the basic equation for the straight-line method of analysis and is an asymptotic expression that expresses the drawdown very closely under the condition that a sufficient time has elapsed since the beginning of pumping.

In Eq. (5-16), Q, T_{xx}, T_{yy}, T_{xy}, and S are constants and t is the only variable. For this reason, Eq. (5-16) may be rewritten as

$$s(x,y,t) = \frac{2.303Q}{4\pi(T_{xx}T_{yy} - T_{xy}^2)^{1/2}} \log\left[\frac{2.25}{S}\left(\frac{T_{xx}T_{yy} - T_{xy}^2}{T_{xx}y^2 + T_{yy}x^2 - 2T_{xy}xy}\right)\right] + \frac{2.303Q}{4\pi(T_{xx}T_{yy} - T_{xy}^2)^{1/2}} \log(t) \tag{5-18}$$

Equation (5-18) is the equation of a straight line on semilogarithmic paper s versus $\log(t)$. An examination of Eq. (5-18) shows that its geometric slope is

$$m = \frac{\Delta s}{\Delta[\log(t)]} = \frac{2.303Q}{4\pi(T_{xx}T_{yy} - T_{xy}^2)^{1/2}} \tag{5-19}$$

The line defined by Eq. (5-19) intercepts the time axis at $s = 0$. Consequently, the interception point has the coordinates $s = 0$ and $t = t_0$. Introducing these values into Eq. (5-16) gives

$$\frac{2.25T_0}{S}\left(\frac{T_{xx}T_{yy} - T_{xy}^2}{T_{xx}y^2 + T_{yy}x^2 - 2T_{xy}xy}\right) = 1 \tag{5-20}$$

or

$$t_0 = \frac{S}{2.25}\left(\frac{T_{xx}y^2 + T_{yy}x^2 - 2T_{xy}xy}{T_{xx}T_{yy} - T_{xy}^2}\right) \tag{5-21}$$

The straight-line method can only be used if all the latter part of the observed drawdown data for all three observation wells fall within the range of time in which Eq. (5-16) is applicable. The steps of the method are given below (Papadopulos, 1965):

Step 1: Select an appropriate x and y Cartesian coordinate system having the pumped well at the origin and determine the x and y coordinates of each observation well.

Step 2: On semilogarithmic paper, plot observed drawdown (s) values against logarithmic time (t) for each observation well.

Step 3: For each well data draw a straight line through observation points.

Step 4: Find the interception point of each straight line with the time axis where $s = 0$. Read the value of t_0 for each line.

Step 5: Determine the geometric slope of each straight line. All three lines should have the same, or approximately the same, geometric slope. If differences exist, determine an average value.

Step 6: Substitute the geometric slope and Q in Eq. (5-19) and determine $(T_{xx}T_{yy} - T_{yy}^2)$.

Step 7: Introduce the value of t_0 of each line and the value of $(T_{xx}T_{yy} - T_{xy}^2)$ into Eq. (5-21) and, using the coordinates of the observation well corresponding to each intercept, solve the resulting three equations for the products ST_{xx}, ST_{yy}, and ST_{xy}.

Step 8: Solve these products for T_{xx}, T_{yy}, and T_{xy} in terms of S. Then, substitute these values into the expression $(T_{xx}T_{yy} - T_{xy}^2)$, whose value is determined in Step 6, and solve it for S.

Step 9: With the known value of S, calculate T_{xx}, T_{yy}, and T_{xy} from the products determined in Step 7.

Step 10: With the values determined from the previous steps, calculate the values of principal transmissivities ($T_{\xi\xi}$ and $T_{\eta\eta}$) and the angle (θ) between the x axis and ξ axis from Eqs. (5-11), (5-12). and (5-13), respectively.

Example 5-2: Using the data of Example 5-1, determine the values of storage coeffficient (S), the components of transmissivity tensor (T_{xx}, T_{yy}, and T_{xy}), principal transmissivities ($T_{\xi\xi}$ and $T_{\eta\eta}$), and the direction of the principal axes by the straight-line method. The geometry of the pumping system is shown in Figure 5-4. The observed drawdown data are given in Table 5-1.

Solution:

Step 1: The coordinate axes are chosen with the x axis passing through OW-1 and the coordinates are determined as: OW-1: $x = 28.3$ m, $y = 0$; OW-2: $x = 9.0$ m, $y = 33.5$ m; and OW-3: $x = -19.3$ m, $y = -5.2$ m.

Step 2: From Table 5-1, the drawdowns are plotted on single logarithmic paper against the time values and are shown in Figure 5-9.

Step 3: Straight lines are drawn through the points in Figure 5-9. The data show that the straight-line method of analysis is applicable.

Step 4: From Figure 5-9, the t intercepts are: $(t_0)_1 = 0.37$ min, $(t_0)_2 = 0.72$ min, and $(t_0)_3 = 0.24$ min.

Step 5: The geometric slopes of the three lines (m) drawn through the latter part of the data points are the same and their value is 1.15 m per log cycle (Figure 5-9).

Step 6: From Eq. (5-19) we obtain

$$T_{xx}T_{yy} - T_{xy}^2 = \left[\frac{(2.303)(12.57 \text{ liters/s})}{(4\pi)(1.15 \text{ m})(1000 \text{ liters/m}^3)}\right]^2 = 4.013 \text{ m}^4/\text{s}^2$$

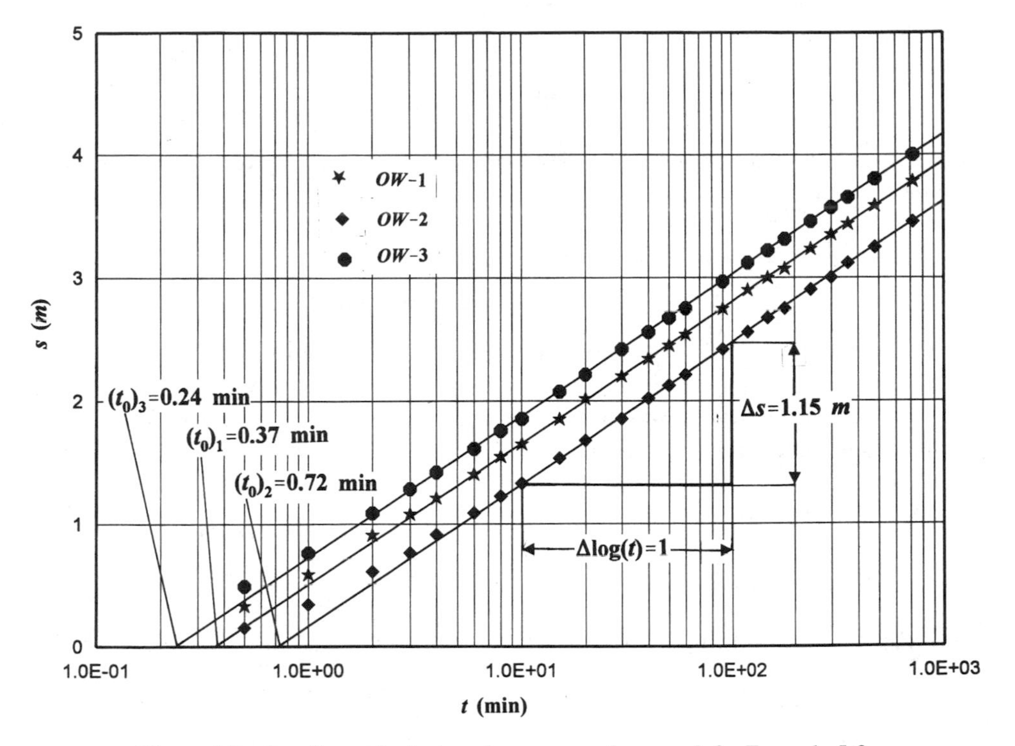

Figure 5-9 Semilogarithmic drawdown versus time graph for Example 5-2.

Step 7: Substitution of the values into Step 1, Step 4, and Step 6 into Eq. (5-21) gives

$$(28.3\ \text{m})^2 ST_{yy} = 2.00 \times 10^{-4}\ \text{m}^4/\text{s}$$

$$(33.5\ \text{m})^2 ST_{xx} + (9.0\ \text{m})^2 ST_{yy} - 2(33.5\ \text{m})(9.0\ \text{m}) ST_{xy} = 3.90 \times 10^{-4}\ \text{m}^4/\text{s}$$

$$(5.2\ \text{m})^2 ST_{xx} + (19.3\ \text{m})^2 ST_{yy} - 2(5.2\ \text{m})(19.3\ \text{m}) ST_{xy} = 1.30 \times 10^{-4}\ \text{m}^4/\text{s}$$

Solving these equations simultaneously gives the following results:

$$T_{xx} = \frac{2.488 \times 10^{-7}}{S}\ \text{m}^2/\text{s}$$

$$T_{yy} = \frac{2.497 \times 10^{-7}}{S}\ \text{m}^2/\text{s}$$

$$T_{xy} = -\frac{1.502 \times 10^{-7}}{S}\ \text{m}^2/\text{s}$$

Step 8: Substituting the values of T_{xx}, T_{yy}, and T_{xy} given in Step 7 into the equation of Step 6 gives the value of S:

$$\frac{(2.488\ \text{m}^2/\text{s})(2.497\ \text{m}^2/\text{s})(10^{-14}) - (-1.502\ \text{m}^2/\text{s})^2(10^{-14})}{S^2}$$

$$= 4.013 \times 10^{-6}\ \text{m}^4/\text{s}^2$$

finally,

$$S = 0.993 \times 10^{-4}$$

Step 9: Substituting the values of S in the equations given in Step 7 gives the transmissivity values:

$$T_{xx} = 2.51 \times 10^{-3}\ \text{m}^2/\text{s} = 25.1\ \text{cm}^2/\text{s}$$
$$T_{yy} = 2.51 \times 10^{-3}\ \text{m}^2/\text{s} = 25.1\ \text{cm}^2/\text{s}$$
$$T_{xy} = -1.51 \times 10^{-3}\ \text{m}^2/\text{s} = -15.1\ \text{cm}^2/\text{s}$$

Step 10: With the known transmissivity tensor components, the principal transmissivities can be determined from Eqs. (5-11) and (5-12) as

$$\begin{aligned} T_{\xi\xi} &= \frac{1}{2}\left\{(2.51\ \text{m}^2/\text{s} + 2.51\ \text{m}^2/\text{s})(10^{-3}) + \left[0 + 4(-1.51\ \text{m}^2/\text{s})^2(10^{-3})^2\right]^{1/2}\right\} \\ &= 4.02 \times 10^{-3}\ \text{m}^2/\text{s} \\ T_{\eta\eta} &= \frac{1}{2}\left\{(2.51\ \text{m}^2/\text{s} + 2.51\ \text{m}^2/\text{s})(10^{-3}) - \left[0 + 4(-1.51\ \text{m}^2/\text{s})^2(10^{-3})^2\right]^{1/2}\right\} \\ &= 1.00 \times 10^{-3}\ \text{m}^2/\text{s} \end{aligned}$$

The angle θ between the x axis and ξ axis can be obtained from Eq. (5-14) as

$$\begin{aligned} \tan(2\theta) &= \frac{2(-1.51 \times 10^{-3}\ \text{m}^2/\text{s})}{(2.51 - 2.51)(10^{-3})\ \text{m}^2/\text{s}} = -\infty \\ 2\theta &= 90^\circ \\ \theta &= 45^\circ \end{aligned}$$

Comparing the values determined from the type-curve method (Example 5-1) with the results of the straight-line method, it can be seen that they are approximately the same.

Three-Well Method. As mentioned previously, the Papadopulos methods for the determination of horizontal anisotropy by means of pumping tests require a minimum of four wells, one for water pumping and three for drawdown observations. Neuman et al. (1984) showed how the same methods can be used for the determination of anisotropy with a minimum of three wells, if at least two of them can be pumped in sequence. Neuman and others also described a method of data analysis by least squares from more than three wells. The details regarding the methods are presented in the following sections.

TYPE-CURVE METHOD EQUATIONS. Consider the case that observation Well 4 in Figure 5-3 is absent and only three wells shown in Figure 5-10 are available. If one pump test with Well 1 pumping at a rate Q_1 is conducted, there are only two time-drawdown data sets, s_{12} and s_{13} (the first subscript, 1, corresponds to the pumping well and the second subscripts, 2 and 3, correspond to the observation wells), which are not sufficient for the determination of the components of transmissivity tensor as given by Eq. (5-1). If at least another pumping

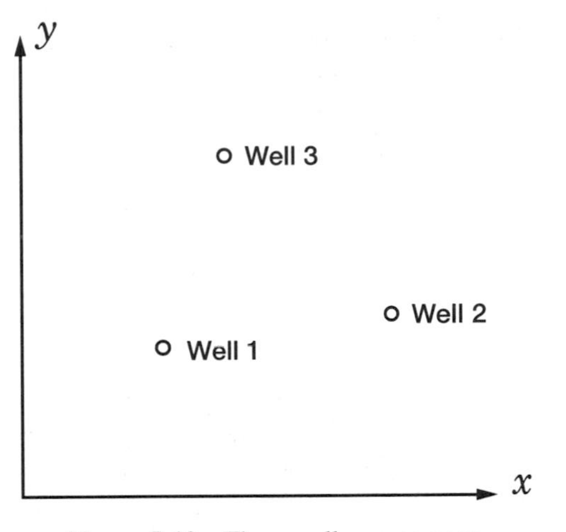

Figure 5-10 Three-well arrangement.

test with Well 2 at a rate Q_2 is conducted and drawdown as a response to pumping Well 2 is observed at least in Well 3, s_{23}, the third set of data can be provided.

From Eq. (5-8) we obtain

$$y^2 T'_{xx} + x^2 T'_{yy} - 2xy T'_{xy} = 4tu_{xy} T_e^2 \tag{5-22}$$

where

$$T'_{xx} = ST_{xx}, \qquad T'_{yy} = ST_{yy}, \qquad T'_{xy} = ST_{xy} \tag{5-23}$$

$$T_e = (T_{xx} T_{yy} - T_{xy}^2)^{1/2} \tag{5-24}$$

$$S = \frac{1}{T_e}(T'_{xx} T'_{yy} - T'_{xy})^{1/2} \tag{5-25}$$

Substituting the three pairs of coordinates of wells into Eq. (5-22) gives

$$y_{12}^2 T'_{xx} + x_{12}^2 T'_{yy} - 2x_{12} y_{12} T'_{xy} = 4t_{12} u_{xy_{12}} T_e^2 \tag{5-26}$$

$$y_{13}^2 T'_{xx} + x_{13}^2 T'_{yy} - 2x_{13} y_{13} T'_{xy} = 4t_{13} u_{xy_{13}} T_e^2 \tag{5-27}$$

$$y_{23}^2 T'_{xx} + x_{23}^2 T'_{yy} - 2x_{23} y_{23} T'_{xy} = 4t_{23} u_{xy_{23}} T_e^2 \tag{5-28}$$

where (x_{12}, y_{12}) and (x_{13}, y_{13}) are the coordinates of Well 2 and Well 3 when the origin of the coordinates is at Well 1; and (x_{23}, y_{23}) are the coordinates of Well 3 when the origin of coordinates is at Well 2.

STRAIGHT-LINE METHOD EQUATIONS. The straight-line methods equations are the modified forms of the equations presented previously. From Eqs. (5-21) and (5-24) we obtain

$$y^2 T'_{xx} + x^2 T'_{yy} - 2xy T'_{xy} = 2.25 T_e^2 t_0 \tag{5-29}$$

where T'_{xx}, T'_{yy}, T'_{xy}, and T_e are defined by Eqs. (5-23) and (5-24). Substituting the three pairs of coordinates into Eq. (5-29) gives

$$y_{12}^2 T'_{xx} + x_{12}^2 T'_{yy} - 2x_{12}y_{12}T'_{xy} = 2.25T_e^2 t_{0_{12}} \quad (5\text{-}30)$$

$$y_{13}^2 T'_{xx} + x_{13}^2 T'_{yy} - 2x_{12}y_{12}T'_{xy} = 2.25T_e^2 t_{0_{13}} \quad (5\text{-}31)$$

$$y_{23}^2 T'_{xx} + x_{23}^2 T'_{yy} - 2x_{23}y_{23}T'_{xy} = 2.25T_e^2 t_{0_{23}} \quad (5\text{-}32)$$

in which the subscript convention is the same as in Eqs. (5-26), (5-27), and (5-28).

IMPORTANT POINTS FOR THE THREE-WELL METHOD. As can be noticed from the type-curve and straight-line equations, for pumping Well 2 the observed drawdown in Well 3 is chosen not in Well 1. This is based on the fact that under ideal conditions the drawdown s_{21} is proportional to s_{12} because of the symmetry of the transmissivity tensor as given by Eq. (5-1). Therefore, the observed drawdown s_{21} in Well 1 as a result of pumping from Well 2 will not provide a linearly independent third equation in Eqs. (5-26), (5-27), and (5-28) or in Eqs. (5-30), (5-31), and (5-32), because the third equation will be merely a multiple of the first equation. As a result, the system cannot be solved uniquely (Neuman et al., 1984).

EQUATIONS FOR THREE PUMPING TESTS FROM THREE WELLS. More reliable results can be determined by conducting up to three pumping tests on those wells shown in Figure 5-10, pumping one well at a time and observing the drawdown in the rest of two wells (Neuman et al., 1984). Each test must be conducted under the conditions that the system has reached equilibrium after the previous pumping test. By this method, from Eqs. (5-22) and (5-29) a system of up to six equations of the form

$$y_{ij}^2 T'_{xx} + x_{ij}^2 T'_{yy} - 2x_{ij}y_{ij}T'_{xy} = b_{ij}, \qquad i, j = 1, 2, 3 \quad (5\text{-}33)$$

where

$$b_{ij} = 4t_{ij}u_{xy_{ij}}T_e^2 \quad (5\text{-}34)$$

in the case of type-curve matching and

$$b_{ij} = 2.25T_e^2 t_{0_{ij}} \quad (5\text{-}35)$$

in the case of straight-line method is used.

Because the number of equations (maximum six) exceeds the number of unknowns (three), the roots of the system of equations can be determined by the method of least squares. The least-squares criterion J is defined as follows:

$$J = \sum_i \sum_j (y_{ij}^2 T'_{xx} + x_{ij}^2 T'_{yy} - 2x_{ij}y_{ij}T'_{xy} - b_{ij})^2 \quad (5\text{-}36)$$

The three unknowns T'_{xx}, T'_{yy}, and T'_{xy} can be determined by minimizing the least-squares criterion (J) with respect to the three unknowns. In Eq. (5-36), the summation is taken over all available pumping wells (i) and observation wells (j). It is well known from elementary calculus that minimization of a function can be achieved by making its first derivative equal to zero and solving for the unknowns. Therefore, by taking the first derivative of J with respect to each unknown and making the result equal to zero gives

$$\left(\sum_i\sum_j y_{ij}^4\right)T'_{xx} + \left(\sum_i\sum_j y_{ij}^2 x_{ij}^2\right)T'_{yy} - 2\left(\sum_i\sum_j x_{ij}y_{ij}^3\right)T'_{xy} = \sum_i\sum_j y_{ij}^2 b_{ij} \tag{5-37}$$

$$\left(\sum_i\sum_j x_{ij}^2 y_{ij}^2\right)T'_{xx} + \left(\sum_i\sum_j x_{ij}^4\right)T'_{yy} - 2\left(\sum_i\sum_j x_{ij}^3 y_{ij}\right)T'_{xy} = \sum_i\sum_j x_{ij}^2 b_{ij} \tag{5-38}$$

$$-\left(\sum_i\sum_j x_{ij}y_{ij}^3\right)T'_{xx} - \left(\sum_i\sum_j x_{ij}^3 y_{ij}\right)T'_{yy} + 2\left(\sum_i\sum_j x_{ij}^2 y_{ij}^2\right)T'_{xy} = -\sum_i\sum_j x_{ij}y_{ij}b_{ij} \tag{5-39}$$

These three equations, which are called *normal equations*, with three unknowns can be solved for T'_{xx}, T'_{yy}, and T'_{xy}. The least-squares procedure for a minimum of four drawdown data sets satisfies the condition that at least three of the lines connecting well i form a triangle. The procedure can also be applied for a case that has more than three wells satisfying the same condition mentioned above.

TYPE-CURVE METHOD. The method is similar to the Theis type-curve method (see Chapter 4, Section 4.2.7.1), and its associated steps are given below (Neuman et al., 1984):

Step 1: Select two different x and y Cartesian coordinate systems having the first and second pumped well as origins and determine the observation wells coordinates for each coordinate system.

Step 2: From the table of the Theis well function (Chapter 4, Table 4-1), prepare a type curve of $W(u_{xy})$ against $1/u_{xy}$ on log–log paper.

Step 3: On another sheet of log–log paper with the same scale as the type curve, plot observed values of the drawdowns (s) against time (t) for each of the three observation wells.

Step 4: Keeping the coordinate axes parallel at all times, superimpose the two sheets of the previous two steps until the best fit of the data curve and the type curve is obtained. Repeat this process for each well. Choose a match point for each well and record the dual coordinates $W(u_{xy})$, s, $1/u_{xy}$, and t of each match point.

Step 5: Using the values of $W(u_{xy})$ and s from Step 4, determine $T_e = (T_{xx}T_{yy} - T_{xy}^2)^{1/2}$ from Eq. (5-7). Differences between the T_e values obtained from different observation wells may often result because of errors in data, type-curve matching, and deviations of the aquifer test conditions. For such cases, the best approach is to determine an average T_e value from the three sets of available data.

Step 6: If there are three time-drawdown data sets, substitute the three pairs (t_{12}, u_{xy12}), (t_{13}, u_{xy13}), and (t_{23}, u_{xy23}) into Eqs. (5-26), (5-27), and (5-28) to a system of three equations in three unknowns, T'_{xx}, T'_{yy}, and T'_{xy} defined by Eq. (5-23). Then solve the resulting three equations for T'_{xx}, T'_{yy}, and T'_{xy}. If there are more than three time-drawdown data sets, develop the equations resulting from the least-squares method as given by Eqs. (5-37), (5-38), and (5-39) with b_{ij} as given by Eq. (5-34).

Step 7: Substitute the values of T'_{xx}, T'_{yy}, T'_{xy}, and T_e (from Step 5) into Eq. (5-25) and determine S.

Step 8: Using S, T'_{xx}, T'_{yy}, and T'_{xy}, determine the components of transmissivity tensor (T_{xx}, T_{yy}, and T_{xy}) from Eq. (5-23).

Step 9: Determine the values of principal transmissivities ($T_{\xi\xi}$ and $T_{\eta\eta}$) and the angle between the x axis and the ξ axis from Eqs. (5-11), (5-12), and (5-13) or (5-14), respectively.

STRAIGHT-LINE METHOD. The straight-line method of analysis can only be used if all the latter part of the observed drawdown data fall within the range of time for which Eq. (5-29) is applicable. The method can be applied step by step as given below (Neuman et al., 1984):

Step 1: Develop two different x and y Cartesian coordinate systems having the first and second pumped well as origins and record the observation wells coordinates for each coordinate system.

Step 2: Plot observed values of drawdowns (s) for each observation well for the first pumped well on semilogarithmic paper with time (t) on the logarithmic scale. Do the same thing for the second pumped well.

Step 3: For each observation well draw a straight line through the points. Determine the geometric slope (m) of each line using Eq. (5-19).

Step 4: From Eq. (5-19) determine $T_e = (T_{xx}T_{yy} - T_{xy}^2)^{1/2}$ using the geometric slope (m) and pumping rate (Q) for each observation well. Differences among the T_e values may be observed because of errors in data and deviations of the aquifer test conditions. Take the arithmetic average of the four values and use it for T_e.

Step 5: Find the intercept point of each straight line with the time axis where $s = 0$. Read the value of t_0 for each line.

Step 6: If there are three time-drawdown data sets, develop Eqs. (5-30), (5-31), and (5-32) to determine T'_{xx}, T'_{yy}, and T'_{xy}. If there are more than three time-drawdown data sets, develop the equations based on the least-squares method as given by Eqs. (5-37), (5-38), and (5-39) with b_{ij} is given by Eq. (5-35).

Step 7: Determine T'_{xx}, T'_{yy}, and T'_{xy} by solving the linear equations system in Step 6.

Step 8: With the known value of T_e from Step 4, T'_{xx}, T'_{yy}, and T'_{xy} (from Step 7), determine the value of S from Eq. (5-25).

Step 9: Using the value of storage coefficient (S) determine the components of transmissivity tensor (T_{xx}, T_{yy}, and T_{xy}) from Eq. (5-23).

Step 10: Determine the values of principal transmissivities ($T_{\xi\xi}$ and $T_{\eta\eta}$) and the angle between x and ξ axes from Eqs. (5-11), (5-12), (5-13), or (5-14), respectively.

Example 5-3: This example is taken from Neuman et al. (1984), who applied the *three-well method* for the determination of anisotropy at the Waste Isolation Pilot Plant (WIPP) site near Carlsbad, New Mexico. The three-well system, called H-6, forms an equilateral triangle horizontally (Figure 5-11) and the length of its each side is 100 ft. Each well is open in the Culebra member of the Rustler Formation which is underlain by the Salado Formation. The Culebra member is composed of silty dolomite with occasional pits, vugs,

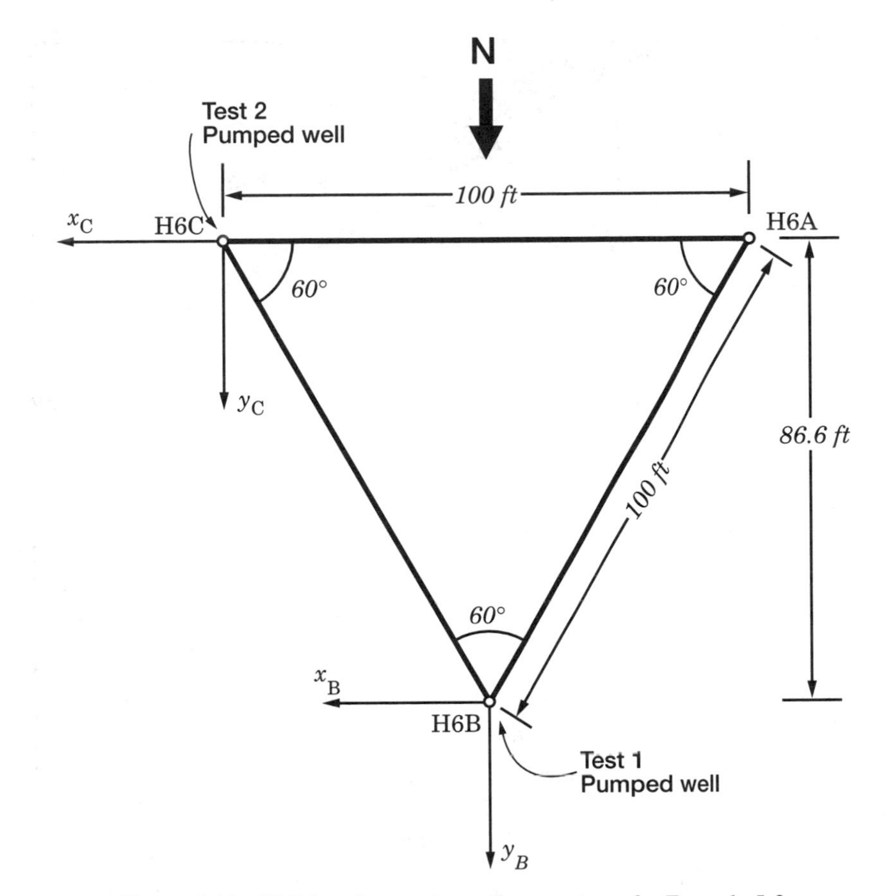

Figure 5-11 Well locations and coordinate systems for Example 5-3.

and fractures and partially filled with selenite (Mercer and Orr, 1979). A total of two pumping tests were conducted. For Test 1 and Test 2, wells H6B and H6C were pumped at constant rates of $Q_B = 4408$ ft^3/day and $Q_C = 3157$ ft^3/day, respectively.

Semilogarithmic observed drawdown versus time curves are presented in Figures 5-12 and 5-13. Using the three-well method with the straight-line method determine the transmissivity tensor components (T_{xx}, T_{yy}, and T_{xy}), principal transmissivities of the aquifer ($T_{\xi\xi}$ and $T_{\eta\eta}$), and storage coefficient (S).

Solution: The applied steps for the straight-line method outlined above are as follows:

Step 1: The two coordinate systems (x_B, y_B) and (x_C, y_C) having their origins at H6B and H6C, respectively, are shown in Figure 5-11. In Figure 5-11, H6B and H6C correspond to Test 1 and Test 2 pumped well locations, respectively. For Test 1 coordinates system, the coordinates of the wells are as follows: Well H6A: $x_{BA} = -50$ ft, $y_{BA} = -86.6$ ft; Well H6B: $x_{BB} = 0$, $y_{BB} = 0$; and Well H6C: $x_{BC} = 50$ ft, $y_{BC} = -86.6$ ft. For Test 2 coordinates system the wells coordinates are: Well H6A: $x_{CA} = -100$ ft, $y_{CA} = 0$; Well H6B: $x_{CB} = -50$ ft, $y_{CB} = 86.6$ ft; and Well H6C: $c_{CC} = 0$, $y_{CC} = 0$.

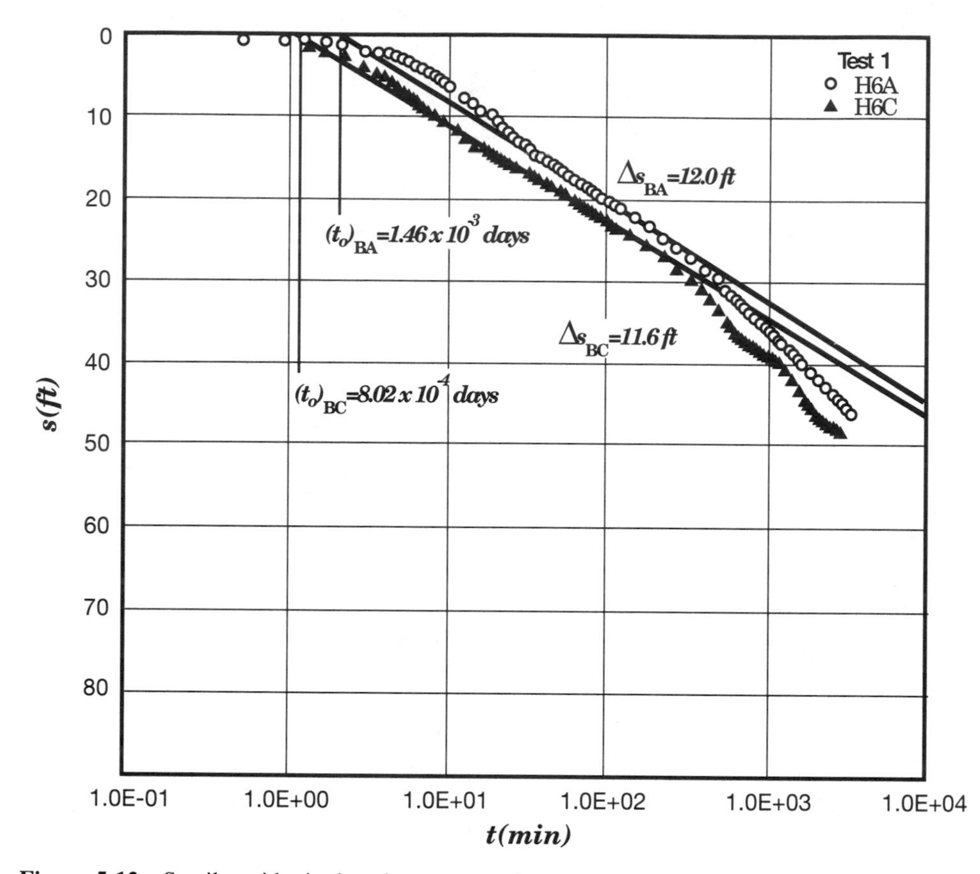

Figure 5-12 Semilogarithmic drawdown versus time graph of Test 1 where H6B is the pumped well. (After Neuman et al., 1984.)

Step 2: For Test 1 and Test 2, the observed values of drawdowns (s) versus time (t on logarithmic scale) are presented in Figures 5-15 and 5-16, respectively.

Step 3: The straight lines through the points are shown in Figures 5-12 and 5-13, respectively. The geometric slopes per log cycle for Test 1 (Figure 5-12) are $\Delta s_{\mathrm{BA}} = 12.0$ ft and $\Delta s_{\mathrm{BC}} = 11.6$ ft. The slopes for Test 2 (Figure 5-13) are $\Delta s_{\mathrm{CA}} = 8.25$ ft and $\Delta s_{\mathrm{CB}} = 8.4$ ft.

Step 4: For Test 1 from Eq. (5-19) we obtain

$$T_e = (T_{xx}T_{yy} - T_{xy}^2)^{1/2} = \frac{2.303}{4\pi}\frac{4408}{\Delta s_{B_i}} \text{ ft}^2\text{/day}, \qquad i = A, C$$

For Test 2 again from Eq. (5-19) we obtain

$$T_e = (T_{xx}T_{yy} - T_{xy}^2) = \frac{2.303}{4\pi}\frac{3157}{\Delta s_{C_i}} \text{ ft}^2\text{/day} \qquad i = A, B$$

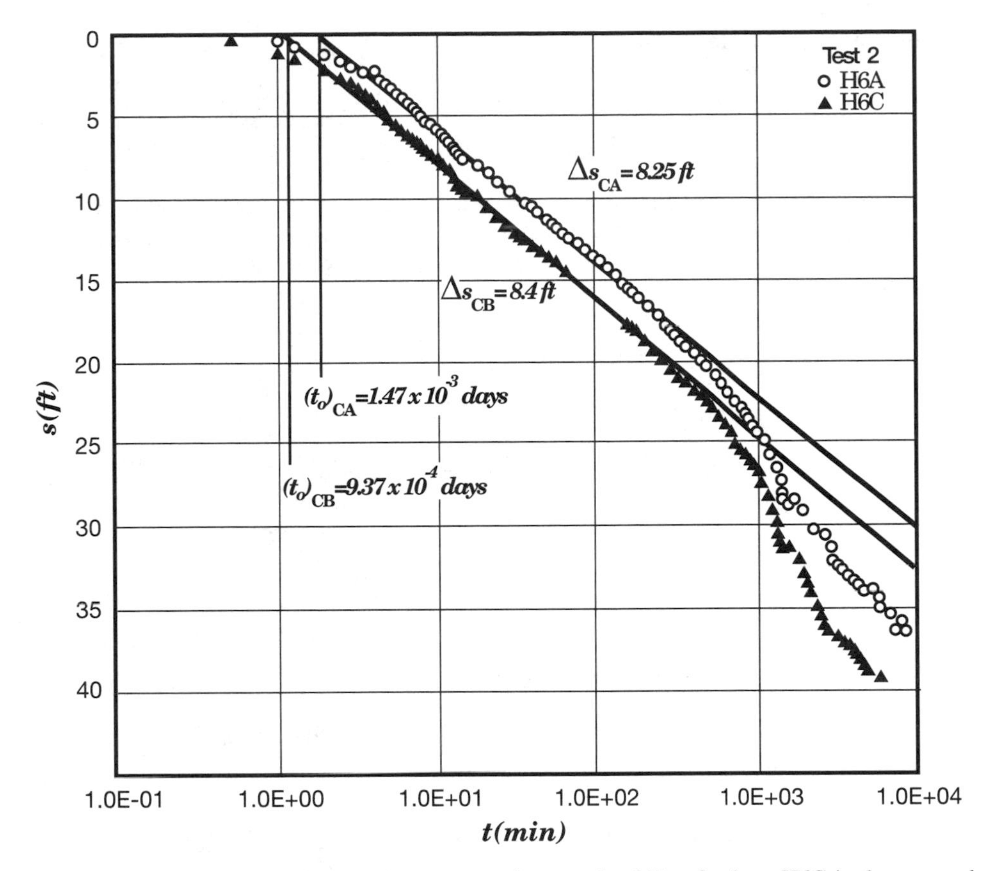

Figure 5-13 Semilogarithmic drawdown versus time graph of Test 2 where H6C is the pumped well. (Neuman et al., 1984.)

Substitution of the Δs values of Step 3 into the above expressions gives four T_e values:

$$T_{e_{BA}} = 67.32 \text{ ft}^2/\text{day}$$
$$T_{e_{BC}} = 69.64 \text{ ft}^2/\text{day}$$
$$T_{e_{CA}} = 71.13 \text{ ft}^2/\text{day}$$
$$T_{e_{CB}} = 68.88 \text{ ft}^2/\text{day}$$

The arithmetic average of the above four values is

$$T_{e_{av}} \cong 69 \text{ ft}^2/\text{day}$$

Step 5: The t intercepts for Test 1 (see Figure 5-12) are

$$(t_0)_{\mathrm{BA}} = 1.46 \times 10^{-3} \text{ days}$$
$$(t_0)_{\mathrm{BC}} = 8.02 \times 10^{-4} \text{ days}$$

The t intercepts for Test 2 (see Figure 5-13) are

$$(t_0)_{CA} = 1.47 \times 10^{-3} \text{ days}$$
$$(t_0)_{CB} = 9.37 \times 10^{-4} \text{ days}$$

Step 6: Since there are four time-drawdown data sets, the appropriate method to compute T'_{xx}, T'_{yy}, and T'_{xy} is by means of the equations based on the least-squares method as given by Eqs. (5-37), (5-38), and (5-39) with b_{ij} given by Eq. (5-35). With these, the normal equations become:

$$1.687 \times 10^8 T'_{xx} + 5.624 \times 10^7 T'_{yy} + 6.494 \times 10^7 T'_{xy} = 2.565 \times 10^5$$
$$5.624 \times 10^7 T'_{xx} + 1.187 \times 10^8 T'_{yy} + 2.165 \times 10^7 T'_{xy} = 2.427 \times 10^5$$
$$6.494 \times 10^7 T'_{xx} + 2.165 \times 10^7 T'_{yy} + 2.249 \times 10^8 T'_{xy} = 2.583 \times 10^4$$

Step 7: The solution of the equations in Step 6 are

$$T'_{xx} = 1.136 \times 10^{-3} \text{ ft}^2/\text{day}$$
$$T'_{yy} = 1.571 \times 10^{-3} \text{ ft}^2/\text{day}$$
$$T'_{xy} = -3.645 \times 10^{-4} \text{ ft}^2/\text{day}$$

Step 8: Substitution the values of T_e (Step 4) and T'_{xx}, T'_{xy} (Step 7) into Eq. (5-25) gives the storage coefficient (S):

$$S = \frac{1}{T_e}(T'_{xx}T'_{yy} - T'_{xy})^{1/2}$$
$$= \frac{1}{69.0 \text{ ft}^2/\text{day}}\Big[(1.136 \times 10^{-3} \text{ ft}^2/\text{day})(1.571 \times 10^{-3} \text{ ft}^2/\text{day}) - (-3.645 \times 10^{-4} \text{ ft}^2/\text{day})^2\Big]^{1/2}$$
$$S = 1.86 \times 10^{-5}$$

Step 9: Introducing the values determined in Step 7 and Step 8 into Eqs. (5-23) gives the components of the transmissivity tensor:

$$T_{xx} = 60.99 \text{ ft}^2/\text{day}$$
$$T_{yy} = 84.34 \text{ ft}^2/\text{day}$$
$$T_{xy} = -19.56 \text{ ft}^2/\text{day}$$

Step 10: With the components of the transmissivity tensor now known, the principal transmissivities ($T_{\xi\xi}$ and $T_{\eta\eta}$) can be determined from Eqs. (5-11) and (5-12), respectively,

$$T_{\xi\xi} = \tfrac{1}{2}\Big\{(60.99 \text{ ft}^2/\text{day} + 84.34 \text{ ft}^2/\text{day}) + [(60.99 \text{ ft}^2/\text{day} - 84.34 \text{ ft}^2/\text{day})^2 + 4(-19.56 \text{ ft}^2/\text{day})^2]^{1/2}\Big\}$$
$$= 95.44 \text{ ft}^2/\text{day}$$

$$
\begin{aligned}
T_{\eta\eta} &= \tfrac{1}{2}\Big\{(60.99\ \text{ft}^2/\text{day} + 84.34\ \text{ft}^2/\text{day}) \\
&\qquad + [(60.99\ \text{ft}^2/\text{day} - 84.34\ \text{ft}^2/\text{day})^2 + 4(-19.56\ \text{ft}^2/\text{day})^2]^{1/2}\Big\} \\
&= 49.88\ \text{ft}^2/\text{day}
\end{aligned}
$$

The angle θ between the positive x axis (east in Figure 5-11) and the direction of maximum principal transmissivity is calculated from Eq. (5-14):

$$
\tan(2\theta) = \frac{2(-19.56\ \text{ft}^2/\text{day})}{60.99\ \text{ft}^2/\text{day} - 84.34\ \text{ft}^2/\text{day}} = 1.675
$$

$$
\theta = 29.58^\circ
$$

$$
\theta = 29^\circ 35'
$$

Thus, the principal ξ coordinate axis has $29^\circ 35'$ from the x axis.

Example 5-4: Determine the possible combinations for the data given in Example 5-3 for the application of three time-drawdown data sets (Neuman et al., 1984).

Solution: Instead of the least-squares solution in Example 5-3, solutions for T_{xx}, T_{yy}, and T_{xy} can also be determined by using the three time-drawdown equations as given by Eqs. (5-30), (5-31), and (5-32). The reader may verify that the least-squares solution in Example 5-3 is very close to those determined by considering only three sets of time-drawdown data at a time, according to either of the following two permissible combinations:

1. Observation wells H6A and H6C from Test 1 and observation well H6A from Test 2.
2. Observation wells H6A and H6B from Test 2 and observation well H6A from Test 1.

The following other two combinations will produce unacceptable results because of the reasons mentioned previously:

1. Observation wells H6A and H6C from Test 1 and observation well H6B from Test 2.
2. Observation wells H6A and H6B from Test 2 and observation well H6C from Test 1.

5.3 HANTUSH'S METHODS

Apart from the methods presented in Section 5.2 for analyzing drawdown data for anisotropic confined aquifers, Hantush (1966a, b) developed methods of drawdown data analysis for anisotropic aquifers based on his original theoretical analysis. In the following sections, Hantush's methods with their associated theory are presented with examples.

5.3.1 Drawdown Equations for Isotropic Confined Aquifers

The pumping tests data analysis methods for confined aquifers are based on the Theis equation (Chapter 4, Section 4.2.4.1):

$$s(r,t) = \frac{Q}{4\pi T} W(u) \tag{5-40}$$

where $s(r,t)$ is the drawdown at the radial distance r and time t, $W(u)$ is the *Theis well function*, and

$$u = \frac{r^2}{4\nu t} \tag{5-41}$$

$$\nu = \frac{T}{S} \tag{5-42}$$

The data analysis methods based on the Theis equation are extensively covered in Chapter 4, and henceforth they will be referred as *isotropic methods of data analysis*.

5.3.2 Drawdown Equations for Anisotropic Confined Aquifers

Hantush (1966a, b) showed that the solutions developed for homogeneous and isotropic aquifers can be used for problems in homogeneous and anisotropic aquifers by using a simple transformation of coordinates. Therefore, the counterparts of Eqs. (5-40), (5-41), and (5-42) are

$$s(r,t) = \frac{Q}{4\pi T_e} W(u') \tag{5-43}$$

$$u' = \frac{r^2}{4\nu' t} \tag{5-44}$$

$$\nu' = \frac{T_r}{S} \tag{5-45}$$

$$T_e = (T_{\xi\xi} T_{\eta\eta})^{1/2} \tag{5-46}$$

in which the coordinate axes ξ and η are parallel the principal directions of anisotropy. In Eq. (5-46), T_e is a constant because $T_{\xi\xi}$ and $T_{\eta\eta}$ are constant. However, in Eq. (5-45), ν' is not a constant but is a function of radial direction, since it depends on T_r that varies with the radial coordinates.

The drawdown equation for homogeneous and anisotropic confined aquifers, Eq. (5-43), is analogous to the drawdown equation for homogeneous and isotropic confined aquifers, Eq. (5-40). The difference is the meaning of the parameters of those equations. Based on this analogy, it is possible to use the isotropic methods of drawdown data analysis for the development of procedures for analyzing similar drawdown data from anisotropic confined aquifers (Hantush, 1966b).

5.3.3 Analysis of Drawdown Equations

5.3.3.1 Applicability of the Isotropic Methods for Anisotropic Confined Aquifers. The isotropic methods of data analysis for confined aquifers are based on Eqs. (5-40), (5-41), and (5-42), in which the parameters T and ν are constants. As mentioned in Section 5.3.2, the aforementioned parameters are not constants for anisotropic aquifers. It is clear from Eq. (5-46) that T_e is constant regardless of the location of the observation point; but, ν', as defined by Eq. (5-45), vary from direction to direction. As a result, the drawdown distribution in anisotropic confined aquifers can be expressed by Eq. (5-43), with constant values for T_e and ν'. Consequently, the isotropic methods of drawdown data analysis for an observation well located on a radial line from the pumped well in an anisotropic aquifer will generate values for T_e and ν' for that direction (Hantush, 1966b).

As can be seen from Eqs. (5-45) and (5-46) the unknowns are $T_{\xi\xi}$, $T_{\eta\eta}$, S, and the principal directions of anisotropy, θ. As a result of this situation, more than one observation well for an anisotropic confined aquifer is needed to determine the aquifer parameters. Details are presented in subsequent sections.

5.3.3.2 Pumping Test Data Analysis Methods for Anisotropic Confined Aquifers. In this section, the equations which are the base of Hantush's data analysis methods for anisotropic aquifers are presented based on Hantush (1966b). Figure 5-14 shows three groups of wells (by definition, one well constitutes a group) located on different radial lines which are called *rays*. The second and third rays (Ray 2 and Ray 3) make the angles α and β, respectively, with Ray 1. The ξ and η coordinate axes are parallel to the principal directions of anisotropy, and the ξ axis makes θ angle with the first line (Ray 1). As a result, the first, second, and third lines of observation wells make the angles θ, $\theta + \alpha$, and $\theta + \beta$ with the ξ axis, respectively. Let the values of the aquifer parameters (ν', T_r), for the first, second, and third observation well groups be denoted by (ν_1', T_{r_1}), (ν_2', T_{r_2}), and (ν_3', T_{r_3}), respectively. The expressions for directional hydraulic conductivities and transmissivities for two-dimensional cases are presented in Section 2.7.1 of Chapter 2.

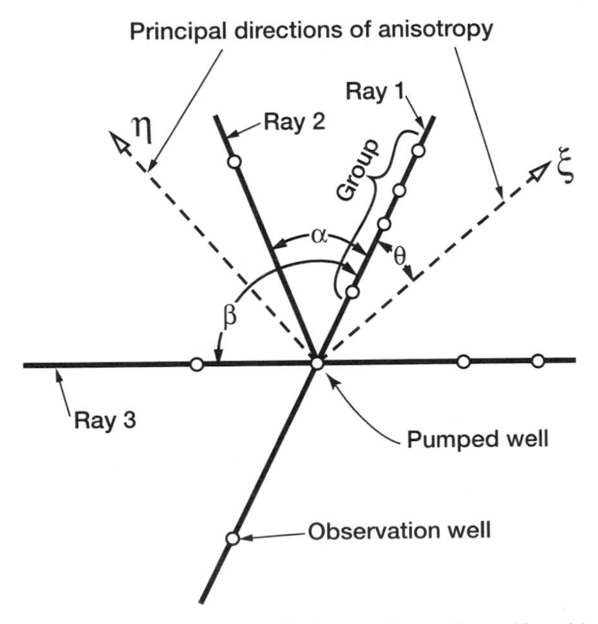

Figure 5-14 Location of three observation wells in an anisotropic aquifer with unknown principal directions of anisotropy. (After Hantush, 1966b.)

From Eq. (2-148) of Chapter 2 we obtain

$$T_{r_1} = \frac{T_{\xi\xi}}{\cos^2\theta + \dfrac{T_{\xi\xi}}{T_{\eta\eta}}\sin^2\theta} \tag{5-47}$$

$$T_{r_2} = \frac{T_{\xi\xi}}{\cos^2(\theta+\alpha) + \dfrac{T_{\xi\xi}}{T_{\eta\eta}}\sin^2(\theta+\alpha)} \tag{5-48}$$

$$T_{r_3} = \frac{T_{\xi\xi}}{\cos^2(\theta+\beta) + \dfrac{T_{\xi\xi}}{T_{\eta\eta}}\sin^2(\theta+\beta)} \tag{5-49}$$

Dividing the corresponding sides of Eqs. (5-47) and (5-48) and Eqs. (5-47) and (5-49) by each other, respectively, gives

$$a = \frac{\cos^2(\theta+\alpha) + n\sin^2(\theta+\alpha)}{\cos^2\theta + n\sin^2\theta} \tag{5-50}$$

$$b = \frac{\cos^2(\theta+\beta) + n\sin^2(\theta+\beta)}{\cos^2\theta + n\sin^2\theta} \tag{5-51}$$

where

$$a = \frac{T_{r_1}}{T_{r_2}} = \frac{\nu_1'}{\nu_2'} \tag{5-52}$$

$$b = \frac{T_{r_1}}{T_{r_3}} = \frac{\nu_1'}{\nu_3'} \tag{5-53}$$

and

$$n = \frac{T_{\xi\xi}}{T_{\eta\eta}} = \left(\frac{T_e}{T_{\eta\eta}}\right)^2 \tag{5-54}$$

5.3.3.3 Analysis of the Equations. The equations that are presented in Section 5.3.3.2 are analyzed in the following sections for the conditions that (1) the principal directions of anisotropy are known and (2) the principal directions of anisotropy are unknown.

Principal Directions of Anisotropy are Known. If the principal directions of anisotropy, which is defined by the angle θ in Figure 5-14, are known, there are only three unknowns ($T_{\xi\xi}$ and $T_{\eta\eta}$, and S) to be determined. If there are at least two groups of observation wells on different radial lines (refer to Figure 5-14), these unknowns can be determined. Let us consider that these lines are Ray 1 and Ray 2 in Figure 5-14. From Eqs. (5-50) and (5-54):

$$T_{\eta\eta} = T_e \left[\frac{\sin^2(\theta+\alpha) - a\sin^2\theta}{a\cos^2\theta - \cos^2(\theta+\alpha)}\right]^{1/2} \tag{5-55}$$

Because $\theta, \alpha, T_e, \nu_1', \nu_2'$, and consequently a are known from the observed drawdown data, the successive solutions of Eqs. (5-55), (5-54), (5-47), (5-48), and

$$S = \frac{T_{r_1}}{\nu_1'}, \qquad S = \frac{T_{r_2}}{\nu_2'} \tag{5-56}$$

will provide values for $T_{\eta\eta}, T_{\xi\xi}, T_{r_1}, T_{r_2}$, and S, respectively.

Principal Directions of Anisotropy Are Unknown. If the principal directions of anisotropy, which is defined by the angle θ in Figure 5-14, are unknown, the number of unknowns become four. If there are at least three groups of observation wells on different radial lines (refer to Figure 5-14), these four unknowns can be determined. Solving Eqs. (5-50) and (5-51) simultaneously for θ and n gives the following equations:

$$\tan(2\theta) = -2\frac{(b-1)\sin^2\alpha - (a-1)\sin^2\beta}{(b-1)\sin(2\alpha) - (a-1)\sin(2\beta)} \tag{5-57}$$

$$n = \frac{\cos^2(\theta+\alpha) - a\cos^2\theta}{a\sin^2\theta - \sin^2(\theta+\alpha)} \tag{5-58}$$

or

$$n = \frac{\cos^2(\theta+\beta) - b\cos^2\theta}{b\sin^2\theta - \sin^2(\theta+\beta)} \tag{5-59}$$

According to the sign convention in Figure 5-14, a negative value of θ indicates that the positive ξ axis is on the left side of the first observation wells line. Equation (5-57) forms a trigonometric equation which has two roots for the unknown 2θ between the angles 0 and 2π in the ξ–η plane. Therefore, if the first root is γ, the other one will be $\pi + \gamma$. As a result, the values of θ will be $\gamma/2$ and $\gamma/2 + \pi/2$. Of these two values of θ, one of them creates $n > 1$ and the other one creates $n < 1$. The ξ axis is assumed to be located along the major axis of anisotropy and $n = T_{\xi\xi}/T_{\eta\eta} > 1$. Therefore, the θ value that makes $n > 1$ locates the major axis of anisotropy. The other value of θ locates the minor axis of anisotropy, the η axis.

The angles α and β can be determined from the geometry of the wells as shown in Figure 5-14. The aquifer parameters T_e, ν_1', ν_2', and ν_3' can be determined from the observed drawdown data using the isotropic methods of analysis. The values of a and b can be determined from Eqs. (5-52) and (5-53), respectively, using the above-mentioned values. Therefore, successive solutions of Eq. (5-57), Eq. (5-58) or (5-59), Eqs. (5-54), (5-47), (5-48), and (5-49), and

$$S = \frac{T_{r_1}}{\nu_1'} \qquad S = \frac{T_{r_2}}{\nu_3'} \qquad S = \frac{T_{r_3}}{\nu_3'} \tag{5-60}$$

will end up values for θ, n, $T_{\eta\eta}$, and then $T_x, T_{r_1}, T_{r_2}, T_{r_3}$, and S, respectively.

5.3.4 Pumping Test Data Analysis Methods

The isotropic drawdown versus time data analysis methods determine the values of ν' and T_e as defined by Eqs. (5-45) and (5-46), respectively. In reference to Figure 5-14, let us consider that the application of the isotropic methods to the first, second, and third group of wells is produced (T_{e_1}, ν_1'), (T_{e_2}, ν_2'), and (T_{e_3}, ν_3'), respectively. From the theoretical standing point of view, the three isotropic transmissivities must be the same. In practice, this requirement should be used as a guide in applying the isotropic methods of analysis. If these three isotropic transmissivities are not at least approximately the same, it means that field conditions are not in agreement with the assumptions of theory and other vitiating conditions must be present.

In the following sections, the data analysis procedures developed by Hantush (1966b) are presented in a stepwise manner under the conditions that the three isotropic transmissivities mentioned above are approximately the same.

5.3.4.1 Procedure I: Principal Directions of Anisotropy Are Known. The principal directions of anisotropy, the ξ and η axes in Figure 5-14, are known. A group of wells (one or more wells) on the same radial line will be referred to as a *ray of observation wells.* The two groups of wells are located on Ray 1 and Ray 2 in Figure 5-14, and Ray 1 makes a θ angle with the ξ axis. Because θ is known, two groups of wells (or two rays) are enough to determine the aquifer parameters. The steps under the conditions that the angles θ and α are known are as follows:

Step 1: Using the isotropic methods of drawdown data analysis for confined aquifers (Chapter 4, Section 4.2.7) determine (T_{e_1}, ν_1') and (T_{e_2}, ν_2').

Step 2: With the known values of ν_1' and ν_2', calculate the value of a from Eq. (5-52).

Step 3: Based on the facts mentioned above, calculate T_e from the following equation:

$$T_e = \tfrac{1}{2}(T_{e_1} + T_{e_2}) \tag{5-61}$$

Step 4: Using the known values of T_e, θ, and α, calculate $T_{\eta\eta}$ from Eq. (5-55).

Step 5: Using the known values of T_e and $T_{\eta\eta}$, calculate $T_{\xi\xi}$ from Eq. (5-54) as

$$T_{\xi\xi} = \frac{T_e^2}{T_{\eta\eta}} \tag{5-62}$$

Step 6: Using the known values of $T_{\xi\xi}$, $T_{\eta\eta}$, θ, and α, calculate T_{r_1} and T_{r_2} from Eqs. (5-47) and (5-48), respectively.

Step 7: Because Eqs. (5-56) provides two S values, calculate the average S value from the following equation:

$$S = \frac{1}{2}\left(\frac{T_{r_1}}{\nu_1'} + \frac{T_{r_2}}{\nu_2'}\right) \tag{5-63}$$

5.3.4.2 Procedure II: Principal Directions of Anisotropy Are Unknown. If the principal directions of anisotropy (the ξ and η axes in Figure 5-14) are not known, three groups of wells are needed to determine the unknowns. Let us assume the three groups of wells are located on Ray 1, Ray 2, and Ray 3 in Figure 5-14. The additional unknown compared with Procedure I (Section 5.3.4.1) is the angle θ between the ξ axis and Ray 1. The angles α and β are the orientation angles of the other group lines (Ray 2 and Ray 3) with the first group line (Ray 1). The unknown principal directions of anisotropy, as well as the aquifer parameters, can be determined in a stepwise manner as follows:

Step 1: Using the isotropic methods of drawdown data analysis for confined aquifers, determine (T_{e_1}, ν_1'), (T_{e_2}, ν_2'), and (T_{e_3}, ν_3') for Ray 1, Ray 2, and Ray 3, respectively.

Step 2: Calculate a from Eq. (5-52) using the known values of ν_1' and ν_2'.

Step 3: Calculate b from Eq. (5-53) using the known values of ν_1' and ν_3'.

Step 4: Determine the value of θ from Eq. (5-57) using the known values of a, b, α, and β. Then, calculate the value of n from Eq. (5-58) or (5-59).

Step 5: As mentioned above, from the theoretical standing point of view, the three isotropic T_e values must be the same. Therefore, calculate the arithmetic average from

$$T_e = \tfrac{1}{3}(T_{e_1} + T_{e_2} + T_{e_3}) \tag{5-64}$$

And based on Eq. (5-54), calculate $T_{\eta\eta}$ from the following equation:

$$T_{\eta\eta} = T_e n^{-1/2} \tag{5-65}$$

Step 6: Calculate the value of $T_{\xi\xi}$ from Eq. (5-62) using the known values of T_e and $T_{\eta\eta}$.

Step 7: Using the known values of $T_{\xi\xi}$, $T_{\eta\eta}$, θ, α, and β, determine the values of T_{r_1}, T_{r_2}, and T_{r_3} from Eqs. (5-47), (5-48), and (5-49), respectively.

Step 8: Calculate the value of storage coefficient from Eq. (5-60) as

$$S = \frac{1}{3}\left(\frac{T_{r_1}}{\nu_1'} + \frac{T_{r_2}}{\nu_2'} + \frac{T_{r_3}}{\nu_3'}\right) \tag{5-66}$$

Step 9: Write the transmissivity equation for any angle θ using Eq. (2-148) of Chapter 2:

$$T_r = \frac{T_{\xi\xi}}{\cos^2\theta + \dfrac{T_{\xi\xi}}{T_{\eta\eta}}\sin^2\theta} \tag{5-67}$$

If more than three groups of observation wells are available, the application of the above-given steps to each combination of three observation wells will provide a set of values for each of the aquifer parameters and principal directions of anisotropy. The average of each set of aquifer parameter values will be considered as the average value of the corresponding aquifer parameter. For example, let us consider that four groups of wells are available. Therefore, there will be four combinations of three observation wells. The application of the method to these combinations will generate four values of θ angles and four values for each of the aquifer parameters. From the theoretical standing point of view, these four set of values should be the same.

Example 5-5: Using Hantush's Procedure II, determine $T_{\xi\xi}$, $T_{\eta\eta}$, and the principal directions of transmissivity of Example 5-1.

Solution: The coordinate system and wells geometry data for Example 5-1 are presented in Figure 5-4. Based on the geometric data given in Figure 5-4, the values of α and β angles and other pertinent data are shown in Figure 5-15. As can be seen from Figure 5-15, there is a total of three observation wells and each one is located on a different ray. Therefore,

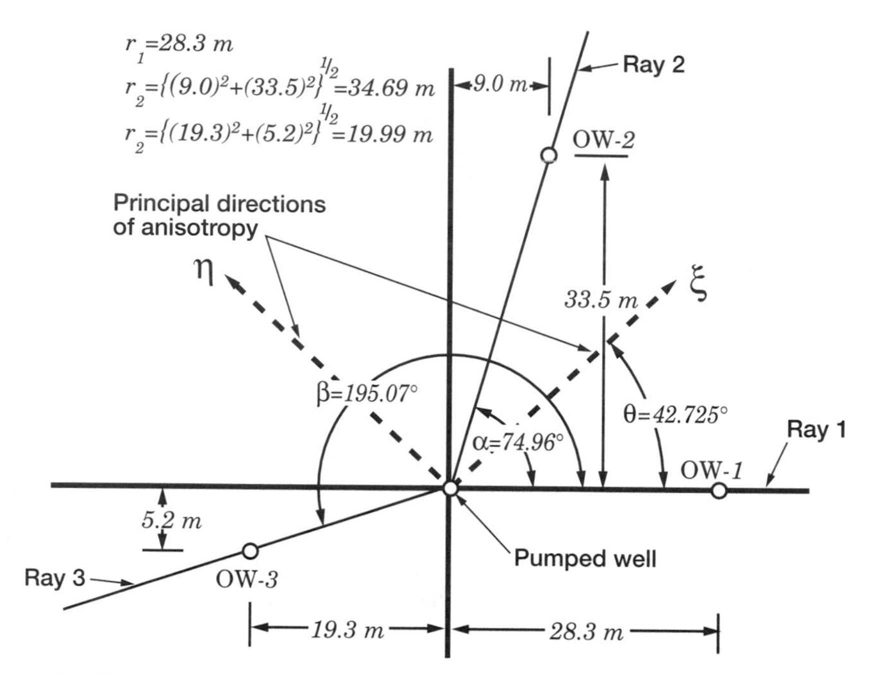

Figure 5-15 Location of the three group lines of observation wells for a pumping test in an anisotropic confined aquifer of Example 5-6.

the aquifer parameters can be determined from the data of these well systems. The stepwise procedure as given in Section 5.3.4.2 is applied as follows:

Step 1: The requirement of this step is to determine T_{e_1}, T_{e_2}, and T_{e_3} using the methods of analysis for isotropic aquifers either by type-curve method or other methods. Here, the results of the Theis type-curve matching method will be used. The match points of observation wells OW-1, OW-2, and OW-3 are shown in Figures 5-6, 5-7, and 5-8, respectively. The Theis type-curve matching coordinates are given in Step 4 of Example 5-1. With these coordinates, Eq. (5-43) gives

$$
\begin{aligned}
T_{e_1} = T_{e_2} = T_{e_3} &= \frac{Q}{4\pi s} W(u') \\
&= \frac{1}{(4\pi)(1.0 \text{ m})} \frac{12.57 \text{ liters/s}}{1000 \text{ liters/m}^3}(2.0) \\
&= 2.0 \times 10^{-3} \text{ m}^2/\text{s}
\end{aligned}
$$

As seen above, all three match points yield the same results because the values of $W(u')$ and s are the same. From Eq. (5-44), using the match-point values of Example 5-1 we obtain

$$
\nu_1' = \frac{(28.3 \text{ m})^2}{(4)(20 \times 60 \text{ s})}(88) = 14.68 \text{ m}^2/\text{s}
$$

$$\nu_2' = \frac{(34.69\text{ m})^2}{(4)(20 \times 60\text{ s})}(48) = 12.03\text{ m}^2/\text{s}$$

$$\nu_3' = \frac{(19.99\text{ m})^2}{(4)(20 \times 60\text{ s})}(135) = 11.24\text{ m}^2/\text{s}$$

Step 2: From Eq. (5-52) we obtain

$$a = \frac{\nu_1'}{\nu_2'} = \frac{14.68\text{ m}^2/\text{s}}{12.03\text{ m}^2/\text{s}} = 1.220$$

Step 3: From Eq. (5-53) we obtain

$$b = \frac{\nu_1'}{\nu_3'} = \frac{14.68\text{ m}^2/\text{s}}{11.24\text{ m}^2/\text{s}} = 1.306$$

Step 4: From Figure 5-16, $\alpha = 74.96°$ and $\beta = 195.07°$. Substitution these values with those in Step 3 and Step 4 into Eq. (5-57) gives

$$\begin{aligned}
\tan(2\theta) &= -2\frac{(b-1)\sin^2\alpha - (a-1)\sin^2\beta}{(b-1)\sin(2\alpha) - (a-1)\sin(2\beta)} \\
&= -2\frac{(1.306-1)\sin^2(74.96°) - (1.220-1)\sin^2(195.07°)}{(1.306-1)\sin(149.92°) - (1.220-1)\sin(390.14°)} \\
&= -12.56 \\
2\theta &= \arctan(-12.56) \\
\theta &= -42.725° = -42°\ 44'
\end{aligned}$$

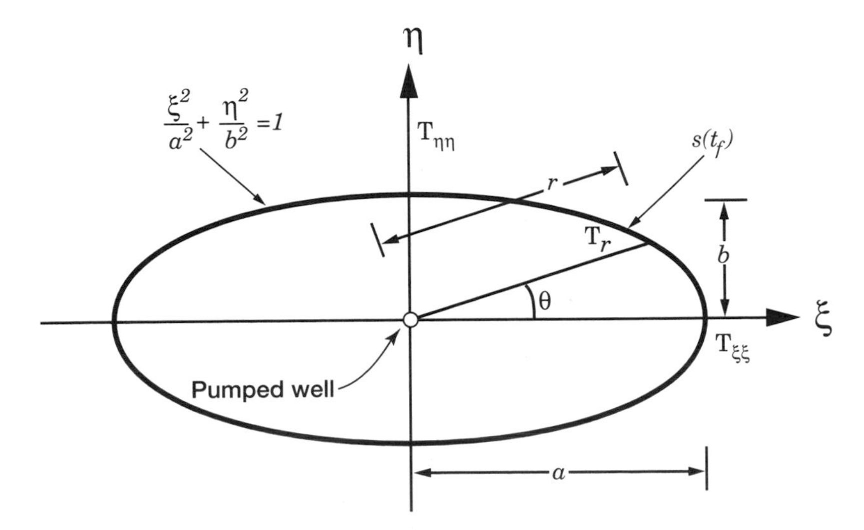

Figure 5-16 Drawdown contour around a well in an anisotropic aquifer.

Based on the sign convention in Figure 5-14, the location of the first principal axis, the ξ axis, is shown in Figure 5-15. The second principal axis, the η axis, makes $\theta + \pi/2$ with ξ axis. The first angle is $\theta = -42.725°$ and the other one is $\theta = -42.725° + 90° = 47.275°$. Substituting the first value into Eq. (5-58) gives

$$n = \frac{\cos^2(\theta + \alpha) - a\cos^2\theta}{a\sin^2\theta - \sin^2(\theta + \alpha)}$$

$$= \frac{\cos^2(-42.725° + 74.96°) - 1.22\cos^2(-42.725°)}{1.22\sin^2(-42.725°) - \sin^2(-42.725° + 74.96°)}$$

$$n = 0.209 < 1$$

Since the ξ axis is assumed to be along the major axis of anisotropy, $n = T_{\xi\xi}/T_{\eta\eta} > 1$. Therefore, the above $n = 0.209$ is not acceptable. Now introducing $\theta = 47.275°$ into Eq. (5-58) gives

$$n = \frac{\cos^2(47.275° + 74.96°) - 1.22\cos^2(47.275°)}{1.22\sin^2(47.275°) - \sin^2(47.275° + 74.96°)}$$

$$= 4.877 > 1$$

This value is acceptable because $n > 1$.

Step 5: From Eq. (5-64) we obtain

$$T_e = \tfrac{1}{3}(T_{e_1} + T_{e_2} + T_{e_3})$$

$$= \tfrac{1}{3}(2.0 \times 10^{-3} + 2.0 \times 10^{-3} + 2.0 \times 10^{-3})\ \text{m}^2/\text{s}$$

$$= 2.0 \times 10^{-3}\ \text{m}^2/\text{s}$$

From Eq. (5-65) we obtain

$$T_{\eta\eta} = T_e n^{-1/2} = (2.0 \times 10^{-3}\ \text{m}^2/\text{s})(4.877)^{-1/2} = 0.91 \times 10^{-3}\ \text{m}^2/\text{s}$$

Step 6: From Eq. (5-62) we obtain

$$T_{\xi\xi} = \frac{T_e^2}{T_{\eta\eta}} = \frac{(2.0 \times 10^{-3}\ \text{m}^2/\text{s})^2}{0.91 \times 10^{-3}\ \text{m}^2/\text{s}} = 4.40 \times 10^{-3}\ \text{m}^2/\text{s}$$

Step 7: From Eq. (5-47) we obtain

$$T_{r_1} = \frac{T_{\xi\xi}}{\sin^2\theta + \dfrac{T_{\xi\xi}}{T_{\eta\eta}}\sin^2\theta}$$

$$= \frac{4.40 \times 10^{-3}\ \text{m}^2/\text{s}}{\cos^2(47.275°) + \dfrac{4.40 \times 10^{-3}\ \text{m}^2/\text{s}}{0.91 \times 10^{-3}\ \text{m}^2/\text{s}}\sin^2(47.275°)}$$

$$T_{r_1} = 1.433 \times 10^{-3}\ \text{m}^2/\text{s}$$

From Eq. (5-48) we obtain

$$T_{r_2} = \frac{T_{\xi\xi}}{\cos^2(\theta + \alpha) + \dfrac{T_{\xi\xi}}{T_{\eta\eta}} \sin^2(\theta + \alpha)}$$

$$= \frac{4.40 \times 10^{-3} \text{ m}^2/\text{s}}{\cos^2(47.275° + 74.96°) + \dfrac{4.40 \times 10^{-3} \text{ m}^2/\text{s}}{0.91 \times 10^{-3} \text{ m}^2/\text{s}} \sin^2(47.275° + 74.96°)}$$

$$T_{r_2} = 1.175 \times 10^{-3} \text{ m}^2/\text{s}$$

From Eq. (5-49) we obtain

$$T_{r_3} = \frac{T_{\xi\xi}}{\cos^2(\theta + \beta) + \dfrac{T_{\xi\xi}}{T_{\eta\eta}} \sin^2(\theta + \beta)}$$

$$= \frac{4.40 \times 10^{-3} \text{ m}^2/\text{s}}{\cos^2(47.275° + 195.07°) + \dfrac{4.40 \times 10^{-3} \text{ m}^2/\text{s}}{0.91 \times 10^{-3} \text{ m}^2/\text{s}} \sin^2(47.275° + 195.07°)}$$

$$T_{r_3} = 1.097 \times 10^{-3} \text{ m}^2/\text{s}$$

Step 8: From Eq. (5-66) we obtain

$$S = \frac{1}{3}\left(\frac{T_{r_1}}{\nu_1'} + \frac{T_{r_2}}{\nu_2'} + \frac{T_{r_3}}{\nu_3}\right)$$

$$= \frac{1}{3}\left(\frac{1.433 \times 10^{-3} \text{ m}^2/\text{s}}{14.68 \text{ m}^2/\text{s}} + \frac{1.175 \times 10^{-3} \text{ m}^2/\text{s}}{12.03 \text{ m}^2/\text{s}} + \frac{1.097 \times 10^{-3} \text{ m}^2/\text{s}}{11.24 \text{ m}^2/\text{s}}\right)$$

$$S = 0.98 \times 10^{-4}$$

Step 9: From Eq. (2-148) of Chapter 2, the transmissivity expression for any direction is

$$T_r = \frac{T_{\xi\xi}}{\cos^2\theta + \dfrac{T_{\xi\xi}}{T_{\eta\eta}} \sin^2\theta} = \frac{4.40 \times 10^{-3} \text{ m}^2/\text{s}}{\cos^2\theta + 4.83 \sin^2\theta}$$

Example 5-6: Compare the results of the Papadopulos method (type-curve and straight line) and the Hantush method (type curve) for the data of Example 5-1.

Solution: As presented in detail in Section 5.2.1.5, the Papadopulos method requires that the aquifer parameters be determined using the isotropic methods of analyses. The results of the Papadopulos method based on the isotropic type-curve and straight-line methods are presented in Example 5-1 and Example 5-2, respectively. The results of Hantush's method (Procedure II) based the isotropic type-curve method as presented in Example 5-2 are presented in Example 5-5. The aquifer parameters based on these methods are presented in

TABLE 5-2 Comparison of Papadopulos and Hantush's Methods for the Data of Example 5-1

Method	$T_{\xi\xi}$ (m^2/s)	$T_{\eta\eta}$ (m^2/s)	θ	S
Papadopulos (Type-curve)	4.42×10^{-3}	0.90×10^{-3}	42° 43′	0.97×10^{-4}
Papadopulos (Straight-line)	4.02×10^{-3}	1.00×10^{-3}	45°	0.99×10^{-4}
Hantush (Procedure II) (Type-curve)	4.40×10^{-3}	0.91×10^{-3}	42° 44′	0.98×10^{-4}

Table 5-2, which shows that the results of different methods are approximately the same. Notice that the results based on the type-curve method are very close as compared with the straight-line method. This is not unexpected because the straight-line method is based on the approximated form of the Theis well function (see Section 4.2.6 of Chapter 4). Minor differences can be attributed to the roundoff errors.

5.4 HANTUSH AND THOMAS METHOD

The methods of Hantush for anisotropic aquifers as presented in Section 5.3 make use of the *isotropic methods*. However, these methods cannot be applied if the isotropic methods do not generate the necessary aquifer parameters for finalizing the analysis. For example, the Theis recovery method (Chapter 4, Section 4.2.7.4) is one of the isotropic methods that does not produce sufficient aquifer parameters to be used in Hantush's anisotropic methods. Drawdown contours under pumping conditions around a well in an anisotropic aquifer are in elliptical form (e.g., Figure 5-2). Hantush and Thomas (1966) developed a method for the determination of transmissivity in any radial direction and storage coefficient of a homogeneous and anisotropic confined aquifer.

If the spatial distribution and number of observation wells are sufficient such that the expected elliptical shape of an equal drawdown curve (or contour) for a certain time under pumping and recovery conditions is known, the principal axes of anisotropy can be determined using the *Hantush and Thomas method*. The method was primarily developed for analyzing recovery data, but it can also be used to analyze data for pumping tests as well.

5.4.1 Equations for the Equal Drawdown Curves

The drawdown equation for anisotropic confined aquifers in term of principal axes (ξ and η) is given by Eq. (5-9). With Eq. (5-46), Eq. (5-9) can be written as

$$s(\xi, \eta, t) = \frac{Q}{4\pi T_e} W(u_{\xi\eta}) \tag{5-68}$$

On the other hand, with $T_r = K_r b$, from Eq. (2-148) of Chapter 2 we obtain

$$T_r = \frac{T_{\xi\xi} T_{\eta\eta}}{T_{\xi\xi} \sin^2 \theta + T_{\eta\eta} \cos^2 \theta} \tag{5-69}$$

From Figure 2-30 of Chapter 2, $\sin\theta = \eta/r$ and $\cos\theta = \xi/r$. With these relations, Eq. (5-69) takes the form

$$T_r = \frac{r^2 T_{\xi\xi} T_{\eta\eta}}{T_{\eta\eta}\xi^2 + T_{\xi\xi}\eta^2} \tag{5-70}$$

Combination of Eqs. (5-10) and (5-70) gives

$$u_{\xi\eta} = \frac{Sr^2}{4T_r t} = \frac{r^2}{4\nu' t}, \qquad \nu' = \frac{T_r}{S} \tag{5-71}$$

Introducing Eq. (5-10) into Eq. (5-68) gives

$$s(\xi, \eta, t) = \frac{Q}{4\pi T_e} W(u_{\xi\eta}) = \frac{Q}{4\pi T_e} W\left[\left(\frac{S}{4t}\right)\left(\frac{\xi^2}{T_{\xi\xi}} + \frac{\eta^2}{T_{\eta\eta}}\right)\right] \tag{5-72}$$

Hantush and Thomas (1966) showed from Eq. (5-72) that the equal drawdown curves expression at any time, t_f, is

$$\frac{\xi^2}{T_{\xi\xi}} + \frac{\eta^2}{T_{\eta\eta}} = \frac{4t_f}{S} W^{-1}\left(\frac{4\pi T_e s}{Q}\right) \tag{5-73}$$

or

$$\frac{\xi^2}{a^2} + \frac{\eta^2}{b^2} = 1 \tag{5-74}$$

where

$$a^2 = \frac{4T_{\xi\xi} t_f}{S} W^{-1}\left(\frac{4\pi T_e s}{Q}\right) \tag{5-75}$$

$$b^2 = \frac{4T_{\eta\eta} t_f}{S} W^{-1}\left(\frac{4\pi T_e s}{Q}\right) \tag{5-76}$$

and $W^{-1}(x)$ is the inverse of the Theis well function. Equation (5-74) means that the equal drawdown curve at any time, t_f, is an ellipse which is shown in Figure 5-16. Dividing the corresponding sides of Eqs. (5-75) and (5-76) to each other gives

$$\frac{a}{b} = \left(\frac{T_{\xi\xi}}{T_{\eta\eta}}\right)^{1/2} \tag{5-77}$$

Combination of Eq. (5-77) with Eq. (5-46) yields

$$\frac{a}{b} = \frac{T_e}{T_{\eta\eta}} = \frac{T_{\xi\xi}}{T_e} \tag{5-78}$$

Equation (5-69) can also be expressed as

$$\frac{1}{T_r} = \frac{1}{T_{\xi\xi}} \cos^2 \theta + \frac{1}{T_{\eta\eta}} \sin^2 \theta \tag{5-79}$$

When we use the trigonometric relations for the θ angle presented earlier, Eq. (5-79) takes the form

$$\frac{r^2}{T_r} = \frac{\xi^2}{T_{\xi\xi}} + \frac{\eta^2}{T_{\eta\eta}} \tag{5-80}$$

and combination with Eq. (5-78) gives

$$\frac{r^2}{T_r} = \frac{ab}{T_e} \left(\frac{\xi^2}{a^2} + \frac{\eta^2}{b^2} \right) \tag{5-81}$$

Substitution of Eq. (5-74) into Eq. (5-81) and solving for the directional transmissivity yields the following equation:

$$T_r = \frac{r^2}{ab} T_e \tag{5-82}$$

5.4.2 Equations for Residual Equal Drawdown Curves

The general theory of the Theis recovery method is given in Section 4.2.7.4 of Chapter 4. The residual drawdown, s', equation for anisotropic aquifers can be expressed from Eq. (4-80) of Chapter 4 as

$$s' = \frac{Q}{4\pi T_e} M_*(\tau, u_0) \tag{5-83}$$

where

$$M_*(\tau, u_0) = W\left(\frac{u_0}{1+\tau} \right) - W\left(\frac{u_0}{\tau} \right) \tag{5-84}$$

$$u_0 = \frac{Sr^2}{4T_r t_0} = \frac{S}{4t_0} \left(\frac{\xi^2}{T_{\xi\xi}} + \frac{\eta^2}{T_{\eta\eta}} \right) \tag{5-85}$$

and

$$\tau = \frac{t'}{t_0} \tag{5-86}$$

where t_0 is the discharging period and t' is the time since pumping stopped. The equal residual drawdown curve at any t'_* since pumping stopped can be expressed as for the pumping case presented Section 5.4.1 (Hantush and Thomas, 1966):

$$\frac{\xi^2}{a_*^2} + \frac{\eta^2}{b_*^2} = 1 \tag{5-87}$$

with

$$a_*^2 = \frac{4T_{\xi\xi}t_0}{S} M_*^{-1}\left(\tau_*, \frac{4\pi T_e s'}{Q}\right) \tag{5-88}$$

$$b_*^2 = \frac{4T_{\eta\eta}t_0}{S} M_*^{-1}\left(\tau_*, \frac{4\pi T_e s'}{Q}\right) \tag{5-89}$$

where $M_*^{-1}(\tau_*, x)$ is the inverse of the function $M_*(\tau_*, u_0)$ for a given value of τ_*—that is, the value of u_0 corresponding to $\tau = \tau_*$ and $M_*(\tau_*, u_0) = \alpha$. The function $M_*(\tau, u_0)$ can be tabulated easily by using the extensively tabulated function $W(u)$ (Table 4-1 of Chapter 4).

Equation (5-87) represents the equal residual drawdown curve at any time t'_* since the pumping stopped. Equations (5-87), (5-88), and (5-89) are analogous to Eqs. (5-74), (5-75), and (5-76) of Section 5.4.1, respectively. Therefore, analysis similar to the one in Section 5.4.1 yields the following equation:

$$T_r = \frac{r^2}{a_* b_*} T_e \tag{5-90}$$

For relatively large time, $\tau > 100u_0$, Eq. (5-83) can be approximated, similar to Eq. (4-82) of Chapter 4, as

$$s' = \frac{2.303Q}{4\pi T_e} \log\left(\frac{t}{t'}\right) \tag{5-91}$$

where

$$t = t_0 + t' \tag{5-92}$$

or

$$s' = \frac{2.303Q}{4\pi T_e} \log\left(\frac{1+\tau}{\tau}\right) \tag{5-93}$$

5.4.3 Pumping Test Data Analysis Methods

As mentioned in Section 5.3.3.2, a group of wells located on a radial line is called *ray of observation wells* (see Figure 5-14). Figure 5-14 shows several rays intersecting at the pumped well location. Based on the theoretical foundations presented in the previous sections, the data analysis methods for pumping and recovery periods proposed by Hantush and Thomas (1966) are presented in the following sections.

5.4.3.1 Analysis During Pumping Period. The steps are as follows (Hantush and Thomas, 1966):

Step 1: Plot drawdown versus time (time on the logarithmic scale) for each observation well including both the pumping and recovery periods.

Step 2: Using the methods of drawdown data analysis for isotropic confined aquifers, determine T_{e_n} and ν'_n for all wells where the subscript n corresponds to the values determined from the nth ray of wells. Theoretically, the values of T_{e_n} should be the same—that is,

$$T_{e_1} = T_{e_2} = \cdots = T_{e_n} = \cdots \tag{5-94}$$

Equation (5-94) should be used as a guide when applying the isotropic methods for the determination of best results. If the values determined from the application of the isotropic methods are significantly different, it means that vitiating conditions other than anisotropy must be present. If Eq. (5-94) is satisfied approximately, take their average to represent the constant value of T_e for the whole system.

Step 3: The value of ν_n' based on any isotropic method of analysis in the direction of a ray of wells should be the same. If the different methods produce approximately equal values of ν_n', take the average of these values.

Step 4: Construct one or more ellipses of equal drawdown using the following general guidelines if needed: With the known values of T_e for the whole flow system and ν_n' for each ray of wells, Eq. (5-72) with Eq. (5-71) can be used to determine drawdown values at any time and any radial distance along the ray of wells. This kind of determination may be required to supplement the observational drawdown data (the semilogarithmic drawdown versus time plots from Step 1) for creating one or more ellipses of equal drawdown. The time corresponding to the equal drawdown curve should preferably be chosen close to the end of the pumping period in order to avoid the usual vitiating conditions related to early flow conditions.

Step 5: From Eq. (5-72) we obtain

$$\frac{4\pi s_f T_e}{Q} = W(u_{\xi\eta}) \tag{5-95}$$

where s_f is the value of particular equal drawdown curve at time t_f. Substitution of Eq. (5-82) into Eq. (5-71) and solving for S gives

$$S = \frac{4T_e t_f u_{\xi\eta}}{ab} \tag{5-96}$$

Because the values of s_f, T_e, and Q are known, the value of Theis well function, $W(u_{\xi\eta})$, can be determined. And using this, from the Theis well function table (Chapter 4, Table 4-1) the corresponding value of $u_{\xi\eta}$ can be read. From the equal drawdown (s_f) ellipse corresponding to t_f, the values of a and b (see Figure 5-16) can be obtained. Introducing all the known values into Eq. (5-96) gives the value of S. If more than one ellipse can be specified, repeat the computation for each ellipse and take the average of all S values.

Step 6: Based on the second expression in Eqs. (5-71), the storage coefficient, S, can also be calculated from the following equation:

$$S = \frac{1}{N}\sum_{n=1}^{N} \frac{T_{r_n}}{\nu_n'} \tag{5-97}$$

where T_{r_n} can be calculated from Eq. (5-82) (see Figure 5-16) and ν_n' is determined in Step 4 for the nth ray and N is the total number of rays.

Step 7: From Eq. (5-78) the expressions for principal transmissivities can be written as

$$T_{\xi\xi} = \frac{a}{b} T_e \tag{5-98}$$

$$T_{\eta\eta} = \frac{b}{a} T_e \tag{5-99}$$

With the known values of a, b, and T_e from the previous steps, calculate the principal transmissivities in ξ and η directions from Eqs. (5-98) and (5-99), respectively.

5.4.3.2 Analysis During Recovery Period. The data analysis steps are as follows (Hantush and Thomas, 1966):

Step 1: If the drawdown versus time can be extrapolated by extending the semilogarithmic curve through the whole recovery period, or a sufficient part of the recovery period, use the procedure that is given in Section 5.4.3.1 to determine the aquifer parameters. For this case, the drawdown (s) and the time (t) should be replaced by ($s - s'$) and t', respectively. Here, s is the extrapolated drawdown had pumping continued, s' is the residual drawdown during recovery, and t' is the time since pumping stopped. If this is not possible, follow the rest of the steps given below.

Step 2: Apply the Theis recovery method (Chapter 4, Section 4.2.7.4) for each observation well to determine T_{e_n}. Theoretically, these values should be the same. If the values are significantly different, conditions other than anisotropy must be present. If the values of T_{e_n} are approximately the same, calculate their average to be used for the constant value of T_e for the whole system.

Step 3: If the values of T_e and ν_n' are available from the pumping test data analysis, use Eq. (5-83) to calculate residual drawdown (s') values at any time and at any distance along the corresponding rays of wells. These residual drawdown values may be necessary when more than one equal residual drawdown curve will be created. For each equal residual drawdown ellipse, determine the values of a_* and b_*.

Step 4: With the known values for one or more ellipses of equal residual drawdown and their half axes lengths, the value of T_r can be determined from Eq. (5-90). The aquifer parameters can be determined from the following equations:

$$T_{\xi\xi} = \frac{a_*}{b_*} T_e \tag{5-100}$$

$$T_{\eta\eta} = \frac{b_*}{a_*} T_e \tag{5-101}$$

$$S = \frac{4 T_e t_0}{a_* b_*} M_*^{-1} \left(\tau_*, \frac{4\pi T_e s'}{Q} \right) \tag{5-102}$$

where s' is the value of residual equal drawdown curve and other values are defined in Section 5.4.2. Equations (5-100), (5-101), and (5-102) can similarly be derived as is done in Step 5 of Section 5.4.3.1.

CHAPTER 6

FULLY PENETRATING PUMPING WELLS IN HOMOGENEOUS AND ISOTROPIC/ANISOTROPIC CONFINED LEAKY AQUIFERS WITHOUT THE STORAGE OF THE CONFINING LAYER

6.1 INTRODUCTION

The Theis model (Chapter 4, Section 4.2) is based on the main assumption that the aquifer is perfectly confined. In other words, this means that the top and bottom boundaries of the confined aquifer are impermeable. As mentioned in Section 3.2.2 of Chapter 3, if the aquifer is not perfectly confined, leakage to the confined aquifer or from the confined aquifer may occur through the underlain or overlain confining layers. The leakage rates from the confined aquifer may be significant under water extraction or injection conditions depending on the magnitude of hydraulic gradients around the well. Hantush and Jacob (1955) extended the Theis model for the condition that the confined aquifer is overlain or underlain by a less permeable formation, called *aquitard* or *confining layer* (see Section 2.2 of Chapter 2) and developed a mathematical model that takes into account the thickness and hydraulic conductivity of the confining layer. Leaky confined aquifers are frequently encountered in subsurface flow systems, being either single or part of a multilayered aquifer system, and the degree of leakage may become significant depending on the thickness and hydraulic conductivity of the confining layer. Under pumping conditions from a leaky confined aquifer, water may be extracted from the overlain and underlain aquifers through the confining layers.

In this chapter the Hantush and Jacob homogeneous and isotropic leaky confined aquifer model (Hantush and Jacob, 1955) will be presented in detail; based on this model, various pumping test data analysis methods will be included with examples. Then, the pumping test data analysis methods for homogeneous and anisotropic leaky confined aquifers will be presented with examples in accordance with those for nonleaky confined aquifers as presented in Chapter 5.

6.2 HANTUSH AND JACOB MODEL FOR HOMOGENEOUS AND ISOTROPIC LEAKY CONFINED AQUIFERS

In the following sections, the theoretical foundations and practical implications of the Hantush and Jacob (1955) model are presented.

6.2.1 Problem Statement and Assumptions

Figure 6-1 shows cross-sectionally and areally a fully penetrating well in a leaky confined aquifer. The confining layer is overlain by an unconfined aquifer, and the main aquifer

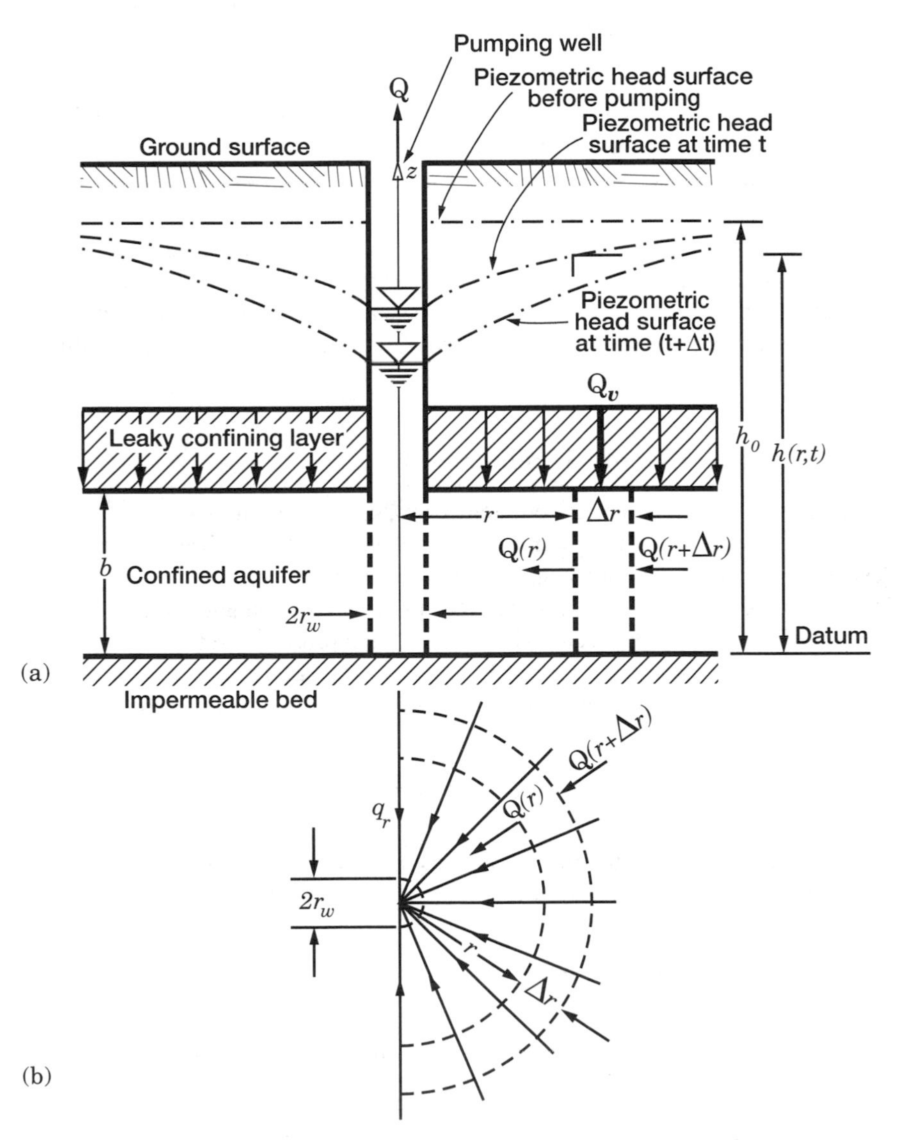

Figure 6-1 Schematic cross section of a leaky confined aquifer

(confined aquifer) is underlain by an impermeable bed. Hantush and Jacob (1955) were the first to develop a mathematical model for the aquifer dynamics under transient pumping conditions with the following assumptions (Reed, 1980):

1. The confined aquifer is homogeneous and isotropic.
2. The extraction rate (Q) of the well is constant.
3. The aquifer is horizontal, has constant thickness (b), and is overlain everywhere by a confining layer having constant vertical hydraulic conductivity (K') and a constant thickness (b').
4. The well penetrates the entire aquifer, and its diameter is infinitesimally small.
5. The confining bed is overlain, or underlain, by an infinite constant-head plane source; that is, the drawdown in the constant-head plane source is zero.
6. The confining layer does not release water from storage.
7. Flow in the confined aquifer is horizontal and in the confining layer is vertical.
8. Initially the water-table elevation of the unconfined aquifer and the potentiometric head elevation of the confined aquifer are the same and equal to h_0.

These assumptions are all self-explanatory with the exception of item 6, which is one of the key assumptions. The assumption given in item 6 means that the hydraulic gradients through the confining layer change instantaneously in accordance with a change in the hydraulic head in the confined aquifer. The models presented in Chapter 7 take into account the storage effects of the confining layers.

6.2.2 Governing Differential Equation

The governing partial differential equation for radial flow in an elastic confined aquifer with linear leakage was originally derived by Jacob (1946). The derivation of this equation is presented in the following paragraphs (Jacob, 1946; Bear, 1979).

The governing partial differential equation will be derived using the continuity principle for the aquifer ring shown in Figure 6-1. The derivation process is similar to that presented in Section 3.2.2.2 of Chapter 3.

The Darcy velocity in the radial direction (see Section 2.5.1.2 of Chapter 2) is

$$q_r = -K_r \frac{\partial h}{\partial r} \tag{6-1}$$

where K_r is the horizontal hydraulic conductivity, h is the hydraulic head, and r is the radial distance. The total flow rate at r for the whole aquifer thickness (b) is

$$Q(r) = -2\pi r b q_r = 2\pi r T \frac{\partial h}{\partial r} \tag{6-2}$$

The hydraulic head at the radius r is $h(r, t)$. After Δt time interval the hydraulic head will be $h(r, t + \Delta t)$ and the head difference will be

$$\Delta h = h(r, t) - h(r, t + \Delta t) \tag{6-3}$$

Applying the continuity principle for the ring shown in Figure 6-1 gives

$$[Q(r) - Q(r + \Delta r)]\Delta t + Q_v \Delta t = 2\pi r \Delta r \Delta h S \tag{6-4}$$

where

$$Q_v = (2\pi r \Delta r) q_v \tag{6-5}$$

in which q_v is the leakage rate and can be expressed from Eq. (2-186) of Chapter 2 as

$$q_v = K' \frac{h_0 - h}{b'} \tag{6-6}$$

where K' is the vertical hydraulic conductivity of the confining layer, and h_0 is the water-table elevation of the unconfined aquifer and the initial potentiometric head elevation of the confined aquifer based on the assumption given by item 8 of Section 6.2.1. The combination of Eqs. (6-4), (6-5), and (6-6) gives

$$[Q(r) - Q(r + \Delta r)]\Delta t + (2\pi r \Delta r) K' \frac{h_0 - h}{b'} \Delta t = 2\pi r \Delta r \Delta h S \tag{6-7}$$

or, as $\Delta r \to 0$ and $\Delta t \to 0$,

$$\frac{\partial Q}{\partial r} + (2\pi r) K' \frac{h_0 - h}{b'} = (2\pi r) S \frac{\partial h}{\partial t} \tag{6-8}$$

and, using Eq. (6-2), gives

$$\frac{1}{r} \frac{\partial}{\partial r} \left(r \frac{\partial h}{\partial r} \right) + K' \frac{h_0 - h}{T b'} = \frac{S}{T} \frac{\partial h}{\partial t} \tag{6-9}$$

With the drawdown, $s = h_0 - h$, Eq. (6-9) takes the form

$$\frac{\partial^2 s}{\partial r^2} + \frac{1}{r} \frac{\partial s}{\partial r} - \frac{K'}{T b'} s = \frac{S}{T} \frac{\partial s}{\partial t} \tag{6-10}$$

where S and T are the storage coefficient and transmissivity of the aquifer, respectively, and

$$B = \left(\frac{T b'}{K'} \right)^{1/2} \tag{6-11}$$

Equation (6-10) takes the form

$$\frac{\partial^2 s}{\partial r^2} + \frac{1}{r} \frac{\partial s}{\partial r} - \frac{s}{B^2} = \frac{S}{T} \frac{\partial s}{\partial t} \tag{6-12}$$

As mentioned in Section 3.2.2.2 of Chapter 3, in the above-given equations, B is the *leakage factor* and $K'/b' = T/B^2$ is the *leakance* or *leakage coefficient* and its numerical values are usually given as a measure of leakage rates.

6.2.3 Initial and Boundary Conditions

The initial condition for head and drawdown, respectively, are

$$\begin{aligned} h(r,0) &= h_0 \qquad \text{for all } r \\ s(r,0) &= 0 \qquad \text{for all } r \end{aligned} \tag{6-13}$$

where h_0 is the constant initial head (see Figure 6-1). This condition states that initially the head is the same everywhere in the confined and unconfined aquifers.

The boundary condition at infinity for head and drawdown, respectively, are

$$\begin{aligned} h(\infty,t) &= h_0 \qquad \text{for all } t \\ s(\infty,t) &= 0 \qquad \text{for all } t \end{aligned} \tag{6-14}$$

This boundary condition means that the drawdown in hydraulic head at infinite distance is zero.

The condition for the discharge rate is

$$\begin{aligned} Q &= 0 \qquad &\text{for } t < 0 \\ Q &= \text{constant} > 0 \qquad &\text{for } t \geq 0 \end{aligned} \tag{6-15}$$

From Eq. (6-2), the constant discharge rate for the head and drawdown at the well, respectively, are

$$\begin{aligned} \lim_{r\to 0}\left(r\frac{\partial h}{\partial r}\right) &= \frac{Q}{2\pi T} \qquad \text{for } t \geq 0 \\ \lim_{r\to 0}\left(r\frac{\partial s}{\partial r}\right) &= -\frac{Q}{2\pi T} \qquad \text{for } t \geq 0 \end{aligned} \tag{6-16}$$

Equation (6-15) states that the discharge of the well is constant throughout the discharging period. Equation (6-16) means that near the discharging well the flow rate towards the well is equal to its discharge.

Before the start of pumping, the piezometric head surface of the confined aquifer coincides with the water table of the unconfined aquifer. This is shown in Figure 6-1.

6.2.4 Solution

6.2.4.1 Hantush and Jacob Equation. Under the initial and boundary conditions described above, the formal solution for drawdown variation for a leaky confined aquifer is given as (Hantush and Jacob, 1955):

$$s(r,t) = \frac{Q}{4\pi T}\left[2K_0\left(\frac{r}{B}\right) - \int_q^{\infty}\frac{1}{y}\exp\left(-y - \frac{r^2}{4B^2 y}\right)\right] dy \tag{6-17}$$

where K_0 is the modified Bessel function of the second kind of zero order and

$$q = \frac{Tt}{SB^2} = \frac{r^2}{4B^2 u} \tag{6-18}$$

The second part of Eq. (6-18) is written using Eq. (4-23) of Chapter 4, $u = r^2S/(4Tt)$. Evaluating the integral in Eq. (6-17) gives (Hantush and Jacob, 1955)

$$s(r,t) = \frac{Q}{2\pi T}K_0\left(\frac{r}{B}\right) - \frac{Q}{4\pi T}I_0\left(\frac{r}{B}\right)\left[-Ei\left(-\frac{r^2}{4B^2}u\right)\right]$$
$$+ \exp\left(-\frac{r^2}{4B^2}u\right)\left\{0.5772 + \ln(u) + [-Ei(-u)]\right.$$
$$- u + u\left[I_0\left(\frac{r}{B}\right) - 1\right]\frac{1}{\frac{r^2}{4B^2}}$$
$$\left. - u^2\sum_{n=1}^{\infty}\sum_{m=1}^{n}\frac{(-1)^{n+m}(n-m+1)!}{[(n+2)!]^2}\left(\frac{r^2}{4B^2}\right)^m u^{n-m}\right\} \tag{6-19}$$

where $I_0(r/B)$ is the modified Bessel function of the first kind and zero order and $-Ei(-r^2/(4B^2)u)$ is the integral-exponential function as defined by Eq. (4-25) of Chapter 4.

Under steady-state conditions—that is, as pumping continues indefinitely—the time, t, in Eq. (6-17) approaches to infinity and the drawdown expression given by Eq. (6-17) takes the form

$$s(r) = \frac{Q}{2\pi T}K_0\left(\frac{r}{B}\right) \tag{6-20}$$

which is exactly the same as Eq. (3-30) of Chapter 3 for a leaky confined aquifer.

Hantush and Jacob (1955) also presented a more concise alternative form of Eq. (6-17) as:

$$s(r,t) = \frac{Q}{4\pi T}\int_u^{\infty}\frac{1}{y}\exp\left(-y - \frac{r^2}{4B^2y}\right)dy \tag{6-21}$$

where

$$u = \frac{r^2S}{4Tt} \tag{6-22}$$

which is the same as Eq. (4-23) of Chapter 4.

If $B \to \infty$ in Eq. (6-21)—that is, when the vertical hydraulic conductivity of the confining layer (K') is very small—the second term in the exponential term of Eq. (6-21) approaches to zero and Eq. (6-21) becomes equivalent to the Theis equation as given by Eq. (4-22) of Chapter 4.

6.2.4.2 Well Function for Leaky Confined Aquifers and Type Curves. In Eq. (6-21), B is defined by Eq. (6-11) and is a function of the aquifer transmissivity (T), thickness of the confining layer (b'), and hydraulic conductivity of the confining layer (K'). Therefore, the ratio between the radial distance (r) and B, r/B, is a parameter representing the aquifer characteristics and y in Eq. (6-21) is a dummy variable. Because of these reasons, Eq. (6-21) can be written as

$$s(r,t) = \frac{Q}{4\pi T} W\left(u, \frac{r}{B}\right) \tag{6-23}$$

where

$$W(u,\beta) = \int_u^\infty \frac{1}{y} \exp\left(-y - \frac{\beta^2}{4y}\right) dy \tag{6-24}$$

where $\beta = r/B$. $W(u,\beta)$ is known as *Hantush and Jacob well function for leaky aquifers*. Table 6-1 lists the values of $W(u,\beta)$ for different values of u and $\beta = r/B$ based on the tables presented by Hantush (1956, 1961b, and 1964). Using the values in Table 6-1, $W(u,\beta)$ versus u or $1/u$ curves for different values of $\beta = r/B$ can be generated (see Figure 6-2).

6.2.4.3 Approximated Forms of the Well Function for Leaky Aquifers. For practical computations the following relations may be used (Hantush, 1964):

$$W(0,\beta) = 2K_0(\beta), \qquad W(u,0) = W(u) \tag{6-25}$$

$$W(u,\beta) = 2K_0(\beta) - W\left(\frac{\beta^2}{4u}, \beta\right) \tag{6-26}$$

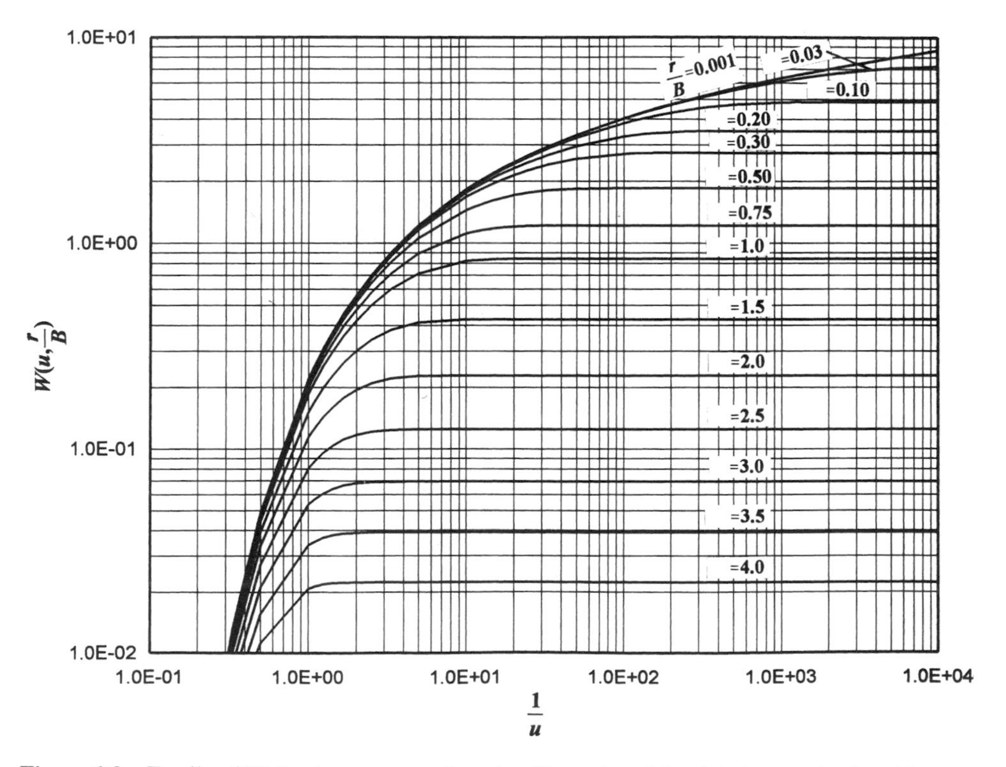

Figure 6-2 Family of Walton's type curves based on Hantush and Jacob leaky confined aquifer well function, $W(u, r/B)$.

TABLE 6-1 Values of the Leaky Confined Aquifer Function, $W(u, r/B)$

$u, r/B$	0	0.001	0.002	0.003	0.004	0.005	0.006	0.007	0.008	0.009	0.01
0	∞	14.0474	12.6611	11.8502	11.2748	10.8286	10.4640	10.1557	9.8887	9.6532	9.4425
.000001	13.2383	13.0031	12.4417	11.8153	11.2711	10.8283	10.4640	10.1557	9.8887		
.000002	12.5451	12.4240	12.1013	11.6716	11.2259	10.8174	10.4619	10.1554	9.8886	9.6532	
.000003	12.1397	12.0581	11.8322	11.5098	11.1462	10.7849	10.4509	10.1523	9.8879	9.6530	9.4425
.000004	11.8520	11.7905	11.6168	11.3597	11.0555	10.7374	10.4291	10.1436	9.8849	9.6521	9.4422
.000005	11.6289	11.5795	11.4384	11.2248	10.9642	10.6822	10.3993	10.1290	9.8786	9.6496	9.4413
.000006	11.4465	11.4053	11.2866	11.1040	10.8764	10.6240	10.3640	10.1094	9.8686	9.6450	9.4394
.000007	11.2924	11.2570	11.1545	10.9951	10.7933	10.5652	10.3255	10.0862	9.8555	9.6382	9.4361
.000008	11.1589	11.1279	11.0377	10.8962	10.7151	10.5072	10.2854	10.0602	9.8398	9.6292	9.4313
.000009	11.0411	11.0135	10.9330	10.8059	10.6416	10.4508	10.2446	10.0324	9.8219	9.6182	9.4251
.00001	10.9357	10.9109	10.8382	10.7228	10.5725	10.3963	10.2038	10.0034	9.8024	9.6059	9.4176
.00002	10.2426	10.2301	10.1932	10.1332	10.0522	9.9530	9.8386	9.7126	9.5781	9.4383	9.2961
.00003	9.8371	9.8288	9.8041	9.7635	9.7081	9.6392	9.5583	9.4671	9.3674	9.2611	9.1499
.00004	9.5495	9.5432	9.5246	9.4940	9.4520	9.3992	9.3366	9.2653	9.1863	9.1009	9.0102
.00005	9.3263	9.3213	9.3064	9.2818	9.2480	9.2052	9.1542	9.0957	9.0304	8.9591	8.8827
.00006	9.1440	9.1398	9.1274	9.1069	9.0785	9.0426	8.9996	8.9500	8.8943	8.8332	8.7673
.00007	8.9899	8.9863	8.9756	8.9580	8.9336	8.9027	8.8654	8.8224	8.7739	8.7204	8.6625
.00008	8.8563	8.8532	8.8439	8.8284	8.8070	8.7798	8.7470	8.7090	8.6661	8.6186	8.5669
.00009	8.7386	8.7358	8.7275	8.7138	8.6947	8.6703	8.6411	8.6071	8.5686	8.5258	8.4792
.0001	8.6332	8.6308	8.6233	8.6109	8.5937	8.5717	8.5453	8.5145	8.4796	8.4407	8.3983
.0002	7.9402	7.9390	7.9352	7.9290	7.9203	7.9092	7.8958	7.8800	7.8619	7.8416	7.8192
.0003	7.5348	7.5340	7.5315	7.5274	7.5216	7.5141	7.5051	7.4945	7.4823	7.4686	7.4534
.0004	7.2472	7.2466	7.2447	7.2416	7.2373	7.2317	7.2249	7.2169	7.2078	7.1974	7.1859
.0005	7.0242	7.0237	7.0222	7.0197	7.0163	7.0118	7.0063	6.9999	6.9926	6.9843	6.9750
.0006	6.8420	6.8416	6.8403	6.8383	6.8353	6.8316	6.8271	6.8218	6.8156	6.8086	6.8009
.0007	6.6879	6.6876	6.6865	6.6848	6.6823	6.6790	6.6752	6.6706	6.6653	6.6594	6.6527
.0008	6.5545	6.5542	6.5532	6.5517	6.5495	6.5467	6.5433	6.5393	6.5347	6.5295	6.5237
.0009	6.4368	6.4365	6.4357	6.4344	6.4324	6.4299	6.4269	6.4233	6.4192	6.4146	6.4094

(continued)

TABLE 6-1 (*continued*)

$u, r/B$	0	0.001	0.002	0.003	0.004	0.005	0.006	0.007	0.008	0.009	0.01
.001	6.3315	6.3313	6.3305	6.3293	6.3276	6.3253	6.3226	6.3194	6.3157	6.3115	6.3069
.002	5.6394	5.6393	5.6389	5.6383	5.6374	5.6363	5.6350	5.6334	5.6315	5.6294	5.6271
.003	5.2349	5.2348	5.2346	5.2342	5.2336	5.2329	5.2320	5.2310	5.2297	5.2283	5.2267
.004	4.9482	4.9482	4.9480	4.9477	4.9472	4.9467	4.9460	4.9453	4.9443	4.9433	4.9421
.005	4.7261	4.7260	4.7259	4.7256	4.7253	4.7249	4.7244	4.7237	4.7230	4.7222	4.7212
.006	4.5448	4.5448	4.5447	4.5444	4.5441	4.5438	4.5433	4.5428	4.5422	4.5415	4.5407
.007	4.3916	4.3916	4.3915	4.3913	4.3910	4.3998	4.3904	4.3899	4.3894	4.3888	4.3882
.008	4.2591	4.2590	4.2590	4.2588	4.2586	4.2583	4.2580	4.2576	4.2572	4.2567	4.2561
.009	4.1423	4.1423	4.1422	4.1420	4.1418	4.1416	4.1413	4.1410	4.1406	4.1401	4.1396
.01	4.0379	4.0379	4.0378	4.0377	4.0375	4.0373	4.0371	4.0368	4.0364	4.0360	4.0356
.02	3.3547	3.3547	3.3547	3.3546	3.3545	3.3544	3.3543	3.3542	3.3540	3.3538	3.3536
.03	2.9591	2.9591	2.9591	2.9590	2.9590	2.9589	2.9589	2.9588	2.9587	2.9585	2.9584
.04	2.6813	2.6812	2.6812	2.6812	2.6812	2.6811	2.6810	2.6810	2.6809	2.6808	2.6807
.05	2.4679	2.4679	2.4679	2.4679	2.4678	2.4678	2.4678	2.4677	2.4676	2.4676	2.4675
.06	2.2953	2.2953	2.2953	2.2953	2.2952	2.2952	2.2952	2.2952	2.2951	2.2950	2.2950
.07	2.1508	2.1508	2.1508	2.1508	2.1508	2.1508	2.1507	2.1507	2.1507	2.1506	2.1506
.08	2.0269	2.0269	2.0269	2.0269	2.0269	2.0269	2.0269	2.0268	2.0268	2.0268	2.0267
.09	1.9187	1.9187	1.9187	1.9187	1.9187	1.9187	1.9187	1.9186	1.9186	1.9186	1.9185

(*continued*)

TABLE 6-1 *(continued)*

$u, r/B$	0	0.001	0.002	0.003	0.004	0.005	0.006	0.007	0.008	0.009	0.01
.1	1.8229	1.8229	1.8229	1.8229	1.8229	1.8229	1.8229	1.8228	1.8228	1.8228	1.8227
.2	1.2227	1.2226	1.2226	1.2226	1.2226	1.2226	1.2226	1.2226	1.2226	1.2226	1.2226
.3	0.9057	0.9057	0.9057	0.9057	0.9057	0.9057	0.9057	0.9057	0.9056	0.9056	0.9056
.4	7024	7024	7024	7024	7024	7024	7024	7024	7024	7024	7024
.5	5598	5598	5598	5598	5598	5598	5598	5598	5598	5598	5598
.6	4544	4544	4544	4544	4544	4544	4544	4544	4544	4544	4544
.7	3738	3738	3738	3738	3738	3738	3738	3738	3738	3738	3738
.8	3106	3106	3106	3106	3106	3106	3106	3106	3106	3106	3106
.9	2602	2602	2602	2602	2602	2602	2602	2602	2602	2602	2602
1.0	0.2194	0.2194	0.2194	0.2194	0.2194	0.2194	0.2194	0.2194	0.2194	0.2194	0.2194
2.0	489	489	489	489	489	489	489	489	489	489	489
3.0	130	130	130	130	130	130	130	130	130	130	130
4.0	38	38	38	38	38	38	38	38	38	38	38
5.0	11	11	11	11	11	11	11	11	11	11	11
6.0	4	4	4	4	4	4	4	4	4	4	4
7.0	1	1	1	1	1	1	1	1	1	1	1
8.0	0	0	0	0	0	0	0	0	0	0	0

(continued)

TABLE 6-1 *(continued)*

$u, r/B$	0.01	0.015	0.02	0.025	0.03	0.035	0.04	0.045	0.05	0.055	0.06	0.065	0.07	0.075	0.08	0.085	0.09	0.095	0.10
0	9.4425	8.6319	8.0569	7.6111	7.2471	6.9394	6.6731	6.4383	6.2285	6.0388	5.8658	5.7067	5.5596	5.4228	5.2950	5.1750	5.0620	4.9553	4.8541
.000001																			
.000002																			
.000003	9.4425																		
.000004	9.4422																		
.000005	9.4413																		
.000006	9.4394																		
.000007	9.4361	8.6319																	
.000008	9.4313	8.6318																	
.000009	9.4251	8.6316																	
.00001	9.4176	8.6313	8.0569																
.00002	9.2961	8.6152	8.0558	7.6111	7.2471														
.00003	9.1499	8.5737	8.0483	7.6101	7.2470														
.00004	9.0102	8.5168	8.0320	7.6069	7.2465	6.9394	6.6731												
.00005	8.8827	8.4533	8.0080	7.6000	7.2450	6.9391	6.6730												
.00006	8.7673	8.3880	7.9786	7.5894	7.2419	6.9384	6.6729	6.4383											
.00007	8.6625	8.3233	7.9456	7.5754	7.2371	6.9370	6.6726	6.4382	6.2285										
.00008	8.5669	8.2603	7.9105	7.5589	7.2305	6.9347	6.6719	6.4381	6.2284										
.00009	8.4792	8.1996	7.8743	7.5402	7.2222	6.9316	6.6709	6.4378	6.2283										
.0001	8.3983	8.1414	7.8375	7.5199	7.2122	6.9273	6.6693	6.4372	6.2282	6.0388	5.8658	5.7067	5.5596	5.4228	5.2950				
.0002	7.8192	7.6780	7.4972	7.2898	7.0685	6.8439	6.6242	6.4143	6.2173	6.0338	5.8637	5.7059	5.5593	5.4227	5.2949	5.1750	5.0620	4.9553	
.0003	7.4534	7.3562	7.2281	7.0759	6.9068	6.7276	6.5444	6.3623	6.1848	6.0145	5.8527	5.6999	5.5562	5.4212	5.2942	5.1747	5.0619	4.9552	4.8541
.0004	7.1859	7.1119	7.0128	6.8929	6.7567	6.6088	6.4538	6.2955	6.1373	5.9818	5.8309	5.6860	5.5476	5.4160	5.2912	5.1730	5.0610	4.9547	4.8539
.0005	6.9750	6.9152	6.8346	6.7357	6.6219	6.4964	6.3626	6.2236	6.0821	5.9406	5.8011	5.6648	5.5330	5.4062	5.2848	5.1689	5.0585	4.9532	4.8530
.0006	6.8009	6.7508	6.6828	6.5988	6.5011	6.3923	6.2748	6.1512	6.0239	5.8948	5.7658	5.6383	5.5134	5.3921	5.2749	5.1621	5.0539	4.9502	4.8510
.0007	6.6527	6.6096	6.5508	6.4777	6.3923	6.2962	6.1917	6.0807	5.9652	5.8468	5.7274	5.6081	5.4902	5.3745	5.2618	5.1526	5.0471	4.9454	4.8478
.0008	6.5237	6.4858	6.4340	6.3695	6.2935	6.2076	6.1136	6.0129	5.9073	5.7982	5.6873	5.5755	5.4642	5.3542	5.2461	5.1406	5.0381	4.9388	4.8430
.0009	6.4094	6.3757	6.2394	6.2716	6.2032	6.1256	6.0401	5.9481	5.8509	5.7500	5.6465	5.5416	5.4364	5.3317	5.2282	5.1266	5.0272	4.9306	4.8368

TABLE 6-1 *(continued)*

$u, r/B$	0.01	0.015	0.02	0.025	0.03	0.035	0.04	0.045	0.05	0.055	0.06	0.065	0.07	0.075	0.08	0.085	0.09	0.095	0.10
.001	6.3069	6.2765	6.2347	6.1823	6.1202	6.0494	5.9711	5.8864	5.7965	5.7026	5.6058	5.5071	5.4075	5.3078	5.2087	5.1109	5.0133	4.9208	4.8292
.002	5.6271	5.6118	5.5207	5.5638	5.5314	5.4939	5.4516	5.4047	5.3538	5.2991	5.2411	5.1803	5.1170	5.0517	4.9848	4.9166	4.8475	4.7778	4.7079
.003	5.2267	5.2166	5.2025	5.1845	5.1627	5.1373	5.1084	5.0762	5.0408	5.0025	4.9615	4.9180	4.8722	4.8243	4.7746	4.7234	4.6707	4.6169	4.5622
.004	4.9421	4.9345	4.9240	4.9105	4.8941	4.8749	4.8530	4.8286	4.8016	4.7722	4.7406	4.7068	4.6710	4.6335	4.5942	4.5533	4.5111	4.4676	4.4230
.005	4.7212	4.7152	4.7068	4.6960	4.6829	4.6675	4.6499	4.6302	4.6084	4.5846	4.5590	4.5314	4.5022	4.4713	4.4389	4.4050	4.3699	4.3335	4.2960
.006	4.5407	4.5357	4.5287	4.5197	4.5088	4.4960	4.4814	4.4649	4.4467	4.4267	4.4051	4.3819	4.3573	4.3311	4.3036	4.2747	4.2446	4.2134	4.1812
.007	4.3882	4.3839	4.3779	4.3702	4.3609	4.3500	4.3374	4.3233	4.3077	4.2905	4.2719	4.2518	4.2305	4.2078	4.1839	4.1588	4.1326	4.1053	4.0771
.008	4.2561	4.2524	4.2471	4.2404	4.2323	4.2228	4.2118	4.1994	4.1857	4.1707	4.1544	4.1368	4.1180	4.0980	4.0769	4.0547	4.0315	4.0073	3.9822
.009	4.1396	4.1363	4.1317	4.1258	4.1186	4.1101	4.1004	4.0894	4.0772	4.0638	4.0493	4.0336	4.0169	3.9991	3.9802	3.9603	3.9395	3.9178	3.8952
.01	4.0356	4.0326	4.0285	4.0231	4.0167	4.0091	4.0003	3.9905	3.9795	3.9675	3.9544	3.9403	3.9252	3.9091	3.8920	3.8741	3.8552	3.8356	3.8150
.02	3.3536	3.3521	3.3502	3.3476	3.3444	3.3408	3.3365	3.3317	3.3264	3.3205	3.3141	3.3071	3.2997	3.2917	3.2832	3.2742	3.2647	3.2547	3.2442
.03	2.9584	2.9575	2.9562	2.9545	2.9523	2.9501	2.9474	2.9444	2.9409	2.9370	2.9329	2.9284	2.9235	2.9183	2.9127	2.9069	2.9007	2.8941	2.8873
.04	2.6807	2.6800	2.6791	2.6779	2.6765	2.6747	2.6727	2.6705	2.6680	2.6652	2.6622	2.6589	2.6553	2.6515	2.6475	2.6432	2.6386	2.6338	2.6288
.05	2.4675	2.4670	2.4662	2.4653	2.4642	2.4628	2.4613	2.4595	2.4576	2.4554	2.4531	2.4505	2.4478	2.4448	2.4416	2.4383	2.4347	2.4310	2.4271
.06	2.2950	2.2945	2.2940	2.2932	2.2923	2.2912	2.2900	2.2885	2.2870	2.2852	2.2833	2.2812	2.2790	2.2766	2.2740	2.2713	2.2684	2.2654	2.2622
.07	2.1506	2.1502	2.1497	2.1491	2.1483	2.1474	2.1464	2.1452	2.1439	2.1424	2.1408	2.1391	2.1372	2.1352	2.1331	2.1308	2.1284	2.1258	2.1232
.08	2.0267	2.0264	2.0260	2.0255	2.0248	2.0240	2.0231	2.0221	2.0210	2.0198	2.0184	2.0169	2.0153	2.0136	2.0118	2.0099	2.0078	2.0056	2.0034
.09	1.9185	1.9183	1.9179	2.9174	1.9169	1.9162	1.9154	1.9146	1.9136	1.9125	1.9114	1.9101	1.9087	1.9072	1.9056	1.9040	1.9022	1.9003	1.8983
.1	1.8227	1.8225	1.8222	1.8218	1.8213	1.8207	1.8200	1.8193	1.8184	1.8175	1.8164	1.8153	1.8141	1.8128	1.8114	1.8099	1.8084	1.8067	1.8050
.2	1.2226	1.2225	1.2224	1.2222	1.2220	1.2218	1.2215	1.2212	1.2209	1.2205	1.2201	1.2196	1.2192	1.2186	1.2181	1.2175	1.2168	1.2162	1.2155
.3	0.9056	0.9056	0.9055	0.9054	0.9053	0.9052	0.9050	0.9049	0.9047	0.9045	0.9043	0.9040	0.9038	0.9035	0.9032	0.9029	0.9025	0.9022	0.9018
.4	7024	7023	7023	7022	7022	7021	7020	7019	7018	7016	7015	7014	7012	7010	7008	7006	7004	7002	7000
.5	5598	5597	5597	5597	5596	5596	5596	5594	5594	5593	5592	5591	5590	5588	5587	5585	5584	5583	5581
.6	4544	4544	4543	4543	4543	4542	4542	4542	4541	4540	4540	4539	4538	4537	4536	4535	4534	4533	4532
.7	3738	3738	3737	3737	3737	3737	3737	3736	3735	3735	3734	3734	3733	3733	3732	3732	3731	3730	3729
.8	3106	3106	3106	3106	3105	3105	3105	3105	3104	3104	3104	3103	3103	3102	3102	3101	3101	3100	3100
.9	2602	2602	2602	2602	2601	2601	2601	2601	2601	2600	2600	2600	2599	2599	2599	2598	2598	2597	2597
1.0	0.2194	0.2194	0.2194	0.2194	0.2193	0.2193	0.2193	0.2193	0.2193	0.2193	0.2192	0.2192	0.2192	0.2191	0.2191	0.2191	0.2191	0.2190	0.2190
2.0	489	489	489	489	489	489	489	489	489	489	489	489	489	489	489	489	489	488	488
3.0	130	130	130	130	130	130	130	130	130	130	130	130	130	130	130	130	130	130	130
4.0	38	38	38	38	38	38	38	38	38	38	38	38	38	38	38	38	38	38	38
5.0	11	11	11	11	11	11	11	11	11	11	11	11	11	11	11	11	11	11	11
6.0	4	4	4	4	4	4	4	4	4	4	4	4	4	4	4	4	4	4	4
7.0	1	1	1	1	1	1	1	1	1	1	1	1	1	1	1	1	1	1	1
8.0	0	0	0	0	0	0	0	0	0	0	0	0	0	0	0	0	0	0	0

TABLE 6-1 *(continued)*

$u, r/B$	0.1	0.15	0.2	0.25	0.3	0.35	0.4	0.45	0.5	0.55	0.6	0.65	0.7	0.75	0.8	0.85	0.9	0.95	1.0
0	4.8541	4.0601	3.5054	3.0830	2.7449	2.4654	2.2291	2.0258	1.8488	1.6931	1.5550	1.4317	1.3210	1.2212	1.1307	1.0485	0.9735	0.9049	0.8420
.0001																			
.0002																			
.0003	4.8541																		
.0004	4.8539																		
.0005	4.8530																		
.0006	4.8510	4.0601																	
.0007	4.8478	4.0600																	
.0008	4.8430	4.0599																	
.0009	4.8368	4.0598																	
.001	4.8292	4.0595	3.5054																
.002	4.7079	4.0435	3.5043	3.0830	2.7449														
.003	4.5622	4.0092	3.4969	3.0821	2.7448														
.004	4.4230	3.9551	3.4806	3.0788	2.7444	2.4654	2.2291												
.005	4.2960	3.8821	3.4567	3.0719	2.7428	2.4651	2.2290												
.006	4.1812	3.8384	3.4274	3.0614	2.7398	2.4644	2.2289	2.0258											
.007	4.0771	3.7529	3.3947	3.0476	2.7350	2.4630	2.2286	2.0257											
.008	3.9822	3.6903	3.3598	3.0311	2.7284	2.4608	2.2279	2.0256	1.8488										
.009	3.8952	3.6302	3.3239	3.0126	2.7202	2.4576	2.2269	2.0253	1.8487										
.01	3.8150	3.5725	3.2875	2.9925	2.7104	2.4434	2.2253	2.0248	1.8486	1.6931	1.5550	1.4317	1.3210	1.2212	1.1307	1.0485			
.02	3.2442	3.1158	2.9521	2.7658	2.5688	2.3713	2.1809	2.0023	1.8379	1.6883	1.5530	1.4309	1.3207	1.2210	1.1306	1.0484	0.9735	0.9049	
.03	2.8873	2.8017	2.6896	2.5571	2.4110	2.2576	2.1031	1.9515	1.8062	1.6695	1.5423	1.4251	1.3177	1.2195	1.1299	1.0481	9733	9048	0.8420
.04	2.6288	2.5655	2.4816	2.3802	2.2661	2.1431	2.0155	1.8869	1.7603	1.6379	1.5213	1.4117	1.3094	1.2146	1.1270	1.0465	9724	9044	8418
.05	2.4271	2.3776	2.3110	2.2299	2.1371	2.0356	1.9283	1.8181	1.7075	1.5985	1.4927	1.3914	1.2955	1.2052	1.1210	1.0426	9700	9029	8409
.06	2.2622	2.2218	2.1673	2.1002	2.0227	1.9369	1.8452	1.7497	1.6524	1.5551	1.4593	1.3663	1.2770	1.1919	1.1116	1.0362	9657	9001	8391
.07	2.1232	2.0894	2.0435	1.9867	1.9206	1.8469	1.7673	1.6835	1.5973	1.5101	1.4232	1.3380	1.2551	1.1754	1.0993	1.0272	9593	8956	8360
.08	2.0034	1.9745	1.9351	1.8861	1.8290	1.7646	1.6947	1.6206	1.5436	1.4650	1.3860	1.3078	1.2310	1.1564	1.0847	1.0161	9510	8895	8316
.09	1.8983	1.8732	1.8389	1.7961	1.7460	1.6892	1.6272	1.5609	1.4918	1.4206	1.3486	1.2766	1.2054	1.1358	1.0682	1.0032	9411	8819	8259

TABLE 6-1 *(continued)*

$u, r/B$	0.1	0.15	0.2	0.25	0.3	0.35	0.4	0.45	0.5	0.55	0.6	0.65	0.7	0.75	0.8	0.85	0.9	0.95	1.0
.1	1.8050	1.7829	1.7527	1.7149	1.6704	1.6198	1.5644	1.5048	1.4422	1.3774	1.3115	1.2451	1.1791	1.1140	1.0505	0.9890	0.9297	0.8730	0.8190
.2	1.2155	1.2066	1.1944	1.1789	1.1602	1.1387	1.1145	1.0879	1.0592	1.0286	0.9964	0.9629	0.9284	0.8932	0.8575	8216	7857	7501	7148
.3	0.9018	0.8969	0.8902	0.8817	0.8713	0.8593	0.8457	0.8306	0.8142	0.7964	7775	7577	7369	7154	6932	6706	6476	6244	6010
.4	7000	6969	6927	6874	6809	6733	6647	6551	6446	6332	6209	6080	5943	5801	5653	5501	5345	5186	5024
.5	5581	5561	5532	5496	5453	5402	5344	5278	5206	5128	5044	4955	4860	4761	4658	4550	4440	4326	4210
.6	4532	4518	4498	4472	4441	4405	4364	4317	4266	4210	4150	4086	4018	3946	3871	3793	3712	3629	3543
.7	3729	3719	3704	3685	3663	3636	3606	3572	3534	3493	3449	3401	3351	3297	3242	3183	3123	3060	2996
.8	3100	3092	3081	3067	3050	3030	3008	2982	2953	2922	2889	2853	2815	2774	2732	2687	2641	2592	2543
.9	2597	2591	2583	2572	2559	2544	2527	2507	2485	2461	2436	2408	2378	2347	2314	2280	2244	2207	2168
1.0	0.2190	0.2186	0.2179	0.2171	0.2161	0.2149	0.2135	0.2120	0.2103	0.2085	0.2065	0.2043	0.2020	0.1995	0.1970	0.1943	0.1914	0.1885	0.1855
2.0	488	488	487	486	485	484	482	480	477	475	473	470	467	463	460	456	452	448	444
3.0	130	130	130	130	130	130	129	129	128	128	127	127	126	125	125	124	123	123	122
4.0	38	38	38	38	38	38	38	37	37	37	37	37	37	37	37	36	36	36	36
5.0	11	11	11	11	11	11	11	11	11	11	11	11	11	11	11	11	11	11	11
6.0	4	4	4	4	4	4	4	4	4	4	4	4	4	4	4	4	4	4	4
7.0	1	1	1	1	1	1	1	1	1	1	1	1	1	1	1	1	1	1	1
8.0	0	0	0	0	0	0	0	0	0	0	0	0	0	0	0	0	0	0	0

(continued)

TABLE 6-1 (*continued*)

u, r/B	1.0	1.5	2.0	2.5	3.0	3.5	4.0	4.5	5.0	6.0	7.0	8.0	9.0
0	0.8420	0.4276	0.2278	0.1247	0.0695	0.0392	0.0223	0.0128	0.0074	0.0025	0.0008	0.0003	0.0001
.01													
.02													
.03	0.8420												
.04	8418												
.05	8409												
.06	8391												
.07	8360	0.4276											
.08	8316	4275											
.09	8259	4274											
.1	0.8190	0.4271	0.2278										
.2	7148	4135	2268	0.1247	0.0695								
.3	6010	3812	2211	1240	694								
.4	5024	3411	2096	1217	691	0.0392							
.5	4210	3007	1944	1174	681	390	0.0223						
.6	3543	2630	1774	1112	664	386	222	0.0128					
.7	2996	2292	1602	1040	639	379	221	127					
.8	2543	1994	1436	961	607	368	218	127	0.0074				
.9	2168	1734	1281	881	572	354	213	125	73				
1.0	0.1855	0.1509	0.1139	0.0803	0.0534	0.0338	0.0207	0.0123	0.0073	0.0025			
2.0	444	394	335	271	210	156	112	77	51	21	0.0008	0.0003	
3.0	122	112	100	86	71	57	45	34	25	12	6	2	
4.0	36	34	31	27	24	20	16	13	10	6	3	2	0.0001
5.0	11	10	10	9	8	7	6	5	4	2	1	1	0
6.0	4	3	3	3	3	2	2	2	2	1	1	0	
7.0	1	1	1	1	1	1	1	1	1	0	0		
8.0	0	0	0	0	0	0	0	0	0				

Source: After Hantush (1956).

and

$$W(u,\beta) \cong W(u) \qquad \text{for } u > 2\beta \tag{6-27}$$

$$W(u,\beta) \cong W(u) \qquad \text{for } u > 5\beta^2 \quad \text{if} \quad \beta < 0.1 \tag{6-28}$$

$$W(u,\beta) \cong 2K_0(\beta) - I_0(\beta)W\left(\frac{\beta^2}{4u}\right) \qquad \text{for} \quad u < \frac{\beta^2}{20} \quad \text{if} \quad u < 1 \tag{6-29}$$

In the above expressions, u is defined by Eq. (6-22) and $W(u)$ is the Theis well function that is tabulated in Table 4-1 of Chapter 4. K_0 is the modified Bessel function of second kind of zero order and is tabulated in Table 3-2 of Chapter 3, and I_0 is the modified Bessel function of the first kind and zero order.

6.2.4.4 Evaluation of the Solution. The assumptions and limitations regarding the foundations of Eq. (6-23) are basically the same as those for the nonleaky confined aqifer solution, the Theis equation, which is presented in Section 4.2 of Chapter 4.

Of those seven assumptions in Section 6.2.1, the sixth assumption neglects the storage effects in the confining layer on delaying the delivery of water. As a result of this situation, the drawdown will be overestimated. Neuman and Witherspoon (1969c, p. 821) investigated this assumption and they concluded that this assumption would not affect the solution if

$$\beta^* = \frac{r}{4b}\left(\frac{K'S_s'}{KS_s}\right)^{1/2} < 0.01 \tag{6-30}$$

where S_s and S_s' are the specific storages of the confined aquifer and confining layer, respectively. Refer to Section 7.4 of Chapter 7 for more of the Neuman and Witherspoon model for wells in multiple aquifer systems.

The seventh assumption in Section 6.2.1 neglects the vertical flow components in the pumped aquifer and horizontal flow components in the confining layer. Hantush and Jacob (1954, p. 917) pointed out that if the hydraulic conductivity of the confined aquifer is much higher than the hydraulic conductivity of the confining layer, the flow may be assumed to be vertical through the confining layer and horizontal in the confined aquifer. The validity of this assumption has been investigated by Neuman and Witherspoon (1969a–c) for the case of two aquifers separated by an aquitard and the details are presented in Chapter 7. Neuman and Witherspoon found that when the hydraulic conductivities of the aquifers are more than two orders of magnitude greater than that of the aquitard, the errors introduced by this assumption are less than 5%. In other words, in aquitard–aquifer systems, with hydraulic conductivity contrasts of two orders of magnitude or more, flow lines tend to become almost horizontal in the aquifers and almost vertical in the aquitard.

According to the fifth assumption of Section 6.2.1, there is no drawdown in the water level in the source bed above the confining layer. This assumption was also investigated by Neuman and Witherspoon (1969b, p. 810), who concluded that the drawdown in the source bed would have negligible effect on drawdown in the pumped aquifer for short times; the required condition is

$$\frac{Tt}{r^2S} < 1.6\frac{\beta^{*^2}}{\left(\frac{r}{B}\right)^4} \tag{6-31}$$

Neuman and Witherspoon (1969b, p.811) also pointed out that neglect of drawdown in the source bed is justified if

$$T_s > 100T \tag{6-32}$$

where T_s is the transmissivity of the source bed.

6.3 PUMPING TEST DATA ANALYSIS METHODS FOR HOMOGENEOUS AND ISOTROPIC AQUIFERS

Several methods have been developed for the analysis of pumping test data for leaky confined aquifers. The Hantush and Jacob formula, as given by Eq. (6-23), is the foundation of these methods. In this section, the type-curve method originally proposed by Walton (1962) will be presented first. Then, Hantush's inflection-point methods and type-curve method (Hantush, 1956) will be presented with examples.

As pointed out in Section 6.2.4.4, the fifth, sixth, and seventh assumptions in Section 6.2.1 are critically important and one should take special care in analyzing field data using the Hantush and Jacob formula.

6.3.1 Walton's Type-Curve Method

As mentioned above, the type-curve method was originally proposed by Walton (1962) and is based on type-curves for $W(u, r/B)$. The method is similar to that of the Theis method as presented in Section 4.2.7.1 of Chapter 4. The type-curve method for leaky confined aquifers are presented below in a stepwise manner:

Step 1: Plot a family of type curve of $1/u$ versus $W(u, r/B)$ on log–log paper using the data in Table 6-1.

Step 2: On another sheet of log–log paper with the same scale as the first, plot the observed drawdown data of s versus t/r^2 (or s versus t for the case of one observation well).

Step 3: Keeping the coordinate axes parallel at all times, superimpose the two sheets until the best fit of data curve and one of the family of type curves is obtained.

Step 4: Select a common point, the match point, arbitrarily chosen on the overlapping part of the curve, or even anywhere on the overlapping portion of the sheets. Read the corresponding values of r/B, $W(u, r/B)$, $1/u$, s, and t/r^2 (or t).

Step 5: Substitute the values of $W(u, r/B)$ and s, along with the known value of Q, into Eq. (6-23) and solve for T.

Step 6: Substitute the values of $1/u$ and t/r^2 (or t) into Eq. (6-22) and solve for S.

Step 7: Substitute the values of T and B into Eq. (6-11) and calculate the hydraulic resistance b'/K'. Then, using the known value of b'/K' and b', determine the value of K', the vertical hydraulic conductivity of the confining layer.

Example 6-1: A confined aquifer is overlain by a leaky confining layer (refer Figure 6-1). A well fully penetrating the aquifer is pumped at a uniform flow rate of 625 m^3/day. The observation well fully penetrating the aquifer completely is located 105 m away from the pumped well and its drawdown data are given in Table 6-2. The thicknesses of the aquifer

TABLE 6-2 Pumping Test Data at r = 105 m for Example 6-1

t (days)	s (m)
0.0	0.0
3.40×10^{-2}	0.055
4.28×10^{-2}	0.062
5.25×10^{-2}	0.066
7.02×10^{-2}	0.073
1.01×10^{-1}	0.090
1.34×10^{-1}	0.098
1.88×10^{-1}	0.108
2.51×10^{-1}	0.116
3.12×10^{-1}	0.122
4.00×10^{-1}	0.135
4.53×10^{-1}	0.136
5.33×10^{-1}	0.137
6.12×10^{-1}	0.138

and overlying confining layer have been estimated to be 80 m and 28 m, respectively. Using the Walton's type-curve method, determine the horizontal hydraulic conductivity value of the confined aquifer (K_h), the vertical hydraulic conductivity of the confining layer (K'), and the storage coefficient of the confined aquifer (S).

Solution: In type-curve matching, one matching is enough. However, matching will be done for two points. The purpose of the selection of two points is to give an idea about the level of potential reading errors.

Step 1: A family of type curves of $W(u, r/B)$ versus $1/u$ on log–log paper for different values of r/B are given in Figure 6-2 using Table 6-1.

Step 2: The graph for the pumping test data is given in Figure 6-3.

Step 3: The superimposed sheets are shown in Figure 6-4. Comparison shows that the plotted points fall along the curve $r/B = 0.3$.

Step 4: In Figure 6-4, A is an arbitrarily chosen point; B is a point on the overlapping part of the curve. The coordinates of A are

$$\left(\frac{1}{u}\right)_A = 10, \qquad W_A\left(u, \frac{r}{B}\right) = 0.88$$

$$s_A = 0.05 \text{ m}, \qquad t_A = 0.086 \text{ days}$$

The coordinates of B are

$$\left(\frac{1}{u}\right)_B = 46, \qquad W_B\left(u, \frac{r}{B}\right) = 2.35$$

$$s_B = 0.13 \text{ m} \qquad t_B = 0.4 \text{ days}$$

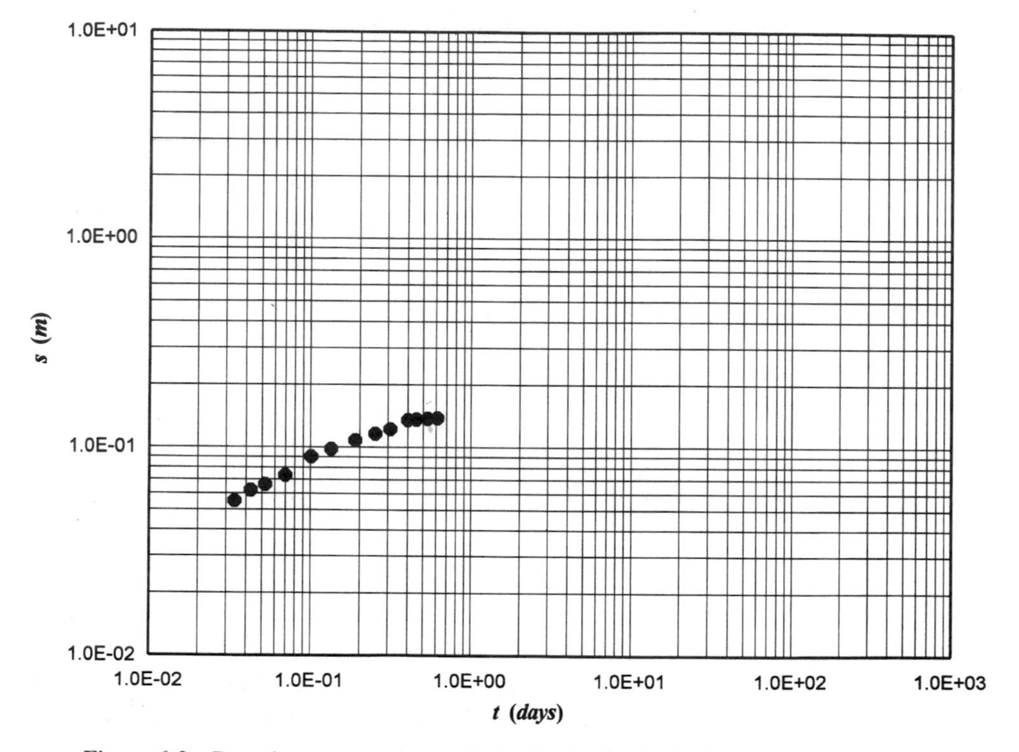

Figure 6-3 Drawdown versus time variation for the data in Table 6-2 for Example 6-1.

Step 5: Using Eq. (6-23), the transmissivity (T) is calculated below. The values for point A gives

$$T = \frac{Q}{4\pi s_A} W_A\left(u, \frac{r}{B}\right) = \frac{625 \text{ m}^3/\text{day}}{(4\pi)(0.05 \text{ m})}(0.88) = 875 \text{ m}^2/\text{day}$$

From the values of point B we obtain

$$T = \frac{Q}{4\pi s_B} W_B\left(u, \frac{r}{B}\right) = \frac{625 \text{ m}^3/\text{day}}{(4\pi)(0.13 \text{ m})}(2.35) = 899 \text{ m}^3/\text{day}$$

Notice that the two points coordinates give approximately the same results. The minor difference can be attributed to the reading errors.

Step 6: Using Eq. (6-22), the storage coefficient (S) is calculated below. The values of point A gives

$$S = \frac{4Tt_A u_A}{r^2} = \frac{(4)(875 \text{ m}^2/\text{day})(0.086 \text{ days})(0.1)}{(105 \text{ m})^2} = 0.0027$$

From the values of point B we obtain

$$S = \frac{4Tt_B u_B}{r^2} = \frac{(4)(899 \text{ m}^2/\text{day})(0.4 \text{ days})\left(\frac{1}{46}\right)}{(105 \text{ m})^2} = 0.0028$$

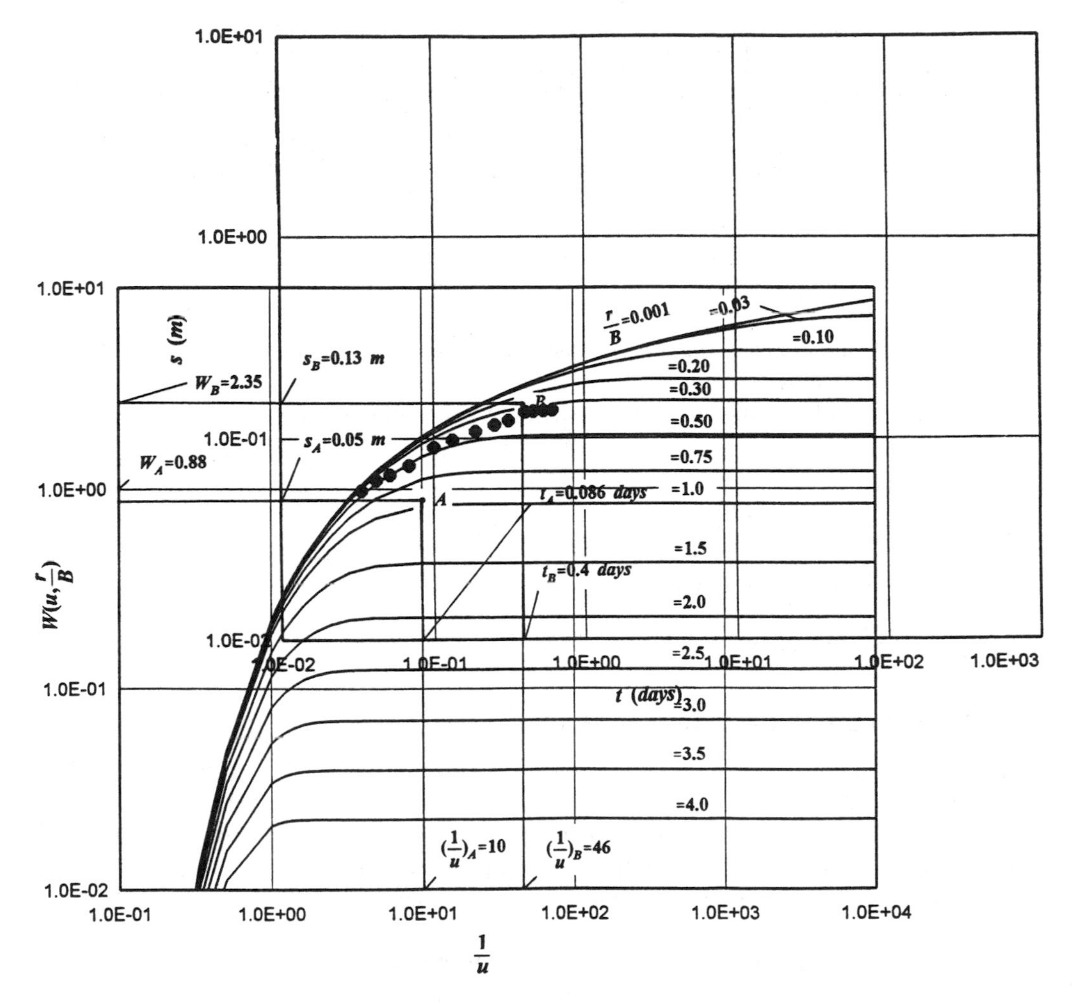

Figure 6-4 Type-curve matching for Example 6-1.

Notice that the storage coefficient values calculated from the two coordinate points are approximately the same. The minor difference can be attributed to the reading errors.

Step 7: From Step 3, $r/B = 0.3$. From Eq. (6-11) we obtain

$$\frac{b'}{K'} = \frac{B^2}{T} = \frac{\left(\frac{r}{0.3}\right)^2}{T}$$

Subtitution of $r = 105$ m and the T values into the above expression gives

$$\frac{b'}{K'} = \frac{\left(\frac{105 \text{ m}}{0.3}\right)^2}{(875 \text{ m}^2/\text{day})} = 140 \text{ days} \qquad \text{for } T = 875 \text{ m}^2/\text{day}$$

$$\frac{b'}{K'} = \frac{\left(\dfrac{105\ \text{m}}{0.3}\right)^2}{(899\ \text{m}^2/\text{day})} = 136\ \text{days} \qquad \text{for } T = 899\ \text{m}^2/\text{day}$$

Using the values of point A, the horizontal hydraulic conductivity of the confined aquifer is

$$K_h = \frac{T}{b} = \frac{875\ \text{m}^2/\text{day}}{80\ \text{m}} = 10.94\ \text{m/day}$$

For point B, the horizontal hydraulic conductivity is 11.24 m/day. As observed, these values are very close.

The vertical hydraulic conductivity values of the confining layer are

$$K' = \frac{28\ \text{m}}{140\ \text{days}} = 0.20\ \text{m/day} \qquad \text{for } T = 875\ \text{m}^2/\text{day}$$

$$K' = \frac{28\ \text{m}}{136\ \text{days}} = 0.21\ \text{m/day} \qquad \text{for } T = 899\ \text{m}^2/\text{day}$$

Again, the values are very close.

6.3.2 Hantush's Inflection-Point Methods

The *inflection-point methods* developed by Hantush (1956) is based on the drawdown equation for leaky confined aquifers originally developed by Hantush and Jacob (1955), and this equation is given by Eq. (6-23). In the following sections the theory of the inflection points method is presented first, and then one observation well method and more than one observation well method are presented in detail, with examples.

6.3.2.1 Theory of the Inflection-Point Methods. Consider a function defined as $y = f(x)$. Making zero the second derivative of this function [i.e., $f''(x) = 0$], define the inflection point of this function. There is always an inflection point between a maximum and a minimum values of the $f(x)$ function. Now, consider the function between the drawdown and time [i.e., $s = f(t)$], whose semilogarithmic variation, t being the logarithmic scale, is shown in Figure 6-5. As shown in Figure 6-5, the drawdown versus time has an inflection point and its properties are given below (Hantush, 1956, 1964).

By differentiating Eq. (6-21) with respect to $\log(t)$, the geometric slope, m, at any point is given by

$$m = \frac{\partial s}{\partial[\log(t)]} = \frac{\Delta s}{\Delta[\log(t)]} = \frac{2.303Q}{4\pi T}\exp\left(-u - \frac{r^2}{4B^2u}\right) \tag{6-33}$$

At the inflection point, the following relations are valid (Hantush, 1956):

(a) First we have

$$u_i = \frac{r^2S}{4Tt_i} = \frac{r}{2B} \tag{6-34}$$

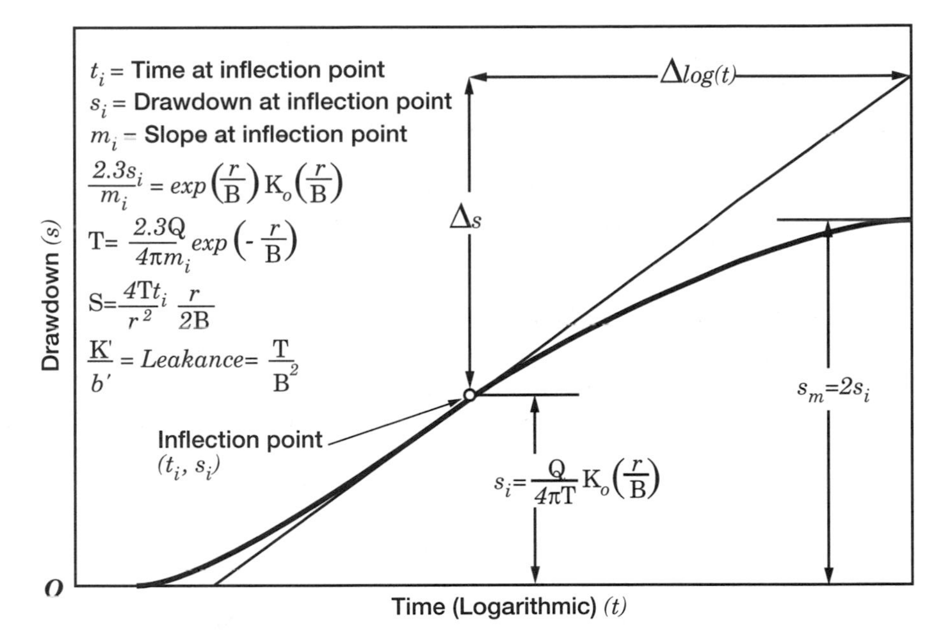

Figure 6-5 Drawdown versus time variation at an observation well of a leaky confined aquifer due to constant rate water extraction.

where the subscript i represents the values of the variables at the inflection point. Equation (6-34) is obtained by making zero the second derivative of s, as given by Eq. (6-21), with respect to $\log(t)$ and solving for u.

(b) The geometric slope of the curve, m_i, at the inflection point is given by

$$m_i = \frac{2.303Q}{4\pi T} \exp\left(-\frac{r}{B}\right) \tag{6-35}$$

and solving for r gives

$$r = 2.303B \left[\log\left(\frac{2.303Q}{4\pi T}\right) - \log(m_i)\right] \tag{6-36}$$

(c) The drawdown at the inflection point, s_i, is given by

$$s_i = \frac{Q}{4\pi T} K_0\left(\frac{r}{B}\right) = \frac{1}{2} s_m \tag{6-37}$$

where s_m is the maximum (steady-state) drawdown. Equation (6-37) is determined as follows: First, s_i determined by substituting $u = r/2B$ into Eq. (6-21). Then, substitution the resulting integral expression into Eq. (6-17) with the value of u_i given by Eq. (6-34) establish the relation given by Eq. (6-37).

(d) Simultaneously solving Eqs. (6-35) and (6-37) for $f(r/B)$ gives the relation between the drawdown and the geometric slope of the curve at the inflection point:

$$f\left(\frac{r}{B}\right) = \exp\left(\frac{r}{B}\right) K_0\left(\frac{r}{B}\right) = 2.303\frac{s_i}{m_i} \tag{6-38}$$

To apply the above theory for the determination of the aquifer parameters, tables of the functions $W(u, r/B)$, $e^x K_0(x)$, and $e^x W(x)$ are necessary. The tabulated form of the first function, $W(u, r/B)$, is given in Table 6-1. The second and third functions are given in Table 3-2 of Chapter 3. In Table 3-2, $-Ei(-x)$ is equivalent to the Theis well function, $W(u)$.

6.3.2.2 One Observation Well Method.

In this method, which is based on Eq. (6-23), the observed drawdowns of a single well are used. One of the main requirements of this method is to determine the maximum drawdown at the observation well. If the pumping test period is long enough so that the maximum drawdown at the observation well can be determined by extrapolation, the following steps can be applied for the determination of the aquifer parameters (Hantush, 1956, 1964; Kruseman and De Ridder, 1991):

Step 1: Plot the drawdown (s) versus time (t) on semilogarithmic paper (t on logarithmic scale) and draw the best fit of the data curve.

Step 2: Using fitted curve determine the value of maximum drawdown (s_m) by extrapolation. This is only possible is the pumping test period is long enough.

Step 3: Using the value of s_m, calculate s_i from Eq. (6-37), that is, $s_i = (1/2)s_m$. Using the value of s_i, determine the inflection point as shown in Figure 6-5.

Step 4: On the s versus t curve with the known s_i value, determine the value of t_i at the inflection point.

Step 5: Determine the geometric slope (m_i) of the s versus t curve at the inflection point. For most cases, m_i can closely be approximated by the geometric slope of the straight-line portion of the curve on which the inflection point is located.

Step 6: Substitute the values of s_i and m_i into Eq. (6-38) and determine by interpolation, if necessary, the value of r/B from Table 3-2 of Chapter 3 of the function $\exp(r/B)K_0(r/B)$.

Step 7: From the known value of r/B and r, calculate the value of B.

Step 8: Using the known values of Q, s_i, m_i, and r/B, calculate the value of T from either Eq. (6-35) or Eq. (6-37).

Step 9: Using the known values of T, t_i, r, and r/B, calculate the value of S from Eq. (6-34).

Step 10: Substitute the values of T and B into the relation $K'/b' = T/B^2$, which is determined from Eq. (6-11), and calculate the value of leakance, K'/b'.

Example 6-2: Using Hantush's one observation well method, determine the values of horizontal hydraulic conductivity of the aquifer (K_h) and vertical hydraulic conductivity of the aquitard (K'), and also determine the storage coefficient of the aquifer (S) for the drawdown data of Example 6-1 (for the aquifer geometry refer to Figure 6-1). Compare the results with with those determined by the Walton's type-curve method.

Solution:

Step 1: From Table 6-2 the values of s are plotted on single logarithmic paper against the values of t. A straight line is drawn through the points as shown in Figure 6-6.

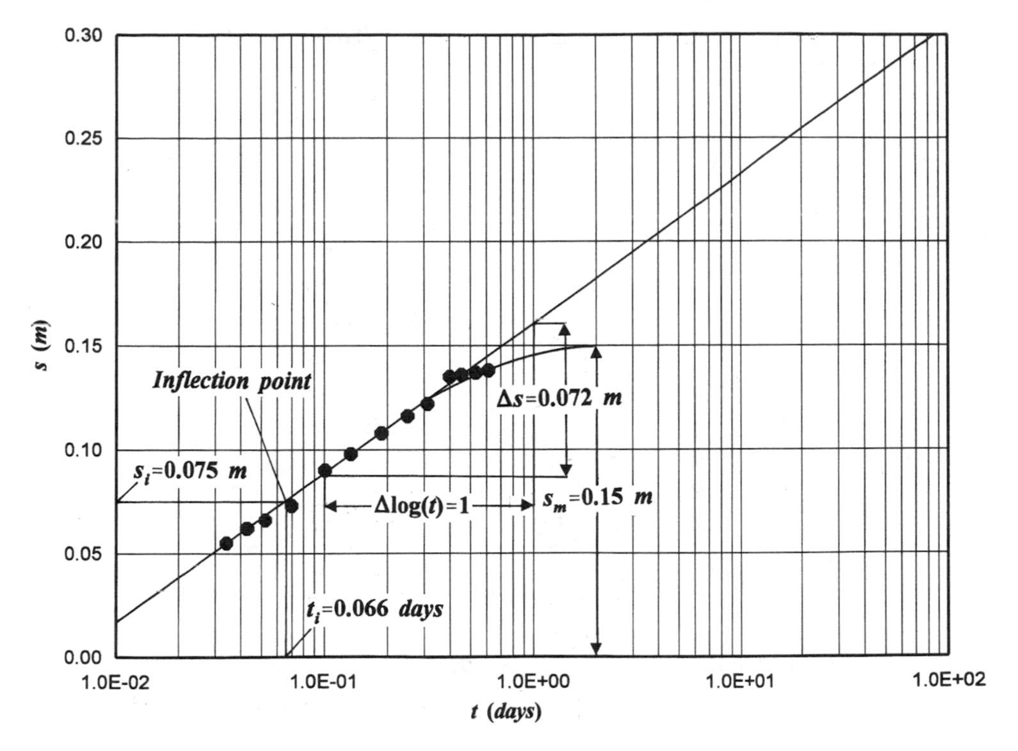

Figure 6-6 Semilogarithmic drawdown versus time variation for Example 6-2.

Step 2: In Figure 6-6 the maximum (or steady-state) drawdown is found by extrapolation, and its value is $s_m = 0.15$ m.

Step 3: From Eq. (6-37) the drawdown at the inflection point is $s_i = (1/2)s_m = (1/2)(0.15 \text{ m}) = 0.075$ m. With this value the inflection point is located and is shown in Figure 6-6.

Step 4: From Figure 6-6 the abscissa of the inflection point can be read as $t_i = 6.6 \times 10^{-2}$ days.

Step 5: From Figure 6-6 the corresponding values for the geometric slope of the straight line portion are $\Delta s = 0.072$ m and $\Delta \log(t) = 1$.

Step 6: From Eq. (6-33) we obtain

$$m_i = \frac{\Delta s}{\Delta \log(t)} = \frac{0.072 \text{ m}}{1} = 0.072 \text{ m}$$

Substitution of the values of s_i and m_i into Eq. (6-38) gives:

$$f\left(\frac{r}{B}\right) = \exp\left(\frac{r}{B}\right) K_0\left(\frac{r}{B}\right) = 2.303\frac{0.075 \text{ m}}{0.072 \text{ m}} = 2.40$$

From Table 3-2 of Chapter 3, the value of r/B can be determined as $r/B = 0.14$.

Step 7: Because $r = 105$ m, $B = 105 \text{ m}/0.14 = 750$ m.

Step 8: Substituting the values $Q = 625\ \text{m}^3/\text{day}$, $m_i = 0.072$ m, and $r/B = 0.14$ into Eq. (6-35) gives

$$T = \frac{2.303Q}{4\pi m_i} \exp\left(-\frac{r}{B}\right) = \frac{(2.303)(625\ \text{m}^3/\text{day})}{(4\pi)(0.072\ \text{m})} \exp(-0.14) = 1382\ \text{m}^2/\text{day}$$

Step 9: From Eq. (6-34) the value of storage coefficient is

$$S = \frac{r}{2B}\frac{4Tt_i}{r^2} = \frac{(105\ \text{m})}{(2)(750\ \text{m})}\frac{(4)(1382\ \text{m}^2/\text{days})(6.6 \times 10^{-2}\text{days})}{(105\ \text{m})^2} = 0.0023$$

Step 10: From Eq. (6-11) we obtain

$$\frac{b'}{K'} = \frac{B^2}{T} = \frac{\left(\dfrac{105\ \text{m}}{0.14\ \text{m}}\right)^2}{1382\ \text{m}^2/\text{days}} = 407\ \text{days}$$

$$K' = \frac{28\ \text{m}}{407\ \text{days}} = 0.069\ \text{m/day}$$

The horizontal hydraulic conductivity of the aquifer is

$$K_h = \frac{T}{b} = \frac{1382\ \text{m}^2/\text{day}}{80\ \text{m}} = 17.28\ \text{m/day}$$

The above results are compared in Table 6-3 with those determined using the Walton's type curve method results. As can be seen from Table 6-3, the horizontal and vertical hydraulic conductivities are close and they are in the same orders of magnitude. The storage coefficients determined by the two methods are close to each other as well. Minor differences can be attributed to reading errors from the graphs and tables and roundoff errors associated with the calculations.

Example 6-3: This example is adapted from Hantush (1956) by changing the units from the English system to those of the metric system.

The drawdown data determined from an observation well under the pumping conditions on the Moutrays farm in the Roswell artesian basin, New Mexico, are given in Table 6-4. Refer to Figure 6-1 for the aquifer geometry. Before the test, water was pumped at a relatively small rate for a period of approximately 30 minutes. The constant drawdown

TABLE 6-3 Comparison Between the Walton's Type-Curve Method and Hantush's One Observation Well Method Results for Example 6-2

Method	T m²/day	K_h m/day	S	b'/K' (days)	K' (m/day)
Walton's type-curve method	875	10.94	0.0027	140	0.20
Hantush's one observation well method (Point A)	1382	17.28	0.0023	407	0.069

TABLE 6-4 Pumping Test Data at r = 323 m for Example 6-3

t (min)	s (m)	t (min)	s (m)
10	1.280	189	3.425
12	1.384	215	3.563
15	1.488	250	3.667
18	1.591	300	3.805
20	1.626	390	3.979
25	1.764	410	4.048
30	1.868	470	4.151
35	2.006	550	4.290
40	2.076	600	4.394
45	2.180	1358	5.086
55	2.318	1470	5.120
65	2.456	1585	5.258
75	2.595	1710	5.328
85	2.698	1850	5.345
95	2.802	2800	5.501
108	2.941	3000	5.535
126	3.044	3500	5.604
136	3.148	4200	5.708
171	3.252		

in the observation well during that 30-minute pumping period was 0.61 m. Probably, the drawdown was reached to this value within the first 20 minutes. After a few minutes from the start of the pumping test, the drawdown was approximately constant in this observation well. At the end of 10 minutes from the start of the pumping test, the water level in the observation well had dropped an additional 0.66 m. As a result, the observed drawdowns during the early part of the test (approximately during the first 40 minutes) were definitely affected by the recovery due to the break of water pumping. The pumping rate was 8147 m^3/day and the observation well was 323 m away from the pumped well. Using Hantush's one observation well method, determine the values of T, b'/K', and S for the leaky aquifer.

Solution:

Step 1: Using the values in Table 6-4, the values of s are plotted on single logarithmic paper against the values of s (Figure 6-7). The curve shown in Figure 6-7 is not adjusted for the early data portion. As mentioned above, the first 40-minute portion does not represent the drawdown resulted from the pumping.

Step 2: From Figure 6-7, s_m is estimated to be 5.7 m.

Step 3: From Eq. (6-37), the drawdown at the inflection point is $s_i = (1/2)s_m = (1/2)(5.7 \text{ m}) = 2.85$ m. With this value, the located inflection point is shown in Figure 6-7.

Step 4: From Figure 6-7, the abscissa of the inflection point can be read as $t_i = 98.0$ min.

Step 5: The corresponding values for the geometric slope are (see Figure 6-7) $\Delta s = 1.9$ m and $\Delta \log(t) = 1$.

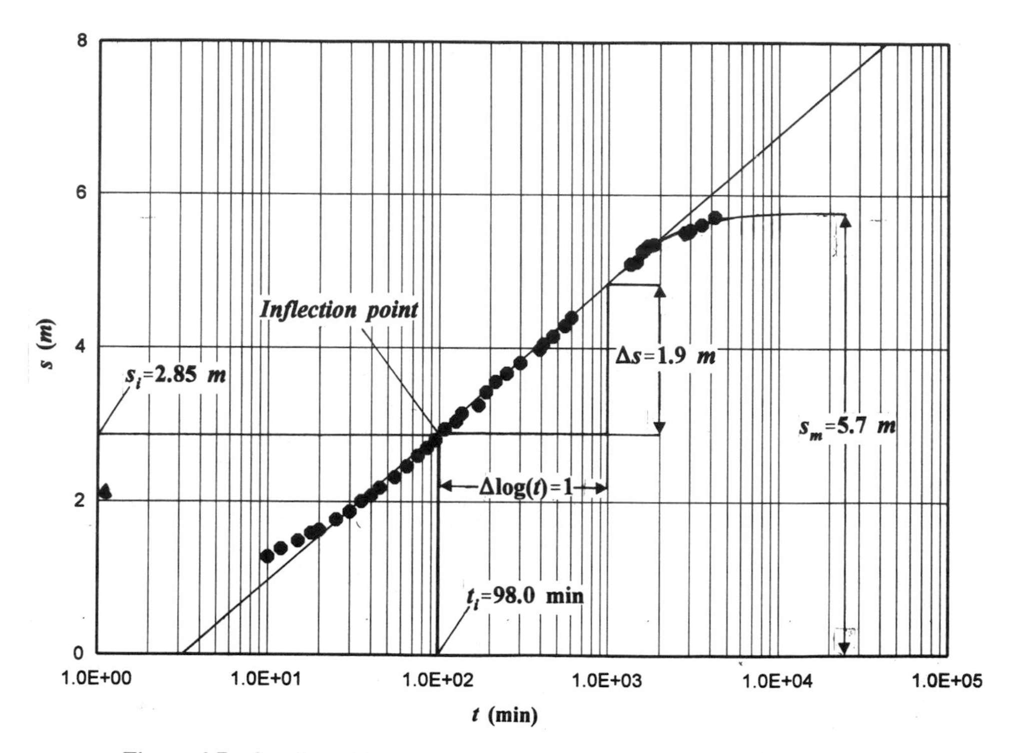

Figure 6-7 Semilogarithmic drawdown versus time variation for Example 6-3.

Step 6: From Eq. (6-33) we obtain

$$m_i = \frac{\Delta s}{\Delta \log(t)} = \frac{1.9 \text{ m}}{1} = 1.9 \text{ m}$$

Introducing the values of s_i and m_i into Eq. (6-38) gives

$$f\left(\frac{r}{B}\right) = \exp\left(\frac{r}{B}\right) K_0\left(\frac{r}{B}\right) = 2.303\frac{2.85 \text{ m}}{1.9 \text{ m}} = 3.45$$

From Table 3-2 of Chapter 3, the value of r/B can be determined as $r/B = 0.041$.

Step 7: Using the known values of r/B and r, $B = 323 \text{ m}/0.041 = 7878$ m.

Step 8: Using $Q = 8147 \text{ m}^3\text{/day}$, $m_i = 1.9$ m, and $r/B = 0.041$, Eq. (6-35) gives

$$T = \frac{2.303Q}{4\pi m_i} \exp\left(-\frac{r}{B}\right) = \frac{(2.303)(8147 \text{ m}^3\text{/day})}{(4\pi)(1.9 \text{ m})} \exp(-0.041)$$
$$= 753 \text{ m}^2\text{/day}$$

Step 9: From Eq. (6-34), with $t_i = 98 \text{ min} = 0.0681$ days and the other values determined from the previous steps gives the value of S:

$$S = \frac{r}{2B}\frac{4Tt_i}{r^2} = \frac{323 \text{ m}}{(2)(7878 \text{ m})}\frac{(4)(753 \text{ m}^2\text{/day})(0.0681 \text{ days})}{(323 \text{ m})^2} S = 0.00004$$

Step 10: From Step 6, $r/B = 0.041$. From Eq. (6-11) we obtain

$$\frac{b'}{K'} = \frac{B^2}{T} = \frac{\left(\frac{r}{0.041}\right)^2}{T} = \frac{(323\ \text{m}/0.041)^2}{753\ \text{m}^2/\text{day}} = 82{,}422\ \text{days}$$

6.3.2.3 More-Than-One-Observation-Well Method. If there are more than one observation wells for a pumping test, the methodology outlined in Section 6.3.2.1 can be applied to each observation well under the condition that the maximum drawdown can be determined by extrapolation from the drawdown versus time data. Then, using the values of each observation well, average aquifer parameter values can be calculated in the vicinity for the pumping test area.

The method for more than one observation well, developed by Hantush (1956), can be used under the prerequisites that (1) data from at least two observation wells are available and (2) the straight-line portion of the semilogarithmic drawdown versus time curves are fully developed. The method is based on the theory of the Hantush inflection-point method presented in Section 6.3.2.1 and Eq. (6-35) is the main equation to be used. After taking the logarithms of the both sides of Eq. (6-35), it can be expressed as

$$r = 2.303B\left[\log\left(\frac{2.303Q}{4\pi T}\right) - \log(m_i)\right] \tag{6-39}$$

A semilogarithmic plot of r versus m_i (m_i is on the logarithmic scale) should be a straight line. If the aforementioned conditions are satisfied, the following steps can be used to determine the aquifer parameters:

Step 1: Plot the drawdown (s) versus time (t) data (t on the logarithmic scale) for all observation wells and draw the best fits of the data curves.

Step 2: Determine the values of the geometric slope (m_i) for the straight-line portion of the curve for each observation well using Eq. (6-33).

Step 3: On another semilogarithmic scale, plot distance (r) versus geometric slope (m_i) data (m_i is on the logarithmic scale), and draw the best-fit straight line through the plotted points. This line is the graphic representation of Eq. (6-39).

Step 4: Find the geometric slope of the line constructed in Step 3.

Step 5: From Eq. (6-39) we obtain

$$\Delta r = 2.303B\Delta \log(m_i) \tag{6-40}$$

or solving for B gives

$$B = \frac{1}{2.303}\frac{\Delta r}{\Delta \log(m_i)} \tag{6-41}$$

Using the known values of Δr and $\Delta \log(m_i)$, calculate the value of B from Eq. (6-41).

Step 6: From the graph in Step 3, determine the value of m_i when r is zero, designated by $(m_i)_0$.

Step 7: Substituting $r = 0$ and $m_i = (m_i)_0$ into Eq. (6-35) gives

$$T = \frac{2.303Q}{4\pi(m_i)_0} \tag{6-42}$$

Calculate the value of T from Eq. (6-42).

Step 8: Using the known values of T and B, calculate the leakance from the relation $K'/b' = T/B^2$, which is determined from Eq. (6-11).

Step 9: From the known values of Q, r, T, and B, calculate the value of s_i for each observation well from Eq. (6-37). Use Table 3-2 of Chapter 3 for the values of $K_0(x)$.

Step 10: Determine the inflection points for each observation well using the known s_i values and read their respective abscissa (t_i) values.

Step 11: Knowing the values of T, r, r/B, and t_i for each observation well, calculate the value of S from Eq. (6-34).

Step 12: Calculate the mean of the S values in Step 11 which represents the average storage coefficient in the well field.

Example 6-4: This example is adapted from Hantush (1956) by changing the units from the English system to the metric system.

A pumping test was conducted on the F. Wortman farm in the Orchard Park area of Roswell artesian basin, New Mexico, and the drawdown data were collected from three different observation wells which were located 738 m, 853 m, and 1219 m away from the pumped well and the data are listed in Tables 6-5, 6-6, and 6-7, respectively. The constant extraction rate was 2936 m^3/day. Using *Hantush's more-than-one-observation well method*, determine the values of T, S, and b'/K' of the aquifer. Refer Figure 6-1 for the aquifer geometry.

Solution:

Step 1: The semilogarithmic drawdown versus time curves for the three observation wells are presented in Figures 6-8, 6-9, and 6-10.

TABLE 6-5 Pumping Test Data at r = 738 m for Example 6-4

t (min)	s (m)	t (min)	s (m)
10	0.044	239	0.376
13	0.053	259	0.371
20	0.066	272	0.415
30	0.084	300	0.433
100	0.230	331	0.460
131	0.265	363	0.486
141	0.279	400	0.517
148	0.287	463	0.544
160	0.301	500	0.566
176	0.340	600	0.601
190	0.362	750	0.654
205	0.380	840	0.676
216	0.389		

TABLE 6-6 Pumping Test Data at r = 853 m for Example 6-4

t (min)	s (m)	t (min)	s (m)
15	0.049	227	0.362
25	0.080	245	0.345
35	0.097	279	0.345
40	0.106	371	0.451
60	0.168	400	0.468
75	0.168	450	0.482
90	0.194	500	0.513
125	0.230	750	0.610
185	0.309	900	0.654

TABLE 6-7 Pumping Test Data at r = 1219 m for Example 6-4

t (min)	s (m)	t (min)	s (m)
10	0.035	136	0.208
20	0.049	152	0.203
25	0.053	164	0.230
30	0.062	185	0.252
35	0.071	190	0.274
40	0.075	210	0.301
45	0.084	221	0.305
50	0.088	245	0.305
55	0.097	272	0.305
60	0.102	300	0.327
65	0.111	331	0.336
70	0.119	363	0.362
75	0.124	400	0.398
80	0.128	441	0.424
85	0.133	476	0.442
90	0.141	600	0.482
105	0.146	725	0.522
108	0.168	850	0.566
125	0.186		

Step 2: From Figures 6-8, 6-9, and 6-10, the geometric slopes of the straight portions of the curves are as follows:

$$m_i = \frac{\Delta s}{\Delta \log(t)} = \frac{0.365 \text{ m}}{0.713} = 0.512 \text{ m} \qquad \text{for } r = 738 \text{ m}$$

$$m_i = \frac{0.330 \text{ m}}{0.638} = 0.517 \text{ m} \qquad \text{for } r = 853 \text{ m}$$

$$m_i = \frac{0.253 \text{ m}}{0.481} = 0.526 \text{ m} \qquad \text{for } r = 1219 \text{ m}$$

Step 3: Using the values in Step 2, the distance versus geometric slope graph is shown in Figure 6-11.

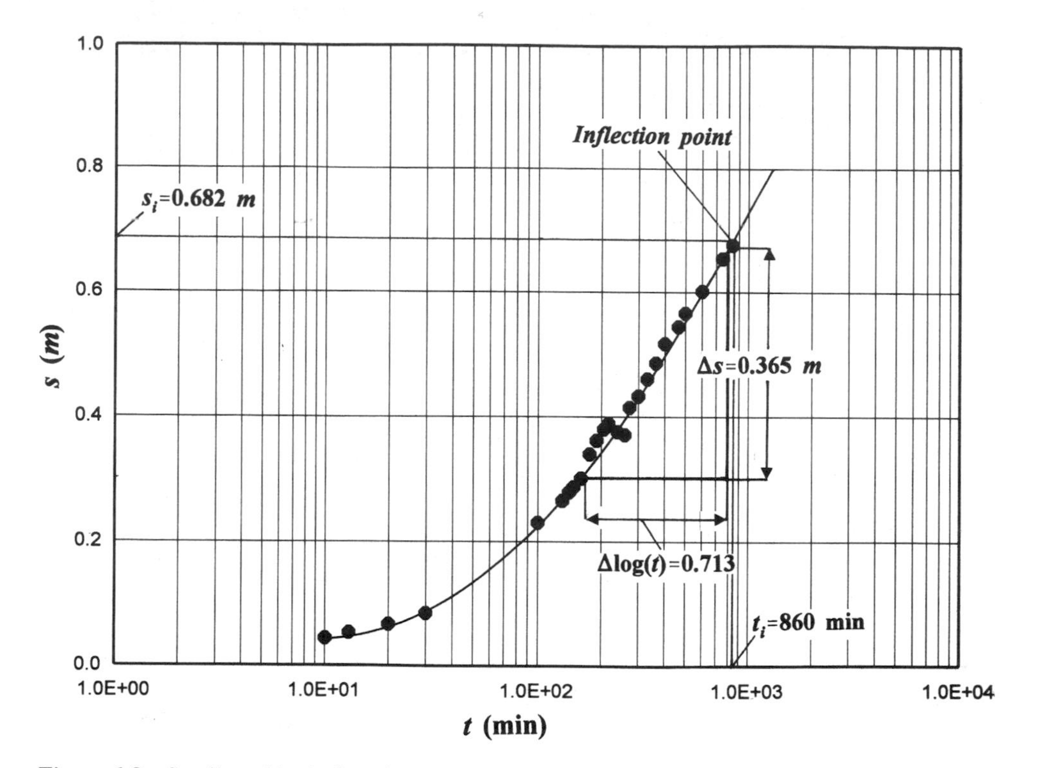

Figure 6-8 Semilogarithmic drawdown versus time variation for the observation well at $r = 738$ m for Example 6-4.

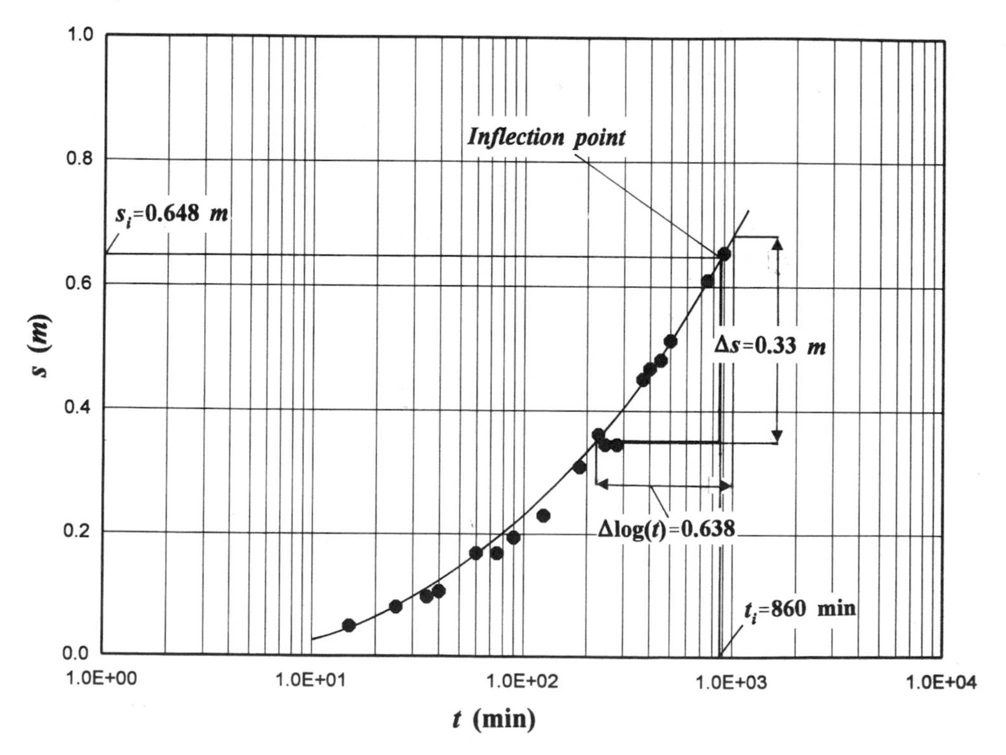

Figure 6-9 Semilogarithmic drawdown versus time variation for the observation well at $r = 853$ m for Example 6-4.

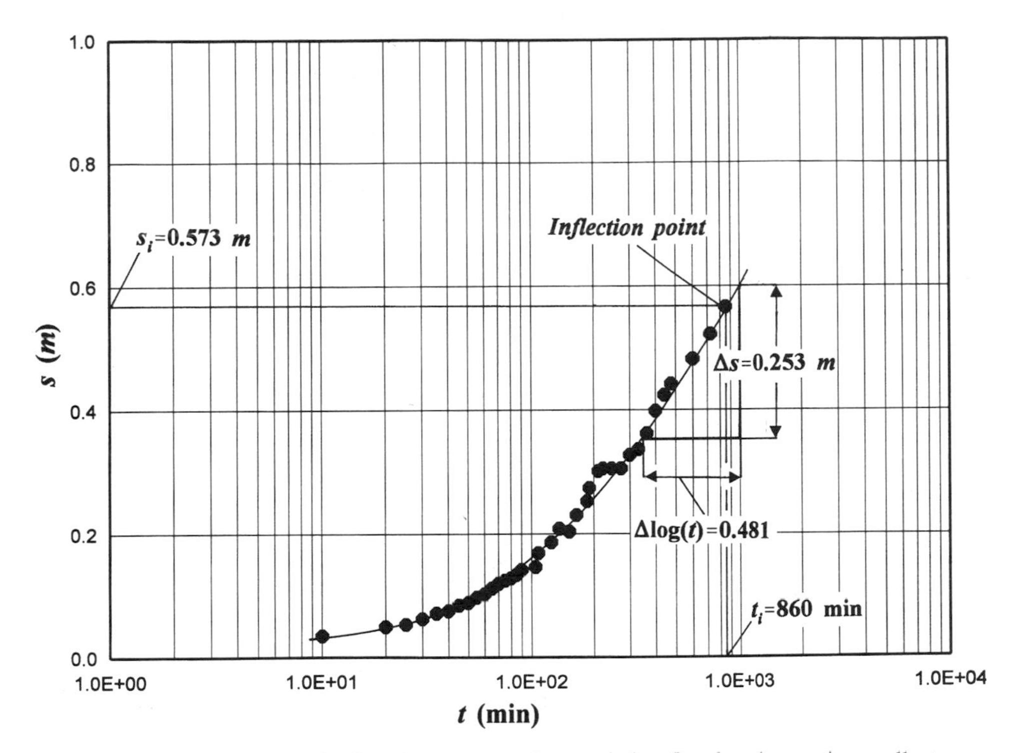

Figure 6-10 Semilogarithmic drawdown versus time variation for the observation well at $r = 1219$ m for Example 6-4.

Step 4: From Figure 6-11, the geometric slope is

$$\frac{\Delta r}{\Delta \log(m_i)} = \frac{1750 \text{ m}}{\log\left(\dfrac{0.547 \text{ m}}{0.490 \text{ m}}\right)} = 36{,}618 \text{ m}$$

Step 5: Substitution the values into Eq. (6-41) gives

$$B = \frac{1}{2.303} \frac{\Delta r}{\Delta \log(m_i)} = \frac{1}{2.303}(36618 \text{ m}) = 15{,}900 \text{ m}$$

Step 6: From Figure 6-11, $(m_i)_0 = 0.49$ m.

Step 7: From Eq. (6-42), with $Q = 2936 \text{ m}^3/\text{day} = 0.034 \text{ m}^3/\text{s}$,

$$T = \frac{2.303Q}{4\pi(m_i)_0} = \frac{(2.303)(0.034 \text{ m}^3/\text{s})}{(4\pi)(0.49 \text{ m})} = 0.0127 \text{ m}^2/\text{s}$$

Step 8: When we use the known values of T and b, the value of leakance is

$$\frac{K'}{b'} = \frac{T}{B^2} = \frac{0.0127 \text{ m}^2/\text{s}}{(15900 \text{ m})^2} = 0.50 \times 10^{-10} \text{ s}^{-1}$$

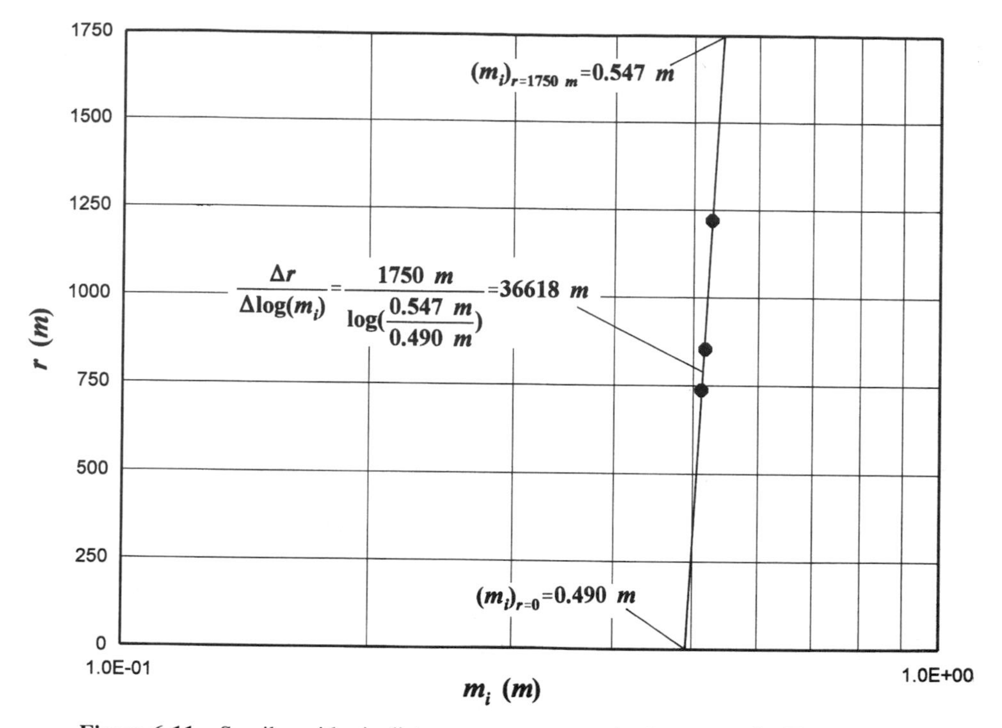

Figure 6-11 Semilogarithmic distance versus geometric slope curve for Example 6-4.

Step 9: From Eq. (6-37) we obtain

$$s_i = \frac{Q}{4\pi T} K_0 \left(\frac{r}{B}\right) = \frac{0.034 \text{ m}^3/\text{s}}{(4\pi)(0.0127 \text{ m}^2/\text{s})} K_0 \left(\frac{738 \text{ m}}{15900 \text{ m}}\right)$$

$$= (0.213 \text{ m})K_0(0.046) = (0.213 \text{ m})(3.2) = 0.682 \text{ m} \qquad \text{for } r = 738 \text{ m}$$

The value of K_0 (0.046) is taken from Table 3-2 of Chapter 3. Similarly, for the other observation wells the following values can be obtained:

$$s_i = (0.213 \text{ m})K_0 \left(\frac{853 \text{ m}}{15900 \text{ m}}\right) = (0.213 \text{ m})K_0(0.054)$$

$$= (0.213 \text{ m})(3.04) = 0.648 \text{ m} \qquad \text{for } r = 853 \text{ m}$$

$$s_i = (0.213 \text{ m})K_0 \left(\frac{1219 \text{ m}}{15900 \text{ m}}\right) = (0.213 \text{ m})K_0(0.077)$$

$$= (0.213 \text{ m})(2.69) = 0.573 \text{ m} \qquad \text{for } r = 1219 \text{ m}$$

Step 10: Using the values of s_i, the following t_i values can be read from Figures 6-8, 6-9, and 6-10:

$$s_i = 0.682 \text{ m}, \quad t_i = 860 \text{ min} \qquad \text{for } r = 738 \text{ m}$$
$$s_i = 0.648 \text{ m}, \quad t_i = 860 \text{ min} \qquad \text{for } r = 853 \text{ m}$$
$$s_i = 0.573 \text{ m}, \quad t_i = 860 \text{ min} \qquad \text{for } r = 1219 \text{ m}$$

TABLE 6-8 Data Summary for Example 6-4 Between Step 1 and Step 11

r (m)	r/B	$K_0(r/B)$	s_i (m) from Eq. (6-37)	t_i (min) from Figures 6-8, 6-9, and 6-10	S from Eq. (6-34)
738	0.046	3.20	0.682	860	0.000112
853	0.054	3.04	0.648	860	0.000097
1219	0.077	2.69	0.573	860	0.000068

Step 11: With $T = 0.0127 \text{ m}^2/\text{s} = 0.762 \text{ m}^2/\text{min}$ and $B = 15{,}900$ m, the values of S are calculated from Eq. (6-34), $S = (2Tt_i)/(Br)$, and are listed in Table 6-8 for different values of the parameters.

Step 12: From the values in Table 6-8, the average value of S is 0.000092.

6.3.3 Hantush's Type-Curve Method

Hantush developed a type-curve method (Hantush, 1964) on the condition that the period of test is long enough so that a sufficient number of observed drawdown data fall within the period $t > 4t_i$ and the maximum drawdown can be determined by extrapolation.

The type-curve method is based on Eq. (6-17), which can also be expressed as

$$s(r,t) = \frac{Q}{4\pi T}\left[2K_0\left(\frac{r}{B}\right) - W\left(q, \frac{r}{B}\right)\right] \tag{6-43}$$

$W(q, r/B)$ represents the integral expression on the right-hand side of Eq. (6-17) and q is defined by Eq. (6-18). Hantush (1964) showed that if $q > 2r/B$, Eq. (6-43) can be approximated by

$$s_m - s = \frac{Q}{4\pi T}W(q) \tag{6-44}$$

with

$$s_m = \frac{Q}{2\pi T}K_0\left(\frac{r}{B}\right) \tag{6-45}$$

where s_m is the maximum or steady-state drawdown. The type curve is a double logarithmic plot of $W(q)$ versus q whose adequate range is $10^{-3} < q < 5$. The observed drawdown data are plotted as $(s_m - s)$ versus t for all available wells, and the values of s_m and t_i are different for different well positions. The maximum drawdown (s_m) values are extrapolated from the semilogarithmic drawdown versus time plot of each observation well. The corresponding values of t_i are estimated by locating the inflection point of the curve using Eq. (6-37), $s_i = (1/2)s_m$.

During the best type-curve matching process, one should bear in mind that the observed drawdown points within the period of $t < 4t_i$, for each observation well, may fall below the type curve which is due to fact that Eq. (6-44) may not be applicable. As a matter of fact, within the period defined above, the drawdown versus time curves for each observation well will be matched to one of the type curves of $W(q, r/B)$ versus q for different values of r/B, if the Walton's type-curve method is applied (see Section 6.3.1).

Under the framework of the above-given theoretical background, the steps of the Hantush's type-curve method are given below:

Step 1: For each observation well, plot s versus t on single logarithmic paper (t on the logarithmic scale) and draw the best fit of the data points.

Step 2: Determine the value of maximum drawdown (s_m) by extrapolation. This is only possible if the period of the test is long enough.

Step 3: Prepare a type curve by plotting $W(q)$ versus q on double logarithmic paper. This curve is identical to $W(u)$ versus u of the Theis type curve. Consequently, Table 4-1 of Chapter 4 can be used for the generation of the type curve.

Step 4: On another sheet of double logarithmic paper with the same scale as the one in Step 3, plot the observed drawdown data curve ($s_m - s$) versus t for all available observation wells. It should be remembered that each observation well has its own value of s_m. According to Eq. (6-18), $q = Tt/(SB^2)$; consequently, q is independent of r. Therefore, the data of all observation wells can be plotted in one graph.

Step 5: Keeping the coordinates axes parallel at all times, superimpose the observed data curve (Step 4) on the type curve (Step 3) until the best fit of the data curve and the type curve is obtained. One should bear in mind that the observed drawdown points within the period $t < 4t_i$ for each of the observation wells may fall below the type curve because in this period Eq. (6-44) may not be applicable.

Step 6: Select a common point, the match point, arbitrarily chosen on the overlapping part of the curve, or even anywhere on the overlapping portion of the sheets. Read the corresponding coordinate values of ($s_m - s$), t, q, and $W(q)$.

Step 7: Using the match-point values and the value of Q, calculate T from Eq. (6-44).

Step 8: Substitute the values of s_m and T into Eq. (6-45) and determine the value of $K_0(r/B)$ and subsequently find r/B from Table 3-2 of Chapter 3.

Step 9: From the known values of r/B and r, calculate the value of B. Then, calculate b'/K' from Eq. (6-11) as $b'/K' = B^2/T$.

Step 10: Substitute the values of T, t, q, and B into Eq. (6-18) and solve for S.

Example 6-5: Determine the aquifer parameters for the data of Example 6-3 (Table 6-4) using the Hantush type-curve method and compare the results with those determined by the inflection-point method of Hantush as presented in Example 6-3.

Solution:

Step 1: The drawdown (s) versus time (t) is given in Figure 6-7 (t on logarithmic scale). As shown, a straight line is drawn through the points.

Step 2: From Figure 6-7, the value of s_m is 5.7 m.

Step 3: The type curve, $W(q)$ versus q, on double logarithmic scale is given in Figure 4-2 of Chapter 4 with the exception that $W(u)$ versus u is plotted (see Figure 6-13).

Step 4: Using the value of s_m in Step 2, t versus $s_m - s$ data are presented in Table 6-9 based on the values in Table 6-4. The drawdown data of ($s - s_m$) versus t on double logarithmic paper having the same scale of the graph of Step 3 is given in Figure 6-12.

Step 5: The superimposed sheets are shown in Figure 6-13. As expected, the early time data ($t < 4t_i$) fall below the type curve.

TABLE 6-9 Pumping Test Data at r = 323 m generated from Table 6-4 with s_m = 5.7 m for Example 6-5

t (days)	$s_m - s$ (m)	t (days)	$s_m - s$ (m)
10	4.420	189	2.275
12	4.316	215	2.137
15	4.212	250	2.033
18	4.109	300	1.895
20	4.074	390	1.721
25	3.936	410	1.652
30	3.832	470	1.549
35	3.694	550	1.410
40	3.624	600	1.306
45	3.520	1358	0.614
55	3.382	1470	0.580
65	3.244	1585	0.442
75	3.105	1710	0.372
85	3.002	1850	0.355
95	2.898	2800	0.199
108	2.759	3000	0.165
126	2.656	3500	0.096
136	2.552		
171	2.448		

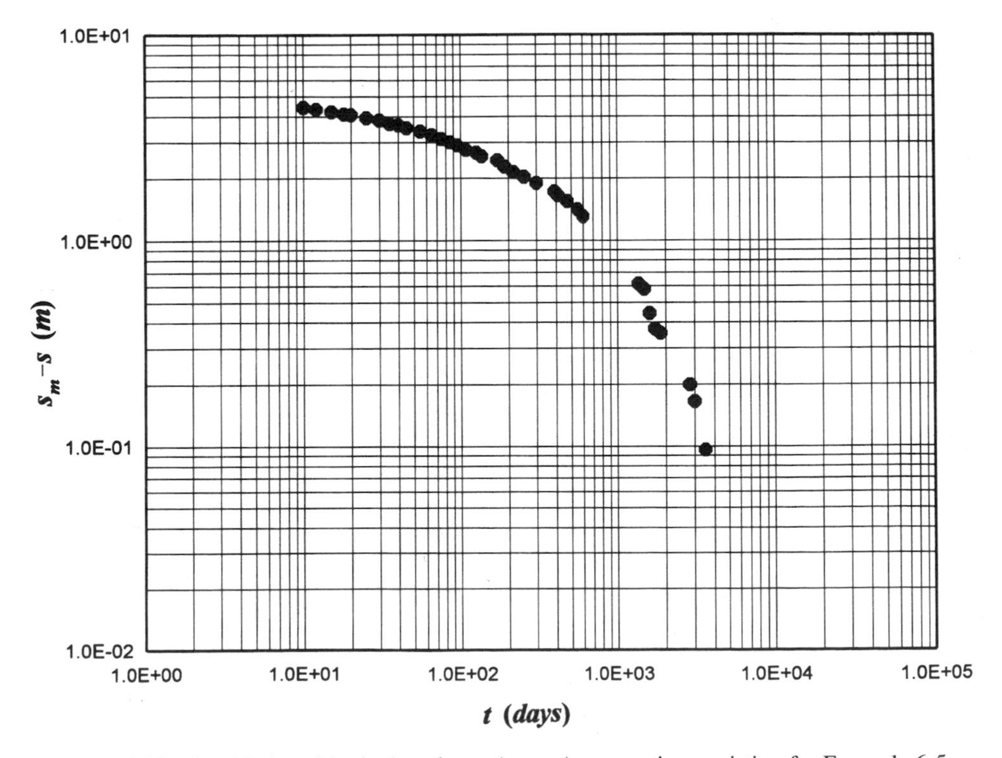

Figure 6-12 Double logarithmic drawdown, $(s - s_m)$, versus time variation for Example 6-5.

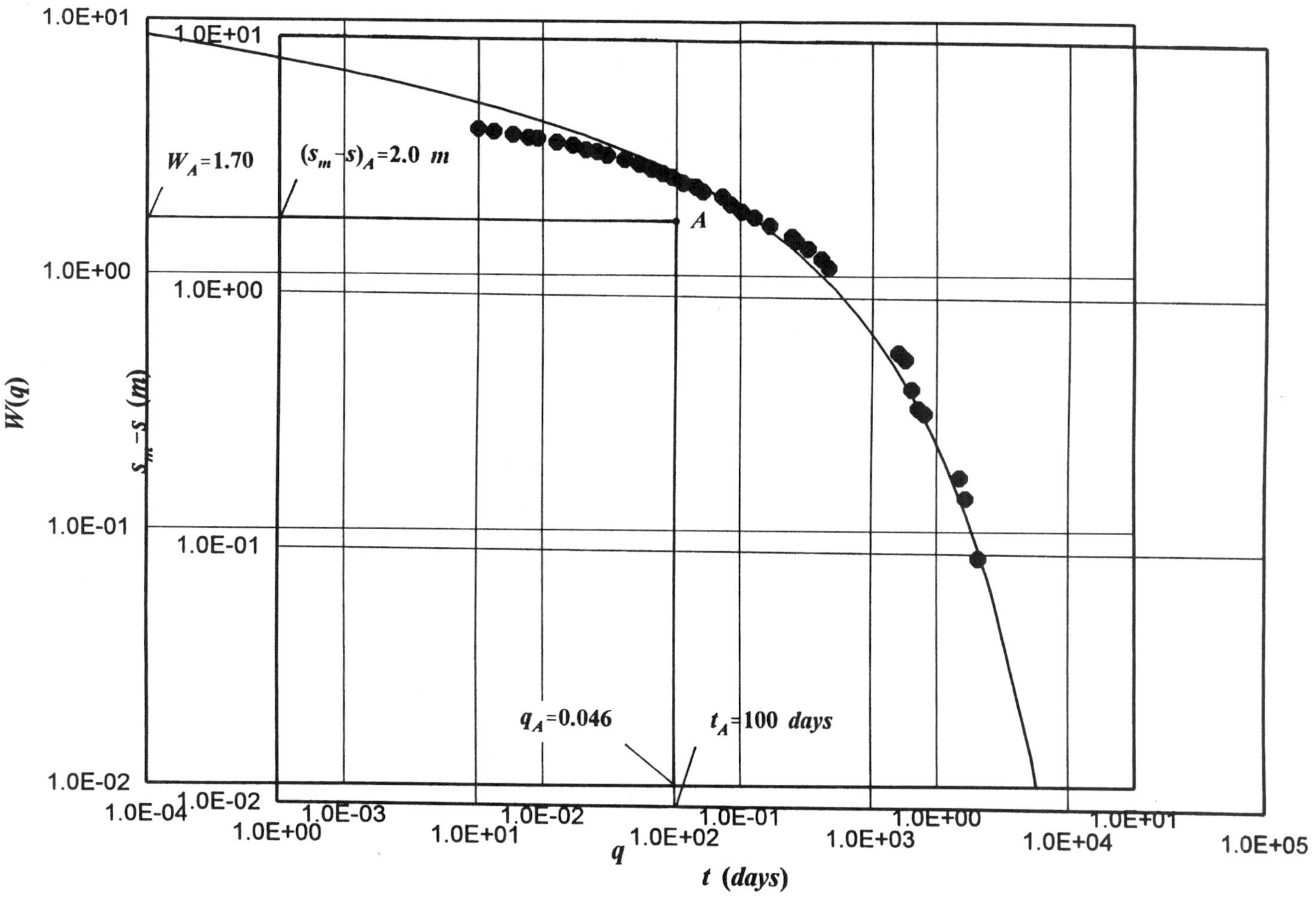

Figure 6-13 Type-curve matching for Example 6-5.

TABLE 6-10 Comparison of the Hantush's Inflection-Point Method and Hantush's Type-Curve Method Results for Example 6-5

Method	T (m^2/day)	b'/K' (days)	S
Hantush's inflection-point method	753	82,422	0.00004
Hantush's type-curve method	551	19,228	0.00008

Step 6: In Figure 6-12, A is an arbitrarily chosen common point and its coordinates are

$$q_A = 4.6 \times 10^{-2}, \qquad W_A = 1.70$$
$$(s_m - s)_A = 2.0 \text{ m}, \qquad t_A = 100 \text{ min}$$

Step 7: From Eq. (6-44), the value of T is

$$T = \frac{Q}{4\pi(s_m - s)_A} W(q)_A = \frac{8147 \text{ m}^3/\text{day}}{(4\pi)(2.0 \text{ m})}(1.70) = 551 \text{ m}^2/\text{day}$$

Step 8: From Eq. (6-45) we obtain

$$K_0\left(\frac{r}{B}\right) = \frac{2\pi T s_m}{Q} = \frac{(2\pi)(551 \text{ m}^2/\text{day})(5.7 \text{ m})}{8147 \text{ m}^3/\text{day}} = 2.42$$

From Table 3-2 of Chapter 3, the corresponding value of r/B is 0.10.

Step 9: Because $r = 323$ m and the value of r/B is known, the calculated value of B is 3230 m. Then, from Eq. (6-11), the value of b'/K' is calculated to be 19,228 days.

Step 10: From Eq. (6-18), the value of S is

$$S = \frac{Tt}{qB^2} = \frac{(551 \text{ m}^2/\text{day})(100 \text{ min}/1440 \text{ min/days})}{(4.6 \times 10^{-2})(3230 \text{ m})^2} = 0.00008$$

The results of the Hantush's inflection-point method and Hantush's type-curve method are compared in Table 6-10. As can be seen, the transmissivity (T) values are in the same order of magnitude and the values of b'/K' and S are reasonably close. Minor differences can be attributed to the reading and roundoff errors.

6.4 PUMPING TEST DATA ANALYSIS METHODS FOR HOMOGENEOUS AND ANISOTROPIC LEAKY CONFINED AQUIFERS

Apart from the methods presented in Section 6.3 for the analysis of drawdown data for isotropic leaky confined aquifers, Hantush (1966a, b) developed methods of drawdown data analysis for anisotropic leaky confined aquifers based on his original theoretical analysis. In Section 5.3 of Chapter 5, the foundations and applications for nonleaky anisotropic confined aquifers theory are presented in detail. Because the general theoretical foundations are the same, only the practical applications regarding leaky confined aquifers cases will be presented here, with some additional equations. Therefore, the reader is advised to review Section 5.3 of Chapter 5, before reading this section.

6.4.1 Equations for Anisotropic Leaky Confined Aquifers

The pumping test data analysis methods for isotropic leaky confined aquifers are based on the Hantush and Jacob equation as given by Eq. (6-23) in Section 6.2.4.1. Hantush (1966a, b) theoretically showed that the solutions developed for homogeneous and isotropic aquifers can also be used for problems in homogeneous and anisotropic aquifers by using a simple transformation of coordinates. Therefore, the counterpart of Eq. (6-23) is

$$s(r,t) = \frac{Q}{4\pi T} W\left(u', \frac{r}{B'}\right) \tag{6-46}$$

in which u' and T_e are defined by Eqs. (5-44) and (5-46) of Chapter 5, respectively. Based on Eq. (6-11), the definition of B' is as follows:

$$B' = \left(\frac{T_r}{\frac{K'}{b}} \right)^{1/2} \tag{6-47}$$

From Eqs. (5-44) through (5-51) of Chapter 5, the modified forms of Eqs. (5-52) and (5-53) of Chapter 5 can, respectively, be expressed as

$$a = \frac{T_{r_1}}{T_{r_2}} = \frac{\nu_1'}{\nu_2'} = \left(\frac{B_1'}{B_2'}\right)^2 \tag{6-48}$$

$$b = \frac{T_{r_1}}{T_{r_3}} = \frac{\nu_1'}{\nu_3'} = \left(\frac{B_1'}{B_3'}\right)^2 \tag{6-49}$$

The equations that are presented above are analyzed in the following sections for the conditions that (1) the principal directions of anisotropy are known and (2) the principal directions of anisotropy are unknown.

6.4.1.1 Principal Directions of Anisotropy Are Known. Details of this case are included in Section 5.3.3.3 of Chapter 5. For leaky confined aquifers case, in addition to Eqs. (5-55) and (5-56) of Chapter 5, the following equations can similarly be derived (Hantush, 1966b):

$$\frac{K'}{b'} = \frac{T_{r_1}}{(B_1')^2}, \qquad \frac{K'}{b'} = \frac{T_{r_2}}{(B_2')^2} \tag{6-50}$$

The above-mentioned expressions along with Eq. (6-50) will produce values for $T_{\eta\eta}$, $T_{\xi\xi}$, T_{r_1}, T_{r_2}, S, and K'/b'.

6.4.1.2 Principal Directions of Anisotropy Are Unknown. Details for this case are included in Section 5.3.3.3 of Chapter 5. For leaky confined aquifers case, in addition to Eqs. (5-57) through (5-60) of Chapter 5, the following equation can similarly be derived (Hantush, 1966b):

$$\frac{K'}{b'} = \frac{T_{r_1}}{(B_1')^2}, \qquad \frac{K'}{b'} = \frac{T_{r_2}}{(B_2')^2}, \qquad \frac{K'}{b'} = \frac{T_{r_3}}{(B_3')^2} \tag{6-51}$$

The above-mentioned expressions along with Eq. (6-51) will produce values for θ, n, $T_{\eta\eta}$, and then $T_{\xi\xi}$, T_{r_1}, T_{r_2}, T_{r_3}, S, and K'/b'.

6.4.2 Hantush's Pumping Test Data Analysis Methods

The isotropic data analysis methods for leaky confined aquifers determine the values of ν', T_e, and B' which are defined by Eqs. (5-45) and (5-46) of Chapter 5 and Eq. (6-47), respectively. In reference to Figure 5-14 of Chapter 5, let us consider that the application of the isotropic methods on the first, second, and third group of wells is produced (T_{e_1}, ν_1', B_1'), (T_{e_2}, ν_2', B_2'), and (T_{e_3}, ν_3', B_3'), respectively. Theoretically, the three isotropic transmissivities must be the same. In practice, this requirement should be used as a guide in applying the isotropic methods of analysis. If T_{e_1}, T_{e_2}, and T_{e_3} are not at least approximately the same, it means that field conditions are not in agreement with the assumptions of theory and that other vitiating conditions must be present. In the following sections, the data analysis procedures originally developed by Hantush (1966b) are presented in a stepwise manner under the conditions that the three transmissivities based on the isotropic ground-water flow theory are approximately the same.

6.4.2.1 *Procedure I: Principal Directions of Anisotropy Are Known.* As mentioned in Section 5.3.4.1 of Chapter 5, the principal directions of anisotropy, the ξ and η axes in Figure 5-14 of Chapter 5, are known. A group of wells (one or more wells) on the same radial line is referred to as *a ray of observation wells*. The two groups of wells are located on Ray 1 and Ray 2 in Figure 5-14 of Chapter 5, and Ray 1 makes θ angle with the ξ axis. Because θ is known, two groups of wells (or two rays) are enough to determine the leaky confined aquifer parameters. The steps, under the conditions that the angles θ and α are known, are as follows:

Step 1: Using the isotropic methods of drawdown data analysis for confined leaky aquifers in Section 6.3, determine (T_{e_1}, ν_1', B_1') and (T_{e_2}, ν_2', B_2').

Step 2: With the known values of ν_1', ν_2', B_1', and B_2', calculate a from

$$a = \frac{1}{2}\left[\frac{\nu_1'}{\nu_2'} + \left(\frac{B_1'}{B_2'}\right)^2\right] \tag{6-52}$$

which is based on Eq. (6-48).

Step 3: Based on the facts mentioned above, calculate T_e from the following equation:

$$T_e = \frac{1}{2}(T_{e_1} + T_{e_2}) \tag{6-53}$$

Step 4: Using the known values of T_e, θ, and α, calculate $T_{\eta\eta}$ from Eq. (5-55) of Chapter 5.

Step 5: Using the known values of T_e and $T_{\eta\eta}$, calculate $T_{\xi\xi}$ from Eq. (5-54) of Chapter 5 as

$$T_{\xi\xi} = \frac{T_e^2}{T_{\eta\eta}} \tag{6-54}$$

Step 6: Using the known values of $T_{\xi\xi}$, $T_{\eta\eta}$, θ, and α, calculate the values of T_{r_1} and T_{r_2} from Eqs. (5-47) and (5-48) of Chapter 5, respectively.

Step 7: Because Eqs. (5-56) of Chapter 5 provides two S values, calculate the average S value from the following equation:

$$S = \frac{1}{2}\left(\frac{T_{r_1}}{\nu_1'} + \frac{T_{r_2}}{\nu_2'}\right) \tag{6-55}$$

Step 8: From Eqs. (5-56) of Chapter 5 and Eqs. (6-51), calculate K'/b' from the following equation:

$$\frac{K'}{b'} = \frac{1}{2}S\left[\frac{\nu_1'}{(B_1')^2} + \frac{\nu_2'}{(B_2')^2}\right] \tag{6-56}$$

6.4.2.2 *Procedure II: Principal Directions of Anisotropy Are Unknown.*

As mentioned in Section 5.3.4.2 of Chapter 5, if the principal directions of anisotropy (the ξ and η axes in Figure 5-14) are not known, three groups of wells are needed for the determination of the unknowns. Let us assume that the three groups of wells are located on Ray 1, Ray 2, and Ray 3 in Figure 5-14 (Chapter 5). The additional unknown compared with Procedure I (Section 6.4.2.1) is the angle θ between the ξ axis and Ray 1. The angles α and β are the orientation angles of the other group lines (Ray 2 and Ray 3) with the first group line (Ray 1). The unknown principal directions of anisotropy, as well as the parameters of a leaky confined aquifer, can be determined in a stepwise manner as follows:

Step 1: Using the isotropic methods of drawdown data analysis for leaky confined aquifers (see Section 6.3), determine (T_{e_1}, ν_1', B_1'), (T_{e_2}, ν_2', B_2'), and (T_{e_3}, ν_3', B_3') for Ray 1, Ray 2, and Ray 3, respectively.

Step 2: Calculate the value of a from Eq. (6-52) using the known values of ν_1', ν_2', B_1', and B_2'.

Step 3: Calculate the value of b from

$$b = \frac{1}{2}\left[\frac{\nu_1'}{\nu_3'} + \left(\frac{B_1'}{B_3'}\right)^2\right] \tag{6-57}$$

which is determined from Eq. (6-49) by taking the arithmetic average.

Step 4: Determine the value of θ from Eq. (5-57) of Chapter 5 using the known values of a, b, α, and β. Then calculate the value of n from Eq. (5-58) or (5-59) of Chapter 5.

Step 5: As mentioned above, theoretically the three isotropic T_e values must be the same. Therefore, calculate the arithmetic average from

$$T_e = \frac{1}{3}(T_{e_1} + T_{e_2} + T_{e_3}) \tag{6-58}$$

And based on Eq. (5-54) of Chapter 5, calculate $T_{\eta\eta}$ from the following equation:

$$T_{\eta\eta} = T_e n^{-1/2} \tag{6-59}$$

Step 6: Calculate the value of $T_{\xi\xi}$ from Eq. (6-54) using the known values of T_e and $T_{\eta\eta}$.

Step 7: Using the known values of $T_{\xi\xi}$, $T_{\eta\eta}$, θ, α, and β, determine the values of T_{r_1}, T_{r_2}, and T_{r_3} from Eqs. (5-47), (5-48), and (5-49) of Chapter 5, respectively.

Step 8: Calculate the value of storage coefficient from Eqs. (5-60) as

$$S = \frac{1}{3}\left(\frac{T_{r_1}}{\nu_1'} + \frac{T_{r_2}}{\nu_2'} + \frac{T_{r_3}}{\nu_3'}\right) \tag{6-60}$$

which is a modified form of Eq. (6-55).

Step 9: Calculate the value of leakance (K'/b') from Eqs. (5-60) and Eqs. (6-51) as

$$\frac{K'}{b'} = \frac{1}{3}S\left[\frac{\nu_1'}{(B_1')^2} + \frac{\nu_2'}{(B_2')^2} + \frac{\nu_3'}{(B_3')^2}\right] \tag{6-61}$$

which is a modified form of Eq. (6-56).

Step 10: Write the transmissivity equation for any θ angle using Eq. (5-67) of Chapter 5.

As mentioned in Section 5.3.4.2 of Chapter 5, if more than three groups of observation wells are available, the application of the method to each combination of three observation wells will provide a set of values for each of the aquifer parameters and principal directions of anisotropy. The average of each set of aquifer parameter values will be considered as the average value of the corresponding aquifer parameter. As an example, consider that four groups of wells are available. Therefore, there will be four combinations of three observation wells. The application of the method to these combinations will generate four values for the θ angle and four values of for each of the aquifer parameters. Theoretically, these four set of values should be the same.

Example 6-6: This example is adapted from Hantush (1966b). Figure 6-14 shows the location of three group lines for observation wells in a leaky confined anisotropic aquifer. The principal directions of anisotropy are not known. The isotropic methods of drawdown data analysis in leaky aquifers (Section 6.3) produced the following results:

$$\begin{aligned} T_{e_1} &= 0.12\ \text{ft}^2/\text{s}, & \nu_1' &= 1150\ \text{ft}^2/\text{s}, & B_1' &= 4800\ \text{ft} \\ T_{e_2} &= 0.10\ \text{ft}^2/\text{s}, & \nu_2' &= 615\ \text{ft}^2/\text{s}, & B_2' &= 3450\ \text{ft} \\ T_{e_3} &= 0.09\ \text{ft}^2/\text{s}, & \nu_3' &= 510\ \text{ft}^2/\text{s}, & B_3' &= 3260\ \text{ft} \end{aligned}$$

Using Hantush's Procedure II, determine $T_{\xi\xi}$, $T_{\eta\eta}$, and the principal directions of anisotropy.

Solution:

Step 1: The aquifer parameters based on the isotropic methods are given.

Step 2: From Eq. (6-52), value of a is

$$a = \frac{1}{2}\left[\frac{\nu_1'}{\nu_2'} + \left(\frac{B_1'}{B_2'}\right)^2\right]$$

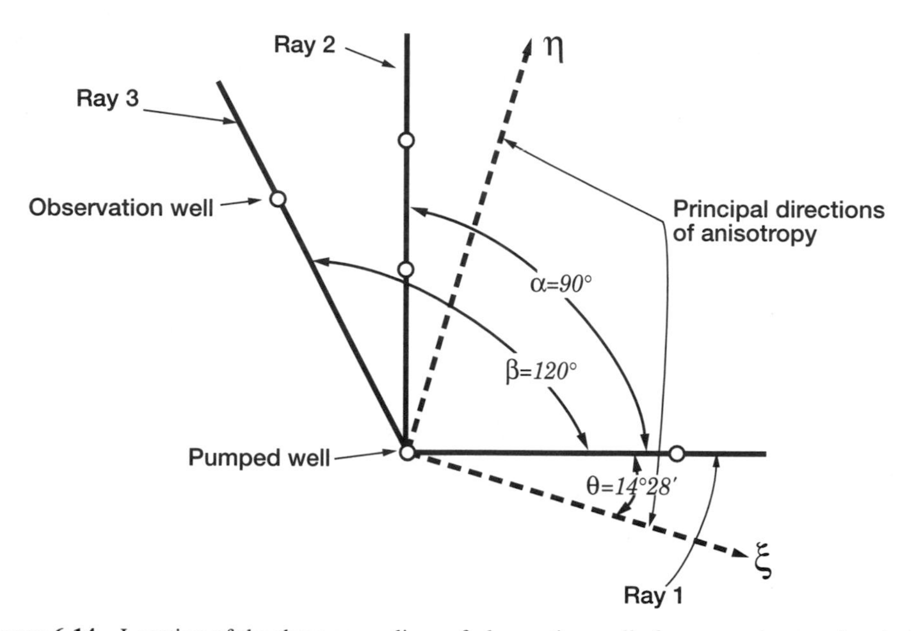

Figure 6-14 Location of the three group lines of observation wells for a pumping test in a leaky anisotropic confined aquifer for Example 6-6.

$$= \frac{1}{2}\left[\frac{1150\ \text{ft}^2/\text{s}}{615\ \text{ft}^2/\text{s}} + \left(\frac{4800\ \text{ft}}{3450\ \text{ft}}\right)^2\right]$$

$$a = 1.903$$

Step 3: From Eq. (6-57) the value of b is

$$b = \frac{1}{2}\left[\frac{\nu_1'}{\nu_3'} + \left(\frac{B_1'}{B_3'}\right)^2\right]$$

$$= \frac{1}{2}\left[\frac{1150\ \text{ft}^2/\text{s}}{510\ \text{ft}^2/\text{s}} + \left(\frac{4800\ \text{ft}}{3260\ \text{ft}}\right)^2\right]$$

$$b = 2.211$$

Step 4: From Eq. (5-57) of Chapter 5, the value of θ is

$$\tan(2\theta) = -2\frac{(b-1)\sin^2\alpha - (a-1)\sin^2\beta}{(b-1)\sin(2\alpha) - (a-1)\sin(2\beta)}$$

$$= -2\frac{(2.211-1)\sin^2(90°) - (1.903)\sin^2(120°)}{(2.211-1)\sin(180°) - (1.903-1)\sin(240°)}$$

$$\tan(2\theta) = 0.552$$

$$2\theta = 28.91° = 28°55'$$

then

$$\theta = 14.46° = 14°28'$$

From Eq. (5-58) of Chapter 5, the value of n is

$$n = \frac{\cos^2(\theta + \alpha) - a\cos^2\theta}{a\sin^2\theta - \sin^2(\theta + \alpha)}$$

$$= \frac{\cos^2(14.46° + 90°) - 1.903\cos^2(14.46°)}{(1.903)\sin^2(14.46°) - \sin^2(14.46° + 90°)}$$

$$n = 2.10 > 1$$

Therefore, the value of θ, which will make $n > 1$, locates the major axis of anisotropy, the ξ axis. The other value locates the minor axis of anisotropy, the η axis.

Step 5: From Eq. (6-58) we obtain

$$T_e = \tfrac{1}{3}(T_{e_1} + T_{e_2} + T_{e_3})$$

$$= \tfrac{1}{3}(0.12 \text{ ft}^2/\text{s} + 0.10 \text{ ft}^2/\text{s} + 0.09 \text{ ft}^2/\text{s})$$

$$T_e = 0.103 \text{ ft}^2/\text{s}$$

and from Eq. (6-59):

$$T_{\eta\eta} = T_e n^{-1/2}$$

$$= (0.103 \text{ ft}^2/\text{s})(2.10)^{-1/2}$$

$$T_{\eta\eta} = 0.071 \text{ ft}^2/\text{s}$$

Step 6: From Eq. (6-54) we obtain

$$T_{\xi\xi} = \frac{T_e^2}{T_{\eta\eta}} = \frac{(0.103 \text{ ft}^2/\text{s})^2}{0.071 \text{ ft}^2/\text{s}} = 0.149 \text{ ft}^2/\text{s}$$

Step 7: From Eq. (5-47) of Chapter 5 we obtain

$$T_{r_1} = \frac{T_{\xi\xi}}{\cos^2\theta + \dfrac{T_{\xi\xi}}{T_{\eta\eta}}\sin^2\theta}$$

$$= \frac{0.149 \text{ ft}^2/\text{s}}{\cos^2(14.46°) + \dfrac{0.149 \text{ ft}^2/\text{s}}{0.071 \text{ ft}^2/\text{s}}\sin^2(14.46°)}$$

$$T_{r_1} = 0.139 \text{ ft}^2/\text{s}$$

From Eq. (5-48) of Chapter 5 we obtain

$$T_{r_2} = \frac{T_{\xi\xi}}{\cos^2(\theta + \alpha) + \dfrac{T_{\xi\xi}}{T_{\eta\eta}} \sin^2(\theta + \alpha)}$$

$$= \frac{0.149 \text{ ft}^2/\text{s}}{\cos^2(14.46° + 90°) + \dfrac{0.149 \text{ ft}^2/\text{s}}{0.071 \text{ ft}^2/\text{s}} \sin^2(14.46° + 90°)}$$

$$T_{r_2} = 0.073 \text{ ft}^2/\text{s}$$

From Eq. (5-49) of Chapter 5 we obtain

$$T_{r_3} = \frac{T_{\xi\xi}}{\cos^2(\theta + \beta) + \dfrac{T_{\xi\xi}}{T_{\eta\eta}} \sin^2(\theta + \beta)}$$

$$= \frac{0.149 \text{ ft}^2/\text{s}}{\cos^2(14.46° + 120°) + \dfrac{0.149 \text{ ft}^2/\text{s}}{0.071 \text{ ft}^2/\text{s}} \sin^2(14.46° + 120°)}$$

$$T_{r_3} = 0.096 \text{ ft}^2/\text{s}$$

Step 8: From Eq. (6-60) we obtain

$$S = \frac{1}{3}\left(\frac{T_{r_1}}{\nu_1'} + \frac{T_{r_2}}{\nu_2'} + \frac{T_{r_3}}{\nu_3'}\right)$$

$$= \frac{1}{3}\left(\frac{0.139 \text{ ft}^2/\text{s}}{1150 \text{ ft}^2/\text{s}} + \frac{0.073 \text{ ft}^2/\text{s}}{615 \text{ ft}^2/\text{s}} + \frac{0.096 \text{ ft}^2/\text{s}}{510 \text{ ft}^2/\text{s}}\right)$$

$$S = 1.4 \times 10^{-4}$$

Step 9: From Eq. (6-61) we obtain

$$\frac{K'}{b'} = \frac{1}{3} S \left[\frac{\nu_1'}{(B_1')^2} + \frac{\nu_2'}{(B_2')^2} + \frac{\nu_3'}{(B_3')^2}\right]$$

$$= \frac{1}{3} 1.4 \times 10^{-4} \left[\frac{1150 \text{ ft}^2/\text{s}}{(4800 \text{ ft})^2} + \frac{615 \text{ ft}^2/\text{s}}{(3450 \text{ ft})^2} + \frac{510 \text{ ft}^2/\text{s}}{(3260 \text{ ft})^2}\right]$$

$$\frac{K'}{b'} = 6.98 \times 10^{-9} \text{ s}^{-1}$$

Step 10: From Eq. (5-67) of Chapter 5, the transmissivity expression for any direction is

$$T_r = \frac{T_{\xi\xi}}{\cos^2\theta + \dfrac{T_{\xi\xi}}{T_{\eta\eta}} \sin^2\theta}$$

$$= \frac{0.149 \text{ ft}^2/\text{s}}{\cos^2\theta + (2.099 \text{ ft}^2/\text{s}) \sin^2\theta}$$

CHAPTER 7

FULLY PENETRATING PUMPING WELLS IN HOMOGENEOUS AND ISOTROPIC CONFINED LEAKY AQUIFERS WITH THE STORAGE OF THE CONFINING LAYERS

7.1 INTRODUCTION

In this chapter, the theoretical foundations and practical implications of the leaky confined aquifer theory will be presented by taking into account the storage effects of the underlain and/or overlain confining layers under transient pumping conditions. This is the main difference of the leaky confined aquifer theory (see Chapter 6) for which the storage effects of the confining layer are neglected. The "no storage in the confining layers" assumption means that at any time during pumping, drawdown varies linearly across confining layers, which is certainly not the case at small values of time. As mentioned in Section 6.2.4.4 of Chapter 6, the neglect of the storage effects of the confining layers in the mathematical model under pumping conditions causes overestimated drawdown around the well.

Hantush (1960b) was the first to develop a mathematical model and derived asymptotic solutions for leaky confined aquifers by taking into account the storage effects of the confining layers. Hantush also developed pumping test data analysis methods based on his original solutions. Later, Neuman and Witherspoon (1969a–c) investigated more or less the same problem and derived a complete solution; they then developed a pumping test data analysis method (Neuman and Witherspoon, 1972) based on the solutions developed by Hantush (1960b) and Neuman and Witherspoon (1969b). The Hantush and Jacob (1955)–model–based pumping test data analysis methods (see Chapter 6) are extensively used by practitioners despite the fact that they are based on more assumptions than the other models are. These assumptions were evaluated extensively by Neuman and Witherspoon (1969c). The aforementioned mathematical models not only made significant contributions from the theoretical point of view to the understanding of the physical mechanisms of leaky confined aquifer systems under pumping conditions, but also contributed extensively for practitioners determining the hydrogeologic parameters of aquifers and confining layers. Therefore, in this chapter, the basic details and practical applications of these models are presented with examples.

7.2 HANTUSH'S ONE CONFINED AQUIFER AND TWO CONFINING LAYERS MODEL: ASYMPTOTIC SOLUTIONS

In 1960, Hantush published a paper entitled "Modification of the Theory of Leaky Confined Aquifers" by modifying the previous theory (Hantush and Jacob, 1955) with the inclusion of the storage effects of the confining layers in the model. In the following sections, Hantush's modified model and its practical applications are presented.

7.2.1 Problem Statement and Assumptions

The problem under question is to determine the drawdown distribution in a confined (or artesian) aquifer bounded by two confining layers under pumping conditions from the aquifer by taking into account the storage effects of the confining layers. The thicknesses of the aquifer, the confining layer overlain the aquifer, and the confining layer underlain the aquifer, are b, b', and b'', respectively. The aquifer is characterized by its horizontal hydraulic conductivity (K_h) and storage coefficient (S). The confining layers are characterized by their vertical hydraulic conductivities (K' and K'') and storage coefficients (S' and S''). Three cases that may exist under natural hydrogeologic conditions are shown in Figures 7-1, 7-2, and 7-3, which are identified as Case 1, Case 2, and Case 3, respectively.

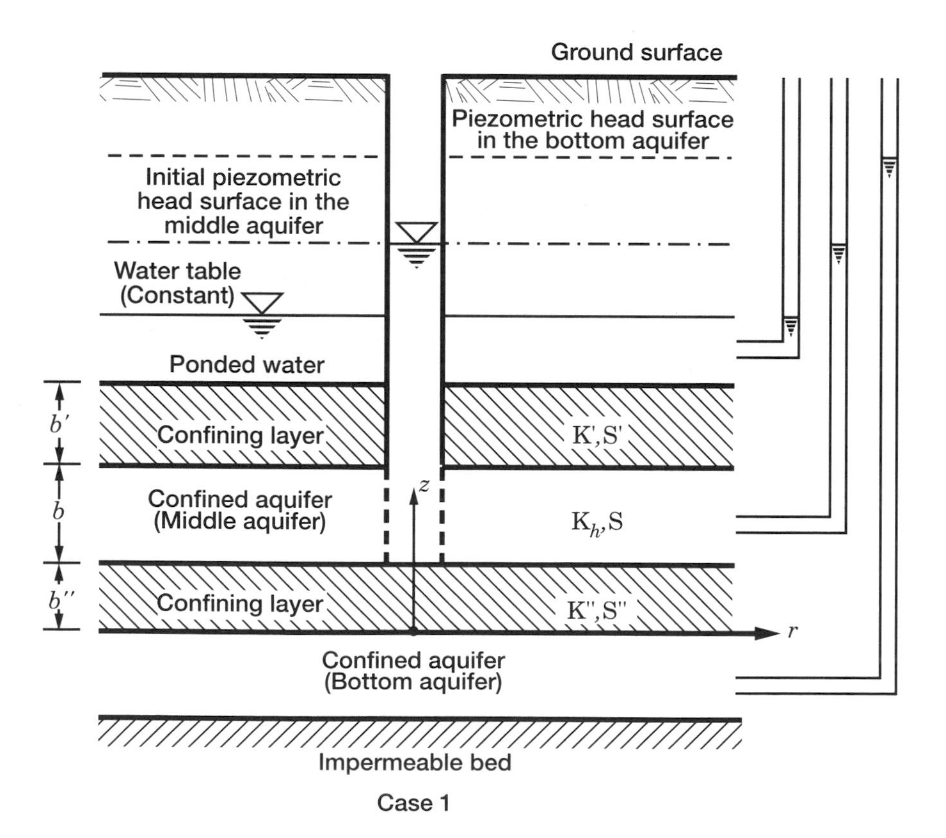

Figure 7-1 Schematic representation of a leaky confined aquifer with constant head plane sources above and below: Case 1. (After Hantush, 1960b.)

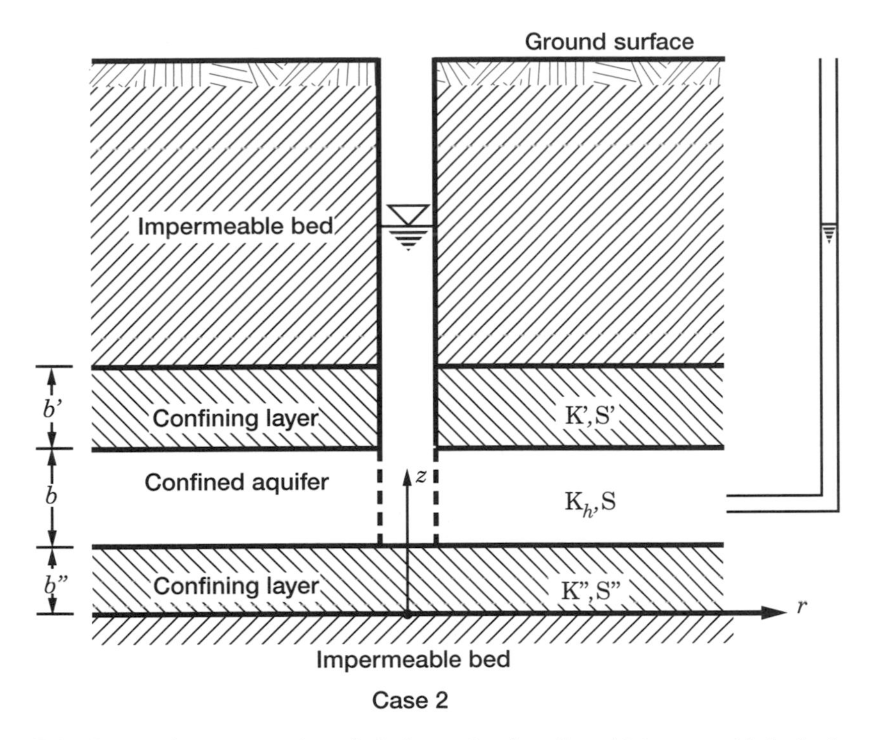

Figure 7-2 Schematic representation of a leaky confined aquifer with impermeable beds above and below: Case 2. (After Hantush, 1960b.)

In Case 1 (Figure 7-1) the confined aquifer is underlain and overlain by two confining layers. The first confining layer (b') is overlain by a water-table aquifer or a ponded water body, and the second confining layer (b'') is underlain by a confined aquifer. As can be seen from Figure 7-2, Case 2 is the same as Case 1 except that the first and second confining layers are bounded by impermeable beds. And the last one, Case 3 (Figure 7-3), is the same as Case 1 except that only the second confining layer is underlain by an impermeable bed.

Under the framework of the above-described statement of the problem, the assumptions regarding the mathematical problem are as follows:

1. The discharge rate (Q) of the well is constant.
2. The well fully penetrates the confined aquifer.
3. The diameter of the well is infinitesimally small.
4. The hydraulic conductivities of the confining layers are very small as compared with those of the confined aquifer. Therefore, the flow is vertical through the confining layers and horizontal in the main confined aquifer.
5. The aquifer and confining layers are horizontal.
6. The hydraulic head in the formations supplying the leakage is constant.

The model of Hantush (1960b) is based on one key assumption: that the direction of flow is vertical in the aquitard and horizontal in the aquifers (the fourth assumption). The validity of this assumption has been investigated by Neuman and Witherspoon (1969a, 1969b, 1969c) based on the finite-element numerical modeling method; these authors found

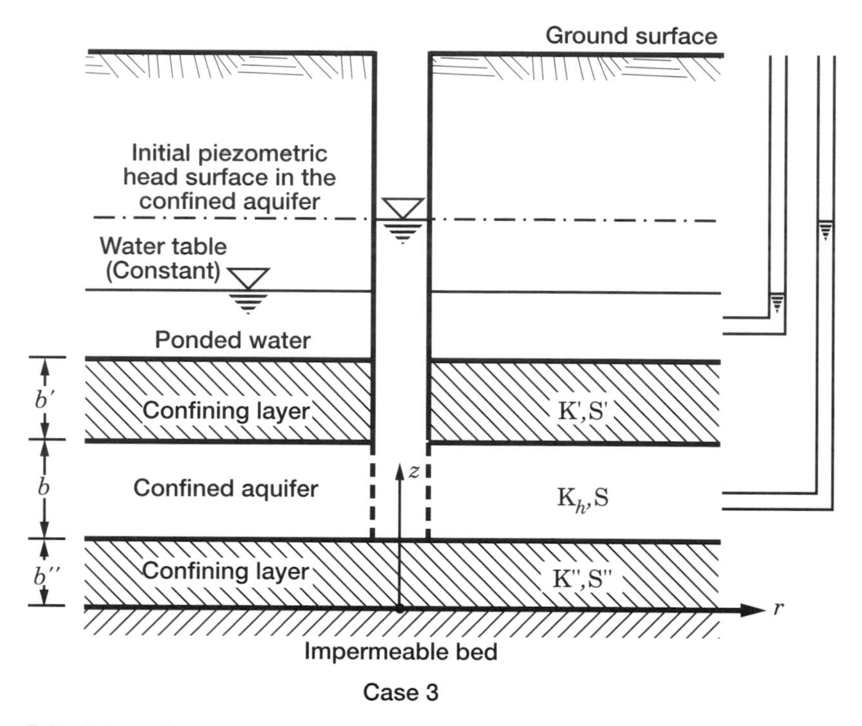

Figure 7-3 Schematic representation of a leaky confined aquifer with constant head plane source above and impermeable bed below: Case 3. (After Hantush, 1960b.)

that when the hydraulic conductivities of the aquifers are two or more orders of magnitude greater than that of the aquitard, the errors introduced by this assumption are usually less than 5%. According to Neuman and Witherspoon (1969b), these errors increase with time, decrease with radial distance from the pumping well, and are smallest in the pumped aquifer and greatest in the aquitard. Since the hydraulic conductivity contrast between aquifer and aquitard is often greater than three orders of magnitude, it would appear that these assumptions can be used safely. The rest of the assumptions given above are all self-explanatory; their adequacy will be evaluated in Section 7.4.

7.2.2 Governing Equations

The governing equations will be formulated in a manner similar to that of Hantush (1960b) (see Chapter 2, Section 2.8 for their derivation). The general differential equation of groundwater flow in homogeneous and isotropic aquifers is given by [Chapter 2, Section 2.8.1.2, Eq. (2-167)]

$$\frac{\partial^2 h_n}{\partial x^2} + \frac{\partial^2 h_n}{\partial y^2} + \frac{\partial^2 h_n}{\partial z^2} = \frac{S_{s_n}}{K_n} \frac{\partial h_n}{\partial t} \tag{7-1}$$

in which S_{s_n} is the specific storage, $K_n = K_x = K_y = K_z$ is the hydraulic conductivity, and $h = h_n(x, y, z, t)$ is the hydraulic head in the nth aquifer layer. The flow equation in a confined aquifer (see Figures 7-1, 7-2, and 7-3) which is underlain and overlain by confining layers of relatively low hydraulic conductivities is approximated (Hantush, 1960b) as

$$\frac{\partial^2 h}{\partial x^2} + \frac{\partial^2 h}{\partial y^2} + \frac{q_z(b'') - q_z(b + b'')}{T} = \frac{S}{T}\frac{\partial h}{\partial t} \tag{7-2}$$

The vertical Darcy velocity is

$$q_z(x, y, z, t) = -K_n \frac{\partial h_n}{\partial z} \tag{7-3}$$

Let $h_0(x, y, t_0)$ and $h_{0_n}(x, y, z, t_0)$ be the initial hydraulic head distributions in the main confined aquifer and confining layers, respectively. Under pumping conditions their respective drawdowns, $s(x, y, t)$ and $s_n(x, y, z, t)$, can be expressed as

$$s = h_0 - h, \qquad s_n = h_{0_n} - h_n \tag{7-4}$$

With these, Eqs. (7-1), (7-2), and (7-3) take the forms, respectively,

$$\frac{\partial^2 s_n}{\partial x^2} + \frac{\partial^2 s_n}{\partial y^2} + \frac{\partial^2 s_n}{\partial z^2} = \frac{S_{s_n}}{K_n}\frac{\partial s_n}{\partial t} \tag{7-5}$$

$$\frac{\partial^2 s}{\partial x^2} + \frac{\partial^2 s}{\partial y^2} + \frac{q_z(b + b'') - q_z b''}{T} = \frac{S}{T} \tag{7-6}$$

$$q_z(x, y, z, t) = K_n \frac{\partial s_n}{\partial z} \tag{7-7}$$

According to the fourth assumption in Section 7.2.1, the flow is vertical in the confining layers and purely horizontal in the main aquifer in which the pumping well is screened. Based on this assumption, the first and second terms on the left-hand side of Eq. (7-1) becomes zero and the differential equations of confining layers become one-dimensional. And expressing the confined aquifer differential equation in radial coordinates, the differential equations for the upper confining layer (b'), the pumped confined aquifer (b), and the lower confining layer (b''), take the following forms, respectively:

$$\frac{\partial^2 s_1}{\partial z^2} = \frac{S'}{K'b'}\frac{\partial s_1}{\partial t} \tag{7-8}$$

$$\frac{\partial^2 s}{\partial r^2} + \frac{1}{r}\frac{\partial s}{\partial r} + \frac{K'}{T}\frac{\partial s_1(r, b'' + b, t)}{\partial z} - \frac{K''}{T}\frac{\partial s_2(r, b'', t)}{\partial z} = \frac{S}{T}\frac{\partial s}{\partial t} \tag{7-9}$$

$$\frac{\partial^2 s_2}{\partial z^2} = \frac{S''}{K''b''}\frac{\partial s_2}{\partial t} \tag{7-10}$$

7.2.3 Initial and Boundary Conditions

Three different cases, called Case 1, Case 2, and Case 3, are identified in Section 7.2.1. The initial and boundary conditions for drawdown for each case are described below:

7.2.3.1 Case 1. The geometry of this case is shown in Figure 7-1. For the upper confining layer, the initial condition for the drawdown, $s_1(r, z, t)$, is

$$s_1(r, z, 0) = 0 \tag{7-11}$$

The boundary condition at the upper boundary layer—that is, $z = z_2$—is

$$s_1(r, z_2, t) = 0 \tag{7-12}$$

The boundary condition at the lower boundary—that is, $z = z_1$—is

$$s_1(r, z_1, t) = s(r, t) \tag{7-13}$$

where $z_1 = b'' + b$ and $z_2 = b'' + b + b' = z_1 + b'$.

For the main confined aquifer, the initial condition for the drawdown, $s(r, t)$, is:

$$s(r, 0) = 0 \tag{7-14}$$

The boundary condition at infinity is

$$s(\infty, t) = 0 \tag{7-15}$$

From Eq. (6-2) of Chapter 6, the constant discharge rate at the well is

$$\lim_{r \to 0} \left(r \frac{\partial s}{\partial r} \right) = -\frac{Q}{2\pi T} \tag{7-16}$$

For the lower confining layer, the initial condition for the drawdown, $s_2(r, z, t)$, is

$$s_2(r, z, 0) = 0 \tag{7-17}$$

The boundary condition at the upper boundary—that is, $z = b''$—is

$$s_2(r, b'', t) = s(r, t) \tag{7-18}$$

The boundary condition at the lower boundary—that is, $z = 0$—is

$$s_2(r, 0, t) = 0 \tag{7-19}$$

7.2.3.2 Case 2. For this case (see Figure 7-2) the initial and boundary conditions are the same as those for Case 1 with the exception of those expressed by Eqs. (7-12) and (7-19); their corresponding forms, respectively, are

$$\frac{\partial s_1(r, z_2, t)}{\partial z} = 0 \tag{7-20}$$

$$\frac{\partial s_2(r, 0, t)}{\partial z} = 0 \tag{7-21}$$

7.2.3.3 Case 3. For this case (see Figure 7-3) the initial and boundary conditions are the same as those for Case 1 with the exception of Eq. (7-19), and its corresponding form is

$$\frac{\partial s_2(r,0,t)}{\partial z} = 0 \tag{7-22}$$

Equations (7-11), (7-14), and (7-17) state that for the three cases initially the drawdown is zero in the confined aquifer and confining layers. Equation (7-12) states that the drawdown is zero at the upper boundary of the upper confining layer which is based on the fact that the hydraulic head of the overlying formation (unconfined aquifer or ponded water) is constant. Equations (7-13) and (7-18) state that the drawdown at the upper and lower boundaries of the confined aquifer is equal to the drawdowns in the confining layers. Equation (7-15) states that the drawdown is zero at infinite distance. Equation (7-16) states that near the well the flow rate towards the well is equal to its discharge. Equation (7-19) states that the drawdown is zero at the lower boundary of the lower confining bed. Equation (7-20) states that the upper boundary of the upper confining layer is an impervious boundary. And finally, Eqs. (7-21) and (7-22) state that the lower boundary of the lower confining layer is an impervious boundary.

7.2.4 Solutions

The boundary value problem defined above was solved by Hantush (1960b) in the Laplace domain, and an exact solution was obtained for the drawdown distribution within the confined aquifer by simultaneously applying the Laplace and Hankel transform techniques to Eqs. (7-8) through (7-22). Hantush (1960b, p. 3722) wrote: "An exact solution that is the inverse Laplace transform of (40) can be obtained. The solution, however, is complicated and is difficult to evaluate numerically. Instead, approximate solutions, one for sufficiently short periods of time and the other sufficiently long periods, will be obtained. These two asymptotic solutions will be used to interpolate graphically for the intermediate range of time. The derivation of these asymptotic solutions follows." In the analysis of Hantush, the drawdown distributions for the confining layers are not included. In order to better explain Hantush's approach for the asymptotic solutions, in the following sections some intermediate details regarding the solution process are also presented.

The final Laplace domain solution for the main confined aquifer [Hantush, 1960b, p. 3722, Eq. (40)] is:

$$\overline{s}(r,p) = \frac{Q}{2\pi T p} K_0(\gamma r) = \frac{Q}{4\pi T p} \int_0^{\infty} \exp\left[-y - \frac{(\gamma r)^2}{4y}\right] \frac{dy}{y} \tag{7-23}$$

where $s(r,p)$ is the Laplace transform of $s(r,t)$ and p is the Laplace transform of t and

$$\gamma = \left[\frac{pS}{T} + \frac{1}{T}\left(\frac{K'S'p}{b'}\right)^{1/2} \coth\left(\frac{b'S'p}{K'}\right)^{1/2} + \frac{1}{T}\left(\frac{K''S''p}{b''}\right)^{1/2} \coth\left(\frac{b''S''p}{K''}\right)^{1/2}\right]^{1/2} \tag{7-24}$$

in which K_0 is the modified Bessel function of the second kind of zero order (Tabulated in Table 3-2 of Chapter 3) and coth is the hyperbolic cotangent.

7.2.4.1 *Solution for Small Values of Time.* The time parameter (t) and its Laplace transform (p) in Eq. (7-23) are inversely related, which means that as the values of t goes up, the values of p goes down. Therefore, the solution for short values of time corresponds to the large values of p in Eq. (7-23).

The well-known definition of a hyperbolic cotangent function is

$$\coth(x) = \frac{e^x + e^{-x}}{e^x - e^{-x}}, \qquad x \neq 0 \tag{7-25}$$

Based on this, the general form of the hyperbolic cotangents in Eq. (7-24) can be expressed as

$$\coth\left(kp^{1/2}\right) = \frac{e^{kp^{1/2}} + e^{-kp^{1/2}}}{e^{kp^{1/2}} - e^{-kp^{1/2}}} \tag{7-26}$$

where k is equal $(b'S'/K')^{1/2}$ or $(b''S''/K'')^{1/2}$. From the inspection of Eq. (7-26), it can be observed that the arguments of the hyperbolic cotangents in Eq. (7-24) approaches to unity as the values of p increases. Hantush (1960b) concluded that under the conditions for the solution in the Laplace domain

$$\frac{b'S'p}{K'} \geq 10, \qquad \frac{b''S''p}{K''} \geq 10 \tag{7-27}$$

the hyperbolic cotangents in Eq. (7-24) may be replaced by unity without appreciably changing the values of Eq. (7-23). Some numerical values of the exponential terms in Eq. (7-26) are presented in Table 7-1. As can be seen from this table, as $x = kp^{1/2}$ increases, the term e^{-x} rapidly decreases and the relative error, defined as e^{-x}/e^x, becomes less than 2% for the x^2 values greater than 4. Hantush's limiting conditions, as given by Eq. (7-27), correspond to $x^2 = 10$ in Table 7-1, and its relative error is 0.18%. However,

TABLE 7-1 Comparisons of the Terms in Eq. (7-25)

x	x^2	e^x	e^{-x}	e^{-x}/e^x
1	1	2.718	1.368	0.1354
1.5	2.25	4.482	0.223	0.0498
2	4	7.389	0.135	0.0183
2.5	6.25	12.183	0.082	0.0067
2.75	7.5625	15.643	0.064	0.0041
2.9	8.41	18.174	0.055	0.0030
3	9	20.086	0.050	0.0025
3.1	9.61	22.198	0.045	0.0020
3.162	10	23.618	0.042	0.0018
3.2	10.24	24.533	0.041	0.0017
3.3	10.89	27.113	0.037	0.0014
3.5	12.25	33.116	0.030	0.0009
4	16	54.6	0.018	0.0003

Hantush (1960b) has not specifically addressed the reasons of selecting Eq. (7-27) as the limiting conditions. In the framework of the values in Table 7-1, perhaps $x^2 = 10$ was selected as a conservative value in order to increase the accuracy of results.

The inverse Laplace transforms of the expressions in Eq. (7-27) are as follows:

$$\frac{b'S'}{K'} \geq 10t, \qquad \frac{b''S''}{K''} \geq 10t \tag{7-28}$$

As will be mentioned in detail in Section 7.4.5.3, Neuman and Witherspoon (1969b) also showed that from a practical standpoint the criterion given by Eq. (7-27) or Eq. (7-28) is overly conservative.

Solutions for the three cases (Case 1, Case 2, and Case 3) for the range defined by Eqs. (7-27) and (7-28) are all the same, and with this approach Eq. (7-23) can be written as

$$\overline{s}(r, p) = \frac{Q}{4\pi T} \int_0^{\infty} \frac{e^{-y}}{y} \overline{g}(p)\, dy \tag{7-29}$$

where

$$\overline{g}(p) = \exp\left(-\frac{r^2 Sp}{4Ty}\right) \frac{1}{p} \exp\left(-\frac{r^2 c^2}{4y} p^{1/2}\right) \tag{7-30}$$

and

$$c^2 = \frac{1}{T} \left[\frac{K'}{\left(\dfrac{K'b'}{S'}\right)^{1/2}} + \frac{K''}{\left(\dfrac{K''b''}{S''}\right)^{1/2}} \right]^{1/2} \tag{7-31}$$

By taking the inverse Laplace transform of Eq. (7-29), finally, Hantush (1960b) obtained the following solution for the drawdown distribution in the main confined aquifer under the conditions expressed by Eqs. (7-27) or (7-28) for the three cases described above:

$$s(r, t) = \frac{Q}{4\pi T} H(u, \beta) \tag{7-32}$$

where

$$u = \frac{r^2 S}{4Tt} \tag{7-33}$$

$$H(u, \beta) = \int_u^{\infty} \frac{e^{-y}}{y} \operatorname{erfc}\left\{ \frac{\beta u^{1/2}}{[y(y-u)]^{1/2}} \right\} dy \tag{7-34}$$

and

$$\beta = \frac{1}{4} r\lambda \tag{7-35}$$

$$\lambda = \left[\frac{\left(\frac{K'}{b'}\right)}{T} \left(\frac{S'}{S}\right) \right]^{1/2} + \left[\frac{\left(\frac{K''}{b''}\right)}{T} \left(\frac{S''}{S}\right) \right]^{1/2} \tag{7-36}$$

If K'' is much smaller than K' and that S' and S'' are not greatly different, or $K'S' \gg K''S''$, Eq. (7-36) takes the form

$$\lambda = \left[\frac{\left(\frac{K'}{b'}\right)}{T} \left(\frac{S'}{S}\right) \right]^{1/2} \tag{7-37}$$

Equation (7-37) corresponds to Figure 6-1 of Chapter 6 for which the storage effects of the confining layer are neglected. In other words, Eqs. (7-32), (7-33), (7-34), (7-35), and (7-37) form the solution for a confined aquifer that is overlain or underlain by a confining layer whose storage effects are taken into account.

The Function $H(u, \beta)$. The function $H(u, \beta)$, defined by Eq. (7-34), has extensively been tabulated by Hantush (1961a, 1964) and is included in Table 7-2. Table 7-2 covers the ranges of u and β for almost all practical purposes. The numbers in parentheses of Table 7-2 are powers of 10 with which other numbers are multiplied. Type curves based on the values in Table 7-2 are presented in Figure 7-4.

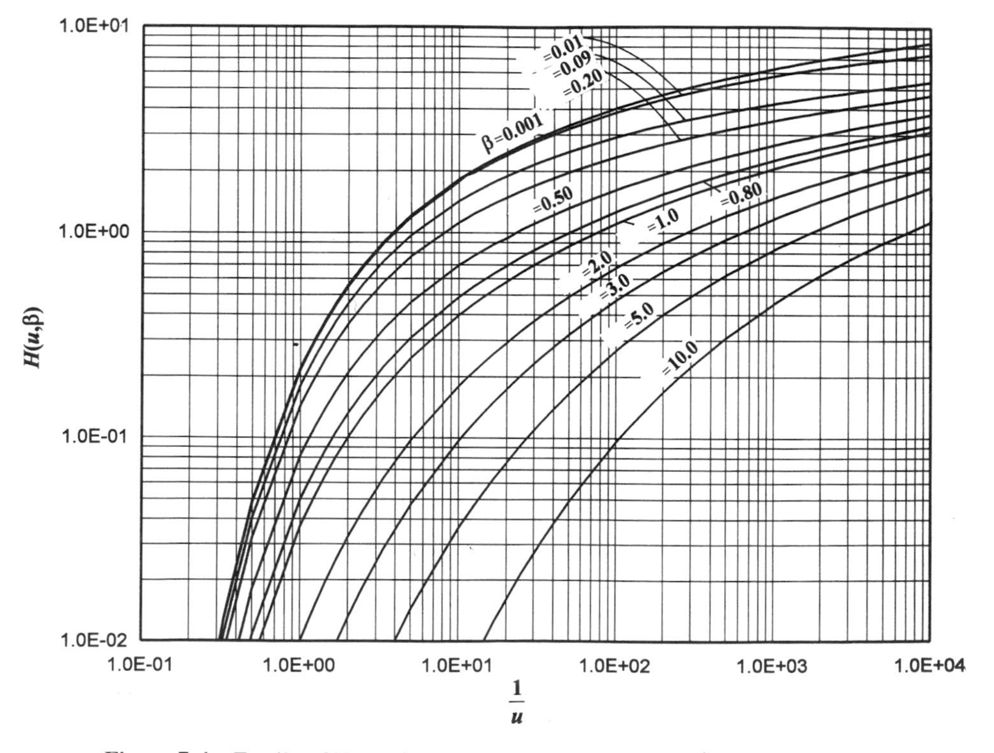

Figure 7-4 Family of Hantush type curves $H(u, \beta)$ versus $1/u$ for values of β.

TABLE 7-2 Values of the Function $H(u, \beta)$

β	[−5]								
u	1	2	3	4	5	6	7	8	9
1[−9]	19.5841	19.2088	18.9309	18.7113	18.5304	18.3767	18.2432	18.1252	18.0196
2[−9]	19.0298	18.7198	18.4778	18.2805	18.1144	17.9713	17.8456	17.7337	17.6329
3[−9]	18.6916	18.4185	18.1989	18.0162	17.8603	17.7246	17.6045	17.4970	17.3997
4[−9]	18.4460	18.1979	17.9943	17.8226	17.6745	17.5446	17.4290	17.3250	17.2306
5[−9]	18.2524	18.0228	17.8317	17.6687	17.5269	17.4018	17.2900	17.1890	17.0970
6[−9]	18.0923	17.8774	17.6963	17.5404	17.4041	17.2831	17.1745	17.0760	16.9861
7[−9]	17.9558	17.7527	17.5800	17.4303	17.2985	17.1811	17.0753	16.9792	16.8911
8[−9]	17.8367	17.6435	17.4779	17.3335	17.2058	17.0915	16.9883	16.8942	16.8079
9[−9]	17.7309	17.5463	17.3869	17.2472	17.1231	17.0116	16.9107	16.8184	16.7336
1[−8]	17.6359	17.4586	17.3047	17.1691	17.0482	16.9394	16.8405	16.7500	16.6666
2[−8]	17.0000	16.8661	16.7457	16.6365	16.5369	16.4454	16.3609	16.2824	16.2092
3[−8]	16.6207	16.5081	16.4050	16.3102	16.2225	16.1411	16.0652	15.9942	15.9274
4[−8]	16.3490	16.2496	16.1577	16.0723	15.9928	15.9184	15.8486	15.7828	15.7207
5[−8]	16.1368	16.0467	15.9629	15.8845	15.8110	15.7418	15.6766	15.6150	15.5565
6[−8]	15.9627	15.8797	15.8019	15.7289	15.6601	15.5951	15.5336	15.4753	15.4198
7[−8]	15.8149	15.7375	15.6646	15.5960	15.5310	15.4695	15.4110	15.3554	15.3024
8[−8]	15.6865	15.6137	15.5449	15.4798	15.4181	15.3594	15.3035	15.2503	15.1994
9[−8]	15.5731	15.5040	15.4386	15.3766	15.3176	15.2614	15.2078	15.1565	15.1075
1[−7]	15.4713	15.4056	15.3431	15.2837	15.2271	15.1731	15.1214	15.0720	15.0245
2[−7]	14.7981	14.7505	14.7047	14.6605	14.6178	14.5766	14.5367	14.4982	14.4609
3[−7]	14.4016	14.3623	14.3242	14.2873	14.2514	14.2166	14.1827	14.1498	14.1177
4[−7]	14.1193	14.0851	14.0517	14.0193	13.9877	13.9568	13.9268	13.8974	13.8688
5[−7]	13.8998	13.8691	13.8390	13.8097	13.7811	13.7531	13.7258	13.6990	13.6728
6[−7]	13.7202	13.6920	13.6645	13.6376	13.6112	13.5853	13.5601	13.5353	13.5110
7[−7]	13.5682	13.5420	13.5164	13.4913	13.4667	13.4426	13.4190	13.3958	13.3730
8[−7]	13.4364	13.4119	13.3878	13.3642	13.3411	13.3184	13.2961	13.2742	13.2527
9[−7]	13.3200	13.2968	13.2741	13.2518	13.2299	13.2083	13.1872	13.1664	13.1459
1[−6]	13.2158	13.1938	13.1722	13.1510	13.1301	13.1096	13.0894	13.0695	13.0499
2[−6]	12.5292	12.5135	12.4981	12.4828	12.4677	12.4528	12.4380	12.4234	12.4090
3[−6]	12.1266	12.1138	12.1011	12.0885	12.0761	12.0637	12.0515	12.0394	12.0275
4[−6]	11.8407	11.8295	11.8185	11.8076	11.7967	11.7860	11.7753	11.7647	11.7542
5[−6]	11.6187	11.6087	11.5988	11.5890	11.5793	11.5696	11.5600	11.5505	11.5410
6[−6]	11.4373	11.4282	11.4191	11.4101	11.4012	11.3923	11.3835	11.3748	11.3661
7[−6]	11.2838	11.2753	11.2670	11.2586	11.2503	11.2421	11.2340	11.2258	11.2178
8[−6]	11.1508	11.1429	11.1351	11.1272	11.1195	11.1118	11.1041	11.0965	11.0889
9[−6]	11.0335	11.0260	11.0186	11.0112	11.0039	10.9966	10.9894	10.9822	10.9750
1[−5]	10.9285	10.9214	10.9144	10.9074	10.9004	10.8935	10.8866	10.8798	10.8730
2[−5]	10.2375	10.2324	10.2274	10.2225	10.2175	10.2126	10.2077	10.2028	10.1979
3[−5]	9.8330	9.8289	9.8248	9.8207	9.8167	9.8126	9.8086	9.8045	9.8005
4[−5]	9.5459	9.5423	9.5388	9.5352	9.5317	9.5282	9.5247	9.5212	9.5177
5[−5]	9.3231	9.3199	9.3167	9.3136	9.3104	9.3073	9.3041	9.3010	9.2979
6[−5]	9.1411	9.1382	9.1353	9.1324	9.1295	9.1266	9.1237	9.1209	9.1180
7[−5]	8.9871	8.9844	8.9818	8.9791	8.9764	8.9738	8.9711	8.9684	8.9658
8[−5]	8.8538	8.8513	8.8488	8.8463	8.8438	8.8413	8.8388	8.8363	8.8338
9[−5]	8.7362	8.7338	8.7314	8.7291	8.7267	8.7243	8.7220	8.7197	8.7173

(*continued*)

TABLE 7-2 (*continued*)

β	[−5]								
u	1	2	3	4	5	6	7	8	9
1[−4]	8.6309	8.6287	8.6264	8.6242	8.6220	8.6197	8.6175	8.6153	8.6130
2[−4]	7.9385	7.9370	7.9354	7.9338	7.9322	7.9306	7.9290	7.9275	7.9259
3[−4]	7.5335	7.5322	7.5309	7.5296	7.5283	7.5270	7.5257	7.5244	7.5231
4[−4]	7.2461	7.2449	7.2438	7.2427	7.2416	7.2405	7.2393	7.2382	7.2371
5[−4]	7.0231	7.0221	7.0211	7.0201	7.0191	7.0181	7.0171	7.0161	7.0151
6[−4]	6.8410	6.8401	6.8392	6.8383	6.8373	6.8364	6.8355	6.8346	6.8337
7[−4]	6.6870	6.6862	6.6853	6.6845	6.6836	6.6828	6.6819	6.6811	6.6803
8[−4]	6.5537	6.5529	6.5521	6.5513	6.5505	6.5497	6.5489	6.5481	6.5473
9[−4]	6.4360	6.4353	6.4345	6.4338	6.4330	6.4323	6.4315	6.4308	6.4300
1[−3]	6.3308	6.3301	6.3294	6.3287	6.3280	6.3272	6.3265	6.3258	6.3251
2[−3]	5.6389	5.6384	5.6379	5.6374	5.6369	5.6364	5.6359	5.6354	5.6349
3[−3]	5.2345	5.2341	5.2337	5.2333	5.2329	5.2325	5.2320	5.2316	5.2312
4[−3]	4.9479	4.9475	4.9472	4.9468	4.9465	4.9461	4.9458	5.9454	5.9451
5[−3]	4.7258	4.7254	4.7251	4.7248	4.7245	4.7242	4.7239	4.7236	4.7232
6[−3]	4.5445	4.5442	4.5439	4.5436	4.5433	4.5430	4.5427	4.5425	4.5422
7[−3]	4.3913	4.3911	4.3908	4.3905	4.3903	4.3900	4.3897	4.3895	4.3892
8[−3]	4.2588	4.2586	4.2583	4.2581	4.2578	4.2576	4.2573	4.2571	4.2568
9[−3]	4.1420	4.1418	4.1416	4.1413	4.1411	4.1409	4.1407	4.1404	4.1402
1[−2]	4.0377	4.0375	4.0372	4.0370	4.0368	4.0366	4.0364	4.0362	4.0359
2[−2]	3.3545	3.3544	3.3542	3.3541	3.3539	3.3538	3.3536	3.3535	3.3533
3[−2]	2.9590	2.9589	2.9587	2.9586	2.9585	2.9584	2.9583	2.9581	2.9580
4[−2]	2.6811	2.6810	2.6809	2.6808	2.6807	2.6806	2.6805	2.6804	2.6803
5[−2]	2.4678	2.4677	2.4676	2.4675	2.4674	2.4673	2.4673	2.4672	2.4671
6[−2]	2.2952	2.2951	2.2951	2.2950	2.2949	2.2948	2.2947	2.2947	2.2946
7[−2]	2.1507	2.1507	2.1506	2.1505	2.1505	2.1504	2.1503	2.1502	2.1502
8[−2]	2.0269	2.0268	2.0267	2.0267	2.0266	2.0265	2.0265	2.0264	2.0263
9[−2]	1.9187	1.9186	1.9185	1.9185	1.9184	1.9184	1.9183	1.9182	1.9182
1[−1]	1.8229	1.8228	1.8227	1.8227	1.8226	1.8226	1.8225	1.8225	1.8224
2[−1]	1.2226	1.2226	1.2225	1.2225	1.2225	1.2224	1.2224	1.2224	1.2223
3[−1]	0.9056	0.9056	0.9056	0.9056	0.9056	0.9055	0.9055	0.9055	0.9055
4[−1]	0.7024	0.7023	0.7023	0.7023	0.7023	0.7023	0.7023	0.7022	0.7022
5[−1]	0.5598	0.5597	0.5597	0.5597	0.5597	0.5597	0.5597	0.5597	0.5596
6[−1]	0.4544	0.4544	0.4543	0.4543	0.4543	0.4543	0.4543	0.4543	0.4543
7[−1]	0.3738	0.3738	0.3737	0.3737	0.3737	0.3737	0.3737	0.3737	0.3737
8[−1]	0.3106	0.3106	0.3106	0.3106	0.3106	0.3106	0.3105	0.3105	0.3105
9[−1]	0.2602	0.2602	0.2602	0.2602	0.2602	0.2601	0.2601	0.2601	0.2601
1	0.2194	0.2194	0.2194	0.2194	0.2194	0.2194	0.2193	0.2193	0.2193
2	489(−4)	489(−4)	489(−4)	489(−4)	489(−4)	489(−4)	489(−4)	489(−4)	489(−4)
3	130(−4)	130(−4)	130(−4)	130(−4)	130(−4)	130(−4)	130(−4)	130(−4)	130(−4)
4	378(−5)	378(−5)	378(−5)	378(−5)	378(−5)	378(−5)	378(−5)	378(−5)	378(−5)
5	115(−5)	115(−5)	115(−5)	115(−5)	115(−5)	115(−5)	115(−5)	115(−5)	115(−5)
6	360(−6)	360(−6)	360(−6)	360(−6)	360(−6)	360(−6)	360(−6)	360(−6)	360(−6)
7	115(−6)	115(−6)	115(−6)	115(−6)	115(−6)	115(−6)	115(−6)	115(−6)	115(−6)
8	377(−7)	377(−7)	377(−7)	377(−7)	377(−7)	377(−7)	377(−7)	377(−7)	377(−7)
9	124(−7)	124(−7)	124(−7)	124(−7)	124(−7)	124(−7)	124(−7)	124(−7)	124(−7)
10	416(−8)	416(−8)	416(−8)	416(−8)	416(−8)	416(−8)	416(−8)	416(−8)	416(−8)

(*continued*)

TABLE 7-2 (*continued*)

β	[−4]									
u	1	2	3	4	5	6	7	8	9	10
1[−9]	17.9240	17.2752	16.8846	16.6043	16.3856	16.2063	16.0542	15.9223	15.8058	15.7014
2[−9]	17.5412	16.9102	16.5257	16.2485	16.0317	15.8535	15.7024	15.5711	15.4551	15.3511
3[−9]	17.3109	16.6934	16.3135	16.0387	15.8233	15.6461	15.4956	15.3648	15.2492	15.1456
4[−9]	17.1441	16.5378	16.1618	15.8889	15.6747	15.4983	15.3484	15.2180	15.1027	14.9993
5[−9]	17.0125	16.4158	16.0432	15.7721	15.5589	15.3832	15.2338	15.1038	14.9888	14.8856
6[−9]	16.9034	16.3153	15.9458	15.6762	15.4639	15.2889	15.1399	15.0103	14.8955	14.7926
7[−9]	16.8100	16.2296	15.8629	15.5948	15.3834	15.2089	15.0603	14.9310	14.8165	14.7138
8[−9]	16.7281	16.1549	15.7908	15.5239	15.3133	15.1394	14.9912	14.8622	14.7479	14.6454
9[−9]	16.6552	16.0885	15.7268	15.4612	15.2514	15.0780	14.9302	14.8014	14.6873	14.5849
1[−8]	16.5894	16.0288	15.6694	15.4050	15.1958	15.0229	14.8754	14.7469	14.6330	14.5308
2[−8]	16.1406	15.6262	15.2846	15.0295	14.8261	14.6569	14.5122	14.3858	14.2735	14.1726
3[−8]	15.8645	15.3815	15.0529	14.8048	14.6056	14.4394	14.2968	14.1719	14.0609	13.9609
4[−8]	15.6619	15.2031	14.8850	14.6425	14.4469	14.2831	14.1422	14.0187	13.9087	13.8096
5[−8]	15.5010	15.0616	14.7524	14.5148	14.3223	14.1606	14.0212	13.8989	13.7898	13.6914
6[−8]	15.3669	14.9440	14.6424	14.4092	14.2194	14.0596	13.9216	13.8003	13.6920	13.5942
7[−8]	15.2517	14.8430	14.5482	14.3189	14.1316	13.9735	13.8367	13.7163	13.6088	13.5117
8[−8]	15.1506	14.7544	14.4657	14.2399	14.0549	13.8983	13.7628	13.6432	13.5364	13.4398
9[−8]	15.0605	14.6752	14.3921	14.1696	13.9867	13.8316	13.6971	13.5784	13.4722	13.3761
1[−7]	14.9790	14.6037	14.3257	14.1062	13.9252	13.7715	13.6380	13.5201	13.4145	13.3189
2[−7]	14.4247	14.1147	13.8727	13.6754	13.5093	13.3661	13.2404	13.1285	13.0277	12.9360
3[−7]	14.0865	13.8133	13.5937	13.4111	13.2551	13.1194	12.9994	12.8919	12.7946	12.7057
4[−7]	13.8408	13.5927	13.3891	13.2174	13.0693	12.9394	12.8239	12.7199	12.6254	12.5389
5[−7]	13.6472	13.4177	13.2265	13.0635	12.9218	12.7967	12.6849	12.5839	12.4918	12.4073
6[−7]	13.4872	13.2722	13.0911	12.9353	12.7989	12.6779	12.5693	12.4709	12.3810	12.2982
7[−7]	13.3506	13.1476	12.9748	12.8251	12.6934	12.5759	12.4702	12.3740	12.2860	12.2048
8[−7]	13.2315	13.0384	12.8728	12.7284	12.6007	12.4864	12.3831	12.2891	12.2027	12.1230
9[−7]	13.1258	12.9412	12.7818	12.6420	12.5179	12.4065	12.3055	12.2133	12.1285	12.0501
1[−6]	13.0307	12.8535	12.6996	12.5640	12.4431	12.3342	12.2353	12.1448	12.0615	11.9842
2[−6]	12.3948	12.2609	12.1405	12.0314	11.9318	11.8403	11.7557	11.6772	11.6040	11.5355
3[−6]	12.0156	11.9029	11.7998	11.7050	11.6174	11.5360	11.4601	11.3890	11.3223	11.2593
4[−6]	11.7438	11.6444	11.5525	11.4672	11.3877	11.3133	11.2434	11.1777	11.1156	11.0568
5[−6]	11.5317	11.4416	11.3577	11.2793	11.2058	11.1367	11.0715	11.0098	10.5149	10.8958
6[−6]	11.3575	11.2745	11.1968	11.1238	11.0550	10.9900	10.9285	10.8701	10.8146	10.7618
7[−6]	11.2098	11.1323	11.0595	10.9908	10.9259	10.8643	10.8059	10.7503	10.6972	10.6466
8[−6]	11.0814	11.0085	10.9397	10.8747	10.8129	10.7543	10.6984	10.6451	10.5942	10.5455
9[−6]	10.9679	10.8989	10.8335	10.7715	10.7125	10.6563	10.6027	10.5514	10.5024	10.4553
1[−5]	10.8662	10.8004	10.7380	10.6786	10.6220	10.5680	10.5163	10.4668	10.4194	10.3739
2[−5]	10.1930	10.1454	10.0996	10.0553	10.0127	9.9715	9.9317	9.8932	9.8558	7.8197
3[−5]	9.7965	9.7573	9.7192	9.6822	9.6464	9.6115	9.5777	9.5447	9.5127	9.4815
4[−5]	9.5142	9.4800	9.4467	9.4142	9.3826	9.3518	9.3217	9.2924	9.2638	9.2358
5[−5]	9.2948	9.2640	9.2340	9.2047	9.1761	9.1481	9.1208	9.0940	9.0679	9.0422
6[−5]	9.1152	9.0870	9.0595	9.0325	9.0062	8.9804	8.9551	8.9303	8.9060	8.8822
7[−5]	8.9631	8.9370	8.9114	8.8863	8.8618	8.8376	8.8140	8.7908	8.7681	8.7457
8[−5]	8.8313	8.8068	8.7828	8.7593	8.7361	8.7134	8.6911	8.6692	8.6477	8.6266
9[−5]	8.7150	8.6919	8.6691	8.6468	8.6249	8.6034	8.5822	8.5614	8.5410	8.5209

(*continued*)

TABLE 7-2 (*continued*)

β	[−4]									
u	1	2	3	4	5	6	7	8	9	10
1[−4]	8.6108	8.5889	8.5673	8.5460	8.5252	8.5046	8.4845	8.4646	8.4450	8.4258
2[−4]	7.9243	7.9087	7.8932	7.8779	7.8628	7.8479	7.8332	7.8187	7.8043	7.7901
3[−4]	7.5218	7.5090	7.4963	7.4838	7.4713	7.4590	7.4468	7.4348	7.4228	7.4110
4[−4]	7.2360	7.2249	7.2139	7.2029	7.1921	7.1814	7.1707	7.1602	7.1497	7.1393
5[−4]	7.0141	7.0042	6.9943	6.9845	6.9748	6.9651	6.9556	6.9461	6.9366	6.9273
6[−4]	6.8328	6.8237	6.8147	6.8057	6.7968	6.7880	6.7792	6.7705	6.7618	6.7533
7[−4]	6.6794	6.6710	6.6626	6.6543	6.6461	6.6379	6.6297	6.6216	6.6136	6.6056
8[−4]	6.5465	6.5386	6.5308	6.5230	6.5153	6.5076	6.5000	6.4924	6.4849	6.4774
9[−4]	6.4293	6.4219	6.4145	6.4071	6.3998	6.3926	6.3854	6.3782	6.3711	6.3640
1[−3]	6.3244	6.3174	6.3104	6.3034	6.2965	6.2896	6.2827	6.2759	6.2691	6.2624
2[−3]	5.6344	5.6294	5.6244	5.6195	5.6146	5.6097	5.6048	5.5999	5.5951	5.5903
3[−3]	5.2308	5.2268	5.2227	5.2187	5.2147	5.2107	5.2067	5.2027	5.1988	5.1948
4[−3]	4.9447	4.9412	4.9377	4.9342	4.9308	4.9273	4.9238	4.9204	4.9170	4.9135
5[−3]	4.7229	4.7198	4.7167	4.7136	4.7105	4.7074	4.7043	4.7012	4.6982	4.6951
6[−3]	4.5419	4.5391	4.5362	4.5334	4.5306	4.5277	4.5249	4.5221	4.5193	4.5165
7[−3]	4.3890	4.3863	4.3837	4.3811	4.3785	4.3759	4.3733	4.3707	4.3681	4.3655
8[−3]	4.2566	4.2542	4.2517	4.2493	4.2468	4.2444	4.2420	4.2396	4.2371	4.2347
9[−3]	4.1400	4.1377	4.1354	4.1331	4.1308	4.1285	4.1262	4.1239	4.1217	4.1194
1[−2]	4.0357	4.0355	4.0314	4.0292	4.0270	4.0249	4.0227	4.0206	4.0184	4.0163
2[−2]	3.3532	3.3517	3.3502	3.3487	3.3472	3.3457	3.3442	3.3427	3.3412	3.3397
3[−2]	2.9579	2.9567	2.9555	2.9543	2.9531	2.9519	2.9507	2.9495	2.9484	2.9472
4[−2]	2.6802	2.6792	2.6782	2.6772	2.6762	2.6752	2.6742	2.6731	2.6721	2.6711
5[−2]	2.4670	2.4661	2.4652	2.4643	2.4634	2.4626	2.4617	2.4608	2.4599	2.4590
6[−2]	2.2945	2.2937	2.2929	2.2921	2.2913	2.2905	2.2897	2.2889	2.2882	2.2874
7[−2]	2.1501	2.1494	2.1487	2.1479	2.1472	2.1465	2.1458	2.1451	2.1443	2.1436
8[−2]	2.0263	2.0256	2.0249	2.0243	2.0236	2.0230	2.0223	2.0216	2.0210	2.0203
9[−2]	1.9181	1.9175	1.9169	1.9163	1.9157	1.9151	1.9145	1.9138	1.9132	1.9126
1[−1]	1.8223	1.8218	1.8212	1.8206	1.8201	1.8195	1.8189	1.8184	1.8178	1.8172
2[−1]	1.2223	1.2220	1.2216	1.2213	1.2209	1.2206	1.2203	1.2199	1.2196	1.2193
3[−1]	0.9054	0.9052	0.9050	0.9047	0.9045	0.9043	0.9040	0.9038	0.9035	0.9033
4[−1]	0.7022	0.7020	0.7018	0.7017	0.7015	0.7013	0.7011	0.7010	0.7008	0.7006
5[−1]	0.5596	0.5595	0.5594	0.5592	0.5591	0.5590	0.5588	0.5587	0.5585	0.5584
6[−1]	0.4543	0.4542	0.4541	0.4539	0.4538	0.4537	0.4536	0.4535	0.4534	0.4533
7[−1]	0.3737	0.3736	0.3735	0.3734	0.3733	0.3732	0.3732	0.3731	0.3730	0.3729
8[−1]	0.3105	0.3105	0.3104	0.3103	0.3102	0.3102	0.3101	0.3100	0.3099	0.3099
9[−1]	0.2601	0.2601	0.2600	0.2599	0.2599	0.2598	0.2598	0.2597	0.2596	0.2596
1	0.2193	0.2193	0.2192	0.2192	0.2191	0.2191	0.2190	0.2190	0.2189	0.2189
2	489(−4)	489(−4)	489(−4)	489(−4)	488(−4)	488(−4)	488(−4)	488(−4)	488(−4)	488(−4)
3	130(−4)	130(−4)	130(−4)	130(−4)	130(−4)	130(−4)	130(−4)	130(−4)	130(−4)	130(−4)
4	378(−5)	378(−5)	378(−5)	378(−5)	378(−5)	377(−5)	377(−5)	377(−5)	377(−5)	377(−5)
5	115(−5)	115(−5)	115(−5)	115(−5)	115(−5)	115(−5)	115(−5)	115(−5)	115(−5)	115(−5)
6	360(−6)	360(−6)	360(−6)	360(−6)	360(−6)	360(−6)	360(−6)	359(−6)	359(−6)	359(−6)
7	115(−6)	115(−6)	115(−6)	115(−6)	115(−6)	115(−6)	115(−6)	115(−6)	115(−6)	115(−6)
8	377(−7)	377(−7)	376(−7)	376(−7)	376(−7)	376(−7)	376(−7)	376(−7)	376(−7)	376(−7)
9	124(−7)	124(−7)	124(−7)	124(−7)	124(−7)	124(−7)	124(−7)	124(−7)	124(−7)	124(−7)
10	416(−8)	416(−8)	415(−8)	411(−8)	416(−8)	415(−8)	415(−8)	415(−8)	415(−8)	415(−8)

(*continued*)

TABLE 7-2 (*continued*)

β / u	[−3] 1	2	3	4	5	6	7	8	9
1[−9]	15.7014	15.0127	14.6087	14.3218	14.0991	13.9171	13.7631	13.6298	13.5121
2[−9]	15.3511	14.6643	14.2609	13.9743	13.7518	13.5699	13.4116	13.2828	13.1651
3[−9]	15.1456	14.4601	14.0573	13.7709	13.5485	13.3667	13.2129	13.0796	12.9621
4[−9]	14.9993	14.3151	13.9126	13.6264	13.4042	13.2225	13.0687	12.9355	12.8180
5[−9]	14.8856	14.2025	13.8003	13.5143	13.2922	13.1105	12.9569	12.8237	12.7062
6[−9]	14.7926	14.1104	13.7085	13.4227	13.2006	13.0190	12.8654	12.7323	12.6148
7[−9]	14.7138	14.0324	13.6309	13.3452	13.1232	12.9417	12.7881	12.6550	12.5375
8[−9]	14.6454	13.9648	13.5636	13.2780	13.0561	12.8746	12.7211	12.5880	12.4706
9[−9]	14.5849	13.9052	13.5042	13.2187	12.9969	12.8155	12.6620	12.5289	12.4115
1[−8]	14.5308	13.8518	13.4510	13.1657	12.9440	12.7626	12.6091	12.4761	12.3587
2[−8]	14.1726	13.4994	13.1006	12.8162	12.5950	12.4141	12.2609	12.1280	12.0108
3[−8]	13.9609	13.2922	12.8948	12.6112	12.3905	12.2098	12.0568	11.9242	11.8071
4[−8]	13.8096	13.1445	12.7485	12.4655	12.2452	12.0647	11.9119	11.7794	11.6624
5[−8]	13.6914	13.0296	12.6347	12.3523	12.1323	11.9521	11.7994	11.6670	11.5501
6[−8]	13.5942	12.9355	12.5415	12.2596	12.0399	11.8599	11.7074	11.5751	11.4583
7[−8]	13.5117	12.8556	12.4626	12.1811	11.9617	11.7819	11.6295	11.4973	11.3806
8[−8]	13.4398	12.7863	12.3941	12.1131	11.8939	11.7143	11.5620	11.4299	11.3133
9[−8]	13.3761	12.7250	12.3336	12.0530	11.8341	11.6546	11.5025	11.3704	11.2538
1[−7]	13.3189	12.6700	12.2794	11.9992	11.7805	11.6012	11.4491	11.3172	11.2006
2[−7]	12.9360	12.3051	11.9206	11.6434	11.4265	11.2484	11.0973	10.9660	10.8500
3[−7]	12.7057	12.0883	11.7084	11.4336	11.2182	11.0410	10.8905	10.7598	10.6441
4[−7]	12.5389	11.9326	11.5567	11.2838	11.0695	10.8932	10.7433	10.6129	10.4976
5[−7]	12.4073	11.8107	11.4381	11.1670	10.9538	10.7781	10.6287	10.4987	10.3837
6[−7]	12.2982	11.7101	11.3406	11.0711	10.8588	10.6838	10.5348	10.4052	10.2905
7[−7]	12.2048	11.6245	11.2578	10.9896	10.7783	10.6038	10.4553	10.3260	10.2115
8[−7]	12.1230	11.5498	11.1856	10.9188	10.7083	10.5343	10.3862	10.2572	10.1429
9[−7]	12.0501	11.4834	11.1217	10.8561	10.6463	10.4729	10.3251	10.1964	10.0823
1[−6]	11.9842	11.4237	11.0643	10.7999	10.5908	10.4178	10.2704	10.1419	10.0280
2[−6]	11.5355	11.0211	10.6795	10.4244	10.2210	10.0519	9.9073	9.7808	9.6686
3[−6]	11.2593	10.7764	10.4479	10.1997	10.0006	9.8344	9.6918	9.5670	9.4560
4[−6]	11.0568	10.5980	10.2799	10.0375	9.8419	9.6782	9.5373	9.4138	9.3038
5[−6]	10.8958	10.4566	10.1473	9.9098	9.7174	9.5557	9.4164	9.2940	9.1849
6[−6]	10.7618	10.3389	10.0374	9.8042	9.6145	9.4547	9.3167	9.1954	9.0872
7[−6]	10.6466	10.2379	9.9432	9.7139	9.5267	9.3686	9.2319	9.1115	9.0040
8[−6]	10.5455	10.1493	9.8607	9.6349	9.4500	9.2935	9.1579	9.0384	8.9316
9[−6]	10.4553	10.0702	9.7872	9.5646	9.3818	9.2267	9.0923	8.9736	8.8674
1[−5]	10.3739	9.9987	9.7207	9.5012	9.3203	9.1667	9.0332	8.9153	8.8097
2[−5]	9.8197	9.5097	9.2678	9.0705	8.9045	8.7614	8.6357	8.5239	8.4231
3[−5]	9.4815	9.2084	8.9889	8.8063	8.6504	8.5147	8.3948	8.2873	8.1901
4[−5]	9.2358	8.9878	8.7843	8.6126	8.4646	8.3348	8.2193	8.1154	8.0210
5[−5]	9.0422	8.8128	8.6217	8.4588	8.3171	8.1921	8.0804	7.9795	7.8875
6[−5]	8.8822	8.6674	8.4864	8.3306	8.1943	8.0734	7.9649	7.8666	7.7767
7[−5]	8.7457	8.5428	8.3701	8.2205	8.0888	7.9715	7.8658	7.7698	7.6818
8[−5]	8.6266	8.4336	8.2681	8.1238	7.9962	7.8820	7.7788	7.6849	7.5986
9[−5]	8.5209	8.3364	8.1771	8.0374	7.9134	7.8021	7.7012	7.6091	7.5244

(*continued*)

TABLE 7-2 (*continued*)

β	[−3]								
u	1	2	3	4	5	6	7	8	9
1[−4]	8.4258	8.2487	8.0949	7.9594	7.8386	7.7299	7.6311	7.5407	7.4574
2[−4]	7.7901	7.6563	7.5361	7.4271	7.3276	7.2363	7.1519	7.0735	7.0005
3[−4]	7.4110	7.2985	7.1956	7.1010	7.0135	6.9323	6.8565	6.7856	6.7191
4[−4]	7.1393	7.0401	6.9484	6.8633	6.7840	6.7098	6.6402	6.5746	6.5127
5[−4]	6.9273	6.8375	6.7538	6.6756	6.6024	6.5334	6.4684	6.4070	6.3487
6[−4]	6.7533	6.6705	6.5930	6.5202	6.4517	6.3869	6.3257	6.2675	6.2123
7[−4]	6.6056	6.5284	6.4559	6.3874	6.3227	6.2615	6.2033	6.1479	6.0951
8[−4]	6.4774	6.4048	6.3363	6.2714	6.2100	6.1516	6.0960	6.0430	6.9923
9[−4]	6.3640	6.2952	6.2301	6.1684	6.1097	6.0538	6.0004	5.9495	5.9007
1[−3]	6.2624	6.1969	6.1348	6.0757	6.0193	5.9656	5.9142	5.8651	5.8179
2[−3]	5.5903	5.5431	5.4976	5.4538	5.4115	5.3707	5.3312	5.2930	5.2561
3[−3]	5.1948	5.1560	5.1184	5.0819	5.0465	5.0121	4.9786	4.9461	4.9145
4[−3]	4.9135	4.8798	4.8470	4.8151	4.7839	4.7536	4.7241	4.6952	4.6771
5[−3]	4.6951	4.6649	4.6355	4.6067	4.5786	4.5512	4.5244	4.4981	4.4725
6[−3]	4.5165	4.4890	4.4620	4.4357	4.4099	4.3846	4.3599	4.3357	4.3120
7[−3]	4.3655	4.3400	4.3150	4.2906	4.2666	4.2431	4.2201	4.1975	4.1753
8[−3]	4.2347	4.2109	4.1875	4.1646	4.1421	4.1200	4.0984	4.0771	4.0562
9[−3]	4.1194	4.0969	4.0749	4.0533	4.0320	4.0112	3.9907	3.9705	3.9507
1[−2]	4.0163	3.9950	3.9741	3.9536	3.9334	3.9136	3.8940	3.8749	3.8530
2[−2]	4.3397	3.3250	3.3104	3.2960	3.2818	3.2678	3.2539	3.2402	3.2267
3[−2]	2.9472	2.9354	2.9237	2.9121	2.9007	2.8894	2.8782	2.8671	2.8561
4[−2]	2.6711	2.6611	2.6512	2.6414	2.6316	2.6220	2.6124	2.6029	2.5935
5[−2]	2.4590	2.4502	2.4415	2.4329	2.4243	2.4159	2.4074	2.3991	2.3908
6[−2]	2.2874	2.2795	2.2717	2.2640	2.2563	2.2487	2.2411	2.2336	2.2262
7[−2]	2.1436	2.1365	2.1294	2.1224	2.1154	2.1085	2.1016	2.0947	2.0880
8[−2]	2.0203	2.0138	2.0073	2.0008	1.9944	1.9880	1.9817	1.9754	1.9692
9[−2]	1.9126	1.9066	1.9005	1.8946	1.8886	1.8827	1.8768	1.8710	1.8652
1[−1]	1.8172	1.8116	1.8060	1.8004	1.7949	1.7893	1.7839	1.7784	1.7730
2[−1]	1.2193	1.2159	1.2125	1.2092	1.2059	1.2026	1.1993	1.1960	1.1928
3[−1]	0.9033	0.9010	0.8986	0.8963	0.8940	0.8917	0.8894	0.8871	0.8848
4[−1]	0.7006	0.6989	0.6971	0.6954	0.6936	0.6919	0.6902	0.6885	0.6867
5[−1]	0.5584	0.5570	0.5557	0.5543	0.5530	0.5516	0.5503	0.5490	0.5476
6[−1]	0.4533	0.4522	0.4511	0.4500	0.4490	0.4479	0.4468	0.4458	0.4447
7[−1]	0.3729	0.3720	0.3711	0.3703	0.3694	0.3685	0.3677	0.3668	0.3659
8[−1]	0.3099	0.3092	0.3084	0.3077	0.3070	0.3063	0.3056	0.3049	0.3042
9[−1]	0.2596	0.2590	0.2584	0.2578	0.2572	0.2566	0.2560	0.2554	0.2549
1	0.2189	0.2184	0.2179	0.2174	0.2169	0.2164	0.2159	0.2154	0.2149
2	488(−4)	487(−4)	486(−4)	485(−4)	484(−4)	483(−4)	482(−4)	481(−4)	480(−4)
3	130(−4)	130(−4)	130(−4)	129(−4)	129(−4)	129(−4)	129(−4)	128(−4)	128(−4)
4	377(−5)	376(−5)	376(−5)	375(−5)	374(−5)	373(−5)	372(−5)	372(−5)	371(−5)
5	115(−5)	114(−5)	114(−5)	114(−5)	114(−5)	113(−5)	113(−5)	113(−5)	113(−5)
6	359(−6)	359(−6)	358(−6)	357(−6)	356(−6)	356(−6)	355(−6)	354(−6)	353(−6)
7	115(−6)	115(−6)	115(−6)	115(−6)	114(−6)	114(−6)	114(−6)	114(−6)	113(−6)
8	376(−7)	375(−7)	374(−7)	374(−7)	373(−7)	372(−7)	371(−7)	371(−7)	370(−7)
9	124(−7)	124(−7)	124(−7)	123(−7)	123(−7)	123(−7)	123(−7)	122(−7)	122(−7)
10	415(−8)	414(−8)	413(−8)	412(−8)	411(−8)	411(−8)	410(−8)	409(−8)	408(−8)

(*continued*)

TABLE 7-2 (*continued*)

β	[−2]									
u	1	2	3	4	5	6	7	8	9	10
1[−9]	13.4069	12.7142	12.3088	12.0212	11.7982	11.6159	11.4617	11.3282	11.2105	11.1051
2[−9]	13.0599	12.3674	11.9622	11.6746	11.4515	11.2692	11.1151	10.9816	10.8639	10.7585
3[−9]	12.8569	12.1645	11.7593	11.4718	11.2487	11.0665	10.9124	10.7789	10.6611	10.5558
4[−9]	12.7128	12.0206	11.6154	11.3279	11.1049	10.9226	10.7685	10.6350	10.5173	10.4119
5[−9]	12.6010	11.9089	11.5038	11.2163	10.9932	10.8110	10.6569	10.5234	10.4057	10.3003
6[−9]	12.5097	11.8176	11.4125	11.1251	10.9021	10.7198	10.5657	10.4322	10.3145	10.2092
7[−9]	12.4324	11.7405	11.3354	11.0480	10.8249	10.6427	10.4886	10.3551	10.2374	10.1321
8[−9]	12.3655	11.6736	11.2686	10.9812	10.7581	10.5759	10.4218	10.2884	10.1706	10.0653
9[−9]	12.3065	11.6147	11.2097	10.9222	10.6992	10.5170	10.3629	10.2295	10.1117	10.0064
1[−8]	12.2536	11.5619	11.1569	10.8695	10.6465	10.4643	10.3012	10.1768	10.0590	9.9538
2[−8]	11.9059	11.2148	10.8100	10.5227	10.2997	10.1176	9.9636	9.8301	9.7124	9.6071
3[−8]	11.7023	11.0116	10.6070	10.3197	10.0968	9.9148	9.7608	9.6273	9.5096	9.4044
4[−8]	11.5577	10.8674	10.4629	10.1757	9.9529	9.7708	9.6168	9.4834	9.3658	9.2605
5[−8]	11.4455	10.7555	10.3511	10.0640	9.8412	9.6592	9.5052	9.3718	9.2541	9.1489
6[−8]	11.3537	10.6640	10.2598	9.9727	9.7500	9.5679	9.4140	9.2806	9.1629	9.0577
7[−8]	11.2761	10.5867	10.1825	9.8955	9.6728	9.4908	9.3368	9.2035	9.0858	8.9806
8[−8]	11.2088	10.5197	10.1156	9.8286	9.6059	9.4239	9.2700	9.1366	9.0190	8.9138
9[−8]	11.1494	10.4605	10.0565	9.7696	9.5470	9.3650	9.2111	9.0777	8.9601	8.8548
1[−7]	11.0963	10.4076	10.0037	9.7169	9.4942	9.3122	9.1583	9.0250	8.9074	8.8021
2[−7]	10.7460	10.0592	9.6560	9.3694	9.1470	8.9651	8.8113	8.6781	8.5605	8.4554
3[−7]	10.5405	9.8552	9.4524	9.1660	8.9437	8.7620	8.6083	8.4751	8.3576	8.2525
4[−7]	10.3943	9.7102	9.3078	9.0216	8.7995	8.6178	8.4642	8.3310	8.2136	8.1085
5[−7]	10.2806	9.5976	9.1955	8.9096	8.6875	8.5059	8.3524	8.2193	8.1018	7.9968
6[−7]	10.1875	9.5055	9.1037	8.8180	8.5960	8.4145	8.2610	8.1279	8.0105	7.9055
7[−7]	10.1087	9.4276	9.0261	8.7405	8.5186	8.3372	8.1837	8.0507	7.9333	7.8283
8[−7]	10.0403	9.3600	8.9588	8.6734	8.4516	8.2702	8.1167	7.9838	7.8664	7.7614
9[−7]	9.9800	9.3003	8.8994	8.6141	8.3924	8.2111	8.0577	7.9247	7.8074	7.7024
1[−6]	9.9259	9.2469	8.8463	8.5611	8.3395	8.1582	8.0048	7.8719	7.7546	7.6497
2[−6]	9.5677	8.8946	8.4960	8.2118	7.9908	7.8099	7.6569	7.5242	7.4071	7.3024
3[−6]	9.3561	8.6875	8.2904	8.0069	7.7864	7.6059	7.4531	7.3206	7.2037	7.0991
4[−6]	9.2047	8.5399	8.1441	7.8613	7.6412	7.4610	7.3084	7.1761	7.0593	6.9547
5[−6]	9.0866	8.4251	8.0304	7.7482	7.5284	7.3485	7.1960	7.0639	6.9472	6.8427
6[−6]	8.9894	8.3310	7.9373	7.6556	7.4362	7.2564	7.1042	6.9721	6.8555	6.7512
7[−6]	8.9069	8.2512	7.8584	7.5772	7.3581	7.1785	7.0264	6.8945	6.7780	6.6737
8[−6]	8.8350	8.1819	7.7900	7.5093	7.2904	7.1110	6.9591	6.8272	6.7108	6.6066
9[−6]	8.7714	8.1206	7.7296	7.4492	7.2306	7.0514	6.8996	6.7679	6.6515	6.5474
1[−5]	8.7142	8.0657	7.6754	7.3955	7.1771	6.9981	6.8464	6.7147	6.5985	6.4944
2[−5]	8.3315	7.7010	7.3170	7.0402	6.8238	6.6461	6.4954	6.3645	6.2489	6.1453
3[−5]	8.1013	7.4844	7.1051	6.8308	6.6159	6.4392	6.2893	6.1590	6.0438	5.9406
4[−5]	7.9346	7.3290	6.9536	6.6813	6.4677	6.2919	6.1425	6.0128	5.8980	5.7951
5[−5]	7.8031	7.2072	6.8353	6.5648	6.3523	6.1772	6.0284	5.8991	5.7847	5.6821
6[−5]	7.6941	7.1068	6.7380	6.4692	6.2576	6.0832	5.9350	5.8060	5.6919	5.5896
7[−5]	7.6007	7.0212	6.6553	6.3880	6.1773	6.0036	5.8558	5.7272	5.6133	5.5113
8[−5]	7.5190	6.9466	6.5834	6.3174	6.1076	5.9344	5.7870	5.6587	5.5452	5.4434
9[−5]	7.4461	6.8804	6.5196	6.2549	6.0459	5.8733	5.7263	5.5983	5.4850	5.3834

(*continued*)

TABLE 7-2 (*continued*)

β	[−2]									
u	1	2	3	4	5	6	7	8	9	10
1[−4]	7.3803	6.8208	6.4623	6.1988	5.9906	5.8185	5.6718	5.5442	5.4311	5.3297
2[−4]	6.9321	6.4190	6.0787	5.8248	5.6226	5.4546	5.3111	5.1857	5.0716	4.9747
3[−4]	6.6563	6.1750	5.8479	5.6012	5.4035	5.2387	5.0974	4.9739	4.8641	4.7655
4[−4]	6.4541	5.9971	5.6807	5.4399	5.2459	5.0837	4.9443	4.8223	4.7138	4.6161
5[−4]	6.2934	5.8561	5.5488	5.3131	5.1223	4.9623	4.8246	4.7039	4.5964	4.4996
6[−4]	6.1596	5.7389	5.4394	5.2082	5.0203	4.8623	4.7262	4.6066	4.5000	4.4040
7[−4]	6.0447	5.6383	5.3458	5.1185	4.9333	4.7772	4.6424	4.5239	4.4182	4.3228
8[−4]	5.9439	5.5501	5.2637	5.0402	4.8573	4.7029	4.5693	4.4518	4.3469	4.2523
9[−4]	5.8539	5.4713	5.1907	4.9704	4.7898	4.6369	4.5046	4.3880	4.2838	4.1898
1[−3]	5.7727	5.4001	5.1247	4.9076	4.7290	4.5776	4.4464	4.3306	4.2272	4.1337
2[−3]	5.2203	4.9139	4.6753	4.4813	4.3184	4.1783	4.0556	3.9467	3.8487	3.7598
3[−3]	4.8837	4.6149	4.3993	4.2205	4.0683	3.9362	3.8197	3.7157	3.6218	3.5363
4[−3]	4.6396	4.3962	4.1972	4.0298	3.8859	3.7601	3.6485	3.5485	3.4578	3.3750
5[−3]	4.4474	4.2231	4.0369	3.8786	3.7415	3.6208	3.5134	3.4166	3.3287	3.2483
6[−3]	4.2888	4.0794	3.9036	3.7528	3.6214	3.5052	3.4013	3.3074	3.2220	3.1436
7[−3]	4.1536	3.9564	3.7893	3.6450	3.5185	3.4062	3.3053	3.2141	3.1308	3.0542
8[−3]	4.0357	3.8488	3.6891	3.5504	3.4282	3.3193	3.2213	3.1324	3.0510	2.9762
9[−3]	3.9313	3.7531	3.5999	3.4661	3.3478	3.2420	3.1465	3.0596	2.9801	2.9068
1[−2]	3.8374	3.6669	3.5195	3.3901	3.2752	3.1722	3.0790	2.9941	2.9162	2.8443
2[−2]	3.2133	3.0880	2.9759	2.8748	2.7829	2.6989	2.6218	2.5505	2.4844	2.4227
3[−2]	2.8452	2.7423	2.6487	2.5631	2.4844	2.4117	2.3443	2.2816	2.2230	2.1680
4[−2]	2.5842	2.4955	2.4140	2.3388	2.2691	2.2043	2.1438	2.0872	2.0341	1.9841
5[−2]	2.3826	2.3040	2.2312	2.1636	2.1007	2.0419	1.9867	1.9349	1.8861	1.8401
6[−2]	2.2188	2.1478	2.0818	2.0203	1.9626	1.9086	1.8577	1.8098	1.7645	1.7217
7[−2]	2.0812	2.0164	1.9558	1.8991	1.8458	1.7957	1.7484	1.7038	1.6614	1.6213
8[−2]	1.9630	1.9031	1.8471	1.7944	1.7448	1.6980	1.6538	1.6119	1.5721	1.5341
9[−2]	1.8595	1.8039	1.7516	1.7024	1.6559	1.6120	1.5704	1.5309	1.4934	1.4577
1[−1]	1.7677	1.7157	1.6667	1.6205	1.5768	1.5354	1.4961	1.4587	1.4232	1.3893
2[−1]	1.1895	1.1579	1.1278	1.0990	1.0714	1.0450	1.0197	0.9954	0.9721	0.9497
3[−1]	0.8825	0.8603	0.8389	0.8184	0.7986	0.7796	0.7613	0.7437	0.7267	0.7103
4[−1]	0.6850	0.6683	0.6523	0.6367	0.6218	0.6073	0.5934	0.5799	0.5669	0.5543
5[−1]	0.5463	0.5333	0.5207	0.5086	0.4969	0.4855	0.4745	0.4639	0.4536	0.4436
6[−1]	0.4437	0.4333	0.4233	0.4135	0.4041	0.3950	0.3862	0.3776	0.3693	0.3613
7[−1]	0.3651	0.3567	0.3485	0.3406	0.3330	0.3255	0.3183	0.3113	0.3046	0.2980
8[−1]	0.3035	0.2966	0.2899	0.2834	0.2770	0.2709	0.2650	0.2592	0.2536	0.2481
9[−1]	0.2543	0.2485	0.2430	0.2376	0.2323	0.2272	0.2223	0.2175	0.2128	0.2082
1	0.2144	0.2097	0.2050	0.2005	0.1961	0.1918	0.1876	0.1836	0.1797	0.1758
2	479(−4)	468(−4)	458(−4)	449(−4)	439(−4)	430(−4)	421(−4)	412(−4)	403(−4)	395(−4)
3	128(−4)	125(−4)	122(−4)	120(−4)	117(−4)	115(−4)	113(−4)	110(−4)	108(−4)	106(−4)
4	370(−5)	362(−5)	355(−5)	348(−5)	340(−5)	333(−5)	327(−5)	320(−5)	313(−5)	307(−5)
5	112(−5)	110(−5)	108(−5)	106(−5)	104(−5)	101(−5)	993(−6)	973(−6)	953(−6)	934(−6)
6	353(−6)	345(−6)	338(−6)	332(−6)	325(−6)	318(−6)	312(−6)	305(−6)	299(−6)	243(−6)
7	113(−6)	111(−6)	109(−6)	106(−6)	104(−6)	102(−6)	100(−6)	980(−7)	960(−7)	941(−7)
8	369(−7)	362(−7)	354(−7)	347(−7)	340(−7)	333(−7)	326(−7)	320(−7)	313(−7)	307(−7)
9	122(−7)	119(−7)	117(−7)	115(−7)	112(−7)	110(−7)	108(−7)	106(−7)	104(−7)	102(−7)
10	407(−8)	399(−8)	391(−8)	383(−8)	375(−8)	368(−8)	360(−8)	353(−8)	346(−8)	339(−8)

TABLE 7-2 *(continued)*

β	[−1]								
u	1	2	3	4	5	6	7	8	9
1[−9]	11.1051	10.4121	10.0066	9.7191	9.4960	9.3137	9.1596	9.0261	8.9083
2[−9]	10.7585	10.0655	9.6602	9.3726	9.1495	8.9672	8.8131	8.6797	8.5619
3[−9]	10.5558	9.8628	9.4575	9.1699	8.9468	8.7646	8.6105	8.4770	8.3593
4[−9]	10.4119	9.7190	9.3137	9.0261	8.8030	8.6208	8.4667	8.3332	8.2155
5[−9]	10.3003	9.6074	9.2021	8.9145	8.6915	8.5092	8.3552	8.2217	8.1040
6[−9]	10.2092	9.5163	9.1109	8.8234	8.6004	8.4181	8.2641	8.1306	8.0129
7[−9]	10.1321	9.4392	9.0339	8.7463	8.5233	8.3411	8.1870	8.0536	7.9359
8[−9]	10.0653	9.3725	8.9671	8.6796	8.4566	8.2744	8.1203	7.9869	7.8692
9[−9]	10.0064	9.3136	8.9083	8.6207	8.3977	8.2155	8.0615	7.9280	7.8103
1[−8]	9.9538	9.2609	8.8556	8.5681	8.3450	8.1628	8.0088	7.8754	7.7577
2[−8]	9.6071	8.9144	8.5091	8.2216	7.9987	7.8165	7.6625	7.5292	7.4115
3[−8]	9.4044	8.7117	8.3065	8.0190	7.7961	7.6140	7.4600	7.3267	7.2091
4[−8]	9.2605	8.5679	8.1627	7.8753	7.6524	7.4703	7.3164	7.1831	7.0655
5[−8]	9.1489	8.4563	8.0512	7.7638	7.5409	7.3589	7.2050	7.0717	6.9541
6[−8]	9.0577	8.3652	7.9601	7.6727	7.4499	7.2678	7.1139	6.9807	6.8631
7[−8]	8.9806	8.2881	7.8830	7.5957	7.3729	7.1909	7.0370	6.9037	6.7862
8[−8]	8.9138	8.2213	7.8163	7.5290	7.3062	7.1242	6.9703	6.8371	6.7196
9[−8]	8.8548	8.1625	7.7575	7.4702	7.2474	7.0654	6.9116	6.7783	6.6608
1[−7]	8.8021	8.1098	7.7048	7.4176	7.1948	7.0128	6.8590	6.7258	6.6083
2[−7]	8.4554	7.7633	7.3585	7.0714	6.8488	6.6669	6.5132	6.3801	6.2628
3[−7]	8.2525	7.5607	7.1560	6.8690	6.6465	6.4647	6.3111	6.1781	6.0608
4[−7]	8.1085	7.4169	7.0124	6.7254	6.5030	6.3213	6.1677	6.0348	5.9176
5[−7]	7.9968	7.3054	6.9009	6.6141	6.3917	6.2101	6.0566	5.9237	5.8066
6[−7]	7.9055	7.2142	6.8099	6.5231	6.3008	6.1192	5.9658	5.8330	5.7159
7[−7]	7.8283	7.1372	6.7329	6.4463	6.2240	6.0425	5.8891	5.7563	5.6393
8[−7]	7.7614	7.0705	6.6663	6.3797	6.1574	5.9760	5.8227	5.6899	5.5729
9[−7]	7.7024	7.0116	6.6075	6.3209	6.0987	5.9173	5.7641	5.6314	5.5144
1[−6]	7.6497	6.9590	6.5549	6.2684	6.0463	5.8649	5.7117	5.5790	5.4621
2[−6]	7.3024	6.6126	6.2091	5.9230	5.7012	5.5202	5.3673	5.2349	5.1183
3[−6]	7.0991	6.4100	6.0069	5.7211	5.4996	5.3188	5.1662	5.0341	4.9176
4[−6]	6.9547	6.2663	5.8635	5.5779	5.3567	5.1761	5.0237	4.8917	4.7755
5[−6]	6.8427	6.1548	5.7523	5.4670	5.2459	5.0655	4.9132	4.7815	4.6654
6[−6]	6.7512	6.0637	5.6615	5.3763	5.1555	4.9753	4.8231	4.6915	4.5755
7[−6]	6.6737	5.9867	5.5847	5.2998	5.0790	4.8990	4.7469	4.6154	4.4996
8[−6]	6.6066	5.9200	5.5182	5.2334	5.0129	4.8329	4.6810	4.5496	4.4339
9[−6]	6.5474	5.8611	5.4596	5.1750	4.9545	4.7747	4.6229	4.4916	4.3760
1[−5]	6.4944	5.8085	5.4071	5.1227	4.9024	4.7227	4.5710	4.4398	4.3243
2[−5]	6.1453	5.4623	5.0624	4.7791	4.5598	4.3810	4.2302	4.0998	3.9851
3[−5]	5.9406	5.2597	4.8610	4.5785	4.3600	4.1818	4.0316	3.9019	3.7877
4[−5]	5.7951	5.1160	4.7182	4.4365	4.2185	4.0409	3.8912	3.7619	3.6482
5[−5]	5.6821	5.0045	4.6075	4.3264	4.1090	3.9319	3.7826	3.6537	3.5404
6[−5]	5.5896	4.9134	4.5172	4.2366	4.0196	3.8429	3.6940	3.5655	3.4526
7[−5]	5.5113	4.8364	4.4408	4.1608	3.9442	3.7679	3.6193	3.4911	3.3785
8[−5]	5.4434	4.7697	4.3747	4.0951	3.8789	3.7029	3.5547	3.4268	3.3145
9[−5]	5.3834	4.7108	4.3164	4.0372	3.4214	3.6458	3.4978	3.3702	3.2581

(continued)

TABLE 7-2 (*continued*)

β	[−1]								
u	1	2	3	4	5	6	7	8	9
1[−4]	5.3297	4.6581	4.2643	3.9855	3.7700	3.5947	3.4470	3.3197	3.2078
2[−4]	4.9747	4.3115	3.9220	3.6463	3.4334	3.2604	3.1149	2.9896	2.8797
3[−4]	4.7655	4.1086	3.7222	3.4489	3.2379	3.0666	2.9227	2.7988	2.6902
4[−4]	4.6161	3.9645	3.5807	3.3093	3.0999	2.9300	2.7873	2.6646	2.5571
5[−4]	4.4996	3.8527	3.4711	3.2013	2.9933	2.8245	2.6829	2.5612	2.4547
6[−4]	4.4040	3.7612	3.3817	3.1133	2.9065	2.7388	2.5982	2.4773	2.3716
7[−4]	4.3228	3.6838	3.3062	3.0391	2.8334	2.6666	2.5268	2.4068	2.3018
8[−4]	4.2523	3.6167	3.2408	2.9749	2.7702	2.6043	2.4653	2.3459	2.2417
9[−4]	4.1898	3.5575	3.1832	2.9185	2.7146	2.5495	2.4112	2.2926	2.1889
1[−3]	4.1337	3.5045	3.1317	2.8680	2.6650	2.5007	2.3630	2.2450	2.1419
2[−3]	3.7598	3.1549	2.7938	2.5383	2.3419	2.1833	2.0508	1.9375	1.8388
3[−3]	3.5363	2.9494	2.5969	2.3475	2.1559	2.0013	1.8725	1.7624	1.6668
4[−3]	3.3750	2.8030	2.4577	2.2130	2.0253	1.8741	1.7481	1.6408	1.5476
5[−3]	3.2483	2.6891	2.3499	2.1094	1.9250	1.7765	1.6531	1.5479	1.4568
6[−3]	3.1436	2.5957	2.2619	2.0252	1.8437	1.6977	1.5763	1.4732	1.3838
7[−3]	3.0542	2.5165	2.1877	1.9543	1.7754	1.6316	1.5122	1.4107	1.3230
8[−3]	2.9762	2.4478	2.1235	1.8932	1.7166	1.5748	1.4572	1.3573	1.2710
9[−3]	2.9068	2.3870	2.0669	1.8394	1.6651	1.5251	1.4091	1.3106	1.2256
1[−2]	2.8443	2.3325	2.0164	1.7915	1.6193	1.4810	1.3664	1.2693	1.1855
2[−2]	2.4227	1.9714	1.6853	1.4807	1.3239	1.1985	1.0951	1.0078	0.9328
3[−2]	2.1680	1.7579	1.4932	1.3029	1.1570	1.0405	0.9446	0.8640	0.7951
4[−2]	1.9841	1.6056	1.3578	1.1789	1.0416	0.9321	0.8421	0.7667	0.7023
5[−2]	1.8401	1.4872	1.2535	1.0841	0.9540	0.8503	0.7652	0.6940	0.6334
6[−2]	1.7217	1.3905	1.1689	1.0077	0.8838	0.7851	0.7042	0.6367	0.5792
7[−2]	1.6213	1.3088	1.0979	0.9439	0.8255	0.7312	0.6540	0.5896	0.5349
8[−2]	1.5343	1.2381	1.0368	0.8893	0.7758	0.6854	0.6116	0.5499	0.4977
9[−2]	1.4577	1.1760	0.9833	0.8417	0.7327	0.6459	0.5749	0.5158	0.4658
1[−1]	1.3893	1.1207	0.9358	0.7996	0.6947	0.6111	0.5429	0.4861	0.4381
2[−1]	0.9497	0.7665	0.6352	0.5367	0.4603	0.3994	0.3498	0.3088	0.2744
3[−1]	0.7103	0.5739	0.4740	0.3981	0.3390	0.2917	0.2533	0.2217	0.1953
4[−1]	0.5543	0.4482	0.3695	0.3092	0.2619	0.2242	0.1935	0.1683	0.1472
5[−1]	0.4436	0.3591	0.2956	0.2467	0.2083	0.1775	0.1525	0.1320	0.1149
6[−1]	0.3613	0.2927	0.2407	0.2005	0.1688	0.1433	0.1227	0.1058	917(−4)
7[−1]	0.2980	0.2415	0.1985	0.1651	0.1386	0.1174	0.1002	861(−4)	744(−4)
8[−1]	0.2481	0.2012	0.1653	0.1373	0.1151	973(−4)	828(−4)	709(−4)	611(−4)
9[−1]	0.2082	0.1690	0.1387	0.1151	963(−4)	812(−4)	690(−4)	590(−4)	507(−4)
1	0.1758	0.1427	0.1172	971(−4)	812(−4)	683(−4)	579(−4)	494(−4)	423(−4)
2	395(−4)	322(−4)	264(−4)	217(−4)	180(−4)	150(−4)	126(−4)	106(−4)	896(−5)
3	106(−4)	862(−5)	707(−5)	582(−5)	481(−5)	399(−5)	332(−5)	278(−5)	233(−5)
4	307(−5)	250(−5)	205(−5)	169(−5)	139(−5)	115(−5)	957(−6)	797(−6)	666(−6)
5	934(−6)	763(−6)	624(−6)	513(−6)	423(−6)	349(−6)	289(−6)	240(−6)	200(−6)
6	293(−6)	239(−6)	196(−6)	161(−6)	133(−6)	109(−6)	905(−7)	750(−7)	624(−7)
7	941(−7)	769(−7)	629(−7)	517(−7)	425(−7)	350(−7)	290(−7)	240(−7)	199(−7)
8	307(−7)	251(−7)	205(−7)	169(−7)	139(−7)	114(−7)	943(−8)	780(−8)	646(−8)
9	102(−7)	830(−8)	679(−8)	557(−8)	458(−8)	377(−8)	311(−8)	257(−8)	213(−8)
10	339(−8)	277(−8)	227(−8)	186(−8)	153(−8)	126(−8)	104(−8)		

(*continued*)

TABLE 7-2 (*continued*)

β	[0]									
u	1	2	3	4	5	6	7	8	9	10
1[−9]	8.8030	8.1102	7.7051	7.4178	7.1950	7.0130	6.8591	6.7259	6.6084	6.5033
2[−9]	8.4566	7.7639	7.3590	7.0717	6.8490	6.6671	6.5134	6.3803	6.2629	6.1579
3[−9]	8.2540	7.5614	7.1565	6.8694	6.6468	6.4650	6.3113	6.1783	6.0610	5.9561
4[−9]	8.1102	7.4178	7.0130	6.7259	6.5033	6.3216	6.1680	6.0350	5.9178	5.8130
5[−9]	7.9987	7.3064	6.9016	6.6146	6.3921	6.2104	6.0569	5.9240	5.8068	5.7020
6[−9]	7.9076	7.2153	6.8106	6.5237	6.3012	6.1196	5.9661	5.8333	5.7161	5.6114
7[−9]	7.8306	7.1384	6.7337	6.4468	6.2244	6.0429	5.8894	5.7566	5.6395	5.5348
8[−9]	7.7639	7.0717	6.6671	6.3803	6.1579	5.9764	5.8230	5.6902	5.5732	5.4685
9[−9]	7.7051	7.0129	6.6084	6.3216	6.0993	5.9178	5.7644	5.6317	5.5147	5.4101
1[−8]	7.6525	6.9604	6.5558	6.2691	6.0468	5.8654	5.7121	5.5793	5.4624	5.3578
2[−8]	7.3063	6.6146	6.2104	5.9240	5.7020	5.5208	5.3678	5.2354	5.1187	5.0145
3[−8]	7.1039	6.4124	6.0085	5.7223	5.5006	5.3196	5.1669	5.0347	4.9182	4.8141
4[−8]	6.9603	6.2691	5.8654	5.5793	5.3578	5.1771	5.0245	4.8924	4.7761	4.6722
5[−8]	6.8490	6.1579	5.7544	5.4685	5.2472	5.0666	4.9141	4.7822	4.6661	4.5623
6[−8]	6.7580	6.0671	5.6617	5.3781	5.1568	4.9764	4.8241	4.6923	4.5763	4.4726
7[−8]	6.6811	5.9904	5.5872	5.3016	5.0805	4.9002	4.7480	4.6164	4.5004	4.3969
8[−8]	6.6145	5.9239	5.5208	5.2354	5.0145	4.8343	4.6821	4.5506	4.4348	4.3314
9[−8]	6.5558	5.8653	5.4624	5.1771	4.9562	4.7761	4.6241	4.4927	4.3770	4.2736
1[−7]	6.5032	5.8129	5.4101	5.1249	4.9041	4.7241	4.5722	4.4409	4.3253	4.2221
2[−7]	6.1578	5.4685	5.0666	4.7822	4.5623	4.3831	4.2319	4.1014	3.9865	3.8839
3[−7]	5.9559	5.2673	4.8661	4.5824	4.3630	4.1844	4.0338	3.9038	3.7894	3.6874
4[−7]	5.8128	5.1248	4.7241	4.4409	4.2220	4.0439	3.8937	3.7641	3.6502	3.5486
5[−7]	5.7018	5.0144	4.6141	4.3313	4.1129	3.9351	3.7854	3.6562	3.5426	3.4413
6[−7]	5.6112	4.9242	4.5244	4.2420	4.0239	3.8465	3.6971	3.5682	3.4549	3.3540
7[−7]	5.5346	4.8481	4.4486	4.1666	3.9488	3.7717	3.6226	3.4940	3.3810	3.2804
8[−7]	5.4683	4.7821	4.3830	4.1013	3.8839	3.7071	3.5583	3.4299	3.3172	3.2168
9[−7]	5.4098	4.7240	4.3252	4.0438	3.8267	3.6501	3.5016	3.3735	3.2610	3.1609
1[−6]	5.3575	4.6721	4.2736	3.9925	3.7756	3.5993	3.4510	3.3231	3.2109	3.1110
2[−6]	5.0141	4.3312	3.9350	3.6561	3.4412	3.2669	3.1205	2.9945	2.8840	2.7857
3[−6]	4.8136	4.1327	3.7382	3.4608	3.2474	3.0745	2.9294	2.8047	2.6955	2.5984
4[−6]	4.6716	3.9922	3.5991	3.3230	3.1109	2.9391	2.7951	2.6714	2.5631	2.4671
5[−6]	4.5617	3.8836	3.4917	3.2167	3.0055	2.8347	2.6916	2.5688	2.4614	2.3661
6[−6]	4.4719	3.7951	3.4042	3.1301	2.9199	2.7499	2.6076	2.4856	2.3789	2.2844
7[−6]	4.3962	3.7204	3.3304	3.0572	2.8478	2.6786	2.5370	2.4157	2.3097	2.2158
8[−6]	4.3306	3.6558	3.2667	2.9943	2.7856	2.6171	2.4762	2.3554	2.2501	2.1568
9[−6]	4.2728	3.5989	3.2106	2.9390	2.7309	2.5631	2.4228	2.3026	2.1978	2.1050
1[−5]	4.2212	3.5481	3.1606	2.8896	2.6822	2.5149	2.3752	2.2556	2.1513	2.0590
2[−5]	3.8827	3.2162	2.8344	2.5686	2.3660	2.2032	2.0677	1.9522	1.8518	1.7632
3[−5]	3.6858	3.0241	2.6464	2.3842	2.1850	2.0254	1.8929	1.7802	1.6825	1.5965
4[−5]	3.5468	2.8889	2.5145	2.2553	2.0588	1.9017	1.7715	1.6610	1.5654	1.4815
5[−5]	3.4394	2.7848	2.4131	2.1564	1.9622	1.8071	1.6790	1.5704	1.4765	1.3943
6[−5]	3.3519	2.7002	2.3309	2.0764	1.8841	1.7310	1.6045	1.4975	1.4052	1.3244
7[−5]	3.2781	2.6290	2.2619	2.0093	1.8189	1.6673	1.5424	1.4369	1.3459	1.2664
8[−5]	3.2143	2.5677	2.2026	1.9517	1.7629	1.6128	1.4893	1.3850	1.2953	1.2169
9[−5]	3.1583	2.5138	2.1505	1.9013	1.7139	1.5652	1.4429	1.3398	1.2512	1.1719

(*continued*)

TABLE 7-2 (*continued*)

β	[0]									
u	1	2	3	4	5	6	7	8	9	10
1[−4]	3.1082	2.4658	2.1042	1.8565	1.6704	1.5230	1.4019	1.2999	1.2121	1.1359
2[−4]	2.7819	2.1549	1.8062	1.5697	1.3937	1.2555	1.1429	1.0488	0.9686	0.8992
3[−4]	2.5937	1.9778	1.6380	1.4092	1.2401	1.1080	1.0011	0.9123	0.8369	0.7721
4[−4]	2.4617	1.8545	1.5217	1.2989	1.1352	1.0079	0.9054	0.8205	0.7489	0.6875
5[−4]	2.3601	1.7604	1.4335	1.2157	1.0564	0.9331	0.8341	0.7525	0.6839	0.6252
6[−4]	2.2778	1.6846	1.3628	1.1493	0.9937	0.8738	0.7778	0.6990	0.6329	0.5765
7[−4]	2.2087	1.6212	1.3039	1.0943	0.9420	0.8250	0.7317	0.6552	0.5913	0.5370
8[−4]	2.1492	1.5670	1.2537	1.0475	0.8982	0.7838	0.6928	0.6185	0.5565	0.5040
9[−4]	2.0971	1.5196	1.2101	1.0070	0.8603	0.7483	0.6594	0.5870	0.5267	0.4758
1[−3]	2.0506	1.4776	1.1715	0.9713	0.8271	0.7172	0.6303	0.5595	0.5008	0.4513
2[−3]	1.7516	1.2116	0.9305	0.7507	0.6238	0.5290	0.4554	0.3965	0.3483	0.3084
3[−3]	1.5825	1.0652	0.8006	0.6340	0.5182	0.4329	0.3673	0.3156	0.2738	0.2394
4[−3]	1.4656	0.9658	0.7139	0.5572	0.4496	0.3711	0.3114	0.2647	0.2274	0.1970
5[−3]	1.3767	0.8915	0.6498	0.5012	0.4001	0.3269	0.2718	0.2290	0.1951	0.1677
6[−3]	1.3054	0.8327	0.5997	0.4578	0.3620	0.2932	0.2419	0.2023	0.1711	0.1460
7[−3]	1.2460	0.7843	0.5589	0.4227	0.3315	0.2665	0.2182	0.1813	0.1523	0.1292
8[−3]	1.1953	0.7435	0.5248	0.3936	0.3064	0.2446	0.1990	0.1643	0.1372	0.1158
9[−3]	1.1512	0.7083	0.4956	0.3689	0.2852	0.2262	0.1830	0.1502	0.1248	0.1047
1[−2]	1.1122	0.6775	0.4702	0.3476	0.2670	0.2106	0.1694	0.1383	0.1144	955(−4)
2[−2]	0.8677	0.4914	0.3214	0.2256	0.1653	0.1248	964(−4)	757(−4)	604(−4)	487(−4)
3[−2]	0.7353	0.3965	0.2491	0.1687	0.1197	877(−4)	659(−4)	504(−4)	392(−4)	308(−4)
4[−2]	0.6467	0.3357	0.2043	0.1346	931(−4)	666(−4)	489(−4)	367(−4)	279(−4)	216(−4)
5[−2]	0.5812	0.2923	0.1733	0.1115	755(−4)	530(−4)	382(−4)	281(−4)	211(−4)	160(−4)
6[−2]	0.5298	0.2593	0.1503	948(−4)	630(−4)	435(−4)	308(−4)	224(−4)	165(−4)	124(−4)
7[−2]	0.4880	0.2332	0.1325	820(−4)	536(−4)	365(−4)	255(−4)	183(−4)	133(−4)	982(−5)
8[−2]	0.4530	0.2119	0.1182	720(−4)	464(−4)	311(−4)	215(−4)	152(−4)	109(−4)	797(−5)
9[−2]	0.4230	0.1941	0.1064	639(−4)	406(−4)	269(−4)	184(−4)	128(−4)	912(−5)	658(−5)
1[−1]	0.3970	0.1789	966(−4)	572(−4)	359(−4)	235(−4)	159(−4)	110(−4)	771(−5)	552(−5)
2[−1]	0.2452	971(−4)	468(−4)	251(−4)	143(−4)	859(−5)	533(−5)	340(−5)	222(−5)	149(−5)
3[−1]	0.1729	629(−4)	281(−4)	140(−4)	752(−5)	424(−5)	248(−5)	151(−5)	936(−6)	592(−6)
4[−1]	0.1296	441(−4)	186(−4)	880(−5)	448(−5)	241(−5)	135(−5)	785(−5)	466(−6)	283(−6)
5[−1]	0.1006	325(−4)	130(−4)	590(−5)	288(−5)	149(−5)	806(−6)	450(−6)	257(−6)	151(−6)
6[−1]	799(−4)	247(−4)	952(−5)	413(−5)	195(−5)	974(−6)	509(−6)	275(−6)	153(−6)	873(−7)
7[−1]	646(−4)	192(−4)	714(−5)	299(−5)	137(−5)	663(−6)	336(−6)	176(−6)	955(−7)	534(−7)
8[−1]	529(−4)	152(−4)	547(−5)	222(−5)	986(−6)	466(−6)	229(−6)	117(−6)	622(−7)	340(−7)
9[−1]	437(−4)	122(−4)	426(−5)	169(−5)	728(−6)	335(−6)	161(−6)	805(−7)	418(−7)	223(−7)
1	365(−4)	993(−5)	337(−5)	130(−5)	547(−6)	246(−6)	115(−6)	566(−7)	288(−7)	151(−7)
2	760(−5)	173(−5)	487(−6)	156(−6)	551(−7)	208(−7)	834(−8)	350(−8)	152(−8)	
3	196(−5)	406(−6)	102(−6)	290(−7)	911(−8)	308(−8)	111(−8)			
4	558(−6)	108(−6)	250(−7)	655(−8)	189(−8)					
5	167(−6)	309(−7)	672(−8)	165(−8)						
6	519(−7)	926(−8)	192(−8)							
7	165(−7)	286(−8)								
8	536(−8)									
9	176(−8)									
10										

(*continued*)

TABLE 7-2 (*continued*)

β	[1]								
u	1	2	3	4	5	6	7	8	9
1[−9]	6.5033	5.8130	5.4101	5.1249	4.9042	4.7241	4.5722	4.4409	4.3253
2[−9]	6.1579	5.4685	5.0666	4.7823	4.5623	4.3831	4.2320	4.1014	3.9865
3[−9]	5.9561	5.2674	4.8661	4.5824	4.3630	4.1844	4.0338	3.9038	3.7894
4[−9]	5.8130	5.1249	4.7241	4.4409	4.2221	4.0439	3.8937	3.7641	3.6502
5[−9]	5.7020	5.0145	4.6142	4.3314	4.1130	3.9352	3.7854	3.6562	3.5426
6[−9]	5.6114	4.9243	4.5244	4.2420	4.0240	3.8465	3.6971	3.5682	3.4550
7[−9]	5.5348	4.8482	4.4487	4.1666	3.9489	3.7718	3.6227	3.4941	3.3811
8[−9]	5.4685	4.7823	4.3831	4.1014	3.8839	3.7051	3.5583	3.4300	3.3172
9[−9]	5.4101	4.7241	4.3253	4.0439	3.8267	3.6502	3.5016	3.3735	3.2611
1[−8]	5.3578	4.6722	4.2737	3.9925	3.7756	3.5993	3.4510	3.3232	3.2109
2[−8]	5.0145	4.3314	3.9352	3.6562	3.4413	3.2670	3.1205	2.9945	2.8840
3[−8]	4.8141	4.1329	3.7383	3.4609	3.2475	3.0746	2.9295	2.8047	2.6955
4[−8]	4.6722	3.9925	3.5993	3.3232	3.1110	2.9392	2.7952	2.6714	2.5632
5[−8]	4.5623	3.8839	3.4919	3.2168	3.0057	2.8348	2.6917	2.5689	2.4615
6[−8]	4.4726	3.7954	3.4044	3.1303	2.9200	2.7500	2.6077	2.4857	2.3790
7[−8]	4.3969	3.7207	3.3307	3.0574	2.8480	2.6787	2.5371	2.4158	2.3098
8[−8]	4.3314	3.6562	3.2670	2.9945	2.7858	2.6172	2.4763	2.3555	2.2502
9[−8]	4.2736	3.5993	3.2109	2.9392	2.7311	2.5632	2.4229	2.3027	2.1979
1[−7]	4.2221	3.5486	3.1609	2.8898	2.6824	2.5151	2.3753	2.2557	2.1514
2[−7]	3.8839	3.2168	2.8348	2.5689	2.3662	2.2034	2.0679	1.9523	1.8519
3[−7]	3.6874	3.0249	2.6469	2.3846	2.1853	2.0257	1.8931	1.7804	1.6826
4[−7]	3.5486	2.8898	2.5151	2.2557	2.0591	1.9019	1.7718	1.6612	1.5656
5[−7]	3.4413	2.7858	2.4137	2.1568	1.9625	1.8074	1.6793	1.5706	1.4767
6[−7]	3.3540	2.7012	2.3316	2.0769	1.8845	1.7313	1.6048	1.4978	1.4054
7[−7]	3.2804	2.6302	2.2627	2.0099	1.8193	1.6677	1.5427	1.4371	1.3461
8[−7]	3.2168	2.5689	2.2034	1.9523	1.7633	1.6132	1.4896	1.3853	1.2955
9[−7]	3.1609	2.5151	2.1514	1.9019	1.7144	1.5656	1.4433	1.3401	1.2514
1[−6]	3.1110	2.4671	2.1051	1.8571	1.6710	1.5234	1.4023	1.3002	1.2125
2[−6]	2.7857	2.1568	1.8074	1.5706	1.3944	1.2561	1.1434	1.0492	0.9689
3[−6]	2.5984	1.9801	1.6395	1.4102	1.2409	1.1087	1.0017	0.9128	0.8374
4[−6]	2.4671	1.8571	1.5234	1.3002	1.1361	1.0087	0.9060	0.8211	0.7494
5[−6]	2.3661	1.7633	1.4354	1.2171	1.0574	0.9339	0.8348	0.7531	0.6844
6[−6]	2.2844	1.6877	1.3648	1.1508	0.9948	0.8747	0.7786	0.6996	0.6334
7[−6]	2.2158	1.6246	1.3061	1.0959	0.9432	0.8260	0.7325	0.6559	0.5919
8[−6]	2.1568	1.5706	1.2560	1.0492	0.8995	0.7849	0.6937	0.6192	0.5571
9[−6]	2.1050	1.5234	1.2125	1.0087	0.8617	0.7494	0.6603	0.5877	0.5273
1[−5]	2.0590	1.4816	1.1741	0.9731	0.8285	0.7183	0.6312	0.5603	0.5015
2[−5]	1.7632	1.2170	0.9339	0.7531	0.6256	0.5304	0.4565	0.3974	0.3491
3[−5]	1.5965	1.0716	0.8046	0.6368	0.5203	0.4344	0.3686	0.3166	0.2746
4[−5]	1.4815	0.9730	0.7183	0.5603	0.4518	0.3728	0.3128	0.2658	0.2283
5[−5]	1.3943	0.8894	0.6546	0.5045	0.4024	0.3287	0.2732	0.2302	0.1961
6[−5]	1.3244	0.8412	0.6048	0.4612	0.3645	0.2951	0.2434	0.2035	0.1720
7[−5]	1.2664	0.7934	0.5643	0.4263	0.3341	0.2685	0.2198	0.1825	0.1533
8[−5]	1.2169	0.7530	0.5304	0.3974	0.3090	0.2466	0.2006	0.1655	0.1382
9[−5]	1.1739	0.7182	0.5014	0.3728	0.2879	0.2283	0.1846	0.1515	0.1258

(*continued*)

TABLE 7-2 (*continued*)

β	[1]								
u	1	2	3	4	5	6	7	8	9
1[−4]	1.1359	0.6878	0.4763	0.3516	0.2698	0.2127	0.1710	0.1396	0.1154
2[−4]	0.8992	0.5044	0.3287	0.2302	0.1684	0.1271	981(−4)	770(−4)	613(−4)
3[−4]	0.7721	0.4111	0.2570	0.1736	0.1229	900(−4)	675(−4)	516(−4)	401(−4)
4[−4]	0.6875	0.3514	0.2126	0.1396	963(−4)	688(−4)	505(−4)	378(−4)	288(−4)
5[−4]	0.6252	0.3089	0.1818	0.1165	787(−4)	551(−4)	397(−4)	292(−4)	219(−4)
6[−4]	0.5765	0.2766	0.1590	998(−4)	661(−4)	455(−4)	323(−4)	234(−4)	172(−4)
7[−4]	0.5370	0.2510	0.1412	870(−4)	567(−4)	385(−4)	269(−4)	192(−4)	140(−4)
8[−4]	0.5040	0.2300	0.1270	770(−4)	494(−4)	330(−4)	228(−4)	161(−4)	115(−4)
9[−4]	0.4758	0.2125	0.1153	688(−4)	436(−4)	288(−4)	196(−4)	137(−4)	970(−5)
1[−3]	0.4513	0.1976	0.1055	621(−4)	388(−4)	253(−4)	170(−4)	117(−4)	825(−5)
2[−3]	0.3084	0.1164	551(−4)	292(−4)	166(−4)	987(−5)	609(−5)	388(−5)	254(−5)
3[−3]	0.2394	816(−4)	355(−4)	175(−4)	927(−5)	518(−5)	303(−5)	183(−5)	113(−5)
4[−3]	0.1970	619(−4)	253(−4)	117(−4)	588(−5)	314(−5)	176(−5)	101(−5)	597(−6)
5[−3]	0.1677	493(−4)	190(−4)	839(−5)	403(−5)	208(−5)	111(−5)	614(−6)	349(−6)
6[−3]	0.1460	404(−4)	149(−4)	628(−5)	292(−5)	145(−5)	745(−6)	398(−6)	221(−6)
7[−3]	0.1292	340(−4)	120(−4)	486(−5)	219(−5)	105(−5)	521(−6)	271(−6)	147(−6)
8[−3]	0.1158	290(−4)	983(−5)	386(−5)	169(−5)	780(−6)	378(−6)	193(−6)	102(−6)
9[−3]	0.1047	252(−6)	821(−5)	313(−5)	133(−5)	595(−6)	282(−6)	141(−6)	729(−7)
1[−2]	955(−4)	221(−4)	695(−5)	258(−5)	106(−5)	464(−6)	215(−6)	106(−6)	532(−7)
2[−2]	487(−4)	831(−5)	205(−5)	609(−6)	203(−6)	751(−7)	291(−7)	118(−7)	505(−8)
3[−2]	308(−4)	426(−5)	888(−6)	224(−6)	662(−7)	210(−7)	717(−8)	264(−8)	102(−8)
4[−2]	216(−4)	253(−5)	456(−6)	104(−6)	269(−7)	767(−8)	242(−8)		
5[−2]	160(−4)	164(−5)	261(−6)	540(−7)	126(−7)	335(−8)			
6[−2]	124(−4)	112(−5)	162(−6)	304(−7)	655(−8)	163(−8)			
7[−2]	982(−5)	799(−6)	106(−6)	182(−7)	368(−8)				
8[−2]	797(−5)	587(−6)	724(−7)	114(−7)	219(−8)				
9[−2]	658(−5)	442(−6)	508(−7)	746(−8)	135(−8)				
1[−1]	552(−5)	340(−6)	365(−7)	505(−8)					
2[−1]	149(−5)	493(−7)	307(−8)						
3[−1]	592(−6)	126(−7)							
4[−1]	283(−6)	424(−8)							
5[−1]	151(−6)	171(−8)							
6[−1]	873(−7)								
7[−1]	534(−7)								
8[−1]	340(−7)								
9[−1]	223(−7)								
1	151(−7)								
2									
3									
4									
5									

(*continued*)

TABLE 7-2 (*continued*)

β	[2]									
u	1	2	3	4	5	6	7	8	9	10
1[−9]	4.2221	3.5486	3.1609	2.8898	2.6824	2.5151	2.3753	2.2557	2.1514	2.0591
2[−9]	3.8839	3.2168	2.8348	2.5689	2.3662	2.2034	2.0679	1.9523	1.8519	1.7633
3[−9]	3.6874	3.0249	2.6469	2.3846	2.1853	2.0257	1.8931	1.7804	1.6826	1.5967
4[−9]	3.5486	2.8898	2.5151	2.2557	2.0591	1.9019	1.7718	1.6612	1.5656	1.4816
5[−9]	3.4413	2.7858	2.4137	2.1568	1.9625	1.8075	1.6793	1.5706	1.4767	1.3944
6[−9]	3.3540	2.7012	2.3316	2.0769	1.8845	1.7313	1.6048	1.4978	1.4054	1.3246
7[−9]	3.2804	2.6302	2.2627	2.0099	1.8193	1.6677	1.5427	1.4371	1.3461	1.2666
8[−9]	3.2168	2.5689	2.2034	1.9523	1.7633	1.6132	1.4896	1.3853	1.2955	1.2171
9[−9]	3.1609	2.5151	2.1514	1.9019	1.7144	1.5656	1.4433	1.3401	1.2514	1.1741
1[−8]	3.1110	2.4671	2.1051	1.8572	1.6710	1.5234	1.4023	1.3002	1.2125	1.1361
2[−8]	2.7858	2.1568	1.8074	1.5706	1.3944	1.2561	1.1434	1.0492	0.9689	0.8995
3[−8]	2.5985	1.9801	1.6395	1.4103	1.2409	1.1087	1.0017	0.9128	0.8374	0.7725
4[−8]	2.4671	1.8572	1.5234	1.3002	1.1361	1.0087	0.9060	0.8211	0.7494	0.6879
5[−8]	2.3662	1.7633	1.4354	1.2171	1.0574	0.9339	0.8348	0.7531	0.6844	0.6257
6[−8]	2.2845	1.6878	1.3648	1.1508	0.9949	0.8747	0.7786	0.6996	0.6334	0.5770
7[−8]	2.2159	1.6246	1.3061	1.0959	0.9432	0.8260	0.7325	0.6559	0.5919	0.5375
8[−8]	2.1568	1.5706	1.2561	1.0492	0.8995	0.7849	0.6937	0.6192	0.5571	0.5045
9[−8]	2.1051	1.5234	1.2125	1.0087	0.8617	0.7494	0.6603	0.5877	0.5274	0.4763
1[−7]	2.0591	1.4816	1.1741	0.9731	0.8285	0.7183	0.6312	0.5603	0.5015	0.4519
2[−7]	1.7633	1.2171	0.9339	0.7531	0.6257	0.5305	0.4565	0.3974	0.3491	0.3091
3[−7]	1.5966	1.0716	0.8046	0.6368	0.5203	0.4345	0.3686	0.3166	0.2746	0.2402
4[−7]	1.4816	0.9731	0.7183	0.5603	0.4519	0.3728	0.3128	0.2659	0.2283	0.1978
5[−7]	1.3944	0.8995	0.6547	0.5045	0.4025	0.3287	0.2732	0.2302	0.1961	0.1685
6[−7]	1.3246	0.8413	0.6049	0.4613	0.3645	0.2952	0.2434	0.2035	0.1720	0.1468
7[−7]	1.2666	0.7935	0.5644	0.4264	0.3341	0.2685	0.2198	0.1825	0.1533	0.1300
8[−7]	1.2171	0.7531	0.5305	0.3974	0.3091	0.2466	0.2006	0.1655	0.1382	0.1166
9[−7]	1.1741	0.7183	0.5015	0.3728	0.2879	0.2283	0.1846	0.1515	0.1258	0.1056
1[−6]	1.1361	0.6879	0.4763	0.3516	0.2698	0.2127	0.1710	0.1396	0.1154	963(−4)
2[−6]	0.8995	0.5045	0.3287	0.2302	0.1685	0.1271	981(−4)	770(−4)	613(−4)	494(−4)
3[−6]	0.7725	0.4113	0.2571	0.1736	0.1230	900(−4)	675(−4)	516(−4)	401(−4)	315(−4)
4[−6]	0.6879	0.3516	0.2127	0.1396	963(−4)	689(−4)	505(−4)	378(−4)	288(−4)	222(−4)
5[−6]	0.6256	0.3091	0.1819	0.1166	787(−4)	552(−4)	397(−4)	292(−4)	219(−4)	166(−4)
6[−6]	0.5770	0.2768	0.1591	999(−4)	662(−4)	456(−4)	323(−4)	234(−4)	172(−4)	129(−4)
7[−6]	0.5375	0.2511	0.1413	871(−4)	568(−4)	385(−4)	269(−4)	192(−4)	140(−4)	103(−4)
8[−6]	0.5045	0.2302	0.1271	770(−4)	494(−4)	331(−4)	228(−4)	161(−4)	115(−4)	842(−5)
9[−6]	0.4763	0.2127	0.1154	689(−4)	436(−4)	288(−4)	196(−4)	137(−4)	970(−5)	700(−5)
1[−5]	0.4519	0.1978	0.1056	621(−4)	388(−4)	253(−4)	171(−4)	118(−4)	826(−5)	590(−5)
2[−5]	0.3091	0.1166	552(−4)	292(−4)	166(−4)	988(−5)	610(−5)	388(−5)	254(−5)	169(−5)
3[−5]	0.2402	818(−4)	356(−4)	175(−4)	929(−5)	519(−5)	304(−5)	184(−5)	114(−5)	715(−6)
4[−5]	0.1978	621(−4)	253(−4)	118(−4)	590(−5)	315(−5)	176(−5)	102(−5)	599(−6)	361(−6)
5[−5]	0.1685	494(−4)	191(−4)	842(−5)	405(−5)	208(−5)	112(−5)	616(−6)	350(−6)	206(−6)
6[−5]	0.1468	406(−4)	149(−4)	631(−5)	293(−5)	145(−5)	748(−6)	399(−6)	221(−6)	127(−6)
7[−5]	0.1300	342(−4)	120(−4)	489(−5)	220(−5)	105(−5)	524(−6)	272(−6)	148(−6)	824(−7)
8[−5]	0.1166	292(−4)	988(−5)	388(−5)	169(−5)	784(−6)	380(−6)	193(−6)	103(−6)	556(−7)
9[−5]	0.1056	253(−4)	826(−5)	315(−5)	133(−5)	599(−6)	283(−6)	142(−6)	733(−7)	386(−7)

(*continued*)

TABLE 7-2 (*continued*)

β / u	[2]									
	1	2	3	4	5	6	7	8	9	10
1[−4]	963(−4)	222(−4)	700(−5)	260(−5)	107(−5)	467(−6)	217(−6)	106(−6)	536(−7)	275(−7)
2[−4]	494(−4)	842(−5)	208(−5)	616(−6)	206(−6)	760(−7)	294(−7)	119(−7)	511(−8)	229(−8)
3[−4]	315(−4)	434(−5)	905(−6)	229(−6)	674(−7)	213(−7)	729(−8)	269(−8)	103(−8)	
4[−4]	222(−4)	260(−5)	467(−6)	106(−6)	275(−7)	785(−8)	248(−8)			
5[−4]	166(−4)	169(−5)	269(−6)	556(−7)	129(−7)	344(−8)				
6[−4]	129(−4)	117(−5)	168(−6)	314(−7)	677(−8)	168(−8)				
7[−4]	103(−4)	834(−6)	111(−6)	189(−7)	383(−8)					
8[−4]	841(−5)	616(−6)	759(−7)	119(−7)	229(−8)					
9[−4]	699(−5)	466(−6)	535(−7)	785(−8)	142(−8)					
1[−3]	590(−5)	361(−6)	386(−7)	535(−8)						
2[−3]	169(−5)	555(−7)	344(−8)							
3[−3]	713(−6)	149(−7)								
4[−3]	360(−6)	534(−8)								
5[−3]	205(−6)	228(−8)								
6[−3]	126(−6)	109(−8)								
7[−3]	821(−7)									
8[−3]	553(−7)									
9[−3]	384(−7)									
1[−2]	274(−7)									
2[−2]	226(−8)									

(*continued*)

TABLE 7-2 *(continued)*

β	[3]								
u	1	2	3	4	5	6	7	8	9
1[−9]	2.0591	1.4816	1.1741	0.9731	0.8285	0.7183	0.6312	0.5603	0.5015
2[−9]	1.7633	1.2171	0.9339	0.7531	0.6257	0.5305	0.4565	0.3974	0.3491
3[−9]	1.5967	1.0716	0.8046	0.6368	0.5203	0.4345	0.3686	0.3166	0.2746
4[−9]	1.4816	0.9731	0.7183	0.5603	0.4519	0.3728	0.3128	0.2659	0.2283
5[−9]	1.3944	0.8995	0.6547	0.5045	0.4025	0.3287	0.2733	0.2302	0.1961
6[−9]	1.3246	0.8413	0.6049	0.4613	0.3645	0.2952	0.2434	0.2035	0.1720
7[−9]	1.2666	0.7935	0.5644	0.4264	0.3341	0.2685	0.2198	0.1825	0.1533
8[−9]	1.2171	0.7531	0.5305	0.3974	0.3091	0.2466	0.2006	0.1655	0.1382
9[−9]	1.1741	0.7183	0.5015	0.3728	0.2879	0.2283	0.1846	0.1515	0.1258
1[−8]	1.1361	0.6879	0.4763	0.3516	0.2698	0.2127	0.1710	0.1396	0.1154
2[−8]	0.8995	0.5045	0.3287	0.2302	0.1685	0.1271	981(−4)	770(−4)	613(−4)
3[−8]	0.7725	0.4113	0.2571	0.1736	0.1230	900(−4)	675(−4)	516(−4)	401(−4)
4[−8]	0.6879	0.3516	0.2127	0.1396	963(−4)	689(−4)	505(−4)	378(−4)	288(−4)
5[−8]	0.6257	0.3091	0.1819	0.1166	787(−4)	552(−4)	397(−4)	292(−4)	219(−4)
6[−8]	0.5770	0.2768	0.1591	999(−4)	662(−4)	456(−4)	323(−4)	234(−4)	172(−4)
7[−8]	0.5375	0.2512	0.1413	871(−4)	568(−4)	385(−4)	269(−4)	192(−4)	140(−4)
8[−8]	0.5045	0.2302	0.1271	770(−4)	494(−4)	331(−4)	228(−4)	161(−4)	115(−4)
9[−8]	0.4763	0.2127	0.1154	689(−4)	436(−4)	288(−4)	196(−4)	137(−4)	970(−5)
1[−7]	0.4519	0.1978	0.1056	621(−4)	388(−4)	253(−4)	171(−4)	118(−4)	826(−5)
2[−7]	0.3091	0.1166	552(−4)	292(−4)	166(−4)	988(−5)	610(−5)	388(−5)	254(−5)
3[−7]	0.2402	818(−4)	356(−4)	175(−4)	929(−5)	519(−5)	304(−5)	184(−5)	114(−5)
4[−7]	0.1978	621(−4)	253(−4)	118(−4)	590(−5)	315(−5)	176(−5)	102(−5)	599(−6)
5[−7]	0.1685	494(−4)	191(−4)	842(−5)	405(−5)	208(−5)	112(−5)	616(−6)	350(−6)
6[−7]	0.1468	406(−4)	149(−4)	631(−5)	293(−5)	145(−5)	748(−6)	399(−6)	221(−6)
7[−7]	0.1300	342(−4)	120(−4)	489(−5)	220(−5)	105(−5)	524(−6)	272(−6)	148(−6)
8[−7]	0.1166	292(−4)	988(−5)	388(−5)	169(−5)	784(−6)	380(−6)	193(−6)	103(−6)
9[−7]	0.1056	253(−4)	826(−5)	315(−5)	133(−5)	599(−6)	284(−6)	142(−6)	733(−7)
1[−6]	963(−4)	222(−4)	700(−5)	260(−5)	107(−5)	467(−6)	217(−6)	106(−6)	536(−7)
2[−6]	494(−4)	842(−5)	208(−5)	616(−6)	206(−6)	760(−7)	294(−7)	119(−7)	511(−8)
3[−6]	315(−4)	434(−5)	905(−6)	229(−6)	674(−7)	213(−7)	729(−8)	269(−8)	103(−8)
4[−6]	222(−4)	260(−5)	467(−6)	106(−6)	275(−7)	785(−8)	248(−8)		
5[−6]	166(−4)	169(−5)	269(−6)	556(−7)	130(−7)	344(−8)			
6[−6]	129(−4)	117(−5)	168(−6)	315(−7)	678(−8)	168(−8)			
7[−6]	103(−4)	835(−6)	111(−6)	189(−7)	383(−8)				
8[−6]	842(−5)	616(−6)	760(−7)	119(−7)	229(−8)				
9[−6]	700(−5)	467(−6)	536(−7)	785(−8)	143(−8)				
1[−5]	590(−5)	361(−6)	387(−7)	535(−8)					
2[−5]	169(−5)	556(−7)	344(−8)						
3[−5]	715(−6)	149(−7)							
4[−5]	361(−6)	535(−8)							
5[−5]	206(−6)	229(−8)							
6[−5]	127(−6)	109(−8)							
7[−5]	824(−7)								
8[−5]	556(−7)								
9[−5]	386(−7)								
1[−4]	275(−7)								
2[−4]	229(−8)								
3[−4]									
4[−4]									
5[−4]									

(continued)

TABLE 7-2 (*continued*)

β	[4]									
u	1	2	3	4	5	6	7	8	9	10
1[−9]	0.4519	0.1978	0.1056	621(−4)	388(−4)	253(−4)	171(−4)	118(−4)	826(−5)	590(−5)
2[−9]	0.3091	0.1166	552(−4)	292(−4)	166(−4)	988(−5)	610(−5)	388(−5)	254(−5)	169(−5)
3[−9]	0.2402	818(−4)	356(−4)	175(−4)	929(−5)	519(−5)	304(−5)	184(−5)	114(−5)	715(−6)
4[−9]	0.1978	621(−4)	253(−4)	118(−4)	590(−5)	315(−5)	176(−5)	102(−5)	599(−6)	361(−6)
5[−9]	0.1685	494(−4)	191(−4)	842(−5)	405(−5)	208(−5)	112(−5)	616(−6)	350(−6)	206(−6)
6[−9]	0.1468	406(−4)	149(−4)	631(−5)	293(−5)	145(−5)	748(−6)	399(−6)	221(−6)	127(−6)
7[−9]	0.1300	342(−4)	120(−4)	489(−5)	220(−5)	105(−5)	524(−6)	272(−6)	148(−6)	824(−7)
8[−9]	0.1166	292(−4)	988(−5)	388(−5)	169(−5)	784(−6)	380(−6)	193(−6)	103(−6)	556(−7)
9[−9]	0.1056	253(−4)	826(−5)	315(−5)	133(−5)	599(−6)	284(−6)	142(−6)	733(−7)	387(−7)
1[−8]	963(−4)	222(−4)	700(−5)	260(−5)	107(−5)	467(−6)	217(−6)	106(−6)	536(−7)	275(−7)
2[−8]	494(−4)	842(−5)	208(−5)	616(−6)	206(−6)	760(−7)	294(−7)	119(−7)	511(−8)	229(−8)
3[−8]	315(−4)	434(−5)	905(−6)	229(−6)	674(−7)	213(−7)	729(−8)	269(−8)	103(−8)	
4[−8]	222(−4)	260(−5)	467(−6)	106(−6)	275(−7)	785(−8)	248(−8)			
5[−8]	166(−4)	169(−5)	269(−6)	556(−7)	130(−7)	344(−8)				
6[−8]	129(−4)	117(−5)	168(−6)	315(−7)	678(−8)	168(−8)				
7[−8]	103(−4)	835(−6)	111(−6)	189(−7)	383(−8)					
8[−8]	842(−5)	616(−6)	760(−7)	119(−7)	229(−8)					
9[−8]	700(−5)	467(−6)	536(−7)	785(−8)	143(−8)					
1[−7]	590(−5)	361(−6)	387(−7)	535(−8)						
2[−7]	169(−5)	556(−7)	344(−8)							
3[−7]	715(−6)	149(−7)								
4[−7]	361(−6)	535(−8)								
5[−7]	206(−6)	229(−8)								
6[−7]	127(−6)	109(−8)								
7[−7]	824(−7)									
8[−7]	556(−7)									
9[−7]	387(−7)									
1[−6]	275(−7)									
2[−6]	229(−8)									

Note: The numbers in parentheses are powers of 10 by which the other numbers are multiplied; for example, 2[−5] = 2.0×10^{-5} and [4] = 10^4.
Source: After Hantush (1961a).

The function $H(u, \beta)$ can be approximated by the following relations (Hantush, 1961a, 1964):

(a) For $u \geq 10^4 \beta^2$

$$H(u, \beta) \cong W(u) - \frac{4\beta}{(\pi u)^{1/2}} [0.258 + 0.693 \exp(-0.5u)] \tag{7-38}$$

(b) For u less than both $10^{-4}/\beta^2$ and $10^{-4}\beta^2$

$$H(u, \beta) \cong \frac{1}{2} \ln \left(\frac{0.044}{\beta^2 u} \right) \tag{7-39}$$

where $W(u)$ is the Theis well function defined by Eq. (4-25) of Chapter 4 and is tabulated in Table 4-1 of Chapter 4.

Rate of Leakage and Total Leakage Volume. For the conditions given by Eqs. (7-27) or (7-28), the rate of leakage $q_L(L^3/T)$ added to the main confined aquifer is given by (Hantush, 1960b)

$$q_L = Q \left\{ 1 - e^{nt} \operatorname{erfc} \left[(nt)^{1/2} \right] \right\} \tag{7-40}$$

where

$$n = \frac{T\lambda^2}{S} \tag{7-41}$$

and erfc(x) is the *complementary error function* and erfc$(x) = 1 - \operatorname{erf}(x)$ in which erf(x) is the *error function* (tabulated in Table 8-5 of Chapter 8). The variation of q_L/Q versus $\log(nt)$ is shown in Figure 7-5.

The total volume of leakage $V_L(L^3)$ from storage under continuous withdrawal conditions from the well within the range of time indicated is determined by integrating q_L, which is given by Eq. (7-40), with respect to time (t) from zero to t and the result is (Hantush, 1960b):

$$V_L = \int_0^t q_L \, dt = V \left[1 - \frac{2}{(n\pi t)^{1/2}} + \frac{q_L}{nQt} \right] \tag{7-42}$$

in which $V = Qt$ is the total volume withdrawn during the pumping test period. The variation of V_L versus $\log(nt)$ is shown in Figure 7-5.

7.2.4.2 Solution for Large Values of Time for Case 1.

The general solution in the Laplace domain as given by Eq. (7-23) remains the same for Case 1 (Figure 7-1). Therefore, its inverse was taken in a manner similar to that for small values of time (Section 7.2.4.1) and the results are presented below based on Hantush (1960b).

Drawdown Variation. For large values of t, which corresponds to small values of p, the hyperbolic cotangent terms in Eq. (7-23) may be approximated by

$$x \coth(x) \cong 1 + \frac{x^2}{3}, \qquad \text{for } 5x^2 < 1 \tag{7-43}$$

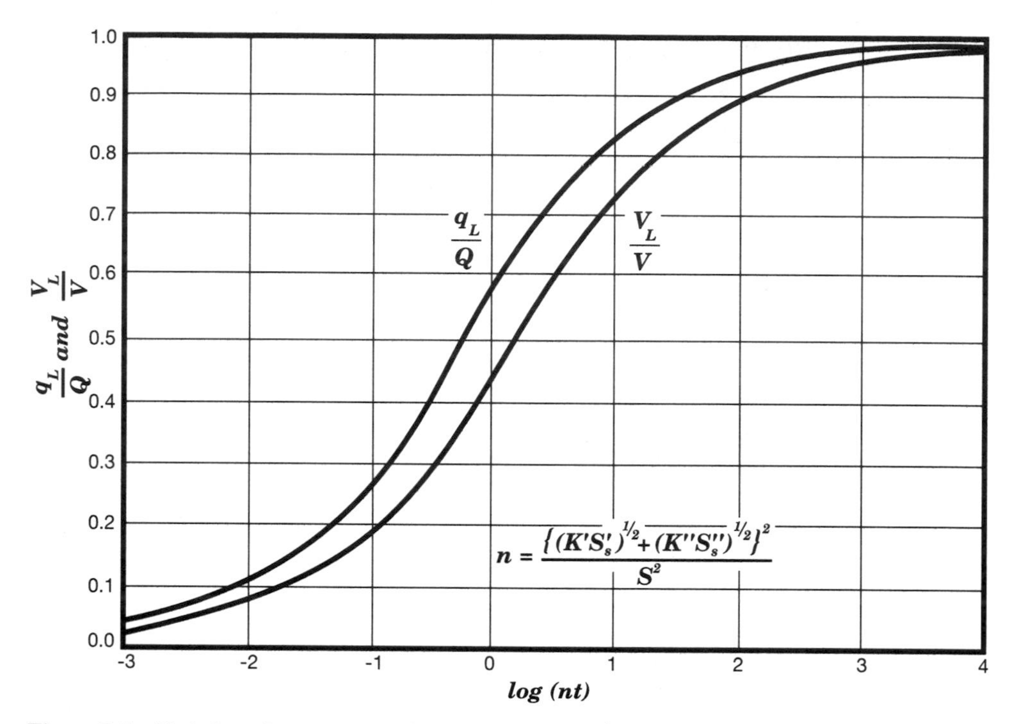

Figure 7-5 Variation of percentages of well discharge (q_L/Q) and total withdrawn volume (V_L/Q) from storage in the confining layers with time and aquifer parameters. (After Hantush, 1960b.)

Therefore, for

$$\frac{5b'S'p}{K'} < 1, \qquad \frac{5b''S''p}{K''} < 1 \tag{7-44}$$

Equation (7-23) may be approximated by

$$\bar{s}(r,p) = \frac{Q}{4\pi T}\int_0^\infty \frac{\overline{g_1}(p)}{y} \exp\left(-y - \frac{\alpha^2}{4y}\right) dy \tag{7-45}$$

where

$$\alpha = r\left(\frac{K'}{b'T} + \frac{K''}{b''T}\right)^{1/2} \tag{7-46}$$

$$\overline{g_1}(p) = \frac{1}{p}\exp\left(-\frac{r^2 S\delta_1}{4Ty}p\right) \tag{7-47}$$

and

$$\delta_1 = 1 + \frac{S' + S''}{3S} \tag{7-48}$$

By taking the inverse Laplace transform of Eq. (7-45), finally Hantush (1960b) determined the following solution for the drawdown distribution:

$$s(r,t) = \frac{Q}{4\pi T} W(u\delta_1, \alpha) \tag{7-49}$$

where

$$W(u\delta_1, \alpha) = \int_{u\delta_1}^{\infty} \frac{1}{y} \exp\left(-y - \frac{\alpha^2}{4y}\right) dy \tag{7-50}$$

and $W(u\delta_1, \alpha)$ is the *Hantush and Jacob well function for leaky aquifers* [see Eq. (6-24) of Chapter 6] and its tabulated values are given in Table 6-1 of Chapter 6. Eq. (7-50) is valid under the following conditions:

$$\frac{5b'S'}{K'} < t, \qquad \frac{5b''S''}{K''} < t \tag{7-51}$$

which are, respectively, the inverse Laplace transforms of the expressions given by Eq. (7-44).

Rate of Leakage and Total Leakage Volume. For the conditions of Eqs. (7-44) or (7-51), the rate of leakage $q_L(L^3/T)$ added to the main confined aquifer is given by (Hantush, 1960b)

$$q_L = Q\left[1 - \frac{1}{\delta_1}\exp\left(-\frac{\alpha^2 tT}{Sr^2\delta_1}\right)\right] \tag{7-52}$$

where α and δ_1 are defined by Eqs. (7-46) and (7-48), respectively.

The total volume of leakage can similarly be determined as in Section 7.2.4.1. Integrating q_L as given by Eq. (7-52) with respect to time from zero to t gives the total volume of leakage:

$$V_L = Q\left\{t + \frac{Sr^2}{T\alpha^2}\left[\exp\left(-\frac{\alpha^2 Tt}{Sr^2\delta_1}\right) - 1\right]\right\} \tag{7-53}$$

7.2.4.3 Solution for Large Values of Time for Case 2. The general solution in the Laplace domain as given by Eq. (7-23) remains the same for Case 2 (Figure 7-2). As a result, its inverse was taken in a manner similar to that for small values of time (Section 7.2.4.1) and the results are presented below based on Hantush (1960b).

Drawdown Variation. Hantush (1960b) has not presented the intermediate steps regarding the inverse Laplace transform of Eq. (7-23) corresponding to Case 2 and only presented the final result for the drawdown variation. In the process of the solution the approximation

$$\tanh(x) \approx x, \qquad \text{for } x^2 < 0.10 \tag{7-54}$$

was made in taking the inverse Laplace transform of Eq. (7-23), and this approximation is valid for the following conditions:

$$\frac{10b'S'p}{K'} < 1, \qquad \frac{10b''S''p}{K''} < 1 \tag{7-55}$$

or

$$\frac{10b'S'}{K'} < t, \qquad \frac{10b''S''}{K''} < t \tag{7-56}$$

and the corresponding solution for drawdown distribution is

$$s(r,t) = \frac{Q}{4\pi T} W(u\delta_2) \tag{7-57}$$

where

$$\delta_2 = 1 + \frac{S' + S''}{S} \tag{7-58}$$

and $W(u\delta_2)$ is the *exponential integral* or *Theis well function* for a single confined aquifer as defined by Eq. (4-25) of Chapter 4 and is tabulated in Table 4-1 of Chapter 4.

Rate of Leakage and Total Leakage Volume. For the conditions of Eqs. (7-55) or (7-56), the rate of leakage $q_L(L^3/T)$ added to the main confined aquifer is given by (Hantush, 1960b)

$$q_L = Q\frac{S' + S''}{S + S' + S''} \tag{7-59}$$

The total volume of leakage can similarly be determined as in Section 7.2.4.1. Integrating q_L as given by Eq. (7-59) with respect to time from zero to t gives the total volume of leakage

$$V_L = Qt\frac{S' + S''}{S + S' + S''} \tag{7-60}$$

7.2.4.4 Solution for Large Values of Time for Case 3. The general solution in the Laplace domain as given by Eq. (7-23) remains the same for Case 3 (Figure 7-3) as well and its inverse was taken in a manner similar to that for small values of time (Section 7.2.4.1) and the results are presented below based on Hantush (1960b). For this case, Hantush proposed that the larger of the two values of $5b'S'/K'$ and $10b''S''/K''$ can be taken as the lower limit of the time range, and the results are presented below.

Drawdown Variation. Hantush (1960b) has not presented the intermediate steps regarding the inverse Laplace transform of Eq. (7-23) corresponding to Case 3 (Figure 7-3). Hantush only presented the final result for the drawdown variation as

$$s(r,t) = \frac{Q}{4\pi T} W\left[u\delta_3, r\left(\frac{K'}{b'T}\right)^{1/2}\right] \tag{7-61}$$

where

$$\gamma = r\left(\frac{K'}{b'T}\right)^{1/2} \tag{7-62}$$

As mentioned in Section 7.2.4.2, $W(u\delta_3, \gamma)$ is the *Hantush and Jacob well function for leaky aquifers* [see Eq. (6-24) of Chapter 6] and its tabulated values are given in Table 6-1 of Chapter 6 and

$$\delta_3 = 1 + \frac{S'}{3S} + \frac{S''}{S} \tag{7-63}$$

Rate of Leakage and Total Leakage Volume. The rate of leakage $q_L(L^3/T)$ added to the main confined aquifer is given by (Hantush, 1960b) as

$$q_L = Q\left[1 - \frac{1}{\delta_3}\exp\left(-\frac{K't}{Sb'\delta_3}\right)\right] \tag{7-64}$$

The total volume of leakage can similarly be determined as in Section 7.2.4.1. Integrating q_L as given by Eq. (7-64) with respect to time from zero to t gives the total volume of leakage

$$V = Q\left\{t + \frac{Sb'}{K'}\left[\exp\left(-\frac{K't}{Sb'\delta_3}\right) - 1\right]\right\} \tag{7-65}$$

7.2.5 Solutions for Special Cases

The drawdown equation for a single confined aquifer (Theis equation) and the drawdown equation for a leaky confined aquifer without storage of the confining layer (Hantush and Jacob equation) can be shown to be special cases of the models presented in the previous sections. These are shown in the following paragraphs.

If $K' = K'' = 0$, the confined aquifers between the confining layers shown in Figures 7-1, 7-2, and 7-3 reduce to single nonleaky confined aquifers. Substituting these values into Eqs. (7-35) and (7-36) gives $\beta = 0$, and with this value the complementary error function in Eq. (7-34) is $\text{erfc}(0) = 1$. Therefore, Eq. (7-34), which is valid for the three cases (Case 1, Case 2, and Case 3) for small values of time, reduces to

$$H(u, 0) = \int_{y=u}^{\infty} \frac{e^{-y}}{y}\, dy \tag{7-66}$$

which exactly the same as the Theis well function as given by Eq. (4-25) of Chapter 4. Substitution of $H(u, \beta) = W(u)$ into Eq. (7-32) gives

$$s(r, t) = \frac{Q}{4\pi T} W(u) \tag{7-67}$$

This is exactly the form of the Theis equation for drawdown distribution in a nonleaky confined aquifer (Theis, 1935) given by Eq. (4-24) of Chapter 4.

It can easily be observed that substitution of $K' = K'' = 0$ into the drawdown equations for large values of time, which are expressed by Eqs. (7-49), (7-57), and (7-61), for Case 1, Case 2, and Case 3, respectively, gives exactly the same form of the Theis equation.

If $K'' = 0$ and $S' = S'' = 0$ in the hydrogeologic system of Case 1 (Figure 7-1) or Case 3 (Figure 7-3), the resulting system reduces to a single leaky confined aquifer. The limit of Eq. (7-49) or (7-61), as K'', S', and S'' approach to zero, will be expressed as (Hantush, 1960b)

$$s(r, t) = \frac{Q}{4\pi T} W\left(u, \frac{r}{B}\right) \tag{7-68}$$

where B is given by Eq. (6-11) of Chapter 6. Equation (7-68) is the Hantush and Jacob formula (Hantush and Jacob, 1955) and given by Eq. (6-23) of Chapter 6.

7.2.6 Evaluation of the Solutions

In the previous sections, Hantush's modified theory for leaky confined aquifers (Hantush, 1960b) is presented in detail. As mentioned above, the main difference with the Hantush and Jacob (1955) model for leaky confined aquifers is that the modified theory takes into account the storage effects of the confining layers and introduces additional parameters in the models.

Hantush (1960b) concluded that if the ratio between the storage coefficients of the confining layers and the main confined aquifer is small ($S'/S \leq 0.01$), the storage effects of the confining layers on the drawdown in the main confined aquifer is very small. Therefore, the limiting drawdown equations, namely Hantush and Jacob (1955) and Theis (1935), can be used to accurately determine the drawdown variation in the main confined aquifer. If S'/S is relatively greater, the discrepancy becomes significant.

In Figure 7-6 (Hantush, 1960b) the dimensionless drawdowns are compared for no-leakage (Curve I), leakage without storage in a finite confining layer (Curve II), leakage with storage in an infinite confining layer (Curve III), and leakage with storage in a finite confining layer (Curve IV). Of these four curves, especially the ones corresponding to finite confining layers (Curve II and Curve IV) from a practical standpoint are important. Strictly speaking, Curve II and Curve IV correspond to Hantush and Jacob (1955) model (HJ55) and Hantush (1960b) model (H60), respectively. As can be seen from Figure 7-6, the H60 model predicts less dimensionless drawdown than HJ55 model does during a certain period of time and afterwards they become approximately the same. From Figure 7-6, the following equations for a certain time (t) can be written:

$$s_{D_{HJ55}} = \frac{4\pi T_{HJ55} s_{HJ55}}{Q} \qquad (7\text{-}69)$$

$$s_{D_{H60}} = \frac{4\pi T_{H60} s_{H60}}{Q} \qquad (7\text{-}70)$$

where $s_{D_{HJ55}}$ and $s_{D_{H60}}$ are the dimensionless drawdowns that correspond to HJ55 and H60 models, respectively. It is clear from Figure 7-6 that $s_{D_{HJ55}}$ is greater than $s_{D_{H60}}$; therefore, from Eqs. (7-69) and (7-70) we obtain

$$T_{HJ55} s_{HJ55} > T_{H60} s_{H60} \qquad (7\text{-}71)$$

In the HJ55 model, the leakage rate through the confining layer is constant because of the neglect of its storage effects. This, obviously, corresponds a steady recharge to the aquifer. During early time periods, the main contribution to the pumping well comes from storage, and in this period the HJ55 model introduces less contribution to the well by storage from the confining layer. In other words, the HJ55 model introduces less recharge through the confining layer than the recharge of the H60 model. Less recharge corresponds more drawdown. In other words, s_{HJ55} becomes greater than s_{H60}. Because of this fact and the condition expressed by Eq. (7-71) we obtain

$$T_{HJ55} > T_{H60} \qquad (7\text{-}72)$$

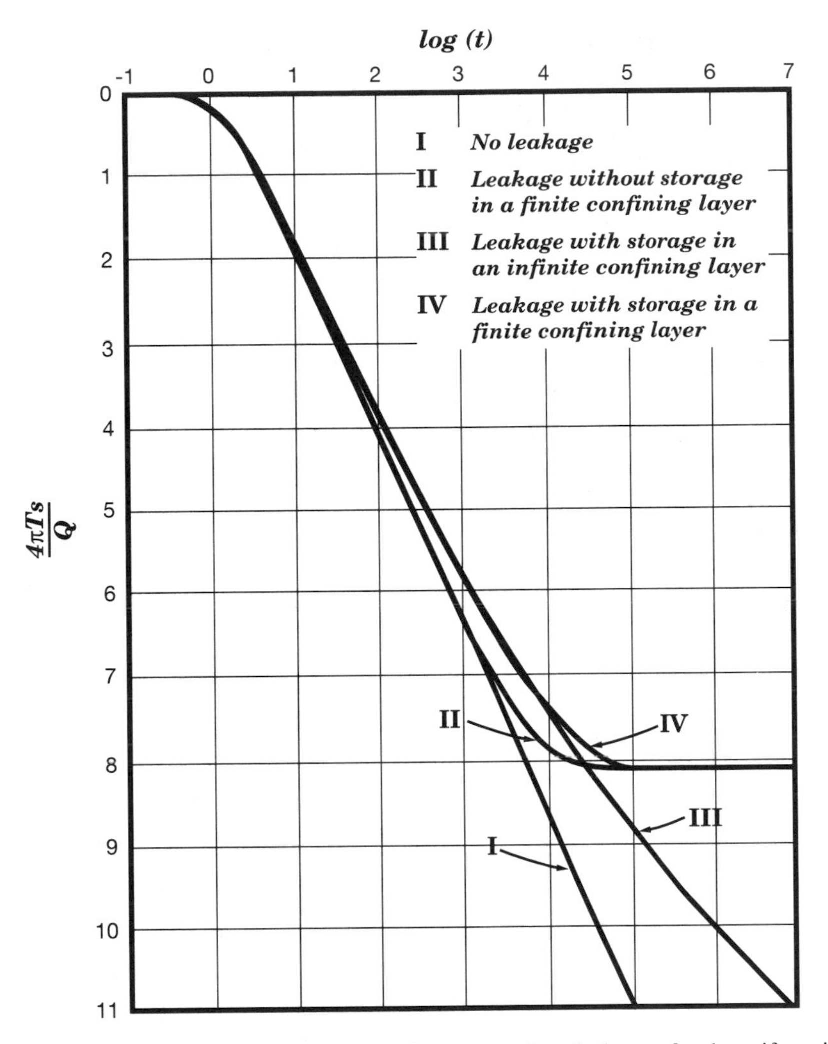

Figure 7-6 Comparison of drawdown versus time curves in a leaky confined aquifer with and without storage in the confining layers. (After Hantush, 1960b.)

Because in both model the aquifer thickness (b) is the same, the HJ55 model predicts greater horizontal hydraulic conductivity (K_h) value than the H60 model does. This important conclusion is shown by an example in Section 7.2.7.1.

7.2.7 Application to Pumping Test Data Analysis and Leakage Rates

The equations derived from Hantush's modified theory for leaky confined aquifers (Hantush, 1960b) have some important potential applications for the evaluation of aquifer and confining layers parameters as well as the determination of the leakage rates through the confining layers, and they are presented in the following sections.

7.2.7.1 Methods for Pumping Test Data Analysis.

The solutions presented in the previous sections can be used for the analysis of pumping test data to determine the parameters of confined aquifers and their confining layers under small and large values of time pumping conditions. These methods are presented in the following sections for both small and large values of time pumping conditions.

Pumping Test Data Analysis for Small Values of Time. The type-curve method can be applied for the determination of hydrogeologic parameters of a main confine aquifer and its confining layers. If the total pumping period (t) is less than $b'S'/(10K')$ and $b''S''/(10K'')$, the analysis methods for the three cases (Case 1, Case 2, and Case 3) are the same. As mentioned above, this is based on the fact that the drawdown equation for the three cases are the same. The details for the solution are presented in Section 7.2.4.1. The type-curve method based on the modified theory of Hantush (Hantush, 1960b) is presented below in a stepwise manner:

Step 1: Plot a family of type curves of $1/u$ versus $H(u, \beta)$ on log–log paper using the data in Table 7-2.

Step 2: On another sheet of log–log paper with the scale as the first, plot the observed drawdown data of s versus t.

Step 3: Keeping the coordinate axes parallel at all times, superimpose the two sheets until the best-fit-of-data curve and one of the family of type curves are obtained. Record the values of β for each well data.

Step 4: Select a common point, the match point, arbitrarily chosen on the overlapping part of the curve, or even anywhere on the overlapping portion of the sheets. Read the corresponding values of $1/u$, $H(u, \beta)$, s, and t.

Step 5: Substitute the values of $H(u, \beta)$ and s, along with the known value of Q, into Eq. (7-32) and solve for T for each observation well. Calculate the average of T values.

Step 6: Substitute the values of $1/u$, t, r, and T (from the previous step) into Eq. (7-33) and solve for S for each observation well. Calculate the average of S values.

Step 7: Substitute the values of β, T, S, b', and b'' into Eq. (7-36) and generate the equations that have the products of $(K'S')$ and $(K''S'')$ as the unknowns and solve them for the same products. If there are more than two pairs, find the unknowns for each pair and then take the arithmetic averages. If there is one confining layer, use Eq. (7-37) for the determination of the product of $(K'S')$.

Step 8: If the hydraulic conductivities of the confining layers (K' and K'') are known from laboratory tests or other pumping test data analysis methods, determine the values of S' and S'' from the known values of the products $(K'S')$ and $(K''S'')$. If the values of consolidation coefficient (c_v) are known, calculate the specific storage coefficients from $S_s = K_v/c_v$ [Eq. (2-71) of Chapter 2 with $K_v = K_z$] and compare them with the previously determined values.

Step 9: Calculate the values of $b'S'/(10K')$ and $b''S''/(10K'')$ and check if these values are higher than the maximum t value of the drawdown versus time data for each well. If these conditions are satisfied, the hydrogeologic parameters determined during the previous steps are acceptable. If these conditions are not satisfied, apply one of the methods presented below that is appropriate to the hydrogeologic system.

Pumping Test Data Analysis for Large Values of Time. As mentioned in Section 7.2.4, for each case there is a unique solution regarding the drawdown equation. Therefore, the pumping test data analysis method for each case depends on the mathematical form of the drawdown solution. The methods are presented in the following sections for each case.

DATA ANALYSIS FOR CASE 1. If the hydrogeologic system under question is the one shown in Figure 7-1, the corresponding large values of time solution can be used under the conditions that the time (t) is greater than $5b'S'/K'$ and $5b''S''/K''$. A method of data analysis similar to that of Walton's type-curve method (Chapter 6, Section 6.3.1) can be used because the function $W(u\delta_1, \alpha)$ is equivalent to the well function for leaky confined aquifers and is known as *Hantush and Jacob well function for leaky aquifers* (see Section 6.2.4.2 of Chapter 6). The steps of the method are presented below.

Step 1: Plot a family of type curves of $1/(u\delta_1)$ versus $W(u\delta_1, \alpha)$ on log–log paper using the data in Table 6-1 of Chapter 6. One should bear in mind that $u\delta_1$ and α correspond to u and r/B in Table 6-1 of Chapter 6, respectively.

Step 2: On another sheet of log–log paper with the same scale as the first, plot the observed drawdown (s) data versus time (t) for each observation well.

Step 3: Keeping the coordinate axes parallel at all times, superimpose the two sheets of each well until the best-fit-of-data curve and one of the family of type curves are obtained.

Step 4: Select a common point on overlapping portion of the sheets and read the corresponding values of α, $W(u\delta_1, \alpha)$, $1/(u\delta_1)$, s, and t.

Step 5: Substitute the values of $W(u\delta_1, \alpha)$ and s, along with the known value of Q, into Eq. (7-49) and solve for T.

Step 6: Multiplying the both sides of Eq. (7-33) by δ_1 and using Eq. (7-48) gives

$$S + \frac{1}{3}S' + \frac{1}{3}S'' = (u\delta_1)\frac{4Tt}{r^2} \tag{7-73}$$

Using the values of $(u\delta_1)$, T, t, and r from the curve matching of each well, calculate the right-hand side of Eq. (7-73) and take the arithmetic average. If the vertical hydraulic conductivities (K' and K'') and consolidation coefficients (c'_v and c''_v) are known, determine the storage coefficients (S' and S'') from Eq. (2-71) of Chapter 2. Substitute the values of S' and S'' into Eq. (7-73) and solve it for S.

Step 7: Equation (7-46) can also be expressed as

$$\frac{K'}{b'} + \frac{K''}{b''} = \frac{\alpha^2 T}{r^2} \tag{7-74}$$

Using the known values of α, T, and r, calculate the right-hand side of Eq. (7-74) for each well and take the arithmetic average of all values. This yields an equation in which K' and K'' are unknown. If K' or K'' is known by laboratory measurements or other methods, calculate K' or K'' from Eq. (7-74). If the values of both (K' and K'') are known from other methods, calculate the left-hand side of Eq. (7-74) and compare it with its right-hand side.

Step 8: Calculate $5b'S'/K'$ and $5b''S''/K''$ and check if the values of these products are higher than the minimum t value of the drawdown versus time data for each well. If these conditions are not satisfied, apply the pumping test data analysis method for small values of time presented previously.

DATA ANALYSIS FOR CASE 2. If the hydrogeologic system under question can be represented with the one shown in Figure 7-2, the corresponding large values of time solution can be used under the conditions that the time (t) is greater than $10b'S'/K'$ and $10b''S''/K''$. A method of data analysis similar to that of Theis' type curve method (Chapter 4, Section 4.2.7.1) can be used because the function $W(u\delta_2)$ is virtually equivalent to the well function for nonleaky confined aquifers and is known as *Theis well function* (see Section 4.2.4.2 of Chapter 4). As can be seen from Section 7.2.4.3, the equations for Case 2 do not include K' and K'' as variables. The steps of the method are presented below.

Step 1: Plot the type curve of $1/(u\delta_2)$ versus $W(u\delta_2)$ on log–log paper using the data in Table 4-1 of Chapter 4. One should bear in mind that $u\delta_1$ corresponds to u in Table 4-1 of Chapter 4.

Step 2: On another sheet of log–log paper with the same scale as the first, plot the observed drawdown (s) data versus t/r^2 for each observation well.

Step 3: Keeping the coordinate axes parallel at all times, superimpose the two sheets of each well until the best-fit-of-data curve and type curve are obtained. Repeat this for each well.

Step 4: Select a common point on overlapping portion of the sheets and read the corresponding values of $1/(u\delta_2)$, $W(u\delta_2)$, s, and t/r^2.

Step 5: Substitute the values of $W(u\delta_2)$ and s, along with the known value of Q, into Eq. (7-57) and solve for T.

Step 6: Multiplying the both sides of Eq. (7-33) by δ_2 and using Eq. (7-58) gives

$$S + S' + S'' = (u\delta_2)\frac{4Tt}{r^2} \tag{7-75}$$

Using the known values of $(u\delta_2)$, T, and t/r^2 for each well, calculate the right-hand side of Eq. (7-75) and take its arithmetic average. If the vertical hydraulic conductivities (K' and K'') and consolidation coefficients (c'_v and c''_v) are known, determine the storage coefficients (S' and S'') from Eq. (2-71) of Chapter 2. Substitute the values of S' and S'' into Eq. (2-75) and solve it for S, the storage coefficient of the main confined aquifer.

Step 7: Calculate $10b'S'/K'$ and $10b''S''/K''$ and check if the values of these products are greater than the minimum t value of the drawdown versus time data for each well. If these conditions are not satisfied, apply the pumping test data analysis method for small values of time presented previously.

DATA ANALYSIS FOR CASE 3. If the hydrogeologic system under question can be represented with the one shown in Figure 7-3, the corresponding large values of time solution can be used under the conditions that the time (t) is greater than $5b'S'/K'$ and $10b''S''/K''$. A method of data analysis similar to that of Walton's type-curve method (Chapter 6, Section 6.3.1) can be used because the function $W(u\delta_3, \gamma)$ is virtually equivalent to the well function for leaky confined aquifers and is known as the *Hantush and Jacob well function for leaky aquifers* (see Section 6.2.4.2 of Chapter 6). As can be seen from Section 7.2.4.4, the equations for Case 3 include only K' and do not include K'' as a variable. The steps of the method are presented below.

Step 1: Plot a family of type curves of $1/(u\delta_3)$ versus $W(u\delta_3, \gamma)$ on log–log paper using the data in Table 6-1 of Chapter 6. One should bear in mind that $u\delta_3$ and γ, expressed by Eq. (7-61), correspond to u and r/B in Table 6-1 of Chapter 6, respectively.

Step 2: On another sheet of log–log paper with the same scale as the first, plot the observed drawdown (s) data versus time (t) for each observation well.

Step 3: Keeping the coordinate axes parallel at all times, superimpose the two sheets of each well until the best-fit-of-data curve and one of the family of type curves is obtained.

Step 4: Select a common point on overlapping portion of the sheets and read the corresponding values of γ, $W(u\delta_3, \gamma)$, $1/(u\delta_3)$, s, and t.

Step 5: Substitute the values of $W(u\delta_3, \gamma)$ and s, along with the known value of Q, into Eq. (7-61) and solve for T.

Step 6: Multiplying both sides of Eq. (7-33) by δ_3 and using Eq. (7-63) gives

$$S + \frac{1}{3}S' + S'' = (u\delta_3)\frac{4Tt}{r^2} \tag{7-76}$$

Using the known values of $(u\delta_3)$, T, t, and r, calculate the right-hand side of Eq. (7-76) for each well and take the arithmetic average of all resulting values. This results an equation in which S, S' and S'' are unknown. If the vertical hydraulic conductivities (K' and K'') and consolidation coefficients (c'_v and c''_v) are known, determine the storage coefficient (S' and S'') from Eq. (2-71) of Chapter 2. Substitute the values of S' and S'' into Eq. (7-76) and solve it for S, the storage coefficient of the main confined aquifer.

Step 7: From Eq. (7-62) we obtain

$$K' = \gamma^2\frac{b'T}{r^2} \tag{7-77}$$

Substitute the values of γ, b', T, and r in Eq. (7-77) and calculate the value of K' for each observation well. Then, take the average of all K' values.

Step 8: Calculate $5b'S'/K'$ and $10b''S''/K''$ and check if the values of these products are greater than the minimum t value of the drawdown versus time data for each well. If these conditions are not satisfied, apply the pumping test data analysis method for small values of time presented previously.

Example 7-1: This example is adapted from Sheahan (1977). A pumping test was conducted for the aquifer system shown in Figure 7-7 to determine the hydrogeologic parameters. The geometry of the well locations is shown in Figure 7-8. The pumped well (PW) was screened in the second confined aquifer (CA2) and its rate was 436 m^3/day. Water levels were monitored in both the first confined aquifer (CA1) and CA2 during the pumping test. The drawdown data obtained from the observation wells TH/OW-1 and TH/OW-2 screened in CA2 are presented in Table 7-3 and Table 7-4, respectively. During the pumping test for CA2, the drawdowns in CA1 were also recorded. The records showed that pumping conditions in CA2 created drawdown in CA1. Previous tests showed that the third aquifer (CA3) was a confined aquifer and was not received drawdown because of pumping in CA2. Using the given drawdown data, do the following:

(a) Evaluate the information gathered during the tests for the aquifers and confining layers.

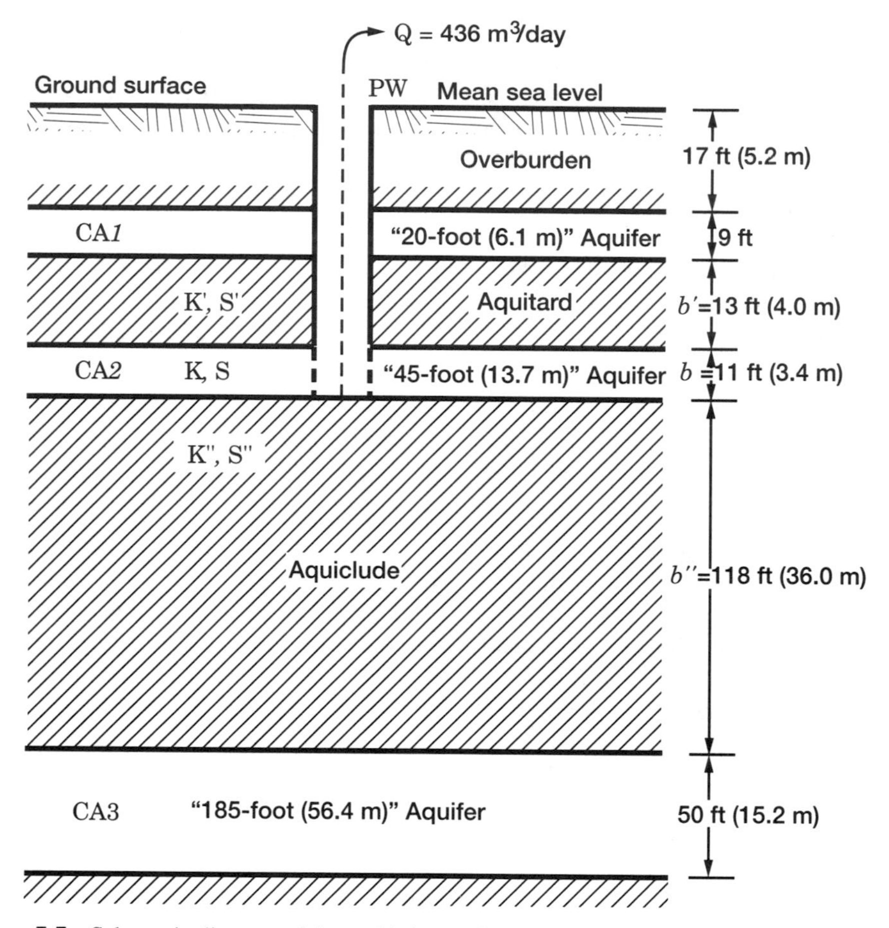

Figure 7-7 Schematic diagram of the multiple aquifers system for Example 7-1. (After Sheahan, 1977.)

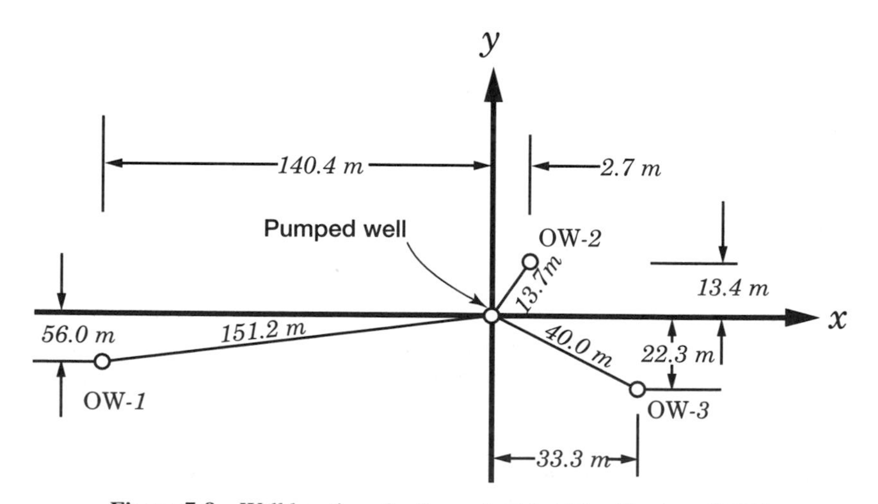

Figure 7-8 Well locations for Example 7-1. (After Sheahan, 1977.)

TABLE 7-3 Drawdown Versus Time Data at r = 151.2 m for TH/OW-1 for Example 7-1

t (min)	s (m)
45	0.05
90	0.12
205	0.25
243	0.28
284	0.29
298	0.31
318	0.32
355	0.34
387	0.34
590	0.43
700	0.45
738	0.46
930	0.49
1330	0.54

TABLE 7-4 Drawdown Versus Time Data at r = 40.0 m for TH/OW-3 for Example 7-1

t (min)	s (m)
1.1	0.15
1.5	0.19
2.0	0.25
3.6	0.49
4.0	0.62
4.5	0.69
5.0	0.74
5.9	0.79
7.0	0.81
8.0	0.84
10.0	0.88
15.0	0.81
20.0	0.90
60.0	1.05
81.0	1.14
118.0	1.17
150.0	1.22
220.0	1.34
265.0	1.33
435.0	1.46
610.0	1.55
675.0	1.54
1073.0	1.62
1400.0	1.68

(b) What are the possible pumping test data analysis methods for this hydrogeologic system and elaborate these methods for the aquifers and confining layers point of view.

(c) Determine the hydrogeologic parameters of CA2 and its confining layer and compare the results based on Hantush (1960b) and Hantush and Jacob (1955) and explain the reasons of differences.

Solution:

(a) The pumping test results can be interpreted as follows: (1) CA2 was receiving recharge from CA1 through the confining layer (or aquitard) above CA2; (2) CA3 was not hydraulically connected to CA1 and CA2; and (3) the formation between CA2 and CA3, called *aquiclude*, is less pervious than the aquitard above CA2. Under these conditions, the bottom of CA2 can be assumed to be impervious and the corresponding model for the leaky aquifer, which is composed of CA2 and its overlain confining layer, is Eq. (7-34) with λ given by Eq. (7-37). This hydrogeologic system also corresponds to the Hantush and Jacob model as given in Section 6.2.4.1 of Chapter 6.

(b) The pumping test data analysis method for small values of time presented above is the method to be used to determine the hydrogeologic parameters of CA2 and its confining layer. However, this method gives only the product $K'S'$ for the confining layer, and an additional method is required to determine the hydraulic conductivity,

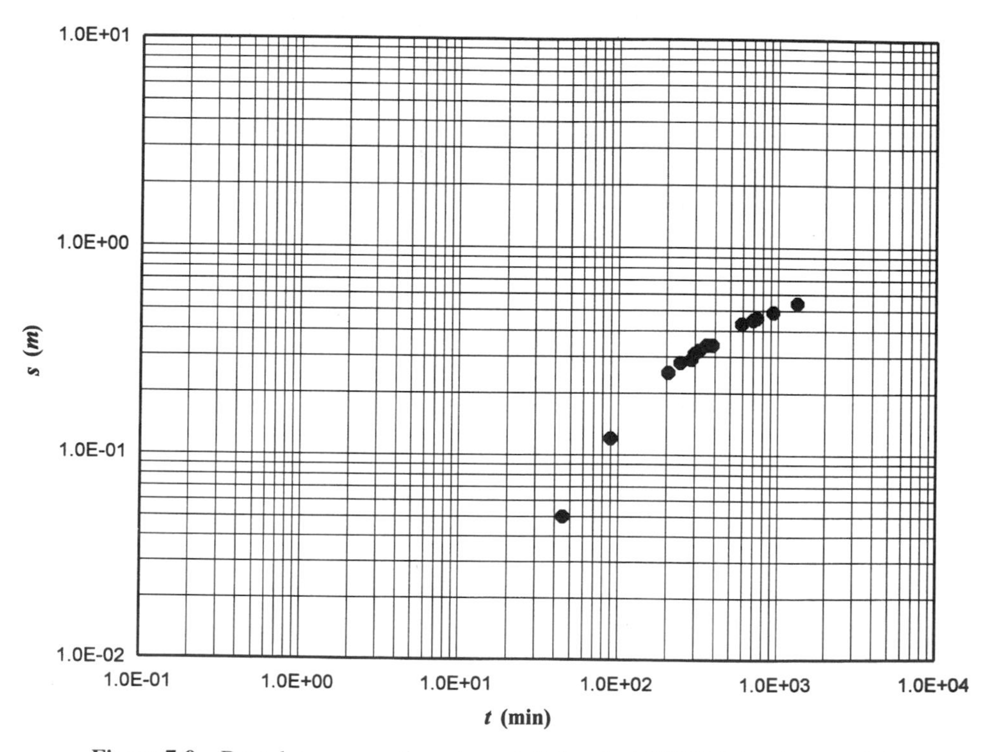

Figure 7-9 Drawdown versus time at $r = 151.2$ m for TH/OW-1 for Example 7-1.

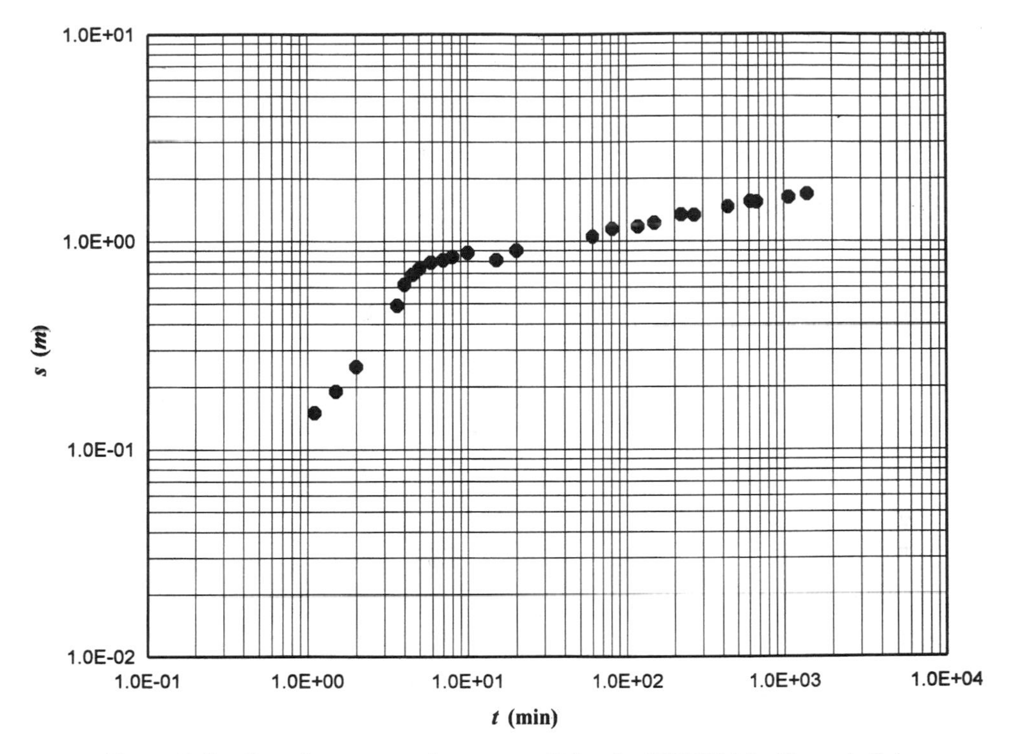

Figure 7-10 Drawdown versus time at $r = 40.0$ m for TH/OW-3 for Example 7-1.

K', of the confining layer because its value is not given. One approach is to use the Walton type-curve method that is based on the Hantush and Jacob model (see Section 6.3.1 of Chapter 6).

(c) Based on the answer of (b), first the Walton type-curve method will be applied to determine the hydrogeologic parameters of CA2 and its confining layer. Then the type-curve method based on Hantush (1960b) will be applied.

Using the values in Table 7-3 and Table 7-4, the log–log drawdown versus time graphs are shown in Figure 7-9 and Figure 7-10 for TH/OW-1 and TH/OW-3, respectively.

Application of Walton Type-Curve Method. The Walton type-curve method will be applied in accordance with the steps given Section 6.3.1 of Chapter 6.

Step 1: A family of type curves of $W(u, r/B)$ versus $1/u$ on log–log paper for different r/B values are given in Figure 6-2 of Chapter 6.

Step 2: The log–log graphs for observation wells TH/OW-1 and TH/OW-3 are given in Figure 7-9 and Figure 7-10, respectively.

Step 3: The superimposed sheets are shown Figure 7-11 and Figure 7-12 for TH/OW-1 and TH/OW-3, respectively. Comparison shows that the plotted points fall along the curves $r/B = 0.3$ and 0.03 for TH/OW-1 and Th/OW-3, respectively.

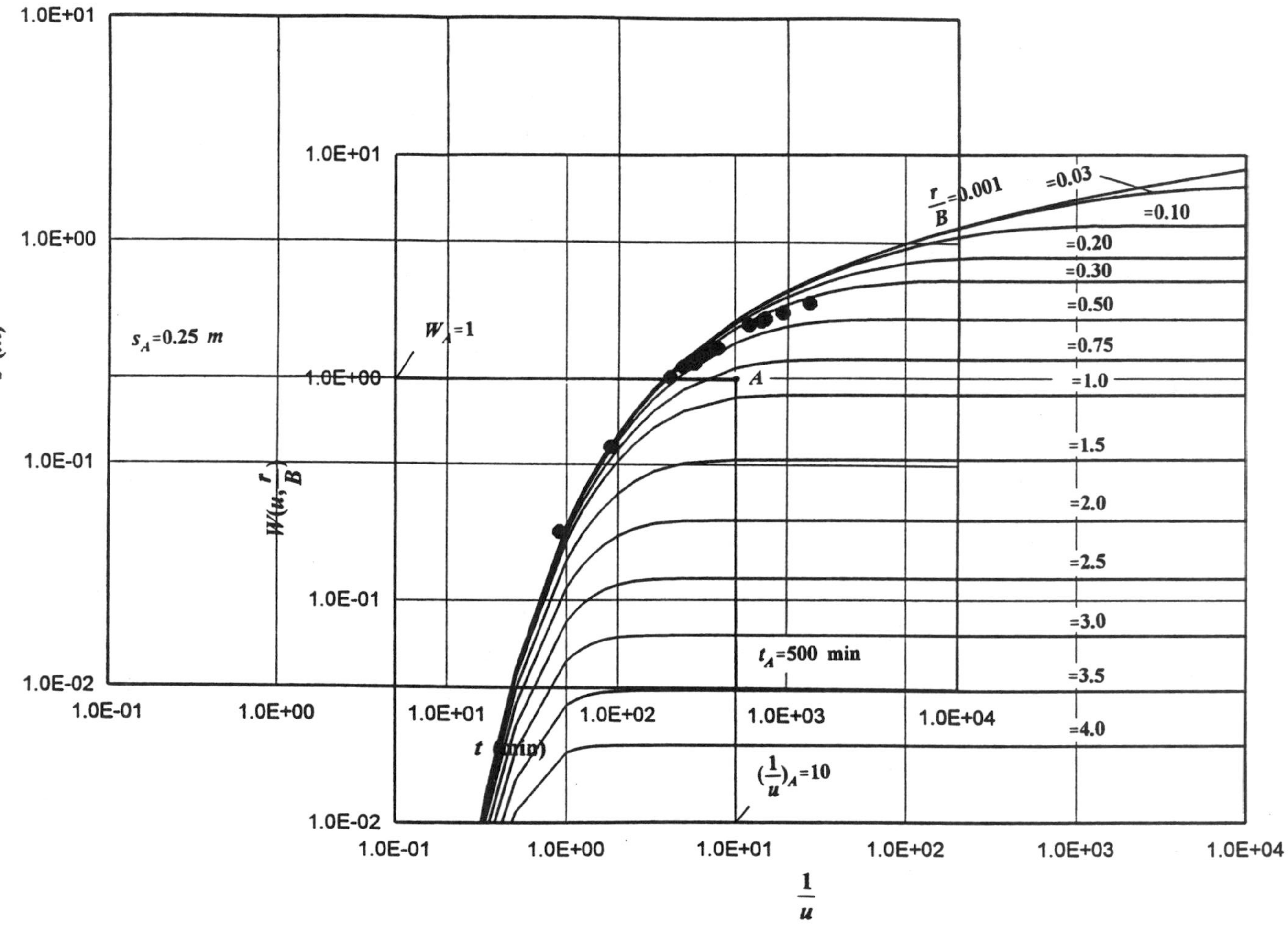

s (m)

1.0E+01
1.0E+00
1.0E-01
1.0E-02
1.0E-01
1.0E+00
1.0E+01
1.0E+02
1.0E+03
1.0E+04

t (min)

s_A=0.25 m

$W\left(u, \frac{r}{B}\right)$

W_A=1

A

t_A=500 min

$\left(\frac{1}{u}\right)_A$=10

$\frac{r}{B}$=0.001
=0.03
=0.10
=0.20
=0.30
=0.50
=0.75
=1.0
=1.5
=2.0
=2.5
=3.0
=3.5
=4.0

1.0E+01
1.0E+00
1.0E-01
1.0E-02
1.0E-01
1.0E+00
1.0E+01
1.0E+02
1.0E+03
1.0E+04

$\frac{1}{u}$

Figure 7-11 Walton type-curve matching for observation well TH/OW-1 for Example 7-1.

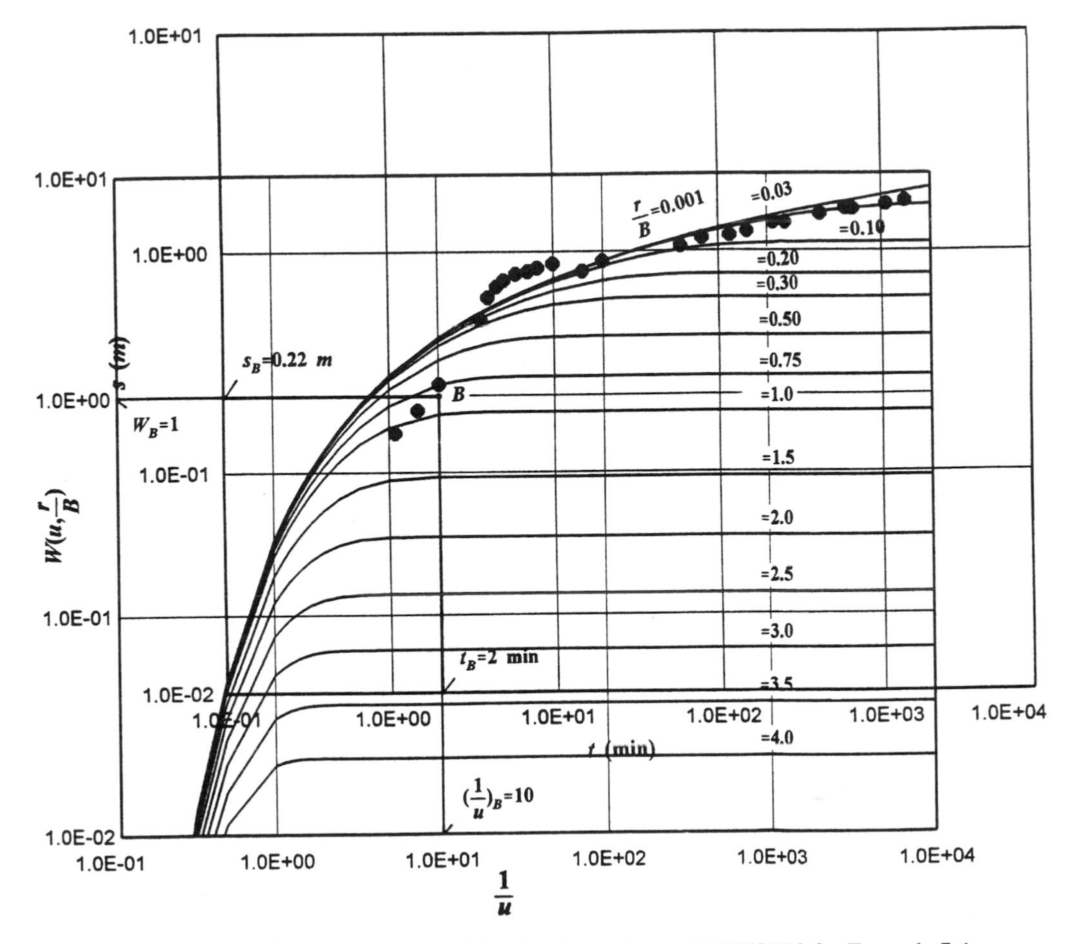

Figure 7-12 Walton type-curve matching for observation well TH/OW-3 for Example 7-1.

Step 4: Match point (A) coordinates in Figure 7-11 for TH/OW-1 are

$$\left(\frac{1}{u}\right)_A = 10, \qquad W_A\left(u, \frac{r}{B}\right) = 1$$

$$s_A = 0.25 \text{ m}, \qquad t_A = 500 \text{ min}$$

Match point (B) coordinates in Figure 7-12 for TH/OW-3 are

$$\left(\frac{1}{u}\right)_B = 10, \qquad W_B\left(u, \frac{r}{B}\right) = 1$$

$$s_B = 0.22 \text{ m}, \qquad t_B = 2 \text{ min}$$

Step 5: From Eq. (6-23) of Chapter 6, the transmissivity values are calculated below. The values for TH/OW-1 gives

$$T = \frac{Q}{4\pi s_A} W_A\left(u, \frac{r}{B}\right) = \frac{436 \text{ m}^3/\text{day}}{(4\pi)(0.25 \text{ m})}(1) = 139 \text{ m}^2/\text{day}$$

For TH/OW-3 we obtain

$$T = \frac{436 \text{ m}^3/\text{day}}{(4\pi)(0.22 \text{ m})}(1) = 158 \text{ m}^2/\text{day}$$

The arithmetic average of the two T values is $T_{av} = 149 \text{ m}^2/\text{day}$.

Step 6: From Eq. (6-22) of Chapter 6, the values of storage coefficient are calculated below. For TH/OW-1 we obtain

$$S = \frac{4Tt_A u_A}{r^2} = \frac{(4)(139 \text{ m}^2/\text{day})\left(\dfrac{500 \text{ min}}{1440 \text{ min/day}}\right)(0.1)}{(151.2 \text{ m})^2} = 8.40 \times 10^{-4}$$

For TH/OW-3 we obtain

$$S = \frac{(4)(158 \text{ m}^2/\text{day})\left(\dfrac{2 \text{ min}}{1440 \text{ min/day}}\right)(0.1)}{(40.0 \text{ m})^2} = 0.55 \times 10^{-4}$$

The arithmetic average of the two S values is $S_{av} = 4.48 \times 10^{-4}$.

Step 7: From Eq. (6-11) of Chapter 6 for TH/OW-1 we obtain

$$\frac{b'}{K'} = \frac{B^2}{T} = \frac{\left(\dfrac{151.2 \text{ m}}{0.3}\right)^2}{139 \text{ m}^2/\text{day}} = 1827 \text{ days}$$

$$K' = \frac{4 \text{ m}}{1827 \text{ days}} = 2.19 \times 10^{-3} \text{ m/day} = 2.53 \times 10^{-6} \text{ cm/s}$$

For TH/OW-3 we obtain

$$\frac{b'}{K'} = \frac{\left(\dfrac{40.0 \text{ m}}{0.03}\right)^2}{158 \text{ m}^2/\text{day}} = 11{,}252 \text{ days}$$

$$K' = \frac{4 \text{ m}}{11{,}252 \text{ days}} = 0.36 \times 10^{-3} \text{ m/day} = 0.41 \times 10^{-6} \text{ cm/day}$$

The arithmetic average of the two K' values is: $K'_{av} = 1.47 \times 10^{-6}$ cm/s.

Application of the Type-Curve Method Based on Hantush (1960b). The type-curve method based on Hantush (1960b) will be applied in accordance with the steps presented previously.

Step 1: A family of type curves of $H(u, \beta)$ versus $1/u$ on log–log paper for different values of β are given in Figure 7-4 based on the values in Table 7-2.

Step 2: The log–log graphs for observation wells TH/OW-1 and TH/OW-3 are given in Figure 7-9 and Figure 7-10, respectively.

Step 3: The superimposed sheets are shown Figure 7-13 and Figure 7-14 for TH/OW-1 and TH/OW-3, respectively. Comparison shows that the plotted points fall along the curves $\beta = 0.20$ and 0.09 for TH/OW-1 and TH/OW-3, respectively.

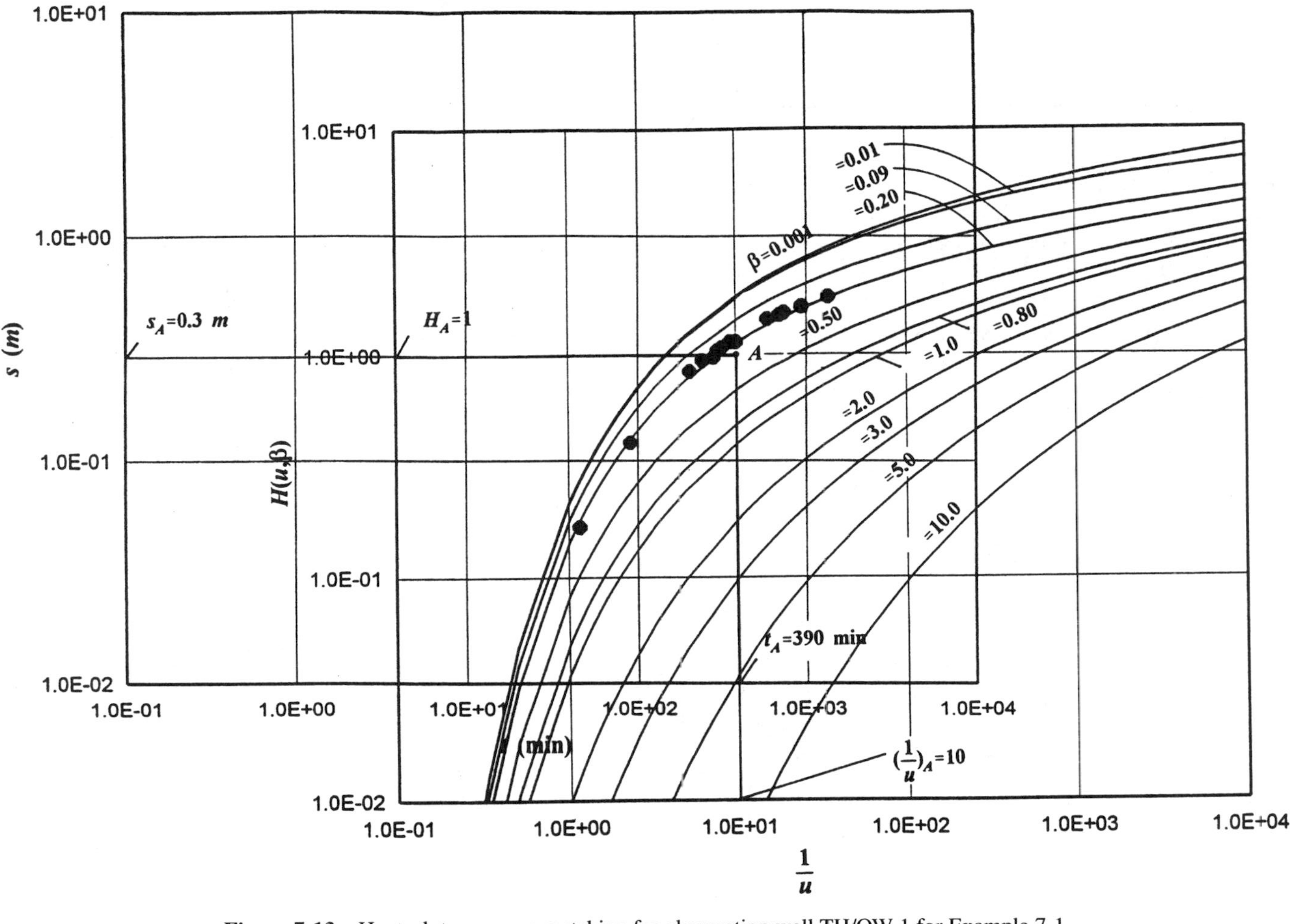

Figure 7-13 Hantush type-curve matching for observation well TH/OW-1 for Example 7-1.

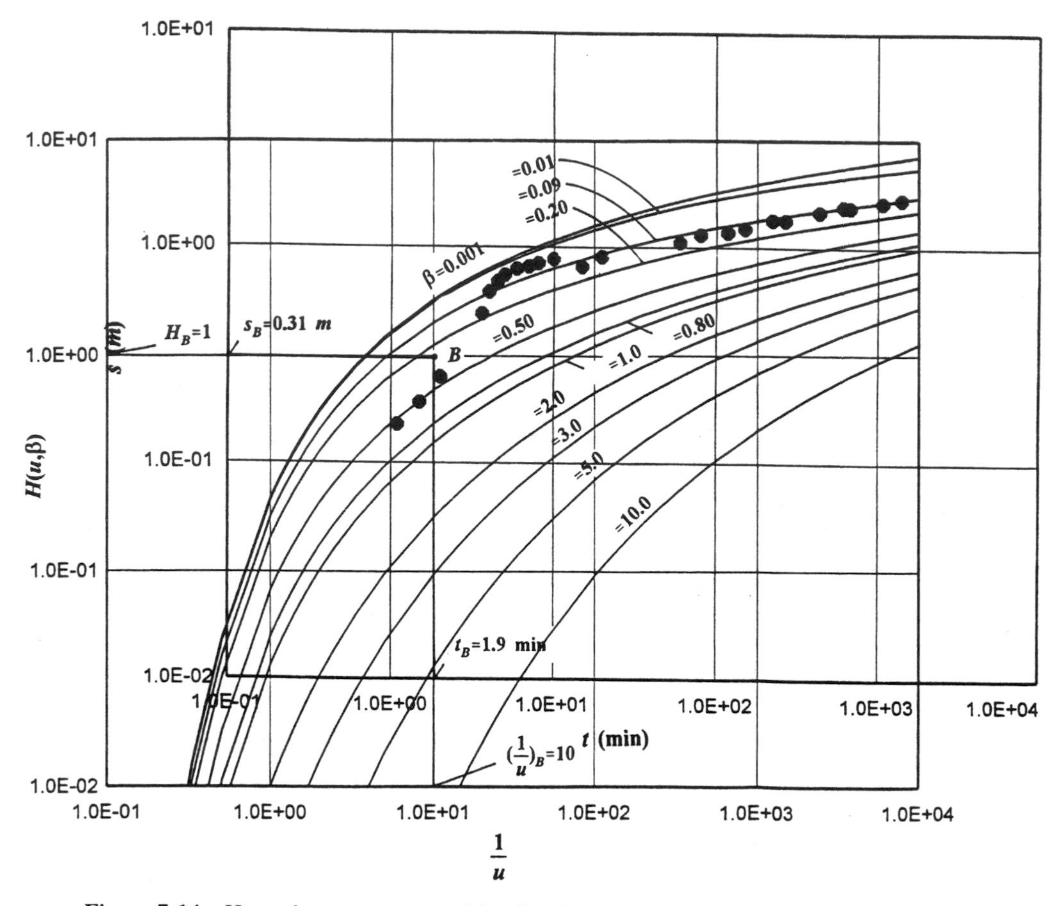

Figure 7-14 Hantush type-curve matching for observation well TH/OW-3 for Example 7-1.

Step 4: Match point (A) coordinates in Figure 7-13 for TH/OW-1 are

$$\left(\frac{1}{u}\right)_A = 10, \qquad H_A(u, \beta) = 1$$

$$s_A = 0.30 \text{ m} \qquad t_A = 390 \text{ min}$$

Match point (B) coordinates in Figure 7-14 for TH/OW-3 are

$$\left(\frac{1}{u}\right)_B = 10, \qquad H_B(u, \beta) = 1$$

$$s_B = 0.31 \text{ m}, \qquad t_B = 1.9 \text{ min}$$

Step 5: From Eq. (7-32), the transmissivity values are calculated below. The values for TH/OW-1 gives

$$T = \frac{Q}{4\pi s_A} H_A(u, \beta) = \frac{436 \text{ m}^3/\text{day}}{(4\pi)(0.30 \text{ m})}(1) = 116 \text{ m}^2/\text{day}$$

For TH/OW-3 we obtain

$$T = \frac{436 \text{ m}^3/\text{day}}{(4\pi)(0.31 \text{ m})}(1) = 112 \text{ m}^2/\text{day}$$

The arithmetic average of the two T values is $T_{av} = 114 \text{ m}^2/\text{day}$.

Step 6: From Eq. (7-33), the values of storage coefficient are calculated below. For TH/OW-1 we obtain

$$S = \frac{4Tt_A u_A}{r^2} = \frac{(4)(116 \text{ m}^2/\text{day})\left(\dfrac{390 \text{ min}}{1440 \text{ min/day}}\right)(0.1)}{(151.2 \text{ m})^2} = 5.50 \times 10^{-4}$$

For TH/OW-3 we obtain

$$S = \frac{(4)(112 \text{ m}^2/\text{day})\left(\dfrac{1.9 \text{ min}}{1440 \text{ min/day}}\right)(0.1)}{(40.0 \text{ m})^2} = 0.37 \times 10^{-4}$$

The arithmetic average of the two S values is: $S_{av} = 2.94 \times 10^{-4}$.

Step 7: For TH/OW-1, substitution of $r = 151.2$ m and $\beta = 0.2$ into Eq. (7-35) with Eq. (7-37) gives

$$0.2 = \frac{151.2 \text{ m}}{4}\left[\frac{\left(\dfrac{K'}{4 \text{ m}}\right)}{116 \text{ m}^2/\text{day}}\left(\frac{S'}{5.5 \times 10^{-4}}\right)\right]^{1/2}$$

and solving for $K'S'$ gives

$$K'S' = 7.14 \times 10^{-6} \text{ m/day}$$

Similarly, for TH/OW-3 using $r = 40.0$ m and $\beta = 0.09$ we obtain

$$0.09 = \frac{40.0 \text{ m}}{4}\left[\frac{\left(\dfrac{K'}{4 \text{ m}}\right)}{112 \text{ m}^2/\text{day}}\left(\frac{S'}{0.37 \times 10^{-4}}\right)\right]^{1/2}$$

and solving for $K'S'$ gives

$$K'S' = 1.34 \times 10^{-6} \text{ m/day}$$

The arithmetic average of the above two values is

$$(K'S')_{av} = 4.24 \times 10^{-6} \text{ m/day}$$

Step 8: Using the K' values determined from the Walton type-curve method the expressions in Step 7 gives the values for storage coefficient, S', of the confining layer:

$$S' = \frac{7.14 \times 10^{-6}\ \text{m/day}}{2.19 \times 10^{-3}\ \text{m/day}} = 3.26 \times 10^{-3} \qquad \text{for TH/OW-1}$$

$$S' = \frac{1.34 \times 10^{-6}\ \text{m/day}}{0.36 \times 10^{-3}\ \text{m/day}} = 3.72 \times 10^{-3} \qquad \text{for TH/OW-3}$$

The arithmetic average of the above two S' values is 3.49×10^{-3}.

Step 9:

$$\frac{b'S'}{10K'} = \frac{(4\ \text{m})(3.26 \times 10^{-3})}{(10)(2.19 \times 10^{-3}\ \text{m/day})} = 0.60\ \text{days} = 864\ \text{min} \qquad \text{for TH/OW-1}$$

$$\frac{b'S'}{10K'} = \frac{(4\ \text{m})(3.72 \times 10^{-3})}{(10)(0.36 \times 10^{-3}\ \text{m/day})} = 4.13\ \text{days} = 5947\ \text{min} \qquad \text{for TH/OW-3}$$

As can be observed from Table 7-3 for TH/OW-1 (the maximum time period is 1330 min), only two measurement points are greater than 864 min and this situation does not affect the curve-matching process for this observation well. For TH/OW-3 the maximum time period is 1400 min (see from Table 7-4) and this value is much smaller than its limiting value (5947 min). Based on the aforementioned analysis, the overall conclusion is that the results are acceptable.

Comparison of the Results. The transmissivity and storage coefficient values for CA2 determined from Hantush and Jacob (1955) (HJ55) and Hantush (1960b) (H60) model are compared in Table 7-5. The transmissivity values in Table 7-5 show that the HJ55 model predicts higher values than the H60 model. This is consistent with the theoretical explanation that is presented in detail in Section 7.2.6.

Because horizontal hydraulic conductivity (K_h) is proportional with transmissivity, the following relations can be written for this example from the values in Table 7-5:

$$\frac{(K_h)_{\text{HJ55}}}{(K_h)_{\text{H60}}} = 1.20 \qquad \text{for TH/OW-1}$$

$$\frac{(K_h)_{\text{HJ55}}}{(K_h)_{\text{H60}}} = 1.41 \qquad \text{for TH/OW-3}$$

The arithmetic average of these values is 1.31. Neuman and Witherspoon (1969c) (NW69) made a similar comparison between their model, which takes into account the storage

TABLE 7-5 Comparison of the Results Determined from Hantush and Jacob (1955) and Hantush (1960b) Models for Example 7-1

	T (m²/day)		S	
Model	TH/OW-1	TH/OW-3	TH/OW-1	TH/OW-3
Hantush and Jacob (1955)	139	158	8.40×10^{-4}	0.55×10^{-4}
Hantush (1960b)	116	112	5.50×10^{-4}	0.37×10^{-4}

effects of the aquitard (see Section 7.4), and the Hantush and Jacob (1955) model and found the following ratio for the horizontal hydraulic conductivity (K_2) of the confined aquifer in their example:

$$\frac{(K_h)_{\mathrm{HJ55}}}{(K_h)_{\mathrm{NW69}}} = 1.34$$

As seen from the above ratios, they are close to each other.

7.2.7.2 Determination of the Leakage Rates Through Confining Layers. The theoretical expressions for the rate of leakage and total leakage volume for small and large values of time are given in detail in Section 7.2.4. These equations have potential practical applications and their application is shown with an example.

Example 7-2: Determine the variation of leakage rate and leakage volume with time for the leaky aquifer in Example 7-1.

Solution: Solutions for small and large values of time for leakage rate and leakage volume are given separately. For short values of time, the expressions for the three cases (Case 1, Case 2, and Case 3) are given in Section 7.2.4.1. The expressions for large values of time are given in Section 7.2.4.4, because the hydrogeologic system in Example 7-1 (see Figure 7-7) corresponds to Case 3.

Hydrogeologic Data Summary. The hydrogeologic data determined from the Hantush (1960b)-model-based type-curve analysis in Example 7-1 are summarized as follows: $T_{\mathrm{av}} = 114\ \mathrm{m}^2/\mathrm{day}$, $S_{\mathrm{av}} = 2.94 \times 10^{-4}$, $(K'S')_{\mathrm{av}} = 4.24 \times 10^{-6}$ m/day, $b' = 4$ m, $S'_{\mathrm{av}} = 3.49 \times 10^{-3}$, $K'_{\mathrm{av}} = (K'S')_{\mathrm{av}}/S'_{\mathrm{av}} = 1.21 \times 10^{-3}$ m/day, and $Q = 436\ \mathrm{m}^3/\mathrm{day}$.

Calculation of the Parameters. Because $K'S' \gg K''S''$, Eq. (7-37) is the equation to be used. Substitution of the above values into Eq. (7-37) gives

$$\lambda = \left(\frac{K'S'}{b'TS}\right)^{1/2} = \left(\frac{4.24 \times 10^{-6}\ \mathrm{m/day}}{(4\ \mathrm{m})(114\ \mathrm{m}^2/\mathrm{day})(2.94 \times 10^{-4})^{1/2}}\right) = 0.00562\ \mathrm{m}^{-1}$$

From Eq. (7-41) we have

$$n = \frac{T\lambda^2}{S} = \frac{(114\ \mathrm{m}^2/\mathrm{day})(0.00562\ \mathrm{m}^{-1})^2}{(2.94 \times 10^{-4})} = 12.25\ \mathrm{day}^{-1}$$

Rate of Leakage and Total Leakage Volume for Small Values of Time. Substitution the above values into the rate of leakage equation given by Eq. (7-40) gives

$$q_L = (436\ \mathrm{m}^3/\mathrm{day})[1 - e^{(12.25\ \mathrm{day}^{-1}t)}]\,\mathrm{erfc}[(12.25\ \mathrm{day}^{-1}t)^{1/2}]$$

in which the units of time (t) and rate of leakage (q_L) are days and $\mathrm{m}^3/\mathrm{day}$, respectively. The complementary error function is connected to the error function by $\mathrm{erfc}(x) = 1 - \mathrm{erf}(x)$ in which $\mathrm{erf}(x)$ is tabulated in Table 8-5. The variation of q_L with time is presented in Figure 7-15. Figure 7-15 shows that the rate of leakage initially increases rapidly and

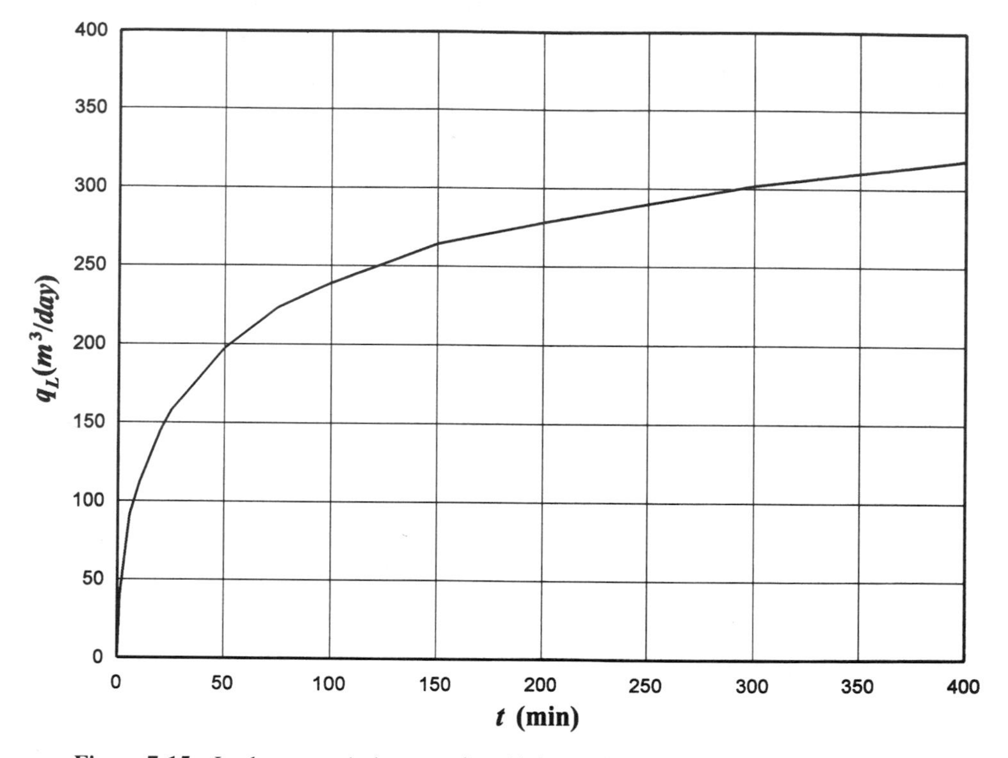

Figure 7-15 Leakage rate (q_L) versus time (t) for small values of time for Example 7-2.

later on asymptotically approaches to a constant value. As can be seen from Figure 7-15 by extrapolation, under almost steady-state conditions the majority of the well discharge comes through the confining layer. The total volume (V_L) withdrawn from leakage is calculated from Eq. (7-42), and its variation with time is shown in Figure 7-16. Figure 7-16 shows that the curves becomes approximately a straight line after approximately 50 min.

Rate of Leakage and Total Leakage Volume for Large Values of Time. As mentioned above, the hydrogeologic system corresponds to Case 3 (see Figure 7-7) and with $K'' = 0$ and $S'' = 0$, Eqs. (7-64) and (7-65) will be used for q_L and V_L, respectively. From Eq. (7-63) we have

$$\delta_3 = 1 + \frac{S'' + \dfrac{S'}{3}}{S} = 1 + \frac{0 + \dfrac{3.49 \times 10^{-3}}{3}}{2.94 \times 10^{-4}} = 4.96$$

and substitution the values of K', S, δ_3, and b' into Eq. (7-64) yields the following equation:

$$q_L = (436 \text{ m}^3/\text{day}) \left\{ 1 - \frac{1}{4.96} e^{-(0.207 \text{ day}^{-1})(t)} \right\}$$

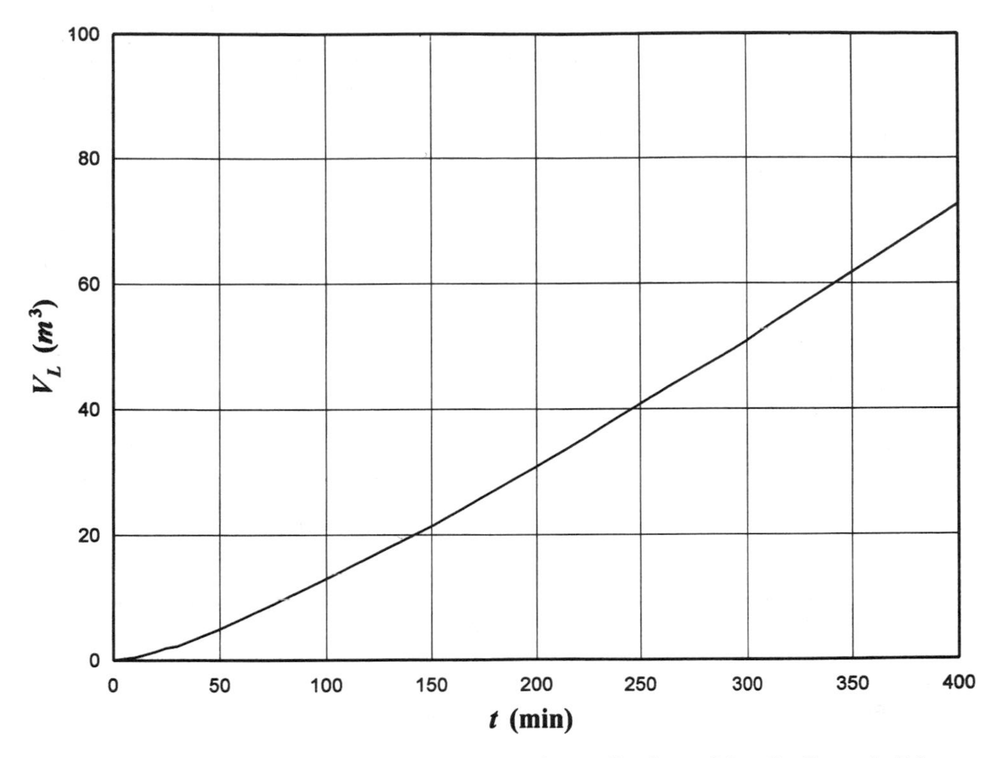

Figure 7-16 Total volume (V_L) versus time (t) for small values of time for Example 7-2.

in which the dimensions of q_L is m^3/day. Introducing the values into Eq. (7-65) gives the variation of V_L (units m^3) with respect to time t (days):

$$V_L = (436 \text{ m}^3/\text{day}) \left\{ t + (0.972 \text{ days})[e^{-(0.207 \text{ day}^{-1})(t)} - 1] \right\}$$

The variation of q_L and V_L with time are shown in Figure 7-17 and Figure 7-18, respectively. As can be seen from Figure 7-17, q_L does not change significantly with time. This is not unexpected because large time is elapsed since the pumping started and the system is close to steady state which means that all hydrogeologic quantities tend to become time-independent. The variation of V_L with time is almost linear (Figure 7-18), as is the case for small values of time (Figure 7-16).

7.3 NEUMAN AND WITHERSPOON'S TWO/ONE CONFINED AQUIFERS AND ONE AQUICLUDE MODELS: COMPLETE SOLUTIONS

An aquiclude is a special forms of confining layers adjacent aquifers. If the vertical hydraulic conductivity of a confining layer is so small that the water flow through it may be negligible, the confining layer is called an *aquiclude*; if the flow rate is appreciable, it is called an *aquitard* (refer to Section 2.2 of Chapter 2 for definitions).

Neuman and Witherspoon (1968) developed an analytical modeling approach to the theory of water flow in an aquiclude adjacent slightly leaky aquifers. According to Neuman

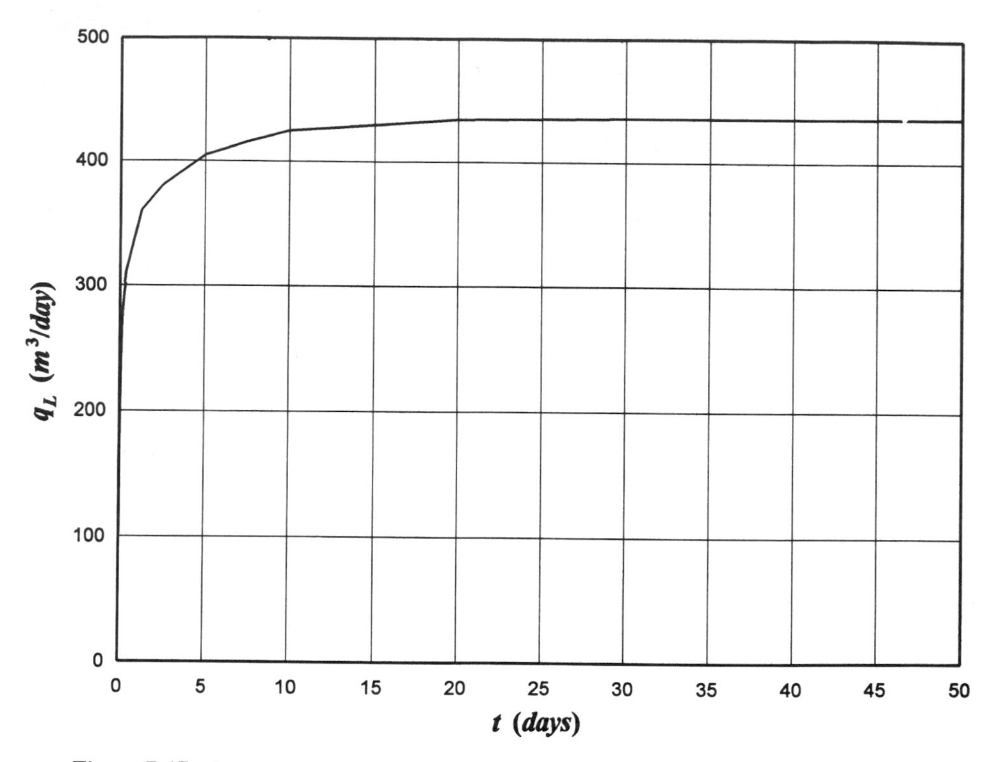

Figure 7-17 Leakage rate (q_L) versus time (t) for large values of time for Example 7-2.

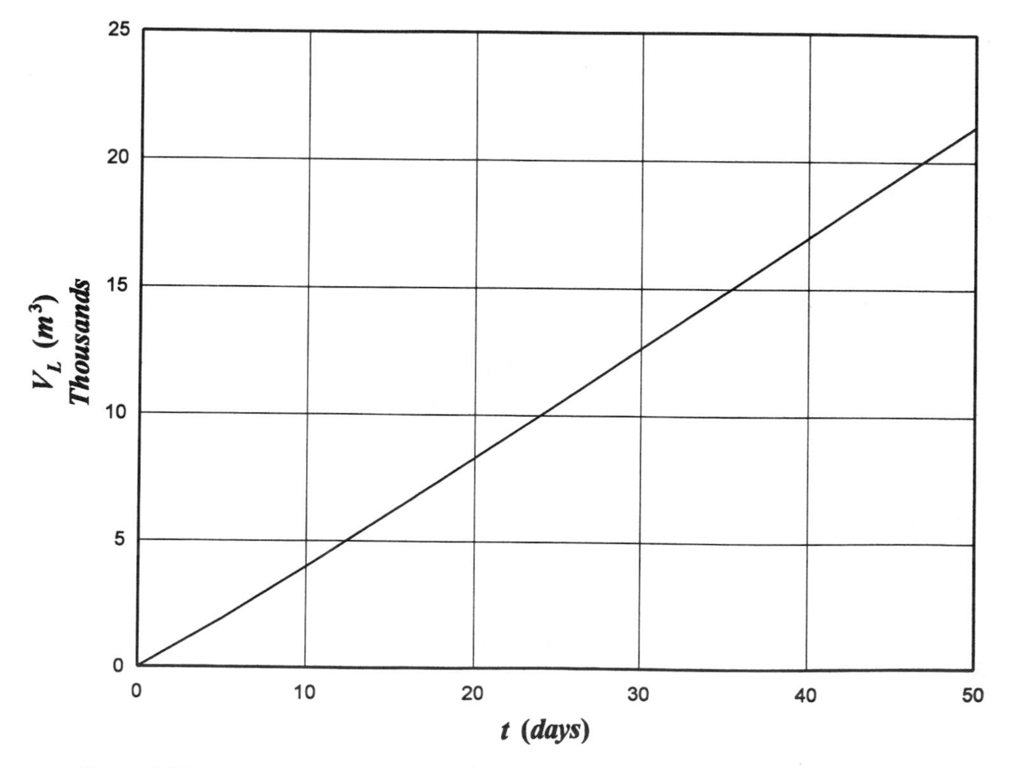

Figure 7-18 Total volume (V_L) versus time (t) for large values of time for Example 7-2.

and Witherspoon, the term "slightly leaky" means that when the aquifer is pumped at constant rate, the drawdowns in the aquifer will be predicted by the Theis equation under the condition that the vertical hydraulic conductivity of the aquiclude is sufficiently small so that the contribution of the aquitard to the total flow reaching the pumping well is not enough to give the usual evidence of a leaky aquifer.

Neuman and Witherspoon (1968) also specified the limits of their theory regarding the flow in aquicludes as follows: From a practical standpoint, a lower limit does not exist. This is due to the fact that after a few minutes of pumping time, the Theis equation holds at all radial distances including the well bore (Witherspoon et al., 1967). The upper limit will depend on the magnitude of error that one is willing to accept for the determination of the properties of the aquiclude. For example, consider that the properties of an aquifer system are known and the drawdown in the adjacent aquiclude will be predicted with an error of less than 10%. It is important to know that the drawdown data for the aquifer follow either a β curve for small values of time (see Section 7.2.4.1) or an r/B curve for large values of time (see Chapter 6, Section 6.2.4.1), as defined by Hantush and Jacob (1955) and Hantush (1960b), but they do not deviate from the Theis solution more than 10%. It can be shown that the vertical hydraulic conductivity of an aquiclude can be significant and still meet the criterion to be a slightly leaky system. Neuman and Witherspoon (1968) also state that, if $r/B \leq 0.01$ in the development by Hantush (1960b), the Theis solution gives acceptable results everywhere in the aquifer except at large values of time.

The models for the vertical flow in aquicludes are relatively simpler as compared with the ones that take into account the horizontal and vertical flows in leaky aquifers systems as presented in Sections 7.2 and 7.4. As will be seen in Section 7.4, the mathematical model for two confined aquifers and one confining layer is a well-verified complete solution, but because of its complex mathematical form its practical usage regarding pumping test data analysis for the determination of hydraulic parameters of confining layers based on type-curve matching method is not a common practice. Instead, some simplified mathematical models are being used to overcome the difficulties that are associated with the aforementioned complete solutions. The models that are presented in detail in the following sections are (1) the two confined aquifers and one aquiclude model and (2) the one confined aquifer and one aquiclude of infinite thickness model.

7.3.1 Two Confined Aquifers and One Aquiclude Model

7.3.1.1 Problem Statement and Assumptions. A finite thickness (b') aquiclude, whose vertical hydraulic conductivity is K', is underlain and overlain by slightly two leaky confined aquifers, each of which is being pumped at constant rates Q_1 and Q_2 (see Figure 7-19). An observation well is located at r_1 and r_2 distances from these two pumping wells. The vertical coordinates (z) are measured from the bottom of the aquiclude. Neuman and Witherspoon (1968) developed a mathematical model for this problem under the following assumptions:

1. Each aquifer is homogeneous and isotropic and of infinite radial extent.
2. The horizontal flow components in the aquiclude are negligibly small compared with the vertical components.
3. The confined system remains saturated at all times and is infinite radial extent.
4. The drawdowns in the aquifers can be predicted by the Theis solution.

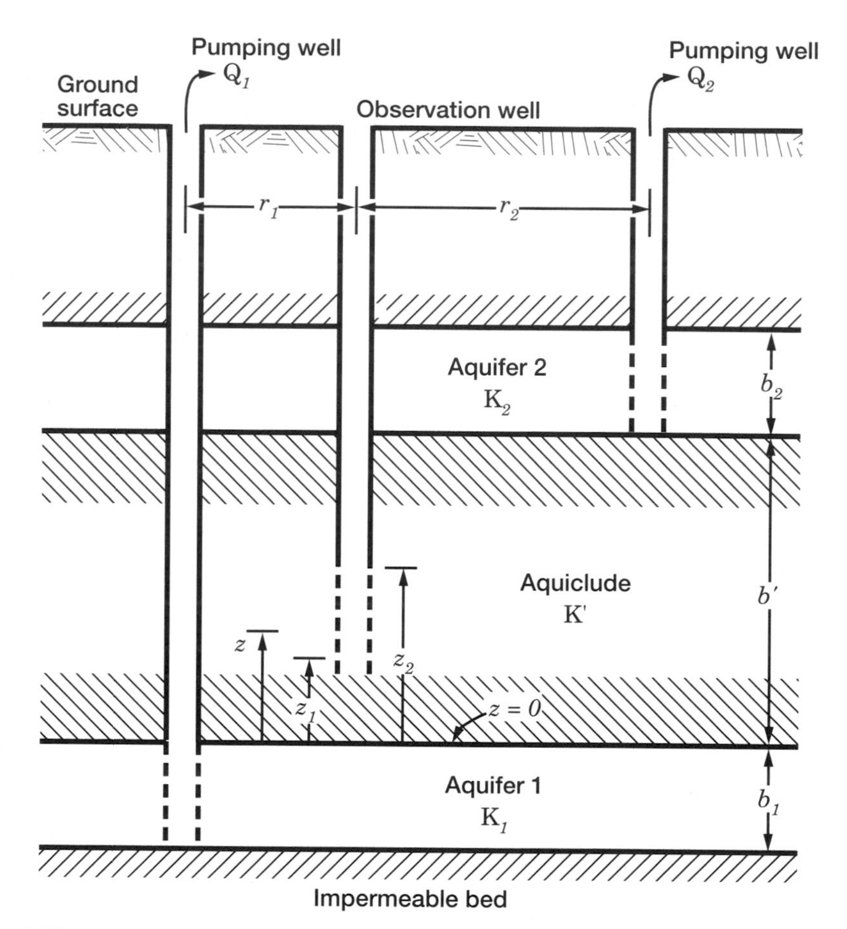

Figure 7-19 Schematic representation of an aquiclude enclosed by two confined aquifers. (After Neuman and Witherspoon, 1968.)

On the basis of the aforementioned assumptions, the horizontal components of flow are neglected and the flow in the aquitard is considered to be one-dimensional. The goal is to determine the drawdown in the aquiclude, $s'(r, z, t)$, on the assumption that the drawdowns at the upper and lower boundaries of the aquiclude can be predicted by the Theis equation.

7.3.1.2 Governing Equations. With the assumptions given above, the flow is one-dimensional in the aquiclude and the governing equation is in the form of either Eq. (7-8) or Eq. (7-10) of Section 7.2:

$$\frac{\partial^2 s'}{\partial z^2} = \frac{S'}{K'b'}\frac{\partial s'}{\partial t} = \frac{1}{\alpha'}\frac{\partial s'}{\partial t} \tag{7-78}$$

where S' and K' are the storage coefficient and vertical hydraulic conductivity of the aquiclude, respectively, and α' is the *hydraulic diffusivity* of the aquiclude and is defined as follows:

$$\alpha' = \frac{K'b'}{S'} = \frac{K'}{S'_s} \tag{7-79}$$

where S'_s is the specific storage of the aquiclude.

7.3.1.3 Initial and Boundary Conditions. The initial condition is

$$s'(r, z, 0) = 0 \tag{7-80}$$

The boundary conditions are

$$s'(r, 0, t) = \frac{Q_1}{4\pi T_1}[-\mathrm{Ei}(-u_1)] = \frac{Q_1}{4\pi T_1} W(u_1) = \phi_1(t) \tag{7-81}$$

$$s'(r, 0, t) = \frac{Q_2}{4\pi T_2}[-\mathrm{Ei}(-u_2)] = \frac{Q_2}{4\pi T_2} W(u_2) = \phi_2(t) \tag{7-82}$$

where

$$u_1 = \frac{r^2}{4\alpha_1 t} = \frac{r^2 S_1}{4T_1 t} = \frac{1}{4t_{D_1}} \tag{7-83}$$

$$u_2 = \frac{r^2}{4\alpha_2 t} = \frac{r^2 S_2}{4T_2 t} = \frac{1}{4t_{D_2}} \tag{7-84}$$

and the hydraulic diffusivities of the aquifers are defined as

$$\alpha_1 = \frac{T_1}{S_1}, \qquad \alpha_2 = \frac{T_2}{S_2} \tag{7-85}$$

where T_1 and T_2 are the transmissivities of the first and second aquifers, respectively, and S_1 and S_2 are their storage coefficients. The function $-\mathrm{Ei}(-u)$ is the *exponential integral* or *Theis well function* as defined by Eq. (4-25) of Chapter 4.

Equation (7-80) states that initially the drawdown in the aquiclude is zero. The boundary conditions given by Eqs. (7-81) and (7-82) are based on the fourth assumption of Section 7.3.1.1, and they state that the drawdown at the boundaries of the aquiclude are only dependent on time and radial distance.

7.3.1.4 Solution. The solution of the aforementioned boundary-value problem was expressed by Neuman and Witherspoon (1968) using the solution of a similar problem in Carslaw and Jaeger (1959). If the boundary conditions, expressed by Eqs. (7-81) and (7-82), are given by the functions $\phi_1(t)$ and $\phi_2(t)$, respectively, Carslaw and Jaeger (1959, p. 103) have shown that the solution of Eq. (7-78) for the conduction of heat under comparable conditions may be obtained using Duhamel's theorem as:

$$s'(r, z, t) = \frac{2\alpha'\pi}{b^2} \sum_{n=1}^{\infty} n \sin\left(\frac{n\pi z}{b'}\right) \int_0^t [\phi_1(\lambda) - (-1)^n \phi_2(\lambda)] \exp\left[-\frac{n^2 \alpha' \pi^2 (t-\lambda)}{b'^2}\right] d\lambda \tag{7-86}$$

By defining a new variable

$$y^2 = \frac{n^2 \alpha' \pi^2 (t-\lambda)}{b'^2} \tag{7-87}$$

and introducing the term

$$u'' = \frac{b'^2}{\pi^2 \alpha' t} = \frac{b' S'}{\pi^2 K' t} \tag{7-88}$$

Eq. (7-86) can be expressed in the following form:

$$\begin{aligned} s'(r,z,t) = {} & \frac{4}{\pi} \sum_{n=1}^{\infty} \frac{1}{n} \sin\left(\frac{n\pi z}{b'}\right) \int_0^{n(u'')^{-1/2}} \phi_1\left(t - \frac{y^2 b'^2}{n^2 \alpha' \pi^2}\right) y e^{-y^2}\, dy \\ & + \frac{4}{\pi} \sum_{n=1}^{\infty} \frac{1}{n} \sin\left[\frac{n\pi(b'-z)}{b'}\right] \int_0^{n(u'')^{-1/2}} \phi_2\left(t - \frac{y^2 b'^2}{n^2 \alpha' \pi^2}\right) y e^{-y^2}\, dy \end{aligned} \tag{7-89}$$

Substitution of $\phi_1(t)$ and $\phi_2(t)$, as given by Eqs. (7-81) and (7-82), respectively, into Eq. (7-89) gives

$$s'(r,z,t) = \frac{Q_1}{4\pi T_1} W\left(u_1, u_1'', \frac{z}{b'}\right) + \frac{Q_2}{4\pi T_2} W\left(u_2, u_2'', 1 - \frac{z}{b'}\right) \tag{7-90}$$

where

$$W\left(u_1, u_1'', \frac{z}{b'}\right) = \frac{4}{\pi} \sum_{n=1}^{\infty} \frac{1}{n} \sin\left(\frac{n\pi z}{b'}\right) \int_0^{n u_1''^{-1/2}} \left[-\mathrm{Ei}\left(-\frac{n^2 u_1}{n^2 - u_1'' y^2} \right) \right] y e^{-y^2}\, dy \tag{7-91}$$

$$W\left(u_2, u_2'', 1 - \frac{z}{b'}\right) = \frac{4}{\pi} \sum_{n=1}^{\infty} \frac{1}{n} \sin\left[\frac{n\pi(b'-z)}{b'}\right] \cdot \int_0^{n u_2''^{-1/2}} \left[-\mathrm{Ei}\left(-\frac{n^2 u_2}{n^2 - u_2'' y^2} \right) \right] y e^{-y^2}\, dy \tag{7-92}$$

The first term on the right side of Eq. (7-90) represents the effect due to pumping conditions in the lower aquifer, and the second term represents the effect due to pumping conditions in the upper aquifer. Because the two wells do not necessarily start pumping at the same time, the term u'' is introduced to allow for this.

Because y in Eq. (7-91) is a dummy variable, it must take on all values between the limits of integration. As a result of this, for the limiting case, when u_1'' becomes sufficiently small, which corresponds large values of pumping time, the exponential term in Eq. (7-91) approaches $-\mathrm{Ei}(-u_1)$; and the remaining integral approaches to $1/2$. Therefore, Eq. (7-91) takes the following form:

$$W\left(u_1, u_1'', \frac{z}{b'}\right) = -\mathrm{Ei}(-u_1) \frac{2}{\pi} \sum_{n=1}^{\infty} \frac{1}{n} \sin\left(\frac{n\pi z}{b'}\right) \tag{7-93}$$

Because the inverse Fourier sine transform of $1/n$ in the above series is simply $1 - z/b'$ (Churchill, 1958), Eq. (7-93) can simply be written as

$$W\left(u_1, u_1'', \frac{z}{b}\right) = -\mathrm{Ei}(-u_1)\left(1 - \frac{z}{b'}\right) \tag{7-94}$$

Similarly, if u_2'' is sufficiently small, Eq. (7-92) can be simplified to

$$W\left(u_2, u_2'', 1 - \frac{z}{b'}\right) = -\operatorname{Ei}(-u_2)\frac{z}{b'} \tag{7-95}$$

Substitution of Eqs. (7-94) and (7-95) into Eq. (7-90) gives

$$s'(r, z, t) = \frac{Q_1}{4\pi T_1}[-\operatorname{Ei}(-u_1)]\left(1 - \frac{z}{b'}\right) + \frac{Q_2}{4\pi T_2}[-\operatorname{Ei}(-u_2)]\left(\frac{z}{b'}\right) \tag{7-96}$$

For the solution for the special case of one pumping well in the lower aquifer under the condition that u_1'' is sufficiently small, Eq. (7-96) takes the form

$$s'(r, z, t) = \frac{Q_1}{4\pi T_1}[-\operatorname{Ei}(-u_1)]\left(1 - \frac{z}{b'}\right) = s(r, t)\left(1 - \frac{z}{b'}\right) \tag{7-97}$$

Equation (7-97) clearly shows that the ratio between the drawdown in the aquiclude at some distance r and the drawdown in the aquifer at the same radial distance from the pumping well at the same pumping time can simply be expressed as

$$\frac{s'(r, z, t)}{s(r, t)} = 1 - \frac{z}{b'} \tag{7-98}$$

Because of the assumption that no drawdown exists in the upper aquifer, there will be a linear variation in the drawdown across the aquiclude as the pseudo-steady state is approached. The same results can be obtained if Aquifer 2 is pumped at constant rate instead of Aquifer 1, and the corresponding form of Eq. (7-98) will be expressed as

$$\frac{s'(r, z, t)}{s(r, t)} = \frac{z}{b'} \tag{7-99}$$

7.3.2 One Confined Aquifer and One Aquiclude of Infinite Thickness Model

7.3.2.1 Problem Statement and Assumptions. In the case of a relatively large aquiclude thickness (b' in Figure 7-19) a somewhat simplified solution can be developed for the drawdown variation in the aquiclude. Mathematically speaking, the aquiclude is considered to be semi-infinite. Neuman and Witherspoon (1968) derived a relatively simple solution for pumping conditions from Aquifer 1 in Figure 7-19 by solving the corresponding boundary-value problem. It is obvious that the solution is applicable for the cases where the aquiclude lies above or below the slightly leaky aquifer.

The assumptions are the same as the ones in Section 7.3.1.1.

7.3.2.2 Governing Equations. The governing equations are given by Eqs. (7-78) and (7-79).

7.3.2.3 Initial and Boundary Conditions. The initial condition is given by

$$s'(r, z, t) = 0 \tag{7-100}$$

The boundary conditions are

$$s'(r,0,t) = \frac{Q_1}{4\pi T_1}[-\,\mathrm{Ei}(-u_1)] = \frac{Q_1}{4\pi T_1}W(u_1) = \phi_1(t) \tag{7-101}$$

$$s'(r,\infty,t) = 0 \tag{7-102}$$

Equation (7-100) states that initially the drawdown in the aquiclude is zero. The boundary condition given by Eq. (7-101) is based on the fourth assumption of Section 7.3.1.1 and states that the drawdown at the boundary of the aquiclude is only dependent on time and radial distance. Equation (7-102) states that at infinite z distance the drawdown in the aquiclude is zero.

7.3.2.4 Solution. The solution of the boundary-value problem defined above was expressed by Neuman and Witherspoon (1968) using the solution of a similar problem in Carslaw and Jaeger (1959). If the boundary condition expressed by Eq. (7-102) is given by the function $\phi_1(t)$, Carslaw and Jaeger (1959, p. 63) have shown that the solution of Eq. (7-78) for the conduction of heat under comparable conditions may be obtained using Duhamel's theorem as

$$s'(r,z,t) = \int_0^t \phi_1(\lambda)\frac{z}{2[\pi(t-\lambda)^3\alpha']^{1/2}}\exp[-\frac{z^2}{4(t-\lambda)\alpha'}]\,d\lambda \tag{7-103}$$

With the introduction of the following variables

$$y^2 = \frac{z^2}{4(t-\lambda)\alpha'} \tag{7-104}$$

$$u' = \frac{z^2}{4\alpha' t} \tag{7-105}$$

Eq. (7-103) takes the form

$$s'(r,z,t) = \frac{2}{\pi^{1/2}}\int_{(u')^{1/2}}^{\infty} \phi_1\left(t - \frac{z^2}{4y^2\alpha'}\right) e^{-y^2}\,dy \tag{7-106}$$

Using Eq. (7-101), the final solution is

$$s'(r,z,t) = \frac{Q_1}{4\pi T_1}W(u_1,u') \tag{7-107}$$

where

$$W(u_1,u') = \frac{2}{\pi^{1/2}}\int_{(u')^{1/2}}^{\infty}\left[-\,\mathrm{Ei}\left(-\frac{u_1 y^2}{y^2-u'}\right)\right]e^{-y^2}\,dy \tag{7-108}$$

$$u' = \frac{z^2}{4\alpha' t} = \frac{S_s' z^2}{4K' t} = \frac{1}{4t_D'} \tag{7-109}$$

and u_1 is given by Eq. (7-83). Therefore, Eq. (7-107) reduces to the following final expression (because of the existence of one aquifer, t_D is written without the subscript):

$$s'(r,z,t) = \frac{Q}{4\pi T}\frac{2}{\pi^{1/2}}\int_{\frac{1}{(4t'_D)^{1/2}}}^{\infty}\left\{-\operatorname{Ei}\left[-\frac{t'_D y^2}{t_D(4t'_D y^2 - 1)}\right]\right\}e^{-y^2}\,dy \tag{7-110}$$

For large values of time, the exponential integral in Eq. (7-107) approaches $-\operatorname{Ei}(-u_1)$, such that

$$\begin{aligned} W(u_1, u') &= -\operatorname{Ei}(-u_1)\frac{2}{\pi^{1/2}}\int_0^{\infty} e^{-y^2}\,dy \\ &= -\operatorname{Ei}(-u_1)\operatorname{erf}(\infty) = -\operatorname{Ei}(-u_1) \end{aligned} \tag{7-111}$$

As a result, Eq. (7-107) reduces to the Theis equation:

$$s'(r,z,t) = \frac{Q_1}{4\pi T_1}[-\operatorname{Ei}(-u_1)] \tag{7-112}$$

The equations presented above show that, for sufficiently large values of time, the drawdown in the aquiclude at some given point must be the same as the drawdown in the aquifer at the same radial distance from the pumping well and at the same pumping time. In other words,

$$\frac{s'(r,z,t)}{s(r,t)} = 1 \tag{7-113}$$

which is in agreement with Eq. (7-98) under the condition that b' is very large ($b' \to \infty$).

7.3.3 Evaluation of the Models and Their Potential Usage

The functions $W(u, u'', z/b')$ and $W(u, u')$ in the above-given solutions require numerical methods for their evaluations. Neuman and Witherspoon (1968) used the Adams–Moulton method to numerically evaluate these functions. In the evaluation process the authors used t_D, dimensionless time in the aquifer, which is defined as

$$t_D = \frac{1}{4u} = \frac{Tt}{r^2 S} \tag{7-114}$$

t'_D, dimensionless time in the semi-infinite aquiclude, which is defined as [from Eq. (7-109)]

$$t'_D = \frac{1}{4u'} = \frac{K't}{S'_s z^2} \tag{7-115}$$

and t''_D, dimensionless time in the finite aquiclude, which is defined as

$$t''_D = \frac{1}{4u''} = \frac{\pi^2 K't}{4b'^2 S'_s} \tag{7-116}$$

Witherspoon et al. (1967) presents an extensive table of results for both $W(u, u'', z/b')$ and $W(u, u')$ that cover a practical range of dimensionless times and values of z/b' (or $1 - z/b'$) ranging from 0 to 0.9. Table 7.6 presents a table for $W(u, u')$.

In the case of the finite aquiclude, the variation of $W(u, u'', z/b')$ at $z/b' = 0.5$ with t''_D for different values of t_D is presented in Figure 7-20. Figure 7-21 is based on the same data of Figure 7-20, but in the form of type curves corresponding to the ratio of t_D/t''_D which is constant for a given pumping test.

In the case of the semi-infinite aquiclude, the variation of $W(u, u')$ with t'_D for different values of t_D/t'_D is shown in Figure 7-22 and is similar to that in Figure 7-21.

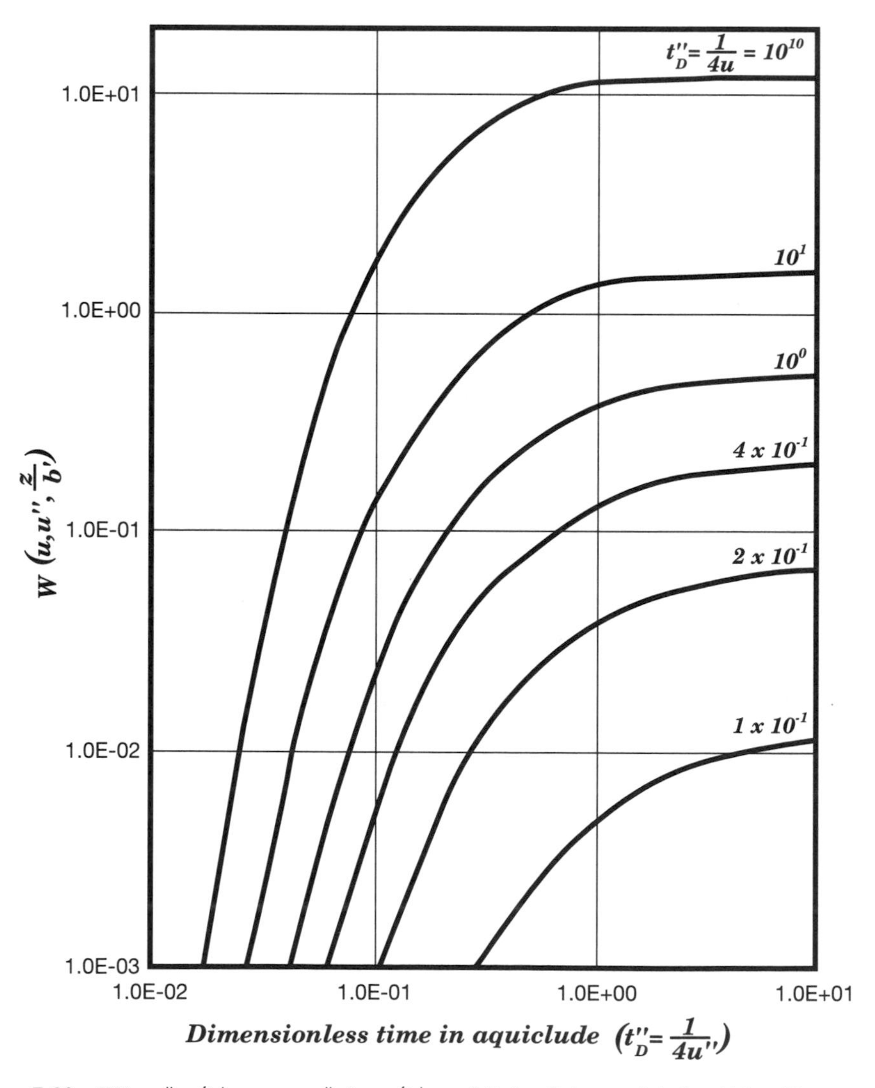

Figure 7-20 $W(u, u'', z/b')$ versus t''_D for $z/b' = 0.5$ for finite aquiclude. (After Neuman and Witherspoon, 1968.)

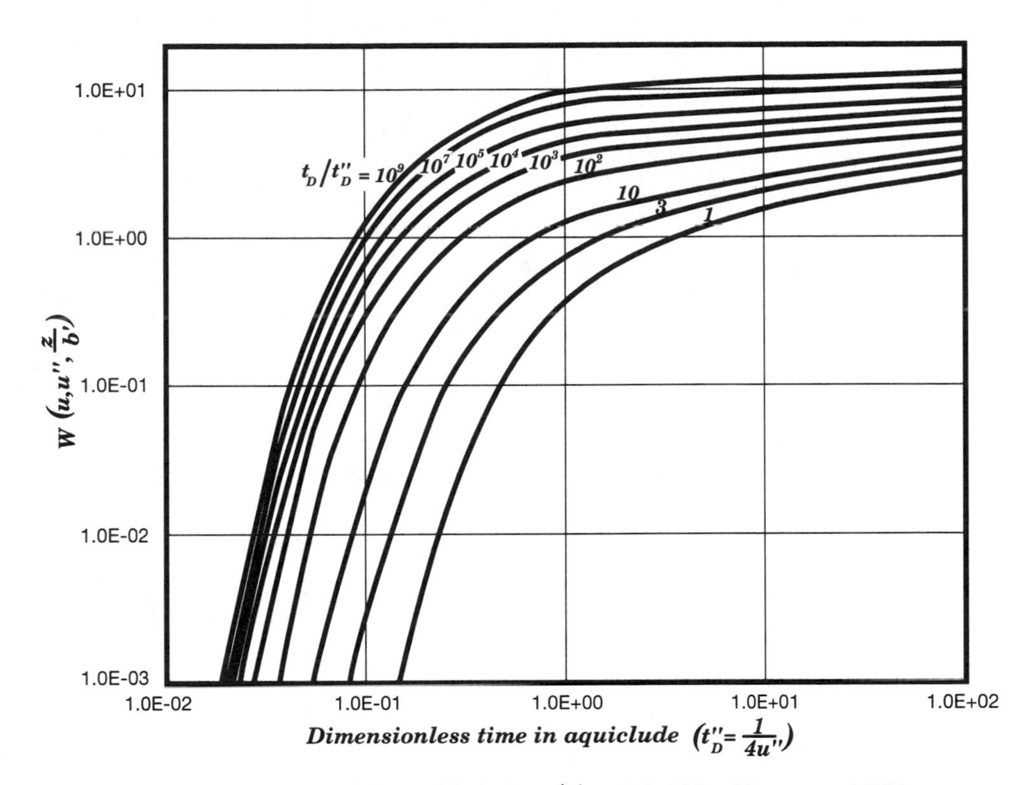

Figure 7-21 Type curves for finite aquiclude for $z/b' = 0.5$. (After Neuman and Witherspoon, 1968.)

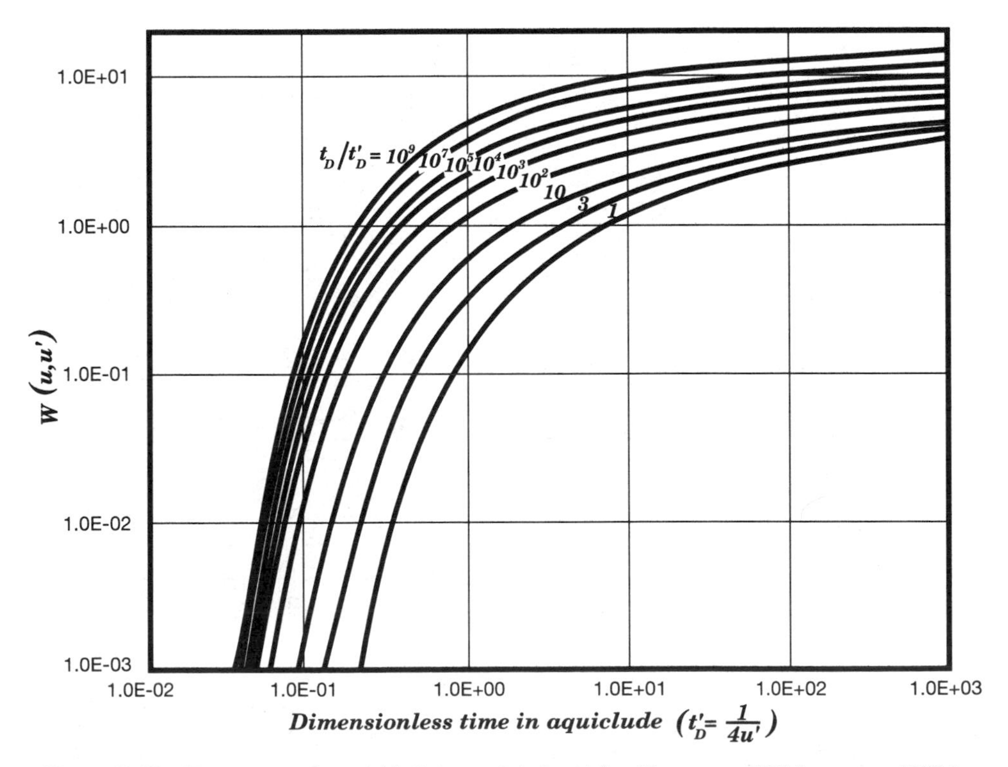

Figure 7-22 Type curves for semi-infinite aquiclude. (After Neuman and Witherspoon, 1968.)

TABLE 7-6 Values of $W(u, u')$ Function for Different values of u and u'

	t_D								
t'_D	0.2	0.4	0.5	0.7	1	2	4	7	10
250	9.07E−01	9.22E−01	9.26E−01	9.31E−01	9.35E−01	9.41E−01	9.46E−01	9.49E−01	9.50E−01
125	8.70E−01	8.91E−01	8.96E−01	9.03E−01	9.09E−01	9.18E−01	9.24E−01	9.28E−01	9.30E−01
83.3333	8.43E−01	8.68E−01	8.74E−01	8.82E−01	8.89E−01	9.00E−01	9.07E−01	9.12E−01	9.14E−01
62.5	8.2E−01	8.48E−01	8.56E−01	8.65E−01	8.73E−01	8.85E−01	8.94E−01	8.99E−01	9.01E−01
50	8.01E−01	8.32E−01	8.40E−01	8.50E−01	8.59E−01	8.72E−01	8.81E−01	8.87E−01	8.90E−01
41.6667	7.84E−01	8.17E−01	8.26E−01	8.36E−01	8.46E−01	8.60E−01	8.71E−01	8.77E−01	8.80E−01
31.25	7.54E−01	7.91E−01	8.01E−01	8.13E−01	8.24E−01	8.40E−01	8.52E−01	8.59E−01	8.62E−01
25	7.29E−01	7.69E−01	7.79E−01	7.92E−01	8.04E−01	8.22E−01	8.35E−01	8.43E−01	8.47E−01
12.5	6.37E−01	6.86E−01	7.00E−01	7.16E−01	7.32E−01	7.55E−01	7.72E−01	7.82E−01	7.87E−01
8.3333	5.73E−01	6.28E−01	6.43E−01	6.62E−01	6.79E−01	7.06E−01	7.25E−01	7.37E−01	7.43E−01
6.25	5.23E−01	5.82E−01	5.98E−01	6.18E−01	6.37E−01	6.66E−01	6.87E−01	7.00E−01	7.07E−01
5	4.82E−01	5.44E−01	5.61E−01	5.82E−01	6.02E−01	6.33E−01	6.55E−01	6.69E−01	6.76E−01
4.1667	4.48E−01	5.11E−01	5.28E−01	5.50E−01	5.71E−01	6.03E−01	6.27E−01	6.42E−01	6.49E−01
3.125	3.92E−01	4.56E−01	4.75E−01	4.98E−01	5.20E−01	5.54E−01	5.79E−01	5.95E−01	6.03E−01
2.5	3.48E−01	4.13E−01	4.31E−01	4.55E−01	4.77E−01	5.12E−01	5.39E−01	5.55E−01	5.64E−01
1.25	2.14E−01	2.73E−01	2.90E−01	3.13E−01	3.36E−01	3.72E−01	3.99E−01	4.17E−01	4.26E−01
0.8333	1.44E−01	1.95E−01	2.10E−01	2.31E−01	2.51E−01	2.85E−01	3.12E−01	3.29E−01	3.38E−01
0.0625	1.02E−01	1.45E−01	1.58E−01	1.76E−01	1.95E−01	2.25E−01	2.50E−01	2.66E−01	2.75E−01
0.5	7.44E−02	1.10E−01	1.22E−01	1.38E−01	1.54E−01	1.82E−01	2.04E−01	2.19E−01	2.27E−01
0.4167	5.55E−02	8.53E−02	9.54E−02	1.09E−01	1.23E−01	1.48E−01	1.68E−01	1.82E−01	1.89E−01
0.3125	3.23E−02	5.33E−02	6.06E−02	7.09E−02	8.18E−02	1.01E−01	1.18E−01	1.29E−01	1.35E−01
0.25	1.96E−02	3.44E−02	3.99E−02	4.75E−02	5.58E−02	7.09E−02	8.40E−02	9.29E−02	9.79E−02
0.125	2.29E−03	5.14E−03	6.34E−03	8.19E−03	1.03E−02	1.46E−02	1.87E−02	2.16E−02	2.33E−02
0.0833	3.35E−04	9.67E−04	1.25E−03	1.72E−03	2.30E−03	3.55E−03	4.81E−03	5.78E−03	6.35E−03
0.0625	6.38E−05	2.03E−04	2.80E−04	4.04E−04	5.60E−04	9.33E−04	1.33E−03	1.65E−03	1.84E−03
0.05	1.24E−05	4.52E−05	6.54E−05	9.91E−05	1.46E−04	2.56E−04	3.84E−04	4.89E−04	5.54E−04
0.0417	4.10E−06	1.08E−05	1.59E−05	2.60E−05	4.06E−05	7.80E−05	1.17E−04	1.50E−04	1.73E−04
0.03125	5.46E−09	6.81E−07	1.06E−06	1.89E−06	3.93E−06	5.73E−06	1.12E−05	1.53E−05	1.78E−05

TABLE 7-6 *(continued)*

	t_D								
t'_D	1E+02	1E+03	1E+04	1E+05	1E+06	1E+07	1E+08	1E+09	1E+10
250	9.56E−01	9.58E−01	9.60E−01	9.60E−01	9.61E−01	9.62E−01	9.62E−01	9.62E−01	9.62E−01
125	9.37E−01	9.41E−01	9.43E−01	9.44E−01	9.45E−01	9.46E−01	9.46E−01	9.47E−01	9.47E−01
83.3333	9.24E−01	9.28E−01	9.30E−01	9.32E−01	9.33E−01	9.34E−01	9.34E−01	9.35E−01	9.35E−01
62.5	9.12E−01	9.17E−01	9.20E−01	9.21E−01	9.23E−01	9.23E−01	9.24E−01	9.25E−01	9.25E−01
50	9.02E−01	9.07E−01	9.10E−01	9.12E−01	9.13E−01	9.14E−01	9.15E−01	9.16E−01	9.16E−01
41.6667	8.93E−01	8.99E−01	9.02E−01	9.04E−01	9.05E−01	9.06E−01	9.07E−01	9.08E−01	9.08E−01
31.25	8.77E−01	8.83E−01	8.87E−01	8.89E−01	8.91E−01	8.92E−01	8.93E−01	8.94E−01	8.94E−01
25	8.63E−01	8.70E−01	8.74E−01	8.76E−01	8.78E−01	8.79E−01	8.80E−01	8.81E−01	8.82E−01
12.5	8.08E−01	8.18E−01	8.23E−01	8.27E−01	8.29E−01	8.31E−01	8.32E−01	8.33E−01	8.34E−01
8.3333	7.68E−01	7.79E−01	7.85E−01	7.89E−01	7.92E−01	7.94E−01	7.95E−01	7.97E−01	7.98E−01
6.25	7.34E−01	7.47E−01	7.54E−01	7.58E−01	7.61E−01	7.63E−01	7.65E−01	7.66E−01	7.67E−01
5	7.05E−01	7.19E−01	7.26E−01	7.31E−01	7.34E−01	7.37E−01	7.39E−01	7.40E−01	7.41E−01
4.1667	6.80E−01	6.94E−01	7.02E−01	7.07E−01	7.11E−01	7.13E−01	7.15E−01	7.17E−01	7.18E−01
3.125	6.36E−01	6.51E−01	6.60E−01	6.65E−01	6.69E−01	6.72E−01	6.74E−01	6.76E−01	6.77E−01
2.5	5.98E−01	6.15E−01	6.24E−01	6.30E−01	6.34E−01	6.36E−01	6.39E−01	6.40E−01	6.42E−01
1.25	4.64E−01	4.83E−01	4.93E−01	4.99E−01	5.04E−01	5.07E−01	5.09E−01	5.11E−01	5.13E−01
0.8333	3.76E−01	3.94E−01	4.04E−01	4.11E−01	4.15E−01	4.18E−01	4.21E−01	4.23E−01	4.24E−01
0.0625	3.11E−01	3.28E−01	3.38E−01	3.44E−01	3.48E−01	3.51E−01	3.54E−01	3.56E−01	3.57E−01
0.5	2.60E−01	2.77E−01	2.86E−01	2.92E−01	2.96E−01	2.99E−01	3.01E−01	3.03E−01	3.04E−01
0.4167	2.20E−01	2.36E−01	2.44E−01	2.50E−01	2.53E−01	2.56E−01	2.58E−01	2.60E−01	2.61E−01
0.3125	1.61E−01	1.74E−01	1.81E−01	1.86E−01	1.89E−01	1.91E−01	1.93E−01	1.94E−01	1.95E−01
0.25	1.19E−01	1.30E−01	1.37E−01	1.40E−01	1.43E−01	1.45E−01	1.46E−01	1.48E−01	1.48E−01
0.125	3.11E−02	3.52E−02	3.76E−02	3.90E−02	4.00E−02	4.08E−02	4.13E−02	4.18E−02	4.21E−02
0.0833	9.07E−03	1.06E−02	1.14E−02	1.19E−02	1.23E−02	1.26E−02	1.28E−02	1.29E−02	1.31E−02
0.0625	2.79E−03	3.32E−03	3.63E−03	3.82E−03	3.95E−03	4.05E−03	4.12E−03	4.18E−03	4.23E−03
0.05	8.85E−04	1.07E−03	1.19E−03	1.26E−03	1.30E−03	1.34E−03	1.37E−03	1.39E−03	1.40E−03
0.0417	2.87E−04	3.55E−04	3.95E−04	4.21E−04	4.38E−04	4.50E−04	4.60E−04	4.68E−04	4.74E−04
0.03125	3.12E−05	4.04E−05	4.55E−05	4.88E−05	5.11E−05	5.27E−05	5.40E−05	5.49E−05	5.57E−05

Source: after Witherspoon et al. (1967) and Kruseman and De Ridder (1991).

7.4 NEUMAN AND WITHERSPOON'S TWO CONFINED AQUIFERS AND ONE AQUITARD MODEL: COMPLETE SOLUTION

One of the main assumptions of the modified theory of leaky aquifers (Hantush, 1960b), as presented in Section 7.2, is that no drawdown occurs in unpumped aquifer. Neuman and Witherspoon (1969a–c, 1972) developed a complete solution with some practical applications without this assumption for a hydrogeologic system which is composed of two confined aquifers and one confining layer. In the following sections, Neuman and Witherspoon's model and its practical applications are presented in detail.

7.4.1 Problem Statement and Assumptions

The schematic representation of the problem that Neuman and Witherspoon (1969a, b) solved is illustrated in Figure 7-23, which shows that two confined aquifers are separated by an aquitard or confining layer with pumping from the lower aquifer (Aquifer 1). The hydrogeologic system shown in Figure 7-23 is a special form of Case 1 of Hantush (1960b) (Figure 7-1), with the exception that the drawdown in the uppermost aquifer is assumed to be zero ($s_2 = 0$) in the Hantush's models.

The assumptions are as follows:

1. The discharge rate (Q_1) of the well is constant.
2. The diameter of the well is infinitesimally small and is completed in the lower confined aquifer.

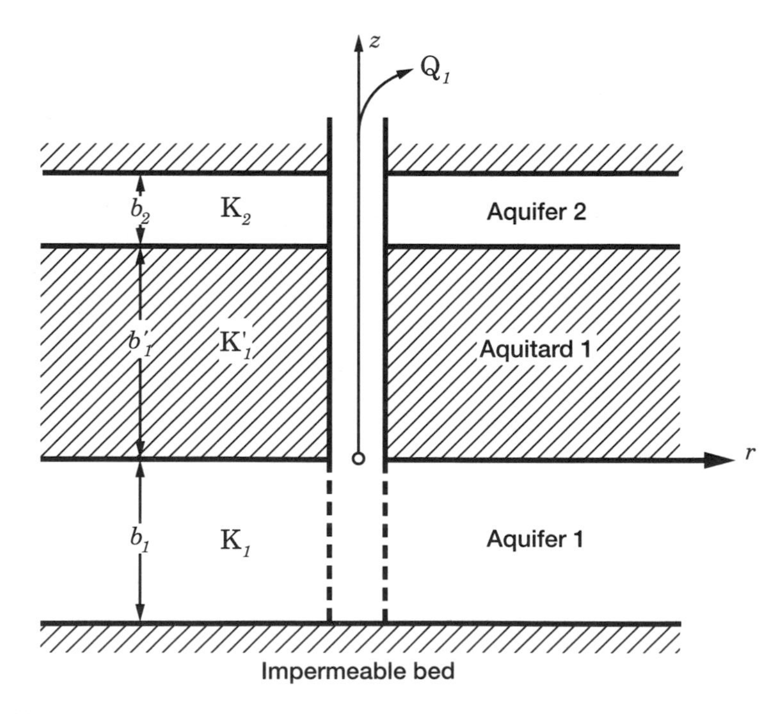

Figure 7-23 Schematic representation of two confined aquifers and one aquitard (After Neuman and Witherspoon, 1969b.)

3. Each layer is homogeneous, isotropic, horizontal, and of infinite radial extent.
4. The system remains saturated at all times, and Darcy's law applies.
5. The direction of flow is vertical in the aquitard or confining layer.
6. The upper and lower limits of the complete system are no-flow boundaries.

As mentioned in Section 7.2.1, the validity of fifth assumption has been investigated by Neuman and Witherspoon (1969a–c) based on the finite element numerical modeling method and they found that when the hydraulic conductivities of the aquifers are two or more orders of magnitude greater than that of the aquitard, the errors introduced by this assumption are usually less than 5%. The rest of assumptions will be evaluated in Sections 7.4.7 and 7.4.8.

7.4.2 Governing Equations

The governing equations for drawdown will be formulated in a manner similar to that in Section 7.2.2 in accordance with the notation of Neuman and Witherspoon (1969a, b). As can be seen from Figure 7-23, the coordinate center is selected at the upper boundary of the lower confined aquifer (Aquifer 1). The notation is the same as in Section 7.2.2.

The governing equation can be written for Aquifer 1 by introducing $K'' = 0$ in Eq. (7-9) because the Neuman and Witherspoon model takes into account only one confining layer. With this, Eq. (7-9) takes the form

$$\frac{\partial^2 s_1}{\partial r^2} + \frac{1}{r}\frac{\partial s_1}{\partial r} + \frac{K_1'}{T_1}\frac{\partial s_1'}{\partial z}\Big|_{z=0} = \frac{S_{s_1}}{K_1}\frac{\partial s_1}{\partial t} \qquad (7\text{-}117)$$

From Eq. (7-8), the governing equation for the confining layer or aquitard is

$$\frac{\partial^2 s_1'}{\partial z^2} = \frac{S_{s_1}'}{K_1'}\frac{\partial s_1'}{\partial t} \qquad (7\text{-}118)$$

Again from Eq. (7-9), the governing equation for Aquifer 2 can be written as

$$\frac{\partial^2 s_2}{\partial r^2} + \frac{1}{r}\frac{\partial s_2}{\partial r} - \frac{K_1'}{T_2}\frac{\partial s_1'}{\partial z}\Big|_{z=b_1'} = \frac{S_{s_2}}{K_2}\frac{\partial s_2}{\partial t} \qquad (7\text{-}119)$$

In Eqs. (7-117) and (7-119) the subscript "1" and "2" refer Aquifer 1 and Aquifer 2, respectively. Because there is only one aquitard the subscript "1" in Eq. (7-118) refers the aquitard separating the aquifers. The positive and negative signs in front of the third terms in Eqs. (7-117) and (7-119), respectively, are the results of the conservation of mass principle applied to each aquifer.

7.4.3 Initial and Boundary Conditions

The initial and boundary conditions are written in a manner similar to that in Section 7.2.3. The initial and boundary conditions for Aquifer 1 are

$$s_1(r, 0) = 0 \qquad (7\text{-}120)$$

$$s_1(\infty, t) = 0 \qquad (7\text{-}121)$$

$$\lim_{r \to 0} \left(r \frac{\partial s_1}{\partial r} \right) = -\frac{Q_1}{2\pi T_1} \tag{7-122}$$

The initial and boundary conditions for the aquitard or confining layer are

$$s_1'(r, z, 0) = 0 \tag{7-123}$$

$$s_1'(r, 0, t) = s_1(r, t) \tag{7-124}$$

$$s_1'(r, b_1', t) = s_2(r, t) \tag{7-125}$$

The initial and boundary conditions for Aquifer 2 are as follows:

$$s_2(r, 0) = 0 \tag{7-126}$$

$$s_2(\infty, t) = 0 \tag{7-127}$$

$$\lim_{r \to 0} \left(\frac{\partial s_2}{\partial r} \right) = 0 \tag{7-128}$$

Equations (7-120), (7-123), and (7-126) state that initially the drawdown is zero in the aquifers and aquitard. Equations (7-121) and (7-127) state that drawdown is zero at large distances in the pumped and unpumped aquifers. Equation (7-122) states that near the pumping well in the pumped aquifer the flow rate is equal to the discharge rate. Equation (7-128) states that in the unpumped aquifer the drawdown near the well does not change with the radial distance. Eqs. (7-124) and (7-125) state that at the upper and lower boundaries of the aquitard, drawdowns in the aquifers are equal to drawdown in the aquitard.

7.4.4 General Solution

The boundary value problem defined above was solved by Neuman and Witherspoon (1969a, b) by applying the Laplace and Hankel transforms similarly as used by Hantush (1960b) to Eqs. (7-8), (7-9), and (7-10) with the associated initial and boundary conditions and determined separate solutions for drawdown in the aquifers and aquitard. In the pumped aquifer (Aquifer 1)

$$s_1(r, t) = \frac{Q_1}{4\pi T_1} \int_0^\infty \left(1 - e^{-y^2 \bar{t}_{D_1}} \right) \{[1 + G(y)] J_0[\omega_1(y)] + [1 - G(y)] J_0[\omega_2(y)]\} \frac{dy}{y} \tag{7-129}$$

In the aquitard

$$s_1'(r, z, t) = \frac{Q_1}{4\pi T_1} \frac{2}{\pi} \sum_{n=1}^{\infty} \frac{1}{n} \sin\left(\frac{n\pi z}{b_1'} \right)$$

$$\cdot \int_0^\infty \left[1 - e^{-n^2 \pi^2 \bar{t}_{D_1}} + \frac{e^{-n^2 \pi^2 \bar{t}_{D_1}} - e^{-y^2 \bar{t}_{D_1}}}{1 - \dfrac{y^2}{n^2 \pi^2}} \right]$$

$$\cdot \left\{ \left[\frac{2\left(\frac{r}{B_{21}}\right)^2 (-1)^n y}{F(y)\sin y} - G(y) - 1 \right] J_0[\omega_1(y)] - \left[\frac{2\left(\frac{r}{B_{21}}\right)^2 (-1)^n y}{F(y)\sin y} - G(y) + 1 \right] J_0[\omega_2(y)] \right\} \frac{dy}{y} \tag{7-130}$$

In the unpumped aquifer (Aquifer 2)

$$s_2(r,t) = \frac{Q_1}{4\pi T_1} \int_0^\infty \left(1 - e^{-y^2 \bar{t}_{D_1}}\right) \frac{2\left(\frac{r}{B_{21}}\right)^2}{F(y)} \{J_0[\omega_1(y)] - J_0[\omega_2(y)]\} \frac{dy}{\sin y} \tag{7-131}$$

where

$$\bar{t}_{D_1} = t_{D_1} \frac{\left(\frac{r}{B_{11}}\right)^4}{(4\beta_{11})^2} \tag{7-132}$$

$$G(y) = \frac{M(y)}{F(y)} \tag{7-133}$$

$$\omega_1^2(y) = \tfrac{1}{2}[N(y) + F(y)] \tag{7-134}$$

$$\omega_2^2(y) = \tfrac{1}{2}[N(y) - F(y)] \tag{7-135}$$

$$F^2(y) = M^2(y) + \left[\frac{2\left(\frac{r}{B_{11}}\right)\left(\frac{r}{B_{21}}\right) y}{\sin y} \right]^2 \tag{7-136}$$

$$M(y) = \left[\frac{\left(\frac{r}{B_{11}}\right)^4}{(4\beta_{11})^2} - \frac{\left(\frac{r}{B_{21}}\right)^4}{(4\beta_{21})^2} \right] y^2 - \left[\left(\frac{r}{B_{11}}\right)^2 - \left(\frac{r}{B_{21}}\right)^2 \right] y \cot(y) \tag{7-137}$$

$$N(y) = \left[\frac{\left(\frac{r}{B_{11}}\right)^4}{(4\beta_{11})^2} + \frac{\left(\frac{r}{B_{21}}\right)^4}{(4\beta_{21})^2} \right] y^2 - \left[\left(\frac{r}{B_{11}}\right)^2 + \left(\frac{r}{B_{21}}\right)^2 \right] y \cot(y) \tag{7-138}$$

$$B_{11} = \left(\frac{K_1 b_1 b_1'}{K_1'} \right)^{1/2} \tag{7-139}$$

$$B_{21} = \left(\frac{K_2 b_2 b_1'}{K_1'} \right)^{1/2} \tag{7-140}$$

$$\beta_{11} = \frac{r}{4b_1} \left(\frac{K_1' S_{s_1}'}{K_1 S_{s_1}} \right)^{1/2} \tag{7-141}$$

$$\beta_{21} = \frac{r}{4b_2} \left(\frac{K_1' S_{s_1}'}{K_2 S_{s_2}} \right)^{1/2} \tag{7-142}$$

$$t_{D_1} = \frac{K_1 t}{S_{s_1} r^2} \tag{7-143}$$

and $J_0(x)$ is a zero-order Bessel function of the first kind. In the above expressions, $J_0[\omega_1(y)]$ and $J_0[\omega_2(y)]$ must be set to zero when $\omega_1^2(y) < 0$ and $\omega_2^2(y) < 0$, respectively.

7.4.5 Solutions for Special Cases

Several special cases are presented based on the general solution of the problem that is presented in Section 7.4.4. Special solutions are presented in the following sections.

7.4.5.1 Special Solution: No Drawdown in the Unpumped Aquifer. As mentioned in Section 7.4.1, the hydrogeologic system shown in Figure 7-23 is a special form of Case 1 of Hantush (1960b) (Figure 7-1) and Case 3 (Figure 7-3) of Hantush (1960b) with the exception that the drawdown in the uppermost aquifer is assumed to be zero in the Hantush's models (see Section 7.2.4). Therefore, it is interesting to derive a complete solution by neglecting the drawdown in the unpumped aquifer to supplement Hantush's asymptotic solutions for small and large values of time.

Mathematically, zero drawdown in the unpumped aquifer can be represented by making the hydraulic conductivity of Aquifer 2 go to infinity. By letting $K_2 \to \infty$, both r/B_{21} and β_{21} approach zero. With these, $G(y) = 1$ because $F(y) = M(y) = N(y)$. When we substitute these expressions into Eqs. (7-129) and (7-130), the following equations can, respectively, be obtained:

$$s_1(r,t) = \frac{Q_1}{4\pi T_1} 2 \int_0^\infty \left(1 - e^{-y^2 \bar{t}_{D_1}}\right) J_0[\omega_1(y)] \frac{dy}{y} \tag{7-144}$$

$$s_1'(r,z,t) = \frac{Q_1}{4\pi T_1} \frac{4}{\pi} \sum_{n=1}^{\infty} \sin\left(\frac{n\pi z}{b_1'}\right) \cdot \int_0^\infty \left[e^{-n^2\pi^2 \bar{t}_{D_1}} - 1 - \frac{e^{-n^2\pi^2 \bar{t}_{D_1}} - e^{-y^2 \bar{t}_{D_1}}}{1 - \dfrac{y^2}{n^2\pi^2}} \right] J_0[\omega_1(y)] \frac{dy}{y} \tag{7-145}$$

where

$$\omega_1^2(y) = \frac{\left(\dfrac{r}{B_{11}}\right)^4}{(4\beta_{11})^2} y^2 - \left(\frac{r}{B_{11}}\right)^2 y \cot(y) \tag{7-146}$$

and as before the general solution case the Bessel function must be set zero when $\omega_1^2(y) < 0$.

7.4.5.2 Special Solution: Reduction to the Theis Solution. It can be shown that if the hydraulic conductivity of the aquitard (K_1') is zero, Eq. (7-129) reduces to the Theis solution. To show this, Neuman and Witherspoon (1969a, b) defined an additional variable as

$$x = \frac{\left(\frac{r}{B_{11}}\right)^2}{4\beta_{11}} y \tag{7-147}$$

When we substitute this into Eq. (7-129) as well as into Eqs. (7-132) through (7-138) and let $K_1' \to 0$, Eq. (7-129) reduces to

$$s_1(r,t) = \frac{Q_1}{2\pi T_1} \int_0^\infty \left(1 - e^{-x^2 t_{D_1}}\right) J_0(x) \frac{dx}{x} \tag{7-148}$$

where

$$t_{D_1} = \frac{K_1 t}{S_{s_1} r^2} = \frac{K_1 b_1 t}{S_{s_1} b_1 r^2} = \frac{T_1 t}{S_1 r^2} \tag{7-149}$$

where T_1 and S_1 are the transmissivity and storage coefficient of the aquifer, respectively. From tables of definite integrals [Gradshteyn and Ryzhik, 1965, p. 717, Eq. (5)] the following expression can be written:

$$\int_0^\infty e^{-ax^2} J_0(x) \frac{dx}{x} = \frac{1}{2} \int_0^{1/4a} e^{-y} \frac{dy}{y} \tag{7-150}$$

where the real part of a is nonnegative. Combination of this expression with Eq. (7-148) yields

$$s_1(r,t) = \frac{Q_1}{4\pi T_1} \int_{\frac{r^2 S_1}{4T_1 t}}^\infty e^{-y} \frac{dy}{y} \tag{7-151}$$

which is the Theis solution as given in Chapter 4 [Section 4.2.4.1, Eq. (4-22)].

7.4.5.3 Special Solutions: Asymptotic Solutions for Small and Large Times. As mentioned above, if the drawdown in the unpumped aquifer remains zero ($s_2 = 0$), the Neuman and Witherspoon general model becomes equivalent to Case 1 of Hantush (1960b) (see Section 7.2.1 and Figure 7-1). The solution for this case is presented in Section 7.4.5.1. Neuman and Witherspoon (1969c) evaluated *Hantush's small- and large-time solutions* (Hantush, 1960b), as presented in Section 7.2.4, using their complete solution. Details of this evaluation are presented below.

When we use Eq. (7-149), Eq. (7-32), *Hantush's small-time solution*, can be expressed as

$$s_1(r,t) = \frac{Q_1}{4\pi T_1} \int_{\frac{1}{4t_{D_1}}}^\infty \exp(-y) \operatorname{erfc}\left\{\frac{\beta_{11}}{[y(4t_{D_1} y - 1)]^{1/2}}\right\} \frac{dy}{y} \tag{7-152}$$

where β_{11} is equal to Hantush's β that is defined by Eq. (7-35) for one aquitard case. Because of the existence of one aquitard (see Figure 7-23), the conditions for small-time solution, as given by Eq. (7-28), takes the following form:

$$t \leq \frac{b_1' S_1'}{10 K_1'} = \frac{b_1' S_{s_1}' b_1'}{10 K_1'} \tag{7-153}$$

Alternatively, by using Eqs. (7-141) and (7-149) one obtains

$$t_{D_1} \leq \frac{1.6\beta_{11}^2}{\left(\dfrac{r}{B_{11}}\right)^4} \tag{7-154}$$

which is equivalent to Eq. (7-28) with the exception of notation. Neuman and Witherspoon (1969c) showed that from a practical standpoint the criterion given by Eq. (7-28) or Eq. (7-154) is overly conservative (see Section 7.2.4.1).

Neuman and Witherspoon (1969a, b) also presented a corresponding short-time solution for drawdown distribution in the aquitard by following in part a procedure previously outlined by Hantush (1960b) (see Section 7.2) for the condition as given by Eq. (7-153). The solution is

$$s_1'(r,z,t) = \frac{Q_1}{4\pi T_1} \int_{\frac{1}{4t_{D_1}}}^{\infty} \frac{e^{-y}}{y} \left[\operatorname{erfc} \left\{ \frac{\beta_{11} + y\left(\dfrac{z}{b_1'}\right)\dfrac{4\beta_{11}}{\left(\dfrac{r}{B_{11}}\right)^2}}{[y(4t_{D_1}y - 1)]^{1/2}} \right\} - \operatorname{erfc} \left\{ \frac{\beta_{11} + y\left(2 - \dfrac{z}{b_1'}\right)\dfrac{4\beta_{11}}{\left(\dfrac{r}{B_{11}}\right)^2}}{[y(4t_{D_1}y - 1)]^{1/2}} \right\} \right] dy \tag{7-155}$$

For the particular case of semi-infinite aquitard where $b_1' \to \infty$, Neuman and Witherspoon (1969c) showed that Eq. (7-155) reduces to

$$s_1'(r,z,t) = \frac{Q_1}{4\pi T_1} \int_{\frac{1}{4t_{D_1}}}^{\infty} \frac{e^{-y}}{y} \operatorname{erfc} \left\{ \frac{\beta_{11} + y\left(\dfrac{t_{D_1}}{t_{D_1'}}\right)^{1/2}}{[y(4t_{D_1}y - 1)]^{1/2}} \right] dy \tag{7-156}$$

where

$$t_{D_1'} = \frac{K_1' t}{S_{s_1}' z^2} \tag{7-157}$$

Hantush's asymptotic solution for large values of time (see Section 7.2.4.2), as given by Eq. (7-49), can be expressed in the following form using the variables defined by Eqs. (7-132), (7-139), and (7-141):

$$s_1(r,t) = \frac{Q_1}{4\pi T_1} \int_{\frac{\delta_1}{4t_{D_1}}}^{\infty} \frac{1}{y} \exp\left[-y - \frac{\left(\frac{r}{B_{11}}\right)^2}{4y} \right] dy \tag{7-158}$$

where the corresponding form of Eq. (7-48) is

$$\delta_1 = 1 + \frac{16\beta_{11}^2}{3\left(\frac{r}{B_{11}}\right)^2} \tag{7-159}$$

and Hantush's criterion for the validity of the solution given by Eq. (7-44) or (7-51) takes the following form:

$$t_{D_1} \geq \frac{80\beta_{11}^2}{\left(\frac{r}{B_{11}}\right)^4} \tag{7-160}$$

Neuman and Witherspoon (1969c) concluded that for given values of β_{11} and r/B_{11}, these solutions cover the entire time domain except for an interval whose span is less than two log cycles.

7.4.6 Evaluation of the Solutions

Neuman and Witherspoon (1969a, b) evaluated the analytical solutions as given by Eqs. (7-129), (7-130), and (7-131) as well as asymptotic solutions given by Eq. (7-152), which is equivalent to Eq. (7-32) of Hantush's small-time solution, and Eq. (7-155) using the Zonnevald adaptation of the Adams–Moulton numerical integration method.

Neuman and Witherspoon (1969a, b) checked the analytical models against the finite element method results by comparing the dimensionless variables for drawdown and time as defined, respectively, as

$$s_D = \frac{4\pi T_1}{Q_1} s, \qquad t_{D_1} = \frac{K_1 t}{S_{s_1} r^2} \tag{7-161}$$

for different numerical values of β_{11}, β_{21}, r/B_{11}, and r/B_{21}. Figure 7-24 presents one of the comparisons which shows that there is excellent agreement between the analytical and numerical solutions. In Figure 7-24, two of the curves show results for the pumped and unpumped aquifers, and the other three curves represent different elevations in the aquitard. The Theis solution is included for reference purposes.

Neuman and Witherspoon (1969a, b) also evaluated the requirement for short-time solutions as given by Eq. (7-28) or Eq. (7-154) and concluded that the properties of the

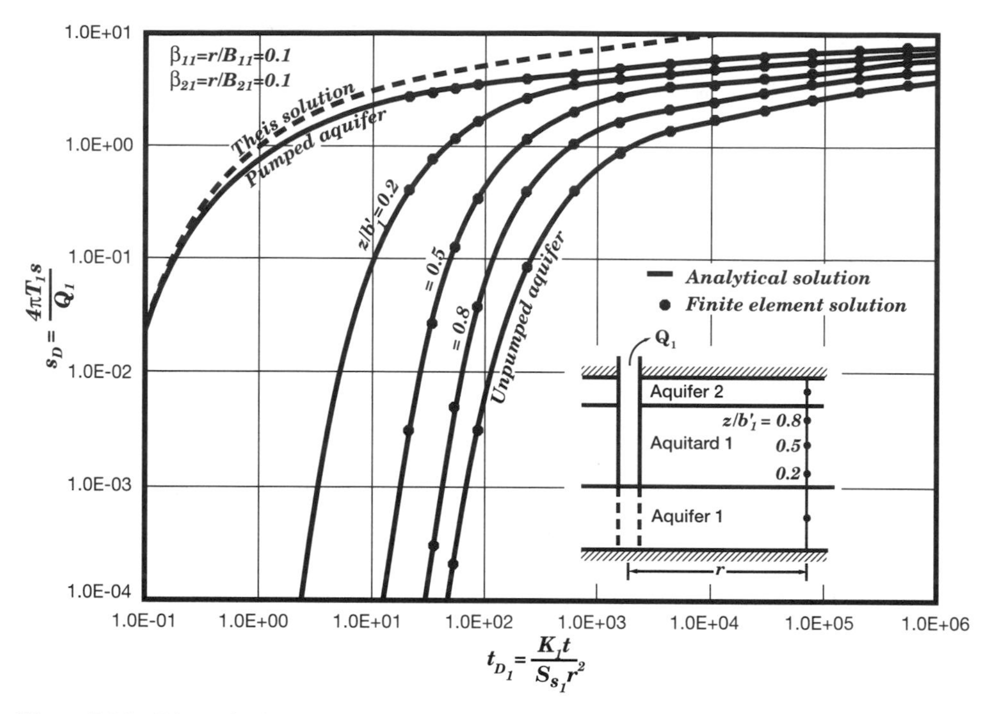

Figure 7-24 Dimensionless drawdown versus dimensionless time for two confined aquifers and one aquitard system for $\beta_{11} = \beta_{21} = r/B_{11} = r/B_{21} = 0.1$. (After Neuman and Witherspoon 1969b.)

unpumped aquifer have a negligible effect on drawdowns in other parts of the system. This means that the asymptotic solutions are independent of the magnitudes of β_{21} and r/B_{21} for the short-time solutions. However, as expected, at larger values of time the solutions everywhere in the system will be affected by the properties of the unpumped aquifer—that is, by β_{21} and r/B_{21} as shown in Figures 7-25 and 7-26. As β_{12} and r/B_{21} range from zero to infinity the unpumped aquifer has infinite and zero transmissivities, respectively. Figure 7-25 show results for $z/b1' = 0.8$ in the aquitard, and Figure 7-26 shows the effects in the pumped aquifer. The limiting values for the dimensionless time are indicated in both figures. The results shown in Figures 7-25 and 7-26 were determined using the finite-element method. The particular curves for $\beta_{21} = r/B_{21} = 0.1$ and 0 were also checked using the analytical solutions as given by Eqs. (7-129) and (7-130). The conclusions reached by Neuman and Witherspoon (1969b) are as follows:

1. As the hydraulic conductivity of the unpumped aquifer approaches zero (i.e., $\beta_{21} = r/B_{21} \to \infty$), the assumption of vertical flow is no longer generally applicable, and consequently the analytical solutions must be used with caution.
2. When the transmissivity of the unpumped aquifer becomes very large (i.e., $T_2 \to \infty$), drawdowns become constant as the pseudo-steady state is reached. In this case, the drawdown changes linearly across the thickness of the aquitard. These results further suggest that when $T_2 > 100T_1$ (i.e., $\beta_{21} < 0.1\beta_{11}$ and $r/B_{21} < 0.1r/B_{11}$), one is probably justified in neglecting drawdown in the unpumped aquifer.

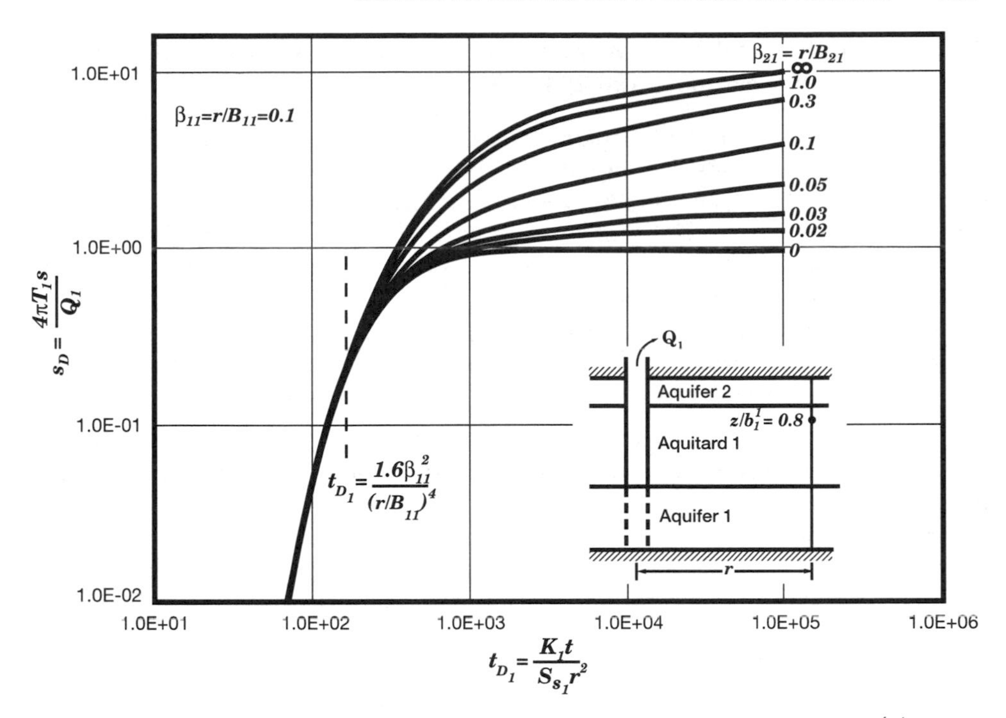

Figure 7-25 Effect of unpumped aquifer on dimensionless drawdown in aquitard at $z/b_1' = 0.8$. (After Neuman and Witherspoon, 1969b.)

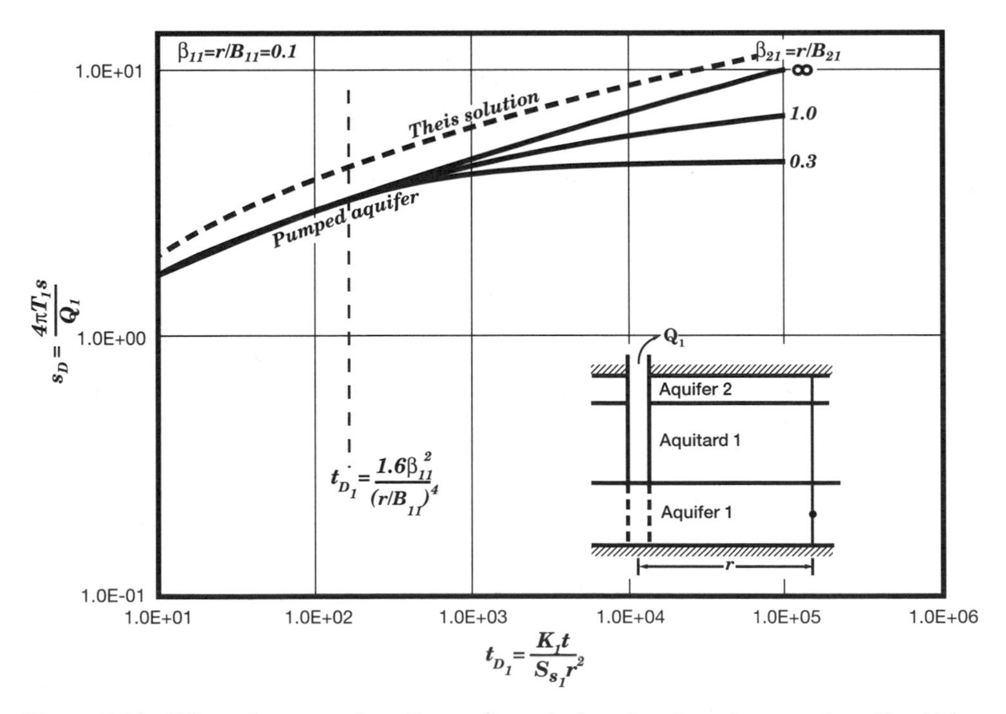

Figure 7-26 Effect of unpumped aquifer on dimensionless drawdown in pumped aquifer. (After Neuman and Witherspoon, 1969b.)

7.4.7 Evaluation of the Two Key Assumptions of the Hantush and Jacob (1955) and Hantush (1960b) Models

Strictly speaking, the actual flow for the hydrogeologic systems which are composed of confined aquifers and aquitards as defined in Section 7.2.1 for Hantush (1960b), Section 7.4.1 for Neuman and Witherspoon (1969a, b) model, and Section 6.2.1 of Chapter 6 for Hantush and Jacob (1955) model is three-dimensional. In order to simplify the mathematics for the determination of analytical solutions, some restrictive assumptions had to be made. Of those assumptions listed in the aforementioned sections, the validity and limitations of the following assumptions need to be known by the user: (1) Storage in the aquitards is negligible and (2) drawdowns in the unpumped aquifers remain zero. In the following sections, the evaluation of these assumptions are presented based on Neuman and Witherspoon (1969c).

7.4.7.1 First Key Assumption: Storage in the Aquitard May Be Neglected.

Hantush and Jacob (1955) presented a solution for the problem of flow for a leaky confined aquifer on the assumption that the storage capacity of the aquitard is negligible (see Chapter 6). The Hantush and Jacob solution, as given by Eq. (6-21) of Chapter 6, can also be expressed in the following form using Neuman and Witherspoon's variables given by Eqs. (7-139) and (7-143):

$$s_1(r,t) = \frac{Q_1}{4\pi T_1} \int_{\frac{1}{4t_{D_1}}}^{\infty} \frac{1}{y} \exp\left[-y - \frac{\left(\frac{r}{B_{11}}\right)^2}{4y} \right] dy \tag{7-162}$$

It can be seen that if the storage is to be neglected, then $\beta_{11} = 0$ in Eq. (7-159), and Eqs. (7-158) and (7-162) become identical. Some key aspects regarding the validity of Hantush and Jacob model, which will be referred to here as *the r/B solution*, are summarized below based on Neuman and Witherspoon's detailed analysis (Neuman and Witherspoon, 1969c).

1. The family of curves for $\beta = 0.01$ shown in Figure 7-27 are almost identical with those of the r/B solution curves (compare with Figure 6-2 of Chapter 6). It can be observed from Figure 7-27 that the Theis solution slowly diverges from the envelope of curves with time instead of coinciding with them. From this, it can be reached to the conclusion that at large values of time for a system, neglecting the storage effects in aquitards the error will be negligible as long as $\beta_{11} \leq 0.01$.
2. As β_{11} increases, the r/B solution becomes less and less representative of the actual behavior in the pumped aquifer. The errors resulted from the r/B solution become significant when β_{11} reaches 0.1, and they are large when β_{11} reaches 1.0. Figure 7-28 presents the superposed r/B solution on the $\beta_{11} = 1.0$ solution and shows that the errors resulted from the r/B solution increase as r/B_{11} decreases. However, these errors decrease with time and disappear altogether for those values of t_{D_1} given by Eq. (7-160). The conclusion is that with large values of β_{11} and zero drawdown in the unpumped aquifer, the r/B solution is subject to significant errors if $t_{D_1} \leq 80\beta_{11}^2/(r/B_{11})^4$.

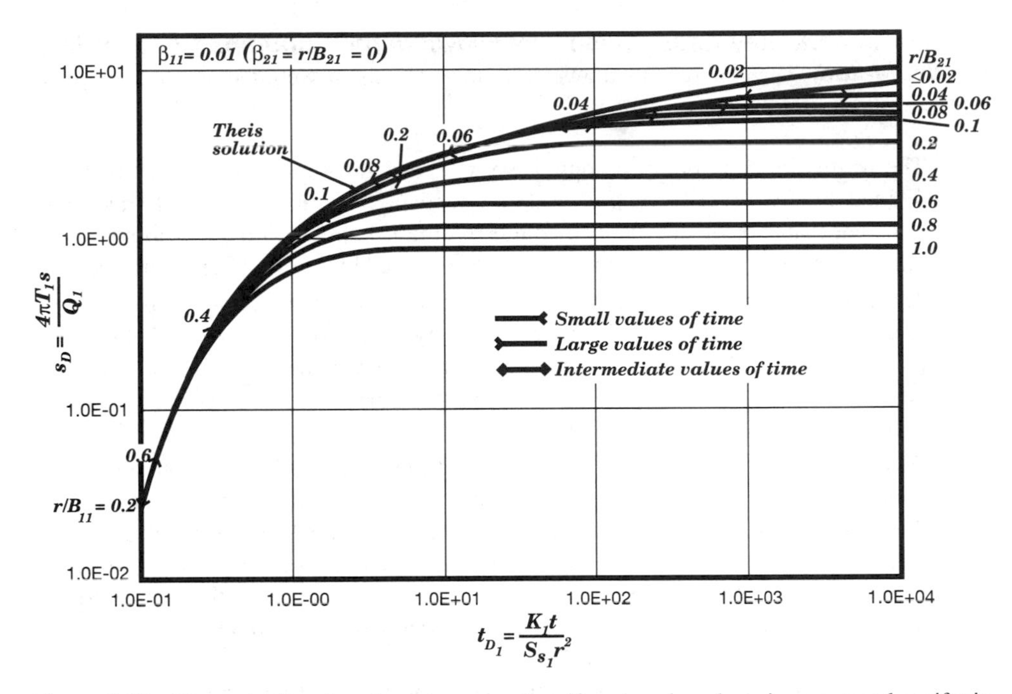

Figure 7-27 Dimensionless drawdown in pumped aquifer when drawdown in unpumped aquifer is zero for $\beta_{11} = 0.01$. (After Neuman and Witherspoon, 1969c.)

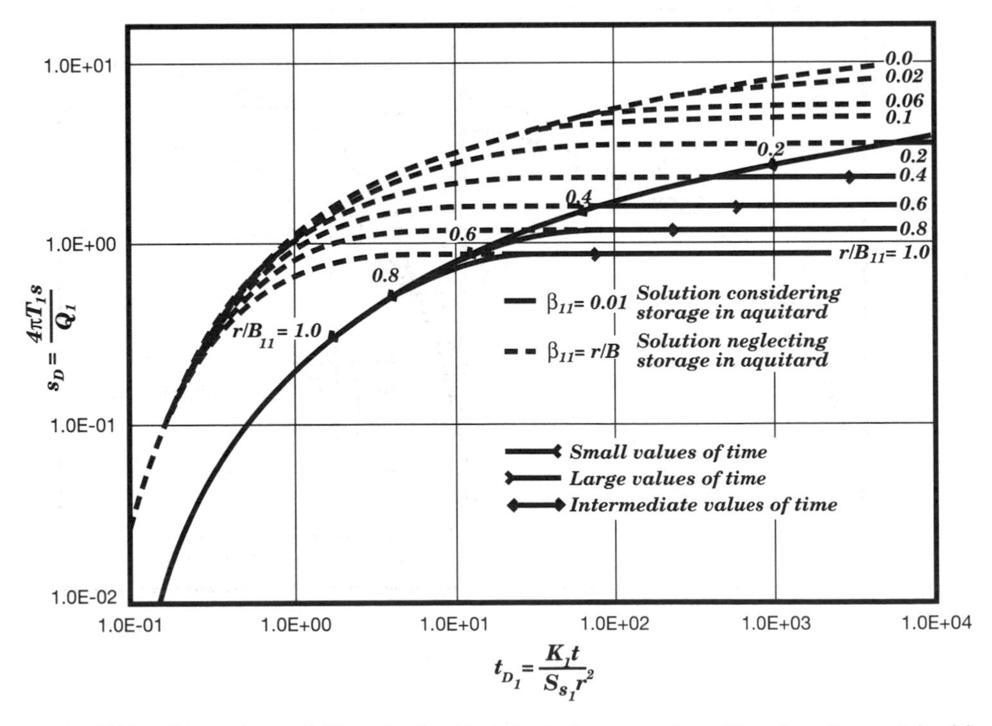

Figure 7-28 Comparison of dimensionless drawdown for pumped aquifer when $\beta_{11} = 1.0$ with Hantush and Jacob's r/B solution. (After Neuman and Witherspoon, 1969c.)

7.4.7.2 Second Key Assumption: Drawdown in the Unpumped Aquifer May Be Neglected. The conclusions drawn from the analysis of Neuman and Witherspoon (1969c) can be summarized as follows:

1. The results in the unpumped aquifer are independent of r/B_{21} and β_{21}, which are defined by Eqs. (7-140) and (7-142), respectively, at small values of time. The transient behavior of the pumped aquifer will not be affected by conditions in the unpumped aquifer as long as the condition expressed by Eq. (7-154) is satisfied.
2. The results in the unpumped aquifer under transient conditions are dependent on r/B_{11}, r/B_{21}, β_{11}, and β_{21}, which are defined by Eqs. (7-139), (7-140), (7-141), and (7-142), respectively. Thus, the assumption (drawdown in the unpumped aquifer may be neglected) is valid at small values of time everywhere in the system, with the exception of the unpumped aquifer itself.
3. When $t_{D_1} > 1.6\beta_{11}^2/(r/B_{11})^4$, the behavior of the unpumped aquifer on drawdown in other parts of the system may be significant. For example, Figure 7-29 shows that the results for the pumped aquifer that are uniquely defined for given values of β_{11} and r/B_{11} at small values of time become separate curves at large values of time, depending on the values of β_{21} and r/B_{21}. As mentioned above, if the transmissivity of the unpumped aquifer is infinitely large, its drawdown becomes zero. Based on Eq. (7-142), $T_2 = \infty$ means that $\beta_{21} = r/B_{21} = 0$. Thus, the lower branch of the curves for the pumped aquifer on Figure 7-29 corresponds to this special case.

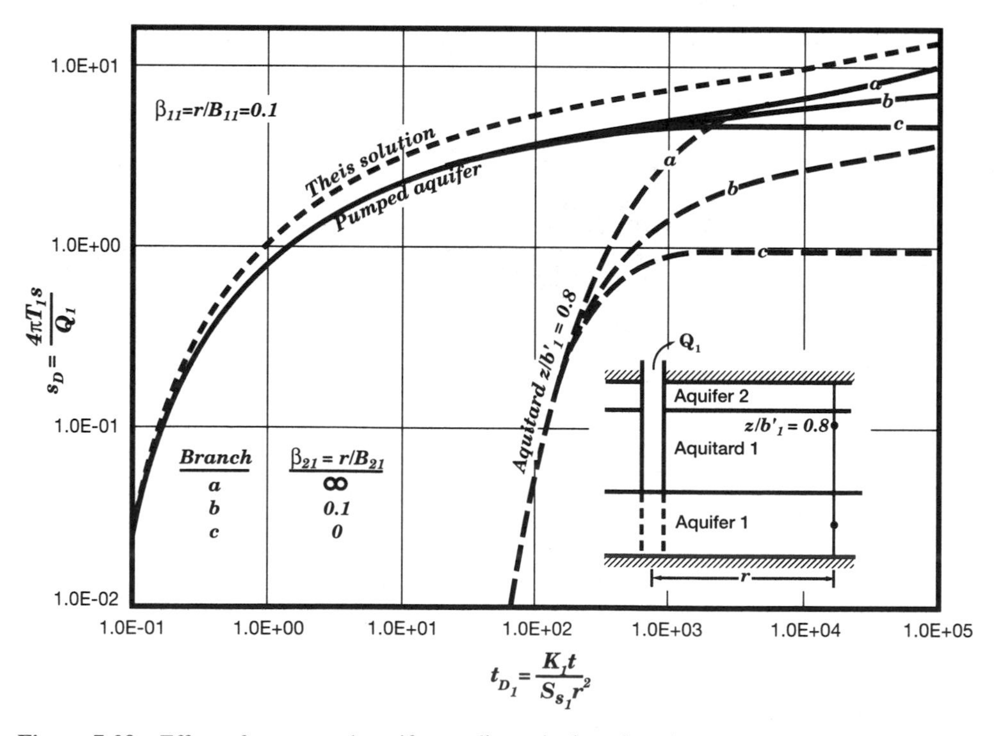

Figure 7-29 Effect of unpumped aquifer on dimensionless drawdown in pumped aquifer and in aquitard at $z/b_1' = 0.8$. (After Neuman and Witherspoon, 1969c.)

7.4.8 Evaluation of the Effect of Neglecting Storage in Aquitard on Drawdown in Pumped and Unpumped Aquifers

Hantush (1967) also analyzed the two aquifers case on the assumption that the storage effects of the aquitard is negligible. Neuman and Witherspoon (1969c) compared their results with the results of Hantush's special case equations for $T_1 = T_2 (r/B_{11} = r/B_{21})$.

For the aforementioned conditions and using Eqs. (7-139) and (7-143), Hantush's equations [Hantush, 1967, Eqs. (17) and (18)] take the following forms:

$$s_1(r,t) = \frac{Q_1}{8\pi T_1} \int_{\frac{1}{4t_{D_1}}}^{\infty} \left\{ \exp(-y) + \exp\left[-y - \frac{2\left(\frac{r}{B_{11}}\right)^2}{4y} \right] \right\} \frac{dy}{y} \tag{7-163}$$

$$s_2(r,t) = \frac{Q_1}{8\pi T_1} \int_{\frac{1}{4t_{D_1}}}^{\infty} \left\{ \exp(-y) - \exp\left[-y - \frac{2\left(\frac{r}{B_{11}}\right)^2}{4y} \right] \right\} \frac{dy}{y} \tag{7-164}$$

Figure 7-30 presents a comparison of dimensionless drawdown for the pumped and unpumped aquifers between Neuman and Witherspoon's solutions given by Eqs. (7-129) and (7-131) and Hantush's solutions given by Eqs. (7-163) and (7-164) for $\beta_{11} = \beta_{21} =$

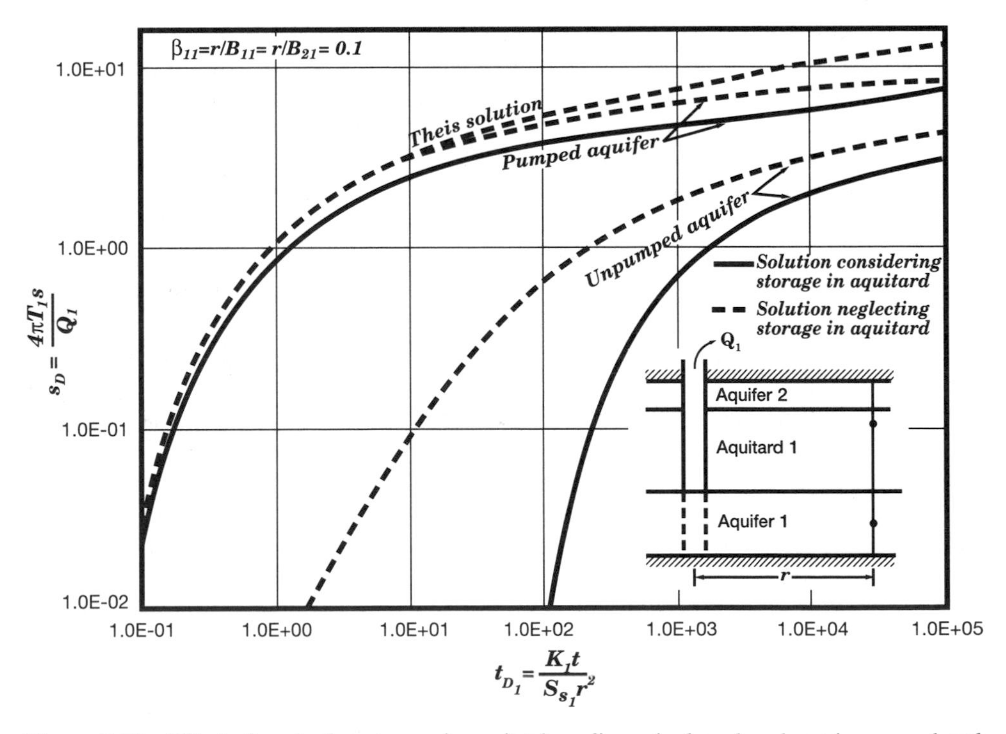

Figure 7-30 Effect of neglecting storage in aquitard on dimensionless drawdown in pumped and unpumped aquifers. (After Neuman and Witherspoon, 1969c.)

$r/B_{11} = r/B_{21} = 0.1$. As observed from Figure 7-30, in the pumped aquifer Neuman and Witherspoon's results are close to Hantush's somewhat overestimated drawdown results. Hantush's solution coincides with the Theis solution at early time and then it diverges from it as time increases but stays between the Neuman and Witherspoon's solution and Theis solution. If $\beta_{11} = \beta_{21} = 1.0$, the differences between the two models would be considerably greater.

For the unpumped aquifer Hantush's model again predicts overestimated drawdowns and the discrepancies between the two models are higher than the ones in the pumped aquifer. As can be seen from Figure 7-30, Hantush's model results an earlier response and as time increases the discrepancies decrease. Prediction of relatively greater drawdown by Hantush's model is due to the fact that in the Neuman and Witherspoon's model the aquitard contributes water from storage to the pumped aquifer. The differences between the two models will be greater if $\beta_{11} = \beta_{21} = 1.0$.

7.4.9 Application to Pumping Test Data Analysis for Leaky Aquifer Conditions: Neuman and Witherspoon's Ratio Method for Aquitards

7.4.9.1 Theory of the Ratio Method. The mathematical model developed by Neuman and Witherspoon for leaky aquifers as given by Eqs. (7-129), (7-130), and (7-131) shows that the behavior of drawdown in each layer is a function of B_{11}, B_{21}, β_{11}, β_{21}, and t_{D_1}, which are defined by Eqs. (7-139), (7-140), (7-141), (7-142), and (7-143), respectively. These quantities depend on hydraulic conductivities, storativities, and thicknesses of the layers. The observation of drawdown in the pumped aquifer alone is not always sufficient for the determination of the values of β and r/B values. In each aquifer in Figure 7-23, the solutions depend on the five dimensionless parameters r/B_{11}, r/B_{21}, β_{11}, β_{21}, and t_{D_1}. The solution for aquitard involves one additional parameter z/b_1'. This relatively large number of dimensionless parameters makes the situation practically impossible to generate a sufficient number of type curves to cover the entire range of values for the analysis of pumping tests conducted in an aquifer shown in Figure 7-23. For the type-curve matching to be efficient, the number of independent dimensionless parameters for a set of type curves should not exceed two.

Neuman and Witherspoon (1972) carried out an analysis for the reduction of the number type curves in restricting the analysis of field data to small values of time based on the fact that at early times the unpumped aquifer does not create any influence on the rest of the hydrogeologic system. This implicitly means that drawdowns are independent of β_{21} and r/B_{21} and the aquitard acts as though it has infinite thickness. Because of this situation, the drawdown is independent of r/B_{11}, defined by Eq. (7-139), and z/b_1'. Therefore, the resulting equations for drawdown will be dependent only on β_{11}, t_{D_1}, and t_{D_1}'.

In the pumped aquifer, the drawdown equation for small values of time is given by Eq. (7-152), which is the same as Eq. (7-32), originally developed by Hantush (1960b). In the aquitard the drawdown solution is given by Eq. (7-156). Theoretically, these equations are limited for the condition expressed by Eq. (7-153) or Eq. (7-154). The dimensionless parameters β_{11}, t_{D_1}, and t_{D_1}' are defined by Eqs. (7-141), (7-143), and (7-157), respectively.

Neuman and Witherspoon investigated the variation of s_1'/s_1, the ratio between the drawdown in the aquitard and the drawdown in the aquifer at the same elapsed time (t) and the same radial distance (r) from the pumping well. As mentioned above, the hydrogeologic system is composed of one aquifer and one aquitard. For this reason, all the subscripts in the aforementioned quantities will be dropped and will be shown by β, t_D, t_D', and s'/s, respectively. The variation of s'/s versus t'_D is shown in Figure 7-31 for different practical range values of t_D and β. As can be seen from Figure 7-31, the curves corresponding to $t_D = 0.2$, $\beta = 0.0, 0.01$, and 1.0 are practically the same. Of these three β values, $\beta = 0.0$ corresponds to the Theis equation as given by Eq. (4-24) of Chapter 4. This situation can be observed from Eq. (7-152) by substituting $\beta = 0.0$ in the same equation. The same situation is valid for $t_D = 10^4$ and for the same values of β. If Eq. (7-107), which is based on Neuman and Witherspoon's slightly leaky aquifer theory, is used instead of Eq. (7-155), and the drawdown in the confined aquifer, s, is determined from the Theis equation, there

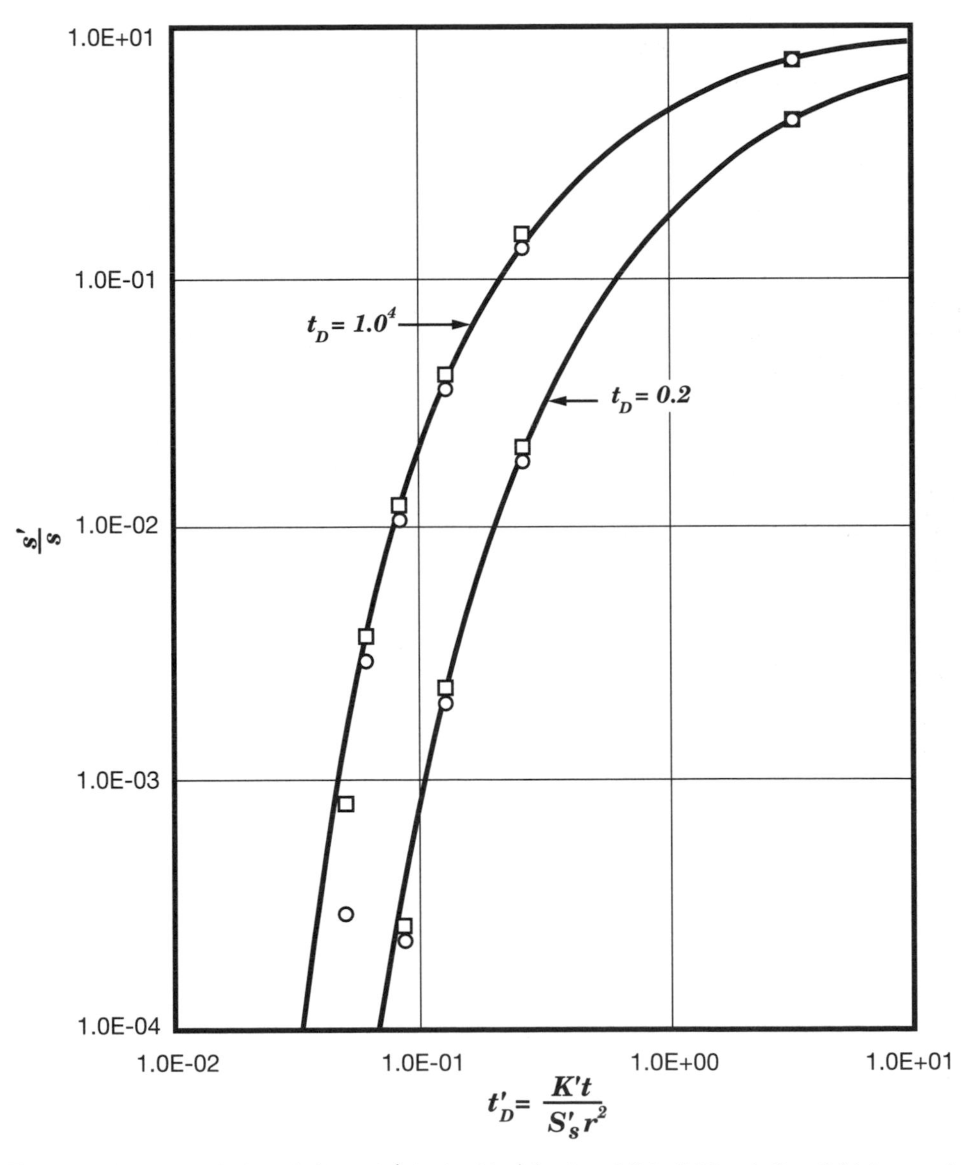

Figure 7-31 The variation of $s'(r, z, t)/s(r, t)$ with t'_D for $\beta = 0.0$ (solid lines), $\beta = 0.01$ (squares), and $\beta = 1.0$ (circles). (After Neuman and Witherspoon, 1972.)

is in effect a special case where $\beta = 0.0$. The two solid lines in Figure 7-31 represent this special case. With modified notation given above, Eq. (7-107) takes the form

$$s'(r, z, t) = \frac{Q}{4\pi T} W(t_D, t'_D) \tag{7-165}$$

where

$$W(u, u') = W(t_D, t'_D) = \frac{2}{\pi^{1/2}} \int_{\frac{1}{(4t'_D)^{1/2}}}^{\infty} \left\{ -\,\mathrm{Ei}\left[-\frac{t'_D y^2}{t_D(4t'_D y^2 - 1)} \right] \right\} e^{-y^2}\, dy \tag{7-166}$$

$$t_D = \frac{1}{4u} = \frac{Kt}{r^2 S_s} = \frac{Tt}{r^2 S} \tag{7-167}$$

and

$$t'_D = \frac{1}{4u'} = \frac{K't}{S'_s z^2} \tag{7-168}$$

For $\beta = 10.0$, the values of s'/s are significantly different than those shown in Figure 7-31. From the above-mentioned facts, Neuman and Witherspoon (1969c) reached the following conclusion: For all practical values of t_D, the ratio s'/s is independent of β if its value does not exceeds 1.0.

Because the ratio s'/s is a key quantity, the method is called as *the ratio method* by Neuman and Witherspoon. The purpose of the ratio method is to determine the hydraulic diffusivity (α'), which is defined as the ratio between the hydraulic conductivity and storage coefficient, of aquitards under arbitrary leaky conditions. From Eq. (7-168) the expression from α' is

$$\alpha' = \frac{K'}{S'_s} = \frac{z^2}{t} t'_D \tag{7-169}$$

Based on the same notation, the Theis equation, as given by Eq. (4-24) of Chapter 4, can be expressed as

$$s(r, t) = \frac{Q}{4\pi T} W(u) \tag{7-170}$$

where

$$W(u) = \int_{\frac{1}{4t_D}}^{\infty} \frac{e^{-y}}{y}\, dy \tag{7-171}$$

The ratio of $s'(r, z, t)$ to $s(r, t)$, which are given by Eqs. (7-165) and (7-170), respectively, is

$$\frac{s'(r, z, t)}{s(r, t)} = \frac{W(t_D, t'_D)}{W(t_D)} = \frac{W(u, u')}{W(u)} \tag{7-172}$$

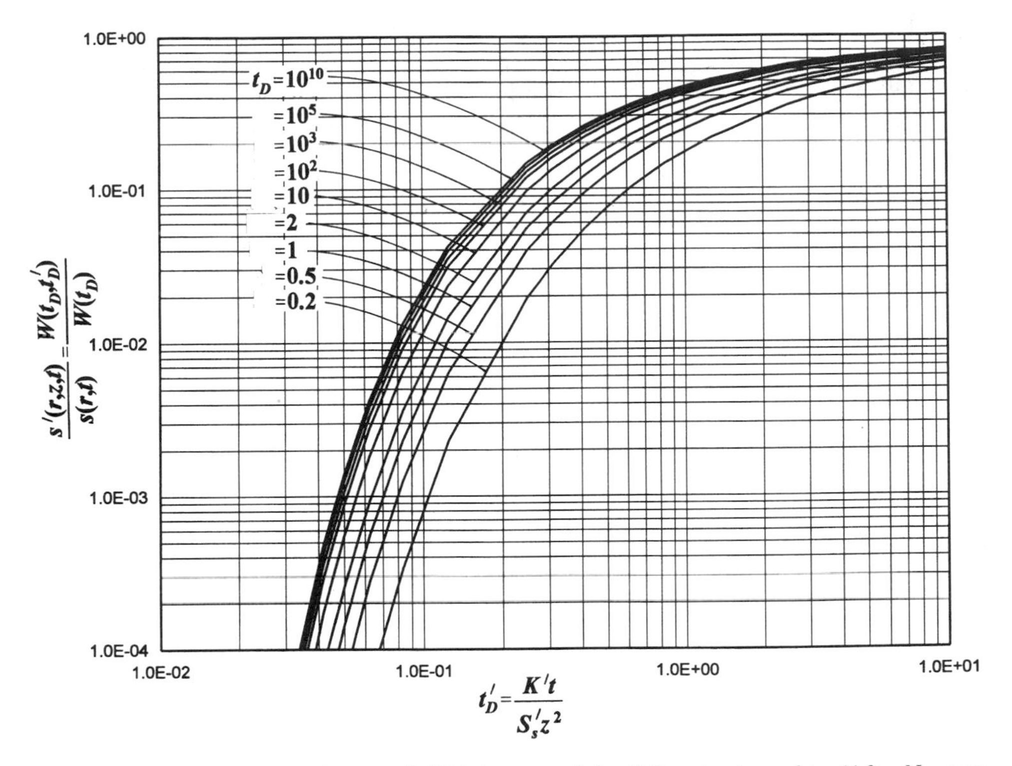

Figure 7-32 The variation of $W(t_D, t'_D)/W(t_D)$ versus t'_D for different values of t_D. (After Neuman and Witherspoon, 1972.)

The variation of $W(t_D, t'_D)/W(t_D)$ versus t'_D for different values of t_D is presented in Figure 7-32, which was prepared from values given by Witherspoon et al. (1967) by Kruseman and De Ridder (1991), and is also presented in Table 7-6.

7.4.9.2 Determination of Hydraulic Conductivity and Specific Storage of an Aquitard from its Hydraulic Diffusivity. To evaluate the hydraulic conductivity (K') and specific storage (S'_s) of an aquitard from its hydraulic diffusivity (α'), either K' or S'_s must first be determined by other methods. In general, the values of K' may vary by several orders of magnitude from one aquitard to another and even from location to location in the same aquitard. For the values of S'_s, the range is relatively much narrow from aquitard to aquitard and location to location in the same aquitard. The potential methods for the determination of K' and S'_s values of an aquitard can be listed as follows:

1. Conducting standard consolidation tests on core samples in laboratory and with the following formula S'_s can be calculated [see Chapter 2, Section 2.4.1.4, Eq. (2-73))]:

$$S'_s = \frac{a_v \gamma_w}{1 + e} \tag{7-173}$$

   where $a_v(LT^2M^{-1})$ is the coefficient of compressibility; $\gamma_w(ML^{-2}T^{-2})$ is the specific weight of water; and e (dimensionless) is the void ratio.

2. The product ($K'S'$) of an aquitard can be determined from the Hantush (1960b) model based on pumping test data analysis methods (see Section 7.2) if the thickness of aquitard (b') is known. Let us show the known quantity by C:

$$K'S' = K'S'_s b' = C \tag{7-174}$$

Then, the combination of Eqs. (7-169) and (7-174) gives

$$S'_s = \left(\frac{C}{\alpha' b'} \right)^{1/2} \tag{7-175}$$

3. If the above-mentioned laboratory and field measured values are not available, S'_s can be estimated by correlating publish results on similar formations. Once the value of S'_s has been estimated, K' can be determined from Eq. (7-169) using the known values α' and S'_s.

7.4.9.3 The Ratio Method for the Determination of Hydraulic Diffusivity (α') of an Aquitard. The ratio method can be applied to any hydrogeologic system that is composed of an aquifer and its underlain and overlain adjacent aquitards (Figure 7-33). Boulton (1963) and Neuman (1972) showed that during early values of time, drawdown in an unconfined aquifer resulting from a steady pumping rate can safely be approximated

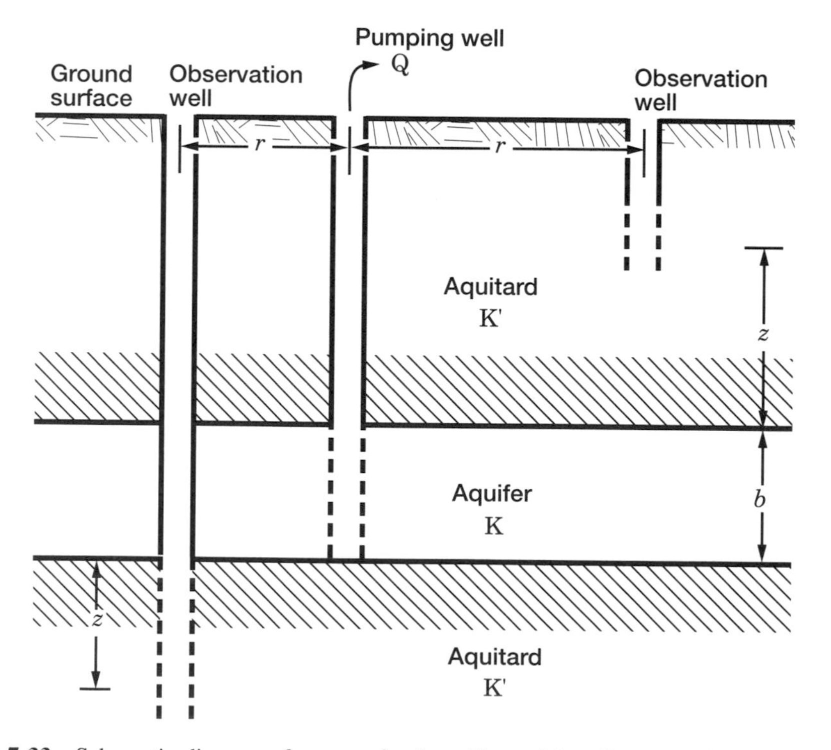

Figure 7-33 Schematic diagram of one confined aquifer and its adjacent aquitards for the ratio method. (After Neuman and Witherspoon, 1972.)

by the Theis equation. Based on this fact, Neuman and Witherspoon (1972) concluded that if the ratio method is applicable to aquitards adjacent to confined leaky aquifers, it should also be applicable to those situations for which the pumped aquifer is unconfined and is underlain by an aquitard. Some key aspects of the ratio method regarding its application can be summarized as follows (Neuman and Witherspoon, 1972):

1. The ratio method can be applied to arbitrary multiple aquifer systems and does not require any graphical curve-matching procedures.
2. The pumped aquifer can be either confined or unconfined.
3. The confining layers or aquitards can be heterogeneous and anisotropic. Under this condition, the ratio method gives the average vertical hydraulic conductivity along the thickness of the aquitard.
4. The method is based on drawdown data collected during the early values of time. As a result, the pumping test period can relatively be small.
5. The drawdown data collected for the unpumped aquifer or in the aquitard provide an *in situ* indication of time limit beyond which the ratio method gives unreliable results.
6. The accuracy of drawdowns measured in the aquitard is not critical because the method is more sensitive to time lag than the magnitude of $s'(r, z, t)/r(r, t)$.
7. The method does not require prior knowledge of the aquitard thickness.

The method is based on a family of curves of $s'(r, z, t)/s(r, t)$, which is given by Eq. (7-172), versus t'_D for different values of t_D. The steps for the ratio method (Neuman and Witherspoon, 1972) are given below.

Step 1: Determine the values of transmissivity (T) and storage coefficient (S) for the confined or unconfined aquifer by any appropriate method as outlined below:

(a) If the pumped aquifer is slightly leaky, determine its T and S values by the methods based on the Theis equation (see Chapter 4, Section 4.2).

(b) If leakage is significant, use the Walton's type-curve method, Hantush's methods based on the Hantush and Jacob (1955) model (Chapter 6, Section 6.3), or the Hantush (1960b) model-based methods (Section 7.2.7.1).

Step 2: At a selected radial distance (r) in the aquifer and aquitard and at a selected vertical distance (z) in the aquitard, determine the ratio $s(r, z, t)/s(r, t)$ at a given early value of time (t). Repeat this process for other values of r, z, and t, if possible.

Step 3: Determine the values of t_D for the particular values of r and t at which the values of $s'(r, z, t)/s(r, t)$ have been obtained. When $t_D < 100$, the curves in Figure 7-32 are sensitive to minor changes of t_D. When $t_D > 100$, the curves in Figure 7-32 are very close to each other and, therefore, they are practically independent of t_D.

Step 4: With the known values of s'/s and t_D from the previous steps, determine the corresponding value of t'_D from Figure 7-32.

Step 5: Using the known values of z, t, and t'_D, determine the diffusivity of the aquitard (α') from Eq. (7-169). One should note from Figure 7-32 that the value of t'_D is not very sensitive to the magnitude of s'/s and, therefore, α' calculated from Eq. (7-169) does not depend significantly on the actual drawdown value in the aquitard. The critical quantity for the determination of α' at a given z elevation is the time period between the beginning of the test and the time when the observation well in the aquitard starts to respond.

Step 6: Determine the hydraulic conductivity (K') and specific storage (S_s') of the aquitard from its hydraulic diffusivity (α') as outlined in Section 7.4.9.2.

Example 7-3: This example is taken from Neuman and Witherspoon (1972).

General Site Description. Neuman and Witherspoon selected the Oxnard aquifer in the city of Oxnard, California. The location of pumped well (22J5) and piezometers are shown in Figure 7-34. The Oxnard aquifer, which is the most important source in the Oxnard basin, is under confined conditions and is underlain and overlain by two aquitards and the upper aquitard is overlain by a semiperched aquifer (Figure 7-35). The lower aquitard is underlain by the Mugu aquifer. The pumped well was screened throughout the full thickness, which is around 93 ft, of the Oxnard aquifer.

Piezometers. For piezometers (22H2, 22H5, 22K2, and 23E2) were available to monitor the hydraulic heads in the Oxnard aquifer at radial distances ranging from 502 ft to 1060 ft. In addition, seven new piezometers were installed at different elevations in the upper, lower aquitards, and semiperched aquifer. The pertinent information for these piezometers are

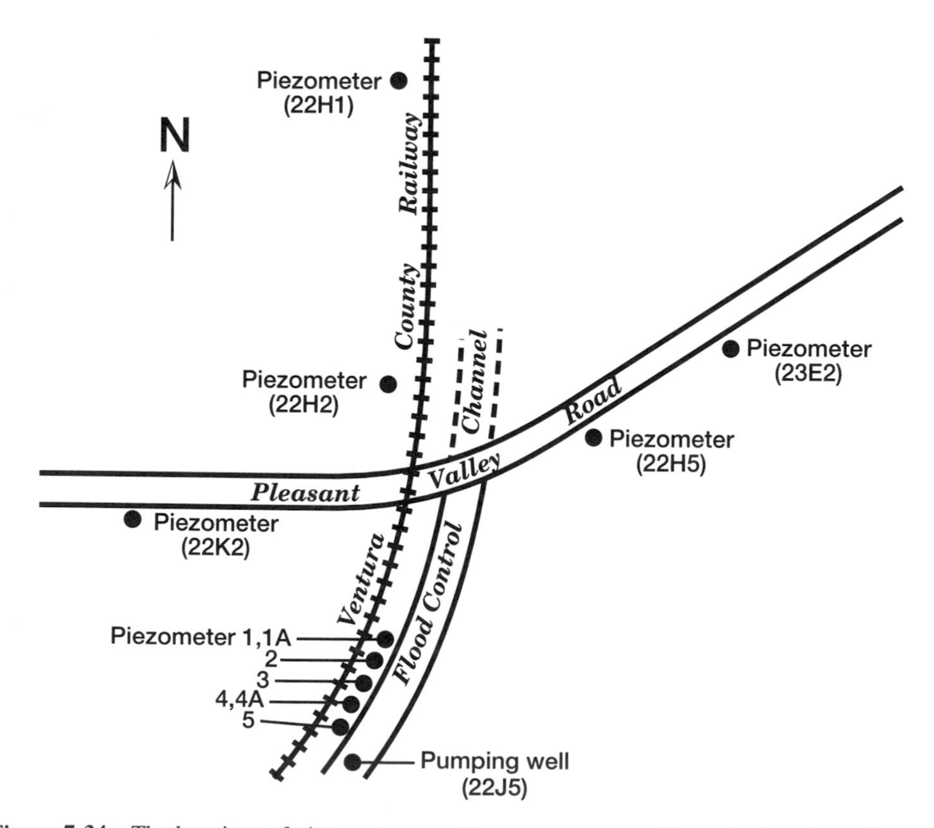

Figure 7-34 The locations of piezometers used in pumping test for Example 7-3. (After Neuman and Witherspoon, 1972.)

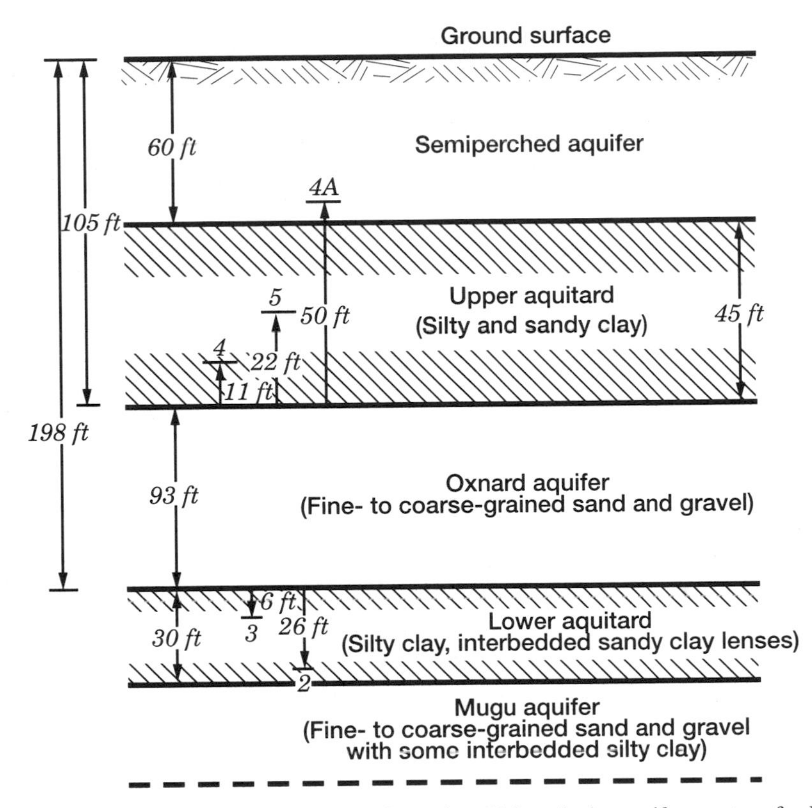

Figure 7-35 Schematic cross section of the Oxnard and Mugu leaky aquifers system for Example 7-3. (Adapted from Neuman and Witherspoon, 1972.)

summarized in Table 7-7. It would be ideal if the seven piezometers had been arranged along a circular arc having its center at the pumped well. Therefore, responses would be determined at different elevations having the same radial distance from the pumped well. However, this arrangement was not possible because of the local conditions.

TABLE 7-7 Location of Piezometers for Oxnard Aquifer Pumping Test for Example 7-3

Piezometer	Distance from 22J5 (ft)	Depth (ft)	Vertical Distance[a] (ft)	Layer
1	100	120	—	Oxnard aquifer
1A	100	239	—	Mugu aquifer
2	91	225	−26	Lower aquitard
3	81	205	− 6	Lower aquitard
4	72	95	+11	Upper aquitard
4A	72	58.5	+50	Semiperched aquifer
5	62	84	+22	Upper aquitard

[a] Positive values are measured from the top of the Oxnard aquifer located at a depth of 105 ft. Negative values are measured from the bottom of the Oxnard aquifer located at a depth of 198 ft (see Figure 7-35).
Source: After Neuman and Witherspoon (1972).

Lithology. A schematic cross section of the site hydrogeologic system is shown in Figure 7-35. The semiperched zone extends until 60 ft depth and is composed of fine- to medium-grained sand with interbedded silty clay lenses. The upper aquitard is 45 ft thick and is made up of predominantly silty and sandy clays, mainly montmorillonite. The Oxnard aquifer is around 93 ft thick and is composed of fine- to coarse-grained sand and gravel. The Oxnard aquifer is underlain by an aquitard (lower aquitard) having 30 ft thickness and is made up of silty clay with some interbedded sandy clay lenses. The lower aquitard is underlain by the Mugu aquifer which is composed of fine- to coarse-grained sand and gravel with some interbedded silty clay.

Responses to Pumping Test. The pumping test period was 31 days and the pumping rate was 1000 gallons/min. The response of the five piezometers (Wells 1, 22H2, 22H5, 22K2, and 23E2) in the Oxnard aquifer in double logarithmic coordinate system is shown in Figure 7-36.

Figure 7-37 shows the response at one particular point in the lower aquitard (Well 3), which is 6 ft below from the lower boundary of the Oxnard aquifer (see Figure 7-35), as well as the responses in the Oxnard aquifer (Well 1) and the Mugu aquifer (Well 1A). Figure 7-38 shows the response at two different elevations in the upper aquitard (Wells 4 and 5) and the response in the overlain semiperched aquifer.

Because piezometer 1 is located farthest from the pumped well, the response was not available in the pumped aquifer directly below the piezometers where drawdowns in the upper aquitard were measured. However, from the distance–drawdown curves in the Oxnard

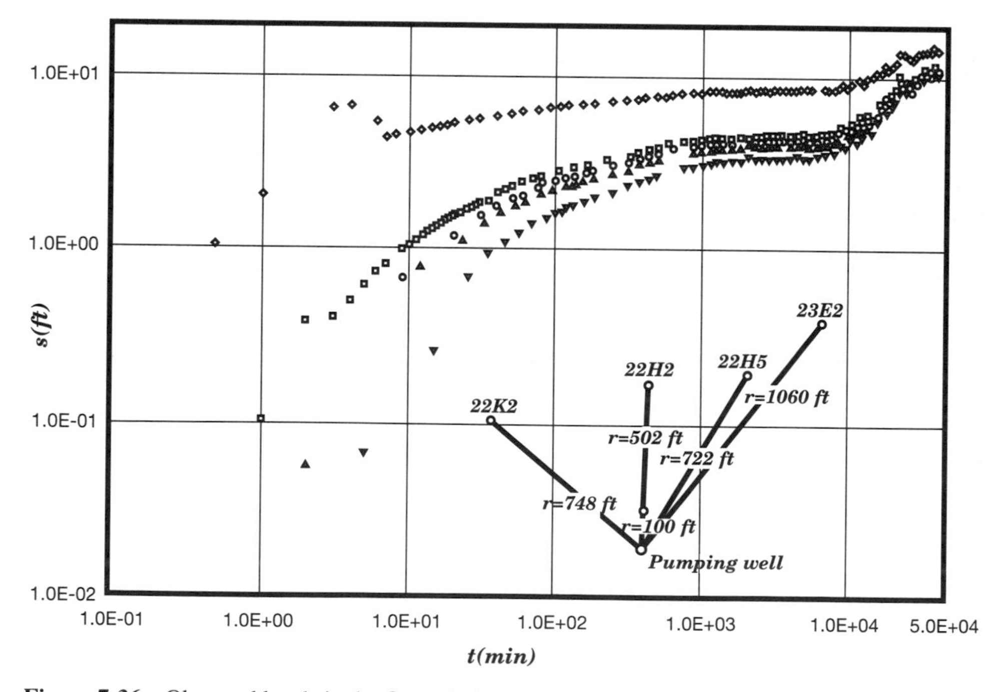

Figure 7-36 Observed heads in the Oxnard piezometers for the pumping test of Example 7-3. The diamonds represent Well 1, the squares represent Well 22H2, the triangles represent Well 22H5, the circles represent Well 22K2, and the inverted triangles represent Well 23E2. (After Neuman and Witherspoon, 1972.)

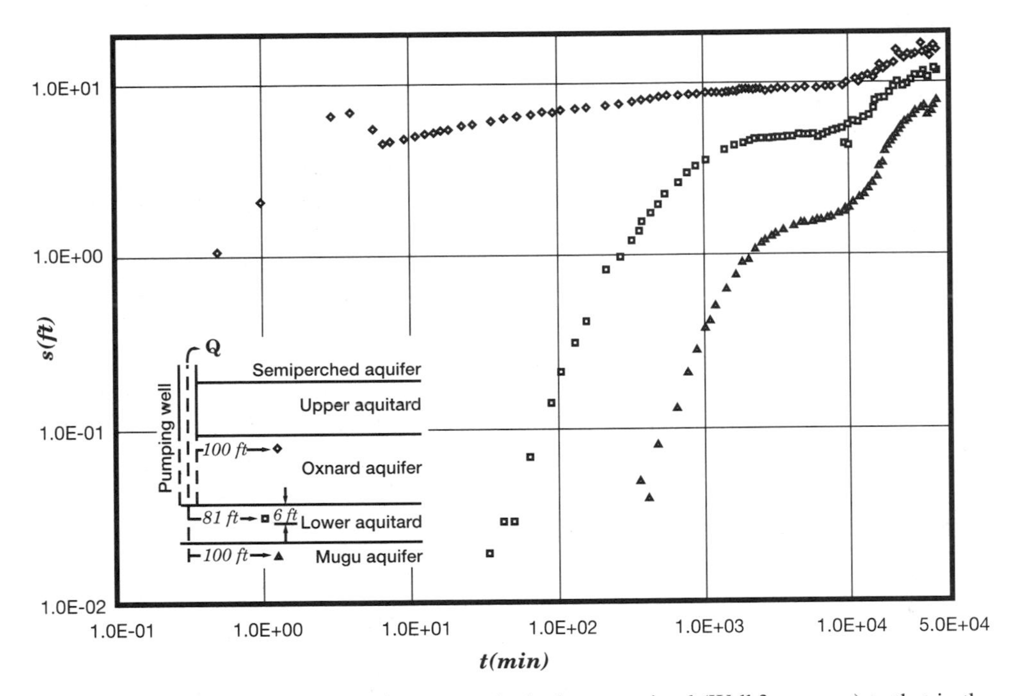

Figure 7-37 The response of the piezometers in the lower aquitard (Well 3, squares) to that in the Oxnard (Well 1, diamonds) and Mugu (Well 1A, triangles) aquifers during the pumping test. (After Neuman and Witherspoon, 1972.)

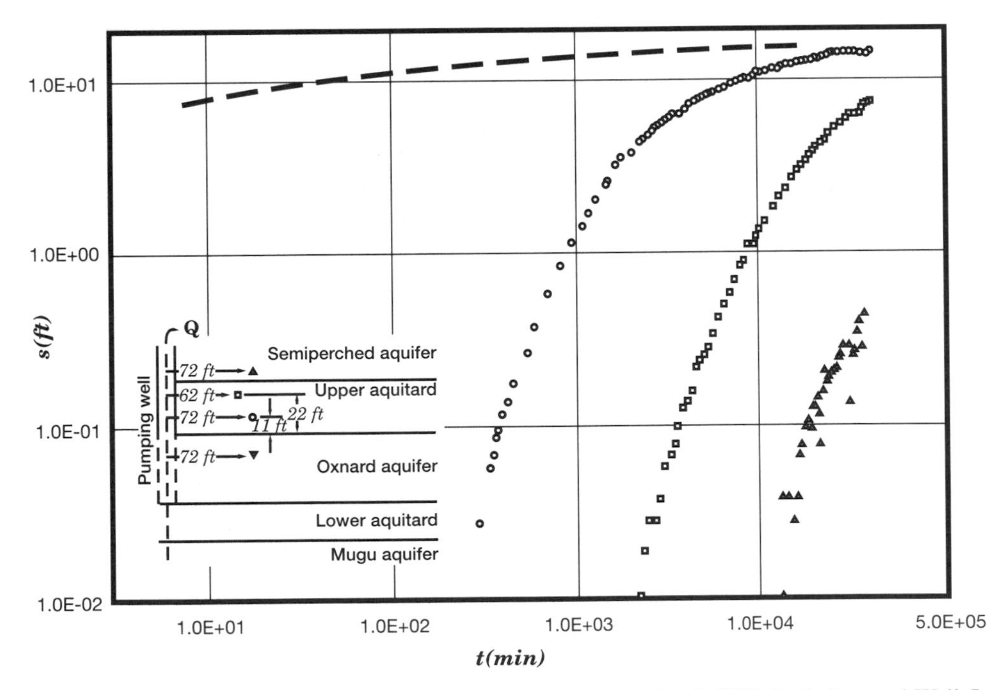

Figure 7-38 The response of the piezometers in the upper aquitard (Well 4, circles; and Well 5, squares) and the semiperched aquifer (Well 4A, triangles) during the pumping test. The broken line indicates the probable response of the Oxnard aquifer at $r = 72$ ft (After Neuman and Witherspoon, 1972.)

aquifer and from the behavior of piezometer 4, Neuman and Witherspoon concluded that the aquifer response was approximately the dashed curve as shown in Figure 7-38. As mentioned above, in Step 5, the ratio method for evaluating aquitards is more sensitive to the time lag than to the actual magnitude of drawdown in the aquifer. As a result, the dashed curve in Figure 7-38 can be considered sufficiently accurate for the purpose of the test. One should note that the shapes of the curves in Figures 7-37 and 7-38 are quite similar to those of Neuman and Witherspoon's theoretical curves as given in Figure 7-24.

From the pumping test data described above, (a) determine the hydraulic diffusivity (α') values of each aquitard using the ratio method, (b) compare the diffusivity, hydraulic conductivity, and specific storage of the aquifer with the ones for the aquitard, and (c) evaluate the values of K' determined from the ratio method with the aid of consolidation tests and from the direct measurement method.

Solution: (a) The pumping test data is analyzed in accordance the steps given above. The steps are as follows:

Step 1: The pumping test data as presented in Figure 7-36 for the Oxnard aquifer was analyzed by Neuman and Witherspoon (1972) using the semilogarithmic method of Cooper and Jacob (see Section 4.2.7.2, Chapter 4), and the values of T and S are presented in Table 7-8.

In general, as shown in Table 7-8, the values of T become higher as r increases, and the reason for this situation can be explained as follows (Neuman and Witherspoon, 1972): Because the Oxnard aquifer is a leaky aquifer (see Figure 7-35), the actual drawdown curve at any given well will be below the Theis curve as shown in Figure 7-39 (see also Figure 7-24). Consider a particular point on the data curve that corresponds to some given value of s and t. If one could match the observed data to the true type curve where β and r/B are not zero, one would determine the true value of s_{D_1} for the chosen point. Because such type curves were not available for this investigation, Neuman and Witherspoon used a method that is the same as the Theis type-curve method as presented in Section 4.2.7.1 of Chapter 4. As a result of this situation, the field data are being shifted upward from their true position, and the chosen point will indicate an apparent value of $s_{D_2} > s_{D_1}$. From the definition of s_D, $s_D = 4\pi Ts/Q$, it is obvious that because s does not change, the value of T is increased. The greater the value of r, the larger β and r/B become, and therefore the larger the difference the true type curve and the Theis type

TABLE 7-8 Values of T and S for the Oxnard Aquifer Determined by the Cooper and Jacob Semilogarithmic Method for Example 7-3

Well	r (ft)	T (gpd/ft)	S
1	100	130,600	1.12×10^{-4}
22H2	502	139,000	3.22×10^{-4}
22H5	722	142,600	3.08×10^{-4}
22K2	748	136,700	2.48×10^{-4}
23E2	1060	157,000	2.53×10^{-4}

Source: After Neuman and Witherspoon (1972).

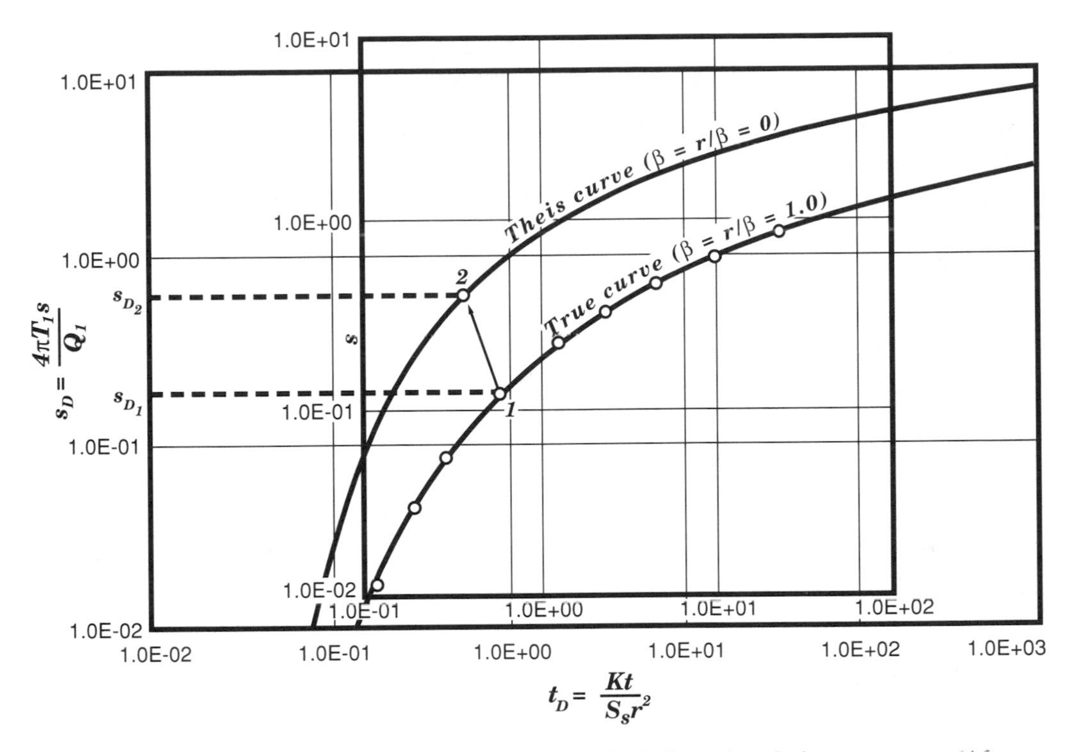

Figure 7-39 A comparison of hypothetical field data with leaky and nonleaky type curves. (After Neuman and Witherspoon, 1972.)

curve is. In other words, as r increases, T should increase as well. This is the case in Table 7-8. Regarding errors in S, the shifting of observed data as indicated in Figure 7-39 may be either to the left or to the right. Therefore, the effect on the calculated values of S in Table 7-8 is not predictable. With this unpredictability in mind, Neuman and Witherspoon decided to select the results from piezometer 1 of $T = 130{,}600$ gpd/ft and $S = 1.12 \times 10^{-4}$ as the most representative values of the Oxnard aquifer in the area of the pumping test.

Step 2: Neuman and Witherspoon (1972) first evaluated piezometer 3 ($r = 81$ ft and $z = 6$ ft, from Table 7-7) in the lower aquitard whose response during the pumping is given in Figure 7-37. In the Oxnard aquifer they used piezometer 1 for which $r = 100$ ft. At two early values of time ($t = 80$ min and $t = 200$ min) the ratio $s'(r, z, t)/s(r, t)$ was determined. According to theory of the ratio method as outlined in Section 7.4.9.1, the ratio between the drawdown in the aquitard and the drawdown in the aquifer at the same elapsed time (t) and the same radial distance (r) from the pumping well must be considered. The above values show that this condition was satisfied approximately because the radial distances of the piezometers in the Oxnard aquifer and lower aquitard to the pumped well are 100 ft and 81 ft, respectively (see Figure 7-37).

At $t = 80$ min, one can read in Figure 7-37 that $s' = 0.078$ ft and $s = 6.6$ ft, and their ratio is $s'/s = 0.078/6.6 = 1.18 \times 10^{-2}$.

Step 3: Substitution of these values with $T = 130{,}600$ gpd/ft $= 12.125$ ft^2/min and $S = 1.12 \times 10^{-4}$ into Eq. (7-167) gives

$$t_D = \frac{Tt}{r^2 S} = \frac{(12.125\ \text{ft}^2/\text{min})(80\ \text{min})}{(81\ \text{ft})^2(1.12 \times 10^{-4})} = 1.32 \times 10^3$$

Step 4: From Figure 7-32, these values of s'/s and t_D correspond to $t_D' = 0.086$.

Step 5: Introducing the values of $z = 6$ ft, $t = 80$ min $= 0.0556$ days, and $t_D' = 0.086$ into Eq. (7-169) gives the value of hydraulic diffusivity:

$$\alpha' = \frac{z^2}{t} t_D' = \frac{(6\ \text{ft})^2}{0.0556\ \text{days}}(0.086)$$

$$= 55.73\ \text{ft}^2/\text{day}$$

$$\alpha' = 4.17 \times 10^2\ \text{gpb/ft}$$

Similarly, one finds that, at $t = 200$ min, $\alpha' = 45.32$ ft^2/day $= 3.39 \times 102$ gpd/ft. Since the ratio method gives more reliable results when t is small, Neuman and Witherspoon adopted $\alpha' = 4.17 \times 10^2$ gpd/ft as the representative value for the top 6 ft of the lower aquitard. The results of the lower and upper aquitards based on similar calculations are summarized in Table 7-9.

Step 6: As mentioned in Section 7.4.9.2, determination of the values of K' require that the values of S_s' should be determined by other methods. Neuman and Witherspoon determined the values of S_s' from laboratory consolidation tests by using Eq. (7-173). These values were then used to calculate K' from Eq. (7-169) as $K' = \alpha' S_s'$, and the results are summarized in Table 7-10.

TABLE 7-9 Results for Hydraulic Diffusivity of Aquitards for Example 7-3

Layer	Section Tested	Hydraulic Diffusivity, $\alpha' = K'/S_s'$ gpd/ft	$\alpha' = K'/S_s'$ cm^2/s
Upper aquitard	Bottom 22 ft	1.02×10^2	1.47×10^{-1}
Upper aquitard	Bottom 11 ft	2.44×10^2	3.51×10^{-1}
Lower aquitard	Top 6 ft	4.17×10^2	5.99×10^{-1}

Source: After Neuman and Witherspoon (1972).

TABLE 7-10 Hydraulic Properties of Aquitard Layers for Example 7-3

Layer	Section Tested	Specific Storage, S_s'		Hydraulic Conductivity, K'	
		ft^{-1}	cm^{-1}	gpd/ft^2	cm/s
Upper aquitard	Bottom 22 ft	2.4×10^{-4}	7.88×10^{-6}	2.45×10^{-2}	1.11×10^{-6}
Upper aquitard	Bottom 11 ft	2.4×10^{-4}	7.88×10^{-6}	5.85×10^{-2}	2.66×10^{-6}
Lower aquitard	Top 6 ft	1.0×10^{-4}	3.28×10^{-6}	4.17×10^{-2}	1.89×10^{-6}

Source: After Neuman and Witherspoon (1972).

(b) The hydraulic diffusivity of the Oxnard aquifer is

$$\alpha = \frac{T}{S} = \frac{130,600 \text{ gpd/ft}}{1.12 \times 10^{-4}} = 1.17 \times 10^9 \text{ gpb/ft}$$

which is more than 1 million times the values obtained for the aquitards. Using $b = 93$ ft as the thickness of the Oxnard aquifer, its hydraulic conductivity and specific storage, respectively, are

$$K = \frac{T}{b} = \frac{130{,}600 \text{ gpd/ft}}{93 \text{ ft}} = 1405 \text{ gpd/ft}^2 = 0.066 \text{ cm/s}$$

$$S_s = \frac{S}{b} = \frac{1.12 \times 10^{-4}}{93 \text{ ft}} = 1.20 \times 10^{-6} \text{ ft}^{-1}$$

The above values show that the hydraulic conductivity of the aquifer exceeds the value for the aquitards by more than four orders of magnitude. However, the specific storage (S_s) of the aquifer is less than S_s' in the lower and upper aquitards by two orders of magnitude. In other words, for the same change in head a unit volume of aquitard material can contribute about 100 times more water from storage than a similar volume of the aquifer can. Based on to this fact, Neuman and Witherspoon (1972) concluded that storage in the aquitards must be considered when one deals with leaky aquifer system.

(c) According to Neuman and Witherspoon (1972), direct measurements performed on undisturbed samples from the same aquitards in the laboratory indicated that the aquitard hydraulic conductivities vary within a range of at least three orders of magnitude. Neuman and Witherspoon also added that the results in Table 7-10 fall on the high side of this range and thus are an indication that the average hydraulic conductivity in the field cannot always be reliably estimated from laboratory experiments.

CHAPTER 8

PARTIALLY PENETRATING PUMPING AND OBSERVATION WELLS IN HOMOGENEOUS AND ANISOTROPIC CONFINED AQUIFERS

8.1 INTRODUCTION

In general, wells partially penetrate the entire thickness of a confined aquifer. If the length of water entry (or the screen interval) in a well is less than the entire thickness of a confined aquifer, the well is called a *partially penetrating well.* The importance of partial penetration of a well increases with the decrease of the ratio between the screen interval and the aquifer thickness. The ground-water flow to or from a partially penetrating well is three-dimensional. As a result, the drawdown around partially penetrating pumping and observation wells will be dependent on their screened lengths and screen location along the thickness of the aquifer.

In this chapter, both the theoretical foundations and the practical applications of the models developed for partially penetrating wells in confined aquifers will be presented in a sequential manner based on Hantush's original works (Hantush, 1957, 1961c, 1961d, and 1964). In 1957, Hantush developed a leaky aquifer model for the condition that the screen of the pumped well in a confined aquifer extends to the top of the aquifer. This model, which is presented in Section 8.2, is important because all the models that Hantush subsequently developed are modified versions of it. As a matter of fact, in 1964, based on this model, Hantush published the general case leaky confined aquifer model for which the screen interval does not extend to the top of the aquifer. This model is presented in Section 8.3. These models are especially important for the determination of hydrogeologic parameters of confined aquifers under a partially penetrating well condition. In this chapter, the details and practical applications of these models are presented, with examples (Section 8.4).

8.2 HANTUSH'S LEAKY AQUIFER MODEL: THE SCREEN OF THE PUMPED WELL EXTENDS TO THE TOP OF THE AQUIFER

As far as the author of this book is aware, Hantush's original paper published in 1957 under the title "Non-steady Flow to a Well Partially Penetrating an Infinite Leaky Aquifer" is the first publication regarding partially penetrating well theory. In the following sections, the mathematical foundations of this model are presented.

8.2.1 Problem Statement and Assumptions

Figure 8-1 illustrates a partially penetrating well in a confined aquifer in which the screened interval extends to the upper boundary of the aquifer. The model is based on the following assumptions (Hantush, 1964; Reed, 1980):

1. The well discharge rate is constant.
2. The well diameter is infinitesimal.
3. The aquifer is overlain, or underlain, everywhere by a confining layer having uniform hydraulic conductivity (K') and thickness (b').
4. The confining layer does not release water from storage.
5. Flow in the confined aquifer is horizontal and in the confining layer is vertical.
6. Initially the water-table elevation of the unconfined aquifer and the potentiometric head elevation of the confined aquifer are the same.

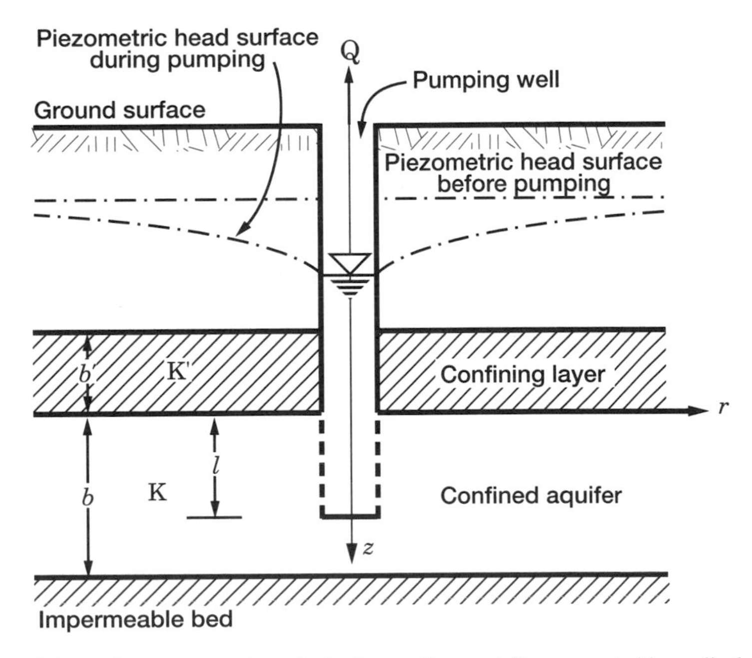

Figure 8-1 Schematic representation of a leaky aquifer partially penetrated by a discharging well whose screen extends to the top of the aquifer.

7. The leakage from the confining bed is assumed to be generated within the aquifer so that in the aquifer no vertical flow results from leakage alone.
8. Horizontal and vertical hydraulic conductivities are the same.

8.2.2 Governing Equations

The governing differential equation of three-dimensional ground-water flow in radial coordinates is [Chapter 2, Eq. (2-174)]:

$$K_r \frac{\partial^2 s}{\partial r^2} + \frac{K_r}{r}\frac{\partial s}{\partial r} + K_z \frac{\partial^2 s}{\partial z^2} = S_s \frac{\partial s}{\partial t} \tag{8-1}$$

By letting $T = K_r b$ and $S = S_s b$, Eq. (6-10) of Chapter 6 can be written as

$$K_r \frac{\partial^2 s}{\partial r^2} + \frac{K_r}{r}\frac{\partial s}{\partial r} - \frac{K'}{bb'} s = S_s \frac{\partial s}{\partial t} \tag{8-2}$$

The last term on the left-hand side of Eq. (8-1) corresponds to the vertical flow component. Adding this term in Eq. (8-2) gives

$$K_r \frac{\partial^2 s}{\partial r^2} + \frac{K_r}{r}\frac{\partial s}{\partial r} + K_z \frac{\partial^2 s}{\partial z^2} - \frac{K'}{bb'} s = S_s \frac{\partial s}{\partial t} \tag{8-3}$$

If $K_r = K_z = K$, Eq. (8-3) takes the form

$$\frac{\partial^2 s}{\partial r^2} + \frac{1}{r}\frac{\partial s}{\partial r} + \frac{\partial^2 s}{\partial z^2} - \frac{s}{B^2} = \frac{S}{T}\frac{\partial s}{\partial t} \tag{8-4}$$

where

$$B = \left(\frac{Tb'}{K'}\right)^{1/2} \tag{8-5}$$

Equation (8-4), which is the equation Hantush used [1957, p. 12, Eq. (2)], describes the governing equation for unsteady radial and vertical flow in a homogeneous confined aquifer with leakage through the confining layer. Equation (8-5) is the same as Eq. (6-11) of Chapter 6.

8.2.3 Initial and Boundary Conditions

The initial condition is

$$s(r, z, t) = 0, \qquad r \geq 0, \quad 0 \leq z \leq b \tag{8-6}$$

The boundary conditions are:

$$s(\infty, z, t) = 0, \quad 0 \leq z \leq b, \quad t \geq 0 \tag{8-7}$$

$$\frac{\partial s(r,0,t)}{\partial z} = 0, \quad r \geq 0, \quad 0 \leq z \leq b, \quad t \geq 0 \tag{8-8}$$

$$\frac{\partial s(r,b,t)}{\partial z} = 0, \quad r \geq 0, \quad 0 \leq z \leq b, \quad t \geq 0 \tag{8-9}$$

The condition for constant discharge at the well itself is

$$\lim_{r \to 0} \left(2\pi K_r \int_0^b \frac{\partial s}{\partial r}\, dz \right) = -Q \tag{8-10}$$

or, for the screening interval in Figure 8-1, gives

$$\lim_{r \to 0} \left(r \frac{\partial s}{\partial r} \right) = -\frac{Q}{2\pi K_r l}, \qquad 0 < z < l \tag{8-11}$$

Equation (8-6) states that initially the drawdown is zero everywhere in the aquifer. Equation (8-7) states that as the distance from the well approaches infinity, the drawdown is zero. Equations (8-8) and (8-9) state that there is no vertical flow at the upper and lower boundaries of the aquifer. Finally, Eq. (8-11) states that near the pumping well the discharge is distributed uniformly over the screening interval of the well and no radial flow occurs below the screened zone.

8.2.4 Solution

Hantush (1957) first applied the Laplace transform method to Eqs. (8-4), (8-6), and (8-7). Due to linearity and homogeneity of Eq. (8-4), Hantush used the method of images (see Chapter 10) to obtain a vanishing flux across the horizons $z = 0$ and $z = b$ and determined the following solution:

$$\begin{aligned} s(r,z,t) = \frac{Q}{4\pi T} \Bigg\{ & W\left(u, \frac{r}{B}\right) + \frac{2b}{\pi l} \sum_{n=1}^{\infty} \frac{1}{n} \cos\left(\frac{n\pi z}{b}\right) \\ & \cdot \sin\left(\frac{n\pi l}{b}\right) W_n \left[u, \left[\left(\frac{r}{B}\right)^2 + \left(\frac{n\pi r}{b}\right)^2 \right]^{1/2} \right] \Bigg\} \end{aligned} \tag{8-12}$$

where

$$u = \frac{r^2 S}{4Tt} \tag{8-13}$$

$$W\left(u, \frac{r}{B}\right) = \int_u^{\infty} \frac{1}{y} \exp\left(-y - \frac{r^2}{4B^2 y} \right) dy \tag{8-14}$$

$$W_n \left[u, \left[\left(\frac{r}{B}\right)^2 + \left(\frac{n\pi r}{b}\right)^2 \right]^{1/2} \right] = \int_u^{\infty} \exp\left[-y - \frac{(r/B)^2 + (n\pi r/b)^2}{4y} \right] \frac{dy}{y} \tag{8-15}$$

In the above equations, $W(u, r/B)$ is the *Hantush and Jacob well function for leaky aquifers* (see Chapter 6, Section 6.2.4.2).

8.2.5 Special Solutions

8.2.5.1 Steady-State Solution. For large time, u goes to zero and steady state is reached. Therefore, as u goes to zero, by making use of the first expression of Eq. (6-25) of Chapter 6 (Section 6.2.4.3), Eq. (8-12) reduces to (Hantush, 1957)

$$s(r,z) = \frac{Q}{2\pi T}\left\{\left[K_0\left(\frac{r}{B}\right) + \frac{2b}{\pi l}\sum_{n=1}^{\infty}\frac{1}{n}\cos\left(\frac{n\pi z}{b}\right)\sin\left(\frac{n\pi l}{b}\right)\right] \cdot K_0\left[\left[\left(\frac{r}{B}\right)^2 + \left(\frac{n\pi r}{b}\right)^2\right]^{1/2}\right]\right\} \tag{8-16}$$

where K_0 is the modified Bessel's function of the second kind of zero order.

8.2.5.2 Unsteady-State Solution for Zero Penetration. This corresponds to a situation that the penetration zone of the well does not exist. In other words, the well extends only to the top of the aquifer ($l = 0$) as shown in Figure 8-1. As l becomes zero, $\sin(n\pi l/b)/(n\pi l/b)$ approaches 1. Therefore, using the first expression of Eq. (6-25) of Chapter 6 (Section 6.2.4.3), Eq. (8-12) reduces to (Hantush, 1957)

$$s(r,z,t) = \frac{Q}{4\pi T}\left\{W\left(r,\frac{r}{B}\right) + 2\sum_{n=1}^{\infty}\cos\left(\frac{n\pi z}{b}\right)W_n\left[u,\left[\left(\frac{r}{B}\right)^2 + \left(\frac{n\pi r}{b}\right)^2\right]^{1/2}\right]\right\} \tag{8-17}$$

8.2.5.3 Steady-State Solution for Zero Penetration. By letting l go to zero in Eq. (8-16), the solution for zero penetration is as follows:

$$s(r,z) = \frac{Q}{2\pi T}\left\{K_0\left(\frac{r}{B}\right) + 2\sum_{n=1}^{\infty}\cos\left(\frac{n\pi z}{b}\right)K_0\left[\left[\left(\frac{r}{B}\right)^2 + \left(\frac{n\pi r}{b}\right)^2\right]^{1/2}\right]\right\} \tag{8-18}$$

8.2.5.4 Solution for Nonleaky Aquifers. The drawdown variation without leakage is obtained from the above solutions by making either K' equals zero or letting B go to infinity. These solutions are given below (Hantush, 1957).

Partially Penetrating Wells

Unsteady-State Solution. By letting B go to infinity, Eq. (8-12) takes the form

$$s(r,z,t) = \frac{Q}{4\pi T}\left[W(u) + \frac{2b}{\pi l}\sum_{n=1}^{\infty}\frac{1}{n}\cos\left(\frac{n\pi z}{b}\right)\sin\left(\frac{n\pi l}{b}\right)W_n\left(u,\frac{n\pi r}{b}\right)\right] \tag{8-19}$$

Steady-State Solution. If the discharge of the well is provided from storage only, which is the case in an infinite nonleaky aquifer, steady-state drawdown will theoretically never be reached because the infinite nonleaky aquifer is a closed reservoir without having flow from other sources. In other words, the flow across the boundary at infinity is zero. However, if the head is assumed to remain uniform at a finite distance, but arbitrarily large distance from the well, steady-state drawdown may be reached. In this case the aquifer is not closed at the outer boundary; and when time becomes infinitely large, the inflow is equal to the discharge of the well. The unsteady-state drawdown variation within practical distances from the well will, for all practical purposes, be the same as that under the condition that the aquifer is infinite.

The steady-state drawdown variation in an effectively infinite nonleaky aquifer was obtained by Hantush (1957). The governing equation for $K_r = K_z$, which is a special form of Eq. (8-3), is

$$\frac{\partial^2 s}{\partial r^2} + \frac{1}{r}\frac{\partial s}{\partial r} + \frac{\partial^2 s}{\partial z^2} = 0 \tag{8-20}$$

The boundary conditions are

$$s(r_e, z) = 0 \tag{8-21}$$

$$\frac{\partial s(r,0)}{\partial z} = \frac{\partial s(r,b)}{\partial z} = 0 \tag{8-22}$$

$$\lim_{r \to 0} \left[2\pi K_r r \int_0^b \frac{\partial s}{\partial r}\, dz \right] = -Q \tag{8-23}$$

Equation (8-21) states that the drawdown at r_e distance from the well is zero. The expressions in Eq. (8-22) state that there is no vertical flow at the upper and lower boundaries of the aquifer. Equation (8-23) states that near the well the flow toward the well is equal to its discharge.

Using the method of separation of variables, Hantush (1957) showed that the solution that satisfies the conditions expressed by Eqs. (8-20) through (8-23) is equal to

$$s(r,z) = \frac{Q}{2\pi T}\left[\ln\left(\frac{r_e}{r}\right) + \frac{2b}{\pi l}\sum_{n=1}^{\infty}\frac{1}{n}\cos\left(\frac{n\pi z}{b}\right)\sin\left(\frac{n\pi l}{b}\right)R_0\left(\frac{n\pi r}{b}\right)\right] \tag{8-24}$$

where

$$R_0\left(\frac{n\pi r}{b}\right) = K_0\left(\frac{n\pi r}{b}\right) - K_0\left(\frac{n\pi r_e}{b}\right)\frac{I_0\left(\frac{n\pi r}{b}\right)}{I_0\left(\frac{n\pi r_e}{b}\right)} \tag{8-25}$$

in which K_0 is the modified Bessel's function of the second kind of zero order and I_0 is the modified Bessel's function of the first kind and zero order. The second term of Eq. (8-25) may be neglected for cases of practical interest.

Wells Just Tapping the Top of the Aquifer

Unsteady-State Solution. By letting B go to infinity, Eq. (8-17) reduces to (Hantush, 1957):

$$s(r,z,t) = \frac{Q}{4\pi T}\left[W(u) + 2\sum_{n=1}^{\infty}\cos\left(\frac{n\pi z}{b}\right)W_n\left(u, \frac{n\pi r}{b}\right)\right] \tag{8-26}$$

Steady-State Solution. By letting l go to zero, Eq. (8-24) takes the following form (Hantush, 1957):

$$s(r,z) = \frac{Q}{2\pi T}\left[\ln\left(\frac{r_e}{r}\right) + 2\sum_{n=1}^{\infty}\cos\left(\frac{n\pi z}{b}\right)R_0\left(\frac{n\pi r}{b}\right)\right] \tag{8-27}$$

where R_0 is defined by Eq. (8-25).

8.2.6 Evaluation of Solutions

The first term in Eqs. (8-16) and (8-18) represents the drawdown under steady-state conditions for fully penetrating well cases [Chapter 3, Eq. (30)], and the function $K_0(r/B)$ has been tabulated for practical ranges of the parameter r/B (Chapter 3, Table 3-2).

The first term in Eqs. (8-12) and (8-17) represents the drawdown under unsteady-state conditions for fully penetrating well cases [Chapter 6, Eq. (6-23)]; and the function $W(u, r/B)$, which is known as *Hantush and Jacob well function for leaky aquifers*, is tabulated in Table 6-1 of Chapter 6.

The infinite sum in each of the solutions given above represents the additional drawdown as a result of partial penetration. The function W_n, which is similar in composition to the function $W(u, r/B)$, can be evaluated from the tabulated values of $W(u, r/B)$. The term r/B in the argument of W_n is very small compared with $(nr\pi/b)$ and, therefore, may be neglected for all practical purposes (Hantush, 1957).

The infinite sums in the above expressions tend to approach zero quite rapidly as n increases. Therefore, very few number of terms are sufficient for the practical ranges of the parameters. In fact, even in the extreme case of zero penetration, the whole series may safely be neglected for $r/b > 2$ [see Eq. (6-17) and Table 6-1 of Chapter 6]. For practical purposes, the series may be neglected for values of $r/b \geq 1.5$. In other words, the equal drawdown lines for zero penetration case rapidly changes to a radial type and can hardly be distinguished from those of a radial system at a distance from the well equal to only 1.5 times the thickness of the aquifer. It is clear that for increased penetration, changing to radial character will take place even more rapidly (Hantush, 1957).

Figure 8-2 illustrates the drawdown variation with time for full and zero penetration for $B = 10{,}000$ ft (3,048 m), $b = 50\pi$ ft (15π m), $r = 100$ ft (30 m), and $z = 0$. The data corresponding to Figure 8-2 are given in Table 1 of Hantush (1957). Figure 8-3 shows the variation of drawdown with distance for $B = 10{,}000$ ft (3,048 m), $b = 50\pi$ ft (15π m), $4Tt/S = 10^6$ ft^2 (92,903 m^2), and $z = 0$. The data corresponding to Figure 8-3 are given in Table 2 of Hantush (1957). The following conclusions can be drawn from Figures 8-2 and 8-3 (Hantush, 1957):

1. At a given distance from the center of the well, the difference between the drawdown induced in the case of partial penetration and that induced in the case of complete penetration increases with the time period of pumping (Figure 8-2). This difference tends to approach to a constant value after a sufficiently long time has elapsed.
2. The constant difference decreases as r/b increases and becomes vanishingly small as r/b approaches 2 (Figure 8-3).
3. For all practical purposes, this difference may be neglected for values of $r/b \geq 1.5$.

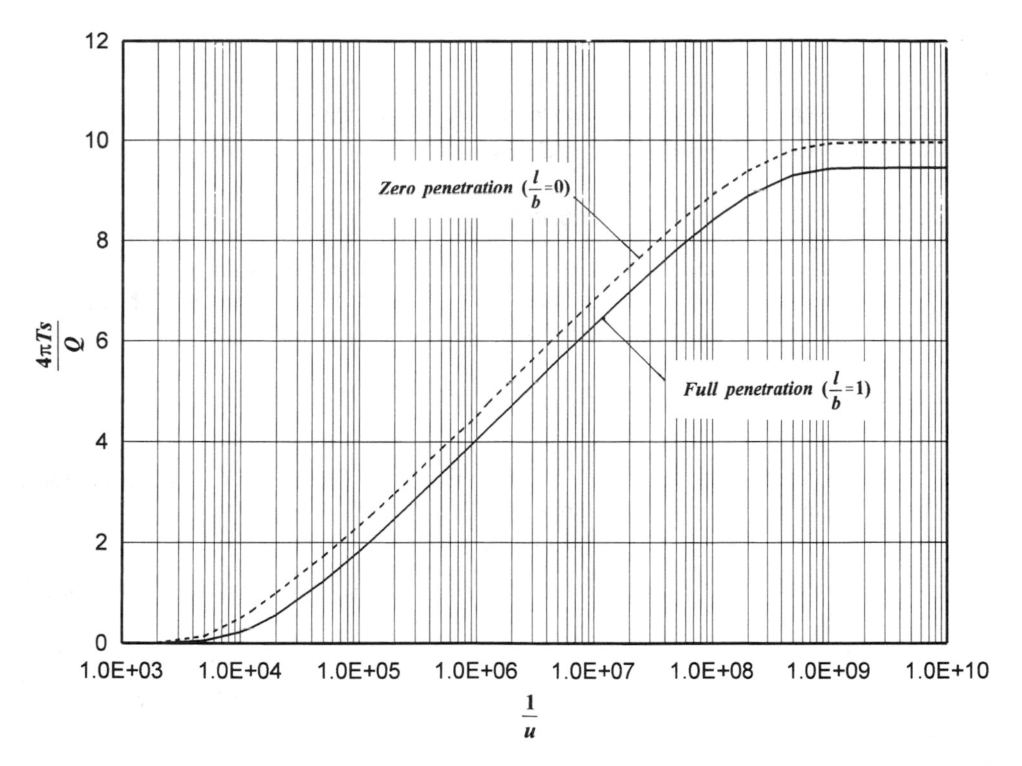

Figure 8-2 Variation of drawdown with time in an infinite leaky aquifer with $B = 10{,}000$ ft (3,048 m), $b = 157$ ft (48 m), $r = 100$ ft (30 m), and $z = 0$.

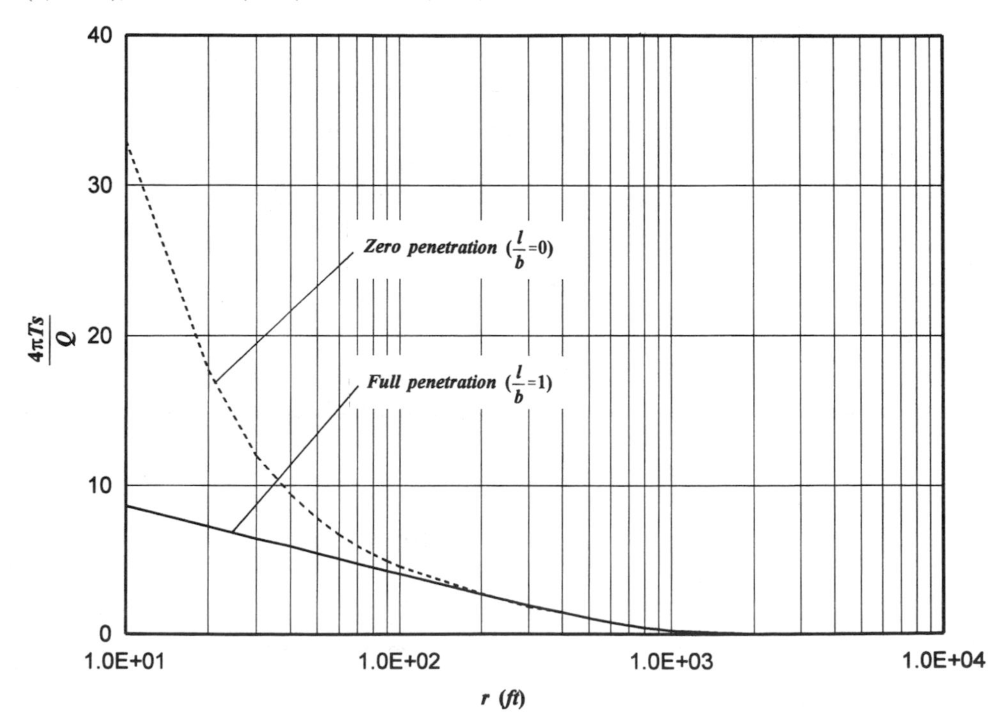

Figure 8-3 Distance versus drawdown in an infinite leaky aquifer with $B = 10{,}000$ ft (3,048 m), $b = 175$ ft (53 m), $4Tt/S = 10^6$ ft^2 (92,903 m^2), and $z = 0$.

8.3 HANTUSH'S LEAKY AQUIFER MODEL: THE SCREEN OF THE PUMPED WELL DOES NOT EXTEND TO THE TOP OF THE AQUIFER (GENERAL CASE)

8.3.1 Problem Statement and Assumptions

Figure 8-4 illustrates a partially penetrating well in a confined aquifer in which the screen interval is located at any place along the aquifer thickness. The assumptions of the model are as follows (Hantush, 1964; Reed, 1980):

1. The well discharge rate (Q) is constant.
2. The well diameter is infinitesimal.
3. The horizontal and vertical hydraulic conductivities (K_r) and (K_z) of the main aquifer are different but have constant values.
4. The leaky system is overlain by a constant-head water body.
5. The confined aquifer is overlain, or underlain, everywhere by a confining bed having uniform hydraulic conductivity (K') and thickness (b').
6. Hydraulic gradient across confining bed changes instantaneously with a change in head in the aquifer. In other words, there is no release of water from storage in the confining bed.
7. Flow is vertical in the confining bed and radial in the aquifer.

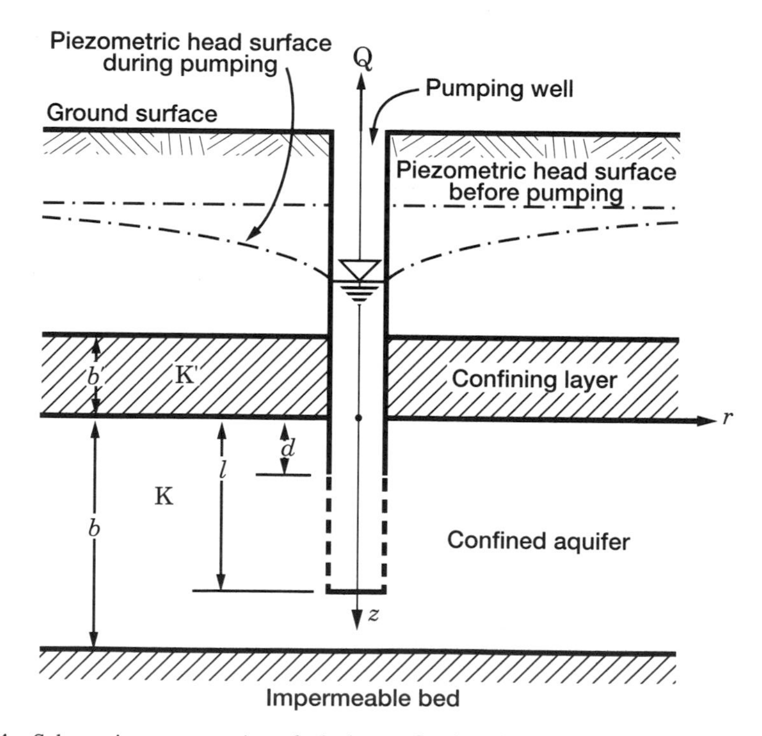

Figure 8-4 Schematic representation of a leaky confined aquifer partially penetrated by a discharging well whose screen does not extend to the top of the aquifer.

8. The leakage from the confining bed is generated within the aquifer, so that in the aquifer no vertical flow results from leakage alone.

8.3.2 Governing Equations

Equation (8-3) in Section 8.2.2 is the governing equation, and with

$$T = K_r b, \qquad a^2 = \frac{K_z}{K_r} \tag{8-28}$$

it takes the following form (Hantush, 1964; Reed, 1980):

$$\frac{\partial^2 s}{\partial r^2} + \frac{1}{r}\frac{\partial s}{\partial r} + a^2 \frac{\partial^2 s}{\partial z^2} - \frac{K'}{Tb'} s = \frac{S}{T}\frac{\partial s}{\partial t} \tag{8-29}$$

Equation (8-29) is the differential equation describing unsteady radial flow in a confined homogeneous aquifer with radial–vertical anisotropy and leakage from the overlain or underlain confining layer.

8.3.3 Initial and Boundary Conditions

The initial condition is

$$s(r, z, 0) = 0, \qquad r \geq 0, \quad 0 \leq z \leq b \tag{8-30}$$

The boundary conditions are

$$s(\infty, z, t) = 0, \qquad 0 \leq z \leq b, \quad t \geq 0 \tag{8-31}$$

$$\frac{\partial s(r, 0, t)}{\partial z} = 0, \qquad r \geq 0, \quad t \geq 0 \tag{8-32}$$

$$\frac{\partial s(r, b, t)}{\partial z} = 0, \qquad r \geq 0, \quad t \geq 0 \tag{8-33}$$

$$\lim_{r \to 0} \left(r \frac{\partial s}{\partial r} \right) = \begin{cases} 0, & 0 < z < d \\ -\dfrac{Q}{2\pi K_r (l - d)}, & d < z < l \\ 0, & l < z < b \end{cases} \tag{8-34}$$

Equation (8-30) states that initially drawdown is zero. Equation (8-31) states that drawdown is zero at large distance from the pumping well. Equations (8-32) and (8-33) state that there is no vertical flow at the upper and lower boundaries of the aquifer. This means that vertical hydraulic gradients in the aquifer are caused by the geometric placement of the pumping well and screen interval and not by leakage through the overlain or underlain confining layer. Equation (8-34) states that near the pumping well the discharge is distributed uniformly over the well screen and that no radial flow occurs above and below the screen interval.

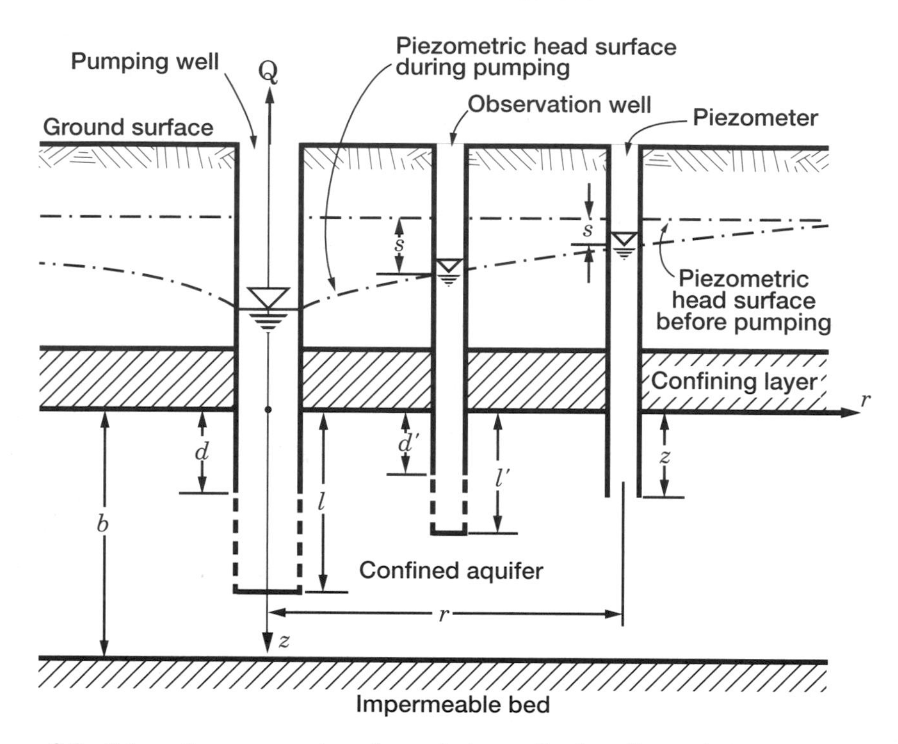

Figure 8-5 Schematic representation of a nonleaky confined aquifer partially penetrated by discharging and observation wells whose screens do not extend to the top of the aquifer.

8.3.4 Solution for Drawdown in Piezometers

A piezometer is a pipe that has small diameter and water enters into it only through its bottom when it is driven into an aquifer (see Figure 8-5). Solutions for drawdown in piezometers with and without leakage are given below.

8.3.4.1 Solution for K_r Is Different than K_z with Leakage. For the drawdown in a piezometer with leakage a solution determined by Hantush (1964, p. 350) is given as (Reed, 1980)

$$s = \frac{Q}{4\pi T}\left\{W(u, \beta_r) + f\left(u, \frac{ar}{b}, \beta_r, \frac{d}{b}, \frac{l}{b}, \frac{z}{b}\right)\right\} \tag{8-35}$$

where

$$W(u, \beta_r) = \int_u^\infty \frac{1}{y} \exp\left(-y - \frac{\beta^2 r}{4y}\right) dy \tag{8-36}$$

$$u = \frac{r^2 S}{4Tt} \tag{8-37}$$

$$\beta_r = r\left(\frac{K'}{Tb'}\right)^{1/2} \tag{8-38}$$

$$a = \left(\frac{K_z}{K_r}\right)^{1/2} \tag{8-39}$$

$$f\left(u, \frac{ar}{b}, \beta_r, \frac{d}{b}, \frac{l}{b}, \frac{z}{b}\right) = \frac{2b}{\pi(l-d)} \sum_{n=1}^{\infty} \frac{1}{n} \left[\sin\left(\frac{n\pi l}{b}\right) - \sin\left(\frac{n\pi d}{b}\right)\right] \cdot \cos\left(\frac{n\pi z}{b}\right) W(u, \kappa_n) \tag{8-40}$$

where

$$\kappa_n = \left[\beta_r^2 + \left(\frac{n\pi ar}{b}\right)^2\right]^{1/2} \tag{8-41}$$

In Eq. (8-36) and Eq. (8-40), $W(u, \beta_r)$ and $W(u, \kappa_n)$, respectively, are the well functions for leaky aquifers [see Chapter 6, Eq. (6-24)].

8.3.4.2 Solution for K_r is Different than K_z without Leakage. The governing equation is Eq. (8-29) with $K' = 0$. For the drawdown in a piezometer, the solution is given by Hantush (1961c, p. 85; 1964, p. 353):

$$s = \frac{Q}{4\pi T}\left[W(u) + f\left(u, \frac{ar}{b}, \frac{l}{b}, \frac{d}{b}, \frac{z}{b}\right)\right] \tag{8-42}$$

where $W(u)$ is defined by Eq. (4-25) of Chapter 4 and is given by

$$W(u) = \int_u^{\infty} \frac{e^{-y}}{y} dy \tag{8-43}$$

and

$$f\left(u, \frac{ar}{b}, \frac{l}{b}, \frac{d}{b}, \frac{z}{b}\right) = \frac{2b}{\pi(l-d)} \sum_{n=1}^{\infty} \frac{1}{n} \cdot \left[\sin\left(\frac{n\pi l}{b}\right) - \sin\left(\frac{n\pi d}{b}\right)\right] \cos\left(\frac{n\pi z}{b}\right) W(u, x_n) \tag{8-44}$$

$$W(u, x_n) = \int_u^{\infty} \frac{1}{y} \exp\left(-y - \frac{x_n^2}{4y}\right) dy \tag{8-45}$$

where u and a are given by Eqs. (8-37) and (8-39), respectively, and

$$x_n = \frac{n\pi ar}{b} \tag{8-46}$$

8.3.4.3 Solution for $K_r = K_z$ without Leakage. This solution corresponds to the case where the horizontal and vertical hydraulic conductivities are equal, in other words, $K = K_r = K_z$ and, therefore, $a = 1$. The geometry of the system is given in Figure 8-5. The solution is given by either of the following two equations (Hantush, 1961c, p. 85; Hantush,

1964, p. 353):

$$s = \frac{Q}{4\pi K b}\left[W(u) + f\left(u, \frac{r}{b}, \frac{l}{b}, \frac{d}{b}, \frac{z}{b}\right)\right] \tag{8-47}$$

or

$$s = \frac{Q}{8\pi K(l-d)}\left[M\left(u, \frac{l+z}{r}\right) + M\left(u, \frac{l-z}{r}\right) + f'\left(u, \frac{b}{r}, \frac{l}{r}, \frac{z}{r}\right)\right.$$
$$\left. - M\left(u, \frac{d+z}{r}\right) - M\left(u, \frac{d-z}{r}\right) - f'\left(u, \frac{b}{r}, \frac{d}{r}, \frac{z}{r}\right)\right] \tag{8-48}$$

$$f = \frac{2b}{\pi(l-d)}\sum_{n=1}^{\infty}\frac{1}{n}\left[\sin\left(\frac{n\pi l}{b}\right) - \sin\left(\frac{n\pi d}{b}\right)\right]\cos\left(\frac{n\pi z}{b}\right) W\left(u, \frac{n\pi r}{b}\right) \tag{8-49}$$

$$f' = \sum_{n=1}^{\infty}\left[M\left(u, \frac{2nb + x + z}{r}\right) - M\left(u, \frac{2nb - x - z}{r}\right)\right.$$
$$\left. + M\left(u, \frac{2nb + x - z}{r}\right) - M\left(u, \frac{2nb - x + z}{r}\right)\right] \tag{8-50}$$

$$u = \frac{r^2 S_s}{4Kt} \tag{8-51}$$

The Function of $M(u, \beta)$. This function is defined by the following infinite integral:

$$M(u, \beta) = \int_u^{\infty} \frac{\exp(-y)}{y} \operatorname{erf}(\beta y^{1/2})\, dy \tag{8-52}$$

where $\operatorname{erf}(x)$ is the *error function* and is defined as

$$\operatorname{erf}(x) = \frac{2}{\pi^{1/2}}\int_0^x \exp(-y^2)\, dy \tag{8-53}$$

The $M(u, \beta)$ function is extensively tabulated by Hantush (1961c, 1964), and its values are presented in Table 8-1. Because $\operatorname{erf}(-x) = -\operatorname{erf}(x)$, it follows that

$$M(u, \beta) = -M(u, \beta) \tag{8-54}$$

The function $M(u, \beta)$ has been approximated with sufficient accuracy as (Hantush, 1964)

$$M(u, \beta) \cong 2\left[\sinh^{-1}(\beta) - 2\beta\left(\frac{u}{\pi}\right)^{1/2}\right] \quad \text{for} \quad u < \frac{0.05}{\beta^2} < 0.01 \tag{8-55}$$

$$M(u, \beta) \cong 2\left[\sinh^{-1}(\beta) - \beta \operatorname{erf}(u^{1/2})\right] \quad \text{for} \quad u < \frac{0.05}{\beta^2} \tag{8-56}$$

$$M(u, \beta) \cong W(u) \quad \text{for} \quad u > \frac{5}{\beta^2} \tag{8-57}$$

in which $\sinh^{-1}(\beta)$ is the inverse hyperbolic sine of β.

TABLE 8-1 Values of the Function $M(u, \beta) = \int_u^\infty \frac{e^{-y}}{y} \operatorname{erf}(\beta \sqrt{y})\, dy$

u	β	0.1	0.2	0.3	0.4	0.5	0.6	0.7	0.8	0.9	1.0	1.2	1.4	1.6	1.8	2.0
	0	0.1997	0.3974	0.5913	0.7801	0.9624	1.1376	1.3053	1.4653	1.6177	1.7627	2.0319	2.2759	2.4979	2.7009	2.8872
	1	0.1994	0.3969	0.5907	0.7792	0.9613	1.1363	1.3037	1.4635	1.6157	1.7605	2.0292	2.2728	2.4943	2.6968	2.8827
	2	0.1993	0.3967	0.5904	0.7788	0.9608	1.1357	1.3031	1.4628	1.6148	1.7595	2.0281	2.2715	2.4929	2.6951	2.8809
	3	0.1993	0.3966	0.5902	0.7785	0.9605	1.1353	1.3026	1.4622	1.6142	1.7588	2.0272	2.2705	2.4917	2.6938	2.8794
	4	0.1992	0.3965	0.5900	0.7783	0.9602	1.1349	1.3022	1.4617	1.6137	1.7582	2.0265	2.2696	2.4907	2.6927	2.8782
10^{-6}	5	0.1992	0.3964	0.5898	0.7780	0.9599	1.1346	1.3018	1.4613	1.6132	1.7577	2.0259	2.2689	2.4899	2.6918	2.8772
	6	0.1991	0.3963	0.5897	0.7779	0.9596	1.1343	1.3014	1.4609	1.6127	1.7572	2.0253	2.2682	2.4891	2.6909	2.8762
	7	0.1991	0.3962	0.5895	0.7777	0.9594	1.1341	1.3011	1.4605	1.6123	1.7568	2.0248	2.2676	2.4884	2.6901	2.8753
	8	0.1990	0.3961	0.5894	0.7775	0.9592	1.1338	1.3009	1.4602	1.6120	1.7563	2.0243	2.2670	2.4877	2.6894	2.8745
	9	0.1990	0.3960	0.5893	0.7774	0.9590	1.1336	1.3006	1.4599	1.6116	1.7560	2.0238	2.2665	2.4871	2.6887	2.8737
	1	0.1989	0.3959	0.5892	0.7772	0.9588	1.1334	1.3003	1.4596	1.6113	1.7556	2.0234	2.2660	2.4865	2.6880	2.8730
	2	0.1987	0.3954	0.5883	0.7760	0.9574	1.1316	1.2983	1.4572	1.6086	1.7526	2.0198	2.2618	2.4818	2.6827	2.8671
	3	0.1984	0.3949	0.5876	0.7751	0.9562	1.1302	1.2967	1.4554	1.6066	1.7504	2.0171	2.2587	2.4782	2.6786	2.8625
	4	0.1982	0.3945	0.5871	0.7744	0.9553	1.1291	1.2953	1.4539	1.6049	1.7485	2.0148	2.2560	2.4751	2.8752	2.8587
10^{-5}	5	0.1981	0.3942	0.5866	0.7737	0.9544	1.1281	1.2941	1.4526	1.6034	1.7468	2.0128	2.2536	2.4724	2.6721	2.8553
	6	0.1979	0.3939	0.5861	0.7731	0.9537	1.1271	1.2931	1.4513	1.6020	1.7452	2.0110	2.2515	2.4700	2.6694	2.8523
	7	0.1978	0.3936	0.5857	0.7725	0.9530	1.1263	1.2921	1.4502	1.6007	1.7438	2.0093	2.2495	2.4677	2.6669	2.8495
	8	0.1976	0.3933	0.5853	0.7720	0.9523	1.1255	1.2912	1.4492	1.5996	1.7425	2.0077	2.2477	2.4657	2.6645	2.8469
	9	0.1975	0.3931	0.5849	0.7715	0.9517	1.1248	1.2903	1.4482	1.5984	1.7413	2.0062	2.2460	2.4637	2.6623	2.8444
	1	0.1974	0.3929	0.5846	0.7710	0.9511	1.1241	1.2895	1.4473	1.5974	1.7402	2.0049	2.2444	2.4619	2.6603	2.8421
	2	0.1965	0.3910	0.5818	0.7673	0.9465	1.1185	1.2830	1.4398	1.5890	1.7308	1.9936	2.2313	2.4469	2.6434	2.8234
	3	0.1958	0.3896	0.5796	0.7644	0.9429	1.1142	1.2780	1.4341	1.5825	1.7236	1.9850	2.2212	2.4354	2.6305	2.8091
	4	0.1952	0.3883	0.5778	0.7620	0.9398	1.1106	1.2737	1.4292	1.5771	1.7176	1.9778	2.2128	2.4258	2.6197	2.7970
10^{-4}	5	0.1946	0.3873	0.5762	0.7599	0.9372	1.1074	1.2700	1.4250	1.5723	1.7123	1.9714	2.2053	2.4172	2.6101	2.7864
	6	0.1941	0.3863	0.5748	0.7580	0.9348	1.1045	1.2666	1.4211	1.5680	1.7075	1.9656	2.1986	2.4095	2.6014	2.7768
	7	0.1937	0.3854	0.5734	0.7562	0.9326	1.1018	1.2635	1.4176	1.5640	1.7030	1.9603	2.1924	2.4025	2.5934	2.7679
	8	0.1933	0.3846	0.5722	0.7545	0.9305	1.0994	1.2607	1.4143	1.5603	1.6989	1.9554	2.1866	2.3959	2.5860	2.7597
	9	0.1929	0.3838	0.5710	0.7530	0.9286	1.0970	1.2579	1.4112	1.5568	1.6951	1.9507	2.1812	2.3897	2.5791	2.7519
	1	0.1925	0.3831	0.5699	0.7515	0.9267	1.0948	1.2554	1.4083	1.5535	1.6914	1.9463	2.1761	2.3838	2.5725	2.7446
	2	0.1896	0.3772	0.5611	0.7397	0.9120	1.0771	1.2347	1.3846	1.5270	1.6619	1.9109	2.1348	2.3367	2.5195	2.6857
	3	0.1873	0.3727	0.5543	0.7307	0.9007	1.0636	1.2189	1.3666	1.5066	1.6393	1.8838	2.1032	2.3006	2.4788	2.6406
	4	0.1854	0.3689	0.5486	0.7231	0.8912	1.0521	1.2056	1.3513	1.4895	1.6203	1.8610	2.0766	2.2702	2.4447	2.6027
10^{-3}	5	0.1837	0.3655	0.5435	0.7163	0.8828	1.0421	1.1938	1.3379	1.4744	1.6035	1.8409	2.0532	2.2434	2.4146	2.5693
	6	0.1822	0.3625	0.5390	0.7103	0.8752	1.0330	1.1832	1.3258	1.4608	1.5884	1.8228	2.0320	2.2193	2.3875	2.5393
	7	0.1808	0.3597	0.5348	0.7047	0.8682	1.0246	1.1735	1.3147	1.4483	1.5745	1.8061	2.0126	2.1972	2.3626	2.5117
	8	0.1795	0.3571	0.5310	0.6995	0.8618	1.0169	1.1645	1.3044	1.4367	1.5616	1.7907	1.9946	2.1766	2.3395	2.4861
	9	0.1783	0.3547	0.5273	0.6947	0.8557	1.0096	1.1560	1.2947	1.4258	1.5495	1.7762	1.9777	2.1573	2.3179	2.4620

TABLE 8-1 *(continued)*

u	β	0.1	0.2	0.3	0.4	0.5	0.6	0.7	0.8	0.9	1.0	1.2	1.4	1.6	1.8	2.0
	1	0.1772	0.3524	0.5239	0.6901	0.8500	1.0027	1.1480	1.2855	1.4155	1.5381	1.7625	1.9617	2.1391	2.2975	2.4394
	2	0.1680	0.3340	0.4962	0.6533	0.8040	0.9476	1.0836	1.2121	1.3329	1.4464	1.6527	1.8340	1.9935	2.1342	2.2587
	3	0.1610	0.3200	0.4753	0.6253	0.7691	0.9057	1.0349	1.1564	1.2703	1.3770	1.5697	1.7376	1.8839	2.0116	2.1233
	4	0.1551	0.3083	0.4578	0.6020	0.7400	0.8708	0.9942	1.1100	1.2183	1.3193	1.5008	1.6577	1.7932	1.9103	2.0117
10^{-2}	5	0.1500	0.2981	0.4425	0.5817	0.7146	0.8404	0.9588	1.0696	1.1730	1.2691	1.4410	1.5884	1.7147	1.8229	1.9156
	6	0.1455	0.2890	0.4289	0.5635	0.6919	0.8132	0.9272	1.0336	1.1326	1.2243	1.3877	1.5268	1.6450	1.7454	1.8307
	7	0.1413	0.2807	0.4164	0.5470	0.6713	0.7885	0.8994	1.0008	1.0958	1.1837	1.3394	1.4711	1.5821	1.6756	1.7543
	8	0.1375	0.2731	0.4050	0.5317	0.6522	0.7658	0.8720	0.9707	1.0621	1.1464	1.2951	1.4200	1.5246	1.6120	1.6848
	9	0.1339	0.2660	0.3943	0.5176	0.6346	0.7447	0.8474	0.9428	1.0308	1.1118	1.2541	1.3729	1.4716	1.5534	1.6210
	1	0.1306	0.2593	0.3844	0.5043	0.6181	0.7249	0.8245	0.9167	1.0016	1.0795	1.2159	1.3290	1.4223	1.4991	1.5619
	2	0.1051	0.2084	0.3081	0.4030	0.4920	0.5744	0.6500	0.7186	0.7806	0.8362	0.9297	1.0029	1.0595	1.1026	1.1352
	3	8.74(−2)	0.1731	0.2554	0.3331	0.4053	0.4713	0.5309	0.5842	0.6313	0.6727	0.7400	0.7899	0.8261	0.8519	0.8699
	4	7.39(−2)	0.1462	0.2153	0.2801	0.3397	0.3935	0.4415	0.4837	0.5203	0.5519	0.6015	0.6363	0.6602	0.6760	0.6863
10^{-1}	5	6.32(−2)	0.1248	0.1835	0.2381	0.2878	0.3323	0.3714	0.4052	0.4341	0.4584	0.4955	0.5203	0.5362	0.5462	0.5521
	6	5.44(−2)	0.1074	0.1575	0.2039	0.2458	0.2828	0.3149	0.3423	0.3652	0.3842	0.4122	0.4300	0.4408	0.4471	0.4506
	7	4.71(−2)	9.29(−2)	0.1360	0.1756	0.2111	0.2421	0.2686	0.2909	0.3093	0.3242	0.3455	0.3583	0.3657	0.3698	0.3719
	8	4.10(−2)	8.06(−2)	0.1179	0.1519	0.1821	0.2082	0.2302	0.2484	0.2632	0.2750	0.2913	0.3007	0.3058	0.3084	0.3096
	9	3.57(−2)	7.03(−2)	0.1026	0.1319	0.1576	0.1797	0.1980	0.2130	0.2250	0.2343	0.2468	0.2537	0.2572	0.2589	0.2597
	1	3.13(−2)	6.14(−2)	8.95(−2)	0.1148	0.1369	0.1555	0.1709	0.1833	0.1929	0.2004	0.2101	0.2151	0.2175	0.2186	0.2191
	2	9.01(−3)	1.75(−2)	2.51(−2)	3.16(−2)	3.67(−2)	4.07(−2)	4.35(−2)	4.55(−2)	4.69(−2)	4.77(−2)	4.85(−2)	4.88(−2)			4.88(−2)
	3	2.82(−3)	5.44(−3)	7.68(−3)	9.47(−3)	1.08(−2)	1.17(−2)	1.23(−2)	1.26(−2)	1.28(−2)	1.30(−2)					1.30(−2)
	4	9.20(−4)	1.76(−3)	2.44(−3)	2.96(−3)	3.31(−3)	3.53(−3)	3.66(−3)	3.72(−3)	3.76(−3)	3.77(−3)					3.77(−3)
1	5	3.07(−4)	5.80(−4)	7.96(−4)	9.49(−4)	1.05(−3)	1.10(−3)	1.13(−3)	1.14(−3)	1.15(−3)						1.15(−3)
	6	1.04(−4)	1.95(−4)	2.64(−4)	3.10(−4)	3.36(−4)	3.50(−4)	3.56(−4)	3.59(−4)	3.60(−4)						3.60(−4)
	7	3.56(−5)	6.61(−5)	8.84(−5)	1.02(−4)	1.10(−4)	1.13(−4)	1.15(−4)	1.15(−4)	1.15(−4)						1.15(−4)
	8	1.23(−5)	2.26(−5)	2.99(−5)	3.42(−5)	3.63(−5)	3.72(−5)	3.75(−5)	3.76(−5)	3.77(−5)						3.77(−5)
	9	4.28(−6)	7.79(−6)	1.02(−5)	1.15(−5)	1.21(−5)	1.23(−5)	1.24(−5)	1.24(−5)							1.24(−5)
	10	1.49(−6)	2.70(−6)	3.48(−6)	3.90(−6)	4.07(−6)	4.13(−6)	4.15(−6)	4.16(−6)							4.16(−6)

Values of $M(u, \beta)$ are equal to $W(u)$ for u greater than 10 and all values of β.
The numbers in parentheses are powers of 10 by which the other numbers are raised; for example, $8.74(-2) = 0.0874$.

TABLE 8-1 *(continued)*

u	β	2.0	2.2	2.4	2.6	2.8	3.0	3.2	3.4	3.6	3.8	4.0	4.2	4.4	4.6	4.8	5.0	5.2	5.4	5.6	5.8	6.0
	0	2.8872	3.0593	3.2188	3.3675	3.5064	3.6369	3.7597	3.8757	3.9856	4.0900	4.1894	4.2842	4.3748	4.4616	4.5448	4.6248	4.7018	4.7760	4.8475	4.9167	4.9835
	1	2.8827	3.0543	3.2134	3.3616	3.5001	3.6301	3.7525	3.8681	3.9775	4.0815	4.1804	4.2747	4.3649	4.4512	4.5340	4.6136	4.6901	4.7638	4.8349	4.9036	4.9700
	2	2.8809	3.0523	3.2112	3.3592	3.4975	3.6273	3.7495	3.8649	3.9742	4.0779	4.1766	4.2708	4.3608	4.4469	4.5295	4.6089	4.6852	4.7588	4.8297	4.8982	4.9644
	3	2.8794	3.0507	3.2095	3.3573	3.4955	3.6251	3.7472	3.8625	3.9716	4.0752	4.1738	4.2678	4.3576	4.4436	4.5261	4.6053	4.6815	4.7549	4.8257	4.8940	4.9601
	4	2.8782	3.0494	3.2080	3.3557	3.4938	3.6233	3.7453	3.8604	3.9694	4.0729	4.1714	4.2653	4.3550	4.4408	4.5232	4.6023	4.6784	4.7516	4.8223	4.8905	4.9565
10^{-6}	5	2.8772	3.0482	3.2067	3.3544	3.4923	3.6217	3.7436	3.8586	3.9675	4.0709	4.1592	4.2630	4.3526	4.4384	4.5206	4.5996	4.6756	4.7488	4.8193	4.8874	4.9533
	6	2.8762	3.0471	3.2056	3.3531	3.4910	3.6203	3.7420	3.8569	3.9658	4.0690	4.1673	4.2610	4.3505	4.4362	4.5183	4.5972	4.6731	4.7462	4.8166	4.8846	4.9504
	7	2.8753	3.0462	3.2045	3.3519	3.4897	3.6190	3.7406	3.8554	3.9642	4.0674	4.1655	4.2591	4.3486	4.4341	4.5162	4.5950	4.6708	4.7438	4.8141	4.8821	4.9477
	8	2.8745	3.0453	3.2035	3.3509	3.4886	3.6177	3.7393	3.8540	3.9627	4.0658	4.1639	4.2574	4.3467	4.4323	4.5142	4.5929	4.6686	4.7415	4.8118	4.8797	4.9452
	9	2.8737	3.0444	3.2026	3.3499	3.4875	3.6166	3.7380	3.8527	3.9613	4.0643	4.1623	4.2558	4.3450	4.4305	4.5124	4.5910	4.6666	4.7394	4.8097	4.8774	4.9429
	1	2.8730	3.0436	3.2017	3.3489	3.4865	3.6155	3.7369	3.8515	3.9600	4.0629	4.1609	4.2542	4.3434	4.4288	4.5106	4.5892	4.6647	4.7375	4.8076	4.8753	4.9407
	2	2.8671	3.0371	3.1946	3.3412	3.4782	3.6066	3.7274	3.8414	3.9493	4.0517	4.1490	4.2418	4.3304	4.4152	4.4964	4.5744	4.6494	4.7215	4.7911	4.8582	4.9230
	3	2.8625	3.0321	3.1892	3.3353	3.4718	3.5998	3.7202	3.8337	3.9412	4.0431	4.1400	4.2323	4.3204	4.4048	4.4855	4.5631	4.6376	4.7093	4.7784	4.8450	4.9094
	4	2.8587	3.0279	3.1846	3.3304	3.4665	3.5941	3.7140	3.8272	3.9343	4.0358	4.1323	4.2243	4.3120	4.3960	4.4764	4.5535	4.6276	4.6989	4.7677	4.8339	4.8979
10^{-5}	5	2.8553	3.0242	3.1806	3.3260	3.4618	3.5890	3.7087	3.8215	3.9282	4.0294	4.1256	4.2172	4.3046	4.3882	4.4683	4.5451	4.6189	4.6899	4.7582	4.8242	4.6878
	6	2.8523	3.0209	3.1769	3.3220	3.4575	3.5844	3.7038	3.8163	3.9227	4.0236	4.1195	4.2108	4.2979	4.3812	4.4610	4.5375	4.6110	4.6816	4.7497	4.8153	4.8787
	7	2.8495	3.0178	3.1735	3.3184	3.4536	3.5802	3.6993	3.8116	3.9177	4.0183	4.1139	4.2049	4.2918	4.3748	4.4543	4.5305	4.6037	4.6741	4.7419	4.8072	4.8703
	8	2.8469	3.0149	3.1704	3.3150	3.4499	3.5763	3.6951	3.8071	3.9130	4.0133	4.1087	4.1994	4.2860	4.3688	4.4480	4.5240	4.5969	4.6670	4.7346	4.7997	4.8625
	9	2.8444	3.0122	3.1675	3.3118	3.4465	3.5727	3.6912	3.8030	3.9086	4.0087	4.1038	4.1943	4.2806	4.3632	4.4421	4.5178	4.5905	4.6604	4.7277	4.7926	4.8551
	1	2.8421	3.0097	3.1647	3.3088	3.4433	3.5692	3.6875	3.7990	3.9044	4.0043	4.0992	4.1894	4.2756	4.3578	4.4356	4.5121	4.5845	4.6542	4.7212	4.7859	4.8482
	2	2.8234	2.9891	3.1423	3.2845	3.4171	3.5412	3.6576	3.7673	3.8708	3.9688	4.0618	4.1502	4.2345	4.3149	4.3918	4.4654	4.5360	4.6038	4.6690	4.7317	4.7922
	3	2.8091	2.9733	3.1251	3.2659	3.3970	3.5197	3.6347	3.7429	3.8450	3.9416	4.0332	4.1202	4.2030	4.2820	4.3574	4.4296	4.4988	4.5652	4.6289	4.6903	4.7493
	4	2.7970	2.9600	3.1106	3.2502	3.3801	3.5015	3.6154	3.7224	3.8233	3.9187	4.0090	4.0948	4.1764	4.2542	4.3285	4.3995	4.4674	4.5326	4.5952	4.6553	4.7132
10^{-4}	5	2.7864	2.9483	3.0978	3.2363	3.3652	3.4856	3.5984	4.7043	3.8042	3.8985	3.9878	4.0725	4.1531	4.2298	4.3030	4.3729	4.4399	4.5040	4.5655	4.6246	4.6814
	6	2.7768	2.9378	3.0863	3.2238	3.3518	3.4712	3.5830	3.6880	3.7869	3.8802	3.9686	4.0524	4.1320	4.2077	4.2800	4.3490	4.4150	4.4781	4.5387	4.5969	4.6527
	7	2.7679	2.9280	3.0757	3.2123	3.3394	3.4579	3.5688	3.6730	3.7710	3.8635	3.9509	4.0338	4.1126	4.1875	4.2588	4.3270	4.3921	4.4544	4.5141	4.5714	4.6264
	8	2.7597	2.9190	3.0658	3.2017	3.3279	3.4456	3.5557	3.6590	3.7562	3.8479	3.9345	4.0166	4.0945	4.1686	4.2392	4.3065	4.3708	4.4323	4.4912	4.5477	4.6019
	9	2.7519	2.9105	3.0565	3.1916	3.3171	3.4340	3.5434	3.6459	3.7423	3.8332	3.9191	4.0004	4.0776	4.1509	4.2207	4.2873	4.3508	4.4116	4.4697	4.5255	4.8789

TABLE 8-1 *(continued)*

u	β	2.0	2.2	2.4	2.6	2.8	3.0	3.2	3.4	3.6	3.8	4.0	4.2	4.4	4.6	4.8	5.0	5.2	5.4	5.6	5.8	6.0
	1	2.7446	2.9024	3.0478	3.1821	3.3069	3.4231	3.5317	3.6335	3.7292	3.8194	3.9046	3.9852	4.0616	4.1342	4.2033	4.2691	4.3320	4.3920	4.4494	4.5045	4.5572
	2	2.6857	2.8377	2.9771	3.1056	3.2245	3.3349	3.4377	3.5337	3.6236	3.7080	3.7874	3.8623	3.9329	3.9998	4.0632	4.1233	4.1805	4.2349	4.2867	4.3360	4.3832
	3	2.6406	2.7881	2.9231	3.0471	3.1616	3.2675	3.3659	3.4575	3.5430	3.6230	3.6981	3.7686	3.8350	3.8975	3.9566	4.0125	4.0654	4.1156	4.1632	4.2084	4.2514
	4	2.6027	2.7464	2.8776	2.9980	3.1087	3.2110	3.3056	3.3936	3.4754	3.5518	3.6233	3.6902	3.7530	3.8120	3.8676	3.9199	3.9694	4.0161	4.0602	4.1020	4.1416
10^{-3}	5	2.5693	2.7098	2.8377	2.9548	3.0623	3.1613	3.2528	3.3375	3.4162	3.4894	3.5577	3.6215	3.6813	3.7372	3.7898	3.8391	3.8856	3.9293	3.9705	4.0094	4.0462
	6	2.5393	2.6767	2.8018	2.9159	3.0205	3.1166	3.2052	3.2871	3.3629	3.4334	3.4989	3.5599	3.6169	3.6702	3.7200	3.7667	3.8105	3.8517	3.8903	3.9267	3.9609
	7	2.5117	2.6464	2.7688	2.8802	2.9822	3.0757	3.1616	3.2409	3.3142	3.3821	3.4451	3.5036	3.5581	3.6090	3.6564	3.7007	3.7422	3.7810	3.8174	3.8515	3.8835
	8	2.4861	2.6183	2.7382	2.8472	2.9466	3.0377	3.1213	3.1982	3.2691	3.3346	3.3953	3.4516	3.5038	3.5524	3.5977	3.6398	3.6792	3.7159	3.7502	3.7823	3.8123
	9	2.4620	2.5920	2.7095	2.8162	2.9134	3.0022	3.0835	3.1582	3.2270	3.2903	3.3489	3.4030	3.4532	3.4998	3.5430	3.5832	3.6206	3.6554	3.6678	3.7181	3.7463
	1	2.4394	2.5671	2.6825	2.7870	2.8820	2.9687	3.0480	3.1206	3.1873	3.2487	3.3052	3.3574	3.4057	3.4503	3.4917	3.5300	3.5656	3.5987	3.6294	3.6580	3.6845
	2	2.2587	2.3692	2.4675	2.5552	2.6337	2.7041	2.7673	2.8243	2.8756	2.9218	2.9637	3.0015	3.0357	3.0666	3.0946	3.1200	3.1430	3.1638	3.1827	3.1998	3.2153
	3	2.1233	2.2212	2.3072	2.3830	2.4498	2.5088	2.5610	2.6073	2.6482	2.6846	2.7168	2.7453	2.7707	2.7932	2.8131	2.8307	2.8464	2.8602	2.8724	2.8832	2.8928
	4	2.0117	2.0996	2.1759	2.2423	2.3000	2.3503	2.3942	2.4324	2.4658	2.4949	2.5202	2.5423	2.5615	2.5782	2.5927	2.6052	2.6161	2.5256	2.6337	2.6408	2.6468
10^{-2}	5	1.9156	1.9951	2.0634	2.1221	2.1725	2.2158	2.2531	2.2851	2.3125	2.3361	2.3563	2.3735	2.3883	2.4009	2.4116	2.4207	2.4284	2.4350	2.4405	2.4452	2.4491
	6	1.8307	1.9031	1.9645	2.0167	2.0610	2.0986	2.1305	2.1574	2.1802	2.1995	2.2157	2.2294	2.2408	2.2504	2.2584	2.2651	2.2706	2.2752	2.2790	2.2821	2.2846
	7	1.7543	1.8205	2.8760	1.9227	1.9618	1.9946	2.0220	2.0449	2.0640	2.0798	2.0929	2.1038	2.1127	2.1201	2.1261	2.1310	2.1350	2.1382	2.1408	2.1429	2.1446
	8	1.6848	1.7455	1.7959	1.8378	1.8725	1.9012	1.9249	1.9444	1.9604	1.9734	1.9841	1.9928	1.9998	2.0055	2.0101	2.0137	2.0166	2.0189	2.0207	2.0221	2.0233
	9	1.6210	1.6767	1.7226	1.7603	1.7912	1.8164	1.8370	1.8536	1.8671	1.8780	1.8867	1.8937	1.8992	1.9036	1.9071	1.9098	1.9119	1.9135	1.9148	1.9158	1.9165
	1	1.5619	1.6133	1.6552	1.6892	1.7167	1.7389	1.7568	1.7711	1.7825	1.7915	1.7987	1.8043	1.8087	1.8121	1.8147	1.8168	1.8183	1.8195	1.8204	1.8211	1.8216
	2	1.1352	1.1596	1.1777	1.1909	1.2004	1.2073	1.2122	1.2156	1.2179	1.2195	1.2206	1.2213	1.2218	1.2221	1.2223	1.2224	1.2225	1.2226	1.2226	1.2226	1.2226
	3	0.8699	0.8823	0.8907	0.8962	0.8998	0.9021	0.9035	0.9044	0.9049	0.9053	0.9054	0.9056	0.9056	0.9056	0.9057						0.9057
	4	0.6863	0.6928	0.6968	0.6992	0.7006	0.7014	0.7019	0.7021	0.7023	0.7023	0.7024										0.7024
10^{-1}	5	0.5521	0.5556	0.5576	0.5587	0.5592	0.5595	0.5597	0.5597	0.5597												0.5597
	6	0.4506	0.4525	0.4535	0.4540	0.4542	0.4543	0.4543														0.4543
	7	0.3719	0.3729	0.3734	0.3736	0.3737	0.3737															0.3737
	8	0.3096	0.3102	0.3104	0.3105																	0.3105
	9	0.2597	0.2600	0.2601	0.2602																	0.2602
	1	0.2191	0.2193	0.2194																		0.2194

Values of $M(u, \beta)$ are equal to $W(u)$ for u greater than 1 and all values of β.

TABLE 8-1 (*continued*)

u	β	6.0	6.2	6.4	6.6	6.8	7.0	7.2	7.4	7.6	7.8	8.0	8.2	8.4	8.6	8.8	9.0	9.2	9.4	9.6	9.8	10
	0	4.9835	5.0482	5.1109	5.1718	5.2308	5.2882	5.3440	5.3983	5.4511	5.5026	5.5529	5.6019	5.6497	5.6965	5.7421	5.7868	5.8305	5.8733	5.9151	5.9562	5.9964
10^{-6}	1	4.9700	5.0343	5.0965	5.1569	5.2155	5.2724	5.3278	5.3816	5.4340	5.4851	5.5349	5.5834	5.6308	5.6771	5.7223	5.7666	5.8098	5.8521	5.8935	5.9341	5.9739
	2	4.9644	5.0285	5.0905	5.1507	5.2091	5.2659	5.3210	5.3747	5.4269	5.4778	5.5274	5.5758	5.6230	5.6691	5.7141	5.7581	5.8012	5.8433	5.8846	5.9250	5.9645
	3	4.9601	5.0240	5.0860	5.1450	5.2043	5.2609	5.3159	5.3694	5.4215	5.4722	5.5217	5.5699	5.6170	5.6629	5.7078	5.7517	5.7946	5.8366	5.8777	5.9179	5.9573
	4	4.9565	5.0203	5.0821	5.1420	5.2002	5.2566	5.3115	5.3649	5.4169	5.4675	5.5168	5.5649	5.6119	5.6577	5.7025	5.7463	5.7890	5.8309	5.8719	5.9120	5.9513
	5	4.9533	5.0170	5.0787	5.1385	5.1965	5.2529	5.3077	5.3610	5.4128	5.4633	5.5126	5.5606	5.6074	5.6531	5.6978	5.7415	5.7841	5.8259	5.8668	5.9068	5.9460
	6	4.9504	5.0140	5.0756	5.1353	5.1933	5.2495	5.3042	5.3574	5.4092	5.4596	5.5087	5.5566	5.6034	5.6490	5.6936	5.7371	5.7797	5.8214	5.8621	5.9021	5.9412
	7	4.9477	5.0112	5.0728	5.1324	5.1902	5.2464	5.3010	5.3541	5.4058	5.4561	5.5052	5.5530	5.5996	5.6452	5.6897	5.7331	5.7756	5.8172	5.8579	5.8977	5.9367
	8	4.9452	5.0087	5.0701	5.1297	5.1874	5.2435	5.2981	5.3511	5.4027	5.4529	5.5019	5.5496	5.5962	5.6416	5.6860	5.7294	5.7718	5.8133	5.8539	5.8937	5.9326
	9	4.9429	5.0063	5.0676	5.1271	5.1848	5.2408	5.2953	5.3482	5.3997	5.4499	5.4988	5.5464	5.5929	5.6383	5.6826	5.7259	5.7683	5.8097	5.8502	5.8899	5.9287
10^{-5}	1	4.9407	5.0040	5.0653	5.1247	5.1823	5.2383	5.2926	5.3455	5.3969	5.4470	5.4958	5.5434	5.5898	5.6352	5.6794	5.7226	5.7649	5.8063	5.8467	5.8863	5.9251
	2	4.9230	4.9857	5.0464	5.1052	5.1622	5.2176	5.2714	5.3236	5.3745	5.4240	5.4722	5.5192	5.5650	5.6097	5.6534	5.6961	5.7377	5.7785	5.8183	5.8573	5.8955
	3	4.9094	4.9716	5.0319	5.0902	5.1468	5.2017	5.2550	5.3069	5.3573	5.4063	5.4541	5.5006	5.5460	5.5903	5.6335	5.6757	5.7169	5.7572	5.7966	5.8351	5.8729
	4	4.8979	4.9598	5.0196	5.0776	5.1338	5.1883	5.2413	5.2927	5.3427	5.3914	5.4388	5.4849	5.5299	5.5738	5.6167	5.6585	5.6993	5.7392	5.7782	5.8164	5.8638
	5	4.8878	4.9493	5.0089	5.0665	5.1224	5.1766	5.2292	5.2803	5.3299	5.3783	5.4253	5.4711	5.5158	5.5594	5.6018	5.6433	5.6838	5.7234	5.7621	5.7999	5.8369
	6	4.8787	4.9399	4.9991	5.0565	5.1120	5.1659	5.2182	5.2690	5.3184	5.3664	5.4131	5.4587	5.5030	5.5463	5.5885	5.6296	5.6698	5.7091	5.7475	5.7850	5.8217
	7	4.8703	4.9312	4.9902	5.0472	5.1025	5.1561	5.2081	5.2586	5.3077	5.3555	5.4020	5.4472	5.4913	5.5342	5.5762	5.6171	5.5570	5.6960	5.7341	5.7713	5.8078
	8	4.8625	4.9232	4.9818	5.0386	5.0937	5.1470	5.1988	5.2490	5.2979	5.3453	5.3915	5.4365	5.4803	5.5231	5.5647	5.6053	5.6450	5.6837	5.7216	5.7586	5.7948
	9	4.8551	4.9156	4.9740	5.0306	5.0853	5.1384	5.1900	5.2400	5.2886	5.3358	5.3818	5.4265	5.4701	5.5126	5.5540	5.5943	5.6338	5.6723	5.7099	5.7466	5.7826
10^{-4}	1	4.8482	4.9084	4.9666	5.0229	5.0775	5.1303	5.1816	5.2314	5.2798	5.3268	5.3725	5.4170	5.4604	5.2026	5.5438	5.5840	5.6231	5.6614	6.6988	5.7353	5.7710
	2	4.7922	4.8506	4.9069	4.9614	5.0141	5.0651	5.1145	5.1624	5.2089	5.2541	5.2980	5.3406	5.3821	5.4225	4.4619	5.5002	5.5375	5.5739	5.6095	5.6441	5.6780
	3	4.7493	4.8062	4.8612	4.9142	4.9655	5.0151	5.0631	5.1096	5.1547	5.1985	5.2409	5.2822	5.3223	5.3612	5.3992	5.4361	5.4720	5.5070	5.5411	5.5744	5.6069
	4	4.7132	4.7689	4.8227	4.8745	4.9246	4.9730	5.0198	5.0652	5.1091	5.1516	5.1929	5.2330	5.2719	5.3097	5.3464	5.3822	5.4169	5.4508	5.4837	5.5158	5.5471
	5	4.3814	4.7361	4.7888	4.8396	4.8886	4.9360	4.9818	5.0261	5.0689	5.1105	5.1507	5.1898	5.2276	5.2644	5.3001	5.3348	5.3686	5.4014	5.4333	5.4644	5.4947
	6	4.6527	4.7065	4.7582	4.8081	4.8562	4.9026	4.9475	4.9908	5.0327	5.0733	5.1127	5.1508	5.1877	5.2236	5.2583	5.2921	5.3249	5.3568	5.3879	5.4180	5.4474
	7	4.6264	4.6793	4.7301	4.7791	4.8264	4.8719	4.9159	4.9584	4.9995	5.0393	5.0777	5.1150	5.1511	5.1861	5.2200	5.2530	5.2850	5.3160	5.3462	5.3755	5.4041
	8	4.6019	4.6540	4.7040	4.7522	4.7987	4.8435	4.8867	4.9284	4.9687	5.0076	5.0453	5.0818	5.1171	5.1513	5.1845	5.2166	5.2478	5.2781	5.3075	5.3361	5.3639
	9	4.5789	4.6302	4.6796	4.7270	4.7727	4.8168	4.8592	4.9002	4.9397	4.9780	5.0149	5.0506	5.0852	5.1187	5.1512	5.1826	5.2131	5.2427	5.2714	5.2992	5.3263

TABLE 8-1 *(continued)*

u	β	6.0	6.2	6.4	6.6	6.8	7.0	7.2	7.4	7.6	7.8	8.0	8.2	8.4	8.6	8.8	9.0	9.2	9.4	9.6	9.8	10
	1	4.5572	4.6078	4.6565	4.7032	4.7482	4.7915	4.8333	4.8736	4.9124	4.9500	4.9862	5.0213	5.0552	5.0880	5.1197	5.1505	5.1803	5.2092	5.2372	5.2644	5.2908
	2	4.3832	4.4282	4.4713	4.5125	4.5519	4.5898	4.6260	4.6609	4.6943	4.7264	4.7573	4.7870	4.8155	4.8430	4.8695	4.8950	4.9193	4.9433	4.9662	4.9882	5.0095
	3	4.2514	4.2923	4.3312	4.3684	4.4038	4.4376	4.4699	4.5007	4.5302	4.5584	4.5854	4.6113	4.5360	4.6597	4.6824	4.7042	4.7251	4.7452	4.7644	4.7829	4.8006
	4	4.1416	4.1792	4.2148	4.2487	4.2808	4.3114	4.3405	4.3682	4.3945	4.4197	4.4436	4.4664	4.4881	4.5089	4.5287	4.5476	4.5356	4.5829	4.5993	4.6150	4.6301
10^{-3}	5	4.0462	4.0809	4.1137	4.1448	4.1742	4.2021	4.2285	4.2535	4.2773	4.2998	4.3212	4.3415	4.3608	4.3792	4.3966	4.4131	4.4288	4.4438	4.4580	4.4715	4.4844
	6	3.9609	3.9932	4.0236	4.0523	4.0793	4.1048	4.1290	4.1517	4.1733	4.1936	4.2129	4.2311	4.2483	4.2646	4.2800	4.2946	4.3084	4.3214	4.3338	4.3455	4.3566
	7	3.8835	3.9136	3.9418	3.9684	3.9934	4.0168	4.0390	4.0598	4.0794	4.0978	4.1152	4.1316	4.1470	4.1616	4.1753	4.1882	4.2004	4.2119	4.2227	4.2329	4.2425
	8	3.8123	3.8404	3.8668	3.8914	3.9146	3.9362	3.9566	3.9756	3.9935	4.0103	4.0261	4.0409	4.0548	4.0678	4.0801	4.0916	4.1024	4.1125	4.1220	4.1309	4.1393
	9	3.7463	3.7726	3.7972	3.8202	3.8416	3.8617	3.8805	3.8980	3.9144	3.9297	3.9440	3.9575	3.9700	3.9817	3.9927	4.0029	4.0125	4.0215	4.0299	4.0377	4.0450
	1	3.6845	3.7093	3.7323	3.7537	3.7737	3.7923	3.8096	3.8258	3.8408	3.8548	3.8679	3.8801	3.8914	3.9020	3.9119	3.9210	3.9296	3.9375	3.9449	3.9518	3.9582
	2	3.2153	3.2293	3.2419	3.2534	3.2638	3.2731	3.2816	3.2892	3.2961	3.3023	3.3079	3.3130	3.3175	3.3215	3.3252	3.3285	3.3314	3.3340	3.3364	3.3385	3.3403
	3	2.8929	2.9012	2.9086	2.9151	2.9209	2.9259	2.9303	2.9342	2.9376	2.9405	2.9431	2.9453	2.9473	2.9489	2.9504	2.9517	2.9528	2.9537	2.9545	2.9552	2.9558
	4	2.6468	2.6520	2.6565	2.6603	2.6636	2.6664	2.6688	2.6708	2.6725	2.6740	2.6752	2.6762	2.6771	2.6778	2.6784	2.6789	2.6793	2.6797	2.6800	2.6802	2.6804
10^{-2}	5	2.4491	2.4523	2.4551	2.4574	2.4593	2.4609	2.4622	2.4632	2.4641	2.4648	2.4654	2.4659	2.4663	2.4666	2.4669	2.4671	2.4673	2.4674	2.4675	2.4676	2.4677
	6	2.2846	2.2867	2.2884	2.2898	2.2909	2.2918	2.2926	2.2931	2.2936	2.2940	2.2943	2.2945	2.2947	2.2948	2.2949	2.2950	2.2951	2.2951	2.2952	2.2952	2.2952
	7	2.1446	2.1460	2.1470	2.1479	2.1486	2.1491	2.1495	2.1498	2.1500	2.1502	2.1504	2.1505	2.1506	2.1506	2.1507	2.1507	2.1508				2.1508
	8	2.0233	2.0241	2.0248	2.0253	2.0257	2.0260	2.0263	2.0264	2.0266	2.0267	2.0267	2.0268	2.0268	2.0269							2.0269
	9	1.9165	1.9171	1.9175	1.9178	1.9181	1.9183	1.9184	1.9185	1.9186	1.9186	1.9186										1.9186
10^{-1}	1	1.8216	1.8219	1.8222	1.8224	1.8226	1.8227	1.8227	1.8228	1.8228	1.8229											1.8229
	2	1.2226	1.2226	1.2226																		1.2226

Values of $M(u, \beta)$ are equal to $W(u)$ for u greater than 2×10^{-1} and all values of β.

TABLE 8-1 *(continued)*

u	β	10	12	14	16	18	20	22	24	26	28	30	32	34	36	38	40	42	44	46	48	50
	0	5.9964	6.3595	6.6668	6.9333	7.1684	7.3789	7.5692	7.7431	7.9030	8.0511	8.1890	8.3180	8.4392	8.5536	8.6615	8.7641	8.8616	8.9546	9.0435	9.1286	9.2102
	1	5.9739	6.3325	6.6353	6.8973	7.1279	7.3339	7.5197	7.6891	7.8445	7.9881	8.1215	8.2460	8.3627	8.4725	8.5761	8.6741	8.7671	8.8556	8.9400	9.0206	9.0977
	2	5.9645	6.3213	6.6223	6.8823	7.1111	7.3152	7.4992	7.6667	7.8202	7.9620	8.0935	8.2161	8.3309	8.4388	8.5406	8.6367	8.7279	8.8145	8.8971	8.9758	9.0510
	3	5.9573	6.3127	6.6122	6.8709	7.0982	7.3008	7.4834	7.6495	7.8016	7.9419	8.0720	8.1932	8.3066	8.4130	8.5133	8.6081	8.6978	8.7830	8.8641	8.9414	9.0152
	4	5.9513	6.3054	6.6038	6.8612	7.0873	7.2887	7.4701	7.6350	7.7859	7.9250	8.0539	8.1739	8.2861	8.3913	8.4904	8.5839	8.6725	8.7565	8.8364	8.9125	8.9851
10^{-6}	5	5.9460	6.2990	6.5963	6.8527	7.0777	7.2781	7.4584	7.6222	7.7721	7.9101	8.0379	8.1569	8.2680	8.3722	8.4702	8.5627	8.6502	8.7331	8.8119	8.8870	8.9586
	6	5.9412	6.2932	6.5896	6.8450	7.0691	7.2685	7.4478	7.6106	7.7596	7.8966	8.0235	8.1415	8.2516	8.3549	8.4519	8.5435	8.6300	8.7120	8.7899	8.8640	8.9346
	7	5.9367	6.2879	6.5834	6.8379	7.0611	7.2596	7.4381	7.6000	7.7481	7.8842	8.0102	8.1273	8.2366	8.3390	8.4352	8.5258	8.6115	8.6926	8.7696	8.8428	8.9126
	8	5.9326	6.2830	6.5776	6.8313	7.0537	7.2514	7.4290	7.5901	7.7374	7.8727	7.9979	8.1142	8.2226	8.3242	8.4196	8.5094	8.5942	8.6745	8.7507	8.8231	8.8921
	9	5.9287	6.2783	6.5722	6.8251	7.0467	7.2436	7.4205	7.5809	7.7273	7.8619	7.9863	8.1018	8.2095	8.3103	8.4049	8.4940	8.5781	8.6576	8.7330	8.8047	8.8729
	1	5.9251	6.2739	6.5671	6.8193	7.0402	7.2363	7.4125	7.5721	7.7178	7.8517	7.9753	8.0901	8.1971	8.2972	8.3910	8.4794	8.5628	8.6416	8.7163	8.7872	8.8547
	2	5.8955	6.2385	6.5257	6.7720	6.9870	7.1773	7.3476	7.5013	7.6412	7.7692	7.8871	7.9960	8.0972	8.1914	8.2795	8.3621	8.4397	8.5127	8.5817	8.6469	8.7087
	3	5.8729	6.2113	6.4940	6.7358	6.9463	7.1321	7.2979	7.4472	7.5826	7.7061	7.8195	7.9240	8.0208	8.1106	8.1943	8.2725	8.3458	8.4145	8.4792	8.5401	8.5976
	4	5.8538	6.1884	6.4673	6.7053	6.9120	7.0940	7.2561	7.4016	7.5332	7.6531	7.7627	7.8636	7.9566	8.0428	8.1229	8.1975	8.2671	8.3322	8.3933	8.4507	8.5047
10^{-5}	5	5.8369	6.1682	6.4438	6.6785	6.8818	7.0606	7.2193	7.3615	7.4899	7.6065	7.7129	7.8105	7.9003	7.9833	8.0602	8.1316	8.1982	8.2602	8.3182	8.3725	8.4235
	6	5.8217	6.1500	6.4225	6.6542	6.8546	7.0303	7.1861	7.3253	7.4508	7.5644	7.6679	7.7626	7.8496	7.9298	8.0038	8.0725	8.1362	8.1955	8.2508	8.3024	8.3507
	7	5.8078	6.1332	6.4030	6.6319	6.8296	7.0025	7.1556	7.2921	7.4149	7.5259	7.6267	7.7188	7.8032	7.8807	7.9522	8.0183	8.0795	8.1364	8.1892	8.2384	8.2843
	8	5.7948	6.1177	6.3848	6.6112	6.8063	6.9767	7.1272	7.2613	7.3815	7.4901	7.5885	7.6781	7.7601	7.8353	7.9044	7.9682	8.0271	8.0817	8.1323	8.1792	8.2229
	9	5.7826	6.1030	6.3678	6.5917	6.7844	6.9525	7.1007	7.2324	7.3503	7.4565	7.5527	7.6401	7.7198	7.7928	7.8597	7.9214	7.9782	8.0306	8.0791	8.1241	8.1658
	1	5.7710	6.0892	6.3517	6.5734	6.7638	6.9296	7.0756	7.2051	7.3208	7.4249	7.5189	7.6042	7.6818	7.7527	7.8177	7.8773	7.9321	7.9826	8.0292	8.0723	8.1122
	2	5.6780	5.9778	6.2221	6.4257	6.5982	6.7463	6.8747	6.9869	7.0856	7.1729	7.2504	7.3194	7.3811	7.4364	7.4861	7.5307	7.5709	7.6072	7.6399	7.6695	7.6962
	3	5.6069	5.8928	6.1233	6.3133	6.4725	6.6075	6.7231	6.8227	6.9091	6.9844	7.0502	7.1080	7.1588	7.2035	7.2429	8.2777	7.3084	7.3556	7.3597	7.3809	7.3997
	4	5.5471	5.8214	6.0406	6.2194	6.3677	6.4920	6.5972	6.6868	6.7635	6.8294	6.8862	6.9353	6.9778	7.0146	7.0465	7.0742	7.0982	7.1191	7.1371	7.1528	7.1663
10^{-4}	5	5.4947	5.7589	5.9681	6.1374	6.2763	6.3915	6.4880	6.5692	6.6378	6.6961	6.7456	6.7878	6.8237	6.8544	6.8805	6.9029	6.9219	6.9380	6.9518	6.9635	6.9734
	6	5.4474	5.7026	5.9031	6.0638	6.1945	6.3019	6.3908	6.4648	6.5266	6.5784	6.6218	6.6583	6.6890	6.7147	6.7363	6.7545	6.7696	6.7823	6.7929	6.8017	6.8090
	7	5.4041	5.6511	5.8437	5.9967	6.1201	6.2205	6.3027	6.3704	6.4263	6.4726	6.5109	6.5427	6.5690	6.5907	6.6087	6.6235	6.6357	6.6457	6.6539	6.6605	6.6661
	8	5.3639	5.6034	5.7887	5.9348	6.0515	6.1456	6.2219	6.2841	6.3349	6.3763	6.4103	6.4380	6.4607	6.4791	6.4942	6.5063	6.5162	6.5241	6.5305	6.5357	6.5397
	9	5.3263	5.5588	5.7374	5.8771	5.9878	6.0762	6.1472	6.2044	6.2506	6.2879	6.3181	6.3424	6.3620	6.3777	6.3903	6.4004	6.4084	6.4147	6.4197	6.4236	6.4267
	1	5.2908	5.5168	5.6892	5.8230	5.9281	6.0113	6.0775	6.1303	6.1724	6.2061	6.2330	6.2543	6.2713	6.2848	6.2954	6.3037	6.3102	6.3153	6.3192	6.3222	6.3246
	2	5.0095	5.1861	5.3123	5.4037	5.4701	5.5184	5.5534	5.5788	5.5970	5.6101	5.6193	5.6257	5.6302	5.6333	5.6354	5.6368	5.6377	5.6383	5.6387	5.6390	5.6391
	3	4.8006	4.9437	5.0402	5.1056	5.1498	5.1795	5.1993	5.2124	5.2208	5.2263	5.2297	5.2318	5.2331	5.2339	5.2343	5.2346	5.2347	5.2348	5.2349		5.2349
	4	4.6301	4.7481	4.8235	4.8714	4.9017	4.9205	4.9320	4.9390	4.9430	4.9454	4.9467	4.9474	4.9478	4.9480	4.9481						4.9481
10^{-3}	5	4.4844	4.5830	4.6426	4.6783	4.6993	4.7114	4.7183	4.7220	4.7241	4.7251	4.7256	4.7259	4.7260	4.7260							4.7260
	6	4.3566	4.4396	4.4872	4.5140	4.5288	4.5367	4.5409	4.5429	4.5439	4.5444	4.5446	4.5447									4.5447
	7	4.2425	4.3128	4.3510	4.3714	4.3819	4.3871	4.3896	4.3908	4.3913	4.3915	4.3916										4.3916
	8	4.1395	4.1991	4.2300	4.2455	4.2530	4.2565	4.2580	4.2587	4.2589	4.2590	4.2591										4.2591
	9	4.0450	4.0961	4.1212	4.1331	4.1385	4.1408	4.1417	4.1421	4.1422	4.1423											4.1423
	1	3.9582	4.0020	4.0224	4.0316	4.0355	4.0370	4.0376	4.0378													4.0378
	2	3.3403	3.3507	3.3537	3.3545	3.3547																3.3547
	3	2.9558	2.9585	2.9590																		2.9590
	4	2.6804	2.6812	2.6812																		2.6812
10^{-2}	5	2.4677	2.4679																			2.4679
	6	2.2952																				2.2952

Values of $M(\mu, \beta)$ are equal to $W(u)$ for u greater than 6×10^{-2} and all values of β.

TABLE 8-1 *(continued)*

u	β	50	52	54	56	58	60	62	64	66	68	70	72	74	76	78	80	82	84	86	88	90
	0	9.2102	9.2886	9.3641	9.4368	9.5069	9.5747	9.6403	9.7037	9.7653	9.8249	9.8829	9.9392	9.9940	10.0473	10.0992	10.1498	10.1992	10.2474	10.2944	10.3404	10.3853
10^{-6}	1	9.0977	9.1716	9.2426	9.3108	9.3765	9.4398	9.5008	9.5598	9.6168	9.6720	9.7255	9.7773	9.8276	9.8764	9.9239	9.9700	10.0148	10.0585	10.1011	10.1425	10.1830
	2	9.0510	9.1231	9.1922	9.2585	9.3223	9.3838	9.4430	9.5001	9.5553	9.6086	9.6602	9.7102	9.7586	9.8056	9.8512	9.8955	9.9385	9.9803	10.0210	10.0606	10.0992
	3	9.0152	0.0859	9.1535	9.2185	9.2809	9.3409	9.3986	9.4543	9.5081	9.5600	9.6102	9.6588	9.7058	9.7513	9.7955	9.8384	9.8800	9.9204	9.9597	9.9979	10.0351
	4	8.9851	9.0545	9.1210	9.1847	9.2459	9.3047	9.3613	9.4158	9.4684	9.5191	9.5681	9.6155	9.6613	9.7057	9.7487	9.7904	9.8308	9.8700	9.9081	9.9452	9.9812
	5	8.9586	9.0269	9.0924	9.1550	9.2152	9.2729	9.3285	9.3819	9.4335	9.4831	9.5311	9.5774	9.6222	9.6655	9.7075	9.7481	9.7875	9.8257	9.8628	9.8988	9.9338
	6	8.9346	9.0020	9.0065	9.1282	9.1874	9.2442	9.2988	9.3513	9.4019	9.4507	9.4977	9.5431	9.5869	9.6293	9.6703	9.7101	9.7485	9.7858	9.8220	9.8571	9.8911
	7	8.9126	8.9791	9.0427	9.1036	9.1619	9.2179	9.2716	9.3232	9.3730	9.4208	9.4670	9.5115	9.5545	9.5961	9.6362	9.6751	9.7127	9.7492	9.7845	9.8187	9.8519
	8	8.8921	8.9578	9.0206	9.0807	9.1382	9.1933	9.2463	9.2971	9.3460	9.3931	9.4385	9.4822	9.5244	9.5652	9.6046	9.6426	9.6795	9.7151	9.7497	9.7831	9.8156
	9	8.8729	8.9378	9.0000	9.0592	9.1160	9.1703	9.2225	9.2726	9.3208	9.3671	9.4118	9.4548	9.4962	9.5362	9.5749	9.6122	9.6483	9.6833	9.7171	9.7498	9.7815
10^{-5}	1	8.8547	8.9190	8.9803	9.0389	9.0949	9.1486	9.2001	9.2495	9.2970	9.3426	9.3865	9.4288	9.4696	9.5089	9.5469	9.5835	9.6189	9.6532	9.6863	9.7183	9.7494
	2	8.7087	8.7673	8.8229	8.8759	8.9263	8.9743	9.0202	9.0640	9.1059	9.1461	9.1845	9.2213	9.2566	9.2905	9.3230	9.3542	9.3843	9.4132	9.4410	9.4677	9.4935
	3	8.5976	8.6519	8.7033	8.7521	8.7983	8.8422	8.8839	8.9237	8.9615	8.9975	9.0319	9.0647	9.0960	9.1260	9.1546	9.1819	9.2081	9.2332	9.2572	9.2802	9.3023
	4	8.5047	8.5555	8.6035	8.6488	8.6916	8.7321	8.7705	8.8069	8.8414	8.8742	8.9053	8.9349	8.9630	8.9898	9.0153	9.0396	9.0628	9.0848	9.1059	9.1260	9.1451
	5	8.4235	8.4713	8.5163	8.5587	8.5986	8.6362	8.6717	8.7053	8.7370	8.7671	8.7955	8.8224	8.8479	8.8721	8.8950	8.9167	8.9374	8.9570	8.9756	8.9932	9.0100
	6	8.3507	8.3959	8.4383	8.4780	8.5154	8.5505	8.5836	8.6147	8.6440	8.6716	8.6977	8.7223	8.7455	8.7675	8.7882	8.8078	8.8263	8.8438	8.8603	8.8760	8.8908
	7	8.2843	8.3271	8.3672	8.4046	8.4397	8.4726	8.5034	8.5324	8.5596	8.5851	8.6091	8.6317	8.6530	8.6730	8.6918	8.7095	8.7262	8.7419	8.7567	8.7706	8.7838
	8	8.2229	8.2636	8.3016	8.3370	8.3700	8.4009	8.4297	8.4568	8.4821	8.5057	8.5279	8.5487	8.5682	8.5865	8.6036	8.6197	8.6348	8.6490	8.6623	8.6747	8.6864
	9	8.1658	8.2045	8.2405	8.2740	8.3052	8.3343	8.3614	8.3866	8.4102	8.4322	8.4528	8.4720	8.4899	8.5067	8.5223	8.5370	8.5507	8.5635	8.5754	8.5866	8.5970
10^{-4}	1	8.1122	8.1491	8.1833	8.2151	8.2446	8.2720	8.2974	8.3211	8.3431	8.3636	8.3827	8.4005	8.4170	8.4324	8.4468	8.4601	8.4726	8.4842	8.4949	8.5050	8.5143
	2	7.6962	7.7203	7.7421	7.7618	7.7797	7.7958	7.8104	7.8236	7.8356	7.8463	7.8560	7.8648	7.8727	7.8798	7.8862	7.8920	7.8972	7.9019	7.9061	7.9098	7.9132
	3	7.3997	7.4163	7.4309	7.4439	7.4553	7.4654	7.4742	7.4820	7.4889	7.4949	7.5002	7.5048	7.5088	7.5124	7.5154	7.5181	7.5204	7.5225	7.5242	7.5258	7.5271
	4	7.1663	7.1780	7.1881	7.1968	7.2043	7.2108	7.2163	7.2211	7.2251	7.2286	7.2315	7.2341	7.2362	7.2380	7.2395	7.2408	7.2419	7.2428	7.2436	7.2442	7.2447
	5	6.9734	6.9818	6.9888	6.9948	6.9998	7.0040	7.0075	7.0105	7.0129	7.0150	7.0167	7.0181	7.0192	7.0202	7.0209	7.0216	7.0221	7.0225	7.0228	7.0231	7.0233
	6	6.8090	6.8151	6.8201	6.8242	6.8276	6.8304	6.8327	6.8345	6.8360	6.8372	6.8382	6.8390	6.8396	6.8401	6.8405	6.8408	6.8411	6.8413	6.8414	6.8416	6.8417
	7	6.6661	6.6705	6.6741	6.6770	6.6793	6.6811	6.6826	6.6838	6.6847	6.6854	6.6860	6.6864	6.6868	6.6871	6.6873	6.6874	6.6875	6.6876	6.6877	6.6878	6.6878
	8	6.5397	6.5430	6.5456	6.5476	6.5492	6.5504	6.5514	6.5521	6.5527	6.5531	6.5535	6.5537	6.5539	6.5541	6.5542	6.5543	6.5543	6.5544	6.5544	6.5544	6.5544
	9	6.4267	6.4291	6.4310	6.4324	6.4335	6.4343	6.4350	6.4354	6.4358	6.4361	6.4363	6.4364	6.4365	6.4366	6.4366	6.4367	6.4367	6.4367	6.4368		6.4368
10^{-3}	1	6.3246	6.3263	6.3277	6.3287	6.3294	6.3300	6.3304	6.3307	6.3310	6.3311	6.3312	6.3313	6.3314	6.3314	6.3315						6.3315
	2	5.6391	5.6392	5.6393	5.6393	5.6393																5.6393

Values of $M(u, \beta)$ are equal to $W(u)$ for u greater than 2×10^{-3} and all values of β.
Source: After Hantush (1961c).

Equation of Drawdown for Relatively Small Values of Time. By virtue of Eq. (8-57), f' terms in Eq. (8-48) can safely be neglected if (Hantush, 1961c)

$$u > 5\left(\frac{r}{2b - l - z}\right)^2 \tag{8-58}$$

For infinite values of b, f' terms approach zero. Therefore, if either

$$t < \frac{(2b - l - z)^2 S_s}{20K} \tag{8-59}$$

or the aquifer is infinitely thick, Eq. (8-48) reduces to [Hantush, 1961c, p. 90, Eq. (7)]

$$s = \frac{Q}{8\pi K(l - d)} E\left(u, \frac{l}{r}, \frac{d}{r}, \frac{z}{r}\right) \tag{8-60}$$

in which

$$E = M\left(u, \frac{l + z}{r}\right) + M\left(u, \frac{l - z}{r}\right) - M\left(u, \frac{d + z}{r}\right) - M\left(u, \frac{d - z}{r}\right) \tag{8-61}$$

Equation (8-60) states that during the initial pumping period, the drawdown around a partially penetrating well will be the same as in the case of infinite aquifer thickness. The length of the initial pumping period depends on the screen interval location, the observation point depth, the formation thickness, and the hydraulic parameters of the formation.

Hantush (1961d, p. 173) points out that Eq. (8-60) may take the following form depending on the geometry of the flow system:

$$s = cM(u, \beta) \tag{8-62}$$

in which case the equation is valid for

$$t < \frac{(2b - r\beta)^2 S_s}{20K} \tag{8-63}$$

and c and β are constants that are dependent on the parameters of the flow system. For example, if $d = 0$ in Figure 8-5 and if the piezometer depth is equal to the depth of the pumped well ($z = l$), Eq. (8-60) will take the form

$$s = \frac{Q}{8\pi Kl} M\left(u, \frac{2l}{r}\right) \tag{8-64}$$

If the penetration depth of the piezometer is zero ($z = 0$), Eq. (8-60) will become

$$s = \frac{Q}{4\pi Kl} M\left(u, \frac{l}{r}\right) \tag{8-65}$$

Equation of Drawdown for Relatively Large Values of Time. Hantush (1961c, p. 90) showed that for

$$u < \frac{1}{20}\left(\frac{\pi r}{b}\right)^2 \tag{8-66}$$

the function $W(u, n\pi r/b)$ can, for all practical purposes, be replaced by $2K_0(n\pi r/b)$, in which the series in Eq. (8-47) become independent of time. Hence, for

$$u < \frac{1}{2}\left(\frac{r}{b}\right)^2 \qquad \text{namely} \quad t > \frac{b^2 S_s}{2K} \tag{8-67}$$

Eq. (8-50) takes the form

$$s = \frac{Q}{4\pi K b}\left[W(u) + f_s\left(\frac{r}{b}, \frac{l}{b}, \frac{d}{b}, \frac{z}{b}\right)\right] \tag{8-68}$$

in which

$$f_s = \frac{4b}{\pi(l-d)} \sum_{n=1}^{\infty} \frac{1}{n} K_0\left(\frac{n\pi r}{b}\right)\left[\sin\left(\frac{n\pi l}{b}\right) - \sin\left(\frac{n\pi d}{b}\right)\right] \cos\left(\frac{n\pi z}{b}\right) \tag{8-69}$$

where K_0 is the zero-order modified Bessel function of the second kind. Equation (8-68) shows that in this time range, the rate of change of drawdown is the same as though the pumped well penetrates the aquifer completely. This means that the effect of partial penetration on the drawdown has attained its maximum value.

8.3.5 Solution for Average Drawdown in Observation Wells

The water level in an observation well (see Figure 8-5) represents the average drawdown in the aquifer profile that is in contact with the well screen or the perforated section of the casing of the observation well. Solutions for drawdown in observation wells with and without leakage are given below.

8.3.5.1 Solution for K_r Is Different than K_z with Leakage. The average drawdown in an observation well which is screened between the depths l' and d' ($l' > d'$) can be obtained by integrating the drawdown equation for piezometers with respect to z between the limits d' and l', and then dividing the result by $(l' - d')$. After performing this operation, the resulting drawdown expression will be as (Reed, 1980, p. 31):

$$s = \frac{Q}{4\pi T}\left[W(u, \beta_r) + \overline{f}\left(u, \frac{ar}{b}, \beta_r, \frac{d}{b}, \frac{l}{b}, \frac{d'}{b}, \frac{l'}{b}\right)\right] \tag{8-70}$$

where $W(u)$ is the Theis well function defined by Eq. (4-25) of Chapter 4 or Eq. (8-43) and

$$\overline{f}\left(u, \frac{ar}{b}, \beta_r, \frac{d}{b}, \frac{l}{b}, \frac{d'}{b}, \frac{l'}{b}\right) = \frac{2b^2}{\pi^2(l-d)(l'-d')} \sum_{n=1}^{\infty} \frac{1}{n^2}\left[\sin\left(\frac{n\pi l}{b}\right) - \sin\left(\frac{n\pi d}{b}\right)\right] \cdot \left[\sin\left(\frac{n\pi l'}{b}\right) - \sin\left(\frac{n\pi d'}{b}\right)\right] W(u, \kappa_n) \tag{8-71}$$

where $W(u, \beta_r)$ and u are defined by Eqs. (8-36) and (8-37), respectively, and κ_n is defined by Eq. (8-41).

8.3.5.2 Solution for K_r Is Different than K_z without Leakage. The geometry of this case is shown in Figure 8-5; the corresponding solutions, based on Hantush's original work (Hantush, 1964, pp. 349–351), is given as (Reed, 1980, pp. 8–10)

$$s = \frac{Q}{4\pi T}\left[W(u) + \overline{f}\left(u, \frac{ar}{b}, \frac{l}{b}, \frac{d}{b}, \frac{l'}{b}, \frac{d'}{b}\right)\right] \quad (8\text{-}72)$$

where $W(u)$ is defined by Eq. (8-43) and

$$\overline{f}\left(u, \frac{ar}{b}, \frac{l}{b}, \frac{d}{b}, \frac{l'}{b}, \frac{d'}{b}\right) = \frac{2b^2}{\pi^2(l-d)(l'-d')}\sum_{n=1}^{\infty}\frac{1}{n^2}\left[\sin\left(\frac{n\pi l}{b}\right) - \sin\left(\frac{n\pi d}{b}\right)\right]$$
$$\cdot\left[\sin\left(\frac{n\pi l'}{b}\right)\right] - \sin\left(\frac{n\pi d'}{b}\right) W\left(u, \frac{n\pi ar}{b}\right) \quad (8\text{-}73)$$

where $W(u, x)$ is defined by Eq. (8-36).

8.3.5.3 Solution for $K_r = K_z$ without Leakage. As mentioned in Section 8.3.5.1, the water level in an observation well (see Figure 8-5) reflects the average drawdown in the aquifer profile that is in contact with the screen or perforated section of the casing of the well. The average drawdown in an observation well, shown in Figure 8-5, screened between the depths l' and d' ($l' > d'$) can be determined by integrating the equation of drawdown in piezometers with respect to z between the limits d' and l' and then dividing the result by $(l' - d')$. After performing this operation on Eq. (8-47), the average drawdown expression will take the following form [Hantush, 1961c, p. 90, Eqs. (9a) and (9b)]:

$$\overline{s} = \frac{Q}{4\pi Kb}\left[W(u) + \overline{f}\left(u, \frac{r}{b}, \frac{l}{b}, \frac{d}{b}, \frac{l'}{b}, \frac{d'}{b}\right)\right] \quad (8\text{-}74)$$

where

$$\overline{f} = \frac{2b^2}{\pi^2(l-d)(l'-d')}\sum_{n=1}^{\infty}\frac{1}{n^2}\left[\sin\left(\frac{n\pi l}{b}\right) - \sin\left(\frac{n\pi d}{b}\right)\right]$$
$$\cdot\left[\sin\left(\frac{n\pi l'}{b}\right) - \sin\left(\frac{n\pi d'}{b}\right)\right] W\left(u, \frac{n\pi r}{b}\right) \quad (8\text{-}75)$$

Hantush (1961c) points out that the result of averaging Eq. (8-48) is rather complicated; and, instead, Hantush presents the corresponding average drawdown equations for relatively small and large values of time, and these solutions are presented in the following sections.

Equation of Average Drawdown for Relatively Small Values of Time. For

$$t < \frac{(2b - l - l')S_s}{20K} \quad (8\text{-}76)$$

the terms f' in Eq. (8-48) can be drop out for all practical purposes, and the result of averaging the remaining four M terms is given by [Hantush, 1961c, p. 91, Eq. (10)]

$$\overline{s} = \frac{Q}{8\pi K(l-d)(l'-d')}\left[F\left(u, \frac{l+l'}{r}, \frac{l-l'}{r}\right) - F\left(u, \frac{d+l'}{r}, \frac{d-l'}{r}\right) + F\left(u, \frac{d+d'}{r}, \frac{d-d'}{r}\right) - F\left(u, \frac{l+d'}{r}, \frac{l-d'}{r}\right)\right] \quad (8\text{-}77)$$

$$F(u, \beta, \alpha) = r\left\{\beta M(u, \beta) - \alpha M(u, \alpha) + 2\left[y^{1/2}\operatorname{erfc}(y^{1/2}u^{1/2}) - x\operatorname{erfc}(x^{1/2}u^{1/2}) + \frac{e^{-xu} - e^{-yu}}{\pi^{1/2}u^{1/2}}\right]\right\} \quad (8\text{-}78)$$

in which

$$x = 1 + \beta^2, \qquad y = 1 + \alpha^2 \quad (8\text{-}79)$$

and M is the function defined by Eq. (8-52). Hantush (1961c) points out that Eq. (8-78) also gives the average drawdown in observation wells tapping an infinite deep aquifer for the whole range of pumping time.

If $l'/l < 2$, a sufficiently accurate form of Eq. (8-77) can be expressed as [Hantush, 1961c, p. 91, Eq. (11)]

$$\overline{s} \cong \frac{Q}{8\pi K(l-d)}\overline{E}\left(u, \frac{l}{r}, \frac{d}{r}, \frac{l'}{r}, \frac{d'}{r}\right) \quad (8\text{-}80)$$

which is valid for

$$t < \frac{\left[2b - \frac{1}{2}(2l + l' + d')\right]^2 S_s}{20K} \quad (8\text{-}81)$$

in which $\overline{E}$ is the function of the function E of Eq. (8-60) in which $z = (l' + d')/2$. In other words, the average drawdown in an observation well screened between the depths l' and d' can be approximated by the average of drawdowns in two piezometers whose penetration depths are l' and d', respectively, provided that $l'/l < 2$. Equation (8-80) is valid under the following condition (Hantush, 1961d, p. 174): If $r/l > 1$ and $l'/l < 1$, the average drawdown in the observation well can, for all practical purposes, be taken as that given by Eq. (8-60), with the value of z arbitrarily chosen between l' and zero. The corresponding equation is [Hantush, 1961c, p. 91, Eq. (12)]:

$$\overline{s} \cong cM(u, \beta) \quad (8\text{-}82)$$

in which case the equation is valid for

$$t < \frac{(2b - r\beta)^2 S_s}{20K} \quad (8\text{-}83)$$

In Eqs. (8-82) and (8-83), c and β are constants whose values depend on the parameters of the flow system under consideration. Hantush (1961c) gives equations for some special cases: If $l = 3d$, a choice of $z = d$ will reduce Eq. (8-60) to

$$\bar{s} \cong \frac{3Q}{16\pi K l} M\left(u, \frac{4}{3}\frac{l}{r}\right) \tag{8-84}$$

Also, if $d = 0$, a choice of $z = 0$ will result in

$$\bar{s} \cong \frac{Q}{4\pi K l} M\left(u, \frac{l}{r}\right) \tag{8-85}$$

whereas a choice of $z = l$ will give

$$\bar{s} \cong \frac{Q}{8\pi K l} M\left(u, \frac{2l}{r}\right) \tag{8-86}$$

Equation of Average Drawdown for Relatively Large Values of Time. Because, for a relatively long period of pumping, that is, for

$$t > \frac{b^2 S_s}{2K} \tag{8-87}$$

the relation

$$W\left(u, \frac{n\pi r}{b}\right) \cong 2K_0\left(\frac{n\pi r}{b}\right) \tag{8-88}$$

is valid, Eq. (8-74) takes the following form (Hantush, 1961c):

$$\bar{s} = \frac{Q}{4\pi K b}\left[W(u) + \overline{f_s}\left(\frac{r}{b}, \frac{l}{b}, \frac{d}{b}, \frac{l'}{b}, \frac{d'}{b}\right)\right] \tag{8-89}$$

in which

$$\overline{f_s} = \frac{4b^2}{\pi^2(l-d)(l'-d')}\sum_{n=1}^{\infty}\frac{1}{n^2}K_0\left(\frac{n\pi r}{b}\right) \cdot \left[\sin\left(\frac{n\pi l}{b}\right) - \sin\left(\frac{n\pi d}{b}\right)\right]\left[\sin\left(\frac{n\pi l'}{b}\right) - \sin\left(\frac{n\pi d'}{b}\right)\right] \tag{8-90}$$

Equation (8-89) shows that in this time range the rate of change of drawdown is the same as though the pumping well fully penetrates the aquifer. In other words, the effect of partial penetration on the drawdown reaches its maximum value.

8.3.6 Drawdown in Piezometers or Wells for $r/b > 1.5$

As mentioned in Section 8.2.6, for relatively large distances—that is, $r/b > 1.5$—Hantush (1957) showed that the equation of drawdown can be expressed as

$$s = \bar{s} = \frac{Q}{4\pi K b} W(u) \tag{8-91}$$

Hantush (1961c) notes that Eq. (8-91) gives sufficiently accurate results for practical purposes even for r/b as small as one, provided that $u < 0.1(r/b)^2$. It is obvious that Eq. (8-91) is the Theis equation [Chapter 4, Eq. (4-24)]. This means that the actual three-dimensional flow pattern becomes radial and hardly distinguishable from that of a radial flow system at a distance from the pumped well equal to or greater than 1.5 times the thickness of the aquifer.

8.3.7 Recovery Equations

The basic theory for recovery is presented in Section 4.2.7.4 of Chapter 4. Here, Hantush's theory (Hantush, 1961c) regarding the recovery equations for partially penetrating wells will be presented.

As mentioned in Section 4.2.7.4 of Chapter 4, if t and t' are the time since pumping started and the time since pumping stopped, respectively, the residual drawdown, s', in a piezometer during recovery is shown to be

$$s' = s(t) - s(t') \tag{8-92}$$

Similarly, the average residual drawdown, $\overline{s'}$, in an observation well can be written as

$$\overline{s'} = \overline{s}(t) - \overline{s}(t') \tag{8-93}$$

where $t = t_0 + t'$ and t_0 is the time at which the pumping has stopped. As a result, the recovery equation corresponding to any of the drawdown equations presented in the previous sections can readily be formulated under the same time criteria.

In the following sections, the recovery equations under the conditions that $K_r = K_z$ and without leakage will be presented.

8.3.7.1 Recovery Equations for Piezometers

Equations for $K_r = K_z$ without Leakage. The drawdown equation for $K_r = K_z$ is given by Eq. (8-47). Applying Eq. (8-92) to Eq. (8-47), the general equation of recovery in a piezometer of penetration depth equal to z can be expressed as (Hantush, 1961c)

$$s' = \frac{Q}{4\pi Kb}\left[W(u) - W(u') + f\left(u, \frac{r}{b}, \frac{l}{b}, \frac{d}{b}, \frac{z}{b}\right) - f\left(u', \frac{r}{b}, \frac{l}{b}, \frac{d}{b}, \frac{z}{b}\right)\right] \tag{8-94}$$

where u' is the value of u after replacing t by t'.

If Eq. (8-48) is used instead of Eq. (8-47) in conjunction with Eq. (8-92), the general recovery equation can be expressed in the following alternative form (Hantush, 1961c):

$$s' = \frac{Q}{8\pi K(l-d)}[(\text{terms of Eq. (8-48)}) - (\text{same terms with } u' \text{ replacing } u)] \tag{8-95}$$

Hantush (1961c) notes that a third form can be obtained by subtracting Eq. (8-48), with u' replacing u, from Eq. (8-47).

Equation (8-94) is suitable for computation when t' is large. Equation (8-95) is suitable for small values of t—that is, $(t_0 + t')$. The third form mentioned above is suitable for computation when t is large and t' is small.

Equations for $K_r = K_z$ without Leakage for Relatively Small Values of Time. Hantush (1961d) also presented the recovery equations for the type of Eq. (8-62), and the result is

$$s' = c\left[M(u,\beta) - M(u',\beta) + f_2\left(u, \frac{b}{r}, \beta\right) - f_2\left(u', \frac{b}{r}, \beta\right)\right] \tag{8-96}$$

in which u' is the value of u after replacing t by t' and

$$f_2\left(u, \frac{b}{r}, \beta\right) = \sum_{n=1}^{\infty}\left\{M\left[u, \left(\frac{2nb}{r} + \beta\right)\right] - M\left[u, \left(\frac{2nb}{r} - \beta\right)\right]\right\} \tag{8-97}$$

For the condition expressed by Eq. (8-63), Eq. (8-96) can be approximated by

$$s' = c[M(u,\beta) - M(u',\beta)] \tag{8-98}$$

For the conditions

$$t_0 > \frac{b^2 S_s}{2K}, \qquad t' < \frac{(2b - r\beta)^2 S_s}{20K} \tag{8-99}$$

Eq. (8-96) can be approximated by

$$s' = \frac{Q}{4\pi K b}[W(u) - f_s] - cM(u',\beta) \tag{8-100}$$

in which f_s is given by Eq. (8-69).

If both t_0 and t' are greater than $b^2 S_s/(2K)$, then Eq. (8-96) and the general recovery equation take the following form (Hantush, 1961d)

$$s' = \frac{Q}{4\pi K b}[W(u) - W(u')] \tag{8-101}$$

8.3.7.2 Recovery Equations for Observation Wells. The equations presented in Section 8.3.7.1 are the recovery equations for piezometers. The recovery equations for observation wells can similarly be determined by using the appropriate drawdown equations. Equation (8-101) gives the recovery for both in piezometers and observation wells (Hantush, 1961d).

8.3.8 Evaluation of Solutions

The effects of partial penetration on drawdown around a pumping well are shown in Figures 8-6 and 8-7. Based on these figures, some key aspects regarding the solutions for partially penetrating wells are given below (Hantush, 1961c):

1. If an observation well is screened throughout the thickness of a confined aquifer, the average drawdown can be represented by the Theis equation as given by Eq. (8-91). The same is also true for the case of a well located at $r/b > 1.5$, regardless of the length and vertical position of its screen. In other words, the average drawdown is not affected by partial penetration. Curve 1 in Figure 8-6 corresponds to this case.

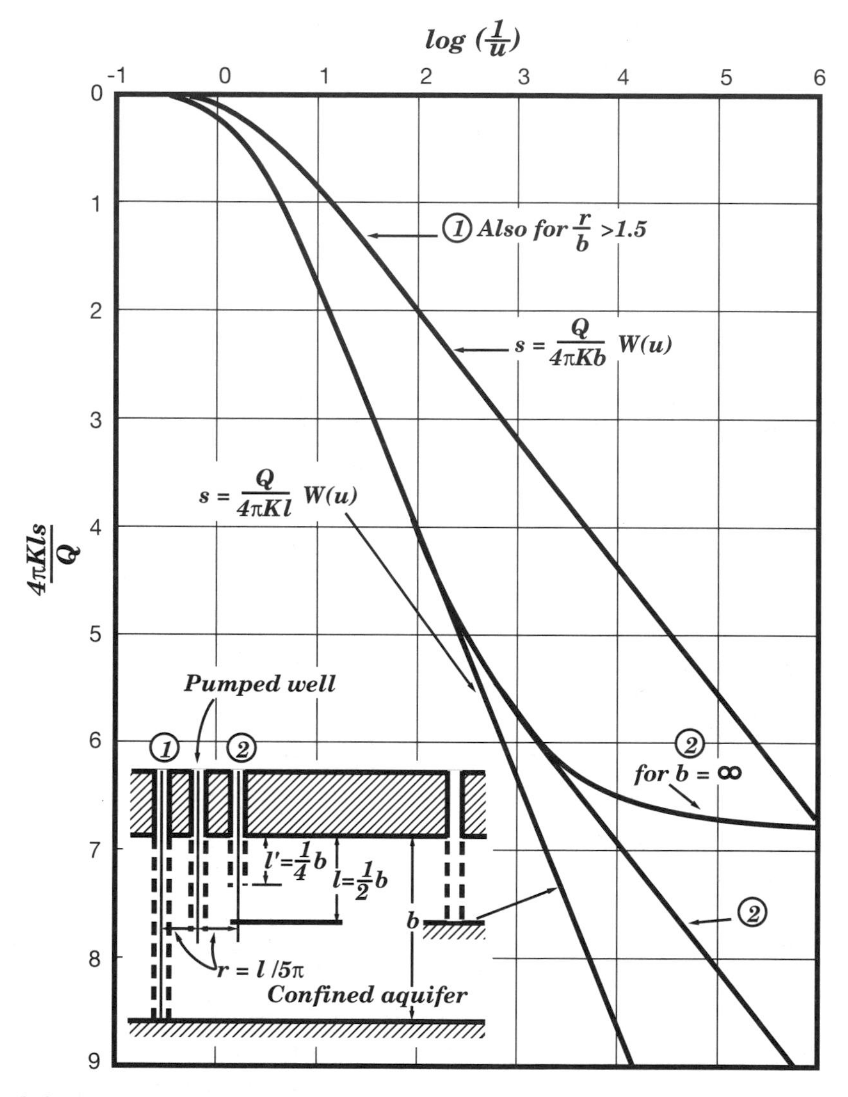

Figure 8-6 Time versus drawdown variation in partially penetrating wells in a confined aquifer. (After Hantush, 1961c.)

2. At relatively large values of time, which satisfy the conditions expressed by Eq. (8-67), the time versus drawdown curves have approximately the same slope regardless of the location of the wells and the vertical position of their screens. This means that the effect of partial penetration has reached to its maximum value (see Curve 1 and Curve 2 in Figure 8-6).

3. If the distance to an observation well is relatively short (i.e., $r/b < 1.5$), or is not screened throughout the aquifer thickness, the drawdown versus time curve will have the trend of Curve 2 in Figure 8-6 and Curves 1 and 2 in Figure 8-7.

4. Figure 8-7 compares the drawdowns at two different observation wells that have equal distances from the pumping well, one with zero penetration and the other

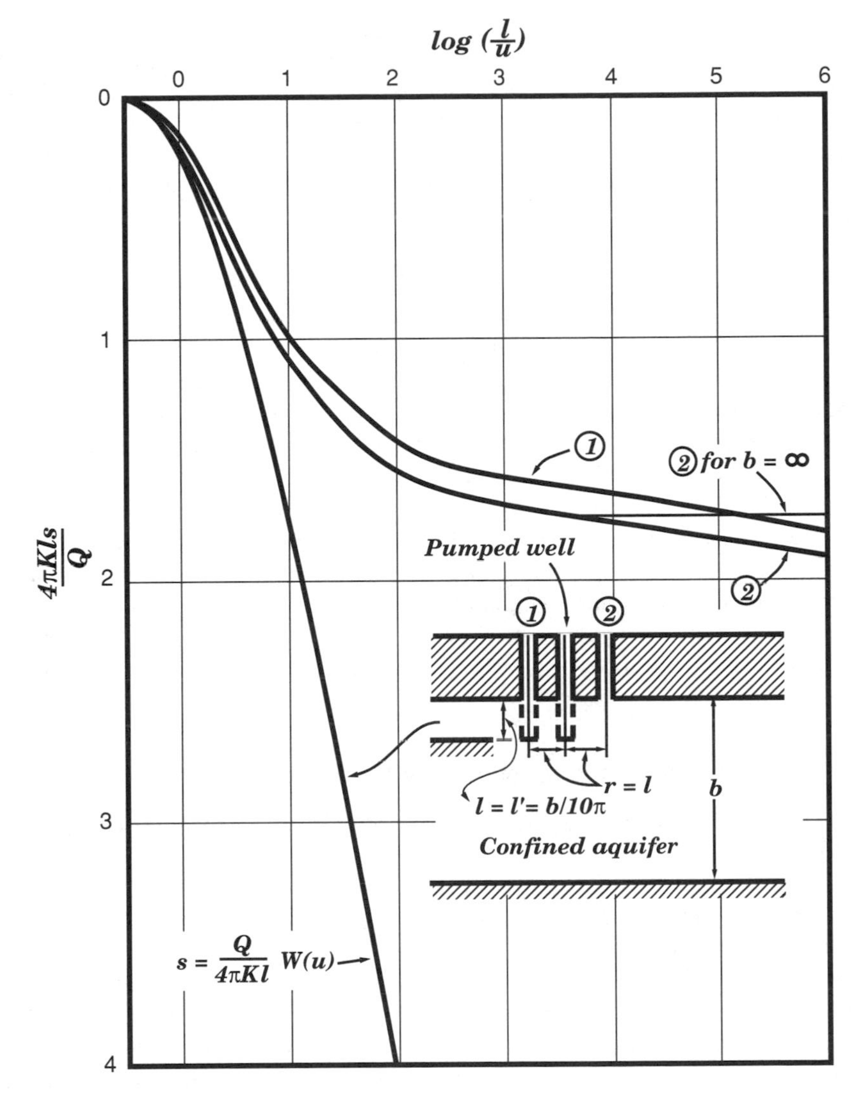

Figure 8-7 Time versus drawdown variation in wells of different penetration depths in a confined aquifer. (After Hantush, 1961c.)

screened throughout the thickness of the aquifer. As can be seen from Figure 8-7, the drawdown curves are significantly different.

8.4 APPLICATION TO PUMPING TEST DATA ANALYSIS

8.4.1 Introduction

In the previous sections, the theoretical foundations of drawdown equations in confined aquifers under partially penetrating well conditions are presented in detail. These solutions are practically important in determining the hydrogeologic parameters of confined aquifers.

As discussed in detail in Chapter 4, the Theis drawdown equation can be used under fully penetrating well conditions for the determination of aquifer parameters. In most cases, both pumping and observation wells partially penetrate the aquifer thickness or the thickness of the aquifer is large or is unknown. Under these conditions, the Theis equation may give erroneous results depending on the degree of partial penetration. The drawdown equations for partially penetrating well cases not only give more realistic aquifer hydraulic parameters, but also provide quantitative information regarding the thickness of a water-bearing formation.

Leaky confined aquifer models under partially penetrating well conditions [Eq. (8-35) for piezometers and Eq. (8-70) for observation wells] involve a large number of variables. Dealing with these relatively large number of variables is not practical or it is impossible to construct a sufficient number of type curves to cover the entire range of values for field application. Therefore, in the following sections, various methods regarding the determination of aquifer parameters based on pumping test data from partially penetrating wells in nonleaky confined aquifers will be presented.

These methods can basically be categorized as (1) the double logarithmic type-curve method and (2) semilogarithmic methods. The first one is the standard type-curve method which needs to generate the type-curve data based on the pumping and observation wells geometry by running a computer program for the corresponding drawdown equation. The second group of methods, which were developed by Hantush (1961d), do not require to run a computer program; instead, the necessary data can be taken from tables originally designed for this type of analyses.

8.4.2 Type-Curve Methods for Nonleaky Confined Aquifers

If the thickness of a confined aquifer is known, the following methods can be used in analyzing data for partially penetrating wells.

8.4.2.1 Type-Curve Method for $r/b > 1.5$. It has been shown in Section 8.3.6 that for relatively large distances (i.e., $r/b > 1.5$) the drawdown equation for piezometers or observation wells under partially penetrating well conditions can be expressed by the Theis equation. Therefore, data collected from observation wells located at $r > 1.5b$ from a partially penetrating pumping well or from wells screened throughout the thickness of the aquifer can be analyzed by the Theis type-curve method (see Chapter 4, Section 4.2.7.1). Additional conditions for the application of the Theis equation are addressed in Section 8.3.6.

8.4.2.2 Type-Curve Method for the General Case. Equation (8-72) in Section 8.3.5.2 gives the average drawdown in an observation well under partially penetrating pumping well conditions in nonleaky confined aquifers. This the general case in the sense that the distance of an observation well from the pumping well does not have any restrictions. In other words, r/b can be smaller than 1.5.

Type-Curve Generation. Equation (8-72) can be expressed in dimensionless form as

$$s_D = \frac{4\pi T s}{Q} = W(u) + \overline{f}\left(u, \frac{ar}{b}, \frac{l}{b}, \frac{d}{b}, \frac{l'}{b}, \frac{d'}{b}\right) \qquad (8\text{-}102)$$

in which $\overline{f}$ is given by Eq. (8-73). The values of the dimensionless drawdown s_D can be determined from a computer program, called TYPCRV, originally developed by the U.S.

Geological Survey (Reed, 1980). A menu-driven modified version of a portion of this program was used to generate Hantush type curves under partially penetrating well conditions in confined aquifers. The modified program is called HTYCPC, and modifications were made by the author of this book. Modifications in HTYCPC are only related with the input data instructions and output structure, and all the rest of part of the original program portion including its algorithm were kept the same. The program generates s_D versus u data for a selected value of the anisotropy factor (a). The number of type curves depends on the number of anisotropy factor expressed by $a = (K_z/K_r)^{1/2}$, as given by Eq. (8-39).

Steps of the Type-Curve Method. The steps of the type curve method are presented as follows:

Step 1: Plot a family of type curves, s_D versus $1/u$, from Eq. (8-102) for different values of K_z/K_r on log–log paper for each pumped well and observation well pair. A suggested set of values of $a^2 = K_z/K_r$ are 0.001, 0.01, 0.02, 0.05, 0.10, 0.25, 0.50, and 1.00. This set can be modified depending on the purpose of test, lithologic structure of the formation, and the data gathered from the pumping test. If the aquifer thickness is unknown and is known to be very thick, generate sets of type curves for more than relatively one large thickness.

Step 2: On another sheet of log–log paper, with the same scale of the first, plot the drawdown (s) versus time (t) for a given observation well at a distance from the pumped well.

Step 3: Keeping the coordinate axes parallel at all times—that is, the s_D axis parallel with the t axis—superimpose the time versus drawdown field data curve of each well on the type curves. Adjust until as many as data points fall on one of the type curves and note the value of $a^2 = K_z/K_r$. If there are more than one set of type curves, repeat the process for each set.

Step 4: Select a common point on overlapping portion of the sheets and read the corresponding values of s_D, $1/u$, s, and t for each set of type curves.

Step 5: Solving Eq. (8-102) for T gives

$$T = \frac{Qs_D}{4\pi s} \tag{8-103}$$

Substitute the known values of Q, s, and s_D and calculate the value of aquifer transmissivity (T) for each set of type curve.

Step 6: Solving Eq. (8-37) for S gives

$$S = \frac{4Tt}{r^2\left(\dfrac{1}{u}\right)} \tag{8-104}$$

Substitute the values of T, t, r, and $1/u$ and calculate the value of storage coefficient (S) for each set of type curves. Calculate the specific storage (S_s) value from $S = S_s b$.

Step 7: Having determined the value of transmissivity (T), calculate the radial horizontal hydraulic conductivity from

$$K_r = \frac{T}{b} \tag{8-105}$$

Repeat this calculation for each value. It must be pointed out that here b is the full thickness of the confined aquifer (see Figure 8-5), not the length of the screen interval (l–d in Figure 8-5) of the pumping well.

Step 8: Solving K_z from the second expression of Eq. (8-28) gives

$$K_z = a^2 K_r \tag{8-106}$$

Substitute the values in Eq. (8-106) and calculate the value of the vertical hydraulic conductivity (K_z).

Example 8-1: The data for this example are taken from Hantush (1961d) by converting the units from the English system to the metric system. A pumping test was conducted near the town of Socorro, New Mexico during September 1952. Both the pumped well and the observation well are perforated throughout their depth of penetration. A schematic representation of the wells is shown in Figure 8-8. The depths from the bottom boundary of the confining layer of the pumped well and observation well are 35.36 m and 18.29 m, respectively. The aquifer is known to be very thick, but its thickness (b) is unknown. The distance of the observation well from the pumped well is 5.18 m. During the test, the pumping rate was 101.09 liters/s (8734 m^3/day). The observed drawdowns at 5.18-m radial distance are presented in Table 8-2. Using the type-curve method outlined above, (a) determine the aquifer parameters and (b) discuss the results.

Solution: (a) The aquifer parameters are determined as follows:

Step 1: Because the relatively large aquifer thickness is unknown, type curves are generated for b = 75 m, 150 m, and 300 m and they are presented in Figures 8-9, 8-10, and 8-11, respectively. Because the curves are too close to each other, only $a^2 = K_z/K_r = 0.001$ and 1 are presented in the figures.

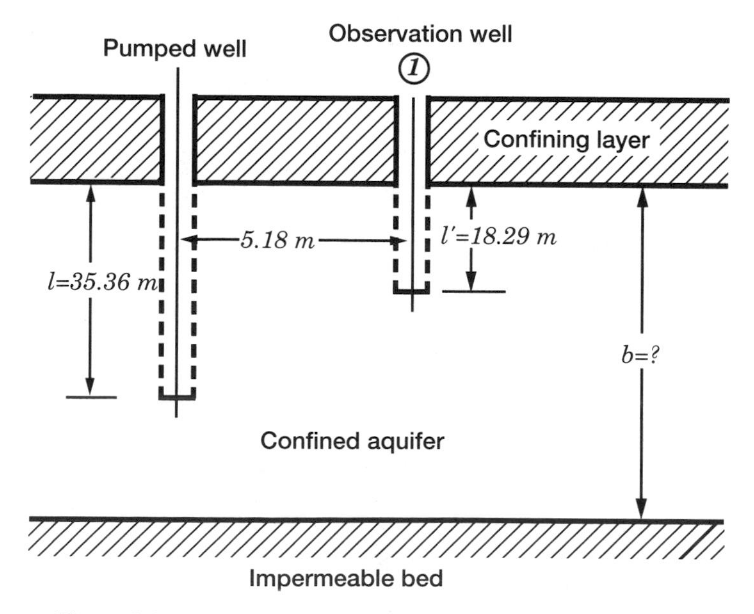

Figure 8-8 Schematic representation of the wells in Example 8-1.

TABLE 8-2 Pumping Test Data for Example 8-1

t (min)	s (m)	t (min)	s (m)
1.14	1.68	24.17	4.42
1.73	1.53	32.72	4.47
2.00	2.19	39.45	4.69
2.67	2.44	43.45	4.63
3.04	2.37	49.43	4.71
3.57	2.59	56.23	4.68
4.67	2.78	66.07	4.74
5.83	2.95	67.14	4.74
7.33	3.04	78.38	4.76
7.57	3.36	91.20	4.82
10.00	3.52	116.14	4.86
10.00	3.69	158.78	4.94
15.64	4.12	223.46	4.94
16.00	3.92	291.07	4.93
18.98	4.27	348.66	4.94
21.59	4.15	582.10	4.96

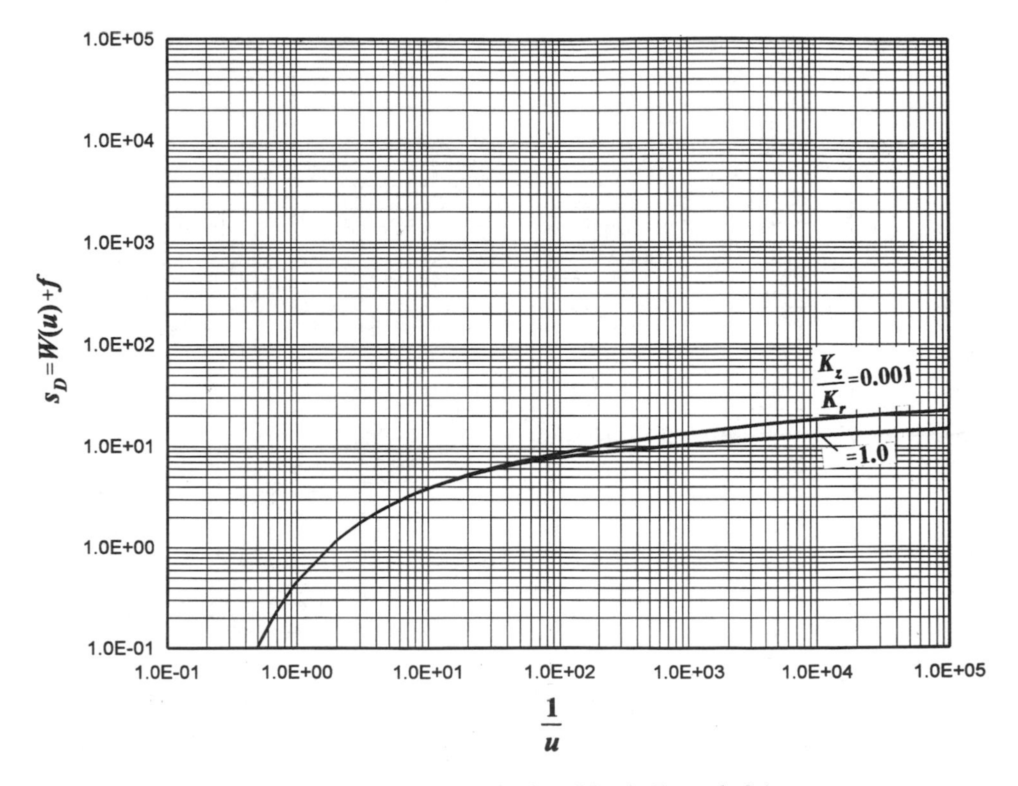

Figure 8-9 Type curves for $b = 75$ m in Example 8-1.

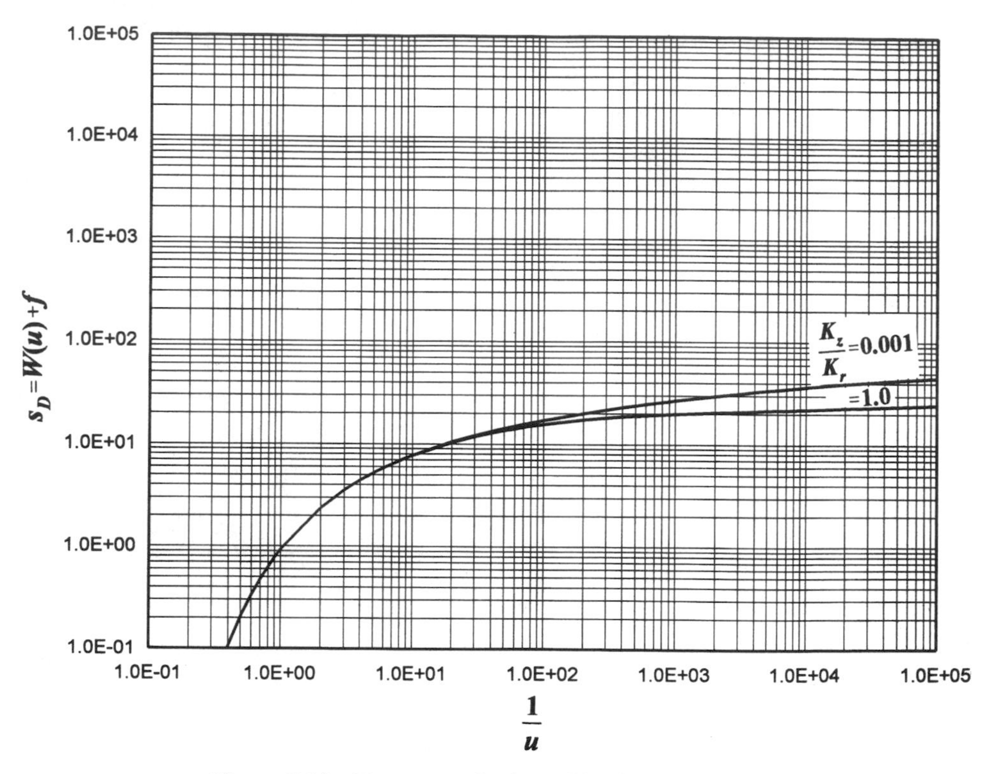

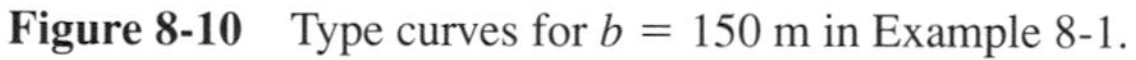

Figure 8-10 Type curves for $b = 150$ m in Example 8-1.

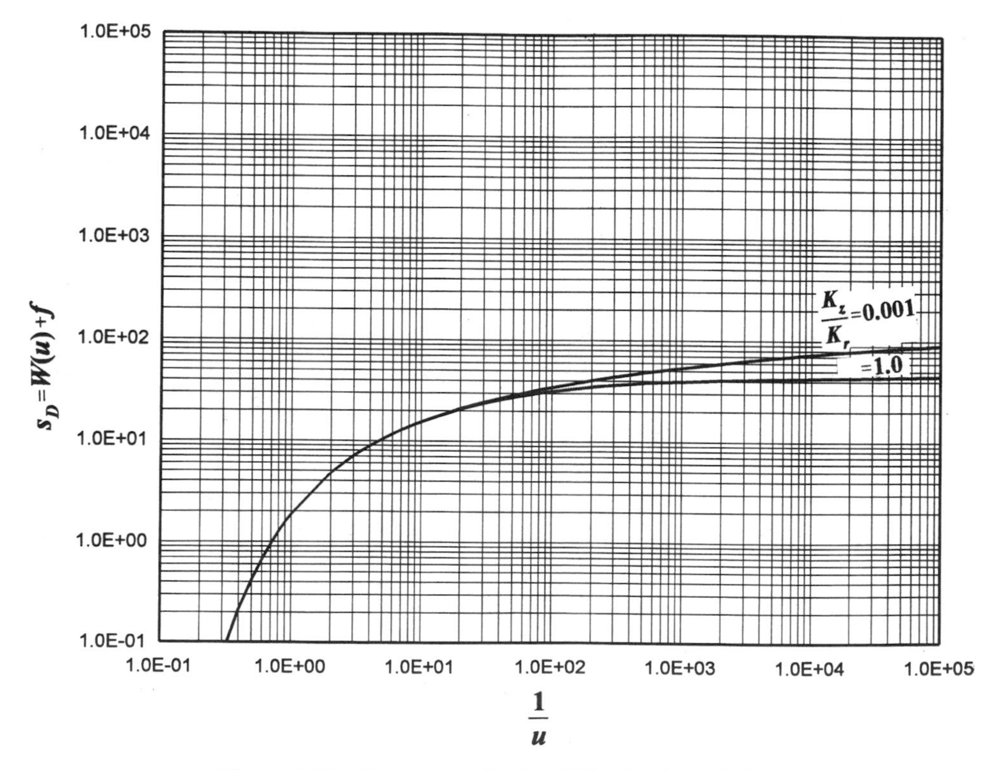

Figure 8-11 Type curves for $b = 300$ m in Example 8-1.

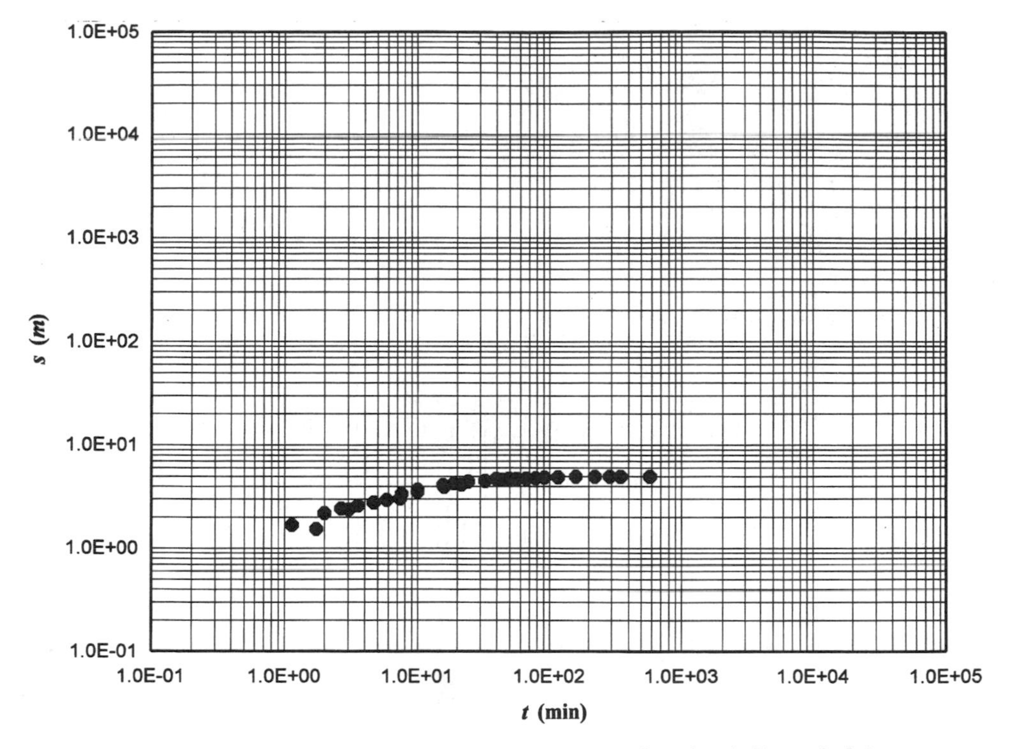

Figure 8-12 Double logarithmic drawdown versus time data in Example 8-1.

Step 2: The observed drawdown versus time data are presented in Figure 8-12.

Step 3: The superimposed graphs for $b = 75$ m, 150 m, and 300 m are shown in Figures 8-13, 8-14, and 8-15, respectively. Because the type curves are too close to each other, the observed data can be fitted to any type curve between $a^2 = K_z/K_r = 0.001$ and 1. Therefore, with the present data only the horizontal hydraulic conductivity (K_r) value can be determined and the vertical hydraulic conductivity (K_z) cannot be determined. For convenience the type-curve matching in each figure corresponds to $a^2 = K_z/K_r = 1$.

Step 4: The type-curve match-point coordinates are presented in Table 8-3, and they are also shown in Figures 8-13, 8-14, and 8-15.

Step 5: Substitution of $Q = 8734$ m^3/day and the corresponding type-curve matching coordinates for $b = 75$ m from Table 8-3 into Eq. (8-103) gives the value of T:

$$T = \frac{Qs_D}{4\pi s} = \frac{(8734 \text{ m}^3/\text{day})(100)}{(4\pi)(45 \text{ m})} = 1545 \text{ m}^2/\text{day}$$

The T values for $b = 150$ m and 300 m are calculated similarly and are listed in Table 8-4.

Step 6: Substitution of the corresponding type-curve matching coordinates for $b = 75$ m from Table 8-3 into Eq. (8-104) gives the value of S:

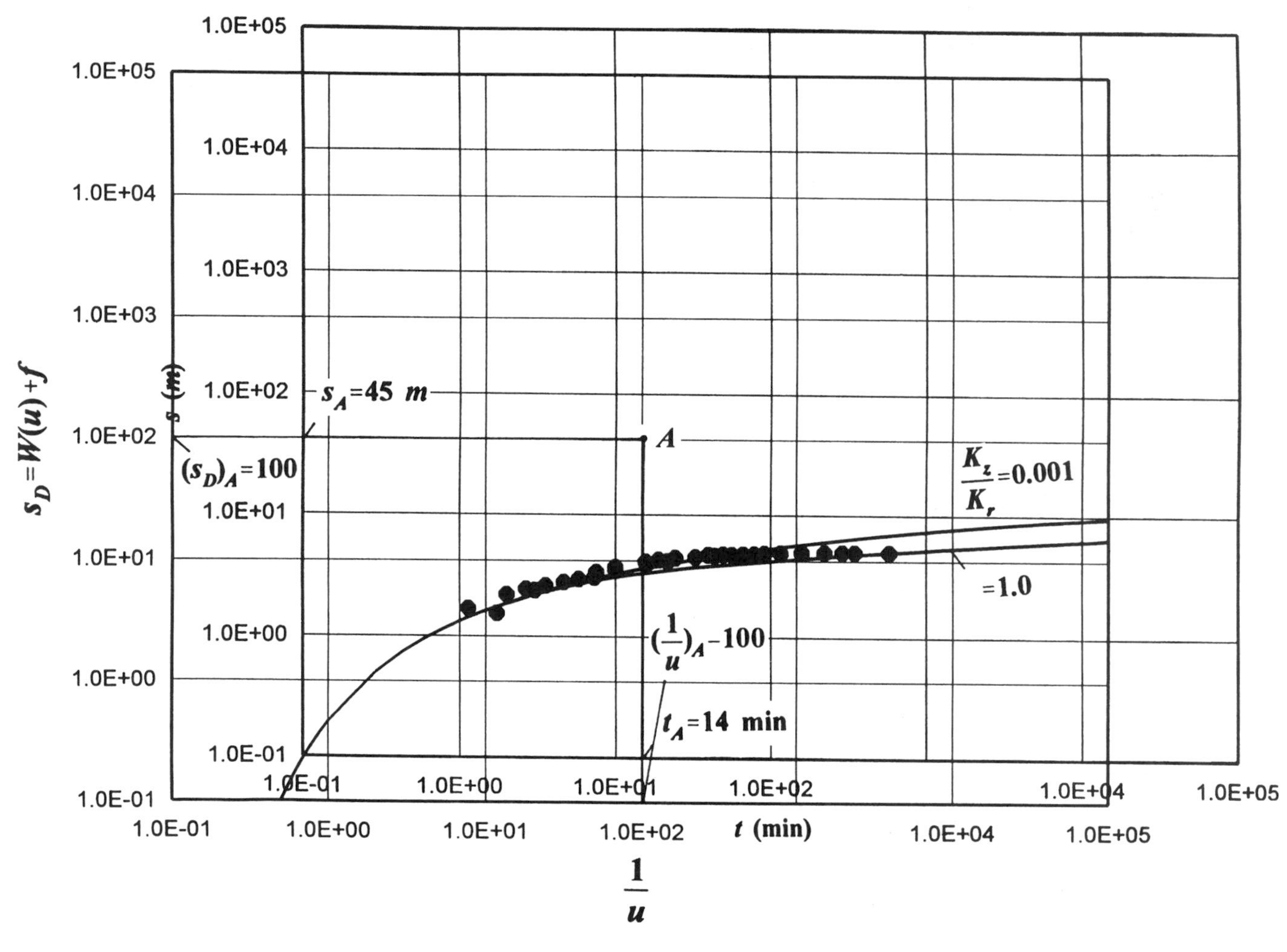

Figure 8-13 Type curve and observed drawdown data matching for $b = 75$ m in Example 8-1.

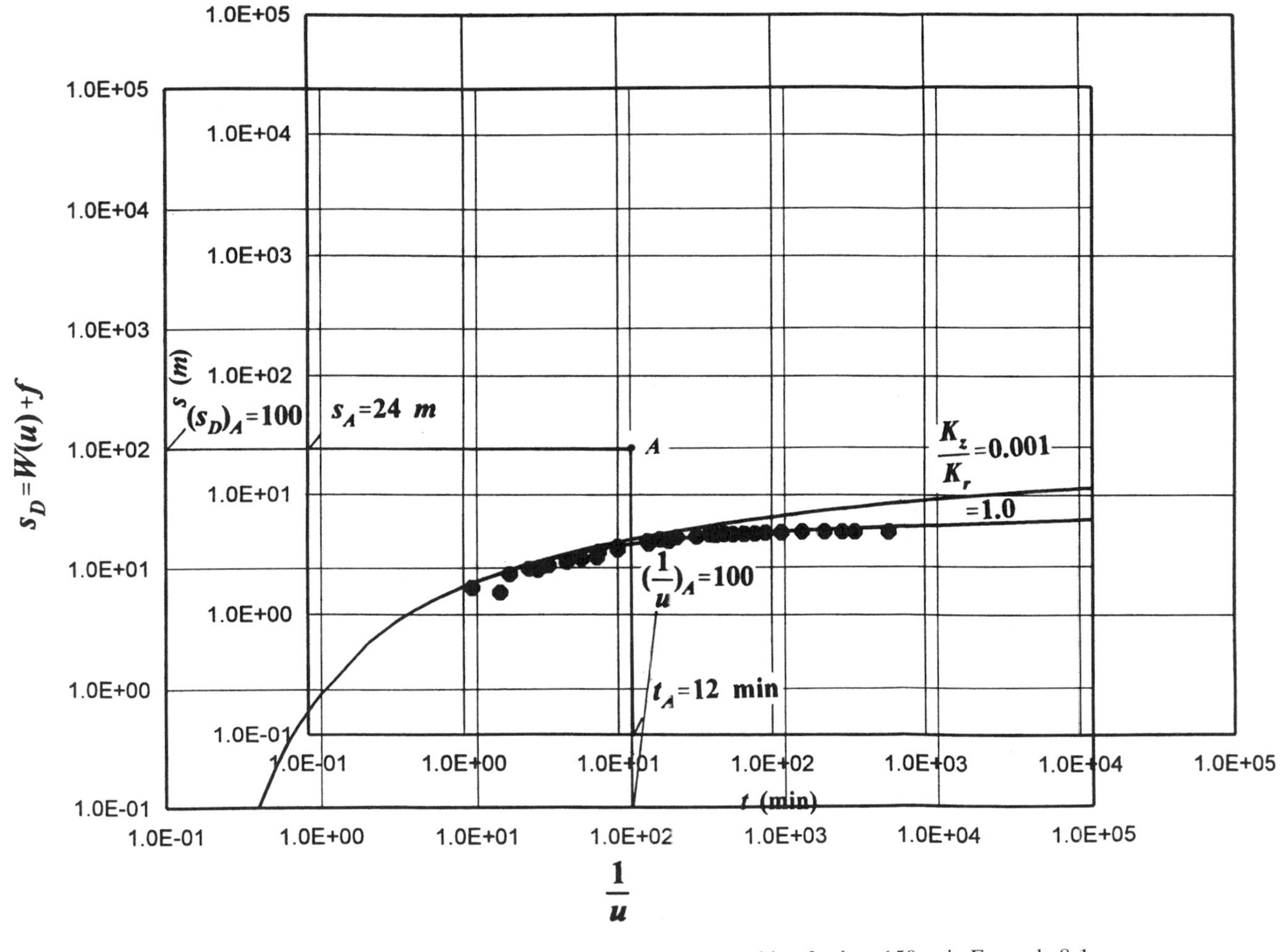

Figure 8-14 Type curve and observed drawdown data matching for $b = 150$ m in Example 8-1.

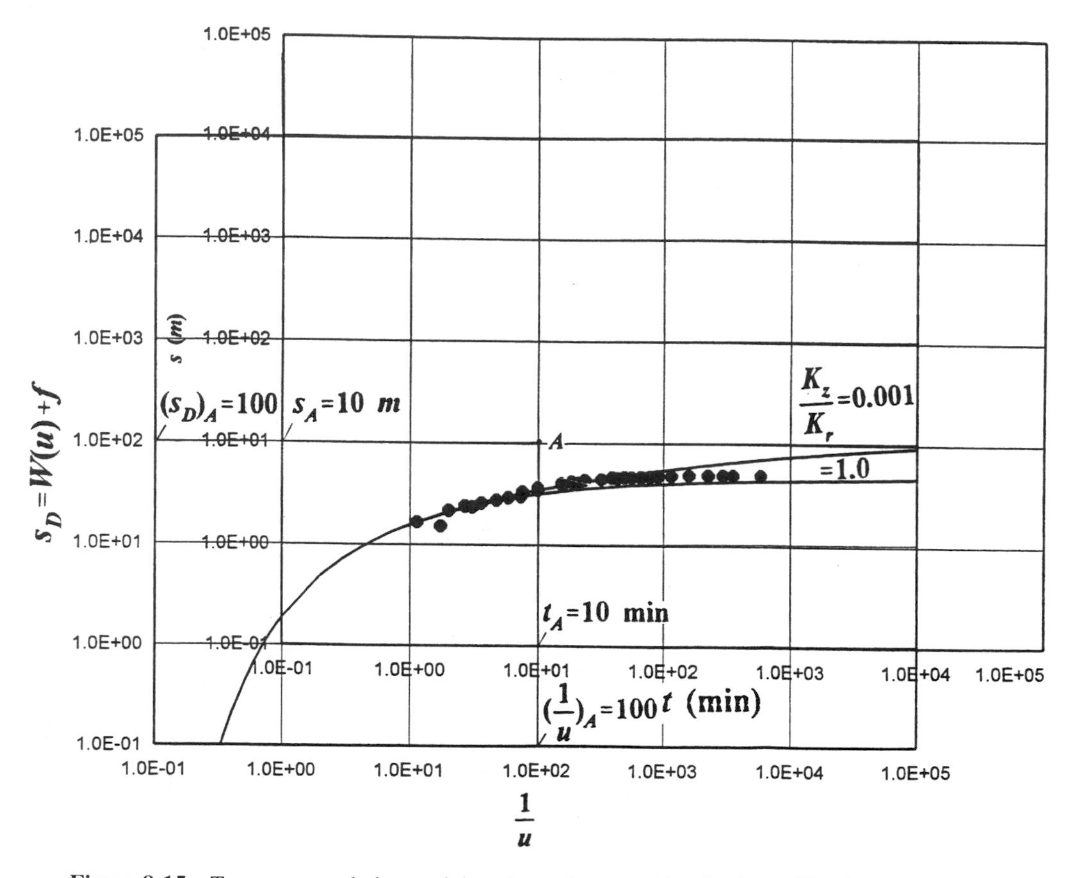

Figure 8-15 Type curve and observed drawdown data matching for b = 300 m in Example 8-1.

TABLE 8-3 Match-Point Coordinates for Example 8-1

b (m)	$1/u$	s_D	t (min)	s (min)
75	100	100	14	45
150	100	100	12	24
300	100	100	10	10

TABLE 8-4 Type-Curve Matching Results for Example 8-1

		K_r			
b (m)	T (m^2/day)	(m/day)	(cm/s)	S	S_s (m^{-1})
75	1545	21.6	2.38×10^{-2}	0.022	2.93×10^{-4}
150	3687	24.6	2.84×10^{-2}	0.046	3.05×10^{-4}
300	8849	29.5	3.41×10^{-2}	0.092	3.07×10^{-4}
	Average values:	25.2	2.92×10^{-2}	0.053	3.02×10^{-4}

$$S = \frac{4Tt}{r^2\left(\frac{1}{u}\right)} = \frac{(4)\left(\dfrac{1545\ \text{m}^2}{24 \times 60\ \text{min}}\right)(14\ \text{min})}{(5.18\ \text{m})^2(100)} = 0.022$$

The corresponding specific storage value is $S_s = S/b = 0.022/(75\ \text{m}) = 2.93 \times 10^{-4}\ \text{m}^{-1}$. The S and S_s values for $b = 150$ m and 300 m are calculated similarly and are listed in Table 8-4.

Step 7: Substitution of the corresponding values for $b = 75$ m from Table 8-3 into Eq. (8-105) gives the value of K_r:

$$K_r = \frac{T}{b} = \frac{1545\ \text{m}^2/\text{day}}{75\ \text{m}} = 21.6\ \text{m/day} = 2.38 \times 10^{-2}\ \text{cm/s}$$

The K_r values for $b = 150$ m and 300 m are calculated similarly and are listed in Table 8-4.

Step 8: Because $a^2 = K_z/K_r$ is unknown, the vertical hydraulic conductivity cannot be determined.

(b) Now let's discuss results: In this pumping test the screen interval length ($l = 35.36$ m in Figure 8-8) is way too long for the determination of the vertical K_z value. This value is the main reason of not having scattered type curves. Because the type curves for the range of values for K_z/K_x mentioned in Step 1 of Section 8.4.2.2 are too close to each other, the identification of K_z/K_x cannot be made. Had the length of the screen interval was much shorter, say between 1 m and 2 m, scattered type curves could be determined.

The data are analyzed under the condition that the aquifer thickness is unknown. The results in Table 8-4 show that different aquifer thicknesses produce approximately the same K_r and S_s values: For $b = 75$ m, 150 m, and 300 m, the type-curve-derived K_r values are 2.38×10^{-2} cm/s, 2.84×10^{-2} cm/s, and 3.41×10^{-2}cm/s, respectively. The corresponding S_s values are $2.93 \times 10^{-4}\ \text{m}^{-1}$, $3.05 \times 10^{-4}\ \text{m}^{-1}$, and $3.07 \times 10^{-4}\ \text{m}^{-1}$, respectively. This approach is also confirmed in Example 8-2 of Section 8.4.3.2.

8.4.3 Straight-Line and Type-Curve Methods for Nonleaky Confined Aquifers

In the following sections, first the properties of the drawdown equation under pumping conditions are presented. Then, under the framework of these properties, several methods are outlined in detail, with examples.

8.4.3.1 Properties of Drawdown Equations. The drawdown equations for confined aquifers under partially penetrating well conditions have some characteristic features which are important in determining the formation hydraulic properties by graphical matching methods. These characteristic features are given below based on Hantush (1961d).

Equations of drawdown in piezometers or observation wells during relatively short periods of pumping [the condition is given by Eq. (8-63) or Eq. (8-81)] are the type of equations given by Eq. (8-62) or (8-82). These equations are called *equations of special cases*. The $M(u, \beta)$ function, which is defined by Eq. (8-52), is tabulated in Table 8-1. In Eq. (8-62), c and β are constants which are dependent on the parameters of the flow system under consideration. The rest of the parameters are all defined in Section 8.3.4.3. Figure 8-16 presents diagrammatic representation of partially penetrating pumping and observation wells and their geometric locations. And Figure 8-17 presents theoretical graphs of drawdown (s) versus the logarithm of time [$\log(t)$] for the equations. The properties of these graphs are given below (Hantush, 1961d):

1. In the period

$$t < \frac{(r\beta)^2 S_s}{20K} \tag{8-107}$$

the drawdown is given by

$$s = cW(u) \tag{8-108}$$

This follows from the value of the function $M(u, \beta)$ for large values of u, that is small values of time as presented by Eq. (8-57) of Section 8.3.4.3. The diagrammatic representation of pumping and observation wells are shown in Figure 8-16. The drawdown (s) versus logarithmic time [$\log(t)$] variation of the equations mentioned above is shown in Figure 8-17. Equation (8-108) indicates that during the initial periods of pumping, the flow behaves as if the aquifer ends at the bottom of the pumping well. The dashed curves identified by I' and I'' in Figure 8-17 present the time versus drawdown variation for wells I and II in Figure 8-16, respectively, on the assumption that the aquifer is assumed to end at the bottom of the pumped well. As can be seen from Figure 8-17, the curves of each pair deviate from each other, and the deviation is greater for larger distances from the pumped well. However, the general trend of variation prior to the inflection point appears to be the same.

2. In general, the curves have an inflection point that develops within the period expressed by Eq. (8-63). The relation regarding this point is

$$\frac{\beta^2}{\pi^{1/2}} = f(x) = xe^{x^2}\operatorname{erf}(x) \tag{8-109}$$

in which

$$x = \beta u_i^{1/2} \tag{8-110}$$

$$u_i = \frac{r^2 S_s}{4Kt_i} \tag{8-111}$$

and t_i is the value of t at the inflection point. The function $f(x)$ is given in Table 8-5 for a wide range of values (Hantush, 1961d). The drawdown s_i at the inflection point is given by

$$s_i = cM(u_i, \beta) \tag{8-112}$$

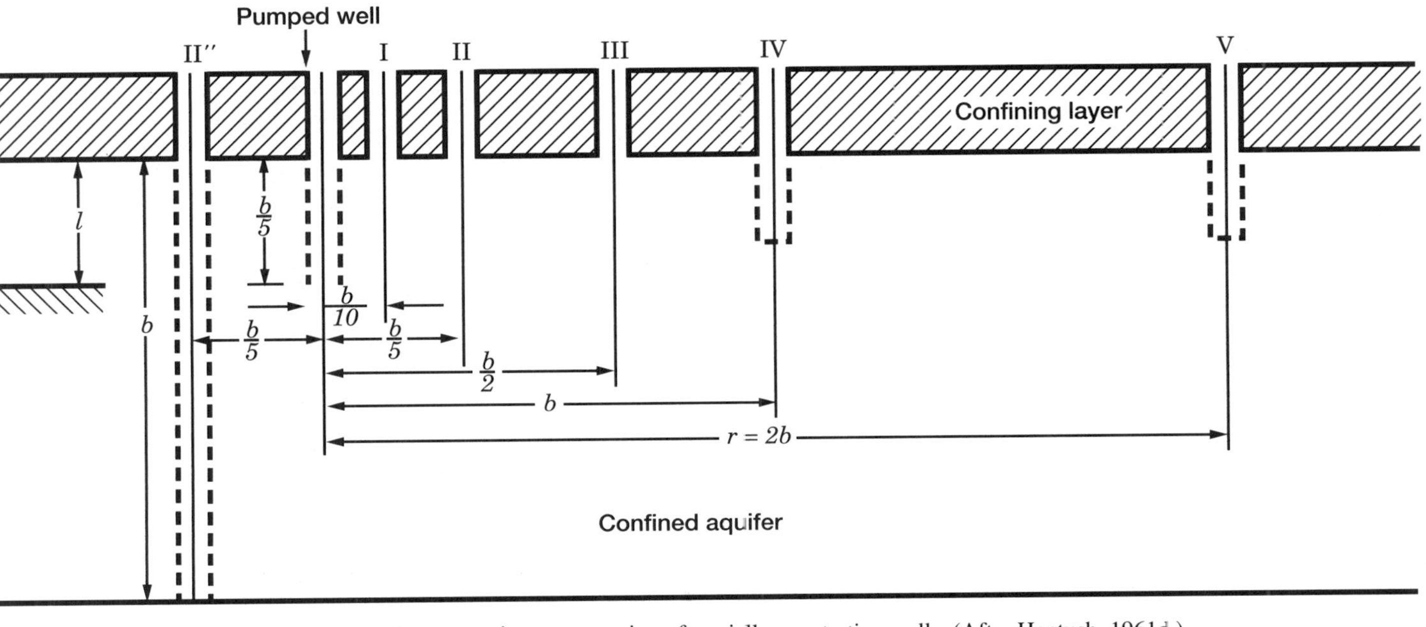

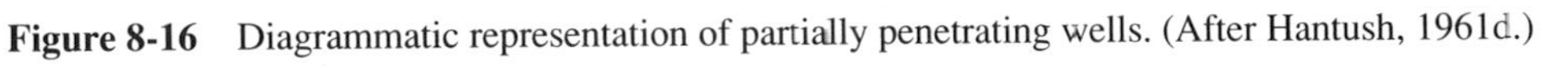

Figure 8-16 Diagrammatic representation of partially penetrating wells. (After Hantush, 1961d.)

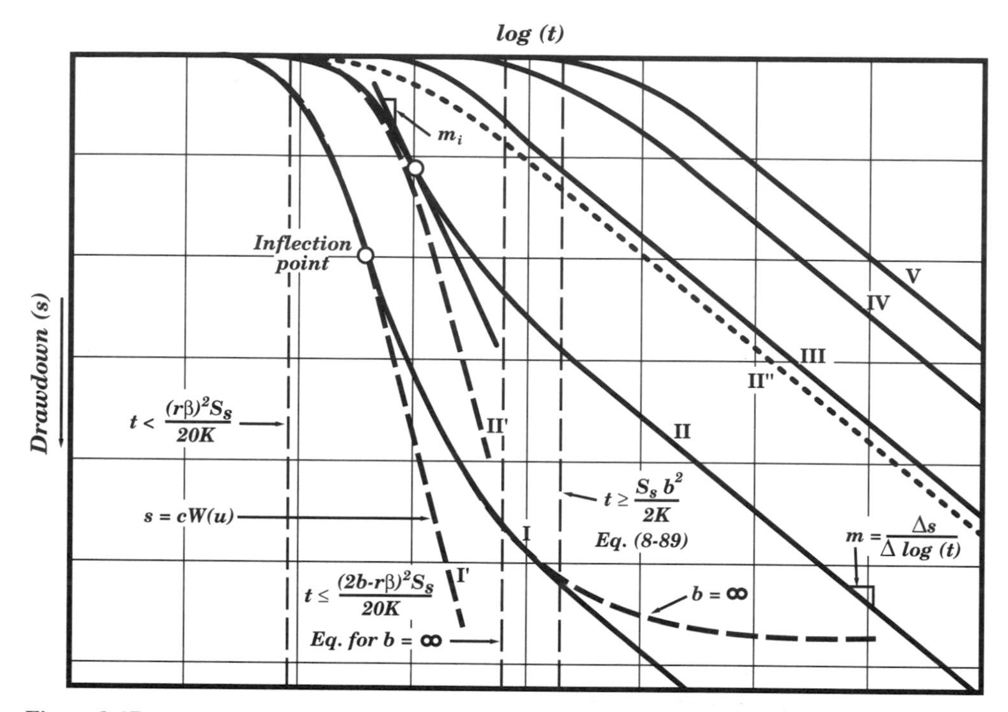

Figure 8-17 Drawdown (s) versus logarithmic time [log(t)] variation for different conditions. (After Hantush, 1961d.)

and the geometric slope m_i at that point is

$$m_i = \frac{\Delta s}{\Delta \log(t)} = 2.303ce^{-u_i}\,\mathrm{erf}(x) \tag{8-113}$$

Equation (8-109) is determined by making zero the second derivative of Eq. (8-62) with respect to log(t), and Eq. (8-112) is determined from Eq. (8-62) by replacing u_i for u. After determining the first derivative of Eq. (8-62) with respect to log(t), u_i is substituted for u to determine Eq. (8-113).

3. The drawdown versus time curves in Figure 8-17 start to deviate from those that correspond to infinitely thick aquifers at an approximate value of t or u given, respectively, by

$$t_d = \frac{(2b - r\beta)^2 S_s}{20K} \tag{8-114}$$

and

$$\frac{1}{u_d} = \frac{(2b - r\beta)^2}{5r^2} \tag{8-115}$$

where t_d and u_d are, respectively, the values of t and u at the point of deviation.

4. Depending on the geometry of the flow system, the inflection disappears (Curves IV and V in Figure 8-17) for

$$\frac{x^2(2b - r\beta)^2}{5(r\beta)^2} \leq 1 \tag{8-116}$$

where x and β are the parameters of Eq. (8-109). Equation (8-116) is determined by noting that the inflection occurs only if

$$t_i < \frac{(2b - r\beta)^2 S_s}{20K} \tag{8-117}$$

5. The drawdown curves tend to become straight lines with the same geometric slope at relatively large values of time, that is $t > b^2 S_s/(2K)$. The geometric slope of these lines is given by

$$m = \frac{\Delta s}{\Delta \log(t)} = \frac{2.303Q}{4\pi Kb} \tag{8-118}$$

which is determined by differentiating Eq. (8-68) with respect to $\log(t)$ and noting that u becomes very small for large values of time.

8.4.3.2 Methods of Pumping Test Data Analysis.

Hantush (1961d) developed several methods for the analysis of pumping test data under partially penetrating wells conditions in confined aquifers. Details of these methods are given in the following sections.

Hantush's Inflection Point Method. If the drawdown equation under partially penetrating well conditions in pumping and observation wells for relatively short periods of pumping is the type of equation expressed by Eq. (8-82), or can be approximated by such equations, the following steps can be followed under the conditions that (a) the observed semilogarithmic time versus drawdown indicates the formation of an inflection point within the data range, and (b) the tangent of this curve at the inflection point can be constructed with approximate accuracy (Hantush, 1961d, p. 183):

Step 1: Plot the observed drawdown data on a single logarithmic paper (t on logarithmic scale) and draw the tangent in the region of the curve inflection and determine its geometric slope from Eq. (8-113).

Step 2: Compute the value of $f(x)$ from Eq. (8-109) as $\beta^2/\pi^{1/2} = f(x)$.

Step 3: Determine the value of x and $\text{erf}(x)$ from Table 8-5 (or from a graph generated from this table).

Step 4: With the known values of x and β, calculate the value of u_i from Eq. (8-110) and then calculate the value of $\exp(-u_i)$.

Step 5: Calculate the value of K from Eq. (8-113) with the appropriate expression for c.

Step 6: Determine the value of $M(u_i, \beta)$ from Table 8-1 using the previously determined values of u_i and β. Then, calculate the value of s_i from Eq. (8-62).

Step 7: From the semilogarithmic graph, determine the value of t_i corresponding to the calculated value of s_i. Then, calculate the value of S_s from Eq. (8-111).

Step 8: If it is discernible within the observed data, plot the ultimate straight line of the curve and determine its geometric slope (m) from Eq. (8-113). Then, calculate the value of the product $Kb(T)$ from Eq. (8-118).

TABLE 8-5 Values of the Function erf(x) and $f(x) = xe^{x^2}$ erf(x)

x	erf(x)	$f(x)$	x	erf(x)	$f(x)$	x	erf(x)	$f(x)$	x	erf(x)	$f(x)$	x	erf(x)	$f(x)$	x	erf(x)	$f(x)$
0.005	0.0056	282(−7)	0.505	0.5249	0.34207	1.005	0.8448	2.3078	1.505	0.9667	14.012	2.005	0.9954	111.17	2.505	0.9997	1330.0
0.010	0.0113	113(−6)	0.510	0.5292	0.35007	1.010	0.8468	2.3721	1.510	0.9673	14.282	2.010	0.9955	113.72	2.510	0.9996	1366.5
0.015	0.0169	254(−6)	0.515	0.5336	0.35825	1.015	0.8488	2.4138	1.515	0.9679	14.554	2.015	0.9956	116.33	2.515	0.9996	1404.0
0.020	0.0226	451(−6)	0.520	0.5379	0.36656	1.020	0.8508	2.4811	1.520	0.9684	14.835	2.020	0.9957	119.01	2.520	0.9996	1442.7
0.025	0.0282	705(−6)	0.525	0.5422	0.37497	1.025	0.8528	2.4995	1.525	0.9690	15.121	2.025	0.9958	121.75	2.525	0.9996	1482.5
0.030	0.0338	0.00102	0.530	0.5465	0.38357	1.030	0.8548	2.5435	1.530	0.9695	15.414	2.030	0.9959	124.57	2.530	0.9997	1523.5
0.035	0.0395	0.00138	0.535	0.5507	0.39227	1.035	0.8567	2.5882	1.535	0.9701	15.711	2.035	0.9960	127.45	2.535	0.9997	1565.7
0.040	0.0451	0.00181	0.540	0.5549	0.40114	1.040	0.8587	2.6338	1.540	0.9706	16.015	2.040	0.9961	130.40	2.540	0.9997	1609.1
0.045	0.0507	0.00204	0.545	0.5591	0.41011	1.045	0.8606	2.6800	1.545	0.9712	16.326	2.045	0.9962	133.43	2.545	0.9997	1653.8
0.050	0.0564	0.00283	0.550	0.5633	0.41929	1.050	0.8624	2.7264	1.550	0.9716	16.642	2.050	0.9963	136.54	2.550	0.9997	1699.8
0.055	0.0620	0.00342	0.555	0.5675	0.42855	1.055	0.8643	2.7752	1.555	0.9721	16.965	2.055	0.9963	139.72	2.555	0.9997	1747.2
0.060	0.0676	0.00407	0.560	0.5716	0.43801	1.060	0.8661	2.8240	1.560	0.9726	17.297	2.060	0.9964	142.98	2.560	0.9997	1796.0
0.065	0.0732	0.00478	0.565	0.5757	0.44763	1.065	0.8680	2.8737	1.565	0.9731	17.634	2.065	0.9965	146.32	2.565	0.9997	1846.2
0.070	0.0789	0.00558	0.570	0.5798	0.45738	1.070	0.8698	2.9242	1.570	0.9736	17.979	2.070	0.9966	149.76	2.570	0.9997	1897.9
0.075	0.0845	0.00637	0.575	0.5839	0.46727	1.075	0.8716	2.9756	1.575	0.9741	18.330	2.075	0.9967	153.27	2.575	0.9997	1951.2
0.080	0.0901	0.00725	0.580	0.5879	0.47735	1.080	0.8733	3.0280	1.580	0.9746	18.691	2.080	0.9967	156.88	2.580	0.9997	2006.0
0.085	0.0957	0.00819	0.585	0.5919	0.48760	1.085	0.8751	3.0813	1.585	0.9750	19.058	2.085	0.9968	160.58	2.585	0.9998	2062.5
0.090	0.1013	0.00919	0.590	0.5959	0.49797	1.090	0.8768	3.1355	1.590	0.9755	19.433	2.090	0.9969	164.37	2.590	0.9998	2120.7
0.095	0.1069	0.01024	0.595	0.5999	0.50858	1.095	0.8785	3.1907	1.595	0.9759	19.815	2.095	0.9970	168.26	2.595	0.9998	2180.6
0.100	0.1125	0.01136	0.600	0.6039	0.51931	1.100	0.8802	3.2470	1.600	0.9764	20.208	2.100	0.9970	172.25	2.600	0.9998	2242.3
0.105	0.1181	0.01253	0.605	0.6078	0.53023	1.105	0.8819	3.3041	1.605	0.9768	20.606	2.105	0.9971	176.34	2.605	0.9998	2305.9
0.110	0.1236	0.01384	0.610	0.6117	0.54132	1.110	0.8835	3.3623	1.610	0.9772	21.015	2.110	0.9972	180.54	2.610	0.9998	2371.4
0.115	0.1292	0.01505	0.615	0.6156	0.55260	1.115	0.8852	3.4215	1.615	0.9776	21.433	2.115	0.9972	184.84	2.615	0.9998	2438.8
0.120	0.1348	0.01640	0.620	0.6194	0.56402	1.120	0.8868	3.4819	1.620	0.9780	21.860	2.120	0.9973	189.25	2.620	0.9998	2508.3
0.125	0.1403	0.01782	0.625	0.6232	0.57568	1.125	0.8884	3.5432	1.625	0.9784	22.293	2.125	0.9973	193.78	2.625	0.9998	2579.9
0.130	0.1459	0.01928	0.630	0.6271	0.58754	1.130	0.8900	3.6058	1.630	0.9788	22.739	2.130	0.9974	198.43	2.630	0.9998	2653.6
0.135	0.1514	0.02082	0.635	0.6308	0.59950	1.135	0.8915	3.6693	1.635	0.9792	23.193	2.135	0.9975	203.17	2.635	0.9998	2713.8
0.140	0.1570	0.02241	0.640	0.6346	0.61173	1.140	0.8931	3.7342	1.640	0.9796	23.658	2.140	0.9975	208.08	2.640	0.9998	2807.8
0.145	0.1625	0.02406	0.645	0.6383	0.62412	1.145	0.8946	3.8002	1.645	0.9800	24.132	2.145	0.9976	213.09	2.645	0.9998	2888.5
0.150	0.1680	0.02577	0.650	0.6420	0.63675	1.150	0.8961	3.8674	1.650	0.9804	24.618	2.150	0.9976	218.24	2.650	0.9998	2971.6
0.155	0.1735	0.02755	0.655	0.6457	0.64952	1.155	0.8976	3.9357	1.655	0.9808	25.113	2.155	0.9977	223.52	2.655	0.9998	3057.3
0.160	0.1790	0.02938	0.660	0.6494	0.66256	1.160	0.8911	4.0055	1.660	0.9811	25.619	2.160	0.9978	228.94	2.660	0.9998	3145.5
0.165	0.1845	0.03128	0.665	0.6530	0.67574	1.165	0.9006	4.0763	1.665	0.9815	26.137	2.165	0.9978	234.49	2.665	0.9998	3236.5
0.170	0.1900	0.03325	0.670	0.6566	0.68921	1.170	0.9020	4.1485	1.670	0.9818	26.666	2.170	0.9979	240.20	2.670	0.9998	3330.2
0.175	0.1955	0.03527	0.675	0.6602	0.70288	1.175	0.9034	4.2220	1.675	0.9822	27.205	2.175	0.9979	246.05	2.675	0.9999	3426.9
0.180	0.2009	0.03736	0.680	0.6638	0.71669	1.180	0.9048	4.2970	1.680	0.9825	27.758	2.180	0.9980	252.06	2.680	0.9999	3526.5
0.185	0.2064	0.03951	0.685	0.6673	0.73079	1.185	0.9062	4.3732	1.685	0.9928	28.322	2.185	0.9980	258.22	2.685	0.9999	3629.1
0.190	0.2118	0.04171	0.690	0.6708	0.74514	1.190	0.9076	4.4509	1.190	0.9832	29.900	2.190	0.9981	264.55	2.690	0.9999	3734.9
0.195	0.2173	0.04401	0.695	0.6743	0.75965	1.195	0.9090	4.5300	1.695	0.9835	29.489	2.195	0.9981	271.04	2.695	0.9999	3844.0
0.200	0.2227	0.04636	0.700	0.6778	0.77446	1.200	0.9103	4.6106	1.700	0.9838	30.092	2.200	0.9981	277.71	2.700	0.9999	3956.5
0.205	0.2281	0.04877	0.705	0.6813	0.78949	1.205	0.9116	4.6925	1.705	0.9841	30.709	2.205	0.9982	284.55	2.705	0.9999	4072.4
0.210	0.2335	0.05125	0.710	0.6847	0.80477	1.210	0.9127	4.7748	1.710	0.9844	31.340	2.210	0.9982	291.58	2.170	0.9999	4191.9
0.215	0.2389	0.05380	0.715	0.6881	0.82025	1.215	0.9143	4.8611	1.715	0.9847	31.983	2.215	0.9983	298.78	2.715	0.9999	4315.1
0.220	0.2443	0.05642	0.720	0.6914	0.83606	1.220	0.9155	4.9481	1.720	0.9850	32.642	2.220	0.9983	306.19	2.720	0.9999	4442.2
0.225	0.2497	0.05909	0.725	0.6948	0.85203	1.225	0.9168	5.0365	1.725	0.9853	33.315	2.225	0.9984	313.78	2.725	0.9999	4573.2
0.230	0.2550	0.06183	0.730	0.6981	0.86827	1.230	0.9181	5.1261	1.730	0.9856	34.004	2.230	0.9984	321.59	2.730	0.9999	4708.3
0.235	0.2604	0.06466	0.735	0.7014	0.88480	1.235	0.9193	5.2180	1.735	0.9859	34.707	2.235	0.9984	330.12	2.735	0.9999	4847.6
0.240	0.2657	0.06755	0.740	0.7047	0.90166	1.240	0.9205	5.3115	1.740	0.9861	35.429	2.240	0.9985	337.82	2.740	0.9999	4991.2
0.245	0.2710	0.07050	0.745	0.7079	0.91869	1.245	0.9217	5.4060	1.745	0.9864	36.164	2.245	0.9985	346.27	2.745	0.9999	5139.4
0.250	0.2763	0.07354	0.750	0.7112	0.93612	1.250	0.9229	5.5033	1.750	0.9867	36.918	2.250	0.9985	354.94	2.750	0.9999	5292.2

0.255	0.2816	0.07664	0.755	0.7144	0.95372	1.255	0.9241	5.6025	1.755	0.9870	37.688	2.255	0.9986	363.85
0.260	0.2869	0.07981	0.760	0.7175	0.97161	1.260	0.9252	5.7031	1.760	0.9872	38.476	2.260	0.9986	373.51
0.265	0.2922	0.08305	0.765	0.7207	0.98985	1.265	0.9264	5.8055	1.765	0.9874	39.279	2.265	0.9986	382.38
0.270	0.2974	0.08637	0.770	0.7238	1.0083	1.270	0.9275	5.9103	1.770	0.9877	40.105	2.270	0.9987	392.03
0.275	0.3027	0.08977	0.775	0.7269	1.0271	1.275	0.9286	6.0159	1.775	0.9879	40.946	2.275	0.9987	401.94
0.280	0.3078	0.09324	0.780	0.7300	1.0464	1.280	0.9297	6.1252	1.780	0.9882	41.809	2.280	0.9987	412.12
0.285	0.3131	0.09678	0.785	0.7331	1.0657	1.285	0.9308	6.2353	1.785	0.9884	42.689	2.285	0.9988	422.57
0.290	0.3183	0.10039	0.790	0.7361	1.0855	1.290	0.9319	6.3481	1.790	0.9886	43.593	2.290	0.9988	433.30
0.295	0.3235	0.10409	0.795	0.7391	1.1055	1.295	0.9330	6.4625	1.795	0.9889	44.514	2.295	0.9988	444.33
0.300	0.3286	0.10788	0.800	0.7421	1.1259	1.300	0.9340	6.5798	1.800	0.9891	45.460	2.300	0.9989	455.66
0.305	0.3338	0.11173	0.805	0.7451	1.1466	1.305	0.9350	6.6991	1.805	0.9893	46.423	2.305	0.9989	467.31
0.310	0.3389	0.11566	0.810	0.7480	1.1677	1.310	0.9361	6.8211	1.810	0.9895	47.413	2.310	0.9989	479.26
0.315	0.3440	0.11967	0.815	0.7509	1.1891	1.315	0.9371	6.9452	1.815	0.9897	48.422	2.315	0.9989	491.55
0.320	0.3491	0.12376	0.820	0.7538	1.2108	1.320	0.9381	7.0717	1.820	0.9899	49.458	2.320	0.9990	504.18
0.325	0.3542	0.12794	0.825	0.7567	1.2330	1.325	0.9391	7.1999	1.825	0.9901	50.514	2.325	0.9990	517.15
0.330	0.3593	0.13221	0.830	0.7595	1.2554	1.330	0.9400	7.3318	1.830	0.9904	51.600	2.330	0.9990	530.48
0.335	0.3643	0.13654	0.835	0.7623	1.2783	1.335	0.9410	7.4658	1.835	0.9906	52.706	2.335	0.9990	544.17
0.340	0.3694	0.14098	0.840	0.7651	1.3015	1.340	0.9419	7.6015	1.840	0.9907	53.843	2.340	0.9991	558.25
0.345	0.3744	0.14549	0.845	0.7679	1.3251	1.345	0.9428	7.7405	1.845	0.9909	55.002	2.345	0.9991	572.71
0.350	0.3794	0.15009	0.850	0.7707	1.3492	1.350	0.9438	7.8827	1.850	0.9911	56.191	2.350	0.9991	590.53
0.355	0.3844	0.15477	0.855	0.7734	1.3722	1.355	0.9447	8.0271	1.855	0.9913	57.405	2.355	0.9991	602.86
0.360	0.3893	0.15956	0.860	0.7761	1.3970	1.360	0.9456	8.1751	1.860	0.9915	58.651	2.360	0.9992	618.56
0.365	0.3943	0.16441	0.865	0.7788	1.4236	1.365	0.9464	8.3252	1.865	0.9917	59.924	2.365	0.9992	634.71
0.370	0.3992	0.16938	0.870	0.7814	1.4492	1.370	0.9473	8.4790	1.870	0.9918	61.229	2.370	0.9992	651.30
0.375	0.4041	0.17443	0.875	0.7841	1.4753	1.375	0.9482	8.6349	1.875	0.9920	62.563	2.375	0.9992	668.36
0.380	0.4090	0.17956	0.880	0.7867	1.5018	1.380	0.9490	8.7948	1.880	0.9922	63.931	2.380	0.9992	685.90
0.385	0.4139	0.18479	0.885	0.7893	1.5286	1.385	0.9499	8.9567	1.885	0.9923	65.328	2.385	0.9993	703.92
0.390	0.4187	0.19014	0.890	0.7918	1.5561	1.390	0.9507	9.1228	1.890	0.9925	66.762	2.390	0.9993	722.46
0.395	0.4236	0.19556	0.895	0.7944	1.5839	1.395	0.9515	9.2925	1.895	0.9926	68.226	2.395	0.9993	741.51
0.400	0.4284	0.20109	0.900	0.7969	1.6122	1.400	0.9523	9.4645	1.900	0.9928	69.729	2.400	0.9993	761.12
0.405	0.4332	0.20671	0.905	0.7994	1.6410	1.405	0.9531	9.6399	1.905	0.9929	71.264	2.405	0.9993	781.25
0.410	0.4380	0.21243	0.910	0.8019	1.6702	1.410	0.9539	9.8203	1.910	0.9931	72.841	2.410	0.9994	801.96
0.415	0.4427	0.21826	0.915	0.8043	1.7000	1.415	0.9546	10.003	1.915	0.9932	74.450	2.415	0.9994	823.27
0.420	0.4475	0.22419	0.920	0.8068	1.7303	1.420	0.9554	10.190	1.920	0.9934	76.103	2.420	0.9994	845.17
0.425	0.4522	0.23021	0.925	0.8092	1.7611	1.425	0.9561	10.380	1.925	0.9935	77.791	2.425	0.9994	867.70
0.430	0.4569	0.23636	0.930	0.8116	1.7924	1.430	0.9569	10.575	1.930	0.9937	79.524	2.430	0.9994	890.86
0.435	0.4616	0.24259	0.935	0.8139	1.8242	1.435	0.9576	10.773	1.935	0.9938	81.295	2.435	0.9994	914.60
0.440	0.4662	0.24896	0.940	0.8163	1.8565	1.440	0.9583	10.975	1.940	0.9939	83.112	2.440	0.9994	939.20
0.445	0.4709	0.25542	0.945	0.8186	1.8894	1.445	0.9590	11.182	1.945	0.9941	84.970	2.445	0.9995	964.41
0.450	0.4755	0.26201	0.950	0.8209	1.9229	1.450	0.9597	11.393	1.950	0.9942	86.916	2.450	0.9995	990.34
0.455	0.4801	0.26868	0.955	0.8232	1.9569	1.455	0.9604	11.605	1.955	0.9943	88.827	2.455	0.9995	1017.0
0.460	0.4847	0.27546	0.960	0.8254	1.9915	1.460	0.9611	11.825	1.960	0.9944	90.828	2.460	0.9995	1044.5
0.465	0.4892	0.28237	0.965	0.8277	2.0267	1.465	0.9617	12.050	1.965	0.9946	92.873	2.465	0.9995	1072.7
0.470	0.4938	0.28943	0.970	0.8299	2.0625	1.470	0.9624	12.273	1.970	0.9947	94.972	2.470	0.9995	1101.7
0.475	0.4983	0.29658	0.975	0.8321	2.0990	1.475	0.9630	12.510	1.975	0.9948	97.118	2.475	0.9995	1131.6
0.480	0.5028	0.30385	0.980	0.8342	2.1360	1.480	0.9637	12.743	1.980	0.9949	99.322	2.480	0.9996	1162.3
0.485	0.5072	0.31124	0.985	0.8368	2.1748	1.485	0.9643	12.990	1.985	0.9950	101.58	2.485	0.9996	1194.0
0.490	0.5117	0.31874	0.990	0.8385	2.2121	1.490	0.9649	13.233	1.990	0.9951	103.89	2.490	0.9996	1226.5
0.495	0.5161	0.32638	0.995	0.8406	2.2509	1.495	0.9655	13.490	1.995	0.9952	106.26	2.495	0.9996	1260.0
0.500	0.5205	0.33416	1.000	0.8427	2.2907	1.500	0.9661	13.750	2.000	0.9953	108.69	2.500	0.9996	1294.5

Note: The numbers in parentheses are powers of 10; for example, $282(-7) = 282 \times 10^{-7}$.

Source: After Hantush (1961d).

Step 9: Calculate the value of b from the known values of K and Kb. Then, calculate the storage coefficient value from $S = S_s b$.

Hantush's Type-Curve Method. If the observed semilogarithmic time versus drawdown curve shows an inflection of the type illustrated in Figure 8-17 and if the number and distribution of the observed points are such that the details of the curve prior to the attainment of the ultimate straight line variation are discernible, a type-curve method originally developed by Hantush (1961d) can be used for the determination of the aquifer parameters and often its thickness as well. The method is essentially the Theis type-curve method as presented in detail in Section 4.2.7.1 of Chapter 4. In the method of Hantush, the type curve is a plot of the E (or $\overline{E}$) function of Eq. (8-60) [or Eq. (8-62)] versus $1/u$ on double logarithmic paper. Equations (8-59), (8-60), and (8-61) are the necessary equations. The observed data curve is a plot of s (or $\overline{s}$) versus t on a double logarithmic paper of the same scale as that of the type curve. The steps of the method are as follows:

Step 1: Plot the type curve for each well, which is the E function (or $\overline{E}$ function) versus $1/u$ on a double logarithmic paper, using the values of the function $M(u, \beta)$ from Table 8-1.

Step 2: On another sheet of log–log paper with the same scale as the first, plot the observed drawdown (s) versus time (t).

Step 3: Superimpose the observed data on the type curve by keeping the coordinate axes parallel and adjust until the best fitting position is obtained. In matching the observed data curve to the type curve, one should keep in mind that the observed data points for relatively large values of time periods may deviate upward from the type curve. This is not unexpected because the type curve is for drawdown values for relatively short periods of pumping.

Step 4: Select an arbitrary matching point anywhere on the two superimposed sheets and record its coordinates—namely E (or $\overline{E}$) and $1/u$ on the type-curve sheet, and s (or $\overline{s}$) and t on the observation data curve sheet.

Step 5: Calculate the value of K from Eq. (8-60) for piezometer data or Eq. (8-80) for observation well data, whichever applies, and calculate S_s from Eq. (8-111) as

$$S_s = \frac{4Kt}{\left(\frac{1}{u}\right) r^2} \tag{8-119}$$

Step 6: If the observed drawdown data curve departs from the type curve, record the value of $1/u$ at the departure point. Then, compute the aquifer thickness (b) value for piezometers from

$$b \cong \frac{1}{2}\left[l + z + r\left(\frac{5}{u_d}\right)^{1/2}\right] \tag{8-120}$$

which is a combination of the time criterion given by Eq. (8-59) and Eq. (8-119). In Eq. (8-120), $1/u_d$ is the recorded value of $1/u$ at the departure point. For observation wells, use the following equation:

$$b \cong \frac{1}{2}\left[l + \frac{1}{2}(l' + d') + r\left(\frac{5}{u_d}\right)^{1/2}\right] \tag{8-121}$$

Equation (8-121) is a combination of Eq. (8-81) and Eq. (8-119). In the case where E (or $\overline{E}$) reduces to the M function as given by Eq. (8-62), use the following equation:

$$b \cong \frac{1}{2}\left[r\beta + r\left(\frac{5}{u_d}\right)^{1/2}\right] \tag{8-122}$$

Equation (8-122) is a combination of Eq. (8-63) and Eq. (8-119).

Step 7: From the computed values of K and b, calculate the transmissivity value from $T = Kb$. Calculate the value of storage coefficient from $S = S_s b$.

Step 8: If the observed drawdown data curve does not depart from the type curve within the observed data range, record the value of $1/u$ of a point in the vicinity of the last observed point. If this value is used in the relations of b mentioned in Step 6, the calculated value of b will be greater than that value in Step 6.

Step 9: Carry out Step 8 of Hantush's inflection point method and determine the value of b, and then using the previously determined value of K, calculate the T value from $T = Kb$, which can be used as a check for the T value determined in Step 7.

Important Points of the Type-Curve Method. Hantush (1961d) has provided some basic guidelines regarding the applicability of the type-curve method. These guidelines are presented below:

1. There is a possibility that the T values determined in Step 7 and Step 9 may not be close to each other. This may be due to the following reasons: (a) The curve fitting of the two superimposed curves is not the best; (b) the ultimate straight-line portion of the curve is drawn with either larger or smaller geometric slope than necessary; or (c) the apparent transmissivity of the aquifer (in case of unconfined aquifers) is not reached to its uniform value within the period of observation. Under these potential reasons, the procedure should be repeated again with the necessary adjustments so that the difference between the two values of T is either eliminated or reduced to a minimum. If the difference under all attempts is still great, it is more likely that the apparent transmissivity has not reached a uniform value within the period of observation.
2. For certain hydrogeologic systems, both the inflection-point method and type-curve method may be applicable, and the results present the opportunity to check each other. In general, the results of the computations may not agree on the first trial. The disagreement may be a result of under- or overestimating the geometric slope at the curve inflection, as well as the geometric slope of the ultimate straight line. It may also be due to a poor choice of the best matching, or to all of these factors. A second trial with the necessary adjustments generally gives a satisfactory answer.

Theis' or Cooper and Jacob's Method or Both. If the semilogarithmic plot of the observed drawdown versus time data curve does not show any inflection even for a long period of pumping, three potential possibilities may exist (Hantush, 1961d, p. 185):

1. The observation well is a fully penetrating well and screened throughout the aquifer thickness.
2. The distance of the observation from the pumped well is relatively great; in other words, $r/b > 1.5$. In order to find out that this is the case, information regarding the approximate thickness of the aquifer is needed. Such information can be provided from previous geological and geophysical investigations, or at least from the range of the thickness given by the relation

$$b < \frac{1}{2}\left(l + l' + 2.4\frac{l}{x}\right) \tag{8-123}$$

where x is the solution of Eq. (8-109).

3. If the first and second possibilities are ruled out, the flow parameters are such that

$$(2b - l - l')^2 \frac{x^2}{5l^2} < 1 \tag{8-124}$$

where x is again the solution of Eq. (8-109).

If the first or second possibility is the case, then the Theis method (see Section 4.2.7.1 of Chapter 4), under the condition that the ultimate straight line is formed, can be used for the determination of the aquifer parameters. If the third possibility is the case, then the type-curve method can be used under the condition that the number and distribution of the observed points on the curved part of the semilogaritmic plot sufficiently define a curve on the logarithmic plot required for the type-curve method. Alternatively, the following procedure can be used under the condition that an estimated value for the aquifer thickness is available (Hantush, 1961d, p. 186).

Cooper and Jacob's Method Adjusted for Partial Penetration. The drawdown equation for relatively large values of time, in which the ultimate semilogarithmic straight line forms, is given by Eq. (8-68) for piezometers and by Eq. (8-89) for observation wells. Because the second term in the bracket on the right-hand side of each of these equations is time-independent, the Cooper and Jacob method (see Section 4.2.7.2 of Chapter 4) can be applied if the numerical value of this constant can be determined. In the following paragraphs, first the equation for storage coefficient (S) is derived, and then the procedure is presented based on the original work of Hantush (Hantush, 1961d).

Equation for Storage Coefficient. A short form of Eq. (8-68) is

$$s = \frac{Q}{4\pi Kb}[W(u) + f_s] \tag{8-125}$$

From Eq. (4-45) of Section 4.5.2.1 of Chapter 4, $W(u)$ can approximately be expressed by the equation

$$W(u) \cong \ln\left(\frac{1}{1.781u}\right) = \ln\left(\frac{2.25Tt}{r^2S}\right) \tag{8-126}$$

Substitution of Eq. (8-126) into Eq. (8-125) gives

$$s = \frac{Q}{4\pi Kb}\left[\ln\left(\frac{2.25Tt}{r^2 S}\right) + f_s\right] \tag{8-127}$$

Equation (8-127) shows a straight line on semilogarithmic paper. The interception point with the time (t) axis has the coordinates $s = 0$ and $t = t_0$. Substitution these values into Eq. (8-127) gives

$$0 = \frac{Q}{4\pi Kb}\left[\ln\left(\frac{2.25Tt_0}{r^2 S}\right) + f_s\right] \tag{8-128}$$

or solving for S we obtain

$$S = \frac{2.25Tt_0 \exp(f_s)}{r^2} \tag{8-129}$$

The steps of the procedure are as follows (Hantush, 1961d):

Step 1: Plot the observed drawdown data on single logarithmic paper (t on logarithmic scale) and construct the ultimate straight line and extend it to the zero-drawdown axis.

Step 2: Determine the geometric slope (m) of the ultimate straight line from the left-hand side of Eq. (8-118) and also determine the interception point with the time axis where $s = 0$. Read the value of t_0.

Step 3: Using the known values of Q and m, calculate the value of the product Kb from the right-hand side of Eq. (8-118).

Step 4: Calculate the value of f_s (or $\overline{f_s}$) from Eq. (8-69) for piezometers and from Eq. (8-90) from observation wells from their appropriate expressions. A few terms of the series involved are generally sufficient.

Step 5: Calculate the value of storage coefficient from Eq. (8-129).

Examples for Pumping Test Data Analysis. In this section, the application of the methods will be shown with examples.

Example 8-2: For the data given in Example 8-1; (a) Analyze the drawdown data; (b) using Hantush's type curve method determine the aquifer parameters; (c) using Hantush's inflection point method, determine the aquifer parameters; and (d) compare the results with those determined by the standard type curve method as presented in Example 8-1.

Solution: As $l'/l = 18.29\text{ m}/35.36\text{ m} < 2$ (see Section 8.3.5.3), the average drawdown $\overline{s}$ in the observed well during relatively short times is given by Eq. (8-80). In Eq. (8-80) the $\overline{E}$ function is the function E of Eq. (8-60) in which $z = (l' + d')/2$. With $d = d' = 0$, $l' = 18.29$ m, $l = 35.36$ m, and $r = 5.18$ m, we obtain $z = (18.29\text{ m} + 0)/2 = 9.15$ m. Substitution of the corresponding values in Eq. (8-61) gives

$$\overline{E} = M\left(u, \frac{35.36\text{ m} + 9.15\text{ m}}{5.18\text{ m}}\right) + M\left(u, \frac{35.36\text{ m} - 9.15\text{ m}}{5.18\text{ m}}\right)$$
$$- M\left(u, \frac{9.15\text{ m}}{5.18\text{ m}}\right) - M\left(u, -\frac{9.15\text{ m}}{5.18\text{ m}}\right)$$

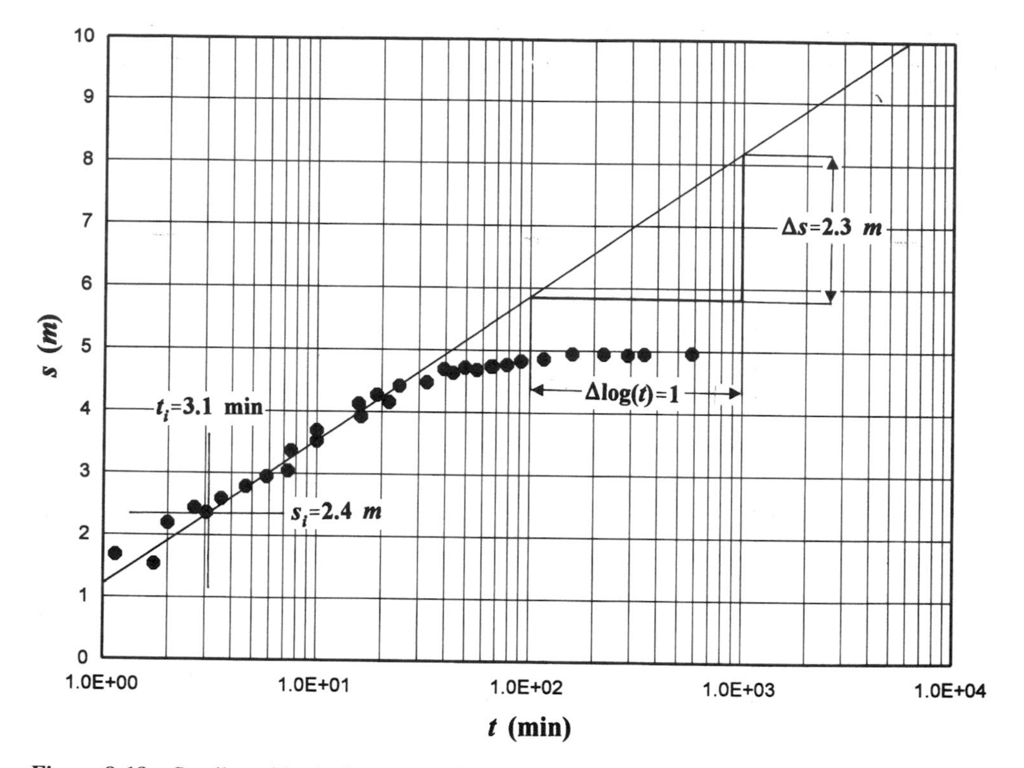

Figure 8-18 Semilogarithmic time versus drawdown variation for the observation well of Example 8-2.

Because of Eq. (8-54), the sum of the last two terms of $\overline{E}$ is zero. Therefore, the result is

$$\overline{s} = \frac{Q}{8\pi K(35.36\ \text{m})}[M(u, 8.6) + M(u, 5.1)]$$

$$= \frac{Q}{\pi K(283\ \text{m})}\overline{E}(u)$$

(a) Analysis of Drawdown Data: The semilogarithmic time versus drawdown variation for the data in Table 8-2 is shown in Figure 8-18, which shows an inflection. As can be seen from Figure 8-18, the time versus drawdown curve exhibits an inflection of the type shown in Figure 8-17 (Curve I). Figure 8-18 also shows that the details of the curve prior the attainment of the ultimate straight-line variation are discernible. Therefore, *the type-curve method* can be used. On the other hand, Figure 8-18 also shows that the tangent to the curve at the inflection point can be drawn with sufficient accuracy. This means that *the inflection-point method* can be used as well. Application of these methods in analyzing the data are given below.

(b) Hantush's Type-Curve Method: The calculation steps in accordance with the previously-presented steps are given below:

Step 1: From the values of $M(u, \beta)$ in Table 8-1 and $\overline{E}(u)$ of the above-given expression a log–log type curve is given in Figure 8-19. Table 8-1 gives $M(u, 8.6)$ directly. For

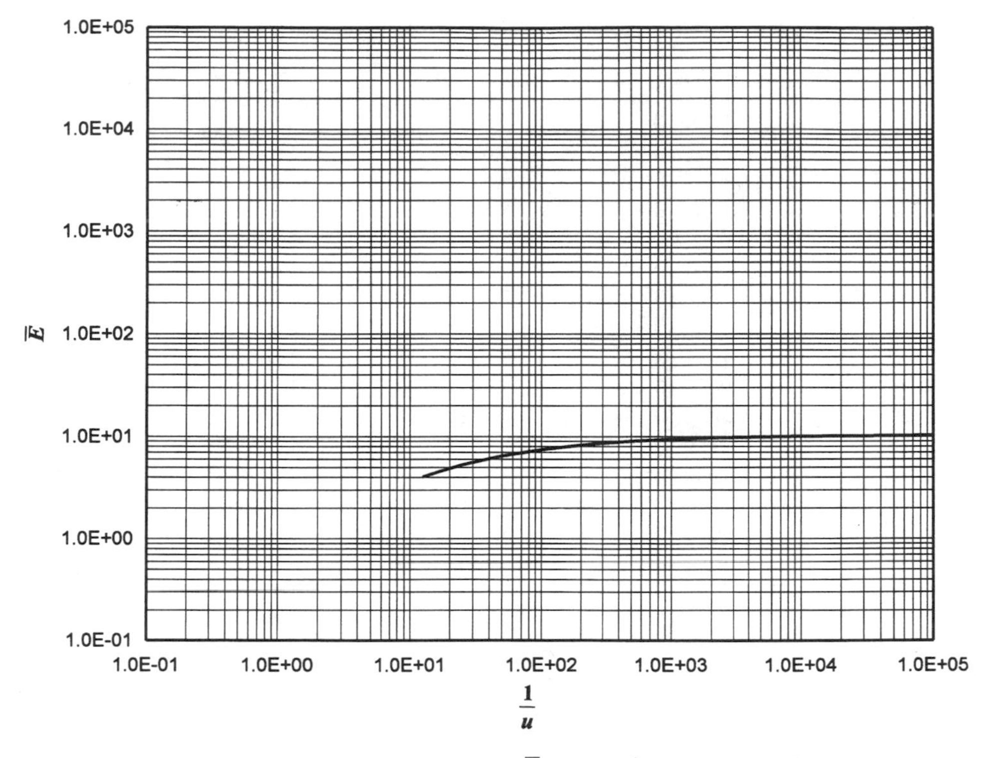

Figure 8-19 Hantush type curve: $\overline{E}$ versus $1/u$ in Example 8-2.

$M(u, 5.1)$, linear interpolation is applied to the $M(u, 5.0)$ and $M(u, 5.2)$ values. The generated type-curve data are given in Table 8-6.

Step 2: The observed data in Table 8-2 are plotted on log–log paper of the same scale and is presented in Figure 8-12.

Step 3: The two curves are superimposed and adjusted to determine the matching position is shown in Figure 8-20.

Step 4: In Figure 8-20, A is an arbitrary chosen point. Its coordinates are

$$\left(\frac{1}{u}\right)_A = 1000, \qquad \overline{E}_A = 100$$

$$t_A = 100 \text{ min}, \qquad \overline{s}_A = 50 \text{ m}$$

Step 5: Substitution of the values in the above-given expression gives the K value:

$$K = \frac{Q}{(283 \text{ m})\pi\overline{s}}\overline{E} = \frac{8734 \text{ m}^3/\text{day}}{(283 \text{ m})(\pi)(50 \text{ m})}(100)$$

$$= 19.65 \text{ m/day} = 2.27 \times 10^{-2} \text{ cm/s}$$

TABLE 8-6 Type-Curve Data, u (or $1/u$) Versus $\overline{E}(u)$ for Example 8-2

u	$1/u$	$M(u, 8.6)$	$M(u, 5.1)$	$\overline{E}(u)$
2.0×10^{-6}	0.50×10^{6}	5.6691	4.6471	10.3162
4.0×10^{-6}	0.25×10^{6}	5.6577	4.6404	10.2981
6.0×10^{-6}	0.17×10^{6}	5.6490	4.6352	10.2842
8.0×10^{-6}	0.13×10^{6}	5.6416	4.6308	10.2724
1.0×10^{-5}	1.00×10^{5}	5.6352	4.6270	10.2622
2.0×10^{-5}	0.50×10^{5}	5.6097	4.6119	10.2216
4.0×10^{-5}	0.25×10^{5}	5.3738	4.5906	10.1644
6.0×10^{-5}	0.17×10^{5}	5.5463	4.5743	10.1206
8.0×10^{-5}	0.13×10^{5}	5.5231	4.5605	10.0836
1.0×10^{-4}	1.00×10^{4}	5.5026	4.5483	10.0509
2.0×10^{-4}	0.50×10^{4}	5.4225	4.5007	9.9232
4.0×10^{-4}	0.25×10^{4}	5.3097	4.4335	9.7432
6.0×10^{-4}	0.17×10^{4}	5.2236	4.3820	9.6056
8.0×10^{-4}	0.13×10^{4}	5.1513	4.3387	9.4900
1.0×10^{-3}	1.00×10^{3}	5.0880	4.3006	9.3886
2.0×10^{-3}	0.50×10^{3}	4.8430	4.1519	8.9949
4.0×10^{-3}	0.25×10^{3}	4.5089	3.9447	8.4536
6.0×10^{-3}	0.17×10^{3}	4.2646	3.7886	8.0532
8.0×10^{-3}	0.13×10^{3}	4.0678	3.6595	7.7273
1.0×10^{-2}	1.00×10^{2}	3.9020	3.5478	7.4498
2.0×10^{-2}	0.50×10^{2}	3.3215	3.1315	6.4530
4.0×10^{-2}	0.25×10^{2}	2.6778	2.6107	5.2885
6.0×10^{-2}	0.17×10^{2}	2.2948	2.2679	4.5627
8.0×10^{-2}	0.13×10^{2}	2.0269	2.0152	4.0421

From Eq. (8-119) the specific storage value is

$$S_s = \frac{4Kt}{\left(\dfrac{1}{u}\right) r^2}$$

$$= \frac{(4)(19.65 \text{ m/day})\left(\dfrac{100 \text{ min}}{1440 \text{ min/day}}\right)}{(1000)(5.18 \text{ m})^2}$$

$$S_s = 2.03 \times 10^{-4} \text{ m}^{-1}$$

Step 6: Within the test period, the observed drawdown curve does not depart from the type curve. This means that the time t_d at the eventual point of departure is greater than the period of observation. As can be seen from Figure 8-20, t_d and $1/u_d$ are greater than 582.10 min and 6000, respectively. Therefore, from Eq. (8-121) the thickness of the aquifer is greater than

$$b = \tfrac{1}{2}\left\{35.36 \text{ m} + \tfrac{1}{2}(18.29 \text{ m} + 0) + (5.18 \text{ m})[(5)(6000)]^{1/2}\right\} = 471 \text{ m}$$

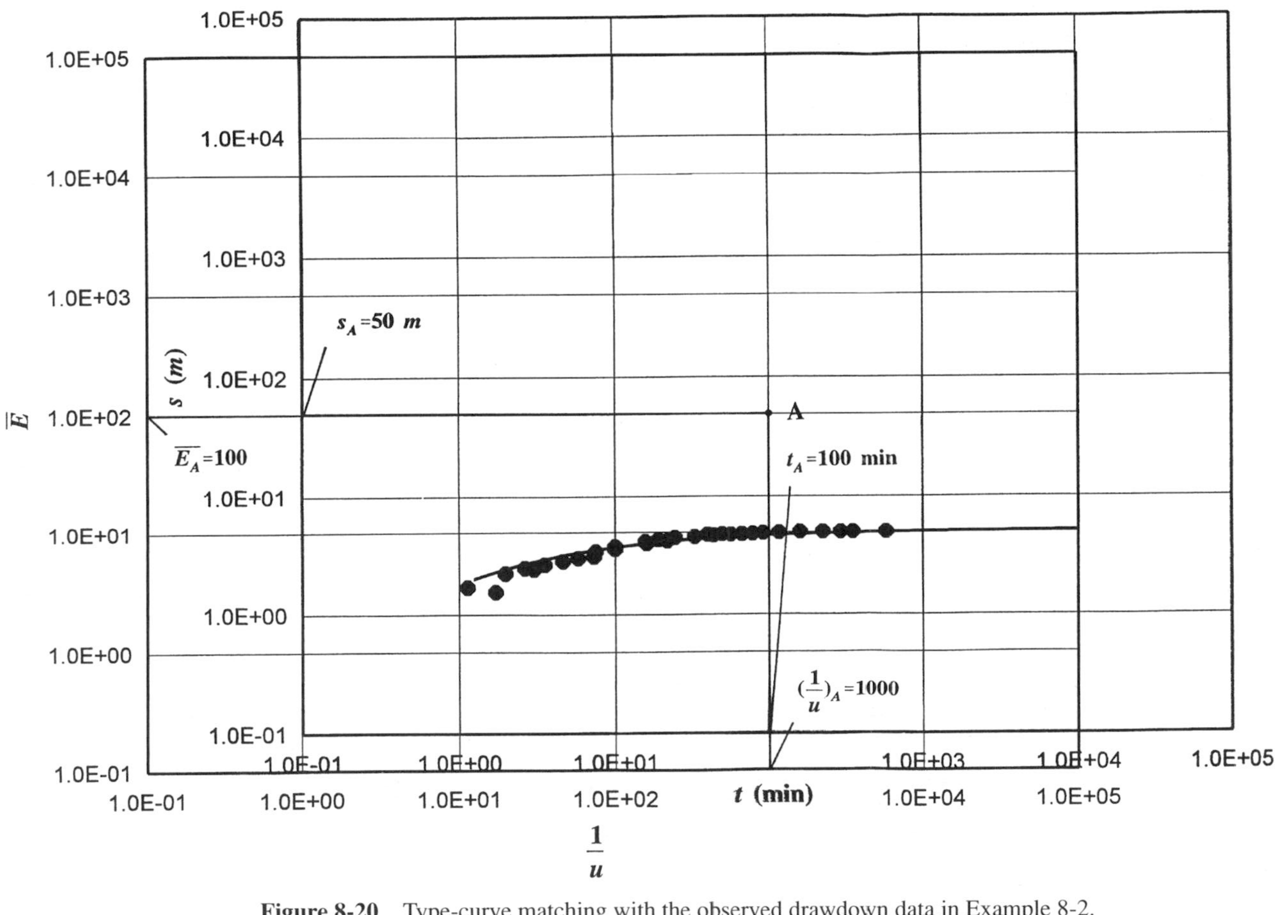

Figure 8-20 Type-curve matching with the observed drawdown data in Example 8-2.

Step 7: The transmissivity and storage coefficient values are greater than

$$T = Kb = (19.65 \text{ m/day})(471 \text{ m}) = 9255 \text{ m}^2/\text{day}$$

$$S = S_s b = (2.03 \times 10^{-4} \text{ m}^{-1})(471 \text{ m}) = 0.096$$

Step 8: The observed drawdown curve does not depart from the type curve within the observed data range. Therefore, the aquifer thickness is greater than $b = 471$ m as determined in Step 6.

Step 9: See Step 8 of the Hantush's inflection point method in (c) as given below.

(c) Inflection-Point Method: The semilogarithmic time versus observed drawdown plot as presented in Figure 8-18 clearly indicates the formation of the inflection point within the observed data. The tangent of this curve at the inflection point can be drawn with sufficient accuracy. As a result, the inflection-point method can be applied. On the other hand, it has been pointed out that the application of the inflection-point method generally gives an approximate solution. This is due to the fact that the actual equation is replaced by an approximate equation of the form of Eq. (8-62). For the inflection-point method case, the drawdown in the observation well is approximated by that in a piezometer of zero penetration ($z = 0$) instead of Eq. (8-80), the actual equation. Introducing $d = 0$ into Eq. (8-61) gives

$$E = M\left(u, \frac{l+z}{r}\right) + M\left(u, \frac{l-z}{r}\right) - M\left(u, \frac{z}{r}\right) - M\left(u, -\frac{z}{r}\right)$$

In the above equation, the sum of the last two terms is zero because of Eq. (8-54). Zero penetration of a piezometer corresponds to $z = 0$ (see Figure 8-5). Therefore, the above expression becomes

$$E = 2M\left(u, \frac{l}{r}\right)$$

And from Eq. (8-80) we obtain

$$s = \frac{Q}{4\pi K(35.36 \text{ m})} M\left(u, \frac{l}{r}\right)$$

$$= \frac{Q}{(141.44 \text{ m})(\pi K)} M(u, 6.82)$$

for which $c = Q/(141.44 \text{ m})(\pi K)$ and $\beta = l/r = (35.36 \text{ m})(5.18 \text{ m}) = 6.82$. Based on this expression, the application of the inflection-point method is given below:

Step 1: The graph of the data is given in Figure 8-18 (t on logarithmic scale). A tangent is drawn in the region of the curve inflection and its geometric slope is

$$m_i = \frac{\Delta s}{\Delta(\log t)} = \frac{2.3 \text{ m}}{1} = 2.3 \text{ m}$$

Step 2: The value of the $f(x)$ function is

$$f(x) = \frac{\beta^2}{\pi^{1/2}} = \frac{(6.82)^2}{\pi^{1/2}} = 26.2$$

Step 3: From Table 8-5, the corresponding value to $f(x) = 26.2$ is $x = 1.666$ and $\operatorname{erf}(x) = 0.982$.

Step 4: From Eq. (8-110) we have

$$u_i = \left(\frac{x}{\beta}\right)^2 = \left(\frac{1.666}{6.82}\right)^2 = 0.06$$

and $\exp(-0.06) = 0.94$.

Step 5: Introducing the above values into Eq. (8-113) gives

$$2.3 \text{ m} = 2.303 \frac{8734 \text{ m}^3/\text{day}}{(141.44 \text{ m})(\pi K)}(0.94)(0.982)$$

and solving it for K gives $K = 18.17$ m/day $= 2.10 \times 10^{-2}$ cms.

Step 6: Interpolation for $M(u, \beta)$ from the values in Table 8-5 gives $M(0.06, 6.82) = 2.291$. Then, from Eq. (8-62) the value of s_i can be calculated as follows:

$$s_i = \frac{8734 \text{ m}^3/\text{day}}{(141.44 \text{ m})(\pi)(18.17 \text{ m/day})}(2.291) = 2.4 \text{ m}$$

Step 7: For $s_i = 2.4$ m, Figure 8-18 shows $t_i = 3.1$ min, and from Eq. (8-111), S_s is calculated as

$$\begin{aligned} S_s &= \frac{4Kt_iu_i}{r^2} \\ &= \frac{(4)(18.17 \text{ m/day})\left(\dfrac{3.1 \text{ min}}{1440 \text{ min/day}}\right)(0.06)}{(5.18 \text{ m})^2} \\ &= 3.50 \times 10^{-4} \text{ m}^{-1} \end{aligned}$$

Steps 8 and 9: As can be seen from Figure 8-18, the straight-line portion of the curve is not discernible within the observed data. Therefore, the transmissivity ($T = Kb$) and, consequently, the thickness of the aquifer cannot be specified from the present drawdown data.

(d) Comparison the Results with the Standard Type-Curve Method Results: The horizontal hydraulic conductivity (K_r) and specific storage (S_s) values by the standard type-curve method (see Example 8-1), Hantush's type-curve method, and Hantush's inflection-point method are compared in Table 8-7. As can be seen from Table 8-7, the three methods produced acceptable close values to each other.

TABLE 8-7 Comparison of the Results by Different Methods for Example 8-2

Method	K_r (cm/sec)	S_s (m^{-1})
Standard type-curve method	2.92×10^{-2}	3.02×10^{-4}
Hantush's type-curve method	2.27×10^{-2}	2.03×10^{-4}
Hantush's inflection-point method	2.10×10^{-2}	3.50×10^{-2}

Example 8-3: This example is taken from Hantush (1961d). The data of this example were provided from the Olson Wells, located about one-half mile north of the Campus of the New Mexico Institute of Mining and Technology, Socorro, New Mexico. The geometry of the wells is shown diagrammatically in Figure 8-21. Both the pumped well and Observation Well No. 1, which are located at 182 ft distance from the pumped well, have a 75 ft of penetration depth and only their 50-ft portion from the bottom are perforated. The pumping rate was 1.6 ft^3/s. The third well, Observation Well No. 2, is 384 ft from the pumped well and is assumed to be zero penetration (actually about 1.1 ft penetration). Drawdowns monitored at Observation Wells No. 1 and No. 2 are given in Table 8-8 and Table 8-9, respectively. Using these data, (a) analyze the drawdown data from Observation Well No. 1 and determine the aquifer parameters, and (b) analyze the drawdown data from Observation Well No. 2 and determine the aquifer parameters.

Solution: **(a) Analysis of Drawdown Data from Observation Well No. 1:** Eq. (8-80) gives the average drawdown in this well because l'/l = (75 ft)/(75 ft) $<$ 1 or = 1. Because of this and r/l = (182 ft)/(75 ft) $>$ 1, the average drawdown can for all practical purposes be expressed by Eq. (8-60), with the value of z arbitrarily chosen between l' and zero (see

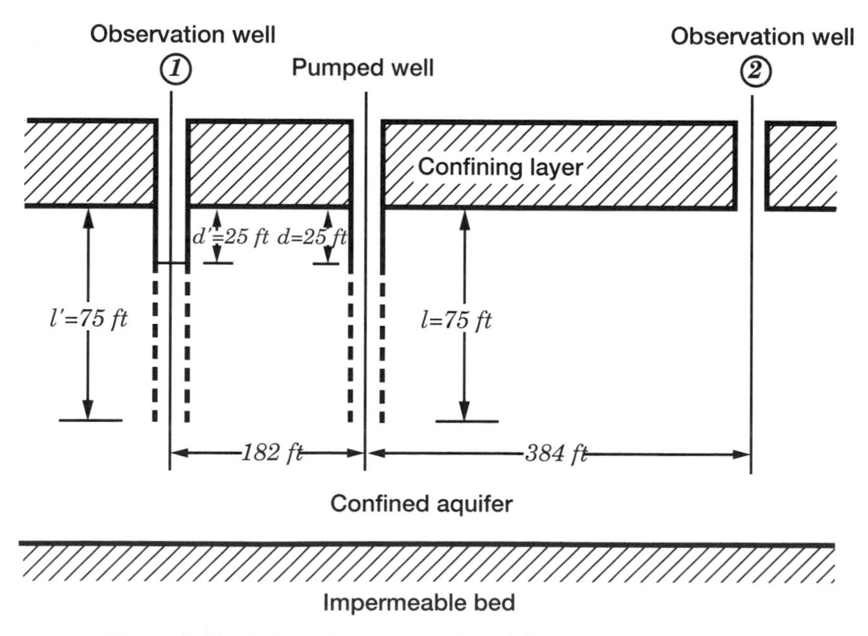

Figure 8-21 Schematic representation of the wells in Example 8-3.

TABLE 8-8 Pumping Test Data for Observation Well No. 1 of Example 8-3

t (min)	s (ft)
75.51	0.41
104.59	0.49
146.89	0.72
206.06	0.87
277.33	1.02
350.75	1.17
394.45	1.26
464.52	1.35
554.63	1.50
671.43	1.61
734.51	1.67
790.68	1.72
1364.58	2.01
1364.60	2.06
1559.55	2.10
1725.84	2.19
1888.00	2.22
1905.46	2.27
1914.26	2.30
2004.47	2.33
2254.24	2.36
2773.32	2.45
3311.31	2.52
3723.92	2.62
4064.43	2.63
4446.31	2.67
4709.77	2.69
5623.41	2.75

Section 8.3.5.3). With $z = 25$ ft (selected value) and $r = 182$ ft, and substitution all values in Eq. (8-60) gives

$$\bar{s} = \frac{Q}{8\pi K(75\text{ ft} - 25\text{ ft})}\left[M\left(u, \frac{75\text{ ft} + 25\text{ ft}}{182\text{ ft}}\right) - M\left(u, \frac{25\text{ ft} + 25\text{ ft}}{182\text{ ft}}\right)\right.$$
$$\left. + M\left(u, \frac{75\text{ ft} - 25\text{ ft}}{182\text{ ft}}\right) - M\left(u, \frac{25\text{ ft} - 25\text{ ft}}{182\text{ ft}}\right)\right]$$

or

$$\bar{s} = \frac{Q}{(400\text{ ft})\pi K}\left[M\left(u, \frac{100\text{ ft}}{182\text{ ft}}\right) - M\left(u, \frac{50\text{ ft}}{182\text{ ft}}\right) + M\left(u, \frac{50\text{ ft}}{182\text{ ft}}\right) - M(u, 0)\right]$$

In the above expression, $M(u, \beta = 0) = 0$, because $\text{erf}(\beta y^{1/2}) = 0$ in Eq. (8-52). Therefore, the above expression takes the form

TABLE 8-9 Pumping Test Data for Observation Well No. 2 of Example 8-3

t (min)	s (ft)
114.29	0.02
200.45	0.03
277.33	0.09
340.41	0.16
416.39	0.22
482.61	0.26
562.73	0.33
661.93	0.39
744.90	0.42
928.97	0.50
1061.70	0.60
1364.58	0.68
1377.21	0.65
1559.55	0.75
1725.84	0.79
1887.99	0.83
2137.96	0.88
2773.32	0.96
3459.39	1.04

$$\overline{s} = \frac{Q}{(400\text{ ft})\pi K} M(u, 0.55) = \frac{Q}{(400\text{ ft})\pi K} \overline{E}(u)$$

which is of the form of Eq. (8-62) with $c = Q/[(400\text{ ft})\pi K]$ and $\beta = 0.55$. Because the inflection of the semilogarithmic drawdown versus time curve is clearly exhibited (see Figure 8-22), both the inflection point method as well as the type curve method can be used in determining the formation parameters. Application of these methods are given below.

Hantush's Type-Curve Method: The steps of the method are presented below in accordance with Section 8.4.3.2.

Step 1: Using the values of $M(u, \beta)$ from Table 8-1, first the $M(u, 0.55)$ function is created for different values of u. Table 8-1 gives $M(u, 0.50)$ and $M(u, 0.60)$. Therefore, linear interpolation is applied to determine the values of $M(u, 0.55)$, and the values are listed in Table 8-10 and the plot is given in Figure 8-23.

Step 2: The observed drawdown data are plotted on the same scale and is given in Figure 8-24.

Step 3: The two curves are superimposed and adjusted to obtain the matching position shown in Figure 8-25.

Step 4: In Figure 8-25, A is an arbitrary chosen point and its coordinates are

$$\left(\frac{1}{u}\right)_A = 100, \qquad \overline{E}_A = 1$$

$$t_A = 14{,}000\text{ min}, \qquad \overline{s}_A = 3.3\text{ ft}$$

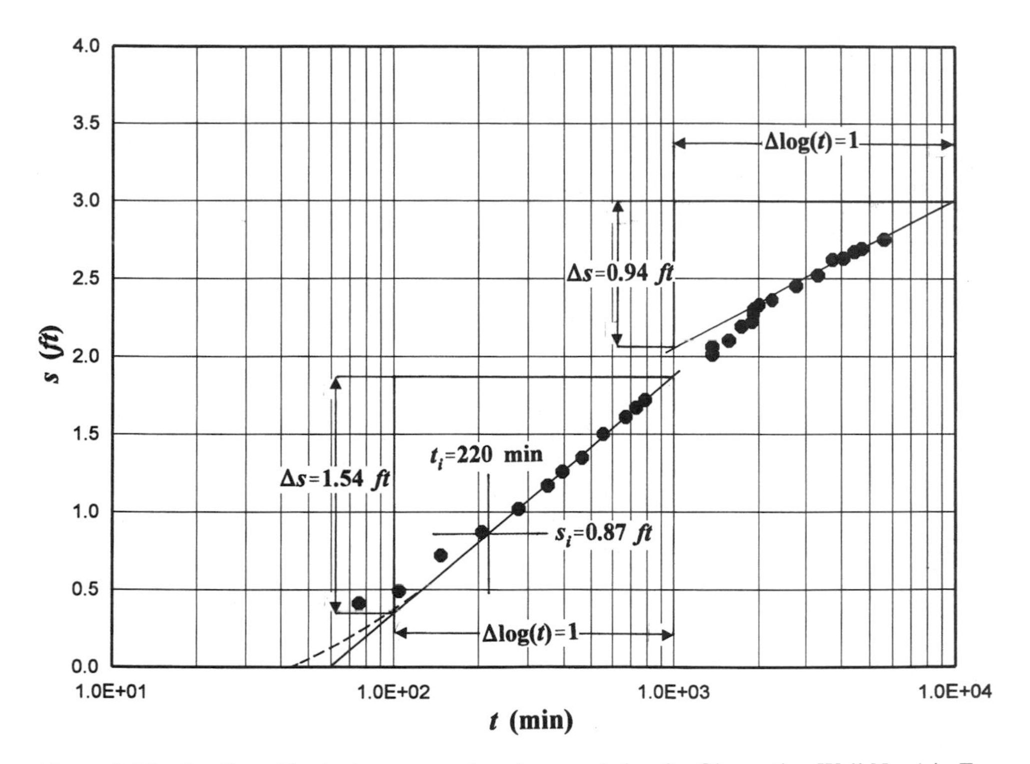

Figure 8-22 Semilogarithmic time versus drawdown variation for Observation Well No. 1 in Example 8-3.

Step 5: From the above expression

$$K = \frac{Q}{(400 \text{ ft})\pi\bar{s}} = \frac{1.6 \text{ ft}^3/\text{s}}{(400 \text{ ft})\pi(3.3 \text{ ft})}(1)$$

$$= 3.86 \times 10^{-4} \text{ ft/s}$$

$$K = 1.18 \times 10^{-2} \text{ cm/s}$$

and from Eq. (8-119) the value of S_s is

$$S_s = \frac{4Kt}{\left(\frac{1}{u}\right) r^2}$$

$$= \frac{(4)(3.86 \times 10^{-4} \text{ ft/s})(14000 \times 60 \text{ s})}{(100)(182 \text{ ft})^2} = 3.91 \times 10^{-4} \text{ ft}^{-1}$$

Step 6: As can be seen from Figure 8-25, the observed drawdown data departure does not occur within the period of the pumping test. Therefore, this step cannot be applied.

Step 7: Because the thickness is not available from Step 6, the transmissivity and storage coefficient values cannot be determined.

TABLE 8-10 Type-Curve Data, u (or $1/u$) versus $\overline{E}(u)$ for Example 8-3

u	$1/u$	$M(u, 0.50)$	$M(u, 0.60)$	$E(u) = M(u, 55)$
2×10^{-6}	0.50×10^{6}	0.9613	1.1353	1.0488
4×10^{-6}	0.25×10^{6}	0.9602	1.1349	1.0476
6×10^{-6}	0.17×10^{6}	0.9596	1.1343	1.0470
8×10^{-6}	0.13×10^{6}	0.9592	1.1338	1.0465
1×10^{-5}	1.00×10^{5}	0.9588	1.1334	1.0461
2×10^{-5}	0.50×10^{5}	0.9574	1.1316	1.0445
4×10^{-5}	0.25×10^{5}	0.9553	1.1291	1.0422
6×10^{-5}	0.17×10^{5}	0.9537	1.1271	1.0404
8×10^{-5}	0.13×10^{5}	0.9523	1.1255	1.0389
1×10^{-4}	1.00×10^{4}	0.9511	1.1241	1.0376
2×10^{-4}	0.50×10^{5}	0.9465	1.1185	1.0325
4×10^{-4}	0.25×10^{4}	0.9398	1.1106	1.0252
5×10^{-4}	0.17×10^{4}	0.9348	1.1045	1.0197
8×10^{-4}	0.13×10^{4}	0.9305	1.0994	1.0150
1×10^{-3}	1.00×10^{3}	0.9267	1.0948	1.0108
2×10^{-3}	0.50×10^{3}	0.9120	1.0771	0.9946
4×10^{-3}	0.25×10^{3}	0.8912	1.0521	0.9717
6×10^{-3}	0.17×10^{3}	0.8752	1.0330	0.9541
8×10^{-3}	0.13×10^{3}	0.8618	1.0169	0.9394
1×10^{-2}	1.00×10^{2}	0.8500	1.0027	0.9264
2×10^{-2}	0.50×10^{2}	0.8040	0.9476	0.8758
4×10^{-2}	0.25×10^{2}	0.7400	0.8708	0.8054
6×10^{-2}	0.17×10^{2}	0.6919	0.8132	0.7526
8×10^{-2}	0.13×10^{2}	0.6522	0.7658	0.7090
9×10^{-2}	0.11×10^{2}	0.6346	0.7447	0.6897
1×10^{-1}	1.00×10^{1}	0.6181	0.7249	0.6715
2×10^{-1}	0.50×10^{1}	0.4920	0.5744	0.5332
4×10^{-1}	0.25×10^{1}	0.3397	0.3935	0.3666
6×10^{-1}	0.17×10^{1}	0.2548	0.2828	0.2643
8×10^{-1}	0.13×10^{1}	0.1821	0.2082	0.1952
9×10^{-1}	0.11×10^{1}	0.1576	0.1797	0.1687
1	1.00	0.1369	0.1555	0.1462
2	0.50	0.0367	0.0407	0.0387
4	0.25	0.0033	0.0035	0.0034
6	0.17	0.000336	0.000350	0.000343
8	0.13	0.0000363	0.0000372	0.00003675
10	0.10	0.00000407	0.00000413	0.00000410

Step 8: The value of $1/u$ in the vicinity of last observed point in Figure 8-25 is around 4. Eq. (8-122) is the equation to be used because $\overline{E}(u)$ reduces to the M function as seen from Table 8-10. Substitution of this value as well as the other values in Eq. (8-122) gives

$$b \cong \frac{1}{2}\left[r\beta + r\left(\frac{5}{u_d}\right)^{1/2}\right]$$

$$= \frac{1}{2}\left\{(182 \text{ ft})(0.55) + (182 \text{ ft})[(5)(40)]^{1/2}\right\} = 1337 \text{ ft}$$

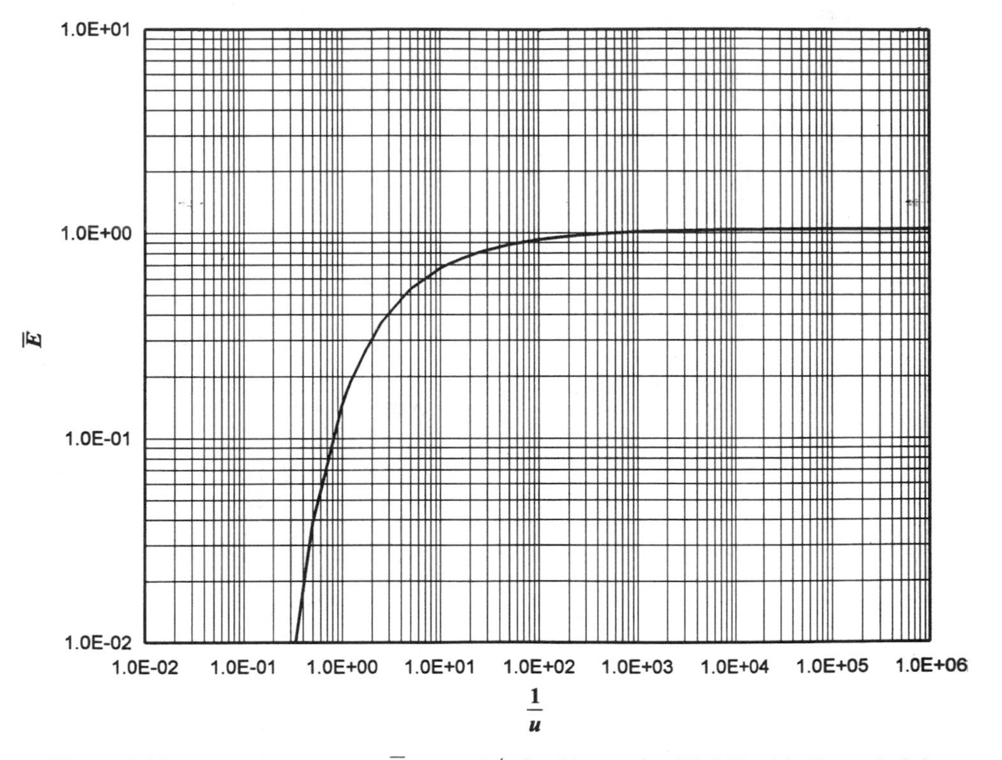

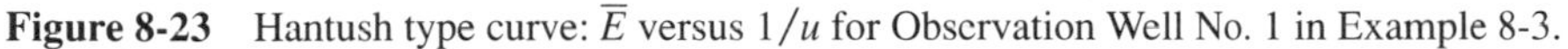

Figure 8-23 Hantush type curve: $\overline{E}$ versus $1/u$ for Observation Well No. 1 in Example 8-3.

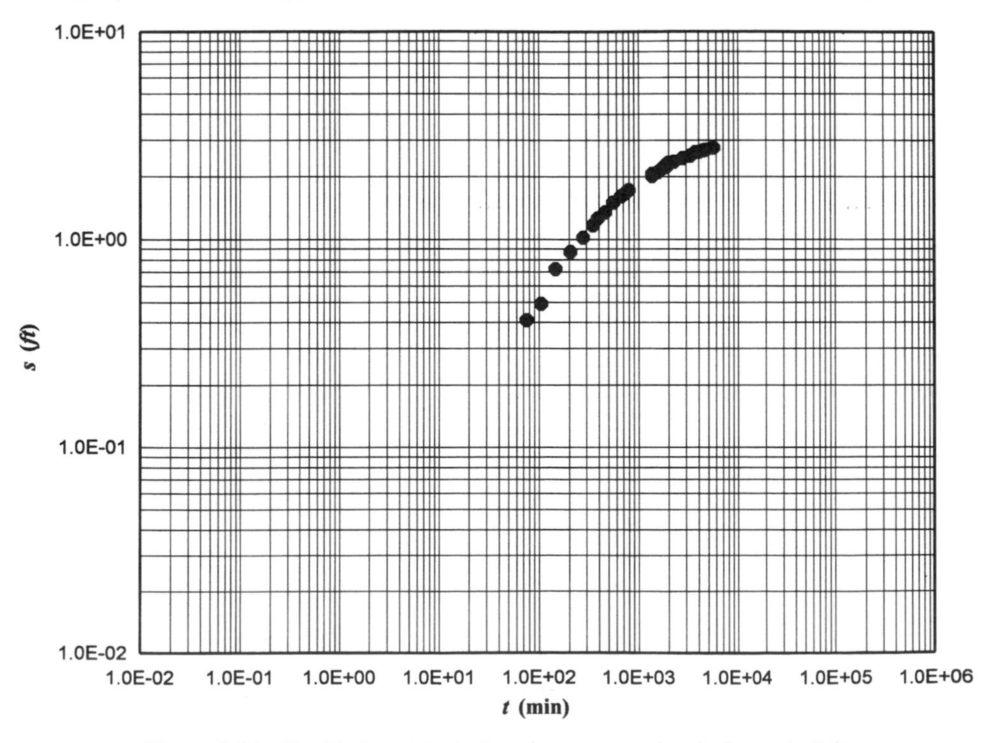

Figure 8-24 Double logarithmic drawdown versus time in Example 8-3.

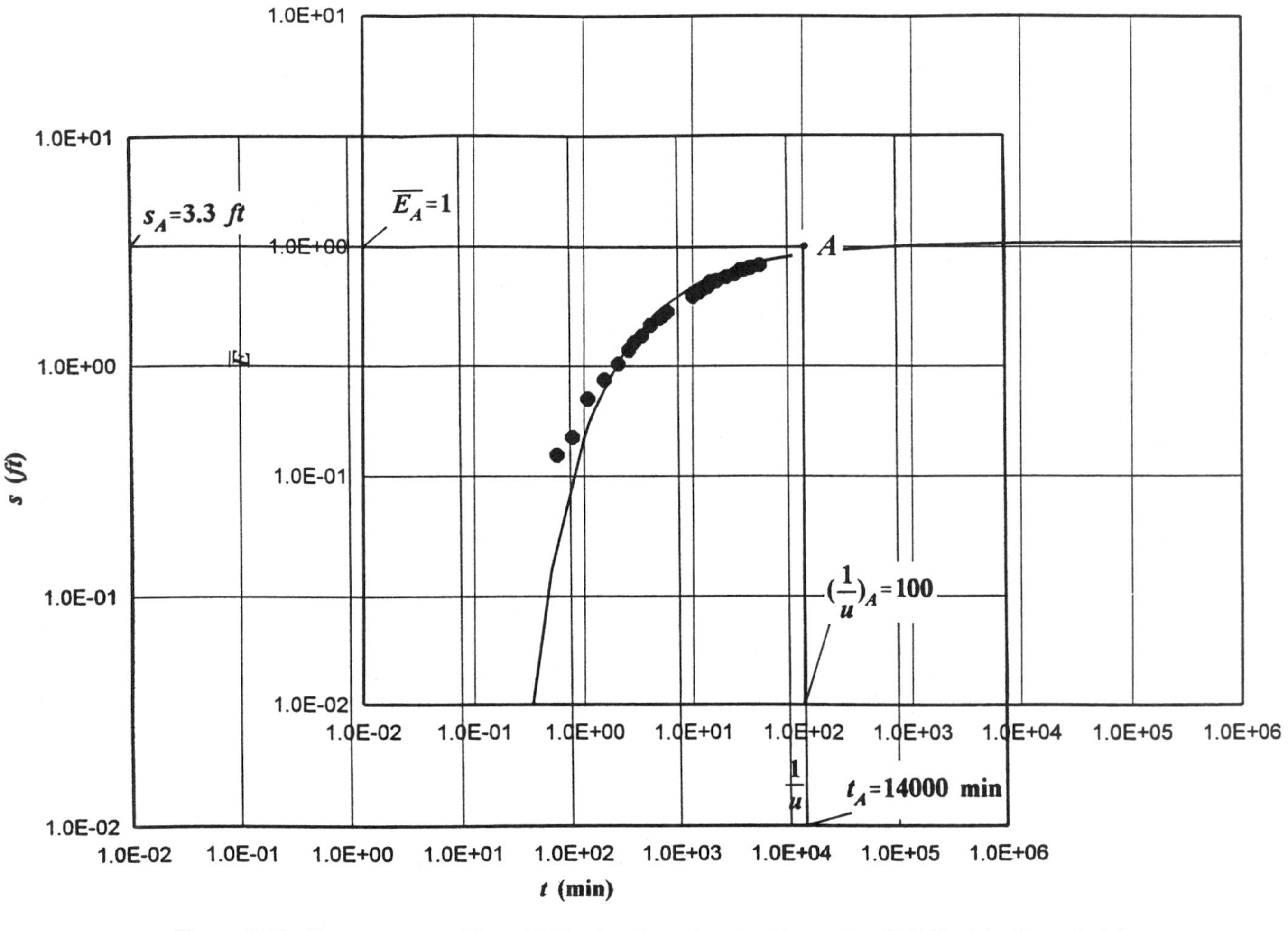

Figure 8-25 Type-curve matching with the drawdown data for Observation Well No. 1 in Example 8-3.

Step 9: From the above values, the transmissivity and storage coefficient values are

$$T = Kb = (3.86 \times 10^{-4} \text{ ft/s})(1337 \text{ ft}) = 0.516 \text{ ft}^2/\text{s}$$

$$S = S_s b = (3.94 \times 10^{-4} \text{ ft}^{-1})(1337 \text{ ft}) = 0.52$$

Hantush's Inflection-Point Method: The observed semilogarithmic plot for Observation Well No. 1 is shown in Figure 8-22. The tangent to the curve in the region of the curve inflection is drawn and the steps of calculations are given below.

Step 1: The geometric slope of the tangent at the inflection point is

$$m_i = \frac{\Delta s}{\Delta(\log t)} = \frac{1.54 \text{ ft}}{1} = 1.54 \text{ ft}$$

Step 2: The value of the $f(x)$ function, as given by Eq. (8-109), for $\beta = 0.55$ is

$$f(x) = \frac{\beta^2}{\pi^{1/2}} = \frac{(0.55)^2}{\pi^{1/2}} = 0.171$$

Step 3: From Table 8-5, the corresponding value to $f(x) = 0.171$ is $x = 0.371$ and $\text{erf}(x) = 0.40$.

Step 4: From Eq. (8-110) the value of u_i and $\exp(-u_i)$ are

$$u_i = \left(\frac{x}{\beta}\right)^2 = \left(\frac{0.371}{0.55}\right)^2 = 0.455$$

$$\exp(-0.455) = 0.634$$

Step 5: Introducing the values in Eq. (8-113)

$$1.54 \text{ ft} = 2.303 \frac{1.6 \text{ ft}^3/\text{s}}{(400 \text{ ft})(\pi K)}(0.634)(0.40)$$

and solving it for K gives $K = 4.82 \times 10^{-4}$ ft/s $= 1.47 \times 10^{-2}$ cm/s.

Step 6: From Table 8-1, the value of the $M(u, \beta)$ function for $u = 0.455$ and $\beta = 0.55$ is $M(0.455, 0.55) = 0.33$. From Eq. (8-62), the value of s_i can be calculated as

$$s_i = \frac{1.6 \text{ ft}^3/\text{s}}{(400 \text{ ft})(\pi)(4.82 \times 10^{-4} \text{ ft/s})}(0.33) = 0.87 \text{ ft}$$

Step 7: From Figure 8-22, the corresponding value for $s_i = 0.87$ ft is $t_i = 220$ min. Then, the value of S_s from Eq. (8-111) is calculated as

$$S_s = \frac{4Kt_iu_i}{r^2}$$

$$= \frac{(4)(4.82 \times 10^{-4} \text{ ft/s})(220 \times 60 \text{ s})(0.455)}{(182 \text{ ft})^2}$$

$$S_s = 3.50 \times 10^{-4} \text{ ft}^{-1}$$

Step 8: The ultimate straight line of the curve in Figure 8-22 is drawn and its geometric slope is

$$m = \frac{\Delta s}{\Delta(\log t)} = \frac{0.94 \text{ ft}}{1} = 0.94 \text{ ft}$$

From Eq. (8-118), the value of transmissivity is

$$Kb = \frac{2.3Q}{4\pi m} = \frac{(2.3)(1.6 \text{ ft}^3/\text{s})}{(4\pi)(0.94 \text{ ft})} = 0.312 \text{ ft}^2/\text{s}$$

Step 9: Using the value of K from Step 5, the thickness of the aquifer (b) is

$$b = \frac{0.312 \text{ ft}^2/\text{s}}{4.82 \times 10^{-4} \text{ ft/s}} = 647 \text{ ft}$$

The storage coefficient is

$$S = S_s b = (3.50 \times 10^{-4} \text{ ft}^{-1})(647 \text{ ft}) = 0.226$$

(b) Analysis of Drawdown Data from Observation Well No. 2: The observed semilogarithmic time versus drawdown curve (see Figure 8-26) for this observation well does not exhibit a curve inflection. It is known that the well has practically zero penetration, and the

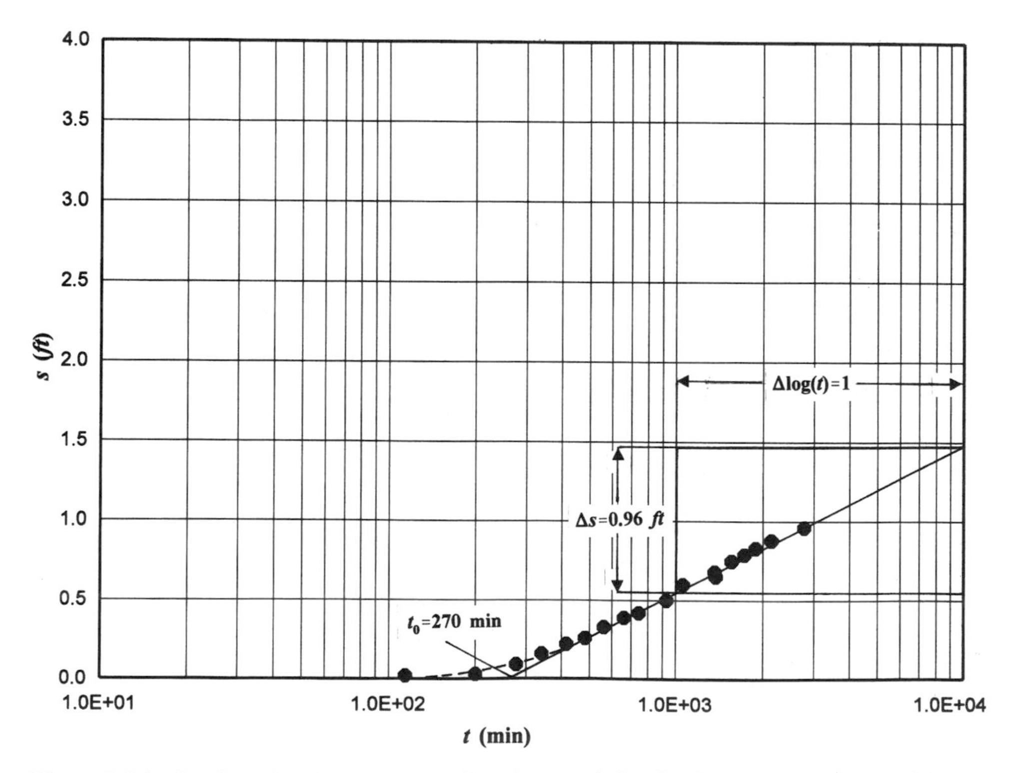

Figure 8-26 Semilogarithmic time versus drawdown variation for Observation Well No. 2 in Example 8-3.

aquifer is known to be too thick. Its estimated thickness from the Observation Well No. 1 analysis is 1337 ft. As a result, $r/b = 384/1337 < 1.5$. This means that the condition given in Section 8.4.2.1 is not valid, and therefore the possibilities for the application of the Theis or Cooper and Jacob method are ruled out, although the general trend of the time versus drawdown variation resembles the one displayed by the Theis equation. Hantush's type-curve method can be applied because there are enough points on the curved part of the observed semilogarithmic curve, and the drawdown equation is Eq. (8-60) with $z = 0$. However, for illustrative purposes, the Cooper and Jacob method adjusted for partial penetration, as presented in Section 8.4.3.2, will be applied (Hantush, 1961d).

Cooper and Jacob's Method Adjusted for Partial Penetration: The steps of this method in accordance to Section 8.4.3.2 are as follows:

Step 1: The observed semilogarithmic time versus drawdown curve for Observation Well No. 2 is shown in Figure 8-26 using the data in Table 8-9.

Step 2: The geometric slope of the straight line portion in Figure 8-26 is

$$m = \frac{\Delta s}{\Delta(\log t)} = \frac{0.96 \text{ ft}}{1} = 0.96 \text{ ft}$$

and $t_0 = 270$ min.

Step 3: From Eq. (8-118) we obtain

$$T = Kb = \frac{2.303Q}{4\pi m} = \frac{(2.303)(1.6 \text{ ft}^3/\text{s})}{(4\pi)(0.96 \text{ ft})} = 0.305 \text{ ft}^2/\text{s}$$

Step 4: This step requires to calculate the value of f_s from its series form from Eq. (8-69). According to Hantush (1961d, p. 192), only five terms of the infinite series are needed. The accuracy of this statement needs to be checked. The calculated value of f_s is 0.81 (Hantush, 1961c).

Step 5: Introducing this value in Eq. (8-129) gives

$$S = \frac{2.25Tt_0 \exp(f_s)}{r^2} = \frac{(2.25)(0.305 \text{ ft}^2/\text{s})(270 \times 60 \text{ s}) \exp(0.81)}{(384 \text{ ft})^2} = 0.17$$

8.4.4 Hantush's Method for Recovery Test Data

8.4.4.1 Properties of Recovery Equations. Some charactersistic features of the recovery equations are important in applying them to the recovery data analysis in confined aquifers. These features are given below based on Hantush (1961d).

Recovery Equations for Special Cases. The equations given in Section 4.2.7.4 of Chapter 4 regarding the Theis recovery data analysis method for fully penetrated confined aquifers can be applied. The drawdown equation during short pumping periods is the type of equation given by Eqs. (8-62) or (8-82). These equations are called *recovery equations for special case*.

Figure 8-27 presents theoretical curves for the residual drawdown (s') versus $\log(t/t')$ based on Eq. (8-96). These curves have three main characteristic aspects, and these aspects are discussed below (Hantush, 1961d).

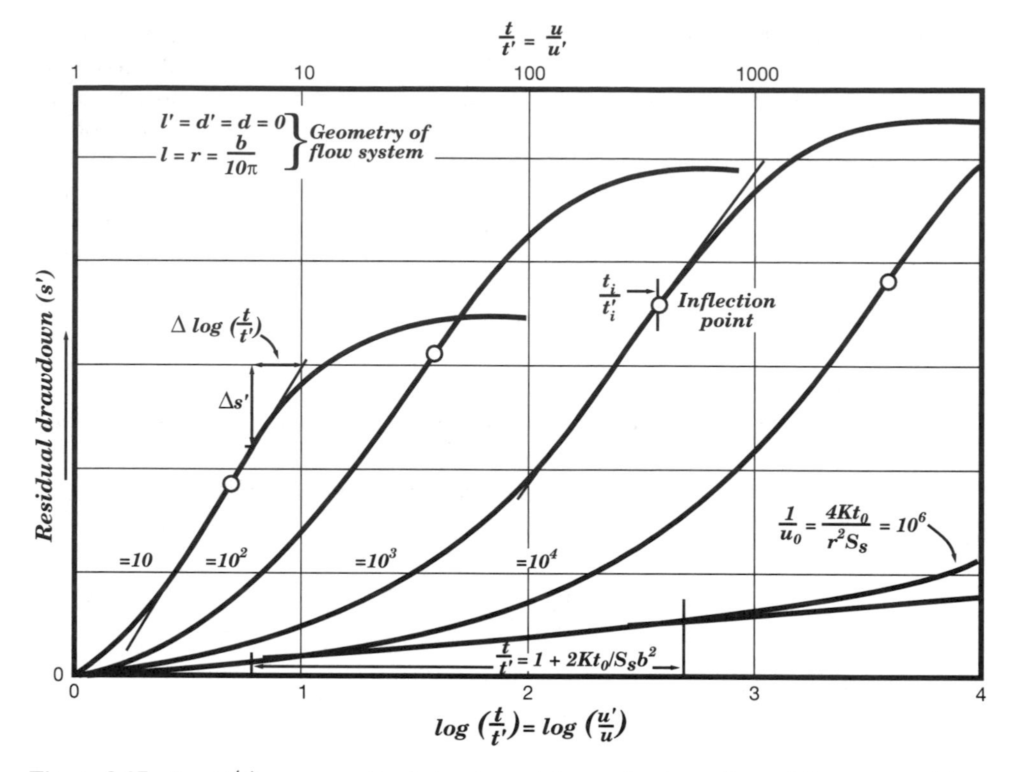

Figure 8-27 Log(t/t') versus residual drawdown (s') variation in wells partially penetrated in a confined aquifer. (After Hantush, 1961d.)

1. The curves have a straight-line portion at their lower left end passing through the point having the coordinates $s' = 0$ and $t/t' = 1$. This line may clearly be defined, as in the curves $1/u_0 = 4Kt_0/(r^2S_s) = 10^6$ and 10^4; barely recognizable, as the curve 10^3; or entirely indistinguishable, as in the curves 10^2 and 10. The range of time in which the line forms depends on the pumping test period (t_0), the formation parameters, and the aquifer thickness. More specifically, if $t/t' < [1 + 2Kt_0/(b^2S_s)]$, the straight line forms. If the straight line is formed, its geometric slope (m) can similarly be derived by noting that both u and u' become very small for larger values of time, as for the case of Eq. (8-118), by differentiating Eq. (8-101) with respect to $\log(t/t')$ gives

$$m = \frac{\Delta s'}{\Delta(\log t)} = \frac{2.303Q}{4\pi Kb} \tag{8-130}$$

2. In general, the curves have an inflection point around which the curve may often be approximated by a straight line. The geometric slope of the straight line, for all practical purposes, is equal the geometric slope of the curve at the inflection point. The inflection point forms, if it does, in the period

$$t' < \left[\frac{(2b - r\beta)^2 S_s}{20K}\right] \tag{8-131}$$

Thus, the recovery equation around the inflection point is given by Eq. (8-96) after dropping out $f^2(u', b/r, \beta)$. By differentiating Eq. (8-96), or any of its equivalent form, twice with respect to $\log(t/t')$ and equating the results to zero, the value of u' at the inflection point—that is, the value of u_i'—can be shown to be given approximately by the relation [Hantush, 1961d, p. 182, Eq. (30)]

$$\frac{\beta^2}{\pi^{1/2}} = f(x) = xe^{x^2}\,\mathrm{erf}(x) \tag{8-132}$$

where

$$x = \beta u_i'^{1/2} \tag{8-133}$$

and

$$u_i' = \frac{r^2 S_s}{4Kt_i'} \tag{8-134}$$

The geometric slope of the curve (m_i) at the inflection point can be obtained from Eq. (8-96) or other equivalent form by differentiating with respect to $\log(t/t')$ and substituting u_i and u_i' for u and u', respectively, in the resulting equation. This slope can be approximated closely by either of the following two relations, depending on the range of (t/t') in which the inflection point occurs (Hantush, 1961d, p. 182):

a. If the inflection point occurs in the range $t/t' < 100$, the geometric slope, for all practical purposes, can be taken as

$$m_i = \frac{\Delta s'}{\Delta(\log t)} = 2.303c\left\{\left(\frac{t_i}{t_0}\right)\exp(-u_i')\,\mathrm{erf}(x) - \left(\frac{t_i'}{t_0}\right)\exp\left[-\left(\frac{t'}{t}\right)_i u_i'\right]\mathrm{erf}\left[x\left(\frac{t'}{t}\right)_i^{1/2}\right]\right\} \tag{8-135}$$

where $(t'/t)_i$ is the value of t'/t at the inflection point.

b. If the inflection point occurs in the range $t/t' > 100$, the second term on the right-hand side in Eq. (8-135) becomes insignificant. As a result, the geometric slope is given by Eq. (8-135) without the second term.

3. If the geometry of the flow system is such that the following relation holds

$$\frac{x^2(2b - r\beta)^2}{5(r\beta)^2} \leq 1 \tag{8-136}$$

the curve under consideration will not show an inflection point. The straight-line portion of the curve that forms through the latter part of data will, however, be the same as that described in the item 1.

8.4.4.2 Analysis of Recovery Data. If the drawdown equation in an observation well during short time periods reduces to the type of Eq. (8-82), the data analysis method is as follows (Hantush, 1961d):

Step 1: Plot the residual drawdown (s') versus $\log(t/t')$ on single logarithmic paper and draw the best-fit curve through the observed points.

Step 2: Draw the tangent to this curve in the curve inflection region and determine its geometric slope from the left-hand side of Eq. (8-130).

Step 3: Locate the inflection point on this tangent by inspection and determine its value of $(t/t')_i$. It should be pointed out that overestimation or underestimation of the inflection-point location will not significantly affect the resulting values, as will be shown subsequently.

Step 4: Compute the following values:

$$\frac{t_0}{t'_i} = \left(\frac{t}{t'}\right)_i - 1 \tag{8-137}$$

and

$$\frac{t_i}{t_0} = \left(\frac{t}{t'}\right)_i \left(\frac{t_0}{t'_i}\right) \tag{8-138}$$

where t_0 is the pumping period.

Step 5: Calculate $f(x)$ from Eq. (8-132) and determine the corresponding value of x and $\text{erf}(x)$ from Table 8-5.

Step 6: Calculate the value of u'_i from Eq. (8-133).

Step 7: There may be two cases:

(a) If $(t/t')_i > 100$, calculate the value of K from Eq. (8-135) by neglecting the second term on the right-hand side.

(b) If $(t/t')_i < 100$, calculate the value of K from Eq. (8-135).

Step 8: From the known values of t_0 and (t'_i/t_0), calculate the value of t'_i.

Step 9: Calculate the value of S_s from Eq. (8-134).

Step 10: If the straight-line variation between s' and $\log(t/t')$ is discernible through the later part of the data, including the origin point, whose coordinates are ($t/t' = 1$, $s' = 0$), then draw this line and measure its geometric slope from $m = \Delta s'/\Delta \log(t)$; then calculate the value of Kb from Eq. (8-130). Then, from the known values of K and (Kb), calculate the value of b. With the values of b and S_s, calculate the storage coefficient value ($S = S_s b$).

Important Points. Some important points regarding the applicability of the method are given below (Hantush, 1961d):

1. If $(t/t')_i > 10$, the second term in Eq. (8-135) is comparatively very small. Therefore, a large error in estimating the value of $(t/t')_i$ will not affect significantly the value of the second term. The largest error, if any, will be in the first term of Eq. (8-135), because of the factor (t_i/t_0). This error, however, rarely exceeds $\pm 3\%$. The error in the value of S_s is approximately directly proportional to the error in estimating the value of t'_i.
2. If the short periods of time drawdown equation is not of the form of Eq. (8-62), an exact graphical solution is not available. The procedure outlined before can be used to make a rough estimation for the formation parameters. However, if the straight

TABLE 8-11 t/t' **Versus Drawdown at the Observation Well No. 1 for Example 8-4**

t/t'	s' (ft)
59.70	2.45
29.79	1.95
16.29	1.59
12.65	1.35
8.13	1.09
7.33	1.03
6.05	0.90
4.44	0.72
4.06	0.70
3.72	0.58
2.94	0.49
2.65	0.45
2.22	0.36
1.97	0.28
1.94	0.26
1.75	0.23
1.60	0.21

line examined in Step 10 forms, the solution for the transmissivity is exact, because in the present case Eq. (8-130) gives the value of Kb.

Example 8-4: This example is adapted from Hantush (1961d). The observed residual drawdown data from Observation Well No. 1 of Example 8-3 are given in Table 8-11. The pumping period was $t_0 = 5623.41$ min. Using Hantush's recovery data analysis method, determine the aquifer parameters.

Solution: The data of this example will be analyzed in accordance the steps as given in Section 8.4.4.2. The steps are as follows:

Step 1: The observed residual drawdown (s') versus $\log(t/t')$ curve is given in Figure 8-28 based on the values in Table 8-11.

Step 2: The tangent in the region of curve inflection is drawn and shown in Figure 8-28. Its geometric slope is

$$m_i = \frac{\Delta s'}{\Delta(\log t)} = \frac{1.74 \text{ ft}}{1} = 1.74 \text{ ft}$$

Step 3: By inspection, the inflection point is located at $(t/t')_i = 30$.

Step 4: From Eqs. (8-137) and (8-138) we obtain

$$\frac{t_0}{t_i'} = \left(\frac{t}{t'}\right)_i - 1 = 30 - 1 = 29$$

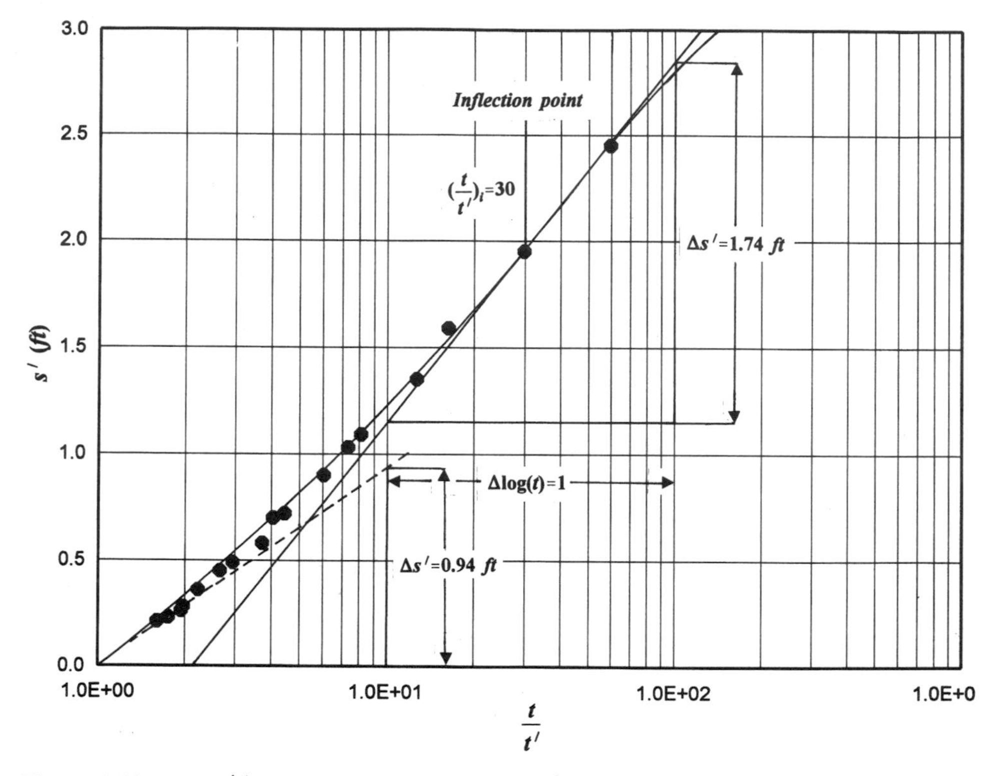

Figure 8-28 Log(t/t') versus residual drawdown (s') variation for Observation Well No. 1 in Example 8-4.

and

$$\frac{t_i}{t_0} = \left(\frac{t}{t'}\right)_i \left(\frac{t_0}{t_i'}\right) = \frac{30}{29} = 1.034$$

Step 5: From (a) of Example 8-3 we have $\beta = 0.55$. Then, from Eq. (8-132) we obtain

$$f(x) = \frac{\beta^2}{\pi^{1/2}} = \frac{(0.55)^2}{\pi^{1/2}} = 0.17$$

for which Table 8-5 gives $x = 0.371$ and $\text{erf}(x) = 0.40$.

Step 6: From Eq. (8-133) we obtain

$$u_i' = \left(\frac{x}{\beta}\right)^2 = \left(\frac{0.371}{0.55}\right)^2 = 0.455$$

Step 7: Because $(t/t')_i < 100$, all terms of Eq. (8-135) must be included. Substituting the values $c = Q/[(400 \text{ ft})\pi K]$, $Q = 1.6 \text{ ft}^3/\text{day}$, and the other values into Eq. (8-135) gives:

$$K = \frac{(2.303)(1.6\ \text{ft}^3/\text{day})}{(400\ \text{ft})(\pi)(1.74\ \text{ft})} \left\{ (1.034)\exp(-0.455)\,\text{erf}(0.371) \right.$$

$$\left. - \frac{1}{29}\exp\left(-\frac{0.455}{30}\right)\text{erf}\left[0.371\left(\frac{1}{30}\right)^{1/2}\right] \right\}$$

$$K = 4.37 \times 10^{-4}\ \text{ft/s} = 1.33 \times 10^{-2}\ \text{cm/s}$$

Step 8: From $t_0/t_i' = 29$ and $t_0 = 5623.41$ min (the pumping period), $t_i' = 11635$ s.
Step 9: From Eq. (8-134), the value of S_s is

$$S_s = \frac{4Kt_i'u_i'}{r^2}$$

$$= \frac{(4)(4.34 \times 10^{-4}\ \text{ft/s})(11{,}635\ \text{s})(0.455)}{(182\ \text{ft})^2}$$

$$S_s = 2.79 \times 10^{-4}\ \text{ft}^{-1}$$

Step 10: The straight-line portion of the lower left-hand portion of the observed residual drawdown curve is not clearly discernible (see Figure 8-28). However, for the purpose of illustrating the procedure, the straight line is drawn by the dashed line in Figure 8-28. And for this straight line, $m = 0.94$ ft. Then, from Eq. (8-130) the product (Kb) is calculated as

$$Kb = \frac{2.303Q}{4\pi m} = \frac{(2.303)(1.6\ \text{ft}^3/\text{s})}{(4\pi)(0.94\ \text{ft})} = 0.312\ \text{ft}^2/\text{s}$$

Finally,

$$b = \frac{(Kb)}{K} = \frac{0.312\ \text{ft}^2/\text{s}}{4.37 \times 10^{-4}\ \text{ft/s}} = 714\ \text{ft}$$

and

$$S = S_s b = (2.79 \times 10^{-4}\ \text{ft}^{-1})(714\ \text{ft}) = 0.20$$

CHAPTER 9

FULLY AND PARTIALLY PENETRATING PUMPING AND OBSERVATION WELLS IN HOMOGENEOUS AND ANISOTROPIC UNCONFINED AQUIFERS

9.1 INTRODUCTION

The responses of unconfined aquifers under fully or partially penetrating pumping well conditions are significantly different from the responses of confined aquifers. There is no doubt that the Theis solution (Theis, 1935), which is valid only for confined aquifers, is an important contribution to the quantitative hydrogeology. The same solution is also used for unconfined aquifers. But in analyzing pumping test data with the Theis solution from unconfined aquifers, one often finds that the drawdown versus time curve for an observation well fails to follow the Theis solution. The typical behavior of the drawdown versus time curve for an observation well under pumping conditions plotted on double logarithmic paper is such that at early times the curve has a steep segment, at intermediate times it has a flat segment, and at later times it has a somewhat steeper segment. Because of this, the physical mechanism of the response for unconfined aquifers has been researched by various investigators since the 1950s. The first mathematical model capable of mimicking the aforementioned behavior was introduced into the literature by Boulton (1954a, 1963). Later, Neuman (1972, 1974, 1975) developed analytical models for the delayed response process characterizing flow to a well in an unconfined aquifer for both fully and partially penetrating pumping and observation well conditions. The main difference between the Boulton and Neuman approaches is the simulation procedure regarding the delayed response of unconfined aquifers.

In this chapter, first because of their importance, the physical mechanisms of flow to a well in an unconfined aquifer will be presented in detail. Then, Boulton's models with and without delayed yield for fully penetrating wells will be presented with examples for pumping test data analysis based on Boulton's type-curve method. Neuman's models will be presented in accordance with their developmental process. As a result, Neuman's first work regarding fully penetrating wells in unconfined aquifers will be presented with its practical implications. Then, Neuman's modified model, taking into account partially penetrating wells, will be presented. Reduction of special cases from Neuman's models will also be included. Based on these solutions, Neuman's type-curve, semilogarithmic, and recovery methods will be presented, with examples.

9.2 PHYSICAL MECHANISM OF FLOW TO A WELL IN AN UNCONFINED AQUIFER

Understanding the physical mechanism of the flow to a well in an unconfined aquifer is critically important in applying the related mathematical models for the determination of aquifer parameters. In the following paragraphs, the physical mechanism of this phenomenon is explained in detail (Boulton, 1954a, 1963; Neuman, 1972, 1974, 1975; Price, 1985).

When water is extracted by pumping from a confined aquifer, only water pressure is reduced and the aquifer usually remains under fully saturated flow conditions. Under extraction conditions from an unconfined aquifer, initially water comes from elastic storage. In other words, initially, little or no water comes from the pores of the unconfined aquifer. The hydraulic head around the well declines rapidly, as is the case in a confined aquifer, but the system remains under saturated flow conditions above the cone of depression curve around the well (see Figure 9-1a). For a short time after extraction starts, the drawdown versus time curve for an observation well follows the drawdown curve predicted by the Theis equation (see Figure 9-2). As time goes on, the water body above the cone of depression in Figure 9-1 starts to contribute water to the water table by the gravity drainage (see Figure 9-1b). This gravity drainage has the recharge effect entering into the cone of depression around the well, and the drawdown becomes less than the drawdown value predicted by the Theis equation. As extraction continues, the cone of depression expands more slowly, and the gravity drainage of the pores inside the cone continues but eventually ceases if the extraction period is long enough (Figure 9-1c).

Based on the foregoing analysis, the complete response of an unconfined aquifer to extraction from a well can be considered to occur in three stages (Price, 1985):

1. Initially, the response is similar to that of a confined aquifer in which water is released from storage.
2. In the second stage, the drawdown decreases because of gravity drainage; this stage is the *delayed yield stage*.
3. The third stage starts when the gravity drainage ceases.

The second stage, which is known as the delayed yield stage, can range from hours to weeks or more depending on the hydrogeologic parameters of the aquifer. Figure 9-2 shows the delayed response of the drawdown for an unconfined aquifer. As can be seen in Figure 9-2, initially the drawdown curve follows the Theis curve, but after a while the curve departs from the Theis curve. When drawdown versus time on double logarithmic paper is plotted, the observed points depict an S-shape curve having three characteristic segments, which correspond to the stages discussed above: (1) a steep portion during early times (Segment 1); (2) a flat portion during intermediate times (Segment 2); and (3) another steep portion during later times (Segment 3). As can be seen from Figure 9-2, along Segment 3 the curve again follows the Theis curve for which the storage coefficient is equal to the specific yield (i.e., $S = S_y$).

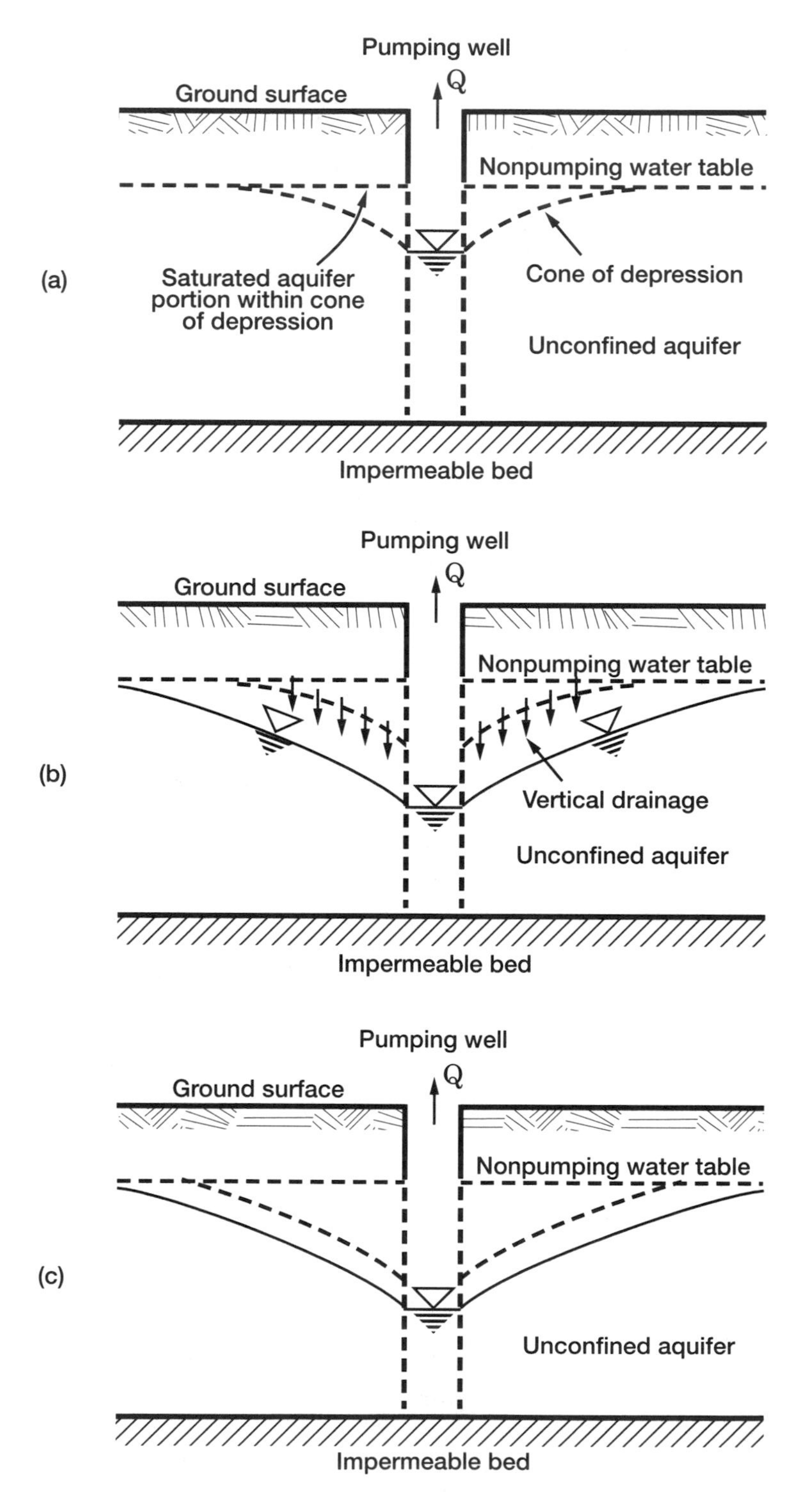

Figure 9-1 The development process of the cone of depression in an unconfined aquifer: (a) initial cone of depression; (b) gravity drainage of the initial cone of depression; and (c) cone of depression under equilibrium conditions.

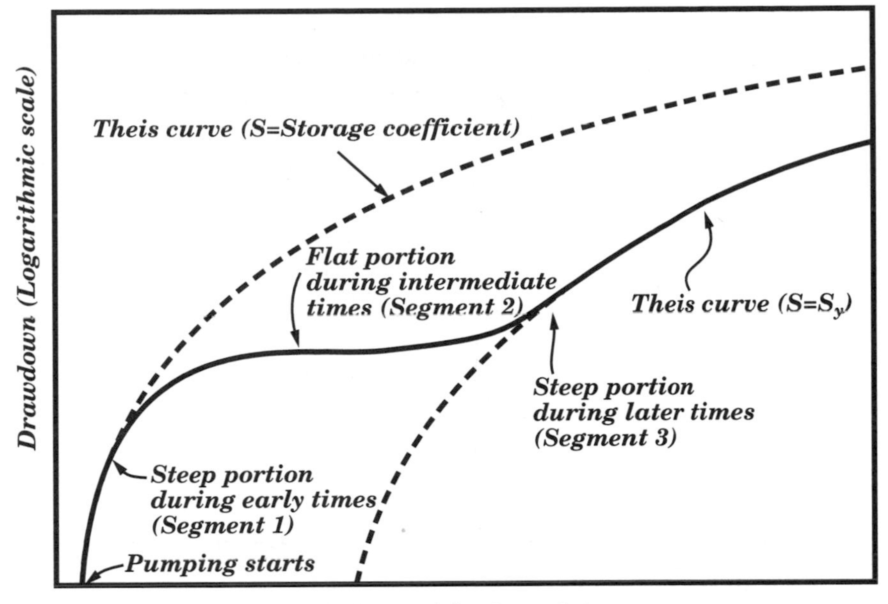

Figure 9-2 Delayed response of the drawdown in an unconfined aquifer.

9.3 BOULTON'S MODEL FOR A FULLY PENETRATING WELL IN AN UNCONFINED AQUIFER WITHOUT DELAYED RESPONSE

9.3.1 Introduction

Boulton (1954a) introduced an equation for the drawdown of the water table near a pumped well before the flow has reached steady state. This solution corresponds to the period after Segment 2 in Figure 9-2; that is, the vertical recharge by gravity drainage ceased by the system is still under transient conditions. In the following sections the mathematical foundations and practical applications of this model are presented.

9.3.2 Problem Statement and Assumptions

The geometry of the problem is illustrated in Figure 9-3, which shows that a well fully penetrates an unconfined aquifer. The assumptions are given below (Boulton, 1954a):

1. The aquifer is homogeneous and anisotropic, is of infinite lateral extent, and is underlain by a horizontal impermeable bed;
2. The well completely penetrates the aquifer and is unlined.
3. The specific yield (S_y) is constant.
4. The flow in the aquifer obeys Darcy's law.
5. The water table is initially horizontal.
6. The well is pumped at a constant rate from the instant $t = 0$.

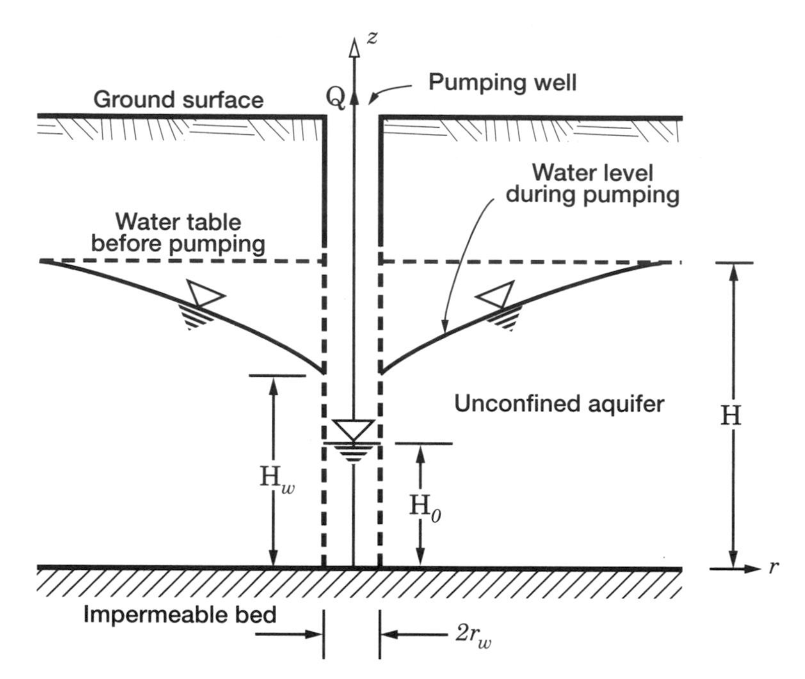

Figure 9-3 Schematic representation of flow to a well in an unconfined aquifer.

7. The contribution to the flow by water and aquifer compression may be neglected, except during the very early period of pumping.
8. The drawdown of the water table is small compared with the saturated thickness of the aquifer.

9.3.3 Governing Equations

The three-dimensional form of the transient ground water flow equation in the Cartesian coordinates system is [Chapter 2, Eq. (2-166)]

$$\frac{\partial^2 h}{\partial x^2} + \frac{\partial^2 h}{\partial y^2} + \frac{\partial^2 h}{\partial z^2} = \frac{S_s}{K} \frac{\partial h}{\partial t} \tag{9-1}$$

In the cylindrical coordinate system, Eq. (9-1) takes the form (see Chapter 2, Section 2.8.2)

$$\frac{\partial^2 h}{\partial r^2} + \frac{1}{r} \frac{\partial h}{\partial r} + \frac{\partial^2 h}{\partial z^2} = \frac{S_s}{K} \frac{\partial h}{\partial t} \tag{9-2}$$

Assumption 7 in Section 9.3.2 is the key assumption for the Boulton's solution. With this assumption, the right-hand side of Eq. (9-2) may be assumed to be zero:

$$\frac{\partial^2 h}{\partial r^2} + \frac{1}{r} \frac{\partial h}{\partial r} + \frac{\partial^2 h}{\partial z^2} = 0 \tag{9-3}$$

The differential equation to be satisfied at the variable free-surface boundary may be obtained in the following way. Because the pressure on this surface is atmospheric (assumed to be zero), the equation to the surface is

$$h(r, z, t) - z = 0 \tag{9-4}$$

A particle on the free surface will stay on the surface, and its mathematical expression is

$$\frac{D}{Dt}(h - z) = 0 \tag{9-5}$$

where D denotes differentiation following the motion of the particle. The equivalent form of D/Dt is

$$\frac{D}{Dt} = \frac{\partial}{\partial t} + \frac{v_r}{S_y}\frac{\partial}{\partial r} + \frac{v_z}{S_y}\frac{\partial}{\partial z} \tag{9-6}$$

where S_y is the specific yield, and v_r/S_y and v_z/S_y are the ground-water velocity components (seepage velocity components). From Eqs. (9-5) and (9-6) it follows that

$$\frac{\partial h}{\partial t} + \frac{v_r}{S_y}\frac{\partial h}{\partial r} + \frac{v_z}{S_y}\left(\frac{\partial h}{\partial z} - 1\right) = 0 \tag{9-7}$$

The Darcy velocity components in the radial and vertical directions, respectively, are

$$v_r = -K\frac{\partial h}{\partial r}, \qquad v_z = -K\frac{\partial h}{\partial z} \tag{9-8}$$

Substituting of these velocity components into Eq. (9-7) gives the free surface boundary equation:

$$\frac{\partial h}{\partial t} = \frac{K}{S_y}\left[\left(\frac{\partial h}{\partial r}\right)^2 + \left(\frac{\partial h}{\partial z}\right)^2 - \frac{\partial h}{\partial z}\right] \tag{9-9}$$

Because Eq. (9-9) is nonlinear, a solution satisfying it cannot easily be determined. However, if the gradients of h are small, the squared terms in Eq. (9-9) may be neglected, and then it reduces to the following linear equation:

$$\frac{\partial h}{\partial t} + \frac{K}{S_y}\frac{\partial h}{\partial z} = 0 \qquad \text{for } z = h \tag{9-10}$$

9.3.4 Initial and Boundary Conditions

Because the initial water table is assumed to be horizontal plane, $h(r, t) = z = H$, it follows that the initial condition is

$$h(r, t) = H \qquad \text{when } t = 0 \tag{9-11}$$

At the surface, the boundary conditions are

$$h(r,t) = H_0 \qquad \text{for} \quad 0 \leq z \leq H_0 \tag{9-12}$$

$$h(r,t) = z \qquad \text{for} \quad H_0 \leq z \leq H_w \tag{9-13}$$

Because the bottom of the aquifer is assumed to be impermeable, its corresponding boundary condition is

$$\frac{\partial h(r,t)}{\partial z} = 0 \qquad \text{for } z = 0 \tag{9-14}$$

Also

$$h(\infty,t) = H \qquad \text{for } 0 \leq z \leq H \tag{9-15}$$

Equation (9-11) states that the head is the same at every point of the water table. Equation (9-12) states that below the water level in the well the head is equal to H_0. Equation (9-13) states that above the water level in the well the head is equal to the geometric elevation (z). Equation (9-14) states that there is no vertical flow beyond the lower boundary of the aquifer; in other words, it is impervious. Equation (9-15) states that as the distance from the well becomes large, the head does not change with the radial distance.

The drawdown of the water table is generally a small fraction of the aquifer thickness (H). If $s_{\max} = H - H_0 < 0.5H$, Eq. (9-10) may be approximated by (Boulton, 1954a)

$$\frac{\partial h}{\partial t} + \frac{K}{S_y}\frac{\partial h}{\partial z} = 0, \qquad \text{for} \quad z = H \tag{9-16}$$

Equation (9-16) states that the lowering of the water table may be replaced by changes in the head (h) along the initial water table position where radial flow components may be neglected.

The exact expression for the discharge of the well is

$$Q = 2\pi r_w K \int_0^{H_w(t)} \frac{\partial h}{\partial r} \Big|_{r=r_w} dz \tag{9-17}$$

which is approximated by (Boulton, 1954a)

$$Q = 2\pi r K H \frac{\partial h}{\partial r} \qquad r \rightarrow 0, \quad 0 \leq z \leq H \tag{9-18}$$

This means that the constant discharge of the well, instead of being received along the depth of saturation at the face of the well, is received along the original depth of saturation.

9.3.5 Solution

The solution of the above-defined problem was solved by Boulton (1954a) by using the Laplace transform, finite Fourier cosine transform, and Hankel transform techniques; and the following solution was found:

$$s(r,t) = H - h(r,t) = \frac{Q}{2\pi T}\int_0^{\infty} \frac{J_0(\beta r)}{\beta}\left\{1 - \frac{\cosh(\beta z)}{\cosh(\beta H)}\exp\left[-\frac{K}{S_y}t\beta\tanh(\beta H)\right]\right\} d\beta \tag{9-19}$$

where $T = KH$ is the transmissivity and J_0 denotes the Bessel function of the first kind of zero order. Boulton (1954a) states that it can be verified that Eq. (9-19) satisfies Eqs. (9-3), (9-11), (9-14), and (9-16), using the properties of the Bessel function. Boulton also states that the exponential term, involving the time, in Eq. (9-19) makes no contribution to the discharge into a well of vanishingly small radius, which means that Eq. (9-18) is satisfied.

Based on the previously stated assumptions, the drawdown of the water table is small. As a result, an approximation can be made by putting $z = H$ in Eq. (9-19). Since the effect of this approximation is to make the predicted drawdown too small, whereas the effect of using Eq. (9-16) instead of Eq. (9-9) is to make the drawdown to large, the errors resulting from these two approximations tend to cancel each other (Boulton, 1954a).

With the following dimensionless variables

$$\rho = \frac{r}{H}, \qquad \tau = \frac{Kt}{S_y H} \tag{9-20}$$

and substitution of $z = H$ and $\lambda = \beta H$, Eq. (9-19) can be expressed as

$$s = \frac{Q}{2\pi T}V(\rho,\tau) \tag{9-21}$$

where

$$V(\rho,\tau) = \int_0^{\infty} \frac{J_0(\lambda\rho)}{\lambda}\{1 - \exp[-\tau\lambda\tanh(\lambda)]\}\, d\lambda \tag{9-22}$$

$V(\rho,\tau)$ is the gravity well function for unconfined aquifers. Values of this function are tabulated in Table 9-1. Table 9-1 is constructed by graphical interpolation from its available form (Stallman, 1961) by Hantush (1964). The approximated forms of Eq. (9-22) are presented in the following section (Section 9.3.6).

9.3.6 Approximate Formulas for $V(\rho, \tau)$

The following approximate formulas may be used for practical purposes:

When the Time Factor τ Is Sufficiently Large. For large values of λ, $\tanh(\lambda)$ may be replaced by λ in Eq. (9-22) without appreciably changing the value of the integral. Under these conditions, the approximate formula is (Boulton 1954a; Hantush, 1964)

$$V(\rho,\tau) \cong \frac{1}{2}W\left(\frac{\rho^2}{4\tau}\right) = \frac{1}{2}W(u) \qquad \text{for } \tau > 5 \tag{9-23}$$

TABLE 9-1 Values of $V(\rho, \tau)$ for Different Values of ρ and τ

		10^{-3}									10^{-2}									
τ	ρ	1	2	3	4	5	6	7	8	9	1	2	3	4	5	6	7	8	9	10
	1	2.99	2.30	1.90	1.64	1.42	1.28	1.15	1.04	0.950	0.875	0.474	0.322	0.240	0.192	0.158	0.135	0.118	0.104	0.093
	2	3.68	2.97	2.58	2.30	2.09	1.92	1.76	1.64	1.52	1.42	0.860	0.610	0.468	0.378	0.316	0.270	0.236	0.210	0.187
	3	4.08	3.40	3.00	2.70	2.46	2.28	2.13	2.00	1.88	1.79	1.18	0.860	0.675	0.555	0.465	0.400	0.350	0.310	0.278
	4	4.35	3.68	3.26	2.98	2.75	2.58	2.42	2.29	2.17	2.06	1.42	1.07	0.850	0.710	0.600	0.525	0.460	0.410	0.368
10^{-2}	5	4.58	3.90	3.49	3.20	2.96	2.79	2.64	2.50	2.38	2.28	1.60	1.24	1.010	0.850	0.725	0.630	0.560	0.500	0.450
	6	4.76	4.06	3.65	3.36	3.15	2.96	2.80	2.68	2.56	2.45	1.78	1.40	1.15	0.970	0.840	0.735	0.650	0.585	0.530
	7	4.92	4.20	3.80	3.51	3.30	3.12	2.96	2.82	2.70	2.60	1.91	1.54	1.28	1.09	0.950	0.835	0.740	0.670	0.610
	8	5.08	4.34	3.94	3.65	3.42	3.24	3.09	2.95	2.84	2.72	2.04	1.65	1.39	1.20	1.04	0.925	0.825	0.750	0.680
	9	5.18	4.47	4.05	3.75	3.54	3.35	3.20	3.05	2.95	2.84	2.14	1.75	1.50	1.29	1.14	1.02	0.910	0.825	0.750
	1	5.24	4.54	4.14	3.85	3.63	3.45	3.30	3.15	3.04	2.94	2.25	1.85	1.58	1.38	1.22	1.09	0.985	0.890	0.815
	2	5.85	5.15	4.78	4.50	4.28	4.10	3.93	3.80	3.66	3.56	2.87	2.46	2.20	1.98	1.80	1.65	1.52	1.42	1.32
	3	6.24	5.50	5.12	4.85	4.61	4.43	4.28	4.14	4.01	3.90	3.24	2.84	2.54	2.32	2.14	1.98	1.85	1.74	1.64
	4	6.45	5.75	5.35	5.08	4.85	4.67	4.50	4.38	4.26	4.15	3.46	3.05	2.76	2.54	2.36	2.20	2.07	1.96	1.86
10^{-1}	5	6.65	6.00	5.58	5.25	5.00	4.85	4.70	4.55	4.45	4.30	3.65	3.24	2.95	2.72	2.52	2.38	2.24	2.14	2.03
	6	6.75	6.10	5.65	5.40	5.15	4.98	4.82	4.68	4.56	4.45	3.76	3.37	3.09	2.85	2.67	2.50	2.38	2.26	2.16
	7	6.88	6.20	5.80	5.50	5.25	5.08	4.92	4.80	4.68	4.55	3.90	3.50	3.20	2.99	2.80	2.64	2.50	2.38	2.28
	8	7.00	6.25	5.85	5.60	5.35	5.20	5.00	4.90	4.80	4.65	3.96	3.55	3.26	3.05	2.86	2.71	2.58	2.46	2.36
	9	7.10	6.35	6.00	5.70	5.50	5.30	5.12	5.00	4.90	4.75	4.05	3.65	3.36	3.15	2.96	2.80	2.66	2.55	2.45
	1	7.14	6.45	6.05	5.75	5.55	5.35	5.20	5.05	4.95	4.83	4.10	3.74	3.45	3.22	3.04	2.90	2.75	2.64	2.54
	2	7.60	6.88	6.45	6.15	5.92	5.75	5.60	5.50	5.35	5.25	4.59	4.18	3.90	3.68	3.50	3.34	3.20	3.09	2.97
	3	7.85	7.15	6.70	6.45	6.20	6.00	5.85	5.75	5.60	5.50	4.82	4.42	4.12	3.90	3.72	3.57	3.45	3.31	3.20
	4	8.00	7.28	6.85	6.58	6.35	6.15	6.00	5.90	5.75	5.70	4.95	4.55	4.26	4.04	3.86	3.70	3.59	3.46	3.36
	5	8.15	7.35	7.00	6.65	6.50	6.25	6.10	6.00	5.85	5.80	5.05	4.68	4.40	4.19	4.00	3.85	3.71	3.60	3.49
1	6	8.20	7.50	7.10	6.75	6.55	6.35	6.20	6.10	5.95	5.85	5.20	4.78	4.50	4.26	4.09	3.92	3.80	3.69	3.59
	7	8.25	7.55	7.15	6.85	6.62	6.40	6.30	6.20	6.05	5.95	5.25	4.85	4.58	4.35	4.18	4.00	3.90	3.78	3.66
	8	8.30	7.60	7.20	6.90	6.70	6.50	6.35	6.25	6.10	6.05	5.30	4.92	4.65	4.40	4.25	4.10	3.95	3.82	3.74
	9	8.32	7.65	7.25	7.00	6.75	6.55	6.40	6.30	6.15	6.10	5.35	5.00	4.70	4.49	4.30	4.15	4.00	3.90	3.80
	10	8.35	7.75	7.35	7.05	6.80	6.60	6.45	6.35	6.20	6.14	5.40	5.02	4.80	4.52	4.35	4.19	4.05	3.92	3.84

TABLE 9-1 (continued)

	ρ	10^{-1}									1				
τ		1	2	3	4	5	6	7	8	9	1	2	3	4	5
	1	0.093	0.0430	0.0264	0.0180	0.0132	0.0100	0.0078	0.0062	0.0049	0.0040	0.00057	0.00015		
	2	0.187	0.0865	0.0530	0.0365	0.0268	0.0205	0.0160	0.0125	0.0100	0.0081	0.00118	0.00020		
	3	0.278	0.130	0.0800	0.0550	0.0405	0.0310	0.0240	0.0190	0.0150	0.0122	0.00184	0.00032		
	4	0.368	0.174	0.107	0.0735	0.0540	0.0415	0.0322	0.0255	0.0202	0.0165	0.00244	0.00043		
10^{-2}	5	0.450	0.215	0.133	0.0920	0.0675	0.0520	0.0400	0.0320	0.0255	0.0206	0.00305	0.00055		
	6	0.530	0.257	0.160	0.110	0.0810	0.0610	0.0478	0.0380	0.0305	0.0250	0.00365	0.00065		
	7	0.610	0.298	0.186	0.130	0.0950	0.0725	0.0565	0.0450	0.0360	0.0292	0.00430	0.00078		
	8	0.680	0.340	0.214	0.148	0.108	0.0825	0.0645	0.0510	0.0412	0.0336	0.00500	0.00090		
	9	0.750	0.378	0.236	0.164	0.122	0.0930	0.0730	0.0585	0.0470	0.0380	0.00570	0.00105		
	1	0.815	0.415	0.260	0.180	0.134	0.103	0.0805	0.0640	0.0515	0.0420	0.00635	0.00118		
	2	1.32	0.750	0.500	0.359	0.268	0.208	0.165	0.132	0.107	0.0880	0.0145	0.00278		
	3	1.64	1.02	0.700	0.515	0.392	0.308	0.246	0.200	0.164	0.135	0.0238	0.00490		
	4	1.86	1.22	0.870	0.650	0.510	0.405	0.328	0.268	0.220	0.182	0.0350	0.00750	0.00160	0.00038
10^{-1}	5	2.03	1.37	1.00	0.770	0.610	0.490	0.400	0.330	0.275	0.230	0.0450	0.0104	0.00240	0.00056
	6	2.16	1.49	1.12	0.875	0.700	0.570	0.468	0.390	0.325	0.276	0.0580	0.0138	0.00320	0.00080
	7	2.28	1.60	1.22	0.965	0.775	0.640	0.525	0.445	0.375	0.320	0.0715	0.0175	0.00425	0.00108
	8	2.36	1.69	1.30	1.04	0.850	0.715	0.600	0.500	0.425	0.364	0.0840	0.0212	0.00525	0.00140
	9	2.45	1.75	1.38	1.11	0.920	0.775	0.650	0.550	0.475	0.404	0.0980	0.0260	0.00630	0.00165
	1	2.54	1.85	1.45	1.18	0.975	0.825	0.700	0.595	0.510	0.444	0.113	0.0310	0.00840	0.00235
	2	2.97	2.29	1.88	1.60	1.38	1.22	1.07	0.950	0.840	0.750	0.259	0.0950	0.0330	0.0115
	3	3.20	2.50	2.10	1.82	1.60	1.42	1.28	1.15	1.05	0.960	0.388	0.165	0.0700	0.0275
	4	3.36	2.66	2.25	1.97	1.75	1.58	1.42	1.30	1.20	1.10	0.495	0.235	0.112	0.0535
	5	3.49	2.78	2.38	2.09	1.87	1.69	1.54	1.42	1.30	1.21	0.580	0.300	0.150	0.0715
1	6	3.59	2.90	2.47	2.18	1.95	1.78	1.65	1.52	1.40	1.30	0.660	0.360	0.195	0.0990
	7	3.66	2.96	2.55	2.25	2.04	1.85	1.70	1.58	1.48	1.38	0.730	0.415	0.230	0.125
	8	3.74	3.00	2.60	2.32	2.11	1.94	1.79	1.66	1.55	1.44	0.790	0.465	0.272	0.155
	9	3.80	3.09	2.67	2.39	2.17	2.00	1.85	1.72	1.60	1.50	0.850	0.515	0.307	0.182
	10	3.84	3.12	2.74	2.45	2.24	2.05	1.90	1.77	1.65	1.55	0.890	0.550	0.340	0.210

For $\tau > 5, V(\tau,\rho) \approx 0.5W\left(\rho^2/4\tau\right)$

For $\tau < 0.01, V(\tau,\rho) \approx \sinh^{-1}(\tau/\rho) - \dfrac{\tau}{(1+\rho^2)^{1/2}}$

For $\tau < 0.01$ and $\tau/\rho > 10, V(\tau,\rho) \approx \sinh^{-1}(1/\rho) - \ln\left(\dfrac{\tau+1}{\tau}\right)$

Source: After Stallman (1961) and Hantush (1964).

where

$$u = \frac{S_y r^2}{4Tt} \tag{9-24}$$

and $W(u)$, as defined by Eq. (4-25) of Chapter 4, is the *Theis well function* and is tabulated in Table 4-1 of Chapter 4.

When the Time Factor τ Is Small. For small values of time ($\tau < 0.05$), the function $V(\rho, \tau)$ is closely given by the following equation (Boulton, 1954a):

$$V(\rho, \tau) \cong \sinh^{-1}\left(\frac{1}{\rho}\right) + \sinh^{-1}\left(\frac{\tau}{\rho}\right) - \sinh\left(\frac{1+\tau}{\rho}\right) \qquad \text{for} \quad \tau < 0 \tag{9-25}$$

The function may also be approximated for the following cases (Hantush, 1964):

$$V(\rho, \tau) \cong \sinh^{-1}\left(\frac{\tau}{\rho}\right) - \frac{\tau}{(1+\rho^2)^{1/2}} \qquad \text{for} \quad \tau < 0.01 \tag{9-26}$$

and

$$V(\rho, \tau) \cong \ln\left(\frac{2\tau}{\rho}\right) \qquad \text{for} \quad \rho < 0.01, \quad \frac{\tau}{\rho} > 10 \tag{9-27}$$

9.3.7 Accuracy and Limitations of the Drawdown Equation: Method of Correction

9.3.7.1 Modified Drawdown Equation for Water Table. To minimize the errors involved in Eq. (9-21), a correction factor C_f is applied (Boulton, 1954a; Hantush, 1964). With this, Eq. (9-21) takes the form

$$s = \frac{Q}{2\pi T}(1 + C_f)V(\rho, \tau) \qquad \text{for} \quad s_{\max} = H - H_0 < 0.5H \tag{9-28}$$

The correction factor, C_f, depends on ρ, τ, r_w/H, and Q/H^2. The value of the correction factor, C_f, ranges from about -0.30 to 0.16. The values of C_f can be determined from Figure 9-4 for $\tau < 0.05$ and from Figure 9-5 for $\tau > 5$. For $0.05 < \tau < 5$, C_f may be assumed to be zero with an error not greater than 6%.

9.3.7.2 Equation of Drawdown in the Pumped Well. If well losses are neglected, the drawdown in the pumped well may be approximated as follows (Boulton, 1954a; Hantush, 1964):

(a) **For $\tau < 0.05$:** The drawdown in the well is given by Eq. (9-28) after replacing H and ρ by H_0 and r_w/H, respectively.
(b) **For $0.05 < \tau < 5$:** The drawdown may be computed from

$$s_w = H - H_0 = \frac{Q}{2\pi T}\left[m + \ln\left(\frac{H}{r_w}\right)\right] \tag{9-29}$$

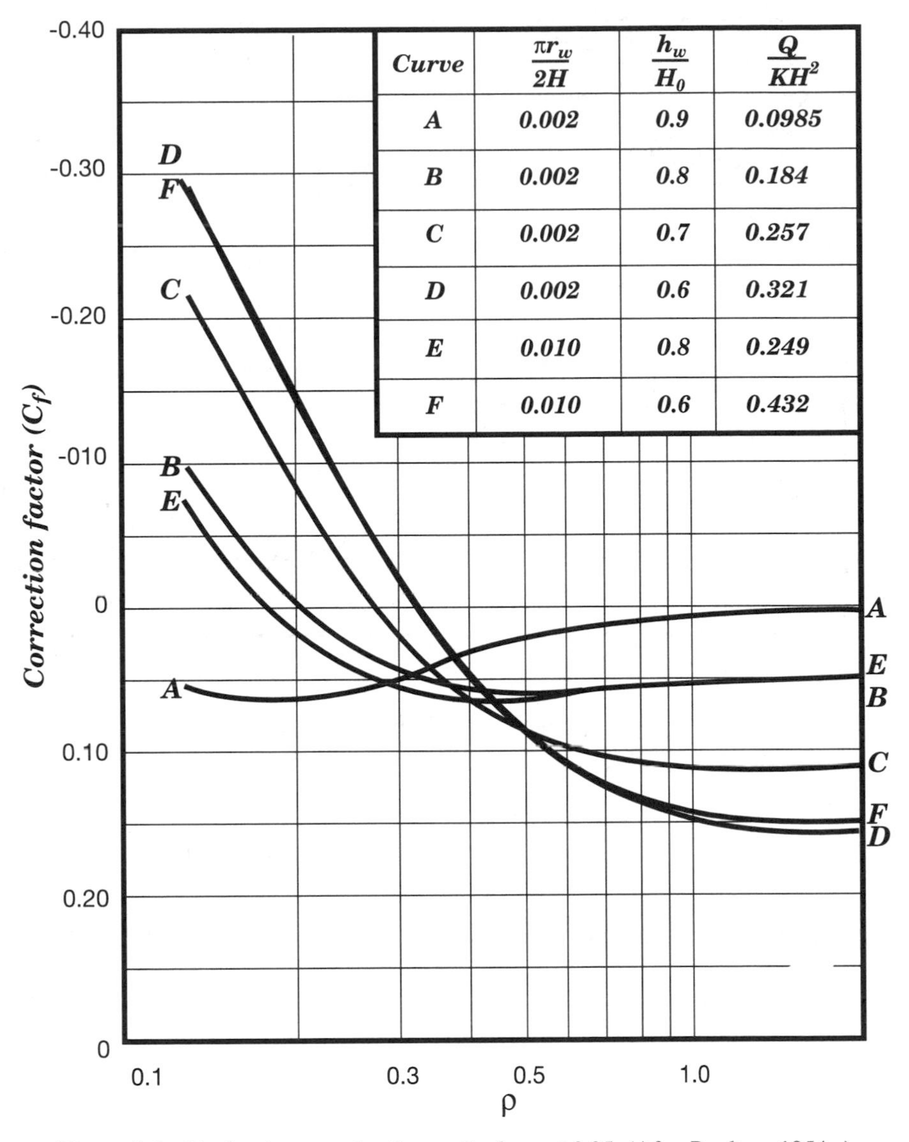

Curve	$\frac{\pi r_w}{2H}$	$\frac{h_w}{H_0}$	$\frac{Q}{KH^2}$
A	0.002	0.9	0.0985
B	0.002	0.8	0.184
C	0.002	0.7	0.257
D	0.002	0.6	0.321
E	0.010	0.8	0.249
F	0.010	0.6	0.432

Figure 9-4 Boulton's correction factor, C_f, for $\tau < 0.05$. (After Boulton, 1954a.)

where m is a function of τ and can be computed from a curve plotted through the points given in Table 9-2 (Boulton, 1954a).

(c) **For $\tau > 5$:** The expression to be used is

$$H^2 - H_0^2 = \frac{Q}{\pi K} \ln \left[\frac{1.5H\tau^{1/2}}{r_w} \right] \tag{9-30}$$

9.3.8 Recovery Equation

If τ and τ' are the time factors, reckoned from the commencement of pumping and from the end, respectively, the drawdown of the water table during recovery can be written from

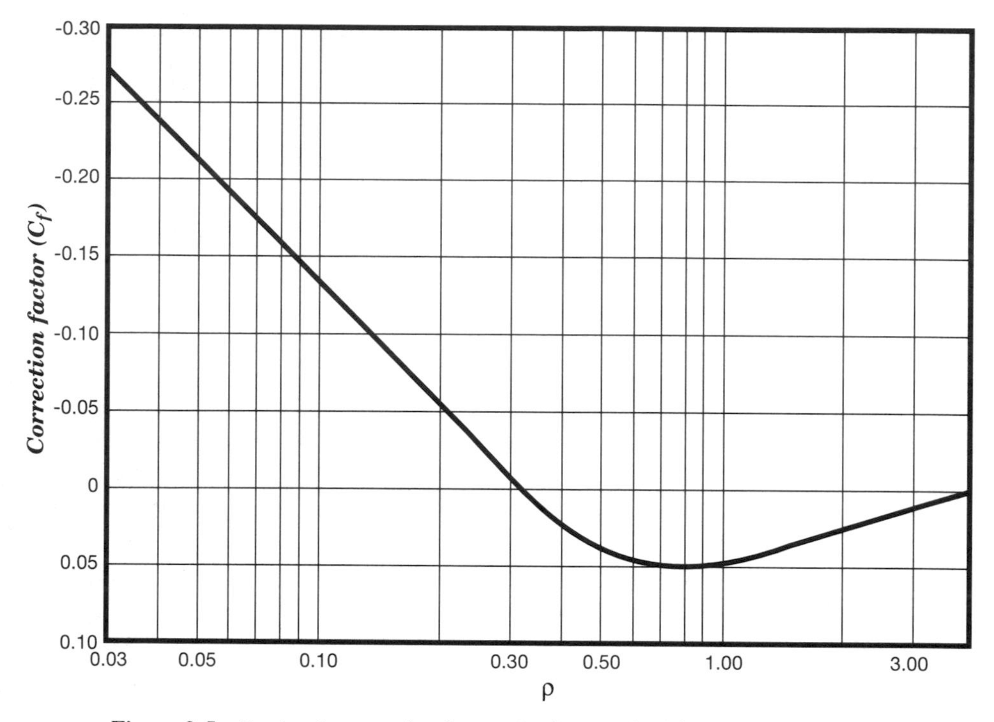

Figure 9-5 Boulton's correction factor, C_f, for $\tau > 5$. (After Boulton, 1954a.)

Eq. (9-21) as (Boulton, 1954a; Hantush, 1964)

$$s = \frac{Q}{2\pi T}[V(\rho, \tau) - V(\rho, \tau')] \tag{9-31}$$

With correction factors, as used in Eq. (9-28), Eq. (9-31) takes the form

$$s = \frac{Q}{2\pi T}[(1 + C_f)V(\rho, \tau) - (1 + C_f')V(\rho, \tau')] \tag{9-32}$$

where C_f' is the correction factor during recovery.

9.3.9 Examples

The usage of the equations presented in the previous sections is shown some examples below.

TABLE 9-2 Values of *m* as a Function of τ

τ	m
0.05	−0.043
0.20	0.087
1.00	0.512
5.00	1.228

Source: After Boulton (1954a).

Example 9-1: The data of this example is adapted from Boulton (1954a). The following data are given for an unconfined aquifer and well: $r_w = 0.30$ m; $H = 244$ m; $K = 0.53$ m/day; $S_y = 0.15$; and $Q = 2450$ m^3/day. Determine the drawdown at $r = 50$ m distance at $t = 3 \times 10^5$ s $= 3.47$ days.

Solution: From Eq. (9-20), the values of ρ and τ are

$$\rho = \frac{r}{H} = \frac{50\text{ m}}{244\text{ m}} = 0.205$$

$$\tau = \frac{Kt}{S_y H} = \frac{(0.53\text{ m/day})(3.47\text{ days})}{(0.15)(244\text{ m})} = 0.05$$

From Table 9-1, $V = 0.215$. Also

$$\frac{\pi r_w}{2H} = \frac{\pi(0.30\text{ m})}{2(244\text{ m})} = 0.0019$$

$$\frac{Q}{KH^2} = \frac{2450\text{ m}^3\text{/day}}{(0.53\text{ m/day})(244\text{ m})^2} = 0.0776$$

Hence, from Curve A of Figure 9-4, $C_f \approx 0.06$. From Eq. (9-28), the drawdown at $r = 50$-m distance is

$$s = \frac{2450\text{ m}^3\text{/day}}{(2\pi)(0.53 \times 244\text{ m}^2\text{/day})}(1 + 0.06)(0.215) = 0.69\text{ m}$$

The drawdown of the pumped well itself can be found from Eq. (9-29) with $m = -0.043$ (Table 9-2):

$$s_w = \frac{2450\text{ m}^3\text{/day}}{(2\pi)(0.53 \times 244\text{ m}^2\text{/day})}\left[-0.043 + \ln\left(\frac{244\text{ m}}{0.30\text{ m}}\right)\right] = 20.08\text{ m}$$

This value is smaller than $0.5H$, which is a requirement of Eq. (9-28).

Example 9-2: Data for a thin and very permeable aquifer are given as $r_w = 0.30$ m, $H = 30$ m, $K = 25$ m/day, $S_y = 0.20$, and $Q = 10{,}000$ m^3/day. Determine the drawdowns at $r = 6$-m distance and at the pumped well itself at $t = 1.16$ days.

Solution: From Eq. (9-20)

$$\rho = \frac{r}{H} = \frac{6\text{ m}}{30\text{ m}} = 0.2$$

$$\tau = \frac{Kt}{S_y H} = \frac{(25\text{ m/day})(1.16\text{ days})}{(0.20)(30\text{ m})} = 4.83$$

From Table 9-1, $V \approx 2.78$. Because $0.05 < \tau = 4.83 < 5$, C_f may be assumed to be zero. Therefore, from Eq. (9-28) the drawdown at $r = 6$-m distance is

$$s = \frac{10{,}000\text{ m}^3\text{/day}}{(2\pi)(25 \times 30\text{ m}^2\text{/day})}(1 + 0.0)(2.78) = 5.9\text{ m}$$

From Eq. (9-29), with $m = 1.228$ (Table 9-2), the drawdown at the pumped well itself is

$$s_w = \frac{10{,}000 \text{ m}^3\text{/day}}{(2\pi)(25 \times 30 \text{ m}^2\text{/day})}\left[1.228 + \ln\left(\frac{30 \text{ m}}{0.30 \text{ m}}\right)\right] = 12.4 \text{ m}$$

9.4 BOULTON'S MODEL FOR A FULLY PENETRATING WELL IN UNCONFINED AQUIFERS WITH THE DELAYED RESPONSE

9.4.1 Problem Statement and Assumptions

As mentioned previously, Boulton (1954b) was the first to introduce the delayed response concept of unconfined aquifers under pumping conditions. Boulton extended the theory of unsteady radial flow to a pumped well in a confined aquifer by including the effect of delayed yield from storage in unconfined aquifers. The geometry of the problem is shown in Figure 9-3. The assumptions are as follows:

1. The assumptions listed in Section 9.3.2.
2. The aquifer is unconfined but shows delayed yield or the aquifer is semi-unconfined.
3. The flow to the well is in unsteady state.
4. The diameter of the well is small; that is, the storage in the well can be neglected.

9.4.2 Governing Equations

Without delayed gravity drainage, the governing equation of flow in cylindrical coordinates under unconfined aquifer conditions is (Boulton, 1954b)

$$T\left(\frac{\partial^2 s}{\partial r^2} + \frac{1}{r}\frac{\partial s}{\partial r}\right) = S\frac{\partial s}{\partial t} \tag{9-33}$$

This equation is the same equation for confined aquifers as given by Eq. (4-10) of Chapter 4.

Boulton (1954b, 1963) assumed that the amount of water derived from storage within an unconfined aquifer, due to an increment of drawdown Δs between times τ and $\tau + \Delta\tau$ since the beginning of pumping, consists of two components:

1. A volume $S\Delta s$ of water instantaneously released from storage per horizontal area, in which S is the storage coefficient under confined conditions.
2. A delayed yield from storage, per unit horizontal area, at any time $t(t > \tau)$ from the start of pumping

   $$\Delta s \alpha S_y e^{-\alpha(t-\tau)}$$

   where e is the base of natural logarithm and $\alpha(T^{-1})$ is an empirical constant. From this, Boulton expressed the total volume of delayed gravity drainage per unit area per unit drawdown by the following equation:

$$\alpha S_y \int_\tau^\infty e^{-\alpha(t-\tau)}\, dt = S_y \tag{9-34}$$

Thus, the total effective storage coefficient is

$$S + S_y = \eta S \tag{9-35}$$

It is convenient to write this expression as

$$\eta = 1 + \frac{S_y}{S} \tag{9-36}$$

From the above expressions, Boulton's partial integrodifferential equation for unsteady-state flow to a fully penetrating well in an unconfined aquifer is (Boulton, 1963)

$$T\left(\frac{\partial^2 s}{\partial r^2} + \frac{1}{r}\frac{\partial s}{\partial r}\right) = S\frac{\partial s}{\partial t} + \alpha S_y \int_0^t \frac{\partial s}{\partial t} e^{-\alpha(t-\tau)}\, dt \tag{9-37}$$

where the last term on the right-hand side expresses the rate of delayed drainage per unit area at time t.

As mentioned above, Boulton identified α as an empirical constant. But its physical meaning remains unclear. This situation is further elaborated by Neuman (1979, p. 899): "Although Boulton's model appears to fit data quite well, it nevertheless fails to provide insight into the physical nature of the delayed yield phenomenon. Since the exponential rate of delay assumed by Boulton is hypothetical, his model is not related explicitly to the actual physics of the delay process. Furthermore, the model does not specify any definite relationship between α (controlling the rate of delay) and other physical characteristics of the aquifer system. This is not to imply that Boulton neglected to speculate about the physical nature of delayed yield. On the contrary, he postulated three possible mechanisms for the process. . . ." The Boulton's empirical constant (α) was also included in a model developed by Moench (1995) by combining the Boulton and Neuman models (see Section 9.5) for flow to a well in an unconfined aquifer. Regarding this constant, Moench (1995, p. 379) states: "No attempt is made to quantitatively relate the empirical constant to physically defined parameters in the zone above or near the water table. Assumptions required to make such an association are deemed unjustified. . . . The inclusion of an empirical constant to partially account for effects of delayed yield is found to improve the match between theoretical type curves and field data. . . ."

9.4.3 Initial and Boundary Conditions

Because the water table is assumed to be horizontal, the initial condition for the drawdown, $s(r, t)$, is

$$s(r, 0) = 0, \qquad r > 0 \tag{9-38}$$

The boundary condition at infinity is

$$s(\infty, t) = 0, \qquad \text{for all } t \tag{9-39}$$

From Eq. (4-14) of Chapter 4, the boundary condition for the constant discharge rate at the well is:

$$\lim_{r \to 0} \left(r \frac{\partial s}{\partial r} \right) = -\frac{Q}{2\pi T} \tag{9-40}$$

Equation (9-38) states that initially the drawdown is zero everywhere in the aquifer. Equation (9-39) states that as the distance from the pumped well approaches infinity, the drawdown is zero. Finally, Eq. (9-40) states that near the pumping well the flow rate toward the well is equal to its discharge rate.

9.4.4 Solution

The integrodifferential equation, as given by Eq. (9-37), with the unusual integral term, was solved with the preceding initial and boundary conditions by Boulton (1963), by using the Laplace transform method along with the application of the Faltung theorem (Carslaw and Jaeger, 1959), which is also known as the *convolution property* (Spiegel, 1971), and the following solution was found:

$$s(r,t) = \frac{Q}{4\pi T} \int_0^\infty \frac{2}{x} \left\{ 1 - e^{-\mu_1} \left[\cosh(\mu_2) + \frac{\alpha t \eta \left(1 - x^2\right)}{2\mu_2} \sinh(\mu_2) \right] \right\} J_0 \left(\frac{r}{\nu D} x \right) dx \tag{9-41}$$

where

$$\mu_1 = \frac{1}{2} \alpha t \eta (1 - x^2) \tag{9-42}$$

$$\mu_2 = \frac{1}{2} \alpha t \left[\eta^2 \left(1 + x^2\right)^2 - 4\eta x^2 \right]^{1/2} \tag{9-43}$$

$$\nu = \left(\frac{\eta - 1}{\eta} \right)^{1/2} = \left(\frac{S_y}{S + S_y} \right)^{1/2} \tag{9-44}$$

$$D = \left(\frac{T}{\alpha S_y} \right)^{1/2} \tag{9-45}$$

and J_0 denotes the Bessel function of the first kind of zero order. Equation (9-41) is the general solution when η has a finite value, and its two limiting cases are presented below.

9.4.4.1 *Solution When η Tends to Infinity.*

For the case when η is large and tends to infinity, Eq. (9-41) can be reduced to (Boulton, 1963)

$$s(r,t) = \frac{Q}{4\pi T} \int_0^\infty 2 J_0 \left(\frac{r}{D} x \right) \left[1 - \frac{1}{x^2 + 1} \exp \left(- \frac{\alpha t x^2}{x^2 + 1} \right) - F \right] \frac{dx}{x} \tag{9-46}$$

where

$$F = \frac{x^2}{x^2 + 1} \exp \left[-\alpha \eta t \left(x^2 + 1\right) \right] \tag{9-47}$$

9.4.4.2 *Solution for Short Times.*

The function F, which is defined by Eq. (9-47), vanishes when $t > 0$, but is finite as t approaches zero and ηt approaches a finite value. As a result, for small values of t, Eq. (9-46) reduces to (Boulton, 1963)

$$s(r,t) = \frac{Q}{4\pi T} \int_0^\infty 2J_0\left(\frac{r}{D}x\right) \frac{x^2}{x^2+1} \left\{1 - \exp\left[-\alpha\eta t\left(x^2+1\right)\right]\right\} \frac{dx}{x} \tag{9-48}$$

Equation (9-48) is identical, with the exception of notation, to the equation of unsteady radial flow in an infinite leaky confined aquifer as given by Eq. (6-17) of Chapter 6, originally derived by Hantush and Jacob (1955). This can be shown as follows: Using the integral relation (Hantush and Jacob, 1955)

$$K_0\left(\frac{r}{b}\right) = \int_0^\infty \frac{\alpha}{\alpha^2+1} J_0\left(\frac{\alpha r}{B}\right) d\alpha \tag{9-49}$$

an alternative form of Eq. (6-17) of Chapter 6 is

$$s(r,t) = \frac{Q}{4\pi T}\left\{2K_0\left(\frac{r}{b}\right) - \int_0^\infty \frac{2\alpha}{\alpha^2+1} J_0\left(\frac{\alpha r}{b}\right) \exp\left[-\frac{Tt}{SB^2}\left(\alpha^2+1\right)\right]\right\} d\alpha \tag{9-50}$$

The combination of Eqs. (9-49) and (9-50) is identical, except for notation, to Eq. (9-48). Therefore, Eq. (9-48) is identical to Eq. (6-17) of Chapter 6.

9.4.5 Boulton's Well Function for Unconfined Aquifers

Equation (9-46) can be expressed as

$$s(r,t) = \frac{Q}{4\pi T} W\left(\alpha t, \frac{r}{D}\right) \tag{9-51}$$

where

$$W\left(\alpha t, \frac{r}{D}\right) = \int_0^\infty 2J_0\left(\frac{r}{D}x\right)\left[1 - \frac{1}{x^2+1}\exp\left(-\frac{\alpha t x^2}{x^2+1}\right) - F\right]\frac{dx}{x} \tag{9-52}$$

From Eq. (9-45) we obtain

$$\frac{r}{D} = r\left(\frac{\alpha S_y}{T}\right)^{1/2} \tag{9-53}$$

Introducing the well-known dimensionless variable, as expressed by Eq. (4-22) of Chapter 4, we obtain

$$u_y = \frac{r^2 S_y}{4Tt} \tag{9-54}$$

and thus Eq. (9-53) takes the form

$$\alpha t = \left(\frac{r}{D}\right)^2 \frac{1}{4u_y} \tag{9-55}$$

Equation (9-55) is applicable for large values of t. A similar variable may be defined for small values of t as

$$u_a = \frac{r^2 S}{4Tt} \tag{9-56}$$

From Eqs. (9-35), (9-54), (9-55), and (9-56) we obtain

$$u_y = u_a(\eta - 1) = \left(\frac{r}{D}\right)^2 \frac{1}{4\alpha t} \tag{9-57}$$

With the equations given above, Eq. (9-51) can be expressed as

$$s(r,t) = \frac{Q}{4\pi T} W\left(u_{ay}, \frac{r}{D}\right) \tag{9-58}$$

$W(u_{ay}, r/D)$ is the *Boulton's well function for fully penetrating unconfined aquifers* for the particular case when η tends to infinity. Values of this function are given in Table 9-3 in terms of the practical range of u_a, u_y, and r/D (Boulton, 1963).

Equation (9-58) and its associated components, as given by Eqs. (9-54) and (9-56), are given in the gallon–day–foot system of units as

$$s(r,t) = \frac{114.6Q}{T} W(u_{ay}, r/D) \tag{9-59}$$

where

$$u_a = \frac{2693 r^2 S}{Tt} \tag{9-60}$$

$$u_y = \frac{2693 r^2 S_y}{Tt} \tag{9-61}$$

$$\frac{r}{D} = 104r \left(\frac{\alpha S_y}{T}\right)^{1/2} \tag{9-62}$$

In Eqs. (9-59) through (9-62), s is expressed in feet, r in feet, Q in gallons per minute (gpm), t in minutes, T in gallons per day per foot (gpd/ft), and α in min^{-1}.

If the delay index, $1/\alpha$, is very large and D goes to infinity, then Eq. (9-58) reduces to the form of the Theis equation as given by Eq. (4-24) of Chapter 4.

9.4.6 Application to Pumping Test Data Analysis

9.4.6.1 Features of the Boulton Solution. Boulton (1963) proposed to use Eq. (9-58) for analyzing pumping test data for unconfined aquifers under transient conditions. Prickett (1965) presented a comprehensive evaluation and application of the Boulton's model. Values of $W(u_{ay}, r/D)$ in Table 9-3 are plotted against values of $1/u_a$ and $1/u_y$ on separate log–log paper as shown in Figures 9-6 and 9-7 for different values of r/D. These curves are called "Type A curves" and "Type B curves," respectively. The Type A curves are essentially the same as the set of leaky confined aquifer type curves,

TABLE 9-3 Values of $W(u_{ay}, r/D)$

$1/u_a = N_n \times 10^n$

$r/D = 0.01$			$r/D = 0.1$			$r/D = 0.2$			$r/D = 0.316$			$r/D = 0.4$			$r/D = 0.6$		
N	n	$W(u_a, r/D)$	N	n	$W(u_a, r/D)$	N	n	$W(u_a, r/D)$	N	n	$W(u_a, r/D)$	N	n	$W(u_a, r/D)$	N	n	$W(u_a, r/D)$
1	1	1.82	1	1	1.80	5	0	1.19	1	0	0.216	1	0	0.213	1	0	0.206
1	2	4.04	5	1	3.24	1	1	1.75	2	0	0.544	2	0	0.534	2	0	0.504
1	3	6.31	1	2	3.81	5	1	2.95	5	0	1.153	5	0	1.114	5	0	0.996
5	3	7.82	2	2	4.30	1	2	3.29	1	1	1.655	1	1	1.564	1	1	1.311
1	4	8.40	5	2	4.71	5	2	3.50	5	1	2.504	5	1	2.181	2	1	1.493
1	5	9.42	1	1	4.83	1	3	3.51	1	2	2.623	1	2	2.225	5	1	1.553
1	6	9.44	1	4	4.85				1	3	2.648	1	3	2.229	1	2	1.555

$r/D = 0.8$			$r/D = 1.0$			$r/D = 1.5$			$r/D = 2.0$			$r/D = 2.5$			$r/D = 3.0$		
N	n	$W(u_a, r/D)$	N	n	$W(u_a, r/D)$	N	n	$W(u_a, r/D)$	N	n	$W(u_a, r/D)$	N	n	$W(u_a, r/D)$	N	n	$W(u_a, r/D)$
5	−1	0.046	5	−1	0.0444	5	−1	0.0394	3.33	−1	0.0100	5	−1	0.0271	5	−1	0.0210
1	0	0.197	1	0	0.1855	1	0	0.1509	5	−1	0.0335	1	0	0.0803	1	0	0.0534
2	0	0.466	2	0	0.421	1.25	0	0.199	1	0	0.114	1.25	0	0.0961	1.25	0	0.0607
5	0	0.857	5	0	0.715	2	0	0.301	1.25	0	0.144	2	0	0.1174	2	0	0.0681
1	1	1.050	1	1	0.819	5	0	0.413	2	0	0.194	5	0	0.1247	5	0	0.0695
2	1	1.121	2	1	0.841	1	1	0.427	5	0	0.227	1	1	0.1247	1	1	0.0695
5	1	1.131	5	1	0.842	2	1	0.428	1	1	0.228						

$1/u_y = N_n \times 10^n$

$r/D = 0.01$			$r/D = 0.1$			$r/D = 0.2$			$r/D = 0.316$			$r/D = 0.4$			$r/D = 0.6$		
N	n	$W(u_y, r/D)$	N	n	$W(u_y, r/D)$	N	n	$W(u_y, r/D)$	N	n	$W(u_y, r/D)$	N	n	$W(u_y, r/D)$	N	n	$W(u_y, r/D)$
4	2	9.45	4	0	4.86	4	−1	3.51	4	−1	2.66	1	−1	2.23	4.44	−1	1.586
4	3	9.54	4	1	4.95	4	0	3.54	4	0	2.74	1	0	2.26	2.22	0	1.707
4	4	10.23	4	2	5.64	1	1	3.69	4	1	3.38	5	0	2.40	4.44	0	1.844
4	5	12.31	4	3	7.72	4	1	3.85	4	2	5.42	1	1	2.55	1.67	1	2.448
4	6	14.61	4	4	10.01	1.5	2	4.55	4	3	7.72	3.75	1	3.20	4.44	1	3.255
						4	2	5.42				1	2	4.05			

$r/D = 0.8$			$r/D = 1.0$			$r/D = 1.5$			$r/D = 2.0$			$r/D = 2.5$			$r/D = 3.0$		
N	n	$W(u_y, r/D)$	N	n	$W(u_y, r/D)$	N	n	$W(u_y, r/D)$	N	n	$W(u_y, r/D)$	N	n	$W(u_y, r/D)$	N	n	$W(u_y, r/D)$
2.5	−2	1.133	4	−2	0.844	7.11	−2	0.444	4	−2	0.239	2.56	−2	0.1321	1.78	−2	0.0743
2.5	−1	1.158	4	−1	0.901	3.55	−1	0.509	2	−1	0.283	1.28	−1	0.1617	8.89	−2	0.0939
1.25	0	1.264	4	0	1.356	7.11	−1	0.587	4	−1	0.337	2.56	−1	0.1988	1.78	−1	0.1189
2.5	0	1.387	4	1	3.140	2.67	0	0.963	1.5	0	0.614	9.6	−1	0.3990	6.67	−1	0.2618
9.37	0	1.938				7.11	0	1.569	4	0	1.111	2.56	0	0.7977	1.78	0	0.5771
2.5	1	2.704															

Source: After Boulton (1963).

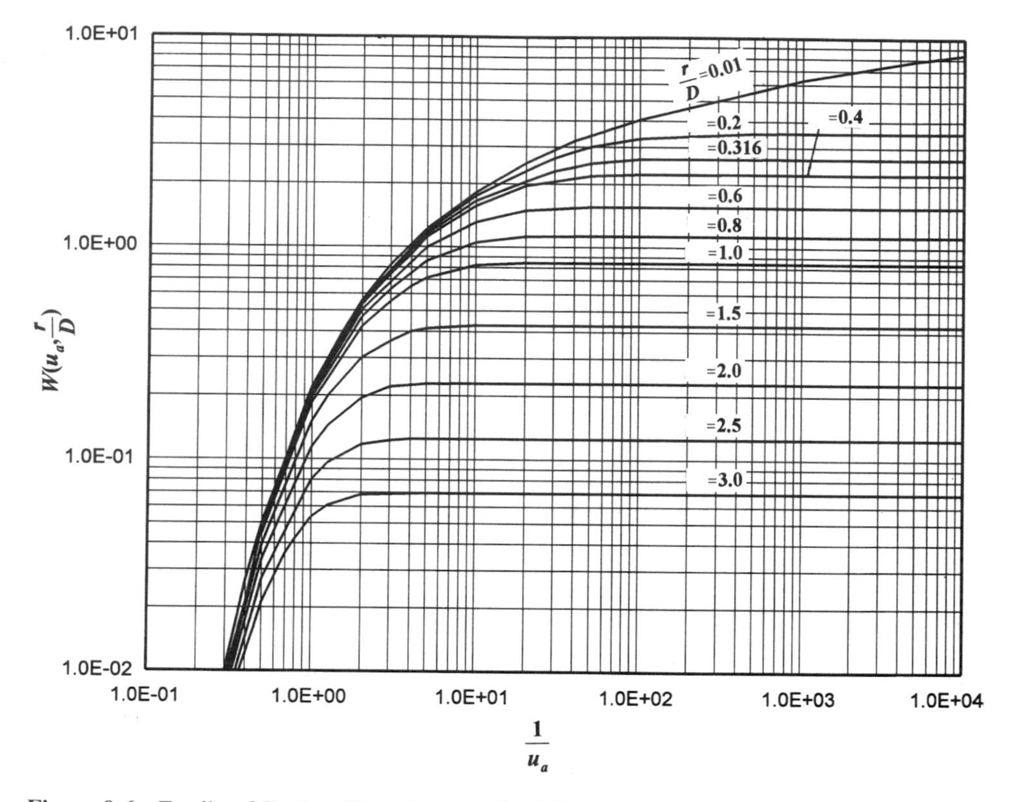

Figure 9-6 Family of Boulton Type A curves for fully penetrating wells in unconfined aquifers: $W(u_a, r/D)$ versus $1/u_a$.

as given by Figure 6-2 of Chapter 6, because of the equivalent form of the corresponding drawdown equations (see Section 9.4.4.2). The Type A curves are used to analyze early drawdown versus time data, whereas the Type B curves are used to analyze late drawdown versus time data.

The right-hand of each Type A curve and the left-hand portion of each Type B curve in Figures 9-6 and 9-7, respectively, approach a horizontal asymptote given by [see Eq. (3-30) of Chapter 3]

$$W\left(u_{ay}, \frac{r}{D}\right) = 2K_0\left(\frac{r}{D}\right) = \frac{4\pi T s}{Q} \tag{9-63}$$

where $K_0(r/D)$ denotes the modified Bessel function of the second kind of zero order (its values are given in Table 3-2 of Chapter 3). Eq. (9-63) represents a steady state drawdown distribution which is identical with that for the leaky confined aquifer solution when t is sufficiently large. One should kept in mind that the definitions of D in Eq. (9-63) and B in Eq. (3-30) of Chapter 3 are different. Equation (9-46) is derived on the assumption that η is infinite, for which case the drawdown of the water table to the value given by Eq. (9-58) would theoretically occur instantaneously at the start of pumping. For the practical case in which η is large but not infinite, type curves for small values of $1/u_y$ cannot be shown

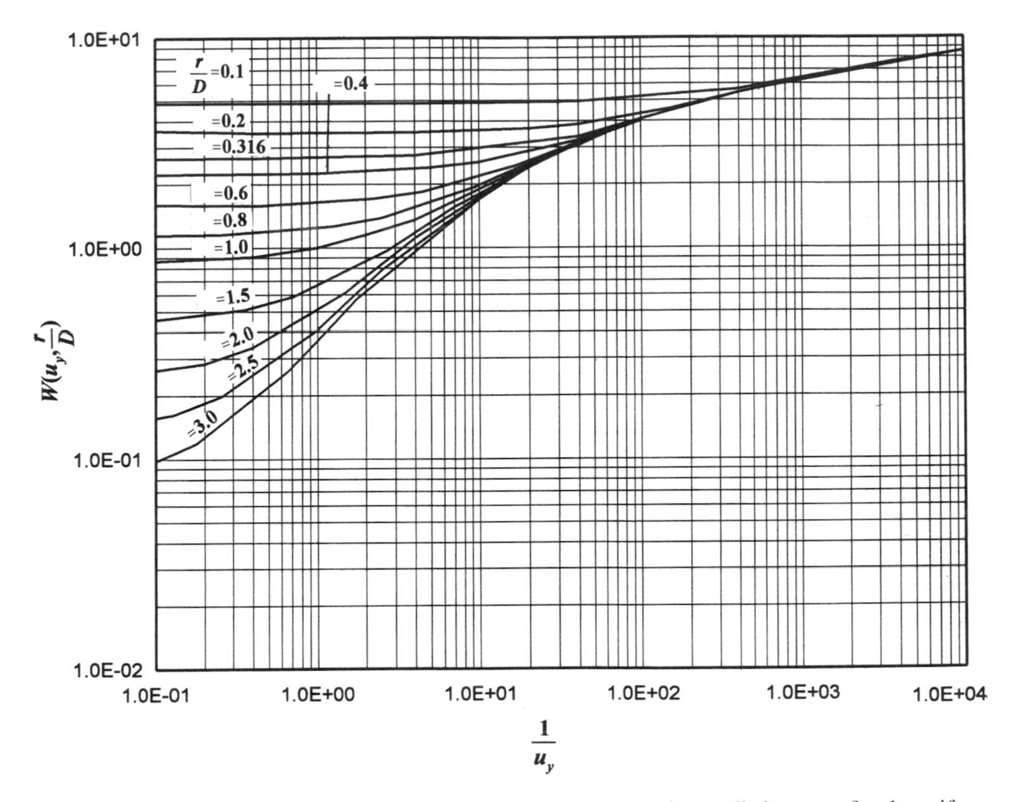

Figure 9-7 Family of Boulton Type B curves for fully penetrating wells in unconfined aquifers: $W(u_y, r/D)$ versus $1/u_y$.

on a base of $1/u_y$ before the value of η is known. Therefore, the early time values of $W(u_{ay}, r/D)$ are shown on a base of $1/u_a = (\eta - 1)/u_y$ determined from Eq. (9-57).

9.4.6.2 *Delay Index.* It is evident from Figure 9-7 that as $1/u_y$ increases the Type B curve for any particular value of r/D becomes indistinguishable from the Theis type curve (the Theis curve almost coalesces with the curve of $r/D = 0.01$). In other words, the Type B curves eventually merge with the lower envelope of the family. The value of $1/u_y$ of Type B curves, being proportional to the time t_{wt} at which the effects of gravity drainage cease to influence the drawdown, can be used to calculate the value of t_{wt} by using the known or assumed values of T and S_y. Boulton (1963) gave a curve for αt_{wt} versus r/D that can be used to calculate the values of t_{wt}. The curve to estimate the time t_{wt} when delayed gravity drainage ceases to influence drawdown is given in Figure 9-8. The values of αt_{wt} are such that the departure parallel to the $W(u_{ay}, r/D)$ axis of any Type B curve from the Theis curve is 0.02, which is negligible for all practical purposes. Boulton (1963) plotted the curve in Figure 9-8 from points $(r/D, \alpha t)$ by determining them from the calculated values of W near the right-hand Theis type curve, and then he applied graphical interpolation to determine the constant departure of 0.02.

The value of $1/\alpha$ (in minutes or hours) is significant when calculating the time at which delayed yield ceases to be effective. Because of this, Boulton (1963) called this quantity

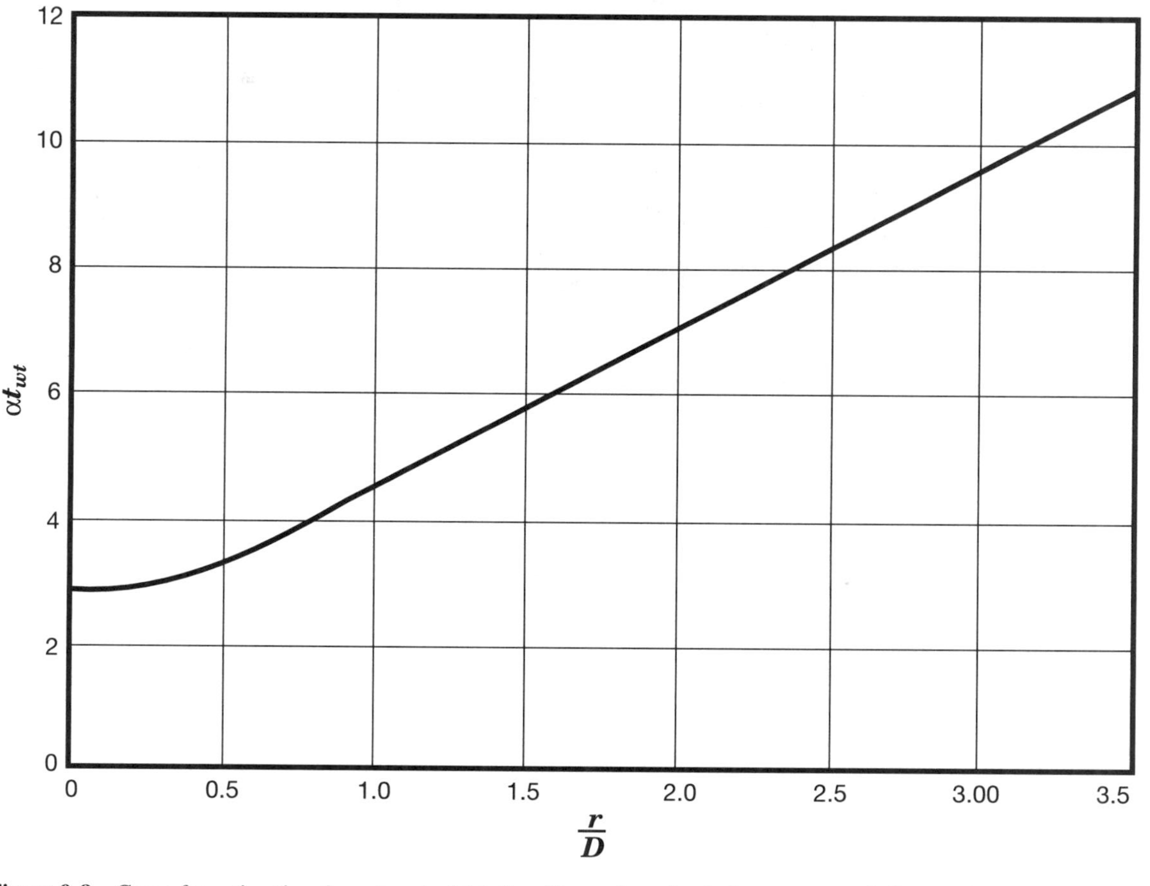

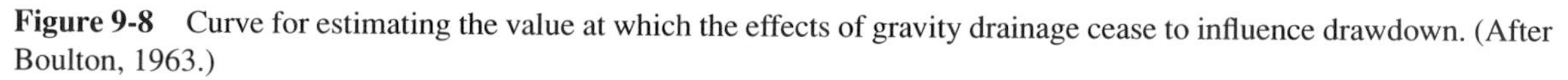

Figure 9-8 Curve for estimating the value at which the effects of gravity drainage cease to influence drawdown. (After Boulton, 1963.)

delay index. It must be pointed out that for the estimation of t_0 the constants T, S_y, and α (or D and α) and the distance r must be known or assumed.

9.4.6.3 Boulton Type-Curve Method. The purpose of the Boulton type-curve method is to find the values of T, S, S_y, η, and α. The steps of the method are given below (Boulton, 1963):

Step 1: Plot the family of Boulton type curves of $W(u_{ay}, r/D)$ versus $1/u_a$ and $1/u_y$ for a practical range of values of r/D separately on log–log papers using the data in Table 9-3.

Step 2: Calculate the ratio of $s_{\max}/b$ for each observation well. Based on this ratio, if the maximum drawdown is a significant portion of the aquifer thickness (b), calculate the corrected drawdowns for only late data using Jacob's drawdown correction (see Section 3.4.3.1 of Chapter 3) scheme as given by

$$s_c = s_{wt} - \frac{s_{wt}^2}{2b} \tag{9-64}$$

which is Eq. (3-64) of Chapter 3. Here, s_c is the corrected drawdown and s_{wt} is the drawdown of the water table (refer also to Section 3.4.3.1 of Chapter 3 for the Jacob's correction scheme). On another sheet of log–log paper with the same scale as the ones in Step 1, plot the values of drawdown s versus time t for a single observation well at a distance r from the pumped well. Repeat this for all other observation wells.

Step 3: Keeping the coordinate axes parallel at all times—that is, $W(u_{ay}, r/D)$ axis parallel with the s axis and the $1/u_a$ axis parallel with the t axis—first superimpose the time versus drawdown data curve on Type A curves. Adjust until as much as possible early data curve fall on one of the Type A curves. Note the r/D value of the selected Type A curve.

Step 4: Select an arbitrary point A on the overlapping of the two sheets of paper, or even anywhere on the overlapping portion of the sheets. Note the values of the coordinates of the match point s, $1/u_a$, $W(u_{ay}, r/D)$, and t.

Step 5: Substitute these values as well as the known Q value, first into Eq. (9-58) and then into Eq. (9-56), and then calculate the T and S values. Then, calculate the horizontal hydraulic conductivity value from $K_r = T/b$, where b is the average aquifer thickness.

Step 6: Move the observed drawdown versus time curve until as much as possible of the late data fall on the Type B curve with the same r/D value for the Type A curve.

Step 7: Select an arbitrary B point on the superimposed curves (this point may also be selected anywhere on the overlapping portion of the sheets) and note the values of s, $1/u_y$, $W(u_{ay}, r/D)$, and t for the match point B.

Step 8: Substitute these values, first into Eq. (9-58) and then into Eq. (9-54), and then calculate the T and S_y values. The calculated value of Step 5 and this step should give approximately the same value for T. Then, calculate the horizontal hydraulic conductivity value from $K_r = T/b$.

Step 9: Substitute the values of S and S_y into Eq. (9-36) and determine the value of η.

Step 10: Calculate the value of α by first determining the value of D from the known value of r/D from Step 3 and the corresponding value of r and subsequently substituting the values of B, S_y, and T into Eq. (9-45). Calculate the reciprocal of α, $1/\alpha$, which is the delay index.

Step 11: As mentioned in Section 9.4.6.2, eventually the effects of delayed gravity drainage become negligibly small, and Type B curve merges with the Theis curve. Determine the merging point of the Type B curve for a particular value of r/D by measuring the value of αt_{wt} for the particular value of r/D on the vertical axis of the *Boulton delay index curve* as given in Figure 9-8. Because $1/\alpha$ is known, calculate the value of t_{wt} from the known value of αt_{wt}.

Step 12: Repeat the procedure for other observation wells. Calculations for T, S, and S_y for different observation wells should give approximately the same results.

Important Points. Some important points about the Boulton type-curve method are given below (Kruseman and De Ridder, 1991):

1. For values $\eta > 100$, the geometric slope of the line joining to the corresponding Type A and Type B curves is essentially zero. For values $10 < \eta < 100$, the geometric slope of this line is small and can be approximated by a line tangent to both curves. The points on the observed data curve that could not be matched with Type A or Type B should fall along this tangent (Boulton, 1963).
2. If delayed yield effects are not apparent, the observed drawdown data curve will fall completely along the left-hand Theis type curve.
3. If sufficient observations are made after the delayed yield has ceased to influence the drawdown versus time curve, the observed data for which $t > t_{wt}$, together with the right-hand Theis curve, can be used to calculate the T and S_y values.
4. If the Boulton type-curve method is applied to pumping tests in semi-unconfined aquifers, it gives no information about the hydrogeologic properties of the underlain or overlain layer, because D is a function of the properties of the unconfined aquifer.

Example 9-3: The data of this example is taken from Kruseman and De Ridder (1991). The drawdown versus time data given in Table 9-4 correspond to an observation well at r = 90-m distance from the pumped well. The steady-state flow rate is $Q = 873\ \text{m}^3/\text{day}$. The average saturated thickness of the aquifer before pumping was 22 m. Using the Boulton's type-curve method, determine the values of transmissivity (T), horizontal hydraulic conductivity (K_r), storage coefficient (S), specific yield (S_y), delay index ($1/\alpha$), time coordinate of the merging point of the drawdown versus time curve (t_{wt}), and η.

Solution:

Step 1: The family of Boulton type curves of $W(u_{ay}, r/D)$ versus $1/u_a$ and $1/u_y$ on log–log papers are given in Figures 9-6 and 9-7, respectively.

Step 2: The observed drawdown versus time data curve is given in Figure 9-9 using the values in Table 9-4. The maximum drawdown ($s_{\max}$) in Table 9-4 is 0.204 m and the initial aquifer thickness (b) is 22 m. Therefore, $s_{\max}/b = 0.204\ \text{m}/22\ \text{m} = 0.009$. This means that, drawdown correction is not required since the maximum drawdown is not a significant fraction of the initial aquifer thickness.

Step 3: The left-hand portion of the observed drawdown versus time data curve as given in Figure 9-9 is adjusted Type A curves (Figure 9-6), and $r/D = 0.6$ established the best match with the observed data. The matched curves are shown in Figure 9-10.

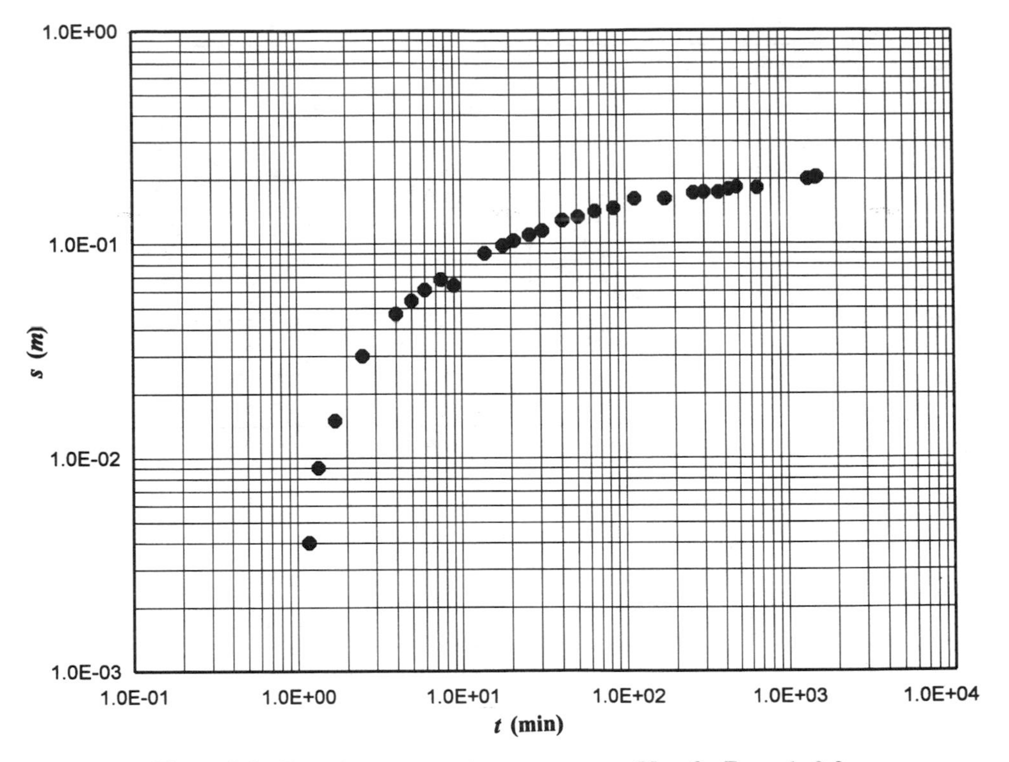

Figure 9-9 Drawdown versus time curve at $r = 90$ m for Example 9-3.

TABLE 9-4 Drawdown Versus Time Data at $r = 90$ m for Example 9-3

t (min)	s (m)	t (min)	s (m)
0.0	0.0	41.0	0.128
1.17	0.004	51.0	0.133
1.34	0.009	65.0	0.141
1.7	0.015	85.0	0.146
2.5	0.030	115.0	0.161
4.0	0.047	175.0	0.161
5.0	0.054	260.0	0.172
6.0	0.061	300.0	0.173
7.5	0.068	370.0	0.173
9.0	0.064	430.0	0.179
14.0	0.090	485.0	0.183
18.0	0.098	665.0	0.182
21.0	0.103	1340.0	0.200
26.0	0.110	1490.0	0.203
31.0	0.115	1520.0	0.204

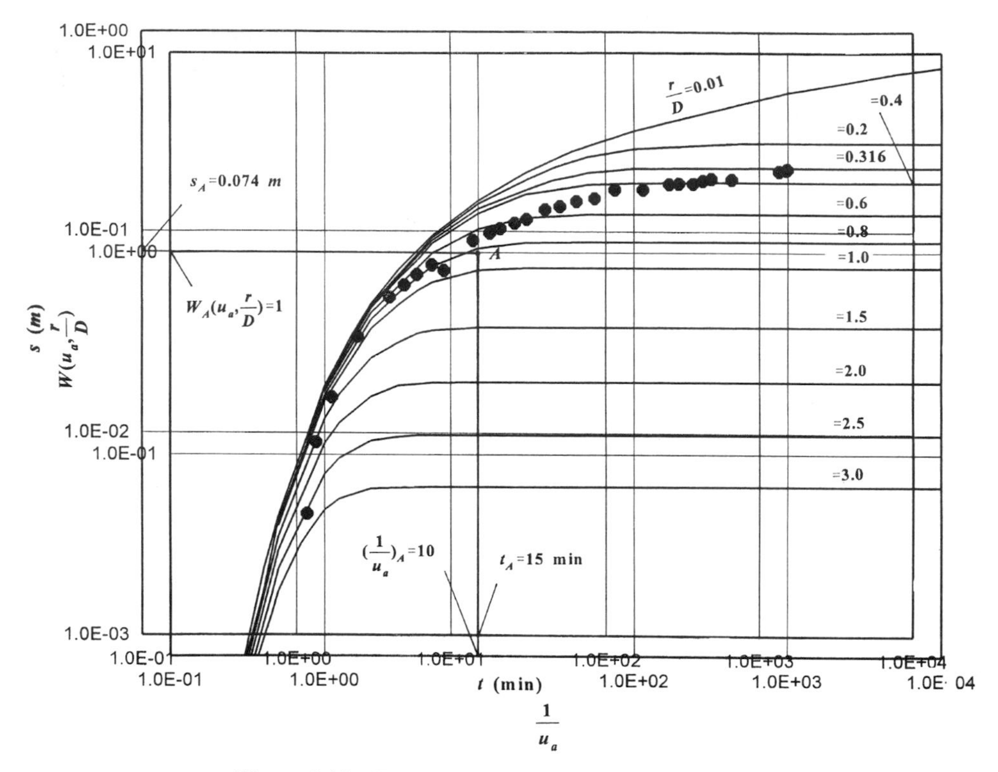

Figure 9-10 Type A curve matching for Example 9-3.

Step 4: A match point (A) is chosen in Figure 9-10. This point is represented by the following coordinates:

$$\left(\frac{1}{u_a}\right)_A = 10, \qquad W_A\left(u_a, \frac{r}{D}\right) = 1$$

$$s_A = 0.074 \text{ m}, \qquad t_A = 15 \text{ min} = 1.04 \times 10^{-2} \text{ days}$$

Step 5: Using Eq. (9-58), the transmissivity (T) value and the K_r value can be calculated as

$$T = \frac{Q}{4\pi s_A} W\left(u_a, \frac{r}{D}\right) = \frac{873 \text{ m}^3/\text{day}}{(4\pi)(0.074 \text{ m})}(1) = 939 \text{ m}^2/\text{day}$$

$$K_r = \frac{T}{b} = \frac{939 \text{ m}^2/\text{day}}{22 \text{ m}} = 42.68 \text{ m/day} = 4.94 \times 10^{-2} \text{ cm/s}$$

From Eq. (9-56), the storage coefficient value is

$$S = \frac{4Tt_A u_{a_A}}{r^2} = \frac{(4)(939 \text{ m}^2/\text{day})\left(1.04 \times 10^{-2} \text{ days}\right)(0.1)}{(90 \text{ m})^2} = 4.8 \times 10^{-4}$$

Step 6: Now, the right-hand portion of the observed drawdown versus time curve is superimposed on the Type B Boulton type curves (Figure 9-7) and adjusted for $r/D = 0.6$. The matched curves are shown in Figure 9-11.

Step 7: A match point (B) is chosen in Figure 9-11. This point is represented by the following coordinates:

$$\left(\frac{1}{u_y}\right)_B = 1, \qquad W_B\left(u_y, \frac{r}{D}\right) = 1$$

$$s_B = 0.1 \text{ m}, \qquad t_B = 230 \text{ min} = 0.16 \text{ days}$$

Step 8: Like before, Eq. (9-58) gives the transmissivity value, and then the horizontal hydraulic conductivity value from $K_r = T/b$ can be calculated as

$$T = \frac{Q}{4\pi s_B} W_B\left(u_y, \frac{r}{D}\right) = \frac{873 \text{ m}^3/\text{day}}{(4\pi)(0.1 \text{ m})}(1) = 695 \text{ m}^2/\text{day}$$

$$K_r = \frac{T}{b} = \frac{695 \text{ m}^2/\text{day}}{22 \text{ m}} = 31.59 \text{ m/day} = 3.66 \times 10^{-2} \text{ cm/s}$$

From Eq. (9-54), the specific yield value is

$$S_y = \frac{4Tt_B u_{y_B}}{r^2} = \frac{(4)\left(695 \text{ m}^2/\text{day}\right)(0.16 \text{ days})}{(90 \text{ m})^2}(1) = 5.5 \times 10^{-2}$$

Step 9: From Eq. (9-36), the value of η is

$$\eta = 1 + \frac{S_y}{S} = 1 + \frac{5.5 \times 10^{-2}}{4.8 \times 10^{-4}} = 116$$

Step 10: From $r/D = 0.6$ we obtain

$$D = \frac{r}{0.6} = \frac{90 \text{ m}}{0.6} = 150 \text{ m}$$

From Eq. (9-45), the value of α is

$$\alpha = \frac{T}{S_y D^2} = \frac{695 \text{ m}^2/\text{day}}{\left(5.5 \times 10^{-2}\right)(150 \text{ m})^2} = 0.56 \text{ day}^{-1}$$

Step 11: Because $r/D = 0.6$, from the Boulton's delay index curve (Figure 9-8) we obtain

$$\alpha t_{wt} = 3.6$$

and, therefore,

$$t_{wt} = \frac{3.6}{\alpha} = \frac{3.6}{0.56 \text{ day}^{-1}} = 6.4 \text{ days}$$

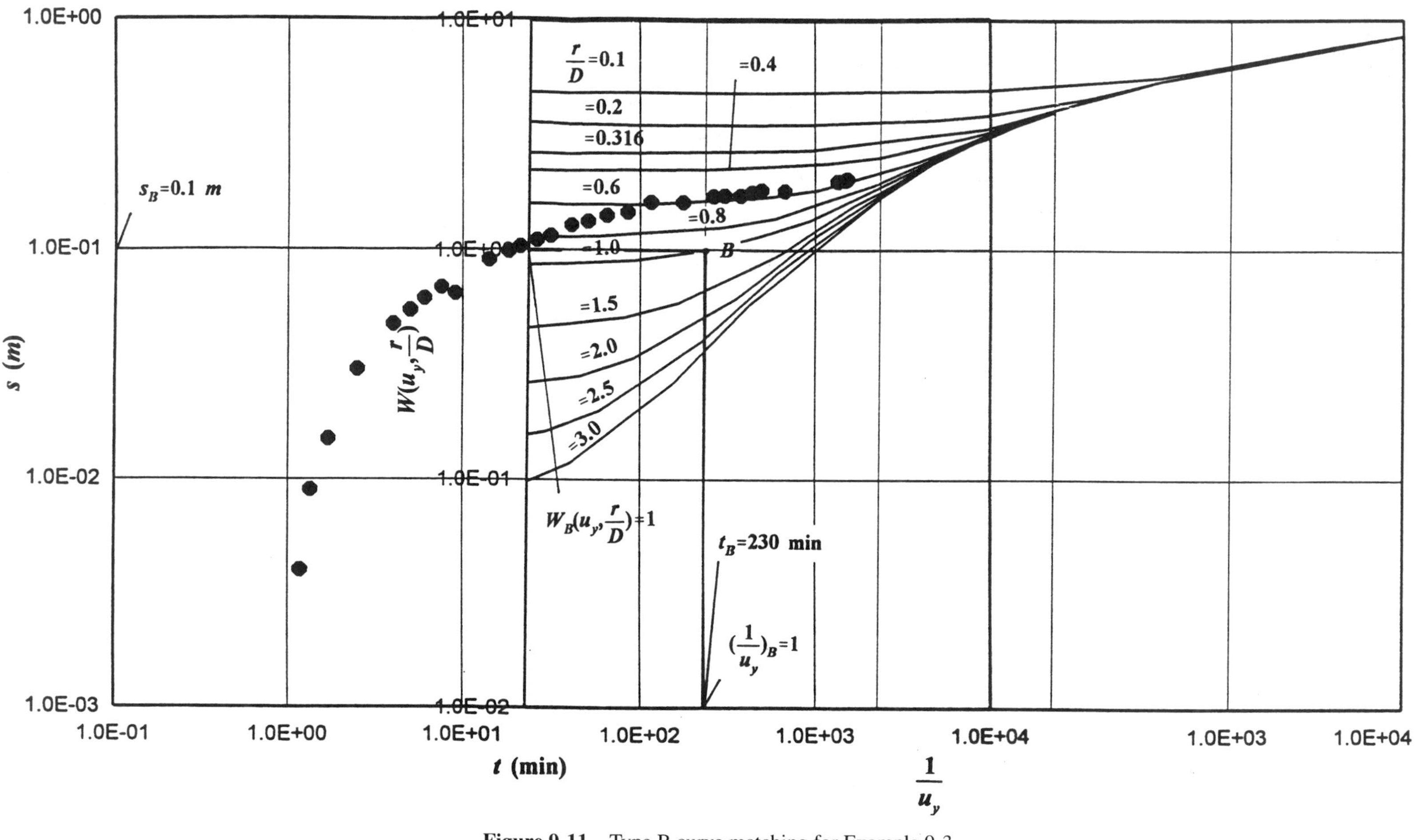

Figure 9-11 Type B curve matching for Example 9-3.

Example 9-4: This example is adapted from Prickett (1965). A pumping test was conducted for an aquifer under water-table (or unconfined aquifer) conditions for a period of about 24 hours at a constant rate of 1000 gpm (gallons per minute) (192,513 ft^3/day). The observed drawdown data given in Table 9-5 corresponds to a well at $r = 200$ ft from the pumped well. The average thickness of the aquifer (b) before pumping was around 80 ft. Using the Boulton's type-curve method, determine the values of transmissivity (T), storage coefficient (S), specific yield (S_y), delay index ($1/\alpha$), the time coordinate of the merging point of the drawdown versus time curve (t_{wt}), and η.

Solution:

Step 1: The family of Boulton type curves of $W(u_{ay}, r/D)$ versus $1/u_a$ and $1/u_y$ on log–log papers are given in Figures 9-6 and 9-7, respectively.

Step 2: The observed drawdown versus time data curve is given in Figure 9-12 using the values in Table 9-5. The maximum drawdown ($s_{\max}$) in Table 9-5 is 1.90 ft and the initial aquifer thickness (b) is 80 ft. Therefore, $s_{\max}/b = 1.90 \text{ ft}/80 \text{ ft} = 0.024$. This means that drawdown correction is not required since the maximum drawdown is not a significant fraction of the initial aquifer thickness.

Step 3: The drawdown versus time data curve (Figure 9-12) is superimposed on the family of Boulton Type A curves (Figure 9-6) and adjusted. The matched curves are shown in

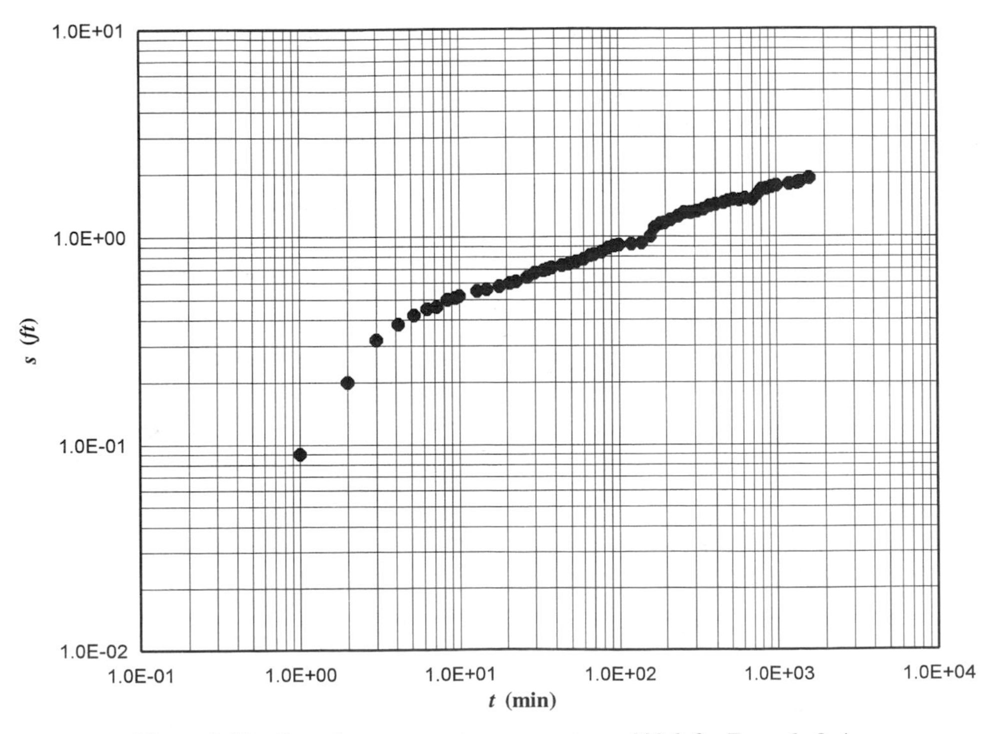

Figure 9-12 Drawdown versus time curve at $r = 200$ ft for Example 9-4.

TABLE 9-5 Drawdown Versus Time Data at r = 200 ft for Example 9-4

t (min)	s (ft)	t (min)	s (ft)
1	0.09	120	0.92
2	0.20	140	0.93
3	0.32	160	1.00
4.15	0.38	170	1.10
5.20	0.42	185	1.15
6.30	0.45	197	1.16
7.20	0.46	215	1.20
8.45	0.50	240	1.25
9.40	0.51	260	1.30
10	0.52	285	1.30
13	0.55	310	1.32
15	0.56	340	1.35
18	0.58	370	1.40
21	0.60	400	1.42
23	0.61	450	1.45
27	0.64	480	1.48
30	0.67	520	1.50
34	0.69	570	1.49
36	0.70	620	1.52
38	0.71	700	1.50
44	0.73	760	1.60
48.5	0.74	800	1.68
54	0.76	870	1.69
60	0.78	930	1.72
66	0.81	1000	1.75
70	0.82	1200	1.78
78	0.84	1350	1.80
86	0.88	1400	1.82
93	0.90	1600	1.90
100	0.91		

Figure 9-13, which shows that an $r/D = 0.8$ type curve establishes a good match with the observed data curve.

Step 4: A match point (A) is chosen in Figure 9-13. This point is represented by the following coordinates:

$$\left(\frac{1}{u_a}\right)_A = 1, \qquad W_A\left(u_a, \frac{r}{D}\right) = 0.2$$

$$s_A = 0.1 \text{ ft}, \qquad t_A = 1 \text{ min} = 6.94 \times 10^{-4} \text{ days}$$

Step 5: Using Eq. (9-58), the transmissivity (T) value and the horizontal hydraulic conductivity (K_r) can be calculated as

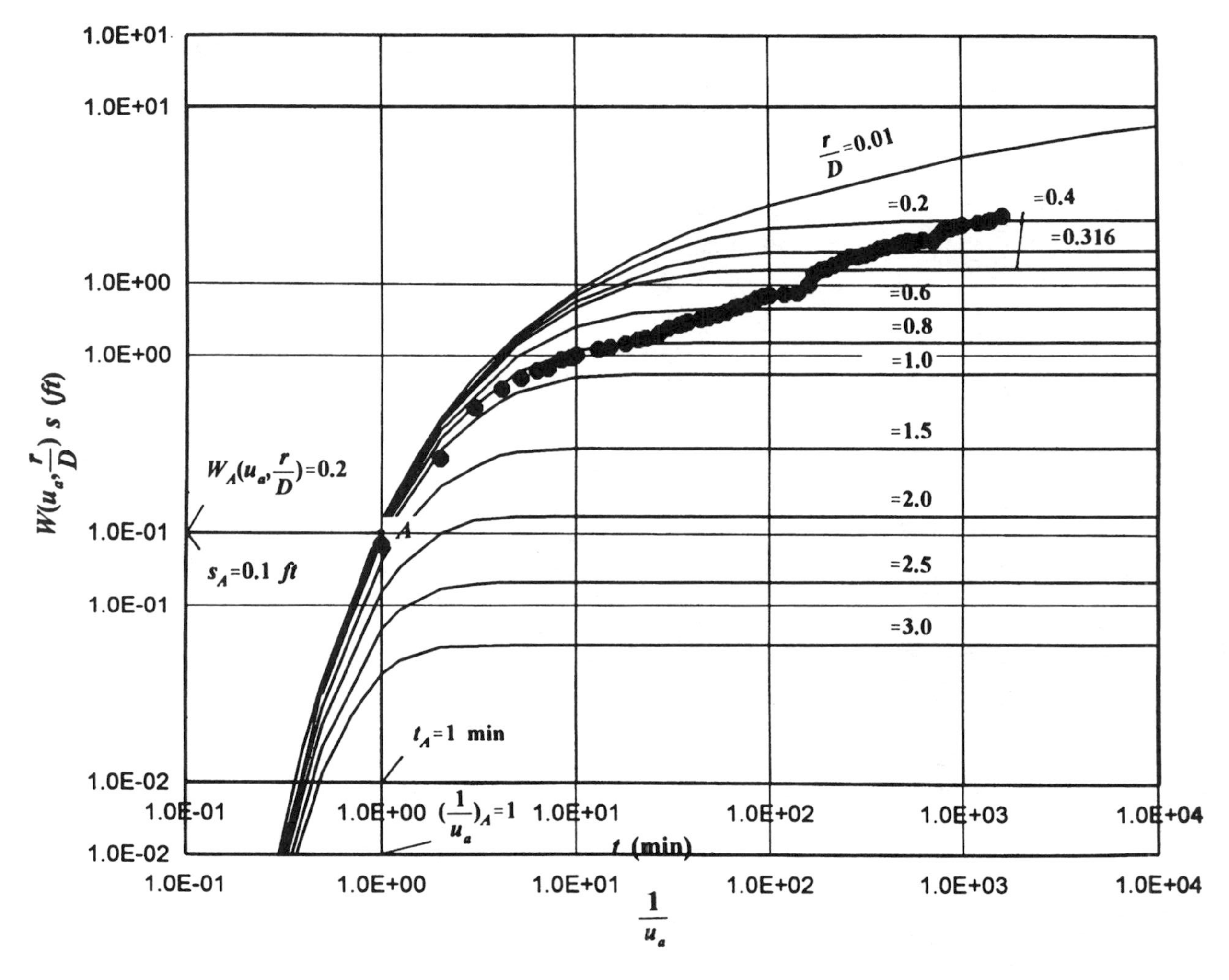

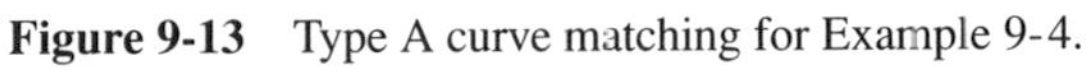

Figure 9-13 Type A curve matching for Example 9-4.

$$T = \frac{Q}{4\pi s_A} W\left(u_a, \frac{r}{D}\right) = \frac{192{,}513 \text{ ft}^3/\text{day}}{(4\pi)(0.1 \text{ ft})}(0.2) = 30{,}369 \text{ ft}^2/\text{day}$$

$$K_r = \frac{T}{b} = \frac{30{,}609 \text{ ft}^2/\text{day}}{80 \text{ ft}} = 382.61 \text{ ft/day} = 0.13 \text{ cm/s}$$

From Eq. (9-56), the storage coefficient value is

$$S = \frac{4Tt_A u_{a_A}}{r^2} = \frac{(4)\left(30{,}369 \text{ ft}^2/\text{day}\right)\left(6.94 \times 10^{-4} \text{ days}\right)(1)}{(200 \text{ ft})^2} = 21.0 \times 10^{-4}$$

Step 6: The right-hand portion of the drawdown versus time curve (Figure 9-12) is superimposed on the Boulton Type B curves (Figure 9-7) for $r/D = 0.8$, and the matched curves are shown in Figure 9-14.

Step 7: A match point (B) is chosen in Figure 9-14 and its coordinates are as follows:

$$\left(\frac{1}{u_y}\right)_B = 1, \qquad W_B\left(u_y, \frac{r}{D}\right) = 0.1$$

$$s_B = 0.038\text{ft}, \qquad t_B = 6 \text{ min} = 4.17 \times 10^{-3} \text{ days}$$

Step 8: Like before, Eq. (9-58) gives the transmissivity value, and then the horizontal hydraulic conductivity value from $K_r = T/b$ can be calculated as

$$T = \frac{Q}{4\pi s_B} W_B\left(u_y, \frac{r}{D}\right) = \frac{192{,}513 \text{ ft}^3/\text{day}}{(4\pi)(0.038 \text{ ft})}(0.1) = 40{,}315 \text{ ft}^2/\text{day}$$

$$K_r = \frac{T}{b} = \frac{40{,}315 \text{ ft}^2/\text{day}}{80 \text{ ft}} = 503.94 \text{ ft/day} = 0.18 \text{ ft/day}$$

From Eq. (9-54), the specific yield value is

$$S_y = \frac{4Tt_B u_{y_B}}{r^2} = \frac{(4)\left(40{,}315 \text{ ft}^2/\text{day}\right)\left(4.17 \times 10^{-3} \text{ days}\right)(1)}{(200 \text{ ft})^2} = 0.017$$

Step 9: From Eq. (9-36), the value of η is

$$\eta = 1 + \frac{S_y}{S} = 1 + \frac{0.017}{0.0021} = 9.1$$

Step 10: From $r/D = 0.8$ we obtain

$$D = \frac{r}{0.8} = \frac{200 \text{ ft}}{0.8} = 250 \text{ ft}$$

From Eq. (9-45), the value of α is

$$\alpha = \frac{T}{S_y D^2} = \frac{30{,}369 \text{ ft}^2/\text{day}}{(0.017)(250 \text{ ft})^2} = 28.58 \text{ day}^{-1} = 1.98 \times 10^{-2} \text{ min}^{-1}$$

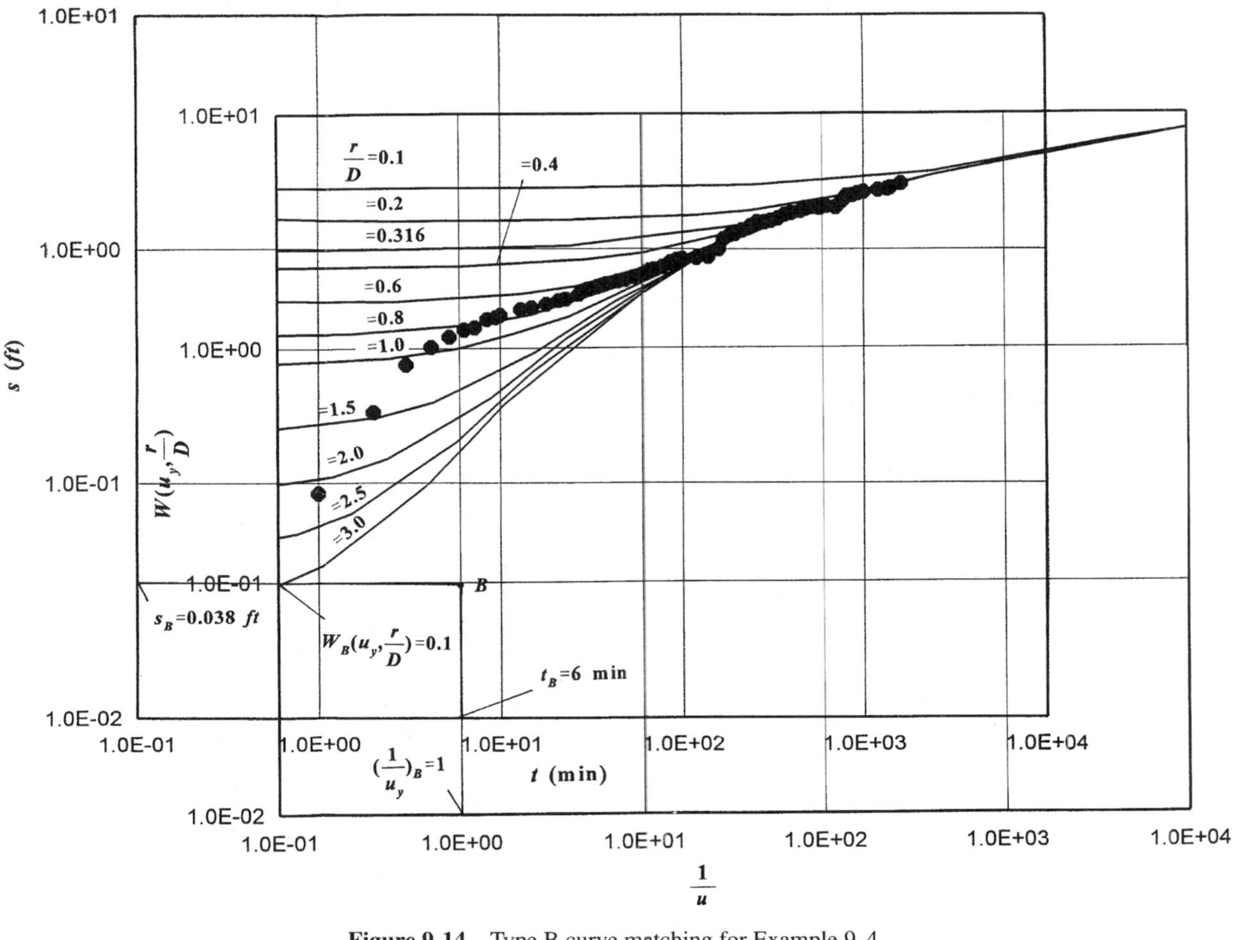

Figure 9-14 Type B curve matching for Example 9-4.

Step 11: Because $r/D = 0.8$, from the Boulton's delay index curve (Figure 9-8) we obtain

$$\alpha t_{wt} = 4.0$$

and, therefore,

$$t_{wt} = \frac{4.0}{\alpha} = \frac{3.6}{1.98 \text{ day}^{-2}} = 202 \text{ min} = 0.14 \text{ days}$$

9.5 NEUMAN'S MODELS FOR FULLY AND PARTIALLY PENETRATING WELLS IN UNCONFINED AQUIFERS WITH THE DELAYED RESPONSE

9.5.1 Introduction

Neuman (1972) used a different approach than Boulton (1954a, 1963), which is presented in Section 9.4, in developing an analytical model for a fully penetrating well in unconfined aquifers. As mentioned in Section 9.4.2, the Boulton model is capable of reproducing all three segments of the delayed yield process, which is described in Section 9.2; but the physical meaning of α is unclear. The main difference between the two models is that the Neuman model is based on well-defined physical parameters of the aquifer system. Later, Neuman (1974) extended the same mathematical approach to partially penetrating wells in unconfined aquifers.

The Neuman approach leads to a solution that can reproduce all three segments of a characteristic drawdown versus time curve in Figure 9-2 without using any empirical constants and without considering the unsaturated zone. This approach allows for the development of vertical gradients and aquifer anisotropy, and it includes the Theis model (Theis, 1935) and Boulton model (Boulton, 1954a) as special cases. Because this approach applies to both extraction of water from wells and injection of water to wells, the *delayed yield* is replaced by a more general concept called *delayed unconfined aquifer response* (Neuman, 1972). In this section, Neuman's fully and partially penetrating well models as well as their application to pumping test data analysis will be presented in detail.

9.5.2 Neuman's Fully Penetrating Wells Model

9.5.2.1 Problem Statement and Assumptions. The geometry of a fully penetrating pumping and observation wells in an unconfined aquifer is given in Figure 9-15. The assumptions are as follows (Neuman, 1972):

1. The pumping rate (Q) of the well is constant.
2. The diameter of the well is infinitesimally small, and the well fully penetrates the aquifer thickness.
3. The aquifer remains saturated at all times, and Darcy's law applies.
4. The unconfined aquifer is infinite lateral extent and rests on an impermeable horizontal layer.
5. The aquifer material is homogeneous but anisotropic, and its principal hydraulic conductivities are oriented parallel to the coordinate axes.

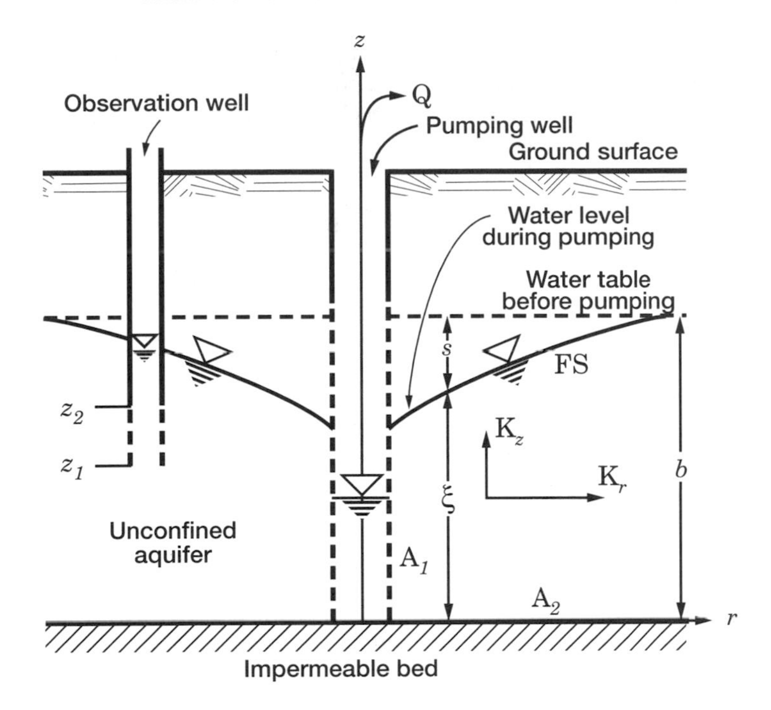

Figure 9-15 Schematic diagram of Neuman's fully-penetrating wells model for unconfined aquifers (Adapted from Neuman, 1972.)

6. Water is released from storage by compaction of the aquifer, expansion of the water, and gravity drainage at the free surface.
7. The well can be treated as a line sink, which means that the seepage face is neglected.
8. The capillarity effects above the water table can be neglected. This means that water is released instanteneously from the unsaturated zone.
9. The drawdown of the water table is small as compared with the saturated thickness of the aquifer.

9.5.2.2 Governing Equations. The governing differential equation is the three-dimensional homogeneous and anisotropic transient ground water flow equation in radial coordinates which is given by Eq. (2-174) of Chapter 2 in terms of the hydraulic head (h). In terms of drawdown, the governing equation is

$$K_r \frac{\partial^2 s}{\partial r^2} + \frac{K_r}{r}\frac{\partial s}{\partial r} + K_z \frac{\partial^2 s}{\partial z^2} = S_s \frac{\partial s}{\partial t}, \qquad 0 < z < \xi \tag{9-65}$$

in which $s = h_0 - h$, where h_0 is the initial head.

9.5.2.3 Initial and Boundary Conditions. The initial and boundary conditions are described based on Neuman and Witherspoon (1970). The position of the free surface of an unconfined aquifer changes in space under transient flow conditions. As a result, it is necessary to treat the free surface as a moving boundary. Under this condition, the boundary of the flow region consists of three complementary parts which are shown by A_1 (prescribed head boundary), A_2 (prescribed flux boundary), and FS (free surface boundary) in Figure 9-15. The other boundaries in the right- and left-hand side boundaries go to infinity. The seepage face is included in A_1.

The capillary effects above the free surface can also be significant in fine-grained sediments, but much less important in coarse sands and gravels. As mentioned in Section 9.5.2.1, the capillary effects are neglected.

The initial conditions for drawdown $s(r, z, t)$ and saturated thickness $\xi(r, t)$, respectively, are

$$s(r, z, 0) = 0 \tag{9-66}$$

$$\xi(r, 0) = b \tag{9-67}$$

The boundary condition at infinity is

$$s(\infty, z, t) = 0 \tag{9-68}$$

The boundary condition at the impervious boundary (A_2) is

$$\frac{\partial s(r, 0, t)}{\partial z} = 0 \tag{9-69}$$

The condition for the constant pumping rate Q at the well is

$$\lim_{r \to 0} \int_0^{\xi} r \frac{\partial s}{\partial r} \, dz = -\frac{Q}{2\pi K_r} \tag{9-70}$$

Two conditions must be satisfied on the free surface. From the definition of $\xi(r, t)$ (Figure 9-15), the first condition can be expressed as

$$\xi(r, t) = b - s(r, \xi, t) \tag{9-71}$$

For the second condition, the movement of an infinitesimal portion of the free surface dA during a dt time interval is considered as shown in Figure 9-16. Let n_i be the unit outer normal and v_i be the average Darcy velocity at the free surface by which the fluid particles are moving. The net average rate of infiltration, which is the rate of fluid added to the free surface from above, is I and S_y is the specific yield of the porous medium. With these, the volume of moving fluid enclosed between dA at t and at $t + dt$ can be expressed by

$$S_y \, dA \, dL = S_y \frac{\partial \xi}{\partial t} n_z \, dA \, dt \tag{9-72}$$

The total inflow during dt is

$$(v_i n_i + I n_z) \, dA \, dt$$

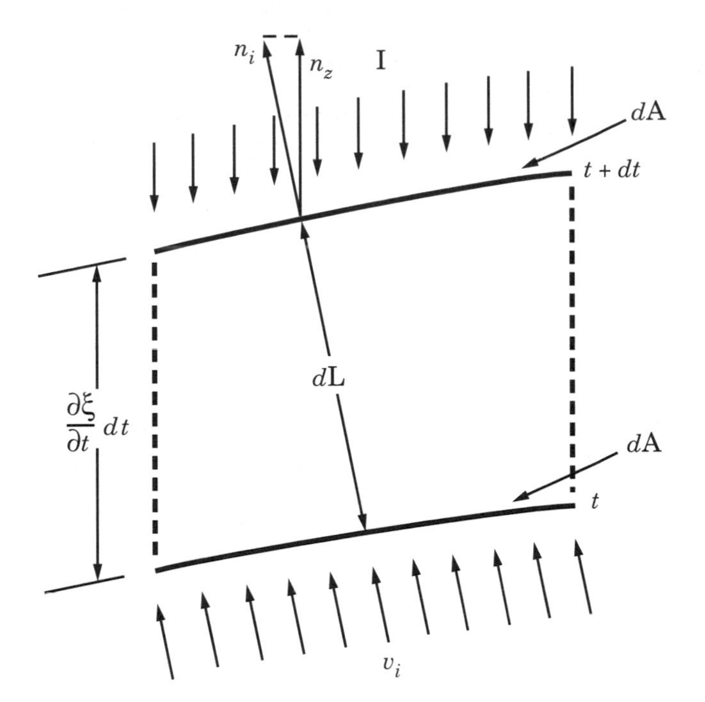

Figure 9-16 An infinitesimal element on the moving free surface of an unconfined aquifer. (Adapted from Neuman and Witherspoon, 1970.)

where n_z is the outer normal in the z direction. Making this equal to Eq. (9-72) and dividing both sides by $dA\,dt$ gives

$$v_i n_i = \left(S_y \frac{\partial \xi}{\partial t} - I \right) n_z \tag{9-73}$$

Then, replacing the Darcy's equation for v_i in radial coordinates gives

$$K_r \frac{\partial s}{\partial r} n_r + K_z \frac{\partial s}{\partial z} n_z = \left(S_y \frac{\partial \xi}{\partial t} - I \right) n_z \qquad \text{at} \quad z = \xi \tag{9-74}$$

Neuman (1972) linearized Eqs. (9-65) through (9-74) by using a perturbation technique similar to that described by Dagan (1964, 1967a,b) on the condition that the aquifer is thick enough and the drawdown (s) remains much smaller than the saturated aquifer thickness (ξ). This technique results a first-order linearized approximation, which is determined by simply shifting the boundary condition from the free surface to the horizontal plane at $z = b$. During the pumping period, the infiltration rate (I) can be neglected, which is the case for all transient drawdown models. This process eliminates ξ from the above equations; finally, Neuman (1972) obtained the following equations for Eqs. (9-65), (9-70), and (9-74), respectively,

$$\frac{\partial^2 s}{\partial r^2} + \frac{1}{r} \frac{\partial s}{\partial r} + K_D \frac{\partial^2 s}{\partial z^2} = \frac{1}{\alpha_s} \frac{\partial s}{\partial t}, \qquad 0 < z < b \tag{9-75}$$

$$\lim_{r \to 0} \int_0^b r \frac{\partial s}{\partial r} dz = -\frac{Q}{2\pi K_r} \tag{9-76}$$

$$\frac{\partial s(r,b,t)}{\partial z} = -\frac{1}{\alpha_y} \frac{\partial s(r,b,t)}{\partial t} \tag{9-77}$$

where

$$K_D = \frac{K_z}{K_r}, \qquad \alpha_s = \frac{K_r}{S_s}, \qquad \alpha_y = \frac{K_z}{S_y} \tag{9-78}$$

Equation (9-66) states that initially the drawdown is zero everywhere in the aquifer. Equation (9-68) states that the drawdown is small at large distances from the pumping well in the aquifer. Equation (9-69) states that the plane at $z = 0$ is an impervious boundary. Equation (9-76) states that near the pumping well the flow rate is equal to the discharge rate. Finally, Equation (9-77) expresses the conditions at the free surface of aquifer.

9.5.2.4 Solution. By applying the Laplace and Hankel transform techniques to Eqs. (9-66), (9-68), (9-69), (9-75), (9-76), and (9-77) and inverting the results, Neuman (1972) obtained the first-order approximation to the original initial and boundary value problem. Later, Neuman (1973) made some supplementary alterations as well as notation changes in the original solution, and these changes are included in Neuman (1975). Here, in order to clearly present the final model of Neuman, both the original and its modified version of his model will be presented.

Neuman's Original Solution. The original solution for $s(r,z,t)$ is (Neuman, 1972)

$$s(r,z,t) = \frac{Q}{4\pi T} \int_0^\infty 4xJ_0[x(K_D)^{1/2}] \left[\omega_0(x) + \sum_{n=1}^{\infty} \omega_n(x) \right] dx \tag{9-79}$$

where J_0 denotes the Bessel function of the first kind of zero order and

$$\omega_0(x) = \frac{\left\{1 - \exp\left[-t_s K_D \left(x^2 - \beta_0^2\right)\right]\right\} \cosh(\beta_0 z_D b_D)}{\left\{x^2 + (1+\sigma)\beta_0^2 - \left[\left(x^2 - \beta_0^2\right)^2 \frac{b_D^2}{\sigma}\right]\right\} \cosh(\beta_0 b_D)} \tag{9-80}$$

$$\omega_n(x) = \frac{\left\{1 - \exp[-t_s K_D (x^2 + \beta_n^2)]\right\} \cos(\beta_n z_D b_D)}{\left\{x^2 - (1+\sigma)\beta_n^2 - \left[(x^2 + \beta_n^2)^2 \frac{b_D^2}{\sigma}\right]\right\} \cos(\beta_n b_D)} \tag{9-81}$$

and

$$t_s = \frac{Tt}{S_s b r^2} = \frac{Tt}{Sr^2}, \qquad t_y = \frac{Tt}{S_y r^2} \tag{9-82}$$

$$b_D = \frac{b}{r}, \qquad z_D = \frac{z}{b}, \qquad \sigma = \frac{S}{S_y} = \frac{t_y}{t_s} \tag{9-83}$$

and K_D in Eq. (9-79) is defined by the first expression of Eq. (9-78). The terms β_0 and β_n are the roots of the equations

$$\frac{\sigma}{b_D}\beta_0 \sinh(\beta_0 b_D) - (x^2 - \beta_0^2)\cosh(\beta_0 b_D) = 0, \qquad \beta_0^2 < x^2 \tag{9-84}$$

$$\frac{\sigma}{b_D}\beta_n \sin(\beta_n b_D) + (x^2 + \beta_n^2)\cos(\beta_n b_D) = 0 \tag{9-85}$$

where

$$(2n-1)\frac{\pi}{2} < \beta_n b_D < n\pi, \qquad n \geq 1 \tag{9-86}$$

Because Eq. (9-79) is the solution of the linear partial differential equation, as given by Eq. (9-75), the principle of superposition can be used to obtain solutions for any number of extraction and injection wells in an unconfined aquifer.

Average Solution Over Aquifer Thickness. The average drawdown in an observation well whose perforations are located between the elevations z_1 and z_2 in Figure 9-15 is (Neuman, 1972)

$$s_{z_1,z_2} = \frac{1}{z_1 - z_2}\int_{z_1}^{z_2} s(r,z,t)\,dz \tag{9-87}$$

Obviously, this average drawdown must always be greater than the actual fall of the water table at the point of observation. If an observation well completely penetrates an unconfined aquifer, the average drawdown can directly be determined from Eq. (9-79) by only redefining Eqs. (9-80) and (9-81), respectively, as (Neuman, 1972)

$$\omega_0(x) = \frac{\left\{1 - \exp\left[-t_s K_D\left(x^2 - \beta_0^2\right)\right]\right\}\tanh(\beta_0 b_D)}{\left\{x^2 + (1+\sigma)\beta_0^2 - \left[\left(x^2 - \beta_0^2\right)^2 \dfrac{b_D^2}{\sigma}\right]\right\} b_D \beta_0} \tag{9-88}$$

and

$$\omega_n(x) = \frac{\left\{1 - \exp\left[-t_s K_D\left(x^2 + \beta_n^2\right)\right]\right\}\tan(\beta_n b_D)}{\left\{x^2 - (1+\sigma)\beta_n^2 - \left[\left(x^2 + \beta_n^2\right)^2 \dfrac{b_D^2}{\sigma}\right]\right\} b_D \beta_n} \tag{9-89}$$

Neuman's Modified Solution. The solution given by Eq. (9-79) is expressed in terms of five dimensionless parameters: σ, z_D, b_D, K_D, and t_s. Neuman (1973) later lumped K_D and b_D into a single dimensionless parameter as

$$\beta = \frac{{K_D}^2}{b_D^2} = \frac{K_z}{K_r}\left(\frac{r}{b}\right)^2 \tag{9-90}$$

Therefore, the total number of parameters is reduced to four. Referring to Eqs. (9-79), (9-84), and (9-85), Neuman (1973) defined a new variable

$$y = xb_D = x\frac{b}{r} \tag{9-91}$$

and a new set of parameters

$$\gamma_0 = \beta_0 b_D = \beta_0 \frac{b}{r} \tag{9-92}$$

$$\gamma_n = \beta_n b_D = \beta_n \frac{b}{r} \tag{9-93}$$

Substitution of these parameters into Eqs. (9-79), (9-84), (9-85), Neuman's original solution takes the following form:

$$s(r,z,t) = \frac{Q}{4\pi T}\int_0^\infty 4yJ_0\left(y\beta^{1/2}\right)\left[\omega_0(y) + \sum_{n=1}^{\infty}\omega_n(y)\right] dy = \frac{Q}{4\pi T} s_D \tag{9-94}$$

$$\omega_0(y) = \frac{\left\{1 - \exp\left[-t_s\beta\left(y^2 - \gamma_0^2\right)\right]\right\}\cosh(\gamma_0 z_D)}{\left\{y^2 + (1+\sigma)\gamma_0^2 - \left[\dfrac{\left(y^2 - \gamma_0^2\right)^2}{\sigma}\right]\right\}\cosh(\gamma_0)} \tag{9-95}$$

$$\omega_n(y) = \frac{\left\{1 - \exp\left[-t_s\beta\left(y^2 + \gamma_n^2\right)\right]\right\}\cos(\gamma_n z_D)}{\left\{y^2 - (1+\sigma)\gamma_n^2 - \left[\dfrac{\left(y^2 + \gamma_n^2\right)^2}{\sigma}\right]\right\}\cos(\gamma_n)} \tag{9-96}$$

and the terms γ_0 and γ_n are the roots of the equations

$$\sigma\gamma_0 \sinh(\gamma_0) - \left(y^2 - \gamma_0^2\right)\cosh(\gamma_0) = 0, \qquad \gamma_0^2 < y^2 \tag{9-97}$$

$$\sigma\gamma_n \sin(\gamma_n) + \left(y^2 + \gamma_n^2\right)\cos(\gamma_n) = 0 \tag{9-98}$$

where

$$(2n-1)\frac{\pi}{2} < \gamma_n < n\pi, \qquad n \geq 1 \tag{9-99}$$

Notice that in Eq. (9-94), the integral is shown by s_D, which is *Neuman's well function for fully penetrating wells*. Also, $s_D = 4\pi Ts/Q$ is called the *dimensionless drawdown*. The average drawdown for an observation well can similarly be obtained from Eqs. (9-87) and (9-94) by redefining $\omega_0(y)$ and $\omega_n(y)$ as

$$\omega_0(y) = \frac{\left\{1 - \exp[-t_s\beta(y^2 - \gamma_0^2)]\right\}\tanh(\gamma_0)}{\left\{y^2 + (1+\sigma)\gamma_0^2 - \left[\dfrac{\left(y^2 - \gamma_0^2\right)^2}{\sigma}\right]\right\}\gamma_0} \tag{9-100}$$

$$\omega_n(y) = \frac{\{1 - \exp[-t_s\beta(y^2 + \gamma_n^2)]\}\tan(\gamma_n)}{\left\{y^2 - (1+\sigma)\gamma_n^2 - \left[\frac{(y^2+\gamma_n^2)^2}{\sigma}\right]\right\}\gamma_n} \tag{9-101}$$

9.5.2.5 Special Cases. Neuman's model for fully penetrating wells in unconfined aquifers has some important special cases. Under certain conditions the Neuman model reduces to the Theis and Boulton equations. Some details regarding these special cases are presented below.

Reduction to the Theis Solution. If gravitational drainage does not exist, the specific yield is zero ($S_y = 0$). According to the last expression of Eq. (9-83), this means that σ approaches to infinity. Therefore, the aquifer is under confined aquifer conditions, and Eq. (9-79) should reduce to the Theis solution with respect to t_s. This can be shown as follows (Neuman, 1972): Multiplying both sides of Eq. (9-84) by β_0 gives

$$\frac{\sigma}{b_D}\beta_0^2\sinh(\beta_0 b_D) - \beta_0(x^2 - \beta_0^2)\cosh(\beta_0 b_D) = 0, \qquad \beta_0^2 < 0 \tag{9-102}$$

In order to have the right-hand side of Eq. (9-102) to be equal zero, the following conditions must be satisfied:

$$\lim_{\sigma\to\infty}\beta_0 = 0 \tag{9-103}$$

$$\lim_{\sigma\to\infty}(\sigma\beta_0^2) = \lim_{\beta_0\to 0}\frac{(x^2-\beta_0^2)\beta_0 b_D\cosh(\beta_0 b_D)}{\sinh(\beta_0 b_D)} = x^2 \tag{9-104}$$

Therefore, Eqs. (9-80) and (9-81) take the form

$$\lim_{\sigma\to\infty}\omega_0(x) = \frac{1-\exp(-t_s K_D x^2)}{2x^2} \tag{9-105}$$

and

$$\lim_{\sigma\to\infty}\omega_n(x) = 0 \qquad \text{for } n = 1, 2, 3, \ldots \tag{9-106}$$

For the above-given expressions the following equation is used:

$$\lim_{x\to 0}\frac{\sinh(x)}{x} = 1 \tag{9-107}$$

Defining a new variable $y = xK_D^{1/2}$ and substituting Eqs. (9-105) and (9-106) into Eq. (9-79) gives

$$s(r,t) = \frac{Q}{4\pi T}\int_0^\infty 2\left[1 - e^{-t_s y^2}\right]J_0(y)\frac{dy}{y} \tag{9-108}$$

From tables of definite integrals [Gradshteyn and Ryzhik, 1965, p. 717, Eq. (5)] we have

$$\int_0^\infty e^{-ax^2}J_0(x)\frac{dx}{x} = \frac{1}{2}\int_0^{1/4a} e^{-y}\frac{dy}{y} \tag{9-109}$$

where the real part of a is nonnegative. Using Eq. (9-109) into Eq. (9-108) gives

$$s(r,t) = \frac{Q}{4\pi T} \int_{\frac{1}{4t_s}}^{\infty} e^{-y} \frac{dy}{y} \tag{9-110}$$

which is the Theis equation derived in Section 4.2.4 of Chapter 4. In Eq. (9-110), t_s is given by the first expression of Eq. (9-82).

Reduction to Boulton's Equation without Delayed Response. Neuman (1972) showed that, at large values of time, drawdown everywhere in the aquifer approaches the asymptote:

$$s(r,z,t) = \frac{Q}{4\pi T} \int_0^{\infty} 2J_0(y\beta^{1/2}) \left\{ 1 - \exp[-t_y \beta y \tanh(y)] \frac{\cosh(yz_D)}{\cosh(y)} \right\} \frac{dy}{y} \tag{9-111}$$

When the aquifer is rigid and the water is considered to be incompressible and $S_s = 0$, Eq. (9-111) holds at all values of time. As a result, one might expect that Eq. (9-111) is identical to Boulton's equation without delayed yield under isotropic aquifer conditions ($K_D = 1$), as given by Eq. (9-19). Neuman (1972) also showed that when the aquifer is very thin, b_D approaches to zero, and Eq. (9-111) reduces to Theis equation with respect to t_y as defined by the second expression of Eqs. (9-82).

9.5.2.6 Evaluation of the Solution. Neuman (1972) extensively evaluated the analytical solution as given by Eq. (9-79) by varying some key parameters in the solution as well as comparing its results with a numerical model. Neuman evaluated the integral in Eq. (9-79) numerically because of its complex nature. The numerical integration was accomplished by using a self-adaptive Simpson scheme with quadratically convergent Newton–Raphson iterative method. In the following section the results of Neuman's evaluation are presented.

Variation of Drawdown Versus Time Curves with σ. In Eq. (9-79), σ, which is defined by the last expression of Eqs. (9-83), is one of the key parameters in controlling the value of drawdown as given by Eq. (9-79). The effect of this parameter on s_D versus t_s and s_D versus t_y curves and their physical meaning are investigated below based on Neuman (1972).

In Figure 9-17, dimensionless drawdown versus dimensionless time t_s for $z_D = z/b = 0$, $b_D = z/r = 1$, and $K_D = K_z/K_r = 1$ and four finite values of $\sigma = S_y/S_s$ are shown. As shown mathematically in Section 9.5.2.5, the Theis curve in Figure 9-17 corresponds to $\sigma = \infty$. Besides this, some typical aspects of Figure 9-17 are as follows (Neuman, 1972): (1) At early values of time, the drawdown versus time curves fall on the Theis curve. This means that during early time periods, water is released from storage only by compaction of the aquifer material and expansion of water. (2) During the second stage, as described in Section 9.2 along with Figure 9-2, gravity drainage becomes more effective. This effect is similar to that of leakage from a nearby source. (3) The smaller the value of σ, the larger the effect of gravity drainage. (4) As time increases, the effect of elastic storage at a point in the aquifer dissipates completely and the drawdown versus time curves once again become parallel to the Theis curve.

In Figure 9-18, dimensionless drawdown versus dimensionless time t_s for $z_D = z/r = 0$, $b_D = b/r = 1$, and $K_D = K_z/K_r = 1$ and six finite values of $\sigma = S_y/S_s$ are shown. Some typical aspects of Figure 9-18 are as follows (Neuman, 1972): (1) As time increases, the curves join the Theis curve. (2) The period of time occupied by the early segment of the time versus drawdown curve becomes less as σ decreases. When σ approaches zero,

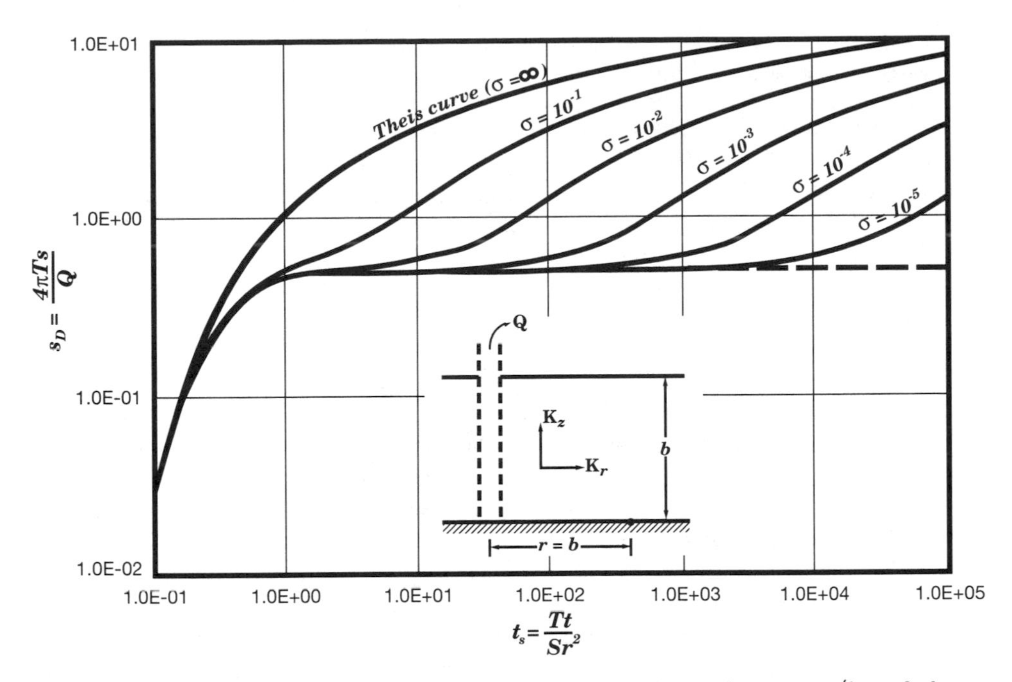

Figure 9-17 Dimensionless drawdown versus dimensionless time t_s for $z_D = z/b = 0$, $b_D = b/r = 1$, and $K_D = K_z/K_r = 1$. (After Neuman, 1972.)

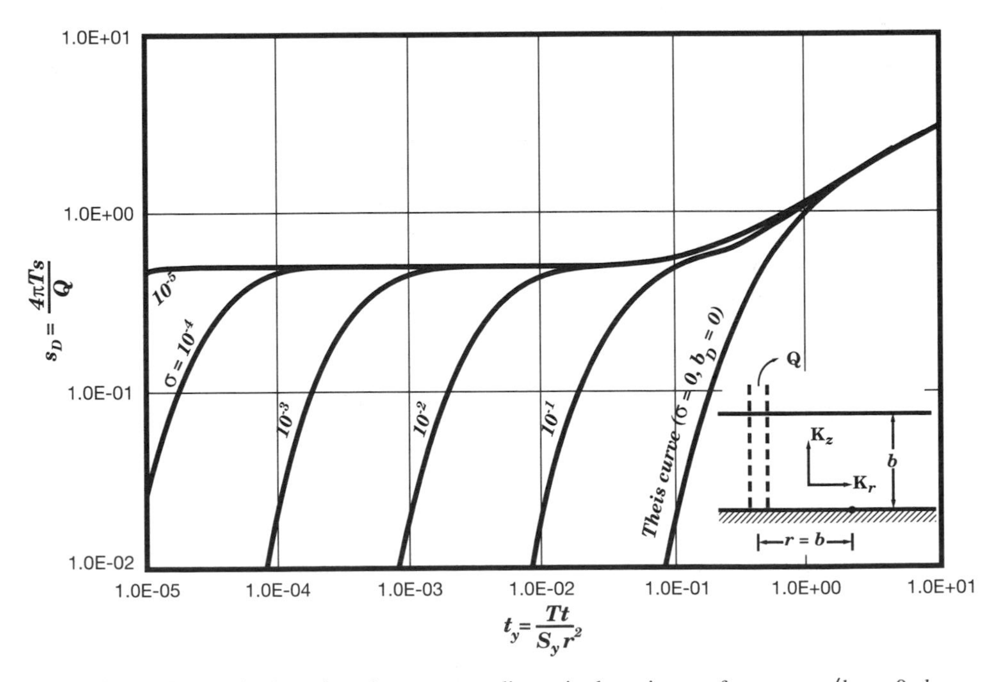

Figure 9-18 Dimensionless drawdown versus dimensionless time t_y for $z_D = z/b = 0$, $b_D = b/r = 1$, and $K_D = K_z/K_r = 1$. (After Neuman, 1972.)

this segment disappears completely, and hydraulic heads everywhere below the water table drop instantaneously after pumping starts.

Effects of Anisotropy on Drawdown. The effect of anisotropy ($K_D = K_z/K_r$) on drawdown is shown in Figure 9-19 for $\sigma = 10^{-2}$, $z_D = z/b = 0$, $b_D = b/r = 1$ (Neuman, 1972) with $K_D = 0.1, 0.5, 1.0, 5.0$, and 10.0. Generally, the horizontal hydraulic conductivity values of an aquifer are higher than its vertical hydraulic conductivity values. However, in fractured rock formations, vertical hydraulic conductivity values may be higher than their horizontal hydraulic conductivity values ($K_D > 1$). Figure 9-19 shows that as K_D becomes smaller the drawdown becomes higher and, therefore, the delayed response becomes more effective.

Comparison with a Numerical Model. As can be seen from Sections 9.4, 9.5.2, and 9.5.3, both Boulton and Neumans models are based on some restrictive assumptions including the linearization of the governing partial differential equation. Therefore, evaluation of the results of these models has a special importance. Neuman (1972) compared the results of Eq. (9-79) with an integrated saturated–unsaturated finite-difference model for flow to a single well in a compressible unconfined aquifer system developed by Cooley (1971). The information on this comparison is as follows: The well fully penetrates the elastic unconfined aquifer; the water table is at an initial elevation of 59.5 ft from the aquifer bottom; the zone above the water table is under capillary pressure, and its properties are

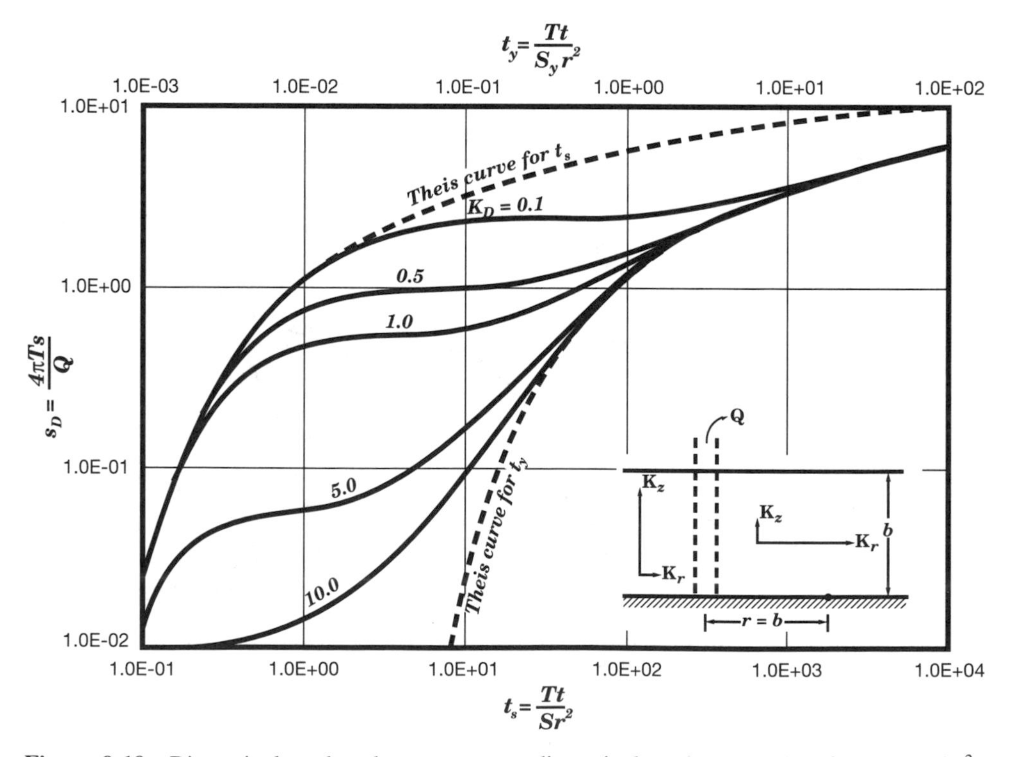

Figure 9-19 Dimensionless drawdown s_D versus dimensionless time t_s and t_y for $\sigma = 10^{-2}$, $z_D = z/b = 0$, and $b_D = b/r = 1$. (After Neuman, 1972.)

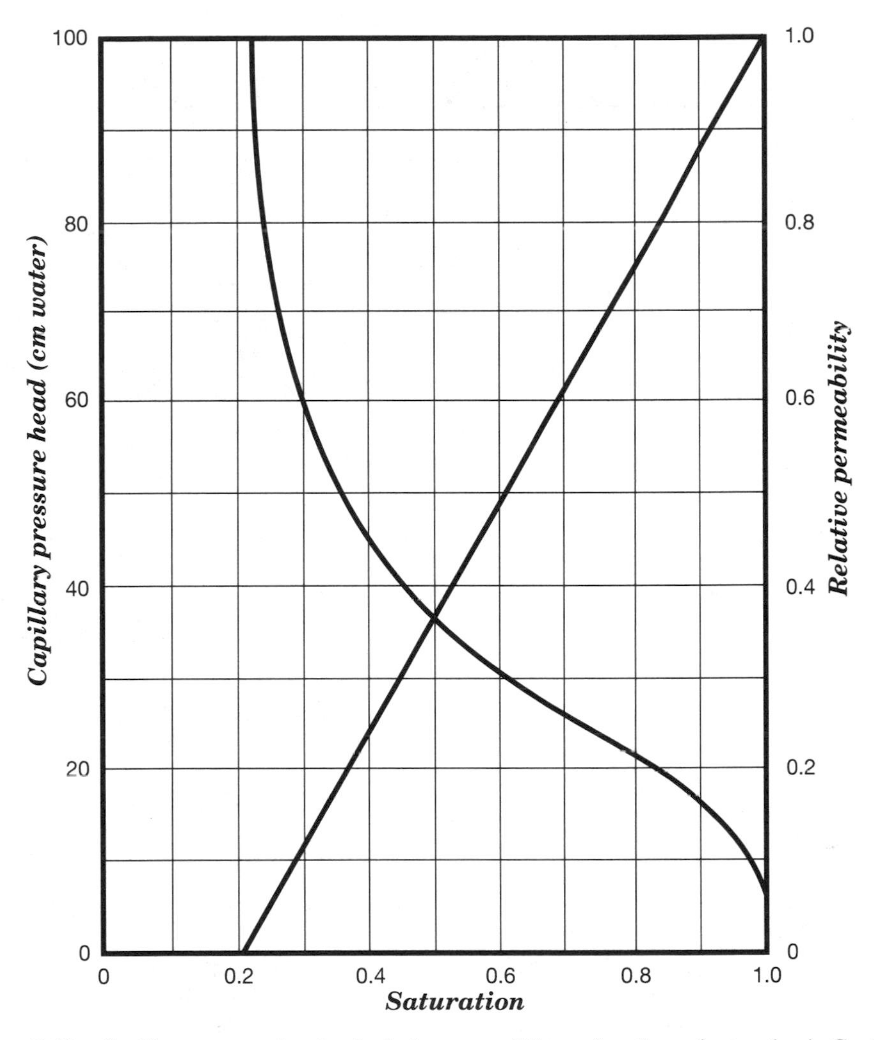

Figure 9-20 Capillary pressure head and relative permeability as functions of saturation in Cooley's (1971) numerical model. (After Neuman, 1972.)

a function of saturation (Figure 9-20). The other numerical values are $Q = 302$ gpm (gallons per minute); effective radius $= 0.53$ ft; $K_r = K_z = 1500$ gal/day/ft^2; $S_y = 0.23$; and $S_s = 0.0006$ ft^{-1}. Figure 9-21 shows a comparison between analytical and numerical models at $r = 41$ ft ($b_D = b/r = 1.45$) and $r = 70$ ft ($b_D = 0.838$), and at $z = 0$ ($z_D = z/b = 0$) and $z = 50$ ft ($z_D = 0.84$).

The analytical model, as given by Eq. (9-79), neglects the unsaturated zone, treats the well as a line sink, and is based on a linearized mathematical approximation. Despite these limitations, as can be seen from Figure 9-21, the comparison with a more comprehensive numerical model is very good.

Moench (1995) developed a semianalytical model by combining the Boulton and Neuman models for flow to a partially penetrating well in an unconfined aquifer by using an expression similar to that given by Eq. (9-34). As Moench (1995) pointed out, the theoretical

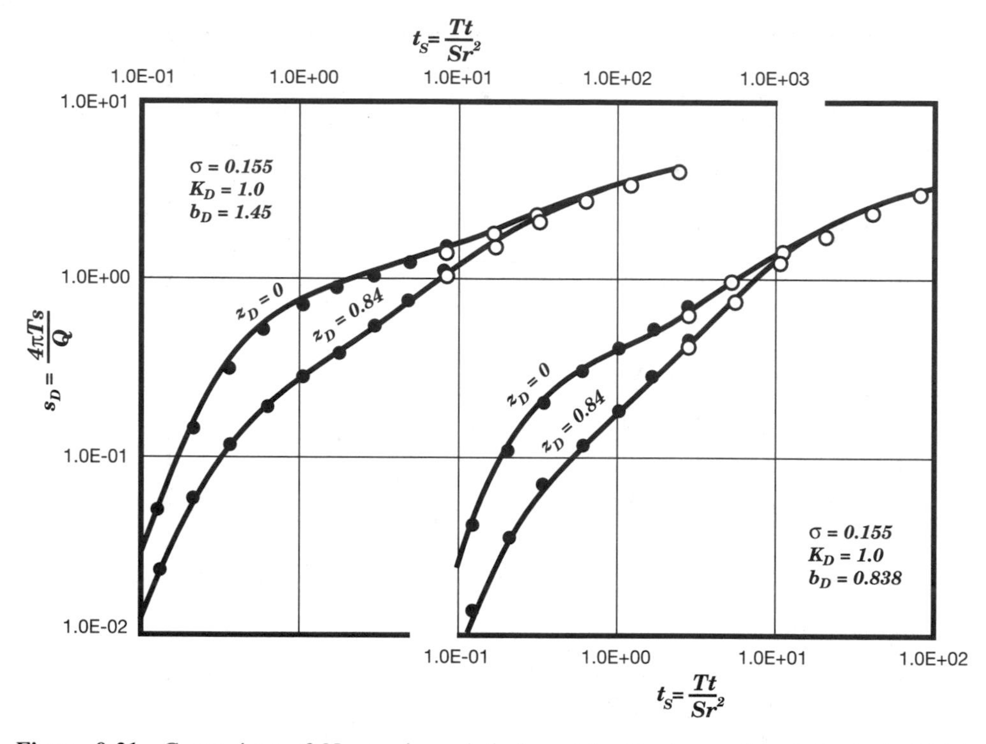

Figure 9-21 Comparison of Neuman's analytical model with the results of Cooley's (1971) saturated–unsaturated numerical model. The solid lines represent the analytical solution. The closed circles are fine mesh; the open circles are coarse mesh. (After Neuman, 1972.)

basis for use of Eq. (9-34) is weak. Moench introduced an additional parameter γ ($\gamma = \infty$ corresponds to the approach of Neuman)—which is a function of the empirical constant α_l [corresponds to α in Eq. (9-34)], vertical hydraulic conductivity, saturated thickness, and specific yield—to be determined by trial-and-error type-curve matching. Moench (1995, p. 378) states: "... Analysis of drawdown in selected piezometers from the published results of two aquifer tests conducted in relatively homogeneous glacial outwash deposits with significantly different hydraulic conductivities reveals improved comparison between the theoretical type curves and the hydraulic head measured in water table piezometers." Moench does not present comparisons between his type-curve and Neuman type-curve results for the hydrogeologic parameters. On the other hand, neither Boulton nor Moench compared their results with the results of a numerical model as, for instance, was done by Neuman. As mentioned above, despite the restrictive assumption for the instantaneous unsaturated flow effects, the comparison of the results of the Neuman model with a more comprehensive saturated-unsaturated numerical model is very good (see Figure 9-21).

9.5.3 Neuman's Partially Penetrating Wells Model

9.5.3.1 Problem Statement and Assumptions. Practicing hydrogeologists frequently face the situation that both pumping and observation wells do not fully penetrate the saturated thickness of unconfined aquifers. For relatively thick unconfined aquifers, partial penetration significantly affects the drawdowns created by a pumping well depending on the length of screen intervals. The models developed by Boulton (1954a,b, 1963) and Neuman (1972, 1973) are based on the assumption that both the pumping and observation wells fully penetrate the saturated thickness of unconfined aquifers. Therefore, they cannot take into account the effects of partial penetration in unconfined aquifers.

Neuman (1974) extended his theory to account for the effects of partial penetration in a pumping well on drawdowns in an unconfined aquifer. The schematic diagram of the unconfined aquifer with a partially penetrating well is shown in Figure 9-22. The well is open to inflow or outflow from a depth d to a depth l beneath the initially static water table.

The assumptions listed in Section 9.5.2.1 for a fully penetrated well in an unconfined aquifer are also valid for a partially penetrating well.

9.5.3.2 Governing Equations The governing differential equation is the three-dimensional homogeneous and anisotropic transient ground-water flow equation in radial coordinates as given by Eq. (9-65) in Section 9.5.2.2.

9.5.3.3 Initial and Boundary Conditions. The initial and boundary conditions expressed by Eqs. (9-66), (9-67), (9-68), (9-69), (9-71), and (9-74) in Section 9.5.2.3 for a fully penetrating well case are also valid for a partially-penetrating well case. The modified forms of the other boundary conditions are given below.

The boundary condition regarding the constant pumping rate Q at the well is

$$\lim_{r \to 0} \int_{b-l}^{\min(b-d,\xi)} r \frac{\partial s}{\partial r}\, dz = -\frac{Q}{2\pi K_r} \tag{9-112}$$

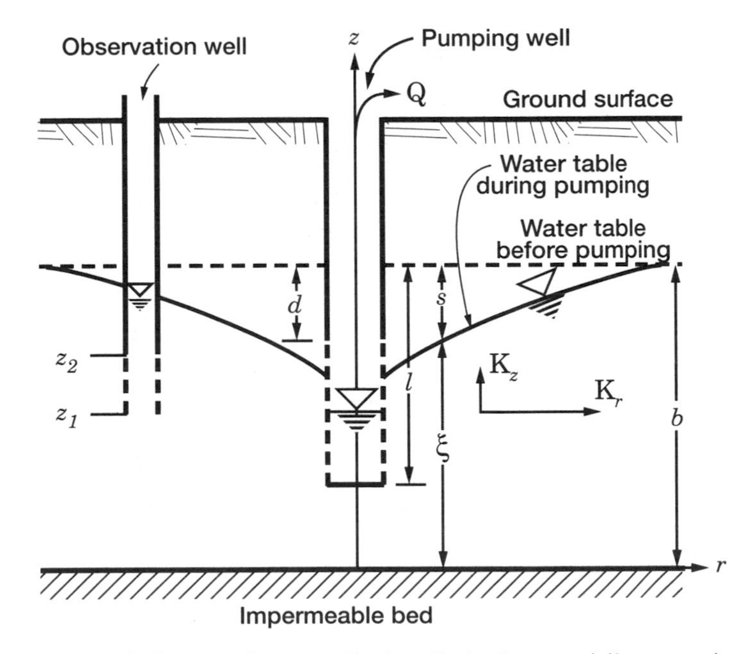

Figure 9-22 Schematic diagram of an unconfined aquifer having a partially penetrating well. (After Neuman, 1974.)

which corresponds to Eq. (9-70) of the fully penetrating well case. The upper limit of the integral in Eq. (9-112) is $b - d$, if the water table around the well is above the upper boundary of the open interval. If the water table around the well is below the upper boundary of the open interval, the upper limit is ξ. The boundary condition for the zones of the well with no inflow or outflow (no perforations) is

$$\frac{\partial s(0, z, t)}{\partial z} = 0, \qquad 0 \leq z \leq b - l, \quad b - d \leq z \leq b \tag{9-113}$$

With the linearization technique outlined in Section 9.5.2.3, the initial- and boundary-value problem can be represented by Eqs. (9-66), (9-68), (9-69), (9-75), (9-77), (9-112), and (9-113). Neuman introduced a further approximation by assuming that flux along the perforated section of the well is uniform. Therefore, Eq. (9-112) takes the form

$$\lim_{r \to 0} \left(r \frac{\partial s}{\partial r} \right) = -\frac{Q}{2\pi K_r (l - d)}, \qquad b - l < z < b - d \tag{9-114}$$

The physical meaning of the initial and boundary conditions are described in Section 9.5.2.3.

9.5.3.4 Solution.

After applying the Laplace and Hankel transforms to Eqs. (9-66), (9-68), (9-69), (9-75), (9-77), (9-112), (9-113), and (9-114) and inverting the results, Neuman (1974) obtained a first-order approximation to the original initial- and boundary-value problem and expressed the final solution in terms of six dimensionless parameters: σ, β, z_D, l_D, d_D, and t_s or t_y. The solution is

$$s(r, z, t) = \frac{Q}{4\pi T} \int_0^\infty 4y J_0 \left(y\beta^{1/2} \right) \left[u_0(y) + \sum_{n=1}^{\infty} u_n(y) \right] dy = \frac{Q}{4\pi T} s_D \tag{9-115}$$

where

$$u_0(y) = \frac{\left\{ 1 - \exp\left[-t_s \beta \left(y^2 - \gamma_0^2 \right) \right] \right\} \cosh\left(\gamma_0 z_D \right)}{\left[y^2 + (1 + \sigma)\gamma_0^2 - \dfrac{\left(y^2 - \gamma_0^2 \right)^2}{\sigma} \right] \cosh\left(\gamma_0 \right)} \cdot \frac{\sinh\left[\gamma_0 \left(1 - d_D \right) \right] - \sinh\left[\gamma_0 \left(1 - l_D \right) \right]}{\left(l_D - d_D \right) \sinh\left(\gamma_0 \right)} \tag{9-116}$$

$$u_n(y) = \frac{\left\{ 1 - \exp\left[-t_s \beta \left(y^2 + \gamma_n^2 \right) \right] \right\} \cos\left(\gamma_n z_D \right)}{\left[y^2 - (1 + \sigma)\gamma_n^2 - \dfrac{\left(y^2 + \gamma_n^2 \right)^2}{\sigma} \right] \cos(\gamma_n)} \cdot \frac{\sin\left[\gamma_n (1 - d_D) \right] - \sin\left[\gamma_n (1 - l_D) \right]}{\left(l_D - d_D \right) \sin(\gamma_n)} \tag{9-117}$$

where J_0 denotes the Bessel function of the first kind of zero order and

$$\beta = K_D \left(\frac{r}{b}\right)^2 = \frac{K_z}{K_r}\left(\frac{r}{b}\right)^2 \tag{9-118}$$

$$t_s = \frac{Tt}{S_s b r^2} = \frac{Tt}{S r^2} \tag{9-119}$$

$$d_D = \frac{d}{b}, \qquad l_D = \frac{l}{b}, \qquad z_D = \frac{z}{b} \tag{9-120}$$

$$\sigma = \frac{S}{S_y} = \frac{t_y}{t_s} \tag{9-121}$$

$$t_y = \frac{Tt}{S_y r^2} \tag{9-122}$$

and γ_0 and γ_n are the roots of the equations

$$\sigma\gamma_0 \sinh(\gamma_0) - \left(y^2 - \gamma_0^2\right)\cosh(\gamma_0) = 0, \qquad \gamma_0^2 < y^2 \tag{9-123}$$

$$\sigma\gamma_n \sin(\gamma_n) + \left(y^2 + \gamma_n^2\right)\cos(\gamma_n) = 0 \tag{9-124}$$

where

$$(2n-1)\frac{\pi}{2} < \gamma_n < n\pi, \qquad n \geq 1 \tag{9-125}$$

Notice that in Eq. (9-115) the integral is shown by s_D, which is *Neuman's well function for partially penetrating wells.* Also, $s_D = 4\pi Ts/Q$ is called the *dimensionless drawdown.* Equation (9-115) is the solution of the linear partial differential equation, as given by Eq. (9-75). Therefore, the principle of superposition can be used to obtain solutions for any number of extraction and injection wells in an unconfined aquifer.

Average Drawdown in Observation Well. The average drawdown in an observation well whose perforations are located between the elevations z_1 and z_2 in Figure 9-22 is simply the average over that vertical distance and is given by Eq. (9-87). This average drawdown is always greater than the fall of the water table at point of observation. Neuman (1974) showed that the average drawdown defined by Eq. (9-87) can be calculated directly from Eq. (9-115) by merely redefining the expressions given by Eqs. (9-116) and (9-117) as

$$u_0(y) = \frac{\left\{1 - \exp\left[-t_s\beta\left(y^2 - \gamma_0^2\right)\right]\right\}\left[\sinh(\gamma_0 z_{2D}) - \sinh(\gamma_0 z_{1D})\right]}{\left[y^2 + (1+\sigma)\gamma_0^2 - \dfrac{\left(y^2 - \gamma_0^2\right)^2}{\sigma}\right]\cosh(\gamma_0)} \cdot \frac{\left\{\sinh\left[\gamma_0(1-d_D)\right] - \sinh\left[\gamma_0(1-l_D)\right]\right\}}{(z_{2D} - z_{1D})\,\gamma_0(l_D - d_D)\sinh(\gamma_0)} \tag{9-126}$$

$$u_n(y) = \frac{\left\{1 - \exp\left[-t_s\beta\left(y^2 + \gamma_n^2\right)\right]\right\}\left[\sin\left(\gamma_n z_{2D}\right) - \sin(\gamma_n z_{1D})\right]}{\left[y^2 - (1+\sigma)\gamma_n^2 - \dfrac{(y^2+\gamma_n^2)^2}{\sigma}\right]\cos(\gamma_n)} \cdot \frac{\{\sin[\gamma_n(1-d_D)] - \sin[\gamma_n(1-l_D)]\}}{(z_{2D} - z_{1D})\gamma_n(l_D - d_D)\sin(\gamma_n)} \quad (9\text{-}127)$$

9.5.3.5 Special Cases.

The partially penetrating wells model developed by Neuman (1974) has some important special cases. Under certain conditions, Neuman's partially penetrating well model reduces to Neuman's fully penetrating well model for unconfined aquifers and Hantush's model (Hantush, 1964) for a confined aquifer. Some details regarding these special cases are presented below.

Reduction to Neuman's Fully Penetrating Well Solution. In the case of a fully penetrating well (Figure 9-15), $d = 0$ and $l = b$. Therefore, from Eq. (9-120) one gets $d_D = 0$ and $l_D = 1$. When we introduce these into Eqs. (9-116) and (9-117), Eq. (9-115) reduces to Eq. (9-94).

Reduction to Hantush's Solution for a Confined Aquifer. If gravity drainage does not exist, which means $S_y = 0$ and, therefore, because of Eq. (9-121), $\sigma = \infty$; the aquifer is under confined flow conditions, and Eq. (9-115) should reduce to Hantush's solution (Hantush, 1961c, 1964), as given by Eq. (8-47) of Chapter 8. This is shown below based on Neuman (1974).

From Eq. (9-123) one has

$$\lim_{\sigma\to\infty} \gamma_0 = 0 \quad (9\text{-}128)$$

and when

$$\lim_{\gamma_0\to 0} \frac{\sinh\left[\gamma_0(1-d_D)\right] - \sinh\left[\gamma_0(1-l_D)\right]}{(l_D - d_D)\sinh(\gamma_0)} = \frac{1}{l_D - d_D}\lim_{\gamma_0\to 0}\left\{\frac{\sinh\left[\gamma_0(1-d_D)\right]}{\sin(\gamma_0)} - \frac{\sinh\left[\gamma_0(1-l_D)\right]}{\sin(\gamma_0)}\right\} = 1 \quad (9\text{-}129)$$

The sum of the terms in the right-hand side of the brackets in Eq. (9-129) is $[(1 - d_D) - (1 - l_D)] = l_D - d_D$, as γ_0 approaches to zero, because of Eq. (9-107). As a result, the final expression becomes equal to one. Because of this, it becomes obvious that

$$\lim_{\sigma\to\infty} u_0(y) = \lim_{\sigma\to\infty} \omega_0(y) \quad (9\text{-}130)$$

where $\omega_0(y)$, which is given by Eq. (9-95), is the equivalent of $u_0(y)$ for full penetration (Figure 9-15). As a result of Eq. (9-130), as well as Eqs. (9-105), (9-106), (9-108), (9-109), and (9-110), the first part of Eq. (9-115) reduces to

$$\lim_{\sigma\to\infty} \frac{Q}{4\pi T}\int_0^\infty 4yJ_0\left(y\beta^{1/2}\right)u_0(y)\,dy = \frac{Q}{4\pi T}\int_{1/4t_s}^\infty \frac{e^{-y}}{y}\,dy \quad (9\text{-}131)$$

which is the Theis equation, as given in Section 4.2.4.1 of Chapter 4, with respect to t_s defined by Eq. (9-119). From Eq. (9-124) one has

$$\lim_{\sigma \to \infty} \gamma_n = n\pi, \qquad n > 1 \tag{9-132}$$

and

$$\begin{aligned} \lim_{\sigma \to \infty} \sigma \gamma_n^2 \sin(\gamma_n) &= \lim_{\gamma_n \to n\pi} \left[-\left(y^2 + \gamma_n^2\right) \gamma_n \cos(\gamma_n) \right] \\ &= -\left(y^2 + n^2\pi^2\right) n\pi \cos(n\pi) \end{aligned} \tag{9-133}$$

which means that

$$\begin{aligned} \lim_{\sigma \to \infty} u_n(y) = & \frac{\left\{1 - \exp\left[-t_s\beta\left(y^2 + n^2\pi^2\right)\right]\right\} \cos(n\pi z_D)}{y^2 + n^2\pi^2} \\ & \cdot \frac{\{\sin[n\pi(1 - d_D)] - \sin[n\pi(1 - l_D)]\}}{n\pi(l_D - d_D)} \end{aligned} \tag{9-134}$$

Neuman [1974, p. 311, Eq. (C8)] showed that

$$\begin{aligned} & \int_0^\infty y J_0\left(y\beta^{1/2}\right) \frac{1 - \exp\left[-t_s\beta\left(y^2 + m^2\pi^2\right)\right]}{y^2 + m^2\pi^2} \, dy \\ & = \frac{1}{2} \int_{1/4t_s}^\infty \exp\left(-y - \frac{\beta m^2\pi^2}{4y}\right) \frac{dy}{y}, \qquad m = 0, 1, 2, \ldots \end{aligned} \tag{9-135}$$

and as a result of this, Eq. (9-115) reduces to

$$\begin{aligned} & \lim_{\sigma \to \infty} \frac{Q}{4\pi T} \int_0^\infty 4y J_0\left(y\beta^{1/2}\right) \sum_{n=1}^\infty u_n(y) \, dy \\ & = \frac{Q}{4\pi T} \frac{2}{\pi(l_D - d_D)} \sum_{n=1}^\infty \frac{1}{n} \cos\left[n\pi(1 - z_D)\right] \\ & \cdot \left[\sin(n\pi l_D) - \sin(n\pi d_D)\right] \int_{1/4t_s}^\infty \exp\left(-y - \frac{\beta^2 n^2\pi^2}{4y}\right) \frac{dy}{y} \end{aligned} \tag{9-136}$$

The sum of Eqs. (9-131) and (9-136) is the equation of Hantush (Hantush, 1961c, 1964), as given by Eq. (8-42) of Chapter 8.

9.5.3.6 Evaluation of the Solutions. Like the evaluation of the fully penetrating well solution (Section 9.5.2.6), Neuman (1974) presented a comprehensive evaluation for Eq. (9-115) along with Eq. (9-87) by a numerical integration procedure as described in Section 9.5.2.6. In the following sections, the results of Neuman's evaluation are presented.

Comparison with Hantush and Dagan Solutions. It is instructive to compare the results of Neuman's partially-penetrating delayed response model for unconfined aquifers with Hantush's partially-penetrating well model for confined elastic aquifers (Hantush, 1961c, 1964), which is given by Eq. (8-42) of Chapter 8, and Dagan's partially penetrating well model for unconfined rigid aquifers (Dagan, 1967a,b). At large values of dimensionless time t_s, the Neuman solution, as given by Eq. (9-115), behaves as if the aquifer were rigid

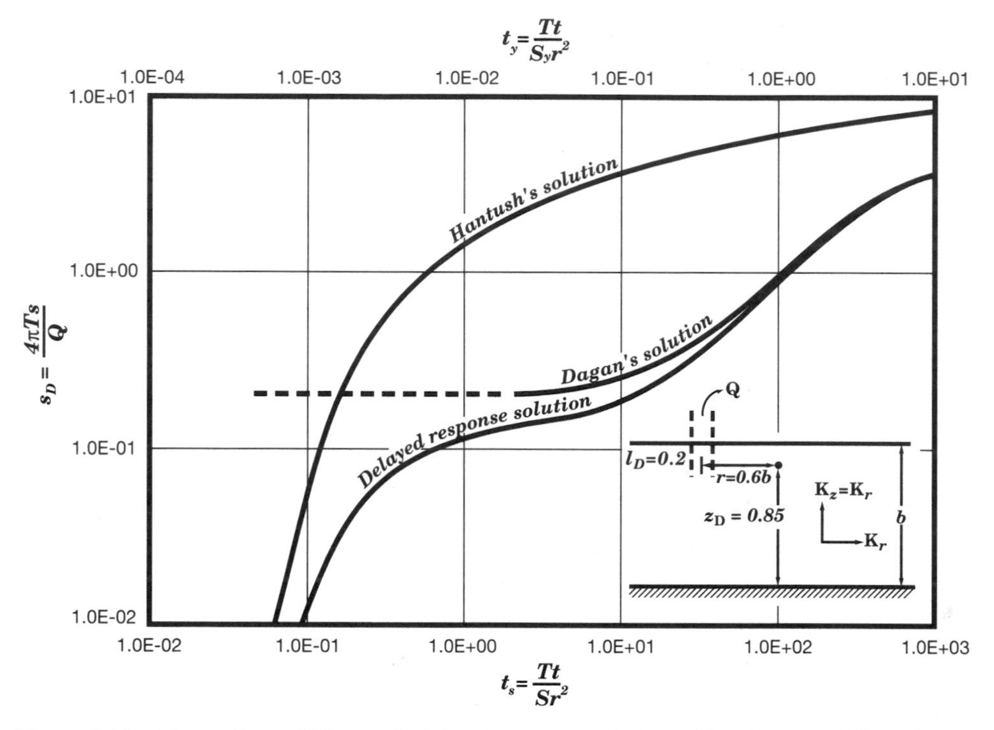

Figure 9-23 Comparison of Neuman's delayed response solution with solutions of Hantush and Dagan for $\sigma = 10^{-2}$, $\beta = 0.36$, $l_D = l/b = 0.2$, $d_D = d/b = 0$, and $z_D = z/b = 0.85$. Hantush's solution: s_D versus t_s. Dagan's solution: s_D versus t_y. (After Neuman, 1974.)

and S_s were zero ($S_s = 0$ corresponds to $t_s \to \infty$). Dagan's solution corresponds to this asymptotic behavior. Eq. (9-87) cannot be reduced analytically to Dagan's solution. As a result, the correspondence between these two solutions at large values of dimensionless time will be demonstrated below numerically.

Comparison of the three models mentioned above is shown in Figure 9-23 which corresponds $\sigma = S/S_y = 10^{-2}$, $\beta = 0.36$, $l_D = l/b = 0.2$, $d_D = d/b = 0$, and $z_D = z/b = 0.85$. Some typical properties of Figure 9-23 are given below (Neuman, 1974):

1. The delayed response curve based on Eq. (9-115) is similar to that shown in Figure 9-2. At early values of time, this curve approaches Hantush's solution. This indicates that water is released from storage primarily by compaction of the aquifer material and expansion of the water.
2. During the second stage the contribution by gravity drainage becomes important. As explained for the fully penetrating well case in Section 9.5.2.6, the smaller the value of σ, the more effective the gravity drainage.
3. As dimensionless time t_s increases, the elastic storage effect at the point under question dissipates completely, and the s_D versus t_s curve approaches Dagan's solution asymptotically. In Figure 9-23, Dagan's solution is expressed in terms of t_y, whereas Hantush's solution is expressed in terms of t_s. The delayed response solution of

Neuman can be expressed in terms of t_s or t_y, since these two parameters are related by Eq. (9-121).

4. Under completely penetrating pumping well conditions, the drawdown curves are bounded by the Theis solution with respect to t_s during the early stage and again by the Theis solution with respect to t_y during the late gravity drainage stage (see Section 9.5.2.6). In the case of partial penetration in unconfined aquifers, Hantush's solution becomes the envelope at early time and Dagan's solution becomes the envelope at later times to the drawdown versus time curve. The behavior of Dagan's solution at $t_y < 10^{-2}$ is indicated by the dashed curve in Figure 9-23.

Effects of Elevation, Radial Distance, and Penetration on Drawdown Versus Time Curves. Figure 9-24 shows the variation of the dimensionless drawdown with elevation and radial distance from the pumping well at $t_s = 1/\beta$ for $K_D = K_z/K_r = 1$ (isotropic aquifer) and for $\sigma = 10^{-2}$, and the well penetrates two-tenths of the upper original saturated thickness. As can be seen from Figure 9-24, for the distances exceeding those corresponding to $\beta = 1$, the drawdown at the water table is less than the drawdown at greater depths. This is the reason why the terms *delayed yield*, *delayed gravity response*, or *delayed response of the water table* are derived and commonly used in ground-water hydrology (Neuman, 1974).

In Figure 9-25, the effect of the depth of penetration to drawdown at the bottom of the aquifer ($z_D = z/b = 0$) for $\sigma = 10^{-2}$, $\beta = 10^{-3}$, $K_r = 10K_z$, and $r = 0.1b$ is shown. Some key aspects are summarized below based on Neuman (1974):

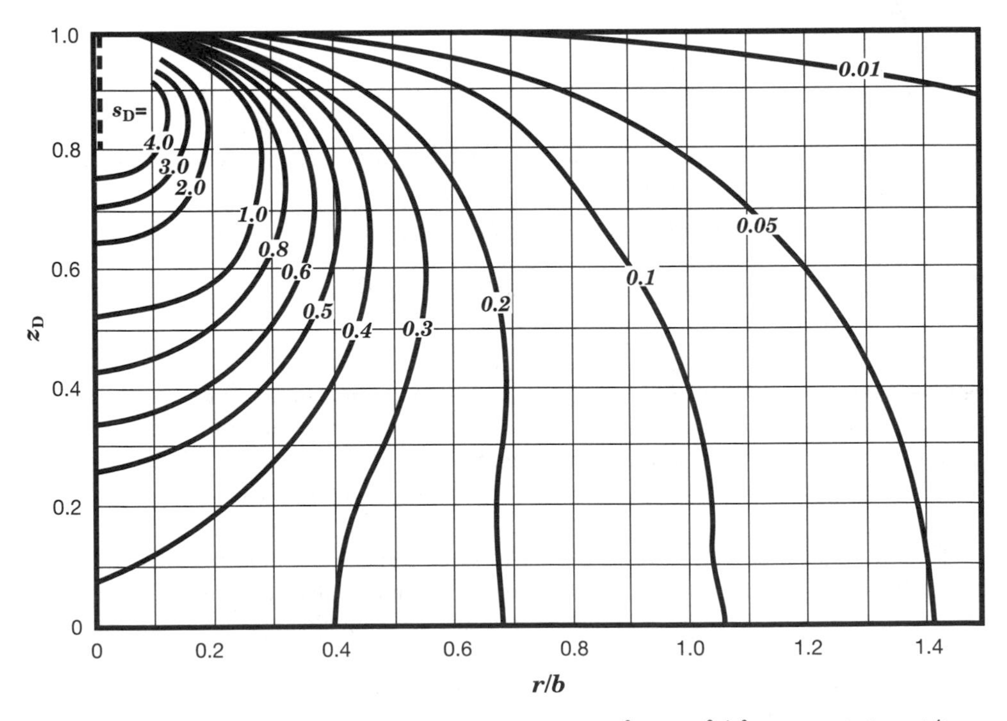

Figure 9-24 Dimensionless drawdown contours for $\sigma = 10^{-2}$, $\beta = r^2/b^2$, $t_s\beta = 1.0$, $l_D = l/b = 0.2$, and $d_D = d/b = 0$. (After Neuman, 1974.)

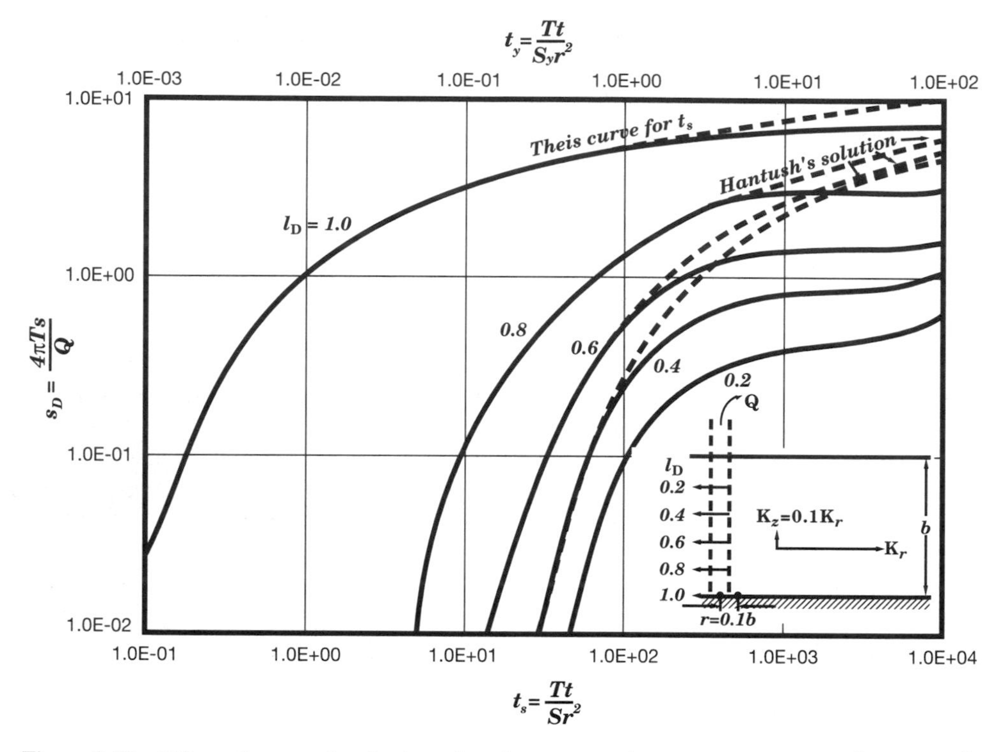

Figure 9-25 Effect of penetration depth on drawdown versus time curves for $\sigma = 10^{-2}$, $\beta = 10^{-3}$, $d_D = d/b = 0$, and $z_D = z/b = 0$. (After Neuman, 1974.)

1. As the well penetrates nearer to the point of observation, an earlier response is observed and the drawdown is greater.
2. As a result of small values of β, the early stage of elastic response is highly pronounced, and the drawdown versus time curves coincide with the Theis curve when $l_D = l/b = 1$ or with Hantush's solution when $l_D < 1$ during this stage.

Effects of β on Time Versus Drawdown Curves. According to Eq. (9-118), β represents both the anisotropy and radial distance because it is the product of $K_D = K_z/K_r$ and r^2/b^2. The effect of β on drawdown in the aquifer at $z_D = z/b = 0$ under fully and partially penetrating well conditions is shown in Figure 9-26 for $\sigma = 10^{-2}$ and for $l_D = l/b = 0.2$ and 1.0. Some important properties of Figure 9-26 are as follows (Neuman 1974):

1. The greater the value of β, the less the effect of partial penetration on the drawdown. The effect of partial penetration decreases with time, with an increasing rate as β becomes greater. When $\beta \geq 1$, this effect disappears completely at dimensionless time values $t_y \geq 10$. This means that flow becomes essentially horizontal and the water is released by only gravity drainage. This is consistent with the Dupuit assumptions (see Chapter 3, Section 3.3).
2. The effect of elastic storage decreases with increasing values of β.

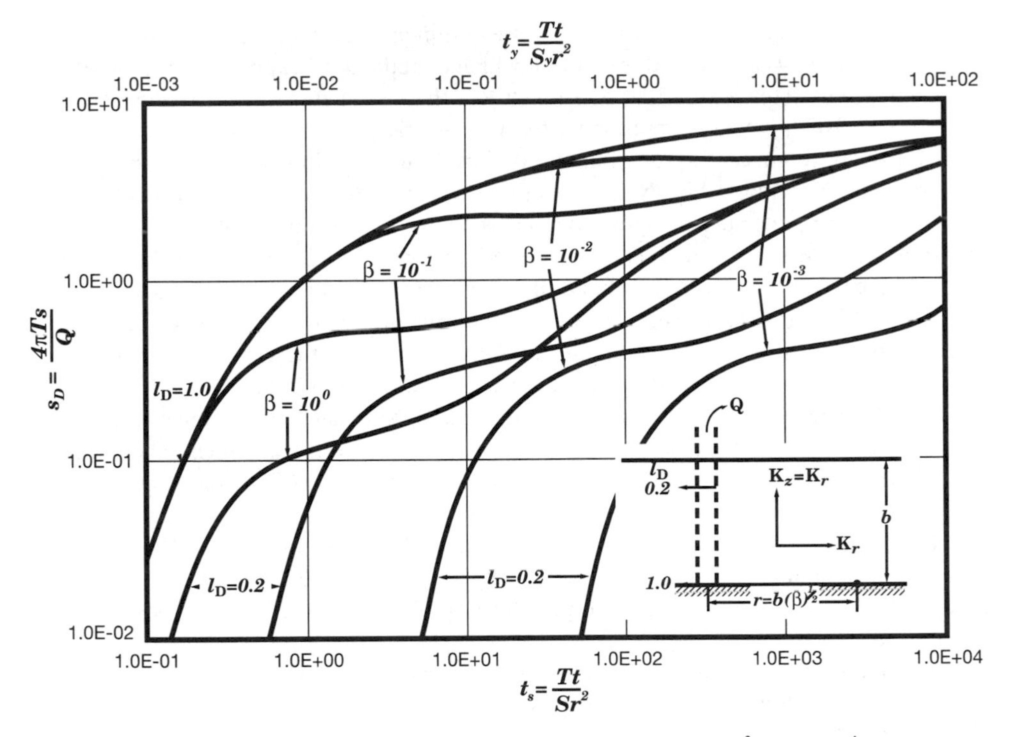

Figure 9-26 Effect of β on drawdown versus time curves for $\sigma = 10^{-2}$, $d_D = d/b = 0$, $z_D = z/b = 0$, and $l_D = l/b = 0.2$ and 1.0. (After Neuman, 1974.)

9.5.4 Application to Pumping Test Data Analysis

The mathematical models of Neuman presented in Section 9.5.2 and Section 9.5.3 for fully penetrating and partially penetrating wells in unconfined aquifers, respectively, have practically important applications in determining aquifer parameters from pumping test data. In the following sections, the data analysis methods will be presented in detail with examples based on Neuman (1975). First, the data analysis methods will be presented for fully penetrating wells. Then, the data analysis methods will be presented for partially penetrating wells.

9.5.4.1 Data Analysis Methods for Fully Penetrating Wells Neuman (1975) developed several methods in analyzing pumping test data for unconfined aquifers. In the following sections, these methods as well as some additional details regarding these methods are presented with examples.

Neuman's Type-Curve Method. The fully penetrating well model, as given by Eq. (9-94), is expressed in terms of three independent parameters: σ, β, t_s, or t_y. The dimensionless time parameters in Eq. (9-94) are related to each other by $t_y = \sigma t_s$, which is the last expression of Eqs. (9-83). It is practically impossible to deal with these three parameters to construct a sufficient number of type curves to analyze a pumping test data. The type curves are normally expressed in terms of not more than two independent parameters.

In order to reduce the number of independent dimensionless parameters from three to two, Neuman (1974) considered the case in which σ in the third expression of Eqs. (9-83) approaches zero. This means that S is much less than S_y. Neuman (1975) suggests that we use $\sigma = 10^{-9}$, which is appropriate for all practical purposes, in generating the type curves. For convenience, the terminology used by Boulton (1963), which is presented in Section 9.4.5, was adopted by Neuman in presenting the Neuman type curves. The type-curve data were generated using a modified version of the original DELAY2 computer program developed by Neuman (1974). The modified program is called DELAY2PC, and modifications were made by the author of this book. Modifications in DELAY2PC are only related to the input data instructions and output structure, and all the rest of part of the original DELAY2 program, including its algorithm, were kept the same. The Type A and Type B type-curve data for fully penetrating wells are given in Table 9-6 and Table 9-7, respectively. These tables are the same as the tables in Neuman (1975, Tables 1a and 1b), with the exception that the t_s and t_y arguments have some additional values. The program requires only t_s as the dimensionless time. Therefore, the values for Type B curves data (Table 9-7) were generated by the program with $t_s = 10^9 t_y$. Based on the data in Table 9-6 and Table 9-7, Type A (for early drawdown data) and Type B (for late drawdown data) curves are presented in Figure 9-27 and Figure 9-28, respectively. As can be seen from Figures 9-27 and 9-28, the scales of Type A and Type B curves are different because σ is an extremely small number.

A methodology similar to the Boulton type-curve method, which is presented in Section 9.4.6.3, will be followed for the Neuman type-curve method. The steps of the method are given below (Neuman, 1975):

Step 1: Plot the family of Neuman type curves of s_D versus t_s and t_y (Type A and Type B family of type curves) for a practical range of values of β separately on log–log paper using the data in Tables 9-6 and 9-7. A suggested range of values for β are 0.001, 0.004, 0.01, 0.03, 0.06, 0.1, 0.2, 0.4, 0.6, 0.8, 1.0, 1.5, 2.0, 2.5, 3.0, 4.0, 5.0, 6.0, and 7.0.

Step 2: Calculate the ratio of $s_{\max}/b$ for each observation well. Based on this ratio, if the maximum drawdown is a significant portion of the aquifer thickness (b), calculate the corrected drawdowns for only late drawdown data using Jacob's correction scheme as given by Eq. (9-64). On another sheet of log–log paper with the same scale as the ones in Step 1, plot the drawdown s against the corresponding time t for a given observation well at a distance r from the pumped well.

Step 3: Keeping the coordinate axes parallel at all times—that is, s_D axis parallel with the s axis and t_y axis parallel with the t axis—first superimpose the drawdown versus field data curve on Type B curves. Adjust until as large as possible an amount of late data fall on one of the Type B curves. Note the β value of the selected Type B curve.

Step 4: Select an arbitrary point B on the overlapping of the two sheets of graph paper, or even anywhere on the overlapping portion of the sheets. Note the values of the coordinates s and s_D along the vertical axis and t and t_y along the horizontal axis.

Step 5: Using the known values of Q, s, and s_D, calculate the transmissivity value from Eq. (9-94):

$$T = \frac{Q}{4\pi} \frac{s_D}{s} \tag{9-137}$$

TABLE 9-6 Values of s_D for t_s and β for Neuman Type A Curves for Fully Penetrating Unconfined Aquifers

	s_D								
t_s	$\beta = 0.001$	$\beta = 0.004$	$\beta = 0.01$	$\beta = 0.03$	$\beta = 0.06$	$\beta = 0.1$	$\beta = 0.2$	$\beta = 0.4$	$\beta = 0.6$
1.0E−01	0.02336	0.02440	0.02403	0.02350	0.02297	0.02240	0.02140	0.01994	0.01880
1.5E−01	0.07708	0.07647	0.07531	0.07326	0.07117	0.06908	0.06525	0.05977	0.05561
2.0E−01	0.14411	0.14235	0.14013	0.13566	0.13127	0.12689	0.11885	0.10758	0.09884
2.5E−01	0.21575	0.21279	0.20908	0.20178	0.19459	0.18744	0.17450	0.15589	0.14161
3.0E−01	0.28790	0.28354	0.27825	0.26778	0.25749	0.24691	0.22809	0.20147	0.18105
3.5E−01	0.35766	0.35187	0.34489	0.33108	0.31750	0.30400	0.27918	0.24374	0.21685
4.0E−01	0.42407	0.41676	0.40807	0.39080	0.37383	0.35696	0.32648	0.28262	0.24906
5.0E−01	0.54938	0.53813	0.52592	0.50160	0.47773	0.45399	0.41036	0.34870	0.30229
6.0E−01	0.66174	0.64836	0.63261	0.60123	0.57042	0.53981	0.48266	0.40323	0.34335
7.0E−01	0.76346	0.74710	0.72784	0.68951	0.65187	0.61447	0.54572	0.44803	0.37587
8.0E−01	0.85606	0.83676	0.81408	0.76891	0.72459	0.68053	0.59956	0.48504	0.40145
9.0E−01	0.93916	0.91697	0.89097	0.84081	0.78995	0.73940	0.64650	0.51580	0.42168
1.0E+00	1.01726	0.99227	0.96298	0.90467	0.84909	0.79219	0.68768	0.54237	0.43779
2.0E+00	1.57138	1.52166	1.46364	1.34798	1.23454	1.12187	0.91770	0.65874	0.49737
3.0E+00	1.92046	1.85021	1.76834	1.60524	1.44180	1.28367	1.00625	0.68571	0.50574
4.0E+00	2.17014	2.08217	1.97965	1.77538	1.57176	1.37588	1.04563	0.69300	0.50711
5.0E+00	2.36518	2.26122	2.14024	1.89908	1.66349	1.43260	1.06454	0.69513	0.50736
6.0E+00	2.52475	2.40628	2.26839	1.99358	1.72624	1.46917	1.07406	0.69579	0.50740
7.0E+00	2.65952	2.52762	2.37407	2.06814	1.77229	1.49348	1.07900	0.69599	0.50741
8.0E+00	2.77604	2.63154	2.46334	2.12836	1.80686	1.51001	1.08162	0.69606	0.50742
9.0E+00	2.87851	2.72210	2.54009	2.17784	1.83326	1.52144	1.08304	0.69608	0.50742
1.0E+01	2.96980	2.80216	2.60704	2.21906	1.85367	1.52946	1.08382	0.69609	0.50742
2.0E+01	3.55857	3.29914	2.99737	2.41285	1.92463	1.54914	1.08480	0.69610	0.50742

TABLE 9-6 *(continued)*

t_s	s_D								
	$\beta = 0.001$	$\beta = 0.004$	$\beta = 0.01$	$\beta = 0.03$	$\beta = 0.06$	$\beta = 0.1$	$\beta = 0.2$	$\beta = 0.4$	$\beta = 0.6$
3.0E+01	3.88887	3.55845	3.17497	2.46644	1.93385	1.55007	1.08480	0.69610	0.50742
4.0E+01	4.11437	3.72399	3.27342	2.48450	1.93532	1.55013	1.08480	0.69610	0.50742
5.0E+01	4.28328	3.83996	3.33309	2.49117	1.93558	1.55013	1.08480	0.69610	0.50742
6.0E+01	4.41682	3.92571	3.37120	2.49377	1.93563	1.55013	1.08480	0.69610	0.50742
7.0E+01	4.52621	3.99139	3.39638	2.49481	1.93564	1.55013	1.08480	0.69610	0.50742
8.0E+01	4.61820	4.04294	3.41342	2.49525	1.93564	1.55013	1.08480	0.69610	0.50742
9.0E+01	4.69702	4.08413	3.42516	2.49543	1.93564	1.55013	1.08480	0.69609	0.50742
1.0E+02	4.76556	4.11753	3.43335	2.49550	1.93564	1.55013	1.08480	0.69610	0.50742
2.0E+02	5.16256	4.25797	3.45340	2.49556	1.93564	1.55013	1.08480	0.69610	0.50742
3.0E+02	5.34090	4.28777	3.45434	2.49556	1.93564	1.55013	1.08480	0.69610	0.50742
4.0E+02	5.44001	4.29561	3.45439	2.49556	1.93564	1.55013	1.08480	0.69610	0.50742
5.0E+02	5.50104	4.29791	3.45440	2.49556	1.93564	1.55013	1.08480	0.69610	0.50742
6.0E+02	5.54054	4.29861	3.54440	2.49556	1.93564	1.55013	1.08480	0.69610	0.50742
7.0E+02	5.56673	4.29884	3.45440	2.49556	1.93564	1.55013	1.08480	0.69610	0.50742
8.0E+02	5.58431	4.29891	3.45440	2.49556	1.93564	1.55013	1.08480	0.68610	0.50742
9.0E+02	5.59622	4.29893	3.45440	2.49556	1.93564	1.55013	1.08480	0.69610	0.50742
1.0E+03	5.60436	4.29894	3.45440	2.49556	1.93564	1.55013	1.08480	0.69610	0.50742
2.0E+03	5.62263	4.29894	3.45440	2.49556	1.93564	1.55013	1.08480	0.69610	0.50742
3.0E+03	5.62340	4.29894	3.45440	2.49556	1.93564	1.55013	1.08480	0.69610	0.50742
4.0E+03	5.62346	4.29894	3.45440	2.49556	1.93564	1.55013	1.08480	0.69610	0.50742
5.0E+03	5.62346	4.29894	3.45440	2.49556	1.93564	1.55013	1.08480	0.69610	0.50742
6.0E+03	5.62346	4.29894	3.45440	2.49556	1.93564	1.55013	1.08481	0.69610	0.50742
7.0E+03	5.62346	4.29894	3.45440	2.49556	1.93564	1.55013	1.08481	0.69610	0.50742
8.0E+03	5.62346	4.29894	3.45440	2.49556	1.93564	1.55013	1.08481	0.69610	0.50742
9.0E+03	5.62346	4.29894	3.45440	2.49557	1.93564	1.55013	1.08481	0.69610	0.50742
1.0E+04	5.62346	4.29894	3.45440	2.49557	1.93564	1.55013	1.08481	0.69610	0.50742

TABLE 9-6 *(continued)*

t_s	s_D									
	$\beta = 0.8$	$\beta = 1.0$	$\beta = 1.5$	$\beta = 2.0$	$\beta = 2.5$	$\beta = 3.0$	$\beta = 4.0$	$\beta = 5.0$	$\beta = 6.0$	$\beta = 7.0$
1.0E−01	0.01786	0.01702	0.01526	0.01376	0.01247	0.01131	0.00933	0.00772	0.00639	0.00530
1.5E−01	0.05210	0.04904	0.04248	0.03700	0.03238	0.02835	0.02184	0.01692	0.01318	0.01031
2.0E−01	0.09148	0.08492	0.07131	0.06031	0.05111	0.04350	0.03172	0.02335	0.01737	0.01306
2.5E−01	0.12944	0.11890	0.09713	0.07978	0.06596	0.05470	0.03817	0.02709	0.01954	0.01430
3.0E−01	0.16411	0.14918	0.11850	0.09516	0.07682	0.06254	0.04217	0.02911	0.02057	0.01484
3.5E−01	0.19435	0.17511	0.13588	0.10671	0.08464	0.06779	0.04454	0.03018	0.02105	0.01505
4.0E−01	0.22084	0.19692	0.14990	0.11546	0.09023	0.07121	0.04592	0.03075	0.02127	0.01514
5.0E−01	0.26382	0.23113	0.16961	0.12683	0.09681	0.07502	0.04720	0.03118	0.02142	0.01519
6.0E−01	0.29552	0.25564	0.18174	0.13313	0.10002	0.07666	0.04760	0.03130	0.02145	0.01520
7.0E−01	0.31914	0.27279	0.18946	0.13675	0.10160	0.07737	0.04774	0.03133	0.02145	0.01520
8.0E−01	0.33682	0.28502	0.19433	0.13869	0.10237	0.07768	0.04779	0.03133	0.02145	0.01520
9.0E−01	0.34965	0.29379	0.19741	0.13978	0.10275	0.07781	0.04781	0.03134	0.02145	0.01520
1.0E+00	0.35974	0.30011	0.19938	0.14039	0.10294	0.07787	0.04782	0.03134	0.02145	0.01520
2.0E+00	0.39076	0.31651	0.20291	0.14120	0.10314	0.07792	0.04782	0.03134	0.02145	0.01520
3.0E+00	0.39340	0.31736	0.20296	0.14120	0.10314	0.07792	0.04782	0.03134	0.02145	0.01520
4.0E+00	0.39367	0.31741	0.20296	0.14120	0.10314	0.07792	0.04782	0.03134	0.02145	0.01520
5.0E+00	0.39370	0.31741	0.20296	0.14120	0.10314	0.07792	0.04782	0.03134	0.02145	0.01520
6.0E+00	0.39370	0.31741	0.20296	0.14120	0.10314	0.07792	0.04782	0.03134	0.02145	0.01520
7.0E+00	0.39370	0.31741	0.20296	0.14120	0.10314	0.07792	0.04782	0.03134	0.02145	0.01520
8.0E+00	0.39370	0.31741	0.20296	0.14120	0.10314	0.07792	0.04782	0.03134	0.02145	0.01520
9.0E+00	0.39370	0.31541	0.20296	0.14120	0.10314	0.07792	0.04782	0.03134	0.02145	0.01520
1.0E+01	0.39370	0.31741	0.20296	0.14120	0.10314	0.07792	0.04782	0.03134	0.02145	0.01520
2.0E+01	0.39370	0.31741	0.20296	0.14120	0.10314	0.07792	0.04782	0.03134	0.02145	0.01520

TABLE 9-6 *(continued)*

t_s	s_D									
	$\beta = 0.8$	$\beta = 1.0$	$\beta = 1.5$	$\beta = 2.0$	$\beta = 2.5$	$\beta = 3.0$	$\beta = 4.0$	$\beta = 5.0$	$\beta = 6.0$	$\beta = 7.0$
3.0E+01	0.39370	0.31741	0.20296	0.14120	0.10314	0.07792	0.04782	0.03134	0.02145	0.01520
4.0E+01	0.39370	0.31741	0.20296	0.14120	0.10314	0.07792	0.04782	0.03134	0.02145	0.01520
5.0E+01	0.39370	0.31741	0.20296	0.14120	0.10314	0.07792	0.04782	0.03134	0.02145	0.01520
6.0E+01	0.39370	0.31741	0.20296	0.14120	0.10314	0.07792	0.04782	0.03134	0.02145	0.01520
7.0E+01	0.39370	0.31741	0.20296	0.14120	0.10314	0.07792	0.04782	0.03134	0.02145	0.01520
8.0E+01	0.39370	0.31741	0.20296	0.14120	0.10314	0.07792	0.04782	0.03134	0.02145	0.01520
9.0E+01	0.39370	0.31741	0.20296	0.14120	0.10314	0.07792	0.04782	0.03134	0.02145	0.01520
1.0E+02	0.39370	0.31741	0.20296	0.14120	0.10314	0.07792	0.04782	0.03134	0.02145	0.01520
2.0E+02	0.39370	0.31741	0.20296	0.14120	0.10314	0.07792	0.04782	0.03134	0.02145	0.01520
3.0E+02	0.39370	0.31741	0.20296	0.14120	0.10314	0.07792	0.04782	0.03134	0.02145	0.01520
4.0E+02	0.39370	0.31741	0.20296	0.14120	0.10314	0.07792	0.04782	0.03134	0.02145	0.01520
5.0E+02	0.39370	0.31741	0.20296	0.14120	0.10314	0.07792	0.04782	0.03134	0.02145	0.01520
6.0E+02	0.39370	0.31741	0.20296	0.14120	0.10314	0.07792	0.04782	0.03134	0.02145	0.01520
7.0E+02	0.39370	0.31741	0.20296	0.14120	0.10314	0.07792	0.04782	0.03134	0.02145	0.01520
8.0E+02	0.39370	0.31741	0.20296	0.14120	0.10314	0.07792	0.04782	0.03134	0.02145	0.01520
9.0E+02	0.39370	0.31741	0.20296	0.14120	0.10314	0.07792	0.04782	0.03134	0.02145	0.01520
1.0E+03	0.39370	0.31741	0.20296	0.14120	0.10314	0.07792	0.04782	0.03134	0.02145	0.01520
2.0E+03	0.39370	0.31741	0.20296	0.14120	0.10314	0.07792	0.04782	0.03134	0.02146	0.01520
3.0E+03	0.39370	0.31741	0.20296	0.14120	0.10314	0.07792	0.04782	0.03134	0.02146	0.01520
4.0E+03	0.39370	0.31742	0.20296	0.14120	0.10314	0.07793	0.04782	0.03134	0.02146	0.01520
5.0E+03	0.39370	0.31742	0.20296	0.14120	0.10314	0.07793	0.04782	0.03134	0.02146	0.01520
6.0E+03	0.39370	0.31742	0.20296	0.14121	0.10314	0.07793	0.04782	0.03134	0.02146	0.01520
7.0E+03	0.39371	0.31742	0.20297	0.14121	0.10314	0.07793	0.04783	0.03134	0.02146	0.01521
8.0E+03	0.39371	0.31742	0.20297	0.14121	0.10314	0.07793	0.04783	0.03134	0.02146	0.01521
9.0E+03	0.39371	0.31742	0.20297	0.14121	0.10315	0.07793	0.04783	0.03134	0.02146	0.01521
1.0E+04	0.39371	0.31742	0.20297	0.14121	0.10315	0.07793	0.04783	0.03135	0.02146	0.01521

TABLE 9-7 Values of s_D for t_y and β for Neuman Type B Curves for Fully Penetrating Unconfined Aquifers

	s_D								
t_y	$\beta = 0.001$	$\beta = 0.004$	$\beta = 0.01$	$\beta = 0.03$	$\beta = 0.06$	$\beta = 0.1$	$\beta = 0.2$	$\beta = 0.4$	$\beta = 0.6$
1.0E−04	5.62346	4.29895	3.45440	2.49558	1.93566	1.55016	1.08485	0.69616	0.50750
1.5E−04	5.62346	4.29895	3.45441	2.49558	1.93567	1.55018	1.08487	0.69620	0.50754
2.0E−04	5.62346	4.29895	3.45441	2.49559	1.93568	1.55019	1.08490	0.69623	0.50758
2.5E−04	5.62346	4.29895	3.45441	2.49559	1.93569	1.55021	1.08492	0.69626	0.50762
3.0E−04	5.62346	4.29895	3.45442	2.49560	1.93570	1.55022	1.08494	0.69630	0.50766
3.5E−04	5.62346	4.29895	3.45442	2.49561	1.93571	1.55024	1.08496	0.69633	0.50770
4.0E−04	5.62346	4.29895	3.45442	2.49561	1.93572	1.55025	1.08499	0.69637	0.50774
5.0E−04	5.62347	4.29896	3.45443	2.49562	1.93574	1.55028	1.08503	0.69643	0.50782
6.0E−04	5.62347	4.29896	3.45443	2.49564	1.93576	1.55031	1.08508	0.69650	0.50790
7.0E−04	5.62347	4.29896	3.45444	2.49565	1.93578	1.55034	1.08513	0.69657	0.50799
8.0E−04	5.62347	4.29896	3.45444	2.49566	1.93580	1.55037	1.08517	0.69664	0.50807
9.0E−04	5.62347	4.29897	3.45445	2.49567	1.93583	1.55040	1.08522	0.69670	0.50815
1.0E−03	5.62347	4.29897	3.45445	2.49569	1.93585	1.55043	1.08527	0.69677	0.50823
2.0E−03	5.62348	4.29899	3.45451	2.49581	1.93606	1.55074	1.08573	0.69745	0.50904
3.0E−03	5.62348	4.29902	3.45456	2.49593	1.93627	1.55104	1.08619	0.69813	0.50986
4.0E−03	5.62349	4.29905	3.45462	2.49605	1.93647	1.55134	1.08666	0.69880	0.51067
5.0E−03	5.62350	4.29907	3.45467	2.49617	1.93668	1.55164	1.08712	0.69948	0.51149
6.0E−03	5.62350	4.29910	3.45473	2.49630	1.93689	1.55195	1.08758	0.70015	0.51230
7.0E−03	5.62351	4.29912	3.45478	2.49642	1.93710	1.55225	1.08804	0.70083	0.51311
8.0E−03	5.62352	4.29915	3.45484	2.49654	1.93731	1.55255	1.08851	0.70151	0.51392
9.0E−03	5.62353	4.29917	3.45489	2.49666	1.93752	1.55285	1.08897	0.70218	0.51474
1.0E−02	5.62353	4.29920	3.45495	2.49678	1.93773	1.55316	1.08943	0.70286	0.51637
2.0E−02	5.62360	4.29946	3.45550	2.49800	1.93982	1.55618	1.09405	0.70961	0.52448
3.0E−02	5.62368	4.29971	3.45605	2.49922	1.94191	1.55919	1.09866	0.71635	0.53257
4.0E−02	5.62375	4.29997	3.45660	2.50043	1.94399	1.56220	1.10327	0.72307	0.54064
5.0E−02	5.62382	4.30023	3.45715	2.50164	1.94608	1.56520	1.10786	0.72978	0.54870
6.0E−02	5.62389	4.30048	3.45769	2.50285	1.94816	1.56820	1.11245	0.73647	0.55673
7.0E−02	5.62396	4.30074	3.45824	2.50407	1.95024	1.57472	1.11703	0.74315	0.56474

TABLE 9-7 (*continued*)

t_y	s_D								
	$\beta = 0.001$	$\beta = 0.004$	$\beta = 0.01$	$\beta = 0.03$	$\beta = 0.06$	$\beta = 0.1$	$\beta = 0.2$	$\beta = 0.4$	$\beta = 0.6$
8.0E−02	5.62403	4.30100	3.45879	2.50528	1.95231	1.57771	1.12161	0.74982	0.57274
9.0E−02	5.62410	4.30125	3.45934	2.50650	1.95439	1.58070	1.12617	0.75647	0.58071
1.0E−01	5.62417	4.30151	3.45989	2.50771	1.95646	1.58369	1.13073	0.76311	0.58866
2.0E−01	5.62488	4.30406	3.46535	2.51982	1.97705	1.61326	1.17584	0.82857	0.66698
3.0E−01	5.62559	4.30661	3.47079	2.53187	1.99743	1.64235	1.22009	0.89086	0.74299
4.0E−01	5.62630	4.30916	3.47620	2.54387	2.01760	1.67099	1.26346	0.95279	0.81653
5.0E−01	5.62701	4.31170	3.48160	2.55580	2.03756	1.69918	1.30594	1.01291	0.88751
6.0E−01	5.62772	4.31423	3.48696	2.56767	2.05731	1.72694	1.34752	1.07119	0.95419
7.0E−01	5.62843	4.31675	3.49230	2.57948	2.07683	1.75426	1.38821	1.12766	1.02000
8.0E−01	5.62913	4.31927	3.49762	2.59122	2.09616	1.78117	1.42801	1.18233	1.08327
9.0E−01	5.62984	4.32178	3.50292	2.60290	2.11527	1.80767	1.46693	1.23522	1.4405
1.0E+00	5.63055	4.32428	3.50816	2.61448	2.13419	1.83377	1.50497	1.28639	1.20244
2.0E+00	5.63757	4.34899	3.55963	2.72657	2.31232	2.07401	1.84507	1.71954	1.68048
3.0E+00	5.64455	4.37309	3.60895	2.83133	2.47181	2.28050	2.11477	2.03791	2.01730
4.0E+00	5.65147	4.39661	3.65626	2.92900	2.61490	2.45931	2.33585	2.28584	2.27371
5.0E+00	5.65833	4.41959	3.70173	3.02006	2.74389	2.61578	2.52144	2.48720	2.47942
6.0E+00	5.66515	4.44205	3.74546	3.10519	2.86088	2.75423	2.68045	2.65605	2.65074
7.0E+00	5.67191	4.46401	3.78760	3.18493	2.96765	2.87800	2.81918	2.80117	2.79735
8.0E+00	5.67856	4.48549	3.82823	3.25984	3.06566	2.98964	2.94195	2.92826	2.92540
9.0E+00	5.68522	4.50652	3.86746	3.33041	3.15612	3.09115	3.05188	3.04123	3.03901
1.0E+01	5.69183	4.52710	3.90541	3.39705	3.24001	3.18411	3.15133	3.14287	3.14111
2.0E+01	5.75539	4.71277	4.22812	3.91226	3.84756	3.83144	3.82221	3.82059	3.82021
3.0E+01	5.81475	4.86999	4.47913	4.26437	4.23259	4.22655	4.22186	4.22138	4.22123
4.0E+01	5.87046	5.00682	4.68454	4.52975	4.51254	4.51013	4.50692	4.50681	4.50674
5.0E+01	5.92298	5.12830	4.85823	4.74177	4.73193	4.73114	4.72857	4.72863	4.72859
6.0E+01	5.97268	5.23766	5.00847	4.91782	4.91208	4.91216	4.90993	4.91007	4.91006
8.0E+01	6.06483	5.42851	5.25874	5.19918	5.19751	5.19834	5.19643	5.19665	5.19665
1.0E+02	6.14883	5.59154	5.46200	5.41938	5.41951	5.42069	5.41891	5.41916	5.41916

TABLE 9-7 *(continued)*

t_y	s_D									
	$\beta = 0.8$	$\beta = 1.0$	$\beta = 1.5$	$\beta = 2.0$	$\beta = 2.5$	$\beta = 3.0$	$\beta = 4.0$	$\beta = 5.0$	$\beta = 6.0$	$\beta = 7.0$
1.0E−04	0.39379	0.31751	0.20306	0.14130	0.10324	0.07802	0.04791	0.03142	0.02152	0.01526
1.5E−04	0.39383	0.31756	0.20311	0.14136	0.10329	0.07807	0.04795	0.03145	0.02156	0.01529
2.0E−04	0.39388	0.31760	0.20316	0.14141	0.10334	0.07812	0.04800	0.03149	0.02159	0.01532
2.5E−04	0.39392	0.31765	0.20322	0.14146	0.10339	0.07817	0.04804	0.03153	0.02163	0.01535
3.0E−04	0.39397	0.31770	0.20327	0.14151	0.10344	0.07822	0.04808	0.03157	0.02166	0.01538
3.5E−04	0.39401	0.31775	0.20332	0.14156	0.10349	0.07826	0.04813	0.03161	0.02169	0.01541
4.0E−04	0.39406	0.31780	0.20337	0.14162	0.10354	0.07831	0.04817	0.03165	0.02173	0.01544
5.0E−04	0.39415	0.31789	0.20347	0.14172	0.10365	0.07841	0.04826	0.03173	0.02180	0.01550
6.0E−04	0.39424	0.31799	0.20358	0.14182	0.10375	0.07851	0.04835	0.03181	0.02187	0.01556
7.0E−04	0.39433	0.31808	0.20368	0.14193	0.10385	0.07861	0.04844	0.03189	0.02193	0.01562
8.0E−04	0.39442	0.31818	0.20378	0.14203	0.10395	0.07871	0.04853	0.03196	0.02200	0.01568
9.0E−04	0.39451	0.31828	0.20389	0.14214	0.10405	0.07880	0.04861	0.03204	0.02207	0.01574
1.0E−03	0.39460	0.31837	0.20399	0.14224	0.10415	0.07890	0.04870	0.03212	0.02214	0.01580
2.0E−03	0.39550	0.31933	0.20502	0.14328	0.10507	0.07988	0.04959	0.03291	0.02284	0.01641
3.0E−03	0.39640	0.32029	0.20605	0.14432	0.10609	0.08087	0.05048	0.03370	0.02353	0.01702
4.0E−03	0.39731	0.32125	0.20708	0.14537	0.10711	0.08185	0.05138	0.03450	0.02425	0.01764
5.0E−03	0.39821	0.32221	0.20812	0.14641	0.10814	0.08284	0.05227	0.03530	0.02496	0.01827
6.0E−03	0.39911	0.32318	0.20915	0.14745	0.10916	0.08383	0.05317	0.03608	0.02568	0.01890
7.0E−03	0.40001	0.32414	0.21018	0.14850	0.11019	0.08482	0.05408	0.03689	0.02640	0.01955
8.0E−03	0.40091	0.32510	0.21121	0.14954	0.11122	0.08582	0.05498	0.03770	0.02712	0.02020
9.0E−03	0.40181	0.32606	0.21225	0.15059	0.11225	0.08689	0.05589	0.03852	0.02784	0.02085
1.0E−02	0.40271	0.32702	0.21328	0.15164	0.11328	0.08789	0.05680	0.03934	0.02857	0.02150
2.0E−02	0.41170	0.33661	0.22362	0.16214	0.12364	0.90797	0.06614	0.04782	0.03624	0.02840
3.0E−02	0.42067	0.34620	0.23398	0.17271	0.13413	0.10822	0.07566	0.05663	0.04438	0.03590
4.0E−02	0.42963	0.35527	0.24436	0.18334	0.14487	0.11852	0.08551	0.06589	0.05293	0.04393
5.0E−02	0.43857	0.36482	0.25508	0.19381	0.15556	0.12907	0.09569	0.07538	0.06201	0.05242
6.0E−02	0.44749	0.37436	0.26549	0.20453	0.16634	0.13977	0.10602	0.08527	0.07144	0.06144
7.0E−02	0.45639	0.38389	0.27590	0.21530	0.17721	0.15074	0.11645	0.09553	0.08118	0.07094

TABLE 9-7 *(continued)*

t_y	s_D									
	$\beta = 0.8$	$\beta = 1.0$	$\beta = 1.5$	$\beta = 2.0$	$\beta = 2.5$	$\beta = 3.0$	$\beta = 4.0$	$\beta = 5.0$	$\beta = 6.0$	$\beta = 7.0$
8.0E−02	0.46527	0.39340	0.28632	0.22610	0.18814	0.16167	0.12719	0.10599	0.09134	0.08074
9.0E−02	0.47413	0.40289	0.29673	0.23694	0.19892	0.17271	0.13811	0.11658	0.10190	0.09099
1.0E−01	0.48297	0.41237	0.30715	0.24780	0.20998	0.18384	0.14935	0.12752	0.11266	0.10166
2.0E−01	0.57103	0.50595	0.41026	0.35730	0.32277	0.29848	0.26628	0.24538	0.23100	0.22035
3.0E−01	0.65558	0.59775	0.51163	0.46457	0.43449	0.41349	0.38596	0.36864	0.35725	0.34852
4.0E−01	0.73721	0.68540	0.61030	0.56966	0.54318	0.52558	0.50293	0.48899	0.47956	0.47277
5.0E−01	0.81572	0.76948	0.70370	0.66892	0.64744	0.63290	0.61451	0.60343	0.59605	0.59080
6.0E−01	0.88931	0.84985	0.79240	0.76274	0.74476	0.73277	0.71786	0.70903	0.70321	0.69911
7.0E−01	0.96140	0.92477	0.87634	0.85105	0.83598	0.82607	0.81390	0.80678	0.80214	0.79888
8.0E−01	1.03033	0.99774	0.95394	0.93231	0.91963	0.91138	0.90135	0.89555	0.89178	0.88915
9.0E−01	1.09621	1.06718	1.02883	1.01028	0.99954	0.99260	0.98427	0.97947	0.97637	0.97420
1.0E+00	1.15917	1.13326	1.09959	1.08358	1.07442	1.06856	1.06155	1.05753	1.05494	1.05313
2.0E+00	1.66297	1.65346	1.63855	1.63369	1.63103	1.62937	1.62739	1.62628	1.62556	1.62506
3.0E+00	2.00874	2.00429	1.99920	1.99702	1.99583	1.99508	1.99419	1.99369	1.99337	1.99314
4.0E+00	2.26887	2.26639	2.26356	2.26235	2.26168	2.26127	2.26077	2.26049	2.26031	2.26018
5.0E+00	2.47638	2.47482	2.47305	2.47228	2.47186	2.47160	2.47128	2.47111	2.47099	2.47091
6.0E+00	2.64868	2.64761	2.64640	2.64588	2.64559	2.64541	2.64520	2.64508	2.64500	2.64494
7.0E+00	2.79586	2.79510	2.79422	2.79385	2.79363	2.79351	2.79335	2.79326	2.79321	2.79316
8.0E+00	2.92427	2.92370	2.92304	2.92275	2.92260	2.92251	2.92239	2.92232	2.92228	2.92225
9.0E+00	3.03814	3.03770	3.03718	3.03696	3.03684	3.03676	3.03668	3.03663	3.03659	3.03657
1.0E+01	3.14041	3.14006	3.13965	3.13947	3.13937	3.13931	3.13925	3.13920	3.13918	3.13916
2.0E+01	3.82006	3.82000	3.81992	3.81988	3.81986	3.81984	3.81984	3.81983	3.81983	3.81983
3.0E+01	4.22118	4.22117	4.22114	4.22113	4.22112	4.22112	4.22113	4.22112	4.22113	4.22113
4.0E+01	4.50672	4.50673	4.50671	4.50670	4.50671	4.50671	4.50672	4.50672	4.50673	4.50673
5.0E+01	4.72859	4.72860	4.72860	4.72859	4.72860	4.72860	4.72861	4.72862	4.72863	4.72863
6.0E+01	4.91006	4.91007	4.91008	4.91007	4.91009	4.91009	4.91010	4.91011	4.91012	4.91012
8.0E+01	5.19665	5.19668	5.19669	5.19669	5.19670	5.19671	5.19672	5.19672	5.19673	5.19674
1.0E+02	5.41916	5.41918	5.41920	5.41920	5.41921	5.41921	5.41922	5.49123	5.41924	5.41924

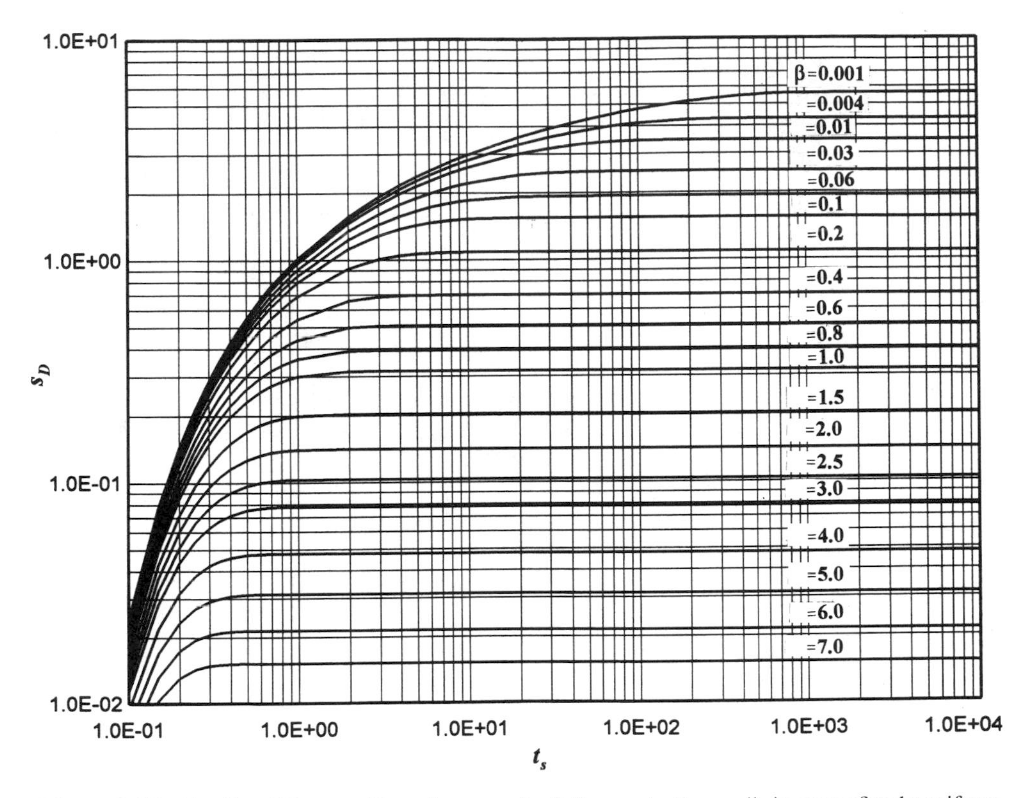

Figure 9-27 Family of Neuman Type A curves for fully penetrating wells in unconfined aquifers: s_D versus t_s.

Also calculate the specific yield value from the second expression of Eq. (9-82):

$$S_y = \frac{Tt}{t_y r^2} \tag{9-138}$$

Equations (9-137) and (9-138) are valid for any set of consistent units. If Q is expressed in gallons per minute (gpm) and the gallon–day–foot system is used for all other quantities, then Eqs. (9-137) and (9-138) take, respectively, the following forms:

$$T = 114.6Q\frac{s_D}{s} \tag{9-139}$$

and

$$S_y = 0.1337\frac{Tt}{t_y r^2} \tag{9-140}$$

where s and r are expressed in feet, Q in gallons per minute (gpm), T in gallons per day per foot (gpd/ft), and t in minutes.

Step 6: Superimpose the drawdown versus time data curve on the Type A curves, keeping the coordinate axes of both graphs parallel to each other and matching as much of the

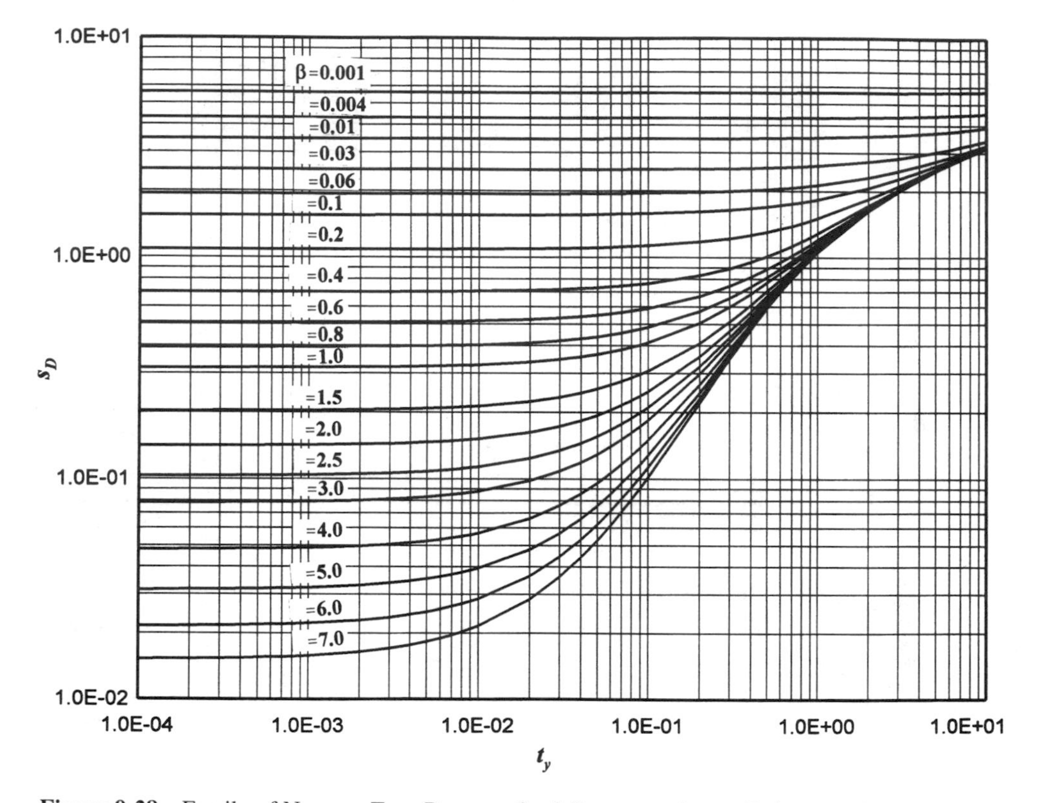

Figure 9-28 Family of Neuman Type B curves for fully penetrating wells in unconfined aquifers: s_D versus t_y.

earliest data curve to a particular type curve as possible. The value of β corresponding to this type curve must be the same as that obtained in Step 3 from the Type B curves.

Step 7: Select an arbitrary point A on the superimposed curves (this point may also be selected anywhere on the overlapping portion of the sheets) and note its coordinates s, s_D, t, and t_s.

Step 8: Substitute these values into Eq. (9-137) and calculate the value of transmissivity (T). This value should be approximately equal to that obtained in Step 5 from the lata drawdown data. Calculate the storage coefficient value from the first expression of Eqs. (9-82):

$$S = \frac{Tt}{t_s r^2} \tag{9-141}$$

For the gallon–foot–day system, as in Eq. (9-140), it takes the following form

$$S = 0.1337 \frac{Tt}{t_s r^2} \tag{9-142}$$

Step 9: From the known transmissivity (T) value of the aquifer, calculate the horizontal hydraulic conductivity (K_r) value from

$$K_r = \frac{T}{b} \tag{9-143}$$

Calculate the degree of anisotropy (K_D) from Eq. (9-90):

$$K_D = \beta \left(\frac{b}{r}\right)^2 \tag{9-144}$$

From the known values of K_D and K_r, calculate the vertical hydraulic conductivity (K_z) value from the first expression of Eqs. (9-78):

$$K_z = K_D K_r \tag{9-145}$$

Step 10: Calculate the parameter σ from the third expression of Eqs. (9-83):

$$\sigma = \frac{S}{S_y} \tag{9-146}$$

Then, calculate the specific storage (S_s) value from

$$S_s = \frac{S}{b} \tag{9-147}$$

Neuman's Semilogarithmic Method. If the data in Tables 9-6 and 9-7 are plotted on semilogarithmic paper, the curves take the form shown in Figure 9-29. As can be seen from Figure 9-29, the late drawdown data tend to fall on a straight line according to Cooper and Jacob (1946). The Cooper and Jacob approximation for the Theis equation is presented by Eq. (4-46) in Section 4.2.7.2 of Chapter 4. In accordance with Eq. (4-46) of Chapter 4, the drawdown equation for the late data is

$$s = \frac{2.303Q}{4\pi T} \log\left(\frac{2.25Tt}{r^2 S_y}\right) \tag{9-148}$$

or, with the dimensionless parameters defined by Eqs. (9-82) and (9-94),

$$s_D = 2.303 \log(2.25 t_y) \tag{9-149}$$

The intermediate data follow approximately a horizontal line (see Figure 9-29), and some of the early data tend to fall around the line

$$s_D = 2.303 \log(2.25 t_s) \tag{9-150}$$

The Relationship Between t_y and β. In order to develop a straight line method, Neuman (1975) first established a relationship between t_y and β. For this purpose, $t_y\beta$ is defined as an additional parameter which is the value of t_y corresponding to the intersection of any horizontal line with the inclined line described by Eq. (9-149). For example, Figure 9-29

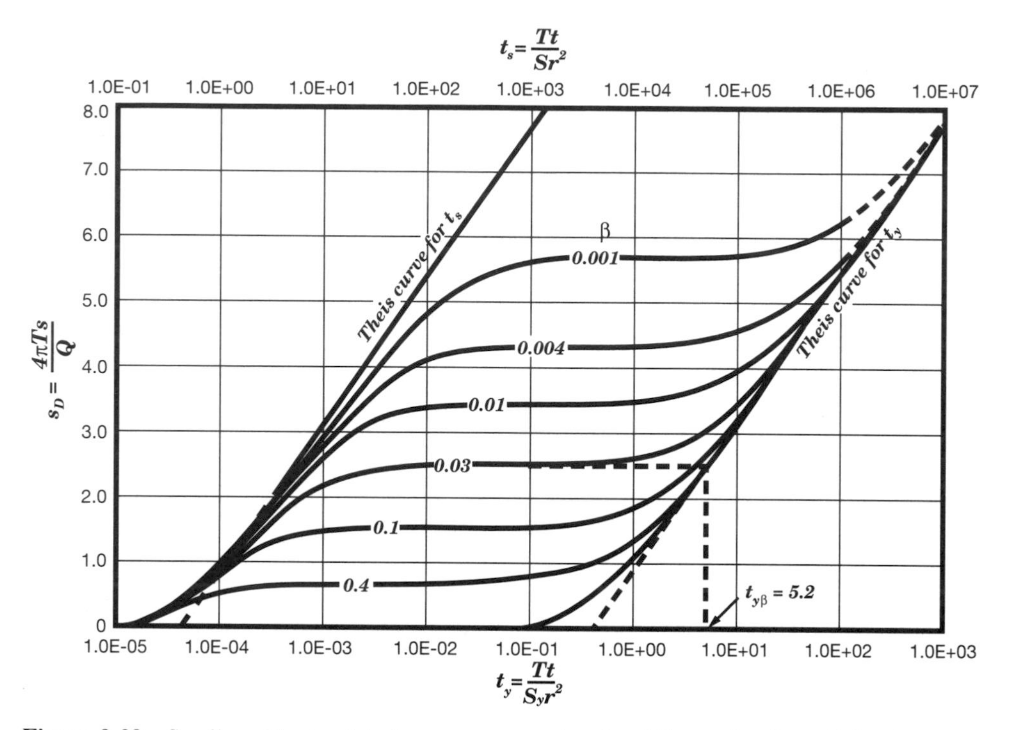

Figure 9-29 Semilogarithmic plot of s_D versus t_s and t_y for different β values. (After Neuman, 1975.)

shows that the value of $t_{y\beta}$ for $\beta = 0.03$ is equal to 5.2. The data for $1/\beta$ versus $t_{y\beta}$ are given in Table 9-8 which was generated by Neuman (1975) with the aid of Tables 9-6 and 9-7, and Eq. (9-149). The graphical display of the values in Table 9-8 on double logarithmic paper is shown in Figure 9-30. A reasonable good approximation for the relationship between β and $t_{y\beta}$ within the range is given by Neuman (1975) as

$$\beta = 0.195 t_{y\beta}^{-1.1053}, \qquad 4.0 \leq t_{y\beta} \leq 100.0 \tag{9-151}$$

This equation is represented by the dashed line in Figure 9-30.

In developing a semilogarithmic method, Neuman (1975) followed a methodology somewhat similar to that previously outlined by Berkaloff (1963) in connection with the theory of Boulton in Section 9.4. The steps of the method are given below based on Neuman (1975):

Step 1: Plot the drawdown s at a given observation well on semilogarithmic paper against the values of time t.

Step 2: Fit a straight line to the late portion of the drawdown versus time data.

Step 3: Find the intersection of this line with the horizontal axis corresponding to $s = 0$. This will be denoted by t_L.

Step 4: Determine the geometric slope of the straight line—that is, the drawdown difference Δs_L per log cycle of time.

TABLE 9-8 Values of $1/\beta$ Versus $t_{y\beta}$ Used in Plotting Figure 9-30

β	$1/\beta$	$t_{y\beta}$
7.0	1.43×10^{-1}	4.52×10^{-1}
6.0	1.67×10^{-1}	4.55×10^{-1}
5.0	2.00×10^{-1}	4.59×10^{-1}
4.0	2.50×10^{-1}	4.67×10^{-1}
3.0	3.33×10^{-1}	4.81×10^{-1}
2.5	4.00×10^{-1}	4.94×10^{-1}
2.0	5.00×10^{-1}	5.13×10^{-1}
1.5	6.67×10^{-1}	5.45×10^{-1}
1.0	1.00×10^{0}	6.11×10^{-1}
0.8	1.25×10^{0}	6.60×10^{-1}
0.6	1.67×10^{0}	7.39×10^{-1}
0.4	2.50×10^{0}	8.93×10^{-1}
0.2	5.00×10^{0}	1.31×10^{0}
0.1	1.00×10^{1}	2.10×10^{0}
0.06	1.67×10^{1}	3.10×10^{0}
0.03	3.33×10^{1}	5.42×10^{0}
0.01	1.00×10^{2}	1.42×10^{1}
0.004	2.50×10^{2}	3.22×10^{1}
0.001	1.00×10^{3}	1.23×10^{2}

Source: After Neuman (1975).

Step 5: Note that in Eq. (9-148), Q, T, and S_y are constants, and only t varies. This equation may be rewritten as

$$s = \frac{2.303Q}{4\pi T} \log\left(\frac{2.25T}{r^2 S_y}\right) + \frac{2.303Q}{4\pi T} \log(t) \tag{9-152}$$

Because Q, T, r, and S_y are constants, the derivative of Eq. (9-152) is

$$\Delta s = \frac{2.303Q}{4\pi T} \Delta[\log(t)] = \Delta s_L \tag{9-153}$$

Therefore, the expression for the transmissivity is

$$T = \frac{2.303Q}{4\pi \Delta s_L} \Delta[\log(t)] \tag{9-154}$$

Calculate the transmissivity value from Eq. (9-154). If the change in s along this line corresponds to a tenfold increase in t, $\Delta[\log(t)] = 1$.

Step 6: Equation (9-152) is the equation of a straight line on semilogarithmic paper s versus $\log(t)$. The geometric slope of the straight line is equal to $2.303Q/(4\pi T)$, and this line intercepts the time axis where $s = 0$. Consequently, the interception point has

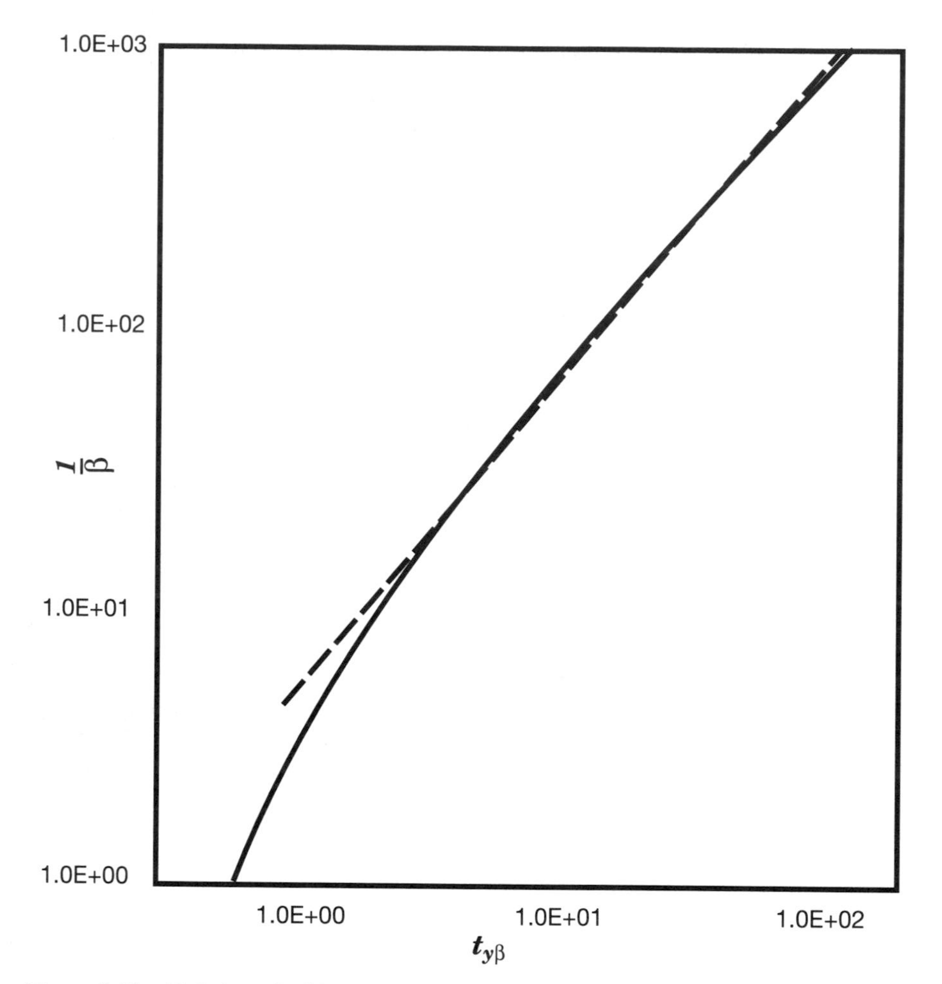

Figure 9-30 Variation of $1/\beta$ versus $t_{y\beta}$ for fully penetrating wells. (After Neuman, 1975.)

the coordinates $s = 0$ and $t = t_L$. Introducing these values into Eq. (9-148) gives

$$0 = \frac{2.303Q}{4\pi T} \log\left(\frac{2.25Tt_L}{r^2 S_y}\right) \tag{9-155}$$

Because $2.303Q/(4\pi T)$ cannot be equal to zero, it follows that

$$\frac{2.25Tt_L}{r^2 S_y} = 1 \tag{9-156}$$

or

$$S_y = \frac{2.25Tt_L}{r^2} \tag{9-157}$$

Calculate S_y from Eq. (9-157).

Step 7: Fit a horizontal line to the intermediate portion of the drawdown versus time data. The value of t corresponding to the intersection of this horizontal line with the straight line passing through the late data is denoted by t_β. Using the known values of T and S_y from Step 5 and Step 6, calculate the dimensionless time $y_{y\beta}$ with the formula

$$t_{y\beta} = \frac{T t_\beta}{S_y r^2} \tag{9-158}$$

The value of β can now be determined directly from the curve in Figure 9-30 or from Eq. (9-151) for a limited range of $t_{y\beta}$ values.

Step 8: Fit a straight line to a portion of the early drawdown versus time data. If the geometric slope of this line differs markedly from that of the line passing through the late data, Step 8 must be skipped, and in this case S must be determined by the Neuman type-curve method presented previously. If the two lines are nearly parallel to each other, the intersection of the early line with the horizontal axis at $s = 0$ is denoted by t_E. Then calculate the transmissivity value from

$$T = \frac{2.303Q}{4\pi \Delta s_E} \Delta[\log(t)] \tag{9-159}$$

which is derived in a manner similar to that in Step 5. If the change in s along this line corresponds to a tenfold increase in t, then $\Delta[\log(t)] = 1$. The T value calculated in this step should be approximately equal to that previously determined in Step 5 from the late drawdown data.

Step 9: Calculate the storage coefficient value from

$$S = \frac{2.25 T t_E}{r^2} \tag{9-160}$$

which is derived in a manner similar to that in Step 6. In Eq. (9-160), t_E corresponds to the intersection of the straight line through early drawdown data with $s = 0$ on semilogarithmic paper. Calculate the values of K_r, K_D, K_z, and S_s from Eqs. (9-143), (9-144), (9-145), and (9-147), respectively.

Step 10: Step 10 is identical to Step 10 of the Neuman type-curve method presented previously, with the exception that σ, as given by the last expression of Eq. (9-83), can also be calculated from

$$\sigma = \frac{t_E}{t_L} \tag{9-161}$$

This equation follows directly from Eqs. (9-157) and (9-160).

Recovery Method. Neuman's analytical model, as given by Eq. (9-94), is derived on the assumption that the delayed response process is not dependent to the capillarity effects above the water table (see Section 9.5.2.1). This assumption is justified by Neuman (see Section 9.5.2.6) by comparing the results with a numerical saturated–unsaturated model. Therefore, the phreatic surface is treated as a moving material boundary. This means that the relationship between the pressure head and water content and its associated hysteresis effects in unsaturated soils during extraction and recovery periods do not play any role in the Neuman model. Therefore, the delayed response process is fully reversible. In other words, Eq. (9-94) is applicable for both falling and rising water-table conditions (Neuman, 1975).

Because of the reasons mentioned above, one can use recovery test data from the pumping well or from the observation wells to determine the aquifer transmissivity value in accordance with Section 4.2.7.4 of Chapter 4. Let t be the time since pumping started and let t_r (or t') be the time the pump was shut off and recovery began. By plotting the residual drawdown on semilogarithmic paper t/t_r (or t/t'), one finds that at large values of t_r, i.e., at small values of t/t_r, these data tend to fall on a straight line. Following the same methodology in Section 4.2.7.4 of Chapter 4, one can easily show that the expression for transmissivity is

$$T = \frac{2.303Q}{4\pi \Delta s_L} \Delta \log \left(\frac{t}{t_r} \right) \tag{9-162}$$

which is the same as Eq. (4-84) of Chapter 4. If Δs_L (or $\Delta s'$) is the residual drawdown corresponding to a tenfold increase in t/t_r along this straight line, then Eq. (9-162) takes the form

$$T = \frac{2.303Q}{4\pi \Delta s_L} \tag{9-163}$$

The steps given in Section 4.2.7.4 of Chapter 4 can also be used for the transmissivity value based on recovery data in unconfined aquifers.

Applicability of Jacob's Correction Scheme. Equation (9-94) is derived on the assumption (see Section 9.5.2.1) that the drawdown of the water table remains small as compared with the saturated thickness of the unconfined aquifer. For cases where the drawdowns do not remain small, Jacob (1944) recommended that before analyzing the pumping test data, the drawdown be corrected according to Eq. (9-64).

Neuman (1972) showed that the Dupuit assumptions do not hold in an unconfined aquifer with delayed gravity response as long as the drawdown data do not fall on the late Theis curve (see Figure 9-19). From this, Neuman concluded that the Jacob's correction scheme is strictly applicable only to the late drawdown data and is not applicable to the early and intermediate data. Therefore, Neuman recommended that Eq. (9-64) be used only in the determination of T and S_y from the late drawdown data, not in the determination of β, T, and S from the early and intermediate data. The application for the steady-state case is given in Section 3.4.3.1 of Chapter 3.

Applicability of Distance Versus Drawdown Analyses. Prickett (1965) claimed that the distance-versus-drawdown-based analysis methods should be applied only after the effects of delayed gravity drainage has dissipated in all observation wells. The reason given by Prickett is that during the period when delayed gravity drainage is influencing drawdown in observation wells, the cone of depression of the pumped well is distorted and this will lead to erroneous results. On the other hand, later Boulton (1970), by basing an observation made by Wenzel (1942), advocated the use of drawdown versus distance analyses for the determination of transmissivity and storage coefficient values in addition to the drawdown-versus-time-based methods.

To check whether Boulton's recommendation is correct, Neuman (1975) brought another view to the application of drawdown versus distance data: From the second expression of Eqs. (9-82) and Eq. (9-90) we obtain

$$\beta t_y = \frac{T K_D t}{S_y b^2} \tag{9-164}$$

which means that if βt_y is kept constant for a given aquifer, t is also constant. As a result, one can develop dimensionless distance versus drawdown curves from the data shown in Figure 9-28 for Type B curves merely by plotting all s_D values for which βt_y is constant against the corresponding values of t_y. Because t is constant for a given value of βt_y, t_y is now directly proportional with $1/r^2$. In Figure 9-31, two distance versus drawdown curves determined from Figure 9-28 by Neuman (1975) are shown for $\beta t_y = 4 \times 10^{-3}$ and 4×10^{-1}. As can be observed from Figure 9-31, the shape of these curves are similar to that of the Theis curve, but substantially different from the shape of Type B curves in Figure 9-28. And based on this, Neuman states that when t is not large enough, analyzing the field data by matching them with the Theis curve may lead to gross errors for the determination of transmissivity and storage coefficient values. Regarding this conclusion, Neuman (1975, p. 335) states: "We are thus led to conclude that one should be cautious in applying distance–drawdown analyses to pumping test data from unconfined aquifers even if these data appear to fall on the Theis curve." Finally, Neuman (1975, p. 335) brings the following explanation regarding Prickett's reasoning for the distorted cone of depression: "It is interesting to mention that Prickett explained his statement by remarking 'the cone of depression is distorted' during the time when delayed gravity response is important. We

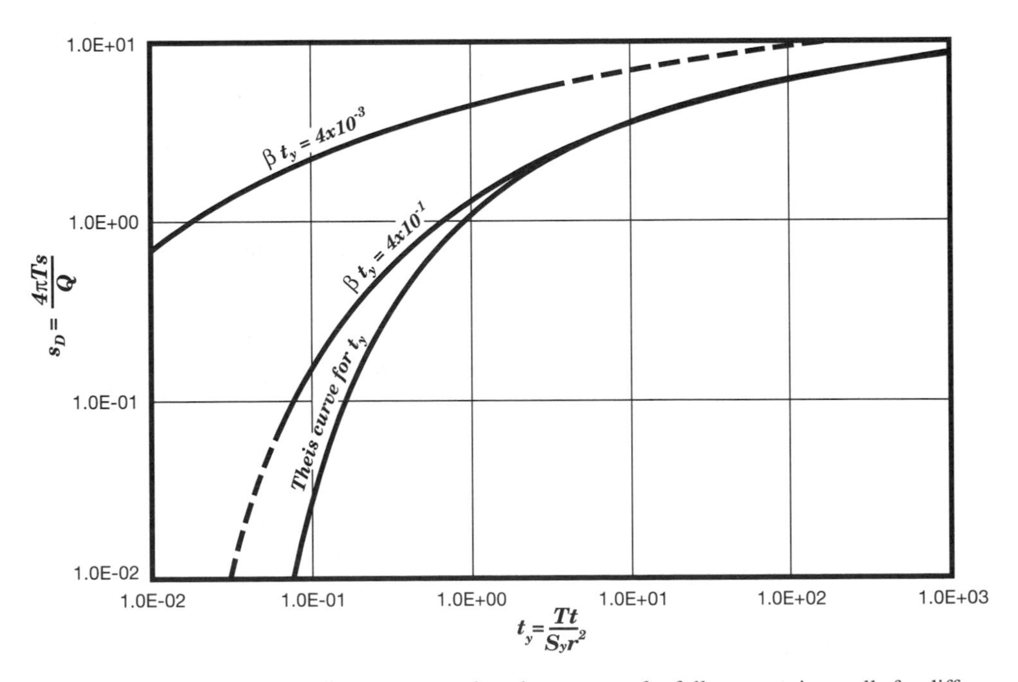

Figure 9-31 Dimensionless distance versus drawdown curves for fully penetrating wells for different values of βt_y = constant. (After Neuman, 1975.)

now know that in fact the cone of depression is not distorted [Neuman, 1972, Figure 7] and its slope varies monotonically with r. The true cause for the errors that may arise from using a distance–drawdown analysis is the difference between the theoretical curves for βt_y constant in Figure 4 and the Theis curve." The figure Neuman referred to above (Figure 4) is presented as Figure 9-31.

Relationship Between Boulton's Delay Index and Aquifer Characteristics. Neuman (1975) obtained the relationship between Boulton's semiempirical quantity α, which is defined in Section 9.4.2, and the physical characteristics of the aquifer. Boulton's type curves are expressed in terms of the dimensionless parameter $r/D = r(rS_y/T)^{1/2}$, which is Eq. (9-53) of Section 9.4.5. And Neuman's type curves are expressed in terms of $\beta = K_z/K_x(r/b)^2$, which is Eq. (9-90) of Section 9.5.2.4. Neuman (1975) first plotted β versus s_D and r/D versus s_D curves (Figure 9-32) by considering the horizontal portions of the Boulton and Neuman type curves as given by Figure 9-7 and Figure 9-28, respectively. Then, Neuman plotted $(r/D)^2/\beta$ versus β for given s_D values on semilogarithmic paper and obtained a set of points as shown in Figure 9-33 and by linear regression found the following straight line:

$$\frac{\left(\frac{r}{D}\right)^2}{\beta} = 3.063 - 0.567\log(\beta) \tag{9-165}$$

The correlation coefficient for this line is very high, $\rho^2 = 0.99$. Neuman (1975) noted that the light scatter of the points around the straight line in Figure 9-33 may be due to extrapolation errors from the associated figures. Substitution of Eqs. (9-53) and (9-90) into Eq. (9-165) gives

$$\alpha = \frac{K_z}{S_y b}\left[3.063 - 0.567\log\left(\frac{K_D r^2}{b^2}\right)\right] \tag{9-166}$$

Based on this equation, Neuman (1975, p. 336) concludes: "This indicates that in a given homogeneous aquifer, α decreases in direct proportion to $\log(r)$, thereby contradicting Boulton's theory in which α is assumed to be a characteristic constant of the aquifer."

Some key aspects regarding Eq. (9-166) and Boulton's and Neuman's type-curve methods are given below based on Neuman (1975):

1. Streltsova (1972) used a finite-difference approximation for the delayed response process and derived the following relationship for the Boulton semiempirical quantity (α) for a homogeneous aquifer:

$$\alpha = \frac{3K_z}{S_y(b - s_{wt})} \tag{9-167}$$

   where s_{wt} is the drawdown of the water table. Because s_{wt} is a function of radial distance (r) and time (t), the α value of Eq. (9-167) decreases with r and increases with t. Neuman's equation, Eq. (9-166), supports Strelsova's conclusion that α decreases with r. But Neuman's equation shows that α is not dependent on time.
2. Equation (9-166) means that the delayed gravity drainage decreases linearly with the logarithm of the radial distance from the pumping well.

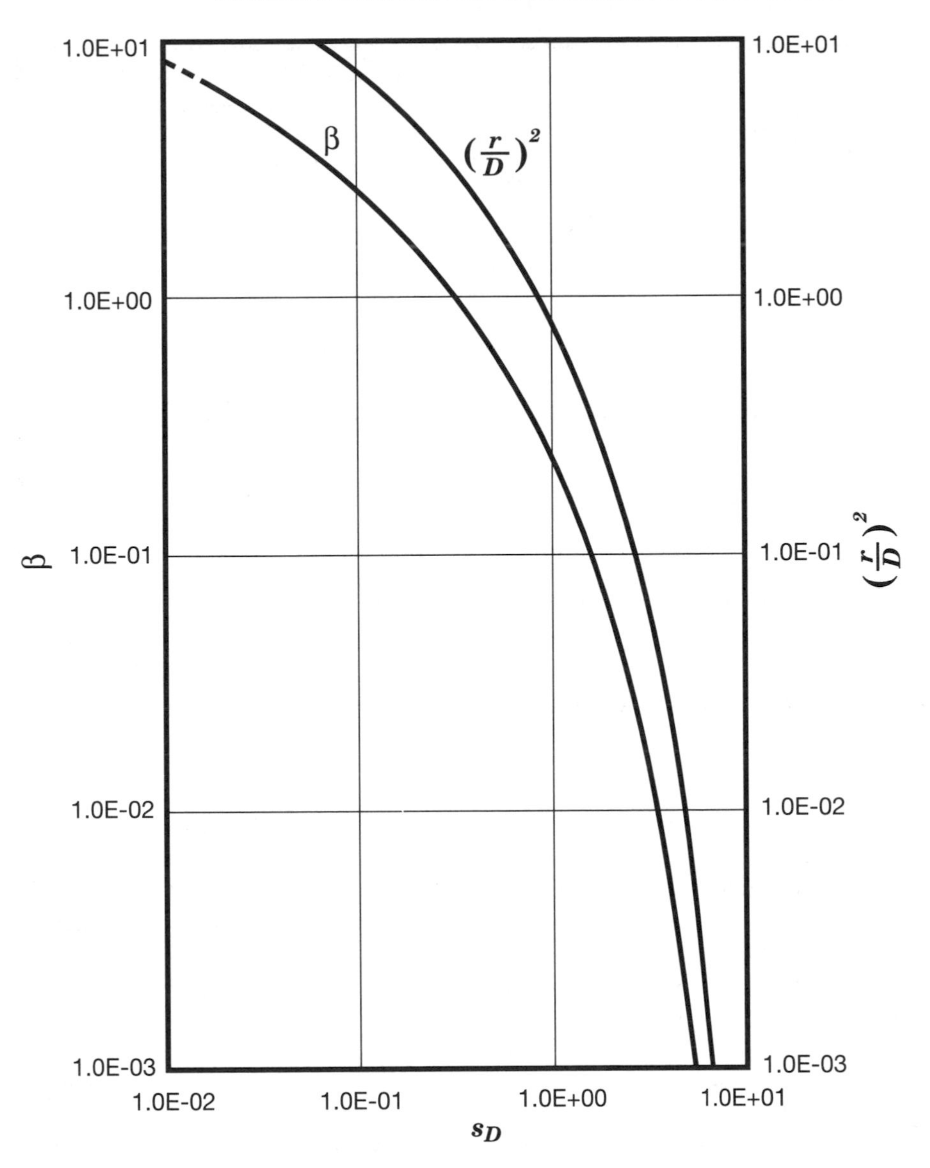

Figure 9-32 Variation of β and r/D versus s_D corresponding to horizontal portion of type curves for fully penetrating wells. (After Neuman, 1975.)

3. The difference between Boulton's and Neuman's models for fully penetrating wells is that Boulton's model can only determine the value of horizontal hydraulic conductivity (K_r), whereas Neuman's model can determine both the horizontal and vertical hydraulic conductivities (K_r and K_z). The storage coefficient (S) and specific yield (S_y) values can be determined by both methods.

Examples. The Neuman type-curve, semilogarithmic, and recovery methods, as presented in the previous sections in detail, will now be described with some examples.

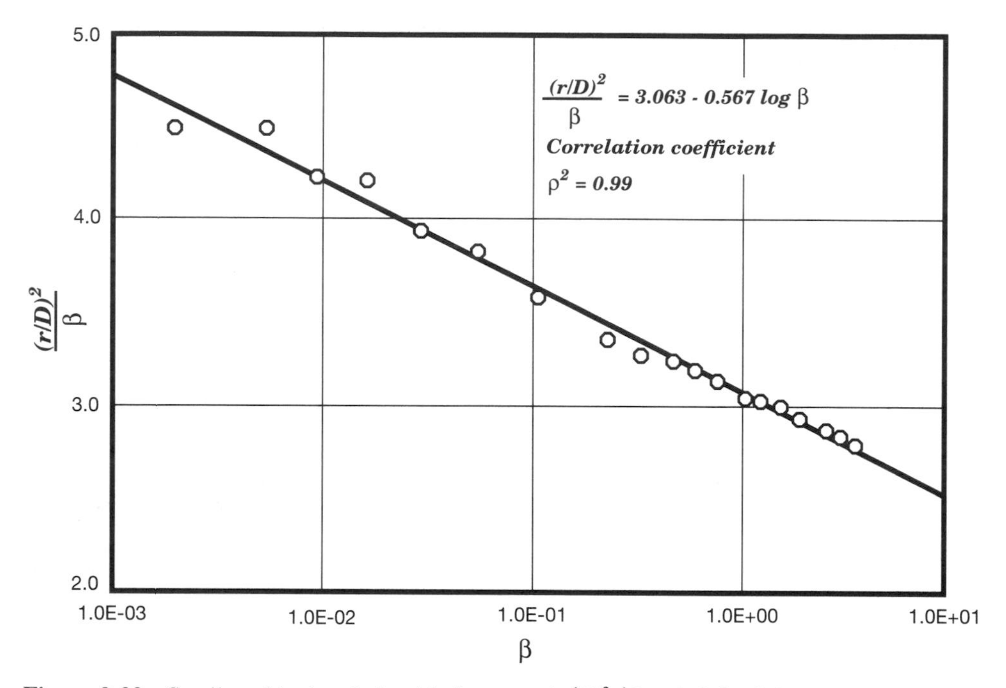

Figure 9-33 Semilogarithmic relationship between $(r/D)^2/\beta$ and β for fully penetrating wells. (After Neuman, 1975.)

Example 9-5: This example is adapted from Neuman (1975). The data for this example are taken from a pumping test performed in 1965 by the French Bureau de Recherches Géologiques et Minières (Bureau of Geologic Research and Minerals of France) at Saint Pardon de Conques, located in the Vallée de la Garonne, Gironde, France (Bonnet et al., 1970; Neuman, 1975).

The aquifer is composed of medium-grained sand with gravel in the deeper part and a clayey zone at shallow depths. The aquifer is underlain by marls having a relatively low hydraulic conductivity. The bottom of the aquifer is at a depth of 13.75 m, and the water table was initially at a depth of 5.51 m. Therefore, the saturated thickness of the aquifer (b) is 8.24 m (see Figure 9-34).

The pumping well is perforated between the depths 7 m and 13.75 m and has a diameter of 0.32 m. The pumping test lasted 48 hours and 50 minutes at a rate (Q) oscillating between 51 m^3/hour (1224 m^3/day) and 54.6 m^3/hour (1310.4 m^3/day) and averaged about 53 m^3/hour (1272 m^3/day). The observed drawdowns at $r = 10$ m and $r = 30$ m radial distances are given in Tables 9-9 and 9-10, respectively.

Using the Neuman type-curve method, presented previously, determine the aquifer parameters.

Solution: The steps of the Neuman type-curve method, presented previously, will be applied to determine the aquifer parameters.

Step 1: The family of Neuman Type A curves (for the use of early drawdown data) and Type B curves (for the use of late drawdown data) are presented in Figures 9-27 and 9-28, respectively, using the data in Table 9-6 and Table 9-7, respectively.

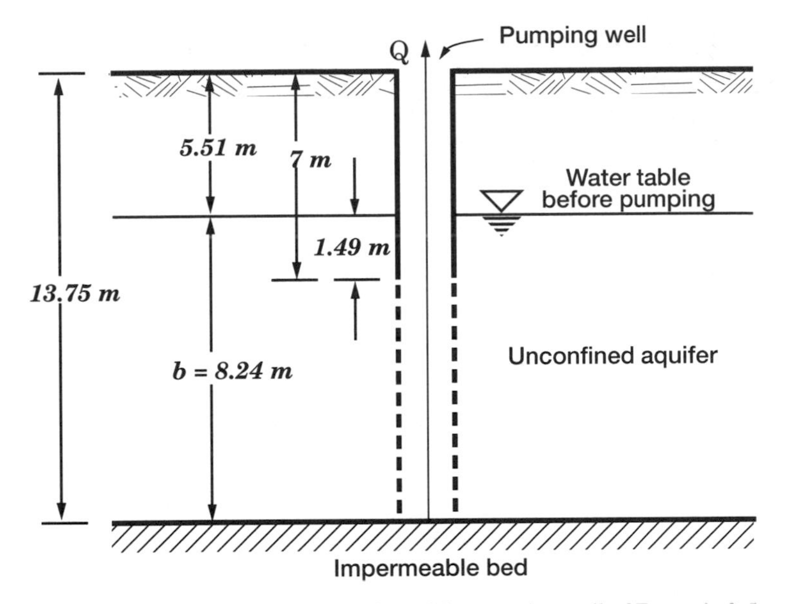

Figure 9-34 Schematic representation of the pumping well of Example 9-5.

TABLE 9-9 Drawdown Versus Time Data at r = 10 m for Example 9-5

t (min)	s (m)	t (min)	s (m)
0.25	0.100	43.87	0.246
0.50	0.121	53.58	0.254
0.73	0.135	73.08	0.265
0.97	0.163	100.00	0.276
1.23	0.172	133.33	0.298
1.45	0.178	155.90	0.309
2.00	0.179	190.48	0.319
2.48	0.185	248.75	0.327
2.93	0.191	303.92	0.349
3.53	0.201	371.23	0.354
4.53	0.216	431.37	0.360
5.53	0.218	544.93	0.366
6.55	0.218	688.25	0.375
7.87	0.218	786.58	0.397
10.00	0.216	987.07	0.403
13.20	0.221	1154.65	0.401
16.12	0.225	1355.03	0.417
19.05	0.229	1585.43	0.430
22.50	0.232	1969.47	0.456
28.72	0.236	2529.58	0.477
35.90	0.242	2939.33	0.506

TABLE 9-10 Drawdown Versus Time Data at r = 30 m for Example 9-5

t (min)	s (m)	t (min)	s (m)
15.08	0.080	544.93	0.181
24.05	0.085	608.35	0.180
37.50	0.085	688.25	0.191
53.58	0.100	786.58	0.211
73.08	0.111	923.32	0.232
97.72	0.121	1105.47	0.239
127.63	0.100	1276.28	0.232
155.90	0.130	1458.30	0.252
190.48	0.148	1585.43	0.256
248.75	0.148	1969.47	0.282
303.92	0.165	2366.18	0.292
371.23	0.176	2703.63	0.301
431.37	0.182	2939.33	0.311
484.90	0.181		

Step 2: The maximum drawdown (s_{max}) in Table 9-9 at $r = 10$ m is 0.506 m and the initial aquifer thickness (b) is 8.24 m. Therefore, $s_{max}/b = 0.506$ m$/8.24$ m $= 0.061$. At $r = 30$ m (Table 9-10), $s_{max}/b = 0.311$ m$/8.24$ m $= 0.038$. These values mean that drawdown correction is not required since the maximum drawdowns are not significant

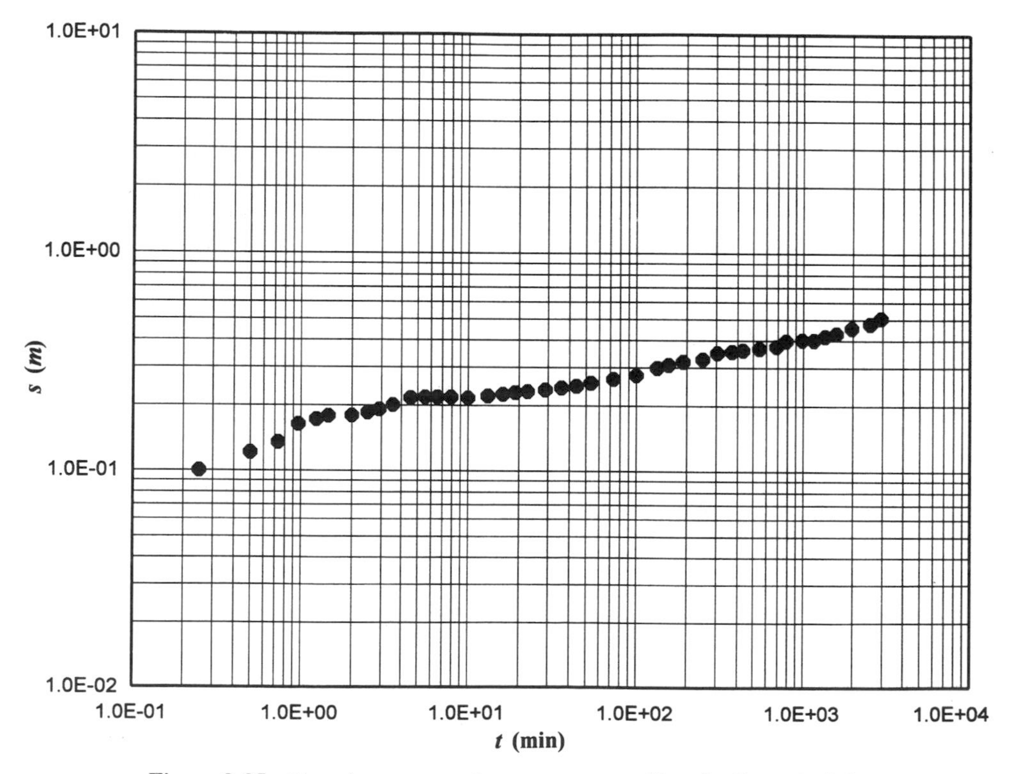

Figure 9-35 Drawdown versus time curve at $r = 10$ m for Example 9-5.

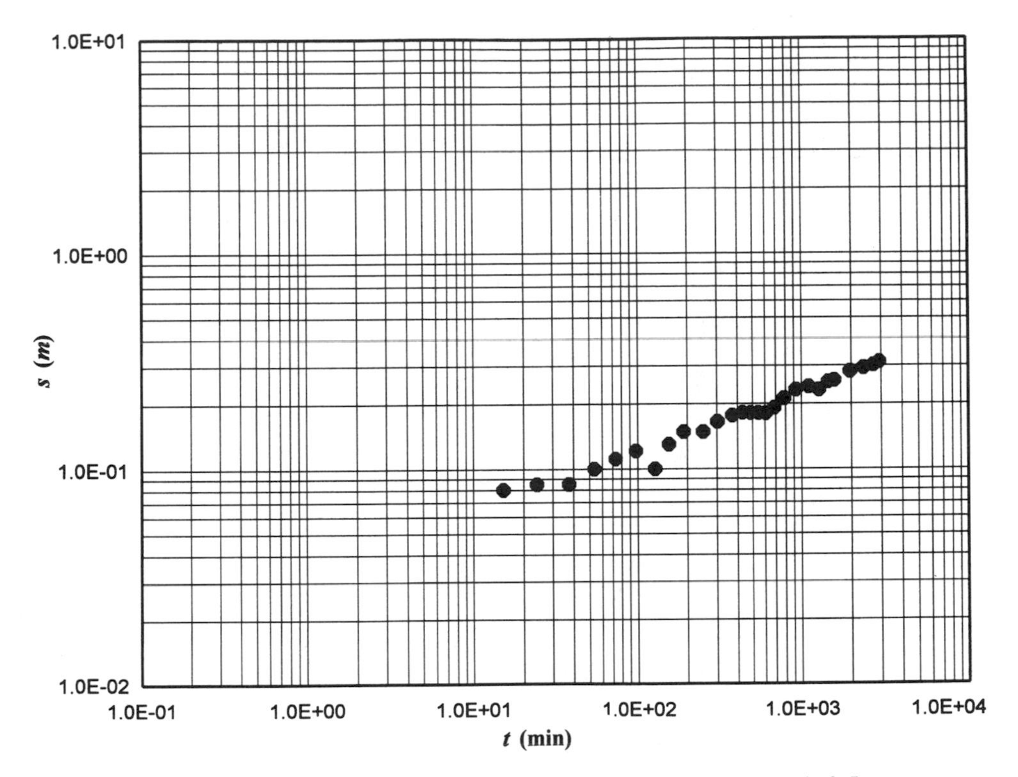

Figure 9-36 Drawdown versus time curve at $r = 30$ m for Example 9-5.

fractions of the initial aquifer thickness. The observed drawdown versus time data in Table 9-9 ($r = 10$ m) and Table 9-10 ($r = 30$ m) on double logarithmic paper with the same scales as those in Step 1 are presented in Figures 9-35 and 9-36, respectively.

Step 3: The drawdown versus time data graphs for $r = 10$ m and $r = 30$ are superimposed on the Type B curves and shown, respectively, in Figures 9-37 and 9-38. As shown in Figure 9-37, for the first observation well ($r = 10$ m), $\beta = 0.01$ establishes a good match with the observed data. The observed data for the second observation well ($r = 30$ m) match with a curve between $\beta = 0.1$ and 0.2 (see Figure 9-38). The data appear to fit a Type B curve for $\beta \approx 0.18$.

Step 4: Match points B_1 and B_2 are chosen in Figures 9-37 and 9-38, respectively. In Figure 9-37, B_1 is represented by the following coordinates:

$$t_1 = 5.3 \text{ min}, \qquad s_1 = 0.065 \text{ m}$$
$$t_{y_1} = 1.0 \qquad s_{D_1} = 1.0$$

The coordinates of B_2 in Figure 9-38 are

$$t_2 = 40 \text{ min}, \qquad s_2 = 0.055 \text{ m}$$
$$t_{y_2} = 1.0, \qquad s_{D_2} = 1.0$$

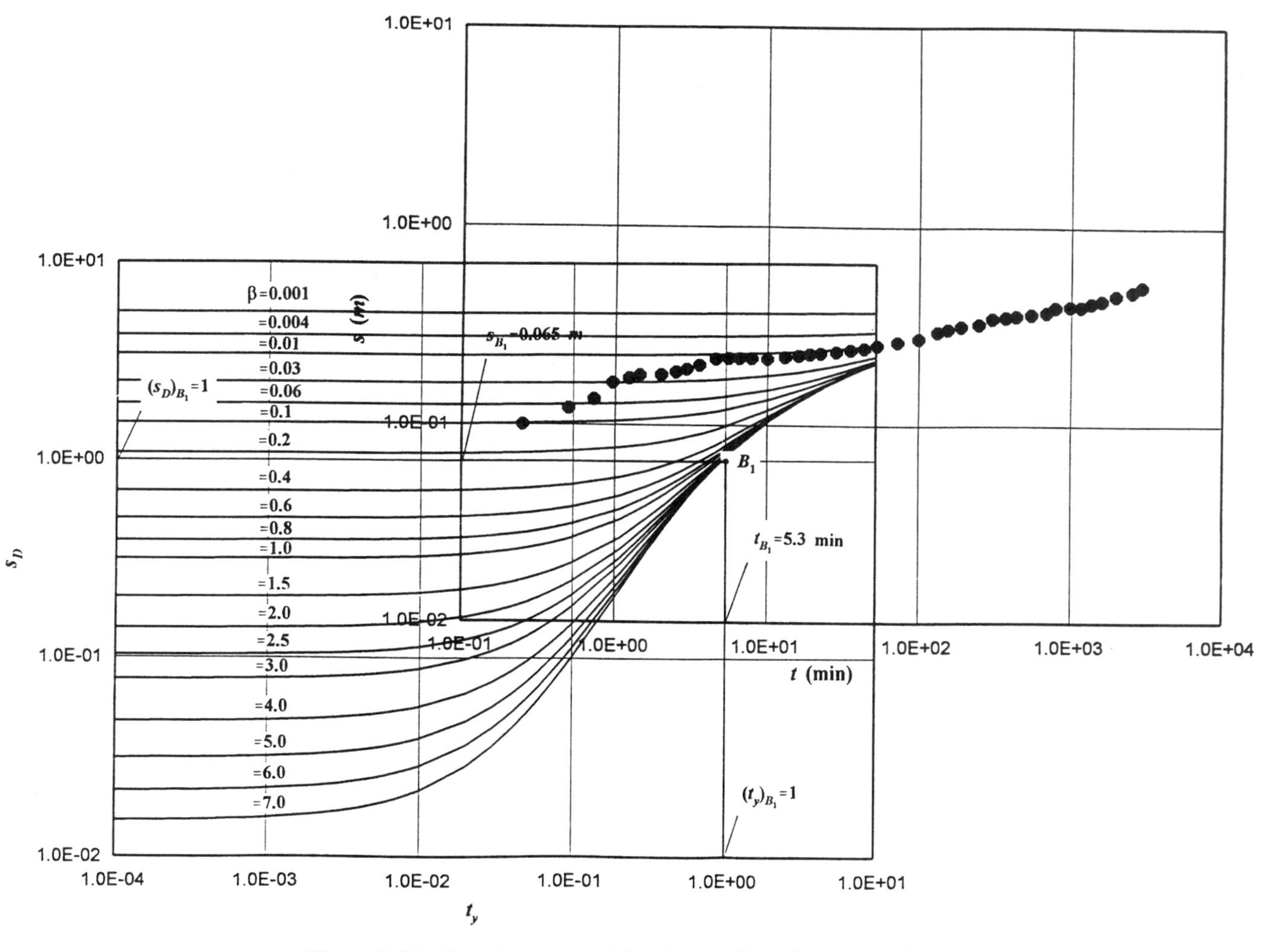

Figure 9-37 Type B curve matching for $r = 10$ m for Example 9-5.

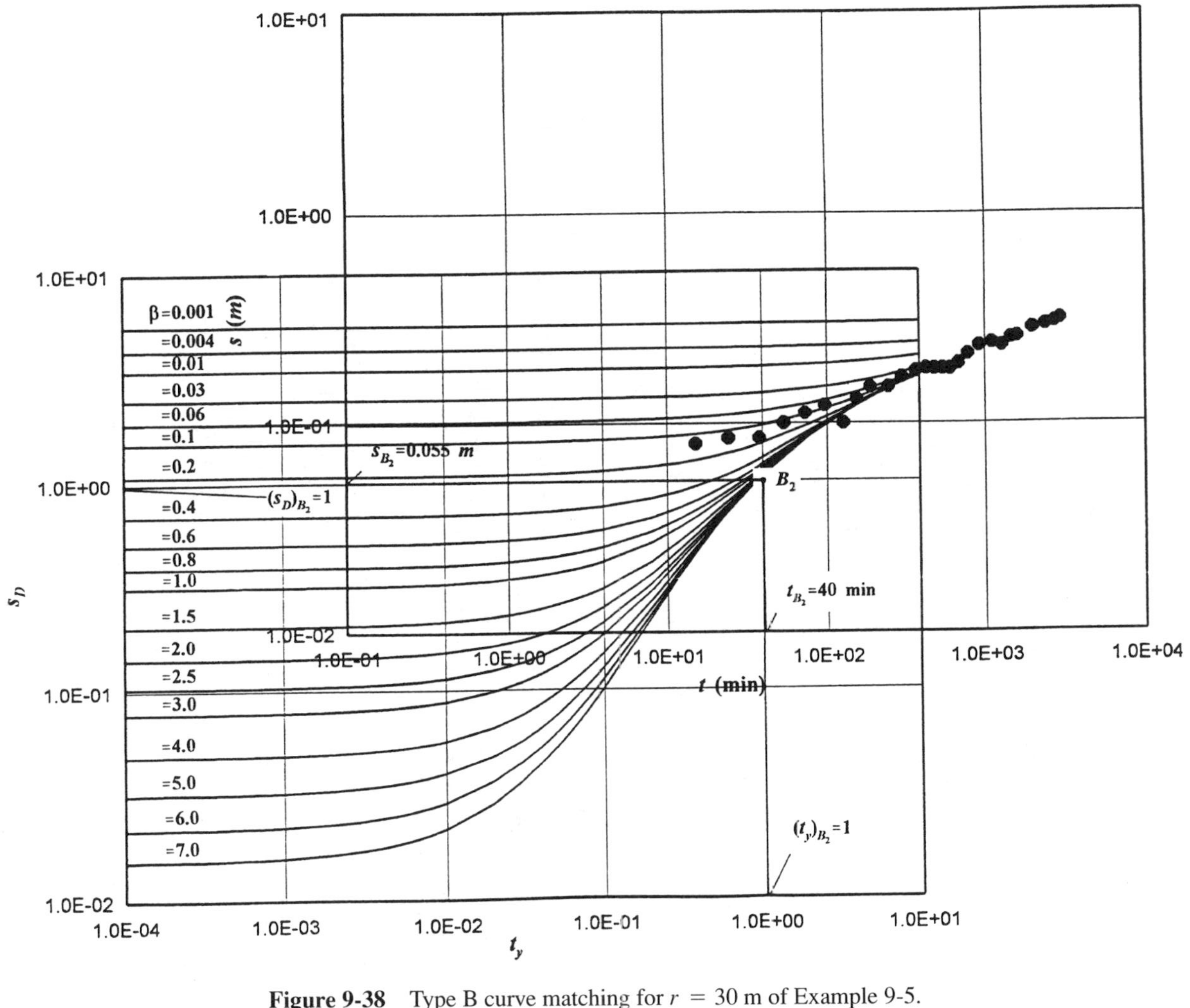

Figure 9-38 Type B curve matching for $r = 30$ m of Example 9-5.

Step 5: Thus, from Eqs. (9-137) and (9-138), the following values can be obtained for $B_1(r = 10 \text{ m})$:

$$T_1 = \frac{Q}{4\pi}\frac{s_{D_1}}{s_1} = \frac{(1272 \text{ m}^3/\text{day})(1.0)}{(4\pi)(0.065 \text{ m})} = 1557.27 \text{ m}^2/\text{day}$$

$$S_{y_1} = \frac{T_1 t_1}{t_{y_1} r_1^2} = \frac{\left(\dfrac{1557.27 \text{ m}^2}{24 \times 60 \text{ min}}\right)(5.3 \text{ min})}{(1.0)(10 \text{ m})^2} = 5.73 \times 10^{-2}$$

Similarly, the following values can be obtained for $B_2(r = 30 \text{ m})$:

$$T_2 = \frac{Q}{4\pi}\frac{s_{D_2}}{s_2} = \frac{(1272 \text{ m}^3/\text{day})(1.0)}{(4\pi)(0.055 \text{ m})} = 1840.41 \text{ m}^2/\text{day}$$

$$S_{y_2} = \frac{T_2 t_2}{t_{y_2} r_2^2} = \frac{\left(\dfrac{1840.41 \text{ m}^2}{24 \times 60 \text{ min}}\right)(40 \text{ min})}{(1.0)(30 \text{ m})^2} = 5.68 \times 10^{-2}$$

It is interesting to note that by using Boulton's type-curve method, which is presented in Section 9.4.6.3, Bonnet et al. (1970) obtained the following values:

$$T_1 = 68.0 \text{ m}^2/\text{h} = 1632.0 \text{ m}^2/\text{day}, \quad S_{y_1} = 4.5 \times 10^{-2} \quad \text{for } r_1 = 10 \text{ m}$$
$$T_2 = 65.0 \text{ m}^2/\text{h} = 1560.0 \text{ m}^2/\text{day}, \quad S_{y_2} = 8.0 \times 10^{-2} \quad \text{for } r_2 = 30 \text{ m}$$

Step 6: The drawdown versus time data graph for $r = 10$ m is superimposed on the Type A curves and is shown in Figure 9-39. The value of β is 0.01 because it must be the same as that obtained in Step 3 from the Type B curves. Type curve matching for $r = 30$ m is not presented because enough data do not exist for the early period.

Step 7: A match point A_1 is chosen in Figure 9-39. This point is represented by the following coordinates:

$$t_1 = 1.33 \text{ min}, \quad s_1 = 0.064 \text{ m}$$
$$t_{s_1} = 10.0, \quad s_{D_1} = 1.0$$

Step 8: From Eqs. (9-137) and (9-141), respectively, the following values can be obtained:

$$T_1 = \frac{Q}{4\pi}\frac{s_{D_1}}{s_1} = \frac{(1272 \text{ m}^3/\text{day})(1.0)}{(4\pi)(0.064 \text{ m})} = 1581.6 \text{ m}^2/\text{day}$$

$$S_1 = \frac{T_1 t_1}{t_{s_1} r^2} = \frac{\left(\dfrac{1581.6 \text{ m}^2}{24 \times 60 \text{ min}}\right)(1.33 \text{ min})}{(10.0)(10 \text{ m})^2} = 1.46 \times 10^{-3}$$

Bonnet et al. (1970) obtained the following values from the Boulton type-curve method:

$$T = 69.0 \text{ m}^2/\text{h} = 1656 \text{ m}^2/\text{day}, \quad S = 1.5 \times 10^{-3}$$

Since the late data give a better fit with the type curves than the early data, the results from the late data appear to be more reliable. Therefore, the arithmetic average of the

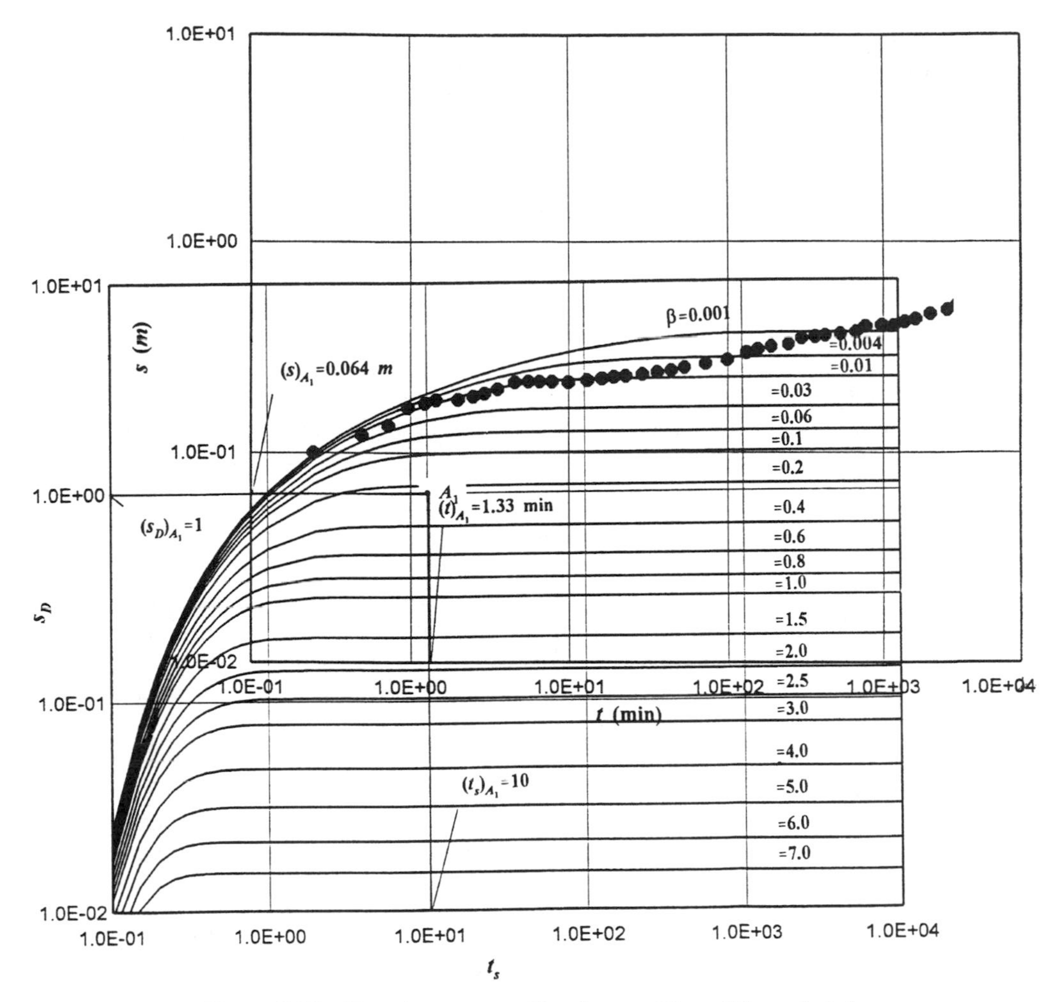

Figure 9-39 Type A curve matching for $r = 10$ m of Example 9-5.

transmissivity values obtained in Step 5 shall be adopted in the calculations in the rest of the steps:

$$T = \frac{1557.27 \text{ m}^2/\text{day} + 1840.41 \text{m}^2/\text{day}}{2} = 1698.84 \text{ m}^2/\text{day}$$

Step 9: From Eq. (9-143), the horizontal hydraulic conductivity value is

$$K_r = \frac{T}{b} = \frac{1698.84 \text{ m}^2/\text{day}}{8.24 \text{ m}} = 206.17 \text{ m/day} = 0.24 \text{ cm/s}$$

From Eq. (9-144), the degree of anisotropy is

$$K_D = \beta \left(\frac{b}{r}\right)^2 = (0.01)\left(\frac{8.24 \text{ m}}{10 \text{ m}}\right)^2 = 6.79 \times 10^{-3}$$

Using the values of K_r and K_D, from Eq. (9-145) the vertical hydraulic conductivity value is

$$K_z = K_D K_r = \left(6.79 \times 10^{-3}\right)(206.17 \text{ m/day}) = 1.40 \text{ m/day} = 1.62 \times 10^{-3} \text{ cm/s}$$

Step 10: From Eq. (9-146) the value of σ, from the S and S_y values of $r = 10$ m, is

$$\sigma = \frac{S}{S_y} = \frac{1.46 \times 10^{-3}}{5.73 \times 10^{-2}} = 2.55 \times 10^{-2}$$

The σ value cannot be determined from the results of $r = 30$ m, because the S value cannot be determined from its early data. From Eq. (9-147), the specific storage value is

$$S_s = \frac{S}{b} = \frac{1.46 \times 10^{-3}}{8.24 \text{ m}} = 1.77 \times 10^{-4} \text{ m}^{-1}$$

Interpretation of the Results. From the results presented above, the horizontal hydraulic conductivity is relatively high ($K_r = 0.24$ cm/s), which corresponds to sand- and gravel-type aquifer materials. In general, the vertical hydraulic conductivity (K_z) of highly permeable materials is close to their horizontal hydraulic conductivity (K_r). Here, the results show that the K_r value is more than a hundred times greater than the K_z value ($K_D = 6.79 \times 10^{-3}$) of the aquifer. This may be attributed to the presence of a clayey formation in the upper part of the aquifer. Furthermore, the values of S_s and σ are relatively high, which is the indication that the compressibility of the aquifer is greater than that usually encountered in deep confined aquifers having similar aquifer materials.

Example 9-6: Using Neuman's semilogarithmic method, determine the aquifer parameters for the aquifer data given in Example 9-5.

Solution: The steps of the Neuman's semilogarithmic method given previously will be applied. The calculational steps are given below:

Step 1: The drawdown versus time data are presented Tables 9-9 and 9-10 for $r = 10$-m and $r = 30$-m distances, respectively. These data are plotted semilogarithmically in Figures 9-40 and 9-41, respectively.

Step 2: The fitted straight lines are shown on Figures 9-40 and 9-41 for $r = 10$-m and $r = 30$-m distances, respectively. At $r = 10$ m, two parallel straight lines can be fitted to the late and early data and a horizontal line can be fitted to the intermediate data. At $r = 30$ m, only a straight line to the late data and a horizontal line to the intermediate data can be fitted.

Step 3: The intersections of the straight lines with the horizontal axes are

$$r = 10 \text{ m}, \qquad t_L = 0.8 \text{ min}$$

$$r = 30 \text{ m}, \qquad t_L = 20 \text{ min}$$

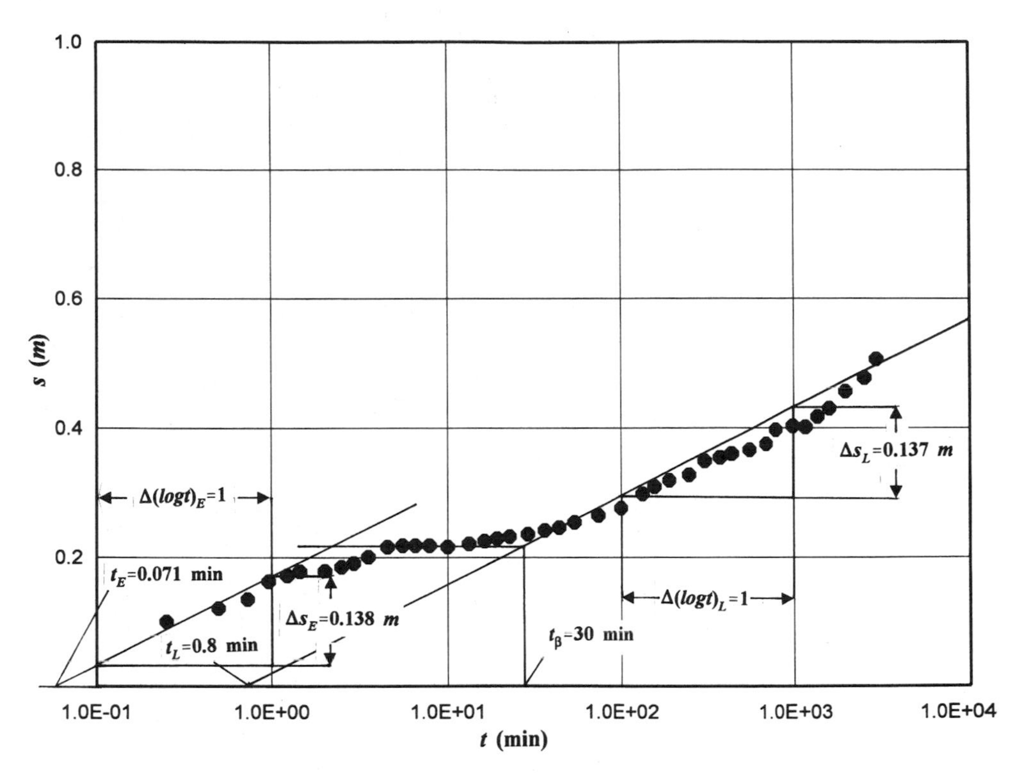

Figure 9-40 Semilogarithmic drawdown versus time curve at $r = 10$ m for Example 9-6.

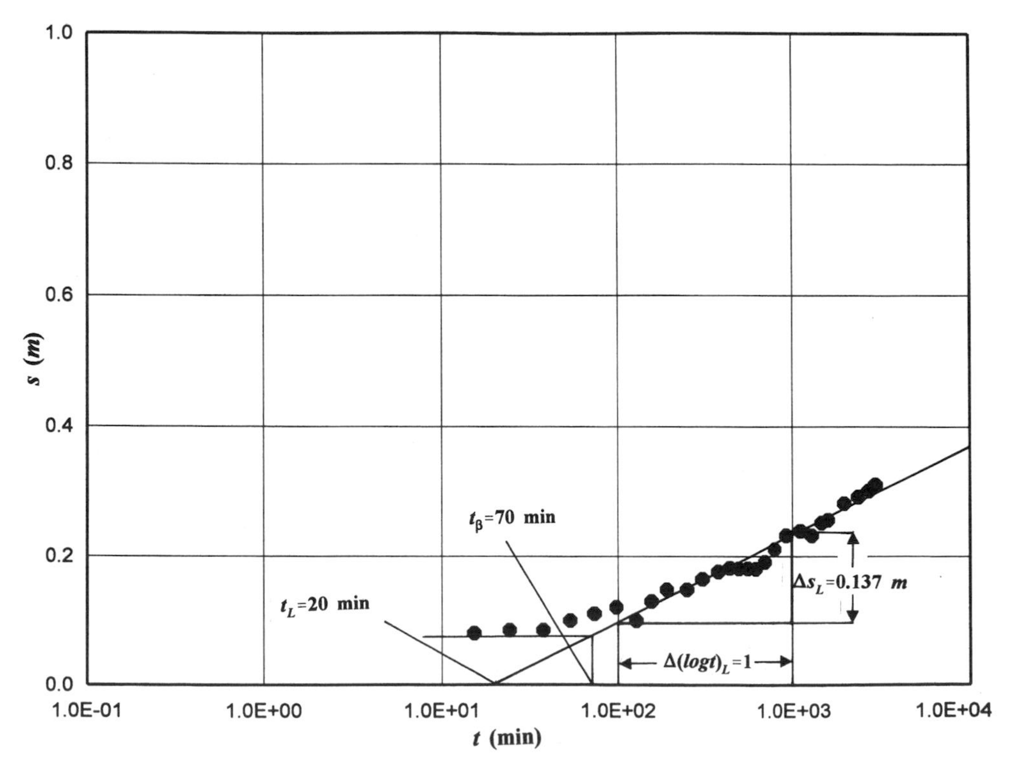

Figure 9-41 Semilogarithmic drawdown versus time curve at $r = 30$ m for Example 9-6.

Step 4: The geometric slopes of the straight lines (the drawdown difference per log cycle of time; see also Figures 9-40 and 9-41) are

$$\Delta s_L = 0.137 \text{ m} \qquad \text{for } r = 10 \text{ m}$$

$$\Delta s_L = 0.137 \text{ m} \qquad \text{for } r = 30 \text{ m}$$

Step 5: Substitution of the $Q = 1272\text{-m}^3/\text{day}$ and $\Delta s_L = 0.137\text{-m}$ values, which are the same for $r = 10$-m and $r = 30$-m observation wells, into Eq. (9-154) gives the T value:

$$T = \frac{2.303Q}{4\pi \Delta s_L}\Delta[\log(t)] = \frac{(2.303)(1272 \text{ m}^3/\text{day})}{(4\pi)(0.137 \text{ m})}(1) = 1701.6 \text{ m}^2/\text{day}$$

This result compares well with those obtained by Bonnet et al. (1970):

$$T = 67.0 \text{ m}^2/\text{h} = 1608.0 \text{ m}^2/\text{day} \qquad \text{for } r = 10 \text{ m}$$

$$T = 63.0 \text{ m}^2/\text{h} = 1512.0 \text{ m}^2/\text{day} \qquad \text{for } r = 30 \text{ m}$$

Step 6: The specific yield (S_y) values can be calculated from Eq. (9-157). For $r = 10$ m we have

$$S_y = \frac{2.25Tt_L}{r^2} = \frac{(2.25)\left(\dfrac{1701.6 \text{ m}^2}{24 \times 60 \text{ min}}\right)(0.8 \text{ min})}{(10 \text{ m})^2} = 2.13 \times 10^{-2}$$

For $r = 30$ m we obtain

$$S_y = \frac{(2.25)\left(\dfrac{1701.6 \text{ m}^2}{24 \times 60 \text{ min}}\right)(20 \text{ min})}{(30 \text{ m})^2} = 5.90 \times 10^{-2}$$

Step 7: The fitted horizontal lines for the intermediate portions of the drawdown versus time data for $r = 10$ m and $r = 30$ m are shown in Figures 9-40 and 9-41, respectively. The t_β values are:

$$t_\beta = 30 \text{ min}, \qquad r = 10 \text{ m}$$
$$t_\beta = 40 \text{ min}, \qquad r = 30 \text{ m}$$

The dimensionless $t_{y\beta}$

$$t_{y\beta} = \frac{Tt_\beta}{S_y r^2} = \frac{\left(\dfrac{1701.60 \text{ m}^2}{24 \times 60 \text{ min}}\right)(30 \text{ min})}{(2.13 \times 10^{-2})(10 \text{ m})^2} = 16.64, \qquad r = 10 \text{ m}$$

$$t_{y\beta} = \frac{\left(\dfrac{1701.60 \text{ m}^2}{24 \times 60 \text{ min}}\right)(70 \text{ min})}{(5.90 \times 10^{-2})(30 \text{ m})^2} = 1.56, \qquad r = 30 \text{ m}$$

The value of $t_{y\beta}$ for $r = 10$ m falls within the range of values for which Eq. (9-151) applies. Therefore, the β value can be determined either by Eq. (9-151) or from Figure 9-30. From Eq. (9-151) we have

$$\beta = \frac{0.195}{t_{y\beta}^{1.1053}} = \frac{0.195}{(16.64)^{1.1053}} = 0.01 \qquad \text{for } r = 10 \text{ m}$$

The value of $t_{y\beta}$ for $r = 30$ m falls outside of the range of values for which Eq. (9-151) is applicable. Therefore, β must be determined from Figure 9-30. According to this figure, $1/\beta = 6.0$, and therefore $\beta = 1/6 = 0.17$.

Step 8: The fitted straight line to the early drawdown versus time data portion for $r = 10$ m is shown on Figure 9-40. The data for $r = 30$ m do not have this kind of portion. Therefore,

$$\Delta s_E = 0.138 \text{ m} \qquad \text{for } r = 10 \text{ m}$$

This value is very close to those obtained in Step 4. Equation (9-159) can be used to calculate the transmissivity value for $r = 10$ m:

$$T = \frac{2.303Q}{4\pi \Delta s_E} = \frac{(2.303)(1272 \text{ m}^3/\text{day})}{(4\pi)(0.138 \text{ m})}(1) = 1687.2 \text{ m}^2/\text{day}$$

The intersection of the early line with the horizontal axis at $s = 0$ is (see also Figure 9-40)

$$t_E = 0.071 \text{ min}$$

Step 9: Eq. (9-160) gives the elastic storage coefficient (S) value:

$$S = \frac{2.25Tt_E}{r^2} = \frac{(2.25)\left(\dfrac{1687.2 \text{ m}^2}{24 \times 60 \text{ min}}\right)(0.071 \text{ min})}{(10 \text{ m})^2} = 1.87 \times 10^{-3}$$

The rest of parameters will be calculated using the arithmetic average of the above three transmissivity values:

$$T = \frac{1701.6 \text{ m}^2/\text{day} + 1701.6 \text{ m}^2/\text{day} + 1687.2 \text{ m}^2/\text{day}}{3} = 1696.8 \text{ m}^2/\text{day}$$

The other parameters can be calculated, respectively, from Eqs. (9-143), (9-144), (9-145), and (9-147) as

$$K_r = \frac{T}{b} = \frac{1696.8 \text{ m}^2/\text{day}}{8.24 \text{ m}} = 205.92 \text{ m/day} = 0.24 \text{ cm/s}$$

$$K_D = \beta \frac{b^2}{r^2} = \frac{(0.01)(8.24 \text{ m})^2}{(10 \text{ m})^2} = 6.79 \times 10^{-3}$$

$$K_z = K_D K_r = (6.79 \times 10^{-3})(205.92 \text{ m/day}) = 1.40 \text{ m/day} = 1.62 \times 10^{-3} \text{ cm/s}$$

$$S_s = \frac{S}{b} = \frac{1.87 \times 10^{-3}}{8.24 \text{ m}} = 2.27 \times 10^{-4} \text{ m}^{-1}$$

Step 10: Using the values of $r = 10$ m, the value of from the third expression of Eqs. (9-83) is

$$\sigma = \frac{S}{S_y} = \frac{1.87 \times 10^{-3}}{2.13 \times 10^{-2}} = 8.78 \times 10^{-2}$$

Eq. (9-161) gives approximately the same value from the values of $r = 10$ m:

$$\sigma = \frac{t_E}{t_L} = \frac{0.071 \text{ min}}{0.8 \text{ min}} = 8.88 \times 10^{-2}$$

9.5.4.2 Data Analysis Methods for Partially Penetrating Wells. As shown in Figure 9-22, when the pumping well or the observation well or both of them are perforated only through a portion of the saturated thickness (b) of the aquifer, Eq. (9-94) is not applicable. As a result, the type curves presented in Figures 9-27 and 9-28 cannot be used to analyze drawdown versus time data for the determination of aquifer parameters. As presented in detail in Section 9.5.3, Neuman (1974) extended the fully penetrating well model of Neuman (1972) to account for the effects of partial penetration on drawdowns in unconfined aquifers and obtained Eq. (9-115) for drawdown as a result of a pumping well. In the following paragraphs, some additional requirements regarding the Neuman type-curve method for partially penetrating wells are presented based on Neuman (1975).

The associated mathematical expressions for drawdown in a piezometer, as given by Eqs. (9-115), (9-116), and (9-117), are expressed in terms of six independent dimensionless parameters: σ, β, l_D, d_D, z_D, and t_s or t_y [recall from Eq. (9-121) that $t_y = \sigma t_s$]. The associated mathematical expressions for drawdown in an observation well, as given by Eqs. (9-115), (9-126), and (9-127), are expressed in terms of seven independent dimensionless parameters: σ, β, l_D, d_D, z_{1D}, z_{2D}, and t_s or t_y. It is practically impossible to generate a sufficient number of type curves to cover the potential range of values to be used for pumping test data analysis in partially penetrating wells. As mentioned in Section 9.5.4.1, the type curves are normally expressed in terms of not more than two independent parameters. The same procedure can be used for partially penetrating well cases under the condition that the parameters regarding the geometry of the wells (l_D, d_D, and z_D, or l_D, d_D, z_{1D}, and z_{2D}) are known. An additional requirement of the procedure for partially penetrating wells is that a special set of type curves (Type A and Type B) must be generated for each observation well in the field. These type curves can be generated using the computer program mentioned in Section 9.5.4.1.

For fully penetrating well cases, the number of independent parameters is reduced from three to two by letting σ approach to zero (see Section 9.5.4.1).

Neuman's Type-Curve Method. The drawdown versus time data analysis steps for fully penetrating wells (see Section 9.5.4.1) are applicable to partially penetrating well cases as well. The steps for partially penetrating well cases are a modified version of the steps in Section 9.5.4.1. The steps are as follows:

Step 1: First, create a diagram and a table that show the z_1 and z_2 values as well as the perforation interval of each observation well. If there are more than one observation well and the number of observation wells is relatively high, generate groups of observation wells whose perforation intervals have approximately the same length and are located approximately at the same place along the thickness of the aquifer. Using the dimensional d and l values of the pumping well and z_1 and z_2 values of each observation well (refer to Figure 9-22 for these variables), calculate the dimensionless parameters by the three expressions in Eq. (9-120). Then, run the computer program to generate s_D versus t_s (Type A curves) and s_D versus t_y (Type B curves) data for a practical range of values of β. A suggested range of values for β are 0.001, 0.004, 0.01, 0.03, 0.06, 0.1, 0.2, 0.4, 0.6, 0.8, 1.0, 1.5, 2.0, 2.5, 3.0, 4.0, 5.0, 6.0, and 7.0. In generating these data, use the instructions for t_s and t_y values in Section 9.5.4.1. For a type curve (Type A or Type B) corresponding to a value of β, l/b, and d/b for the pumping well, z/b for a piezometer or z_1/b and z_2/b for an observation well are required as input data. The runs need to be repeated for each β value. For example, if one uses the above-given suggested range of 19 β values, 19 runs need to be conducted. Using the aforementioned data, plot the family of Neuman Type A and Type B curves for each group of observation wells.

Step 2 Through Step 10: These steps are the same as Step 2 through Step 10 for fully penetrating wells as presented in Section 9.5.4.1.

Example 9-7: This example is adapted from Mock and Merz (1990), in which the original data are presented in the English system. Here, the data are converted to the metric system.

The site is located around Tucson, Arizona, in the northwest corner of the Upper Santa Cruz Basin and Range Province. A 72-hour pumping test was conducted at the Sweetwater Recharge Site for the determination of the aquifer parameters and provide a better understanding of ground water flow. Description of the test area and the test itself is presented below based on Mock and Merz (1990).

The lithology of the saturated sediments in the test area appears to be relatively uniform and homogeneous to a depth of approximately 183.0 m below the ground surface. The saturated sediments between 30.6-m and 183.0-m depths are composed of volcanic, poorly sorted, moderately weathered sandy gravels and gravelly sands with occasional lenses of cobbles, boulders, and silty sandy clay. A fine-grained unit is present below 183.0 m beneath the test area, and the top of this unit is assumed to be impervious. The saturated thickness (b) of the aquifer was 152.4 m. Mock and Merz (1990) used six observation wells for the pumping test. In this example, only one observation well (WR-62) data are analyzed. Figure 9-42 shows the arrangement and completions of the pumped well (EW-1) and the observation well (WR-62).

The discharge rate from EW-1 was 11,011 m^3/day for 72 hours, and water was discharged to the nearby Santa Cruz River. The observed drawdown versus time data are given in Table 9-11. The radial distance (r) from the pumped well (EW-1) to the observation well (WR-62) was 39.3 m.

Using the Neuman type-curve method for partially penetrating wells (presented above) determine the aquifer parameters.

Solution: The steps of the Neuman type-curve method for partially penetrating wells, which are presented above, will be applied to determine the aquifer parameters.

Step 1: As can be seen from Figure 9-42, there is one observation well. Using the dimensional d and l values of the pumped well and z_1 and z_2 values of the observation well, the dimensionless parameters needed for the computer program for the Neuman type curves data generation are listed in Table 9-12. The suggested range of β values in Step 1 are

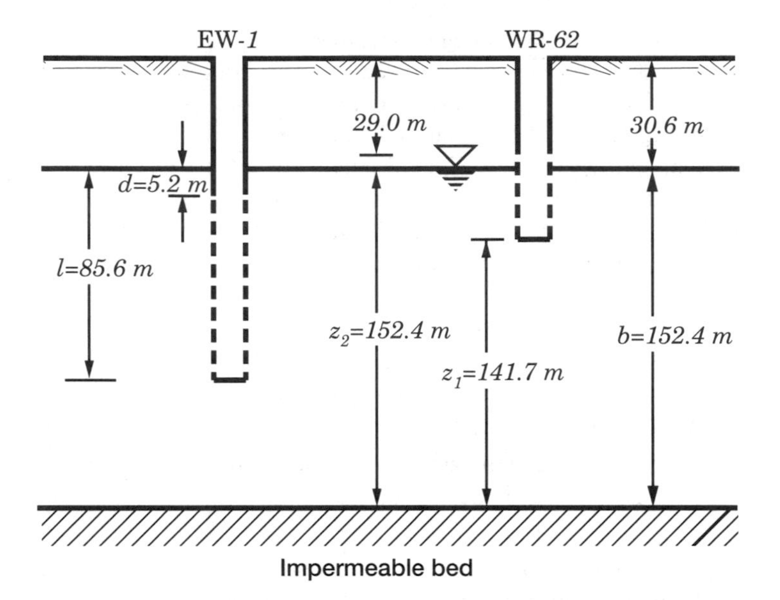

Figure 9-42 Completion interval data of the pumped and observation wells for Example 9-7. (Adapted from Mock and Merz, 1990.)

used. The Type A and Type B curves data are generated by running the DELAY2PC program using the data given in Table 9-12 and $\sigma = 10^{-9}$. The family of Type A and Type B curves are given in Figures 9-43 and 9-44, respectively.

Step 2: The observed drawdown versus time data as given (Table 9-11) at WR-62 on double logarithmic papers with the same scales as those in Step 1 is presented in Figure 9-45. The maximum drawdown during the test period is 2.16 m (see Table 9-11). Since this value is only a small fraction of the initial saturated thickness (1.4%), drawdown correction is not required.

Step 3: The drawdown versus time data graph (Figure 9-45) is superimposed on the Type B curves (Figure 9-44) and shown in Figure 9-47. The data appear to fit a Type B curve for $\beta = 0.004$.

Step 4: Match point B is chosen on Figure 9-47 and is represented by the following coordinates:

$$t = 3.0 \text{ min}, \qquad s = 0.70 \text{ m}$$
$$t_y = 0.01, \qquad s_D = 1$$

Step 5: Thus, from Eqs. (9-137) and (9-138), the following values can be obtained:

$$T = \frac{Q}{4\pi}\frac{s_D}{s} = \frac{(11{,}011 \text{ m}^3/\text{day})\,(1.0)}{(4\pi)(0.70 \text{ m})} = 1251.8 \text{ m}^2/\text{day}$$

$$S_y = \frac{Tt}{t_y r^2} = \frac{\left(\dfrac{1251.8 \text{ m}^2}{24 \times 60 \text{ min}}\right)(3.0 \text{ min})}{(0.01)(39.3 \text{ m})^2} = 0.17$$

TABLE 9-11 Drawdown Versus Time Data at WR-62 for Example 9-7

t (min)	s (m)	t (min)	s (m)	t (min)	s (m)
1	0.36	423	1.31	2004	1.80
2	0.44	453	1.31	2063	1.81
3	0.52	482	1.33	2125	1.83
4	0.55	515	1.34	2181	1.83
5	0.59	546	1.36	2245	1.85
6	0.62	574	1.37	2310	1.86
7	0.65	604	1.39	2369	1.87
8	0.67	634	1.40	2430	1.88
9	0.69	665	1.41	2486	1.89
10	0.71	696	1.43	2453	1.90
12	0.74	727	1.44	2603	1.92
14	0.77	755	1.44	2665	1.93
16	0.80	786	1.46	2736	1.94
18	0.82	822	1.47	2789	1.95
20	0.84	851	1.48	2855	1.96
25	0.89	880	1.49	2914	1.97
30	0.92	909	1.49	2965	1.98
35	0.95	940	1.51	3026	1.98
40	0.97	967	1.51	3086	1.98
45	0.99	1000	1.52	3147	2.00
50	1.01	1027	1.53	3204	2.00
55	1.02	1056	1.54	3259	2.01
60	1.03	1087	1.55	3324	2.04
70	1.05	1116	1.56	3384	2.05
80	1.07	1145	1.57	3444	2.06
90	1.09	1180	1.58	3505	2.08
100	1.10	1210	1.59	3563	2.08
110	1.12	1240	1.60	3622	2.08
125	1.13	1291	1.61	3682	2.09
140	1.14	1346	1.63	3742	2.09
155	1.15	1409	1.65	3806	2.10
170	1.16	1465	1.66	3868	2.11
185	1.17	1524	1.66	3923	2.12
200	1.18	1585	1.68	3987	2.13
220	1.19	1645	1.69	4048	2.14
240	1.20	1702	1.70	4102	2.15
260	1.21	1762	1.72	4166	2.16
326	1.24	1825	1.73	4198	2.16
363	1.26	1885	1.75	4224	2.16
390	1.28	1941	1.76	4284	2.16

TABLE 9-12 Type Curves Generation Input Data for Example 9-7

Pumped Well	Observation Well	l/b	d/b	z/b	z_1/b	z_2/b
EW-1	WR-62	0.5617	0.0341	0.0	0.93	1.00

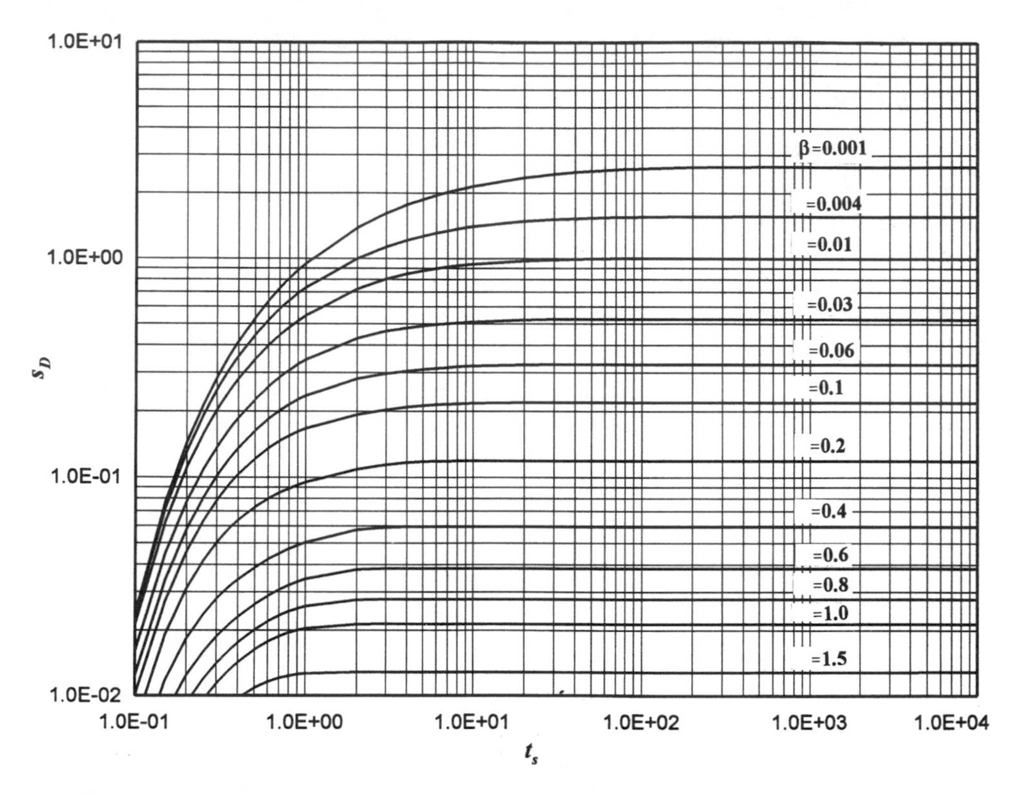

Figure 9-43 Neuman Type A curves for the partially penetrating wells of Example 9-7.

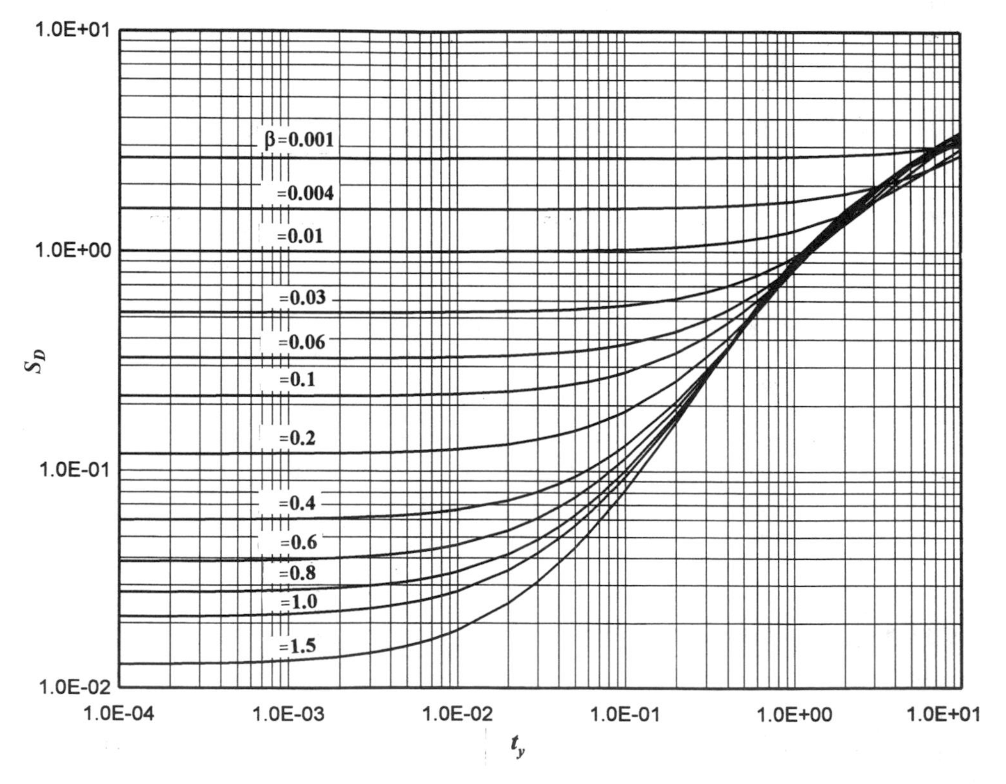

Figure 9-44 Neuman Type B curves for the partially penetrating wells of Example 9-7.

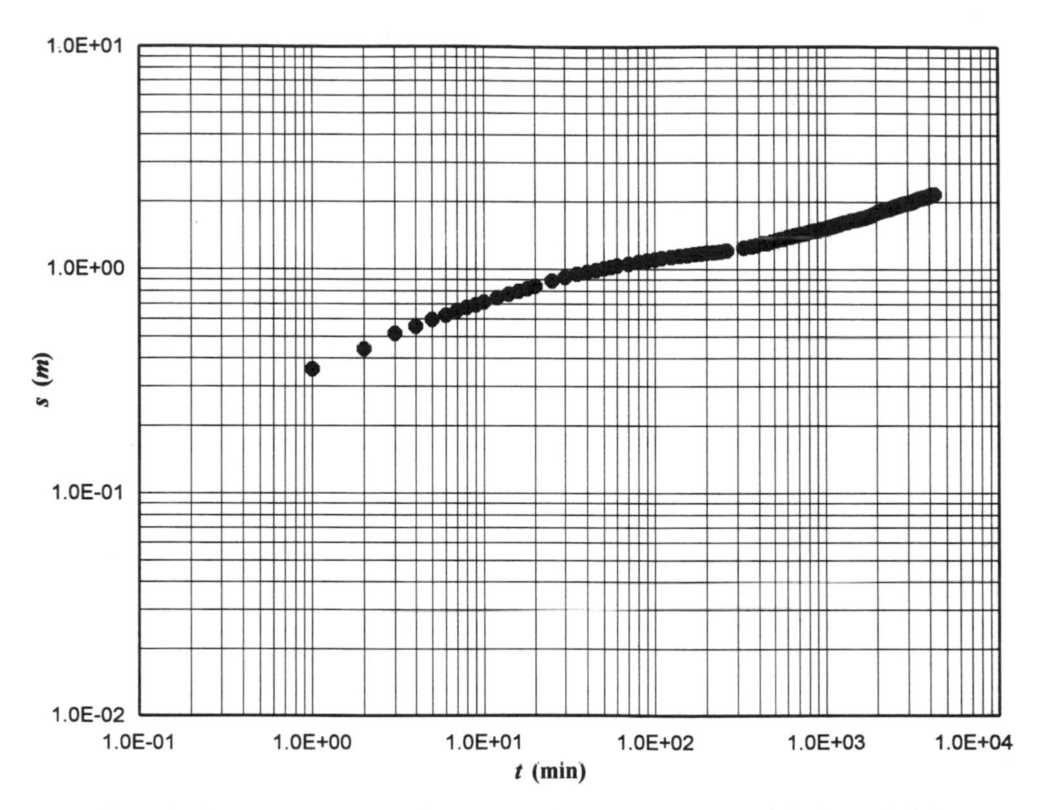

Figure 9-45 Drawdown versus time curve at observation well WR-62 for Example 9-7.

Step 6: The drawdown versus time data graph (Figure 9-45) is superimposed on the Type A curves and is shown in Figure 9-46. The value of β is 0.004 because it must be the same as that obtained in Step 3 from the Type B curves.

Step 7: A match point A is chosen on Figure 9-46 and is represented by the following coordinates:

$$t = 9.0 \text{ min}, \qquad s = 0.45 \text{ m}$$
$$t_s = 10.0, \qquad s_D = 1.0$$

Step 8: From Eqs. (9-137) and (9-141), respectively, the following values can be obtained:

$$T = \frac{Q}{4\pi}\frac{s_D}{s} = \frac{(11{,}011 \text{ m}^3/\text{day})(1.0)}{(4\pi)(0.45 \text{ m})} = 1947.2 \text{ m}^2/\text{day}$$

$$S = \frac{Tt}{t_s r^2} = \frac{\left(\dfrac{1947.2 \text{ m}^2}{24 \times 60 \text{ min}}\right)(9.0 \text{ min})}{(10.0)(39.3 \text{ m})^2} = 7.9 \times 10^{-4}$$

Step 9: The arithmetic average of the transmissivity values obtained in the previous steps shall be adopted in the calculations in the rest of the steps:

$$T = \frac{1251.8 \text{ m}^2/\text{day} + 1947.2 \text{ m}^2/\text{day}}{2} = 1599.5 \text{ m}^2/\text{day}$$

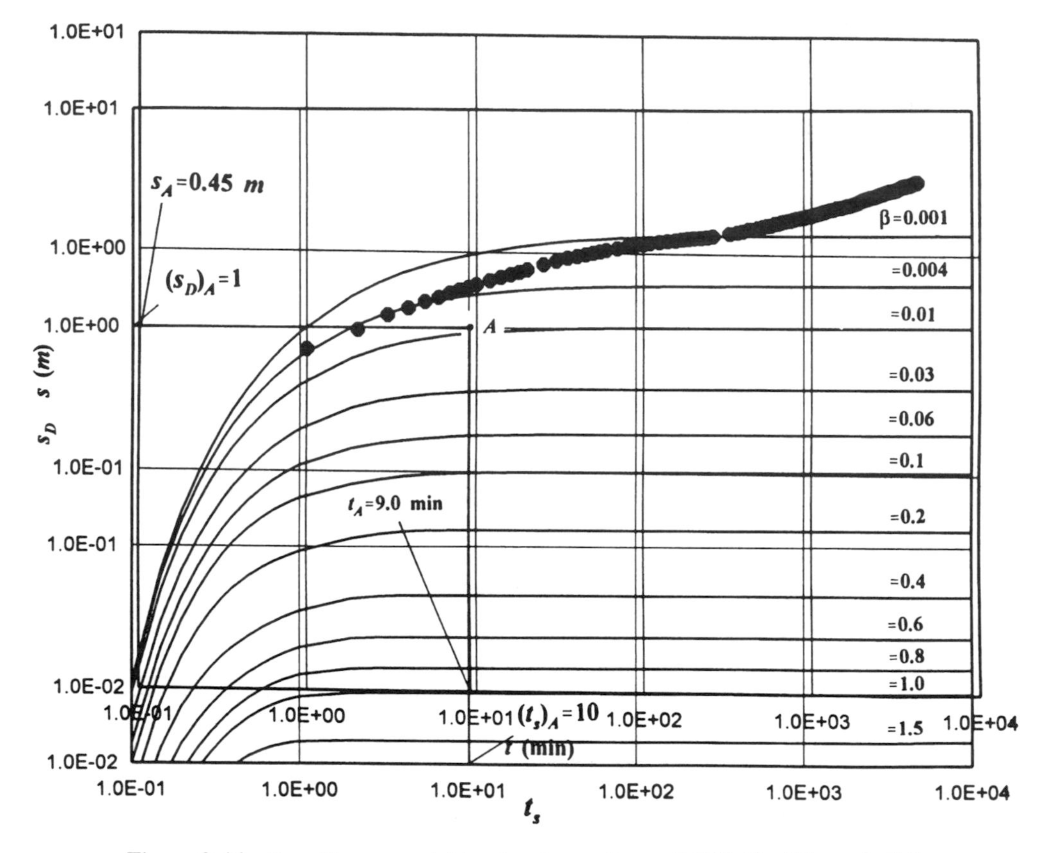

Figure 9-46 Type B curve matching for observation well WR-62 of Example 9-7.

From Eq. (9-143), the horizontal hydraulic conductivity value is

$$K_r = \frac{T}{b} = \frac{1599.5 \text{ m}^2/\text{day}}{152.4\text{m}} = 10.50 \text{ m/day} = 1.21 \times 10^{-2} \text{ cm/s}$$

From Eq. (9-144), the degree of an isotropy is

$$K_D = \beta \left(\frac{b}{r}\right)^2 = (0.004) \left(\frac{152.4 \text{ m}}{39.3 \text{ m}}\right)^2 = 0.06$$

Using the values of K_r and K_D, from Eq. (9-145) the vertical hydraulic conductivity value is

$$K_z = K_D K_r = (0.06)(10.88 \text{ m/day}) = 0.63 \text{ m/day} = 7.29 \times 10^{-4} \text{ cm/s}$$

Step 10: From Eq. (9-146), the value of σ, from the S and S_y values, is

$$\sigma = \frac{S}{S_y} = \frac{7.9 \times 10^{-4}}{0.17} = 4.6 \times 10^{-3}$$

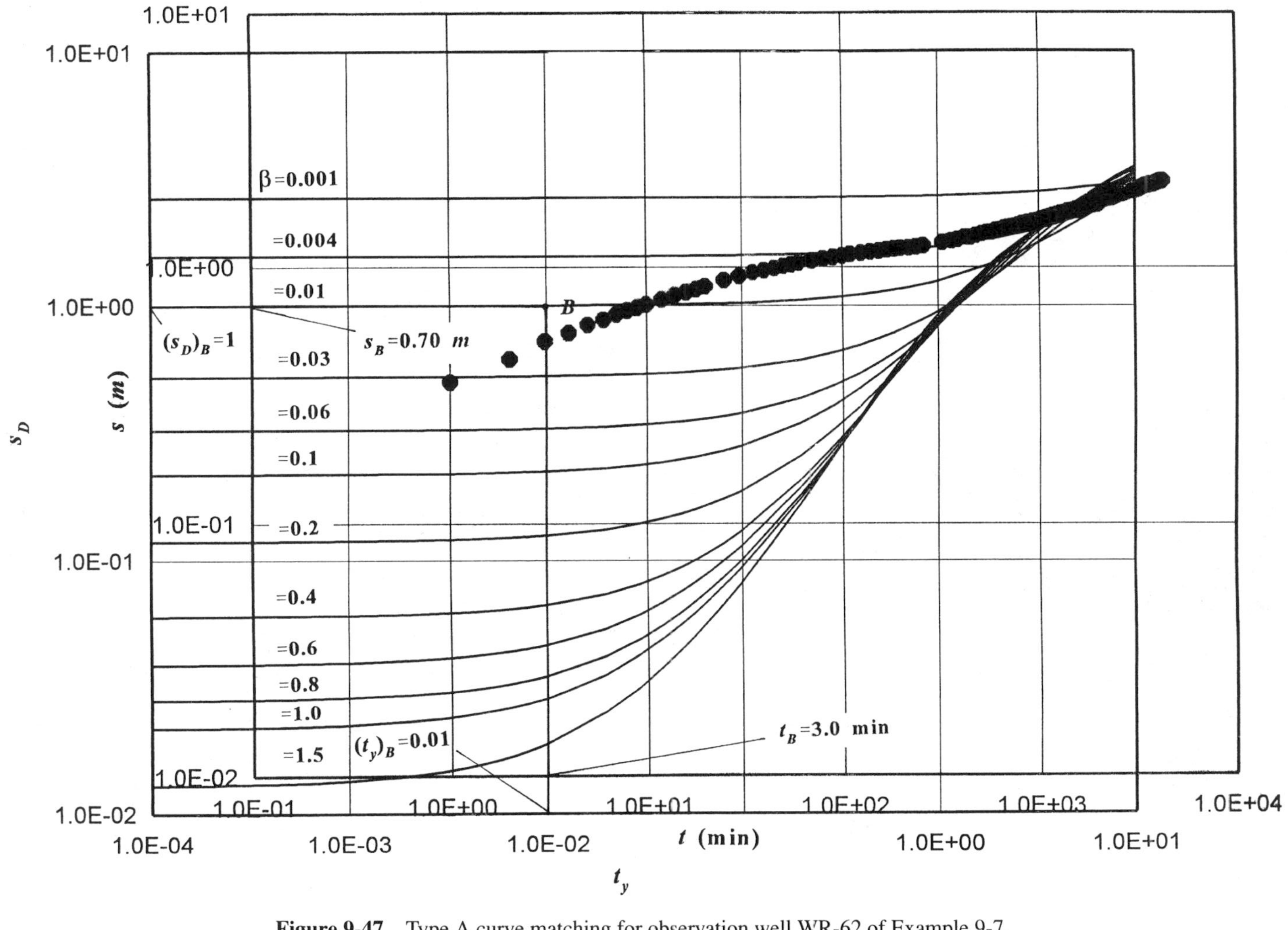

Figure 9-47 Type A curve matching for observation well WR-62 of Example 9-7.

Neuman's Semilogarithmic Method. Neuman's semilogarithmic method, which is presented in detail in Section 9.5.4.1, for fully penetrating wells can also be applied to partially penetrating well cases with some modifications. But one must bear in mind that Eq. (9-151) or Figure 9-30, which are based on the values in Table 9-7, cannot be used because they belong to fully penetrating well cases.

Step 1 through Step 6 of Section 9.5.4.1 can be used as they are. For each group of observation wells, a table corresponding to Table 9-8 is required in order to generate a formula like Eq. (9-151) and a figure like Figure 9-30 using the corresponding Type A and Type B type-curve data. This seems to be a somewhat time-consuming and impractical process.

Recovery Method. The recovery method can be applied for partially penetrating wells as described in Section 9.5.4.1.

CHAPTER 10

FULLY PENETRATING PUMPING WELLS IN HOMOGENEOUS AND ISOTROPIC BOUNDED NONLEAKY CONFINED AQUIFERS

10.1 INTRODUCTION

Analytical mathematical models regarding extraction or injection wells in aquifers are mostly based on the assumption that the medium is infinite in lateral extent. In practice, the location of the zone of influence boundary of a well, beyond which the drawdowns are zero, in an aquifer depends on the extraction or injection rate, the time period, and the hydrogeologic parameters of the aquifer. For a homogeneous and isotropic aquifer the zone of influence boundary of a well has a circular shape, whereas in homogeneous and anisotropic aquifers it has an elliptic shape (see Figure 5-2 of Chapter 5). If the zone of influence boundary of a well does not reach to the nearby impervious boundaries and surface water bodies (streams, canals, lakes, seas, etc.) during the extraction or injection period, the infinite lateral extent assumption-based models can still safely be used for all practical purposes, including pumping test data analyses.

If the zone of influence boundary of a well interferes with the impervious boundaries and surface water bodies near the well, the infinite areal extent assumption no longer holds. The method of images (Jacob, 1950; Ferris et al., 1962; Todd, 1980; Chapius, 1994), which is also an important tool in heat conduction in solids, electricity, and some other similar fields, emerged as a tool after the 1950s in the evaluation of the influence of the aforementioned boundaries on the flow around a well. The method of images is based on the principle of *image wells* and uses the analytical well models based on the infinite areal extent assumption. Although the principle is general in the sense that the analytical models for confined and unconfined aquifers under fully and partially penetrating well conditions can be used for the cases mentioned above, the method of images application is mostly limited in the literature to the Theis equation (Theis, 1935) and its modified form, the Cooper and Jacob equation (Cooper and Jacob, 1946), which are presented in Chapter 4.

In this chapter, after the general foundations of the method of images are presented, the solutions for one and two impervious boundaries as well as their practical applications regarding the detection of buried impervious boundaries will be presented based on the

Cooper and Jacob equation for confined aquifers. Then, the solutions of one recharge boundary and their practical applications regarding pumping test data analysis near a river will be presented based on the Theis and the Cooper and Jacob equations for confined aquifers.

10.2 ONE OR TWO IMPERVIOUS BOUNDARIES

10.2.1 Theory with the Application of the Theis Equation

10.2.1.1 Drawdown Equations. First, the theoretical foundations of the effects on drawdown of one straight impervious boundary around a fully penetrating well in a confined aquifer will be presented (Ferris et al., 1962).

Figure 10-1a shows a discharging well in a confined aquifer bounded on the right by an impermeable vertical planar barrier. The hydraulic condition imposed by the vertical boundary is that there can be no ground-water flow across the boundary. In other words, the impermeable material cannot contribute water to the pumped well. In Figure 10-2, a generalized flow net for streamlines and equipotential lines for the system corresponding to Figure 10-1 is presented. The problem is to obtain a solution for the drawdown variation around the well by taking into account the vertical impermeable barrier. In Figure 10-1b, the real well and its imaginary well with the same rate, located symmetrically according to a fictitious vertical plane corresponding to the plane of the impermeable vertical boundary in Figure 10-1a, are shown in an infinite confined aquifer. The real and imaginary wells are assumed to be located on a common line perpendicular to the fictitious vertical plane. As a result of the geometry, the drawdown cones for the real and imaginary wells will be symmetrical and will create a ground-water divide at every point along the fictitious vertical plane in Figure 10-1b. Because there can be no flow across a ground-water divide, it acts exactly like an impervious boundary. This means that the image well system satisfies the boundary condition of the real problem in Figure 10-1a. Therefore, to take into account the effects of a vertical impervious barrier near a pumping well, one should consider an image well, with the same rate of the real well, symmetrically located according to a vertical plane exactly in the place of the vertical impervious barrier in an infinite aquifer.

Because the partial differential equation for drawdown in a confined aquifer is linear [Eq. (4-10) of Chapter 4], the superposition principle can be applied. Therefore, the resultant drawdown at any point on the cone of depression in the real region is the sum of the drawdowns created at that point by the real well and its image well (see also Figure 10-1b). If an observation well is at distances r_r and r_i from the real well and imaginary well, respectively, the drawdown from Eq. (4-24) of Chapter 4 is

$$s = s_r + s_i = \frac{Q}{4\pi T}[W(u_r) + W(u_i)] \tag{10-1}$$

where

$$u_r = \frac{r_r^2 S}{4Tt}, \qquad u_i = \frac{r_i^2 S}{4Tt} \tag{10-2}$$

where $W(u)$ is the Theis well function for fully penetrating wells in confined aquifers [Eq. (4-25) of Chapter 4], Q is the discharge rate, T is transmissivity, S is the storage

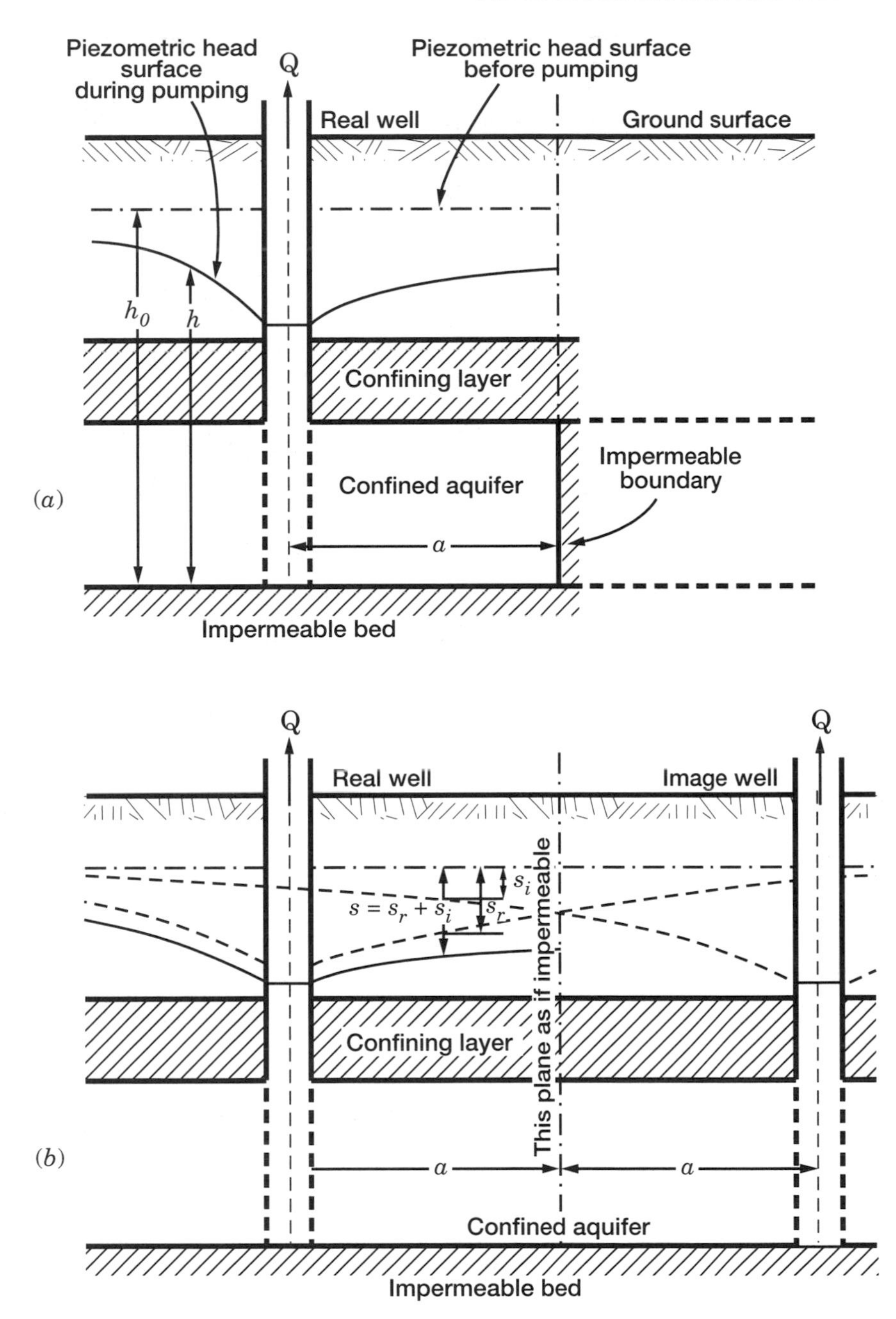

Figure 10-1 Cross sections of an idealized semi-infinite confined aquifer having an impervious vertical plane boundary: (a) Discharging well near an impermeable plane boundary; (b) mathematically equivalent system in a confined aquifer of infinite areal extent. (Adapted from Ferris et al., 1962; Chapius, 1994.)

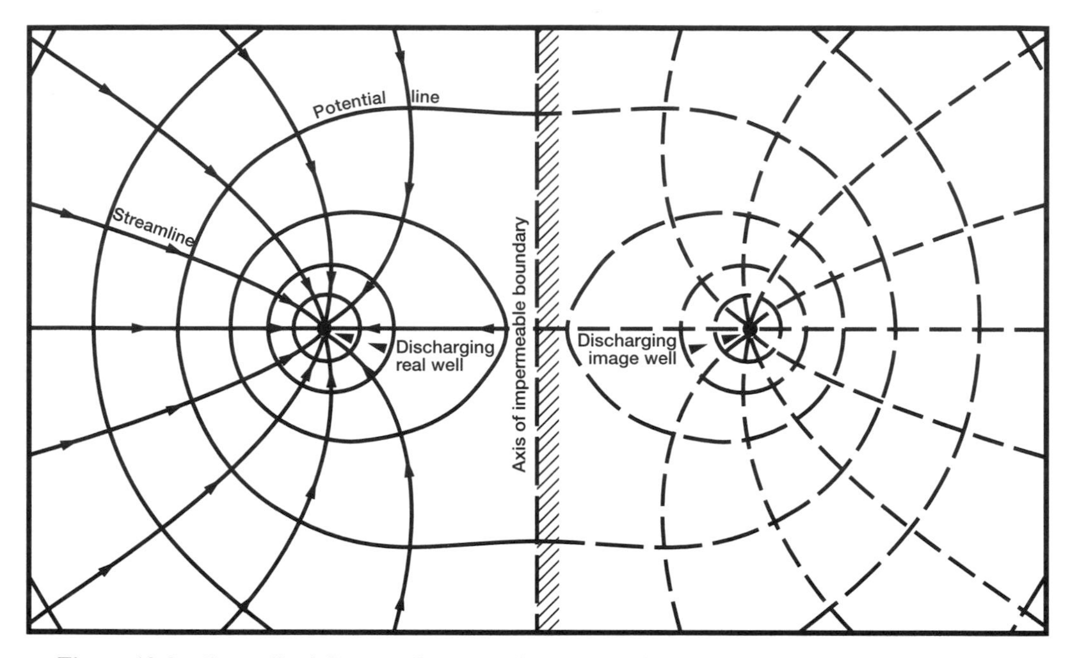

Figure 10-2 Generalized flow net for streamlines and equipotentiallines in the vicinity of a discharging well near an impermeable vertical plane in a confined aquifer. (After Ferris et al., 1962.)

coefficient, and t is time. Using the Cooper and Jacob approximation (Cooper and Jacob, 1946), as given by Eq. (4-46) of Chapter 4, Eq. (10-1) takes the form

$$s = \frac{2.303Q}{4\pi T}\left[\log\left(\frac{2.25Tt}{r_r^2 S}\right) + \log\left(\frac{2.25Tt}{r_i^2 S}\right)\right] \tag{10-3}$$

One should bear in mind that the Cooper and Jacob approximation is valid for $u < 0.01$ (see Section 4.2.7.2 of Chapter 4).

10.2.1.2 Relationship Between Time and Radial Distance

(a) Classical Equation. The equal drawdowns condition created by the real and imaginary wells can be derived from the expressions in Eq. (10-2). In order the drawdowns to be equal, u_r and u_i in Eq. (10-2) must be equal. Equal drawdowns generated by the real and imaginary wells at any point cannot occur at the same time with the exception of the impervious vertical plane. Since $S/4T$ is constant, one must have

$$\frac{r_r^2}{t_r} = \frac{r_i^2}{t_i} \tag{10-4}$$

where t_r and t_i are the corresponding times for the real and imaginary wells.

(b) Equation Proposed by Castany (1982) and Raghunath (1987). Another equation similar to Eq. (10-4) is given by Castany (1982) and Raghunath (1987) and relates the time

t_I (see Figure 10-3), at which the geometric slope is doubled, to the time t_0 at $s = 0$:

$$\frac{r_r^2}{t_0} = \frac{r_i^2}{t_I} \tag{10-5}$$

The relationship between the storage coefficient S and t_0 is

$$S = \frac{2.25Tt_0}{r^2} \tag{10-6}$$

which is Eq. (4-50) in of Chapter 4. Equation (10-5) may be viewed as a special case of Eq. (10-4).

(c) Equation Proposed by Chapius (1994). Chapius (1994) proposed an equation similar to those given above, and its derivational details are given below.

From the u_r and u_i expressions in Eq. (10-2) one can write

$$\frac{u_i}{u_r} = \left(\frac{r_i}{r_r}\right)^2 = \beta^2 \tag{10-7}$$

From Eqs. (10-3), (10-5), (10-6), and (10-7) one can write

$$s = \frac{2.303Q}{4\pi T}\left[\log\left(\frac{t}{t_0}\right) + \log\left(\frac{t}{\beta^2 t_0}\right)\right] \tag{10-8}$$

On the other hand, from Eq. (4-52) of Chapter 4, Δs for one log cycle can be expressed as

$$\Delta s = \frac{2.303Q}{4\pi T} \tag{10-9}$$

and with this, Eq. (10-8) takes the following form:

$$s = \Delta s \log\left(\frac{t^2}{\beta^2 t_0^2}\right) = 2\Delta s \log\left(\frac{t}{\beta t_0}\right) \tag{10-10}$$

Equation (10-10) indicates a straight-line relationship between s and $\log(t)$ having $2\Delta s$ geometric slope per time log cycle and a time intercept

$$t_0^* = \beta t_0 \tag{10-11}$$

when drawdown becomes zero. According to Eq. (10-10), the geometric slope of the drawdown versus logarithmic time will be doubled once the influence of the impervious boundary is detected. The distance from the observation well to the imaginary well, r_i, may then be determined by extrapolating the second straight-line portion of the time intercept, t_0^* (see Figure 10-3). From Eqs. (10-7) and (10-11) one gets

$$\frac{r_i}{r_r} = \beta = \frac{t_0^*}{t_0} \tag{10-12}$$

which was proposed by Chapius (1994).

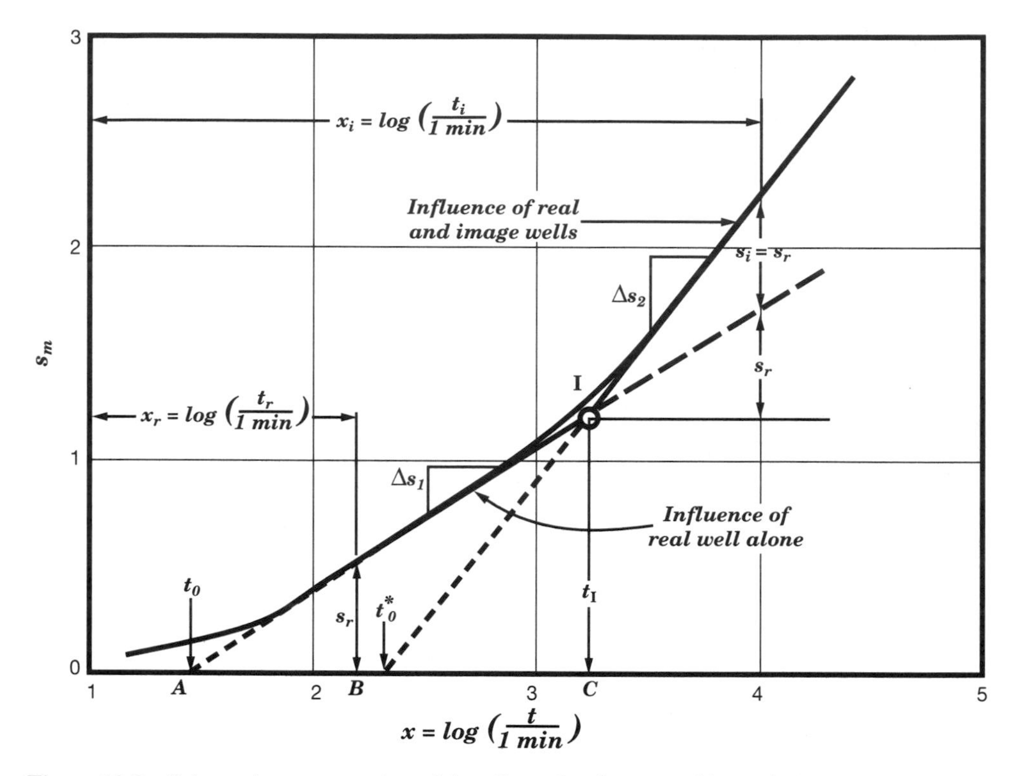

Figure 10-3 Schematic representation of the effect of an impermeable vertical plane boundary in the drawdown versus time curve. (After Chapius, 1994.)

From Eq. (10-8), the required time t_r for the real pumping well to create the drawdown s_r is

$$s_r = \frac{2.303Q}{4\pi T} \log\left(\frac{t_r}{t_0}\right) \tag{10-13}$$

Similarly, again from Eq. (10-8), the required time t_i for imaginary pumping well to create the drawdown s_i is

$$s_i = \frac{2.303Q}{4\pi T} \log\left(\frac{t_i}{\beta^2 t_0}\right) \tag{10-14}$$

If t_r and t_i are selected in Figure 10-3 such as $s_i = s_r$, Eqs. (10-13) and (10-14) gives

$$t_r = \frac{t_i}{\beta^2} \tag{10-15}$$

Combination of Eqs. (10-12) and (10-15) yields the previously given Eq. (10-4). In Figure 10-3, the point I is the intersection of Eqs. (10-13) and (10-14) and I is the starting point where the geometric slope becomes double that of the geometric slope of the first line. At the point I, the time is t_I and $s_r = s_I$, which gives

$$\Delta s \log\left(\frac{t_I}{t_0}\right) = 2\Delta s \log\left(\frac{t_I}{\beta t_0}\right) \tag{10-16}$$

which yields the previously given Eq. (10-5). Equation (10-16) can also be expressed as

$$\log(t_I) - \log(t_0) = 2\left[\log(t_I) - \log(\beta t_0)\right] = 2\left[\log(t_I) - \log(t_0^*)\right] \tag{10-17}$$

The left-hand side of Eq. (10-17) represents the AC distance in Figure 10-3. This distance is twice the distance BC corresponding to the brackets in the right-hand side of Eq. (10-17).

10.2.2 Boundary Detection Methods for Confined Aquifers

The boundary detection methods for confined aquifers are based on the Cooper and Jacob equation (Cooper and Jacob, 1946), which is a modified form of the Theis equation (Theis, 1935). The details of these equations are presented above. In the following sections several methods regarding boundary detection for confined aquifers will be presented.

10.2.2.1 Method for One Impervious Boundary. The classical boundary detection method requires at least three observation wells to locate the imaginary well under the condition that all other assumptions are satisfied. The method and its conditions are presented below (Todd, 1980; Chapius, 1994).

Governing Equations and Methodology. Recharge boundaries around a pumping well in an aquifer (streams, canals, lakes, seas, etc.) would normally be visible. However, impermeable or less permeable subsurface boundaries (localized rocky and clayey formations, manmade barriers, etc.) may not be known. Where these kinds of barriers are encountered at some distance from the pumping well, the drawdown versus time curves for points around the well may respond differently as compared with those without barriers. If an impermeable boundary exists near the pumped well, the geometric slope of the drawdown versus time curve will double under the combined influence of the real and imaginary pumping well as shown in Figure 10-3. This is due to the fact that the geometric slope of the straight-line portion depends only on T and Q.

To determine the location of the image well, follow the following steps:

Step 1: Plot the drawdown versus time curves for at least three observation wells on semilogarithmic papers against the values of time (time on the logarithmic axis).

Step 2: Fit straight lines through the two portions of each observed drawdown graph as shown schematically in Figure 10-3 and determine their geometric slopes.

Step 3: Select an arbitrary drawdown s_r under the influence of the real well in one of the drawdown versus time graphs and measure the corresponding time t_r as shown in Figure 10-3. Repeat this process for at least two additional observation wells.

Step 4: Then, determine the time t_i corresponding to the same selected drawdown ($s_i = s_r$) for the image well from the second straight-line portion as shown in Figure 10-3. Repeat this process for the other observation wells.

Step 5: From the graphs, determine the values of t_0, t_0^*, and t_I for each well.

Step 6: Using the known values of r_r, t_r, t_i, t_0, t_0^*, and t_I for each observation well, calculate the r_i distances between the imaginary and observation wells from Eqs. (10-4), (10-5), and (10-12), which are r_{ia}, r_{ib}, and r_{ic}, respectively. If the discrepancies between these values are great, reevaluate especially the second straight-line fit with the data points and make the necessary adjustments using the graphical property mentioned earlier ($AB = BC$ in Figure 10-3) in Section 10.2.1.2. After the adjustments, Eqs. (10-4), (10-5), and (10-12) should give the same r_i values.

Repeat the above process for the rest of observation wells. For three observation wells case, these distances are shown by r_{i1}, r_{i2}, and r_{i3} in Figure 10-4. The distance r_{i1} only specifies the radius of a circle on which the imaginary well is located. As mentioned above, at least three observation wells are required to locate the imaginary well because two circles of radii r_{i1} and r_{i2} have two intersections, and therefore the third circle of radius r_{i3} will determine the correct intersection (see Figure 10-4).

Example 10-1: This example is taken from Chapius (1994) which is based on the data in Castany (1982). The pumping test was conducted in the confined aquifer of Manga which is located in the southwest of Madacascar. The drawdown versus time curve for a piezometer at a distance $r_r = 5.90$ m from the pumped well at a constant rate $Q = 5 \times 10^{-3}$ m^3/s is given in Figure 10-5. Castany (1982), as presented also in Chapius (1994), gave the following values from the straight-line data analysis:

$$\Delta s_1 = 0.046 \text{ m}, \quad \Delta s_2 = 0.086 \qquad \text{per time log cycle}$$

$$t_0 = 9.3 \text{ s}, \quad t_I = 2.5 \times 10^4 \text{ s}$$

Using the method presented above, evaluate the given data whether there is an impervious boundary or not.

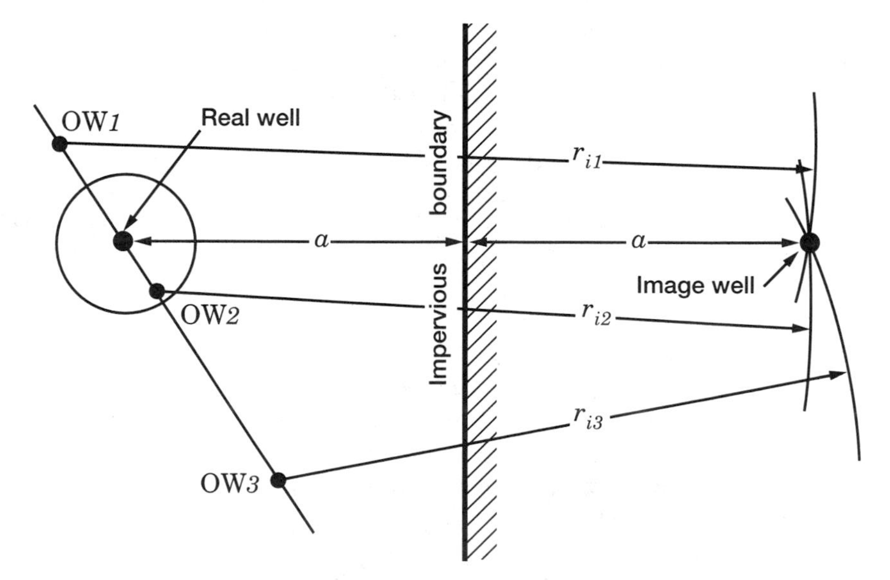

Figure 10-4 Determination of the location of the imaginary well with the drawdown versus time data of three observation well. (After Chapius, 1994.)

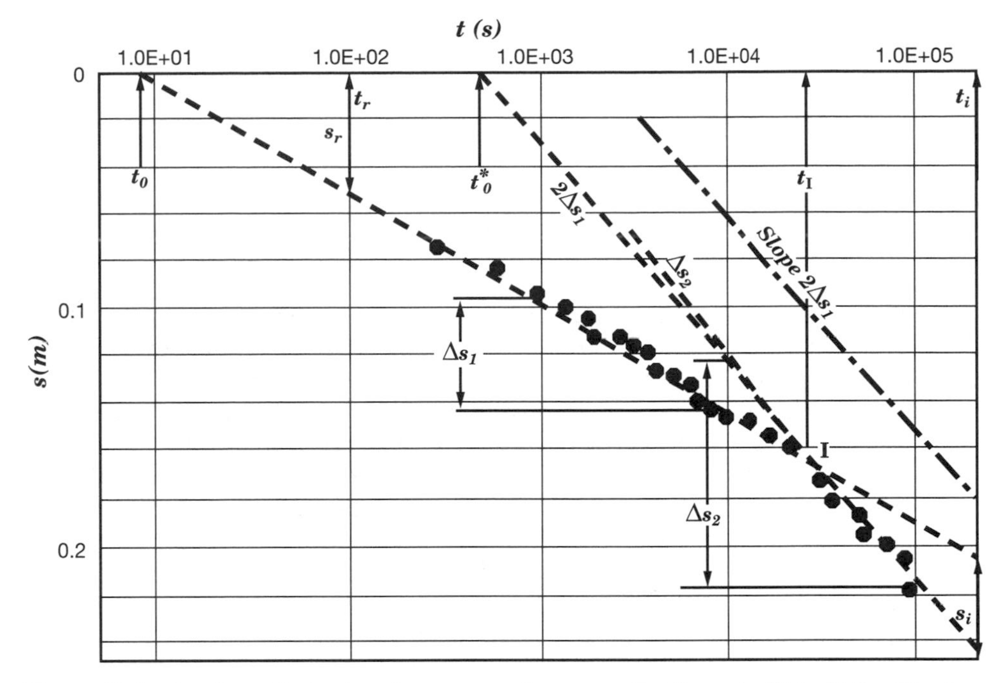

Figure 10-5 Semilogarithmic drawdown versus time curve for Example 10-1. (After Chapius, 1994.)

Solution: As mentioned above, at least three observation wells are required to determine the location of the imaginary well. Obviously, with one observation well this cannot be done. Only the circle on which the imaginary well is located can be specified. The analysis is given below in accordance with the steps given above.

Step 1: The semilogarithmic drawdown versus time curve is given in Figure 10-5.

Step 2: The straight lines through the two portions are shown in Figure 10-5, and their geometric slopes per time log cycle are given above. The values $\Delta s_1 = 0.046$ m and $\Delta s_2 = 0.086$ m show that the geometric slope is approximately doubled. Therefore, there is a good possibility that an impervious boundary exists.

Step 3: The arbitrarily selected drawdown and its corresponding time on the first straight-line are $s_r = 0.049$ m and $t_r = 1.0 \times 10^2$ s.

Step 4: For $s_i = s_r = 0.049$ m on the second straight line, $t_i = 2.0 \times 10^5$ s.

Step 5: From Figure 10-5 we have $t_0 = 9.3$ s, $t_0^* = 570$ s, and $t_I = 2.5 \times 10^4$ s.

Step 6: From Eq. (10-4) we obtain

$$r_i = r_{ia} = r_r \left(\frac{t_i}{t_r} \right)^{1/2} = (5.9 \text{ m}) \left(\frac{2.0 \times 10^5 \text{ s}}{1.0 \times 10^2 \text{ s}} \right)^{1/2} = 263.9 \text{ m}$$

From Eq. (10-5) we have

$$r_i = r_{ib} = r_r \left(\frac{t_I}{t_0} \right)^{1/2}$$

$$= (5.9 \text{ m}) \left(\frac{2.5 \times 10^4 \text{ s}}{9.3 \text{ s}} \right)^{1/2} = 305.9 \text{ m}$$

From Eq. (10-12) we obtain

$$r_i = r_{ic} = r_r \left(\frac{t_0^*}{t_0} \right) = (5.9 \text{ m}) \left(\frac{570 \text{ s}}{9.3 \text{ s}} \right) = 361.6 \text{ m}$$

The three r_i values calculated above from the three different formulas have large differences. The most likely reason of these discrepancies is that the measure of Δs_2 is not accurate. On the other hand, the data points show that the measure of Δs_1 is sufficiently accurate. Therefore, the graphical property mentioned earlier ($AB = BC$ in Figure 10-3) in Section 10.2.1.2 may be used to modify slightly the second straight-line. Since Δs_1 is well-defined, a straight line of geometric slope $2\Delta s_1$ is drawn on the graph. Then, a parallel line is moved until a visual best match with the observed data points is obtained. This fit gives $t_0^* = 480$ s and $t_I = 2.5 \times 10^4$ s for $\Delta s_2 = 2\Delta s_1$. For $s_i = s_r = 0.049$ m, from the second straight line the corresponding time t_i equals 2.7×10^5 s (this value is not shown on Figure 10-5). With these new values of t_i and t_0^*, the previous calculations are repeated as follows: From Eq. (10-4) we obtain

$$r_i = r_{ia} = r \left(\frac{t_i}{t_r} \right)^{1/2} = (5.9 \text{ m}) \left(\frac{2.7 \times 10^5 \text{ s}}{1.0 \times 10^2 \text{ s}} \right)^{1/2} = 306.6 \text{ m}$$

From Eq. (10-5) we have

$$r_i = r_{ib} = r \left(\frac{t_I}{t_0} \right)^{1/2} = (5.9 \text{ m}) \left(\frac{2.5 \times 10^4 \text{ s}}{9.3 \text{ s}} \right)^{1/2} = 306.0 \text{ m}$$

From Eq. (10-12) we obtain

$$r_i = r_{ic} = r_r \left(\frac{t_0^*}{t_0} \right) = (5.9 \text{ m}) \left(\frac{480 \text{ s}}{9.3 \text{ s}} \right) = 304.5 \text{ m}$$

The above values which are obtained from the three equations are approximately the same. Therefore, $r_{ia} = r_{ib} = r_{ic} = 306$ m. In conclusion, there is good possibility that there is an impervious boundary near the test area. The imaginary well is located on a circle that has a 306 m radius. Since there is only one observation well, the exact location of the imaginary well cannot be specified.

Example 10-2: This example is taken from Chapius (1994), which is based on the data in Castany (1982). The pumping test was conducted in a confined aquifer of Niger. The drawdown versus time curve for a piezometer at a distance $r_r = 20$ m from the pumped well at a constant rate $Q = 13 \times 10^{-3}$ m^3/s is given in Figure 10-6. Castany (1982), as presented also in Chapius (1994), gave the following values from the straight-line data analysis:

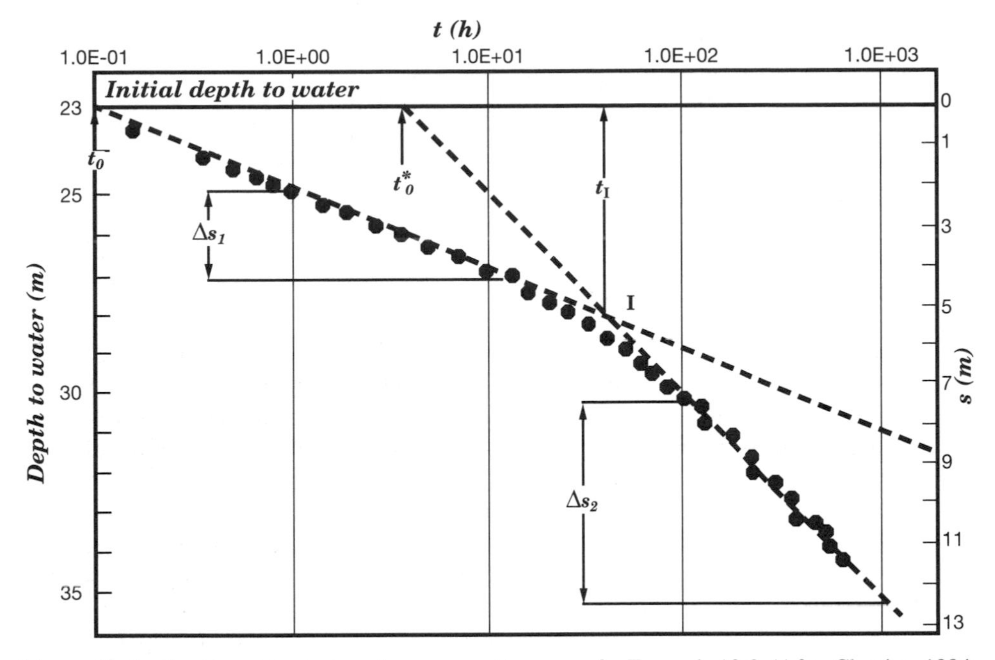

Figure 10-6 Semilogarithmic drawdown versus time curve for Example 10-2. (After Chapius, 1994; as presented in Castany, 1982.)

$$\Delta s_1 = 2.10 \text{ m}, \quad \Delta s_2 = 5.0, \qquad \text{per time log cycle}$$

$$t_0 = 0.10 \text{ h}, \quad t_I = 36 \text{ h}$$

Using the method presented above, evaluate the given data whether there is an impervious boundary or not.

Solution:

Step 1: The semilogarithmic drawdown versus time curve is given in Figure 10-6.

Step 2: The straight lines through the two portions are shown in Figure 10-6, and their geometric slopes per time log cycle are given above. The values $\Delta s_1 = 2.10$ m and $\Delta s_2 = 5.0$ m show that the geometric slope is more than doubled, and two intersecting straight lines can well be defined . However, it may be questioned whether an impervious boundary like the one in Figure 10-1 will be detected.

Step 3: The arbitrarily selected drawdown and its corresponding time on the first straight line are: $s_r = 4.0$ m and $t_r = 10$ h.

Step 4: For $s_i = s_r = 4.0$ m on the second straight line, $t_i = 900$ h.

Step 5: From Figure 10-6 we have $t_0 = 0.1$ h, $t_0^* = 3.6$ h, and $t_I = 36$ h.

Step 6: From Eq. (10-4) we obtain

$$r_i = r_{ia} = r_r \left(\frac{t_i}{t_r} \right)^{1/2} = (20 \text{ m}) \left(\frac{900 \text{ h}}{10 \text{ h}} \right)^{1/2} = 189.7 \text{ m}$$

From Eq. (10-5) we have

$$r_i = r_{ib} = r_r \left(\frac{t_I}{t_0} \right)^{1/2} = (20 \text{ m}) \left(\frac{36 \text{ h}}{0.1 \text{ h}} \right)^{1/2} = 379.5 \text{ m}$$

From Eq. (10-12) we obtain

$$r_i = r_{ic} = r_r \left(\frac{t_0^*}{t_0} \right) = (20 \text{ m}) \left(\frac{3.6 \text{ h}}{0.1 \text{ h}} \right) = 720.0 \text{ m}$$

The differences between calculated values $r_{ia} = 189.7$ m, $r_{ib} = 379.5$ m, and $r_{ic} = 720.0$ m are large. The interpretation of Chapius (1994) is that the detected impervious boundary is not a simple straight plane as assumed in the theory (Figures 10-1 and 10-2). On the other hand, the interpretation of Castany (1982) is that the aquifer is wedge-shaped and vanishes at a distance of 100–200 m from the pumped well.

Conditions for the Detection of an Impervious Boundary. As mentioned in Section 10.2.1.1, the Cooper and Jacob approximation for the Theis equation is valid for $u < 0.01$ (see also Section 4.2.7.2 of Chapter 4). Under the framework of this requirement, some conditions for the detection of impervious boundaries are presented below (Chapius, 1994).

First, consider a pumping well in an infinite confined aquifer without any boundary. Based on the above-mentioned requirement, a minimum time (or a maximum u) is required for the drawdown versus time data to match on the straight line of the Cooper and Jacob equation. In general, this minimum time is calculated for each observation well data to check whether the condition is satisfied or not. Chapius (1994) suggests that we use the condition for t/t_0 instead of u as given above. From $u = r^2 S/(4Tt)$, $u < 0.01$, and Eq. (10-6), the following condition can be derived:

$$\frac{t}{t_0} > 56.1 \tag{10-18}$$

Equation (10-18) can be more easily used in the drawdown versus time graph. In order to use this equation, the differences between the Theis equation and its approximated form (the Cooper and Jacob equation) as a function of u must be known (see Table 10-1). As can be seen from this table, the first straight-line portion of the semilogarithmic drawdown versus time curve may be correctly defined if the effects of the imaginary well are negligible

TABLE 10-1 Relative Error Between the Theis Well Function $W(u)$ and Its Approximated Form of the Cooper and Jacob Equation

u	t/t_0	Error (%)
0.10	5.6	5.4
0.09	6.2	4.6
0.05	11.2	2.0
0.02	28.1	0.59
0.01	56.1	0.24

Source: Chapius (1994).

when u_r drops from 0.1 to 0.01. This means that the second straight-line portion must start after $u_r = 10^{-2}$, and it starts for $u_i = 1$ approximately. For $u_r = 10^{-2}$, from Table 4-1 of Chapter 4, $W(u_r) = 4.04$, and when $u_i < 1$, $W(u_i) < 0.22$, which means that the influence of the imaginary well is less than 5.4%. From the expressions of Eq. (10-2) one can write

$$\frac{u_i}{u_r} = \frac{r_i^2}{r_r^2} = \beta^2 \tag{10-19}$$

which shows that u_r and u_i are proportional to r_r^2 and r_i^2, respectively, and the ratio β^2 must be higher than 100 to properly detect the two straight-line portions. By showing $u_r = u$, the combination of Eqs. (10-1) and (10-19) gives

$$s_D = \frac{4\pi Ts}{Q} = W(u) + W(\beta^2 u) \tag{10-20}$$

where s_D is the dimensionless drawdown. As can be seen from Figure 10-7, $r_i/r_r = \beta > 10$ is required to adequately draw the first straight-line portion. If $r_i/r_r = \beta < 5$, detection of the impervious boundary becomes impossible.

Finally, Chapius (1994) reaches the following conditions regarding the values of time: (1) The time t must be approximately higher than $6t_I$ to reduce the error less than 5% at the beginning of the second straight-line portion. Then a time log cycle is required to adequately draw the second straight-line portion. (2) The total pumping period required for a good detection of an impervious boundary is approximately $10^4 t_0$. This means that four time cycles, two for each straight-line portion, are necessary as shown in Figure 10-7.

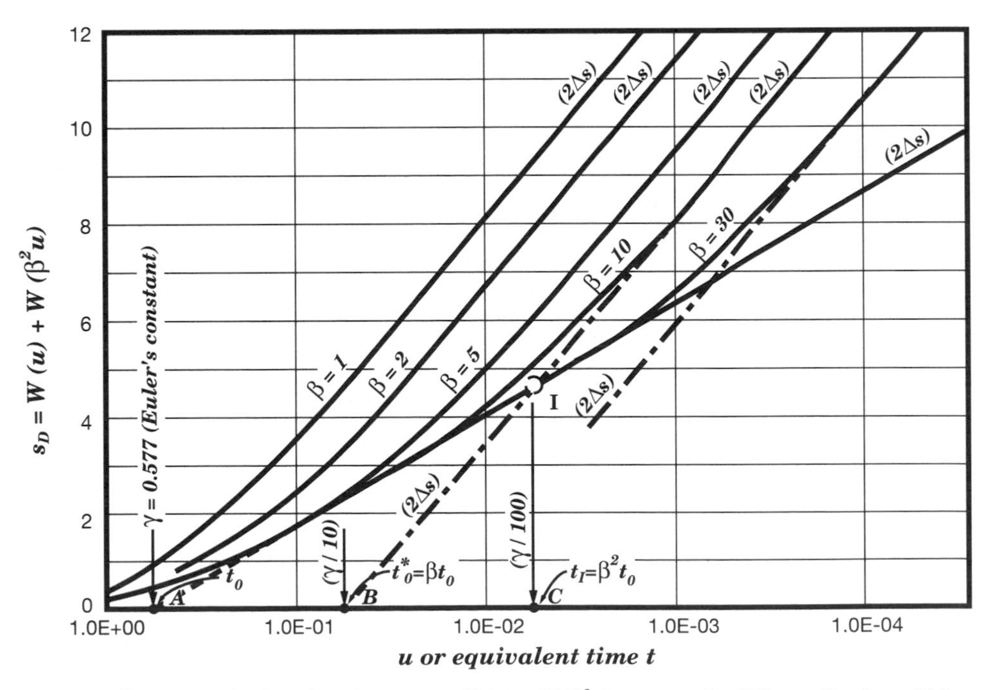

Figure 10-7 Dimensionless drawdown $s_D = W(u) + W(\beta^2 u)$ versus u for different β values. (After Chapius, 1994.)

10.2.2.2 Method for Two Intersecting Boundaries. The approach presented in Section 10.2.2.1 can be extended to the case of two straight vertical impervious plane boundaries along with the application of the method of images. In the following sections, three different cases are presented (Ferris et al., 1962; Chapius, 1994).

Case 1: Two Impervious Boundaries Form a $2\pi/4$ Angle. The imaginary well system for a discharging well in a confined aquifer bounded by two vertical impermeable vertical planes intersecting at $2\pi/4$ (90°) angle is shown in Figure 10-8. The principle is the same as the one used for one vertical barrier case which is shown in Figure 10-1. As shown in Figure 10-8, the imaginary wells I_1 and I_2 are necessary, but they are not enough to represent the effects of the intersecting impermeable barriers. The third image well, I_3, is necessary in order to make the system symmetrical with respect to the barriers, thus creating ground-water divides along these planes. Since the ground-water divides are impervious, the image system becomes hydraulically complete. As a result, the problem has been simplified by considering of four discharging wells in an infinite confined aquifer.

Since there are four wells, one real and three imaginary wells, each well discharges one-fourth of the infinite confined aquifer. If u_r, u_{i1}, u_{i2}, and u_{i3} are small enough and satisfy the necessary conditions, mentioned in Section 10.2.1.1, the Cooper and Jacob approximation may be used for the four wells. From Eqs. (10-8) and (10-9), the drawdown equation for

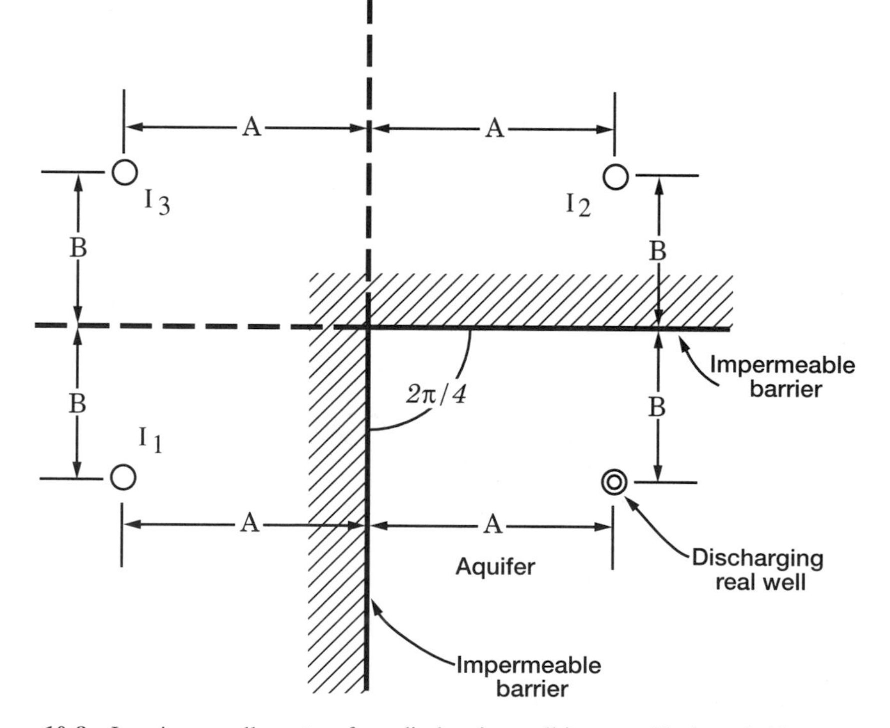

Figure 10-8 Imaginary wells system for a discharging well in an aquifer bounded by two vertical impermeable barriers intersecting at a $2\pi/4$ angle: Image wells, I, are numbered in the sequence in which they were considered and located, and open circles signify discharging wells. (After Ferris et al., 1962.)

four wells can similarly be written as (Chapius, 1994)

$$s = \Delta s \left[\log\left(\frac{t}{t_0}\right) + \log\left(\frac{t}{\beta_1^2 t_0}\right) + \log\left(\frac{t}{\beta_2^2 t_0}\right) + \log\left(\frac{t}{\beta_3^2 t_0}\right) \right] \tag{10-21}$$

or

$$s = 4\Delta s \log\left(\frac{t}{\beta_1^{1/2}\beta_2^{1/2}\beta_3^{1/2} t_0}\right) \tag{10-22}$$

This equation is a straight-line relationship between the drawdown s and $\log(t)$ whose geometric slope is $4\Delta s$ per time log cycle and whose time intercept equals $\beta_1^{1/2}\beta_2^{1/2}\beta_3^{1/2} t_0$.

Case 2: Two Impervious Boundaries Form a $2\pi/3$ Angle. If the two impervious vertical barriers form a $2\pi/3$ angle, by applying the similar methodology to the one presented above, one can find that one real and two imaginary discharging wells satisfy the hydraulic conditions for the system. Since there are three wells, one real and two imaginary wells, each well discharges one-third of the infinite confined aquifer. If u_r, u_{i1}, and, u_{i2} are small enough and satisfy the necessary conditions, mentioned in Section 10.2.1.1, the Cooper and Jacob approximation may be used for the three wells. From Eqs. (10-8) and (10-9), the drawdown equation for three wells can similarly be written as (Chapius, 1994)

$$s = \Delta s \left[\log\left(\frac{t}{t_0}\right) + \log\left(\frac{t}{\beta_1^2 t_0}\right) + \log\left(\frac{t}{\beta_2^2 t_0}\right) \right] \tag{10-23}$$

or

$$s = 3\Delta s \left(\frac{t}{\beta_1^{2/3}\beta_2^{2/3} t_0}\right) \tag{10-24}$$

This equation is a straight-line relationship between the drawdown s and $\log(t)$ whose geometric slope is $3\Delta s$ per time log cycle and time intercept equals $\beta_1^{2/3}\beta_2^{2/3} t_0$.

Case 3: Two Impervious Boundaries Form a $2\pi/6$ Angle. If the two impervious vertical barriers form a $2\pi/6$ angle, by applying the similar methodology to the one in a previous section, one can find that one real and five imaginary discharging wells satisfy the hydraulic conditions for the system. Since there are six wells, one real and five imaginary wells, each well discharges one-sixth of the infinite confined aquifer. If u_r, u_{i1}, u_{i2}, u_{i3}, u_{i4}, and u_{i5} are small enough and satisfy the necessary conditions, mentioned in Section 10.2.1.1, the Cooper and Jacob approximation may be used for the four wells. From Eqs. (10-8) and (10-9), the drawdown equation for three wells can similarly be written as (Chapius, 1994)

$$\begin{aligned} s = \Delta s \left[\log\left(\frac{t}{t_0}\right) + \log\left(\frac{t}{\beta_1^2 t_0}\right) + \log\left(\frac{t}{\beta_2^2 t_0}\right) \right. \\ \left. + \log\left(\frac{t}{\beta_3^2 t_0}\right) + \log\left(\frac{t}{\beta_4^2 t_0}\right) + \log\left(\frac{t}{\beta_5^2 t_0}\right) \right] \end{aligned} \tag{10-25}$$

or

$$s = 6\Delta s \left(\frac{t}{\beta_1^{1/3}\beta_2^{1/3}\beta_3^{1/3}\beta_4^{1/3}\beta_5^{1/2}t_0} \right) \tag{10-26}$$

This equation is a straight-line relationship between the drawdown s and $\log(t)$ whose geometric slope is $6\Delta s$ per time log cycle and whose time intercept equals $\beta_1^{1/3}\beta_2^{1/3}\beta_3^{1/3} \times \beta_4^{1/3}\beta_5^{1/3}t_0$.

10.3 ONE RECHARGE BOUNDARY

There may be cases where pumping tests need to be conducted near a river or stream. If the drawdown versus time curves are affected by the river, the pumping test can no longer be analyzed by the methods based on the infinite areal extent assumption. In the following sections, application of the theory of the method of images for recharge boundaries as well as data analysis methods will be presented in detail. In the following sections, the method of images application for recharge boundaries will be presented by using the Theis equation (Theis, 1935). Even though almost all models for a single well (fully and partially penetrating) based on the infinite areal extent assumption can be used for the same purpose, unfortunately the data analysis methods near a river in the pertinent literature are mostly limited with the application of the Theis equation (Hantush, 1959).

10.3.1 Theory

First, the theoretical foundations of the effects of a perennial river around a fully penetrating well in a confined aquifer will be presented on the assumption that the river fully penetrates the aquifer (Ferris et al., 1962; Hantush, 1959).

Figure 10-9a shows the cross section through a discharging well in a confined aquifer which is bounded on the right by a river. If the river stage is not lowered by the well, the drawdown along the river will be zero. In other words, the hydraulic head along the river will be constant throughout the time. The problem is to obtain a solution for drawdown variation around the well by taking into account the constant-head (or Dirichlet) boundary condition along the river. In Figure 10-9b, the real discharging well and its imaginary recharge well with the same rate, located symmetrically according to a fictitious vertical plane corresponding to the left boundary of the river in Figure 10-9a, which will be called the *plane source* hereafter, are shown in an infinite aquifer. Both the real discharging well and the imaginary recharge well are located on a common line perpendicular to the river. As shown in Figure 10-9b, the imaginary recharge well creates a mounding (s_i) everywhere along the position of the plane source that is equal to and cancels the drawdown (s_r) generated by the real discharging well. Since $s_r = s_i$ on the plane source, the water level in the river remains the same and this satisfies the aforementioned constant-head boundary condition for the river. The resultant drawdown at any point on the cone of depression in the real region is the algebraic sum of the drawdown ($+s_r$) generated by the real well and the mounding ($-s_i$) generated by the imaginary well:

$$s = s_r - s_i \tag{10-27}$$

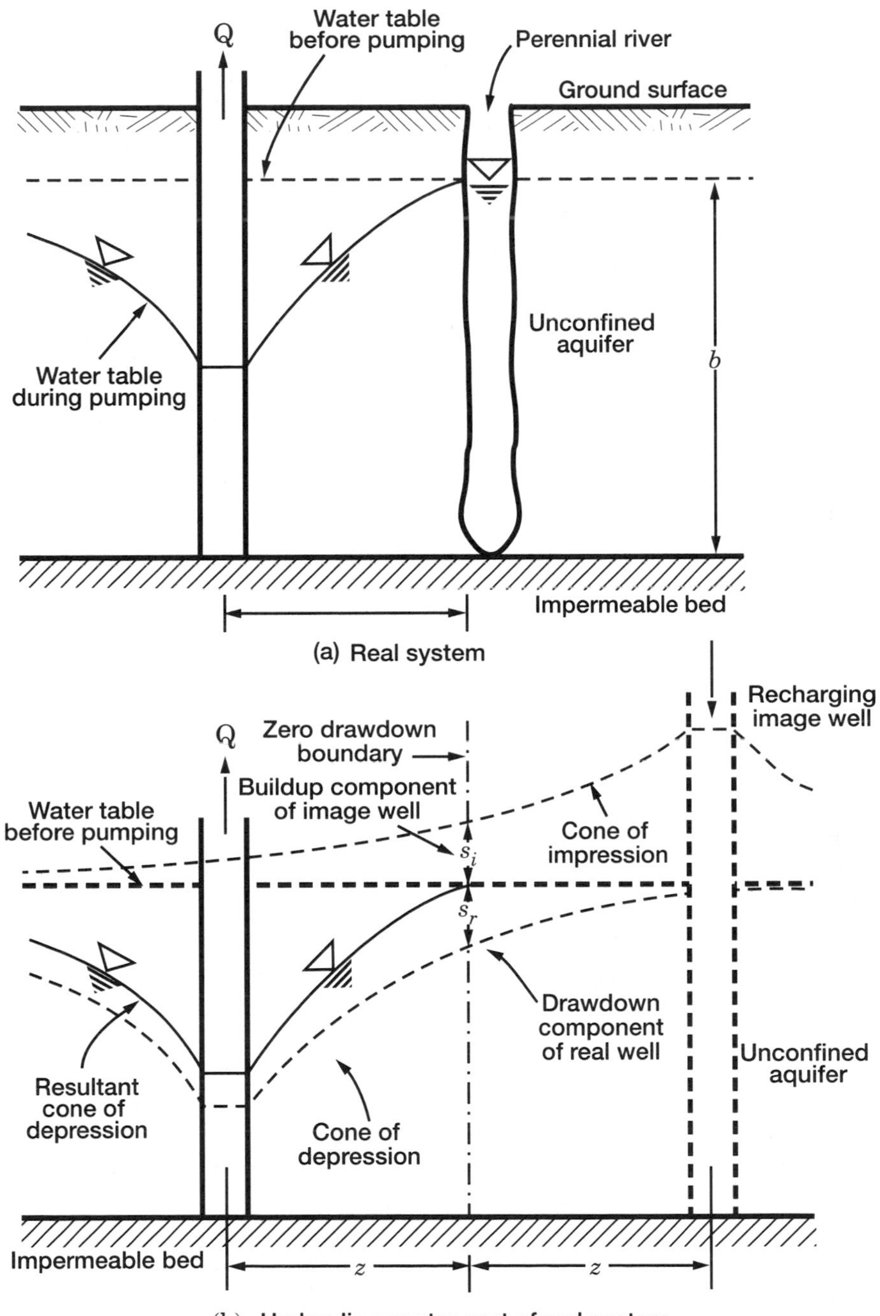

Figure 10-9 Cross sections of an idealized semi-infinite confined aquifer: (a) Discharging well near a perennial river; (b) mathematically equivalent system in a confined aquifer of infinite areal extent. (After Ferris et al., 1962.)

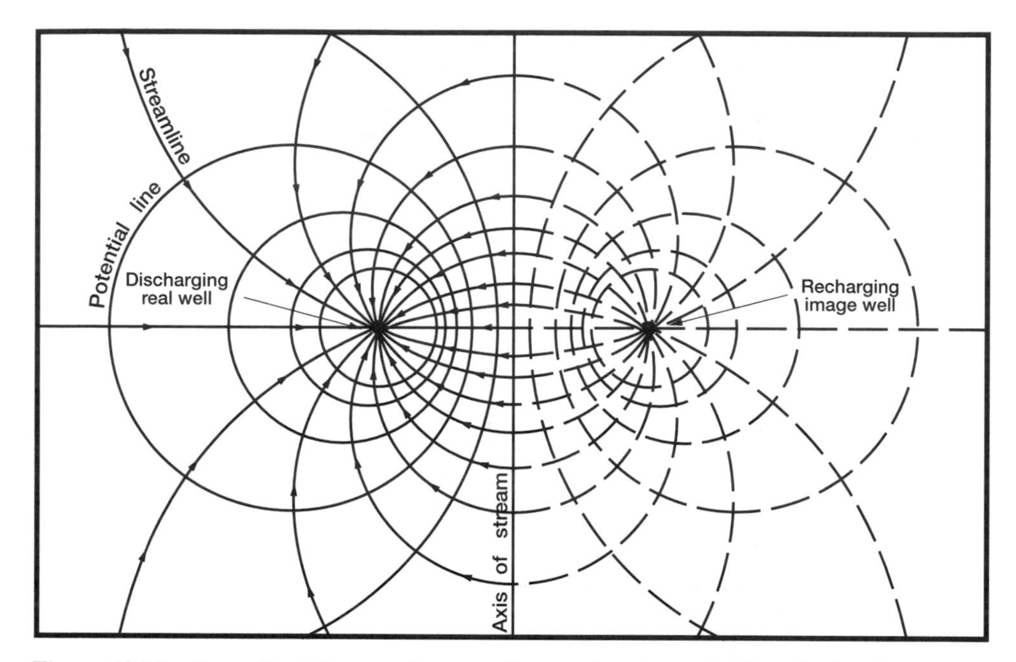

Figure 10-10 Generalized flow net for streamlines and equipotential lines in the vicinity of a discharging well near a river. (After Ferris et al., 1962.)

In Figure 10-10, a generalized flow net for streamlines and equipotential lines for the system corresponding to Figure 10-9 is presented.

10.3.2 Hantush's Methods for One Recharge Boundary

10.3.2.1 Theis' Solution-Based Equations. Hantush (1959) developed several methods based on the Theis equation to analyze drawdown versus time data for pumping tests near a river. In the following paragraphs, the governing equations developed by Hantush are presented.

Basically, the Theis equation is for a well that fully penetrates a confined aquifer (see Chapter 4, Section 4.2). However, the drawdown equations for confined aquifers under both steady- and unsteady-state conditions can be used for unconfined aquifer conditions by using the Jacob's drawdown correction scheme [Chapter 3, Eq. (3-64)] if the maximum drawdown is a significant portion of the saturated thickness.

Based on the general theory outlined in Section 10.3.1, the diagrammatic representation of a discharging well near a vertical plane of constant head is shown in Figure 10-11. Using the Theis drawdown equation [Chapter 4, Eq. (4-24)] in Eq. (10-27), one can write (Hantush, 1959)

$$s = \frac{Q}{4\pi T} M(u, \beta) \tag{10-28}$$

where

$$M(u, \beta) = W(u) - W(\beta^2 u) \tag{10-29}$$

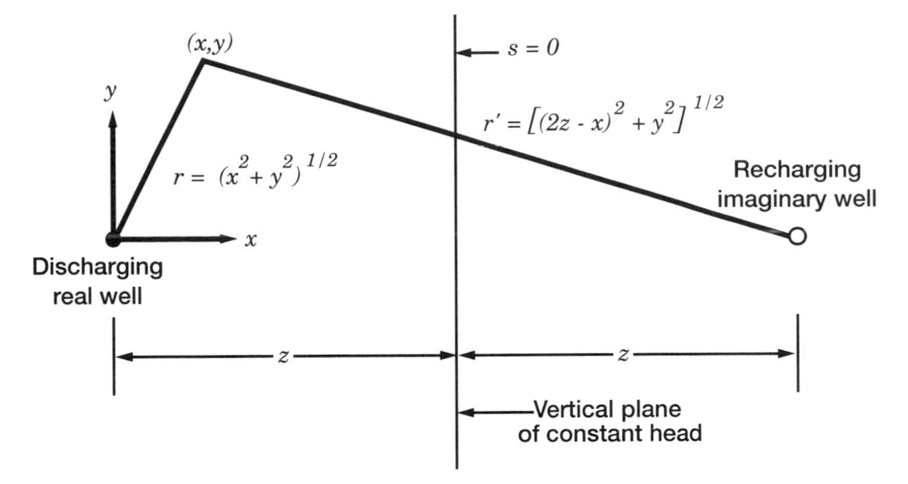

Figure 10-11 Diagrammatic representation of a discharging well near a vertical plane of constant head. (After Hantush, 1959.)

$$u = \frac{\alpha}{t} \tag{10-30}$$

$$\alpha = \frac{r^2 S}{4T} \tag{10-31}$$

$$\beta = \frac{r'}{r} \tag{10-32}$$

where s is the resultant drawdown at any point on the cone of depression at any time t in the real region since the start of pumping and any distance

$$r = \left(x^2 + y^2\right)^{1/2} \tag{10-33}$$

In Eq. (10-32)

$$r' = \left[(2z - x)^2 + y^2\right]^{1/2} \tag{10-34}$$

is the distance from the imaginary well to the point in question. In the above expressions, x and y are the Cartesian coordinates of any point in the aquifer with respect to the center of the discharging well, and z is the effective half distance between the discharging real well and recharging imaginary well.

From Figure 10-11 one can write the following equations:

$$(2z - x)^2 + y^2 = r'^2 \tag{10-35}$$

$$x^2 + y^2 = r^2 \tag{10-36}$$

Combining of Eqs. (10-35) and (10-36) gives

$$4z^2 - 4zx - r^2\left(\beta^2 - 1\right) = 0 \tag{10-37}$$

where β is defined by Eq. (10-32).

Using the series form of the Theis equation [Chapter 4, Eq. (4-26)], Eq. (10-28) can be written in an expanded convergent series form as

$$\begin{aligned} s &= \frac{Q}{4\pi T}\left[2\ln(\beta) + \sum_{n=1}^{\infty}(-1)^n\left(\beta^{2n}-1\right)\frac{\alpha^n}{n\cdot n!}\right] \\ &= \frac{Q}{4\pi T}\left[2\ln(\beta) - \left(\beta^2-1\right)\frac{\alpha}{t} + \left(\beta^4-1\right)\frac{\alpha^2}{2\cdot 2!t^2} - \cdots\right] \end{aligned} \tag{10-38}$$

For small values of $(\beta^2 u)$, that is large relative values of time $(\beta^2 u \leq 0.10)$, Eq. (10-38) may be approximated by (Hantush, 1959)

$$s \cong \frac{Q}{4\pi T}\left[2\ln(\beta) - \left(\beta^2-1\right)\frac{\alpha}{t} + \left(\beta^4-1\right)\frac{\alpha^2}{2\cdot 2!t^2}\right] \tag{10-39}$$

and, for values of $\beta^2 u < 0.05$, by

$$s \cong \frac{Q}{4\pi T}\left[2\ln(\beta) - \left(\beta^2-1\right)\frac{\alpha}{t}\right] \tag{10-40}$$

10.3.2.2 Properties of the Solution. Some properties of Eq. (10-28) are given below (Hantush, 1959).

(a) Properties of the Theoretical Curve for Drawdown (s) Versus Logarithmic Time [log(t)]. The theoretical drawdown (s) versus logarithmic time [$\log(t)$] shown in Figure 10-12 has the following properties:

1. The curve has an inflection point whose abscissa is defined by

$$u_i = \frac{\alpha}{t_i} = \frac{2\ln(\beta)}{\beta^2-1} \tag{10-41}$$

where the subscript i corresponds to values of the variables at the inflection point. Equation (10-41) is obtained by making zero the second derivative of s with respect to $\ln(t)$ and solving for u.

2. The geometric slope of the curve (m) at any point is given by

$$m = \frac{ds}{d[\log(t)]} = \frac{2.303Q}{4\pi T}\left(e^{-u} - e^{-\beta^2 u}\right) \tag{10-42}$$

which is derived by differentiating s in Eq. (10-28) with respect to $\log(t)$.

3. The geometric slope of the curve (m_i) at the inflection point is

$$m_i = \frac{2.303Q}{4\pi T}\left(e^{-u_i} - e^{-\beta^2 u_i}\right) \tag{10-43}$$

which is obtained by making $u = u_i$ in Eq. (10-42).

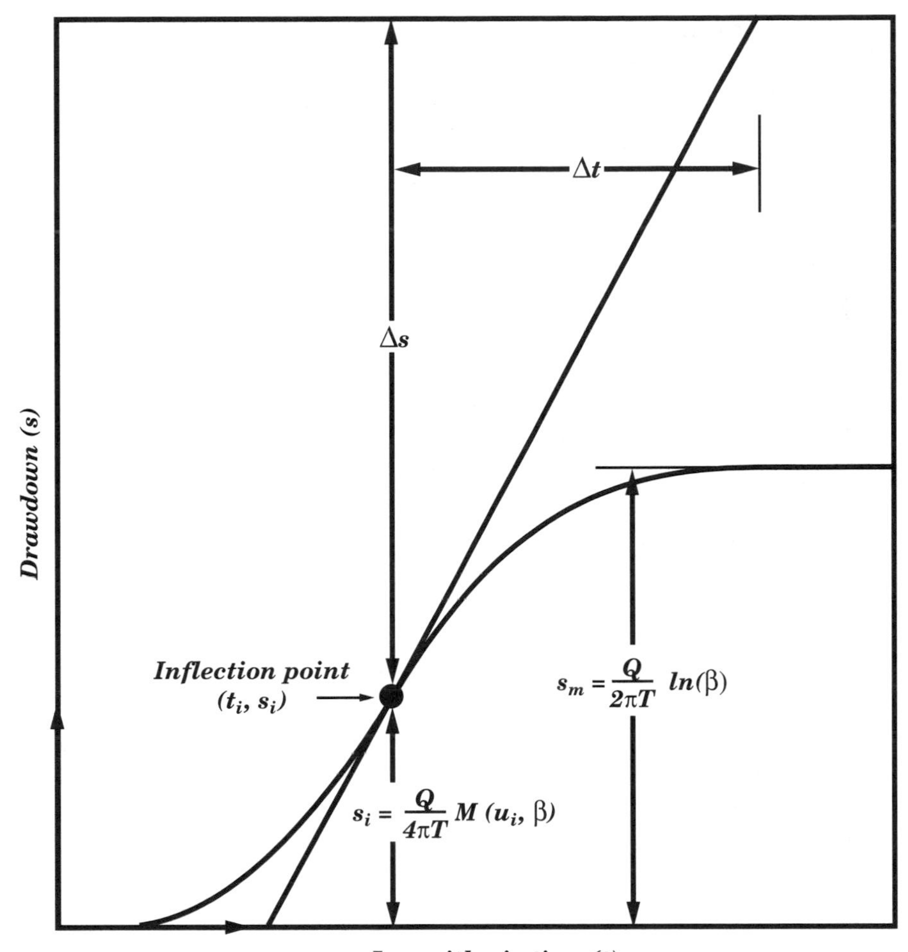

Figure 10-12 The theoretical drawdown (s) versus logarithmic time [log(t)] variation near a vertical plane of constant head. (After Hantush, 1959.)

4. The drawdown at the inflection point (s_i) is obtained by letting $u = u_i$ in Eq. (10-28):

$$s_i = \frac{Q}{4\pi T} M(u_i, \beta) \tag{10-44}$$

5. As t increases, the drawdown increases and eventually approaches to an asymptotic maximum value (s_m) which corresponds to steady-state conditions. By letting t become infinite in Eq. (10-28) or Eq. (10-38), the expression for maximum drawdown takes the form

$$s_m = \frac{Q}{2\pi T} \ln(\beta) \tag{10-45}$$

6. From Eqs. (10-43) and (10-45), the ratio s_m/m_i is

$$\frac{s_m}{m_i} = \frac{2\log(\beta)}{e^{-u_i} - e^{-\beta^2 u_i}} = f(\beta) \tag{10-46}$$

which is only the function of β since u_i, as given by Eq. (10-41), is only the function of β as well.

(b) Properties of the Theoretical Curve for Drawdown (s) Versus Reciprocal of Time (1/t). The theoretical drawdown (s) versus reciprocal of time ($1/t$) shown in Figure 10-13 has the following properties:

1. The curve approaches asymptotically to the ($1/t$) axis.
2. The curve is approximately the parabola given by Eq. (10-39) in the range where $\beta^2 u < 0.10$ and the equation of the parabola is

$$s = s_m - \frac{m_t}{t} + \frac{c}{t^2} \tag{10-47}$$

where s_m, m_t, and c, whose dimensions are [L], [LT], and [LT^2], respectively; represent the corresponding coefficients of ($1/t$) given by Eq. (10-38). And from Eq. (10-39) the expression for c is

$$c = \frac{Q}{4\pi T}\left(\beta^4 - 1\right)\frac{\alpha^2}{4} \tag{10-48}$$

3. The curve is approximately the straight line given by Eq. (10-40) in the range $\beta^2 u < 0.05$, or

$$s = s_m - \frac{m_t}{t} \tag{10-49}$$

4. The curve has an s intercept that is equal to the maximum drawdown s_m.
5. The geometric slope at the s intercept is equal to $-m_t$, which is given as

$$m_t = \frac{Q}{4\pi T}\left(\beta^2 - 1\right)\alpha \tag{10-50}$$

which is obtained by differentiating s in Eq. (10-38) with respect to ($1/t$).

Equation (10-49) is the equation of the tangent line to the curve at the intercept point on the s-axis. The intersection of the tangent line with the ($1/t$) axis can be obtained by making s zero in Eq. (10-49):

$$t = \frac{m_t}{s_m} \tag{10-51}$$

where s_m and m_t are given by Eqs. (10-45) and (10-50), respectively. Therefore, Eq. (10-51) can be written as

$$\frac{\alpha}{t} = u = \frac{2\ln(\beta)}{\beta^2 - 1} \tag{10-52}$$

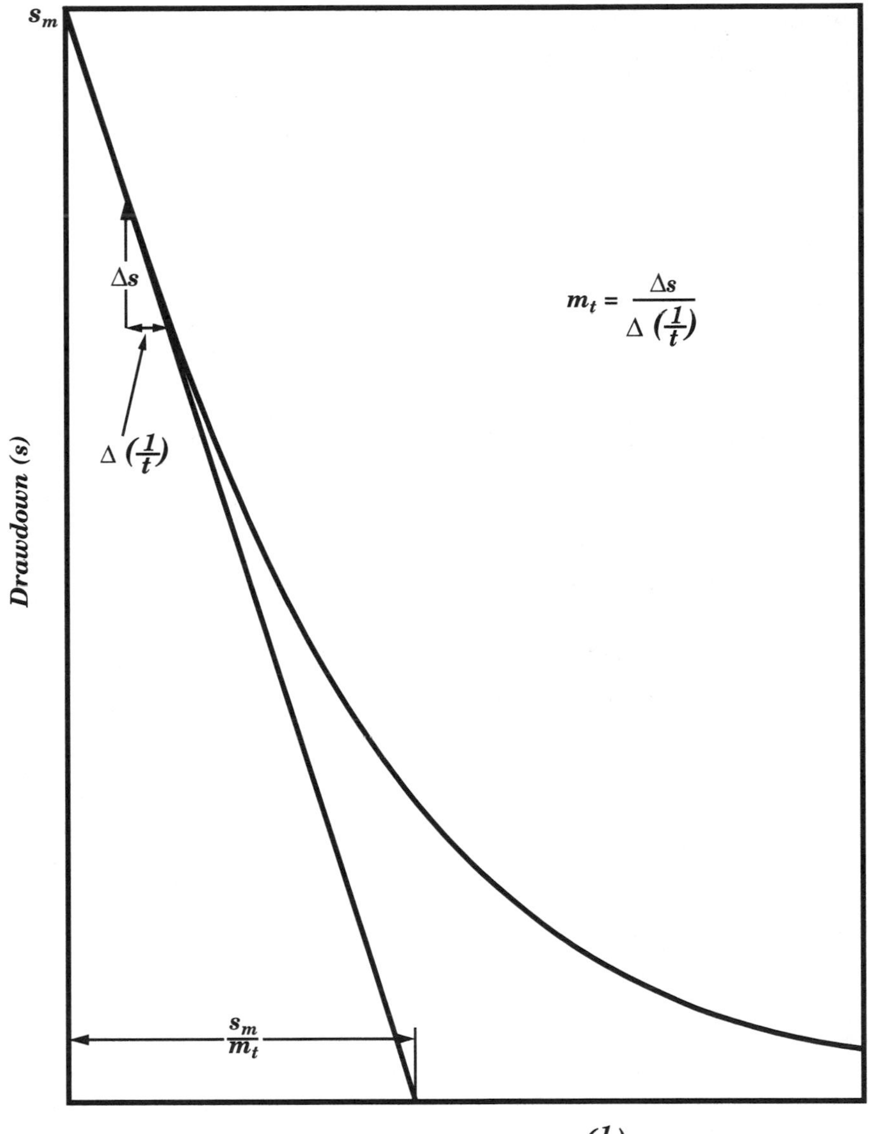

Figure 10-13 The theoretical drawdown (s) versus ($1/t$) variation near a vertical plane of constant head. (After Hantush, 1959.)

From the comparison of Eqs. (10-41) and (10-52), it can be seen that u of Eq. (10-52) is equal u_i of Eq. (10-41). As a result, the value of t, as obtained by Eq. (10-51), is the value of the time t_i at which the inflection point of the semilogarithmic time versus drawdown curve occurs. Thus, the value of t_i can be obtained from Eq. (10-51).

6. By eliminating α and $Q/(4\pi T)$ from Eqs. (10-45), (10-48), and (10-50), one obtains the expression relating s_m, m_t, and c to β as

$$F(\beta) = \frac{\beta^2 + 1}{\beta^2 - 1} \log \beta = 0.868 \frac{c s_m}{m_t^2} \tag{10-53}$$

(c) Properties of Eq. (10-50). Introducing the expressions for α and β [given, respectively, by Eqs. (10-31) and (10-32)] into Eq. (10-50) and using the expressions for r and r' [given, respectively by Eqs. (10-35) and (10-36)], followed by simplifications, we obtain

$$m_t = \frac{QSz}{4\pi T^2}(z - x) \tag{10-54}$$

Geometrically, Eq. (10-54) means that a plot of m_t versus x is a straight line and its slope $(-m_x)$ is given by

$$m_x = \frac{QSz}{4\pi T^2} \tag{10-55}$$

and its x intercept is equal to z, which is the effective half distance between the real discharging well and imaginary recharging well shown in Figure 10-11.

(d) Properties of Eq. (10-45). Equation (10-45) can also be expressed as

$$s_m = \frac{2.303Q}{2\pi T} \log(\beta) \tag{10-56}$$

A semilogarithmic plot of s_m versus β (β on logarithmic scale) is a straight line passing through the point ($s_m = 0$, $\beta = 1$) and having a geometric slope equal to

$$m_m = \frac{2.303Q}{2\pi T} \tag{10-57}$$

10.3.2.3 Tables of Functions. Hantush (1959) presented tables for the functions $M(u_i, \beta)$, $F(\beta)$, $f(\beta)$, and u_i for the determination of different parameters in order to apply the above theory. The $M(u_i, \beta)$ function, given by Eq. (10-29), is only dependent to the Theis well function which is tabulated in Table 4-1 of Chapter 4. Values of the aforementioned functions are given in Tables 10-2 and 10-3 for different values of β. These functions may be approximated by the following formulas for $\beta > 100$ (Hantush, 1959):

$$f(\beta) \cong 2 \log(\beta) \tag{10-58}$$

$$F(\beta) \cong \log(\beta) \tag{10-59}$$

$$M(u_i, \beta) \cong 2.303 \log\left(\frac{0.562}{u_i}\right) \tag{10-60}$$

10.3.2.4 Determination of the Coefficients. The parabolic drawdown equation, as given by Eq. (10-47), has three unknown coefficients: s_m, m_t, and c. These unknown coefficients can be found from the application of the method of least squares under the

TABLE 10-2 Values of the Functions u_i, $M(u_i, \beta)$, and $f(\beta)$

β	u_i	$M(u_i, \beta)$	$f(\beta)$	β	u_i	$M(u_i, \beta)$	$f(\beta)$
1.0	1.000	0.000	1.179	10	0.0466	2.534	2.115
1.1	0.909	0.070	1.183	11	0.0400	2.680	2.188
1.2	0.830	0.135	1.188	12	0.0348	2.815	2.251
1.3	0.761	0.195	1.194	13	0.0306	2.940	2.312
1.4	0.702	0.252	1.203	14	0.0271	3.057	2.367
1.5	0.649	0.306	1.214	15	0.0241	3.172	2.423
1.6	0.603	0.357	1.223	16	0.0218	3.271	2.472
1.7	0.562	0.407	1.235	17	0.0203	3.342	2.520
1.8	0.525	0.456	1.247	18	0.0179	3.462	2.564
1.9	0.492	0.502	1.262	19	0.0164	3.551	2.609
2.0	0.462	0.548	1.273	20	0.0150	3.637	2.647
2.2	0.411	0.635	1.301	21	0.0138	3.716	2.687
2.4	0.368	0.717	1.329	22	0.0128	3.793	2.725
2.6	0.332	0.796	1.357	23	0.0119	3.867	2.761
2.8	0.301	0.872	1.385	24	0.0111	3.938	2.796
3.0	0.275	0.945	1.413	25	0.0103	4.007	2.837
3.2	0.252	1.016	1.435	26	0.00966	4.072	2.862
3.4	0.232	1.083	1.467	27	0.00906	4.135	2.893
3.6	0.214	1.149	1.493	28	0.00852	4.196	2.923
3.8	0.199	1.212	1.500	29	0.00803	4.256	2.952
4.0	0.185	1.273	1.545	30	0.00757	4.313	2.980
4.2	0.173	1.333	1.571	31	0.00716	4.369	3.008
4.4	0.162	1.390	1.597	32	0.00678	4.423	3.034
4.6	0.152	1.447	1.619	33	0.00643	4.475	3.059
4.8	0.142	1.500	1.642	34	0.00611	4.526	3.085
5.0	0.134	1.553	1.667	35	0.00582	4.576	3.109
5.2	0.127	1.604	1.688	36	0.00554	4.624	3.134
5.4	0.120	1.653	1.710	37	0.00528	4.671	3.155
5.6	0.114	1.703	1.731	38	0.00505	4.717	3.178
5.8	0.108	1.750	1.752	39	0.00483	4.761	3.199
6.0	0.102	1.796	1.770	40	0.00462	4.805	3.221
6.2	0.0976	1.840	1.794	41	0.00443	4.847	3.242
6.4	0.0930	1.988	1.814	42	0.00424	4.889	3.262
6.6	0.0888	1.927	1.833	43	0.00407	4.930	3.282
6.8	0.0848	1.969	1.852	44	0.00391	4.969	3.301
7.0	0.0812	2.010	1.871	45	0.00376	5.008	3.321
7.2	0.0777	2.050	1.889	46	0.00362	5.046	3.339
7.4	0.0745	2.089	1.908	47	0.00349	5.084	3.357
7.6	0.0715	2.127	1.925	48	0.00336	5.120	3.375
7.8	0.0687	2.165	1.943	49	0.00325	5.156	3.393
8.0	0.0661	2.202	1.960	50	0.00313	5.191	3.410
8.2	0.0636	2.238	1.977	55	0.00265	5.358	3.491
8.4	0.0613	2.273	1.994	60	0.00228	5.510	3.565
8.6	0.0590	2.308	2.010	65	0.00198	5.650	3.634
8.8	0.0570	2.342	2.026	70	0.00174	5.781	3.697
9.0	0.0550	2.376	2.041	75	0.00154	5.903	3.757
9.2	0.0531	2.408	2.057	80	0.00137	6.017	3.812
9.4	0.0513	2.441	2.072	85	0.00123	6.124	3.864
9.6	0.0497	2.472	2.087	90	0.00111	6.226	3.913
9.8	0.0481	2.503	2.102	95	0.00102	6.311	3.960
				100	0.00092	6.412	4.004

Source: After Hantush (1959).

TABLE 10-3 Values of the Function $F(\beta) = (\beta^2 + 1)/(\beta^2 - 1)\log(\beta)$

β	0	0.1	0.2	0.3	0.4	0.5	0.6	0.7	0.8	0.9	1.0
1.0	0.434	0.436	0.439	0.444	0.451	0.458	0.466	0.474	0.483	0.492	0.502
2.0	0.502	0.511	0.521	0.530	0.540	0.550	0.560	0.568	0.578	0.587	0.600
3.0	0.600	0.605	0.614	0.623	0.632	0.641	0.650	0.658	0.666	0.674	0.682
4.0	0.682	0.690	0.698	0.706	0.713	0.720	0.728	0.736	0.743	0.750	0.757
5.0	0.757	0.764	0.771	0.778	0.784	0.791	0.797	0.804	0.810	0.816	0.823
6.0	0.823	0.829	0.835	0.841	0.846	0.852	0.858	0.863	0.869	0.874	0.880
7.0	0.880	0.885	0.891	0.896	0.902	0.907	0.912	0.917	0.922	0.927	0.932
8.0	0.932	0.936	0.941	0.946	0.951	0.955	0.960	0.964	0.969	0.973	0.978
9.0	0.978	0.982	0.987	0.991	0.996	1.000	1.004	1.008	1.012	1.016	1.020
10.0	1.020	1.024	1.028	1.032	1.036	1.040	1.044	1.047	1.051	1.055	1.059
11.0	1.059	1.062	1.066	1.069	1.073	1.077	1.081	1.084	1.087	1.090	1.094
12.0	1.094	1.097	1.101	1.104	1.108	1.111	1.114	1.117	1.121	1.124	1.127
13.0	1.127	1.130	1.133	1.136	1.140	1.143	1.146	1.149	1.152	1.155	1.158
14.0	1.158	1.161	1.164	1.167	1.170	1.172	1.175	1.178	1.181	1.184	1.187
	0	1	2	3	4	5	6	7	8	9	10
10	1.020	1.059	1.094	1.127	1.158	1.187	1.214	1.239	1.263	1.286	1.308
20	1.308	1.328	1.348	1.367	1.385	1.402	1.419	1.437	1.451	1.466	1.480
30	1.480	1.494	1.508	1.521	1.534	1.547	1.559	1.570	1.582	1.593	1.604
40	1.604	1.615	1.625	1.635	1.645	1.655	1.664	1.674	1.683	1.692	1.700
50	1.700	1.709	1.717	1.725	1.734	1.741	1.749	1.757	1.764	1.772	1.779
60	1.779	1.786	1.793	1.800	1.807	1.814	1.820	1.827	1.833	1.840	1.846
70	1.846	1.850	1.858	1.864	1.870	1.876	1.881	1.887	1.892	1.898	1.904
80	1.904	1.909	1.914	1.919	1.925	1.930	1.935	1.940	1.945	1.950	1.955
90	1.955	1.959	1.964	1.969	1.974	1.978	1.983	1.987	1.992	1.996	2.000

Source: After Hantush (1959).

condition that the observed drawdown versus time data fall within the period during which the drawdown is approximated by Eq. (10-47). Because there are three unknowns in Eq. (10-47), there must be three normal equations to determine the values of the three unknowns. The normal equations are (Hantush, 1959)

$$\sum s = ns_m - m_t \sum \left(\frac{1}{t}\right) + c \sum \left(\frac{1}{t}\right)^2 \tag{10-61}$$

$$\sum \left(\frac{1}{t}\right) s = s_m \sum \left(\frac{1}{t}\right) - m_t \sum \left(\frac{1}{t}\right)^2 + c \sum \left(\frac{1}{t}\right)^3 \tag{10-62}$$

$$\sum \left(\frac{1}{t}\right)^2 s = s_m \sum \left(\frac{1}{t}\right)^2 - m_t \sum \left(\frac{1}{t}\right)^3 + c \sum \left(\frac{1}{t}\right)^4 \tag{10-63}$$

Kazmann (1960) presented a critical discussion of the theory proposed by Hantush. Both Kazmann's discussion and Hantush's reply (Hantush, 1960a) provide answers to some potential questions regarding the theory and its application to pumping tests.

10.3.2.5 Hantush's Pumping Test Data Analysis Methods. Hantush (1959) developed several graphical methods for determining the hydraulic parameters of aquifers based on his theory presented previously. The methods are based on the assumption that the maximum drawdown in an observation well can be extrapolated.

Hantush basically developed three procedures. As mentioned above, each procedure requires an estimate of maximum drawdown (s_m). However, an approximate value of s_m can be obtained (a desirable but not necessary step) by using the method of least squares in fitting the parabola of Eq. (10-47) to the observed s versus $1/t$ data. Depending on the number of observation wells available, one of these procedures may be used. The values of c and m_t may be in great error because of the uncertainty for the determination of the time after which the drawdown versus $1/t$ curve becomes parabolic.

Hantush's methods are (1) the one observation well method and (2) the two and three observation wells method. For the first method there are two procedures (Procedure I and Procedure II). In the following sections, these data analysis methods are presented in detail with some examples.

One Observation Well Method

PROCEDURE I. In this procedure, the observed drawdowns for a single well are used. The steps of the procedure are as follows (Hantush, 1959):

Step 1: If the aquifer is an unconfined aquifer calculate the ratio of $s_{\max}/b$ for each observation well. Based on this ratio, if the maximum drawdown is a significant portion of the aquifer thickness (b), calculate the corrected drawdowns using the Jacob's correction scheme [Chapter 3, Eq. (3-64)]:

$$s_c = s_w - \frac{s_{wt}^2}{2b} \tag{10-64}$$

where s_{wt} is the drawdown of the water table and b is the initial saturated thickness. Using the corrected drawdown data, plot the drawdown (s) versus time (t) data on semilogarithmic paper (t on logarithmic scale). If the approximate value of s_m is obtained by the least squares method, indicate this value on the graph. Draw the best fit of curve through the data points.

Step 2: Find the value of the maximum drawdown (s_m) by extrapolation or by the least-squares method.

Step 3: Draw the tangent line to the curve at the inflection point and determine its geometric slope (m_i). Because the location of the inflection point is not known, this tangent line is usually approximated by the straight portion of the curve.

Step 4: Using the known values of s_m and m_i from the previous steps, calculate the ratio of s_m/m_i, which is equal to $f(\beta)$ according to Eq. (10-46). Then, determine the corresponding value of β from Table 10-3.

Step 5: Using the known values of s_m, Q, and β, calculate the value of T from Eq. (10-45).

Step 6: Determine the values of u_i and $M(u_i, \beta)$ using the values of β from Table 10-2. Then, determine the corresponding value of s_i from Eq. (10-44).

Step 7: With the known value of s_i, locate the inflection point on the curve and read t_i.

Step 8: From Eqs. (10-30) and (10-31) we obtain

$$S = \frac{4u_i t_i T}{r^2} \tag{10-65}$$

Calculate the value of S from Eq. (10-65) using the values of u_i, t_i, T, and r.

Step 9: Calculate the value of z from

$$z = \frac{x}{2} \pm \frac{1}{2}\left[x^2 + r^2\left(\beta^2 - 1\right)\right]^{1/2} \tag{10-66}$$

which are the roots of the quadratic equation given by Eq. (10-37). Here, x and r are defined in Figure 10-11 and β is given by Eq. (10-32).

Important Points for the Method. Hantush (1959) provided some important points regarding the application of the method:

1. Theoretically, the drawdowns predicted by Eq. (10-28) with the calculated parameters and with assigned values of (t) should fall on the observed drawdown versus time curve. This may not be the case for the early portion of the observed drawdown curve because of the variation of the hydrogeologic parameters during the early pumping stages. Drawdowns in the latter part of the curve should compare fairly well with the observed values.
2. Sometimes, the extrapolated maximum drawdown (s_m) is either overestimated or underestimated, or the tangent line at the inflection point is drawn either flatter or steeper than necessary. In such cases, the calculated drawdown versus time curve obtained from Eq. (10-28) will deviate from the observed curve. As a result, adjustment needs to be made for the extrapolated s_m value and/or the geometric slope of the curve at the inflection point. Then, the above steps need to be repeated until a satisfactory match is established.

PROCEDURE II. In this procedure, like the previous one, the observed drawdowns for a single well are used. The steps of the procedure are as follows (Hantush, 1959):

Step 1: If the aquifer is unconfined, calculate the ratio of s_{max}/b for each observation well. Based on this ratio, if the maximum drawdown is a significant portion of the aquifer thickness (b), calculate the corrected drawdowns using the Jacob's correction scheme given by Eq. (10-64). Then, plot the drawdown (s) versus time (t) data on semilogarithmic paper (t on logarithmic scale).

Step 2: Eliminate the observed drawdown points that deviate significantly from the general trend of the curve.

Step 3: Using the remaining data points, apply the least-squares method to determine the parameters s_m, m_t, and c of Eq. (10-47) by taking s and $1/t$ as the dependent and the independent variables, respectively.

Step 4: Using the values of the parameters from the previous step, calculate the value of $F(\beta)$ from Eq. (10-53) and then determine the corresponding value of β from Table 10-3.

Step 5: Calculate the value of T from Eq. (10-45). Then, calculate the value of S from

$$S = \frac{16\pi m_t T^2}{Q\left(\beta^2 - 1\right) r^2} \quad (10\text{-}67)$$

which is the combination of Eqs. (10-31) and (10-50).

Step 6: Calculate the value of z from Eq. (10-66).

Important Points For the Method. Hantush (1959) provided some important points regarding the application of the method:

1. This method may not produce good results unless care and precision in data collection have been exercised, and unless field conditions approach very closely the assumptions in the theoretical developments.
2. Although the method has potential application in determining hydrogelogic parameters, it may be more applicable to problems in heat conduction for which the requirements for data collection as well as the theoretical assumptions may be satisfied with higher precision.
3. This method gives reasonable results if the drawdown versus time data fall to the parabola given by Eq. (10-47). If some of the selected observed data points, from which the coefficients in Eq. (10-47) are to be determined, do not fall in the region where the approximation can be made, the resulting parabola equation may deviate significantly from the normally expected equation. The greatest difference is in the last term (c) of Eq. (10-47). The difference is somewhat less in the second term (m_t). The least affected term is the first one (s_m). If one uses a fourth-degree equation approximation from Eq. (10-38), a better representation of the actual curve may be obtained.

Two or More Observation Wells Method. If there are at least two observation wells, this method may be used. The steps of the method are as follows (Hantush, 1959):

Step 1: If the aquifer is unconfined, calculate the ratio of $s_{\max}/b$ for each observation well. Based on this ratio, if the maximum drawdown is a significant portion of the aquifer thickness (b), calculate the corrected drawdowns using the Jacob's correction scheme given by Eq. (10-64). Then, plot the drawdown (s) versus time (t) data on semilogarithmic paper (t on logarithmic scale).

Step 2: Plot the drawdown versus $(1/t)$ curve using all observational data that fall within the range in which the recharge is effective as indicated by the curve in Step 1. In plotting this curve, use the properties of the theoretical curve as presented in Section 10.3.2.2(b).

Step 3: Using a larger scale for $(1/t)$ than that used in Step 2, plot the curve for most of the latter part of the data by making use of the general trend of the curve plotted in Step 2.

Step 4: Draw the tangent line to the curve at the s intercept and determine its geometric slope ($-m_t$). Determine also the value of the s intercept.

Step 5: Using the known values of m_t and x (position of the observation well), plot m_t versus x on linear scales and draw the best-fit straight line through the points. The equation of this line is given by Eq. (10-54).

Step 6: Determine the geometric slope of this line and its x intercept. This intercept is the value of the distance from the pumped well to the vertical recharge plane, which is equal to z. There may be wide scattering for points of the m_t versus x graph. In general, this may be attributed (1) to the nonuniformity of the formation coefficients and (2) to the errors in estimating the values of s_m and m_t. In such cases, adjust the values of s_m or m_t, or both by reexamining the fit of the drawdown versus $(1/t)$ curve. Several trials may be required before a satisfactory fit is obtained. If, after adjustments the scattering of the points still persists, it means that the hydrogeologic parameters vary significantly in the pumping test area.

Step 7: Using the values of r and r', calculate the β values for each observation well from Eq. (10-32).

Step 8: Then, plot s_m versus β data on semilogarithmic paper (β on logarithmic scale). Draw the best straight-line fit through the point ($s_m = 0, \beta = 1$) and the data points, and determine the geometric slope (m_m) of this line. If a fairly good straight line cannot be made, it is the indication that nonuniform transmissivity exists in the pumping test area.

Step 9: With the known values of m_m and Q, calculate the T value from Eq. (10-57).

Step 10: Then, compute the value of S from Eq. (10-55).

Examples. The Hantush one observation well method (Procedure I and Procedure II) and two or more observation wells method will now be shown with some examples.

Example 10-3: This example is adapted from Hantush (1959). The data belong to a pumping test conducted in the unconfined aquifer, which is composed of alluvial deposits, at the Norbert Irsik site near the Arkansas River, in the Ingalls area, Kansas (Stramel et al., 1958). The location of the wells and river are shown in Figure 10-14. The pumped well is 29 ft deep, 16 in. in diameter, 135 ft from the Arkansas River, and it fully penetrates the unconfined aquifer. The observation wells partially penetrate the aquifer. The saturated thickness of the aquifer was 22 ft before the pumping test. The discharge rate from the pumped well was 1.55 ft^3/s. The drawdown data for the seven observation wells are given in Table 10-4. Using Hantush's one observation well method (Procedure I), determine the hydrogeologic parameters of the aquifer.

Solution: The steps given previously for Procedure I are as follows:

Step 1: The maximum drawdown (s_{max}) in the observation wells have values between 3.13 ft (1S) and 0.73 ft (3N) as given in Table 10-4. Therefore, with $b = 22$ ft, the s_{max}/b ratio varies from 3.3% to 14.2%. The drawdown (s_{wt}) data given in Table 10-4 are corrected using Eq. (10-64), and the corrected drawdown (s_c) data are given in Table 10-5. Using these data, the semilogarithmic s_c versus time curves for the seven observation wells are shown in Figures 10-15 and 10-16. As can be seen from these figures, the drawdown versus time curves during early periods of pumping do not follow the trend indicated by the theory. As a result, the data points in Figures 10-15 and 10-16 before to the period of recharge from the river are neglected.

Step 2: An approximate value of the maximum drawdown for each observation is obtained by the least-squares method by fitting the parabola of Eq. (10-47) through the points s_c versus $(1/t)$ during the recharge period as given in Table 10-5. Equations (10-61), (10-62), and (10-63) are used in determining the parameters in Eq. (10-47), and the results are included in Table 10-6.

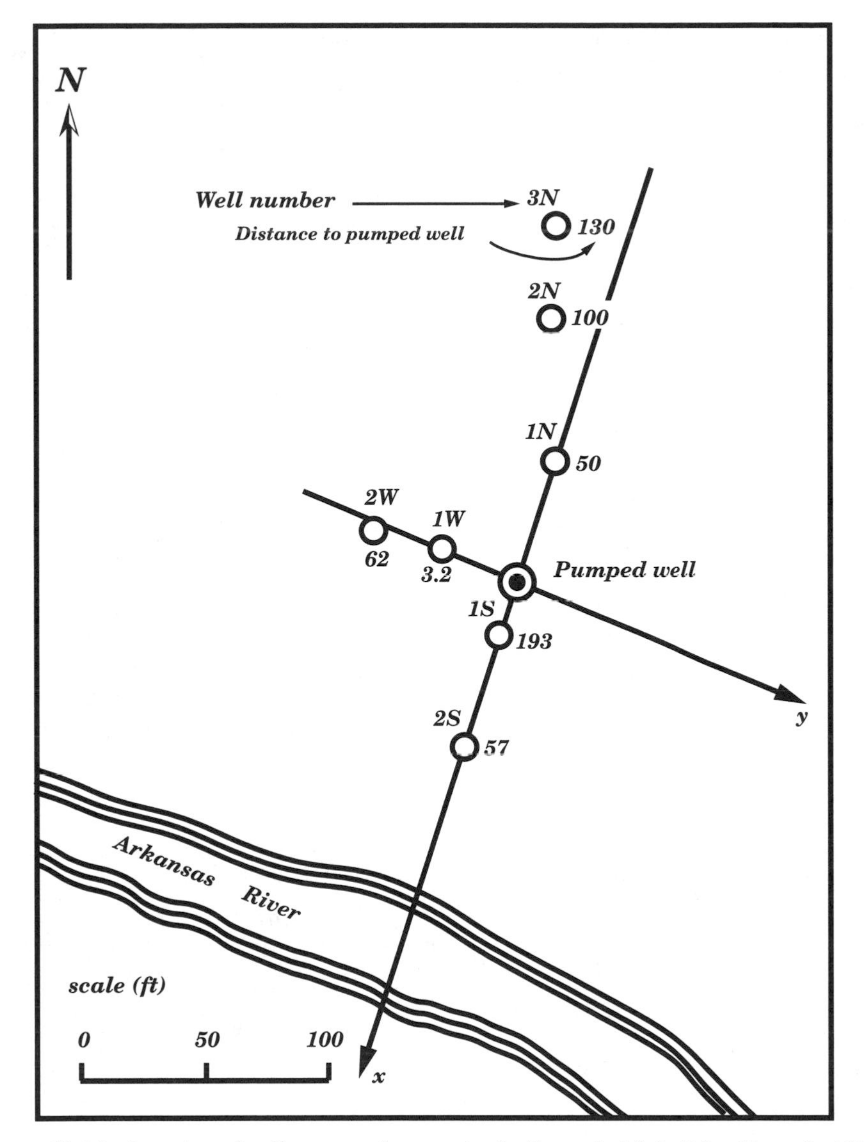

Figure 10-14 Location of wells at pumping test site for Example 10-3. (After Hantush, 1959.)

Step 3: The tangent lines are approximated by the straight portion of the curves in Figures 10-15 and 10-16, and the values of the geometric slope (m_i) at the inflection points are listed in Table 10-7.

Step 4: The calculated values of $f(\beta)$ and the corresponding values of β are listed in Table 10-7 by applying linear interpolation when necessary.

Step 5: Using the values of s_{c_m}, β in Table 10-7, and $Q = 1.55$ ft^3/s in Eq. (10-45), the calculated values of T are listed in Table 10-7.

Step 6: The values of u_i and $M(u_i, \beta)$ corresponding to the β values are listed in Table 10-7 by applying linear interpolation when necessary.

TABLE 10-4 Observed Drawdown Data for Example 10-3

	s_{wt} (ft)						
t (min)	1S	1W	1N	2W	2S	2N	3N
17.54	2.48	—	—	—	0.68	—	—
27.48	2.54	—	—	—	0.76	—	—
41.50	2.60	—	—	—	0.82	—	—
50.00	2.66	—	—	—	0.85	—	—
60.26	2.70	—	—	—	0.90	—	—
75.51	2.73	1.81	1.22	0.95	0.95	0.45	—
90.00	2.78	1.83	1.27	0.97	1.00	0.49	0.39
109.90	2.82	1.89	1.30	1.02	1.02	0.51	0.41
132.43	2.86	1.95	1.34	1.04	1.13	0.55	0.45
151.01	2.88	1.97	1.39	1.07	1.14	0.57	0.46
181.97	2.92	2.03	1.42	1.11	1.20	0.61	0.51
211.35	2.96	2.08	1.47	1.16	1.25	0.65	0.52
236.59	3.01	2.10	1.49	1.19	1.28	0.68	0.54
300.00	3.03	2.14	1.55	1.21	1.30	0.71	0.59
400.00	3.07	2.22	1.59	1.28	1.37	0.76	0.63
490.91	3.10	2.24	1.63	1.35	1.39	0.80	0.67
600.00	3.12	2.27	1.66	1.36	1.42	0.84	0.70
727.78	3.14	2.29	1.68	1.39	1.43	0.86	0.71
900.00	3.13	2.31	1.68	1.44	1.44	0.89	0.73

TABLE 10-5 Corrected Drawdown Data for Example 10-3

		s_c (ft)						
t (min)	$1/t$ (min^{-1})	1S	1W	1N	2W	2S	2N	3N
17.54	0.057013	2.34	—	—	—	0.67	—	—
27.48	0.036390	2.39	—	—	—	0.75	—	—
41.50	0.024096	2.45	—	—	—	0.80	—	—
50.00	0.020000	2.50	—	—	—	0.83	—	—
60.26	0.016595	2.53	—	—	—	0.88	—	—
75.51	0.013243	2.56	1.74	1.19	0.93	0.93	0.45	—
90.00	0.011111	2.60	1.75	1.23	0.95	0.98	0.48	0.39
109.90	0.009099	2.64	1.81	1.26	1.00	1.00	0.50	0.41
132.43	0.007551	2.67	1.86	1.30	1.02	1.10	0.54	0.45
151.01	0.006622	2.69	1.88	1.35	1.04	1.11	0.56	0.46
181.97	0.005495	2.73	1.94	1.37	1.08	1.17	0.60	0.50
211.35	0.004731	2.76	1.98	1.42	1.13	1.21	0.64	0.51
236.59	0.004227	2.80	2.00	1.44	1.16	1.24	0.67	0.53
300.00	0.003333	2.82	2.04	1.50	1.18	1.26	0.70	0.58
400.00	0.002500	2.86	2.11	1.53	1.24	1.33	0.75	0.62
490.91	0.002037	2.88	2.13	1.57	1.31	1.35	0.79	0.66
600.00	0.001667	2.90	2.15	1.60	1.32	1.37	0.82	0.69
727.28	0.001375	2.92	2.17	1.62	1.35	1.38	0.84	0.70
900.00	0.001111	2.91	2.19	1.62	1.39	1.39	0.87	0.72

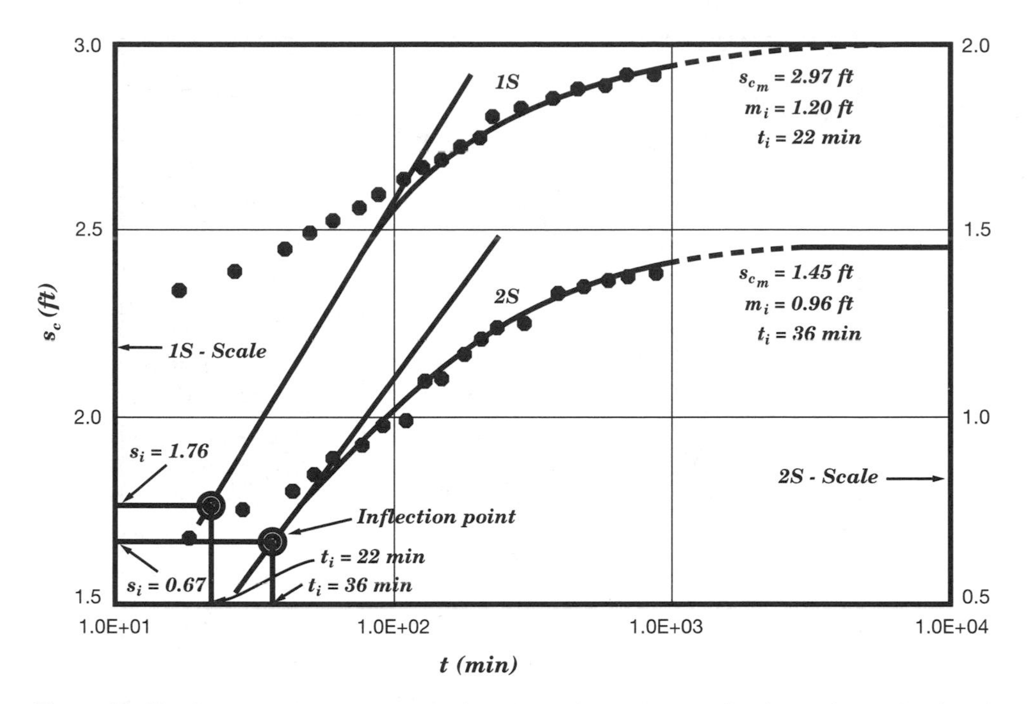

Figure 10-15 Corrected drawdown versus time data points and curves for observation wells 1S and 2S of Example 10-3. (After Hantush, 1959.)

Step 7: With the known values of s_i in Table 10-7, the inflection points are located in Figures 10-15 and 10-16 and the read values for t_i are listed in Table 10-7.

Step 8: With the known values of u_i, t_i, T, and r, the calculated values of S from Eq. (10-65) are listed in Table 10-7.

Step 9: The x coordinates of the observation wells, r, and β are needed for calculation of z from Eq. (10-66). The values of x coordinates of the observation wells in Figure 10-14 are given in Table 10-8. The values of r and β are listed in Table 10-7. From these values, the calculated z values are listed in Table 10-7 as well.

Comments on the Results. Some comments on the results are given below (Hantush, 1959):

1. Consider the drawdown versus time curves for observation wells 1S and 2S in Figure 10-15. The data points for the early period of pumping are included in the figures in order to show their effects on the aquifer parameters. For example, if all data points are included for the application of the theory on observation well 1S, the transmissivity would be approximately three times the 0.231-ft^2/day value in Table 10-7. And the distance between the pumped well and vertical constant head boundary (z) would be approximately 5000 ft instead of the listed value of 166 ft in Table 10-7. Obviously, this is an extremely unrealistic value.
2. The z values determined by this method are supposed to be close to each other. However, as can be seen from Table 10-7, the z values for observation wells 2N and 3N deviate significantly from the average value of the rest of wells. This may be the

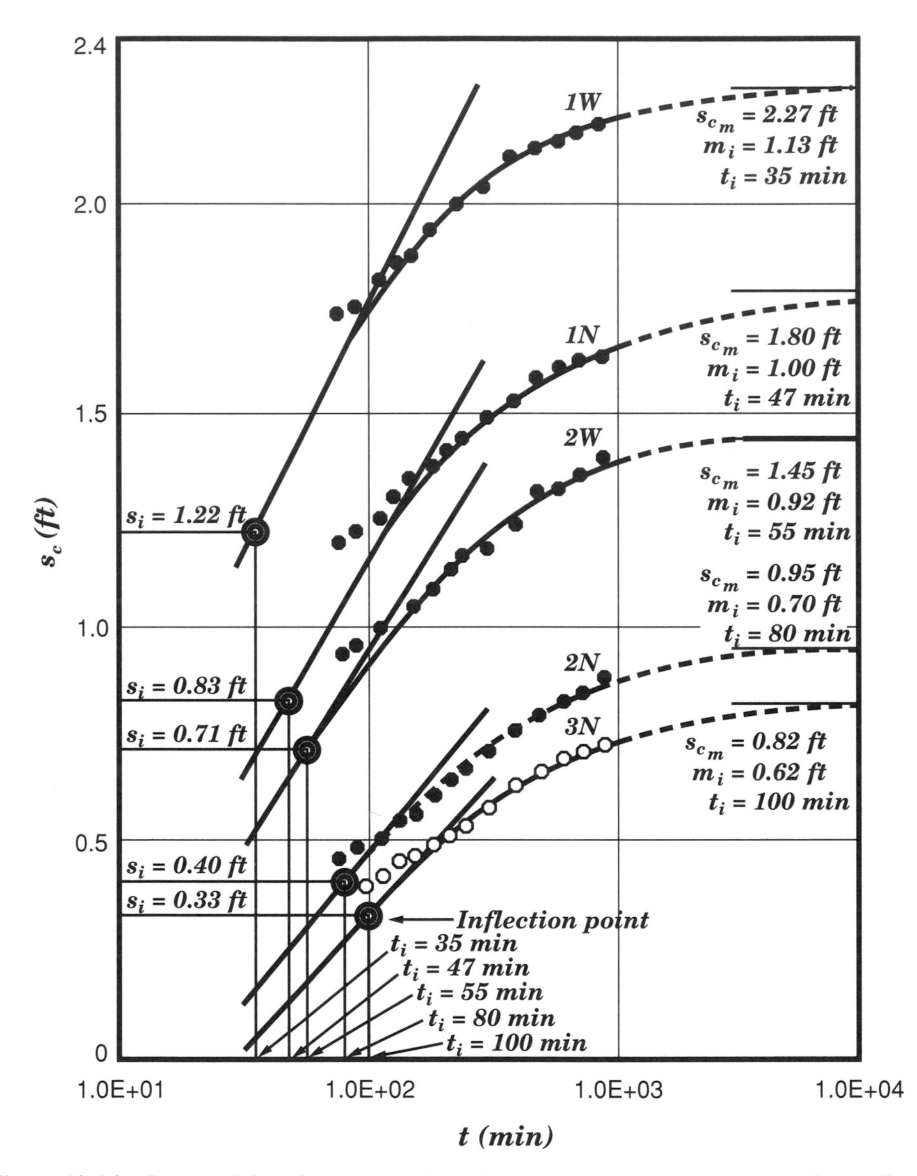

Figure 10-16 Corrected drawdown versus time data points and curves for observation wells 1N, 2N, 3N, 1W, and 3W of Example 10-3. (After Hantush, 1959.)

indication that the amount of induced recharge could not have been obtained from the Arkansas River only. Perhaps local recharge played a role as well.

Example 10-4: Using Hantush's one observation well method (Procedure II), determine the hydrogeologic parameters of the aquifer described in Example 10-3 and interpret the results.

Solution: The steps given previously for Procedure II are as follows:

TABLE 10-6 Least-Squares Curve-Fitting Coefficients of Eq. (10-47) for Examples 10-3 and 10-4

Well No.	Number of Data Points	Curve-Fitting Parameters s_{c_m} (ft)	m_t (ft · min)	c (ft · min^2)	$F(\beta)$
1S	10	2.97	43.46	147.77	0.202
1W	12	2.27	75.78	2747.91	0.943
1N	9	1.70	64.23	918.92	0.329
2W	10	1.49	109.08	6299.00	0.685
2S	14	1.47	64.87	1802.65	0.547
2N	12	0.95	85.34	4004.03	0.453
3N	8	0.81	84.08	4296.42	0.427

Step 1: The maximum drawdown ($s_{\max}$) in the observation wells have values between 3.13 ft (1S) and 0.73 ft (3N) as given in Table 10-4. Therefore, with $b = 22$ ft, the $s_{\max}/b$ ratio varies from 3.3% to 14.2%. The drawdown (s_{wt}) data given in Table 10-4 are corrected using Eq. (10-64), and the corrected drawdown (s_c) data are given in Table 10-5. Using these data, the semilogarithmic s_c versus time curves for the seven observation wells are shown in Figures 10-15 and 10-16.

Step 2: As can be seen from Figures 10-15 and 10-16, the drawdown versus time curves during early periods of pumping do not follow the trend indicated by the theory. As a result, the data points in Figures 10-15 and 10-16 before the period of recharge from the river are neglected.

Step 3: An approximate value of the maximum drawdown for each observation well is obtained by the least-squares method by fitting the parabola of Eq. (10-47) through the points s_c versus $(1/t)$ during the recharge period as given in Table 10-5. Equations (10-61), (10-62), and (10-63) are used in determining the parameters in Eq. (10-47), and the results are included in Table 10-6.

Step 4: The calculated values of $F(\beta)$ from Eq. (10-53) using the values of s_{c_m} (here s_{c_m} is equal to s_m), m_t, and c are included in Table 10-6. And the corresponding values of β are taken from Table 10-3.

Step 5: Using the known values of s_m, Q, and β, the calculated values of T from Eq. (10-45) are given in Table 10-9. Notice from this table that the β values for observation wells 1S, 1N, and 3N are less than 1. For $\beta < 1$, transmissivity (T) values cannot be calculated because the $\ln(\beta)$ values in Eq. (10-45) will be negative. Using the m_t, T, β, and r values for each observation well from Tables 10-6, 10-7, and 10-9, along with $Q = 1.55$ ft^3/s, the calculated S values are listed in Table 10-9.

Step 6: Using the x coordinates, radial distances, and β values of the observation wells, the calculated z values are listed in Table 10-9.

COMMENTS ON THE RESULTS. As can be seen from Table 10-9, only the data for observation wells 1W, 2W, 2S, and 2N produced results. The rest of observation wells (1S, 1N, and 3N) could not produce results because their corresponding β values are less than 1. This situation, obviously, yields negative transmissivity values from Eq. (10-45). The T values are more scattered than the ones resulted from Procedure I (see Table 10-7). The S values compare reasonably well with the ones in Table 10-7. Finally, the z values are more scattered

TABLE 10-7 Calculation of the Hydrogeologic Parameters Using Hantush's First Method (Procedure I) for Example 10-3

Well No.	r (ft)	s_{c_m} (ft)	m_i	$f(\beta)$	β	$\log(\beta)$	u_i	$M(u_i, \beta)$	$Q/(4\pi T)$ (ft)	s_i (ft)	t_i (min)	z (ft)	T (ft^2/s)	S
1S	19.3	2.97	1.20	2.48	16.2	1.21	0.022	3.29	0.534	1.76	22	166	0.231	0.072
1W	32.0	2.27	1.13	2.01	8.6	0.93	0.059	2.31	0.527	1.22	35	137	0.234	0.113
1N	50.0	1.70	1.00	1.70	5.3	0.72	0.123	1.63	0.510	0.83	47	108	0.242	0.131
2W	62.0	1.49	0.92	1.62	4.6	0.66	0.152	1.45	0.487	0.71	55	143	0.253	0.132
2S	57.0	1.47	0.96	1.53	3.9	0.59	0.192	1.24	0.541	0.67	36	140	0.228	0.116
2N	100.0	0.95	0.70	1.36	2.6	0.41	0.332	0.80	0.497	0.40	80	81[a]	0.248	0.158
3N	130.0	0.81	0.62	1.31	2.3	0.36	0.390	0.68	0.485	0.33	100	85[a]	0.254	0.141
										Average Parameters		139[b]	0.241	0.123

[a]Indicating additional source of recharge (may be due to increase in formation thickness.)
[b]Arithmetic average excluding wells 2N and 3N.

TABLE 10-8 The x Coordinates of the Observation Wells for Example 10-3

Well No.	x (ft)
1S	19.3
1W	0.0
1N	−50.0
2W	0.0
2S	57.0
2N	−97.0[a]
3N	−127.0[a]

[a]Estimated value.

TABLE 10-9 Results for the Hydrogeologic Parameters for Example 10-4

Well No.	$F(\beta)$	β	T (ft^2/s)	S	z (ft)
1S	0.202	< 1	—	—	—
1W	0.943	8.25	0.229	0.113	262
1N	0.329	< 1	—	—	—
2W	0.685	4.05	0.232	0.193	243
2S	0.547	2.45	0.150	0.175	168
2N	0.453	1.42	0.091	0.135	91
3N	0.427	< 1	—	—	—

and different than the values in Table 10-3. The overall results are consistent with the general evaluation of Procedure II presented previously.

Example 10-5: Using Hantush's two or more observation wells method, determine the hydrogeologic parameters of the aquifer whose pumping test data are given in Example 10-3.

Solution: The steps given previously are as follows:

Step 1: The corrected drawdown (s_c) versus time (t) curves on semilogarithmic paper are given in Figures 10-15 and 10-16.

Step 2: The corrected drawdown (s_c) versus ($1/t$) curves are presented in Figures 10-17 and 10-18 using the data given in Table 10-5.

Step 3: This step is not needed because Step 2 has provided satisfactory results.

Step 4: The tangents to the curves at the s_c intercept are drawn and their geometric slopes (m_t) as well as their measured s_{c_m} values are presented in Table 10-10.

Step 5: The values of m_t versus the x coordinates are presented in Figure 10-19. With the exception of the data points of observation wells 2N and 3N, the points seem to approximate a straight line. Through these points, the straight lines are drawn: (a) the best-fit line; (b) the line of steepest slope; and (3) the line of flattest slope.

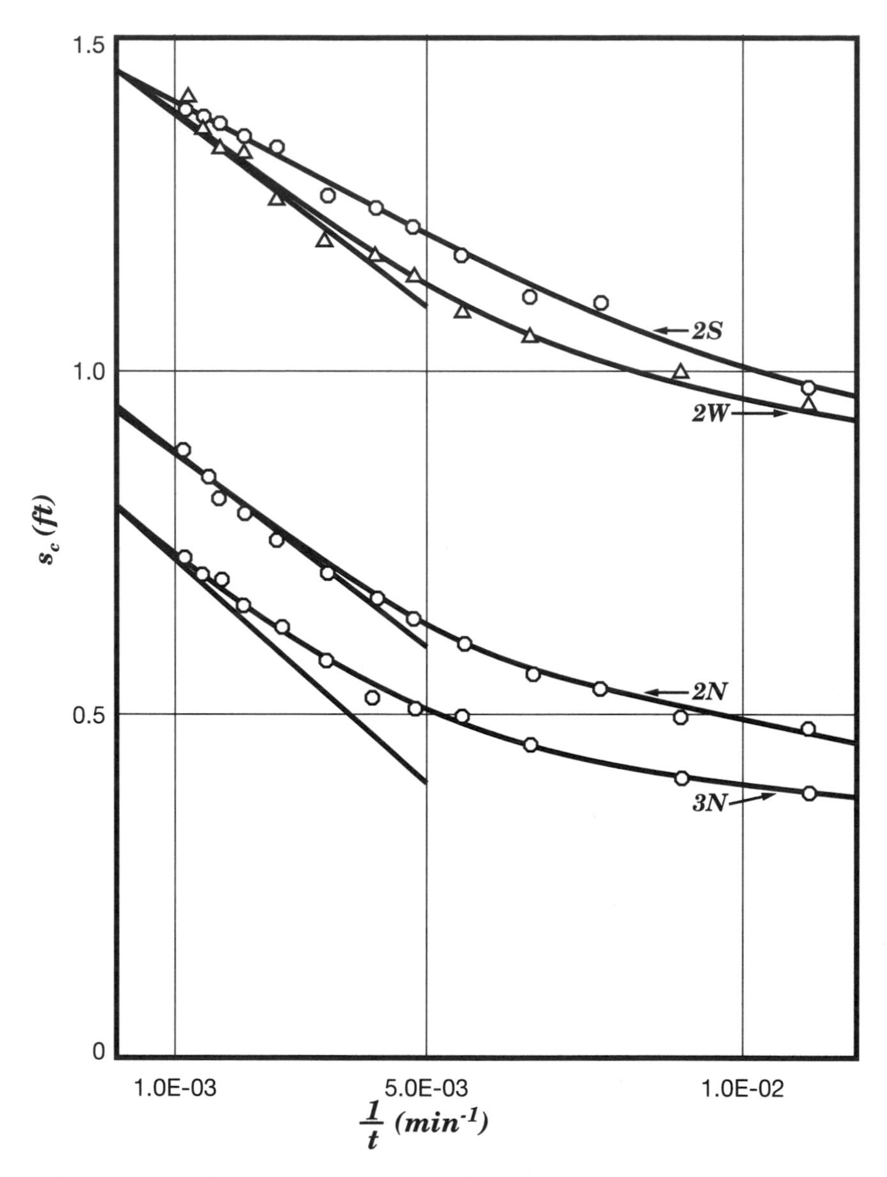

Figure 10-17 Corrected drawdown (s_c) versus ($1/t$) variation for observation wells 2S, 2W, 2N, and 3N for Example 10-5. (After Hantush, 1959.)

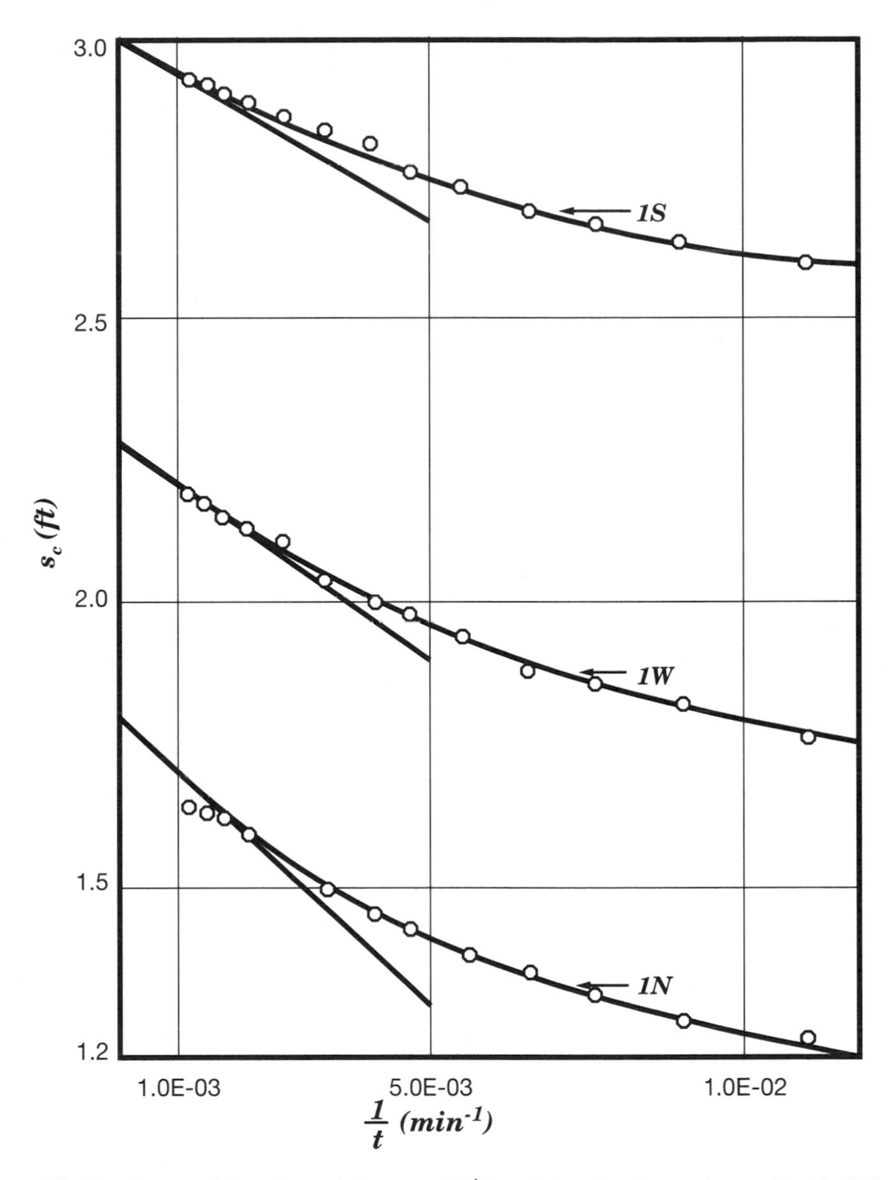

Figure 10-18 Corrected drawdown (s_c) versus ($1/t$) variation for observation wells 1S, 1W, and 1N for Example 10-5. (After Hantush, 1959.)

TABLE 10-10 Calculation of the Hydrogeologic Parameters Using Hantush's Two or Observation Wells Method for Example 10-5

Well No.	s_{Cm} (ft)	m_t (ft · min)	r (ft)	x (ft)	z (ft)[a]	r' (ft)	m_x (min)[a]	β	m_m (ft)[a]	T (ft^2/s)[a]	S
1S	3.00	65	19.3	19.3	140	261	0.53	13.6	2.32	0.245	0.11
1W	2.28	76	32.0	0.0	140	280	0.53	8.8			
1N	1.80	102	50.0	−50.0	140	330	0.53	6.6			
2W	1.45	72	62.0	0.0	140	286	0.53	4.6			
2S	1.45	49	57.0	57.0	140	223	0.53	3.9			
2N	0.95	70	100.0	−97.0[b]	80[c]	260	0.40	2.6			
3N	0.82	82	130.0	−127.0[b]	80[c]	290	0.40	2.2			

[a] Average value for the group indicated.
[b] Estimated value.
[c] Indicating additional source of recharge (may be due to increase in formation thickness).

Step 6: The x intercepts give, respectively, the average value of $z = 140$ ft, the minimum value of $z = 120$ ft, and the maximum value of $z = 180$ ft. The two points of 2N and 3N, taken together, give a value of $z = 80$ ft. The measured geometric slopes of these lines are given in Table 10-10.

Step 7: Using the average value of $z = 140$ ft, the value of β for each well is computed using Eq. (10-32) and given in Table 10-10.

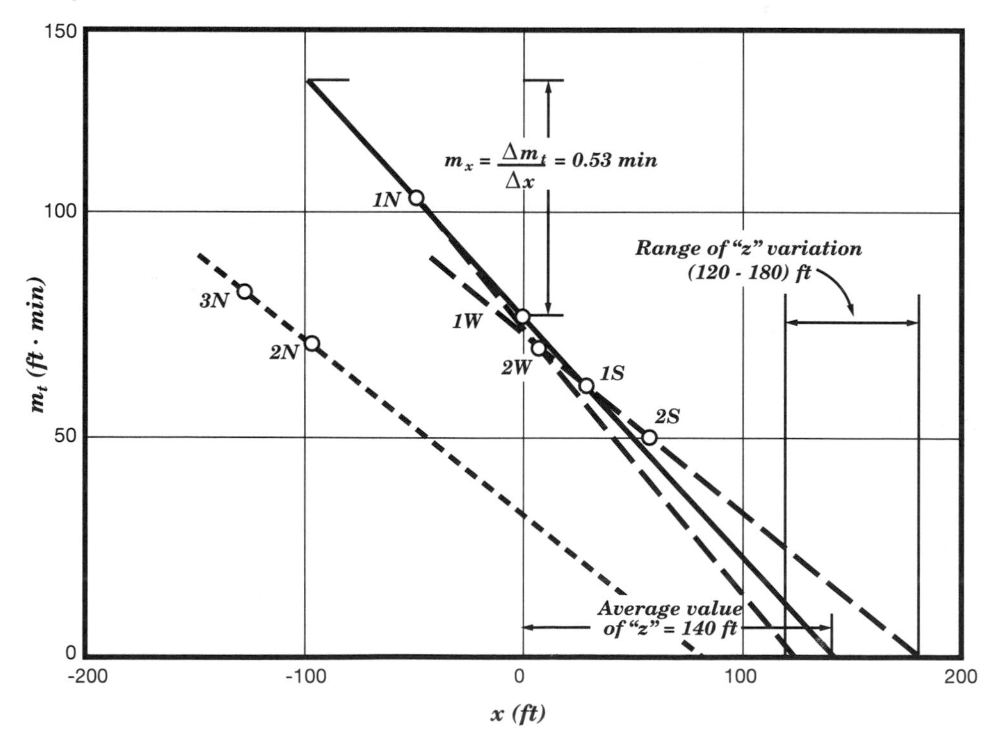

Figure 10-19 Variation of m_t with the x coordinate of the observation wells for Example 10-5. (After Hantush, 1959.)

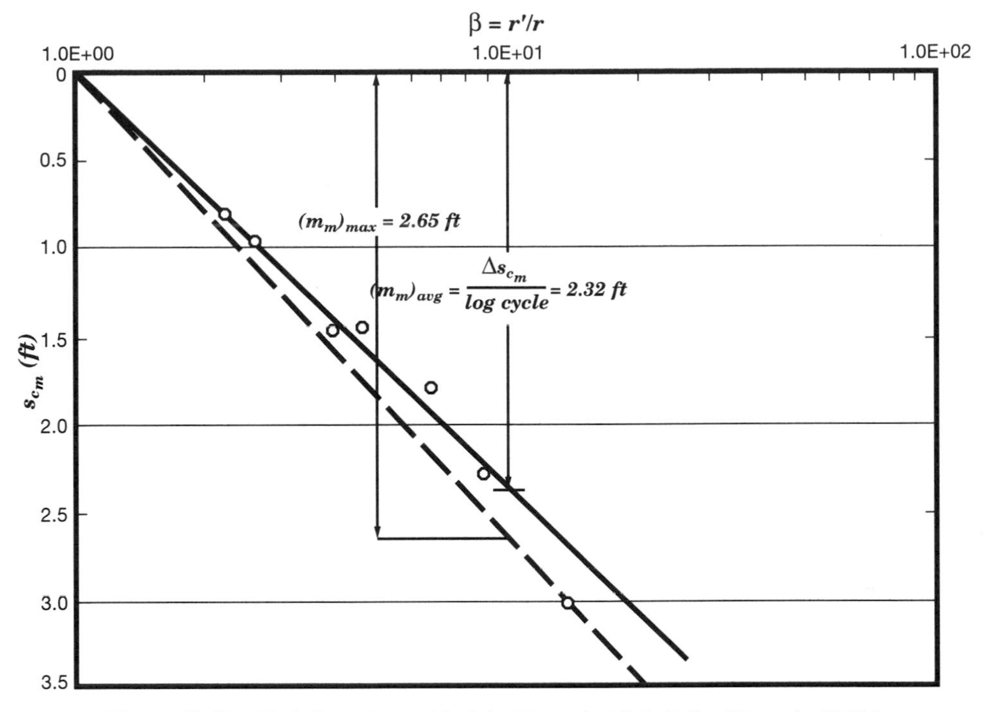

Figure 10-20 Variation of s_{c_m} with β for Example 10-5. (After Hantush, 1959.)

Step 8: The values of β versus s_{c_m} on semilogarithmic paper are shown in Figure 10-20. The geometric slope of the best fit straight line per log cycle is 2.32 ft. The line through the origin of steepest slope is (2.65 ft per log cycle) also shown in Figure 10-20.

Step 9: From Eq. (10-57), the average and minimum values of T are as follows:

$$T_{\text{avg}} = \frac{2.303Q}{2\pi m_m} = \frac{(2.303)\left(1.55\ \text{ft}^3/\text{s}\right)}{(2\pi)(2.32\ \text{ft})} = 0.245\ \text{ft}^2/\text{s}$$

$$T_{\text{min}} = \frac{(2.303)\left(1.55\ \text{ft}^3/\text{s}\right)}{(2\pi)(2.65\ \text{ft})} = 0.214\ \text{ft}^2/\text{s}$$

Step 10: From Eq. (10-55), the value of S for T_{avg} is

$$S = \frac{4\pi T^2 m_x}{Qz} = \frac{(4\pi)\left(0.245\ \text{ft}^2/\text{s}\right)^2 (0.53 \times 60\ \text{s})}{\left(1.55\ \text{ft}^3/\text{s}\right)(140\ \text{ft})} = 0.11$$

The value of S for T_{min} is

$$S = \frac{(4\pi)\left(0.214\ \text{ft}^2/\text{s}\right)^2 (0.53 \times 60\ \text{s})}{\left(1.55\ \text{ft}^3/\text{s}\right)(140\ \text{ft})} = 0.084$$

PART IV

WELL EFFICIENCY AND HYDROGEOLOGIC DATA ANALYSIS METHODS

CHAPTER 11

FULLY PENETRATING PUMPING WELLS IN HOMOGENEOUS AND ISOTROPIC NONLEAKY CONFINED AQUIFERS

11.1 INTRODUCTION

A production well requires certain design aspects to properly extract water from an aquifer. The screen interval and the gravel envelope around it certainly affect the yield of the well in any given aquifer. As a result, after the construction of a production well is complete, the well needs to be tested to evaluate its productivity. The productivity of a well is generally evaluated by *constant rate tests* or *variable rate tests*, which are also called *step drawdown tests*. These tests are required because of the difficulty of expressing analytically the drawdown in and near the well after the well completion due to the fact that the well development and gravel envelope may change the permeability of the formation next to the well. In the constant rate test, the well is pumped at a constant rate for a period of at least 8 hours, and water levels in the well are measured at frequent intervals. In the variable rate or step drawdown tests, the well is pumped during successive periods usually about 1 hour in duration at constant fractions of full capacity, and water levels in the production well are measured at frequent intervals. The productivity of a well is generally expressed by the term *specific capacity*, which is defined as the ratio of the extraction rate to the drawdown of the well (Jacob, 1946, 1950; Walton, 1962, 1970; Hantush, 1964).

In this chapter, the methods regarding the evaluation of well characteristics will be presented with examples. The methods will be presented using Jacob's drawdown equations (Jacob, 1946, 1950) as well as the Cooper and Jacob approximation (Cooper and Jacob, 1946) for the Theis equation (Theis, 1935). Even though, from a theoretical point of view, these equations are valid for fully penetrating wells in confined aquifers, they can also be used for wells in unconfined aquifers for certain conditions.

11.2 DRAWDOWN IN DISCHARGING WELLS

The drawdown in a pumping well (s_w) is composed of the head loss resulting from the laminar flow (s_l) and the head loss resulting from the turbulent flow of water through the screens and the pump intake (s_t). As shown in Figure 11-1, s_l is the drawdown in the aquifer and s_t is the drawdown that occurs as water moves from the aquifer into the well and up the well bore to the pump intake. As a result, the drawdown in most pumping wells is greater than the drawdown in the aquifer at the radius of the well.

The total drawdown (s_w) in a pumping well (see Figure 11-1) can be expressed in the form

$$s_w = s_l + s_t \tag{11-1}$$

where s_l is the drawdown in the aquifer at the effective radius of the pumping well and s_t is the well loss. The second drawdown component (s_t) in Eq. (11-1), which is the well loss caused by flow through the well screen and flow inside of the well to the pump intake, cannot be determined with the porous media flow theory, because it is associated with the turbulent flow inside the well. Because of these difficulties, Jacob (1947) suggested that the effects on the drawdown of turbulent flow may be accounted for approximately by introducing the concept of the *effective well radius*. If this hypothetical, empirically determined radius is substituted in the drawdown equation of that well, the actual drawdown outside the screen of the well in a confined aquifer will be yielded. This radius will also yield the actual water level in a well in an unconfined aquifer if well losses are negligible. The effective radius may be greater or smaller than the actual well radius depending on the hydraulic characteristics of the gravel envelope and the magnitude of the head loss caused by the turbulent flow near

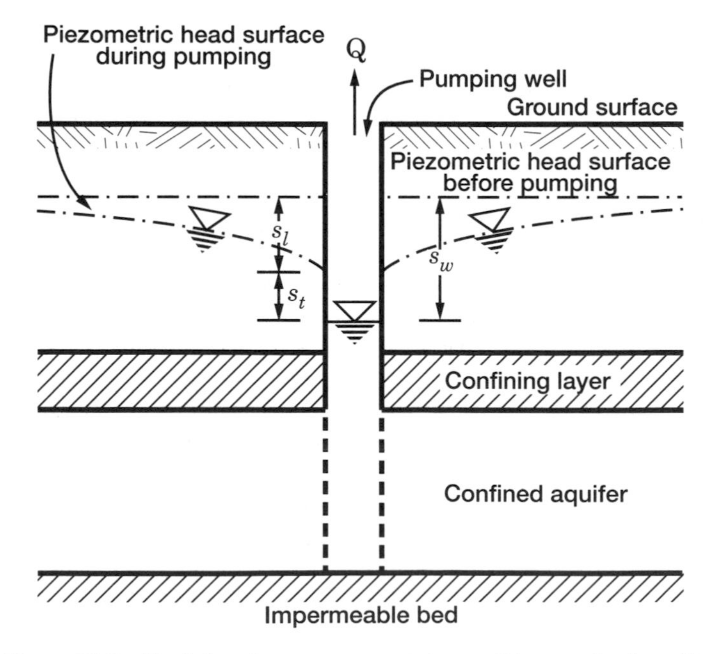

Figure 11-1 Total drawdown components in a well in a confined aquifer.

the well. Under the framework of these theoretical considerations, Jacob (1947) proposed some drawdown equations to determine the drawdown in a well. These equations and their practical applications will be included in detail in the next sections.

11.3 SPECIFIC CAPACITY OF WELLS

The specific capacity of a well is defined as the ratio of its discharge (Q) to its drawdown (s_w) (Jacob, 1950):

$$\text{Specific capacity} = \frac{Q}{s_w} \tag{11-2}$$

The specific capacity of a well is a measure of its productivity. The higher the specific capacity, the better the well. In the following sections, the theoretical specific capacity of a discharging well at a constant rate in a homogeneous, isotropic, nonleaky confined aquifer under steady and unsteady conditions will be presented in detail.

11.3.1 Specific Capacity of Wells Under Steady-State Conditions

The drawdown in a well under steady-state conditions is the difference between the pumping level in the well and the static level in the aquifer.

11.3.1.1 Ideal Well Without Screen or Casing. If a well is drilled in sandstone or dolomitic formations, the well bore may not require any casings, screen, or gravel packs (see Figure 11-2a). For this type of uncased well or rock boring, the head is distributed logarithmically in accordance with the Thiem equation (Thiem, 1906)

$$h_R - h = \frac{Q}{2\pi T} \ln\left(\frac{R}{r}\right) \tag{11-3}$$

which is derived in Section 3.2.1.4 of Chapter 3 [Eq. (3-10)]. In Eq. (11-3), r is the radial distance from the axis of the well and R is the effective outer radius at which the drawdown is zero. As $r \rightarrow r_w$, the drawdown approaches its value in the well, or (Jacob, 1947, 1950)

$$s_w = h_R - h_w = \frac{Q}{2\pi T} \ln\left(\frac{R}{r_w}\right) = BQ \tag{11-4}$$

where B is a constant. From Eqs. (11-2) and (11-4), the specific capacity of a well is

$$\frac{Q}{s_w} = \frac{2\pi T}{\ln\left(\frac{R}{r_w}\right)} \tag{11-5}$$

which is constant because of the fact that T, R, and r_w are all constants.

Example 11-1: This example is taken from Jacob (1950). Determine the relative efficiency of a well in a confined aquifer for different r_w/R values.

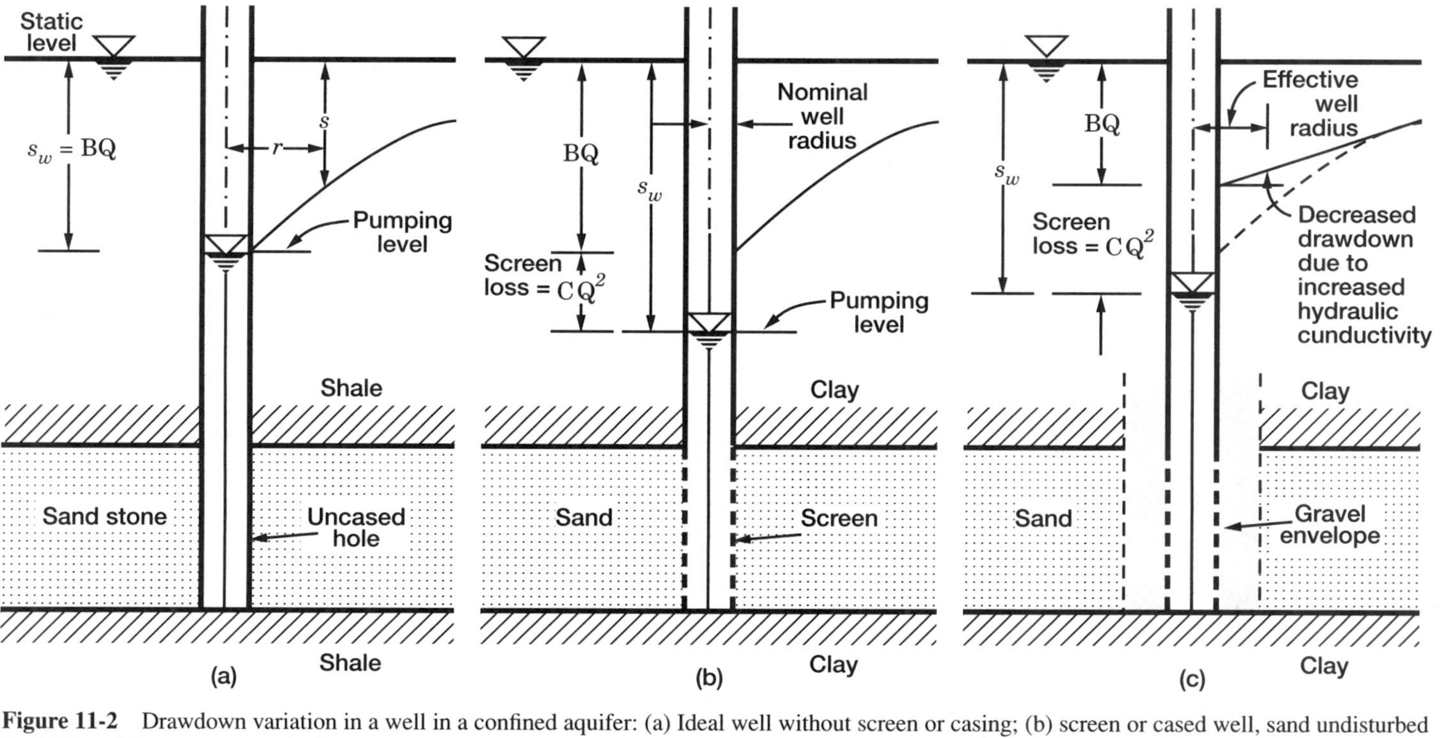

Figure 11-2 Drawdown variation in a well in a confined aquifer: (a) Ideal well without screen or casing; (b) screen or cased well, sand undisturbed by drilling; (c) screen or cased with gravel enveloped well (effective well radius increased). (Adapted from Jacob, 1950.)

Solution: Equation (11-5) can also be expressed as

$$\frac{2.303Q}{2\pi T s_w} = \frac{1}{\log\left(\dfrac{R}{r_w}\right)}$$

which means that the specific capacity, Q/s_w, is multiplied by the constant $2.303/(2\pi T)$. Using this equation, the desired values of the relative efficiencies can be obtained. The relative efficiency is calculated with respect to the first value of $1/\log(R/r_w)$ and the values are presented in Table 11-1. All values in the third column are divided by 0.25 to obtain the values in the fourth column. Consider a well 0.08-m-diameter ($2r_w$) whose effective radius (R) is 400 m. The value of r_w/R is 0.0001. A 0.16-m-diameter well for the same R distance would have $r_w/R = 0.0002$, but would only be 1.08 times as efficient. A 0.40-m-diameter well at the same R distance again would be 1.21 times as efficient as the 0.08-m-diameter well.

11.3.1.2 Screened or Cased Wells. Figure 11-2b shows a screened well without gravel envelope in a confined aquifer. The nominal radius of the hole (or screen) is approximately equal to the radius of the well (r_w) because of the inexistence of the gravel pack. Under pumping conditions from the well, screen losses (or well losses) are produced by friction during water entry into the well. Since the velocity of water increases as the well approached, the flow will become turbulent. Jacob (1950) expressed the well loss as CQ^2, where Q is the discharge rate of the well and C is a constant, referred to as the *well loss constant.*

The well shown in Figure 11-2c has a gravel envelope which changes (generally increases) the hydraulic conductivity of the aquifer around the well. With the increased hydraulic conductivity, the drawdown in and around the well will be less and the effective radius of the well will be increased as shown in Figure 11-2c. The dashed curve in Figure 11-2c represents the drawdown variation that would occur if the formation were left undisturbed with uniform hydraulic conductivity. It is the same drawdown curve as in Figures 11-2a and 11-2b.

The total drawdown in Figure 11-2c is somewhat less than in Figure 11-2b because of the increased effective radius of the well. By assuming that the well loss is the same in both

TABLE 11-1 Calculation of the Relative Efficiency of a Well in a Confined Aquifer Under Ideal Conditions for Example 11-1

r_w/R	$\log(R/r_w)$	$1/\log(R/r_w)$	Relative Efficiency
0.0001	4.000	0.250	1.00
0.0002	3.699	0.270	1.08
0.0005	3.301	0.303	1.21
0.0010	3.000	0.333	1.33
0.0020	2.699	0.371	1.48
0.0050	2.301	0.435	1.74
0.0100	2.000	0.500	2.00
0.0200	1.699	0.589	2.36
0.0500	1.301	0.769	3.08
0.1000	1.000	1.000	4.00

cases and adding this well loss to the formation loss, Jacob (1950) expressed the drawdown in the well as

$$s_w \cong BQ + CQ^2 \tag{11-6}$$

From Eqs. (11-2) and (11-6), the specific capacity of the well in Figure 11-2c is

$$\frac{Q}{s_w} \cong \frac{1}{B + CQ} \tag{11-7}$$

where C is a constant. From Eq. (11-4), the expression for B is

$$B = \frac{1}{2\pi T} \ln\left(\frac{R}{r_w}\right) \tag{11-8}$$

where R and r_w are the effective outer radius and effective well radius, respectively.

11.3.2 Specific Capacity of Wells Under Unsteady-State Conditions

The specific capacity of a well in a confined aquifer varies not only with its discharge but also with time as well. As can be seen from Eq. (11-6) and Figure 11-2, the drawdown at the well has two components, BQ and BQ^2. The first component, BQ, is the drawdown portion that can be predicted by the theoretical drawdown equations. For a confined aquifer case, BQ is the quantity that can be predicted by the Theis equation [Chapter 4, Section 4.2.4.2, Eq. (4-24)]. As mentioned in Section 4.2.7.2 of Chapter 4, for relatively large values of time (generally u less than 0.01) the Cooper and Jacob approximation as given by Eq. (4-46) of Chapter 4 can safely be used. Thus, Eq. (11-6) takes the form

$$s_w = BQ + CQ^2 = \frac{2.303Q}{4\pi T} \log\left(\frac{2.25Tt}{r_w^2 S}\right) + CQ^2 \tag{11-9}$$

or

$$\frac{Q}{s_w} = \left[\frac{2.303}{4\pi T} \log\left(\frac{2.25Tt}{r_w^2 S}\right) + CQ\right]^{-1} \tag{11-10}$$

Equations (11-9) and (11-10) replace Eqs. (11-6) and (11-7), respectively, which apply to steady-state cases.

Rorabaugh (1953) proposed the well loss term to be CQ^n instead of CQ^2, where n is a constant greater than 1, and pointed out that n can deviate significantly from 2 and its value needs to be determined individually for each well. As a result, the drawdown equation given by Eq. (11-6) will have the following form:

$$s_w \cong BQ + CQ^n \tag{11-11}$$

Observations (Rorabaugh, 1953; Hantush, 1964) have shown that well drawdowns determined from Jacob's and Rorabaugh's equations within the range of operating well discharges agreed very closely with observed data, but they differed appreciably for excess discharges. For large well discharges, Rorabaugh (1953) suggested that a margin of safety can be provided using the two equations. Rorabaugh also noted that Jacob's equation predicts closer values to the true head variation outside the well and should be used for designing radii of wells.

With Eq. (11-11), Eq. (11-10) takes the form

$$\frac{Q}{s_w} = \left[\frac{2.303}{4\pi T}\log\left(\frac{2.25Tt}{r_w^2 S}\right) + CQ^{n-1}\right]^{-1} \tag{11-12}$$

From this equation, or Eq. (11-10), the following conclusions can be drawn (Hantush, 1964):

1. The specific capacity, Q/s_w, decreases with time and discharge. Therefore, the discharge is directly proportional with drawdown, which implies a constant specific capacity and may create significant errors.
2. The effect of the well radius (r_w) on the specific capacity (Q/s_w) depends on the magnitude of the discharge (Q) as well as the empirical constant C.
3. The radius of the well (r_w) is important in designing wells, especially for relatively large discharges.
4. If the well losses resulted from turbulent flow near and inside the well are relatively small, the influence of the well radius on the specific capacity may not be important.

Example 11-2: This example is adapted from Jacob (1950). The hydrogeologic parameters of an aquifer are given as: $T = 0.24\ \text{ft}^2/\text{s}$, $S = 2.6 \times 10^{-4}$, $r_w = 0.75$ ft, and $C = 3.2\ \text{s}^2/\text{ft}^5$. Determine the variation of the specific capacity as function of time.

Solution: Substitution of the values in Eq. (11-10) gives

$$\frac{Q}{s_w} = \left\{\frac{2.303}{(4\pi)\left(0.24\ \text{ft}^2/\text{s}\right)}\log\left[\frac{(2.25)\left(0.24\ \text{ft}^2/\text{s}\right)t}{(0.75\ \text{ft})^2\left(2.6\times 10^{-4}\right)}\right] + (3.2\text{s}^2/\text{ft}^5)Q\right\}^{-1}$$

or

$$\frac{Q}{s_w} = \left[0.763\log(3692.31t) + 3.2Q\right]^{-1}$$

where the units of Q and t are ft^3/s and s, respectively. Using this equation, the specific capacity values up to 1000 days for $Q = 1, 2,$ and $3\ \text{ft}^3$/day are presented in Table 11-2 and their graphical forms are shown in Figure 11-3. The specific capacity of a well is constant under steady-state conditions because B, C, and n are constants. Under unsteady-state conditions, Q/s_w decreases with time because B depends on the time of pumping. As can be seen from Table 11-2 and Figure 11-3, the specific capacity decreases with increasing time and discharge rate.

11.3.3 Properties of the Well Loss Constant

The well loss constant (C) can be determined by analyzing the step drawdown test data (see Section 11.4) on the assumption that the well is stable and C remains constant during

TABLE 11-2 Calculated Specific Capacity (Q/s_w) Values for Different Discharge Rate (Q) and Time (t) Values for Example 11-2

	Q/s_w (ft³/s)		
t (days)	$Q = 1$ ft³/s	$Q = 2$ ft³/s	$Q = 3$ ft³/s
1	0.1032	0.0776	0.0622
2	0.1008	0.0762	0.0613
3	0.0995	0.0755	0.0608
5	0.0978	0.0745	0.0602
10	0.0957	0.0733	0.0593
20	0.0934	0.0720	0.0585
30	0.0925	0.0713	0.0581
50	0.0910	0.0705	0.0575
100	0.0892	0.0694	0.0568
200	0.0874	0.0683	0.0560
300	0.0864	0.0677	0.0556
400	0.0857	0.0672	0.0553
500	0.0851	0.0669	0.0551
700	0.0843	0.0664	0.0548
1000	0.0835	0.0659	0.0544

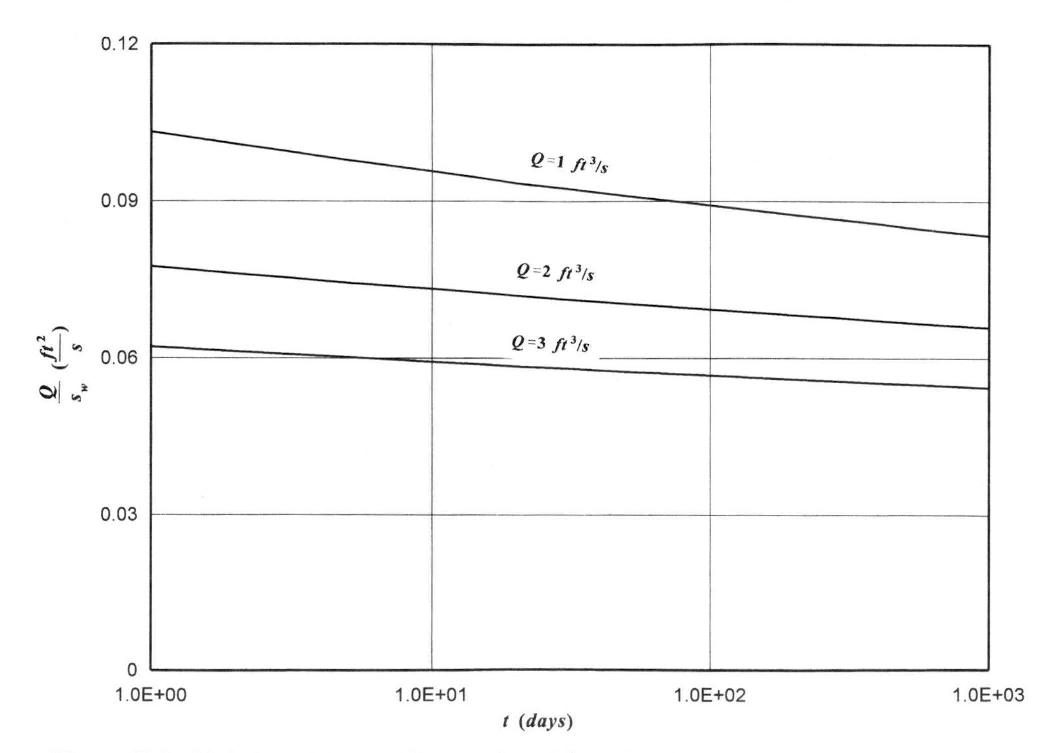

Figure 11-3 Variation of the specific capacity (Q/s_w) with discharge rate (Q) and time (t) for Example 11-2.

TABLE 11-3 Range of Well Loss Coefficient (*C*) Values with the Well Conditions

	Well Loss Coefficient (*C*)	
Wall Condition	s^2/ft^5	s^2/m^5
Properly developed and designed well	< 5	< 1900
Mild deterioration	5–10	1900–3800
Clogging is severe	> 10	> 3800
Difficult to restore well to original capacity	> 40	$> 15{,}200$

Source: After Walton (1962).

the test. There are some potential factors that affect the values of C, and these are presented below (Walton, 1962).

During the development process of well, drilling activities often clog the openings of the well face and well screen. The well yield cannot be increased unless the clogging materials are removed from the well face and screen. The effectiveness of development can be determined by analyzing the data of a step-drawdown test. If the power of n in Eq. (11-11) is 2.0 as in Eq. (11-6), the dimensions of C will be T^2L^{-5} or specifically s^2/ft^5 (s^2/m^5). Based on field experience, Walton (1962) suggested criteria for the well loss coefficient in Eqs. (11-6) and (11-11), and these values are included in Table 11-3. As can be seen from this table, the value of C of a properly developed and designed well is generally less than 5 s^2/ft^5 (1900 s^2/m^5). It is difficult or sometimes impossible to restore the original capacity by one of several rehabilitation methods. The success of rehabilitation activities can be appraised by analyzing the step-drawdown tests data generated prior to and after the improvement work on the well.

11.4 STEP-DRAWDOWN TEST DATA ANALYSIS METHODS

The values of effective well radius (r_w), well loss coefficient (C), and exponent n in Eq. (11-11) may be determined (Jacob, 1950; Hantush, 1964) by analyzing the data of a step-drawdown pumping test. In a step-drawdown test the well is operated during successive periods (usually about 1 hour in duration) at constant fractions of full discharge capacity (three or more), and the drawdowns in the well are measured at frequent intervals. The final data sets will be $Q_1, Q_2, \ldots, Q_n$, and with the corresponding well drawdowns s_{w_1}, $s_{w_2}, \ldots, s_{w_n}$. In addition, observations must be made in outlying wells to determine the value of S; otherwise the value of r_w cannot be determined.

In the following sections, several methods will be presented for step-drawdown tests data analysis to determine the above-mentioned parameters.

11.4.1 Straight-Line Method

The straight-line method can be applied to Eq. (11-6), which corresponds to $n = 2$. Equation (11-6) can also be expressed as (Todd, 1980; Raghunath, 1987)

$$\frac{s_w}{Q} = B + CQ \tag{11-13}$$

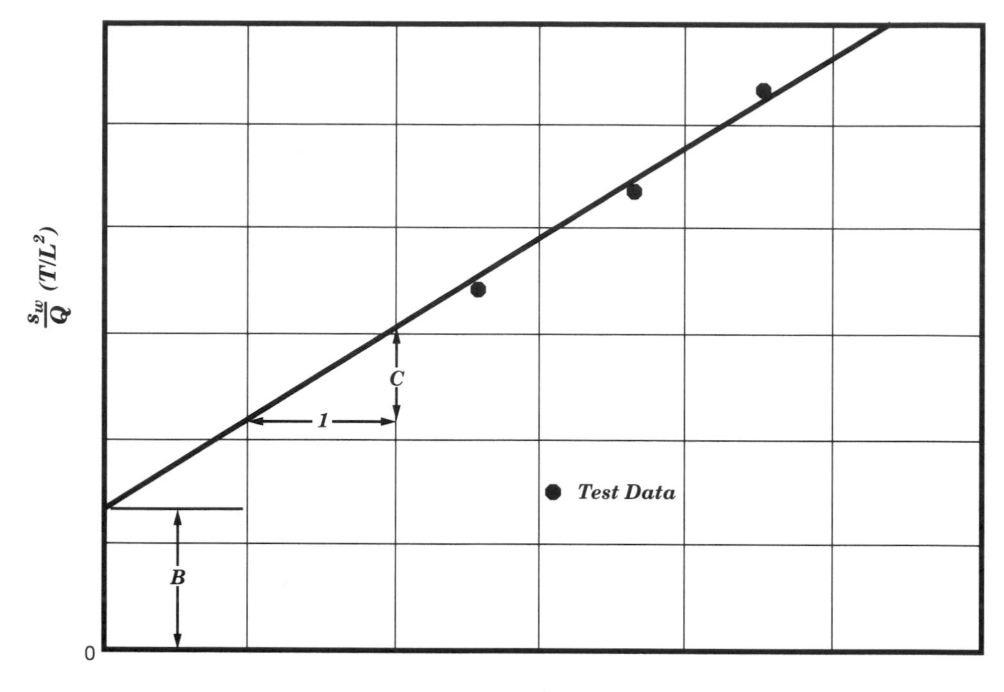

Figure 11-4 Schematic representation of the straight-line method for the determination of well loss coefficients B and C.

which indicates that a plot of s_w/Q versus Q represents a straight line (see Figure 11-4) whose geometric slope corresponds to the well loss coefficient C, and the intercept on the s_w/Q axis gives the value of B.

Example 11-3: The data of this example are taken from Jacob (1950). The data for a step-drawdown test on a 20-in.-diameter well are given in Table 11-4. The aquifer parameters are $T = 0.78$ ft^2/s and $S = 4.8 \times 10^{-3}$. Using the straight-line method, determine the values of B and C as well as the effective radius (r_w) after 1 hr.

Solution: The s_w/Q values are given in the third column of Table 11-4. Using the data in Table 11-4, the specific drawdown (s_w/Q) versus discharge rate (Q) curve is presented in

TABLE 11-4 Input Data for Example 11-3

Discharge Rate, (Q) (ft^3/s)	Drawdown, (s_w) (ft)	s_w/Q (s/ft^2)
2.88	13.0	4.51
2.10	7.8	3.71
0.84	2.5	2.98
0.00	0.3	—

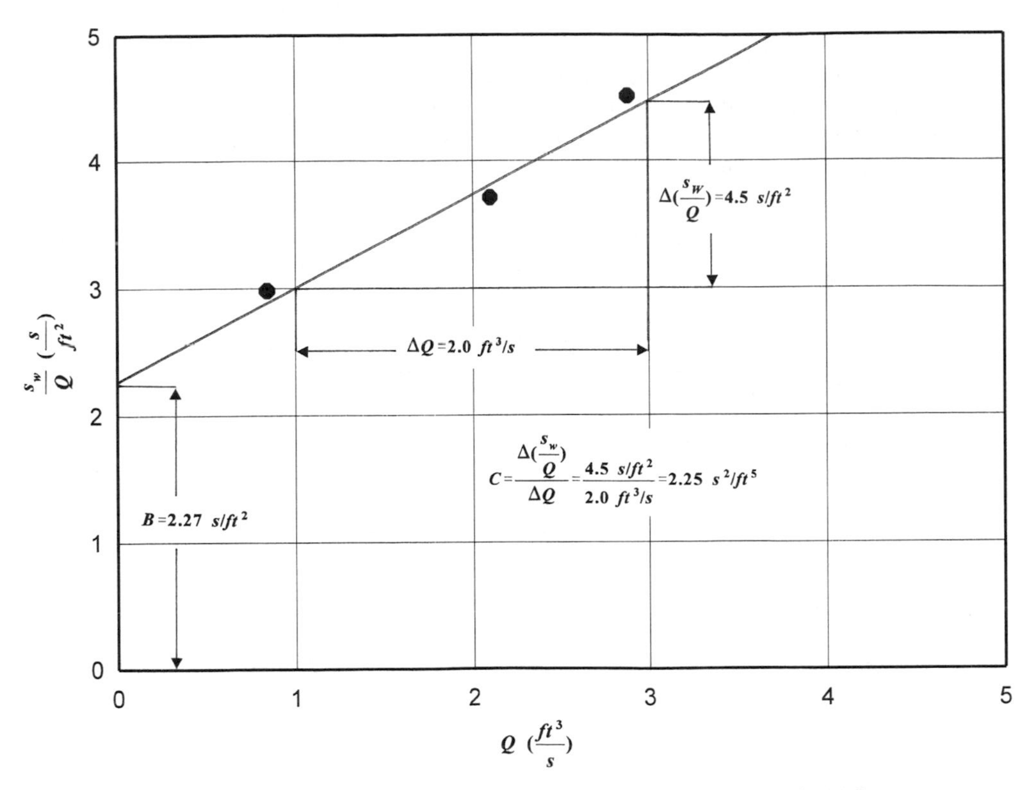

Figure 11-5 Specific drawdown versus discharge rate for Example 11-3.

Figure 11-5. From this graph, $B = 2.27\ \text{s/ft}^2$ and $C = 2.25\ \text{s}^2/\text{ft}^5$, where C is the geometric slope of the straight line and B is the s_w/Q intercept. From Eq. (11-9) we obtain

$$B = \frac{2.303}{4\pi T} \log\left(\frac{2.25Tt}{r_w^2 S}\right)$$

or

$$\log\left(\frac{2.25Tt}{r_w^2 S}\right) = \frac{4\pi TB}{2.303} = \frac{(4\pi)\left(0.78\ \text{ft}^2/\text{s}\right)\left(2.27\ \text{s/ft}^2\right)}{2.303} = 9.66$$

and solving for r_w^2 gives

$$r_w^2 = \frac{2.25Tt}{10^{9.66}S} = \frac{(2.25)\left(0.78\ \text{ft}^2/\text{s}\right)(3600\ \text{s})}{\left(10^{9.66}\right)\left(4.8 \times 10^{-3}\right)} = 0.28 \times 10^{-4}$$

and finally

$$r_w = 1.70 \times 10^{-2}\ \text{ft}$$

Example 11-4: This example is adapted from Raghunath (1982). A production well fully penetrates a confined aquifer. The aquifer parameters are $T = 800\ \text{m}^2/\text{day}$ and $S =$

6.0×10^{-4}. The effective radius and well loss constant are known from a step-drawdown test data analysis as $r_w = 0.12$ m and $C = 2200$ s^2/m^5. Using these data, (a) construct specific capacity curves for the well for $Q = 480$, 960, and 1200 liters/min discharge rates; and (b) calculate the specific capacities and drawdowns for each discharge rate at $t = 1$ hr, 1 day, 1 month, and 1 year.

Solution: (a) Eq. (11-10) is the specific capacity equation to be used. With the values of $T = 800$ m^2/day $= 9.26 \times 10^{-3}$ m^2/s, $r_w = 0.12$ m, and $C = 2200$ s^2/m^5 we have

$$\frac{Q}{s_w} = \left\{ \frac{2.303}{(4\pi)\left(9.26 \times 10^{-3}\ \text{m}^2/\text{s}\right)} \log\left[\frac{(2.25)\left(9.26 \times 10^{-3}\ \text{m}^2/\text{s}\right) t}{(0.12\ \text{m})^2\left(6.0 \times 10^{-4}\right)} \right] + (2200\text{s}^2/\text{m}^5)Q \right\}^{-1}$$

or

$$\frac{Q}{s_w} = \left[19.767 \log(2411.27t) + 2200Q\right]^{-1}$$

For $Q = 480$ liters/min $= 0.008$ m^3/s we obtain

$$\frac{Q}{s_w} = \left[19.767 \log(2411.27t) + 17.6\right]^{-1}$$

For $Q = 960$ liters/min $= 0.016$ m^3/s we have

$$\frac{Q}{s_w} = \left[19.767 \log(2411.27t) + 35.2\right]^{-1}$$

For $Q = 1200$ liters/min $= 0.02$ m^3/s we have

$$\frac{Q}{s_w} = \left[19.767 \log(2411.27t) + 44.0\right]^{-1}$$

Using the formulas given above, the specific capacity versus time curves for the discharge rates are given in Figure 11-6, and their corresponding values are given in Table 11-5. As can be seen from this figure, the specific capacity decreases with the increasing discharge rate.

(b) Specific capacities can be determined from either the corresponding equations or Figure 11-6 at different times. The calculated specific capacity (Q/s_w) and well drawdown (s_w) values are given in Table 11-6.

11.4.2 Rorabaugh's Method

As mentioned in Section 11.3.2, Rorabaugh (1953) suggested that the well loss resulting from turbulence can be expressed as CQ^n as shown in Eq. (11-11). Rorabaugh developed a graphical method for the determination of laminar and turbulent well loss components based on Eqs. (11-6) and (11-11). In the following paragraphs, Rorabaugh's graphical method is presented with examples.

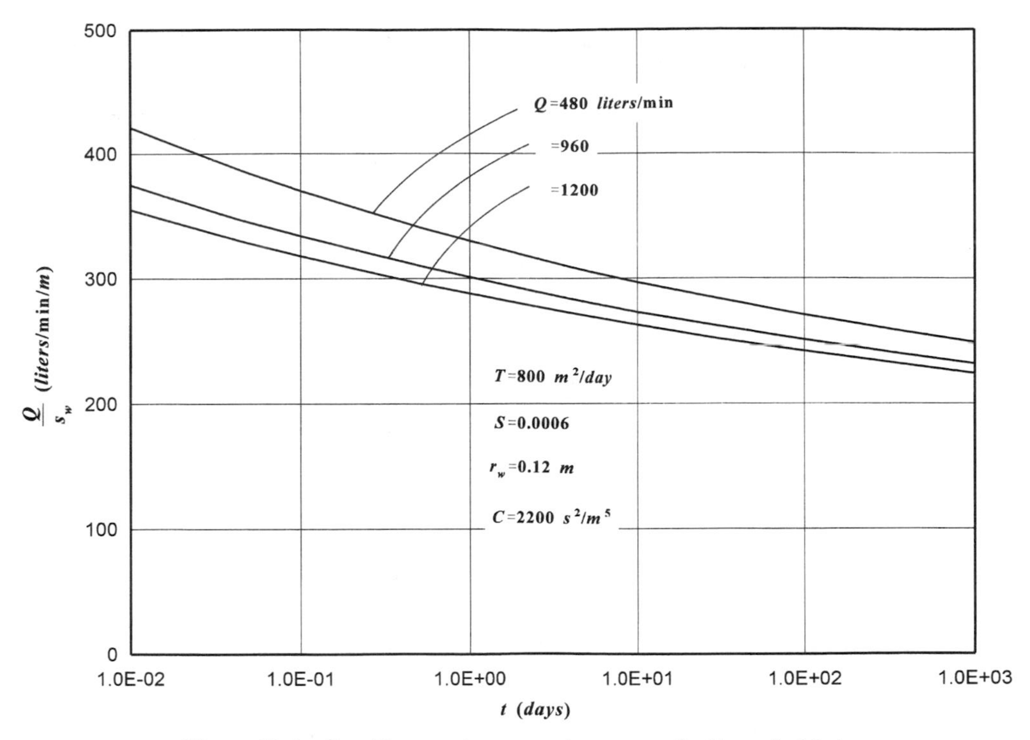

Figure 11-6 Specific capacity versus time curves for Example 11-4.

TABLE 11-5 Input Data for Example 11-4

t	Q/s_w (liters/min/m)		
(days)	Q = 480 liters/min	Q = 960 liters/min	Q = 1200 liters/min
0.01	421	375	355
0.0417	388	348	331
0.05	384	345	328
0.1	370	334	318
0.5	341	310	296
1.0	330	301	288
5.0	306	281	270
10.0	297	273	263
30.0	284	262	252
100.0	271	251	242
365.0	258	240	232
1000.0	249	232	224

TABLE 11-6 Specific Capacity and Drawdown Values for Different Discharge Rates and at Different Times for Example 11-4

	$Q = 480$ liters/min		$Q = 960$ liters/min		$Q = 1200$ liters/min	
t	Q/s_w (liters/min/m)	s_w (m)	Q/s_w (liters/min/m)	s_w (m)	Q/s_w (liters/min/m)	s_w (m)
1 hr	388	1.24	348	2.76	331	3.63
1 day	330	1.45	301	3.19	288	4.17
1 month	284	1.69	262	3.66	252	4.76
1 year	258	1.86	240	4.00	232	5.17

Jacob's equation, as given by Eq. (11-6), can be expressed in the form

$$\frac{s_w}{Q} - B = CQ \tag{11-14}$$

or, taking the logarithms of both sides,

$$\log\left(\frac{s_w}{Q} - B\right) = \log C + \log Q \tag{11-15}$$

Equation (11-14) is in the form of a straight-line equation when Q is plotted against $(s_w/Q - B)$ on double logarithmic paper. It is clear from Eq. (11-15) that for $Q = 1$ we obtain $C = s_w/Q - B$, which means that the geometric slope of the line is unit.

If the exponent of the turbulent well loss component is an unknown constant n as shown in Eq. (11-11), it can be rearranged as

$$\frac{s_w}{Q} - B = CQ^{n-1} \tag{11-16}$$

or

$$\log\left(\frac{s_w}{Q} - B\right) = \log C + (n-1)\log Q \tag{11-17}$$

which is in the form of a straight-line equation when Q is plotted against $(s_w/Q - B)$ on double logarithmic paper. The geometric slope of this line is $(n - 1)$. When $Q = 1$, we obtain $C = s_w/Q - B$.

In Rorabaugh's method, values of B need to be assumed. The steps are as follows: (1) For each value of B, plot $(s_w/Q - B)$ versus Q values on a double logarithmic paper. (2) Repeat this process until an approximate straight line is obtained. Determine the C value from the intercept—that is, where $Q = 1, C = s_w/Q - B$. (3) Determine the n value from the known value of $(n - 1)$, which is the geometric slope of the straight line.

Example 11-5: This example is adapted from Rorabaugh (1953). The step-drawdown test results of a production well are given in Table 11-7. Using Rorabaugh's method, (a) determine the B and C values in Jacob's equation ($n = 2$), (b) determine the B, C, and n values in the modified Jacob's equation, and (c) compare the results of the two cases.

TABLE 11-7 Discharge Rate and Drawdown Data for Example 11-5

Step No.	Discharge Rate, (Q) (ft^3/s)	Drawdown, (s_w) (ft)
1	1.21	25.4
2	2.28	50.4
3	3.12	72.3
4	3.39	80.3

Solution: With the assumed values of B, Table 11-8 is prepared and these values are presented graphically in Figure 11-7.

(a) Of those data sets in Table 11-8, the ones corresponding to $B = 19$ s/ft^2 fall approximately on a straight line at (shown by L1 in Figure 11-7) a 45° angle (geometric slope = 1) which corresponds to $n = 2$. As can be seen from Figure 11-7, points 2, 3, and 4 tend to be located on L1 and point 1 is somewhat off from the line. For $Q = 1$ ft^3/s, it can be read from the graph that $s_w/Q - B = C = 1.30$ s^2/ft^5.

(b) As can be seen from Figure 11-7, $B = 20.4$ s/ft^2 in Table 11-8 produces a straight line (L2 in Figure 11-7) among its other values. Curves are also shown in the same figure for other trials of B values. From Figure 11-7, $n - 1 = 1.68$. Therefore, $n = 2.68$. From this straight line, for $Q = 1$ ft^3/s, $C = 0.44$ s^n/ft^{3n-1} (units of CQ^n must be feet, so that the units of C are s^n/ft^{3n-1}).

TABLE 11-8 Values of ($s_w/Q - B$) for Different Values of B for Example 11-5

B (s/ft^2)	Q (ft^3/s)	$s_w/Q - B$ (s/ft^2)
18	1.21	3.0
	2.28	4.1
	3.12	5.2
	3.39	5.7
19	1.21	2.0
	2.28	3.1
	3.12	4.2
	3.39	4.7
20	1.21	1.0
	2.28	2.1
	3.12	3.2
	3.39	3.7
20.4	1.21	0.6
	2.28	1.7
	3.12	2.8
	3.39	3.3
20.9	1.21	0.1
	2.28	1.2
	3.12	2.3
	3.39	2.8

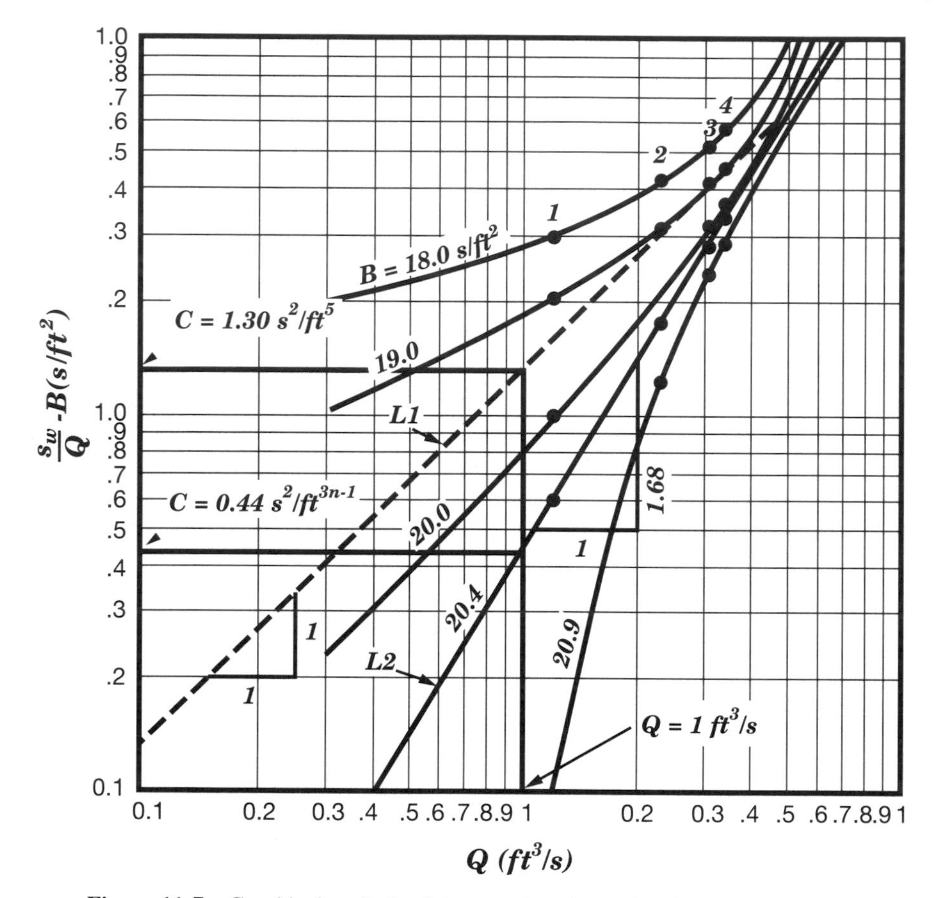

Figure 11-7 Graphical analysis of the step-drawdown data for Example 11-5.

(c) Table 11-9 presents the original drawdown data and drawdowns calculated by both methods. Drawdowns for higher discharge rates are given as well. From the inspection of Figure 11-7 and Table 11-9, the following conclusions can be drawn for the graphical method in using Eqs. (11-6) and (11-11) (Rorabaugh, 1953):

1. Well drawdowns calculated by Eqs. (11-6) and (11-11) are very close to the observed values. However, the problem is to fit a curve to the four observed data points that are in a relatively narrow range for which any number of different curves could be applied. A margin of safety can be provided by using Eqs. (11-6) and (11-11), using the least favorable solution for the purpose.
2. Comparison of the first drawdown component (BQ) based on Eqs. (11-6) and (11-11) shows that Eq. (11-11) assigns a larger portion of the total drawdown to BQ.
3. Equation (11-11) requires accurate data. Small errors in discharge or drawdown measurements will cause relatively large errors in B, C, and n.
4. Rorabaugh's method can be applied by treating the exponent as an unknown constant. The method has also the potential application of accepting the exponent as square, and the best fit of a 45° line will give a solution.

TABLE 11-9 Comparison of Observed Drawdowns and Drawdowns Calculated by the Straight-Line and Rorabaugh's Method for Example 11-5

			Jacob's Equation (n = 2)			Rorabaugh's Modification (n is a parameter)		
Step No.	Q (ft^3/s)	s_w (ft)	BQ (ft)	CQ^2 (ft)	s_w (ft)	BQ (ft)	CQ^n (ft)	s_w (ft)
1	1.21	25.4	23.0	1.9	24.9	24.7	0.7	25.4
2	2.28	50.4	43.3	6.8	50.1	46.5	4.0	50.5
3	3.12	72.3	59.3	12.7	72.0	63.6	9.3	72.9
4	3.39	80.3	64.4	14.9	79.3	69.2	11.6	80.8
5	4.0	—	76.0	20.8	96.8	81.6	18.1	99.7
6	5.0	—	95.0	32.5	127.5	102.0	32.9	134.9
7	6.0	—	114.0	46.8	160.8	122.4	53.6	176.0

Example 11-6: A production well fully penetrates a confined aquifer and the parameters are given as: $T = 450\ \text{m}^2/\text{day}$, $S = 5.0 \times 10^{-4}$, and $Q = 25$ liters/s. The effective radius and well loss constants are known from a step-drawdown test as: $r_w = 0.15$ m, $C = 2050\ \text{s}^2/\text{m}^5$, and $n = 2.55$. Determine the specific capacity and drawdown in the pumped well after 1 year of pumping.

Solution: Equation (11-12) is the equation to be used. With the given values $T = 450\ \text{m}^2/\text{day} = 5.21 \times 10^{-3}\text{m}^2/\text{s}$, $Q = 25$ liters/s $= 0.025\ \text{m}^3/\text{s}$, and the other values, Eq. (11-12) gives

$$\frac{Q}{s_w} = \left\{ \frac{2.303}{(4\pi)\left(5.21 \times 10^{-3}\ \text{m}^2/\text{s}\right)} \log \left[\frac{(2.25)\left(5.21 \times 10^{-3}\ \text{m}^2/\text{s}\right)(365 \times 86{,}400\ \text{s})}{(0.15\ \text{m})^2 \left(5.0 \times 10^{-4}\right)} \right] + \left(2050\ \text{s}^2/\text{m}^5\right)\left(0.025\ \text{m}^3/\text{s}\right)^{2.55-1} \right\}^{-1} = 2.66 \times 10^{-3}\ \text{m}^2/\text{s}$$

and using the value of Q, after one year the drawdown is $s_w = 9.4$ m.

11.4.3 Miller and Weber Method

As mentioned in the above sections, the goal of a step-drawdown test data analysis is to determine the values of B, C, and n in Eq. (11-11). Miller and Weber (1983) developed a method for the evaluation of these parameters based on some derived analytical expressions. In the following sections, after presenting the analytical expressions the Miller and Weber method is presented with examples.

11.4.3.1 Governing Equations. Determination of three parameters, B, C, and n requires three independent equations. Equation (11-11) can be used for this purpose because these parameters are independent of discharge rate. Therefore, from Eq. (11-11), the three required equations may be written as

$$s_{w_1} = BQ_1 + CQ_1^n \tag{11-18}$$

$$s_{w_2} = BQ_2 + CQ_2^n \tag{11-19}$$

$$s_{w_3} = BQ_3 + CQ_3^n \tag{11-20}$$

where $Q_1 < Q_2 < Q_3$ are distinct discharge rates and $s_{w_1} < s_{w_2} < s_{w_3}$ are the corresponding resulting drawdowns.

Equations (11-18), (11-19), and (11-20) represent a highly nonlinear system of equations if the exponent n is greater than one. As a result, traditional solution methods such as the Newton–Raphson, fixed-point iteration, and bisection often fail to result in a uniformly convergent solution (Miller and Weber, 1983). As a result, Miller and Weber proposed the following manipulations for the aforementioned equations as a preliminary step for their method.

The elimination of B between Eqs. (11-18) and (11-19) gives

$$s_{w_1} - \frac{Q_1}{Q_2} s_{w_2} = C\left(Q_1^n - \frac{Q_1}{Q_2} Q_2^n\right) \tag{11-21}$$

Similarly, eliminating B between Eqs. (11-19) and (11-20) yields

$$s_{w_3} - \frac{Q_3}{Q_2} s_{w_2} = C\left(Q_3^n - \frac{Q_3}{Q_2} Q_2^n\right) \tag{11-22}$$

Miller and Weber (1983) defined the following group variables:

$$k_1 = s_{w_1} - \frac{Q_1}{Q_2} s_{w_2} \tag{11-23}$$

$$k_2 = s_{w_3} - \frac{Q_3}{Q_2} s_{w_2} \tag{11-24}$$

Dividing Eqs. (11-21) and (11-22), respectively, by the variables defined in Eqs. (11-23) and (11-24) and equating the resulting ratio gives

$$\frac{Q_1^n}{k_1} - \frac{Q_1}{k_1 Q_2} Q_2^n = \frac{Q_3^n}{k_2} - \frac{Q_3}{k_2 Q_2} Q_2^n \tag{11-25}$$

Rearranging and simplifying yields

$$\frac{k_2}{k_1}\left(\frac{Q_1}{Q_2}\right)^n - \frac{k_2}{k_1}\left(\frac{Q_1}{Q_2}\right) - \left(\frac{Q_3}{Q_2}\right)^n + \frac{Q_3}{Q_2} = 0 \tag{11-26}$$

A residual group is then defined by the following expression:

$$\epsilon = \frac{k_2}{k_1}\left(\frac{Q_1}{Q_2}\right)^n \tag{11-27}$$

As the magnitude of n increases, the residual ϵ becomes smaller because $Q_1 < Q_2$. Under limiting conditions

$$\lim_{n \to \infty} \epsilon = 0 \tag{11-28}$$

Substitution of Eq. (11-27) into Eq. (11-26) yields

$$\epsilon - \frac{k_2}{k_1}\left(\frac{Q_1}{Q_2}\right) - \left(\frac{Q_3}{Q_2}\right)^n + \frac{Q_3}{Q_2} = 0 \tag{11-29}$$

Solving Eq. (11-29) for n gives the following expression:

$$n = \frac{\log\left(\epsilon + \dfrac{Q_3}{Q_2} - \dfrac{k_2}{k_1}\dfrac{Q_1}{Q_2}\right)}{\log\left(\dfrac{Q_3}{Q_2}\right)} \tag{11-30}$$

11.4.3.2 Numerical Solution Procedure. Equation (11-30) is explicitly solvable as n approaches to infinity. As a result, for large values of n (> 5), which is a rare occurrence in practice, a simple approximate solution to Eqs. (11-18), (11-19), and (11-20) can be obtained. In general, n has values between 2 and 3 (Rorabaugh, 1953), and the impact of the values in this range becomes greater and may not be neglected. Despite this fact, a relatively simple solution is still available by coupling Eqs. (11-27) and (11-30) as (Miller and Weber, 1983)

$$\epsilon_i = \frac{k_2}{k_1}\left(\frac{Q_1}{Q_2}\right)^{n_{i-1}} \tag{11-31}$$

$$n_i = \frac{\log\left(\epsilon_i + \dfrac{Q_3}{Q_2} - \dfrac{k_2}{k_1}\dfrac{Q_1}{Q_2}\right)}{\log\left(\dfrac{Q_3}{Q_2}\right)} \tag{11-32}$$

where i is iteration level resulting from the repeated alternate solution of Eqs. (11-31) and (11-32). For each calculation, the latest values of ϵ and n are used. The value of n can be obtained after a few iterations. From the known value of n, the values of B and C can easily be determined. The application of the Miller and Weber method is shown with the following examples.

Example 11-7: This example is adapted from Miller and Weber (1983). The data for a step-drawdown test for a production well are given in Table 11-10. Using the Miller and Weber method, determine the values of B, C, and n.

Solution: Since three equations are needed to solve for the three unknowns, the first three set of values in Table 11-10 are selected as the representative points for the solution. First, k_1 and k_2 need to be calculated from Eqs. (11-23) and (11-24), respectively:

TABLE 11-10 Discharge and Drawdown Data for Example 11-7

Step No.	Discharge Rate, Q (gpm)	Drawdown, s_w (ft)
1	100	10.40
2	200	22.77
3	300	38.63
4	400	58.75

$$k_1 = s_{w_1} - \frac{Q_1}{Q_2} s_{w_2} = 10.40 \text{ ft} - \frac{100 \text{ gpm}}{200 \text{ gpm}}(22.77 \text{ ft}) = -0.985 \text{ ft}$$

$$k_2 = s_{w_3} - \frac{Q_3}{Q_2} s_{w_2} = 38.63 \text{ ft} - \frac{300 \text{ gpm}}{200 \text{ gpm}}(22.77 \text{ ft}) = 4.475 \text{ ft}$$

Introducing $\epsilon_1 = 0$ in Eq. (11-32) gives $n_1 = 3.274$. With this value, Eqs. (11-31) and (11-32) will yield $\epsilon_2 = -0.470$ ft and $n_2 = 2.946$. Similarly, the rest of values can be obtained as shown in Table 11-11. As can be seen from this table, n converges on a value of 2.810. Therefore, from Eq. (11-22)

$$C = \frac{s_{w_3} - \frac{Q_3}{Q_2} s_{w_2}}{Q_3^n - \frac{Q_3}{Q_2} Q_2^n} = \frac{38.63 \text{ ft} - \frac{300 \text{ gpm}}{200 \text{ gpm}}(22.77 \text{ ft})}{\left[(300 \text{ gpm})^{2.810} - \frac{300 \text{ gpm}}{200 \text{ gpm}}(200 \text{ gpm})^{2.810}\right]}$$

$$= 0.942 \times 10^{-6} \text{ min}^n/\text{ft}^{3n-1}$$

Substitution of the values of n and C into Eq. (11-18) gives

$$10.40 \text{ ft} = (100 \text{ gpm})B + \left(0.942 \times 10^{-6} \frac{\text{min}^n}{\text{ft}^{2n-1}}\right)(100 \text{ gpm})^{2.810}$$

or

$$B = 0.100 \text{ min/ft}^2$$

TABLE 11-11 Intermediate Results for the Solution Process of Example 11-7

i	ϵ_i	n_i
1	0	3.274
2	−0.470	2.946
3	−0.590	2.854
4	−0.628	2.825
5	−0.641	2.815
6	−0.646	2.811
7	−0.647	2.810
8	−0.648	2.810

TABLE 11-12 Comparison of the Observed Drawdowns with the Results of Miller and Weber Method for Example 11-7

Step No.	Discharge Rate, Q (gpm)	Observed Drawdown, s_w (ft)	Calculated Drawdown, s_w, by the Miller and Weber Method (ft)
1	100	10.40	10.39
2	200	22.77	22.75
3	300	38.63	38.61
4	400	58.75	59.31

Finally, substitution of the values of n, B, and C into Eq. (11-11) gives the following relationship:

$$s_w = 0.100Q + 0.942 \times 10^{-6} Q^{2.810}$$

where the units of s_w and Q are ft and gpm (gallons per minute), respectively. The results calculated from this equation are listed in Table 11-12. As observed from this table, the results determined by the Miller and Weber method are very close to the observed drawdown values.

Example 11-8: Using the Miller and Weber method, determine the values of n, B, and C for the data of Example 11-5 as given in Table 11-7. Compare the results with the ones determined by the Rorabaugh's method in Example 11-5.

Solution: Since three equations are needed to solve for the three unknowns (B, C, and n), three representative points are chosen from Table 11-7. Let us select the first three data points. First, k_1 and k_2 need to be calculated from Eqs. (11-23) and (11-24), respectively:

$$k_1 = s_{w_1} - \frac{Q_1}{Q_2} s_{w_2} = 25.4 \text{ ft} - \frac{1.21 \text{ ft}^3/\text{s}}{2.28 \text{ ft}^3/\text{s}}(50.4 \text{ ft}) = -1.347 \text{ ft}$$

$$k_2 = s_{w_3} - \frac{Q_3}{Q_2} s_{w_2} = 72.3 \text{ ft} - \frac{3.12 \text{ ft}^3/\text{s}}{2.28 \text{ ft}^3/\text{s}}(50.4 \text{ ft}) = 3.332 \text{ ft}$$

Introducing $\epsilon_1 = 0$ in Eq. (11-32) gives $n_1 = 3.145$. With this value, Eqs. (11-31) and (11-32) will yield $\epsilon_2 = -0.338$ ft and $n_2 = 2.715$. Similarly, the rest of values can be obtained as shown in Table 11-13. As can be seen from this table, n converges on a value of 2.451. Therefore, from Eq. (11-22) we obtain

$$C = \frac{s_{w_3} - \dfrac{Q_3}{Q_2} s_{w_2}}{Q_3^n - \dfrac{Q_3}{Q_2} Q_2^n} = \frac{72.3 \text{ ft} - \dfrac{3.12 \text{ ft}^3/\text{s}}{2.28 \text{ ft}^3/\text{s}}(50.4 \text{ ft})}{\left(3.12 \text{ ft}^3/\text{s}\right)^{2.451} - \dfrac{3.12 \text{ ft}^3/\text{s}}{2.28 \text{ ft}^3/\text{s}}\left(2.28 \text{ ft}^3/\text{s}\right)^{2.451}} = 0.560 \text{ s}^n/\text{ft}^{3n-1}$$

Substitution of the values of n and C into Eq. (11-18) gives

$$25.4 \text{ ft} = \left(1.21 \text{ ft}^3/\text{s}\right) B + \left(0.560 \text{ s}^n/\text{ft}^{2n-1}\right)\left(1.21 \text{ ft}^3/\text{s}\right)^{2.451}$$

TABLE 11-13 Intermediate Results for the Solution Process of Example 11-8

i	ϵ_i	n_i
1	0	3.145
2	−0.338	2.715
3	−0.444	2.567
4	−0.487	2.505
5	−0.507	2.476
6	−0.516	2.462
7	−0.521	2.456
8	−0.523	2.453
9	−0.524	2.451
10	−0.524	2.451

or

$$B = 20.253 \text{ s/ft}^2$$

Finally, substitution of the values of n, B, and C into Eq. (11-11) gives the following relationship:

$$s_w = 20.253Q + 0.560Q^{2.451}$$

where the units of s_w and Q are ft and ft^3/s, respectively.

The results of Rorabaugh's method are determined in Example 11-5. The values of n, B, and C determined by the Miller and Weber's and Rorabaugh's methods are presented in Table 11-14. The values compare favorable and the minor differences can be attributed to the graphical readings of Rorabaugh's method. Also, the predicted drawdown values by the two methods are presented in Table 11-15 which shows that the results are very close.

11.4.3.3 Extension of the Method to More Than Three Steps. As mentioned in Section 11.4.3.2, the Miller and Weber method based on the assumption that the step-drawdown relationship is defined by three steps. However, in practice, four or five, or more steps may be used in conducting a step-drawdown test. In order to apply the method for more than three steps cases, Miller and Weber (1983) suggests that we construct an approximate best-fit curve for the data points and then select three representative data points among other points and use the method described Section 11.4.3.2. Miller and Weber also

TABLE 11-14 Comparison of the Values for n, B, and C Obtained by Miller and Weber's and Rorabaugh's Methods for Example 11-8

Parameter	Miller and Weber Method	Rorabaugh Method
n	2.45	2.68
B (s/ft^2)	20.25	20.40
C ($\text{s}^n/\text{ft}^{3n-1}$)	0.56	0.44

TABLE 11-15 Comparison of the Predicted Drawdown Values Obtained by Miller and Weber's and Rorabaugh's Methods for Example 11-8

	Observed Values		Predicted Drawdown, s_w (ft)	
Step No., i	Q (ft^3/s)	s_w (ft)	Miller and Weber Method	Rorabaugh Method
1	1.21	25.4	25.400	25.417
2	2.28	50.4	50.399	50.518
3	3.12	72.3	72.296	72.933
4	3.39	80.3	79.819	80.754

suggests that we select different combinations of three points and solve each for n, B, and C values. Wide variation would indicate a deviation from the assumptions of the method.

11.4.4 Labadie and Helweg's Computer Method

An alternative to the previous approaches is a least-squares curve-fitting analysis for determining the values of B, C, and n in Eq. (11-11). The goal is to find those particular values of B, C, and n such that the actual drawdown observed in the field from the step-drawdown test is as close as possible to the drawdown calculated from Eq. (11-11). For this purpose, Labadie and Helweg (1975) developed a relatively simple computer algorithm for the analysis of step-drawdown test data, and this method as well as its application are presented below.

11.4.4.1 Governing Equations. Evaluation of the values of B, C, and n can be formulated as an optimization problem in the following form (Labadie and Helweg, 1975):

$$\underset{B,C,n}{\text{minimize}}\, E(B,C,n) = \min_{B,C,n} \sum_{i=1}^{N} \left[\left(BQ_i + CQ_i^n\right) - s_i\right]^2 \qquad (11\text{-}33)$$

where N is the total number of steps, Q_i is the discharge rate during step i of the test; s_i is the drawdown observed after step i of the test, and $E(B, C, n)$ is the squared fitting error as a function of chosen B, C, and n for given step-drawdown test data.

There are three conventional approaches to solve the problem described above. These approaches are discussed in Labadie and Helweg (1975). One of these approaches is based on the application of any of several standard nonlinear programming algorithms for solving unconstrained optimization problems. This approach uses methods that require an initial guess as to the best values for B, C, and n. A systematic search procedure then attempts to determine the best values that minimize $E(B, C, n)$, based on the trial guess given. According to Labadie and Helweg, for many problems this works fine. Therefore, this method is the one to be used.

It is assumed that the value of n is known. The remaining problem of determining B and C would then be easy. For the given n, Eq. (11-33) takes the form

$$E(n) = \min_{B,C} \sum_{i=1}^{N} \left[\left(BQ_i + CQ_i^n\right) - s_i\right]^2 \qquad (11\text{-}34)$$

To satisfy the minimization condition, one must have the following:

$$\frac{\partial E(n)}{\partial B} = \sum_{i=1}^{N} \left[\left(BQ_i + CQ_i^n \right) - s_i \right] Q_i = 0 \tag{11-35}$$

$$\frac{\partial E(n)}{\partial C} = \sum_{i=1}^{N} \left[\left(BQ_i + CQ_i^n \right) - s_i \right] nCQ_i^{n-1} = 0 \tag{11-36}$$

These are two simultaneous linear equations and the Cramer's rule can be applied, yielding $B^*(n)$ and $C^*(n)$. These are listed as a function of n since they are based on some given n, and will vary if n is changed. Eq. (11-34) can now be written as

$$E(n) = \sum_{i=1}^{N} \left[B^*(n)Q_i + C^*(n)Q_i^n - s_i \right]^2 \tag{11-37}$$

Initially, it is assumed that n is given. The next step is to find the minimum n value, namely,

$$\min_n E(n) \tag{11-38}$$

TABLE 11-16 Step-Drawdown Test Data for Example 11-9

Test Identification	Discharge Rate, Q	Drawdown, s_w
Case 1A:	400 gpm	8.6 ft
Sheahan (1971)	700	20.0
	1200	56.5
	1800	141.0
Case 1B:	25.2 liters/s	2.62 m
Sheahan (1971)	44.2	6.10
	75.7	17.22
	113.6	42.98
Case 2:	1.21 ft^3/s	25.4 ft
Rorabaugh (1953)	2.28	50.4
	3.12	72.2
	3.39	80.3
Case 3:	100 gpm	2.5 ft
Bierschenk (1963)	200	6.5
	400	22.2
	500	34.9
	550	46.5
Case 4:	100 gpm	10.40 ft
Miller and Weber (1983)	200	22.77
	300	38.63
	400	58.75

TABLE 11-17 Comparison of the Values of *n*, *B*, and *C* Values Obtained by Labadie and Helweg's, Rorabaugh's, and Miller and Weber's Methods for Example 11-9

Test Identification	Rorabaugh (1953)	Labadie and Helweg (1975)	Miller and Weber (1983)
Case 1A[a]:	$n = 2.8$	$n = 2.759$	—
Sheahan (1971)	$B = 0.017$	$B = 0.017$	—
	$C = 8.2 \times 10^{-8}$	$C = 11.52 \times 10^{-8}$	—
Case 2[b]:	$n = 2.68$	$n = 2.901$	$n = 2.45$
Rorabaugh (1953)	$B = 20.40$	$B = 20.62$	$B = 20.25$
	$C = 0.44$	$C = 0.298$	$C = 0.56$
Case 4[c]:	—	$n = 2.658$	$n = 2.810$
Miller and	—	$B = 0.099$	$B = 0.1$
Weber (1983)	—	$C = 0.233 \times 10^{-5}$	$C = 0.94 \times 10^{-6}$

[a]The results for Rorabaugh's method are given in Labadie and Helweg (1975). [b]The results for Rorabaugh's and for Miller and Weber's methods are given in Example 11-8 (Table 11-14). [c]The results for Miller and Weber's method are given in Example 11-7.

11.4.4.2 Numerical Solution Procedure. Labadie and Helweg (1975) applied a simple one-dimensional grid-search technique for solving this problem in order to determine the optimal n^* value. For each n specified in the search procedure, Eqs. (11-35) and (11-36) are solved and Eq. (11-37) is evaluated. An important advantage is that n will generally lie in a narrow range (between 1 and 4 at the most) and this restricts the range over which the search is carried out. Labadie and Helweg concluded that the efficiency of the search technique is dependent upon the structure of $E(n)$ and the function of $E(n)$ is smooth and convex for all cases.

Labadie and Helweg (1975) developed a computer program, called FASTEP. The listing of this program is given in Labadie and Helweg (1975). In this program, an initial value $n = 1.1$ is chosen and n is iteratively increased by Δn until $E(n)$ begins to increase. At this point the direction is reversed and Δn is set to 0.1, and so on in this manner, until a desired order of accuracy for n^* is determined. The program requires data for Q_i and s_i, $i = 1, \ldots, N$.

The results of the following example were generated using a modified version of the FASTEP program. The modified program is called FASTEPPC, and modifications were made by the author of this book. Modifications in FASTEPPC are only related with the input data instructions and output structure; all the rest of part of the original FASTEP program including its algorithm were kept the same.

Example 11-9: Using the Labadie and Helweg's computer method, determine the values of n, B, and C for the data given in Table 11-16. Compare the results with the ones obtained by the methods of Rorabaugh, and of Miller and Weber.

Solution: The results in Table 11-17 are obtained by running a modified version of the FASTEP computer program using the data given in Table 11-16. The values of n, B, and C obtained by the methods of Labadie and Helweg, Rorabaugh, and Miller and Weber agree reasonably well.

PART V

HYDRAULICS OF SLUG TESTS AND HYDROGEOLOGIC DATA ANALYSIS METHODS

CHAPTER 12

FULLY AND PARTIALLY PENETRATING WELLS IN AQUIFERS

12.1 INTRODUCTION

The purpose of slug tests is to measure the horizontal hydraulic conductivity and storage coefficients of aquifers under field conditions. Slug tests are a quick and relatively inexpensive way of evaluating hydraulic conductivity values compared with pumping tests because using observation wells and pumping the well are not necessary. This is why slug tests became so popular in aquifer evaluation studies.

Slug tests may be categorized as *rising head tests* and *falling head tests*. Both tests are conducted by causing a sudden change in the water level in a well or piezometer by removing or introducing a known volume of water. For rising head tests, a known volume or slug water is suddenly removed from the well, after which the water level in the well is measured as a function of time. For falling head tests, a known volume of water or slug is suddenly injected into the well, and the fall of the water level is measured as a function of time.

Since the 1950s, several two-dimensional analytical/semianalytical mathematical models have been developed for the analysis of slug test data. The three models and their associated data analysis methods, which are known as the Hvorslev method (Hvorslev, 1951), the Cooper, Bredehoeft, and Papadopulos method (Cooper et al., 1967), and the Bouwer and Rice method (Bouwer and Rice, 1976), are widely used in environmental, hydrogeological, and geotechnical investigations. One of the main limitations of these two-dimensional analytical models is that they do not take into account partially penetrating well conditions. In other words, the horizontal planes passing through the upper and lower ends of the screen interval are assumed to be impervious, and therefore the flow is purely horizontal. Since the 1990s, this important assumption as well as other assumptions have been investigated by some researchers using three-dimensional analytical and numerical models for the purpose of establishing some guidelines for practitioners. The results of these studies showed that the distance between the water table (or the upper boundary of a confined aquifer) and the upper boundary of the screen interval, the distance between the lower boundary of the screen interval and the impermeable lower boundary of the aquifer, the length of the screen interval, the permeability of skin around the screen interval, and anisotropy may result in underpredicted or overpredicted formation hydraulic conductivities by the aforementioned two-dimensional models. These studies undoubtedly increased the potential usage of the two-dimensional analytical models because the errors resulting from their usage can be estimated based on the results of the three-dimensional models.

In this chapter the mathematical theories and practical applications of the above-mentioned three analytical models, along with the results of their evaluations based on the three-dimensional models, will be presented in detail.

12.2 BOUWER AND RICE SLUG TEST MODEL

12.2.1 Theory

In the following sections the theory and application of the Bouwer and Rice method are presented in detail (Bouwer and Rice, 1976; Bouwer, 1989).

Geometry and symbols of a well in an unconfined aquifer are shown in Figure 12-1. The theory of the Bouwer and Rice slug test analysis method is based on the following assumptions:

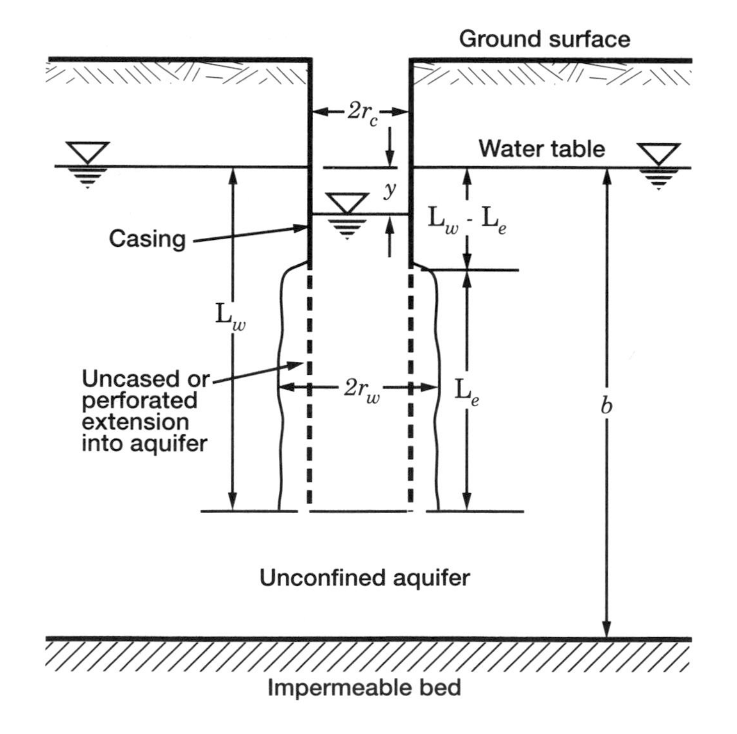

Figure 12-1 Geometry and symbols of Bouwer and Rice slug test model. (After Bouwer and Rice, 1976.)

1. Water is removed from the piezometer or added to the piezometer instantaneously.
2. The aquifer is homogeneous and isotropic.
3. Darcy's law is valid.
4. The aquifer extends to infinity in all directions.
5. The position of the water table does not change with time.
6. Flow above the water table (in the capillary fringe) can be ignored.
7. The aquifer is uniform with depth.
8. Head losses as water enters the well (well losses) are negligible.
9. The specific storage of the formation is negligible.
10. The formation is isotropic with respect to hydraulic conductivity.
11. A zone of disturbance created by drilling or development around the well does not exist.
12. Flow toward the screen interval is horizontal.
13. The horizontal planes passing through the upper and lower ends of the screen interval are impermeable.

The rate of ground-water flow into the well when the water level in the well is a distance y lower than the static ground-water level (Figure 12-1) around the well is calculated with the Thiem equation (Thiem, 1906; see also section 3.2.1) as

$$Q = 2\pi K_r L_e \frac{y}{\ln\left(\dfrac{R_e}{r_w}\right)} \tag{12-1}$$

where Q is flow rate into the well, K_r is the horizontal hydraulic conductivity in the radial direction from the well, L_e is the length of screened, perforated, or otherwise open section of the well, y is the vertical difference between the water level inside the well and the static water table outside the well, R_e is the effective radial distance over which y is dissipated, and r_w is the radial distance of undisturbed portion of aquifer from centerline.

The value of R_e depends on the geometry of the flow system, and its values were determined with an electrical resistance network analog for different values of r_w, L_e, L_w, and H (Bouwer and Rice, 1976). Refer to Figure 12-1 for geometrical meaning of the symbols. The value of r_w is the radius of the screened or open section of the well plus the thickness of a sand gravel pack. Therefore, r_w is the radial distance from the center of well to the edge of the undisturbed aquifer. The thickness of the developed zone is almost never known. For this reason, the tendency is to ignore it and take only gravel or sand packs into account.

Bouwer and Rice (1976) derived the equations for the withdrawal case. As will be discussed later, the same equations can also be used for the adding case. For rising head tests, the rate of rise, dy/dt, of the water level in the well after suddenly removing a slug of water can be related to the inflow Q by the equation

$$\frac{dy}{dt} = -\frac{Q}{\pi r_c^2} \tag{12-2}$$

where r_c is the radius of the casing or other section of the well where the rise of the water level is measured. The minus sign in Eq. (12-2) is introduced because y decreases as t increases.

If the water level rises in the screened or open section of the well with a gravel pack around it, the thickness and porosity of the gravel envelope should be taken into account to calculate the equivalent value of r_c for the rising water level. This calculation is based on the total free-surface area in the well and sand or gravel pack:

$$\pi r_{C_{eq}}^2 = \pi r_c^2 + \pi \left(r_w^2 - r_c^2 \right) n \tag{12-3}$$

or the equivalent radius ($r_{C_{eq}}$) is

$$r_{C_{eq}} = \left[(1-n) r_c^2 + n r_w^2 \right]^{1/2} \tag{12-4}$$

where n is the porosity.

Combining Eqs. (12-1) and (12-2) yields the governing differential equation:

$$\frac{dy}{y} = - \frac{2 K_r L_e}{r_c^2 \ln \left(\dfrac{R_e}{r_w} \right)} \tag{12-5}$$

which can be integrated to

$$\ln(y) = - \frac{2 K_r L_e t}{r_c^2 \ln \left(\dfrac{R_e}{r_w} \right)} + C \tag{12-6}$$

where C is a constant. The initial condition is

$$y = y_0, \qquad \text{at } t = 0 \tag{12-7}$$

and

$$y = y_t, \qquad \text{at } t = t \tag{12-8}$$

From Eqs. (12-6), (12-7), and (12-8), the following equation can be obtained:

$$K_r = \frac{r_c^2}{2 L_e} \ln \left(\frac{R_e}{r_w} \right) \frac{1}{t} \ln \left(\frac{y_0}{y_t} \right) \tag{12-9}$$

Equation (12-9) enables for calculation of K_r from the rise of the water level in the well after suddenly removing a slug of water from the well or from the fall of the water level in the well after suddenly adding a slug of water to the well. Because K_r, r_c, r_w, R_e, and L_e in Eq. (12-9) are constants, $(1/t)\ln(y_0/y_t)$ must also be a constant. Because y_t and t are the only variables in Eq. (12-9), a plot of $\ln(y_t)$ versus t must show a straight line. Thus, instead of calculating K_r on the basis of two measurements of y and t (y_0 at $t = 0$ and y_t at t), a number of y and t measurements can be taken and $[\ln(y_0/y_t)]/t$ determined as the slope of the best-fitting line through the y versus t points on semilogarithmic paper. The straight line through the data points can also be used to select two values of y, namely y_0 and y_t, along with the time interval t between them for substitution into Eq. (12-9). Eq. (12-9) can also be expressed as

$$m = \frac{1}{t} \ln \left(\frac{y_0}{y_t} \right) = \frac{2K_r L_e}{r_c^2 \ln \left(\frac{R_e}{r_w} \right)} \tag{12-10}$$

in which m is a constant. From Eq. (12-10) one can write

$$\ln(y_t) = -mt + \ln(y_0) \tag{12-11}$$

Therefore, as mentioned above, m can be determined graphically from the y_t versus t points on semilogarithmic paper.

The transmissivity (T) of the aquifer can be determined by multiplying the both sides of Eq. (12-9) by the thickness of the aquifer (b):

$$T = \frac{br_c^2}{2L_e} \ln \left(\frac{R_e}{r_w} \right) \frac{1}{t} \ln \left(\frac{y_0}{y_t} \right) \tag{12-12}$$

Equations (12-9) and (12-12) are all dimensionally correct. Thus, K_r and T are expressed in the same units as the length and time parameters in the equations. Values of R_e were experimentally determined for different values of r_w, L_e, L_w, and H (Bouwer and Rice, 1976), and the results are presented in the next section.

12.2.2 Evaluation of R_e

Using an electrical analog model, Bouwer and Rice (1976) provided a convenient set of curves relating the effective radial distance (R_e) to other known well dimensions. Values of R_e, expressed as $\ln(R_e/r_w)$, were determined for different values of r_w, L_e, and D (see Figure 12-1), using the same assumptions as those for Eq. (12-1), the Thiem equation. For a partially penetrating well ($L_w < b$) Bouwer and Rice obtained

$$\ln \left(\frac{R_e}{r_w} \right) = \left\{ \frac{1.1}{\ln \left(\frac{R_e}{r_w} \right)} + \frac{A + B \ln \left[\frac{(b - L_w)}{r_w} \right]}{\frac{L_e}{r_w}} \right\}^{-1} \tag{12-13}$$

where A and B are dimensionless coefficients that are functions of L_e/r_w as shown in Figure 12-2. The experiments indicated the effective upper limit of $\ln[(H - L_w)/r_w]$ is 6. For a fully penetrating well ($L_w = b$)

$$\ln \left(\frac{R_e}{r_w} \right) = \left[\frac{1.1}{\ln \left(\frac{L_w}{r_w} \right)} + \frac{C}{\frac{L_e}{r_w}} \right]^{-1} \tag{12-14}$$

where C is dimensionless coefficient plotted in Figure 12-2 as a function of L_e/r_w.

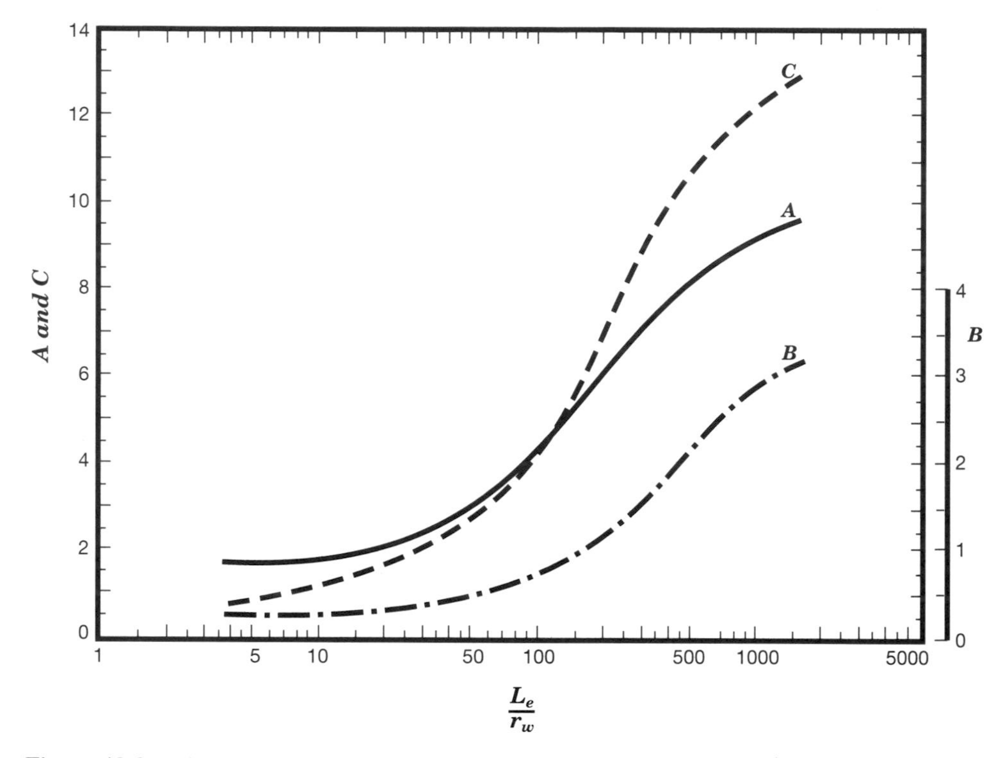

Figure 12-2 Dimensionless parameters A, B, and C as a function of L_e/r_w for calculation of $\ln(R_e/r_w)$. (After Bouwer and Rice, 1976.)

12.2.3 Data Analysis Methodology

Bouwer and Rice (1976) developed a method for the determination of horizontal hydraulic conductivity using the measured values of head difference (y) as a function of time (t). The theoretical background is presented in Section 12.2.1. Here, the methodology of data analysis will be presented in a stepwise manner for both rising and falling head tests. The steps are given below:

Step 1: Plot the measured data of y_t versus t for each well on a semilogarithmic paper as having y on the logarithmic axis.

Step 2: Because y_t and t are the only variables in Eq. (12-9), the plot of Step 1 must show a straight line. In other words, the straight-line portion is the valid part of the readings. Determine the intersection point (y_0) with the $\log(y_t)$ axis. If there is *double straight-line effect*, follow the guidelines described in Section 12.2.4. Compare this y_0 with the measured one.

Step 3: Using an arbitrarily selected t value, read the corresponding y_t value from the plot generated in the previous steps.

Step 4: Using the known values of t, y_0, and y_t, calculate $1/t[\ln(y_0/y_t)]$ in Eq. (12-9).

Step 5: Based on the values of L_w and H in Figure 12-1, select the equation to be used for $\ln(R_e/r_w)$ from Eqs. (12-13) and (12-14).

Step 6: If Eq. (12-13) is to be used, determine the values of A and B from Figure 12-2 using the value of the ratio L_e/r_w (partially penetrating well case). If Eq. (12-14) is to be used, determine the value of C from Figure 12-2 using the value of the ratio L_e/r_w (fully penetrating well case).

Step 7: Calculate the value of $\ln(R_e/r_w)$ using Eq. (12-13) or (12-14) with the values of Step 6.

Step 8: Check if the water level rises in the screened or open section of the well. If this is the case, use Eq. (12-4) to calculate $r_{c_{eq}}$. Use $r_{c_{eq}}$ instead of r_c in Eq. (12-9).

Step 9: Calculate the value of hydraulic conductivity from Eq. (12-9) using the values of r_c, L_e, $\ln(R_e/r_w)$, and $1/t \ln(y_0/y_t)$.

12.2.4 Guidelines for the Applicability of the Bouwer and Rice Method

The slug test data analysis method developed by Bouwer and Rice permits the evaluation of saturated horizontal hydraulic conductivity of aquifers with a single well. Furthermore, Bouwer (1989) has discussed some critical aspects with regard to applicability of the method. These are all presented in the following sections.

12.2.4.1 Double Straight-Line Effect. When plotting $\log(y_t)$ versus t, the points may produce a double straight line as shown schematically in Figure 12-3. The straight lines are shown as AB and BC. The first part (AB) is straight and deep, whereas the next part (BC) is straight and less steep. At point C, the points start to deviate from the straight line as the drawdown around the hole becomes significant compared with y_t. The first straight line (AB) is probably due to a highly permeable zone around the well (gravel pack or developed zone), which quickly sends water into the well immediately after the water level in the well has been lowered (Figure 12-4a). If the ground-water level is above the screened or open section of the hole, and the water level in the hole does not drop below the top of the open section (Figure 12-4b), the gravel pack zone around the open section cannot drain. Under this condition, the double straight-line effect should not occur. The second straight line (BC) is more indicative of the flow from the undisturbed aquifer into the well. Therefore, segment BC should be used in calculating K_r of the aquifer from Eq. (12-9).

12.2.4.2 Falling Head Test. The method can also be applied for a falling head test provided that the equilibrium water level is above the screened or open section of the borehole (Figure 12-4b). Falling water level tests can be performed by instantaneously adding a known volume of water to the hole and measuring the subsequent fall of water level as a function of time. In this case, the outflow from the well due to falling water levels occurs only through the screened or open section of the well, and the aquifer is a true reverse of the flow system for the rising water level after a slug of water has been removed. Thus, Eqs. (12-9), (12-13), and (12-14) are also applicable for a falling head test as well.

If the equilibrium water level in the borehole is below the top of the screen or open section (Figure 12-5) and water is added (hatched section in Figure 12-5), the subsequent flow of water into the aquifer due to the falling water level travels not only through the screen or perforations below the original water table, but also through the vadose zone above the original water table (arrows in Figure 12-5). This increases the rate of fall of the water level in the borehole beyond that caused by inflow into the aquifer and results an overestimated hydraulic conductivity value. The grater the ratio of y/L, the more the slug test will overestimate hydraulic conductivity.

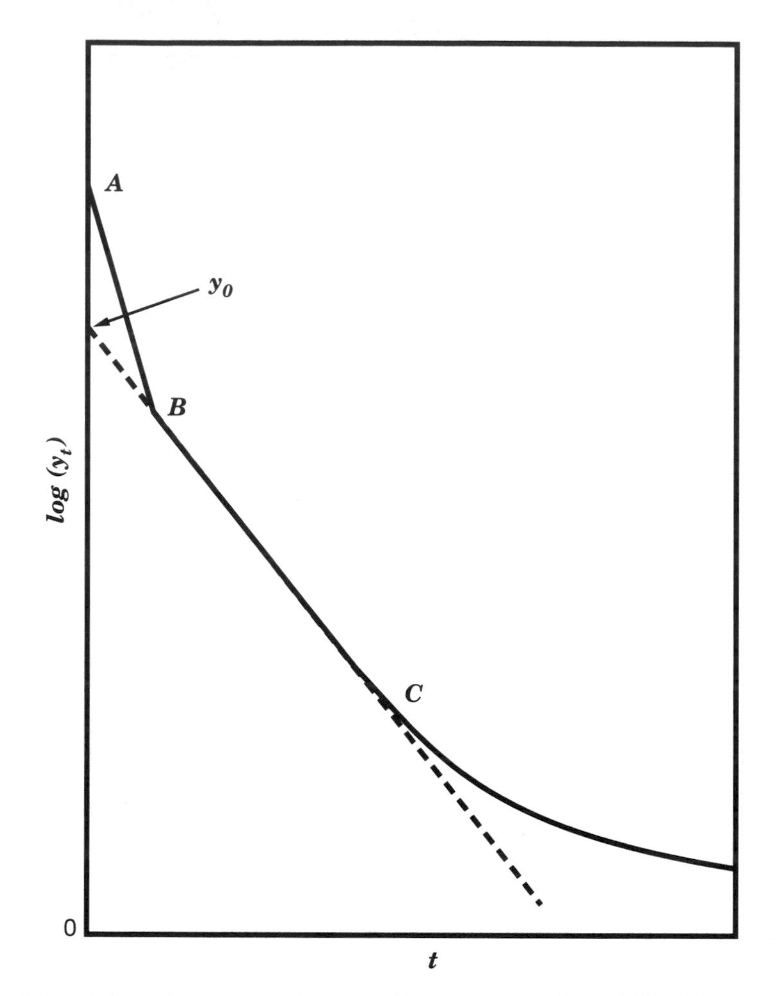

Figure 12-3 Schematic representation of double straight line effect. (After Bouwer, 1989.)

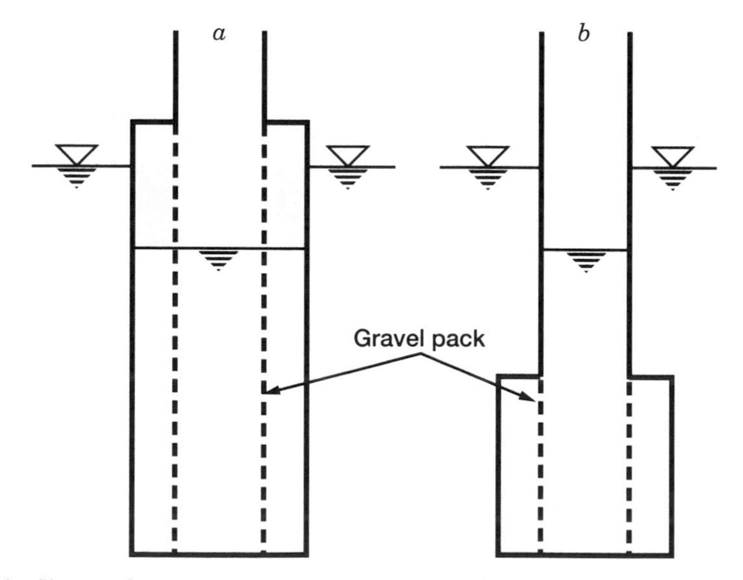

Figure 12-4 Slug test for (a) the well water level below top of screen or perforated section and (b) the well water level above the top of the screen or perforated section. (After Bouwer, 1989.)

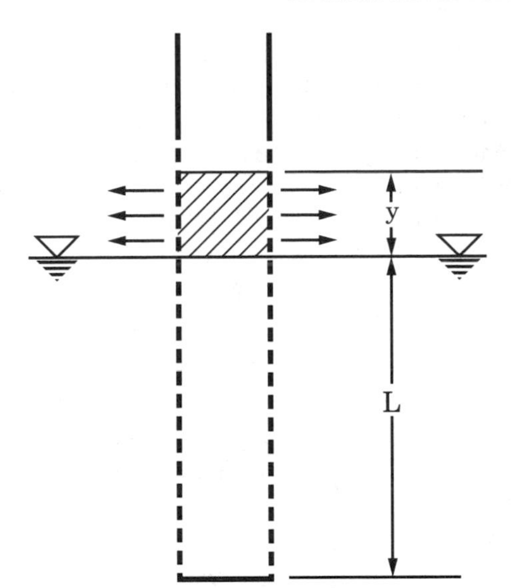

Figure 12-5 Slug test with the equilibrium water level below the top of the screen or perforated section with flow through the unsaturated zone. (After Bouwer, 1989.)

12.2.4.3 Application to Confined Aquifers. Bouwer (1989) has extended the application of slug test to confined aquifers as well. Theoretically, the slug test applies to aquifers where the upper boundary is a plane source (rising head test) or sink (falling-head test), as in an unconfined aquifer. Because most of the head difference (y) between the static water table and the water level in the well is dissipated in the vicinity of the well around the screen or perforated section, the method should also be applicable to situations where the upper boundary of the aquifer is an impermeable or semipermeable plane—that is, an impermeable or semipermeable upper confining layer. From this logic, Bouwer (1989) concludes that the slug test should also give reasonable values for hydraulic conductivity in confined, semiconfined, or stratified aquifers. Theoretically, the larger the distance between the top of the screened or open section of the well and the upper confining layer ($L_w - L_e$ in Figure 12-1), the more accurate the resulting values of K_r will be.

12.2.4.4 Effect of Well Diameter. Theoretically, the Bouwer and Rice method can be applied to any diameter of the borehole. However, from a practical standpoint, the hole dimensions should be selected such that the geometrical parameters are covered by Figure 12-2. The larger the values of r_w and L_e (Figure 12-1), the larger the portion of the aquifer on which K_r is determined. For layered aquifers, smaller values of L_e may sometimes be preferable because they give more resolution about the vertical distribution of K_r when the slug test is carried out at different depths. Very small hole diameters—for example, 5 cm.—should still give accurate values of K_r. But the values apply to only a small region around the well and, therefore, are more sensitive to spatial variability. Also, inaccuracies in the estimates of the thickness of gravel envelopes and developed zones have a greater effect on the calculated values of K_r.

12.2.4.5 Estimating the Rate of Rise or Fall of Water Level in a Well. If the water level in a well rises or falls at a relatively slow rate, which is the case for relatively low permeability materials, the measurement procedure is relatively simple. On the other hand, if the formation has relatively high permeability, the water level in the well moves relatively fast and the process requires some special attention. For these kinds of cases, it may be helpful to get some idea regarding the rate of water level movement before starting the test. The rate of rise or fall of water level in a well can be estimated from Eq. (12-9) by solving it for t (Bouwer and Rice, 1976; Bouwer, 1989):

$$t = \frac{r_c^2}{2K_r L_e} \ln\left(\frac{R_e}{r_w}\right) \ln\left(\frac{y_0}{y_t}\right) \tag{12-15}$$

Consider that $t_{90\%}$, which is the time required for the water level in the well to rise or fall 90% of the initial lowering or raising, respectively, needs to be known. According to Figure 12-1, this means that $y_t = 0.1y_0$. Substitution of this expression into Eq. (12-15) gives

$$t_{90\%} = \frac{r_c^2}{2K_r L_e} \ln\left(\frac{R_e}{r_w}\right) \ln(10) \tag{12-16}$$

or

$$t_{90\%} = 1.15\frac{r_c^2}{K_r L_e} \ln\left(\frac{R_e}{r_w}\right) \tag{12-17}$$

where K_r must be taken as the estimated or expected value of K_r of the aquifer.

12.2.4.6 Pressure Distribution Around the Screened Section. The Bouwer and Rice method assumes that the Darcy flux through the screened or perforated section of the pipe (refer to Figure 12-1) is constant and does not change with depth. This situation can easily be seen from Eq. (12-1), which is the foundation of the method. The relative magnitude of the initial pressure at the upper end of the screened portion is critically important. For a falling head test, $y/(L_w - L_e)$ is a measure for this pressure. For a rising head test, $y/(L_w - L_e - y)$ is the corresponding value. Theoretically, the larger the $(L_w - L_e)$ value, the more accurate the resulting values of K_r will be. Special attention needs to be paid while constructing the well in order to maximize the $(L_w - L_e)$ value.

12.2.4.7 Application

Example 12-1: This example is taken from Bouwer and Rice (1976). A slug test was performed in alluvial deposits of the Salt River bed west of Phoenix, Arizona. The geometry of the aquifer and well is shown in Figure 12-6. A solid cylinder with a volume equivalent to a 0.32-m change in water level in the well was also placed below the water level. When the water level had returned to equilibrium, the cylinder was quickly removed. The measured values of y_t versus time (t) are given in Table 12-1. The theoretical value of y_0 calculated from the displacement of the submerged cylinder is 0.32 m. Applying the Bouwer and Rice slug test data analysis method, determine the horizontal hydraulic conductivity (K_r) of the aquifer.

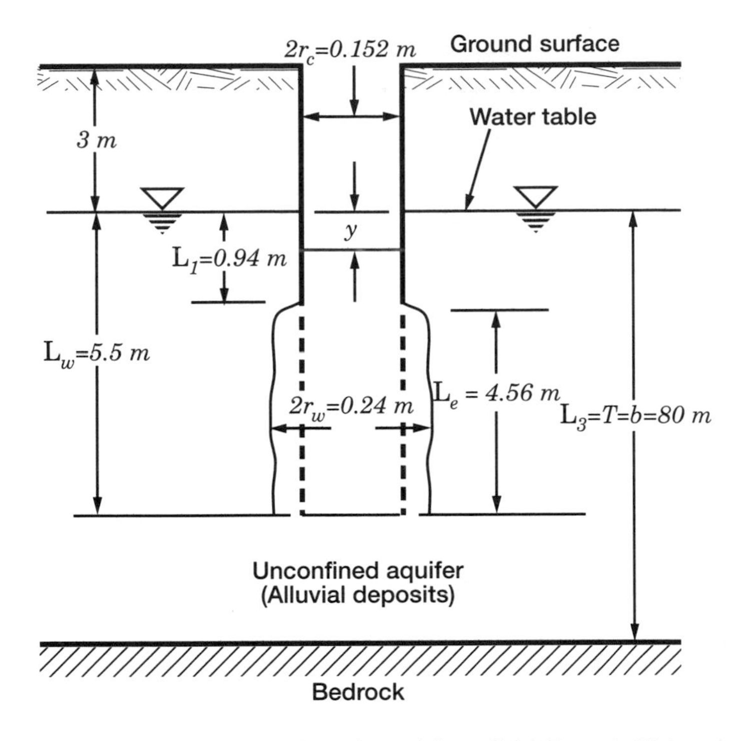

Figure 12-6 Geometrical dimensions of the well for Example 12-1.

Solution:

Step 1: The graph y_t versus t is given in Figure 12-7.

Step 2: From Figure 12-7, the coordinate of the intersection point (y_0) with the log y_t axis is 0.29 m. Figure 12-7 does not show double straight-line effect.

TABLE 12-1 Rising-Head Test Data for Example 12-1

t (s)	y_t (m)	y_t/y_0
0.0	0.290	1.000
0.7	0.238	0.821
1.4	0.190	0.655
2.8	0.150	0.517
3.7	0.117	0.403
6.0	0.072	0.248
8.8	0.030	0.103
12.8	0.012	0.041
18.6	0.005	0.017
28.6	0.002	0.007
38.6	0.001	0.003

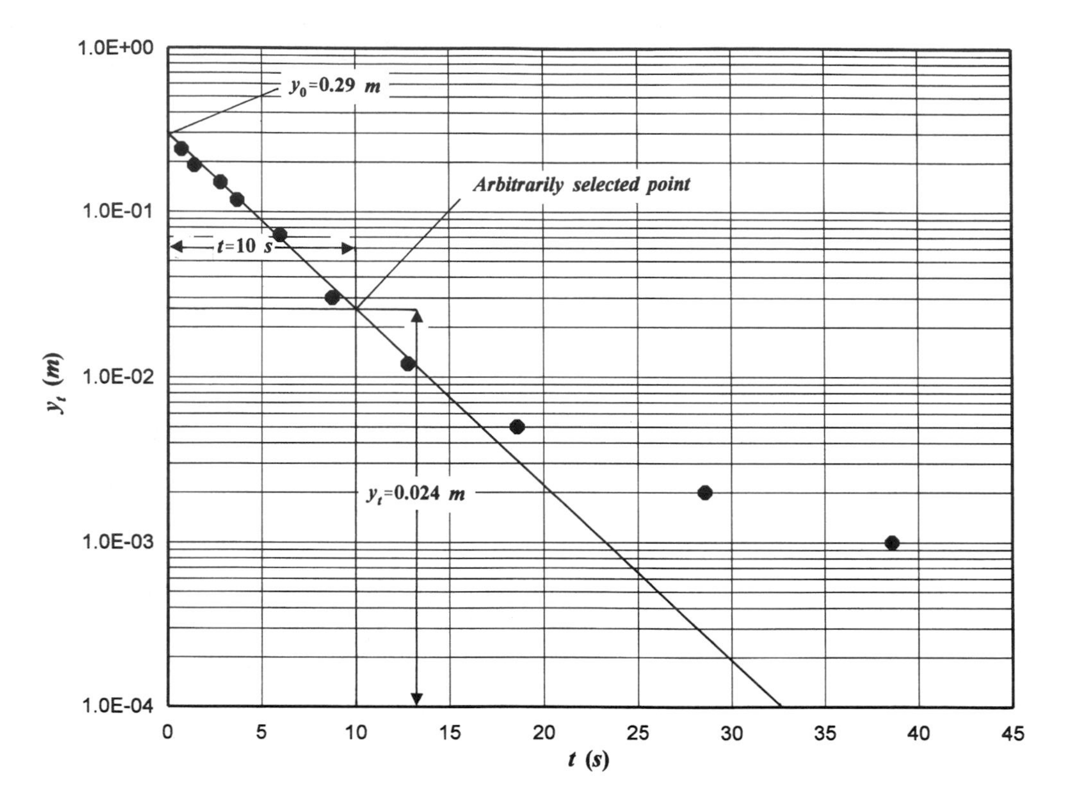

Figure 12-7 Semilogarithmic plot of y_t versus t for Example 12-1.

Step 3: The coordinates of the arbitrarily selected point are

$$t = 10 \text{ s}, \qquad y_t = 0.024 \text{ m}$$

Step 4: From the values of the previous steps we obtain

$$\frac{1}{t} \ln\left(\frac{y_0}{y_t}\right) = \frac{1}{10 \text{ s}} \ln\left(\frac{0.29 \text{ m}}{0.024 \text{ m}}\right) = 0.249 \text{ s}^{-1}$$

Step 5: Because $L_w = 5.5$ m $< H = 80$ m, Eq. (12-13) should be used for $\ln(R_e/r_w)$.

Step 6: $L_e/r_w = 4.56 \text{ m}/0.12 \text{ m} = 38$. Therefore, from Figure 12-2, $A = 2.6$ and $B = 0.42$.

Step 7: Substituting the values into Eq. (12-13) gives

$$\ln\left(\frac{R_e}{r_w}\right) = \left\{ \frac{1.1}{\ln\left(\dfrac{5.5 \text{ m}}{0.12 \text{ m}}\right)} + \frac{2.6 + 0.42 \ln\left[\dfrac{(80 \text{ m} - 5.5 \text{ m})}{0.12 \text{ m}}\right]}{\dfrac{4.56 \text{ m}}{0.12 \text{ m}}} \right\}^{-1} = 2.34$$

Step 8: The water level does not rise in the screened section of the well. Therefore, Eq. (12-4) is not required if we want to calculate an equivalent radius ($r_{C_{eq}}$).

Step 9: Introducing the values of the previous steps into Eq. (12-9) gives the value of hydraulic conductivity:

$$K_r = \frac{(0.076\ \text{m})^2}{2(4.56\ \text{m})}(2.34)(0.249\ \text{s}^{-1}) = 0.00037\ \text{m/s} = 32\ \text{m/day}$$

This value agrees with the K_r values of 10 and 53 m/day obtained with the tube method on two nearby observation wells (Bouwer and Rice, 1976).

12.2.5 Evaluation of the Bouwer and Rice Model with Other Models and Cautions

The assumptions expressed in the items 6, 9, 10, 11, 12, and 13 in Section 12.2.1 are critically important for the Bouwer and Rice model. As a result, the model was evaluated by a different group of investigators during the 1990s to quantify the error that is introduced into the horizontal hydraulic conductivity estimate from a given slug test data based on the Bouwer and Rice method. The first investigation was conducted by Hyder and Butler (1995) based on a three-dimensional semianalytical model developed by Hyder et al. (1994). The second investigation was conducted by Brown and Narasimhan (1995) using a three-dimensional numerical ground-water flow model based on an integral finite-difference algorithm.

In the following sections, after presenting the foundations of the three-dimensional semianalytical and numerical models, the results of three evaluation cases of the Bouwer and Rice model are presented.

12.2.5.1 Description of the Models Used in the Evaluation Process

Description of the Three-Dimensional Semianalytical Model. As mentioned above, Hyder and Butler (1995) evaluated the Bouwer and Rice model using the three-dimensional semianalytical model developed by Hyder et al. (1994). In order to better understand the evaluation results, the governing equation and initial and boundary conditions of the authors' model (Figure 12-8) are presented below.

The governing equation in radial coordinates for $h(r, z, t)$, which is also given by Eq. (2-174) of Chapter 2, is

$$\frac{\partial^2 h}{\partial r^2} + \frac{1}{r}\frac{\partial h}{\partial r} + \left(\frac{K_z}{K_r}\right)\frac{\partial^2 h}{\partial z^2} = \frac{S_s}{K_r}\frac{\partial h}{\partial t} \tag{12-18}$$

where the z coordinate increases downward direction and is equal to zero at the water table. The initial conditions are

$$h(r, z, 0) = 0, \qquad r_w < r < \infty, \quad 0 < z < b \tag{12-19}$$

$$H(0) = H_0 \tag{12-20}$$

The boundary conditions are

$$h(\infty, z, t) = 0, \qquad t > 0, \quad 0 \le z \le b \tag{12-21}$$

$$h(r, 0, t) = 0, \qquad r_w < r < \infty, \quad t > 0 \tag{12-22}$$

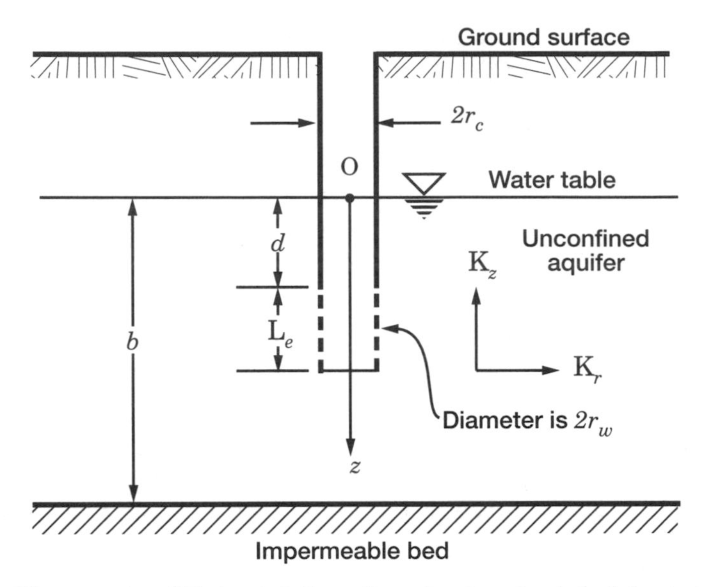

Figure 12-8 The geometry of Hyder et al. three-dimensional semianalytical slug test model. (After Hyder et al., 1994.)

$$\frac{\partial h(r,b,t)}{\partial z} = 0, \qquad r_w < r < \infty, \quad t > 0 \tag{12-23}$$

$$\frac{1}{L_e}\int_d^{d+L_e} h(r_w, z, t)\, dz = H(t), \qquad t > 0 \tag{12-24}$$

$$2\pi r_w K_r \frac{\partial h(r_w, z, t)}{\partial r} = \frac{\pi r_c^2}{L_e}\frac{\partial H(t)}{\partial t}, \qquad t > 0, \quad d \le z \le d + L_e \tag{12-25}$$

The physical meaning of the above-given initial and boundary conditions are presented below (refer also Section 2.9.3 of Chapter 2): Equation (12-19) states that initially the hydraulic head (h) in the formation is zero (water table is the datum). Equation (12-20) states that initial water level above the water table in the casing is equal to H_0. Equation (12-21) states that at large distances from the well the head (h) is zero. Equation (12-22) expresses the condition that the water table is a constant-head boundary (Dirichlet condition). Equation (12-23) states that the lower boundary of the formation is an impermeable boundary. Equation (12-24) expresses the assumption that over the screen interval, $H(t)$ does not change with the vertical coordinate z. Finally, Eq. (12-25) states that the rate of flow of water into the aquifer or out of the aquifer through the screen interval is equal to the rate of decrease or to the rate of increase in the volume of water inside the well.

The evaluation studies with the three-dimensional analytical model were conducted using dimensionless parameters. For this purpose, Hyder and Butler (1995) defined the following dimensionless parameters:

$$\alpha = \frac{2r_w^2 S_s L_e}{r_c^2} \tag{12-26}$$

$$\beta = \frac{L_e}{r_w} \tag{12-27}$$

$$\gamma = \frac{K_{BR}}{K_r} \tag{12-28}$$

where α is the dimensionless storage parameter, β is the aspect ratio, K_{BR} is the estimated horizontal hydraulic conductivity by the Bouwer and Rice method, and K_r is the actual horizontal hydraulic conductivity.

Description of the Three-Dimensional Numerical Model. As mentioned above, Brown and Narasimhan (1995) used a three-dimensional numerical ground-water flow model based on an integral finite-difference algorithm. For the evaluation purpose, Brown and Narasimhan solved several hypothetical slug test problems using known values of K_r and S_s; they also generated the $y(t)$ as a function of t, which represents the change in water level in a well during a slug test. Then, the authors used the simulated $y(t)$ values to estimate the K_r values using the Bouwer and Rice method.

Except where varied in the parametric studies, the authors' base case parameters are as follows (refer to Figures 12-1 and 12-8 for the parameters): $K_r = 1.0 \times 10^{-6}$ m/s, $S_s = 1.0 \times 10^{-3}$, $r_c = r_w = 0.08$ m, $L_e = 0.50$ m, $b = 62.0$ m, $L_w = 5.0$ m, $d = 4.5$ m, and $y_0 = 0.50$ m. Therefore, $L_w - L_e = 4.5$ m, which means that the water level in the well remained above the top of the screen in all cases. The system was considered to be homogeneous and isotropic, and therefore, the hydraulic conductivity values $K_r = K_z = 1.0 \times 10^{-6}$ m/s were used in the model. The estimated K_{BR} value from the Bouwer and Rice method was compared with the K_r value used in the numerical model, and the authors calculated the error according to the relation

$$E = \frac{K_{BR} - K_r^{\prime}}{K_r^{\prime}} \tag{12-29}$$

where E is the percent error and $K_r^{\prime}$ is the K_r specified in the numerical model.

Based on the above-mentioned semianalytical and numerical models, the results of the three evaluation cases of the Bouwer and Rice model are presented in the following paragraphs.

12.2.5.2 Case I: Thick Unconfined Aquifer Having the Screen Interval at Its Center.

This is the case that the screen interval of the well is located at the center of a relatively thick isotropic unconfined aquifer. The results of Hyder and Butler (1995) showed that the measure of the error, K_{BR}/K_r, basically depends on the aspect ratio, anisotropy, and storage parameter, and the results of each are presented in the following sections.

Dependence on Aspect Ratio (L_e/r_w). The aspect ratio (L_e/r_w) versus K_{BR}/K_r for different values of α, having values from 0.1 to 1.0×10^{-7}, are shown in Figure 12-9. As can be seen from Figure 12-9, for typical aspect ratios encountered in practice, K_{BR}/K_r has values around 0.75 when α is less than 0.001. For relatively large values of α, which correspond low permeable clay-dominated formations, K_{BR}/K_r can have values as high as 4 for the same aspect ratio values. As a conclusion, it can be said that the Bouwer and Rice model does not appear appropriate for these types of formations (Hyder and Butler, 1995).

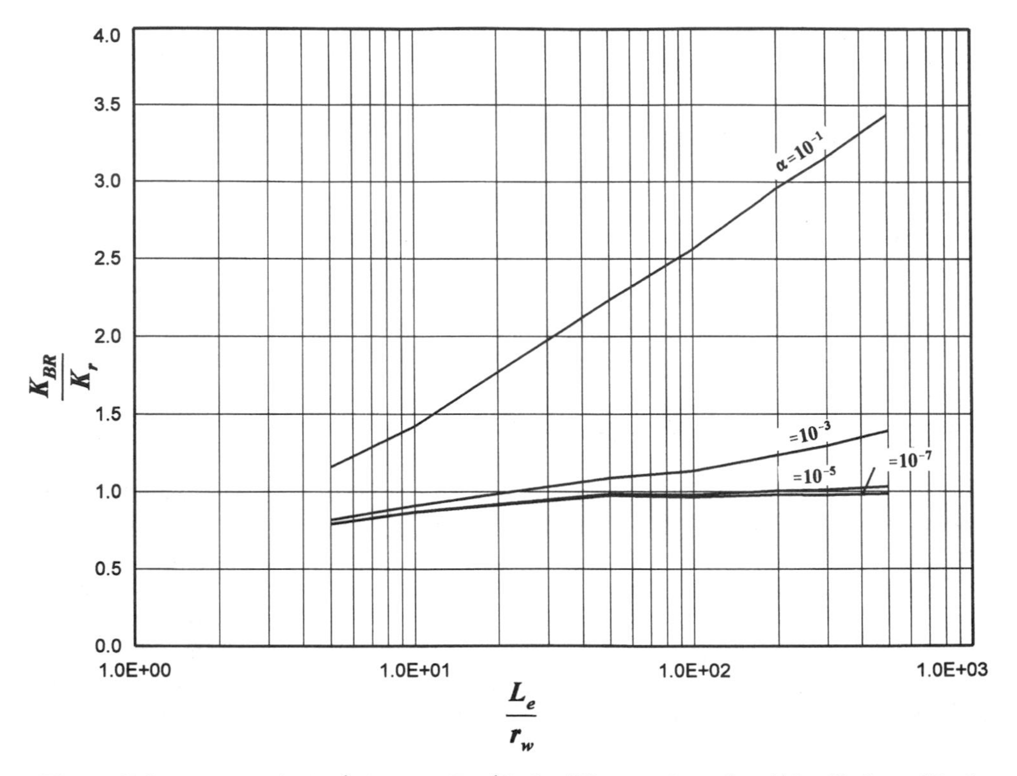

Figure 12-9 Aspect ratio (L_e/r_w) versus K_{BR}/K_r for different values of α. (After Hyder and Butler, 1995.)

Dependence on Anisotropy (K_z/K_r). Natural formations are generally anisotropic in hydraulic conductivity ($K_r > K_z$). The anisotropy ratio (K_z/K_r) versus K_{BR}/K_r for different values of aspect ratio (L_e/r_w) are shown in Figure 12-10. The results show that the Bouwer and Rice model considerably underpredicts the horizontal hydraulic conductivity values as the degree of anisotropy increases. This effect increases at smaller aspect ratios because the horizontal flow assumption through the screen interval will be violated significantly (see Section 12.2.1). As a result, Hyder and Butler (1995) state that the Bouwer and Rice data analysis method must be used with caution for slug tests conducted in formations that are expected to exhibit a considerable degree of anisotropy. If a significant degree of anisotropy is expected and the aspect ratio is moderate or large (i.e., $L_e/r_w > 50$), the authors state that the Cooper et al. (1967) model results, as presented in Section 12.4, should result in horizontal hydraulic conductivity values with greater reliability than the Bouwer and Rice method. [The Cooper et al. (1967) model results are designated as CBP in Figure 12-10.]

Dependence on the Storage Parameter ($2r_w^2 S_s L_e/r_c^2$). The Bouwer and Rice model is based on the assumption that the storage effects can be neglected (see Section 12.2.1). The storage parameter (α) versus K_{BR}/K_r for different aspect ratio (L_e/r_w) values are shown in Figure 12-11. As can be seen from Figure 12-11, K_{BR}/K_r has values between 0.80 and 1.40 for α values less than 0.001. But, the error increases significantly for α values greater than 0.001 for moderate to large aspect ratios because K_{BR}/K_r values can be as high as 4. Under these conditions, Hyder et al. (1994) state that the Cooper et al. (1967) model, as presented in Section 12.4, should be preferred for slug test data analysis.

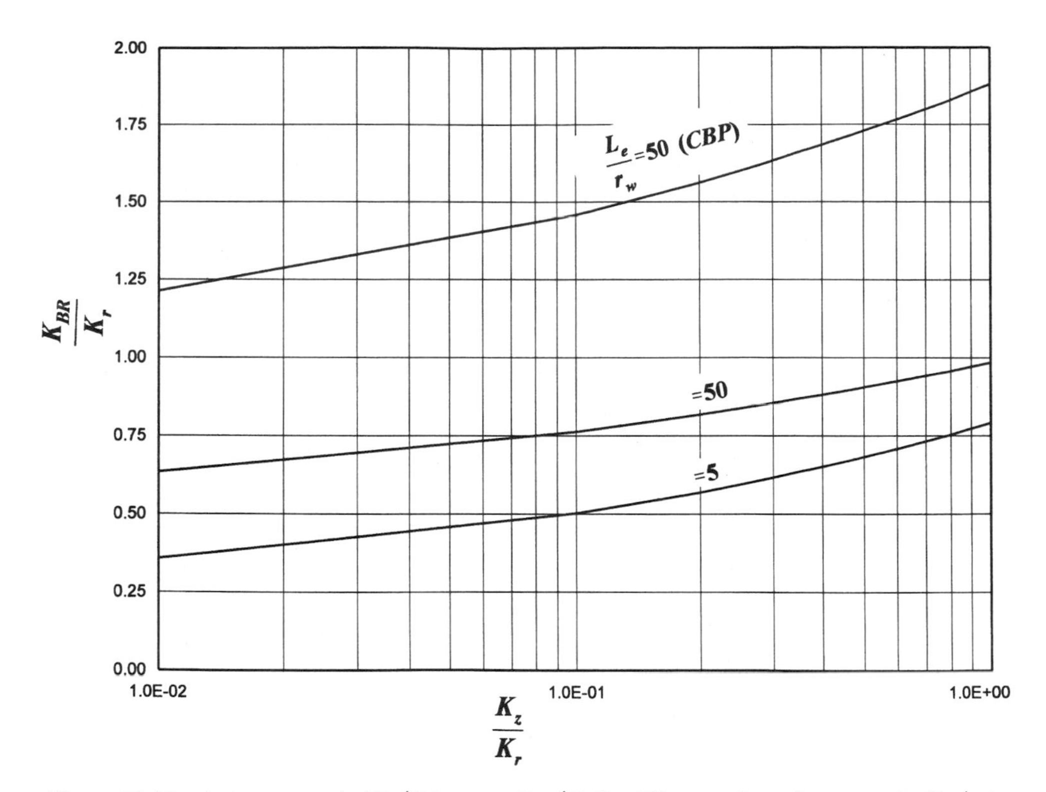

Figure 12-10 Anisotropy ratio (K_z/K_r) versus K_{BR}/K_r for different values of aspect ratio (L_e/r_w). (After Hyder and Butler, 1995.) CBP: Cooper, Bredehoeft, and Papadopulos (1967) model.

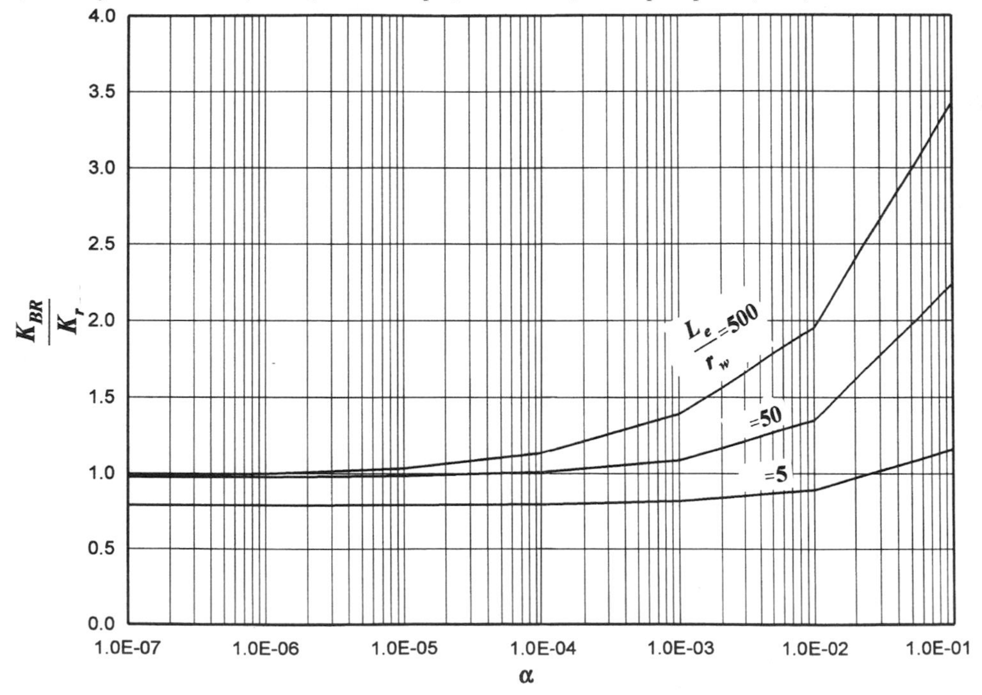

Figure 12-11 Storage parameter (α) versus K_{BR}/K_r for different values of aspect ratio (L_e/r_w). (After Hyder and Butler, 1995.)

12.2.5.3 Case II: Screen Interval Close to the Water Table or a Lower Impermeable Boundary. In many field situations, a well may be screened close to the water table or to a lower impermeable boundary. For this kind of cases, by using the semi-analytical and numerical models mentioned above, Hyder and Butler (1995) and Brown and Narasimhan (1995) investigated the effects of the distance from water table, the screen length, the distance to a lower boundary, and the specific storage on the horizontal hydraulic conductivity derived from the Bouwer and Rice method. The results of these investigations are presented below.

Dependence on Distance from Water Table and Screen Length. The distance below water table (d/L_e) versus K_{BR}/K_r for different aspect ratio (L_e/r_w) values are shown in Figure 12-12 (Hyder and Butler, 1995). Figure 12-12 shows that as d/L_e becomes smaller, the Bouwer and Rice model underpredicts horizontal hydraulic conductivity values and as d/L_e increases K_{BR}/K_r approaches 1 depending on the value of L_e/r_w.

Brown and Narasimhan (1995), whose numerical model is described in Section 12.2.5.1, used $L_e = 0.5$ m and 3.0 m and the base values given in Section 12.2.5.1 in the three-dimensional numerical model, and they found the K_{BR} values to be 8.1×10^{-7} m/s and 9.0×10^{-7} m/s, respectively. The corresponding errors according to Eq. (12-29) are $E =$ 19% and 10%, respectively. Substitution of these values into Eq. (12-29) yields $K_{BR}/K_r' =$ 1.19 and 1.10, respectively. These values will be compared with those derived from the results of Hyder and Butler (1995) which are presented in Figure 12-13. The aspect ratios

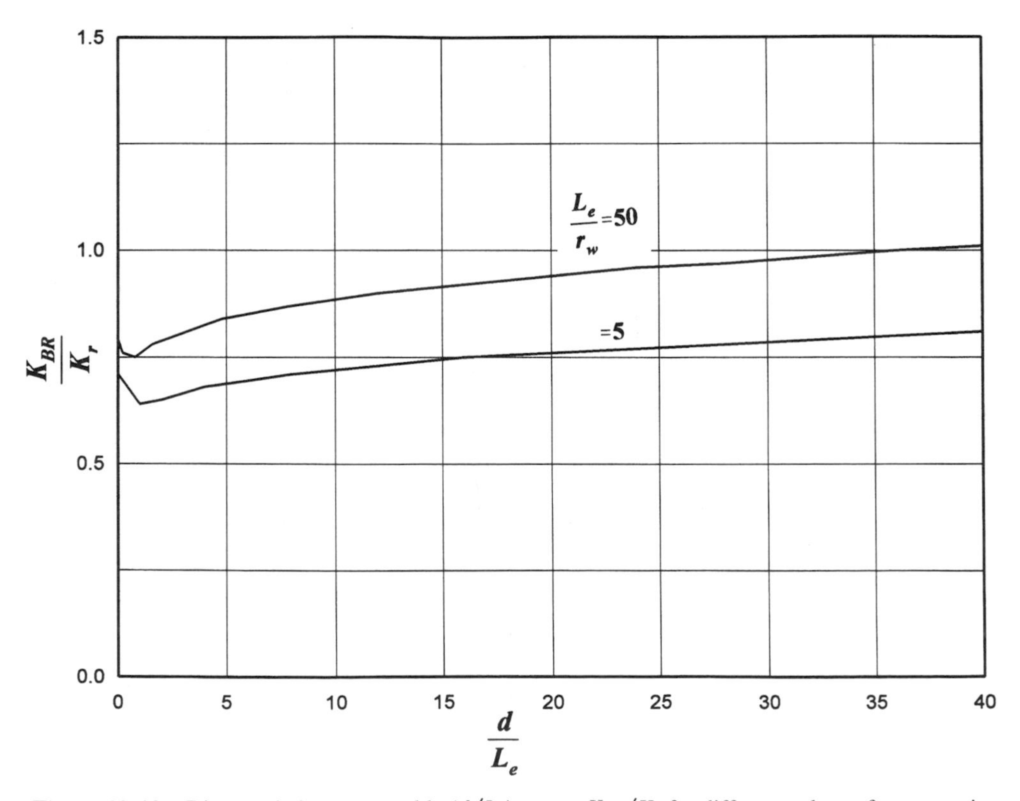

Figure 12-12 Distance below water table (d/L_e) versus K_{BR}/K_r for different values of aspect ratio (L_e/r_w). (After Hyder and Butler, 1995.)

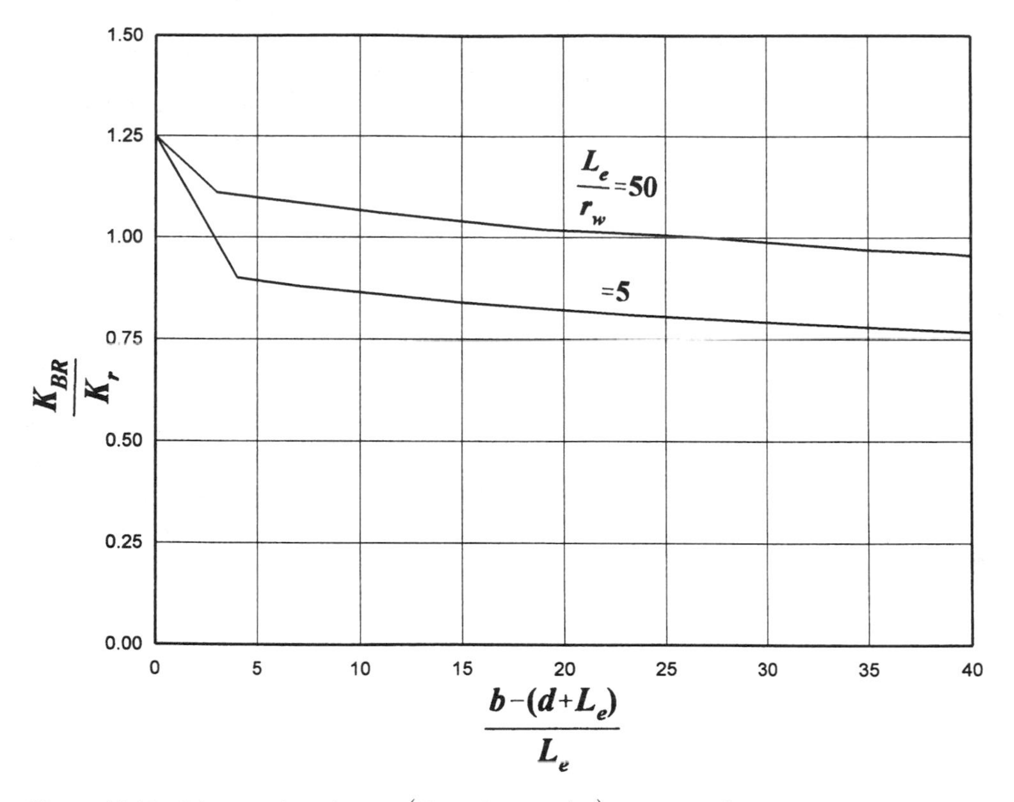

Figure 12-13 Distance above bottom $\left([b-(d+L_e)]/L_e\right)$ versus K_{BR}/K_r for different aspect ratio (L_e/r_w). (After Hyder and Butler, 1995.)

are $L_e/r_w = (0.5 \text{ m})/(0.08 \text{ m}) = 6.25$ and $L_e/r_w = (3.0 \text{ m})/(0.08 \text{ m}) = 37.5$ and, $d/L_e = (4.5 \text{ m})/(0.5 \text{ m}) = 9$ and $d/L_e = (2.0 \text{ m})/(3.0 \text{ m}) = 0.67$ for $L_e = 0.5$ m and $L_e = 3.0$ m respectively. Using these values, with an approximate evaluation, it can be seen from Figure 12-12 that K_{BR}/K_r has values between 0.70 and 0.80. These values are not too much off from the values derived from the numerical model despite the fact that the numerical values of Hyder and Butler (1995) and Brown and Narasimhan (1995) are not exactly the same. As a conclusion, it can be said that the results of the two different groups of authors are in agreement with each other.

Dependence on Distance from Lower Impermeable Boundary. There may be field situations for which a well may be screened quite close to a lower impermeable boundary such as bedrocks or less pervious formations. The distance from lower impermeable boundary $([b-(d+L_e)]/L_e)$ versus K_{BR}/K_r for different aspect ratio (L_e/r_w) values are shown in Figure 12-13 (Hyder and Butler, 1995). Figure 12-13 shows that as the distance becomes smaller, the values of K_{BR} and K_r approach each other regardless of the values of L_e/r_w. Also, as the distance increases, still they do not deviate significantly.

Dependence on Specific Storage. Brown and Narasimhan (1995) evaluated the effects of specific storage on the estimated horizontal hydraulic conductivity values from the Bouwer and Rice method using the numerical model and its parameters described in Section 12.2.5.1. The authors used specific storage values 10^{-5} m^{-1}, 10^{-4} m^{-1}, and

10^{-3} m^{-1} and reported that K_{BR} values are 7.9×10^{-7} m/s, 8.1×10^{-7} m/s, and 8.1×10^{-7} m/s, respectively. The corresponding errors are 19%, 19%, and 21%, respectively. These results show that the specific storage does not have significant effect on the K_{BR} results when the well screen interval is close to a lower impermeable boundary.

12.2.5.4 Case III: Screen Interval at the Center of a Thick Formation and Surrounded by a Skin of Different Permeability.

The term *skin effect* is used to describe the changes in permeability near the well bore as a result of drilling activities. A well bore skin can have either lower or higher permeability than the formation itself, depending on the drilling method, the well emplacement procedure, and the characteristics of the formation. Potential skin effects must be well understood to determine representative hydraulic conductivity values based on slug test data analysis methods. For this purpose, studies were carried out by different investigators for the determination of the skin effects. In the following sections the results of Hyder and Butler (1995) based on a three-dimensional semianalytical model and Brown and Narasimhan (1995) based on a three-dimensional numerical model are presented.

Dependence on Low-Permeability Skin. For most cases, the permeability of the formation near the well bore may be reduced as a result of drilling and completion practices. Hyder and Butler (1995) evaluated the effects of low-permeability skin on the results produced by the Bouwer and Rice model, and their results are presented in Figure 12-14. Figure 12-14 presents the aspect ratio (L_e/r_w) versus K_{BR}/K_r for $K_{\text{skin}}/K_r = 0.1$ and 0.01, where K_{skin} is the skin hydraulic conductivity. As can be seen from Figure 12-14, the Bouwer and Rice

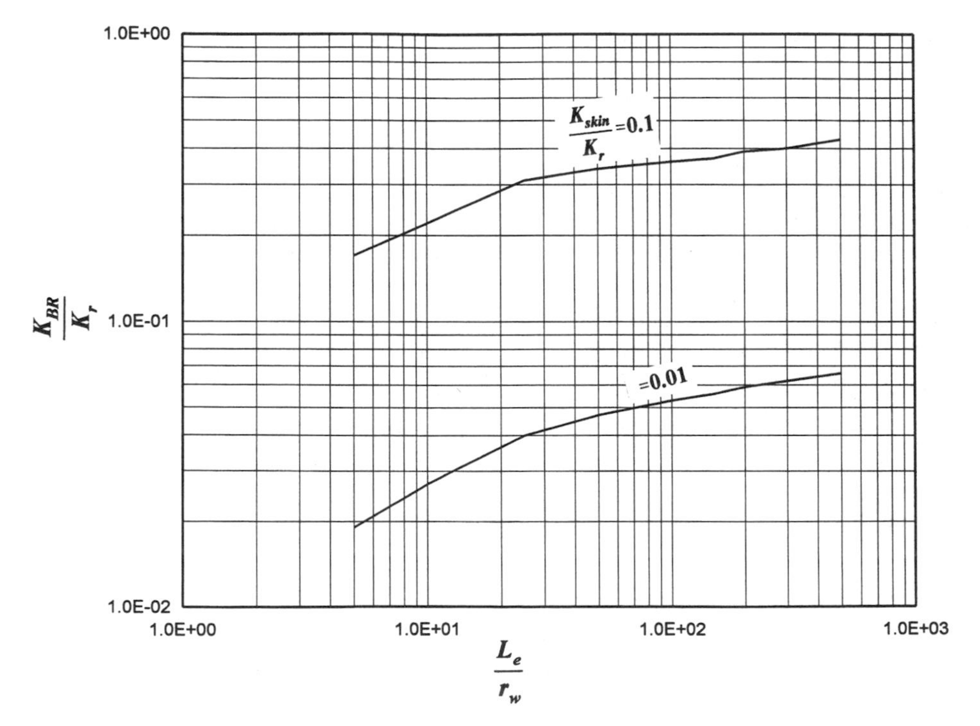

Figure 12-14 Aspect ratio (L_e/r_w) versus K_{BR}/K_r for low-permeability skin case for different values of K_{skin}/K_r. (After Hyder and Butler, 1995.)

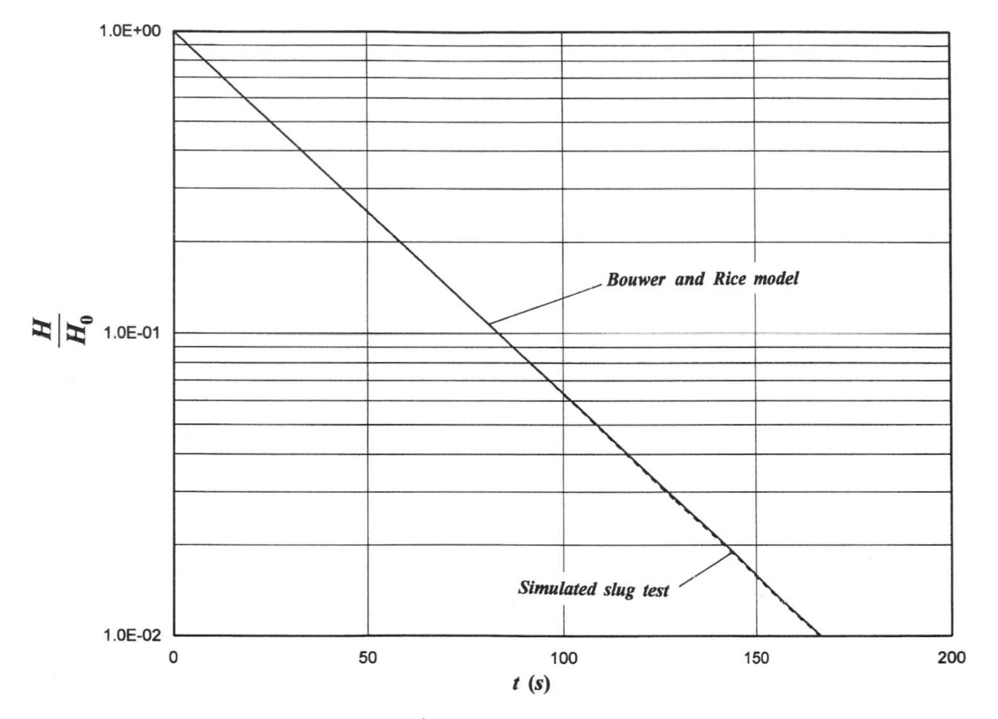

Figure 12-15 Time versus simulated H/H_0 and the best-fit Bouwer and Rice model for the case of a skin two orders of magnitude less permeable than the formation ($L_e/r_w = 50$, $r_w = r_c = 0.05$ m, $r_{\text{skin}} = 0.10$ m, $K_r = K_z = 0.001$ m/s, $K_{\text{skin}} = 0.01K_r$, $S_s = 1.0 \times 10^{-5}$ m^{-1}; well screened at the center of a thick formation). (After Hyder and Butler, 1995.)

hydraulic conductivity values are heavily weighted toward the values of the skin. However, the value of K_{BR}/K_r increases with the increasing aspect ratio, which is due to the fact that at higher aspect ratios the importance of the vertical flow becomes less. The results of Brown and Narasimhan (1995) based on a three-dimensional numerical model described in Section 12.2.5.1 are in agreement with these conclusions.

Additional analysis conducted by Hyder and Butler (1995) showed that an excellent Bouwer and Rice model fit to a set of slug test data is not always a definitive proof that the assumptions of the Bouwer and Rice model are fulfilled. The authors' results are shown in Figure 12-15, which is representative of all low-permeability skin cases shown in Figure 12-14. As can be seen from Figure 12-15, the Bouwer and Rice model generates an excellent fit to the simulated data. Based on some additional simulations, the authors state that the Bouwer and Rice fit for the low-permeability skin case is almost always better than that for the homogeneous formation case. Finally, the authors suggest that the possibility that the behavior is reflective of a low-permeability skin must first be discounted prior to acceptance of the validity of the estimated hydraulic conductivity values.

Dependence of High-Permeability Skin. In Figure 12-16 Hyder and Butler (1995) presents the aspect ratio (L_e/r_w) versus K_{BR}/K_r for the case of well skin is one and two orders of magnitude more permeable than the formation. The authors conducted two series of analyses on the same set of simulated responses. In the first series, the screen radius is assumed to the nominal radius of the well screen. In the second series, the authors performed

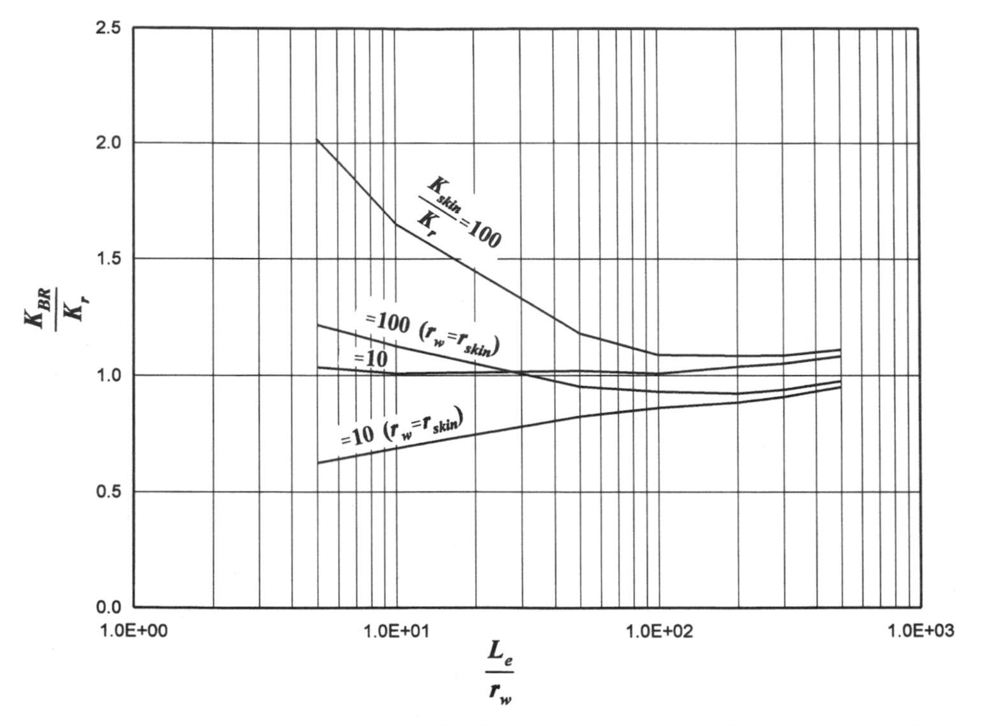

Figure 12-16 Aspect ratio (L_e/r_w) versus K_{BR}/K_r for the high-permeability skin case for different values of K_{skin}/K_r. (After Hyder and Butler, 1995.)

the analysis by assuming that the screen radius is equal to the radius of the skin, which is recommended by Bouwer and Rice (1976). Figure 12-16 shows that the Bouwer and Rice model provides reasonable values with the exception of wells with small aspect ratios. The results of Brown and Narasimhan (1995) based on a three-dimensional numerical model described in Section 12.2.5.1 are in agreement with these conclusions.

12.3 HVORSLEV SLUG TEST MODEL

Hvorslev (1951) was the first to point out that a slug test can be used to determine *in situ* hydraulic conductivity. The Hvorslev method can be used for both rising head and falling head tests. As mentioned in Section 12.1, for rising head tests, a known volume or slug water is suddenly removed from the well and the water level in the well is measured as function of time. For falling head tests, a known volume of water or slug is suddenly injected into the well, and the fall of water is measured as function of time. In the following sections, the theory and application of the Hvorslev method are presented.

12.3.1 Theory for a Fully-Penetrated Well in a Confined Aquifer

Geometry and symbols of a slug tested well are shown in Figure 12-17. The derivation of governing equations in the Hvorslev (1951) paper proceeds quite differently than that

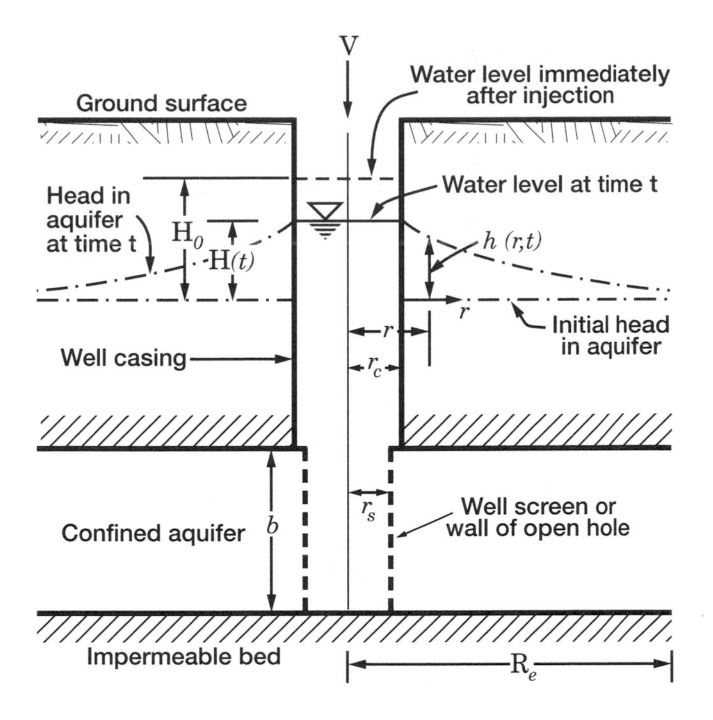

Figure 12-17 Schematic representation of Hvorslev (1951) and Cooper et al. (1995) slug test models.

following, which is based on Chirlin (1989). The theory of the Hvorslev slug test analysis method is based on the following assumptions:

1. Water is removed from the piezometer or added to the piezometer instantaneously.
2. The medium is homogeneous and isotropic.
3. Both soil and water are incompressible (this is equivalent in assuming no aquifer storage).
4. The medium extends to infinity in all directions.
5. The position of the water table or piezometric head of the confined aquifer (equilibrium level) does not change with time.
6. Flow above the water table (in the capillary fringe) can be ignored.
7. Head losses as water enters the well are negligible.
8. The horizontal planes passing through the upper and lower ends of the screen interval are impermeable.

Hvorslev (1951) presented slug test models for a number of different well and aquifer geometries. The slug test shown in Figure 12-17 has wide application in practice and is similar in many respects to the Bouwer and Rice slug test method (Section 12.2) and Cooper et al. slug test method (Section 12.4). The equations corresponding to Figure 12-17 are presented below based on Chirlin (1989).

The governing differential equation for steady-state ground-water flow in radial coordinates is (see Section 2.8.2 of Chapter 2)

$$\frac{\partial (r q_r)}{r \partial r} = 0, \qquad r > r_s, \quad t > 0 \tag{12-30}$$

where r is the radial coordinate and the Darcy flux (q_r) in the radial direction is

$$q_r = -K_r \frac{\partial h}{\partial r}, \qquad r > r_s, \quad t > 0 \tag{12-31}$$

The boundary conditions are

$$h(r,t) = H(t), \qquad r = r_s, \quad t > 0 \tag{12-32}$$

$$\left(K_r \frac{\partial h}{\partial r} \right) (2\pi r b) = \pi r_c^2 \frac{dH(t)}{dt}, \qquad r = r_s, \quad t > 0 \tag{12-33}$$

$$h(r,t) = 0, \qquad r = R_e, \quad t > 0 \tag{12-34}$$

$$h(r,t) = 0, \qquad r > r_s, \quad t = 0 \tag{12-35}$$

where q_r is Darcy velocity in the radial direction, K_r is aquifer horizontal hydraulic conductivity, h is the aquifer piezometric head, $H(t)$ is the water elevation in the well, H_0 is the water elevation in the well at $t = 0$, r is the radial coordinate, r_c is the radius of the well casing within which water level changes occur, R_e is the effective radial distance over which the aquifer head is dissipated, r_s is the radius of well screen, and t is time. The initial head level in Figure 12-17 is considered to be as the datum.

Equation (12-32) states that after the first instant the piezometric head in the aquifer at the face of the well is equal to that in the well. Equation (12-33) expresses the fact that the rate of flow of water into (or out of) an aquifer is equal to the rate of decrease (or increase) in the volume of water within the well. Equation (12-34) states that the head in the aquifer beyond the effective radial distance (R_e) is zero. Equation (12-35) states that initially the piezometric head is zero at every point in the aquifer.

Combinations of Eqs. (12-30) and (12-31) give

$$\frac{\partial^2 h}{\partial r^2} + \frac{1}{r} \frac{\partial h}{\partial r} = 0 \tag{12-36}$$

Equation (12-36) means that the storage effects are zero, which is an assumption of the theory as given above. The physical implication is that at any specific time, the total flux across an imaginary cylinder centered on the well is the same for cylinders of every radius $r_s \leq r \leq R_e$. From Eq. (12-33)

$$2\pi b K_r r \frac{\partial h}{\partial r} = \pi r_c^2 \frac{dH(t)}{dt}, \qquad r > r_s, \quad t > 0 \tag{12-37}$$

Note that Eq. (12-37) is written for $r > r_s$ and $t > 0$ for the assumption given above. Making the substitution

$$r \frac{\partial h}{\partial r} = \frac{\partial h}{\partial (\ln r)} \tag{12-38}$$

and using the independence of this factor from r, one can rewrite without approximation that

$$\frac{\partial h}{\partial(\ln r)} = \frac{\Delta h}{\Delta(\ln r)} = \frac{h(R_e) - h(r_s)}{\ln(R_e) - \ln(r_s)} = \frac{0 - H(t)}{\ln\left(\frac{R_e}{r_s}\right)} \tag{12-39}$$

where the slope is evaluated for the particular pair of points at r_s and R_e and Eqs. (12-32) and (12-34) are invoked. Then Eqs. (12-37) and (12-39) give

$$-\frac{2\pi b K_r H(t)}{\ln\left(\frac{R_e}{r_s}\right)} = \pi r_c^2 \frac{dH(t)}{dt} \tag{12-40}$$

After rearrangement we obtain

$$-\int_0^t \frac{2\pi b K_r}{\ln\left(\frac{R_e}{r_s}\right)} dt = \int_0^{H_0} \pi r_c^2 \frac{dH(t)}{H(t)} \tag{12-41}$$

or

$$-\frac{F K_r t}{\pi r_c^2} = \ln\left[\frac{H(t)}{H_0}\right] \tag{12-42}$$

where F is the *theoretical shape factor* and is defined by

$$F = \frac{2\pi b}{\ln\left(\frac{R_e}{r_s}\right)} \tag{12-43}$$

Combination of Eqs. (12-42) and (12-43) gives

$$K_r = \frac{r_c^2}{2b} \ln\left(\frac{R_e}{r_s}\right) \frac{1}{t} \ln\left[\frac{H_0}{H(t)}\right] \tag{12-44}$$

Equation (12-44) is exactly the same as the Bouwer and Rice equation, Eq. (12-9), with the exception of notation difference. The only difference is the way of evaluating the term $\ln(R_e/r_s)$ in both equations. Hvorslev's approach for this term is presented in the following section.

12.3.2 Hvorslev Equations

Hvorslev (1951) presented formulas for a wide variety of shape factors. A modified form of the portion of Hvorslev equations are included in a document prepared by the U.S. Department of the Navy (USDN) (1982, Table 15). Lambe and Whitman (1969), Cedergren (1977), and Das (1983) also include the same table.

Condition F of Table 15 in the aforementioned USDN document includes an equation that corresponds to the theoretically derived equation, Eq. (12-44), in Section 12.3.1.

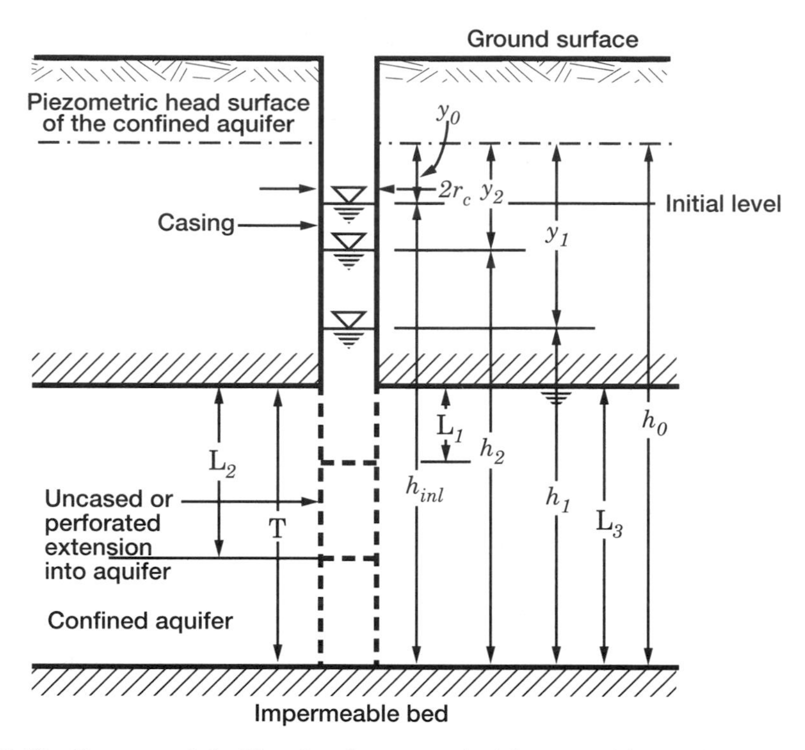

Figure 12-18 Geometry of the Hvorslev slug test method for a cased hole, with an uncased or perforated extension into an aquifer of finite thickness. (After U.S. Department of the Navy, 1982.)

Figure 12-18 presents a schematic representation of Condition F in the USDN document. These equations are included below for the purpose of comparison with the theoretically derived equation as presented in Section 12.3.1. Based on the notation in Figure 12-18, the equations for Condition F are presented below:

Case 1: $L_1/T \leq 0.2$

$$K_r = \frac{\pi r_c}{C_s(t_2 - t_1)} \ln\left(\frac{y_1}{y_2}\right), \qquad F = C_s r_c \tag{12-45}$$

The functional distribution of C_s versus uncased length/radius is given in Figure 12-19.

Case 2: $0.2 < L_2/T < 0.85$

$$K_r = \frac{r_c^2 \ln\left(\dfrac{L_2}{r_c}\right)}{2L_2(t_2 - t_1)} \ln\left(\frac{y_1}{y_2}\right), \qquad F = \frac{2\pi L_2}{\ln\left(\dfrac{L_2}{r_c}\right)} \tag{12-46}$$

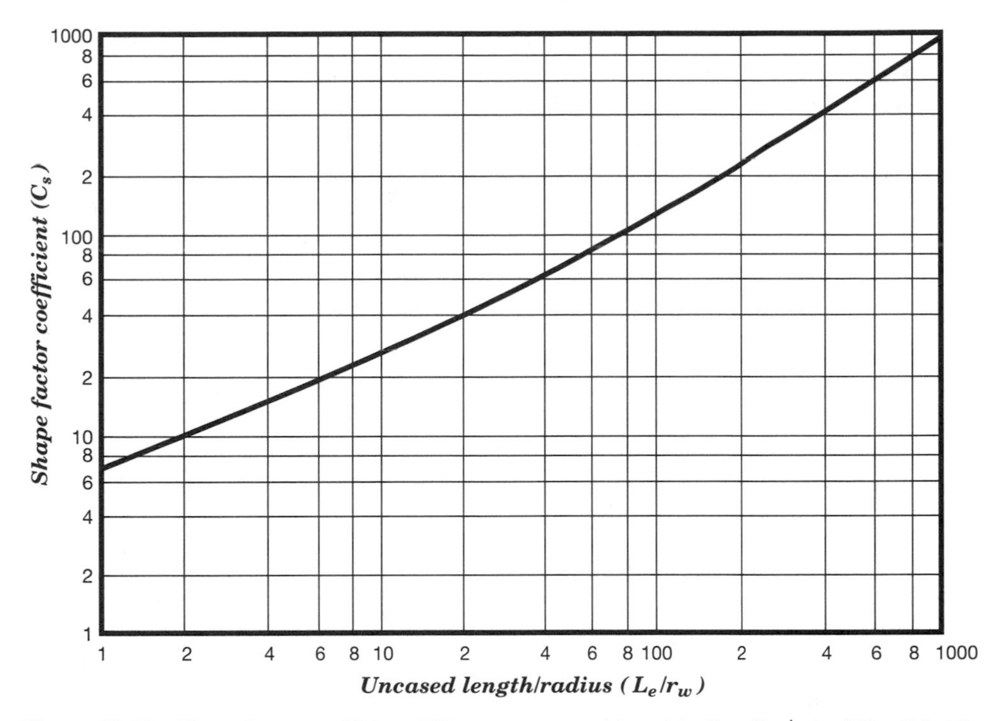

Figure 12-19 Shape factor coefficient (C_s) versus uncased length/radius (L_e/r_w) of Eq. (12-45). (After U.S. Department of the Navy, 1982.)

Case 3: $L_3/T = 1.00$

$$K_r = \frac{r_c^2 \ln\left(\dfrac{R_e}{r_c}\right)}{2L_3(t_2 - t_1)} \ln\left(\frac{y_1}{y_2}\right), \qquad F = \frac{2\pi L_3}{\ln\left(\dfrac{R_e}{r_c}\right)} \tag{12-47}$$

where F is valid for $L_2/r_c > 8$.

12.3.3 Comparison of the Hvorslev Equations with the Theoretically Derived Equation

Hvorslev (1951) does not present derivations of the equations for the three cases presented in Section 12.3.2. The theoretically derived equation, Eq. (12-44), and the Hvorslev equations are compared below:

1. The structure of Case 1 equation, Eq. (12-45), is similar to Eq. (12-44). However, its shape factor is very different than the theoretical shape factor as given by Eq. (12-43).
2. If in the first logarithmic term $L_2 = R_e$ and $r_c = r_s$, and $t_1 = 0, t_2 = t, y_1 = H_0, y_2 = H(t)$, and $L_2 = b$, the first expression in Eqs. (12-46) for Case 2, turns out to be exactly the same as the theoretically derived equation, Eq. (12-44). Similarly, if in the logarithmic term of the second expression of Eqs. (12-46) $L_2 = R_e$ and $r_c = r_s$, and

$L_2 = b$, the shape factor expression F becomes exactly the same as the theoretically derived shape factor in Eq. (12-43). Also using the notation in Eq. (12-9), the first expression of Eqs. (12-46) takes the form

$$K_r = \frac{r_c^2}{2L_2} \ln\left(\frac{L_2}{r}\right) \frac{1}{t} \ln\left(\frac{y_0}{y_t}\right) \tag{12-48}$$

Eq. (12-48) is almost equivalent to the Bouwer and Rice equation as given by Eq. (12-9).

3. Similarly, it can be shown that Hvorslev's K_r and F expressions given by Eqs. (12-47) turn out to be exactly the same theoretically derived expressions given by Eq. (12-43) and Eq. (12-44), respectively.

12.3.4 Data Analysis Methodology

Hvorslev (1951) developed a method for the determination of horizontal hydraulic conductivity using measured values of head difference (y) versus time (t). The method is similar in many respects to the Bouwer and Rice method. The methodology of data analysis is presented in a stepwise manner for both rising and falling head tests. The steps are given below:

Step 1: Plot y_t/y_0 (logarithmic) versus t on semilogarithmic paper. Here, y_0 is the initial head difference and is equal to $h_0 - h_{\text{inl}}$ (see Figure 12-18).

Step 2: Because y_t and t are the only variables in the equations, the plot must show a straight line. In other words, the straight-line portion is the valid part of the readings. If there is a *double straight-line effect*, follow the guidelines described in Section 12.2.4.1.

Step 3: Check if the water level rises in the screened or open section of the well with a gravel pack around it. If this is the case, Use Eq. (12-4) to calculate $r_{C_{eq}}$. Select the equation to be used from the equations in Section 12.3.2.

Step 4: Select two points on the straight-line portion of the curve and record their (t_1, y_1) and (t_2, y_2) coordinates. Record the other parameters.

Step 5: Using the corresponding equation, calculate horizontal hydraulic conductivity (K_r).

12.3.5 Guidelines for the Applicability of the Hvorslev Method

The slug test analysis method developed by Hvorslev permits the evaluation of saturated horizontal hydraulic conductivity of aquifers with a single well. The guidelines that are presented for the Bouwer and Rice method in Section 12.2.4 are also valid for the Hvorslev method.

12.3.6 Application

Example 12-2: Falling-head and rising-head slug tests were performed in an unconfined aquifer. Well dimensions are given in Figure 12-20. The initial water level in the well is

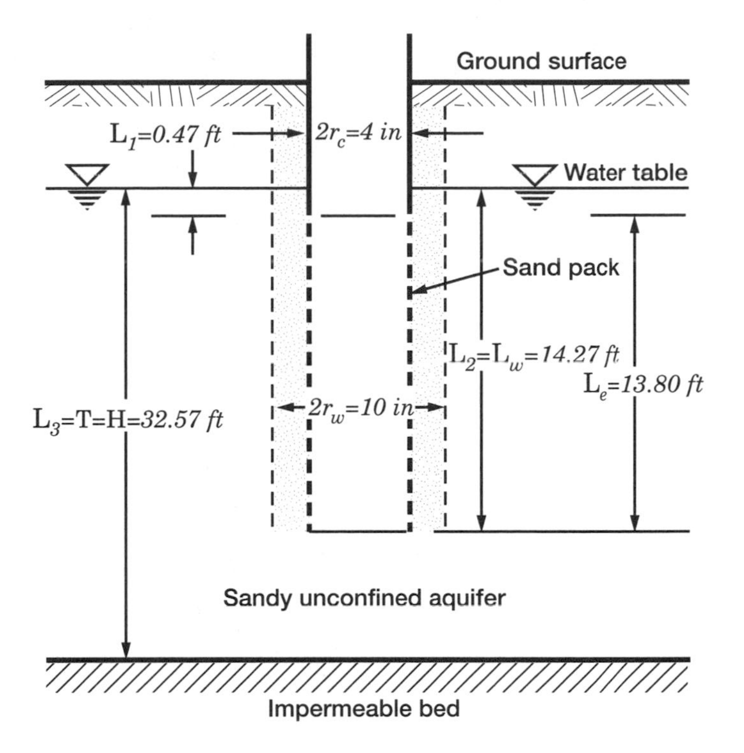

Figure 12-20 Geometrical dimensions of the well for Example 12-2.

10 ft below the measuring point. Depths of water level from the measuring point (MP) as a function of time are listed in Table 12-2a and Table 12-2b for falling-head and rising-head tests, respectively. Applying the Hvorslev method, determine the horizontal hydraulic conductivity values for each test data.

Solution: First of all, the head difference (y_t) needs to be calculated from the formula $y_t = 10 \text{ ft} - d_{\text{MP}}$ for the falling-head test and $y_t = d_{\text{MP}} - 10 \text{ ft}$ for rising-head test. These values are listed in the third columns of Table 12-2a and Table 12-2b.

Step 1: From Tables 12-2a and 12-2b, the initial head differences are

$$y_0 = h_0 - h_{\text{inl}} = 1.48 \text{ ft} \quad \text{for falling-head test}$$
$$= 1.51 \text{ ft} \quad \text{for rising-head test}$$

y_t/y_0 versus t values are presented in the third columns of Table 12-2a and Table 12-2b. Using these tables, the graphs are presented in Figure 12-21 and Figure 12-22 for falling-head and rising-head data, respectively.

Step 2: The drawn straight lines are shown in Figure 12-21 and Figure 12-22 for falling-head and rising-head tests, respectively. The guidelines described in Section 12.2.4.1 were used in drawing the lines.

TABLE 12-2a Falling-Head Test Data for Example 12-2

t (s)	d_{MP} (ft)	y_t (ft)[a]	y_t/y_0
0	8.52	1.48	1.00
2	8.56	1.44	0.97
4	8.59	1.41	0.95
6	8.67	1.33	0.90
8	8.77	1.23	0.83
10	8.73	1.27	0.86
20	8.83	1.17	0.79
30	8.91	1.09	0.73
40	8.99	1.01	0.68
50	9.05	0.95	0.64
60	9.11	0.89	0.60
70	9.17	0.83	0.56
86	9.25	0.75	0.50
101	9.31	0.69	0.46
111	9.37	0.63	0.43
121	9.39	0.61	0.41
136	9.45	0.55	0.37
151	9.49	0.51	0.34
166	9.53	0.47	0.32
186	9.58	0.42	0.28
201	9.60	0.40	0.27
221	9.64	0.36	0.24
231	9.67	0.33	0.22
246	9.68	0.32	0.21
256	9.71	0.29	0.20
271	9.72	0.28	0.19
301	9.75	0.25	0.17
311	9.76	0.24	0.16
326	9.79	0.21	0.14

[a]Here $y_t = 10 \text{ ft} - d_{MP}$.

TABLE 12-2b Rising-Head Test Data for Example 12-2

t (s)	d_{MP} (ft)[a]	y_t (ft)	y_t/y_0
0	11.51	1.51	1.00
2	11.44	1.44	0.95
4	11.40	1.40	0.93
10	11.30	1.30	0.86
20	11.19	1.19	0.78
30	11.08	1.08	0.71
40	10.97	0.97	0.64
50	10.90	0.90	0.59
70	10.74	0.74	0.49
90	10.63	0.63	0.42
100	10.56	0.56	0.37
120	10.47	0.47	0.31
140	10.39	0.39	0.26
160	10.32	0.32	0.21
180	10.26	0.26	0.17
200	10.22	0.22	0.14
220	10.17	0.17	0.11
240	10.13	0.13	0.09
260	10.10	0.10	0.07
280	10.08	0.08	0.05
300	10.06	0.06	0.04
320	10.04	0.04	0.03
340	10.03	0.03	0.02

[a]Here $y_t = d_{MP} - 10$ ft.

Step 3: Geometrical dimensions of the well are given in Figure 12-20. Because $L_2/T = 13.80/32.57 = 0.42$, Eq. (12-46) should be used. The upper end of the screened interval is 0.47 ft below the water table. Initially, $y_t = 1.51$ ft (see Table 12-2b or Figure 12-22). Therefore, the water rises through the screened interval. Assuming a porosity of 0.40 for the gravel pack and that $r_c = 2$ in. $= 0.17$ ft and $r_w = 5$ in. $= 0.42$ ft, the equivalent radius from Eq. (12-4) is

$$r_{c_{eq}} = \left[(1 - 0.40)(0.17 \text{ ft})^2 + (0.40)(0.42 \text{ ft})^2\right]^{1/2} = 0.30 \text{ ft}$$

Step 4: The selected points are shown on Figures 12-21 and 12-22. Their coordinates are as follows: For the falling-head test, Figure 12-21 and Table 12-2a gives

$$t_1 = 50 \text{ s}, \qquad y_1 = 0.95 \text{ ft}$$

$$t_2 = 186 \text{ s}, \qquad y_2 = 0.42 \text{ ft}$$

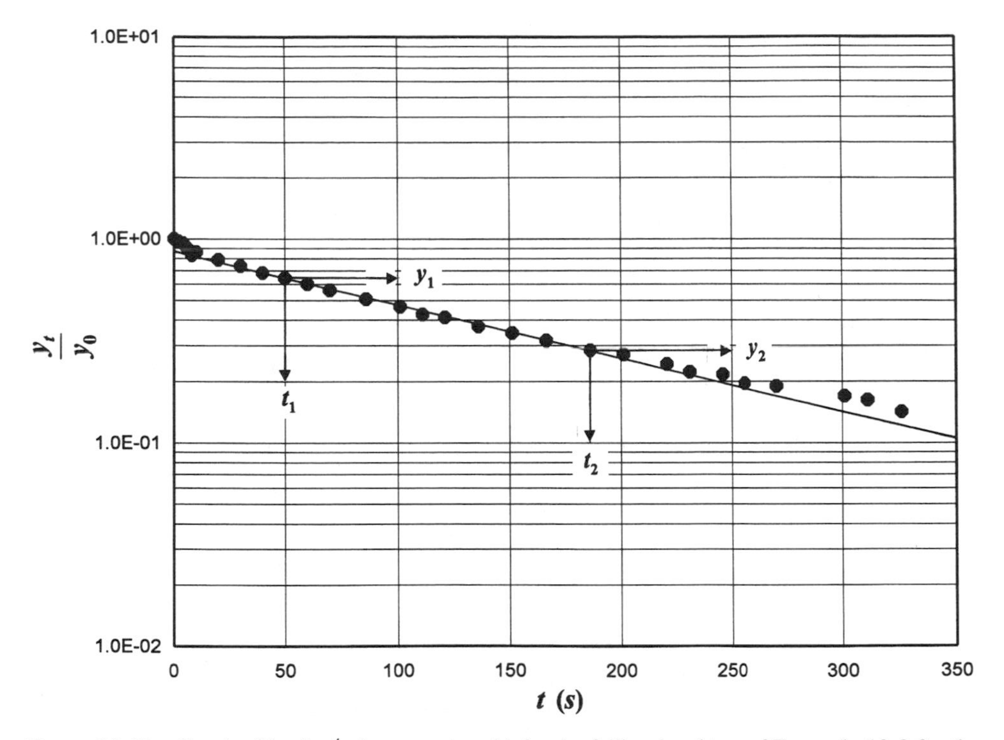

Figure 12-21 Graph of $\log(y_t/y_0)$ versus time (t) for the falling-head test of Example 12-2 for the Hvorslev method.

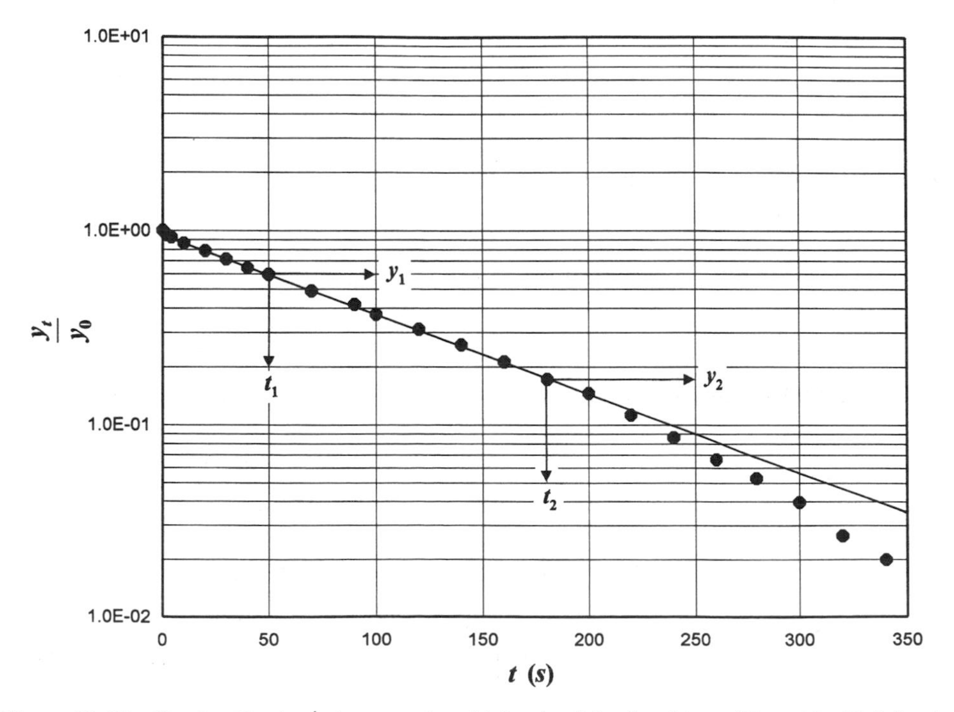

Figure 12-22 Graph of $\log(y_t/y_0)$ versus time (t) for the rising-head test of Example 12-2 for the Hvorslev method.

For the rising-head test, Figure 12-22 and Table 12-2b gives

$$t_1 = 50 \text{ s}, \qquad y_1 = 0.90 \text{ ft}$$

$$t_2 = 180 \text{ s}, \qquad y_2 = 0.26 \text{ ft}$$

Step 5: $L_2 = 13.80$ ft is the length of the screen interval. From Eq. (12-46) for the falling-head test ($r_c = 2$ in. $= 0.17$ ft) we obtain

$$K_r = \frac{(0.17 \text{ ft})^2 \ln\left(\dfrac{13.80 \text{ ft}}{0.17 \text{ ft}}\right)}{2(13.80 \text{ ft})(186 \text{ s} - 50 \text{ s})} \ln\left(\frac{0.95 \text{ ft}}{0.42 \text{ ft}}\right) = 2.76 \times 10^{-5} \text{ ft/s} = 8.42 \times 10^{-4} \text{ cm/s}$$

From Eq. (12-46) for the rising-head test ($r_c = r_{C_{eq}} = 0.30$ ft) we obtain

$$K_r = \frac{(0.30 \text{ ft})^2 \ln\left(\dfrac{13.80 \text{ ft}}{0.30 \text{ ft}}\right)}{2(13.80 \text{ ft})(180 \text{ s} - 50 \text{ s})} \ln\left(\frac{0.90 \text{ ft}}{0.26 \text{ ft}}\right) = 1.19 \times 10^{-4} \text{ ft/s} = 3.63 \times 10^{-3} \text{ cm/s}$$

Example 12-3: Determine the horizontal hydraulic conductivity value for the aquifer described in Example 12-2 using the Bouwer and Rice slug test data analysis method, which is outlined in Section 12.2.3. Compare the results with the ones obtained from the Hvorslev method.

Solution:

Step 1: The Bouwer and Rice graphs for falling-head and rising-head tests are given in Figure 12-23 and Figure 12-24, respectively, using the data in Table 12-2a and Table 12-2b.

Step 2: From Figures 12-23 and 12-24, the coordinates of the intersection points (y_0) with the $\log(y_t)$ axis are

$$y_0 = 1.28 \text{ ft} \qquad \text{for falling-head data}$$

$$y_0 = 1.40 \text{ ft} \qquad \text{for rising-head data}$$

Step 3: The coordinates of the arbitrarily selected points are

$$t = 100 \text{ s}, \qquad y_t = 0.68 \text{ ft} \qquad \text{for falling-head data}$$

$$t = 100 \text{ s}, \qquad y_t = 0.56 \text{ ft} \qquad \text{for rising-head data}$$

Step 4: From the values of the previous steps

$$\frac{1}{t} \ln\left(\frac{y_0}{y_t}\right) = \frac{1}{(100 \text{ s})} \ln\left(\frac{1.28 \text{ ft}}{0.68 \text{ ft}}\right) = 6.33 \times 10^{-3} \text{ s}^{-1} \qquad \text{for falling-head data}$$

$$\frac{1}{t} \ln\left(\frac{y_0}{y_t}\right) = \frac{1}{(100 \text{ s})} \ln\left(\frac{1.40 \text{ ft}}{0.56 \text{ ft}}\right) = 9.16 \times 10^{-3} \text{ s}^{-1} \qquad \text{for rising-head data}$$

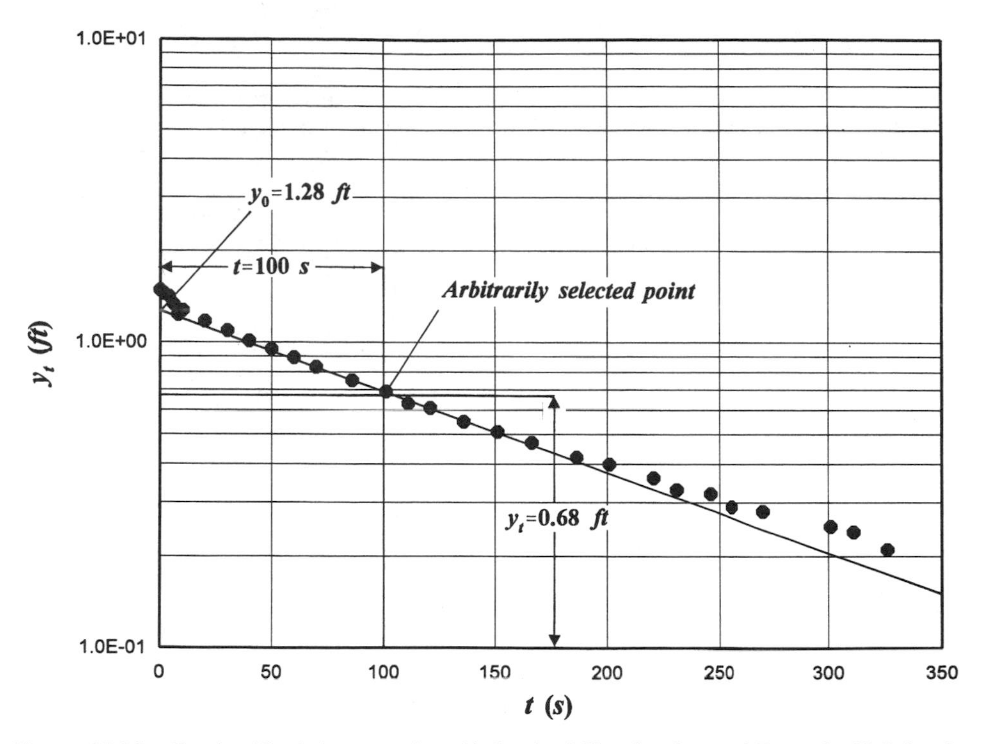

Figure 12-23 Graph of log(y_t) versus time (t) for the falling-head test of Example 12-3 for the Bouwer and Rice method.

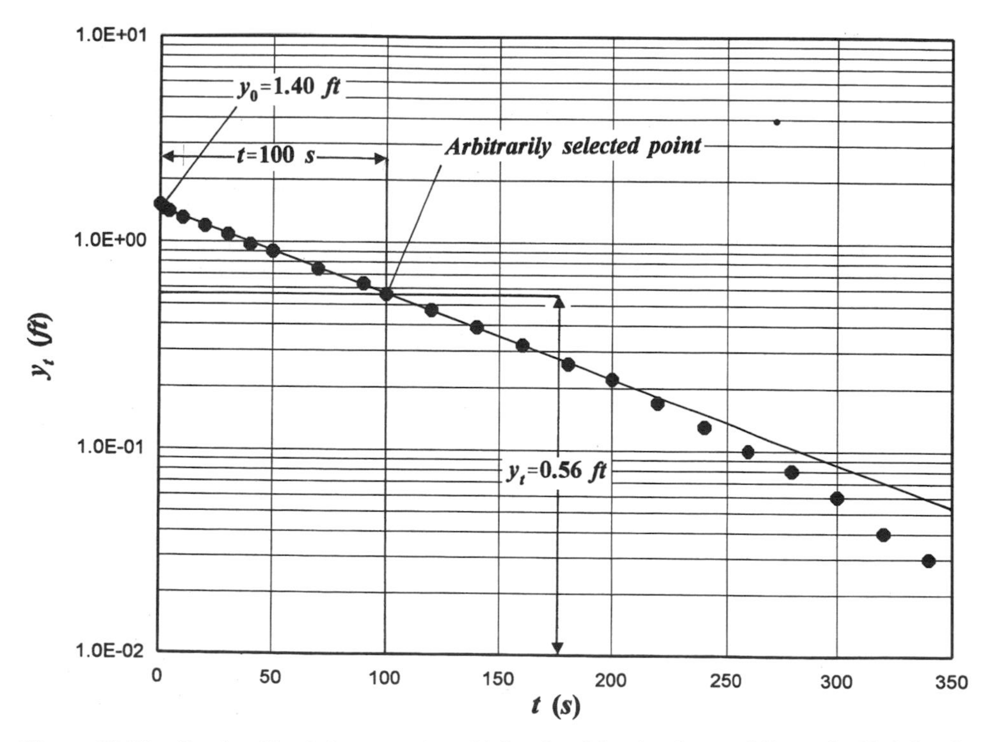

Figure 12-24 Graph of log(y_t) versus time (t) for the rising-head test of Example 12-3 for the Bouwer and Rice method.

TABLE 12-3 Comparison of the K_r Values of Bouwer and Rice, Hvorslev, and Cooper, Bredehoeft, and Papadopulos Methods for Example 12-3 and Example 12-6

	K_r (cm/s)		
Test Type	Bouwer and Rice Method	Hvorslev Method	Cooper, Bredehoeft, and Papadopulos Method
Falling-Head	4.45×10^{-4}	8.42×10^{-4}	9.44×10^{-4}
Rising-Head	6.45×10^{-4}	3.63×10^{-3}	1.23×10^{-3}

Step 5: Because $L_w = 14.27$ ft $< H = 32.57$ ft (see Figure 12-20), Eq. (12-13) should be used for $\ln(R_e/r_w)$.

Step 6:

$$\frac{L_e}{r_w} = \frac{13.8 \text{ ft}}{\left(\frac{5}{12} \text{ ft}\right)} = 33.12$$

Therefore, from Figure 12-2 we have $A = 2.60$ and $B = 0.40$.

Step 7: Substitution the values into Eq. (12-13) gives

$$\ln\left(\frac{R_e}{r_w}\right) = \left\{ \frac{1.1}{\ln\left(\frac{14.27 \text{ ft}}{\frac{5}{12} \text{ ft}}\right)} + \frac{2.60 + 0.40 \ln\left[\frac{(32.57 \text{ ft} - 14.27 \text{ ft})}{\frac{5}{12} \text{ ft}}\right]}{\frac{13.80 \text{ ft}}{\frac{5}{12} \text{ ft}}} \right\}^{-1} = 2.294$$

Step 8: As can be seen from Figure 12-20 and Table 12-2a, the water level in the well does not fall in the open section of the well. Therefore, $r_c = 2$ in. will be used for the falling-head data. On the other hand, Figure 12-20 and Table 12-2b show that after 120 s, the water level rises in the closed section of the casing. Total time period is approximately 350 s to reach equilibrium. Under these conditions, $r_c = 2$ in. will be used for this case as well.

Step 9: Substitution of the values for the falling-head test in the previous steps into Eq. (12-9) gives the value of horizontal hydraulic conductivity:

$$K_r = \frac{\left(\frac{2}{12} \text{ ft}\right)^2}{2(13.80 \text{ ft})}(2.294)\left(6.33 \times 10^{-3} \text{ s}^{-1}\right) = 14.61 \times 10^{-6} \text{ ft/s} = 4.45 \times 10^{-4} \text{ cm/s}$$

Similarly, for the rising-head test we obtain

$$K_r = \frac{\left(\frac{2}{12} \text{ ft}\right)^2}{(2)(13.80 \text{ ft})}(2.294)\left(9.16 \times 10^{-3} \text{ s}^{-1}\right) = 21.15 \times 10^{-6} \text{ ft/s} = 6.45 \times 10^{-4} \text{ cm/s}$$

The Bouwer and Rice, and Hvorslev methods results are presented in Table 12-3.

Example 12-4: Determine the horizontal hydraulic conductivity (K_r) value for the data of Example 12-1 (Table 12-1) using the Hvorslev slug test data analysis method, which is outlined in Section 12.3.4. Compare the value with the one obtained from the Bouwer and Rice method. Refer to Figures 12-6 and 12-18.

Solution:

Step 1: From Figure 12-7 or Table 12-1, $y_0 = h_0 - h_{\text{inl}} = 0.29$ m. The values of y_t/y_0 versus t are presented in Table 12-1 and their graph is shown in Figure 12-25.

Step 2: The drawn straight line is shown in Figure 12-25. The guidelines described in Section 12.2.4 are used in drawing the line.

Step 3: As can be seen from Table 12-1, the maximum head difference (the initial head difference) is 0.29 m. The upper end of the screen interval is 0.94 m below the water table. Therefore, water rises beyond the screened section of the well and a modified r_c need not be used. From the values of Figure 12-6, $L_1/T = 0.94\ \text{m}/80\ \text{m} = 0.01$. Therefore, Eq. (12-45) should be used. From Figure 12-6, $r_w = 0.12$ m and $L_e = 4.56$ m. Therefore, $L_e/r_w = 38$; and from Figure 12-19, C_s is approximately 60.

Step 4: The selected points are shown in Figure 12-25. Their coordinates are as follows:

$$t_1 = 5\ \text{s}, \qquad y_1 = (0.29\ \text{m})(0.30) = 0.087\ \text{m}$$

$$t_2 = 10\ \text{s}, \qquad y_2 = (0.29\ \text{m})(0.090) = 0.026\ \text{m}$$

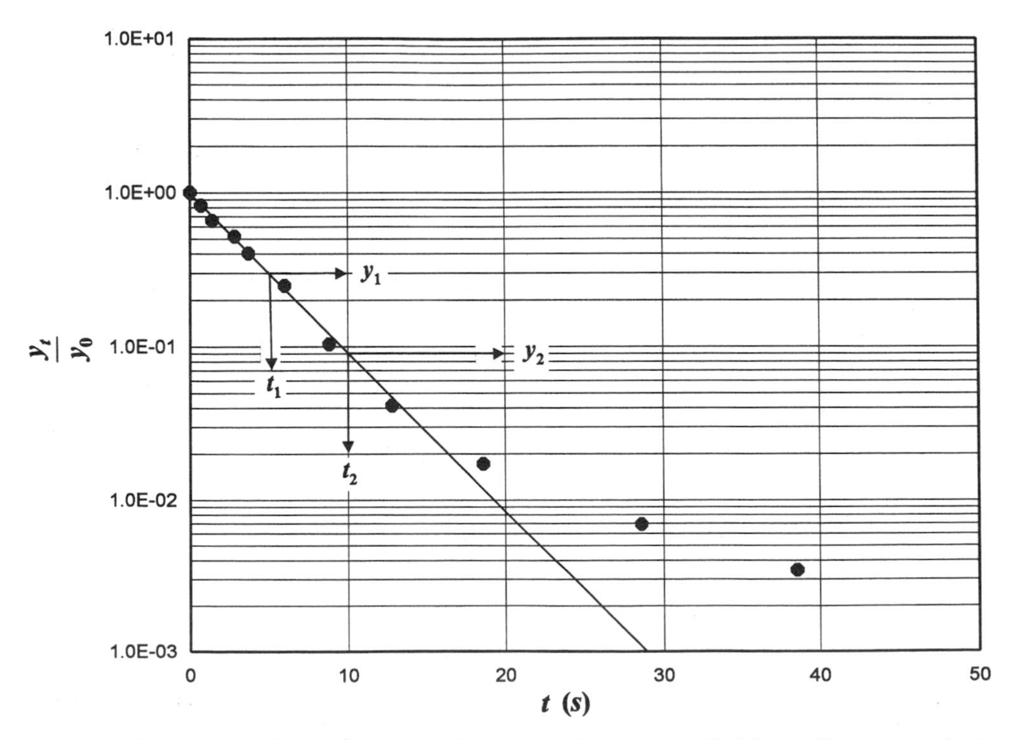

Figure 12-25 Graph of $\log(y_t/y_0)$ versus time (t) data for Example 12-4 for the Hvorslev method.

Step 5: Substitution of the above values into Eq. (12-45) gives

$$K_r = \frac{\pi(0.076 \text{ m})}{(60)(10 \text{ s} - 5 \text{ s})} \ln\left(\frac{0.087 \text{ m}}{0.026 \text{ m}}\right) = 9.61 \times 10^{-4} \text{ m/s} = 83 \text{ m/day}$$

This value is almost three times higher than the value, 32 m/day, of the Bouwer and Rice method (see Example 12-1).

Because $L_2/T = (5.5 \text{ m})/(80 \text{ m}) = 0.07$, Eq. (12-46) is not selected. Despite this fact, if one selects Eq. (12-46), the result is

$$K_r = \frac{(0.076 \text{ m})^2 \ln\left(\dfrac{5.5 \text{ m}}{0.076 \text{ m}}\right)}{2(5.5 \text{ m})(10 \text{ s} - 5 \text{ s})} \ln\left(\frac{0.087 \text{ m}}{0.026 \text{ m}}\right) = 5.93 \times 10^{-4} \text{ m/s} = 47 \text{ m/day}$$

This value is closer to the value determined by the Bouwer and Rice method (32 m/day) which is given in Example 12-1.

12.4 COOPER, BREDEHOEFT, AND PAPADOPULOS SLUG TEST MODEL

The Cooper et al. (1967) slug test method was developed to measure aquifer horizontal hydraulic conductivity and storage coefficient around boreholes including production, monitoring, or test wells. The wells can be screened or open. The slug test is based on a volume V of water that is instantaneously injected to the well, and the subsequent fall of water level in the well is measured. The method can be used for confined and unconfined aquifers. In the following sections, the theory and application of the method are presented and discussed in detail.

12.4.1 Theory

Geometry and symbols of a slug tested well are shown in Figure 12-17. The objective is to find a solution for $h(r, t)$ and $H(t)$. The theoretical analysis is based on the following assumptions:

1. A known volume of water (V) is instantaneously injected in the well or is discharged from the well when $t = 0$.
2. The piezometric head in the aquifer is initially uniform and constant.
3. The well fully penetrates the aquifer and its diameter is finite.
4. The inertia of column of water in the well is negligible.
5. The medium extends to infinity in all directions.
6. The medium is homogeneous and isotropic.
7. Darcy's law is valid.

The governing differential equation is the well-known transient two-dimensional groundwater flow equation for a confined aquifer in the radial coordinates [Chapter 2, Eq. (2-175)]:

$$\frac{\partial^2 h}{\partial r^2} + \frac{1}{r}\frac{\partial h}{\partial r} = \frac{S}{T}\frac{\partial h}{\partial t}, \qquad r > r_s \tag{12-49}$$

where h is the piezometric head, r is the radial coordinate, S is the storage coefficient, and T is the transmissivity.

The initial and boundary conditions may be described as follows (Cooper et al., 1967):

$$h(r_s, t) = H(t), \qquad t > 0 \tag{12-50}$$

$$h(r = \infty, t) = 0, \qquad t > 0 \tag{12-51}$$

$$2\pi r_s T \frac{\partial h(r_s, t)}{\partial r} = \pi r_c^2 \frac{\partial H(t)}{\partial t}, \qquad t > 0 \tag{12-52}$$

$$h(r, 0) = 0, \qquad r > r_s \tag{12-53}$$

$$H(0) = H_0 = \frac{V}{\pi r_c^2} \tag{12-54}$$

Equation (12-50) states that the head in the aquifer at the face of the well is equal to the head in the well at any time. Equation (12-51) states that the head in the aquifer approaches zero as distance from the well approaches infinity. Equation (12-52) states that the rate of flow of water into the aquifer or out of the aquifer is equal to the rate of decrease or to the rate of increase in volume of water inside the well. Finally, Eqs. (12-53) and (12-54) state that initially the head change is zero everywhere outside the well and equal to H_0 inside the well.

By applying the Laplace transform techniques to the above-defined problem, Cooper et al. (1967) obtained the following solution:

$$h(r,t) = \frac{2H_0}{\pi} \int_0^\infty \exp\left(-\frac{\beta u^2}{\alpha}\right) \left\{ J_0\left(\frac{ur}{r_s}\right) [uY_0(u) - 2\alpha Y_1(u)] - Y_0\left(\frac{ur}{r_s}\right) [uJ_0(u) - 2\alpha J_1(u)] \right\} \frac{du}{\Delta(u)} \tag{12-55}$$

where

$$\alpha = \frac{r_s^2 S}{r_c^2} \tag{12-56}$$

$$\beta = \frac{Tt}{r_c^2} \tag{12-57}$$

and

$$\Delta(u) = [uJ_0(u) - 2\alpha J_1(u)]^2 + [uY_0(u) - 2\alpha Y_1(u)]^2 \tag{12-58}$$

J_0 and Y_0, J_1 and Y_1, are zero-order and first-order Bessel functions of the first and second kind, respectively.

The head $H(t)$ inside the well, obtained by substituting $r = r_s$ in Eq. (12-55), is

$$\frac{H}{H_0} = F(\alpha, \beta) \tag{12-59}$$

where

$$F(\alpha, \beta) = \frac{8\alpha}{\pi^2} \int_0^\infty \exp\left(-\frac{\beta u^2}{\alpha}\right) \frac{1}{u\Delta(u)}\, du \tag{12-60}$$

12.4.2 Type-Curve Method

Numerical values of $H(t)/H_0$ for different parameters are given in Table 12-4 (Cooper et al., 1967) and Table 12-5 (Papadopulos et al., 1973). Based on the values in Table 12-4 and Table 12-5, a family of type curves are presented in Figure 12-26 with Tt/r_c^2 on the logarithmic scale and $H(t)/H_0$ on the arithmetic scale. The authors developed a type-curve method which is similar to the Theis type-curve method. The steps of the method are given below:

Step 1: Determine the initial head difference (H_0). If initially ($t = 0$) the value of H_0 is recorded, use that value. If the volume of water (V) is known, the following formula may be used to determine H_0:

$$H_0 = \frac{V}{\pi r_c^2} \tag{12-61}$$

TABLE 12-4 Values of $H(t)/H_0$ Computed from Eq. (12-59)

	H/H_0				
Tt/r_c^2	$\alpha = 10^{-1}$	$\alpha = 10^{-2}$	$\alpha = 10^{-3}$	$\alpha = 10^{-4}$	$\alpha = 10^{-5}$
1.00×10^{-3}	0.9771	0.9920	0.9969	0.9985	0.9992
2.15×10^{-3}	0.9658	0.9876	0.9949	0.9974	0.9985
4.64×10^{-3}	0.9490	0.9807	0.9914	0.9954	0.9970
1.00×10^{-2}	0.9238	0.9693	0.9853	0.9915	0.9942
2.15×10^{-2}	0.8860	0.9505	0.9744	0.9841	0.9888
4.64×10^{-2}	0.8293	0.9187	0.9545	0.9701	0.9781
1.00×10^{-1}	0.7460	0.8655	0.9183	0.9434	0.9572
2.15×10^{-1}	0.6289	0.7782	0.8538	0.8935	0.9167
4.64×10^{-1}	0.4782	0.6436	0.7436	0.8031	0.8410
1.00×10^{0}	0.3117	0.4598	0.5729	0.6520	0.7080
2.15×10^{0}	0.1665	0.2597	0.3543	0.4364	0.5038
4.64×10^{0}	0.07415	0.1086	0.1554	0.2082	0.2620
7.00×10^{0}	0.04625	0.06204	0.08519	0.1161	0.1521
1.00×10^{1}	0.03065	0.03780	0.04821	0.06355	0.08378
1.40×10^{1}	0.02092	0.02414	0.02844	0.03492	0.04426
2.15×10^{1}	0.01297	0.01414	0.01545	0.01723	0.01999
3.00×10^{1}	0.009070	0.009615	0.01016	0.01083	0.01169
4.64×10^{1}	0.005711	0.005919	0.006111	0.006319	0.006554
7.00×10^{1}	0.003722	0.003809	0.003884	0.003962	0.004046
1.00×10^{2}	0.002577	0.002618	0.002653	0.002688	0.002725
2.15×10^{2}	0.001179	0.001187	0.001194	0.001201	0.001208

Source: From Cooper et al. (1967).

TABLE 12-5 Values of $H(t)/H_0$ Computed from Eq. (12-59)

Tt/r_c^2	$\alpha = 10^{-6}$	$\alpha = 10^{-7}$	$\alpha = 10^{-8}$	$\alpha = 10^{-9}$	$\alpha = 10^{-10}$
0.001	0.9994	0.9996	0.9996	0.9997	0.9997
0.002	0.9989	0.9992	0.9993	0.9994	0.9995
0.004	0.9980	0.9985	0.9987	0.9989	0.9991
0.006	0.9972	0.9978	0.9982	0.9984	0.9986
0.008	0.9964	0.9971	0.9976	0.9980	0.9982
0.01	0.9956	0.9965	0.9971	0.9975	0.9978
0.02	0.9919	0.9934	0.9944	0.9952	0.9958
0.04	0.9848	0.9875	0.9894	0.9908	0.9919
0.06	0.9782	0.9819	0.9846	0.9866	0.9881
0.08	0.9718	0.9765	0.9799	0.9824	0.9844
0.1	0.9655	0.9712	0.9753	0.9784	0.9807
0.2	0.9361	0.9459	0.9532	0.9587	0.9631
0.4	0.8828	0.8995	0.9122	0.9220	0.9298
0.6	0.8345	0.8569	0.8741	0.8875	0.8984
0.8	0.7901	0.8173	0.8383	0.8550	0.8686
1.0	0.7489	0.7801	0.8045	0.8240	0.8401
2.0	0.5800	0.6235	0.6591	0.6889	0.7139
3.0	0.4554	0.5033	0.5442	0.5792	0.6096
4.0	0.3613	0.4093	0.4517	0.4891	0.5222
5.0	0.2893	0.3351	0.3768	0.4146	0.4487
6.0	0.2337	0.2759	0.3157	0.3525	0.3865
7.0	0.1903	0.2285	0.2655	0.3007	0.3337
8.0	0.1562	0.1903	0.2243	0.2573	0.2888
9.0	0.1292	0.1594	0.1902	0.2208	0.2505
10.0	0.1078	0.1343	0.1620	0.1900	0.2178
20.0	0.02720	0.03343	0.04129	0.05071	0.06149
30.0	0.01286	0.01448	0.01667	0.01956	0.02320
40.0	0.008337	0.008898	0.009637	0.01062	0.01190
50.0	0.006209	0.006470	0.006789	0.007192	0.007709
60.0	0.004961	0.005111	0.005283	0.005487	0.005735
80.0	0.003547	0.003617	0.003691	0.003773	0.003863
100.0	0.002763	0.002803	0.002845	0.002890	0.002938
200.0	0.001313	0.001322	0.001330	0.001339	0.001348

Source: From Papadopulos et al. (1973).

The other method is to use the log$[H(t)]$ versus t curve. Compare the H_0 value with the one obtained from Eq. (12-61). Then, create a t versus $H(t)/H_0$ data table.

Step 2: Plot $H(t)/H_0$ versus t data on semilogarithmic paper (t is logarithmic) of the same scale as the type curves in Figure 12-26.

Step 3: Keeping the logarithmic axes (t and Tt/r_c^2 axes) coincident, translate the data plot horizontally to a position until the best fit of the data curve and one of the family of type curves is obtained. Record the value of α.

Step 4: Select a common point, the match point, arbitrarily chosen on the overlapping part of the curve, or even anywhere on the overlapping portion of the sheets. Read the corresponding values of t and $\beta = Tt/r_c^2$.

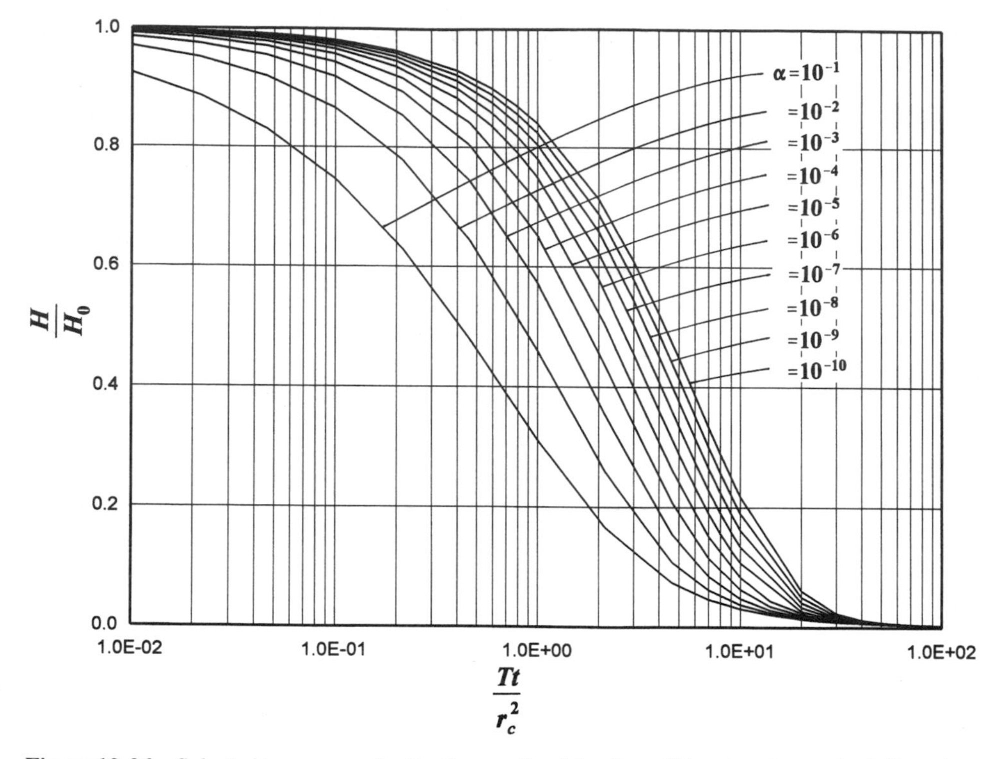

Figure 12-26 Selected type curves for the Cooper, Bredehoeft, and Papadopulos method. (Papadopulos et al., 1973.)

Step 5: Using the known values of β, t, and r_c, calculate the transmissivity value (T) from Eq. (12-57).

Step 6: Using the known values of α, r_s, and r_c, calculate the storage coefficient (S) from Eq. (12-56).

12.4.3 Guidelines for the Type-Curve Method

Cooper et al. (1967) and Papadopulos et al. (1973) established some key guidelines in using the method. These guidelines are as follows:

1. The type curves in Figure 12-26 are very similar in shape. For this reason, it may be difficult to obtain a unique match between the plot for measurement and type curves. However, the determination of transmissivity (T) is not so sensitive to the choice of the curves to be matched.
2. A determination of storage coefficient (S) by this method has questionable reliability because the matching of the data curve to the type curves differ only slightly when α differs by an order of magnitude.
3. Papadopulos et al. (1973) showed that if $\alpha < 10^{-5}$, an error of two orders of magnitude in α will cause an error of less than about 30% of the calculated transmissivity.

12.4.4 Application

Example 12-5: This example is taken from Cooper et al. (1967). A rising-head test was performed in a well cased to 24 m with 15.2-cm (6-inch) casing and drilled as a 15.2-cm open hole to a depth of 122 m. The water levels in the well were recorded after a sudden withdrawal of a long weighted float from the well. The weight of the float was 10.16 kg (kilograms). The measured values of $H(t)$ versus time (t) are given in Table 12-6. Applying the Cooper et al. type-curve method, determine the transmissivity (T) and storage coefficient (S) of the aquifer.

Solution:

Step 1: The float had displaced a volume of 0.01016 m^3 of water because of 10.16 kg of weight. Therefore, the volume of the negative charge was $V = 0.01016\ m^3$. From Eq. (12-61) we obtain

$$H_0 = \frac{V}{\pi r_c^2} = \frac{0.01016\ \mathrm{m}^3}{\pi (0.076\ \mathrm{m})^2} = 0.560\ \mathrm{m}$$

Using this value, $H(t)/H_0$ values are listed in Table 12-6.

TABLE 12-6 Rise of Water Level Versus Time After Instantaneous Withdrawal of Weighted Float of Example 12-5

t (s)	$H(t)$ (m)	$H(t)/H_0$
0	0.560	1.000
3	0.547	0.816
6	0.392	0.700
9	0.345	0.616
12	0.308	0.550
15	0.280	0.500
18	0.252	0.450
21	0.224	0.400
24	0.205	0.366
27	0.187	0.334
30	0.168	0.300
33	0.149	0.266
36	0.140	0.250
39	0.131	0.234
42	0.112	0.200
45	0.108	0.193
48	0.093	0.166
51	0.089	0.159
54	0.082	0.146
57	0.075	0.134
60	0.071	0.127
63	0.065	0.116

Source: After Cooper et al. (1967).

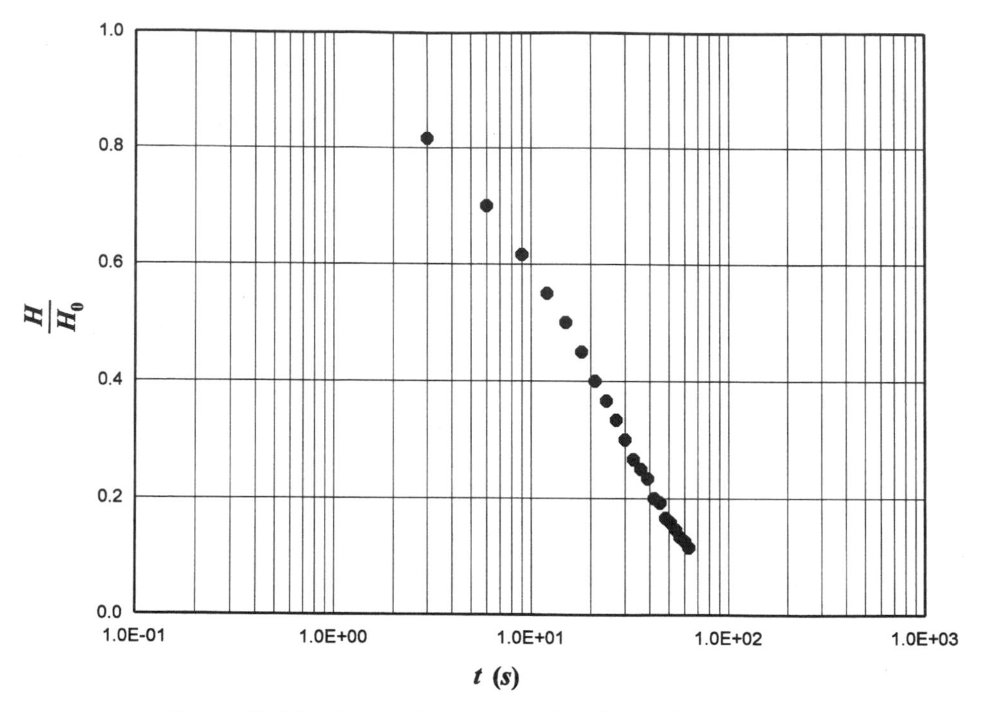

Figure 12-27 Graph of log(t) versus $H(t)/H_0$ for Example 12-5.

Step 2: The data in Table 12-6 are plotted in Figure 12-27 on semilogarithmic paper of the scale as the type curves in Figure 12-26.

Step 3: The superimposed sheets are shown in Figure 12-28. A comparison shows that the plotted points fall along the curve $\alpha = 10^{-3}$.

Step 4: In Figure 12-28, A is an arbitrarily chosen point and its coordinates are

$$t_A = 11.0 \text{ s}, \qquad \beta_A = \left(\frac{Tt}{r_c^2}\right)_A = 1.0$$

Step 5: From the values of the previous step, the value of transmissivity from Eq. (12-57) is

$$T = \frac{1.0 r_c^2}{t_A} = \frac{(1.0)(0.076 \text{ m})^2}{11.0 \text{ s}} = 5.25 \times 10^{-4} \text{ m}^2/\text{s}$$

Step 6: From Eq. (12-56) the storage coefficient is

$$S = \alpha \frac{r_c^2}{r_s^2} = 10^{-3} \frac{(0.076 \text{ m})^2}{(0.076 \text{ m})^2} = 10^{-3}$$

Example 12-6: Determine the values of horizontal hydraulic conductivity (K_r) and storage coefficient (S) for the data of Example 12-2 (Section 12.3.6) using the Cooper, Bredehoeft, and Papadopulos method. Compare the results with the ones obtained from the Bouwer and Rice method and the Hvorslev method.

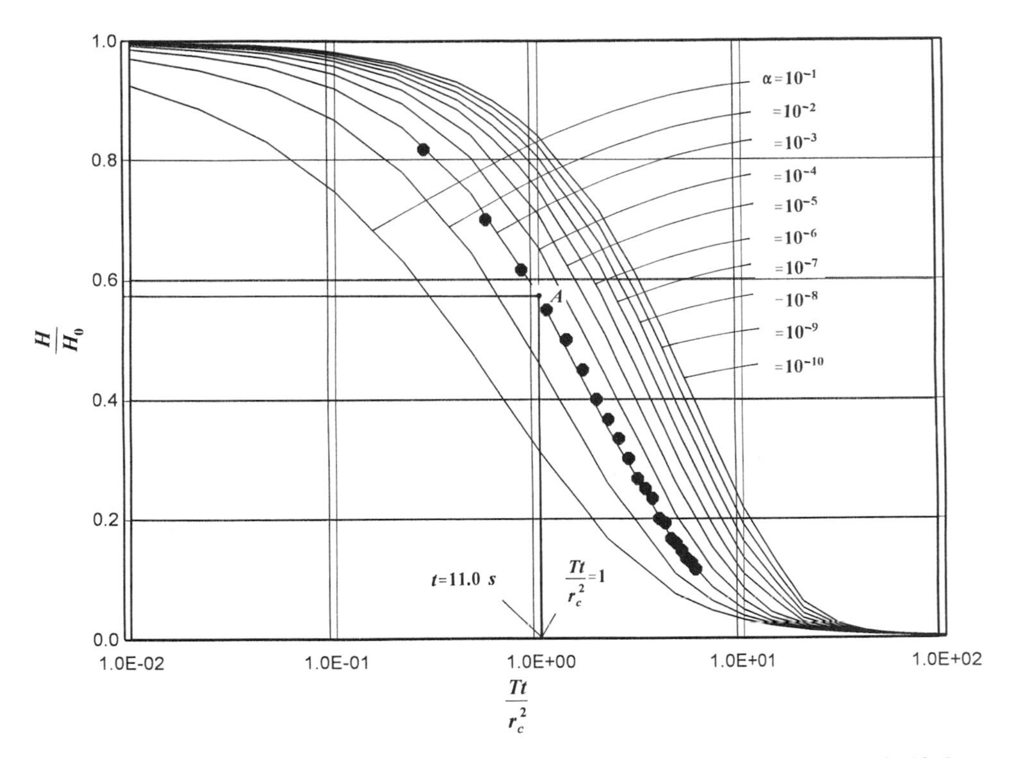

Figure 12-28 Cooper, Bredehoeft, and Papadopulos type-curve matching for Example 12-5.

Solution:

Step 1: From Tables 12-2a and 12-2b, the values of H_0 are

$$H_0 = 1.48 \text{ ft} \qquad \text{for falling-head test}$$

$$H_0 = 1.51 \text{ m} \qquad \text{for rising-head test}$$

With these values, $H(t)/H_0$ (y_t/y_0 in Tables 12-2a and 12-2b) versus t are listed in Tables 12-2a and 12-2b.

Step 2: The data in Table 12-2a and Table 12-2b are plotted in Figure 12-29 and Figure 12-30 for rising-head and falling-head tests, respectively, on semilogarithmic paper of the same scale as the type curves in Figure 12-26.

Step 3: The superimposed sheets for falling-head and rising-head tests are shown in Figure 12-31 and Figure 12-32, respectively. Comparison shows that for both cases the plotted points fall along the type curve $\alpha = 10^{-3}$.

Step 4: For the falling-head test (Figure 12-31), A is an arbitrarily chosen point and its coordinates are

$$t_A = 65.0 \text{ s}, \qquad \beta_A = \left(\frac{Tt}{r_c^2}\right)_A = 1.0$$

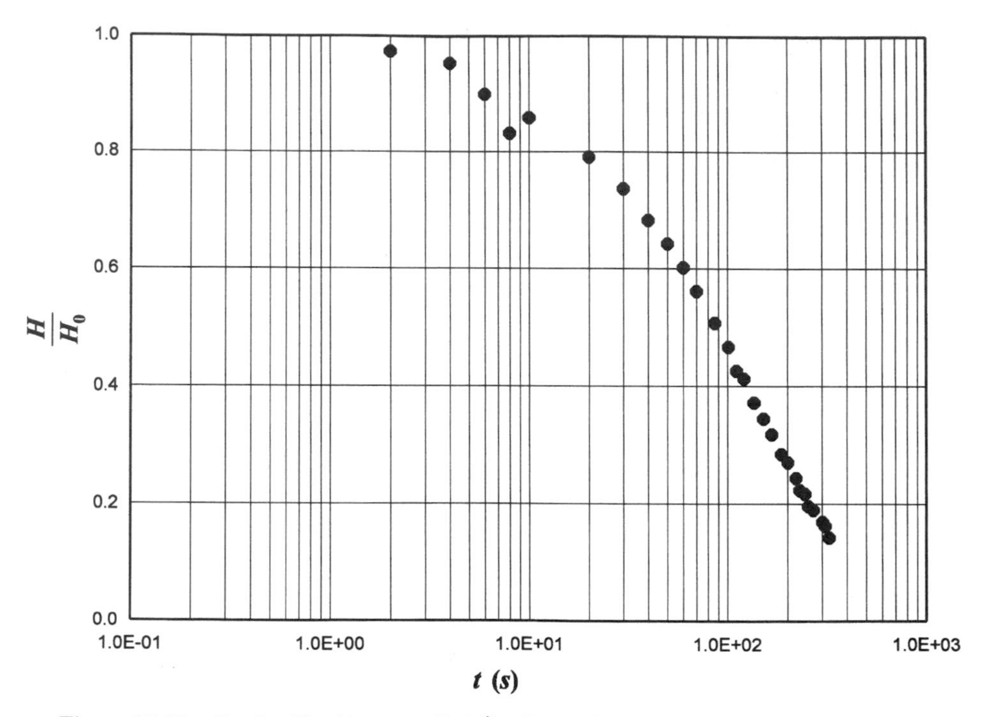

Figure 12-29 Graph of log(t) versus $H(t)/H_0$ for the falling-head test for Example 12-6.

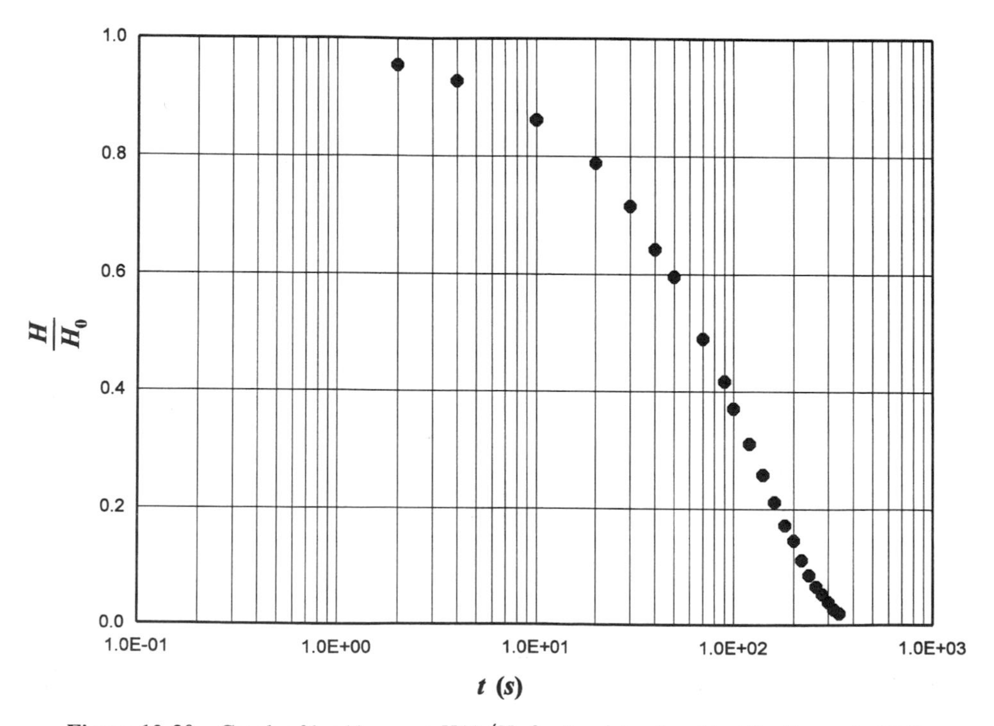

Figure 12-30 Graph of log(t) versus $H(t)/H_0$ for the rising-head test for Example 12-6.

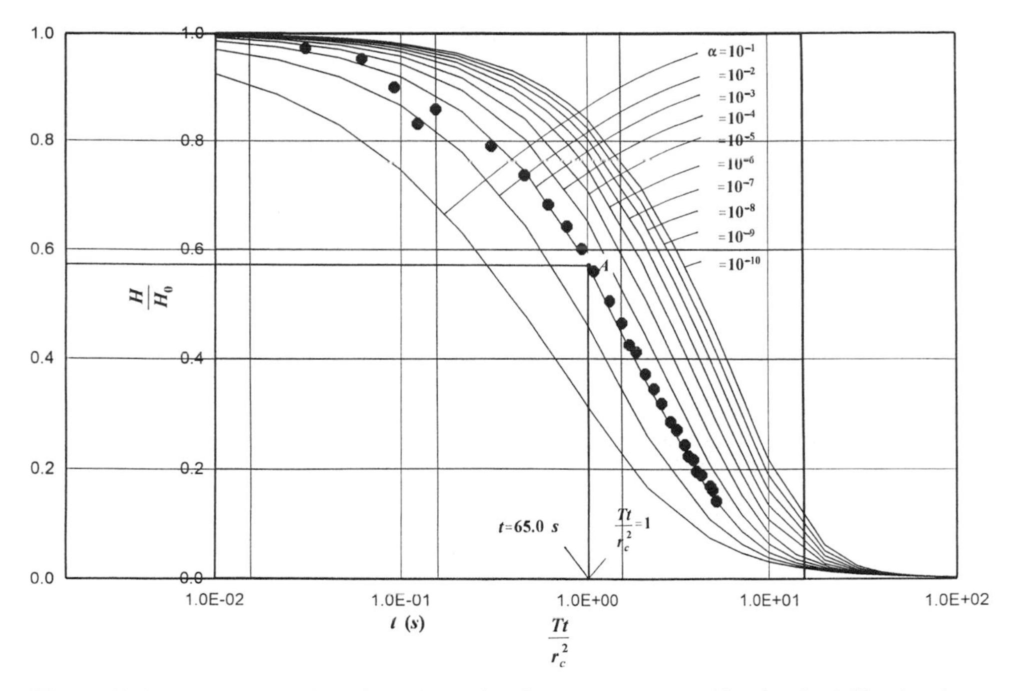

Figure 12-31 Cooper, Bredehoeft, and Papadopulos type-curve matching for the falling-head test for Example 12-6.

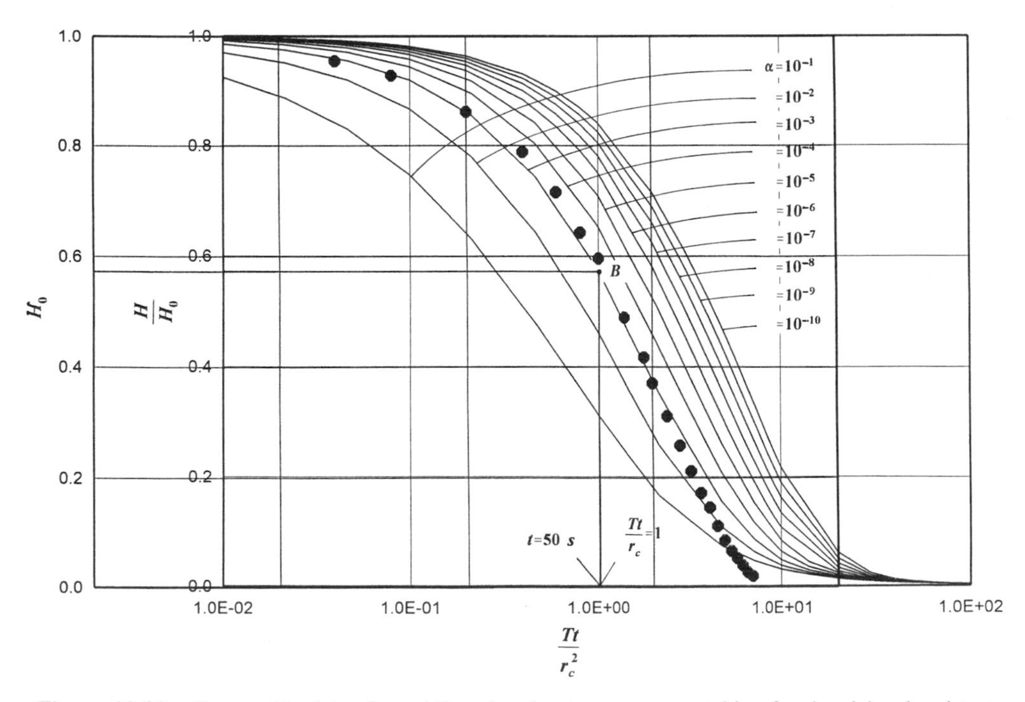

Figure 12-32 Cooper, Bredehoeft, and Papadopulos type-curve matching for the rising-head test for Example 12-6.

For the rising-head test (Figure 12-32), B is an arbitrarily chosen point and its coordinates are

$$t_B = 50.0 \text{ s}, \qquad \beta_B = \left(\frac{Tt}{r_c^2}\right) = 1.0$$

Step 5: For the falling-head test, from Eq. (12-57), the values of T and K_r are as follows:

$$T = \frac{\beta_A r_c^2}{t_A} = \frac{(1.0)\left(\dfrac{2}{12}\text{ ft}\right)^2}{65.0 \text{ s}} = 4.27 \times 10^{-4} \text{ ft}^2/\text{s} = 36.92 \text{ ft}^2/\text{day}$$

$$K_r = \frac{36.92 \text{ ft}^2/\text{day}}{(14.27 \text{ ft} - 0.47 \text{ ft})} = \frac{36.92 \text{ ft}^2/\text{day}}{13.80 \text{ ft}} = 2.68 \text{ ft/day} = 9.44 \times 10^{-4} \text{ cm/s}$$

Similarly, for the rising-head test the following values can be obtained:

$$T = \frac{\beta_B r_c^2}{t_B} = \frac{(1.0)\left(\dfrac{2}{12}\text{ ft}\right)^2}{50.0 \text{ s}} = 5.55 \times 10^{-4} \text{ ft}^2/\text{s} = 48.0 \text{ ft}^2/\text{day}$$

$$K_r = \frac{48.0 \text{ ft}^2/\text{day}}{13.80 \text{ ft}} = 3.48 \text{ ft/day} = 1.23 \times 10^{-3} \text{ cm/s}$$

Step 6: From Eq. (12-56), the value of storage coefficient (S) for the falling-head and rising-head tests is

$$S = \alpha \frac{r_c^2}{r_s^2} = (10^{-3}) \frac{\left(\dfrac{2}{12}\text{ ft}\right)^2}{\left(\dfrac{2}{12}\text{ ft}\right)^2} = 10^{-3}$$

The results obtained from the Bouwer and Rice, the Hvorslev, and the Cooper, Bredehoeft, and Papadopulos methods are compared in Table 12-3.

PART VI

HYDRAULICS OF PRESSURE PULSE AND CONSTANT HEAD INJECTION TESTS FOR TIGHT FORMATIONS AND HYDROGEOLOGIC DATA ANALYSIS METHODS

CHAPTER 13

FULLY PENETRATING WELLS IN CONFINED AQUIFERS

13.1 INTRODUCTION

Rocks and compacted clays are relatively rigid compared with other porous materials such as sandy materials. Therefore, these types of formations are frequently called *tight formations*, and their horizontal or vertical hydraulic conductivity values are generally less than 1.0×10^{-6} cm/s. For some rocks and compacted clays, this value may be as low as 1.0×10^{-10} cm/s. If fractures are generally vertically oriented, the vertical hydraulic conductivity may be greater than the horizontal hydraulic conductivity. From a theoretical standpoint, the other test procedures, such as pumping and slug tests, can also be applied. However, because of their relatively low permeability, a productive test period may take days, weeks, months, or years depending on the type of formation. Most projects require relatively short periods of time because of deadlines and budgetary constraints. As a result, the test of a tight formation needs to be carried out relatively quickly. To achieve this goal, the pressure along the screen interval of a well in a tight formation is controlled manually and the response is recorded as a function of time.

During the past half century, several methods have been developed for measuring horizontal hydraulic conductivity and the storage coefficient of tight formations using two-dimensional analytical models. One important limitation of these methods is that they are based on the assumption that the formation exists only between the horizontal planes passing through the upper and lower boundaries of the screen interval. In other words, the horizontal planes passing through the bottom and top boundaries of the screen interval are assumed to be impervious.

There are basically two kinds of test procedures currently used in different hydrogeologic studies. The first test procedure is called the *pressure pulse test method*; it was developed by Bredehoeft and Papadopulos (1980) based on a modified transient analytical model originally developed by Cooper et al. (1967). In this test, the well is filled with water to the surface and is suddenly pressurized with an additional amount of water. The well is then shut-in, and the decay of pressure (or pressure head) is measured. Decay of this pressure is relatively fast depending on the magnitude of pressure and the type of formation. The other test procedure is called the *constant head injection test method*; and this method is based on the models developed by Thiem (1906) under steady-state flow conditions and by Jacob and Lohman (1952) under transient flow conditions. In this test, the test interval of a tight formation is instantaneously pressurized and kept constant throughout the test period, and the variation of injected flow rate versus time is recorded. Initially, the flow rate is at a maximum; as time increases, its value decreases and eventually approaches a constant value if the test period is long enough. This is the steady-state condition; reaching steady state depends on the degree of the formation tightness.

13.2 PRESSURE PULSE TEST METHOD

13.2.1 Introduction

Pumping and slug tests can also be used to measure the hydrogeologic parameters of tight formations. To create a significant response in an observation well under pumping conditions in a tight formation, the test period may become extremely long as compared with a typical pumping test. Therefore, it can be said that pumping tests are not feasible for tight formations. The slug test method presented by Cooper et al. (1967) has also been presented as a test procedure to measure the horizontal hydraulic conductivity of tight formations (Papadopulos et al., 1973). The period of this test can be reduced by a standpipe of small radius. However, Bredehoeft and Papadopulos (1980) showed that when the horizontal hydraulic conductivity of the formation is extremely small, the standpipe radius required to conduct the test within a reasonable time period is too small to be practical.

13.2.2 Bredehoeft and Papadopulos Method

Bredehoeft and Papadopulos (1980) presented a test procedure and a two-dimensional analytical model that is a modified version of the model developed by Cooper et al. (1967). Later, Neuzil (1982) modified the test procedure to reduce some potential errors. In the following sections, both test procedures, with some examples, are presented in detail.

13.2.2.1 Bredehoeft and Papadopulos' Test Procedure. Figure 13-1 shows the arrangement of the test proposed by Bredehoeft and Papadopulos (1980). The authors' procedure is as follows: The initial hydraulic head in the test interval may or may not be known, depending on the time that has elapsed since the well drilling activities were started. Thus, before the test, the water level in the well could be stable, or it could be rising or falling. However, any rise or fall of the water level is at a very slow rate because of relatively low horizontal hydraulic conductivity. After all preparatory work for the well is done, the well is filled with water; after a reasonable time period of observation of the water level, the system is suddenly pressurized by injecting an additional amount of water with a high-pressure pump. Then, the system is shut-in. The initial excess pressure head, H_0, is measured and then allowed to decay. The schematic head variation in the system is shown in Figure 13-2.

13.2.2.2 Neuzil's Modified Test Procedure. Neuzil (1982) critically evaluated the test procedure of Bredehoeft and Papadopulos and concluded that the authors' test procedure

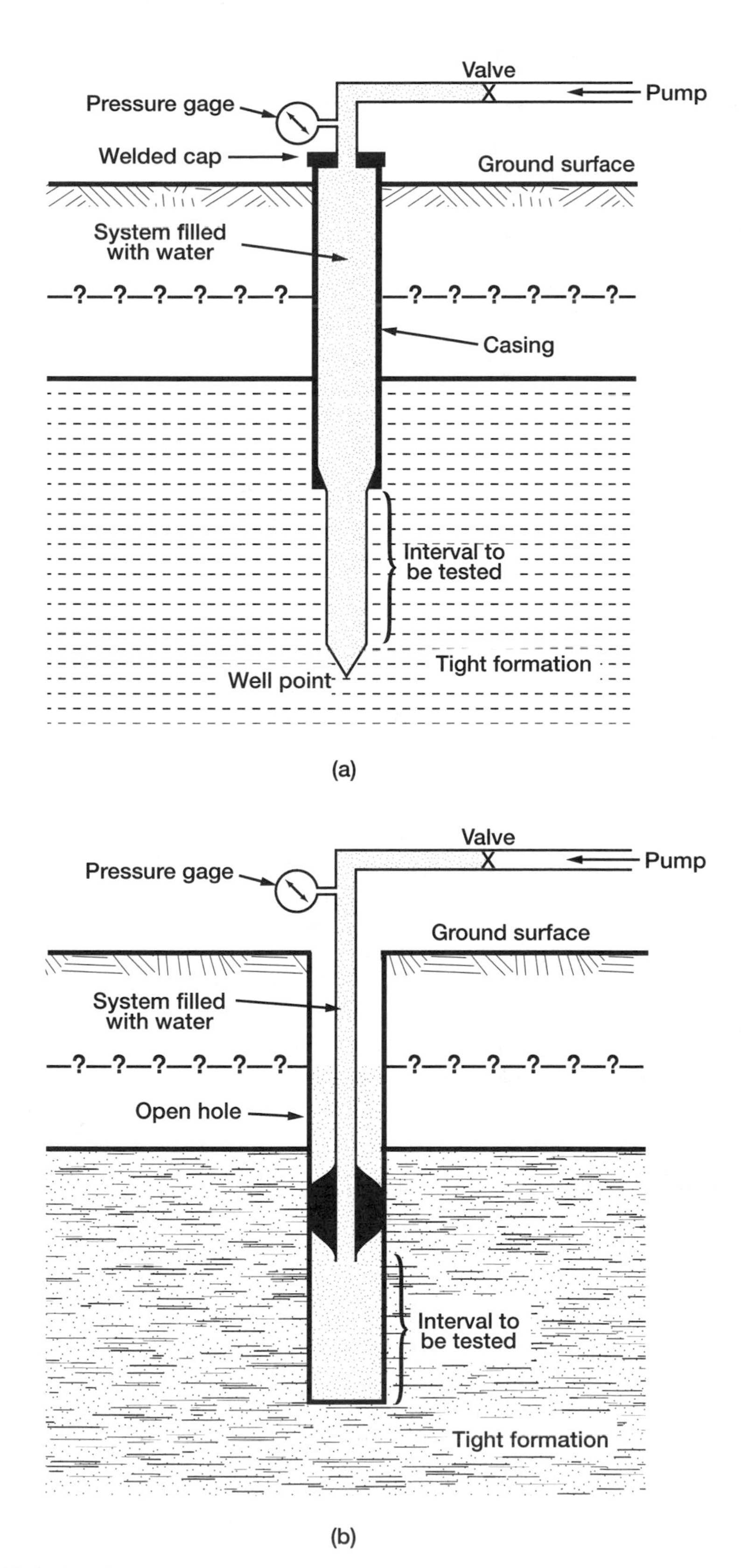

Figure 13-1 Possible arrangements for conducting pressure pulse test: (a) in unconsolidated formations and (b) in consolidated formations. (After Bredehoeft and Papadopulos, 1980.)

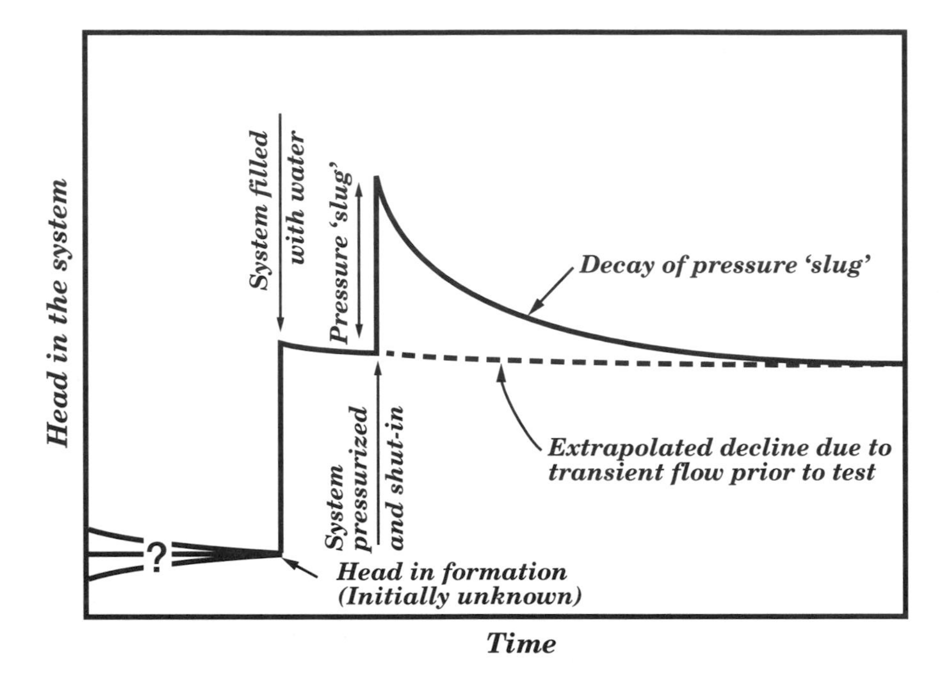

Figure 13-2 Bredehoeft and Papadopulos' schematic head variation in well system before and after pulse pressure application. (After Bredehoeft and Papadopulos, 1980.)

had to be modified. Neuzil states that the authors' procedure does not ensure approximate compliance with the initial condition of equal hydraulic head in the well and formation to be measured. As a result, calculated values of the hydrogeologic parameters could be too large or too small. For example, Bredehoeft and Papadopulos (1980) indicated that the compressibility of water, C_w, should be used to account for the change in storage in the shut-in well. Neuzil states that the compressibility in the shut-in well can be larger than the compressibility of water, and the horizontal hydraulic conductivity and storage coefficient based on C_w may become too small. Because of its importance, the theoretical background of Neuzil's modified test is presented below with an example.

Conditions Before the Test. Before the test, the borehole is filled with water and transient flow may become significant, depending on the permeability of formation. Transient flow in the well may be either a rising or a falling water level in the well. If the water level in the well is falling before shut-in, the apparent decay of the pressure will have a relatively higher rate. This will create an artificially high horizontal hydraulic conductivity value. On the contrary, a rising water level in the well before the start of the test will lead to a horizontal hydraulic conductivity value that is artificially small. Schematic variation in hydraulic head in the system before and after pressurization and shut-in is shown in Figure 13-3. To avoid these artificial conditions on the calculated horizontal hydraulic conductivity, approximate equilibrium conditions must be established before applying the slug pressure head.

Suggested Test Procedure. Figure 13-4 shows the arrangement suggested by Neuzil (1982). The borehole is filled with water, and two packers are set near one another. Two pressure transducers are used to monitor pressures. One transducer is placed between the packers and the other one is placed below the lower packer. After closing the valve, the

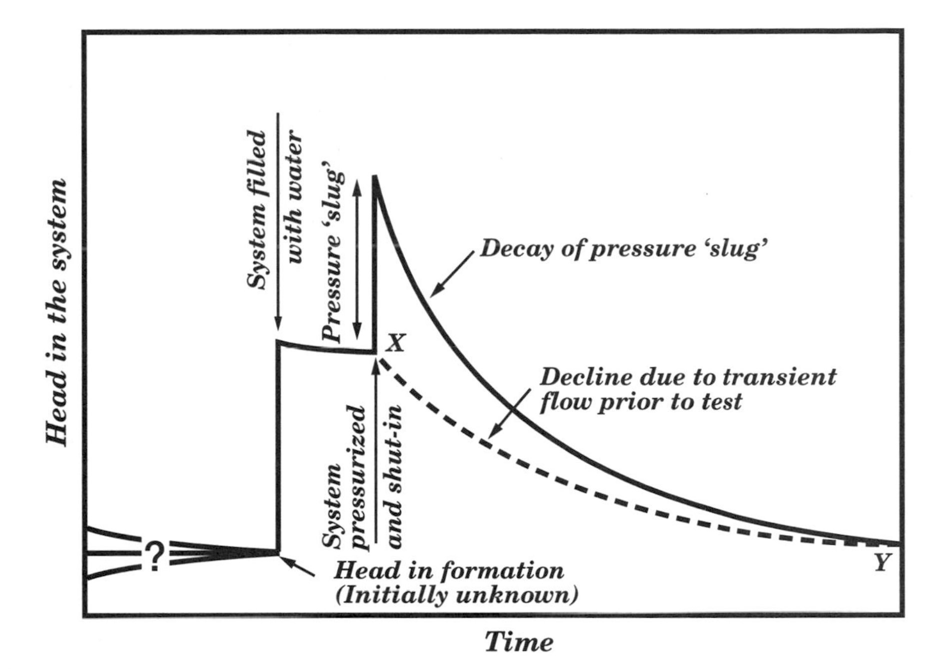

Figure 13-3 Schematic variation in hydraulic head in the system before and after pressurization and shut-in. (After Neuzil, 1982.)

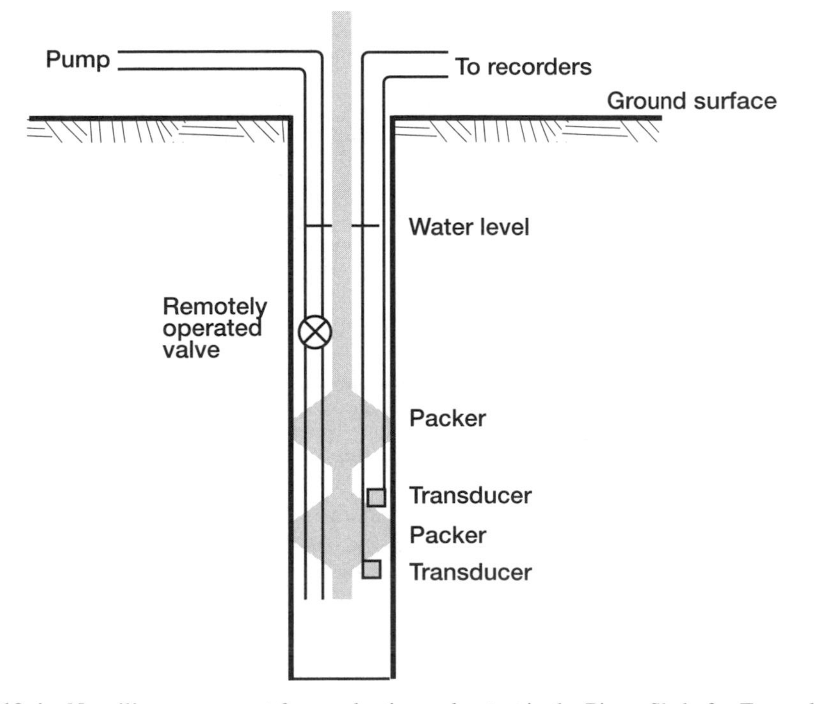

Figure 13-4 Neuzil's arrangement for conducting a slug test in the Pierre Shale for Example 13-1. (After Neuzil, 1982.)

pressures in both sections are monitored until they vary very slowly, which indicates that approximate equilibrium conditions are established. This slow head change under shut-in conditions can be extrapolated for the test period (curve X–Y in Figure 13-3).

Observed Compressibility (C_{obs}). In order to facilitate the calculation of compressibility in the well, the volume of water injected into the well by slug pressure must be known. Measuring the amount of water pumped from a container can provide the necessary data. Then, the compressibility of water in the well can be calculated by using the shut-in volume and the magnitude of the pressure change. From the definition of compressibility [Chapter 2, Section 2.4.1.3, Eq. (2.55)] one can write

$$C_{obs} = \frac{\left|\dfrac{\Delta V}{V_w}\right|}{\Delta p} \tag{13-1}$$

where ΔV is the volume of water pumped in the well, V_w is the volume of water in the shut-in segment of the well, and Δp is the magnitude of the pressure change. In Eq. (13-1), C_{obs} is equivalent to β in Eq. (2-55) of Chapter 2.

Example 13-1: This example is adapted from Neuzil (1982). Figure 13-4 shows the arrangement used for a test conducted in the Pierre Shale. The well was pressurized using a hand-operated piston pump and was shut-in by closing a valve. The shut-in pressure recorded during the test is shown in Figure 13-5. Approximately 20 pump strokes were required to pressurize the well. Later measurement showed that each pump stroke injected approximately 100 cm^3 of water. The shut-in volume in the test was 1.59 m^3 and the pressure slug was 0.462 MPa (67 lb/in^2). Calculate the observed compressibility in the shut-in well (C_{obs}) and compare it with the compressibility of water (C_w).

Solution: Substitution of $V_w = 1.59\ m^3$, $\Delta V = 20 \times 100\ cm^3 = 20 \times 100(0.01\ m)^3 = 0.002\ m^3$, and $\Delta p = 0.462\ MPa = 0.462 \times 106\ Pa$ into Eq. (13-1) gives

$$C_{obs} = \frac{\dfrac{0.002\ m^3}{1.59\ m^3}}{0.462 \times 10^6\ Pa} = 2.723 \times 10^{-9}\ Pa^{-1}$$

The compressibility of water, C_w, at 10°C (50°F) is $4.76 \times 10^{-10}\ Pa^{-1}$ (m^2/N) [Chapter 2, Table 2-11]. Thus,

$$\frac{C_{obs}}{C_w} = \frac{2.723 \times 10^{-9}\ Pa^{-1}}{4.76 \times 10^{-10}\ Pa^{-1}} = 5.72$$

This value indicates that the compressibility in the shut-in well was approximately 6 times higher than the compressibility of water. If C_w is used instead of C_{obs}, the resulting horizontal hydraulic conductivity and storage coefficient will be proportionally smaller. In other words, C_w causes underestimated hydrogeologic parameters.

13.2.2.3 Bredehoeft and Papadopulos' Model. Under the framework of the above-described pulse test procedure, Bredehoeft and Papadopulos (1980) developed an analytical two-dimensional model based on the two-dimensional analytical slug test model of Cooper

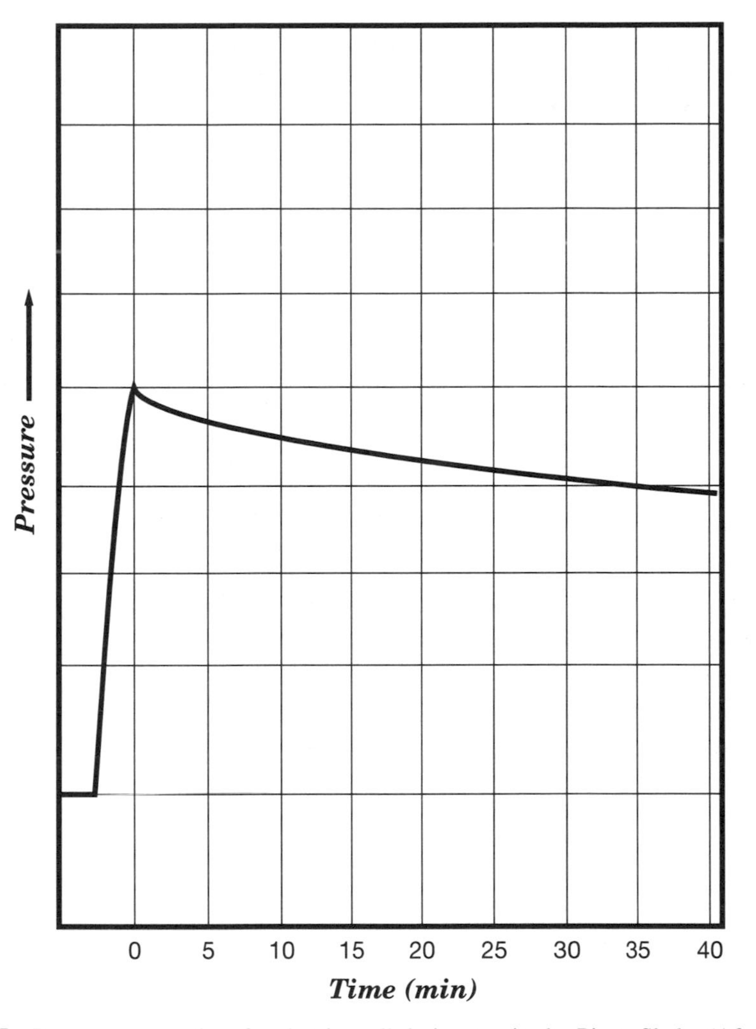

Figure 13-5 Pressure versus time for shut-in well during test in the Pierre Shale. (After Neuzil, 1982.)

et al. (1967). In the following sections the theoretical foundations as well as practical implications of this model are presented in detail.

Assumptions, Governing Equations, and Initial and Boundary Conditions. The geometry of the well system is shown in Figure 13-1. As indicated in the figure, the initial hydraulic head in the interval to be tested may or may not be known. The status of the initial head depends on the time that has elapsed since the well has been drilled. As a result, before the test the water level in the well could be stable, rising, or falling. Any rise or fall of the water level will be at a relatively very slow rate depending on the permeability of formation. The test procedure is explained in Section 13.2.2.1 and the mathematical model is based on the following assumptions:

1. After filling of the system with water, the water-level decline is negligible for the period of the pressurization.

2. The flow in the tested interval is primarily radial, and the hydraulic characteristics of the formation remain constant throughout the test.
3. The volume of water within the pressurized section of the system (V_w) remains constant.
4. The volumetric changes due to expansion and contraction of the components of the system are negligible.

For tight formations, the water-level decline after filling the system with water is at such a slow rate that either it is negligible during the pressurization period or it can be projected at the end of the pressurization period. As a result, the associated errors for assumption 1 are not significant. Assumption 2 means that the horizontal planes passing through the upper and lower ends of the pressurized section are impervious.

The two-dimensional governing equation in radial coordinates is [Chapter 2, Eq. (2-175)]

$$\frac{\partial^2 h}{\partial r^2} + \frac{1}{r}\frac{\partial h}{\partial r} = \frac{S}{T}\frac{\partial h}{\partial t} \tag{13-2}$$

where $h(L)$ is the head change in the tested interval of the formation due to pressurization, S (dimensionless) is the storage coefficient of the tested interval, $T(L^2/T)$ is transmissivity of the tested interval, and $r(L)$ is the radial distance from the center of the well. The initial and boundary conditions are presented below.

Initial Condition for $h(r, t)$. The initial condition for $h(r, t)$ is

$$h(r, 0) = 0 \tag{13-3}$$

Equation (13-3) states that initially the head change is zero everywhere outside the well.

Boundary Conditions at Well for $h(r, t)$. The boundary conditions for $h(r, t)$ are

$$h(\infty, t) = 0 \tag{13-4}$$

$$h(r_s, t) = H(t) \tag{13-5}$$

$$H(0) = H_0 \tag{13-6}$$

Equation (13-4) states that the head in the formation approaches zero as distance from the well approaches infinity. Equation (13-5) states that the head in the formation at the face of the well is equal to the head in the well at any time. Equation (13-6) states that the initial head change in the well is equal to H_0.

Boundary Condition at Well for Flux. Bredehoeft and Papadopulos (1980) only present the final equation of the corresponding boundary condition. Derivation of this equation is presented below:

As mentioned above, V_w is the volume of water introduced in the well to pressurize it, and it is assumed to remain constant throughout the test. In other words, it is the volume of the shut-in segment of the well which is initially fully occupied by water. The introduction of the additional water increases the mass of water occupying this volume and, therefore, causes the pressure to increase in order to compress the initial water mass in the same volume. As the water mass begins to leave the well through the well screen, the pressure begins to drop. The rate at which the mass leaving the well, as expressed by applying Darcy's law at the well face, is equal to the rate at which the mass is changing within the shut-in segment of the well; that is,

$$2\pi r_s T \frac{\partial h(r_s, t)}{\partial r} \Delta t \gamma_w = \Delta M(t) \tag{13-7}$$

or in the limiting case it takes the following form:

$$2\pi r_s T \frac{\partial h(r_s, t)}{\partial r} \gamma_w = \frac{\partial M(t)}{\partial t} \tag{13-8}$$

The compressibility of water is defined as [Chapter 2, Eq. (2-55)]

$$C_w = \frac{1}{E_w} = \frac{\left| \dfrac{\Delta V}{V_0} \right|}{\Delta p} = \frac{\left| \dfrac{\Delta M}{M_0} \right|}{\Delta p} \tag{13-9}$$

where M_0 is the initial mass prior to pressurization within the shut-in segment and is given by

$$M_0 = \gamma_w V_0 \tag{13-10}$$

In Eq. (13-9), C_w is equivalent to β in Eq. (2-55) of Chapter 2. Substitution of Eq. (13-10) into Eq. (13-9) and letting $\Delta p = \gamma_w \Delta H$ gives

$$\Delta M = C_w \gamma_w^2 V_w \Delta H \tag{13-11}$$

and dividing both sides by Δt for the limiting case one can write

$$\frac{\partial M}{\partial t} = C_w \gamma_w^2 V_w \frac{\partial H}{\partial t} \tag{13-12}$$

When we substitute Eq. (13-12) into Eq. (13-8) and let $\gamma_w = \rho_w g$, finally the following equation can be obtained:

$$2\pi r_s T \frac{\partial h(r_s, t)}{\partial r} = C_w V_w \rho_w g \frac{\partial H(t)}{\partial t} \tag{13-13}$$

Equation (13-13) is exactly the same as Eq. (6) of Bredehoeft and Papadopulos (1980). Equation (13-13) states that the rate at which water flows from the well into formation is equal to the rate at which the volume of water stored within the pressurized system expands as the pressure head within the system declines.

Solution. The solution of the above-defined initial and boundary value problem was determined by Bredehoeft and Papadopulos (1980) using the similarity between the slug test solution of Cooper et al. (1967), which is given in detail in Section 12.4 of Chapter 12, and the pressure pulse model. As a matter of fact, with the exception of the form of Eq. (13-13), the boundary value problem for the pressure pulse test is identical to that of the Cooper et al. (1967) conventional slug test model. The corresponding form of Eq. (13-13) for the conventional slug test of Cooper et al. (1967) is given by Eq. (12-52) of Chapter 12. Comparing Eq. (13-13) and Eq. (12-52) of Chapter 12, it can be seen that the expression for a conventional slug test of Cooper et al. (1967) becomes the solution to the problem described above if

$$\pi r_c^2 = C_w V_w \rho_w g \tag{13-14}$$

or

$$r_c = \left(\frac{C_w V_w \rho_w g}{\pi} \right)^{1/2} \tag{13-15}$$

This means that the expression given for r_c by Eq. (13-15) must be replaced everywhere in the solution as given by Eq. (12-60) of Chapter 12. Therefore, the expressions for α and β, as given by Eqs. (12-56) and (12-57) of Chapter 12, respectively, take the following forms:

$$\alpha = \frac{\pi r_s^2 S}{C_w V_w \rho_w g} \tag{13-16}$$

$$\beta = \frac{\pi T t}{C_w V_w \rho_w g} \tag{13-17}$$

Therefore, Eq. (12-59) of Chapter 12 remains the same:

$$\frac{H}{H_0} = F(\alpha, \beta) \tag{13-18}$$

where $F(\alpha, \beta)$ is given by Eq. (12-60) of Chapter 12.

Tables for $F(\alpha, \beta)$ against β are given in Tables 12-4 and 12-5 of Chapter 12 for 10 orders of α, $\alpha = 10^{-10}$ to 10^{-1}. And the type curves based on these tables are presented in Figure 12-26 of Chapter 12. However, α in Eq. (13-18) may have values that are much higher than 10^{-1}. As a result, Bredehoeft and Papadopulos (1980) presented additional values of the function for several α, in the range $10^{-1} \leq \alpha \leq 10$, and are given in Table 13-1; their graphical form is given in Figure 13-6.

Approximate Equation. Bredehoeft and Papadopulos (1980) showed that for small values of the ratio β/α—that is, for small β or large α—the $F(\alpha, \beta)$ can be approximated by the following expression:

$$F(\alpha, \beta) \cong e^{4\alpha\beta} \operatorname{erfc} \left[2(\alpha\beta)^{1/2} \right] \tag{13-19}$$

When this approximation is valid, the function is no longer to be dependent on both α and β and is only dependent on the product parameter of $\alpha\beta$. Table 13-2 presents the values of this approximation, and their graphical representation is presented in Figure 13-6. Plots based on $F(\alpha, \beta)$, as defined by Eq. (12-60) of Chapter 12, and the approximation given by Eq. (13-19) against the product parameter $\alpha\beta$, are presented in Figure 13-7. As can be seen

TABLE 13-1 Values of the Function $F(\alpha, \beta)$

	$F(\alpha, \beta)$						
β	$\alpha = 0.1$	$\alpha = 0.2$	$\alpha = 0.5$	$\alpha = 1$	$\alpha = 2$	$\alpha = 5$	$\alpha = 10$
0.000001	0.9993	0.9990	0.9984	0.9977	0.9968	0.9948	0.9923
0.000002	0.9990	0.9986	0.9977	0.9968	0.9955	0.9927	0.9894
0.000004	0.9986	0.9980	0.9968	0.9955	0.9936	0.9898	0.9853
0.000006	0.9982	0.9975	0.9961	0.9945	0.9922	0.9876	0.9822
0.000008	0.9980	0.9971	0.9955	0.9936	0.9910	0.9857	0.9796
0.00001	0.9977	0.9968	0.9949	0.9929	0.9900	0.9841	0.9773
0.00002	0.9968	0.9955	0.9929	0.9900	0.9858	0.9776	0.9683
0.00004	0.9955	0.9936	0.9899	0.9858	0.9801	0.9687	0.9558
0.00006	0.9944	0.9922	0.9877	0.9827	0.9757	0.9619	0.9464
0.00008	0.9936	0.9909	0.9858	0.9800	0.9720	0.9562	0.9387
0.0001	0.9928	0.9899	0.9841	0.9777	0.9688	0.9512	0.9318
0.0002	0.9898	0.9857	0.9776	0.9687	0.9562	0.9321	0.9059
0.0004	0.9855	0.9797	0.9685	0.9560	0.9389	0.9061	0.8711
0.0006	0.9822	0.9752	0.9615	0.9465	0.9258	0.8869	0.8458
0.0008	0.9794	0.9713	0.9557	0.9385	0.9151	0.8711	0.8253
0.001	0.9769	0.9679	0.9505	0.9315	0.9057	0.8576	0.8079
0.002	0.9670	0.9546	0.9307	0.9048	0.8702	0.8075	0.7450
0.004	0.9528	0.9357	0.9031	0.8686	0.8232	0.7439	0.6684
0.006	0.9417	0.9211	0.8825	0.8419	0.7896	0.7001	0.6178
0.008	0.9322	0.9089	0.8654	0.8202	0.7626	0.6662	0.5797
0.01	0.9238	0.8982	0.8505	0.8017	0.7400	0.6384	0.5492
0.02	0.8904	0.8562	0.7947	0.7336	0.6595	0.5450	0.4517
0.04	0.8421	0.7980	0.7214	0.6489	0.5654	0.4454	0.3556
0.06	0.8048	0.7546	0.6697	0.5919	0.5055	0.3872	0.3030
0.08	0.7734	0.7190	0.6289	0.5486	0.4618	0.3469	0.2682
0.1	0.7459	0.6885	0.5951	0.5137	0.4276	0.3168	0.2428
0.2	0.6418	0.5774	0.4799	0.4010	0.3234	0.2313	0.1740
0.4	0.5095	0.4458	0.3566	0.2902	0.2292	0.1612	0.1207
0.6	0.4227	0.3642	0.2864	0.2311	0.1817	0.1280	0.09616
0.8	0.3598	0.3072	0.2397	0.1931	0.1521	0.1077	0.08134
1.	0.3117	0.2648	0.2061	0.1663	0.1315	0.09375	0.07120
2.	0.1786	0.1519	0.1202	0.09912	0.08044	0.05940	0.04620
4.	0.08761	0.07698	0.06420	0.05521	0.04668	0.03621	0.02908
6.	0.05527	0.04999	0.04331	0.03830	0.03326	0.02663	0.02185
8.	0.03963	0.03658	0.03254	0.02933	0.02594	0.02125	0.01771
10.	0.03065	0.02870	0.02600	0.02376	0.02130	0.01776	0.01499
20.	0.01408	0.01361	0.01288	0.01219	0.01133	0.009943	0.008716
40.	0.006680	0.006568	0.006374	0.006171	0.005897	0.005395	0.004898
60.	0.004367	0.004318	0.004229	0.004132	0.003994	0.003726	0.003445
80.	0.003242	0.003214	0.003163	0.003105	0.003022	0.002853	0.002668
100.	0.002577	0.002559	0.002526	0.002487	0.002431	0.002313	0.002181
200.	0.001271	0.001266	0.001258	0.001247	0.001230	0.001194	0.001149
400.	0.0006307	0.0006295	0.0006272	0.0006242	0.0006195	0.0006085	0.0005944
600.	0.0004193	0.0004188	0.0004177	0.0004163	0.0004141	0.0004087	0.0004016
800.	0.0003140	0.0003137	0.0003131	0.0003123	0.0003110	0.0003078	0.0003035
1000.	0.0002510	0.0002508	0.0002504	0.0002499	0.0002490	0.0002469	0.0002440

Source: After Bredehoeft and Papadopulos (1980).

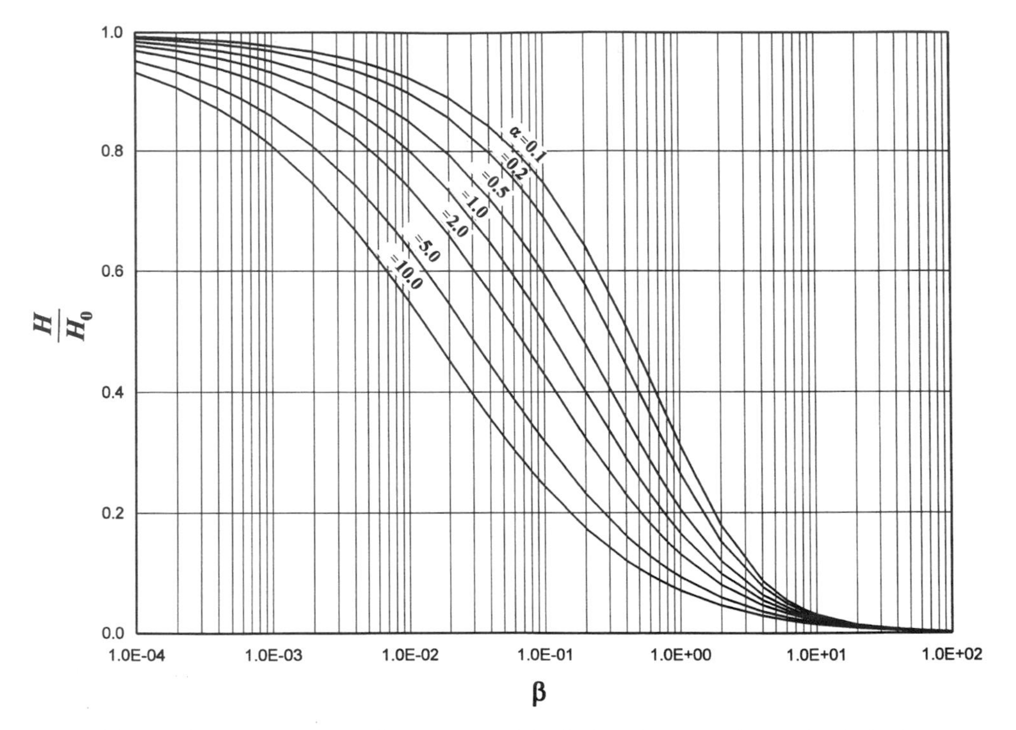

Figure 13-6 Type curves of the function $F(\alpha, \beta)$ based on the values in Table 13-1.

from Figure 13-7, for the practical purpose of analyzing pulse test data, the aforementioned approximation can be used whenever α is greater than 10.

13.2.2.4 Bredehoeft and Papadopulos' Type-Curve Method. Based on the analytical model presented above, Bredehoeft and Papadopulos (1980) developed a type-curve method for the determination of the formation hydrogeologic parameters. The authors' method of analyzing pressure head or decay head data depends on whether the parameter, α, is greater or smaller than 0.1. The applicable methods for $\alpha \leq 0.1$ and for $\alpha > 0.1$ are given in the following paragraphs. The magnitude of α would not be known in advance. First, an attempt should be made for the data analysis by the method applicable for $\alpha \leq 0.1$. If this analysis indicates that α is greater than 0.1, then the method for $\alpha > 0.1$ should be used.

Analysis of Data for $\alpha \leq 0.1$. For $\alpha \leq 0.1$, the data are analyzed in a manner similar to that of Cooper et al. (1967), which is presented in detail in Section 12.4.2 of Chapter 12, for the analysis of conventional slug test data. The steps of the method are given below:

Step 1: Using the pressure pulse (Δp), calculate the initial head change (H_0) in the well from

$$H_0 = \frac{\Delta p}{\gamma_w} \tag{13-20}$$

TABLE 13-2 Values of $F(\alpha, \beta)$ Function Based on the Approximation Given by Eq. (13-19)

$\alpha\beta$	Eq. (13-19)	$\alpha\beta$	Eq. (13-19)
0.000001	0.9977	0.6	0.3137
0.000002	0.9968	0.8	0.2800
0.000004	0.9955	1.	0.2554
0.000006	0.9945	2.	0.1888
0.000008	0.9936	4.	0.1370
0.00001	0.9929	6.	0.1129
0.00002	0.9900	8.	0.09825
0.00004	0.9859	10.	0.08813
0.00006	0.9828	20.	0.06269
0.00008	0.9801	40.	0.04446
0.0001	0.9778	60.	0.03634
0.0002	0.9689	80.	0.03149
0.0004	0.9564	100.	0.02817
0.0006	0.9470	200.	0.01993
0.0008	0.9392	400.	0.01410
0.001	0.9325	600.	0.01151
0.002	0.9066	800.	0.009972
0.004	0.8719	1000.	0.008920
0.006	0.8467	2000.	0.006307
0.008	0.8263	4000.	0.004460
0.01	0.8090	6000.	0.003642
0.02	0.7466	8000.	0.003154
0.04	0.6708	10000.	0.002821
0.06	0.6209	20000.	0.001995
0.08	0.5835	40000.	0.001410
0.1	0.5536	60000.	0.001152
0.2	0.4582	80000.	0.0009974
0.4	0.3647	100000.	0.0008921

Source: After Bredehoeft and Papadopulos (1980).

Step 2: Using the values in Tables 12-4 and 12-5 of Chapter 12, plot a family of type curves similar to those in Figure 12-26 of Chapter 12, one for each α, of $F(\alpha, \beta)$ against β on semilogarithmic paper (β is logarithmic).

Step 3: Plot $H(t)/H_0$ versus t data on semilogarithmic paper (t is logarithmic) of the same scale as the type curves, as given in Figure 13-6.

Step 4: Keeping the logarithmic axes (t and β axes) coincident, translate the data plot horizontally to a position until the best fit of the data curve and one of the family of type curves is obtained. Record the value of α.

Step 5: Select a common point, the match point, arbitrarily chosen on the overlapping part of the curve, or even anywhere on the overlapping portion of the sheets. Read the corresponding values of t and β.

Step 6: Using the observed values of ΔV, V_w, and Δp, calculate the value of C_{obs} from Eq. (13-1).

Step 7: Using the values determined from the previous steps, calculate the value of transmissivity (T) from Eq. (13-17). Then, calculate the horizontal hydraulic conductivity

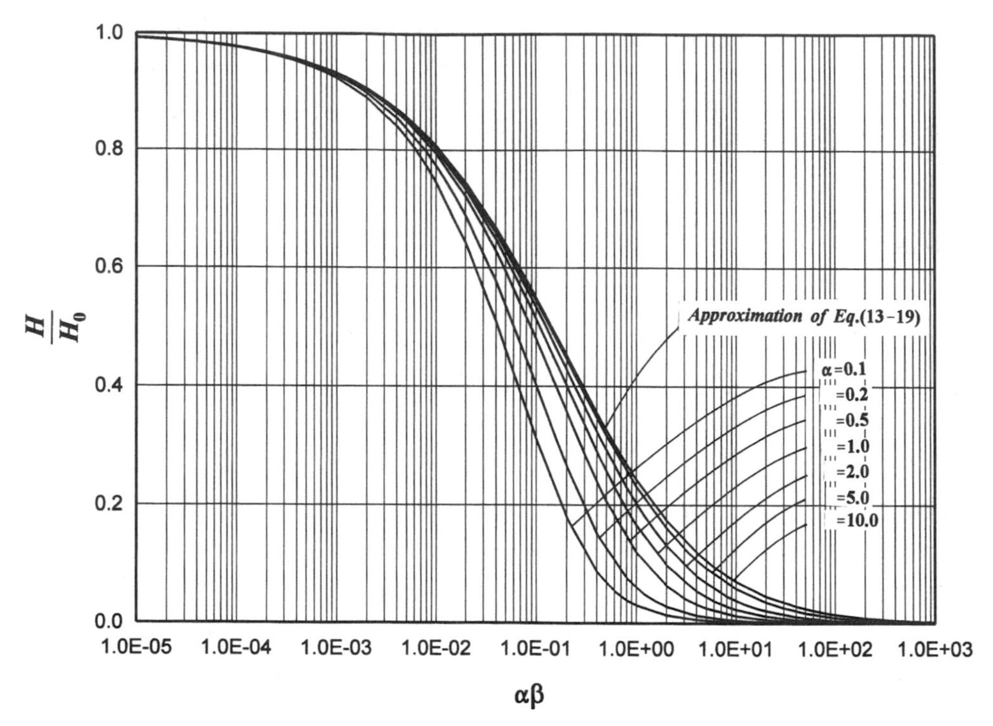

Figure 13-7 Type curves of the function $F(\alpha, \beta)$ against the product parameter $\alpha\beta$. (After Bredehoeft and Papadopulos, 1980.)

value of the tested interval from

$$K_r = \frac{T}{b} \tag{13-21}$$

where b is the length of the tested interval.

Step 8: Likewise, using the values in the previous steps, calculate the value of storage coefficient (S) from Eq. (13-16).

CAUTIONS AND LIMITATIONS OF THE METHOD. There are some cautions and limitations in applying the method for $\alpha \leq 0.1$, and they are given below based on Bredehoeft and Papadopulos (1980):

1. For $\alpha \leq 0.1$, the shape of the type curves of $F(\alpha, \beta)$ versus β are somewhat similar to each other. But the horizontal shift from one type curve to the next is small and becomes smaller as α becomes smaller. As a result, if the variation of the test data is such that a clear distinction cannot be made between two adjacent type curves, the error in the determined storage coefficient would be as large as the error in α. However, the error in the transmissivity value would be small because of the small horizontal shift in the type curves.
2. Analysis on various data sets showed that even an error of two orders of magnitude in α would result in an error of less than 30% in the transmissivity value (Papadopulos et al., 1973).

3. Based on the above facts, the use of the aforementioned type curves yields good estimates for the transmissivity value when $\alpha \leq 0.1$. However, the reliability of the storage coefficient could be questionable as also stated in Section 12.4.3 of Chapter 12 based on Cooper et al. (1967) and Papadopulos et al. (1973).

Analysis of Data for $\alpha > 0.1$. The data analysis method described above is not suitable for $\alpha > 0.1$. Some critical points regarding this situation are given below (Bredehoeft and Papadopulos, 1980):

1. As the value of α becomes large, the similarity in the shape of the type curves increases, as does the horizontal separation from one curve to the next one; the type curves tend to become parallel to each other as shown in Figure 13-6. Because of this, the horizontal separation for these type curves having one order of magnitude different α values becomes approximately one log cycle. As a result, not only the likelihood of having incorrect type-curve matching increases, but also the error in the resulted transmissivity value, caused by the incorrect choice of type curve, becomes as large as in the storage coefficient value.
2. The aforementioned situation is a result of the fact that for large α values a major part of the type curve is defined by the approximation given by Eq. (13-19) and is dependent on the product $\alpha\beta$ (see Figure 13-7). As can be seen from Figure 13-7, the upper part of the curves for different α values coincides with the curve for the approximation as given by Eq. (13-19). For the larger the α the greater the portion of the curve that coincides with the approximation.
3. Because

$$\alpha\beta = \frac{\pi^2 r_s^2 TSt}{(V_w C_w \rho_w g)^2} \tag{13-22}$$

the analysis of pulse test data falling in the range of approximation would yield only the product of transmissivity and storage coefficient, TS.

Example 13-2: The data of this example were taken from C. E. Neuzil of U.S. Geological Survey (Neuzil, 1995, personal communication). The data corresponds to the test described in Example 13-1. The test section was from 368 ft (112.17 m) to 604 ft (184.10 m) depths. Therefore, the length of the screen interval was 236 ft (71.93 m). As mentioned in Example 13-1, the volume shut-in (V_0) was 1.59 m^3 and the injected volume (ΔV) was 2000 cm^3. Pressure pulse was 0.462 MPa (67 psi). The radius of the well (r_s) was 0.067 m. The recorded gauge pressure head (h) versus time are given in Table 13-3. Using the Bredehoeft and Papadopulos method, determine the horizontal hydraulic conductivity and storage coefficient of the formation.

Solution:

Step 1: From the given values, and using $\gamma_w = 9.8k\text{N/m}^3$ (see Table 2-11 of Chapter 2), Eq. (13-20) gives

$$H_0 = \frac{\Delta p}{\gamma_w} = \frac{0.462 \times 10^6 \text{ N/m}^2}{9800 \text{ N/m}^3} = 47.14 \text{ m}$$

Step 2: The type curves are given in Figure 12-26 in which Tt/r_c^2 should be replaced by β.

TABLE 13-3 Pressure Head Versus Time for Example 13-2

Time, t (s)	Pressure, H (m)	H/H_0 (H_0 = 47.14 m)
60	45.97	0.975
120	45.43	0.964
180	45.17	0.958
300	44.63	0.947
480	43.55	0.924
600	43.28	0.918
900	41.94	0.890
1200	40.73	0.864
1800	38.17	0.810
2400	36.83	0.781
3000	35.22	0.747
4200	33.74	0.716
6000	31.18	0.661
8400	28.23	0.599
10200	26.08	0.553
12000	24.06	0.510
14400	22.04	0.468
17400	19.09	0.405
21000	16.94	0.359
27000	15.32	0.325

Step 3: The data in Table 13-3 are plotted in Figure 13-8 on semilogarithmic paper of the same scale of the type curves.

Step 4: The superimposed sheets are shown in Figure 13-9. A comparison shows that the plotted points fall along the curve $\alpha = 10^{-2}$.

Step 5: In Figure 13-9, A is an arbitrarily chosen point and its coordinates are

$$t_A = 2700 \text{ s}, \qquad \beta_A = 0.2$$

Step 6: The value of C_{obs} is calculated in Example 13-1 and its value is

$$C_{obs} = 2.723 \times 10^{-9} \text{ m}^2/\text{N (Pa}^{-1})$$

Step 7: Using the values in the previous steps, Eq. (13-17) with $C_w = C_{obs}$ gives

$$T = \frac{\beta C_{obs} V_w \gamma_w}{\pi t} = \frac{(0.2)(2.723 \times 10^{-9} \text{ m}^2/\text{N})(1.59 \text{ m}^3)(9.8 \times 10^3 \text{ N/m}^3)}{(\pi)(2700 \text{ s})}$$

$$= 1.00 \times 10^{-9} \text{ m}^2/\text{s}$$

And from Eq. (13-21), the horizonal hydraulic conductivity value is

$$K_r = \frac{T}{b} = \frac{1.00 \times 10^{-9} \text{ m}^2/\text{s}}{71.93 \text{ m}} = 1.39 \times 10^{-11} \text{ m/s}$$

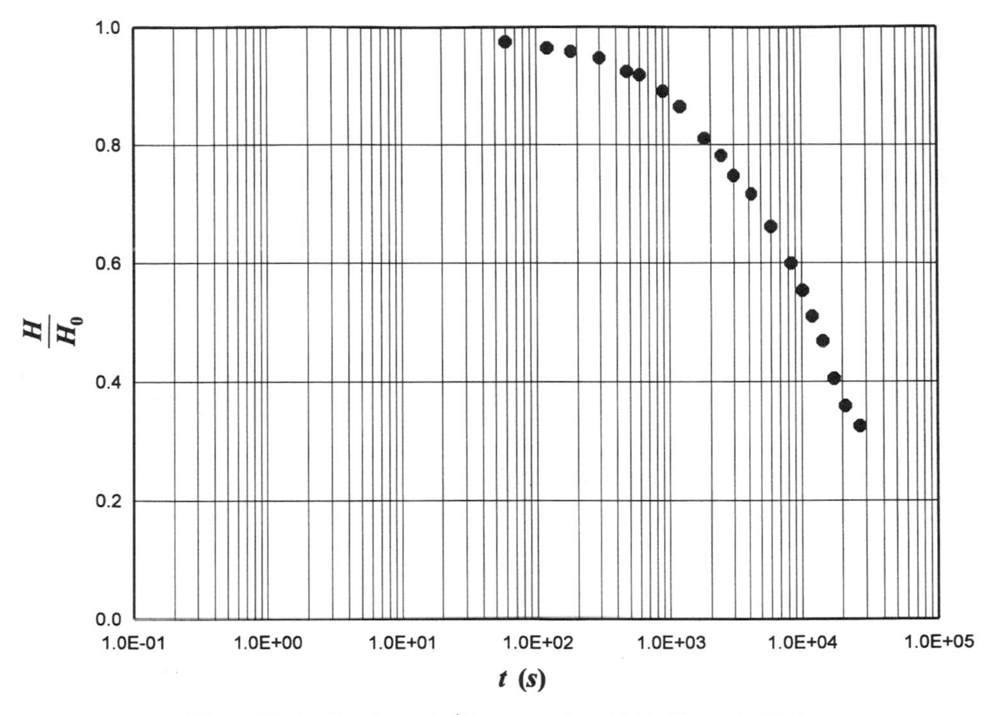

Figure 13-8 Graphs of H/H_0 versus time (t) for Example 13-2.

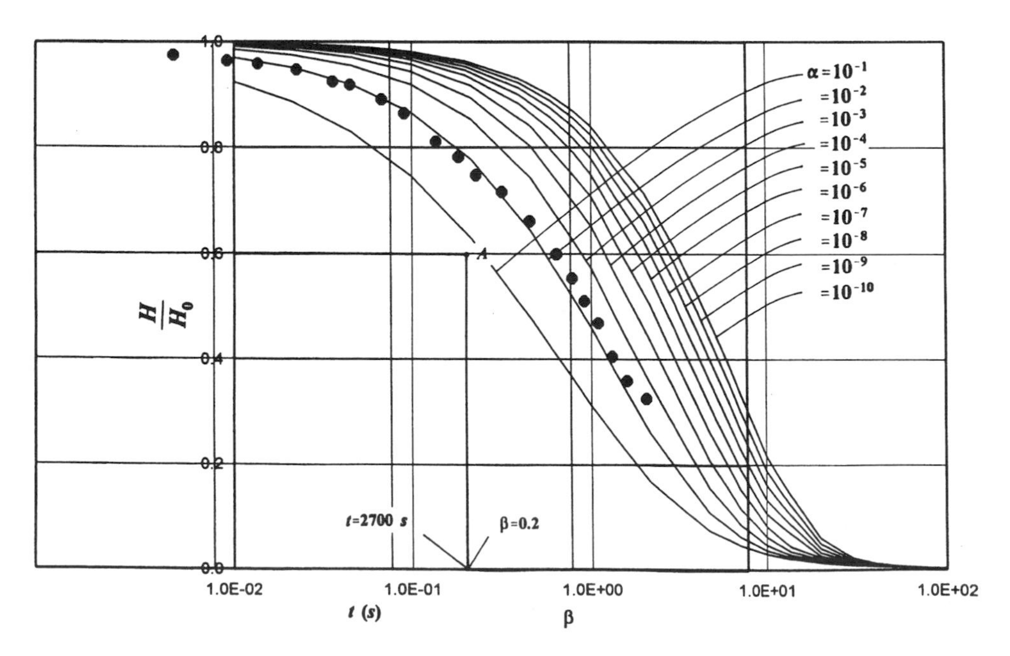

Figure 13-9 Bredehoeft and Papadopulos type-curve matching for Example 13-2.

Step 8: From Eq. (13-16) with $C_w = C_{\text{obs}}$, the value of storage coefficient is

$$S = \frac{\alpha C_{\text{obs}} V_w \gamma_w}{\pi r_s^2} = \frac{\left(10^{-2}\right)\left(2.723 \times 10^{-9}\ \text{m}^2/\text{N}\right)\left(1.59\ \text{m}^3\right)\left(9.8 \times 10^3 \text{N/m}^3\right)}{\pi (0.067\ \text{m})^2}$$

$$= 3.0 \times 10^{-5}$$

13.3 CONSTANT HEAD INJECTION TESTS

13.3.1 Introduction

Constant head injection tests are routinely used for the determination of hydraulic characteristics of tight formations, especially low-permeability rocks and compacted clays. More specifically, constant head injection tests are used for the determination of the values of transmissivity (T) and storage coefficient. The test interval in a tight formation is instantaneously pressurized and kept constant throughout the test period, and the variation of injected flow rate versus time is recorded. Initially, the flow rate is at a maximum; and as time increases, its value decreases and eventually approaches a constant value if the test period is long enough. This is the steady-state condition, and reaching steady state depends on the degree of the formation tightness.

During the past century, several methods were developed for the analysis of flow rate versus time data to determine the hydrogeologic parameters of a given formation. One important limitation of these methods is that they are all based on the assumption that the formation only exists along the screen interval of the well. In other words, the horizontal planes passing through the bottom and top of the screen interval are assumed to be impermeable.

Under steady-state conditions, which can be reached in relatively long time periods in tight formations, the Thiem equation (Thiem, 1906) can be used. One problem associated with the Thiem equation is that the zone of influence radius is unknown and a realistic guess needs to be made for the determination of horizontal hydraulic conductivity using the constant flow rate and the radius of the well. The equations presented by the U.S. Bureau of Reclamation (1968) are based on the Thiem equation by the inclusion of some empirical parameters. Under transient conditions, the data analysis methods are mostly based on the fully penetrating well model developed by Jacob and Lohman (1952) for a nonleaky confined aquifer, and this model has been adopted as a standard test method by the American Society of Testing Materials (1986).

13.3.2 Steady-State Flow Model

Thiem (1906) was the first who developed a mathematical model for a well in a confined aquifer under steady-state conditions. The mathematical foundations of the Thiem equation are given in Section 3.2.1 of Chapter 3. In the following sections, the general aspects of the Thiem equation as well as its application to constant head injection tests under steady state flow conditions are given.

13.3.2.1 Thiem's Equation. The Thiem equation is [Chapter 3, Eq. (3-9)]

$$h(r_2) - h(r_1) = \frac{Q}{2\pi T} \ln\left(\frac{r_2}{r_1}\right) \tag{13-23}$$

As mentioned in Section 3.2.1.4 of Chapter 3, in Eq. (13-23) the piezometric head increases indefinitely with increasing radial distance (r_2). Steady radial flow does not exist in an areally extensive aquifer because the piezometric surface cannot rise above $h(R)$. It is also important to note that Eq. (13-23) is valid only in the close proximity of a well where steady flow has been established. For $r_1 = r$ and $r_2 = R$, Eq. (13-23) takes the following form, which is Eq. (3-10) of Chapter 3:

$$h(R) - h(r) = \frac{Q}{2\pi T} \ln\left(\frac{R}{r}\right) \tag{13-24}$$

As mentioned in Section 3.2.1.4 of Chapter 3, some important aspects of Eq. (3-24) are as follows: (1) The distance R, for which the drawdown is zero, is the *influence radius of the well*; (2) the parameter R has to be estimated before the prediction of drawdowns; (3) in Eq. (13-24), R is in the form of $\ln(R)$. For this reason, even a large error in estimating R does not significantly affect the drawdown determined by Eq. (13-24).

13.3.2.2 Application to Constant Head Injection Tests Data Analysis. If the flow rate to a well under constant head injection conditions does not change significantly after a certain period of time, the Thiem equation, as given by Eq. (13-24), can be used in determining the transmissivity of a tight formation. From Eq. (13-24)

$$T = \frac{Q}{2\pi H} \ln\left(\frac{R}{r_w}\right) \tag{13-25}$$

where $H = h(R) - h(r)$ is the constant head change imposed on the interval and r_w is the radius of the well. The radius to the boundary; in other words, the radius of influence, R, is never known precisely, and because of this the accuracy of the estimate is limited. However, because the term appears as a logarithmic variable, large variations in R result in only small changes in T. The equations presented by the U.S. Bureau of Reclamation (1968) are based on the Thiem equation by the inclusion of some empirical parameters.

Example 13-3: A constant head injection test was conducted in a tight formation with $H = 10$-m constant head. Under steady-state conditions it was recorded that Q is equal 1.0×10^{-5} liters/s (1.0×10^{-8} m^3/s). The length of the screen interval (b) and radius of the well (r_w) were 1 m and 0.1 m, respectively. Determine the formation horizontal hydraulic conductivity by conducting a sensitivity analysis for different values of the radius of influence (R).

Solution: As mentioned above, the value of the radius of influence (R) is unknown. Therefore, the hydraulic conductivity value must be determined by evaluating the sensitivity of R on the calculated hydraulic conductivity values. Substitution of the values in Eq. (13-25) gives

$$T = \frac{1.0 \times 10^{-8}\ \text{m}^3/\text{s}}{(2\pi)(10\text{m})} \ln\left(\frac{R}{r_w}\right) = 1.59 \times 10^{-10} \ln\left(\frac{R}{r_w}\right)\ \text{m}^2/\text{s}$$

TABLE 13-4 Hydraulic Conductivity Values for Different Values of R for Example 13-1

R (m)	$\ln(R/r_w)$	T (m²/s)	K_r (cm/s)
0.5	1.61	2.56×10^{-10}	2.56×10^{-8}
1	2.30	3.66×10^{-10}	3.66×10^{-8}
2	3.00	4.77×10^{-10}	4.77×10^{-8}
4	3.69	5.87×10^{-10}	5.87×10^{-8}
5	3.91	6.22×10^{-10}	6.22×10^{-8}

The calculated values of hydraulic conductivity for different values of R are listed in Table 13-4 from $T = K_r b$. The values of R ranges from 0.5 m to 5 m. As can be seen from Table 13-4, the horizontal hydraulic conductivity for $R = 0.5$ m is close to the half of the hydraulic conductivity for $R = 5$ m.

13.3.3 Unsteady-State Flow Model: The Jacob And Lohman Equation.

Jacob and Lohman (1952) developed a model for a well in a confined aquifer for constant head injection under transient flow conditions. In the following sections, the mathematical foundations of the Jacob and Lohman model and its practical applications in determining horizontal hydraulic conductivity and storage coefficient of tight formations are presented.

13.3.3.1 Problem Statement and Assumptions. Jacob and Lohman (1952) developed a mathematical model for the determination of the values of transmissivity (T) and storage coefficient (S) from tests in which the screen interval is applied a constant head and the discharge varied with time. Figure 13-10 shows a cross section through a well with constant head (s_w) in a nonleaky confined formation. Figure 13-11 conceptually illustrates the flow rate versus time for a well in confined aquifer under constant pressure head. The assumptions regarding this problem are as follows:

1. The difference between the water level in the well and the head of the confined aquifer is s_w is constant throughout the test period.
2. The well has a finite diameter and fully penetrates the formation.
3. Darcy's law is applicable.

According to the second assumption, the well fully penetrates the formation. This means that the horizontal planes passing through the upper and lower ends of the screen interval are impermeable. As a result, the flow is assumed to be horizontal toward the screen interval, and the vertical flow components are neglected. This is an important assumption, and its adequacy has not been evaluated nor has an extended model been introduced into the literature yet by taking into account the effects of partial penetration.

13.3.3.2 Governing Equations. Because the well fully penetrates the formation, the excess pressure head (s) is constant at any vertical section. Therefore, the governing partial differential equation of the system is [Chapter 2, Eq. (2-175)]

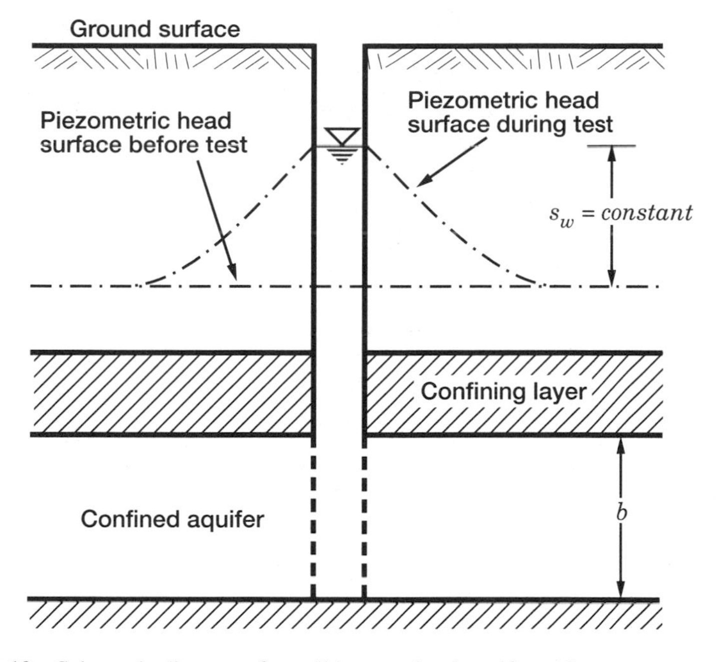

Figure 13-10 Schematic diagram of a well in a confined aquifer with constant pressure head.

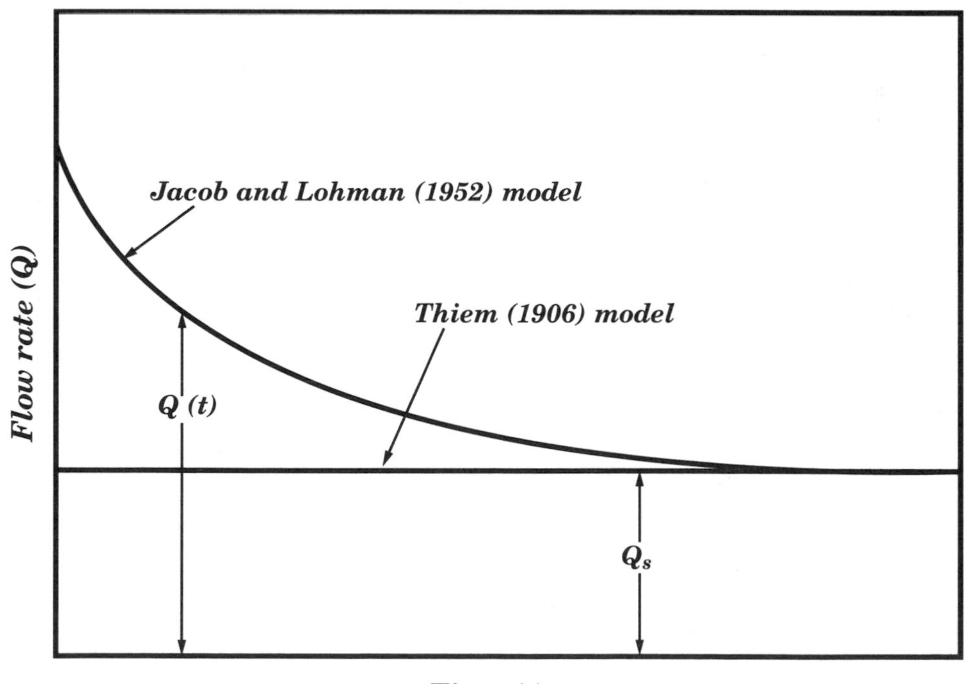

Figure 13-11 Conceptual flow rate versus time for a well in a confined aquifer under constant pressure head.

$$\frac{\partial^2 s}{\partial r^2} + \frac{1}{r}\frac{\partial s}{\partial r} = \frac{S}{T}\frac{\partial s}{\partial t} \tag{13-26}$$

This is the differential equation describing nonsteady radial flow in a homogeneous and isotropic confined aquifer.

13.3.3.3 Initial and Boundary Conditions. The initial and boundary conditions are

$$s(r, 0) = 0, \qquad r_w \leq r \leq \infty \tag{13-27}$$

$$\begin{aligned} s(r_w, t) &= 0, & t &< 0 \\ s(r_w, t) &= s_w = \text{constant}, & t &\geq 0 \end{aligned} \tag{13-28}$$

$$s(\infty, t) = 0, \qquad t \geq 0 \tag{13-29}$$

Equation (13-27) states that initially the excess pressure head is zero everywhere in the formation. Equation (13-28) states that the excess pressure head at the well is constant and the discharge varies with time. Equation (13-29) states that the excess pressure head approaches zero as the distance from the well approaches to infinity.

13.3.3.4 Solution. The solution for the well discharge is (Jacob and Lohman, 1952)

$$Q(\alpha) = 2\pi T s_w G(\alpha) \tag{13-30}$$

where

$$G(\alpha) = \frac{4\alpha}{\pi}\int_0^{\infty} x \exp\left(-\alpha x^2\right)\left\{\frac{\pi}{2} + \tan^{-1}\left[\frac{Y_0(x)}{J_0(x)}\right]\right\} dx \tag{13-31}$$

and

$$\alpha = \frac{Tt}{Sr_w^2} \tag{13-32}$$

and $J_0(x)$ and $Y_0(x)$ are the Bessel functions of zero order of the first and second kinds, respectively. Jacob and Lohman (1952) tabulated the $G(\alpha)$ function for various values of α, and these values are given in Table 13-5.

13.3.3.5 Asymptotic Solutions. Jacob and Lohman (1952) noted that for small values of α (or small relative values of t), $G(\alpha)$ appear to approach the line plotted for $1/(\pi\alpha)^{1/2}$, as it should give the discharge from an infinite aquifer bounded by a plane when r_w approaches to infinity.

For large values of α (or large values of t), the G function approaches to $2/W(1/4\alpha)$, where W stands for the Theis well function (see Section 4.2.4.2 of Chapter 4). This is also as it should be because the drawdown in a well of radius r_w discharging at the rate Q for a period of time t is (Jacob and Lohman, 1952, p. 561)

$$s_w = \frac{Q}{4\pi T} W\left(\frac{1}{4\alpha}\right) = \frac{Q}{2\pi T}\frac{W\left(\dfrac{r_w^2 S}{4Tt}\right)}{2} \tag{13-33}$$

TABLE 13-5 Values of $G(\alpha)$ for Various Values of α

α	$\alpha \times 10^{-4}$	10^{-3}	10^{-2}	10^{-1}	1	10	10^2	10^3	10^4	10^5
1	56.9	18.34	6.13	2.249	0.985	0.534	0.346	0.251	0.1964	0.1608
2	40.4	13.11	4.47	1.716	.803	.461	.311	.232	.1841	.1524
3	33.1	10.79	3.74	1.477	.719	.427	.294	.222	.1777	.1479
4	28.7	9.41	3.30	1.333	.667	.405	.283	.215	.1733	.1449
5	25.7	8.47	3.00	1.234	.630	.389	.274	.210	.1701	.1426
6	23.5	7.77	2.78	1.160	.602	.377	.268	.206	.1675	.1408
7	21.8	7.23	2.60	1.103	.580	.367	.263	.203	.1654	.1393
8	20.4	6.79	2.46	1.057	.562	.359	.258	.200	.1636	.1380
9	19.3	6.43	2.35	1.018	.547	.352	.254	.198	.1621	.1369
α	$\alpha \times 10^{6}$	10^{7}	10^{8}	10^{9}	10^{10}	10^{11}	10^{12}	10^{13}	10^{14}	10^{15}
1	0.1360	0.1177	0.1037	0.0927	0.0838	0.0764	0.0704	0.0651	0.0605	0.0566
2	.1299	.1131	.1002	.0899	.0814	.0744	.0686	.0636	.0593	.0555
3	.1266	.1106	.0982	.0883	.0801	.0733	.0677	.0628	.0586	.0549
4	.1244	.1089	.0968	.0872	.0792	.0726	.0671	.0622	.0581	
5	.1227	.1076	.0958	.0864	.0785	.0720	.0666	.0618	.0577	
6	.1213	.1066	.0950	.0857	.0779	.0716	.0662	.0615	.0574	
7	.1202	.1057	.0943	.0851	.0774	.0712	.0658	.0612	.0572	
8	.1192	.1049	.0937	.0846	.0770	.0709	.0655	.0609	.0569	
9	.1184	.1043	.0932	.0842	.0767	.0706	.0653	.0607	.0567	

Source: After Jacob and Lohman (1952) and Reed (1980).

By comparing Eqs. (13-30) and (13-33), it can be seen that $2/W$ in the latter corresponds to G in the former. Jacob and Lohman (1952, p. 561) reach the following conclusions: "Inasmuch as the two approach each other asymptotically, this is another way of saying that the ratios of discharge to drawdown in two isolated wells—one with constant discharge, the other with constant drawdown—approach equality at sufficiently large relative values of t."

Further Approximation of $G(\alpha)$. For large relative values of t the following equation can approximately be written from Eq. (4-46) as

$$W\left(\frac{r_w^2 S}{4Tt}\right) \cong (2.303)\log\left(\frac{2.25Tt}{r_w^2 S}\right) \tag{13-34}$$

This is equivalent to the Cooper and Jacob approximation (Cooper and Jacob, 1946), and its details are given in Section 4.2.7.2 of Chapter 4. Jacob and Lohman (1952) state that the difference is only about 0.02 percent of W at $\alpha = 200$, where the difference between G and $2/W$ is five percent of G. As a result, the authors conclude that G can be replaced by $2/2.303\log(2.25Tt/r_w^2 S)$ instead of $2/W$, and the discharge can be determined from

$$Q = \frac{4\pi T s_w}{(2.303)\log\left(\dfrac{2.25Tt}{r_w^2 S}\right)} \tag{13-35}$$

Equation (13-35) can also be written in inverted form as given below:

$$\frac{4\pi T s_w}{Q} = 2.303 \log\left(\frac{2.25Tt}{r_w^2 S}\right) \tag{13-36}$$

From the above expressions, the expression for transmissivity (T) and storage coefficient (S) can be derived to be (Lohman, 1972)

$$T = \frac{2.303}{4\pi \dfrac{\Delta\left(\dfrac{s_w}{Q}\right)}{\Delta\left[\log\left(\dfrac{t}{r_w^2}\right)\right]}} \tag{13-37}$$

$$S = \frac{\dfrac{2.25Tt}{r_w^2}}{\log^{-1}\left[\dfrac{\left(\dfrac{s_w}{Q}\right)}{\Delta\left(\dfrac{s_w}{Q}\right)}\right]} \tag{13-38}$$

Equation (13-38) can also be expressed as

$$S = \frac{\dfrac{2.25Tt}{r_w^2}}{10^{\left[\dfrac{\dfrac{s_w}{Q}}{\Delta\left(\dfrac{s_w}{Q}\right)}\right]}} \tag{13-39}$$

For the above approximation, Jacob and Lohman (1952, p. 562) make clear that this approximation is similar to the Cooper and Jacob (1946) approximation for the Theis well function. But they only state that their straight-line approximation is valid for large α values of time without giving a criterion for the dimensionless time parameter (α).

13.3.4 Data Analysis Methods

13.3.4.1 Type-Curve Method. The steps of the type curve method are as follows:

Step 1: Plot the type curve $G(\alpha)$ versus α on log–log paper using the type-curve data in Table 13-5.

Step 2: On another sheet of log–log paper with the same scale as in Step 1, plot the observed flow rate data versus time.

Step 3: Keeping the coordinates axes at all times parallel, superimpose the two sheets until the best fit of data curve and as much as the earliest part of the type curve is obtained.

Step 4: Select a common point, the match point, arbitrarily chosen on the overlapping portion of the curve, or even anywhere on the overlapping portion of the sheets. Read the corresponding α, $G(\alpha)$, Q, and t coordinates.

TABLE 13-6 Flow Rate Versus Time Data for Example 13-4

Time Since Flow Started (min)	Rate of Flow (gpm)	(m³/day)	s_w/Q (m/m³/day)	t/r_w^2 (min/m²)
1	7.28	39.68	0.709	141.72
2	6.94	37.82	0.744	283.45
3	6.88	37.50	0.750	425.17
4	6.28	34.23	0.822	566.89
5	6.22	33.90	0.830	708.62
6	6.22	33.90	0.830	850.34
8	5.95	32.43	0.868	1133.79
11	5.85	31.88	0.883	1558.96
16	5.66	30.85	0.912	2267.57
21	5.50	29.98	0.939	2976.19
26	5.34	29.10	0.967	3684.81
31	5.34	29.10	0.967	4393.42
41.5	5.22	28.45	0.989	5881.52
51	5.14	28.01	1.000	7227.89
61	5.11	27.85	1.010	8645.12
76	5.05	27.52	1.020	10770.98
91	5.00	27.25	1.030	12896.83
103	4.92	26.81	1.050	14597.51
113	4.88	26.60	1.060	16014.74

[a] After Lohman (1972).

Step 5: Substitute the values of Q, $G(\alpha)$, and s_w in Eq. (13-30) and calculate the value of T.

Step 6: Substitute the corresponding values in Eq. (13-32) and calculate the value of S.

Step 7: Calculate the value of horizontal hydraulic conductivity (K_r) from $K_r = T/b$ in which b is the length of the screen interval.

Example 13-4: The data of this example are taken from Lohman (1972) by converting them to metric units. The test was conducted on the Artesia Heights well near Grand Junction, Colorado on September 28, 1948 (Lohman, 1965, Tables 6 and 7, Well 28). After the well was shut in for a period of several days, the head just prior to the test was 92.33 ft (28.14 m). The radius of the well (r_w) was 0.084 m. The data are given in Table 13-6. The length of screen interval is not given. Using the type-curve method given in Section 13.3.4.1, determine the transmissivity and storage coefficient values.

Solution:

Step 1: Two segments of the Jacob and Lohman type curve, $G(\alpha)$ versus α, on log–log paper using the data in Table 13-5 are given in Figures 13-12 and 13-13. Note from Table 13-5 that the range α is quite large, from 1.0×10^{-4} to 3.0×10^{15}.

Step 2: The observed flow rate data versus time on log–log paper are presented in Figure 13-14 using the first and third column values in Table 13-6.

Step 3: The type-curve matching is presented in Figure 13-15.

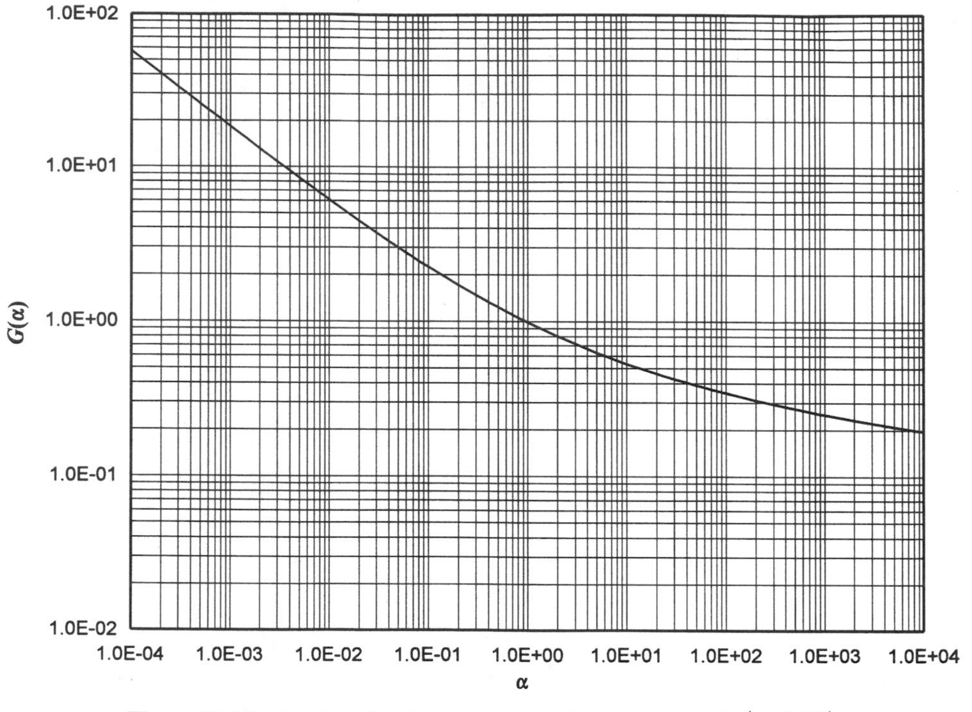

Figure 13-12 Jacob and Lohman type curve between $\alpha = 10^{-4}$ and 10^4.

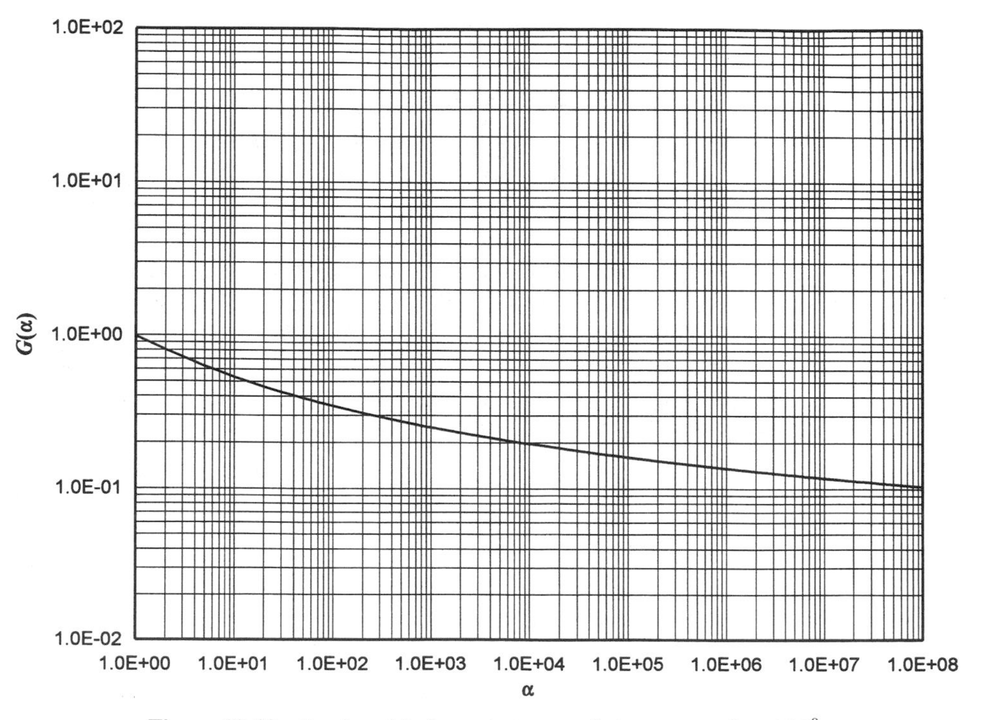

Figure 13-13 Jacob and Lohman type curve between $\alpha = 1$ and 10^8.

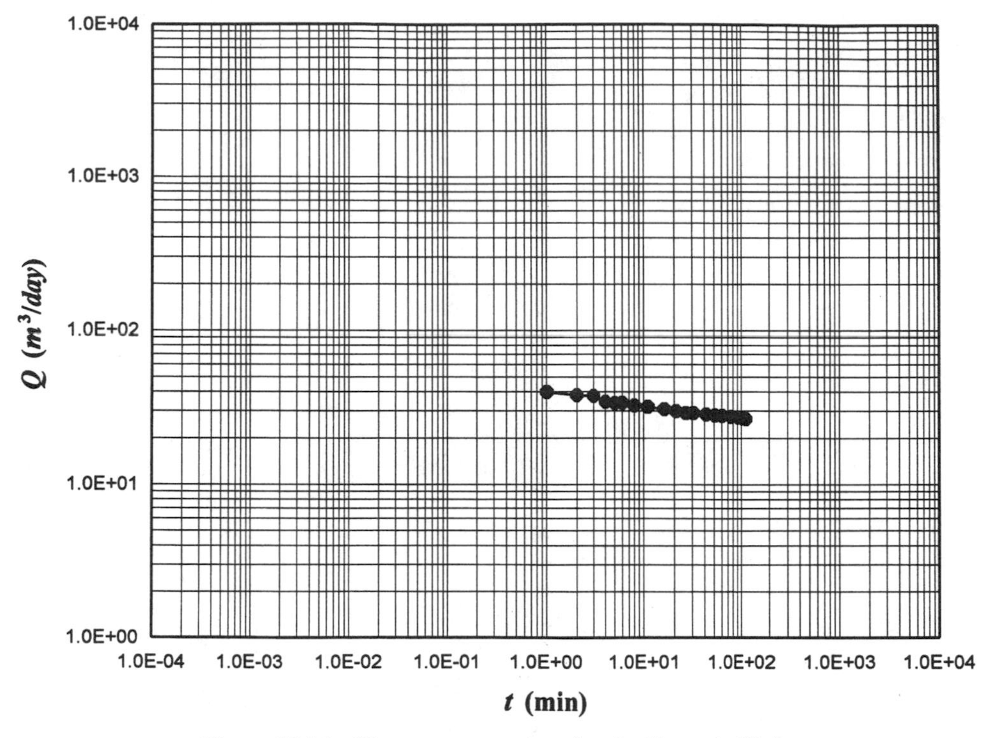

Figure 13-14 Flow rate versus time data for Example 13-4.

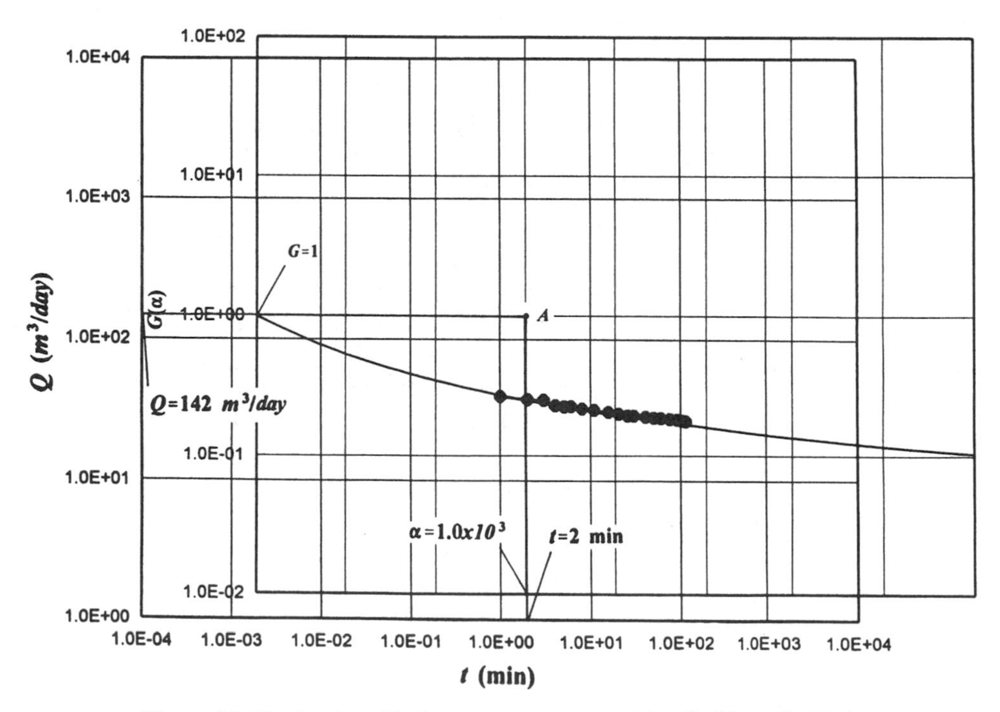

Figure 13-15 Jacob and Lohman type-curve matching for Example 13-4.

Step 4: The coordinates of the match point (A) from Figure 13-15 are

$$t_A = 2 \text{ min}, \qquad Q_A = 142 \text{ m}^3/\text{day}$$

$$\alpha_A = 1.0 \times 10^3, \qquad G_A(\alpha) = 1.0$$

Step 5: Substitution of the values of Q, $G(\alpha)$, and s_w in Eq. (13-30) gives

$$T = \frac{Q}{2\pi G(\alpha)s_w} = \frac{142 \text{ m}^3/\text{day}}{(2\pi)(1.0)(28.14 \text{ m})} = 0.803 \text{ m}^2/\text{day}$$

Step 6: Substituting the corresponding values in Eq. (13-32) gives the value of S

$$S = \frac{Tt}{\alpha r_w^2} = \frac{(0.803 \text{ m}^2/\text{day})\,(2/1440 \text{ days})}{(10^3)\,(0.084 \text{ m})^2} = 1.58 \times 10^{-4}$$

Step 7: Because the length of the screen interval is not given, the horizontal hydraulic conductivity cannot be determined.

13.3.4.2 *Straight-Line Method.* The steps of the straight-line method are as follows:

Step 1: Plot the s_w/Q versus t/r_w^2 data on single logarithmic paper (t/r_w^2 on logarithmic scale) and draw a line through the points.

Step 2: Determine the geometric slope of the line, $\Delta(s_w/Q)/\Delta(\log t/r_w^2)$, and introduce it into Eq. (13-37) and calculate the value of T.

Step 3: Select a point on the straight line and read its t/r_w^2 and s_w/Q coordinate values.

Step 4: Substitute these coordinates as well as the other values in Eq. (13-38) or (13-39) and calculate the value of S.

Step 5: Calculate the horizontal hydraulic conductivity value from $K_r = T/b$ where b is the length of the screen interval.

Cautions and Limitations of the Method.

1. As stated above, the Jacob and Lohman straight-line approximation is similar to the Cooper and Jacob (1946) approximation for the Theis well function. The criterion for the Cooper and Jacob approximation is $u < 0.01$ (see Section 4.2.7.2 of Chapter 4). A similar criterion is not given for the Jacob and Lohman straight-line method.
2. Jacob and Lohman (1952) state that their straight-line approximation is valid for large values of time. This, obviously, creates certain difficulties in applying the method.
3. Under the circumstances mentioned above, evaluation of the results by other methods, if available, become important.

Example 13-5: Using the Jacob and Lohman straight-line method, determine the hydrogeologic parameters of the formation described in Example 13-4.

Step 1: Using the data in Table 13-6, an s_w/Q versus t/r_w^2 graph with the drawn straight line is given in Figure 13-16.

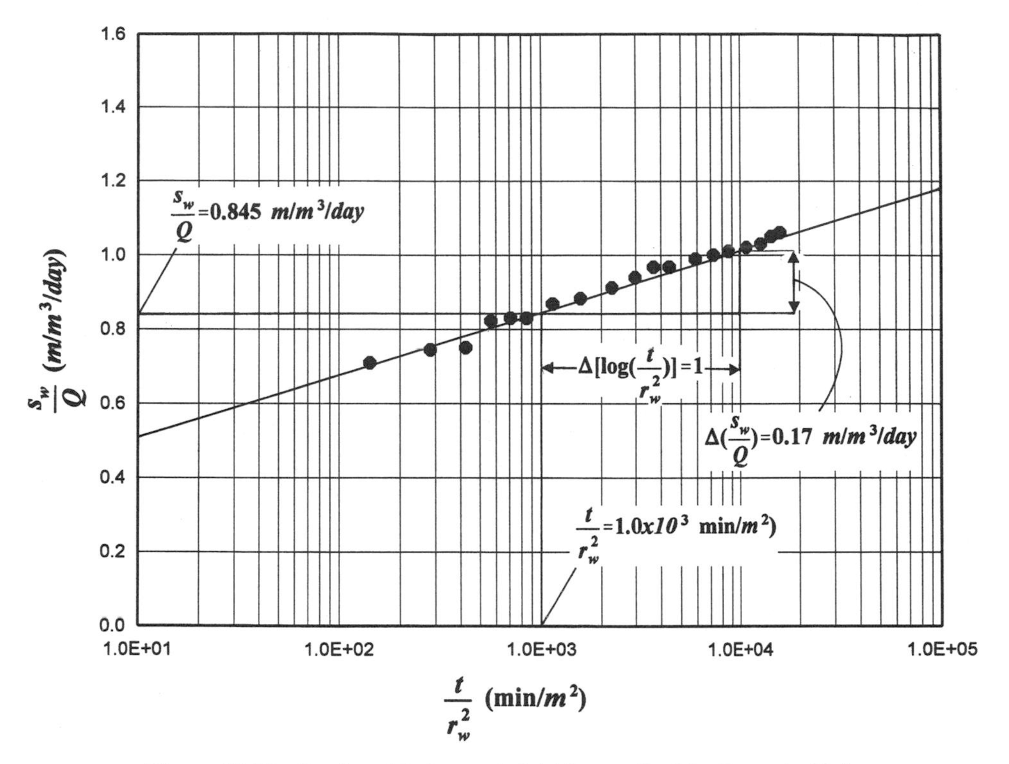

Figure 13-16 Jacob and Lohman straight-line method for Example 13-5.

Step 2: From Figure 13-16, from the geometric slope of the straight line one gets

$$\Delta\left(\frac{s_w}{Q}\right) = 0.17 \text{ m/m}^3\text{/day}$$

From Eq. (13-37) for one log cycle one can write

$$T = \frac{2.303}{(4\pi)\Delta\left(\frac{s_w}{Q}\right)} = \frac{2.303}{(4\pi)\left(0.17 \text{ m/m}^3\text{/day}\right)} = 1.078 \text{ m}^2\text{/day} = 7.486\times10^{-4} \text{ m}^2\text{/min}$$

Step 3: The coordinates of the selected points are

$$\frac{t}{r_w^2} = 1.0 \times 10^3 \text{ min/m}^2, \qquad \frac{s_w}{Q} = 0.845 \text{ m/m}^3\text{/day}$$

Step 4: Substitution the above values in Eqs. (13-38) or (13-39) gives

$$S = \frac{(2.25)\left(4.786 \times 10^{-4} \text{ m}^2\text{/min}\right)\left(10^3\text{min/m}^2\right)}{10^{\left(\frac{0.845 \text{ m/m}^3\text{/day}}{0.17 \text{ m/m}^3\text{/day}}\right)}} = 1.15 \times 10^{-5}$$

Step 5: Because the length of the screen interval is not given, the horizontal hydraulic conductivity cannot be determined.

REFERENCES

Alyamani, M. S., and Z. Şen, "Determination of Hydraulic Conductivity from Complete Grain-Size Distribution Curves," *Ground Water*, Vol. 31, No. 4, pp. 551–555, July–August 1993.

American Society of Testing Materials (ASTM), "Standard Test Method for Determining Transmissivity and Storativity of Low-Permeable Rocks by In Situ Measurements Using the Constant Head Injection Test," *Designation: D 4630-86*, pp. 207–212, Philadelphia, Pennsylvania, 1986.

Babbitt, H. E., and D. H. Caldwell, "The Free Surface Around and Interference Between Gravity Wells," *University of Illinois Bulletin 30*, Series 374, Vol. 45, January 1948.

Batu, V., "A Finite Element Dual Mesh Method to Calculate Nodal Darcy Velocities in Nonhomogeneous and Anisotropic Aquifers," *Water Resources Research*, Vol. 20, No. 11, pp. 1705–1717, November 1984.

Bear, J., *Dynamics of Fluids in Porous Media*, American Elsevier, New York, 764 pp., 1972.

Bear, J., *Hydraulics of Groundwater*, McGraw-Hill, New York, 569 pp., 1979.

Berkaloff, E., "Essai de Puits—Interprétation—Nappe Libre Ave Strata Conductrice d'Eau Privilégée," *Bureau de Recherches Géologique et Miniéres Report DS 63 A 18*, Orleans, France, 1963.

Bierschenk, W. H., "Determining Well Efficiency by Multiple Step-Drawdown Tests," International Association of Scientific Hydrology, No. 64, pp. 493–507, Berkeley, California, 1963.

Beyer, W. H., *Standard Mathematical Tables*, sixth edition, CRC Press, Boca Raton, Florida, 615 pp., 1986.

Bonnet M., J. Forasiewicz, and P. Peaudecerf, "Méthodes d'Intérpretation de Pompages d'Essai en Nappe Libre," *Bureau de Recherches Géologique et Miniéres Report 70 SGN 359 HYD* (in French), Orleans, France, 1970.

Boulton, N. S., "The Flow Pattern Near a Gravity Well in a Uniform Water Bearing Medium," *Journal of the Institution of Civil Engineer*, London, Great Britain, Vol. 36, p. 534–550, December 1951.

Boulton, N. S., "The Drawdown of the Water-Table Under Non-Steady Conditions Near a Pumped Well in an Unconfined Formation," *Proceedings of the Institution of Civil Engineers*, London, Great Britain, Part III, pp. 564–579, 1954a.

Boulton, N. S., "Unsteady Radial Flow to a Pumped Well Allowing for Delayed Yield from Storage," *International Association of Scientific Hydrology*, Vol. II, pp. 472–477, 1954b.

Boulton, N. S., "Analysis of Data from Non-equilibrium Pumping Tests Allowing for Delayed Yield from Storage," *Proceedins of the Institution of Civil Engineers*, London, Great Britain, Vol. 26, pp. 469–482, 1963.

Boulton, N. S., "Analysis of Data from Pumping Tests in Unconfined Anisotropic Aquifers," *Journal of Hydrology*, Vol. 10, pp. 369–378, 1970.

Bouwer, H., *Groundwater Hydrology*, McGraw-Hill, 480 pp., New York, 1978.

Bouwer, H., "The Bouwer and Rice Slug Test—An Update," *Ground Water*, Vol. 27, No. 3, pp. 304–309, 1989.

Bouwer, H., and R. C. Rice, "A Slug Test for Determining Hydraulic Conductivity of Unconfined Aquifers with Completely or Partially Penetrating Wells," *Water Resources Research*, Vol. 12, No. 3, pp. 423–438, 1976.

Bredeheoft, J. D., and S. S. Papadopulos, "A Method for Determining the Hydraulic Properties of Tight Formations," *Water Resources Research*, Vol. 16, No. 1, pp. 233–238, February 1980.

Brown, D. L., and T. N. Narasimhan, "An Evaluation of the Bouwer and Rice Method of Slug Test Analysis," *Water Resources Research*, Vol. 31, No. 5, pp. 1239–1246, 1995.

Carman P. C., "Fluid Flow Through a Granular Bed," *Transaction of the Institutions of Chemical Engineers*, Vol. 15, pp. 150–156, London, Great Britain, 1937.

Carman, P. C., *Flow of Gases Through Porous Media*, Butterworths, London, Great Britain, 1956.

Carslaw, H. S., and J. C. Jaeger, *Conduction of Heat in Solids*, second edition, Oxford University Press, London, Great Britain, 510 pp., 1959.

Castany, G., *Principles et Méthodes de L'Hydrogéologie* (in French), Dunod Université, Paris, France, 1982.

Cedergren, H. R., *Seepage, Drainage, and Flow Nets*, John Wiley & Sons, second edition, New York, 534 pp., 1977.

Chapius, R. P., "Discussion of 'Estimation of Storativity from Recovery Data,' " by P. N. Bullukraya and K. K. Sharma," *Ground Water*, Vol. 30, No. 2, pp. 269–272, March–April 1992a.

Chapius, R. P., "Using Cooper–Jacob Approximation to Take Account of Pumping Well Pipe Storage Effects in Early Drawdown Data of a Confined Aquifer," *Ground Water*, Vol. 30, No. 3, pp. 331–337, May–June, 1992b.

Chapius, R. P., "Assessment of Methods and Conditions to Locate Boundaries: I. One or Two Straight Impervious Boundaries," *Ground Water*, Vol. 32, No. 4, pp. 576–582, July–August 1994.

Chen, C. S., "A Reinvestigation of the Analytical Solution for Drawdown Distributions in a Finite Confined Aquifer," *Water Resources Research*, Vol. 20, No. 10, pp. 1466–1468, October 1984.

Chirlin, G. R., "A Critique of the Hvorslev Method for Slug Test Analysis: The Fully Penetrating Well," *Ground Water Monitoring Review*, Spring, pp. 130–138, 1989.

Chow, V. T., "On the Determination of Transmissibility and Storage Coefficients from Pumping Test Data," *Transactions, American Geophysical Union*, Vol. 33, pp. 397–404, 1952.

Churchill, R. V., *Operational Mathematics*, McGraw-Hill, New York, 1958.

Clark, W. E., "Computing the Barometric Efficiency of a Well," *Journal of Hydraulics Division*, Proceedings of the American Society of Civil Engineers, Vol. 93, HY4, pp. 93–98, 1967.

Cooley, R. L., "A Finite Difference Method for Unsteady Flow in Variably Saturated Porous Media: Application to a Single Pumping Well," *Water Resources Research*, Vol. 7, No. 6, pp. 1607–1625, 1971.

Cooper, H. H., Jr., "The Equation of Groundwater Flow in Fixed and Deforming Coordinates," *Journal of Geophysical Research*, Vol. 71, pp. 4785–4790, 1966.

Cooper H. H., Jr., and C. E. Jacob, "A Generalized Graphical Method for Evaluation Formation Constants and Summarizing Well-Field History," *Transactions, American Geophysical Union*, Vol. 27, No. IV, pp. 526–534, August 1946.

Cooper, H. H., Jr., J. D. Bredehoeft, and I. S. Papadopulos, "Response of a Finite-Diameter Well to an Instantaneous Charge of Water," *Water Resources Research*, Vol. 3, No. 1, pp. 263–269, 1967.

Crawford, L. A., "Water Level and Gradient Responses to Barometric Loading at the Savannah River Site," Master of Science Thesis, University of Georgia, Athens, Georgia, 140 pp., 1994.

Dagan, G., "Second Order Linearized Theory of Free Surface Flow in Porous Media," *La Houille Blanche*, Vol. 8, pp. 901–910, 1964.

Dagan, G., "A Method of Determining the Permeability and Effective Porosity of Unconfined Anisotropic Aquifers," *Hydraulics Laboratory Report P.N. 1/1967*, Technion Israel Institute of Technology, Haifa, Israel, 1967a.

Dagan, G., "A Method of Determining the Permeability and Effective Porosity of Unconfined Anisotropic Aquifers," *Water Resources Research*, Vol. 3, No. 4, pp. 1059–1071, 1967b.

Darcy, H., "Les Fontaines Publiques de la Ville Dijon," (in French) Dalmont, Paris, France, 1856.

Das, B. J., *Advanced Soil Mechanics*, Hemisphere Publishing Corporation, New York, 511 pp., 1983.

Daubré, A., "Les Aux Souterraines à L'époque Actuelle," (in French) Vol. 1, p. 19, Paris, France, 1887.

Davis, S. N., "Porosity and Permeability of Natural Materials," *Flow Through Porous Media*, edited by R. J. M. De Wiest, Academic Press, New York, pp. 54–89, 1969.

Davis, S. N., and R. J. M. DeWiest, *Hydrogeology*, John Wiley & Sons, 463 pp., New York, 1966.

Davis, A. R., and T. C. Rasmussen, "A Comparision of Linear Regression with Clark's Method for Estimating Barometric Efficiency of Confined Aquifers," *Water Resources Research*, Vol. 29, No. 6, pp. 1849–1854, June 1993.

Dawson, K. J., and J. D. Istok, *Aquifer Testing: Design and Analysis of Pumping and Slug Tests*, Lewis Publishers, 344 pp., Chelsea, Michigan, 1991.

De Glee, G. J., "Over Grondwaterstromingen Bij Wateronttrekking Door Middel Van Putten," Thesis (in Dutch), J. Waltman, Delft, The Netherlands, 175 pp., 1930.

de Marsily, G., *Quantitative Hydrogeology - Groundwater Hydrology for Engineers*, translated from French by Gunilla de Marsily, Academic Press, 440 pp., Orlando, Florida, 1986.

Domenico, P. A., *Concepts and Models in Groundwater Hydrology*, McGraw-Hill, New York, 405 pp., 1972.

Domenico, P. A., "Determination of Bulk Rock Properties from Ground-Water Level Fluctuations," *Bulletin of the Association of Engineering Geologists*, Vol. 20, No. 3, pp. 283–287, 1983.

Domenico, P. A., and M. D. Mifflin, "Water from Low-Permeability Sediments and Land Subsidence," *Water Resources Research*, Vol. 1, No. 4, pp. 563–576, Fourth Quarter, 1965.

Domenico, P. A., and F. W. Schwartz, *Physical and Chemical Hydrogeology*, John Wiley & Sons, 824 pp., New York, 1990.

Driscoll, F. G., *Groundwater and Wells*, second edition, Johnson Filtration Systems Inc., St. Paul, Minnesota, 1089 pp., 1986.

Dupuit, J., *Etudes Théoriques et Pratiques sur le Mouvement des Eaux dans les Canaux Découverts et á Travers les Terrains Perméables* (in French), second edition, Dunod, Paris, France, 304 pp., 1863.

Fair, G. M., and L. P. Hatch, "Fundamental Factors Governing the Streamline Flow of Water Through Sand," *Journal of American Water Works Association*, Vol. 25, pp. 1551–1565, 1933.

Ferris, J. G., D. B. Knowles, R. H. Brown, and R. W. Stallman, "Theory of Aquifer Tests," U.S. Geological Survey Water-Supply Paper 1536-E, pp. 69–174, 1962.

Fetter, C. W., *Applied Hydrogeology*, third edition, Prentice-Hall, Englewood Cliffs, New Jersey, 691 pp., 1994.

Franke, O. L., T. E. Reilly, and G. D. Bennett, *Definition of Boundary and Initial Conditions in the Analysis of Saturated Ground-Water Flow Systems, An Introduction*, U.S. Geological Survey Open-File Report 84-458, 26 pp., Reston, Virginia, 1984.

Freeze R. A., and J. A. Cherry, *Groundwater*, Prentice-Hall, Englewood Cliffs, New Jersey, 604 pp., 1979.

Fuller, M. L., "Significance of the Term 'Artesian,' " U.S. Geological Survey Water-Supply Paper 160, pp. 9–15, 1906.

Gelhar, L. W., *Stochastic Subsurface Hydrology*, Prentice-Hall, Englewood Cliffs, New Jersey, 390 pp., 1993.

Gradshteyn, I. S., and I. M. Ryzhik, *Table of Integrals, Series, and Products*, Translated from the Russian by Scripta Technica, Inc., Translation edited by A. Jeffrey, Academic Press, New York, 1086 pp., 1965.

Ground Water Publishing Company, "Conversions of Hydraulic Conductivity, Intrinsic Permeability, and Transmissivity," *Ground Water*, Vol. 34, No. 5, p. 955, September–October 1996.

Hall, P., and J. Chen, *Water Well and Aquifer Test Analysis*, Water Resources Publications, Highlands Ranch, Colorado, 412 pp., 1996.

Hantush, M. S., "Analysis of Data from Pumping Tests in Leaky Aquifers," *Transactions, American Geophysical Union*, Vol. 37, No. 6, pp. 702–714, 1956.

Hantush, M. S., "Non-steady Flow to a Well Partially Penetrating an Infinite Leaky Aquifer," *Proceedings, Iraqi Scientific Society*, Vol. 1, pp. 1–19, Baghdad, Iraq, 1957.

Hantush, M. S., "Analysis of Data from Pumping Wells Near a River," *Journal of Geophysical Research*, Vol. 64, No. 11, pp. 1921–1932, November 1959.

Hantush, M. S., "Author's Reply to the Preceding Discussion," *Journal of Geophysical Research*, Vol. 65, No. 5, pp. 1627–1629, May 1960a.

Hantush, M. S., "Modification of the Theory of Leaky Aquifers," *Journal of Geophysical Research*, Vol. 65, No. 11, pp. 3713–3725, November 1960b.

Hantush, M. S., "Tables of the Function $H(u, \beta) = \int_0^\infty \frac{e^{-y}}{y} \operatorname{erfc}\left\{\frac{\beta u^{1/2}}{[y(y-u)]^{1/2}}\right\} dy$," New Mexico Institute of Mining and Technology, *Professional Paper 103*, 14 pp., Socorro, New Mexico, 1961a.

Hantush, M. S., "Tables of the Function $W(u, \beta) = \int_0^\infty \frac{e^{-y-\frac{\beta^2}{4y}}}{y} dy$," New Mexico Institute of Mining and Technology, *Professional Paper 104*, 13 pp., Socorro, New Mexico, 1961b.

Hantush, M. S., "Drawdown Around a Partially Penetrating Well," *Journal of the Hydraulics Division*, American Society of Civil Engineers, Vol. 87, No. 4, pp. 83–98, July 1961c.

Hantush, M. S., "Aquifer Tests on Partially Penetrating Wells," *Journal of the Hydraulics Division*, American Society of Civil Engineers, Vol. 87, No. 5, pp. 171–195, September 1961d.

Hantush, M. S., "On the Validity of the Dupuit–Forchheimer Well-Discharge Formula," *Journal of Geophysical Research*, Vol. 67, No. 6, pp. 2417–2420, June 1962.

Hantush, M. S., "Hydraulics of Wells," *Advances in Hydroscience*, edited by Ven Te Chow, Academic Press, New York, pp. 281–442, 1964.

Hantush, M. S., "Wells in Homogeneous Anisotropic Aquifers," *Water Resources Research*, Vol. 2, No. 2, pp. 273–279, Second Quarter, 1966a.

Hantush, M. S., "Analysis of Data from Pumping Tests in Anisotropic Aquifers," *Journal of Geophysical Research*, Vol. 71, No. 2, pp. 421–426, January 1966b.

Hantush, M. S., "Flow to Wells in Aquifers Separated by a Semipervious Layer," *Journal of Geophysical Research*, Vol. 72, No. 6, pp. 1709–1720, 1967.

Hantush, M. S., and C. E. Jacob, "Plane Potential Flow of Ground Water with Linear Leakage," *Transactions, American Geophysical Union*, Vol. 35, No. 6, pp. 917–936, 1954.

Hantush, M. S., and C. E. Jacob, "Non-Steady Radial Flow in an Infinite Leaky Aquifer," *Transactions, American Geophysical Union*, Vol. 36, No. 1, pp. 95–100, 1955.

Hantush, M. S., and R. G. Thomas, "A Method for Analyzing a Drawdown Test in Anisotropic Aquifers," *Water Resources Research*, Vol. 2, No. 2, pp. 281–285, Second Quarter, 1966.

Hazen, A., "Some Physical Properties of Sands and Gravels," Massachusetts State Board of Health, *24th Annual Report*, pp. 539–556, 1892.

Highdon, A., E. H. Ohlsen, W. B. Stiles, J. A. Weese, and W. F. Riley, *Mechanics of Materials*, John Wiley & Sons, 756 pp., New York, 1976.

Hvorslev, M. J., *Time Lag and Soil Permeability in Groundwater Observations*, Bulletin No. 36, U.S. Army Corps of Engineers, Waterways Experiment Station, Vicksburg, Mississippi, 49 pp., 1951.

Hyder, Z., and J. J. Butler, Jr., "Slug Tests in Unconfined Formations: An Assessment of the Bouwer and Rice Technique," *Ground Water*, Vol. 33, No. 1, pp. 16–22, January–February 1995.

Hyder, Z., J. J. Butler, Jr., C. D. McElwee, and W. Liu, "Slug Tests in Partially Penetrating Wells," *Water Resources Research*, Vol. 30, No. 11, pp. 2945–2957, November 1994.

Jacob, C. E., "On the Flow of Water in an Elastic Artesian Aquifer," *Transactions, American Geophysical Union*, Vol. 21, pp. 574–586, 1940.

Jacob, C. E., "Notes on Determining Permeability by Pumping Tests Under Water-Table Conditions," U.S. Geological Survey Mimeo Rep., 1944.

Jacob, C. E., "Radial Flow in a Leaky Artisian Aquifer," *Transactions, American Geophysical Union*, Vol. 27, pp. 198–205, 1946.

Jacob, C. E., "Drawdown Test to Determine Effective Radius of Artesian Wells," *Transactions, American Society of Civil Engineers*, Vol. 112, pp. 1047–1070, 1947.

Jacob, C. E., "Flow of Ground Water," in *Engineering Hydraulics*, edited by H. Rouse, John Wiley & Sons, New York, pp. 321–386, 1950.

Jacob, C. E., and S. W. Lohman, "Nonseady Flow to a Well of Constant Drawdown in an Extensive Aquifer," *Transactions, American Geophysical Union*, Vol. 33, pp. 559–569, 1952.

Jaeger, C., *Engineering Fluid Mechanics*, translated from the German by P. O. Wolf, Blackie & Son Ltd., London, 529 pp., 1956.

Jorgensen, D. G., "Relationships Between Basic Soils-Engineering Equations and Basic Ground-Water Equations," *Geological Survey Water-Supply Paper 2064*, U.S. Government Printing Office, 40 pp., Washington, D.C., 1980.

Kashef, A. I., *Groundwater Engineering*, McGraw-Hill, New York, 512 pp., 1986.

Kazmann, R. G., "Discussion of Paper by Mahdi S. Hantush, 'Analysis of Data from Pumping Wells Near a River,' " *Journal of Geophysical Research*, Vol. 65, pp. 1625–1626, May 1960.

Kézdi, A., *Handbook of Soil Mechanics, Soil Physics*, Vol. 1, Elsevier Scientific Publishing Company, New York, 294 pp., 1974.

Kozeny, J., "Über Kapillare Leitung des Wassers im Boden" (in German), *Sitzungsber. Akad. Wiss.*, Vol. 136, pp. 271–306, Wien, Austria, 1927.

Kruseman, G. P., and N. A. De Ridder, *Analysis and Evaluation of Pumping Test Data*, second edition, International Institute for Land Reclamation and Improvement, Wageningen, The Netherlands, Publication 47, 377 pp., 1991.

Labadie, J. W., and O. J. Helweg, "Step-Drawdown Test Analysis by Computer," *Ground Water*, Vol. 13, No. 5, pp. 438–444, September–October 1975.

Lambe, T. W., and R. V. Whitman, *Soil Mechanics*, John Wiley & Sons, New York, 553 pp., 1969.

Lohman, S. W.,*Geology and Artesian Water Supply of the Grand Junction Area, Colorado*, U.S. Geological Survey Professional Paper 451, 149 pp., 1965.

Lohman, S. W., *Ground-Water Hydraulics*, U.S. Geological Survey Professional Paper 708, U.S. Government Printing Office, Washington, D.C., 70 pp., 1972.

Maasland, M., "Theory of Fluid Flow Through Anisotropic Media," Section II, *Drainage of Agricultural Lands*, edited by J. N. Luthin, pp. 216–236, American Society of Agronomy, Madison, Wisconsin, 1957.

Mariño, M. A., and J. N. Luthin, *Seepage and Groundwater*, Elsevier Scientific Publishing Company, New York, 489 pp., 1982.

McCarthy, D. F., *Essenticals of Soil Mechanics and Foundations: Basic Geotechnics*, third edition, Prentice-Hall, Englewood Cliffs, New Jersey, 614 pp., 1988.

Meinzer, O. E., "Outline of Ground-Water Hydrology with Definitions," *Geological Survey Water-Supply Paper 494*, U.S. Government Printing Office, Washington, D.C., 71 pp., 1923.

Meinzer, O. E., "Compressibility and Elasticity of Artesian Aquifers," *Economic Geology*, Vol. 23, pp. 263–291, 1928.

Mercer, J. W., and B. R. Orr, *Interim Data Report on the Geohydrology of the Proposed Waste Isolation Pilot Plant (WIPP) Site*, Southeastern New Mexico, U.S. Geological Survey, Water Resources Investigations 79–98, 1979.

Mercer, J. W., S. D. Thomas, and B. Ross, *Parameters and Variables Appearing in Repository Models*, U.S. Nuclear Regulatory Commission, NUREG/CR-3066, Washington, D.C., 1982.

Miller T. M., and W. J. Weber, Jr., "Rapid Solution of the Nonlinear Step-Drawdown Equation," *Ground Water*, Vol. 21, No. 5, pp. 584–588, September–October 1983.

Mock, P., and J. Merz, "Observation of Delayed Gravity Response in Partially Penetrating Wells," *Ground Water*, Vol. 28, No. 1, pp. 11–16, January–February 1990.

Moench, A. F., "Transient Flow to a Large-Diameter Well in a Confined Aquifer with Storativity Semiconfining Layers," *Water Resources Research*, Vol. 21, No. 8, pp. 1121–1131, August 1985.

Moench, A. F., "Combining the Neuman and Boulton Models for Flow to a Well in an Unconfined Aquifer," *Ground Water*, Vol. 33, No. 3, pp. 378–384, May–June, 1995.

Morris, D. A., and A. I. Johnson, "Summary of Hydrologic and Physical Properties of Rock and Soil Materials as Analyzed by the Hydrologic Laboratory of the U.S. Geological Survey," *U.S. Geological Survey Water Supply Paper 1839-D*, 1967.

Neuman, S. P., "Theory of Flow in Unconfined Aquifers Considering Delayed Response of the Water Table," *Water Resources Research*, Vol. 8, No. 4, pp. 1031–1045, August, 1972.

Neuman, S. P., "Supplementary Comments on 'Theory of Flow in Unconfined Aquifers Considering Delayed Response of the Water Table,' " *Water Resources Research*, Vol. 9, No. 4, pp. 1102–1103, August, 1973.

Neuman, S. P., "Effect of Partial Penetration on Flow in Unconfined Aquifers Considering Delayed Gravity Response," *Water Resources Research*, Vol. 10, No. 2, pp. 303–312, April, 1974.

Neuman, S. P., "Analysis of Pumping Test Data from Anisotropic Unconfined Aquifers Considering Delayed Gravity Response," *Water Resources Research*, Vol. 11, No. 2, pp. 329–342, April, 1975.

Neuman, S. P., "Perspective on Delayed Yield," *Water Resources Research*, Vol. 15, No. 4, pp. 899–908, October, 1978.

Neuman, S. P., G. R. Walter, H. W. Bentley, J. J. Ward, and D. D. Gonzalez, "Determination of Horizontal Anisotropy with Three Wells," *Ground Water*, Vol. 22, No. 1, pp. 66–72, January–February 1984.

Neuman, S. P., and P. A. Witherspoon, "Theory of Flow in Aquicludes Adjacent to Slightly Leaky Aquifers," *Water Resources Research*, Vol. 4, No. 1, pp. 103–112, February 1968.

Neuman, S. P., and P. A. Witherspoon, *Transient Flow of Ground Water to Wells in Multiple Aquifer Systems*, Geotechnical Engineering Report 69-1, University of California, Berkeley, California, 182 pp., January, 1969a.

Neuman, S. P., and P. A. Witherspoon, "Theory of Flow in a Confined Two Aquifer System," *Water Resources Research*, Vol. 5, No. 4, pp. 803–816, August 1969b.

Neuman, S. P., and P. A. Witherspoon, "Applicability of Current Theories of Flow in Leaky Aquifers," *Water Resources Research*, Vol. 5, No. 4, pp. 817–829, August 1969c.

Neuman, S. P., and P. A. Witherspoon, "Variational Principles for Confined and Unconfined Flow of Ground Water," *Water Resources Research*, Vol. 6, No. 5, pp. 1376–1382, September, 1970.

Neuman, S. P., and P. A. Witherspoon, "Field Determination of the Hydraulic Properties of Leaky Multiple Aquifer Systems," *Water Resources Research*, Vol. 8, No. 5, pp. 1284–1298, October 1972.

Neuzil, C. E., "On Conducting the Modified 'Slug Test' in Tight Formations," *Water Resources Research*, Vol. 18, No. 2, pp. 439–441, April, 1982.

Norton, W. H., "Artesian Wells of Iowa," *Iowa Geological Survey*, Vol. 6, p. 130, 1897.

Papadopulos, I. S., "Nonsteady Flow to a Well in an Infinite Anisotropic Aquifer," *Proceedings of Dubrovnik Symposium on the Hydrology of Fractured Rocks*, International Association of Scientific Hydrology, Dubrovnik, Yugoslavia, pp. 21–31, 1965.

Papadopulos, I. S., "Drawdown Distribution Around a Large-Diameter Well," *National Symposium on Ground-Water Hydrology*, San Francisco, California, pp. 157–168, November 6–8, 1967.

Papadopulos, I. S., and H. H. Cooper, Jr., "Drawdown in a Well of Large Diameter," *Water Resources Research*, Vol. 3, No. 1, pp. 241–244, First Quarter, 1967.

Papadopulos, S. S., and J. D. Bredehoeft, and H. H. Cooper, Jr., "On the Analysis of 'Slug Test' Data," *Water Resources Research*, Vol. 9, No. 4, pp. 1087–1089, August, 1973.

Polubarinova–Kochina, P. Ya., *Theory of Ground Water Movement*, translated from the Russian by R. J. M. De Wiest, Princeton University Press, Princeton, New Jersey, 613 pp., 1962.

Price, M., *Introducing Groundwater*, George Allen & Unwin, London, Great Britain, 195 pp., 1985.

Prickett, T. A., "Type-Curve Solution to Aquifer Tests Under Water-Table Conditions," *Ground Water*, Vol. 3, No. 3, pp. 5–14, 1965.

Quiñones-Aponte, V., "Horizontal Anisotropy of the Principal Ground-Water Flow Zone in the Salinas Alluvial Fan, Puerto Rico," *Ground Water*, Vol. 27, No. 4, pp. 491–500, July–August 1989.

Raghunath, H. M., *Ground Water*, Wiley Eastern Ltd., New Delhi, India, 563 pp., 1987.

Reed, J. D., *Type Curves for Selected Problems of Flow to Wells in Confined Aquifers*, U.S. Geological Survey, Book 3, Chapter B3, U.S. Government Printing Office, Washington, D.C., 106 pp., 1980.

Robinson, E. S., and R. T. Bell, "Tides in Confined Well-Aquifer System," *Journal of Geophysical Research*, Vol. 76, No. 8, pp. 1857–1869, March 1971.

Rojstaczer, S., and D. C. Agnew, "The Influence of Formation Material Properties on the Response of Water Levels in Wells to Earth Tides and Atmospheric Loading," *Journal of Geophycial Research*, Vol. 94, No. 12, pp. 12403–12411, 1989.

Rorabaugh, M. I., "Graphical and Theoretical Analysis of Step-Drawdown Test of Artesian Wells," *Transactions, American Society of Civil Engineers*, Vol. 79, pp. 362-1–362-23, 1953.

Rushton, K. R., and S. C. Redshaw, *Seepage and Groundwater Flow*, John Wiley & Sons, Ltd., Chichester, Great Britain, 339 pp., 1979.

Şen, Z., *Applied Hydrogeology for Scientists and Engineers*, CRC Lewis Publishers, 444 pp., New York, 1995.

Sheahan, N. T., "Injection/Extraction Well System—A Unique Seawater Intrusion Barrier," *Ground Water*, Vol. 15, No. 1, pp. 32–49, January–February, 1977.

Sheahan, N. T., "Type-Curve Solution of Step-Drawdown Test," *Ground Water*, Vol. 9, No. 1, pp. 25–29, January–February 1971.

Shepherd, R. G., "Correlations of Permeability and Grain Size," *Ground Water*, Vol. 27, No. 5, pp. 633–638, September–October 1989.

Sperry, M. S., and J. J. Peirce, "A Model for Estimating the Hydraulic Conductivity of Granular Material Based on Grain Shape, Grain Size, and Porosity," *Ground Water*, Vol. 33, No. 6, pp. 892–898, November–December 1995.

Spiegel, M. R., *Advanced Mathematics for Engineers and Scientists*, Shaum's Outline Series in Mathematics, McGraw-Hill, New York, 407 pp., 1971.

Stallman, R. W., "Boulton's Integral for Pumping Tests Analysis," *U.S. Geological Survey Professional Papers 400-C*, Geological Survey Research 1961, Short Papers in Geological Hydrological Science, Articles 147–149, C-24 to C-29, 1961.

Stramel, G. J., C. W. Lane, and W. G. Hodson, "Geology and Ground-Water Hydrology of the Ingalls Area, Kansas," *State Geological Survey of Kansas Bulletin 132*, pp. 45–51, 1958.

Streltsova, T. D., "Unsteady Radial Flow in an Unconfined Aquifer," *Water Resources Research*, Vol. 8, No. 4, pp. 1059–1066, August, 1972.

Terzaghi, K., *Erdbaummechanic auf Bodenphysikalishe Grundlage* (in German), Franz Deticke, Leipzig, Germany, 390 pp., 1925.

Terzaghi, K., *Theoretical Soil Mechanics*, John Wiley & Sons, New York, 510 pp., 1943.

Terzaghi, K., and R. B. Peck, *Soil Mechanics in Engineering Practice*, second edition, John Wiley & Sons, New York, 729 pp., 1967.

Theis, C. V., "The Relation Between the Lowering of the Piezometric Surface and the Rate and Duration of Discharge of a Well Using Ground-Water Storage," *Transactions, American Geophysical Union*, Vol. 16, pp. 519–524, August 1935.

Thiem, G., *Hydrologische Methoden* (in German), J. M. Gebhardt, Leipzig, Germany, 56 pp., 1906.

Todd, D. K., *Ground Water Hydrology*, John Wiley & Sons, New York, 535 pp., 1980.

U.S. Bureau of Reclamation, *Earth Manual*, U.S. Government Printing Office, Washington, D.C., 783 pp., 1968.

U.S. Department of the Interior, *Ground Water Manual*, Revised reprint 1981, U.S. Goverment Printing Office, 480 pp., Denver, Colorado, 1981.

U.S. Department of the Navy, *Soil Mechanics Design Manual 7.1*, Department of the Navy, Naval Facilities and Engineering Command, Alexandria, Virginina, 1982.

Vennard, J. K., and R. L. Street, *Elementary Flud Mechanics*, sixth edition, John Wiley & Sons, New York, 689 pp., 1982.

Verruijt, A., *Theory of Groundwater Flow*, Gordon and Breach Science Publishers, 190 pp., New York, 1970.

Walton, W. C., "Selected Analytical Methods for Well and Aquifer Evaluation," Illinois State Water Survey, *Bulletin No. 49*, 81 pp., 1962.

Walton, W. C., *Groundwater Resource Evaluation*, McGraw-Hill, 664 pp., New York, 1970.

Ward, D. S., M. Reeves, and L. E. Duda, *Verification and Field Comparison of the Sandia Waste-Isolation Flow and Transport Model (SWIFT)*, NUREG/CR-3316, SAND83-1154, Albuquerque, New Mexico, 1984.

Weeks, E. P., "Barometric Fluctuations in Wells Tapping Deep Unconfined Aquifers," *Water Resources Research*, Vol. 15, No. 5, pp. 1167–1176, October, 1979.

Wenzel, L. K., *Methods for Determining Permeability of Water-Bearing Materials, with Special Reference to Discharging Well Methods*, U.S. Geological Survey Water-Supply Paper 887, 192 pp., 1942.

Wikramaratna, R. S., "A New Type Curve Method for the Analysis of Pumping Tests in Large-Diameter Wells," *Water Resources Research*, Vol. 21, No. 2, pp. 261–264, February 1985.

Witherspoon, P. A., I. Javandel, S. P. Neuman, and R. A. Freeze, *Interpretation of Aquifer Gas Storage from Water Pumping Tests*, American Gas Association, 237 pp., New York, 1967.

ABOUT THE AUTHOR

Dr. Vedat Batu received his Dipl. Ing. in civil engineering and his Ph.D. in hydraulic engineering from the Civil Engineering Faculty of Istanbul Technical University, Istanbul, Turkey. He also received his Doçent (University Associate Professor) Degree (gained with an additional approved thesis and by passing some examinations) in the hydraulic engineering area by the Interuniversities Institution of Turkey; his thesis was based on his research work carried out at the University of Wisconsin–Madison. All the theses of Dr. Batu are on the subject of ground-water hydrology, and he is intensely involved in different ground-water-related problems, including research and practice on saturated and unsaturated flows, aquifer hydraulics, contaminant transport, and analytical and numerical modeling techniques. Dr. Batu served as a faculty member and head of the Department of Civil Engineering, Karedeniz Technical University, Trabzon, Turkey. He has had visiting positions at several universities in the United States and United Kingdom, and for more than a decade he has been working with private corporations in the United States, mainly on aquifers and ground-water-related projects.

Dr. Batu's own research and publications have continued throughout his professional life in the academy and industry. He has published in internationally recognized journals such as *Ground Water*, *Journal of Hydraulic Engineering*, *Journal of Irrigation and Drainage Engineering*, *Journal of Hydrology*, *Soil Science Society of America Journal*, and *Water Resources Research* and in numerous symposia proceedings. He is serving as an associate editor of the *Journal of Hydrologic Engineering* of the American Society of Civil Engineers. Dr. Batu has developed numerous analytical and numerical models and developed new concepts and methods relating to ground-water flow and contaminant transport in aquifers.

Dr. Batu is a registered Professional Engineer. Currently, he is a Senior Consultant with Rust Environment & Infrastructure, Inc., in Oak Brook, Illinois.

ABOUT THE DISK

INTRODUCTION

The enclosed disk contains sixteen type curves created in Lotus 1-2-3 Release 5 and three computer programs, HTYCPC, DELAY2PC, and FASTEPPC. The three programs are DOS-based programs written in FORTRAN. HTYCPC generates type curve data for partially penetrating wells in confined aquifers. DELAY2PC generates type curve data for partially penetrating wells in unconfined aquifers. FASTEPPC calculates the well characteristics for step-drawdown tests.

For further information, please refer to the readme files in each directory. The readme files contain detailed information about each file and program.

DISK TABLE OF CONTENTS

Directory	Description
TCURVES	16 type curves generated in Lotus 1-2-3 Release 5
HTYCPC	HTYCPC program and 2 sample files
DELAY2PC	DELAY2PC program and 4 sample files
FASTEPPC	FASTEPPC program and 1 sample file

MINIMUM SYSTEM REQUIREMENTS

- IBM PC or compatible computer with 386 or higher processor (with math coprocessor)
- 4 MB RAM
- 3.5″ floppy disk drive

- To use computer programs: DOS 6.22 or higher
- To read type curves: Windows 3.1 or higher and Lotus 1-2-3 Release 5 or higher (or other spreadsheet program capable of reading Lotus 1-2-3 Release 5 files such as Microsoft Excel.)

HOW TO INSTALL THE SOFTWARE ONTO YOUR COMPUTER

If you would like to copy the files from the disk to your computer's hard drive, run the installation program by doing the following:

1. Insert the enclosed disk into the floppy disk drive of your computer.
2. Windows 3.1 and NT 3.51: From the Program Manager, choose File, Run.
   Windows 95 and NT 4.0: From the Start Menu, choose Run.
3. Type **A:\INSTALL** (where A is the letter of your floppy disk drive). Press Enter.
4. The opening screen of the installation program will appear. Press Enter to continue.
5. The default destination directory is C:\BATU. If you wish to change the default destination, you may do so now. Follow the instructions on the screen.
6. The installation program will copy the files onto your hard drive in the C:\BATU or user-designated directory.

USING THE SOFTWARE

To read the type curves files in the TCURVES directory, launch Lotus 1-2-3 or other spreadsheet software. Go to File, Open and choose C:\BATU\TCURVES or A:\TCURVES if you are using the files off the floppy disk. Double click on the file you want to open.

To use the programs, go to the DOS prompt and go to the appropriate drive and program subdirectory (type **cd C:\BATU\HTYCPC**). Then type the name of the program (e.g. **HTYCPC.EXE**) and Press Enter.

For more detailed instructions, refer to the readme files in each subdirectory.

USER ASSISTANCE

If you need basic assistance or have a damaged disk, please contact Wiley Technical Support at:

Phone: (212) 850-6753
Fax: (212) 850-6800 (Attention: Wiley Technical Support)
Email: techhelp@wiley.com

To place additional orders or to request information about Wiley products, please call (800) 225-5945.

AUTHOR INDEX

Agnew, D.C., 67
Alyamani, M.S., 42, 43, 46
American Society of Testing Materials (ASTM), 692

Babbitt, H.E., 126
Batu, V., 82, 107, 109
Bear, J., 5, 34, 40, 42, 51, 53, 80, 83, 84, 87, 90, 91, 92, 107, 124, 147, 156, 250
Bell, R.T., 78
Bennett, G.D., 99, 100, 102
Bentley, H.W., 206, 211, 222, 224, 225, 226, 228, 229, 231
Berkaloff, E., 526
Beyer, W.H., 118, 150
Bierschenk, W.H., 622
Bonnet, M., 534, 540
Boulton, N.S., 8, 126, 376, 458, 459, 461, 462, 464, 465, 468, 469, 470, 471, 472, 473, 474, 475, 476, 477, 478, 479, 480, 481, 482, 485, 487, 490, 492, 500, 503, 505, 514, 526, 530, 531, 532, 540
Bouwer, H., 5, 8, 627, 628, 629, 631, 632, 633, 634, 635, 636, 639, 641, 642, 644, 647, 648, 651, 654, 658, 659, 660, 662, 668, 672
Bredehoeft, J.D., 9, 627, 642, 643, 649, 660, 662, 663, 664, 665, 666, 667, 668, 669, 671, 672, 675, 676, 677, 678, 680, 681, 682, 683, 684, 685, 686, 687, 688, 689, 691
Brown, D.L., 639, 641, 644, 645, 646, 647, 648
Brown, R.H., 62, 63, 65, 66, 68, 150, 153, 158, 159, 160, 165, 176, 177, 178, 555, 556, 557, 558, 568, 570, 571, 572
Butler, J.J., Jr., 639, 640, 641, 642, 643, 644, 645, 646, 647, 648

Caldwell, D.H., 126
Castany, G., 558, 562, 564, 565, 566
Carman, P.C., 42, 43, 44, 46
Carslaw, H.S., 149, 349, 352, 474
Cedergren, H.R., 651
Chapius, R.P., 178, 180, 182, 555, 557, 559, 560, 561, 562, 563, 564, 565, 566, 567, 568, 569
Chen, C.S., 154, 155
Chen J., 5
Cherry, J.A., 5, 34, 36, 40, 47, 80
Chirlin, G.R., 649
Chow, V.T., 171, 173, 174
Churchill, R.V., 350
Clark, W.E., 73, 76, 77, 78
Cooley, R.L., 502, 503
Cooper, H.H., Jr., 7, 9, 16, 53, 145, 150, 156, 163, 164, 176, 182, 187, 188, 189, 190, 195, 196, 197, 200, 201, 202, 203, 204, 211, 219, 382, 433, 434, 451, 525, 555, 556, 558, 561, 566, 568, 599, 604, 627, 642, 643, 649, 660, 662, 663, 664, 665,

Cooper, H.H., Jr. (*cont'd*)
666, 667, 668, 669, 671, 672, 675, 676, 680, 683, 684, 686, 688, 689, 697, 698, 702
Crawford, L.A., 67

Dagan, G., 495, 509, 510, 511
Darcy, H., 22, 29
Das, B.J., 651
Daubré, A., 22
Davis, A.R., 73, 76, 78
Davis, S.N., 5, 34, 36, 73, 76, 78
Dawson, K.J., 5
DeGlee, G.J., 7, 25, 122, 131
De Ridder, N.A., 5, 122, 131, 165, 177, 178, 270, 357, 482
de Marsily, G., 5
DeWiest, R.J.M., 5
Domenico, P.A., 5, 59, 60, 61, 67, 83
Driscoll, F.G., 5, 156, 178
Dupuit, J., 7, 124
Duda, L.E., 209

Fair, G.M., 42, 45, 46
Ferris, J.G., 62, 63, 65, 66, 68, 150, 153, 158, 159, 160, 165, 176, 177, 178, 555, 556, 557, 558, 568, 570, 571, 572
Fetter, C.W., 5
Forkasiewicz, J., 534, 540
Franke, O.L., 99, 100, 102
Freeze, R.A., 5, 34, 36, 40, 47, 80, 347, 354, 357
Fuller, M.L., 24

Gelhar, L.W., 5
Gonzalez, D.D., 206, 211, 222, 224, 225, 226, 228, 229, 231
Gradshteyn, I.S., 363, 499
Ground Water Publishing Company, 37, 38

Hall, P., 5
Hantush, M.S., 7, 8, 25, 53, 64, 71, 78, 80, 92, 106, 121, 122, 126, 131, 160, 187, 206, 231, 232, 233, 236, 237, 241, 242, 244, 245, 247, 248, 249, 250, 252, 253, 254, 262, 264, 268, 270, 272, 275, 276, 281, 282, 285, 286, 287, 289, 293, 294, 295, 296, 297, 299, 300, 301, 302, 320, 321, 322, 323, 324, 325, 326, 327, 328, 329, 330, 334, 335, 338, 339, 340, 342, 343, 347, 349, 358, 360, 362, 363, 364, 365, 368, 371, 372, 376, 377, 386, 387, 388, 389, 390, 391, 392, 394, 395, 396, 397, 398, 406, 407, 409, 410, 411, 412, 413, 414, 415, 416, 418, 425, 426, 427, 428, 429, 431, 432, 433, 434, 435, 436, 437, 441, 442, 447, 449, 451, 452, 453, 454, 455, 465, 467, 468, 475, 508, 509, 510, 512, 570, 572, 573, 574, 575, 577, 578, 579, 580, 581, 582, 583, 584, 585, 587, 588, 590, 591, 592, 593, 594, 595, 599, 604, 607
Hatch, L.P., 42, 45, 46
Hazen, A., 40, 43, 44, 46
Helweg, O.J., 8, 621, 623
Highdon, A., 87
Hodson, W.G., 584
Hvorslev, M.J., 9, 627, 648, 649, 651, 653, 654, 657, 660, 661, 668, 672
Hyder, Z., 639, 640, 641, 642, 643, 644, 645, 646, 647, 648

Istok, J.D., 5

Jaeger, C., 34
Jaeger, J.C., 149, 349, 352, 474
Jacob, C.E., 7, 9, 16, 25, 36, 53, 56, 67, 69, 71, 78, 122, 138, 147, 151, 156, 160, 163, 164, 176, 177, 182, 211, 219, 248, 249, 250, 252, 253, 254, 264, 268, 286, 293, 294, 323, 325, 326, 329, 330, 334, 342, 347, 349, 368, 377, 382, 389, 392, 433, 434, 451, 475, 525, 530, 555, 556, 558, 561, 566, 568, 572, 599, 600, 601, 602, 603, 604, 605, 607, 608, 612, 615, 676, 692, 694, 695, 696, 697, 698, 699, 700, 701, 702, 703
Javandel, I., 347, 354, 357
Jorgensen, D.G., 64, 65
Johnson, A.I., 27, 28, 35, 67

Knowles, D.B., 62, 63, 65, 66, 68, 150, 153, 158, 159, 160, 165, 176, 177, 178, 555, 556, 557, 558, 568, 570, 571, 572
Kashef, A.I., 5
Kazmann, R.G., 580
Kézdi, A., 59
Kozeny, J., 42, 43, 44, 46
Kruseman, G.P., 5, 122, 131, 165, 177, 178, 270, 357, 482

Labadie, J.W., 8, 621, 623
Lambe, T.W., 55, 59, 60, 651
Lane, C.W., 584
Liu, W., 639, 640, 642

Lohman, S.W., 9, 52, 66, 139, 140, 676, 692, 694, 695, 696, 697, 698, 699, 700, 701, 702, 703
Luthin, J.N., 5, 119

Maasland, M., 90, 91
Mariño, M.A., 5, 119
McElwee, C.D., 639, 640, 642
McCarthy, D.F., 44
Meinzer, O.E., 22, 24, 53, 65, 66
Mercer, J.W., 227
Mercer, J.W., 28, 35, 59, 67
Merz, J., 547, 548
Mifflin, M.D., 60, 61
Miller, T.M., 8, 615, 616, 617, 619, 620, 621, 622, 623
Mock, P., 547, 548
Moench, A.F., 187, 473, 503, 504
Morris, D.A., 27, 28, 34, 67

Narasimhan, T.N., 639, 641, 644, 645, 646, 647, 648
Neuman, S.P., 7, 8, 25, 104, 206, 211, 222, 224, 225, 226, 228, 229, 231, 263, 264, 293, 295, 296, 301, 345, 347, 349, 351, 352, 353, 354, 355, 357, 358, 359, 360, 363, 364, 365, 366, 367, 368, 369, 370, 371, 372, 373, 374, 375, 376, 377, 378, 379, 380, 381, 382, 383, 384, 385, 458, 459, 473, 492, 493, 495, 496, 497, 498, 499, 500, 501, 502, 503, 504, 505, 506, 507, 508, 510, 511, 512, 513, 514, 515, 519, 523, 524, 525, 526, 527, 528, 529, 530, 531, 532, 533, 534, 542, 546, 547, 550, 554
Neuzil, C.E., 676, 678, 679, 680, 681, 689
Norton, W.H., 24

Ohlsen, E.H., 87
Orr, B.R., 227

Papadopulos, S.S. (or I.S.), 7, 9, 145, 150, 187, 188, 189, 190, 194, 195, 196, 197, 200, 201, 202, 203, 204, 206, 207, 208, 209, 210, 211, 212, 219, 222, 241, 242, 627, 642, 643, 649, 660, 662, 663, 664, 665, 666, 667, 668, 669, 671, 672, 675, 676, 677, 678, 680, 681, 682, 683, 684, 685, 686, 687, 688, 689, 691
Peaudecerf, P., 534, 540
Peck, R.B., 55
Peirce, J.J., 43

Polubarinova-Kochina, P., Ya., 5, 34, 80
Price, M., 459
Prickett, T.A., 476, 487, 530, 531

Quiñoncs-Aponte, V., 207

Raghunath, H.M., 5, 558, 607, 609
Rasmussen, T.C., 73, 76, 78
Redshaw, S.C., 5
Reed, J.D., 8, 189, 194, 250, 387, 394, 395, 396, 408, 409, 417, 697
Reeves, M., 209
Reilly, T.E., 99, 100, 102
Rice, R.C., 8, 627, 628, 629, 631, 632, 633, 636, 639, 641, 642, 644, 647, 648, 651, 654, 658, 659, 660, 662, 668, 672
Riley, W.F., 87
Robinson, E.S., 78
Rojstaczer, S., 67
Rorabaugh, M.I., 604, 605, 610, 612, 614, 615, 617, 619, 620, 621, 622, 623
Ross, B., 28, 35, 59, 67
Rushton, K.R., 5
Ryzhik, I.M., 363, 499

Schwartz, F.W., 5, 83
Şen, Z., 5, 42, 43, 46
Sheahan, N.T., 331, 332, 622, 623
Shepherd, R.G., 41, 46
Sperry, M.S., 43
Spiegel, M.R., 92, 93, 474
Stallmann, R.W., 62, 63, 65, 66, 68, 150, 153, 158, 159, 160, 165, 176, 177, 178, 465, 467, 555, 556, 557, 558, 568, 570, 571, 572
Stiles, W.B., 87
Stramel, G.J., 584
Street, R.L., 39, 57
Streltsova, T.D., 532
Street, R.L.,

Terzaghi, K., 53, 55, 87
Theis, C.V., 7, 16, 25, 145, 149, 150, 158, 176, 190, 209, 211, 232, 242, 243, 244, 246, 248, 253, 263, 264, 270, 282, 325, 324, 326, 330, 347, 348, 349, 350, 351, 355, 373, 374, 377, 382, 413, 433, 451, 458, 468, 479, 492, 500, 509, 511, 525, 555, 561, 566, 570, 572, 599
Thiem, G., 7, 9, 113, 128, 601, 629, 676, 692, 693, 695
Thomas, R.G., 7, 206, 242, 244, 245, 247
Thomas, S.D., 28, 35, 59, 67

Todd, D.K., 5, 40, 51, 52, 68, 79, 183, 555, 561, 607

U.S. Bureau of Reclamation, 692, 693
U.S. Department of the Interior, 73, 182
U.S. Department of the Navy, 651, 652, 653

Vennard, J.K., 39, 57
Verruijt, A., 87

Walter, G.R., 206, 211, 222, 224, 225, 226, 228, 229, 231
Walton, W.C., 5, 264, 265, 270, 272, 281, 336, 337, 599, 607
Ward, D.S., 209
Ward, J.J., 206, 211, 222, 224, 225, 226, 228, 229, 231
Weber, W.J., Jr., 8, 615, 616, 617, 619, 620, 621, 622, 623
Weeks, E.P., 68, 69, 70
Weese, J.A., 87, 96
Wenzel, L.K., 150, 530
Whitman, R.V., 55, 59, 60, 651
Witherspoon, P.A., 7, 25, 104, 263, 264, 293, 295, 296, 301, 345, 347, 349, 351, 352, 353, 354, 355, 357, 358, 359, 360, 363, 364, 365, 366, 367, 368, 369, 370, 371, 372, 373, 374, 375, 376, 377, 378, 379, 380, 381, 382, 383, 384, 385, 493, 495
Wikramaratna, R.S., 196, 198, 205

SUBJECT INDEX

Adhesion, 51
Alyamani and Şen equation, for estimation of hydraulic conductivity, 42–43, 46
Aquifer:
anisotropic, 47
artesian (see confined)
classification of, 47–50
compressibility and elasticity of, 53–80
confined, 24
equations of motion in, 80–84
heteregeneous and anisotropic, 50, 94–95
heteregeneous and isotropic, 48–49, 94
homogeneous and isotropic, 47–48, 94, 96
isotropic, 47
leaky, 24
multiple, 24–25
nonleaky, 24
perched, 22
principal directions of anisotropy, 47
thick unconfined, 138–139
thin, unconfined, 139
transversely isotropic, 50, 94
types of, 22–25
unconfined, 22
Aquiclude, 22, 345
Aquifuge, 22
Aquitard, 22, 248, 345
Artesian aquifer (see Confined aquifer)
Aspect ratio, 641

Barometric efficiency, definition of, 67–68
Barometric efficiency, estimation of, 72–78
Barometric fluctuations in wells, confined aquifers, 68–72
Barometric fluctuations in wells, unconfined aquifers, 68
Barometric pressure, 67
Bernoulli equation, 31
Boulton's models, unconfined aquifers, full penetration, 461–470, 472–481
Boulton's well function, unconfined aquifers, full penetration, 475–476
Boundary conditions, 97–109
Boundary condition, types of:
Cauchy (or third-type) 104
constant head, 98–100
Dirichlet (or first-type), 98, 115, 125, 570
first-type (see Dirichlet)
free surface, 106–107
head-dependent flux (see also Cauchy or third-type), 103–104
impermeable boundary layer, 101–102
interface, 104–106
no-flow, 101–102
prescribed head, 101

Boundary condition, types of (*cont'd*)
seepage face, 102–103
specified flux, 101
specified head (see prescribed head)
third-type (see Cauchy)
water divide, 102
Boundary detection method, confined aquifer:
one impervious boundary, 561–567
two intersecting boundaries, 568–570
Boundary value problem, 97
Bouwer and Rice slug test model, 628–633
Bredehoeft and Papadopulos pressure pulse test model, 680–686
Bredehoeft and Papadopulos type curve method, pressure pulse test data analysis, 686–692

Capillarity, 51
Capillary fringe, 24, 51–52
Capillary rise, 51–52
Capillary rise, quantification of, 51–53
Chow method, pumping test data analysis, confined aquifers, full penetration, 171–175
Clark's method for barometric efficiency, 76–78
Coefficient of volume change, 55
Cohesion, 51
Compressibility coefficient of soils, 55
Compressibility values of water, 57
Compressibility, vertical, of the soil skeleton, 55
Confined aquifer:
definition of, 24
infinite assumption criterion, 154–155
leaky, homogeneous, anisotropic, 285
leaky, homogeneous, isotropic, 116, 249
Confined aquifer model, full penetration:
anisotropic, transient, 207–209
anisotropic, transient, approximated, 210
isotropic, transient, 145–150
large diameter well, transient, 187–195
leaky, without storage of confining layer, unsteady-sate, 249–263
leaky, with storage of confining layers, unsteady-state, 293–326
two/one confined aquifer and one aquiclude, unsteady-state, 345–357
two confined aquifers and one aquitard, unsteady-state, 358–365
Confined aquifer model, partial penetration, steady state:
Hantush, leaky, screen extends to upper boundary (Zero penetration), 390
Hantush, nonleaky, screen extends to upper boundary (Zero penetration), 392
Confined aquifer model, partial penetration, transient:
Hantush, leaky, screen extends to upper boundary (Zero penetration), 389
Hantush, leaky, general case for piezometers, 396–397
Hantush, leaky, general case for observation wells, 408
Hantush, nonleaky, general case for observation wells, 409
Hantush, nonleaky, general case for observation wells ($K_r = K_z$), 409
Hantush, nonleaky, general case for piezometers, 396–397
Hantush, nonleaky, general case for piezometers ($K_r = K_z$), 397–408
Hantush, nonleaky, general case for recovery, observation wells, 413
Hantush, nonleaky, general case for recovery, piezometers, 412
Hantush, nonleaky, screen extends to upper boundary (Zero penetration), 390
Confining layer, 22, 25, 248–250, 294–296
Consolidation, 58–61
Consolidation coefficient, 59
Consolidation tests, 375
Consolodation, theory of, 59
Constant head injection test, 692
Constant head injection data analysis method:
steady state, 693–694
unsteady state, 698–703
Constant head injection test model:
steady state (Thiem), 692–693
unsteady state (Jacob and Lohman), 694–698
Cooper, Bredehoeft, and Papadopulos slug test model, 662–664
Cooper and Jacob approximation, 156, 163–165, 558, 599, 604
Cooper and Jacob method for pumping test data analysis, 163–171

Darcy experiment, 32–33
Darcy (units), 38, 40
Darcy flux, 32
Darcy's law, 29–31
Darcy's law, generalization of, 80–84
Darcy's law, range of validity, 33–34
Darcy velocity, 32

Darcy velocities:
in two-dimensional Cartesian coordinates, 80–82
in three-dimensional Cartesian coordinates, 80–82
in two-dimensional cylindrical coordinates, 82–83
in three-dimensional cylindrical coordinates, 82–83
De Glee equation, 122
Delay index, Boulton's, 479–481, 532
Delayed gravity response, 511
Delayed unconfined aquifer response, 492
Delay yield, 492, 511
Delayed yield stage, 459
Density of water, 39, 57
Differential equations of flow in aquifers under unsteady-state conditions:
in polar coordinates, 95
in two dimensions, 96
in three dimensions, 92–95
Differential equations of flow in aquifers under steady state conditions, 96
Differential equation of free surface, 106–107
Dimensionless storage parameter for slug test model, 641
Discharge velocity, 28–29
Double-straight line effect in slug test analysis, 633
Drawdown, in observation well, 408–411, 497, 507–508
Drawdown, in piezometer, 396–408, 411–413, 496–497, 506–507
Drawdown, definition of, 27
Drawdown correction (see Jacob's drawdown correction scheme)
Drawdown, projected, 180
Drawdown, residual, 176–177
Dug well (see Large diameter well)
Dupuit assumptions, 124, 126
Dupuit-Forchheimer well discharge formula, 113, 123–127

Effective stress concept, 53
Effective well radius, 600
Equations of motion in aquifers, 80–84

Fair and Hatch equation, for estimation of hydraulic conductivity, 42
Falling head test (see Slug test)
Flow equation, two-dimensional, 93
Flow equation, three-dimensional, 92–93
Flowing well, 24
Fractured media flow theory, 5

Geometric elevation, 30
Gradient of piezometric head, 31
Grain diameter, effective, 42
Grain diameter, representative, 42
Grain-size distribution, 40
Ground water velocity, 28, 29

Hantush's generalized leaky confined aquifer model with storage of confining layers, 294–326
Hantush and Jacob model for leaky confined aquifer without storage of confining layer, 248–263
Hantush and Jacob well function, leaky confined aquifers, 253–263, 323, 389, 392
Hazen equation, for estimation of hydraulic conductivity, 40, 44
Head, 26–27
Horizontal well methods, 21
Hydraulic conductivity:
conversions of, 37
definition of, 32
directional, 88–92
ellipse of, 90
ellipsoid of, 91–92
estimation from grain size, 40–46
horizontal, 21
principal, 84–88
tensor, 93
values of, 35–36
Hvorslev slug test model, 648–651
Hydraulic charactersitics of aquifers, 25
Hydrogeologic characteristics of aquifers, 25
Hydraulic diffusivity, 348, 374, 377
Hydraulic head (see Head)
Hydraulic resistance, 118

Image wells, 555
Inflection point methods, theory of, 268–270, 425–429
Inflection point methods for leaky confined aquifers:
one observation well method, 270–275
more than one observation well method, 275–281
Inflection point method for nonleaky confined aquifers, partial penetration, 440–441
Influence radius, 116, 156, 693
Initial conditions, 97

Isotropic aquifer, 47
Isotropic methods of data analysis, 232

Jacob and Lohman constant head injection tests model, 694–698
Jacob's drawdown correction scheme, 138, 139, 530

Kozeny-Carman equation for estimation of hydraulic conductivity, 42, 44–45

Labadie and Helweg method for step drawdown test data analysis, 621–623
Large diameter well, 187
Leakage coefficient, 251
Leakage factor, 251
Leakage rate, 321, 323, 324, 325
Leaky aquifer, 24
Leakance, 251

Match line, 196, 198, 199
Method of images, 555
Miller and Weber method for step drawdown test data analysis, 615–621
Mohr circle for principal directions of hydraulic conductivity and transmissivity, 87–88
Modulus of elasticity, definition of, 54
Modulus of elasticity values of porous materials, 61
Modulus of elasticity values of water, 57
Multiple aquifers, 24–25

Neuman model for unconfined aquifers, full penetration, 492–500
Neuman model for unconfined aquifers, partial penetration, 504–509
Neuman and Witherspoon's two/one confined aquifer and one aquiclude model, 345–357
Neuman and Witherspoon's two confined aquifers and one aquitard model, 358–365
Nonleaky aquifer, 24

Observed compressibility, 680

Papadopulos model, confined aquifer, full penetration, anisotropic, 206–209
Papadopulos and Cooper model, confined aquifer, full penetration large diameter wells, 187–195
Partially penetrating well, 386

Permeability:
  coefficient of, 35, 38
  intrinsic, 38
  skin of, 646–648
  units of, 34–35
Phreatic aquifer (see Unconfined aquifer)
Piezometric head, 26–27, 30
Piezometric surface, 24
Porosity:
  definition of, 27
  effective, 27
  range of values, 28
Porous media flow theory, 5
Potentiometric surface, 24
Pressure aquifer (see Confined aquifer)
Pressure head, 27
Pressure pulse test method, definition of, 675–680
Pressure pulse test data analysis method, 686–692
Pressure pulse test model:
  Bredehoeft and Papadopulos, 680–684
  Bredehoeft and Papadopulos, approximate equation, 684–686
Pressure pulse test procedures:
  Bredehoeft and Papadopulos, 676
  Neuzil, 676–680
Projected drawdown, 180
Principal directions of hydraulic conductivity, 84
Pumping test data analysis methods, full penetration, steady state:
  confined, nonleaky aquifer (Thiem), 128–131
  straight-line, confined leaky aquifer, 135–137
  straight-line, unconfined aquifer, 137–141
  type-curve, confined leaky aquifer, 131–135
Pumping test data analysis methods, full penetration, transient:
  confined aquifer, isotropic (Chow), 171–175
  confined aquifer, anisotropic, nonleaky, Procedure I (Hantush), 236
  confined aquifer, anisotropic, nonleaky, Procedure II (Hantush), 236–242
  confined aquifer, anisotropic, nonleaky (Hantush and Thomas), 245–246
  confined aquifer, anisotropic, leaky, Procedure I (Hantush), 287–288
  confined aquifer, anisotropic, leaky, Procedure II (Hantush), 288–292
  inflection point, leaky confined aquifer, one observation well (Hantush), 270–275

inflection point, leaky confined aquifer, more than one observation well (Hantush), 275–281
ratio method (Neuman and Witherspoon), 372–385
recovery, nonleaky confined aquifer, isotropic (Theis), 176–186
recovery, nonleaky confined aquifer, anisotropic (Hantush and Thomas), 244–245, 247
straight-line, anisotropic, nonleaky confined aquifer, four wells (Papadopulos) 219–222
straight-line, anisotropic, nonleaky confined aquifer, three wells (Neuman and others) 226–231
straight-line, isotropic, confined (Cooper and Jacob), 165–171
type-curve, anisotropic, nonleaky confined aquifer, four wells (Papadopulos) 211–219
type-curve, anisotropic, nonleaky confined aquifer, three wells (Neuman and others) 225–226
type-curve, confined, isotropic, leaky, with storage of confining layers (Hantush), 328–343
type-curve, confined nonleaky aquifer, isotropic (Theis), 158–163
type-curve, confined aquifer, large diameter well (Papadopulos and Cooper), 196, 199–205
type-curve, confined aquifer, large diameter well (Wikramaratna), 196–205
type-curve, confined leaky aquifer, without storage of confining layers (Walton), 264–268
type-curve, leaky confined aquifer, without storage of confining layer (Hantush), 281–285

Pumping test data analysis method, full penetration, recharge boundary:
one observation well method, Procedure I (Hantush), 581–582, 584–588
one observation well method, Procedure II (Hantush), 582–583, 588–595
two or more observation wells method (Hantush), 583–584, 591–595

Pumping test data analysis methods, confined nonleaky aquifer, partial penetration, transient:
adjusted (Cooper and Jacob), 434–435, 451
inflection point (Hantush), 429–432, 440–441, 449–450
recovery (Hantush), 451–457
type-curve, 416–425
type-curve (Hantush), 432–434, 436–440, 444–449
type-curve or straight line (Theis' or Cooper and Jacob or both), 433–434

Pumping test data analysis method, unconfined, full penetration, transient:
recovery, 529–530
semilogarithmic, 525–529, 542–546
type-curve (Boulton), 481–492
type-curve (Neuman), 513–525, 534–542

Pumping test data analysis method, unconfined, partial penetration, transient:
recovery, 554
semilogarithmic, 554
type-curve (Neuman), 546–553

Radius of influence (see Influence radius)
Ratio method (Neuman and Witherspoon)(see Pumping test data analysis methods)
Ratio method, theory of (Neuman and Witherspoon), 372–375
Ray of observation wells, 233, 287
Recharge boundary, confined aquifer, theory of, 570–580
Recovery equations, confined aquifer, full penetration, 177, 180–183
Recovery equations, confined aquifer, partial penetration:
observation well, 413
piezometer, 412–413
Recovery equation, unconfined aquifer, full penetration, 469–470
Remediation, 4
Remedial investigation and feasibility study, 4
Reynolds number, 33
Rising head test (see Slug test)
Rorabaugh's method for step drawdown data analysis, 610–615

Saturated hydraulic conductivity (see Hydraulic conductivity)
Scientific notation, 10
Seepage velocity (see Velocity)
Sepherd equations for estimation of hydraulic conductivity, 41–42, 44
S.I. units, 10, 56, 61
Skin effect, 646–648
Slug test:
double straight-line effect, 633
effects of well diameter, 635

Slug test (*cont'd*)
estimation rate of rise or fall, 635–636
falling head test, 627, 633
pressure distribution around the screened section, 636
rising head test, 627
Slug test data analysis method:
Bouwer and Rice, 632–633, 636–639
Hvorslev, 654–662
Cooper, Bredehoeft, and Papadopulos, 664–672
Slug test model:
Bouwer and Rice, 628–631
Hvorslev, 628–631
Cooper, Bredehoeft, and Papadopulos, 662–664
Specific capacity, definition of, 156, 601
Specific capacity equations, 156–157
Specific capacity, steady-state, 601–604
Specific capacity, unsteady-state, 604–605
Specific discharge, 32
Specific retention, 66–67
Specific storage, 21, 55, 58
Specific storage equation, 58
Specific storage equations, based on consolidation coefficient, 58–61
Specific storage, range of values, 59, 61
Specific storage of natural sediments, 61
Soil-water zone, 51
Specific yield, 65–66
Specific yield, values, 67
Specific weight, values of water, 39, 57
Step drawdown test:
data analysis methods, 607–623
definition, 599
Straight-line method for pumping test data analysis (see Cooper and Jacob Method)
Straight-line method for step drawdown, 607–610
Storage of aquitard, 368
Storage coefficient, 61–62
Storage coefficient, effective, 65, 473
Storage coefficient of water, 62
Storage coefficient of soil, 62
Storage coefficient of layered systems, 65
Storativity (see Storage coefficient)
Stress-strain relationships, 54–55
Superposition principle, 556
Surface tension, 52

Tangent law, 106
Theis equation (see Theis model)
Theis model, 145–150
Theis model, approximated form (Cooper and Jacob), 156, 164
Theis recovery method (see Pumping test data analysis methods)
Theis solution (see Theis model)
Theis-type curve, 150
Theis type-curve method (see Pumping test data analysis methods)

Theis well function, 150, 209, 321, 324, 349, 468, 556
Thiem equation, 116, 601, 629, 693
Tidal efficiency, 78–80
Tidal fluctuation in aquifers, 78
Tight formations, 675
Total leakage volume, 321, 323, 324, 325
Transient (see also Unsteady-state)
Transmissivity, conversions, 38
Transmissivity, definition of, 35
Transmissivity, directional, 89–90
Transmissivity, ellipse, 90
Transmissivity, estimation, 156–157
Transmissivities, principal, 86, 87–88, 207, 210
Transmissivity tensor, 207
Turbulent flow in porous media, 34

Unconfined aquifer model, full penetration:
Average solution over aquifer thickness, 497
Boulton, without delayed response, 461–465, 500
Boulton, with delayed response, 472–476
Neuman, with delay yield, 492–499
Recovery, 469–470, 529–530
Unconfined aquifer model, partial penetration:
Neuman, with delay yield, 504–509
recovery, 554
Unsaturated zone, 22, 51–52
Unsteady-state (see also Transient)

Vadose zone, 51–52
Vadose zone, intermediate, 51–52
Variable rate test (see Step drawdown test)
Vertical well methods, 21
Void ratio, 55
Velocity:
discharge, 25, 28, 29
ground water (or seepage), 25, 28, 29
Velocity head, 30

Viscosity:
dynamic, 33–35
dynamic, values of water, 39
kinematic, 34
kinematic, values of water, 39
Volume elasticity of aquifers, 53
Water divide, 102
Water table, 24
Water table aquifer (see Unconfined aquifer)
Well loss constant, 603
Well loss constant, properties of, 605–607

CUSTOMER NOTE: IF THIS BOOK IS ACCOMPANIED BY SOFTWARE, PLEASE READ THE FOLLOWING BEFORE OPENING THE PACKAGE.

This software contains files to help you utilize the models described in the accompanying book. By opening the package, you are agreeing to be bound by the following agreement.

This software product is protected by copyright and all rights are reserved by the author, John Wiley & Sons, Inc., or their licensors. You are licensed to use this software on a single computer. Copying the software to another medium or format for use on a single computer does not violate the U.S. Copyright Law. Copying the software for any other purpose is a violation of the U.S. Copyright Law.

This software product is sold as is without warranty of any kind, either express or implied, including but not limited to the implied warranty of merchantability and fitness for a particular purpose. Neither Wiley nor its dealers or distributors assumes any liability for any alleged or actual damages arising from the use of or the inability to use this software. (Some states do not allow the exclusion of implied warranties, so the exclusion may not apply to you.)